| | | |
|---|---|---|
| Thermal resistance | 1 K/W | = 0.52750 °F/h · Btu |
| Viscosity (dynamic) | 1 N · s/m$^2$ | = 1 kg/s · m |
| | | = 2419.1 lb/ft · h |
| | | = 5.8016 × 10$^{-6}$ lbf · h/ft$^2$ |
| | | = 2.089 × 10$^{-2}$ lbf · s/ft$^2$ |
| Volume | 1 m$^3$ | = 6.1023 × 10$^4$ in.$^3$ |
| | | = 35.314 ft$^3$ |
| | | = 264.17 gal |
| | | = 10$^3$ L |
| | 1 gal | = 0.13368 ft$^3$ |
| Volume flow rate | 1 m$^3$/s | = 2.1188 × 10$^3$ ft$^3$/min |
| | | = 1.5850 × 10$^4$ gal/min |

# Physical Constants

Universal Gas Constant:

$\bar{R} = 8.314$ kJ/kmol · K
$= 8314$ N · m/kmol · K
$= 1545$ ft · lbf/lbmol · °R
$= 1.986$ Btu/lbmol · °R

Stefan-Boltzmann Constant:

$\sigma = 5.670 \times 10^{-8}$ W/m$^2$ · K$^4$
$= 0.1714 \times 10^{-8}$ Btu/h · ft$^2$ · °R$^4$

Blackbody Radiation Constants:

$C_1 = 3.7420 \times 10^8$ W · μm$^4$/m$^2$
$= 1.187 \times 10^8$ Btu · μm$^4$/h · ft$^2$
$C_2 = 1.4388 \times 10^4$ μm · K
$= 2.5897 \times 10^4$ μm · °R
$C_3 = 2897.8$ μm · K
$= 5215.6$ μm · °R

Gravitational Acceleration (Sea Level):

$g = 9.807$ m/s$^2$ $= 32.174$ ft/s$^2$

Standard Atmospheric Pressure:

$p = 1.01325$ bar $= 101{,}325$ N/m$^2$ $= 14.696$ lbf/in.$^2$

# Introduction to Thermal Systems Engineering:

## Thermodynamics, Fluid Mechanics, and Heat Transfer


**Michael J. Moran**
*The Ohio State University*

**Howard N. Shapiro**
*Iowa State University of Science and Technology*

**Bruce R. Munson**
*Iowa State University of Science and Technology*

**David P. DeWitt**
*Purdue University*

John Wiley & Sons, Inc.

| | |
|---|---|
| Acquisitions Editor | Joseph Hayton |
| Production Manager | Jeanine Furino |
| Production Editor | Sandra Russell |
| Senior Marketing Manager | Katherine Hepburn |
| Senior Designer | Harold Nolan |
| Production Management Services | Suzanne Ingrao |
| Cover Design | Howard Grossman |
| Cover Photograph | © Larry Fleming. All rights reserved. |

This book was typeset in 10/12 Times Roman by TechBooks, Inc. and printed and bound by R. R. Donnelley and Sons (Willard). The cover was printed by The Lehigh Press.

The paper in this book was manufactured by a mill whose forest management programs include sustained yield harvesting of its timberlands. Sustained yield harvesting principles ensure that the number of trees cut each year does not exceed the amount of new growth.

This book is printed on acid-free paper. ♾



ISBN 0-471-20490-0

Printed in the United States of America.

10 9 8 7 6 5 4 3 2 1

# Preface

Our objective is to provide an integrated introductory presentation of thermodynamics, fluid mechanics, and heat transfer. The unifying theme is the application of these principles in *thermal systems engineering*. Thermal systems involve the storage, transfer, and conversion of energy. Thermal systems engineering is concerned with how energy is utilized to accomplish beneficial functions in industry, transportation, the home, and so on.

*Introduction to Thermal Systems Engineering: Thermodynamics, Fluid Mechanics, and Heat Transfer* is intended for a three- or four-credit hour course in thermodynamics, fluid mechanics, and heat transfer that could be taught in the second or third year of an engineering curriculum to students with appropriate background in elementary physics and calculus. Sufficient material also is included for a two-course sequence in the thermal sciences. The book is suitable for self-study, including reference use in engineering practice and preparation for professional engineering examinations. SI units are featured but other commonly employed engineering units also are used.

The book has been developed in recognition of the team-oriented, interdisciplinary nature of engineering practice, and in recognition of trends in the engineering curriculum, including the move to reduce credit hours and the ABET-inspired objective of introducing students to the *common themes* of the thermal sciences. In conceiving this new presentation, we identified those critical subject areas needed to form the basis for the engineering analysis of thermal systems and have provided those subjects within a book of manageable size.

Thermodynamics, fluid mechanics, and heat transfer are presented following a traditional approach that is familiar to faculty, and crafted to allow students to master fundamentals before moving on to more challenging topics. This has been achieved with a more integrated presentation than available in any other text. Examples of integration include: unified notation (symbols and definitions); engaging case-oriented introduction to thermodynamics, fluid mechanics, and heat transfer engineering; *mechanical energy* and *thermal energy* equations developed from thermodynamic principles; *thermal boundary layer* concept as an extension of *hydrodynamic boundary layer* principles; and more.

**Features especially useful for students are:**

- Readable, highly accessible, and largely self-instructive presentation with a strong emphasis on engineering applications. Fundamentals *and* applications provided at a *digestible* level for an introductory course.
- An engaging, case-oriented introduction to thermal systems engineering provided in Chapter 1. The chapter describes thermal systems engineering generally and shows the interrelated roles of thermodynamics, fluid mechanics, and heat transfer for analyzing thermal systems.
- Generous collection of detailed examples featuring a structured problem-solving approach that encourages systematic thinking.
- Numerous realistic applications and homework problems. End-of-chapter problems classified by topic.
- Student study tools (summarized in Sec. 1.4) include chapter introductions giving a clear statement of the objective, chapter summary and study guides, and key terms provided in the margins and coordinated with the text presentation.
- A CD-ROM with hyperlinks providing the full print text plus additional content, answers to selected end-of-chapter problems, short fluid flow video clips, and software for solving problems in thermodynamics and in heat transfer.
- Access to a website with additional learning resources: http://www.wiley.com/college/moran

**Features especially useful for faculty are:**

- Proven content and student-centered pedagogy adapted from leading textbooks in the respective disciplines:

  M.J. Moran and H.N. Shapiro, *Fundamentals of Engineering Thermodynamics,* 4th edition, 2000.

  B.R. Munson, D.F. Young, and T.H. Okiishi, *Fundamentals of Fluid Mechanics,* 4th edition, 2002.

  F.P. Incropera and D.P. DeWitt, *Fundamentals of Heat and Mass Transfer,* 5th edition, 2002.

- Concise presentation and flexible approach readily tailored to individual instructional needs. Topics are carefully structured to allow faculty wide latitude in choosing the coverage they provide to students—with no loss in continuity. The accompanying CD-ROM provides additional content that allows faculty further opportunities to customize their courses and/or develop two-semester courses.

- Highly integrated presentation. The authors have worked closely as a team to ensure the material is presented seamlessly and works well as a whole. Special attention has been given to smooth transitions between the three core areas. Links between the core areas have been inserted throughout.
- Instructor's Manual containing complete, detailed solutions to all the end-of-chapter problems to assist with course planning.

## A Note on the Creative Process

How did four experienced authors come together to develop this book? It began with a face-to-face meeting in Chicago sponsored by our Publisher. It was there that we developed the broad outline of the book and the unifying thermal systems engineering theme. At first we believed it would be a straightforward task to achieve our objectives by identifying the core topics in the respective subject areas and adapting material from our previous books to provide them concisely. We quickly found that it was easier to agree on overall objectives than to achieve them. Since we come from the somewhat different technical *cultures* of thermodynamics, fluid mechanics, and heat transfer, it might be expected that challenges would be encountered as the author team reached for a common vision of an integrated book, and this was the case.

Considerable effort was required to harmonize different viewpoints and writing styles, as well as to agree on the breadth and depth of topic coverage. Building on the good will generated at our Chicago meeting, collaboration among the authors has been extraordinary as we have taken a problem-solving approach to this project. Authors have been open and mutually supportive, and have shared common goals. Concepts were honed and issues resolved in weekly telephone conferences, countless e-mail exchanges, and frequent one-to-one telephone conversations. A common vision evolved as written material was exchanged between authors and critically evaluated. By such teamwork, overlapping concepts were clarified, links between the three disciplines strengthened, and a single voice achieved. This process has paralleled the engineering design process we describe in Chapter 1. We are pleased with the outcome.

We believe that we have developed a unique, user-friendly text that clearly focuses on the essential aspects of the subject matter. We hope that this new, concise introduction to thermodynamics, fluid mechanics, and heat transfer will appeal to both students and faculty. Your suggestions for improvement are most welcome.

## Acknowledgments

Many individuals have contributed to making this book better than it might have been without their participation. Thanks are due to the following for their thoughtful comments on specific sections and/or chapters of the book: Fan-Bill Cheung (Pennsylvania State University), Kirk Christensen (University of Missouri-Rolla), Prateen V. DeSai (Georgia Institute of Technology), Mark J. Holowach (Pennsylvania State University), Ron Mathews (University of Texas-Austin), S. A. Sherif (University of Florida). Organization and topical coverage also benefited from survey results of faculty currently teaching thermal sciences courses.

Thanks are also due to many individuals in the John Wiley & Sons, Inc., organization who have contributed their talents and efforts to this book. We pay special recognition to Joseph Hayton, our editor, who brought the author team together, encouraged its work, and provided resources in support of the project.

April 2002

*Michael J. Moran*
*Howard N. Shapiro*
*Bruce R. Munson*
*David P. DeWitt*

# Contents

## THERMO

# WHAT IS THERMAL SYSTEMS ENGINEERING?

## *Introduction...*

The ***objective*** of this chapter is to introduce you to thermal systems engineering using several contemporary applications. Our discussions use certain terms that we assume are familiar from your background in physics and chemistry. The roles of thermodynamics, fluid mechanics, and heat transfer in thermal systems engineering and their relationship to one another also are described. The presentation concludes with tips on the effective use of the book.

***chapter objective***

## 1.1 Getting Started

Thermal systems engineering is concerned with how energy is utilized to accomplish beneficial functions in industry, transportation, and the home, and also the role energy plays in the study of human, animal, and plant life. In industry, thermal systems are found in electric power generating plants, chemical processing plants, and in manufacturing facilities. Our transportation needs are met by various types of engines, power converters, and cooling equipment. In the home, appliances such as ovens, refrigerators, and furnaces represent thermal systems. Ice rinks, snow-making machines, and other recreational uses involve thermal systems. In living things, the respiratory and circulatory systems are thermal systems, as are equipment for life support and surgical procedures.

Thermal systems involve the *storage, transfer,* and *conversion* of energy. Energy can be *stored* within a system in different forms, such as kinetic energy and gravitational potential energy. Energy also can be stored within the matter making up the system. Energy can be *transferred* between a system and its surroundings by *work, heat transfer,* and the *flow* of hot or cold streams of matter. Energy also can be *converted* from one form to another. For example, energy stored in the chemical bonds of fuels can be converted to electrical or mechanical power in fuel cells and internal combustion engines.

The sunflowers shown on the cover of this book can be thought of as thermal systems. Solar energy aids the production of chemical substances within the plant required for life (*photosynthesis*). Plants also draw in water and nutrients through their root system. Plants interact with their environments in other ways as well.

Selected areas of application that involve the engineering of thermal systems are listed in Fig. 1.1, along with six specific illustrations. The *turbojet engine, jet ski,* and *electrical power plant* represent thermal systems involving conversion of energy in fossil fuels to achieve a desired outcome. Components of these systems also involve work and heat transfer. For life support on the *International Space Station,* solar energy is converted to electrical energy and provides energy for plant growth experimentation and other purposes. Semiconductor manufacturing processes such as *high temperature annealing of silicon wafers* involve energy conversion and significant heat transfer effects. The *human cardiovascular*

Prime movers: internal-combustion engines, turbines
Fluid machinery: pumps, compressors
Fossil- and nuclear-fueled power stations
Alternative energy systems
  Fuel cells
  Solar heating, cooling and power generation
Heating, ventilating, and air-conditioning equipment
Biomedical applications
  Life support and surgical equipment
  Artificial organs
Air and water pollution control equipment
Aerodynamics: airplanes, automobiles, buildings
Pipe flow: distribution networks, chemical plants
Cooling of electronic equipment
Materials processing: metals, plastics, semiconductors
Manufacturing: machining, joining, laser cutting
Thermal control of spacecraft

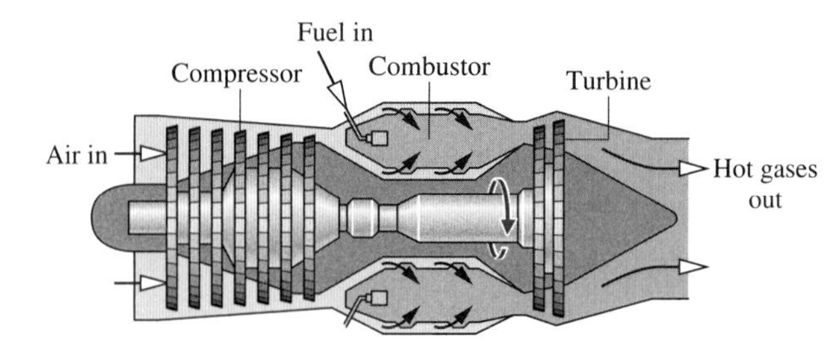

Turbojet engine

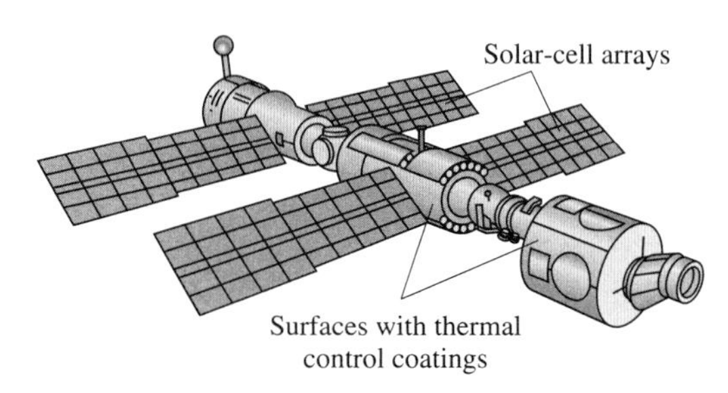

International Space Station

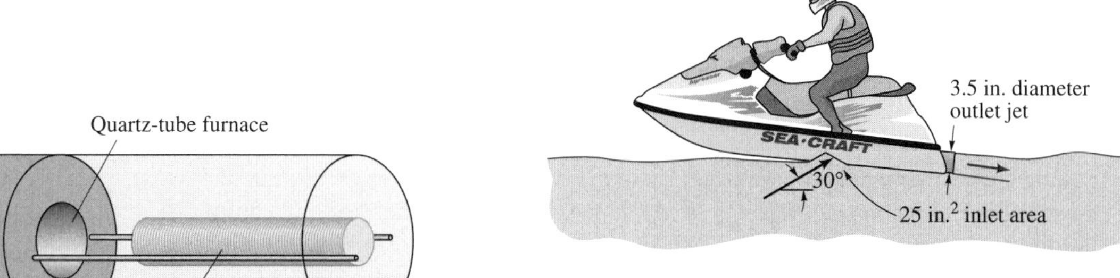

Jet ski water =-pump propulsion

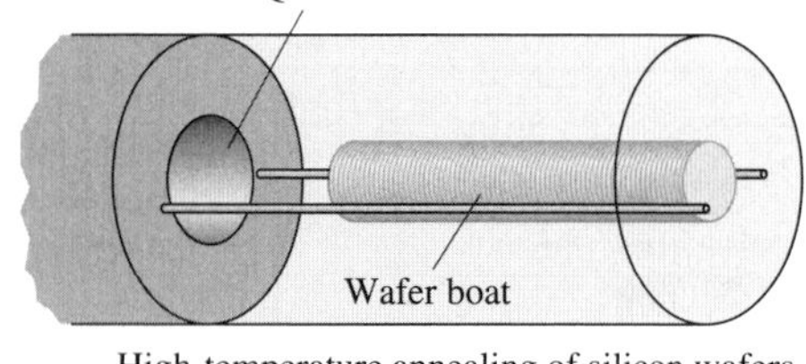

High-temperature annealing of silicon wafers

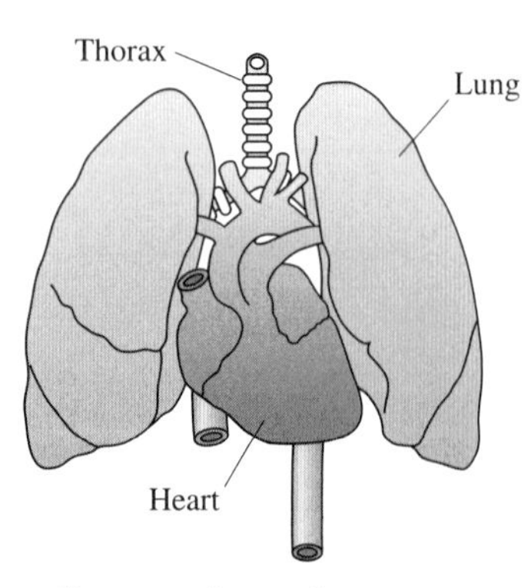

Human cardiovascular system

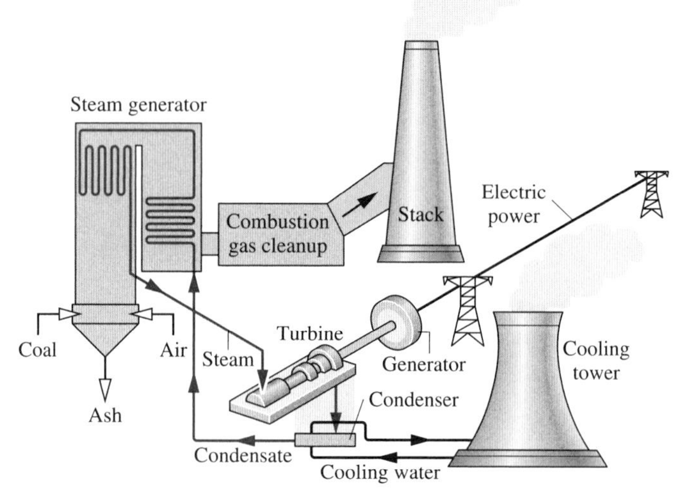

Electrical power plant

*Figure 1.1* Selected areas of applications for thermal systems engineering.

*system* is a complex combination of fluid flow and heat transfer components that regulates the flow of blood and air to within the relatively narrow range of conditions required to maintain life.

In the next section, three case studies are discussed that bring out important features of thermal systems engineering. The case studies also suggest the breadth of this field.

## 1.2 Thermal System Case Studies

Three cases are now considered to provide you with background for your study of thermal systems engineering. In each case, the message is the same: Thermal systems typically consist of a combination of components that function together as a whole. The components themselves and the overall system can be analyzed using principles drawn from three disciplines: thermodynamics, fluid mechanics, and heat transfer. The nature of an analysis depends on what needs to be understood to evaluate system performance or to design or upgrade a system. Engineers who perform such work need to learn thermal systems principles and how they are applied in different situations.

### 1.2.1 Domestic Hot Water Supply

The installation that provides hot water for your shower is an everyday example of a thermal system. As illustrated schematically in Fig. 1.2*a*, a typical system includes:

- a water supply
- a hot-water heater
- hot-water and cold-water delivery pipes
- a faucet and a shower head

The function of the system is to deliver a water stream with the desired flow rate and temperature.

Clearly the temperature of the water changes from when it enters your house until it exits the shower head. Cold water enters from the supply pipe with a pressure greater than the atmosphere, at low velocity and an elevation below ground level. Water exits the shower head at atmospheric pressure, with higher velocity and elevation, and it is comfortably hot. The increase in temperature from inlet to outlet depends on energy added to the water by heating elements (electrical or gas) in the hot water heater. The energy added can be evaluated using principles from thermodynamics and heat transfer. The relationships among the values of pressure, velocity, and elevation are affected by the pipe sizes, pipe lengths, and the types of fittings used. Such relationships can be evaluated using fluid mechanics principles.

Water heaters are designed to achieve appropriate heat transfer characteristics so that the energy supplied is transferred to the water in the tank rather than lost to the surrounding air. The hot water also must be maintained at the desired temperature, ready to be used on demand. Accordingly, appropriate insulation on the tank is required to reduce energy losses to the surroundings. Also required is a thermostat to call for further heating when necessary. When there are long lengths of pipe between the hot water heater and the shower head, it also may be advantageous to insulate the pipes.

The flow from the supply pipe to the shower head involves several fluid mechanics principles. The pipe diameter must be sized to provide the proper flow rate—too small a diameter and there will not be enough water for a comfortable shower; too large a diameter and the material costs will be too high. The flow rate also depends on the length of the pipes and

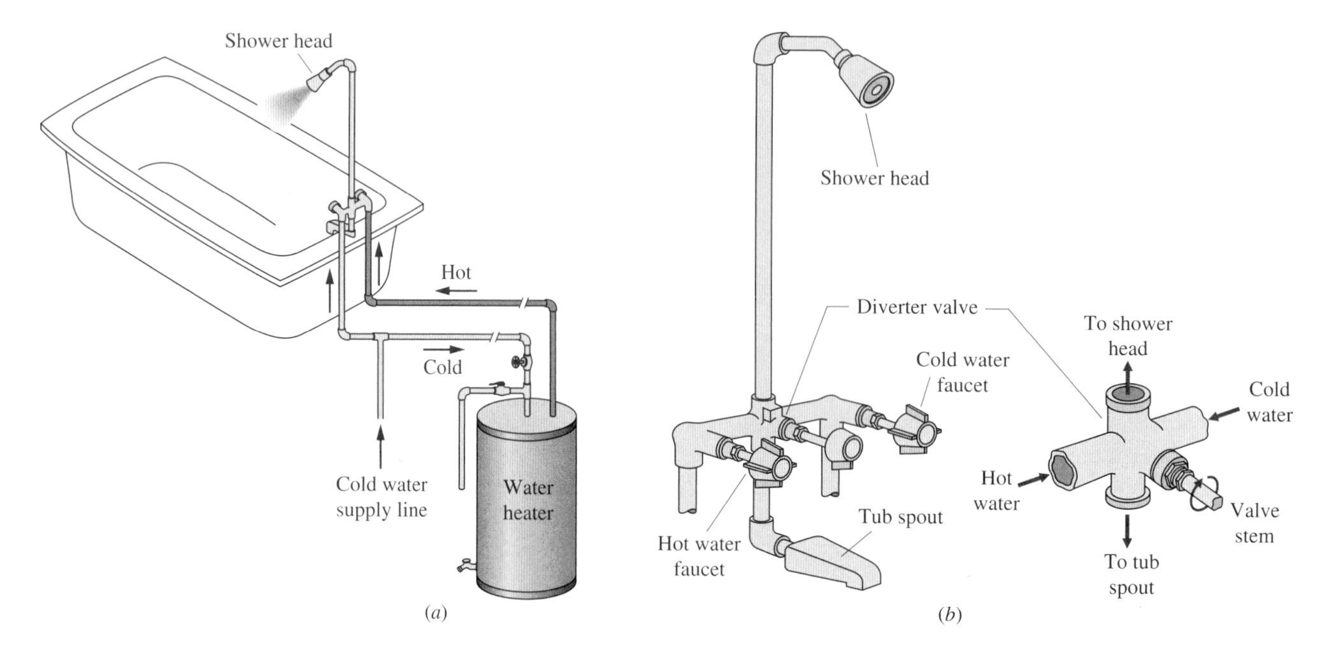


*Figure 1.2* Home hot water supply. (*a*) Overview. (*b*) Faucet and shower head.

the number of valves, elbows, and other fittings required. As shown in Fig. 1.2*b*, the faucet and the shower head must be designed to provide the desired flow rate while mixing hot and cold water appropriately.

From this example we see some important ideas relating to the analysis and design of thermal systems. The everyday system that delivers hot water for your shower is composed of various components. Yet their individual features and the way they work together as a whole involve a broad spectrum of thermodynamics, fluid mechanics, and heat transfer principles.

### 1.2.2 Hybrid Electric Vehicle

Automobile manufacturers are producing hybrid cars that utilize two or more sources of power within a single vehicle to achieve fuel economy up to 60–70 miles per gallon. Illustrated in Fig. 1.3*a* is a *hybrid electric vehicle* (HEV) that combines a gasoline-fueled engine with a set of batteries that power an electric motor. The gasoline engine and the electric motor are each connected to the transmission and are capable of running the car by themselves or in combination depending on which is more effective in powering the vehicle. What makes this type of hybrid particularly fuel efficient is the inclusion of several features in the design:

- the ability to recover energy during braking and to store it in the electric batteries,
- the ability to shut off the gasoline engine when stopped in traffic and meet power needs by the battery alone,
- special design to reduce *aerodynamic drag* and the use of tires that have very low rolling resistance (friction), and
- the use of lightweight composite materials such as carbon fiber and the increased use of lightweight metals such as aluminum and magnesium.

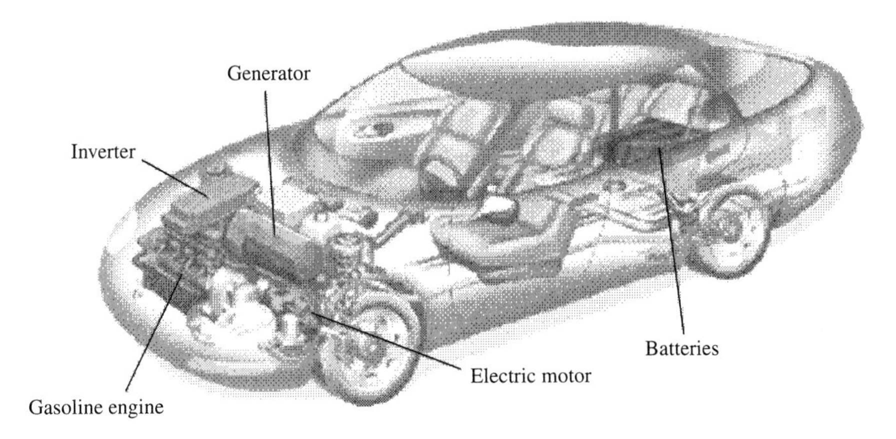


(*a*) Overview of the vehicle showing key thermal systems

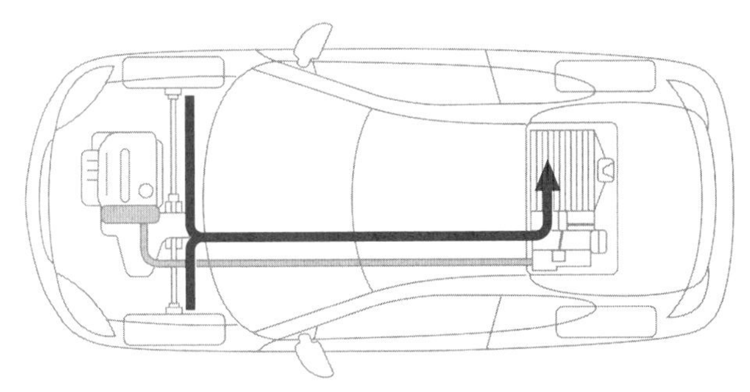

(*b*) Regenerative braking mode with energy flow from wheels to battery

*Figure 1.3* Hybrid electric vehicle combining gasoline-fueled engine, storage batteries, and electric motor. (Illustrations by George Retseck.)

The energy source for such hybrid vehicles is gasoline burned in the engine. Because of the ability to store energy in the batteries and use that energy to run the electric motor, the gasoline engine does not have to operate continuously. Some HEVs use only the electric motor to accelerate from rest up to about 15 miles per hour, and then switch to the gasoline engine. A specially designed transmission provides the optimal power split between the gasoline engine and the electric motor to keep the fuel use to a minimum and still provide the needed power.

Most HEVs use *regenerative braking*, as shown in Fig. 1.3*b*. In conventional cars, stepping on the brakes to slow down or stop dissipates the kinetic energy of motion through the frictional action of the brake. Starting again requires fuel to re-establish the kinetic energy of the vehicle. The hybrid car allows *some* of the kinetic energy to be converted during braking to electricity that is stored in the batteries. This is accomplished by the electric motor serving as a generator during the braking process. The net result is a significant improvement in fuel economy and the ability to use a smaller-sized gasoline engine than would be possible to achieve comparable performance in a conventional vehicle.

The overall energy notions considered thus far are important aspects of thermodynamics, which deals with *energy conversion*, *energy accounting*, and the *limitations* on how energy is converted from one form to another. In addition, there are numerous examples of fluid mechanics and heat transfer applications in a hybrid vehicle. Within the engine, air, fuel, engine coolant, and oil are circulated through passageways, hoses, ducts, and manifolds. These must be designed to ensure that adequate flow is obtained. The fuel pump and water pump also must be designed to achieve the desired fluid flows. Heat transfer principles guide the design of the cooling system, the braking system, the lubrication system, and numerous other aspects of the vehicle. Coolant circulating through passageways in the engine block must absorb energy transferred from hot combustion gases to the cylinder surfaces so those surfaces do not become too hot. Engine oil and other viscous fluids in the transmission and braking systems also can reach high temperatures and thus must be carefully managed.

Hybrid electric vehicles provide examples of complex thermal systems. As in the case of hot water systems, the principles of thermodynamics, fluid mechanics, and heat transfer apply to the analysis and design of individual parts, components, and to the entire vehicle.

### 1.2.3 Microelectronics Manufacturing: Soldering Printed-Circuit Boards

Printed-circuit boards (PCBs) found in computers, cell phones, and many other products, are composed of integrated circuits and electronic devices mounted on epoxy-filled fiberglass boards. The boards have been metallized to provide interconnections, as illustrated in Fig. 1.4*a*. The pins of the integrated circuits and electronic devices are fitted into holes, and a droplet of powdered solder and flux in paste form is applied to the pin-pad region, Fig. 1.4*b*. To achieve reliable mechanical and electrical connections, the PCB is heated in an oven to a temperature above the solder melting temperature; this is known as the *reflow* process. The

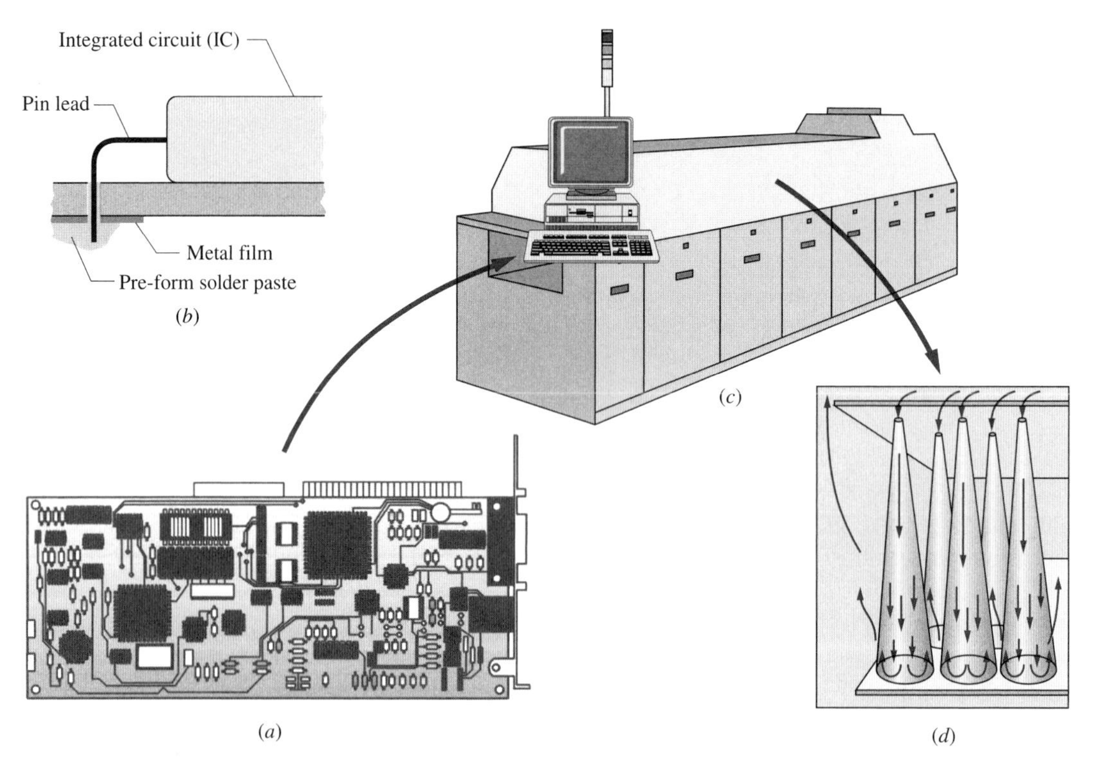


*Figure 1.4* Soldering printed-circuit boards (*a*) with pre-form solder paste applied to integrated circuit pins and terminal pads (*b*) enter the solder-reflow oven (*c*) on a conveyor and are heated to the solder melting temperature by impinging hot air jets (*d*).

PCB and its components must be gradually and uniformly heated to avoid inducing thermal stresses and localized overheating. The PCB is then cooled to near-room temperature for subsequent safe handling.

The PCB prepared for soldering is placed on a conveyor belt and enters the first zone within the solder reflow oven, Fig. 1.4*c*. In passing through this zone, the temperature of the PCB is increased by exposure to hot air jets heated by electrical resistance elements, Fig. 1.4*d*. In the final zone of the oven, the PCB passes through a cooling section where its temperature is reduced by exposure to air that has been cooled by passing through a water-cooled heat exchanger.

From the foregoing discussion, we recognize that there are many aspects of this manufacturing process that involve electric power, flow of fluids, air-handling equipment, heat transfer, and thermal aspects of material behavior. In thermal systems engineering, we perform analyses on *systems* such as the solder-reflow oven to evaluate system performance or to design or upgrade the system. For example, suppose you were the operations manager of a factory concerned with providing electrical power and chilled water for an oven that a vendor claims will meet your requirements. What information would you ask of the vendor? Or, suppose you were the oven designer seeking to maximize the production of PCBs. You might be interested in determining what air flow patterns and heating element arrangements would allow the fastest flow of product through the oven while maintaining necessary uniformity of heating. How would you approach obtaining such information? Through your study of thermodynamics, fluid mechanics, and heat transfer you will learn how to deal with questions such as these.

## 1.3 Analysis of Thermal Systems

In this section, we introduce the basic laws that govern the analysis of thermal systems of all kinds, including the three cases considered in Sec. 1.2. We also consider further the roles of thermodynamics, fluid mechanics, and heat transfer in thermal systems engineering and their relationship to one another.

Important engineering functions are to design and analyze things intended to meet human needs. Engineering *design* is a decision-making process in which principles drawn from engineering and other fields such as economics and statistics are applied to devise a system, system component, or process. Fundamental elements of design include establishing objectives, analysis, synthesis, construction, testing, and evaluation.

Engineering *analysis* frequently aims at developing an *engineering model* to obtain a simplified mathematical representation of system behavior that is sufficiently faithful to reality, even if some aspects exhibited by the actual system are not considered. For example, idealizations often used in mechanics to simplify an analysis include the assumptions of point masses, frictionless pulleys, and rigid beams. Satisfactory modeling takes experience and is a part of the *art* of engineering. Engineering analysis is featured in this book.

The first step in analysis is the identification of the system and how it interacts with its surroundings. Attention then turns to the pertinent *physical laws* and relationships that allow system behavior to be described. Analysis of thermal systems uses, directly or indirectly, one or more of four basic laws:

- *Conservation of mass*
- *Conservation of energy*
- *Conservation of momentum*
- *Second law of thermodynamics*

In your earlier studies in physics and chemistry, you were introduced to these laws. In this book, we place the laws in forms especially well suited for use in thermal systems engineering and help you learn how to apply them.

### 1.3.1 The Three Thermal Science Disciplines

As we have observed, thermal systems engineering typically requires the use of three thermal science disciplines: thermodynamics, fluid mechanics, and heat transfer. Figure 1.5 shows the roles of these disciplines in thermal system engineering and their relationship to one another. Associated with each discipline is a list of principles featured in the part of the book devoted to that discipline.

*Thermodynamics* provides the foundation for analysis of thermal systems through the conservation of mass and conservation of energy principles, the second law of thermodynamics, and property relations. *Fluid mechanics* and *heat transfer* provide additional concepts, including the empirical laws necessary to specify, for instance, material choices, component sizing, and fluid medium characteristics. For example, thermodynamic analysis can tell you the final temperature of a hot workpiece quenched in an oil, but the *rate* at which it will cool is predicted using a heat transfer analysis.

*Fluid mechanics* is concerned with the behavior of fluids at rest or in motion. As shown in Fig. 1.5, two fundamentals that play central roles in our discussion of fluid mechanics are the *conservation of momentum principle* that stems from Newton's second law of motion and the *mechanical energy equation*. Principles of fluid mechanics allow the study of fluids flowing inside pipes (internal flows) and over surfaces (external flows) with consideration of frictional

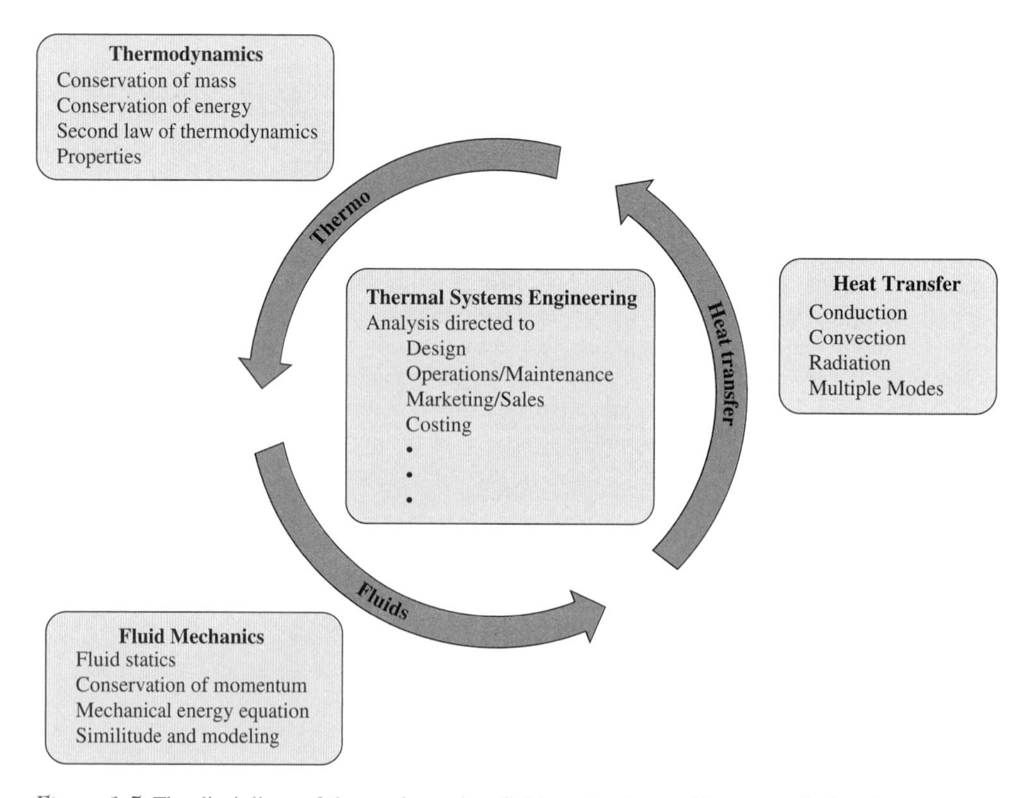


*Figure 1.5* The disciplines of thermodynamics, fluid mechanics, and heat transfer involve fundamentals and principles essential for the practice of thermal systems engineering.

effects and lift/drag forces. The concept of *similitude* is used extensively in scaling measurements on laboratory-sized *models* to full-scale systems.

*Heat transfer* is concerned with energy transfer as a consequence of a temperature difference. As shown in Fig. 1.5, there are three *modes* of heat transfer. *Conduction* refers to heat transfer through a medium across which a temperature difference exists. *Convection* refers to heat transfer between a surface and a moving or still fluid having a different temperature. The third mode of heat transfer is termed thermal *radiation* and represents the net exchange of energy between surfaces at different temperatures by electromagnetic waves independent of any intervening medium. For these modes, the heat transfer rates depend on the *transport properties* of substances, geometrical parameters, and temperatures. Many applications involve more than one of these modes; this is called *multimode* heat transfer.

Returning again to Fig. 1.5, in the thermal systems engineering box we have identified some application areas involving analysis. Earlier we mentioned that *design* requires analysis. Engineers also perform analysis for many other reasons, as for example in the *operation* of systems and determining when systems require *maintenance*. Because of the complexity of many thermal systems, engineers who provide *marketing* and *sales* services need analysis skills to determine whether their product will meet a customer's specifications. As engineers, we are always challenged to optimize the use of financial resources, which frequently requires *costing* analyses to justify our recommendations.

### 1.3.2 The Practice of Thermal Systems Engineering

Seldom do practical applications involve only one aspect of the three thermal sciences disciplines. Practicing engineers usually are required to combine the basic concepts, laws, and principles. Accordingly, as you proceed through this text, you should recognize that thermodynamics, fluid mechanics, and heat transfer provide powerful analysis tools that are complementary. Thermal systems engineering is interdisciplinary in nature, not only for this reason, but because of ties to other important issues such as controls, manufacturing, vibration, and materials that are likely to be present in real-world situations.

Thermal systems engineering not only has played an important role in the development of a wide range of products and services that touch our lives daily, it also has become an enabling technology for evolving fields such as nanotechnology, biotechnology, food processing, health services, and bioengineering. This textbook will prepare you to work in both traditional and emerging energy-related fields.

Your background should enable you to

- contribute to teams working on thermal systems applications.
- specify equipment to meet prescribed needs.
- implement energy policy.
- perform economic assessments involving energy.
- manage technical operations.

This textbook also will prepare you for further study in thermodynamics, fluid mechanics, and heat transfer to strengthen your understanding of fundamentals and to acquire more experience in model building and solving applications-driven problems.

## 1.4 How to Use This Book Effectively

This book has several features and learning resources that facilitate study and contribute further to understanding.

## Core Study Features

***Examples and Problems . . .***

- Numerous annotated solved examples are provided that feature the *solution methodology* presented in Sec. 2.6, and illustrated initially in Example 2.1. We encourage you to study these examples, including the accompanying comments.
- Less formal examples are given throughout the text. They open with the words ***For Example…*** and close with the symbol ▲. These examples also should be studied.
- A large number of end-of-chapter problems are provided. The problems are sequenced to coordinate with the subject matter and are listed in increasing order of difficulty. The problems are classified under headings to expedite the process of selecting review problems to solve.

***Other Study Aids . . .***

- Each chapter begins with an introduction stating the chapter objective and concludes with a summary and study guide.
- Key words are listed in the margins and coordinated with the text material at those locations.
- Key equations are set off by a double horizontal bar.

- *Methodology Update* in the margin identifies where we refine our problem-solving methodology, introduce conventions, or sharpen our understanding of specific concepts.
- For quick reference, conversion factors and important constants are provided on the inside front cover and facing page.
- A list of symbols is provided on the inside back cover and facing page.
- (CD-ROM) directs you to the accompanying CD where *supplemental* text material and learning resources are provided.

***Icons . . .***

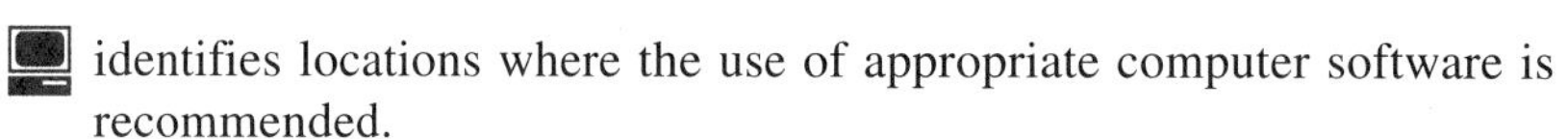

identifies locations where the use of appropriate computer software is recommended.

directs you to short fluid mechanics video segments.

## Enhanced Study Features

***Computer Software . . .***

To allow you to retrieve appropriate data electronically and model and solve complex thermal engineering problems, instructional material and computer-type problems are provided on the CD for *Interactive Thermodynamics (IT)* and *Interactive Heat Transfer (IHT)*. These programs are built around equation solvers enhanced with property data and other valuable features. With the IT and IHT software you can obtain a single numerical solution or vary parameters to investigate their effects. You also can obtain graphical output, and the Windows-based format allows you to use any Windows word-processing software or spreadsheet to generate reports. Tutorials are available from the 'Help' menu, and both programs include several worked examples.

***Accompanying CD . . .***

The CD contains the entire print version of the book plus the following additional content and resources:

- answers to selected end-of-chapter problems
- additional text material not included in the print version of the book

- the computer software *Interactive Thermodynamics (IT)* and *Interactive Heat Transfer (IHT)*, including a directory entitled *Things You Should Know About IT and IHT* that contains helpful information for using the software with this book.
- short video segments that illustrate fluid mechanics principles
- built-in hyperlinks to show connections between topics

***Special Note:*** Content provided on the CD may involve equations, figures, and examples that are not included in the print version of the book.

## Problems

**1.1** List thermal systems that you might encounter in everyday activities such as cooking, heating or cooling a house, and operating an automobile.

**1.2** Using the Internet, obtain information about the operation of a thermal system of your choice or one of those listed or shown in Fig. 1.1. Obtain sufficient information to provide a description to your class on the function of the system and relevant thermodynamics, fluid mechanics, and heat transfer aspects.

**1.3** Referring to the thermal systems of Fig. 1.1, in cases assigned by your instructor or selected by you, explain how energy is *converted* from one form to another and how energy is *stored*.

**1.4** Consider a rocket leaving its launch pad. Briefly discuss the conversion of energy stored in the rocket's fuel tanks into other forms as the rocket lifts off.

*Figure P1.4*

**1.5** Referring to the U.S. patent office Website, obtain a copy of a patent granted in the last five years for a thermal system. Describe the function of the thermal system and explain the claims presented in the patent that relate to thermodynamics, fluid mechanics, and heat transfer.

**1.6** Contact your local utility for the amount you pay for electricity, in cents per kilowatt-hour. What are the major contributors to this cost?

**1.7** A newspaper article lists solar, wind, hydroelectric, geothermal, and biomass as important *renewable* energy resources. What is meant by renewable? List some energy resources that are *not* considered renewable.

**1.8** Reconsider the energy resources of Problem 1.7. Give specific examples of how each is used to meet human needs.

**1.9** Our energy needs are met today primarily by use of *fossil fuels*. What fossil fuels are most commonly used for (a) transportation, (b) home heating, and (c) electricity generation?

**1.10** List some of the roles that coal, natural gas, and petroleum play in our lives. In a memorandum, discuss environmental, political, and social concerns regarding the continued use of these *fossil* fuels. Repeat for *nuclear* energy.

**1.11** A utility advertises that it is less expensive to heat water for domestic use with natural gas than with electricity. Determine if this claim is correct in your locale. What issues determine the relative costs?

**1.12** A news report speaks of *greenhouse* gases. What is meant by greenhouse in this context? What are some of the most prevalent greenhouse gases and why have many observers expressed concern about those gases being emitted into the atmosphere?

**1.13** Consider the following household appliances: desktop computer, toaster, and hair dryer. For each, what is its function and what is the typical power requirement, in Watts? Can it be considered a thermal system? Explain.

*Figure P1.13*

**1.14** A person adjusts the faucet of a shower as shown in Figure P1.14 to a desired water temperature. Part way through the shower the dishwasher in the kitchen is turned on and the temperature of the shower becomes too cold. Why?

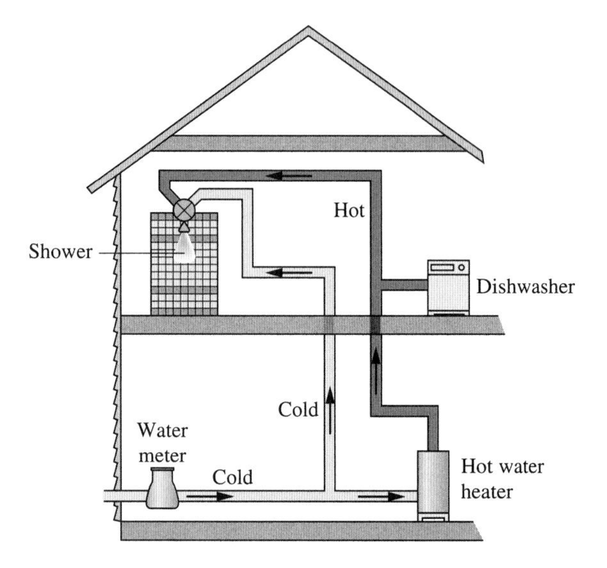


*Figure P1.14*

**1.15** The everyday operation of your car involves the use of various gases or liquids. Make a list of such fluids and indicate how they are used in your car.

**1.16** Your car contains various fans or pumps, including the radiator fan, the heater fan, the water pump, the power steering pump, and the windshield washer pump. Obtain approximate values for the power (horsepower or kilowatts) required to operate each of these fans or pumps.

**1.17** When a hybrid electric vehicle such as the one described in Section 1.2.2 is braked to rest, *only a fraction* of the vehicle's kinetic energy is stored chemically in the batteries. Why only a *fraction*?

**1.18** Discuss how a person's driving habits would affect the fuel economy of an automobile in stop-and-go traffic and on a freeway.

**1.19** The solder-reflow oven considered in Section 1.2.3 operates with the *conveyer speed* and *hot air supply parameters* adjusted so that the PCB soldering process is performed slightly above the solder melting temperature as required for quality joints. The PCB also is cooled to a *safe temperature* by the time it reaches the oven exit. The operations manager wants to increase the rate per unit time that PCBs pass through the oven. How might this be accomplished?

**1.20** In the discussion of the soldering process in Section 1.2.3, we introduced the requirement that the PCB and its components be *gradually* and *uniformly* heated to avoid thermal stresses and localized overheating. Give examples from your personal experience where detrimental effects have been caused to objects heated too rapidly, or very nonuniformly.

**1.21** Automobile designers have worked to reduce the aerodynamic drag and rolling resistance of cars, thereby increasing the fuel economy, especially at highway speeds. Compare the sketch of the 1920s car shown in Figure P1.21 with the appearance of present-day automobiles. Discuss any differences that have contributed to the increased fuel economy of modern cars.

*Figure P1.21*

**1.22** Considering the hot water supply, hybrid electric vehicle, and solder-reflow applications of Sec. 1.2; give examples of conduction, convection, and radiation modes of heat transfer.

**1.23** A central furnace or air conditioner in a building uses a fan to distribute air through a duct system to each room as shown in Fig. P1.23. List some reasons why the temperatures might vary significantly from room to room, even though each room is provided with conditioned air.

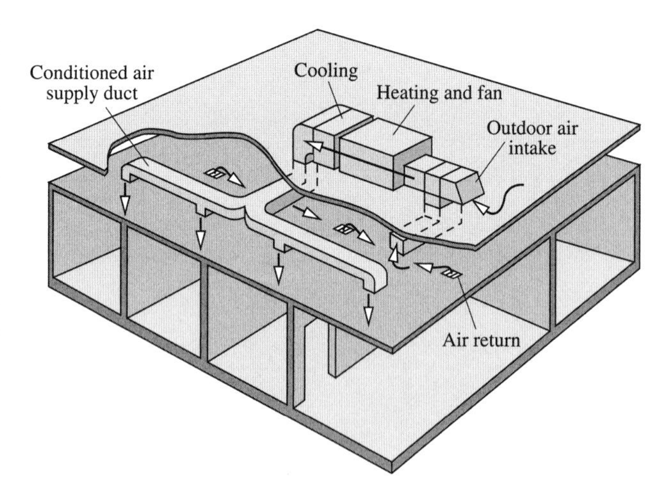


*Figure P1.23*

**1.24** Figure P1.24 shows a wind turbine-electric generator mounted atop a tower. Wind blows steadily across the turbine blades, and electricity is generated. The electrical output of the generator is fed to a storage battery. For the overall thermal system consisting of the wind-turbine generator and storage battery, list the sequence of processes that convert the energy of the wind to energy stored in the battery.

Figure P1.24

**1.25** A plastic workpiece in the form of a thin, square, flat plate removed from a hot injection molding press at 150°C must be cooled to a safe-to-handle temperature. Figure P1.25 shows two arrangements for the cooling process: The workpiece is suspended vertically from an overhead support, or positioned horizontally on a wire rack, each in the presence of ambient air. Calling on your experience and physical intuition, answer the following:

**(a)** Will the workpiece cool more quickly in the vertical or horizontal arrangement if the only air motion that occurs is due to buoyancy of the air near the hot surfaces of the workpiece (referred to as *free* or *natural* convection)?

**(b)** If a fan blows air over the workpiece (referred to as *forced* convection), would you expect the cooling rate to increase or decrease? Why?

**1.26** An automobile engine normally has a coolant circulating through passageways in the engine block and then through a finned-tube *radiator*. Lawn mower engines normally have finned surfaces directly attached to the engine block, with no radiator, in order to achieve the required cooling. Why might the cooling strategies be different in these two applications?

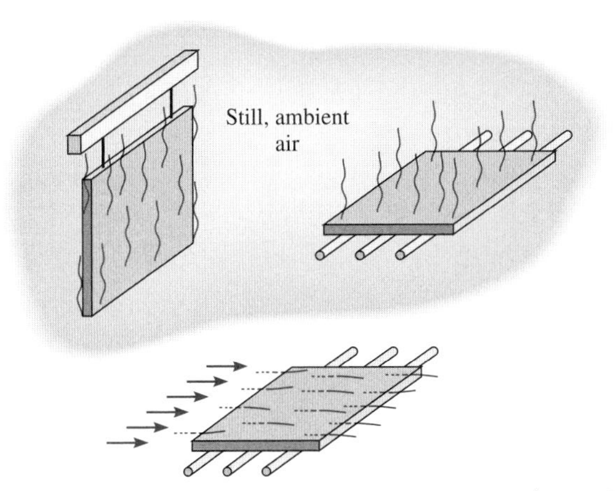


Figure P1.25

Figure P1.26

# GETTING STARTED IN THERMODYNAMICS: INTRODUCTORY CONCEPTS AND DEFINITIONS

## *Introduction...*

The word thermodynamics stems from the Greek words *therme* (heat) and *dynamis* (force). Although various aspects of what is now known as thermodynamics have been of interest since antiquity, the formal study of thermodynamics began in the early nineteenth century through consideration of the motive power of *heat:* the capacity of hot bodies to produce *work.* Today the scope is larger, dealing generally with *energy* and with relationships among the *properties* of matter.

*chapter objective*

The ***objective*** of this chapter is to introduce you to some of the fundamental concepts and definitions that are used in our study of thermodynamics. In most instances the introduction is brief, and further elaboration is provided in subsequent chapters.

## 2.1 Defining Systems

An important step in any engineering analysis is to describe precisely what is being studied. In mechanics, if the motion of a body is to be determined, normally the first step is to define a *free body* and identify all the forces exerted on it by other bodies. Newton's second law of motion is then applied. In thermal systems engineering, the term *system* is used to identify the subject of the analysis. Once the system is defined and the relevant interactions with other systems are identified, one or more physical laws or relations are applied.

*system*

The ***system*** is whatever we want to study. It may be as simple as a free body or as complex as an entire chemical refinery. We may want to study a quantity of matter contained within a closed, rigid-walled tank, or we may want to consider something such as a pipeline through which natural gas flows. The composition of the matter inside the system may be fixed or may be changing through chemical or nuclear reactions. The shape or volume of the system being analyzed is not necessarily constant, as when a gas in a cylinder is compressed by a piston or a balloon is inflated.

*surroundings*
*boundary*

Everything external to the system is considered to be part of the system's ***surroundings.*** The system is distinguished from its surroundings by a specified ***boundary,*** which may be at rest or in motion. You will see that the interactions between a system and its surroundings, which take place across the boundary, play an important part in thermal systems engineering. It is essential for the boundary to be delineated carefully before proceeding with an analysis. However, the same physical phenomena often can be analyzed in terms of alternative choices of the system, boundary, and surroundings. The choice of a particular boundary defining a particular system is governed by the convenience it allows in the subsequent analysis.

### Types of Systems

Two basic kinds of systems are distinguished in this book. These are referred to, respectively, as *closed systems* and *control volumes.* A closed system refers to a fixed quantity of matter, whereas a control volume is a region of space through which mass may flow.

*closed system*

A ***closed system*** is defined when a particular quantity of matter is under study. A closed system always contains the same matter. There can be no transfer of mass across its boundary. A special type of closed system that does not interact in any way with its surroundings is called an ***isolated system.***

*isolated system*

Figure 2.1 shows a gas in a piston–cylinder assembly. When the valves are closed, we can consider the gas to be a closed system. The boundary lies just inside the piston and cylinder walls, as shown by the dashed lines on the figure. The portion of the boundary between the gas and the piston moves with the piston. No mass would cross this or any other part of the boundary.

In subsequent sections of this book, analyses are made of devices such as turbines and pumps through which mass flows. These analyses can be conducted in principle by studying a particular quantity of matter, a closed system, as it passes through the device. In most cases it is simpler to think instead in terms of a given region of space through which mass flows. With this approach, a *region* within a prescribed boundary is studied. The region is called a ***control volume.*** Mass may cross the boundary of a control volume.

*control volume*

A diagram of an engine is shown in Fig. 2.2*a*. The dashed line defines a control volume that surrounds the engine. Observe that air, fuel, and exhaust gases cross the boundary. A schematic such as in Fig. 2.2*b* often suffices for engineering analysis.

The term *control mass* is sometimes used in place of closed system, and the term *open system* is used interchangeably with control volume. When the terms control mass and control volume are used, the system boundary is often referred to as a *control surface.*

In general, the choice of system boundary is governed by two considerations: (1) what is known about a possible system, particularly at its boundaries, and (2) the objective of the analysis.

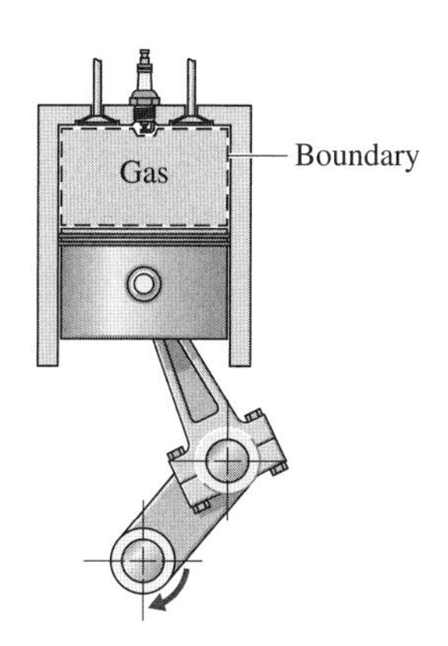


*Figure 2.1* Closed system: A gas in a piston–cylinder assembly.

***For Example...*** Figure 2.3 shows a sketch of an air compressor connected to a storage tank. The system boundary shown on the figure encloses the compressor, tank, and all of the piping. This boundary might be selected if the electrical power input were known, and the objective of the analysis were to determine how long the compressor must operate for the pressure in the tank to rise to a specified value. Since mass crosses the boundary, the system would be a control volume. A control volume enclosing only the compressor might be chosen if the condition of the air entering and exiting the compressor were known, and the objective were to determine the electric power input. ▲

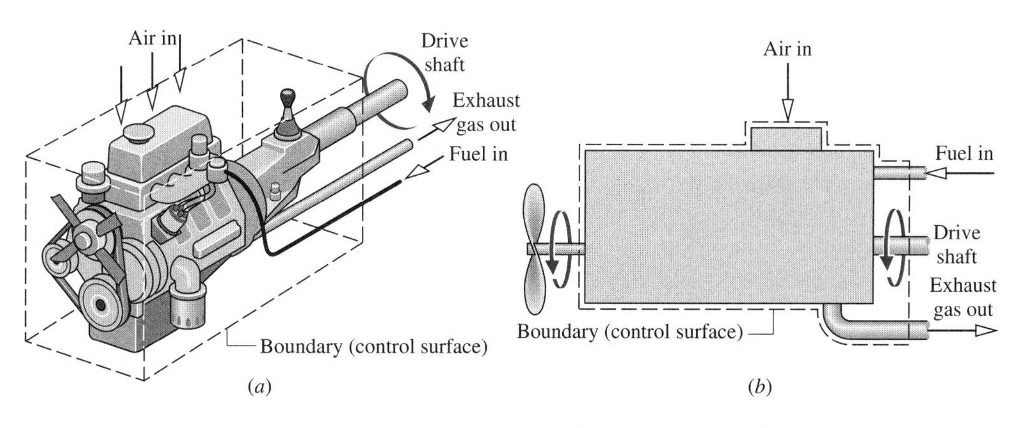


*Figure 2.2* Example of a control volume (open system): An automobile engine.

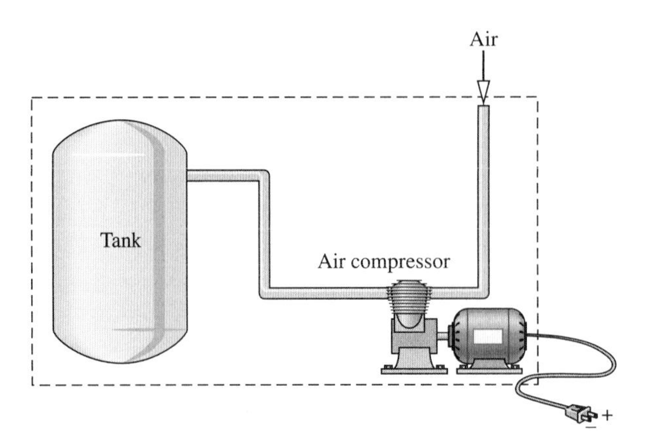


*Figure 2.3* Air compressor and storage tank.

## 2.2 Describing Systems and Their Behavior

Engineers are interested in studying systems and how they interact with their surroundings. In this section, we introduce several terms and concepts used to describe systems and how they behave.

### Macroscopic and Microscopic Approaches

Systems can be studied from a *macroscopic* or a *microscopic* point of view. The macroscopic approach is concerned with the gross or overall behavior of matter. No model of the structure of matter at the molecular, atomic, and subatomic levels is directly used. Although the behavior of systems *is* affected by molecular structure, the macroscopic approach allows important aspects of system behavior to be evaluated from observations of the overall system. The microscopic approach is concerned directly with the structure of matter. The objective is to characterize by statistical means the average behavior of the particles making up a system of interest and relate this information to the observed macroscopic behavior of the system. For the great majority of thermal systems applications, the macroscopic approach not only provides a more direct means for analysis and design but also requires far fewer mathematical complications. For these reasons the macroscopic approach is the one adopted in this book.

### Property, State, and Process

*property*

To describe a system and predict its behavior requires knowledge of its properties and how those properties are related. A ***property*** is a macroscopic characteristic of a system such as mass, volume, energy, pressure, and temperature to which a numerical value can be assigned at a given time without knowledge of the previous behavior (*history*) of the system. Many other properties are considered during the course of our study.

*state*

The word ***state*** refers to the condition of a system as described by its properties. Since there are normally relations among the properties of a system, the state often can be specified by providing the values of a subset of the properties. All other properties can be determined in terms of these few.

*process*

*steady state*

*thermodynamic cycle*

When any of the properties of a system change, the state changes and the system is said to have undergone a ***process***. A process is a transformation from one state to another. However, if a system exhibits the same values of its properties at two different times, it is in the same state at these times. A system is said to be at ***steady state*** if none of its properties changes with time.

A ***thermodynamic cycle*** is a sequence of processes that begins and ends at the same state. At the conclusion of a cycle all properties have the same values they had at the beginning.

Consequently, over the cycle the system experiences no *net* change of state. Cycles that are repeated periodically play prominent roles in many areas of application. For example, steam circulating through an electrical power plant executes a cycle.

At a given state each property has a definite value that can be assigned without knowledge of how the system arrived at that state. Therefore, the change in value of a property as the system is altered from one state to another is determined solely by the two end states and is independent of the particular way the change of state occurred. That is, the change is independent of the details of the process. It follows that if the value of a particular quantity depends on the details of the process, and not solely on the end states, that quantity cannot be a property.

## Extensive and Intensive Properties

Thermodynamic properties can be placed in two general classes: extensive and intensive. A property is called ***extensive*** if its value for an overall system is the sum of its values for the parts into which the system is divided. Mass, volume, energy, and several other properties introduced later are extensive. Extensive properties depend on the size or extent of a system. The extensive properties of a system can change with time,

*extensive property*

***Intensive*** properties are not additive in the sense previously considered. Their values are independent of the size or extent of a system and may vary from place to place within the system at any moment. Thus, intensive properties may be functions of both position and time, whereas extensive properties vary at most with time. Specific volume (Sec. 2.4.1), pressure, and temperature are important intensive properties; several other intensive properties are introduced in subsequent chapters.

*intensive property*

***For Example...*** to illustrate the difference between extensive and intensive properties, consider an amount of matter that is uniform in temperature, and imagine that it is composed of several parts, as illustrated in Fig. 2.4. The mass of the whole is the sum of the masses of the parts, and the overall volume is the sum of the volumes of the parts. However, the temperature of the whole is not the sum of the temperatures of the parts; it is the same for each part. Mass and volume are extensive, but temperature is intensive. ▲

## Phase and Pure Substance

The term ***phase*** refers to a quantity of matter that is homogeneous throughout in both chemical composition and physical structure. Homogeneity in physical structure means that the matter is all *solid,* or all *liquid,* or all *vapor* (or equivalently all *gas*). A system can contain one or more phases. For example, a system of liquid water and water vapor (steam) contains *two* phases. When more than one phase is present, the phases are separated by *phase boundaries.*

*phase*

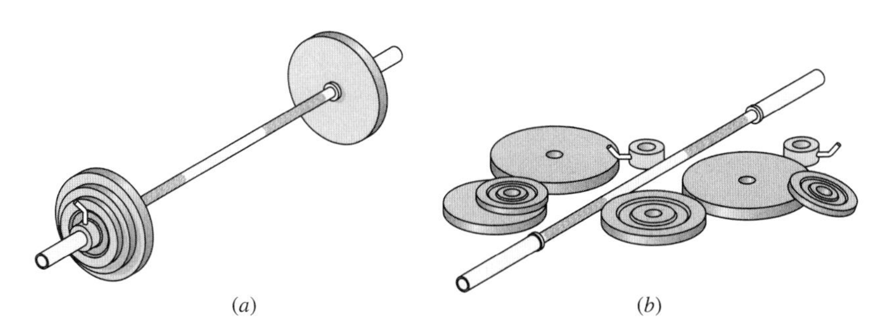


*Figure 2.4* Figure used to discuss the extensive property concept.

*pure substance*

A ***pure substance*** is one that is uniform and invariable in chemical composition. A pure substance can exist in more than one phase, but its chemical composition must be the same in each phase. For example, if liquid water and water vapor form a system with two phases, the system can be regarded as a pure substance because each phase has the same composition. A uniform mixture of gases can be regarded as a pure substance provided it remains a gas and does not react chemically.

## Equilibrium

*equilibrium*

Thermodynamics places primary emphasis on equilibrium states and changes from one equilibrium state to another. Thus, the concept of ***equilibrium*** is fundamental. In mechanics, equilibrium means a condition of balance maintained by an equality of opposing forces. In thermodynamics, the concept is more far-reaching, including not only a balance of forces but also a balance of other influences. Each kind of influence refers to a particular aspect of thermodynamic, or complete, equilibrium. Accordingly, several types of equilibrium must exist individually to fulfill the condition of complete equilibrium; among these are mechanical, thermal, phase, and chemical equilibrium.

*equilibrium state*

We may think of testing to see if a system is in thermodynamic equilibrium by the following procedure: Isolate the system from its surroundings and watch for changes in its observable properties. If there are no changes, we conclude that the system was in equilibrium at the moment it was isolated. The system can be said to be at an ***equilibrium state.***

When a system is isolated, it cannot interact with its surroundings; however, its state can change as a consequence of spontaneous events occurring internally as its intensive properties, such as temperature and pressure, tend toward uniform values. When all such changes cease, the system is in equilibrium. Hence, for a system to be in equilibrium it must be a single phase or consist of a number of phases that have no tendency to change their conditions when the overall system is isolated from its surroundings. At equilibrium, temperature is uniform throughout the system. Also, pressure can be regarded as uniform throughout as long as the effect of gravity is not significant; otherwise, a pressure variation can exist, as in a vertical column of liquid.

## Actual and Quasiequilibrium Processes

There is no requirement that a system undergoing an actual process be in equilibrium *during* the process. Some or all of the intervening states may be nonequilibrium states. For many such processes we are limited to knowing the state before the process occurs and the state after the process is completed. However, even if the intervening states of the system are not known, it is often possible to evaluate certain *overall* effects that occur during the process. Examples are provided in the next chapter in the discussions of *work* and *heat*. Typically, nonequilibrium states exhibit spatial variations in intensive properties at a given time. Also, at a specified position intensive properties may vary with time, sometimes chaotically.

*quasiequilibrium process*

Processes are sometimes modeled as an idealized type of process called a ***quasiequilibrium (or quasistatic) process.*** A quasiequilibrium process is one in which the departure from thermodynamic equilibrium is at most infinitesimal. All states through which the system passes in a quasiequilibrium process may be considered equilibrium states. Because nonequilibrium effects are inevitably present during actual processes, systems of engineering interest can at best approach, but never realize, a quasiequilibrium process.

Our interest in the quasiequilibrium process concept stems mainly from two considerations: (1) Simple thermodynamic models giving at least *qualitative* information about the behavior of actual systems of interest often can be developed using the quasiequilibrium process concept. This is akin to the use of idealizations such as the point mass or the frictionless pulley in mechanics for the purpose of simplifying an analysis. (2) The quasiequilibrium process concept is instrumental in deducing relationships that exist among the properties of systems at equilibrium.

## 2.3 Units and Dimensions

When engineering calculations are performed, it is necessary to be concerned with the *units* of the physical quantities involved. A unit is any specified amount of a quantity by comparison with which any other quantity of the same kind is measured. For example, meters, centimeters, kilometers, feet, inches, and miles are all *units of length.* Seconds, minutes, and hours are alternative *time units.*

Because physical quantities are related by definitions and laws, a relatively small number of them suffice to conceive of and measure all others. These may be called *primary* (or *basic*) *dimensions.* The others may be measured in terms of the primary dimensions and are called *secondary.*

Four primary dimensions suffice in thermodynamics, fluid mechanics, and heat transfer. They are mass (M), length (L), time (t), and temperature (T). Alternatively, force (F) can be used in place of mass (M). These are known, respectively, as the **MLtT** and **FLtT** dimensional systems.

**MLtT, FLtT**

Once a set of primary dimensions is adopted, a ***base unit*** for each primary dimension is specified. Units for all other quantities are then derived in terms of the base units. Let us illustrate these ideas by first considering SI units for mass, length, time, and force, and then considering other units for these quantities commonly encountered in thermal systems engineering.

***base unit***

### 2.3.1 SI Units for Mass, Length, Time, and Force

In the present discussion we consider the SI system of units. SI is the abbreviation for Système International d'Unités (International System of Units), which is the legally accepted system in most countries. The conventions of the SI are published and controlled by an international treaty organization. The ***SI base units*** for mass, length, and time are listed in Table 2.1. They are, respectively, the kilogram (kg), meter (m), and second (s). The SI base unit for temperature is the kelvin (K). (Units for temperature are discussed in Sec. 2.5.) The SI unit of force, called the newton, is defined in terms of the base units for mass, length, and time, as discussed next.

***SI base units***

Newton's second law of motion states that the net force acting on a body is proportional to the product of the mass and the acceleration, written $F \propto ma$. The newton is defined so that the proportionality constant in the expression is equal to unity. That is, Newton's second law is expressed as the equality

$$F = ma \tag{2.1}$$

The newton, N, is the force required to accelerate a mass of 1 kilogram at the rate of 1 meter per second per second. With Eq. 2.1

$$1\text{ N} = (1\text{ kg})(1\text{ m/s}^2) = 1\text{ kg}\cdot\text{m/s}^2 \tag{2.2}$$

***For Example...*** to illustrate the use of the SI units introduced thus far, let us determine the weight in newtons of an object whose mass is 1000 kg, at a place on the earth's surface where the acceleration due to gravity equals a *standard* value defined as 9.80665 m/s$^2$. Recalling

**Table 2.1** SI Units for Mass, Length, Time, and Force

| Quantity | Unit | Symbol |
|---|---|---|
| mass | kilogram | kg |
| length | meter | m |
| time | second | s |
| force | newton<br>($= 1\text{ kg}\cdot\text{m/s}^2$) | N |

that the weight of an object refers to the force of gravity, and is calculated using the mass of the object, $m$, and the local acceleration of gravity, $g$, with Eq. 2.1 we get

$$\begin{aligned} F &= mg \\ &= (1000\ \text{kg})(9.80665\ \text{m/s}^2) = 9806.65\ \text{kg} \cdot \text{m/s}^2 \end{aligned}$$

This force can be expressed in terms of the newton by using Eq. 2.2 as a *unit conversion factor*. That is

$$F = \left(9806.65 \frac{\text{kg} \cdot \text{m}}{\text{s}^2}\right) \left| \frac{1\ \text{N}}{1\ \text{kg} \cdot \text{m/s}^2} \right| = 9806.65\ \text{N} \quad \blacktriangle$$

Observe that in the above calculation of force the unit conversion factor is set off by a pair of vertical lines. This device is used throughout the text to identify unit conversions.

SI units for other physical quantities also are derived in terms of the SI base units. Some of the derived units occur so frequently that they are given special names and symbols, such as the newton. Since it is frequently necessary to work with extremely large or small values when using the SI unit system, a set of standard prefixes is provided in Table 2.2 to simplify matters. For example, km denotes kilometer, that is, $10^3$ m.

**Table 2.2** SI Unit Prefixes

| Factor | Prefix | Symbol |
|---|---|---|
| $10^{12}$ | tera | T |
| $10^{9}$ | giga | G |
| $10^{6}$ | mega | M |
| $10^{3}$ | kilo | k |
| $10^{2}$ | hecto | h |
| $10^{-2}$ | centi | c |
| $10^{-3}$ | milli | m |
| $10^{-6}$ | micro | μ |
| $10^{-9}$ | nano | n |
| $10^{-12}$ | pico | p |

### 2.3.2 Other Units for Mass, Length, Time, and Force

Although SI units are the worldwide standard, at the present time many segments of the engineering community in the United States regularly use some other units. A large portion of America's stock of tools and industrial machines and much valuable engineering data utilize units other than SI units. For many years to come, engineers in the United States will have to be conversant with a variety of units. Accordingly, in this section we consider the alternative units for mass, length, time, and force listed in Table 2.3.

In Table 2.3, the first unit of mass listed is the pound mass, lb, defined in terms of the kilogram as

$$1\ \text{lb} = 0.45359237\ \text{kg} \tag{2.3}$$

The unit for length is the foot, ft, defined in terms of the meter as

$$1\ \text{ft} = 0.3048\ \text{m} \tag{2.4}$$

The inch, in., is defined in terms of the foot

$$12\ \text{in.} = 1\ \text{ft}$$

One inch equals 2.54 cm. Although units such as the minute and the hour are often used in engineering, it is convenient to select the second as the preferred unit for time.

For the choice of pound mass, foot, and second as the units for mass, length, and time, respectively, a force unit can be defined, as for the newton, using Newton's second law written as Eq. 2.1. From this viewpoint, the unit of force, the pound force, lbf, is the force required

**Table 2.3** Other Units for Mass, Length, Time, and Force

| Quantity | Unit | Symbol |
|---|---|---|
| mass | pound mass | lb |
| | slug | slug |
| length | foot | ft |
| time | second | s |
| force | pound force<br>$(= 32.1740\ \text{lb} \cdot \text{ft/s}^2$<br>$= 1\ \text{slug} \cdot \text{ft/s}^2)$ | lbf |

to accelerate one pound mass at 32.1740 $ft/s^2$, which is the standard acceleration of gravity. Substituting values into Eq. 2.1

$$1 \text{ lbf} = (1 \text{ lb})(32.1740 \text{ ft/s}^2) = 32.1740 \text{ lb} \cdot \text{ft/s}^2 \tag{2.5}$$

The pound force, lbf, is not equal to the pound mass, lb. Force and mass are fundamentally different, as are their units. The double use of the word "pound" can be confusing, however, and care must be taken to avoid error.

***For Example...*** to show the use of these units in a single calculation, let us determine the weight of an object whose mass is 1000 lb at a location where the local acceleration of gravity is 32.0 $ft/s^2$. By inserting values into Eq. 2.1 and using Eq. 2.5 as a *unit conversion* factor

$$F = mg = (1000 \text{ lb})\left(32.0 \frac{\text{ft}}{\text{s}^2}\right)\left|\frac{1 \text{ lbf}}{32.1740 \text{ lb} \cdot \text{ft/s}^2}\right| = 994.59 \text{ lbf}$$

This calculation illustrates that the pound force is a unit of force distinct from the pound mass, a unit of mass. ▲

Another mass unit is listed in Table 2.3. This is the *slug*, which is defined as the amount of mass that would be accelerated at a rate of 1 $ft/s^2$ when acted on by a force of 1 lbf. With Newton's second law, Eq. 2.1, we get

$$1 \text{ lbf} = (1 \text{ slug})(1 \text{ ft/s}^2) = 1 \text{ slug} \cdot \text{ft/s}^2 \tag{2.6}$$

Comparing Eqs. 2.5 and 2.6, the relationship between the slug and pound mass is

$$1 \text{ slug} = 32.1740 \text{ lb} \tag{2.7}$$

***For Example...*** to show the use of the slug, let us determine the weight, in lbf, of an object whose mass is 10 slug at a location where the acceleration of gravity is 32.0 $ft/s^2$. Inserting values into Eq. 2.1 and using Eq. 2.6 as a *unit conversion* factor, we get

$$F = mg = (10 \text{ slug})\left(32.0 \frac{\text{ft}}{\text{s}^2}\right)\left|\frac{1 \text{ lbf}}{1 \text{ slug} \cdot \text{ft/s}^2}\right| = 320 \text{ lbf}$$ ▲

METHODOLOGY UPDATE

Because of its global acceptance and intrinsic convenience, the SI system is used throughout this book. In addition, recognizing common practice in the United States, the units listed in Table 2.3 also are used selectively. In particular, the pound mass is used in the thermodynamics portion of the book (Chaps. 2–10) and the slug is used in the fluid mechanics portion (Chaps. 11–14). When the pound mass is the preferred mass unit, the entries of Table 2.3 are called *English* units. When the slug is the preferred mass unit, the entries of Table 2.3 are called *British Gravitational* units. Such terms are part of the jargon of thermal systems engineering with which you should become familiar.

## 2.4 Two Measurable Properties: Specific Volume and Pressure

Three intensive properties that are particularly important in thermal systems engineering are specific volume, pressure, and temperature. In this section specific volume and pressure are considered. Temperature is the subject of Sec. 2.5.

### 2.4.1 Specific Volume

From the macroscopic perspective, the description of matter is simplified by considering matter to be distributed continuously throughout a region. This idealization, known as the *continuum* hypothesis, is used throughout the book.

When substances can be treated as continua, it is possible to speak of their intensive thermodynamic properties "at a point." Thus, at any instant the density ρ at a point is defined as

$$\rho = \lim_{V \to V'} \left( \frac{m}{V} \right) \tag{2.8}$$

where $V'$ is the smallest volume for which a definite value of the ratio exists. The volume $V'$ contains enough particles for statistical averages to be significant. It is the smallest volume for which the matter can be considered a continuum and is normally small enough that it can be considered a "point." With density defined by Eq. 2.8, density can be described mathematically as a continuous function of position and time.

The density, or local mass per unit volume, is an intensive property that may vary from point to point within a system. Thus, the mass associated with a particular volume $V$ is determined in principle by integration

$$m = \int_V \rho \, dV \tag{2.9}$$

and *not* simply as the product of density and volume.

*specific volume*

The ***specific volume*** $v$ is defined as the reciprocal of the density, $v = 1/\rho$. It is the volume per unit mass. Like density, specific volume is an intensive property and may vary from point to point. SI units for density and specific volume are $kg/m^3$ and $m^3/kg$, respectively. However, they are also often expressed, respectively, as $g/cm^3$ and $cm^3/g$. Other units used for density and specific volume in this text are $lb/ft^3$ and $ft^3/lb$, respectively. In the fluid mechanics part of the book, density also is given in $slug/ft^3$.

*molar basis*

In certain applications it is convenient to express properties such as a specific volume on a molar basis rather than on a mass basis. The amount of a substance can be given on a ***molar basis*** in terms of the kilomole (kmol) or the pound mole (lbmol), as appropriate. In either case we use

$$n = \frac{m}{M} \tag{2.10}$$

The number of kilomoles of a substance, $n$, is obtained by dividing the mass, $m$, in kilograms by the molecular weight, $M$, in kg/kmol. Similarly, the number of pound moles, $n$, is obtained by dividing the mass, $m$, in pound mass by the molecular weight, $M$, in lb/lbmol. Appendix Tables T-1 and T-1E provide molecular weights for several substances.

In thermodynamics, we signal that a property is on a molar basis by placing a bar over its symbol. Thus, $\bar{v}$ signifies the volume per kmol or lbmol, as appropriate. In this text the units used for $\bar{v}$ are $m^3/kmol$ and $ft^3/lbmol$. With Eq. 2.10, the relationship between $\bar{v}$ and $v$ is

$$\bar{v} = Mv \tag{2.11}$$

where $M$ is the molecular weight in kg/kmol or lb/lbmol, as appropriate.

### 2.4.2 Pressure

Next, we introduce the concept of pressure from the continuum viewpoint. Let us begin by considering a small area A passing through a point in a fluid at rest. The fluid on one side of the area exerts a compressive force on it that is normal to the area, $F_{normal}$. An equal but oppositely directed force is exerted on the area by the fluid on the other side. For a fluid at rest, no other forces than these act on the area. The ***pressure*** $p$ at the specified point is defined as the limit

*pressure*

$$p = \lim_{A \to A'} \left( \frac{F_{normal}}{A} \right) \tag{2.12}$$

where A′ is the area at the "point" in the same limiting sense as used in the definition of density. The pressure is the same for all orientations of A′ around the point. This is a consequence of the equilibrium of forces acting on an element of volume surrounding the point. However, the pressure can vary from point to point within a fluid at rest; examples are the variation of atmospheric pressure with elevation and the pressure variation with depth in oceans, lakes, and other bodies of water.

### Pressure Units

The SI unit of pressure is the pascal.

$$1 \text{ pascal} = 1 \text{ N/m}^2$$

However, in this text it is convenient to work with multiples of the pascal: the kPa, the bar, and the MPa.

$$1 \text{ kPa} = 10^3 \text{ N/m}^2$$
$$1 \text{ bar} = 10^5 \text{ N/m}^2$$
$$1 \text{ MPa} = 10^6 \text{ N/m}^2$$

Other commonly used units for pressure are pounds force per square foot, $\text{lbf/ft}^2$, and pounds force per square inch, $\text{lbf/in.}^2$ Although atmospheric pressure varies with location on the earth, a standard reference value can be defined and used to express other pressures:

$$1 \text{ standard atmosphere (atm)} = \begin{cases} 1.01325 \times 10^5 \text{ N/m}^2 \\ 14.696 \text{ lbf/in.}^2 \end{cases} \tag{2.13}$$

Pressure as discussed above is called ***absolute pressure.*** In thermodynamics the term pressure refers to absolute pressure unless explicitly stated otherwise. For further discussion of pressure, including pressure measurement devices, see Chap. 11.

***absolute pressure***

## 2.5 Measuring Temperature

In this section the intensive property temperature is considered along with means for measuring it. Like force, a concept of temperature originates with our sense perceptions. It is rooted in the notion of the "hotness" or "coldness" of a body. We use our sense of touch to distinguish hot bodies from cold bodies and to arrange bodies in their order of "hotness," deciding that 1 is hotter than 2, 2 hotter than 3, and so on. But however sensitive the human body may be, we are unable to gauge this quality precisely. Accordingly, thermometers and temperature scales have been devised to measure it.

### 2.5.1 Thermal Equilibrium and Temperature

A definition of temperature in terms of concepts that are independently defined or accepted as primitive is difficult to give. However, it is possible to arrive at an objective understanding of *equality* of temperature by using the fact that when the temperature of a body changes, other properties also change.

To illustrate this, consider two copper blocks, and suppose that our senses tell us that one is warmer than the other. If the blocks were brought into contact and isolated from their surroundings, they would interact in a way that can be described as a *heat interaction.* During this interaction, it would be observed that the volume of the warmer block decreases somewhat with time, while the volume of the colder block increases with time. Eventually, no further changes in volume would be observed, and the blocks would feel equally warm. Similarly, we would be able to observe that the electrical resistance of the warmer block decreases with time, and that of the colder block increases with time; eventually the electrical resistances would become constant also. When all changes in such observable properties cease, the interaction is at an end. The two blocks are then in ***thermal equilibrium.*** Considerations such as these lead us to infer that the blocks have a physical property that determines whether they will be in thermal equilibrium. This property is called ***temperature,*** and we may postulate that when the two blocks are in thermal equilibrium, their temperatures are equal. A process occurring at constant temperature is an ***isothermal process.***

*thermal equilibrium*

*temperature*

*isothermal process*

## 2.5.2 Thermometers

Any body with at least one measurable property that changes as its temperature changes can be used as a thermometer. Such a property is called a ***thermometric property.*** The particular substance that exhibits changes in the thermometric property is known as a *thermometric* substance.

*thermometric property*

A familiar device for temperature measurement is the liquid-in-glass thermometer pictured in Fig. 2.5, which consists of a glass capillary tube connected to a bulb filled with a liquid such as alcohol and sealed at the other end. The space above the liquid is occupied by the vapor of the liquid or an inert gas. As temperature increases, the liquid expands in volume and rises in the capillary. The length $L$ of the liquid in the capillary depends on the temperature. Accordingly, the liquid is the thermometric substance and $L$ is the thermometric property. Although this type of thermometer is commonly used for ordinary temperature measurements, it is not well suited for applications where extreme accuracy is required. Various other types of thermometers have been devised to give accurate temperature measurements.

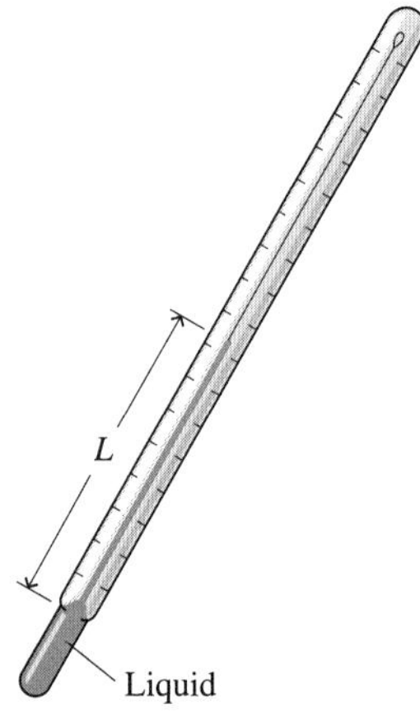


*Figure 2.5* Liquid-in-glass thermometer.

Sensors known as *thermocouples* are based on the principle that when two dissimilar metals are joined, an electromotive force (emf) that is primarily a function of temperature will exist in a circuit. In certain thermocouples, one thermocouple wire is platinum of a specified purity and the other is an alloy of platinum and rhodium. Thermocouples also utilize copper and constantan (an alloy of copper and nickel), iron and constantan, as well as several other pairs of materials. Electrical-resistance sensors are another important class of temperature measurement devices. These sensors are based on the fact that the electrical resistance of various materials changes in a predictable manner with temperature. The materials used for this purpose are normally conductors (such as platinum, nickel, or copper) or semiconductors. Devices using conductors are known as *resistance temperature detectors,* and semiconductor types are called *thermistors.* A variety of instruments measure temperature by sensing radiation. They are known by terms such as *radiation thermometers* and *optical pyrometers.* This type of thermometer differs from those previously considered in that it does not actually come in contact with the body whose temperature is to be determined, an advantage when dealing with moving objects or bodies at extremely high temperatures. All of these temperature sensors can be used together with automatic data acquisition.

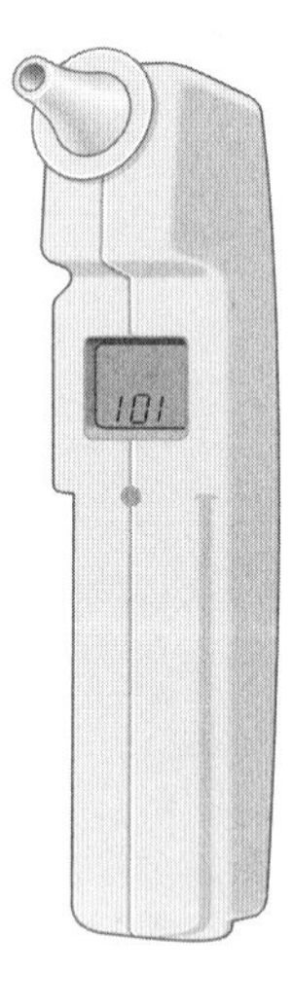


## 2.5.3 Kelvin Scale

Empirical means of measuring temperature such as considered in Sec. 2.5.2 have inherent limitations. ***For Example...*** the tendency of the liquid in a liquid-in-glass thermometer to freeze at low temperatures imposes a lower limit on the range of temperatures that can be measured. At high temperatures liquids vaporize, and therefore these temperatures also cannot be determined by a liquid-in-glass thermometer. Accordingly, several *different* thermometers might be required to cover a wide temperature interval. ▲

In view of the limitations of empirical means for measuring temperature, it is desirable to have a procedure for assigning temperature values that does not depend on the properties of any particular substance or class of substances. Such a scale is called a *thermodynamic* temperature scale. The ***Kelvin scale*** is an absolute thermodynamic temperature scale that provides a continuous definition of temperature, valid over all ranges of temperature. Empirical measures of temperature, with different thermometers, can be related to the Kelvin scale.

*Kelvin scale*

To develop the Kelvin scale, it is necessary to use the conservation of energy principle and the second law of thermodynamics; therefore, further discussion is deferred to Sec. 6.4.1 after these principles have been introduced. However, we note here that the Kelvin scale has a zero of 0 K, and lower temperatures than this are not defined.

### 2.5.4 Celsius, Rankine, and Fahrenheit Scales

Temperature scales are defined by the numerical value assigned to a *standard fixed point.* By international agreement the standard fixed point is the easily reproducible ***triple point of water:*** the state of equilibrium between steam, ice, and liquid water (Sec. 4.2). As a matter of convenience, the temperature at this standard fixed point is defined as 273.16 kelvins, abbreviated as 273.16 K. This makes the temperature interval from the *ice point*[1] (273.15 K) to the *steam point*[2] equal to 100 K and thus in agreement over the interval with the Celsius scale that assigns 100 Celsius degrees to it.

*triple point*

The ***Celsius temperature scale*** (formerly called the centigrade scale) uses the unit degree Celsius (°C), which has the same magnitude as the kelvin. Thus, temperature *differences* are identical on both scales. However, the zero point on the Celsius scale is shifted to 273.15 K, as shown by the following relationship between the Celsius temperature and the Kelvin temperature:

*Celsius scale*

$$T(°\text{C}) = T(\text{K}) - 273.15 \tag{2.14}$$

From this it can be seen that on the Celsius scale the triple point of water is 0.01°C and that 0 K corresponds to −273.15°C.

Two other temperature scales are in common use in engineering in the United States. By definition, the ***Rankine scale,*** the unit of which is the degree rankine (°R), is proportional to the Kelvin temperature according to

*Rankine scale*

$$T(°\text{R}) = 1.8T(\text{K}) \tag{2.15}$$

As evidenced by Eq. 2.15, the Rankine scale is also an absolute thermodynamic scale with an absolute zero that coincides with the absolute zero of the Kelvin scale. In thermodynamic relationships, temperature is always in terms of the Kelvin or Rankine scale unless specifically stated otherwise.

A degree of the same size as that on the Rankine scale is used in the ***Fahrenheit scale,*** but the zero point is shifted according to the relation

*Fahrenheit scale*

$$T(°\text{F}) = T(°\text{R}) - 459.67 \tag{2.16}$$

Substituting Eqs. 2.14 and 2.15 into Eq. 2.16, it follows that

$$T(°\text{F}) = 1.8T(°\text{C}) + 32 \tag{2.17}$$

This equation shows that the Fahrenheit temperature of the ice point (0°C) is 32°F and of the steam point (100°C) is 212°F. The 100 Celsius or Kelvin degrees between the ice point and steam point correspond to 180 Fahrenheit or Rankine degrees.

When making engineering calculations, it is common to round off the last numbers in Eqs. 2.14 and 2.16 to 273 and 460, respectively. This is frequently done in subsequent sections of the text.

**METHODOLOGY UPDATE**

[1] The state of equilibrium between ice and air-saturated water at a pressure of 1 atm.

[2] The state of equilibrium between steam and liquid water at a pressure of 1 atm.

## 2.6 Methodology for Solving Problems

A major goal of this textbook is to help you learn how to solve engineering problems that involve thermal systems engineering principles. To this end, numerous solved examples and end-of-chapter problems are provided. It is extremely important for you to study the examples *and* solve problems, for mastery of the fundamentals comes only through practice.

To maximize the results of your efforts, it is necessary to develop a systematic approach. You must think carefully about your solutions and avoid the temptation of starting problems *in the middle* by selecting some seemingly appropriate equation, substituting in numbers, and quickly "punching up" a result on your calculator. Such a haphazard problem-solving approach can lead to difficulties as problems become more complicated. Accordingly, we strongly recommend that problem solutions be organized using the steps in the box below, as appropriate. The solved examples of this text illustrate this step-wise approach.

***Known:*** State briefly in your own words what is known. This requires that you read the problem carefully *and* think about it.

***Find:*** State concisely in your own words what is to be determined.

***Schematic and Given Data:*** Draw a sketch of the system to be considered. Decide whether a closed system or control volume is appropriate for the analysis, and then carefully identify the boundary. Label the diagram with relevant information from the problem statement. Record all property values you are given. When appropriate, sketch property diagrams (see Sec. 4.2), locating key state points and indicating, if possible, the processes executed by the system.

The importance of good sketches of the system and property diagrams cannot be overemphasized. They are often instrumental in enabling you to think clearly about the problem.

***Assumptions:*** To form a record of how you *model* the problem, list all simplifying assumptions and idealizations made to reduce it to one that is manageable. Sometimes this information also can be noted on the sketches of the previous step.

***Properties:*** Compile property values you anticipate will be needed for subsequent calculations and identify the source from which they are obtained.

***Analysis:*** Using your assumptions and idealizations, reduce the appropriate governing equations and relationships to forms that will produce the desired results.

It is advisable to work with equations in symbol form as long as possible before substituting numerical data. When the equations are reduced to final forms, consider them to determine what additional data may be required. Identify the tables, charts, or property equations that provide the required values.

When all equations and data are in hand, substitute numerical values into the equations. Carefully check that a consistent and appropriate set of units is being employed. Then perform the needed calculations. Finally, consider whether the magnitudes of the numerical values are reasonable and the algebraic signs associated with the numerical values are correct.

***Comments:*** The solved examples provided in the book are frequently annotated with various comments intended to assist learning, including commenting on what was learned, and identifying key aspects of the solution. You are urged to comment on your results. Such a discussion may include a summary of key conclusions, a critique of the original assumptions, and an inference of trends obtained by performing additional *what-if* and *parameter sensitivity* calculations.

The importance of following these steps should not be underestimated. They provide a useful guide to thinking about a problem before effecting its solution. Of course, as a particular solution evolves, you may have to return to an earlier step and revise it in light of a better understanding of the problem. For example, it might be necessary to add or delete an assumption, revise a sketch, determine additional property data, and so on.

The example to follow illustrates the use of this solution methodology together with important concepts introduced previously.

### *Example 2.1* Identifying System Interactions

A wind turbine–electric generator is mounted atop a tower. As wind blows steadily across the turbine blades, electricity is generated. The electrical output of the generator is fed to a storage battery.
(a) Considering only the wind turbine–electric generator as the system, identify locations on the system boundary where the system interacts with the surroundings. Describe changes occurring within the system with time.
(b) Repeat for a system that includes only the storage battery.

### Solution

***Known:*** A wind turbine-electric generator provides electricity to a storage battery.
***Find:*** For a system consisting of (a) the wind turbine–electric generator, (b) the storage battery, identify locations where the system interacts with its surroundings, and describe changes occurring within the system with time.

***Schematic and Given Data:***

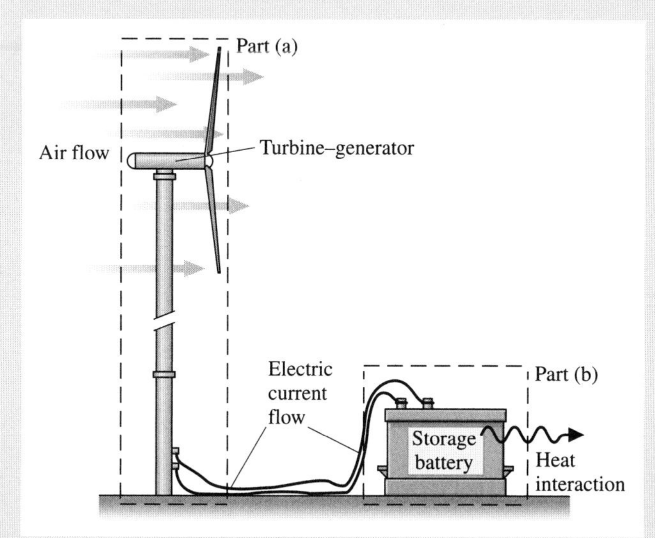


***Assumptions:***
1. In part (a), the system is the control volume shown by the dashed line on the figure.
2. In part (b), the system is the closed system shown by the dashed line on the figure.
3. The wind is steady.

*Figure E2.1*

❶ ***Analysis:*** (a) In this case, there is air flowing across the boundary of the control volume. Another principal interaction between the system and surroundings is the electric current passing through the wires. From the macroscopic perspective, such an interaction is not considered a mass transfer, however. With a steady wind, the turbine–generator is likely to reach steady-state operation, where the rotational speed of the blades is constant and a steady electric current is generated. An interaction also occurs between the turbine-generator tower and the ground: a force and moment are required to keep the tower upright.

(b) The principal interaction between the system and its surroundings is the electric current passing into the battery through the wires. As noted in part (a), this interaction is not considered a mass transfer. The system is a closed system. As the battery is charged and chemical reactions occur within it, the temperature of the battery surface may become somewhat elevated and a heat interaction might occur between the battery and its surroundings. This interaction is likely to be of secondary importance.

---

❶ Using terms from Chap. 1, the system of part (a) involves the *conversion* of kinetic energy to electricity, whereas the system of part (b) involves energy *storage* within the battery.

## 2.7 Chapter Summary and Study Guide

In this chapter, we have introduced some of the fundamental concepts and definitions used in thermodynamics, fluid mechanics, and heat transfer. An important aspect of engineering analysis is to identify appropriate closed systems and control volumes, and to describe system behavior in terms of properties and processes. Three important properties discussed in this chapter are specific volume, pressure, and temperature.

*closed system*
*control volume*
*boundary*
*surroundings*
*property*
*extensive property*
*intensive property*
*state*
*process*
*thermodynamic cycle*
*phase*
*pure substance*
*equilibrium*
*pressure*
*specific volume*
*temperature*
*isothermal process*
*Kelvin scale*
*Rankine scale*

In this book, we consider systems at equilibrium states and systems undergoing processes. We study processes during which the intervening states are not equilibrium states as well as quasiequilibrium processes during which the departure from equilibrium is negligible.

In Tables 2.1 and 2.3, we have introduced both SI and other units for mass, length, time, and force. You will need to be familiar with such units as you use this book. The chapter concludes with a discussion of how to solve problems systematically.

The following checklist provides a study guide for this chapter. When your study of the text and the end-of-chapter exercises has been completed you should be able to

- write out the meanings of the terms listed in the margin throughout the chapter and understand each of the related concepts. The subset of key terms listed here in the margin is particularly important in subsequent chapters.
- identify an appropriate system boundary and describe the interactions between the system and its surroundings.
- use appropriate units for mass, length, time, force, and temperature and apply appropriately Newton's second law and Eqs. 2.14–2.17.
- work on a molar basis using Eqs. 2.10 and 2.11.
- apply the methodology for problem solving discussed in Sec. 2.6.

# Problems

**Exploring System Concepts**

**2.1** Referring to Figs. 2.1 and 2.2, identify locations on the boundary of each system where there are interactions with the surroundings.

**2.2** As illustrated in Fig. P2.2, electric current from a storage battery runs an electric motor. The shaft of the motor is connected to a pulley–mass assembly that raises a mass. Considering the motor as a system, identify locations on the system boundary where the system interacts with its surroundings and describe changes that occur within the system with time. Repeat for an enlarged system that also includes the battery and pulley–mass assembly.

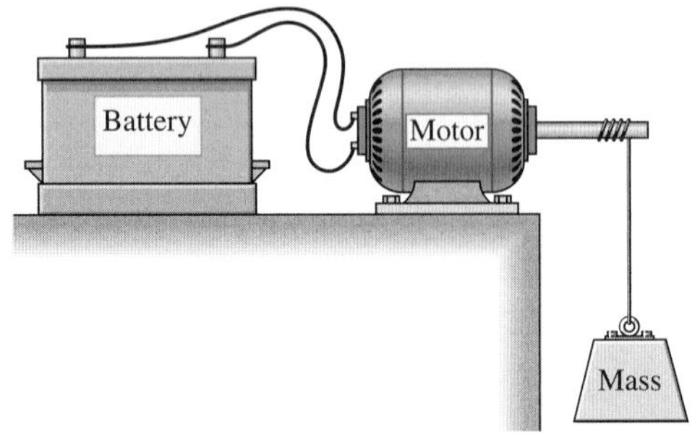


*Figure P2.2*

**2.3** As illustrated in Fig. P2.3, water circulates between a storage tank and a solar collector. Heated water from the tank is used for domestic purposes. Considering the solar collector as a system, identify locations on the system boundary where the system interacts with its surroundings and describe events that occur within the system. Repeat for an enlarged system that includes the storage tank and the interconnecting piping.

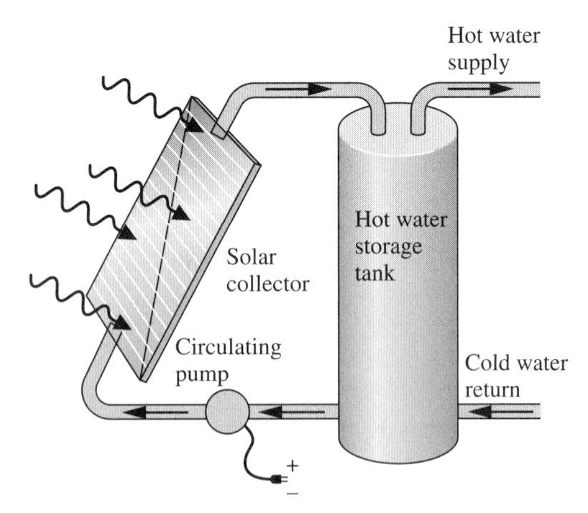


*Figure P2.3*

**2.4** As illustrated in Fig. P2.4, steam flows through a valve and turbine in series. The turbine drives an electric generator. Considering the valve and turbine as a system, identify locations on the system boundary where the system interacts with its surroundings and describe events occurring within the system. Repeat for an enlarged system that includes the generator.

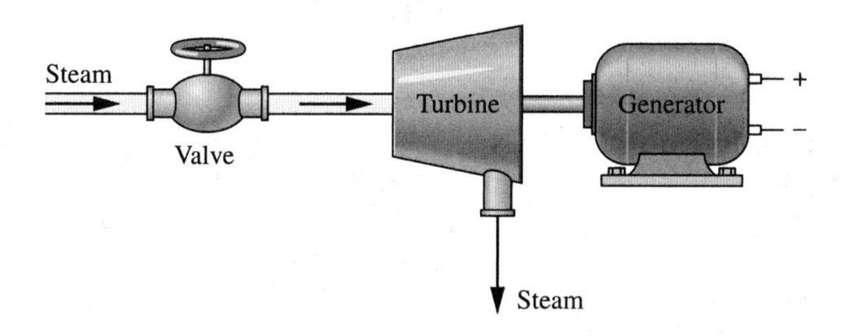


*Figure P2.4*

2.5 As illustrated in Fig. P2.5, water for a fire hose is drawn from a pond by a gasoline engine-driven pump. Considering the engine-driven pump as a system, identify locations on the system boundary where the system interacts with its surroundings and describe events occurring within the system. Repeat for an enlarged system that includes the hose and the nozzle.

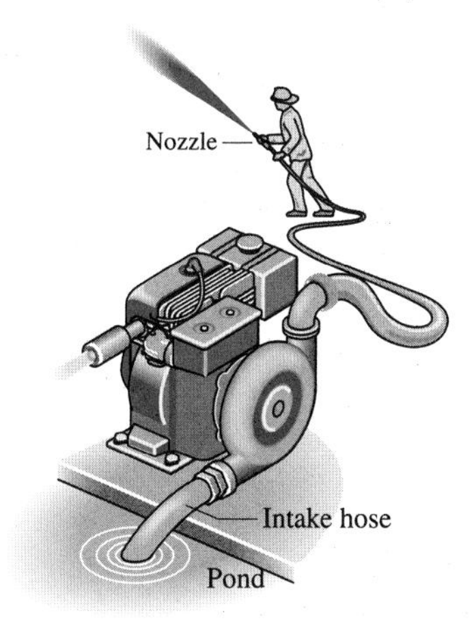


*Figure P2.5*

2.6 A system consists of liquid water in equilibrium with a gaseous mixture of air and water vapor. How many phases are present? Does the system consist of a pure substance? Explain. Repeat for a system consisting of ice and liquid water in equilibrium with a gaseous mixture of air and water vapor.

2.7 A system consists of liquid oxygen in equilibrium with oxygen vapor. How many phases are present? The system undergoes a process during which some of the liquid is vaporized. Can the system be viewed as being a pure substance during the process? Explain.

2.8 A system consisting of liquid water undergoes a process. At the end of the process, some of the liquid water has frozen, and the system contains liquid water and ice. Can the system be viewed as being a pure substance during the process? Explain.

2.9 A dish of liquid water is placed on a table in a room. After a while, all of the water evaporates. Taking the water and the air in the room to be a closed system, can the system be regarded as a pure substance *during* the process? *After* the process is completed? Discuss.

### Force and Mass

2.10 An object has a mass of 20 kg. Determine its weight, in N, at a location where the acceleration of gravity is 9.78 $m/s^2$.

2.11 An object weighs 10 lbf at a location where the acceleration of gravity is 30.0 $ft/s^2$. Determine its mass, in lb and slug.

2.12 An object whose mass is 10 kg weighs 95 N. Determine
(a) the local acceleration of gravity, in $m/s^2$.
(b) the mass, in kg, and the weight, in N, of the object at a location where $g = 9.81$ $m/s^2$.

2.13 An object whose mass is 10 lb weighs 9.6 lbf. Determine
(a) the local acceleration of gravity, in $ft/s^2$.
(b) the mass, in lb and slug, and the weight, in lbf, of the object at a location where $g = 32.2$ $ft/s^2$.

2.14 A gas occupying a volume of 25 $ft^3$ weighs 3.5 lbf on the moon, where the acceleration of gravity is 5.47 $ft/s^2$. Determine its weight, in lbf, and density, in $lb/ft^3$, on Mars, where $g = 12.86$ $ft/s^2$.

2.15 Atomic and molecular weights of some common substances are listed in Appendix Tables T-1 and T-1E. Using data from the appropriate table, determine
(a) the mass, in kg, of 20 kmol of each of the following: air, C, $H_2O$, $CO_2$.
(b) the number of lbmol in 50 lb of each of the following: $H_2$, $N_2$, $NH_3$, $C_3H_8$.

2.16 A simple instrument for measuring the acceleration of gravity employs a *linear* spring from which a mass is suspended. At a location on earth where the acceleration of gravity is 32.174 $ft/s^2$, the spring extends 0.291 in. If the spring extends 0.116 in. when the instrument is on Mars, what is the Martian acceleration of gravity? How much would the spring extend on the moon, where $g = 5.471$ $ft/s^2$?

2.17 A closed system consists of 0.5 lbmol of liquid water and occupies a volume of 0.145 $ft^3$. Determine the weight of the system, in lbf, and the average density, in $lb/ft^3$ and $slug/ft^3$, at a location where the acceleration of gravity is $g = 30.5$ $ft/s^2$.

2.18 The weight of an object on an orbiting space vehicle is measured to be 42 N based on an artificial gravitational acceleration of 6 $m/s^2$. What is the weight of the object, in N, on earth, where $g = 9.81$ $m/s^2$?

2.19 The storage tank of a water tower is nearly spherical in shape with a radius of 30 ft. If the density of the water is 62.4 $lb/ft^3$, what is the mass of water stored in the tower, in lb, when the tank is full? What is the weight, in lbf, of the water if the local acceleration of gravity is 32.1 $ft/s^2$?

### Specific Volume, Pressure

2.20 A spherical balloon has a diameter of 10 ft. The average specific volume of the air inside is 15.1 $ft^3/lb$. Determine the weight of the air, in lbf, at a location where $g = 31.0$ $ft/s^2$.

2.21 Five kg of methane gas is fed to a cylinder having a volume of 20 $m^3$ and initially containing 25 kg of methane at a pressure of 10 bar. Determine the specific volume, in $m^3/kg$, of the methane in the cylinder initially. Repeat for the methane in the cylinder after the 5 kg has been added.

**2.22** A closed system consisting of 2 kg of a gas undergoes a process during which the relationship between pressure and specific volume is $pv^{1.3}$ = constant. The process begins with $p_1 = 1$ bar, $v_1 = 0.5$ m$^3$/kg and ends with $p_2 = 0.25$ bar. Determine the final volume, in m$^3$, and plot the process on a graph of pressure versus specific volume.

**2.23** A closed system consisting of 1 lb of a gas undergoes a process during which the relation between the pressure and volume is $pV^n$ = constant. The process begins with $p_1 = 20$ lbf/in.$^2$, $V_1 = 10$ ft$^3$ and ends with $p_2 = 100$ lbf/in.$^2$ Determine the final volume, in ft$^3$, for each of the following values of the constant $n$: 1, 1.2, 1.3, and 1.4. Plot each of the processes on a graph of pressure versus volume.

**2.24** A system consists of air in a piston–cylinder assembly, initially at $p_1 = 20$ lbf/in.$^2$, and occupying a volume of 1.5 ft$^3$. The air is compressed to $p_2 = 100$ lbf/in.$^2$ and a final volume of 0.5 ft$^3$. During the process, the relation between pressure and volume is linear. Determine the pressure, in lbf/in.$^2$, at an intermediate state where the volume is 1.2 ft$^3$, and sketch the process on a graph of pressure versus volume.

**2.25** A gas initially at $p_1 = 1$ bar and occupying a volume of 1 liter is compressed within a piston–cylinder assembly to a final pressure $p_2 = 4$ bar.

**(a)** If the relationship between pressure and volume during the compression is $pV$ = constant, determine the volume, in liters, at a pressure of 3 bar. Also plot the overall process on a graph of pressure versus volume.

**(b)** Repeat for a linear pressure–volume relationship between the same end states.

**2.26** A gas contained within a piston–cylinder assembly undergoes a thermodynamic cycle consisting of three processes:

***Process 1–2:*** Compression with $pV$ = constant from $p_1 = 1$ bar, $V_1 = 1.0$ m$^3$ to $V_2 = 0.2$ m$^3$

***Process 2–3:*** Constant-pressure expansion to $V_3 = 1.0$ m$^3$

***Process 3–1:*** Constant volume

Sketch the cycle on a $p$–$V$ diagram labeled with pressure and volume values at each numbered state.

### Temperature

**2.27** Convert the following temperatures from °C to °F: **(a)** 21°C, **(b)** −17.78°C, **(c)** −50°C, **(d)** 300°C, **(e)** 100°C, **(f)** −273.15°C. Convert each temperature to °R.

**2.28** Convert the following temperatures from °F to °C: **(a)** 212°F, **(b)** 68°F, **(c)** 32°F, **(d)** 0°F, **(e)** −40°F, **(f)** −459.67°F. Convert each temperature to K.

**2.29** Two temperature measurements are taken with a thermometer marked with the Celsius scale. Show that the *difference* between the two readings would be the same if the temperatures were converted to the Kelvin scale.

**2.30** On a day in January, a household digital thermometer gives the same outdoor temperature reading in °C as in °F. What is that reading? Express the reading in K and °R.

**2.31** A new absolute temperature scale is proposed. On this scale the ice point of water is 150°S and the steam point is 300°S. Determine the temperatures in °C that correspond to 100° and 400°S, respectively. What is the ratio of the size of the °S to the kelvin?

**2.32** As shown in Fig. P2.32, a small-diameter water pipe passes through the 6-in.-thick exterior wall of a dwelling. Assuming that temperature varies linearly with position $x$ through the wall from 68°F to 20°F, would the water in the pipe freeze?

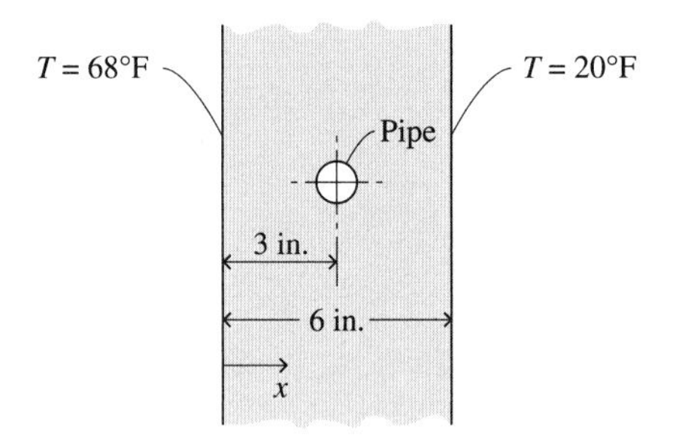


*Figure P2.32*

# 3 thermo USING ENERGY AND THE FIRST LAW OF THERMODYNAMICS

*Introduction...*

Energy is a fundamental concept of thermodynamics and one of the most significant aspects of engineering analysis. In this chapter we discuss energy and develop equations for applying the principle of conservation of energy. The current presentation is limited to closed systems. In Chap. 5 the discussion is extended to control volumes.

Energy is a familiar notion, and you already know a great deal about it. In the present chapter several important aspects of the energy concept are developed. Some of these we have encountered in Chap. 1. A basic idea is that energy can be *stored* within systems in various forms. Energy also can be *converted* from one form to another and *transferred* between systems. For closed systems, energy can be transferred by *work* and *heat transfer.* The total amount of energy is *conserved* in all transformations and transfers.

***chapter objective***

The ***objective*** of this chapter is to organize these ideas about energy into forms suitable for engineering analysis. The presentation begins with a review of energy concepts from mechanics. The thermodynamic concept of energy is then introduced as an extension of the concept of energy in mechanics.

## 3.1 Reviewing Mechanical Concepts of Energy

Building on the contributions of Galileo and others, Newton formulated a general description of the motions of objects under the influence of applied forces. Newton's laws of motion, which provide the basis for classical mechanics, led to the concepts of *work, kinetic energy,* and *potential energy,* and these led eventually to a broadened concept of energy. In the present section, we review mechanical concepts of energy.

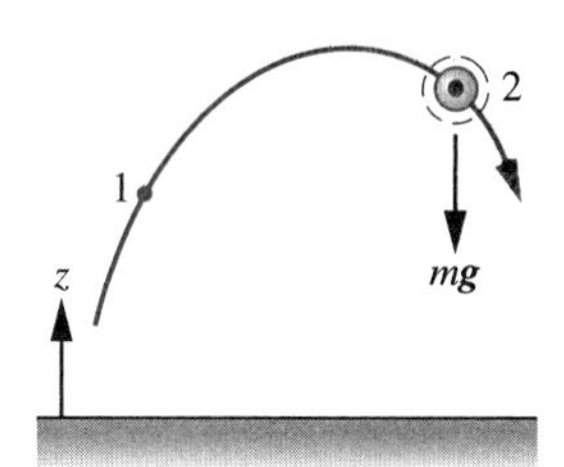


### 3.1.1 Kinetic and Potential Energy

Consider a body of mass $m$ that moves from a position where the magnitude of its velocity is $V_1$ and its elevation is $z_1$ to another where its velocity is $V_2$ and elevation is $z_2$, each relative to a specified coordinate frame such as the surface of the earth. The quantity $\frac{1}{2}mV^2$ is the ***kinetic energy,*** KE, of the body. The *change* in kinetic energy, ΔKE, of the body is

***kinetic energy***

$$\Delta KE = KE_2 - KE_1 = \frac{1}{2}m(V_2^2 - V_1^2) \tag{3.1}$$

Kinetic energy can be assigned a value knowing only the mass of the body and the magnitude of its instantaneous velocity relative to a specified coordinate frame, without regard for how this velocity was attained. Hence, *kinetic energy is a property* of the body. Since kinetic energy is associated with the body as a whole, it is an *extensive* property.

*gravitational potential energy*

The quantity $mgz$ is the ***gravitational potential energy,*** PE. The *change* in gravitational potential energy, $\Delta$PE, is

$$\Delta \text{PE} = \text{PE}_2 - \text{PE}_1 = mg(z_2 - z_1) \tag{3.2}$$

Potential energy is associated with the force of gravity (Sec. 2.3) and is therefore an attribute of a system consisting of the body and the earth together. However, evaluating the force of gravity as $mg$ enables the gravitational potential energy to be determined for a specified value of $g$ knowing only the mass of the body and its elevation. With this view, potential energy is regarded as an *extensive property* of the body.

To assign a value to the kinetic energy or the potential energy of a system, it is necessary to assume a datum and specify a value for the quantity at the datum. Values of kinetic and potential energy are then determined relative to this arbitrary choice of datum and reference value. However, since only *changes* in kinetic and potential energy between two states are required, these arbitrary reference specifications cancel.

*Units.* In SI, the energy unit is the newton-meter, N · m, called the joule, J. In this book it is convenient to use the kilojoule, kJ. Other commonly used units for energy are the foot-pound force, ft · lbf, and the British thermal unit, Btu.

When a system undergoes a process where there are changes in kinetic and potential energy, special care is required to obtain a consistent set of units.

***For Example...*** to illustrate the proper use of units in the calculation of such terms, consider a system having a mass of 1 kg whose velocity increases from 15 m/s to 30 m/s while its elevation decreases by 10 m at a location where $g = 9.7 \text{ m/s}^2$. Then

$$\begin{aligned}
\Delta \text{KE} &= \frac{1}{2} m(\text{V}_2^2 - \text{V}_1^2) \\
&= \frac{1}{2}(1 \text{ kg})\left[\left(30 \frac{\text{m}}{\text{s}}\right)^2 - \left(15 \frac{\text{m}}{\text{s}}\right)^2\right]\left|\frac{1 \text{ N}}{1 \text{ kg} \cdot \text{m/s}^2}\right|\left|\frac{1 \text{ kJ}}{10^3 \text{ N} \cdot \text{m}}\right| \\
&= 0.34 \text{ kJ} \\
\Delta \text{PE} &= mg(z_2 - z_1) \\
&= (1 \text{ kg})\left(9.7 \frac{\text{m}}{\text{s}^2}\right)(-10 \text{ m})\left|\frac{1 \text{ N}}{1 \text{ kg} \cdot \text{m/s}^2}\right|\left|\frac{1 \text{ kJ}}{10^3 \text{ N} \cdot \text{m}}\right| \\
&= -0.10 \text{ kJ}
\end{aligned}$$

For a system having a mass of 1 lb whose velocity increases from 50 ft/s to 100 ft/s while its elevation decreases by 40 ft at a location where $g = 32.0 \text{ ft/s}^2$, we have

$$\begin{aligned}
\Delta \text{KE} &= \frac{1}{2}(1 \text{ lb})\left[\left(100 \frac{\text{ft}}{\text{s}}\right)^2 - \left(50 \frac{\text{ft}}{\text{s}}\right)^2\right]\left|\frac{1 \text{ lbf}}{32.2 \text{ lb} \cdot \text{ft/s}^2}\right|\left|\frac{1 \text{ Btu}}{778 \text{ ft} \cdot \text{lbf}}\right| \\
&= 0.15 \text{ Btu} \\
\Delta \text{PE} &= (1 \text{ lb})\left(32.0 \frac{\text{ft}}{\text{s}^2}\right)(-40 \text{ ft})\left|\frac{1 \text{ lbf}}{32.2 \text{ lb} \cdot \text{ft/s}^2}\right|\left|\frac{1 \text{ Btu}}{778 \text{ ft} \cdot \text{lbf}}\right| \\
&= -0.05 \text{ Btu} \ \blacktriangle
\end{aligned}$$

### 3.1.2 Work in Mechanics

In mechanics, when a body moving along a path is acted on by a resultant force that may vary in magnitude from position to position along the path, the work of the force is written as the scalar product (dot product) of the force vector **F** and the displacement vector of the body along the path $d\mathbf{s}$. That is

$$Work = \int_1^2 \mathbf{F} \cdot d\mathbf{s} \tag{3.3}$$

When the resultant force causes the elevation to be increased, the body to be accelerated, or both, the work done by the force can be considered a *transfer* of energy *to* the body, where it is *stored* as gravitational potential energy and/or kinetic energy. The notion that *energy is conserved* underlies this interpretation.

### 3.1.3 Closure

The presentation thus far has centered on systems for which applied forces affect only their overall velocity and position. However, systems of engineering interest normally interact with their surroundings in more complicated ways, with changes in other properties as well. To analyze such systems, the concepts of kinetic and potential energy alone do not suffice, nor does the rudimentary conservation of energy principle introduced above. In thermodynamics the concept of energy is broadened to account for other observed changes, and the principle of ***conservation of energy*** is extended to include other ways in which systems interact with their surroundings. The basis for such generalizations is experimental evidence. These extensions of the concept of energy are developed in the remainder of the chapter, beginning in the next section with a fuller discussion of work.

*conservation of energy*

## 3.2 Broadening Our Understanding of Work

The work done by, or on, a system evaluated in terms of forces and displacements is given by Eq. 3.3. This relationship is important in thermodynamics, and is used later in the present section. It is also used in Sec. 3.3 to evaluate the work done in the compression or expansion of a gas (or liquid). However, thermodynamics also deals with phenomena not included within the scope of mechanics, so it is necessary to adopt a broader interpretation of work, as follows.

A particular interaction is categorized as a work interaction if it satisfies the following criterion, which can be considered the ***thermodynamic definition of work:*** *Work is done by a system on its surroundings if the sole effect on everything external to the system could have been the raising of a weight.* Notice that the raising of a weight is, in effect, a force acting through a distance, so the concept of work in thermodynamics is an extension of the concept of work in mechanics. However, the test of whether a work interaction has taken place is not that the elevation of a weight has actually taken place, or that a force has actually acted through a distance, but that the sole effect *could have been* an increase in the elevation of a weight.

*thermodynamic definition of work*

***For Example...*** consider Fig. 3.1 showing two systems labeled A and B. In system A, a gas is stirred by a paddle wheel: the paddle wheel does work on the gas. In principle, the work could be evaluated in terms of the forces and motions at the boundary between the paddle wheel and the gas. Such an evaluation of work is consistent with Eq. 3.3, where work is the product of force and displacement. By contrast, consider system B, which includes only the battery. At the boundary of system B, forces and motions are not evident. Rather, there is an electric current $i$ driven by an electrical potential difference existing across the terminals a and b. That this type of interaction at the boundary can be classified as work follows from the thermodynamic definition of work given previously: We can imagine the current is supplied to a *hypothetical* electric motor that lifts a weight in the surroundings. ▲

Work is a means for transferring energy. Accordingly, the term work does not refer to what is being transferred between systems or to what is stored within systems. Energy is transferred and stored when work is done.

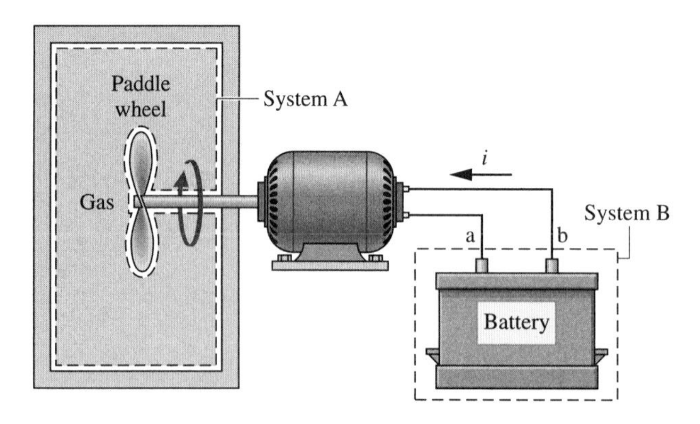


*Figure 3.1* Two examples of work.

### 3.2.1 Sign Convention and Notation

Engineering thermodynamics is frequently concerned with devices such as internal combustion engines and turbines whose purpose is to do work. Hence, it is often convenient to consider such work as positive. That is,

$W > 0$: work done *by* the system
$W < 0$: work done *on* the system

*sign convention for work*

**METHODOLOGY UPDATE**

This ***sign convention*** is used throughout the book. In certain instances, however, it is convenient to regard the work done *on* the system to be positive. To reduce the possibility of misunderstanding in any such case, the direction of energy transfer is shown by an arrow on a sketch of the system, and work is regarded as positive in the direction of the arrow.

*work is not a property*

Returning briefly to Eq. 3.3, to evaluate the integral it is necessary to know how the force varies with the displacement. This brings out an important idea about work: The value of $W$ depends on the details of the interactions taking place between the system and surroundings during a process and not just the initial and final states of the system. It follows that ***work is not a property*** of the system or the surroundings. In addition, the limits on the integral of Eq. 3.3 mean "from state 1 to state 2" and cannot be interpreted as the *values* of work at these states. The notion of work at a state *has no meaning,* so the value of this integral should never be indicated as $W_2 - W_1$.

The differential of work, $\delta W$, is said to be *inexact* because, in general, the following integral cannot be evaluated without specifying the details of the process

$$\int_1^2 \delta W = W$$

On the other hand, the differential of a property is said to be *exact* because the change in a property between two particular states depends in no way on the details of the process linking the two states. For example, the change in volume between two states can be determined by integrating the differential $dV$, without regard for the details of the process, as follows

$$\int_{V_1}^{V_2} dV = V_2 - V_1$$

where $V_1$ is the volume *at* state 1 and $V_2$ is the volume *at* state 2. The differential of every property is exact. Exact differentials are written, as above, using the symbol *d.* To stress the difference between exact and inexact differentials, the differential of work is written as $\delta W$. The symbol $\delta$ is also used to identify other inexact differentials encountered later.

### 3.2.2 Power

Many thermodynamic analyses are concerned with the time rate at which energy transfer occurs. The rate of energy transfer by work is called ***power*** and is denoted by $\dot{W}$. When a work interaction involves an observable force, the rate of energy transfer by work is equal to the product of the force and the velocity at the point of application of the force

*power*

$$\dot{W} = \mathbf{F} \cdot \mathbf{V} \tag{3.4}$$

A dot appearing over a symbol, as in $\dot{W}$, is used to indicate a time rate. In principle, Eq. 3.4 can be integrated from time $t_1$ to time $t_2$ to get the total work done during the time interval

$$W = \int_{t_1}^{t_2} \dot{W}\, dt = \int_{t_1}^{t_2} \mathbf{F} \cdot \mathbf{V}\, dt$$

The same sign convention applies for $\dot{W}$ as for $W$. Since power is a time rate of doing work, it can be expressed in terms of any units for energy and time. In SI, the unit for power is J/s, called the watt. In this book the kilowatt, kW, is generally used. Other commonly used units for power are ft · lbf/s, Btu/h, and horsepower, hp.

***For Example...*** to illustrate the use of Eq. 3.4, let us evaluate the power required for a bicyclist traveling at 20 miles per hour to overcome the drag force imposed by the surrounding air. This *aerodynamic drag* force, discussed in Sec. 14.9, is given by

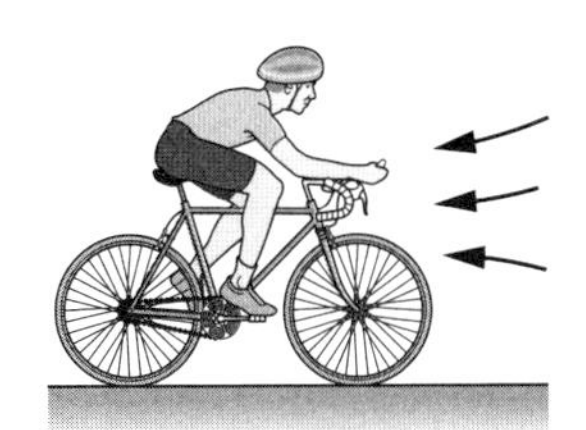

$$F_D = \tfrac{1}{2} C_D \mathrm{A} \rho \mathrm{V}^2$$

where $C_D$ is a constant called the *drag coefficient,* A is the frontal area of the bicycle and rider, and $\rho$ is the air density. By Eq. 3.4 the required power is $\mathbf{F}_D \cdot \mathbf{V}$ or

$$\begin{aligned}\dot{W} &= (\tfrac{1}{2} C_D \mathrm{A} \rho \mathrm{V}^2)\mathrm{V} \\ &= \tfrac{1}{2} C_D \mathrm{A} \rho \mathrm{V}^3\end{aligned}$$

Using typical values: $C_D = 0.88$, $\mathrm{A} = 3.9\ \mathrm{ft}^2$, and $\rho = 0.075\ \mathrm{lb/ft}^3$ together with $\mathrm{V} = 20$ mi/h = 29.33 ft/s, and also converting units to horsepower, the power required is

$$\begin{aligned}\dot{W} &= \frac{1}{2}(0.88)(3.9\ \mathrm{ft}^2)\left(0.075\ \frac{\mathrm{lb}}{\mathrm{ft}^3}\right)\left(29.33\ \frac{\mathrm{ft}}{\mathrm{s}}\right)^3 \left|\frac{1\ \mathrm{lbf}}{32.2\ \mathrm{lb \cdot ft/s^2}}\right| \left|\frac{1\ \mathrm{hp}}{550\ \mathrm{ft \cdot lbf/s}}\right| \\ &= 0.183\ \mathrm{hp}\ \blacktriangle\end{aligned}$$

***Power Transmitted by a Shaft.*** A rotating shaft is a commonly encountered machine element. Consider a shaft rotating with angular velocity $\omega$ and exerting a torque $\mathcal{T}$ on its surroundings. Let the torque be expressed in terms of a tangential force $F_t$ and radius $R$: $\mathcal{T} = F_t R$. The velocity at the point of application of the force is $\mathrm{V} = R\omega$, where $\omega$ is in radians per unit time. Using these relations with Eq. 3.4, we obtain an expression for the *power* transmitted from the shaft to the surroundings

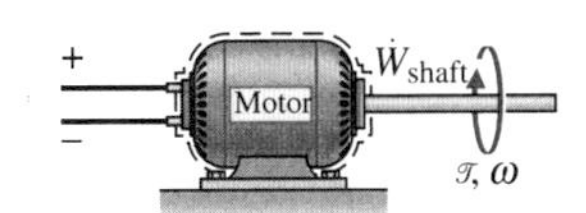


$$\dot{W} = F_t \mathrm{V} = (\mathcal{T}/R)(R\omega) = \mathcal{T}\omega \tag{3.5}$$

A related case involving a gas stirred by a paddle wheel is considered in the discussion of Fig. 3.1.

***Electric Power.*** Shown in Fig. 3.1 is a system consisting of a battery connected to an external circuit through which an electric current, $i$, is flowing. The current is driven by the electrical potential difference $\mathscr{E}$ existing across the terminals labeled a and b. That this type of interaction can be classed as work is considered in the discussion of Fig. 3.1.

The rate of energy transfer by work, or the power, is

$$\dot{W} = -\mathscr{E}i \tag{3.6}$$

The minus sign is required to be in accord with our previously stated sign convention for power. When the power is evaluated in terms of the watt, and the unit of current is the ampere (an SI base unit), the unit of electric potential is the volt, defined as 1 watt per ampere.

## 3.3 Modeling Expansion or Compression Work

Let us evaluate the work done by the closed system shown in Fig. 3.2 consisting of a gas (or liquid) contained in a piston-cylinder assembly as the gas expands. During the process the gas pressure exerts a normal force on the piston. Let $p$ denote the pressure acting at the interface between the gas and the piston. The force exerted by the gas on the piston is simply the product $p\mathrm{A}$, where A is the area of the piston face. The work done by the system as the piston is displaced a distance $dx$ is

$$\delta W = p\mathrm{A}\, dx \tag{3.7}$$

The product A $dx$ in Eq. 3.7 equals the change in volume of the system, $dV$. Thus, the work expression can be written as

$$\delta W = p\, dV \tag{3.8}$$

Since $dV$ is positive when volume increases, the work at the moving boundary is positive when the gas expands. For a compression, $dV$ is negative, and so is work found from Eq. 3.8. These signs are in agreement with the previously stated sign convention for work.

For a change in volume from $V_1$ to $V_2$, the work is obtained by integrating Eq. 3.8

$$W = \int_{V_1}^{V_2} p\, dV \tag{3.9}$$

Although Eq. 3.9 is derived for the case of a gas (or liquid) in a piston-cylinder assembly, it is applicable to systems of *any* shape provided the pressure is uniform with position over the moving boundary.

### Actual Expansion or Compression Processes

To perform the integral of Eq. 3.9 requires a relationship between the gas pressure *at the moving boundary* and the system volume, but this relationship may be difficult, or even impossible, to obtain for actual compressions and expansions. In the cylinder of an automobile engine, for example, combustion and other nonequilibrium effects give rise to nonuniformities throughout the cylinder. Accordingly, if a pressure transducer were mounted on the cylinder head, the recorded output might provide only an approximation for the pressure at the

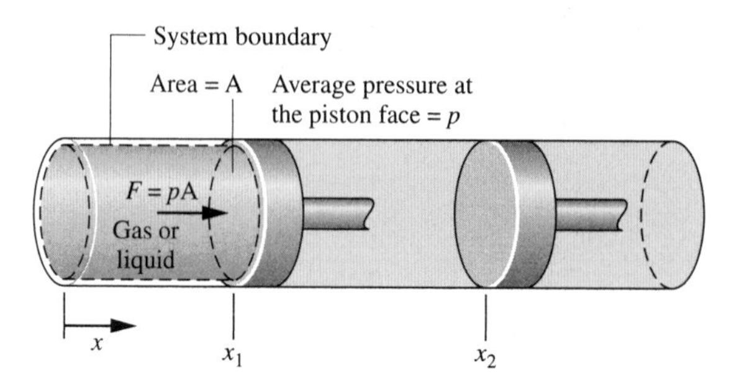


*Figure 3.2* Expansion or compression of a gas or liquid.

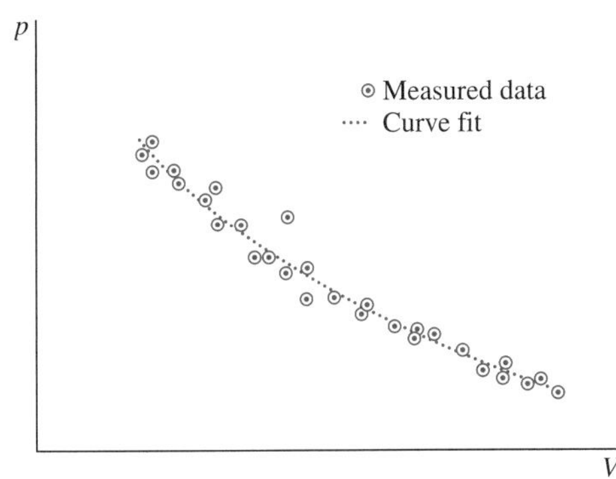


*Figure 3.3* Pressure–volume data.

piston face required by Eq. 3.9. Moreover, even when the measured pressure is essentially equal to that at the piston face, scatter might exist in the pressure–volume data, as illustrated in Fig. 3.3. We will see later that in some cases where lack of the required pressure–volume relationship keeps us from evaluating the work from Eq. 3.9, the work can be determined alternatively from an *energy balance* (Sec. 3.6).

## Quasiequilibrium Expansion or Compression Processes

An idealized type of process called a *quasiequilibrium* process is introduced in Sec. 2.2. A ***quasiequilibrium process*** is one in which all states through which the system passes may be considered equilibrium states. A particularly important aspect of the quasiequilibrium process concept is that the values of the intensive properties are uniform throughout the system, or every phase present in the system, at each state visited.

*quasiequilibrium process*

To consider how a gas (or liquid) might be expanded or compressed in a quasiequilibrium fashion, refer to Fig. 3.4, which shows a system consisting of a gas initially at an equilibrium state. As shown in the figure, the gas pressure is maintained uniform throughout by a number of small masses resting on the freely moving piston. Imagine that one of the masses is removed, allowing the piston to move upward as the gas expands slightly. During such an expansion the state of the gas would depart only slightly from equilibrium. The system would eventually come to a new equilibrium state, where the pressure and all other intensive properties would again be uniform in value. Moreover, were the mass replaced, the gas would be restored to its initial state, while again the departure from equilibrium would be slight. If several of the masses were removed one after another, the gas would pass through a sequence of equilibrium states without ever being far from equilibrium. In the limit as the increments of mass are made vanishingly small, the gas would undergo a quasiequilibrium expansion process. A quasiequilibrium compression can be visualized with similar considerations.

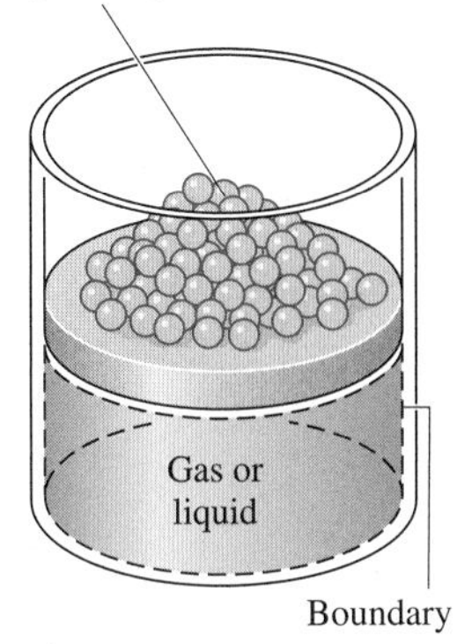


*Figure 3.4* Illustration of a quasiequilibrium expansion or compression.

Equation 3.9 can be applied to evaluate the work in quasiequilibrium expansion or compression processes. For such idealized processes the pressure $p$ in the equation is the pressure of the entire quantity of gas (or liquid) undergoing the process, and not just the pressure at the moving boundary. The relationship between the pressure and volume may be graphical or analytical. Let us first consider a graphical relationship.

A graphical relationship is shown in the pressure-volume diagram ($p$-$V$ diagram) of Fig. 3.5. Initially, the piston face is at position $x_1$, and the gas pressure is $p_1$; at the conclusion of a quasiequilibrium expansion process the piston face is at position $x_2$, and the pressure is reduced to $p_2$. At *each* intervening piston position, the uniform pressure throughout the gas is shown as a point on the diagram. The curve, or *path,* connecting states 1 and 2 on the diagram represents the equilibrium states through which the system has passed during the process. The work done by the gas on the piston during the expansion is given by $\int p\,dV$, which can be interpreted as the area under the curve of pressure versus volume. Thus, the shaded area on Fig. 3.5 is equal to the work for the process. Had the gas been *compressed* from 2 to 1 along the same path on

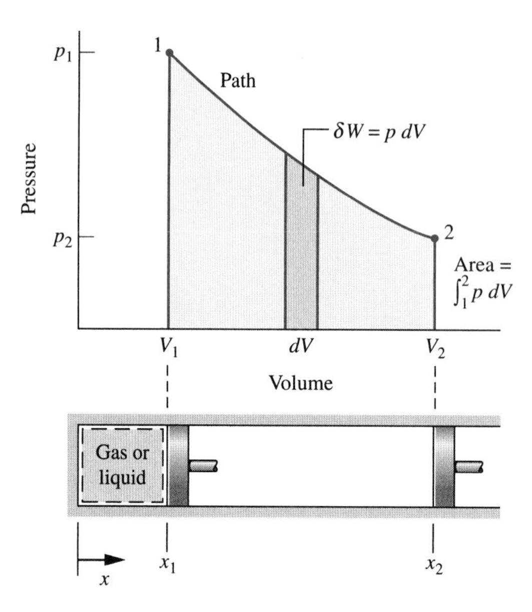


*Figure 3.5* Work of a quasiequilibrium expansion or compression process.

the $p$–$V$ diagram, the *magnitude* of the work would be the same, but the sign would be negative, indicating that for the compression the energy transfer was from the piston to the gas.

The area interpretation of work in a quasiequilibrium expansion or compression process allows a simple demonstration of the idea that work depends on the process. This can be brought out by referring to Fig. 3.6. Suppose the gas in a piston–cylinder assembly goes from an initial equilibrium state 1 to a final equilibrium state 2 along two different paths, labeled A and B on Fig. 3.6. Since the area beneath each path represents the work for that process, the work depends on the details of the process as defined by the particular curve and not just on the end states. Recalling the discussion of property given in Sec. 2.2, we can conclude that *work is not a property.* The value of work depends on the nature of the process between the end states.

The relationship between pressure and volume during an expansion or compression process also can be described analytically. An example is provided by the expression $pV^n = constant$, where the value of $n$ is a constant for the particular process. A quasiequilibrium process described by such an expression is called a ***polytropic process.*** Additional analytical forms for the pressure–volume relationship also may be considered.

*polytropic process*

The example to follow illustrates the application of Eq. 3.9 when the relationship between pressure and volume during an expansion is described analytically as $pV^n = constant$.

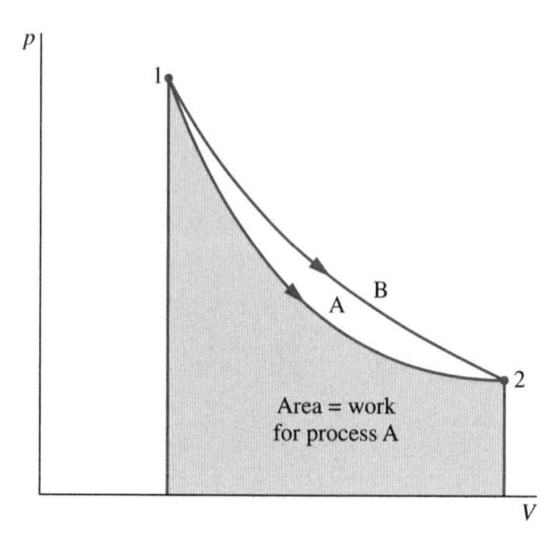


*Figure 3.6* Illustration that work depends on the process.

## *Example 3.1* Evaluating Expansion Work

A gas in a piston–cylinder assembly undergoes an expansion process for which the relationship between pressure and volume is given by

$$pV^n = constant$$

The initial pressure is 3 bar, the initial volume is 0.1 m$^3$, and the final volume is 0.2 m$^3$. Determine the work for the process, in kJ, if (a) $n = 1.5$, (b) $n = 1.0$, and (c) $n = 0$.

### Solution

***Known:*** A gas in a piston–cylinder assembly undergoes an expansion for which $pV^n = constant$.

***Find:*** Evaluate the work if (a) $n = 1.5$, (b) $n = 1.0$, (c) $n = 0$.

***Schematic and Given Data:*** The given $p$–$V$ relationship and the given data for pressure and volume can be used to construct the accompanying pressure–volume diagram of the process.

❶

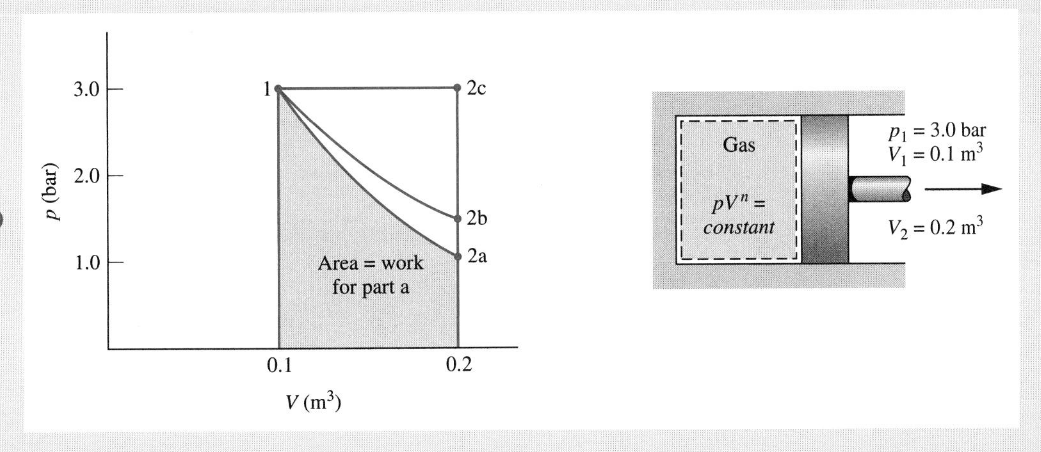


*Figure E3.1*

***Assumptions:***

1. The gas is a closed system.
2. The moving boundary is the only work mode.
3. ❷ The expansion is a polytropic process.

***Analysis:*** The required values for the work are obtained by integration of Eq. 3.9 using the given pressure–volume relation.

(a) Introducing the relationship $p = constant/V^n$ into Eq. 3.9 and performing the integration

$$W = \int_{V_1}^{V_2} p\,dV = \int_{V_1}^{V_2} \frac{constant}{V^n}\,dV$$

$$= \frac{(constant)\,V_2^{1-n} - (constant)\,V_1^{1-n}}{1-n}$$

The constant in this expression can be evaluated at either end state: $constant = p_1V_1^n = p_2V_2^n$. The work expression then becomes

$$W = \frac{(p_2V_2^n)\,V_2^{1-n} - (p_1V_1^n)\,V_1^{1-n}}{1-n} = \frac{p_2V_2 - p_1V_1}{1-n} \tag{1}$$

This expression is valid for all values of $n$ except $n = 1.0$. The case $n = 1.0$ is taken up in part (b).

To evaluate $W$, the pressure at state 2 is required. This can be found by using $p_1V_1^n = p_2V_2^n$, which on rearrangement yields.

$$p_2 = p_1\left(\frac{V_1}{V_2}\right)^n = (3\text{ bar})\left(\frac{0.1}{0.2}\right)^{1.5} = 1.06\text{ bar}$$

Accordingly

❸
$$W = \left(\frac{(1.06\text{ bar})(0.2\text{ m}^3) - (3)(0.1)}{1 - 1.5}\right)\left|\frac{10^5\text{ N/m}^2}{1\text{ bar}}\right|\left|\frac{1\text{ kJ}}{10^3\text{ N}\cdot\text{m}}\right|$$
$$= +17.6\text{ kJ} \triangleleft$$

(b) For $n = 1.0$, the pressure–volume relationship is $pV = constant$ or $p = constant/V$. The work is

$$W = constant \int_{V_1}^{V_2} \frac{dV}{V} = (constant)\ln\frac{V_2}{V_1} = (p_1V_1)\ln\frac{V_2}{V_1} \tag{2}$$

Substituting values

$$W = (3\text{ bar})(0.1\text{ m}^3)\left|\frac{10^5\text{ N/m}^2}{1\text{ bar}}\right|\left|\frac{1\text{ kJ}}{10^3\text{ N}\cdot\text{m}}\right|\ln\left(\frac{0.2}{0.1}\right) = +20.79\text{ kJ} \triangleleft$$

❹ (c) For $n = 0$, the pressure–volume relation reduces to $p = constant$, and the integral becomes $W = p(V_2 - V_1)$, which is a special case of the expression found in part (a). Substituting values and converting units as above, $W = +30$ kJ.

---

❶ In each case, the work for the process can be interpreted as the area under the curve representing the process on the accompanying $p$–$V$ diagram. Note that the relative areas are in agreement with the numerical results.

❷ The assumption of a polytropic process is significant. If the given pressure–volume relationship were obtained as a fit to experimental pressure–volume data, the value of $\int p\,dV$ would provide a plausible estimate of the work only when the measured pressure is essentially equal to that exerted at the piston face.

❸ Observe the use of unit conversion factors here and in part (b).

❹ It is not necessary to identify the gas (or liquid) contained within the piston–cylinder assembly. The calculated values for $W$ are determined by the process path and the end states. However, if it is desired to evaluate other properties such as temperature, both the nature and amount of the substance must be provided because appropriate relations among the properties of the particular substance would then be required.

## 3.4 Broadening Our Understanding of Energy

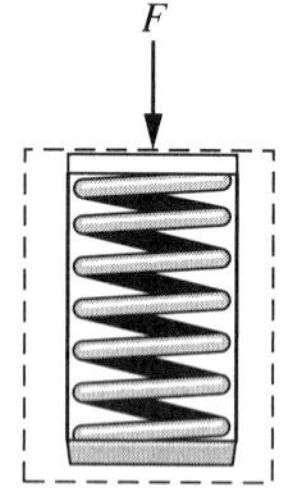


The objective in this section is to use our deeper understanding of work developed in Secs. 3.2 and 3.3 to broaden our understanding of the energy of a system. In particular, we consider the *total* energy of a system, which includes kinetic energy, gravitational potential energy, and other forms of energy. The examples to follow illustrate some of these forms of energy. Many other examples could be provided that enlarge on the same idea.

When work is done to compress a spring, energy is stored within the spring. When a battery is charged, the energy stored within it is increased. And when a gas (or liquid) initially at an equilibrium state in a closed, insulated vessel is stirred vigorously and allowed to come to a final equilibrium state, the energy of the gas is increased in the process. In each of these examples the change in system energy cannot be attributed to changes in the system's kinetic or gravitational potential energy. The change in energy can be accounted for in terms of *internal energy,* as considered next.

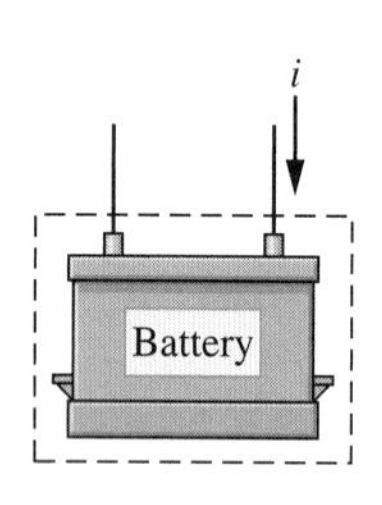


*internal energy*

In engineering thermodynamics the change in the total energy of a system is considered to be made up of three *macroscopic* contributions. One is the change in kinetic energy, associated with the motion of the system *as a whole* relative to an external coordinate frame. Another is the change in gravitational potential energy, associated with the position of the system *as a whole* in the earth's gravitational field. All other energy changes are lumped together in the ***internal energy*** of the system. Like kinetic energy and gravitational potential energy, *internal energy is an extensive property* of the system, as is the total energy.

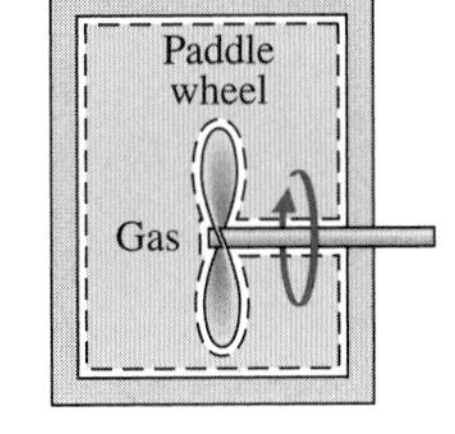


Internal energy is represented by the symbol $U$, and the change in internal energy in a process is $U_2 - U_1$. The specific internal energy is symbolized by $u$ or $\bar{u}$, respectively, depending on whether it is expressed on a unit mass or per mole basis.

The change in the total energy of a system is

$$E_2 - E_1 = (\text{KE}_2 - \text{KE}_1) + (\text{PE}_2 - \text{PE}_1) + (U_2 - U_1)$$

or

$$\Delta E = \Delta\text{KE} + \Delta\text{PE} + \Delta U \tag{3.10}$$

All quantities in Eq. 3.10 are expressed in terms of the energy units previously introduced.

The identification of internal energy as a macroscopic form of energy is a significant step in the present development, for it sets the concept of energy in thermodynamics apart from that of mechanics. In Chap. 4 we will learn how to evaluate changes in internal energy for practically important cases involving gases, liquids, and solids by using empirical data.

To further our understanding of internal energy, consider a system we will often encounter in subsequent sections of the book, a system consisting of a gas contained in a tank. Let us develop a ***microscopic interpretation of internal energy*** by thinking of the energy attributed to the motions and configurations of the individual molecules, atoms, and subatomic particles making up the matter in the system. Gas molecules move about, encountering other molecules or the walls of the container. Part of the internal energy of the gas is the *translational* kinetic energy of the molecules. Other contributions to the internal energy include the kinetic energy due to *rotation* of the molecules relative to their centers of mass and the kinetic energy associated with *vibrational* motions within the molecules. In addition, energy is stored in the chemical bonds between the atoms that make up the molecules. Energy storage on the atomic level includes energy associated with electron orbital states, nuclear spin, and binding forces in the nucleus. In dense gases, liquids, and solids, intermolecular forces play an important role in affecting the internal energy.

*microscopic interpretation of internal energy for a gas*

## 3.5 Energy Transfer by Heat

Thus far, we have considered quantitatively only those interactions between a system and its surroundings that can be classed as work. However, closed systems also can interact with their surroundings in a way that cannot be categorized as work. An example is provided by a gas in a container undergoing a process while in contact with a flame at a temperature greater than that of the gas. This type of interaction is called an ***energy transfer by heat.***

*energy transfer by heat*

On the basis of experiment, beginning with the work of Joule in the early part of the nineteenth century, we know that energy transfers by heat are induced only as a result of a temperature difference between the system and its surroundings and occur only in the direction of decreasing temperature. Because the underlying concept is so important in thermal systems engineering, this section is devoted to a further consideration of energy transfer by heat.

### 3.5.1 Sign Convention and Notation

The symbol $Q$ denotes an amount of energy transferred across the boundary of a system in a heat interaction with the system's surroundings. Heat transfer *into* a system is taken to be *positive,* and heat transfer *from* a system is taken as *negative.*

$Q > 0$: heat transfer *to* the system
$Q < 0$: heat transfer *from* the system

This ***sign convention*** is used throughout the book. However, as was indicated for work, it is sometimes convenient to show the direction of energy transfer by an arrow on a sketch of

*sign convention for heat transfer*

the system. Then the heat transfer is regarded as positive in the direction of the arrow. In an *adiabatic process* there is no energy transfer by heat.

*adiabatic process*

The sign convention for heat transfer is just the *reverse* of the one adopted for work, where a positive value for $W$ signifies an energy transfer *from* the system to the surroundings. These signs for heat and work are a legacy from engineers and scientists who were concerned mainly with steam engines and other devices that develop a work output from an energy input by heat transfer. For such applications, it was convenient to regard both the work developed and the energy input by heat transfer as positive quantities.

The value of a heat transfer depends on the details of a process and not just the end states. Thus, like work, *heat is not a property*, and its differential is written as $\delta Q$. The amount of energy transfer by heat for a process is given by the integral

*heat is not a property*

$$Q = \int_1^2 \delta Q$$

where the limits mean "from state 1 to state 2" and do not refer to the values of heat at those states. As for work, the notion of "heat" at a state has no meaning, and the integral should *never* be evaluated as $Q_2 - Q_1$.

Methods based on experiment are available for evaluating energy transfer by heat. We refer to the different types of heat transfer processes as *modes*. There are three primary modes: conduction, convection, and radiation. *Conduction* refers to energy transfer by heat through a medium across which a temperature difference exists. *Convection* refers to energy transfer between a surface and a moving or still fluid having a different temperature. The third mode is termed thermal *radiation* and represents the net exchange of energy between surfaces at different temperatures by electromagnetic waves independent of any intervening medium. For these modes, the rate of energy transfer depends on the properties of the substances involved, geometrical parameters and temperatures. The physical origins and rate equations for these modes are introduced in Section 15.1.

*Units.* The units for $Q$ and the heat transfer rate $\dot{Q}$ are the same as those introduced previously for $W$ and $\dot{W}$, respectively.

### 3.5.2 Closure

The first step in a thermodynamic analysis is to define the system. It is only after the system boundary has been specified that possible heat interactions with the surroundings are considered, for these are *always* evaluated at the system boundary. In ordinary conversation, the term *heat* is often used when the word *energy* would be more correct thermodynamically. For example, one might hear, "Please close the door or 'heat' will be lost." In *thermodynamics,* heat refers only to a particular means whereby energy is transferred. It does not refer to what is being transferred between systems or to what is stored within systems. Energy is transferred and stored, not heat.

Sometimes the heat transfer of energy to, or from, a system can be neglected. This might occur for several reasons related to the mechanisms for heat transfer discussed in Sec. 15.1. One might be that the materials surrounding the system are good insulators, or heat transfer might not be significant because there is a small temperature difference between the system and its surroundings. A third reason is that there might not be enough surface area to allow significant heat transfer to occur. When heat transfer is neglected, it is because one or more of these considerations apply.

In the discussions to follow, the value of $Q$ is provided or it is an unknown in the analysis. When $Q$ is provided, it can be assumed that the value has been determined by the methods introduced in Sec. 15.1. When $Q$ is the unknown, its value is usually found by using the *energy balance,* discussed next.

## 3.6 Energy Acounting: Energy Balance for Closed Systems

As our previous discussions indicate, the *only ways* the energy of a closed system can be changed is through transfer of energy by work or by heat. Further, a fundamental aspect of the energy concept is that energy is conserved. This is the ***first law of thermodynamics.*** These considerations are summarized in words as follows:

*first law of thermodynamics*

$$\begin{bmatrix} \textit{change} \text{ in the amount} \\ \text{of energy contained} \\ \text{within the system} \\ \text{during some time} \\ \text{interval} \end{bmatrix} = \begin{bmatrix} \textit{net} \text{ amount of energy} \\ \text{transferred } \textit{in} \text{ across} \\ \text{the system boundary by} \\ \textit{heat} \text{ transfer during} \\ \text{the time interval} \end{bmatrix} - \begin{bmatrix} \textit{net} \text{ amount of energy} \\ \text{transferred } \textit{out} \text{ across} \\ \text{the system boundary} \\ \text{by } \textit{work} \text{ during the} \\ \text{time interval} \end{bmatrix}$$

This word statement is just an accounting balance for energy, an energy balance. It requires that in any process of a closed system the energy of the system increases or decreases by an amount equal to the net amount of energy transferred across its boundary.

The phrase *net amount* used in the word statement of the energy balance must be carefully interpreted, for there may be heat or work transfers of energy at many different places on the boundary of a system. At some locations the energy transfers may be into the system, whereas at others they are out of the system. The two terms on the right side account for the *net* results of all the energy transfers by heat and work, respectively, taking place during the time interval under consideration.

The ***energy balance*** can be expressed in symbols as

$$E_2 - E_1 = Q - W \tag{3.11a}$$

Introducing Eq. 3.10 an alternative form is

*energy balance*

$$\Delta \text{KE} + \Delta \text{PE} + \Delta U = Q - W \tag{3.11b}$$

which shows that an energy transfer across the system boundary results in a change in one or more of the macroscopic energy forms: kinetic energy, gravitational potential energy, and internal energy. All previous references to energy as a conserved quantity are included as special cases of Eqs. 3.11.

Note that the algebraic signs before the heat and work terms of Eqs. 3.11 are different. This follows from the sign conventions previously adopted. A minus sign appears before $W$ because energy transfer by work *from* the system *to* the surroundings is taken to be positive. A plus sign appears before $Q$ because it is regarded to be positive when the heat transfer of energy is *into* the system *from* the surroundings.

### Other Forms of the Energy Balance

Various special forms of the energy balance can be written. For example, the energy balance in differential form is

$$dE = \delta Q - \delta W \tag{3.12}$$

where $dE$ is the differential of energy, a property. Since $Q$ and $W$ are not properties, their differentials are written as $\delta Q$ and $\delta W$, respectively.

The instantaneous ***time rate form of the energy balance*** is

$$\frac{dE}{dt} = \dot{Q} - \dot{W} \tag{3.13}$$

*time rate form of the energy balance*

The rate form of the energy balance expressed in words is

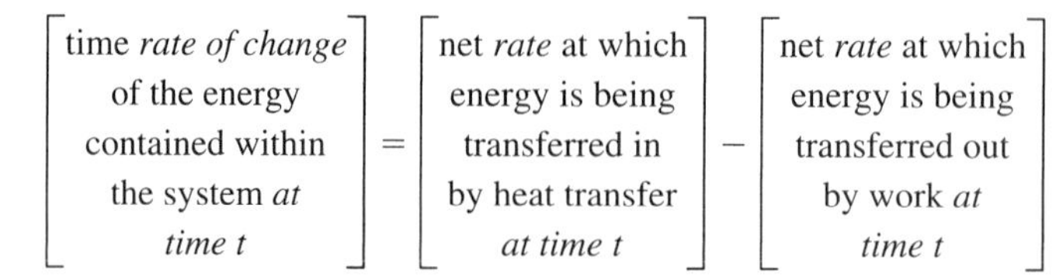

$$\begin{bmatrix}\text{time } \textit{rate of change}\\ \text{of the energy}\\ \text{contained within}\\ \text{the system } \textit{at}\\ \textit{time } t\end{bmatrix} = \begin{bmatrix}\text{net } \textit{rate} \text{ at which}\\ \text{energy is being}\\ \text{transferred in}\\ \text{by heat transfer}\\ \textit{at time } t\end{bmatrix} - \begin{bmatrix}\text{net } \textit{rate} \text{ at which}\\ \text{energy is being}\\ \text{transferred out}\\ \text{by work } \textit{at}\\ \textit{time } t\end{bmatrix}$$

Equations 3.11 through 3.13 provide alternative forms of the energy balance that may be convenient starting points when applying the principle of conservation of energy to closed systems. In Chap. 5 the conservation of energy principle is expressed in forms suitable for the analysis of control volumes. When applying the energy balance in *any* of its forms, it is important to be careful about signs and units and to distinguish carefully between rates and amounts. In addition, it is important to recognize that the location of the system boundary can be relevant in determining whether a particular energy transfer is regarded as heat or work.

***For Example...*** consider Fig. 3.7, in which three alternative systems are shown that include a quantity of a gas (or liquid) in a rigid, well-insulated container. In Fig. 3.7*a*, the gas itself is the system. As current flows through the copper plate, there is an energy transfer from the copper plate to the gas. Since this energy transfer occurs as a result of the temperature difference between the plate and the gas, it is classified as a heat transfer. Next, refer to Fig. 3.7*b*, where the boundary is drawn to include the copper plate. It follows from the thermodynamic definition of work that the energy transfer that occurs as current crosses the boundary of this system must be regarded as work. Finally, in Fig. 3.7*c*, the boundary is located so that no energy is transferred across it by heat or work. ▲

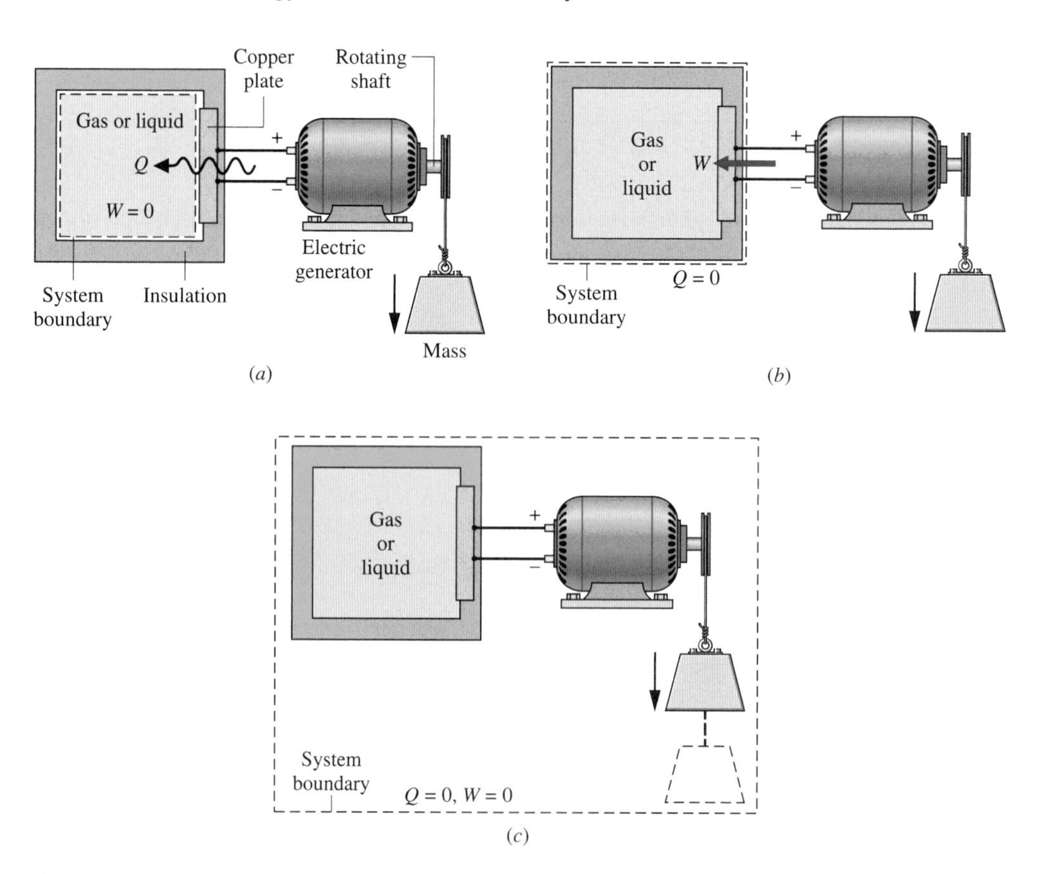


*Figure 3.7* Alternative choices for system boundaries.

*Closing Comment.* Thus far, we have been careful to emphasize that the quantities symbolized by $W$ and $Q$ in the foregoing equations account for transfers of *energy* and not transfers of work and heat, respectively. The terms work and heat denote different *means* whereby energy is transferred and not *what* is transferred. However, to achieve economy of expression in subsequent discussions, $W$ and $Q$ are often referred to simply as work and heat transfer, respectively. This less formal manner of speaking is commonly used in engineering practice.

## Illustrations

The examples to follow bring out many important ideas about energy and the energy balance. They should be studied carefully, and similar approaches should be used when solving the end-of-chapter problems.

In this text, most applications of the energy balance will not involve significant kinetic or potential energy changes. Thus, to expedite the solutions of many subsequent examples and end-of-chapter problems, we indicate in the problem statement that such changes can be neglected. If this is not made explicit in a problem statement, you should decide on the basis of the problem at hand how best to handle the kinetic and potential energy terms of the energy balance.

*Processes of Closed Systems.* The next two examples illustrate the use of the energy balance for processes of closed systems. In these examples, internal energy data are provided. In Chap. 4, we learn how to obtain thermodynamic property data using tables, graphs, equations, and computer software.

### *Example 3.2* Cooling a Gas in a Piston-Cylinder

Four kilograms of a certain gas is contained within a piston–cylinder assembly. The gas undergoes a process for which the pressure–volume relationship is

$$pV^{1.5} = constant$$

The initial pressure is 3 bar, the initial volume is 0.1 m$^3$, and the final volume is 0.2 m$^3$. The change in specific internal energy of the gas in the process is $u_2 - u_1 = -4.6$ kJ/kg. There are no significant changes in kinetic or potential energy. Determine the net heat transfer for the process, in kJ.

## Solution

***Known:*** A gas within a piston–cylinder assembly undergoes an expansion process for which the pressure–volume relation and the change in specific internal energy are specified.

***Find:*** Determine the net heat transfer for the process.

***Schematic and Given Data:***

❶

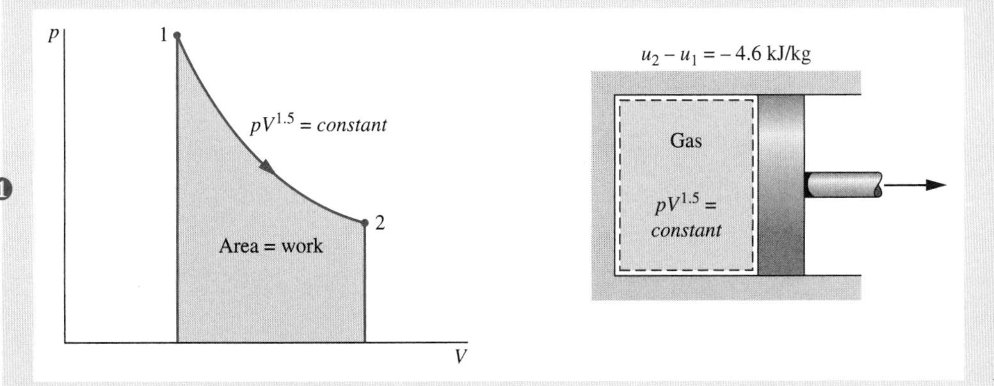


*Figure E3.2*

***Assumptions:***
1. The gas is a closed system.
2. The process is described by $pV^{1.5} = constant$.
3. There is no change in the kinetic or potential energy of the system.

***Analysis:*** An energy balance for the closed system takes the form

$$\cancelto{0}{\Delta KE} + \cancelto{0}{\Delta PE} + \Delta U = Q - W$$

where the kinetic and potential energy terms drop out by assumption 3. Then, writing $\Delta U$ in terms of specific internal energies, the energy balance becomes

$$m(u_2 - u_1) = Q - W$$

where $m$ is the system mass. Solving for $Q$

$$Q = m(u_2 - u_1) + W$$

The value of the work for this process is determined in the solution to part (a) of Example 3.1: $W = +17.6$ kJ. The change in internal energy is obtained using given data as

$$m(u_2 - u_1) = 4 \text{ kg}\left(-4.6 \frac{\text{kJ}}{\text{kg}}\right) = -18.4 \text{ kJ}$$

Substituting values

❷ $$Q = -18.4 + 17.6 = -0.8 \text{ kJ} \triangleleft$$

---

❶ The given relationship between pressure and volume allows the process to be represented by the path shown on the accompanying diagram. The area under the curve represents the work. Since they are not properties, the values of the work and heat transfer depend on the details of the process and cannot be determined from the end states only.

❷ The minus sign for the value of $Q$ means that a net amount of energy has been transferred from the system to its surroundings by heat transfer.

In the next example, we follow up the discussion of Fig. 3.7 by considering two alternative systems. This example highlights the need to account correctly for the heat and work interactions occurring on the boundary as well as the energy change.

## *Example 3.3* Considering Alternative Systems

Air is contained in a vertical piston–cylinder assembly fitted with an electrical resistor. The atmosphere exerts a pressure of 14.7 lbf/in.$^2$ on the top of the piston, which has a mass of 100 lb and a face area of 1 ft$^2$. Electric current passes through the resistor, and the volume of the air slowly increases by 1.6 ft$^3$ while its pressure remains constant. The mass of the air is 0.6 lb, and its specific internal energy increases by 18 Btu/lb. The air and piston are at rest initially and finally. The piston–cylinder material is a ceramic composite and thus a good insulator. Friction between the piston and cylinder wall can be ignored, and the local acceleration of gravity is $g = 32.0$ ft/s$^2$. Determine the heat transfer from the resistor to the air, in Btu, for a system consisting of (a) the air alone, (b) the air and the piston.

### Solution

***Known:*** Data are provided for air contained in a vertical piston–cylinder fitted with an electrical resistor.

***Find:*** Considering each of two alternative systems, determine the heat transfer from the resistor to the air.

***Schematic and Given Data:***

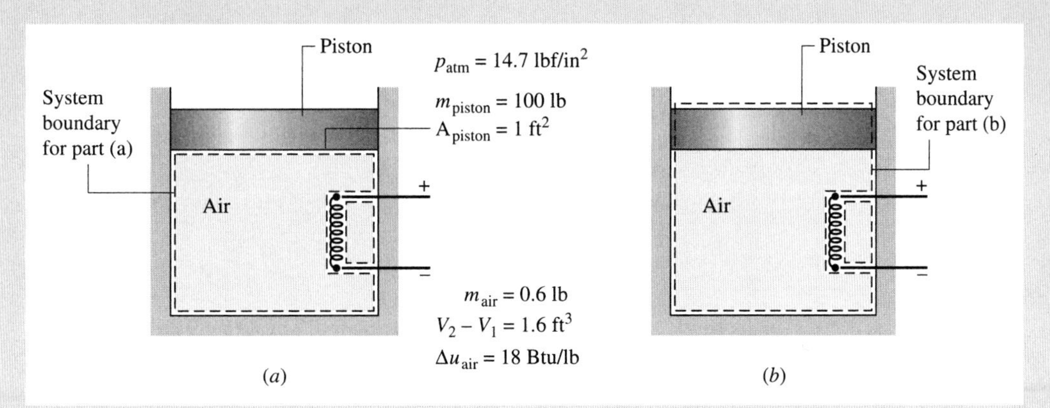


Figure E3.3

***Assumptions:***

1. Two closed systems are under consideration, as shown in the schematic.
2. The only significant heat transfer is from the resistor to the air, during which the air expands slowly and its pressure remains constant.
3. There is no net change in kinetic energy; the change in potential energy of the air is negligible; and since the piston material is a good insulator, the internal energy of the piston is not affected by the heat transfer. ❶
4. Friction between the piston and cylinder wall is negligible.
5. The acceleration of gravity is constant; $g = 32.0$ ft/s$^2$.

***Analysis:*** (a) Taking the air as the system, the energy balance, Eq. 3.11b, reduces with assumption 3 to

$$(\cancel{\Delta KE}^{0} + \cancel{\Delta PE}^{0} + \Delta U)_{air} = Q - W$$

Or, solving for $Q$

$$Q = W + \Delta U_{air}$$

For this system, work is done by the force of the pressure $p$ acting on the *bottom* of the piston as the air expands. With Eq. 3.9 and the assumption of constant pressure

$$W = \int_{V_1}^{V_2} p\, dV = p(V_2 - V_1)$$

To determine the pressure $p$, we use a force balance on the slowly moving, frictionless piston. The upward force exerted by the air on the *bottom* of the piston equals the weight of the piston plus the downward force of the atmosphere acting on the *top* of the piston. In symbols

$$pA_{piston} = m_{piston}\, g + p_{atm}A_{piston}$$

Solving for $p$ and inserting values

$$\begin{aligned} p &= \frac{m_{piston}\, g}{A_{piston}} + p_{atm} \\ &= \frac{(100 \text{ lb})(32.0 \text{ ft/s}^2)}{1 \text{ ft}^2} \left| \frac{1 \text{ lbf}}{32.2 \text{ lb} \cdot \text{ft/s}^2} \right| \left| \frac{1 \text{ ft}^2}{144 \text{ in.}^2} \right| + 14.7 \frac{\text{lbf}}{\text{in.}^2} = 15.4 \frac{\text{lbf}}{\text{in.}^2} \end{aligned}$$

Thus, the work is

$$\begin{aligned} W &= p(V_2 - V_1) \\ &= \left(15.4 \frac{\text{lbf}}{\text{in.}^2}\right)(1.6 \text{ ft}^3) \left| \frac{144 \text{ in.}^2}{1 \text{ ft}^2} \right| \left| \frac{1 \text{ Btu}}{778 \text{ ft} \cdot \text{lbf}} \right| = 4.56 \text{ Btu} \end{aligned}$$

With $\Delta U_{\text{air}} = m_{\text{air}}(\Delta u_{\text{air}})$, the heat transfer is

$$Q = W + m_{\text{air}}(\Delta u_{\text{air}})$$
$$= 4.56 \text{ Btu} + (0.6 \text{ lb})\left(18 \frac{\text{Btu}}{\text{lb}}\right) = 15.4 \text{ Btu} \triangleleft$$

**(b)** Consider next a system consisting of the air and the piston. The energy change of the overall system is the sum of the energy changes of the air and the piston. Thus, the energy balance, Eq. 3.11*b*, reads

$$(\cancel{\Delta KE}^{0} + \cancel{\Delta PE}^{0} + \Delta U)_{\text{air}} + (\cancel{\Delta KE}^{0} + \Delta PE + \cancel{\Delta U}^{0})_{\text{piston}} = Q - W$$

where the indicated terms drop out by assumption 3. Solving for $Q$

$$Q = W + (\Delta PE)_{\text{piston}} + (\Delta U)_{\text{air}}$$

For this system, work is done at the *top* of the piston as it pushes aside the surrounding atmosphere. Applying Eq. 3.9

$$W = \int_{V_1}^{V_2} p\, dV = p_{\text{atm}}(V_2 - V_1)$$
$$= \left(14.7 \frac{\text{lbf}}{\text{in.}^2}\right)(1.6 \text{ ft}^3)\left|\frac{144 \text{ in.}^2}{1 \text{ ft}^2}\right|\left|\frac{1 \text{ Btu}}{778 \text{ ft} \cdot \text{lbf}}\right| = 4.35 \text{ Btu}$$

The elevation change, $\Delta z$, required to evaluate the potential energy change of the piston can be found from the volume change of the air and the area of the piston face as

$$\Delta z = \frac{V_2 - V_1}{A_{\text{piston}}} = \frac{1.6 \text{ ft}^3}{1 \text{ ft}^2} = 1.6 \text{ ft}$$

Thus, the potential energy change of the piston is

$$(\Delta PE)_{\text{piston}} = m_{\text{piston}}\, g \Delta z$$
$$= (100 \text{ lb})\left(32.0 \frac{\text{ft}}{\text{s}^2}\right)(1.6 \text{ ft})\left|\frac{1 \text{ lbf}}{32.2 \text{ lb} \cdot \text{ft/s}^2}\right|\left|\frac{1 \text{ Btu}}{778 \text{ ft} \cdot \text{lbf}}\right| = 0.2 \text{ Btu}$$

Finally

$$Q = W + (\Delta PE)_{\text{piston}} + m_{\text{air}}\Delta u_{\text{air}}$$
$$= 4.35 \text{ Btu} + 0.2 \text{ Btu} + (0.6 \text{ lb})\left(18 \frac{\text{Btu}}{\text{lb}}\right) = 15.4 \text{ Btu} \triangleleft$$

❷ which agrees with the result of part (a).

---

❶ Using the change in elevation $\Delta z$ determined in the analysis, the change in potential energy of the air is about $10^{-3}$ Btu, which is negligible in the present case. The calculation is left as an exercise.

❷ Although the value of $Q$ is the same for each system, observe that the values for $W$ differ. Also, observe that the energy changes differ, depending on whether the air alone or the air and the piston is the system.

***Steady-State Operation.*** A system is at steady state if none of its properties change with time (Sec. 2.2). Many devices operate at steady state or nearly at steady state, meaning that property variations with time are small enough to ignore. The two examples to follow illustrate the application of the energy rate equation to closed systems at steady state.

## *Example 3.4* Gearbox at Steady State

During steady-state operation, a gearbox receives 60 kW through the input shaft and delivers power through the output shaft. For the gearbox as the system, the rate of energy transfer by heat is

$$\dot{Q} = -\text{h}A(T_b - T_f)$$

where h is a constant, h = 0.171 kW/m$^2 \cdot$ K, A = 1.0 m$^2$ is the outer surface area of the gearbox, $T_b$ = 300 K (27°C) is the temperature at the outer surface, and $T_f$ = 293 K (20°C) is the temperature of the surrounding air away from the immediate vicinity of the gearbox. For the gearbox, evaluate the heat transfer rate and the power delivered through the output shaft, each in kW.

## Solution

***Known:*** A gearbox operates at steady state with a known power input. An expression for the heat transfer rate from the outer surface is also known.

***Find:*** Determine the heat transfer rate and the power delivered through the output shaft, each in kW.

***Schematic and Given Data:***

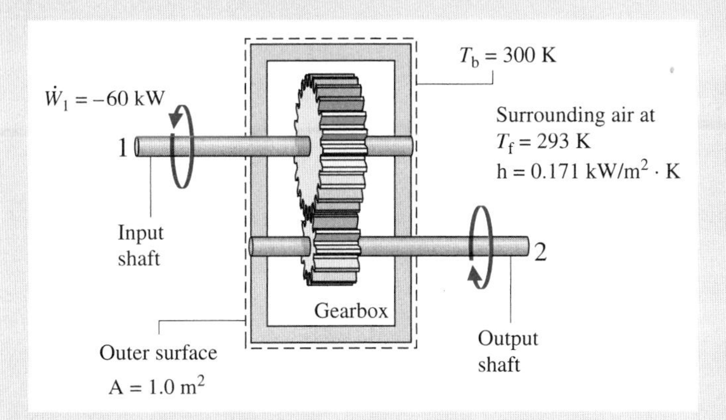


Figure E3.4

***Assumption:*** The gearbox is a closed system at steady state.

***Analysis:*** Using the given expression for $\dot{Q}$ together with known data, the rate of energy transfer by heat is

❶
$$\begin{aligned}\dot{Q} &= -\text{hA}(T_b - T_f) \\ &= -\left(0.171 \frac{\text{kW}}{\text{m}^2 \cdot \text{K}}\right)(1.0\ \text{m}^2)(300 - 293)\text{K} \\ &= -1.2\ \text{kW} \triangleleft\end{aligned}$$

The minus sign for $\dot{Q}$ signals that energy is carried *out* of the gearbox by heat transfer.

The energy rate balance, Eq. 3.13, reduces at steady state to

❷
$$\frac{dE}{dt}^{0} = \dot{Q} - \dot{W} \quad \text{or} \quad \dot{W} = \dot{Q}$$

The symbol $\dot{W}$ represents the *net* power from the system. The net power is the sum of $\dot{W}_1$ and the output power $\dot{W}_2$

$$\dot{W} = \dot{W}_1 + \dot{W}_2$$

With this expression for $\dot{W}$, the energy rate balance becomes

$$\dot{W}_1 + \dot{W}_2 = \dot{Q}$$

Solving for $\dot{W}_2$, inserting $\dot{Q}$ = −1.2 kW, and $\dot{W}_1$ = −60 kW, where the minus sign is required because the input shaft brings energy *into* the system, we have

❸
$$\begin{aligned}\dot{W}_2 &= \dot{Q} - \dot{W}_1 \\ &= (-1.2\ \text{kW}) - (-60\ \text{kW}) \\ &= +58.8\ \text{kW} \triangleleft\end{aligned}$$

❹ The positive sign for $\dot{W}_2$ indicates that energy is transferred from the system through the output shaft, as expected.

---

❶ This expression accounts for heat transfer by convection (Sec. 15.1). It is written to be in accord with the sign convention for the heat transfer rate in the energy rate balance (Eq. 3.13): $\dot{Q}$ is negative when $T_b$ is greater than $T_f$.

❷ Properties of a system at steady state do not change with time. Energy $E$ is a property, but heat transfer and work are not properties.

❸ For this system energy transfer by work occurs at two different locations, and the signs associated with their values differ.

❹ At steady state, the rate of heat transfer from the gear box accounts for the difference between the input and output power. This can be summarized by the following energy rate "balance sheet" in terms of *magnitudes:*

| | Input | Output |
|---|---|---|
| | 60 kW (input shaft) | 58.8 kW (output shaft) |
| | | 1.2 kW (heat transfer) |
| Total: | 60 kW | 60 kW |

## *Example 3.5* Silicon Chip at Steady State

A silicon chip measuring 5 mm on a side and 1 mm in thickness is embedded in a ceramic substrate. At steady state, the chip has an electrical power input of 0.225 W. The top surface of the chip is exposed to a coolant whose temperature is 20°C. The rate of energy transfer by heat between the chip and the coolant is given by $\dot{Q} = -\text{hA}(T_b - T_f)$, where $T_b$ and $T_f$ are the surface and coolant temperatures, respectively, A is the surface area, and $\text{h} = 150 \text{ W/m}^2 \cdot \text{K}$. If heat transfer between the chip and the substrate is negligible, determine the surface temperature of the chip, in °C.

### Solution

***Known:*** A silicon chip of known dimensions is exposed on its top surface to a coolant. The electrical power input and other data are known.

***Find:*** Determine the surface temperature of the chip at steady state.

***Schematic and Given Data:***

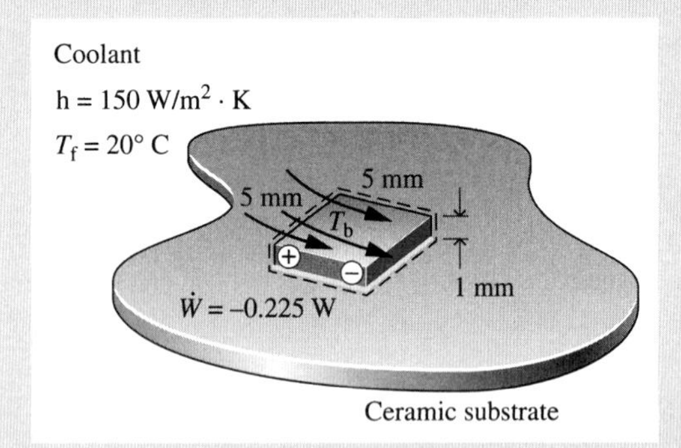


***Assumptions:***

1. The chip is a closed system at steady state.
2. There is no heat transfer between the chip and the substrate.

*Figure E3.5*

***Analysis:*** The surface temperature of the chip, $T_b$, can be determined using the energy rate balance, Eq. 3.13, which at steady state reduces as follows

❶
$$\cancelto{0}{\frac{dE}{dt}} = \dot{Q} - \dot{W}$$

With assumption 2, the only heat transfer is to the coolant, and is given by

❷
$$\dot{Q} = -\text{hA}(T_b - T_f)$$

Collecting results

$$0 = -\text{hA}(T_b - T_f) - \dot{W}$$

Solving for $T_b$

$$T_b = \frac{-\dot{W}}{\text{hA}} + T_f$$

In this expression, $\dot{W} = -0.225$ W, $A = 25 \times 10^{-6}$ m², $h = 150$ W/m²·K, and $T_f = 293$ K, giving

$$T_b = \frac{-(-0.225 \text{ W})}{(150 \text{ W/m}^2 \cdot \text{K})(25 \times 10^{-6} \text{ m}^2)} + 293 \text{ K}$$
$$= 353 \text{ K } (80°\text{C}) \triangleleft$$

❶ Properties of a system at steady state do not change with time. Energy $E$ is a property, but heat transfer and work are not properties.

❷ This expression accounts for heat transfer by convection (Sec. 15.1). It is written to be in accord with the sign convention for heat transfer in the energy rate balance (Eq. 3.13): $\dot{Q}$ is negative when $T_b$ is greater than $T_f$.

***Transient Operation.*** Many devices undergo periods of transient operation where the state changes with time. This is observed during startup and shutdown periods. The next example illustrates the application of the energy rate balance to an electric motor during startup. The example also involves both electrical work and power transmitted by a shaft.

### *Example 3.6* Transient Operation of a Motor

The rate of heat transfer between a certain electric motor and its surroundings varies with time as

$$\dot{Q} = -0.2[1 - e^{(-0.05t)}]$$

where $t$ is in seconds and $\dot{Q}$ is in kW. The shaft of the motor rotates at a constant speed of $\omega = 100$ rad/s (about 955 revolutions per minute, or RPM) and applies a constant torque of $\mathcal{T} = 18$ N·m to an external load. The motor draws a constant electric power input equal to 2.0 kW. For the motor, plot $\dot{Q}$ and $\dot{W}$, each in kW, and the change in energy $\Delta E$, in kJ, as functions of time from $t = 0$ to $t = 120$ s. Discuss.

**Solution (CD-ROM)**

## 3.7 Energy Analysis of Cycles

In this section the energy concepts developed thus far are illustrated further by application to systems undergoing thermodynamic cycles. Recall from Sec. 2.2 that when a system at a given initial state goes through a sequence of processes and finally returns to that state, the system has executed a thermodynamic cycle. The study of systems undergoing cycles has played an important role in the development of the subject of engineering thermodynamics. Both the first and second laws of thermodynamics have roots in the study of cycles. In addition, there are many important practical applications involving power generation, vehicle propulsion, and refrigeration for which an understanding of thermodynamic cycles is necessary. In this section, cycles are considered from the perspective of the conservation of energy principle. Cycles are studied in greater detail in subsequent chapters, using both the conservation of energy principle and the second law of thermodynamics.

### 3.7.1 Cycle Energy Balance

The energy balance for any system undergoing a thermodynamic cycle takes the form

$$\Delta E_{\text{cycle}} = Q_{\text{cycle}} - W_{\text{cycle}} \tag{3.14}$$

where $Q_{\text{cycle}}$ and $W_{\text{cycle}}$ represent *net* amounts of energy transfer by heat and work, respectively, for the cycle. Since the system is returned to its initial state after the cycle, there is no *net* change

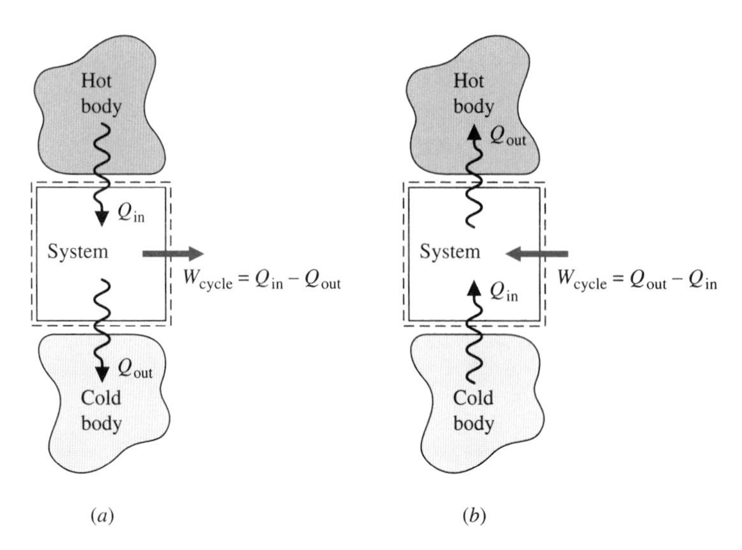


*Figure 3.8* Schematic diagrams of two important classes of cycles. (*a*) Power cycles. (*b*) Refrigeration and heat pump cycles.

in its energy. Therefore, the left side of Eq. 3.14 equals zero, and the equation reduces to

$$W_{cycle} = Q_{cycle} \tag{3.15}$$

Equation 3.15 is an expression of the conservation of energy principle that must be satisfied by *every* thermodynamic cycle, regardless of the sequence of processes followed by the system undergoing the cycle or the nature of the substances making up the system.

Figure 3.8 provides simplified schematics of two general classes of cycles considered in this book: power cycles and refrigeration and heat pump cycles. In each case pictured, a system undergoes a cycle while communicating thermally with two bodies, one hot and the other cold. These bodies are systems located in the surroundings of the system undergoing the cycle. During each cycle there is also a net amount of energy exchanged with the surroundings by work. Carefully observe that in using the symbols $Q_{in}$ and $Q_{out}$ on Fig. 3.8 we have departed from the previously stated sign convention for heat transfer. In this section it is advantageous to regard $Q_{in}$ and $Q_{out}$ as transfers of energy in the *directions indicated by the arrows*. The direction of the net work of the cycle, $W_{cycle}$, is *also indicated by an arrow.* Finally, note that the directions of the energy transfers shown in Fig. 3.8*b* are opposite to those of Fig. 3.8*a*.

## 3.7.2 Power Cycles

*power cycle*

Systems undergoing cycles of the type shown in Fig. 3.8*a* deliver a net work transfer of energy to their surroundings during each cycle. Any such cycle is called a *power cycle.* From Eq. 3.15, the net work output equals the net heat transfer to the cycle, or

$$W_{cycle} = Q_{in} - Q_{out} \quad \text{(power cycle)} \tag{3.16}$$

where $Q_{in}$ represents the heat transfer of energy *into* the system from the hot body, and $Q_{out}$ represents heat transfer *out* of the system to the cold body. From Eq. 3.16 it is clear that $Q_{in}$ must be greater than $Q_{out}$ for a *power* cycle. The energy supplied by heat transfer to a system undergoing a power cycle is normally derived from the combustion of fuel or a moderated nuclear reaction; it can also be obtained from solar radiation. The energy $Q_{out}$ is generally discharged to the surrounding atmosphere or a nearby body of water.

The performance of a system undergoing a *power cycle* can be described in terms of the extent to which the energy added by heat, $Q_{in}$, is *converted* to a net work output, $W_{cycle}$. The extent of the energy conversion from heat to work is expressed by the following ratio, commonly called the ***thermal efficiency:***

$$\eta = \frac{W_{cycle}}{Q_{in}} \quad \text{(power cycle)} \tag{3.17a}$$

*thermal efficiency*

Introducing Eq. 3.16, an alternative form is obtained as

$$\eta = \frac{Q_{in} - Q_{out}}{Q_{in}} = 1 - \frac{Q_{out}}{Q_{in}} \quad \text{(power cycle)} \tag{3.17b}$$

Since energy is conserved, it follows that the thermal efficiency can never be greater than unity (100%). However, experience with *actual* power cycles shows that the value of thermal efficiency is invariably *less* than unity. That is, not all the energy added to the system by heat transfer is converted to work; a portion is discharged to the cold body by heat transfer. Using the second law of thermodynamics, we will show in Chap. 6 that the conversion from heat to work cannot be fully accomplished by any power cycle. The thermal efficiency of *every* power cycle must be less than unity: $\eta < 1$.

## 3.7.3 Refrigeration and Heat Pump Cycles

*refrigeration and heat pump cycles*

Next, consider the ***refrigeration and heat pump cycles*** shown in Fig. 3.8*b*. For cycles of this type, $Q_{in}$ is the energy transferred by heat *into* the system undergoing the cycle *from* the cold body, and $Q_{out}$ is the energy discharged by heat transfer *from* the system *to* the hot body. To accomplish these energy transfers requires a net work *input*, $W_{cycle}$. The quantities $Q_{in}$, $Q_{out}$, and $W_{cycle}$ are related by the energy balance, which for refrigeration and heat pump cycles takes the form

$$W_{cycle} = Q_{out} - Q_{in} \quad \text{(refrigeration and heat pump cycles)} \tag{3.18}$$

Since $W_{cycle}$ is positive in this equation, it follows that $Q_{out}$ is greater than $Q_{in}$.

Although we have treated them as the same to this point, refrigeration and heat pump cycles actually have different objectives. The objective of a refrigeration cycle is to cool a refrigerated space or to maintain the temperature within a dwelling or other building *below* that of the surroundings. The objective of a heat pump is to maintain the temperature within a dwelling or other building *above* that of the surroundings or to provide heating for certain industrial processes that occur at elevated temperatures.

Since refrigeration and heat pump cycles have different objectives, their performance parameters, called *coefficients of performance,* are defined differently. These coefficients of performance are considered next.

### Refrigeration Cycles

The performance of *refrigeration cycles* can be described as the ratio of the amount of energy received by the system undergoing the cycle from the cold body, $Q_{in}$, to the net work into the system to accomplish this effect, $W_{cycle}$. Thus, the ***coefficient of performance,*** $\beta$, is

$$\beta = \frac{Q_{in}}{W_{cycle}} \quad \text{(refrigeration cycle)} \tag{3.19a}$$

*coefficient of performance*

Introducing Eq. 3.18, an alternative expression for $\beta$ is obtained as

$$\beta = \frac{Q_{\text{in}}}{Q_{\text{out}} - Q_{\text{in}}} \quad \text{(refrigeration cycle)} \tag{3.19b}$$

For a household refrigerator, $Q_{\text{out}}$ is discharged to the space in which the refrigerator is located. $W_{\text{cycle}}$ is usually provided in the form of electricity to run the motor that drives the refrigerator.

***For Example...*** in a refrigerator the inside compartment acts as the cold body and the ambient air surrounding the refrigerator is the hot body. Energy $Q_{\text{in}}$ passes to the circulating refrigerant *from* the food and other contents of the inside compartment. For this heat transfer to occur, the refrigerant temperature is necessarily below that of the refrigerator contents. Energy $Q_{\text{out}}$ passes *from* the refrigerant *to* the surrounding air. For this heat transfer to occur, the temperature of the circulating refrigerant must necessarily be above that of the surrounding air. To achieve these effects, a work *input* is required. For a refrigerator, $W_{\text{cycle}}$ is provided in the form of electricity. ▲

### Heat Pump Cycles

The performance of *heat pumps* can be described as the ratio of the amount of energy discharged from the system undergoing the cycle to the hot body, $Q_{\text{out}}$, to the net work into the system to accomplish this effect, $W_{\text{cycle}}$. Thus, the ***coefficient of performance,*** $\gamma$, is

*coefficient of performance*

$$\gamma = \frac{Q_{\text{out}}}{W_{\text{cycle}}} \quad \text{(heat pump cycle)} \tag{3.20a}$$

Introducing Eq. 3.18, an alternative expression for this coefficient of performance is obtained as

$$\gamma = \frac{Q_{\text{out}}}{Q_{\text{out}} - Q_{\text{in}}} \quad \text{(heat pump cycle)} \tag{3.20b}$$

From this equation it can be seen that the value of $\gamma$ is never less than unity. For residential heat pumps, the energy quantity $Q_{\text{in}}$ is normally drawn from the surrounding atmosphere, the ground, or a nearby body of water. $W_{\text{cycle}}$ is usually provided by electricity.

The coefficients of performance $\beta$ and $\gamma$ are defined as ratios of the desired heat transfer effect to the cost in terms of work to accomplish that effect. Based on the definitions, it is desirable thermodynamically that these coefficients have values that are as large as possible. However, as discussed in Chap. 6, coefficients of performance must satisfy restrictions imposed by the second law of thermodynamics.

## 3.8 Chapter Summary and Study Guide

In this chapter, we have considered the concept of energy from an engineering perspective and have introduced energy balances for applying the conservation of energy principle to closed systems. A basic idea is that energy can be stored within systems in three macroscopic forms: internal energy, kinetic energy, and gravitational potential energy. Energy also can be transferred to and from systems.

Energy can be transferred to and from closed systems by two means only: work and heat transfer. Work and heat transfer are identified at the system boundary and are not properties. In mechanics, work is energy transfer associated with forces and displacements at the system boundary. The thermodynamic definition of work introduced in this chapter extends the notion of work from mechanics to include other types of work. Energy transfer by heat is due to a temperature difference between the system and its surroundings, and occurs in the direction of decreasing temperature. Heat transfer modes include conduction, radiation, and convection. These sign conventions are used for work and heat transfer:

- $W, \dot{W} \begin{cases} > 0: \text{work done by the system} \\ < 0: \text{work done on the system} \end{cases}$
- $Q, \dot{Q} \begin{cases} > 0: \text{heat transfer to the system} \\ < 0: \text{heat transfer from the system} \end{cases}$

Energy is an extensive property of a system. Only changes in the energy of a system have significance. Energy changes are accounted for by the energy balance. The energy balance for a process of a closed system is Eq. 3.11 and an accompanying time rate form is Eq. 3.13. Equation 3.15 is a special form of the energy balance for a system undergoing a thermodynamic cycle.

The following checklist provides a study guide for this chapter. When your study of the text and end-of-chapter exercises has been completed, you should be able to

- write out the meanings of the terms listed in the margins throughout the chapter and understand each of the related concepts. The subset of key terms listed here in the margin is particularly important in subsequent chapters.
- evaluate these energy quantities

  –kinetic and potential energy changes using Eqs. 3.1 and 3.2, respectively.
  –work and power using Eqs. 3.3 and 3.4, respectively.
  –expansion or compression work using Eq. 3.9
- apply closed system energy balances in each of several alternative forms, appropriately modeling the case at hand, correctly observing sign conventions for work and heat transfer, and carefully applying SI and other units.
- conduct energy analyses for systems undergoing thermodynamic cycles using Eq. 3.15, and evaluating, as appropriate, the thermal efficiencies of power cycles and coefficients of performance of refrigeration and heat pump cycles.

*internal energy*
*kinetic energy*
*potential energy*
*work*
*power*
*heat transfer*
*adiabatic process*
*energy balance*
*power cycle*
*refrigeration cycle*
*heat pump cycle*

# Problems

## Energy Concepts from Mechanics

**3.1** An automobile has a mass of 1200 kg. What is its kinetic energy, in kJ, relative to the road when traveling at a velocity of 50 km/h? If the vehicle accelerates to 100 km/h, what is the change in kinetic energy, in kJ?

**3.2** An object of weight 40 kN is located at an elevation of 30 m above the surface of the earth. For $g = 9.78 \text{ m/s}^2$, determine the gravitational potential energy of the object, in kJ, relative to the surface of the earth.

**3.3** (CD-ROM)

**3.4** A body whose volume is 1.5 ft$^3$ and whose density is 3 lb/ft$^3$ experiences a decrease in gravitational potential energy of 500 ft · lbf. For $g = 31.0 \text{ ft/s}^2$, determine the change in elevation, in ft.

**3.5** What is the change in potential energy, in ft · lbf, of an automobile weighing 2600 lbf at sea level when it travels from sea level to an elevation of 2000 ft? Assume the acceleration of gravity is constant.

**3.6** An object of mass 10 kg, initially having a velocity of 500 m/s, decelerates to a final velocity of 100 m/s. What is the change in kinetic energy of the object, in kJ?

**3.7** (CD-ROM)

**3.8** (CD-ROM)

## Work and Power

**3.9** The drag force, $F_D$, imposed by the surrounding air on a vehicle moving with velocity V is given by

$$F_D = C_D A \tfrac{1}{2} \rho V^2$$

where $C_D$ is a constant called the drag coefficient, A is the projected frontal area of the vehicle, and $\rho$ is the air density. Determine the power, in kW, required to overcome aerodynamic drag for a truck moving at 110 km/h, if $C_D = 0.65$, $A = 10 \text{ m}^2$, and $\rho = 1.1 \text{ kg/m}^3$.

**3.10** A major force opposing the motion of a vehicle is the rolling resistance of the tires, $F_r$, given by

$$F_r = f\mathcal{W}$$

where $f$ is a constant called the rolling resistance coefficient and $\mathcal{W}$ is the vehicle weight. Determine the power, in kW, required to overcome rolling resistance for a truck weighing 322.5 kN that is moving at 110 km/h. Let $f = 0.0069$.

**3.11** (CD-ROM)

**3.12** Measured data for pressure versus volume during the compression of a refrigerant within the cylinder of a refrigeration compressor are given in the table below. Using data from the table, complete the following:

(a) Determine a value of $n$ such that the data are fit by an equation of the form $pV^n = constant$.

(b) Evaluate analytically the work done on the refrigerant, in Btu, using Eq. 3.9 along with the result of part (a).

| Data Point | $p$ (lbf/in.$^2$) | $V$ (in.$^3$) |
|---|---|---|
| 1 | 112 | 13.0 |
| 2 | 131 | 11.0 |
| 3 | 157 | 9.0 |
| 4 | 197 | 7.0 |
| 5 | 270 | 5.0 |
| 6 | 424 | 3.0 |

**3.13** (CD-ROM)

**3.14** One-half kg of a gas contained within a piston–cylinder assembly undergoes a constant-pressure process at 4 bar beginning at $v_1 = 0.72$ m$^3$/kg. For the gas as the system, the work is $-84$ kJ. Determine the final volume of the gas, in m$^3$.

**3.15** (CD-ROM)

**3.16** A gas is compressed from $V_1 = 0.09$ m$^3$, $p_1 = 1$ bar to $V_2 = 0.03$ m$^3$, $p_2 = 3$ bar. Pressure and volume are related linearly during the process. For the gas, find the work, in kJ.

**3.17** Carbon dioxide gas in a piston–cylinder assembly expands from an initial state where $p_1 = 60$ lbf/in.$^2$, $V_1 = 1.78$ ft$^3$ to a final pressure of $p_2 = 20$ lbf/in.$^2$ The relationship between pressure and volume during the process is $pV^{1.3} = constant$. For the gas, calculate the work done, in lb · lbf. Convert your answer to Btu.

**3.18** A gas expands from an initial state where $p_1 = 500$ kPa and $V_1 = 0.1$ m$^3$ to a final state where $p_2 = 100$ kPa. The relationship between pressure and volume during the process is $pV = constant$. Sketch the process on a $p$–$V$ diagram and determine the work, in kJ.

**3.19** A closed system consisting of 0.5 lbmol of air undergoes a polytropic process from $p_1 = 20$ lbf/in.$^2$, $v_1 = 9.26$ ft$^3$/lb to a final state where $p_2 = 60$ lbf/in.$^2$, $v_2 = 3.98$ ft$^3$/lb. Determine the amount of energy transfer by work, in Btu, for the process.

**3.20** Air undergoes two processes in series:

***Process 1–2:*** polytropic compression, with $n = 1.3$, from $p_1 = 100$ kPa, $v_1 = 0.04$ m$^3$/kg to $v_2 = 0.02$ m$^3$/kg

***Process 2–3:*** constant–pressure process to $v_3 = v_1$

Sketch the processes on a $p$–$v$ diagram and determine the work per unit mass of air, in kJ/kg.

**3.21** A gas undergoes three processes in series that complete a cycle:

***Process 1–2:*** compression from $p_1 = 10$ lbf/in.$^2$, $V_1 = 4.0$ ft$^3$ to $p_2 = 50$ lbf/in.$^2$ during which the pressure–volume relationship is $pv = constant$

***Process 2–3:*** constant volume to $p_3 = p_1$

***Process 3–1:*** constant pressure

Sketch the cycle on a $p$–$V$ diagram and determine the *net* work for the cycle, in Btu.

**3.22** (CD-ROM)

**3.23** The driveshaft of a building's air-handling fan is turned at 300 RPM by a belt running on a 0.3-m-diameter pulley. The net force applied circumferentially by the belt on the pulley is 2000 N. Determine the torque applied by the belt on the pulley, in N · m, and the power transmitted, in kW.

**3.24** Figure P 3.24 shows an object whose mass is 50 lb attached to a rope wound around a pulley. The radius of the pulley is 3 in. If the mass falls at a constant velocity of 3 ft/s, determine the power transmitted to the pulley, in horsepower, and the rotational speed of the pulley, in RPM. The acceleration of gravity is $g = 32.0$ ft/s$^2$.

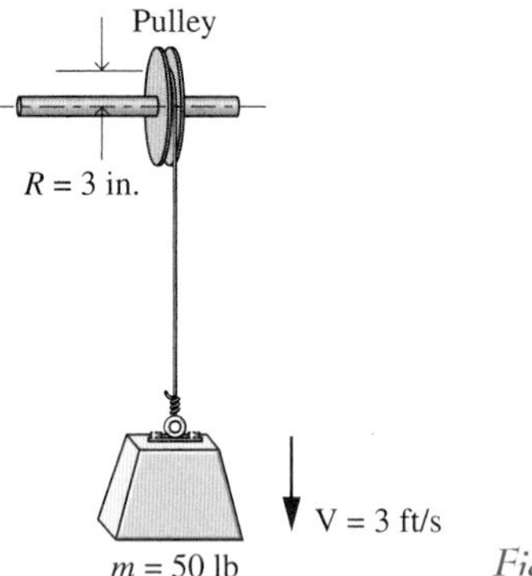


*Figure P3.24*

**3.25** An electric motor draws a current of 10 amp with a voltage of 110 V. The output shaft develops a torque of 10.2 N · m and a rotational speed of 1000 RPM. For operation at steady state, determine

(a) the electric power required by the motor and the power developed by the output shaft, each in kW.

(b) the net power input to the motor, in kW.

(c) the amount of energy transferred to the motor by electrical work and the amount of energy transferred out of the motor by the shaft, in kW · h during 2 h of operation.

**3.26** A 12-V automotive storage battery is charged with a constant current of 2 amp for 24 h. If electricity costs \$0.08 per kW · h, determine the cost of recharging the battery.

**3.27** (CD-ROM)

### Energy Balance

**3.28** Each line in the following table gives information about a process of a closed system. Every entry has the same energy units. Fill in the blank spaces in the table.

| Process | $Q$ | $W$ | $E_1$ | $E_2$ | $\Delta E$ |
|---|---|---|---|---|---|
| a | +50 | −20 | | +50 | |
| b | +50 | +20 | +20 | | |
| c | −40 | | | +60 | +20 |
| d | | −90 | | +50 | 0 |
| e | +50 | | +20 | | −100 |

**3.29** Each line in the following table gives information about a process of a closed system. Every entry has the same energy units. Fill in the blank spaces in the table.

| Process | $Q$ | $W$ | $E_1$ | $E_2$ | $\Delta E$ |
|---|---|---|---|---|---|
| a | +1000 | | +100 | +800 | |
| b | | −500 | +200 | +300 | |
| c | −200 | +300 | | +1000 | |
| d | | −400 | +400 | | +600 |
| e | −400 | | | +800 | −400 |

**3.30** A closed system of mass 2 kg undergoes a process in which there is heat transfer of magnitude 25 kJ from the system to the surroundings. The elevation of the system increases by 700 m during the process. The specific internal energy of the system *decreases* by 15 kJ/kg and there is no change in kinetic energy of the system. The acceleration of gravity is constant at $g = 9.6\ \text{m/s}^2$. Determine the work, in kJ.

**3.31** A closed system of mass 3 kg undergoes a process in which there is a heat transfer of 150 kJ from the system to the surroundings. The work done on the system is 75 kJ. If the initial specific internal energy of the system is 450 kJ/kg, what is the final specific internal energy, in kJ/kg? Neglect changes in kinetic and potential energy.

**3.32** (CD-ROM)

**3.33** A closed system of mass 2 lb undergoes two processes in series:

***Process 1–2:*** $v_1 = v_2 = 4.434\ \text{ft}^3/\text{lb}$, $p_1 = 100\ \text{lbf/in.}^2$, $u_1 = 1105.8\ \text{Btu/lb}$, $Q_{12} = -581.36\ \text{Btu}$

***Process 2–3:*** $p_2 = p_3 = 60\ \text{lbf/in.}^2$, $v_3 = 7.82\ \text{ft}^3/\text{lb}$, $u_3 = 1121.4\ \text{Btu/lb}$

Kinetic and potential energy effects can be neglected. Determine the work and heat transfer for process 2–3, each in Btu.

**3.34** An electric generator coupled to a windmill produces an average electric power output of 15 kW. The power is used to charge a storage battery. Heat transfer from the battery to the surroundings occurs at a constant rate of 1.8 kW. Determine, for 8 h of operation

**(a)** the total amount of energy stored in the battery, in kJ.
**(b)** the value of the stored energy, in \$, if electricity is valued at \$0.08 per kW · h.

**3.35** (CD-ROM)

**3.36** A closed system undergoes a process during which there is energy transfer *from* the system by heat at a constant rate of 10 kW, and the power varies with time according to

$$\dot{W} = \begin{cases} -8t & 0 < t \leq 1\ \text{h} \\ -8 & t > 1\ \text{h} \end{cases}$$

where $t$ is time, in h, and $\dot{W}$ is in kW.

**(a)** What is the time rate of change of system energy at $t = 0.6$ h, in kW?
**(b)** Determine the change in system energy after 2 h, in kJ.

**3.37** (CD-ROM)

**3.38** A gas expands in a piston–cylinder assembly from $p_1 = 8.2$ bar, $V_1 = 0.0136\ \text{m}^3$ to $p_2 = 3.4$ bar in a process during which the relation between pressure and volume is $pV^{1.2} = constant$. The mass of the gas is 0.183 kg. If the specific internal energy of the gas *decreases* by 29.8 kJ/kg during the process, determine the heat transfer, in kJ. Kinetic and potential energy effects are negligible.

**3.39** Air is contained in a rigid well-insulated tank with a volume of $0.6\ \text{m}^3$. The tank is fitted with a paddle wheel that transfers energy to the air at a constant rate of 4 W for 1 h. The initial density of the air is $1.2\ \text{kg/m}^3$. If no changes in kinetic or potential energy occur, determine

**(a)** the specific volume at the final state, in $\text{m}^3/\text{kg}$
**(b)** the energy transfer by work, in kJ.
**(c)** the change in specific internal energy of the air, in kJ/kg.

**3.40** A gas is contained in a closed rigid tank. An electric resistor in the tank transfers energy *to* the gas at a constant rate of 1000 W. Heat transfer between the gas and the surroundings occurs at a rate of $\dot{Q} = -50t$, where $\dot{Q}$ is in watts, and $t$ is time, in min.

**(a)** Plot the time rate of change of energy of the gas for $0 \leq t \leq 20$ min, in watts.
**(b)** Determine the net change in energy of the gas after 20 min, in kJ.
**(c)** If electricity is valued at \$0.08 per kW · h, what is the cost of the electrical input to the resistor for 20 min of operation?

**3.41** Steam in a piston–cylinder assembly undergoes a polytropic process, with $n = 2$, from an initial state where $p_1 = 500\ \text{lbf/in.}^2$, $v_1 = 1.701\ \text{ft}^3/\text{lb}$, $u_1 = 1363.3\ \text{Btu/lb}$ to a final state where $u_2 = 990.58\ \text{Btu/lb}$. During the process, there is a heat transfer from the steam of magnitude 342.9 Btu. The mass of steam is 1.2 lb. Neglecting changes in kinetic and potential energy, determine the work, in Btu, and the final specific volume, in $\text{ft}^3/\text{lb}$.

**3.42** A gas is contained in a vertical piston–cylinder assembly by a piston weighing 675 lbf and having a face area of 8 in.$^2$ The atmosphere exerts a pressure of 14.7 lbf/in.$^2$ on the top of the piston. An electrical resistor transfers energy to the gas in the amount of 3 Btu. The internal energy of the gas increases by 1 Btu, which is the only significant internal energy change of any component present. The piston and cylinder are poor thermal conductors and friction can be neglected. Determine the change in elevation of the piston, in ft.

**3.43** Air is contained in a vertical piston–cylinder assembly by a piston of mass 50 kg and having a face area of 0.01 $m^2$. The mass of the air is 4 g, and initially the air occupies a volume of 5 liters. The atmosphere exerts a pressure of 100 kPa on the top of the piston. Heat transfer of magnitude 1.41 kJ occurs slowly from the air to the surroundings, and the volume of the air decreases to 0.0025 $m^3$. Neglecting friction between the piston and the cylinder wall, determine the change in specific internal energy of the air, in kJ/kg.

**3.44** (CD-ROM)

### Thermodynamic Cycles

**3.45** The following table gives data, in kJ, for a system undergoing a thermodynamic cycle consisting of four processes in series. For the cycle, kinetic and potential energy effects can be neglected. Determine

(a) the missing table entries, each in kJ.
(b) whether the cycle is a power cycle or a refrigeration cycle.

| Process | $\Delta U$ | $Q$ | $W$ |
|---|---|---|---|
| 1–2 | | | −610 |
| 2–3 | 670 | | 230 |
| 3–4 | | 0 | 920 |
| 4–1 | −360 | | 0 |

**3.46** The following table gives data, in Btu, for a system undergoing a thermodynamic cycle consisting of four processes in series. Determine

(a) the missing table entries, each in Btu.
(b) whether the cycle is a power cycle or a refrigeration cycle.

| Process | $\Delta U$ | $\Delta KE$ | $\Delta PE$ | $\Delta E$ | $Q$ | $W$ |
|---|---|---|---|---|---|---|
| 1 | 950 | 50 | 0 | | 1000 | |
| 2 | | 0 | 50 | −450 | | 450 |
| 3 | −650 | | 0 | −600 | | 0 |
| 4 | 200 | −100 | −50 | | 0 | |

**3.47** A gas undergoes a thermodynamic cycle consisting of three processes:

*Process 1–2:* compression with $pV = constant$, from $p_1 = 1$ bar, $V_1 = 1.6\ m^3$ to $V_2 = 0.2\ m^3$, $U_2 - U_1 = 0$

*Process 2–3:* constant pressure to $V_3 = V_1$

*Process 3–1:* constant volume, $U_1 - U_3 = -3549$ kJ

There are no significant changes in kinetic or potential energy. Determine the heat transfer and work for Process 2–3, in kJ. Is this a power cycle or a refrigeration cycle?

**3.48** (CD-ROM)

**3.49** A closed system undergoes a thermodynamic cycle consisting of the following processes:

*Process 1–2:* adiabatic compression with $pV^{1.4} = constant$ from $p_1 = 50$ lbf/in.$^2$, $V_1 = 3\ ft^3$ to $V_2 = 1\ ft^3$

*Process 2–3:* constant volume

*Process 3–1:* constant pressure, $U_1 - U_3 = 46.7$ Btu

There are no significant changes in kinetic or potential energy.

(a) Sketch the cycle on a $p$–$V$ diagram.
(b) Calculate the net work for the cycle, in Btu.
(c) Calculate the heat transfer for process 2–3, in Btu.

**3.50** For a power cycle operating as in Fig. 3.8*a*, the heat transfers are $Q_{in} = 25{,}000$ kJ and $Q_{out} = 15{,}000$ kJ. Determine the net work, in kJ, and the thermal efficiency.

**3.51** (CD-ROM)

**3.52** The net work of a power cycle operating as in Fig. 3.8*a* is $8 \times 10^6$ Btu, and the heat transfer $Q_{out}$ is $12 \times 10^6$ Btu. What is the thermal efficiency of the power cycle?

**3.53** (CD-ROM)

**3.54** A power cycle receives energy by heat transfer from the combustion of fuel at a rate of 300 MW. The thermal efficiency of the cycle is 33.3%.

(a) Determine the net rate power is developed, in MW.
(b) For 8000 hours of operation annually, determine the net work output, in kW · h per year.
(c) Evaluating the net work output at $0.08 per kW · h, determine the value of the net work, in $/year.

**3.55** (CD-ROM)

**3.56** For each of the following, what plays the roles of the hot body and the cold body of the appropriate Fig. 3.8 schematic?

(a) Window air conditioner
(b) Nuclear submarine power plant
(c) Ground-source heat pump

**3.57** A refrigeration cycle operating as shown in Fig. 3.8*b* has heat transfer $Q_{out} = 3200$ Btu and net work of $W_{cycle} = 1200$ Btu. Determine the coefficient of performance for the cycle.

**3.58** (CD-ROM)

**3.59** A refrigeration cycle removes energy from the refrigerated space at a rate of 12,000 Btu/h. For a coefficient of performance of 2.6, determine the net power required, in Btu/h. Convert your answer to horsepower.

**3.60** A heat pump cycle whose coefficient of performance is 2.5 delivers energy by heat transfer to a dwelling at a rate of 20 kW.

(a) Determine the net power required to operate the heat pump, in kW.
(b) Evaluating electricity at $0.08 per kW · h, determine the cost of electricity in a month when the heat pump operates for 200 hours.

**3.61** (CD-ROM)

**3.62** A household refrigerator with a coefficient of performance of 2.4 removes energy from the refrigerated space at a rate of 600 Btu/h. Evaluating electricity at $0.08 per kW · h, determine the cost of electricity in a month when the refrigerator operates for 360 hours.

# EVALUATING PROPERTIES

## *Introduction...*

To apply the energy balance to a system of interest requires knowledge of the properties of the system and how the properties are related. The *objective* of this chapter is to introduce property relations relevant to engineering thermodynamics. As part of the presentation, several examples are provided that illustrate the use of the closed system energy balance introduced in Chap. 3 together with the property relations considered in this chapter.

*chapter objective*

## 4.1 Fixing the State

The state of a closed system at equilibrium is its condition as described by the values of its thermodynamic properties. From observation of many systems, it is known that not all of these properties are independent of one another, and the state can be uniquely determined by giving the values of the *independent* properties. Values for all other thermodynamic properties can be determined once this independent subset is specified. A general rule known as the *state principle* has been developed as a guide in determining the number of independent properties required to fix the state of a system.

*state principle*

For most applications, we are interested in what the state principle says about the *intensive* states of systems. Of particular interest are systems of commonly encountered substances, such as water or a uniform mixture of nonreacting gases. These systems are classed as *simple compressible systems.* Experience shows that the simple compressible systems model is useful for a wide range of engineering applications. For such systems, the state principle indicates that the number of independent intensive properties is *two.*

*simple compressible systems*

*For Example...* in the case of a gas, temperature and another intensive property such as specific volume might be selected as the two independent properties. The state principle then affirms that pressure, specific internal energy, and all other pertinent *intensive* properties could be determined as functions of $T$ and $v$: $p = p(T, v)$, $u = u(T, v)$, and so on. The functional relations would be developed using experimental data and would depend explicitly on the particular chemical identity of the substances making up the system. ▲

Intensive properties such as velocity and elevation that are assigned values relative to datums *outside* the system are excluded from present considerations. Also, as suggested by the name, changes in volume can have a significant influence on the energy of *simple compressible systems.* The only mode of energy transfer by work that can occur as a simple compressible system undergoes *quasiequilibrium* processes, is associated with volume change and is given by $\int p\, dV$.

# Evaluating Properties: General Considerations

This part of the chapter is concerned generally with the thermodynamic properties of simple compressible systems consisting of *pure* substances. A pure substance is one of uniform and invariable chemical composition. Property relations for systems in which composition changes by chemical reaction are not considered in this book. In the second part of this chapter, we consider property evaluation using the *ideal gas model.*

## 4.2 $p$–$v$–$T$ Relation

We begin our study of the properties of pure, simple compressible substances and the relations among these properties with pressure, specific volume, and temperature. From experiment it is known that temperature and specific volume can be regarded as independent and pressure determined as a function of these two: $p = p(T, v)$. The graph of such a function is a *surface,* the $p$–$v$–$T$ surface.

### 4.2.1 $p$–$v$–$T$ Surface

Figure 4.1 is the $p$–$v$–$T$ surface of water. Since similarities exist in the $p$–$v$–$T$ behavior of most pure substances, Fig. 4.1 can be regarded as representative. The coordinates of a point on the $p$–$v$–$T$ surface represents the values that pressure, specific volume, and temperature would assume when the substance is at equilibrium.

There are regions on the $p$–$v$–$T$ surface of Fig. 4.1 labeled *solid, liquid,* and *vapor*. In these *single-phase* regions, the state is fixed by *any* two of the properties: pressure, specific volume, and temperature, since all of these are independent when there is a single phase present. Located between the single-phase regions are ***two-phase regions*** where two phases exist in equilibrium: liquid–vapor, solid–liquid, and solid–vapor. Two phases can coexist during changes in phase such as vaporization, melting, and sublimation. Within the two-phase regions, pressure and temperature are not independent; one cannot be changed without changing the other. In these regions the state cannot be fixed by temperature and pressure alone; however, the state can be fixed by specific volume and either pressure or temperature. Three phases can exist in equilibrium along the line labeled ***triple line.***

*two-phase regions*

*triple line*

*saturation state*

*vapor dome*

A state at which a phase change begins or ends is called a ***saturation state.*** The dome-shaped region composed of the two-phase liquid–vapor states is called the ***vapor dome.*** The lines bordering the vapor dome are called saturated liquid and saturated vapor lines. At the top of the dome, where the saturated liquid and saturated vapor lines meet, is the ***critical point.*** The *critical temperature* $T_c$ of a pure substance is the maximum temperature at which liquid and vapor phases can coexist in equilibrium. The pressure at the critical point is called the *critical pressure,* $p_c$. The specific volume at this state is the *critical specific volume*. Values of the critical point properties for a number of substances are given in Tables T-1 and T-1E located in the Appendix.

*critical point*

The three-dimensional $p$–$v$–$T$ surface is useful for bringing out the general relationships among the three phases of matter normally under consideration. However, it is often more convenient to work with two-dimensional projections of the surface. These projections are considered next.

### 4.2.2 Projections of the $p$–$v$–$T$ Surface

#### The Phase Diagram

If the $p$–$v$–$T$ surface is projected onto the pressure–temperature plane, a property diagram known as a ***phase diagram*** results. As illustrated by Fig. 4.1$b$, when the surface is projected

*phase diagram*

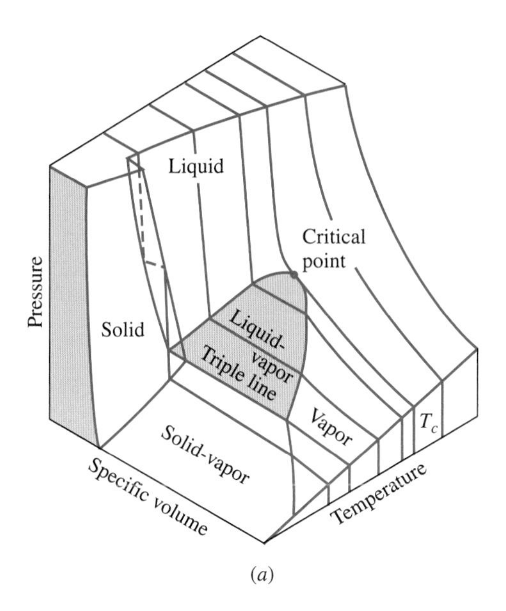


(*a*)

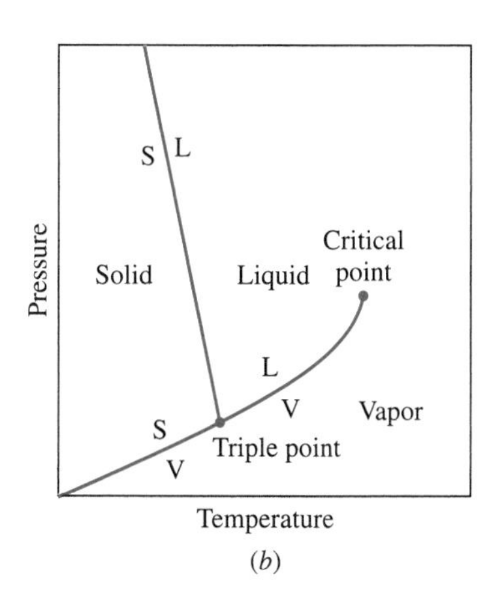


(*b*)

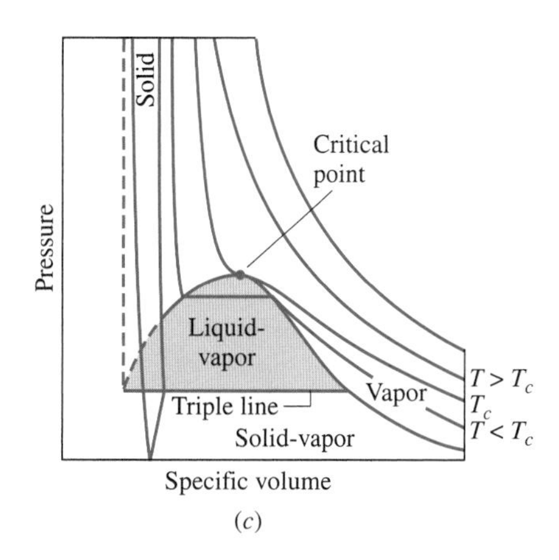


(*c*)

*Figure 4.1* $p$–$v$–$T$ surface and projections for water (not to scale). (*a*) Three-dimensional view. (*b*) Phase diagram. (*c*) $p$–$v$ diagram.

in this way, the two-phase *regions* reduce to *lines*. A point on any of these lines represents all two-phase mixtures at that particular temperature and pressure.

*saturation temperature*

*saturation pressure*

The term ***saturation temperature*** designates the temperature at which a phase change takes place at a given pressure, and this pressure is called the ***saturation pressure*** for the given temperature. It is apparent from the phase diagrams that for each saturation pressure there is a unique saturation temperature, and conversely.

*triple point*

The triple *line* of the three-dimensional $p$–$v$–$T$ surface projects onto a *point* on the phase diagram. This is called the ***triple point.*** Recall that the triple point of water is used as a reference in defining temperature scales (Sec. 2.5.4). By agreement, the temperature *assigned* to the triple point of water is 273.16 K (491.69°R). The *measured* pressure at the triple point of water is 0.6113 kPa (0.00602 atm).

The line representing the two-phase solid–liquid region on the phase diagram, Fig. 4.1*b*, slopes to the left for substances such as water that expand on freezing and to the right for those that contract. Although a single solid phase region is shown on the phase diagram,

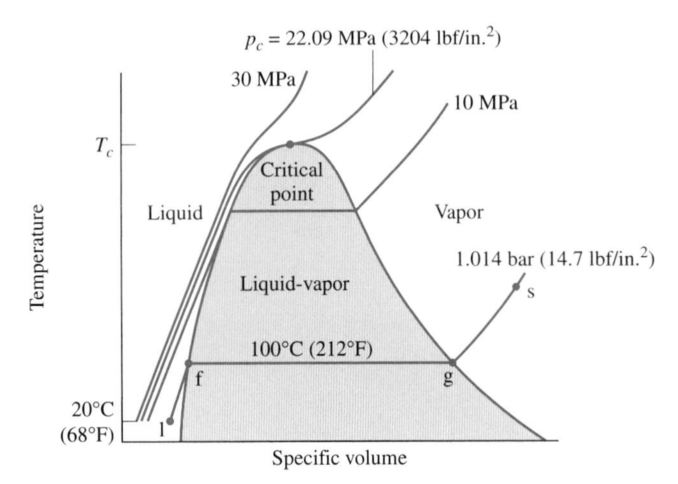


*Figure 4.2* Sketch of a temperature–specific volume diagram for water showing the liquid, two-phase liquid–vapor, and vapor regions (not to scale).

solids can exist in different solid phases. For example, seven different crystalline forms have been identified for water as a solid (ice).

### $p$–$v$ Diagram

Projecting the $p$–$v$–$T$ surface onto the pressure–specific volume plane results in a $p$–$v$ diagram, as shown by Fig. 4.1*c*. The figure is labeled with terms that have already been introduced.

When solving problems, a sketch of the $p$–$v$ diagram is frequently convenient. To facilitate the use of such a sketch, note the appearance of constant-temperature lines (isotherms). By inspection of Fig. 4.1*c*, it can be seen that for any specified temperature *less than* the critical temperature, pressure remains constant as the two-phase liquid–vapor region is traversed, but in the single-phase liquid and vapor regions the pressure decreases at fixed temperature as specific volume increases. For temperatures greater than or equal to the critical temperature, pressure decreases continuously at fixed temperature as specific volume increases. There is no passage across the two-phase liquid–vapor region. The critical isotherm passes through a point of inflection at the critical point and the slope is zero there.

### $T$–$v$ Diagram

Projecting the liquid, two-phase liquid–vapor, and vapor regions of the $p$–$v$–$T$ surface onto the temperature–specific volume plane results in a $T$–$v$ diagram as in Fig. 4.2.

As for the $p$–$v$ diagram, a sketch of the $T$–$v$ diagram is often convenient for problem solving. To facilitate the use of such a sketch, note the appearance of constant-pressure lines (isobars). For pressures *less than* the critical pressure, such as the 10 MPa isobar on Fig. 4.2, the pressure remains constant with temperature as the two-phase region is traversed. In the single-phase liquid and vapor regions, the temperature increases at fixed pressure as the specific volume increases. For pressures greater than or equal to the critical pressure, such as the one marked 30 MPa on Fig. 4.2, temperature increases continuously at fixed pressure as the specific volume increases. There is no passage across the two-phase liquid–vapor region.

The projections of the $p$–$v$–$T$ surface used in this book to illustrate processes are not generally drawn to scale. A similar comment applies to other property diagrams introduced later.

## 4.2.3 Studying Phase Change

It is instructive to study the events that occur as a pure substance undergoes a phase change. To begin, consider a closed system consisting of a unit mass (1 kg or 1 lb) of liquid water

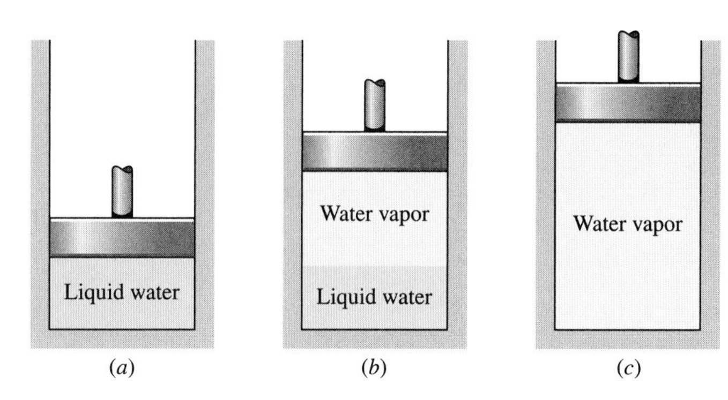


*Figure 4.3* Illustration of constant-pressure change from liquid to vapor for water.

at 20°C (68°F) contained within a piston–cylinder assembly, as illustrated in Fig. 4.3*a*. This state is represented by point l on Fig. 4.2. Suppose the water is slowly heated while its pressure is kept constant and uniform throughout at 1.014 bar (14.7 lbf/in.$^2$).

## Liquid States

As the system is heated at constant pressure, the temperature increases considerably while the specific volume increases slightly. Eventually, the system is brought to the state represented by f on Fig. 4.2. This is the saturated liquid state corresponding to the specified pressure. For water at 1.014 bar (14.7 lbf/in.$^2$) the saturation temperature is 100°C (212°F). The liquid states along the line segment l–f of Fig. 4.2 are sometimes referred to as *subcooled liquid* states because the temperature at these states is less than the saturation temperature at the given pressure. These states are also referred to as *compressed liquid* states because the pressure at each state is higher than the saturation pressure corresponding to the temperature at the state. The names liquid, subcooled liquid, and compressed liquid are used interchangeably.

*subcooled liquid*

*compressed liquid*

## Two-Phase, Liquid–Vapor Mixture

When the system is at the saturated liquid state (state f of Fig. 4.2), additional heat transfer at fixed pressure results in the formation of vapor without any change in temperature but with a considerable increase in specific volume. As shown in Fig. 4.3*b*, the system would now consist of a two-phase liquid–vapor mixture. When a mixture of liquid and vapor exists in equilibrium, the liquid phase is a saturated liquid and the vapor phase is a saturated vapor. If the system is heated further until the last bit of liquid has vaporized, it is brought to point g on Fig. 4.2, the saturated vapor state. The intervening *two-phase liquid–vapor mixtures* can be distinguished from one another by the *quality*, an intensive property.

*two-phase liquid–vapor mixture*

For a two-phase liquid–vapor mixture, the ratio of the mass of vapor present to the total mass of the mixture is its *quality*, $x$. In symbols,

*quality*

$$x = \frac{m_{\text{vapor}}}{m_{\text{liquid}} + m_{\text{vapor}}} \tag{4.1}$$

The value of the quality ranges from zero to unity: at saturated liquid states, $x = 0$, and at saturated vapor states, $x = 1.0$. Although defined as a ratio, the quality is frequently given as a percentage. Examples illustrating the use of quality are provided in Sec. 4.3. Similar parameters can be defined for two-phase solid–vapor and two-phase solid–liquid mixtures.

### Vapor States

Let us return to a consideration of Figs. 4.2 and 4.3. When the system is at the saturated vapor state (state g on Fig. 4.2), further heating at fixed pressure results in increases in both temperature and specific volume. The condition of the system would now be as shown in Fig. 4.3*c*. The state labeled s on Fig. 4.2 is representative of the states that would be attained by further heating while keeping the pressure constant. A state such as s is often referred to as a *superheated vapor* state because the system would be at a temperature greater than the saturation temperature corresponding to the given pressure.

*superheated vapor*

Consider next the same thought experiment at the other constant pressures labeled on Fig. 4.2, 10 MPa (1450 lbf/in.$^2$), 22.09 MPa (3204 lbf/in.$^2$), and 30 MPa (4351 lbf/in.$^2$). The first of these pressures is less than the critical pressure of water, the second is the critical pressure, and the third is greater than the critical pressure. As before, let the system initially contain a liquid at 20°C (68°F). First, let us study the system if it were heated slowly at 10 MPa (1450 lbf/in.$^2$). At this pressure, vapor would form at a higher temperature than in the previous example, because the saturation pressure is higher (refer to Fig. 4.2). In addition, there would be somewhat less of an increase in specific volume from saturated liquid to vapor, as evidenced by the narrowing of the vapor dome. Apart from this, the general behavior would be the same as before. Next, consider the behavior of the system were it heated at the critical pressure, or higher. As seen by following the critical isobar on Fig. 4.2, there would be no change in phase from liquid to vapor. At all states there would be only one phase. Vaporization (and the inverse process of condensation) can occur only when the pressure is less than the critical pressure. Thus, at states where pressure is greater than the critical pressure, the terms liquid and vapor tend to lose their significance. Still, for ease of reference to states where the pressure is greater than the critical pressure, we use the term liquid when the temperature is less than the critical temperature and vapor when the temperature is greater than the critical temperature.

## 4.3 Retrieving Thermodynamic Properties

Thermodynamic property data can be retrieved in various ways, including tables, graphs, equations, and computer software. The emphasis of the present section is on the use of *tables* of thermodynamic properties, which are commonly available for pure, simple compressible substances of engineering interest. The use of these tables is an important skill. The ability to locate states on property diagrams is an important related skill. The software *Interactive Thermodynamics: IT* available with this text is also used selectively in examples and end-of-chapter problems included on the CD. Skillful use of tables and property diagrams is prerequisite for the effective use of software to retrieve thermodynamic property data.

Since tables for different substances are frequently set up in the same general format, the present discussion centers mainly on Tables T-2 through T-5 giving the properties of water, commonly referred to as the *steam tables,* and Tables T-6 through T-8 for Refrigerant 134a. These tables are provided in the Appendix and on the CD. Similar tables are provided only on the CD for Refrigerant 22, ammonia, and propane. We provide all tables in SI units and in English units (see Sec. 2.3.2). Tables in English units are designated with a letter E. For example, the steam tables in English units are Tables T-2E through T-5E.

*steam tables*

### 4.3.1 Evaluating Pressure, Specific Volume, and Temperature

#### Vapor and Liquid Tables

The properties of water vapor are listed in Tables T-4 and of liquid water in Tables T-5. These are often referred to as the *superheated* vapor tables and *compressed* liquid tables, respectively. The sketch of the phase diagram shown in Fig. 4.4 brings out the structure of

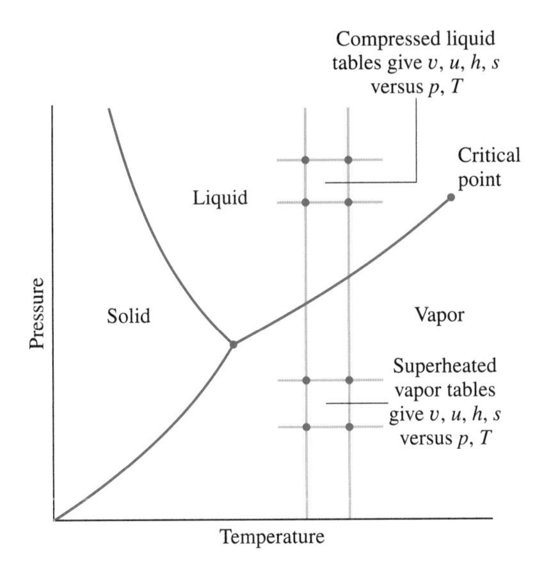


*Figure 4.4* Sketch of the phase diagram for water used to discuss the structure of the superheated vapor and compressed liquid tables (not to scale).

these tables. Since pressure and temperature are independent properties in the single-phase liquid and vapor regions, they can be used to fix the state in these regions. Accordingly, Tables T-4 and T-5 are set up to give values of several properties as functions of pressure and temperature. The first property listed is specific volume. The remaining properties are discussed in subsequent sections.

For each pressure listed, the values given in the superheated vapor table (Tables T-4) *begin* with the saturated vapor state and then proceed to higher temperatures. The data in the compressed liquid table (Tables T-5) *end* with saturated liquid states. That is, for a given pressure the property values are given as the temperature increases to the saturation temperature. In these tables, the value shown in parentheses after the pressure in the table heading is the corresponding saturation temperature.

***For Example...*** in Tables T-4 and T-5, at a pressure of 10.0 MPa, the saturation temperature is listed as 311.06°C. In Tables T-4E and T-5E, at a pressure of 500 lbf/in.$^2$, the saturation temperature is listed as 467.1°F. ▲

***For Example...*** to gain more experience with Tables T-4 and T-5 verify the following: Table T-4 gives the specific volume of water vapor at 10.0 MPa and 600°C as 0.03837 m$^3$/kg. At 10.0 MPa and 100°C, Table T-5 gives the specific volume of liquid water as $1.0385 \times 10^{-3}$ m$^3$/kg. Table T-4E gives the specific volume of water vapor at 500 lbf/in.$^2$ and 600°F as 1.158 ft$^3$/lb. At 500 lbf/in.$^2$ and 100°F, Table T-5E gives the specific volume of liquid water as 0.016106 ft$^3$/lb. ▲

The states encountered when solving problems often do not fall exactly on the grid of values provided by property tables. *Interpolation* between adjacent table entries then becomes necessary. Care always must be exercised when interpolating table values. The tables provided in the Appendix are extracted from more extensive tables that are set up so that *linear interpolation,* illustrated in the following example, can be used with acceptable accuracy. Linear interpolation is assumed to remain valid when using the abridged tables of the text for the solved examples and end-of-chapter problems.

*linear interpolation*

***For Example...*** let us determine the specific volume of water vapor at a state where $p = 10$ bar and $T = 215$°C. Shown in Fig. 4.5 is a sampling of data from Table T-4. At a

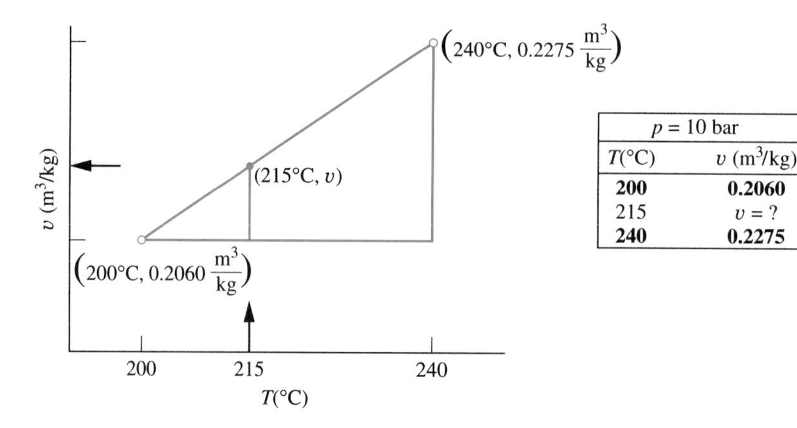


| $p = 10$ bar | |
|---|---|
| $T$(°C) | $v$ (m³/kg) |
| **200** | **0.2060** |
| 215 | $v = ?$ |
| **240** | **0.2275** |

*Figure 4.5* Illustration of linear interpolation.

pressure of 10 bar, the specified temperature of 215°C falls between the table values of 200 and 240°C, which are shown in bold face. The corresponding specific volume values are also shown in bold face. To determine the specific volume $v$ corresponding to 215°C, we may think of the *slope* of a straight line joining the adjacent table states, as follows

$$\textit{slope} = \frac{(0.2275 - 0.2060)\ \text{m}^3/\text{kg}}{(240 - 200)°\text{C}} = \frac{(v - 0.2060)\ \text{m}^3/\text{kg}}{(215 - 200)°\text{C}}$$

Solving for $v$, the result is $v = 0.2141\ \text{m}^3/\text{kg}$. ▲

## Saturation Tables

The saturation tables, Tables T-2 and T-3, list property values for the saturated liquid and vapor states. The property values at these states are denoted by the subscripts f and g, respectively. Table T-2 is called the *temperature table,* because temperatures are listed in the first column in convenient increments. The second column gives the corresponding saturation pressures. The next two columns give, respectively, the specific volume of saturated liquid, $v_f$, and the specific volume of saturated vapor, $v_g$. Table T-3 is called the *pressure table,* because pressures are listed in the first column in convenient increments. The corresponding saturation temperatures are given in the second column. The next two columns give $v_f$ and $v_g$, respectively.

The specific volume of a two-phase liquid–vapor mixture can be determined by using the saturation tables and the definition of quality given by Eq. 4.1 as follows. The total volume of the mixture is the sum of the volumes of the liquid and vapor phases

$$V = V_{\text{liq}} + V_{\text{vap}}$$

Dividing by the total mass of the mixture, $m$, an *average* specific volume for the mixture is obtained

$$v = \frac{V}{m} = \frac{V_{\text{liq}}}{m} + \frac{V_{\text{vap}}}{m}$$

Since the liquid phase is a saturated liquid and the vapor phase is a saturated vapor, $V_{\text{liq}} = m_{\text{liq}} v_f$ and $V_{\text{vap}} = m_{\text{vap}} v_g$, so

$$v = \left(\frac{m_{\text{liq}}}{m}\right) v_f + \left(\frac{m_{\text{vap}}}{m}\right) v_g$$

Introducing the definition of quality, $x = m_{vap}/m$, and noting that $m_{liq}/m = 1 - x$, the above expression becomes

$$v = (1 - x)v_f + xv_g = v_f + x(v_g - v_f) \tag{4.2}$$

The increase in specific volume on vaporization $(v_g - v_f)$ is also denoted by $v_{fg}$.

*For Example...* consider a system consisting of a two-phase liquid–vapor mixture of water at 100°C and a quality of 0.9. From Table T-2 at 100°C, $v_f = 1.0435 \times 10^{-3}$ m³/kg and $v_g = 1.673$ m³/kg. The specific volume of the mixture is

$$\begin{aligned} v &= v_f + x(v_g - v_f) = 1.0435 \times 10^{-3} + (0.9)(1.673 - 1.0435 \times 10^{-3}) \\ &= 1.506 \text{ m}^3/\text{kg} \end{aligned}$$

Similarly, the specific volume of a two-phase liquid–vapor mixture of water at 212°F and a quality of 0.9 is

$$\begin{aligned} v &= v_f + x(v_g - v_f) = 0.01672 + (0.9)(26.80 - 0.01672) \\ &= 24.12 \text{ ft}^3/\text{lb} \end{aligned}$$

where the $v_f$ and $v_g$ values are obtained from Table T-2E. ▲

To facilitate locating states in the tables, it is often convenient to use values from the saturation tables together with a sketch of a $T$–$v$ or $p$–$v$ diagram. For example, if the specific volume $v$ and temperature $T$ are known, refer to the appropriate temperature table, Table T-2 or T-2E, and determine the values of $v_f$ and $v_g$. A $T$–$v$ diagram illustrating these data is given in Fig. 4.6. If the given specific volume falls between $v_f$ and $v_g$, the system consists of a two-phase liquid–vapor mixture, and the pressure is the saturation pressure corresponding to the given temperature. The quality can be found by solving Eq. 4.2. If the given specific volume is greater than $v_g$, the state is in the superheated vapor region. Then, by interpolating in Table T-4 or T-4E, the pressure and other properties listed can be determined. If the given specific volume is less than $v_f$, Table T-5 or T-5E would be used to determine the pressure and other properties.

*For Example...* let us determine the pressure of water at each of three states defined by a temperature of 100°C and specific volumes, respectively, of $v_1 = 2.434$ m³/kg, $v_2 = 1.0$ m³/kg, and $v_3 = 1.0423 \times 10^{-3}$ m³/kg. Using the known temperature, Table T-2 provides the values of $v_f$ and $v_g$: $v_f = 1.0435 \times 10^{-3}$ m³/kg, $v_g = 1.673$ m³/kg. Since $v_1$ is greater

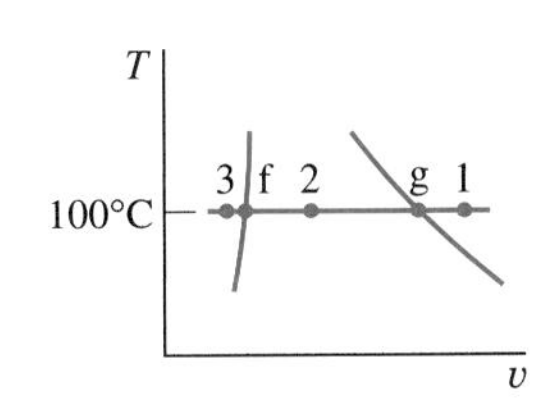


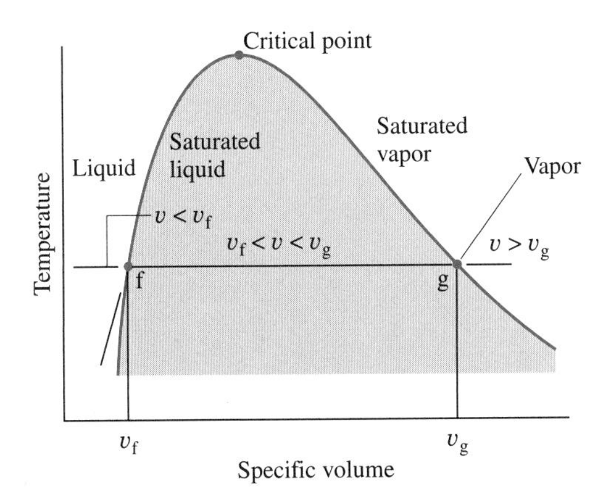


*Figure 4.6* Sketch of a $T$–$v$ diagram for water used to discuss locating states in the tables.

than $v_g$, state 1 is in the vapor region. Table T-4 gives the pressure as 0.70 bar. Next, since $v_2$ falls between $v_f$ and $v_g$, the pressure is the saturation pressure corresponding to 100°C, which is 1.014 bar. Finally, since $v_3$ is less than $v_f$, state 3 is in the liquid region. Table T-5 gives the pressure as 25 bar. ▲

### Examples

The following two examples feature the use of sketches of $p$–$v$ and $T$–$v$ diagrams in conjunction with tabular data to fix the end states of processes. In accord with the state principle, two independent intensive properties must be known to fix the state of the systems under consideration.

## *Example 4.1* Heating Water at Constant Volume

A closed, rigid container of volume 0.5 m$^3$ is placed on a hot plate. Initially, the container holds a two-phase mixture of saturated liquid water and saturated water vapor at $p_1 = 1$ bar with a quality of 0.5. After heating, the pressure in the container is $p_2 = 1.5$ bar. Indicate the initial and final states on a $T$–$v$ diagram, and determine

(a) the temperature, in °C, at each state.
(b) the mass of vapor present at each state, in kg.
(c) If heating continued, determine the pressure, in bar, when the container holds only saturated vapor.

### Solution

***Known:*** A two-phase liquid–vapor mixture of water in a closed, rigid container is heated on a hot plate. The initial pressure and quality and the final pressure are known.

***Find:*** Indicate the initial and final states on a $T$–$v$ diagram and determine at each state the temperature and the mass of water vapor present. Also, if heating continued, determine the pressure when the container holds only saturated vapor.

***Schematic and Given Data:***

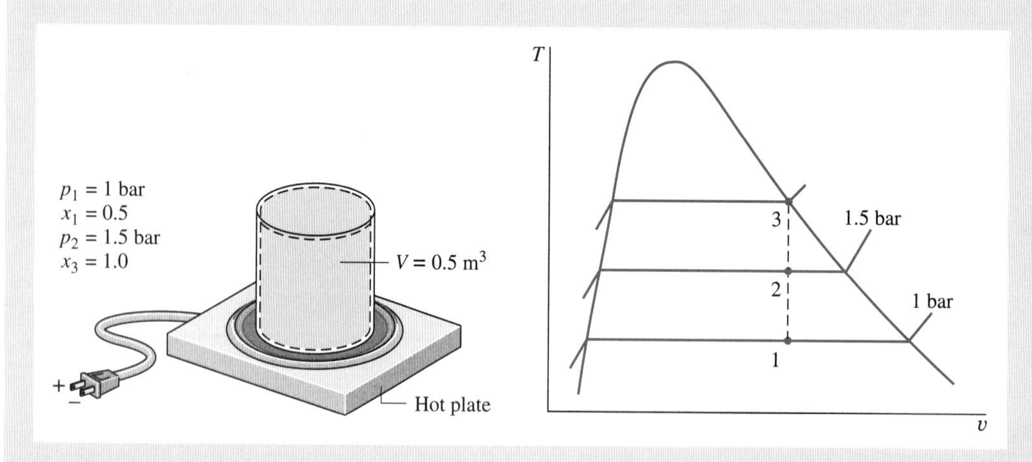


*Figure E4.1*

***Assumptions:***

1. The water in the container is a closed system.
2. States 1, 2, and 3 are equilibrium states.
3. The volume of the container remains constant.

***Analysis:*** Two independent properties are required to fix states 1 and 2. At the initial state, the pressure and quality are known. As these are independent, the state is fixed. State 1 is shown on the $T$–$v$ diagram in the two-phase region. The specific volume at state 1 is found using the given quality and Eq. 4.2. That is

$$v_1 = v_{f1} + x(v_{g1} - v_{f1})$$

From Table T-3 at $p_1 = 1$ bar, $v_{f1} = 1.0432 \times 10^{-3}$ m³/kg and $v_{g1} = 1.694$ m³/kg. Thus

$$v_1 = 1.0432 \times 10^{-3} + 0.5(1.694 - 1.0432 \times 10^{-3}) = 0.8475 \text{ m}^3/\text{kg}$$

At state 2, the pressure is known. The other property required to fix the state is the specific volume $v_2$. Volume and mass are each constant, so $v_2 = v_1 = 0.8475$ m³/kg. For $p_2 = 1.5$ bar, Table T-3 gives $v_{f2} = 1.0582 \times 10^{-3}$ and $v_{g2} = 1.159$ m³/kg. Since

❶ $$v_{f2} < v_2 < v_{g2}$$

❷ state 2 must be in the two-phase region as well. State 2 is also shown on the $T$–$v$ diagram above.

(a) Since states 1 and 2 are in the two-phase liquid–vapor region, the temperatures correspond to the saturation temperatures for the given pressures. Table T-3 gives

$$T_1 = 99.63°\text{C} \quad \text{and} \quad T_2 = 111.4°\text{C} \quad \triangleleft$$

(b) To find the mass of water vapor present, we first use the volume and the specific volume to find the *total* mass, $m$. That is

$$m = \frac{V}{v} = \frac{0.5 \text{ m}^3}{0.8475 \text{ m}^3/\text{kg}} = 0.59 \text{ kg}$$

Then, with Eq. 4.1 and the given value of quality, the mass of vapor at state 1 is

$$m_{g1} = x_1 m = 0.5(0.59 \text{ kg}) = 0.295 \text{ kg} \quad \triangleleft$$

The mass of vapor at state 2 is found similarly using the quality $x_2$. To determine $x_2$, solve Eq. 4.2 for quality and insert specific volume data from Table T-3 at a pressure of 1.5 bar, along with the known value of $v$, as follows

$$x_2 = \frac{v - v_{f2}}{v_{g2} - v_{f2}} = \frac{0.8475 - 1.0528 \times 10^{-3}}{1.159 - 1.0528 \times 10^{-3}} = 0.731$$

Then, with Eq. 4.1

$$m_{g2} = 0.731(0.59 \text{ kg}) = 0.431 \text{ kg} \quad \triangleleft$$

❸ (c) If heating continued, state 3 would be on the saturated vapor line, as shown on the $T$–$v$ diagram above. Thus, the pressure would be the corresponding saturation pressure. Interpolating in Table T-3 at $v_g = 0.8475$ m³/kg, $p_3 = 2.11$ bar. $\triangleleft$

❶ The procedure for fixing state 2 is the same as illustrated in the discussion of Fig. 4.6.

❷ Since the process occurs at constant specific volume, the states lie along a vertical line.

❸ If heating continued at constant volume past state 3, the final state would be in the superheated vapor region, and property data would then be found in Table T-4. As an exercise, verify that for a final pressure of 3 bar, the temperature would be approximately 282°C.

## *Example 4.2* Heating Refrigerant 134a at Constant Pressure

A vertical piston–cylinder assembly containing 0.1 lb of Refrigerant 134a, initially a saturated vapor, is placed on a hot plate. Due to the weight of the piston and the surrounding atmospheric pressure, the pressure of the refrigerant is 20 lbf/in.² Heating occurs slowly, and the refrigerant expands at constant pressure until the final temperature is 65°F. Show the initial and final states on $T$–$v$ and $p$–$v$ diagrams, and determine

(a) the volume occupied by the refrigerant at each state, in ft³.

(b) the work for the process, in Btu.

## Solution

***Known:*** Refrigerant 134a is heated at constant pressure in a vertical piston–cylinder assembly from the saturated vapor state to a known final temperature.

***Find:*** Show the initial and final states on $T$–$v$ and $p$–$v$ diagrams, and determine the volume at each state and the work for the process.

***Schematic and Given Data:***

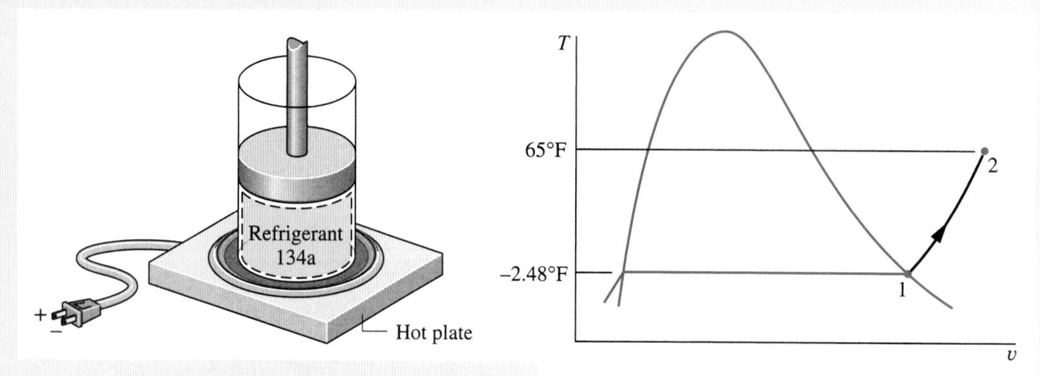


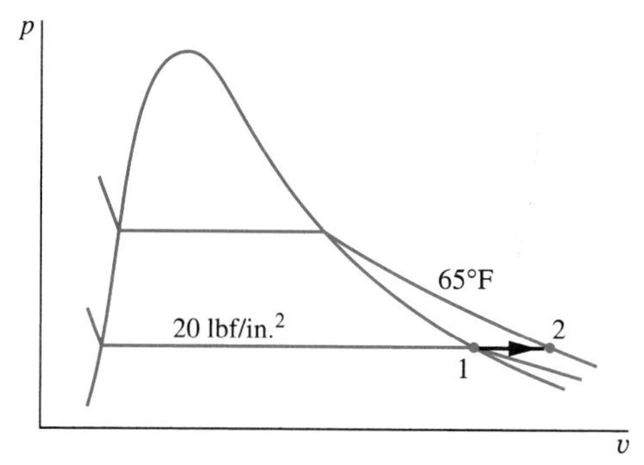


***Assumptions:***

1. The refrigerant is a closed system.
2. States 1 and 2 are equilibrium states.
3. The process occurs at constant pressure.

*Figure E4.2*

***Analysis:*** The initial state is a saturated vapor condition at 20 lbf/in.$^2$ Since the process occurs at constant pressure, the final state is in the superheated vapor region and is fixed by $p_2 = 20$ lbf/in.$^2$ and $T_2 = 65$°F. The initial and final states are shown on the $T$–$v$ and $p$–$v$ diagrams above.

(a) The volumes occupied by the refrigerant at states 1 and 2 are obtained using the given mass and the respective specific volumes. From Table T-7E at $p_1 = 20$ lbf/in.$^2$, we get $v_1 = v_{g1} = 2.2661$ ft$^3$/lb. Thus

$$V_1 = mv_1 = (0.1 \text{ lb})(2.2661 \text{ ft}^3/\text{lb}) = 0.2266 \text{ ft}^3 \triangleleft$$

Interpolating in Table T-8E at $p_2 = 20$ lbf/in.$^2$ and $T_2 = 65$°F, we get $v_2 = 2.6704$ ft$^3$/lb. Thus

$$V_2 = mv_2 = (0.1 \text{ lb})(2.6704 \text{ ft}^3/\text{lb}) = 0.2670 \text{ ft}^3 \triangleleft$$

(b) In this case, the work can be evaluated using Eq. 3.9. Since the pressure is constant

$$W = \int_{V_1}^{V_2} p\, dV = p(V_2 - V_1)$$

Inserting values

❶
$$W = (20 \text{ lbf/in.}^2)(0.2670 - 0.2266)\text{ft}^3 \left|\frac{144 \text{ in.}^2}{1 \text{ ft}^2}\right| \left|\frac{1 \text{ Btu}}{778 \text{ ft} \cdot \text{lbf}}\right| = 0.1496 \text{ Btu} \triangleleft$$

❶ Note the use of conversion factors in this calculation.

### 4.3.2 Evaluating Specific Internal Energy and Enthalpy

In many thermodynamic analyses the sum of the internal energy $U$ and the product of pressure $p$ and volume $V$ appears. Because the sum $U + pV$ occurs so frequently in subsequent discussions, it is convenient to give the combination a name, ***enthalpy,*** and a distinct symbol, $H$. By definition

*enthalpy*

$$H = U + pV \tag{4.3}$$

Since $U$, $p$, and $V$ are all properties, this combination is also a property. Enthalpy can be expressed on a unit mass basis

$$h = u + pv \tag{4.4}$$

and per mole

$$\bar{h} = \bar{u} + p\bar{v} \tag{4.5}$$

Units for enthalpy are the same as those for internal energy.

The property tables introduced in Sec. 4.3.1 giving pressure, specific volume, and temperature also provide values of specific internal energy $u$, enthalpy $h$, and entropy $s$. Use of these tables to evaluate $u$ and $h$ is described in the present section; the consideration of entropy is deferred until it is introduced in Chap. 7.

Data for specific internal energy $u$ and enthalpy $h$ are retrieved from the property tables in the same way as for specific volume. For saturation states, the values of $u_f$ and $u_g$, as well as $h_f$ and $h_g$, are tabulated versus both saturation pressure and saturation temperature. The specific internal energy for a two-phase liquid–vapor mixture is calculated for a given quality in the same way the specific volume is calculated

$$u = (1 - x)u_f + xu_g = u_f + x(u_g - u_f) \tag{4.6}$$

The increase in specific internal energy on vaporization $(u_g - u_f)$ is often denoted by $u_{fg}$. Similarly, the specific enthalpy for a two-phase liquid–vapor mixture is given in terms of the quality by

$$h = (1 - x)h_f + xh_g = h_f + x(h_g - h_f) \tag{4.7}$$

The increase in enthalpy during vaporization $(h_g - h_f)$ is often tabulated for convenience under the heading $h_{fg}$.

***For Example…*** to illustrate the use of Eqs. 4.6 and 4.7, we determine the specific enthalpy of Refrigerant 134a when its temperature is 12°C and its specific internal energy is 132.95 kJ/kg. Referring to Table T-6, the given internal energy value falls between $u_f$ and $u_g$ at 12°C, so the state is a two-phase liquid–vapor mixture. The quality of the mixture is found by using Eq. 4.6 and data from Table T-6 as follows:

$$x = \frac{u - u_f}{u_g - u_f} = \frac{132.95 - 65.83}{233.63 - 65.83} = 0.40$$

Then, with values from Table T-6, Eq. 4.7 gives

$$\begin{aligned} h &= (1 - x)h_f + xh_g \\ &= (1 - 0.4)(66.18) + 0.4(254.03) = 141.32 \text{ kJ/kg} \quad \blacktriangle \end{aligned}$$

In the superheated vapor tables, $u$ and $h$ are tabulated along with $v$ as functions of temperature and pressure.

***For Example...*** let us evaluate $T$, $v$, and $h$ for water at 0.10 MPa and a specific internal energy of 2537.3 kJ/kg. Turning to Table T-3, note that the given value of $u$ is greater than $u_g$ at 0.1 MPa ($u_g$ = 2506.1 kJ/kg). This suggests that the state lies in the superheated vapor region. From Table T-4 it is found that $T$ = 120°C, $v$ = 1.793 m³/kg, and $h$ = 2716.6 kJ/kg. Alternatively, $h$ and $u$ are related by the definition of $h$

$$
\begin{aligned}
h &= u + pv \\
&= 2537.3 \frac{\text{kJ}}{\text{kg}} + \left(10^5 \frac{\text{N}}{\text{m}^2}\right)\left(1.793 \frac{\text{m}^3}{\text{kg}}\right)\left|\frac{1 \text{ kJ}}{10^3 \text{ N}\cdot\text{m}}\right| \\
&= 2537.3 + 179.3 = 2716.6 \text{ kJ/kg}
\end{aligned}
$$

As another illustration, consider water at a state fixed by a pressure equal to 14.7 lbf/in.² and a temperature of 250°F. From Table T-4E, $v$ = 28.42 ft³/lb, $u$ = 1091.5 Btu/lb, and $h$ = 1168.8 Btu/lb. As above, $h$ may be calculated from $u$. Thus

$$
\begin{aligned}
h &= u + pv \\
&= 1091.5 \frac{\text{Btu}}{\text{lb}} + \left(14.7 \frac{\text{lbf}}{\text{in.}^2}\right)\left(28.42 \frac{\text{ft}^3}{\text{lb}}\right)\left|\frac{144 \text{ in.}^2}{1 \text{ ft}^2}\right|\left|\frac{1 \text{ Btu}}{778 \text{ ft}\cdot\text{lbf}}\right| \\
&= 1091.5 + 77.3 = 1168.8 \text{ Btu/lb} \quad \blacktriangle
\end{aligned}
$$

Specific internal energy and enthalpy data for liquid states of water are presented in Tables T-5. The format of these tables is the same as that of the superheated vapor tables considered previously. Accordingly, property values for liquid states are retrieved in the same manner as those of vapor states.

## Reference States and Reference Values

*reference states*
*reference values*

The values of $u$ and $h$ given in the property tables are not obtained by direct measurement but are calculated from other data that can be more readily determined experimentally. Because $u$ and $h$ are calculated, the matter of ***reference states*** and ***reference values*** becomes important and is briefly considered next.

When applying the energy balance, it is *differences* in internal, kinetic, and potential energy between two states that are important, and *not* the values of these energy quantities at each of the two states. ***For Example...*** consider the case of potential energy. The numerical value of potential energy determined relative to the surface of the earth is different from the value relative to the top of a flagpole at the same location. However, the difference in potential energy between any two elevations is precisely the same regardless of the datum selected, because the datum cancels in the calculation. ▲

Similarly, values can be assigned to specific internal energy and enthalpy relative to arbitrary reference values at arbitrary reference states. For example, the reference state in the steam tables is saturated liquid at the triple point temperature: 0.01°C. At this state, the specific internal energy is set to zero, as shown in Table T-2. Values of the specific enthalpy are calculated from $h = u + pv$, using the tabulated values for $p$, $v$, and $u$. As for the case of potential energy considered above, the use of values of a particular property determined relative to an arbitrary reference is unambiguous as long as the calculations being performed involve only differences in that property, for then the reference value cancels.

### 4.3.3 Evaluating Properties Using Computer Software (CD-ROM)

### 4.3.4 Examples

In the following examples, closed systems undergoing processes are analyzed using the energy balance. In each case, sketches of $p$–$v$ and/or $T$–$v$ diagrams are used in conjunction with appropriate tables to obtain the required property data. Using property diagrams and table data introduces an additional level of complexity compared to similar problems in Chap. 3.

## *Example 4.3* Stirring Water at Constant Volume

A well-insulated rigid tank having a volume of 10 ft$^3$ contains saturated water vapor at 212°F. The water is rapidly stirred until the pressure is 20 lbf/in.$^2$ Determine the temperature at the final state, in °F, and the work during the process, in Btu.

### Solution

***Known:*** By rapid stirring, water vapor in a well-insulated rigid tank is brought from the saturated vapor state at 212°F to a pressure of 20 lbf/in.$^2$

***Find:*** Determine the temperature at the final state and the work.

***Schematic and Given Data:***

***Assumptions:***

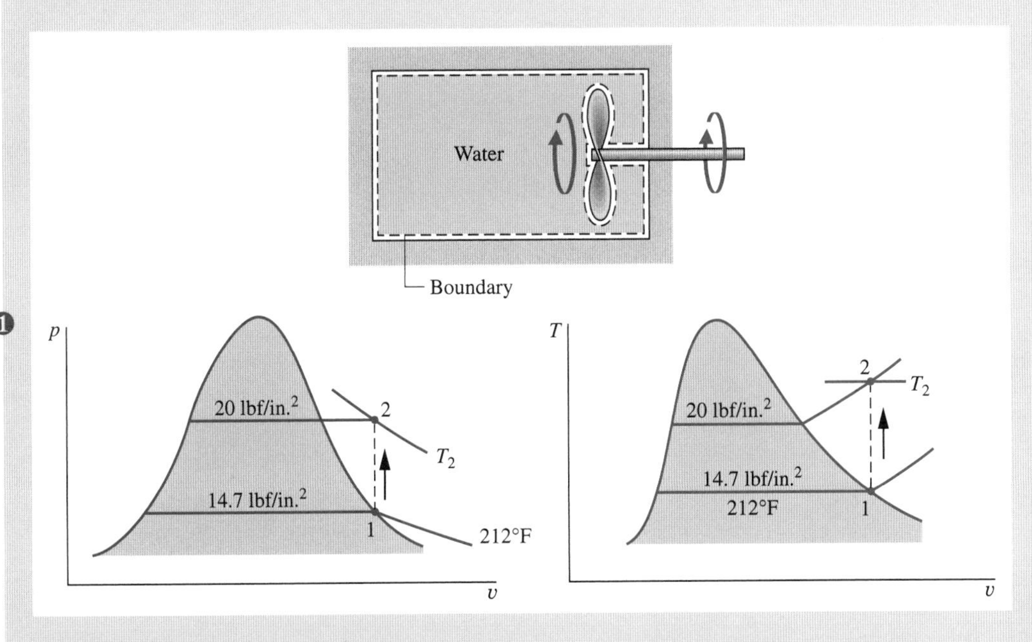


➊

*Figure E4.3*

1. The water is a closed system.
2. The initial and final states are at equilibrium. There is no net change in kinetic or potential energy.
3. There is no heat transfer with the surroundings.
4. The tank volume remains constant.

***Analysis:*** To determine the final equilibrium state, the values of two independent intensive properties are required. One of these is pressure, $p_2 = 20$ lbf/in.$^2$, and the other is the specific volume: $v_2 = v_1$. The initial and final specific volumes are equal because the total mass and total volume are unchanged in the process. The initial and final states are located on the accompanying $T$–$v$ and $p$–$v$ diagrams.

From Table T-2E, $v_1 = v_g(212°F) = 26.80$ ft$^3$/lb, $u_1 = u_g(212°F) = 1077.6$ Btu/lb. By using $v_2 = v_1$ and interpolating in Table T-4E at $p_2 = 20$ lbf/in.$^2$

$$u_2 = 1161.6 \text{ Btu/lb}, \qquad T_2 = 445°\text{F} \triangleleft$$

Next, with assumptions 2 and 3 an energy balance for the system reduces to

$$\Delta U + \cancel{\Delta \text{KE}}^{0} + \cancel{\Delta \text{PE}}^{0} = \cancel{Q}^{0} - W$$

On rearrangement

$$W = -(U_2 - U_1) = -m(u_2 - u_1)$$

To evaluate $W$ requires the system mass. This can be determined from the volume and specific volume

$$m = \frac{V}{v_1} = \left(\frac{10 \text{ ft}^3}{26.8 \text{ ft}^3/\text{lb}}\right) = 0.373 \text{ lb}$$

Finally, by inserting values into the expression for $W$

$$W = -(0.373 \text{ lb})(1161.6 - 1077.6) \text{ Btu/lb} = -31.3 \text{ Btu} \triangleleft$$

where the minus sign signifies that the energy transfer by work is to the system.

---

❶ Although the initial and final states are equilibrium states, the intervening states are not at equilibrium. To emphasize this, the process has been indicated on the $T$–$v$ and $p$–$v$ diagrams by a dashed line. Solid lines on property diagrams are reserved for processes that pass through equilibrium states only (quasiequilibrium processes). The analysis illustrates the importance of carefully sketched property diagrams as an adjunct to problem solving.

## *Example 4.4* Analyzing Two Processes in Series

Water contained in a piston–cylinder assembly undergoes two processes in series from an initial state where the pressure is 10 bar and the temperature is 400°C.

***Process 1–2:*** The water is cooled as it is compressed at a constant pressure of 10 bar to the saturated vapor state.

***Process 2–3:*** The water is cooled at constant volume to 150°C.

(a) Sketch both processes on $T$–$v$ and $p$–$v$ diagrams.
(b) For the overall process determine the work, in kJ/kg.
(c) For the overall process determine the heat transfer, in kJ/kg.

### Solution

***Known:*** Water contained in a piston–cylinder assembly undergoes two processes: It is cooled and compressed while keeping the pressure constant, and then cooled at constant volume.

***Find:*** Sketch both processes on $T$–$v$ and $p$–$v$ diagrams. Determine the net work and the net heat transfer for the overall process per unit of mass contained within the piston–cylinder assembly.

***Schematic and Given Data:***

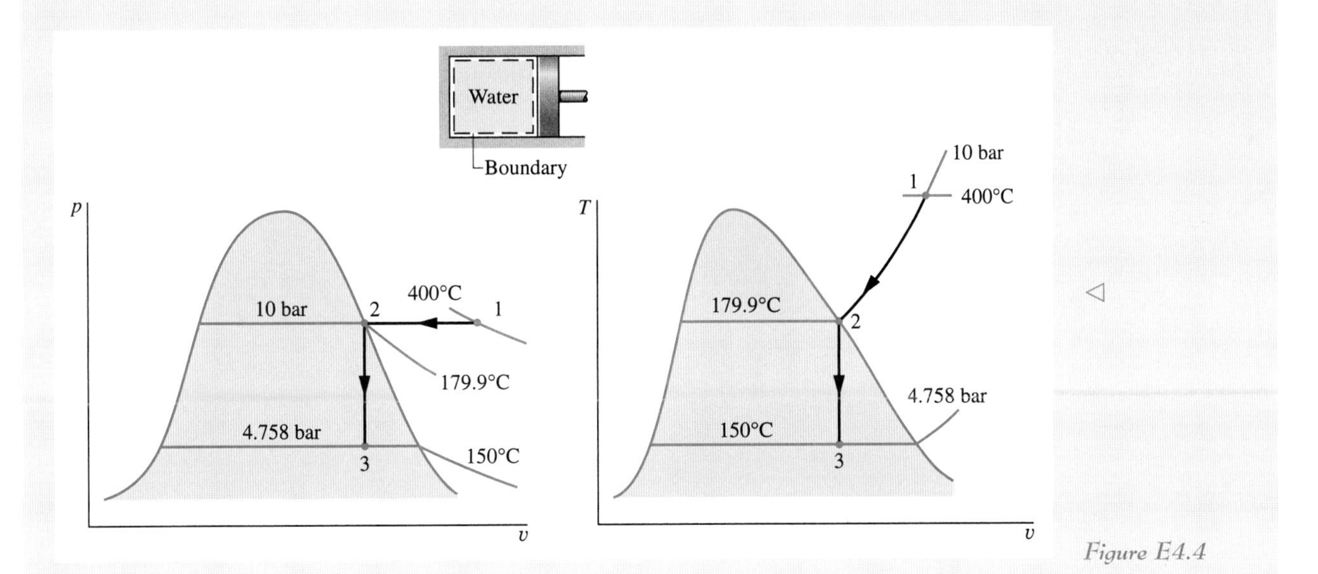


*Figure E4.4*

***Assumptions:***

1. The water is a closed system.
2. The piston is the only work mode.
3. There are no changes in kinetic or potential energy.

***Analysis:*** (a) The accompanying $T-v$ and $p-v$ diagrams show the two processes. Since the temperature at state 1, $T_1$ = 400°C, is greater than the saturation temperature corresponding to $p_1$ = 10 bar: 179.9°C, state 1 is located in the superheat region.

(b) Since the piston is the only work mechanism

$$W = \int_1^3 p\, dV = \int_1^2 p\, dV + \cancelto{0}{\int_2^3 p\, dV}$$

The second integral vanishes because the volume is constant in Process 2–3. Dividing by the mass and noting that the pressure is constant for Process 1–2

$$\frac{W}{m} = p(v_2 - v_1)$$

The specific volume at state 1 is found from Table T-4 using $p_1$ = 10 bar and $T_1$ = 400°C: $v_1$ = 0.3066 m$^3$/kg. Also, $u_1$ = 2957.3 kJ/kg. The specific volume at state 2 is the saturated vapor value at 10 bar: $v_2$ = 0.1944 m$^3$/kg, from Table T-3. Hence

$$\frac{W}{m} = (10 \text{ bar})(0.1944 - 0.3066)\left(\frac{\text{m}^3}{\text{kg}}\right)\left|\frac{10^5 \text{ N/m}^2}{1 \text{ bar}}\right|\left|\frac{1 \text{ kJ}}{10^3 \text{ N}\cdot\text{m}}\right|$$

$$= -112.2 \text{ kJ/kg} \triangleleft$$

The minus sign indicates that work is done *on* the water vapor by the piston.

(c) An energy balance for the *overall* process reduces to

$$m(u_3 - u_1) = Q - W$$

By rearranging

$$\frac{Q}{m} = (u_3 - u_1) + \frac{W}{m}$$

To evaluate the heat transfer requires $u_3$, the specific internal energy at state 3. Since $T_3$ is given and $v_3 = v_2$, two independent intensive properties are known that together fix state 3. To find $u_3$, first solve for the quality

$$x_3 = \frac{v_3 - v_{f3}}{v_{g3} - v_{f3}} = \frac{0.1944 - 1.0905 \times 10^{-3}}{0.3928 - 1.0905 \times 10^{-3}} = 0.494$$

where $v_{f3}$ and $v_{g3}$ are from Table T-2 at 150°C. Then

$$u_3 = u_{f3} + x_3(u_{g3} - u_{f3}) = 631.68 + 0.494(2559.5 - 631.98)$$
$$= 1583.9 \text{ kJ/kg}$$

where $u_{f3}$ and $u_{g3}$ are from Table T-2 at 150°C.

Substituting values into the energy balance

$$\frac{Q}{m} = 1583.9 - 2957.3 + (-112.2) = -1485.6 \text{ kJ/kg}$$

The minus sign shows that energy is transferred *out* by heat transfer. ◁

The next example illustrates the use of *Interactive Thermodynamics: IT* for solving problems. In this case, the software evaluates the property data, calculates the results, and displays the results graphically.

### *Example 4.5* Plotting Thermodynamic Data Using Software

For the system of Example 4.1, plot the heat transfer, in kJ, and the mass of saturated vapor present, in kg, each versus pressure at state 2 ranging from 1 to 2 bar. Discuss the results.

**Solution (CD-ROM)**

### 4.3.5 Evaluating Specific Heats $c_v$ and $c_p$

Several properties related to internal energy are important in thermodynamics. One of these is the property enthalpy introduced in Sec. 4.3.2. Two others, known as *specific heats,* are considered in this section. The specific heats are particularly useful for thermodynamic calculations involving the *ideal gas model* introduced in Sec. 4.5.

The intensive properties $c_v$ and $c_p$ are defined for pure, simple compressible substances as partial derivatives of the functions $u(T, v)$ and $h(T, p)$, respectively

$$c_v = \left.\frac{\partial u}{\partial T}\right)_v \tag{4.8}$$

$$c_p = \left.\frac{\partial h}{\partial T}\right)_p \tag{4.9}$$

where the subscripts $v$ and $p$ denote, respectively, the variables held fixed during differentiation. Values for $c_v$ and $c_p$ can be obtained using the microscopic approach to thermodynamics together with *spectroscopic* measurements. They also can be determined macroscopically using other exacting property measurements. Since $u$ and $h$ can be expressed either on a unit mass basis or per mole, values of the specific heats can be similarly expressed. SI units are kJ/kg · K or kJ/kmol · K. Other units are Btu/lb · °R or Btu/lbmol · °R.

The property $k$, called the *specific heat ratio,* is simply the ratio

$$k = \frac{c_p}{c_v} \tag{4.10}$$

The properties $c_v$ and $c_p$ are referred to as ***specific heats*** (or *heat capacities*) because under certain *special conditions* they relate the temperature change of a system to the amount of energy added by heat transfer. However, it is generally preferable to think of $c_v$ and $c_p$ in terms of their definitions, Eqs. 4.8 and 4.9, and not with reference to this limited interpretation involving heat transfer.

*specific heats*

In general, $c_v$ is a function of $v$ and $T$ (or $p$ and $T$), and $c_p$ depends on both $p$ and $T$ (or $v$ and $T$). Specific heat data are available for common gases, liquids, and solids. Data for gases are introduced in Sec. 4.5 as a part of the discussion of the ideal gas model. Specific heat values for some common liquids and solids are introduced in Sec. 4.3.6 as a part of the discussion of the incompressible substance model.

## 4.3.6 Evaluating Properties of Liquids and Solids

Special methods often can be used to evaluate properties of liquids and solids. These methods provide simple, yet accurate, approximations that do not require exact compilations like the compressed liquid tables for water, Tables T-5. Two such special methods are discussed next: approximations using saturated liquid data and the incompressible substance model.

### Approximations for Liquids Using Saturated Liquid Data

Approximate values for $v$, $u$, and $h$ at liquid states can be obtained using saturated liquid data. To illustrate, refer to the compressed liquid tables, Tables T-5. These tables show that the specific volume and specific internal energy change very little with pressure *at a fixed temperature*. Because the values of $v$ and $u$ vary only gradually as pressure changes at fixed temperature, the following approximations are reasonable for many engineering calculations:

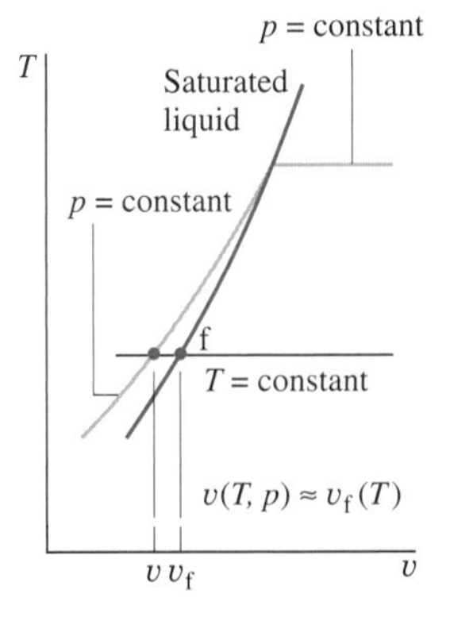


$$v(T, p) \approx v_f(T) \tag{4.11}$$

$$u(T, p) \approx u_f(T) \tag{4.12}$$

That is, for liquids $v$ and $u$ may be evaluated at the saturated liquid state corresponding to the temperature at the given state.

An approximate value of $h$ at liquid states can be obtained by using Eqs. 4.11 and 4.12 in the definition $h = u + pv$; thus

$$h(T, p) \approx u_f(T) + pv_f(T)$$

This can be expressed alternatively as

$$h(T, p) \approx h_f(T) + \underline{v_f(T)[p - p_{sat}(T)]} \tag{4.13}$$

where $p_{sat}$ denotes the saturation pressure at the given temperature. When the contribution of the underlined term of Eq. 4.13 is small, the specific enthalpy can be approximated by the saturated liquid value, as for $v$ and $u$. That is

$$h(T, p) \approx h_f(T) \tag{4.14}$$

Although the approximations given here have been presented with reference to liquid water, they also provide plausible approximations for other substances *when the only liquid data available are for saturated liquid states.* In this text, compressed liquid data are presented only for water (Tables T-5). Also note that *Interactive Thermodynamics: IT* does not provide compressed liquid data for *any* substance, but uses Eqs. 4.11, 4.12, and 4.14 to return liquid values for $v$, $u$, and $h$, respectively. When greater accuracy is required than provided by these approximations, other data sources should be consulted for more complete property compilations for the substance under consideration.

## Incompressible Substance Model

As noted above, there are regions where the specific volume of liquid water varies little and the specific internal energy varies mainly with temperature. The same general behavior is exhibited by the liquid phases of other substances and by solids. The approximations of Eqs. 4.11–4.14 are based on these observations, as is the ***incompressible substance model*** under present consideration.

*incompressible substance model*

To simplify evaluations involving liquids or solids, the specific volume (density) is often assumed to be constant and the specific internal energy assumed to vary only with temperature. A substance idealized in this way is called *incompressible.*

Since the specific internal energy of a substance modeled as incompressible depends only on temperature, the specific heat $c_v$ is also a function of temperature alone

$$c_v(T) = \frac{du}{dT} \qquad \text{(incompressible)} \tag{4.15}$$

This is expressed as an ordinary derivative because $u$ depends only on $T$.

Although the specific volume is constant and internal energy depends on temperature only, enthalpy varies with both pressure and temperature according to

$$h(T, p) = u(T) + pv \qquad \text{(incompressible)} \tag{4.16}$$

For a substance modeled as incompressible, the specific heats $c_v$ and $c_p$ are equal. This is seen by differentiating Eq. 4.16 with respect to temperature while holding pressure fixed to obtain

$$\left.\frac{\partial h}{\partial T}\right)_p = \frac{du}{dT}$$

The left side of this expression is $c_p$ by definition (Eq. 4.9), so using Eq. 4.15 on the right side gives

$$c_p = c_v \qquad \text{(incompressible)} \tag{4.17}$$

Thus, for an incompressible substance it is unnecessary to distinguish between $c_p$ and $c_v$, and both can be represented by the same symbol, $c$. Specific heats of some common liquids and solids are given versus temperature in Appendix Tables HT-1, 2, 4, and 5. Over limited temperature intervals the variation of $c$ with temperature can be small. In such instances, the specific heat $c$ can be treated as constant without a serious loss of accuracy.

Using Eqs. 4.15 and 4.16, the changes in specific internal energy and specific enthalpy between two states are given, respectively, by

$$u_2 - u_1 = \int_{T_1}^{T_2} c(T)\, dT \qquad \text{(incompressible)} \tag{4.18}$$

$$\begin{aligned} h_2 - h_1 &= u_2 - u_1 + v(p_2 - p_1) \\ &= \int_{T_1}^{T_2} c(T)\, dT + v(p_2 - p_1) \qquad \text{(incompressible)} \end{aligned} \tag{4.19}$$

If the specific heat $c$ is taken as constant, Eq. 4.18 becomes

$$u_2 - u_1 = c(T_2 - T_1) \qquad \text{(incompressible, constant } c\text{)} \tag{4.20}$$

When $c$ is constant, Eq. 4.19 reads

$$h_2 - h_1 = c(T_2 - T_1) + \underline{v(p_2 - p_1)} \qquad \text{(incompressible, constant } c) \tag{4.21a}$$

In Eq. 4.21a, the underlined term is often small relative to the first term on the right side; the equation then reduces to the same form as Eq. 4.20

$$h_2 - h_1 \approx c(T_2 - T_1) \qquad \text{(incompressible, constant } c) \tag{4.21b}$$

## 4.4 $p$–$v$–$T$ Relations for Gases

The object of the present section is to gain a better understanding of the relationship among pressure, specific volume, and temperature of gases. This is important not only for understanding gas behavior but also for the discussions of the second part of the chapter, where the *ideal gas model* is introduced. The current presentation is conducted in terms of the *compressibility factor* and begins with the introduction of the *universal gas constant.*

### Universal Gas Constant, $\overline{R}$

Let a gas be confined in a cylinder by a piston and the entire assembly held at a constant temperature. The piston can be moved to various positions so that a series of equilibrium states at constant temperature can be visited. Suppose the pressure and specific volume are measured at each state and the value of the ratio $p\overline{v}/T$ ($\overline{v}$ is volume per mole) determined. These ratios can then be plotted versus pressure at constant temperature. The results for several temperatures are sketched in Fig. 4.7. When the ratios are extrapolated to zero pressure, *precisely the same limiting value is obtained* for each curve. That is,

$$\lim_{p \to 0} \frac{p\overline{v}}{T} = \overline{R} \tag{4.22}$$

where $\overline{R}$ denotes the common limit for all temperatures. If this procedure were repeated for other gases, it would be found in every instance that the limit of the ratio $p\overline{v}/T$ as $p$ tends to zero at fixed temperature is the same, namely $\overline{R}$. Since the same limiting value is exhibited by all gases, $\overline{R}$ is called the ***universal gas constant.*** Its value as determined experimentally is

*universal gas constant*

$$\overline{R} = \begin{cases} 8.314 \text{ kJ/kmol} \cdot \text{K} \\ 1.986 \text{ Btu/lbmol} \cdot {}^\circ\text{R} \\ 1545 \text{ ft} \cdot \text{lbf/lbmol} \cdot {}^\circ\text{R} \end{cases} \tag{4.23}$$

Having introduced the universal gas constant, we turn next to the compressibility factor.

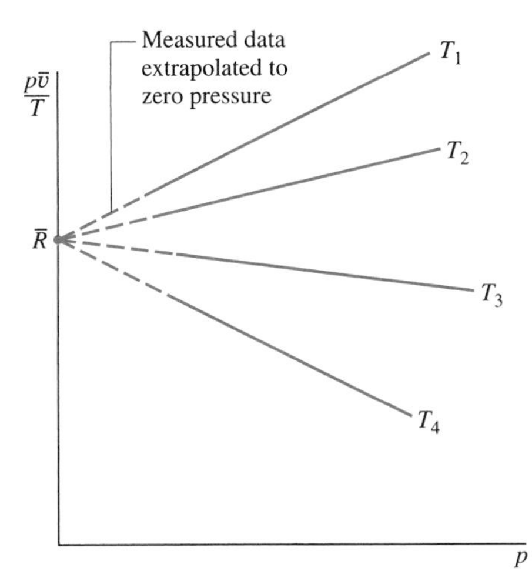


*Figure 4.7* Sketch of $p\overline{v}/T$ versus pressure for a gas at several specified values of temperature.

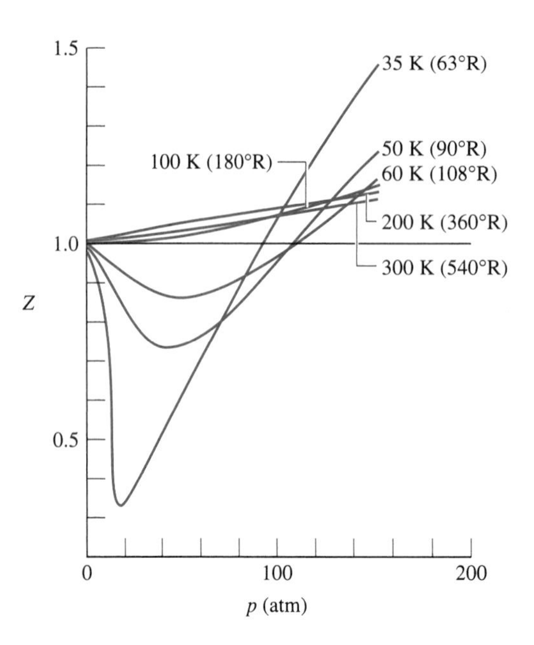


*Figure 4.8* Variation of the compressibility factor of hydrogen with pressure at constant temperature.

### Compressibility Factor, Z

*compressibility factor*

The *dimensionless* ratio $p\bar{v}/\bar{R}T$ is called the ***compressibility factor*** and is denoted by $Z$. That is,

$$Z = \frac{p\bar{v}}{\bar{R}T} \tag{4.24}$$

When values for $p$, $\bar{v}$, $\bar{R}$, and $T$ are used in consistent units, $Z$ is unitless.

With $\bar{v} = Mv$ (Eq. 2.11), where $M$ is the atomic or molecular weight, the compressibility factor can be expressed alternatively as

$$Z = \frac{pv}{RT} \tag{4.25}$$

where

$$R = \frac{\bar{R}}{M} \tag{4.26}$$

$R$ is a constant for the particular gas whose molecular weight is $M$. Alternative units for $R$ are kJ/kg · K, Btu/lb · °R, and ft · lbf/lb · °R.

Equation 4.22 can be expressed in terms of the compressibility factor as

$$\lim_{p \to 0} Z = 1 \tag{4.27}$$

That is, the compressibility factor $Z$ tends to unity as pressure tends to zero at fixed temperature. This is illustrated by Fig. 4.8, which shows $Z$ for hydrogen plotted versus pressure at a number of different temperatures. In general, at states of a gas where pressure is small relative to the critical pressure of the gas, $Z$ approaches 1.

### Generalized Compressibility Data (CD-ROM)

**Special Note:** Content provided on the accompanying CD-ROM may involve equations, figures, and examples that are not included in the print version of the book. In the present case, Fig. 4.9, Eqs. 4.28 and 4.29, and Example 4.6 are found **only** on the CD.

## Evaluating Properties Using the Ideal Gas Model

The discussion of Sec. 4.4 shows that the compressibility factor $Z = pv/RT$ tends to unity as pressure decreases at fixed temperature. For gases generally, we find that at states where the pressure is small relative to the critical pressure $p_c$, the compressibility factor is approximately 1. At such states, we can assume with reasonable accuracy that $Z = 1$, or

$$pv = RT \tag{4.30}$$

Known as the ***ideal gas equation of state,*** Eq. 4.30 underlies the second part of this chapter dealing with the ideal gas model.

*ideal gas equation of state*

Alternative forms of the same basic relationship among pressure, specific volume, and temperature are obtained as follows. With $v = V/m$, Eq. 4.30 can be expressed as

$$pV = mRT \tag{4.31}$$

In addition, since $v = \bar{v}/M$ and $R = \bar{R}/M$, where $M$ is the atomic or molecular weight, Eq. 4.30 can be expressed as

$$p\bar{v} = \bar{R}T \tag{4.32}$$

or, with $\bar{v} = V/n$, as

$$pV = n\bar{R}T \tag{4.33}$$

# 4.5 Ideal Gas Model

For any gas whose equation of state is given *exactly* by $pv = RT$, the specific internal energy depends on temperature *only*. This conclusion is supported by experimental observations, beginning with the work of Joule, who showed in 1843 that the internal energy of air at low density depends primarily on temperature. The specific enthalpy of a gas described by $pv = RT$ also depends on temperature only, as can be shown by combining the definition of enthalpy, $h = u + pv$, with $u = u(T)$ and the ideal gas equation of state to obtain $h = u(T) + RT$. Taken together, these specifications constitute the ***ideal gas model,*** summarized as follows

*ideal gas model*

$$pv = RT \tag{4.30}$$

$$u = u(T) \tag{4.34}$$

$$h = h(T) = u(T) + RT \tag{4.35}$$

The specific internal energy and enthalpy of gases generally depend on two independent properties, not just temperature as presumed by the ideal gas model. Moreover, the ideal gas equation of state does not provide an acceptable approximation at all states. Accordingly, whether the ideal gas model is used depends on the error acceptable in a given calculation. Still, gases often do *approach* ideal gas behavior, and a particularly simplified description is obtained with the ideal gas model.

To expedite the solutions of subsequent examples and end-of-chapter problems involving air, oxygen ($O_2$), nitrogen ($N_2$), carbon dioxide ($CO_2$), carbon monoxide (CO), hydrogen ($H_2$), and other common gases, we assume the ideal gas model is valid. The suitability of this assumption could be verified by reference to appropriate data, including compressibility data such as shown in Fig. 4.8.

The next example illustrates the use of the ideal gas equation of state and reinforces the use of property diagrams to locate principal states during processes.

## *Example 4.7* Air as an Ideal Gas Undergoing a Cycle

One pound of air undergoes a thermodynamic cycle consisting of three processes.

***Process 1–2:*** constant specific volume

***Process 2–3:*** constant-temperature expansion

***Process 3–1:*** constant-pressure compression

At state 1, the temperature is 540°R, and the pressure is 1 atm. At state 2, the pressure is 2 atm. Employing the ideal gas equation of state,
(a) sketch the cycle on $p$–$v$ coordinates.
(b) determine the temperature at state 2, in °R.
(c) determine the specific volume at state 3, in ft³/lb.

## Solution

***Known:*** Air executes a thermodynamic cycle consisting of three processes: Process 1–2, $v$ = constant; Process 2–3, $T$ = constant; Process 3–1, $p$ = constant. Values are given for $T_1$, $p_1$, and $p_2$.
***Find:*** Sketch the cycle on $p$–$v$ coordinates and determine $T_2$ and $v_3$.

***Schematic and Given Data:***

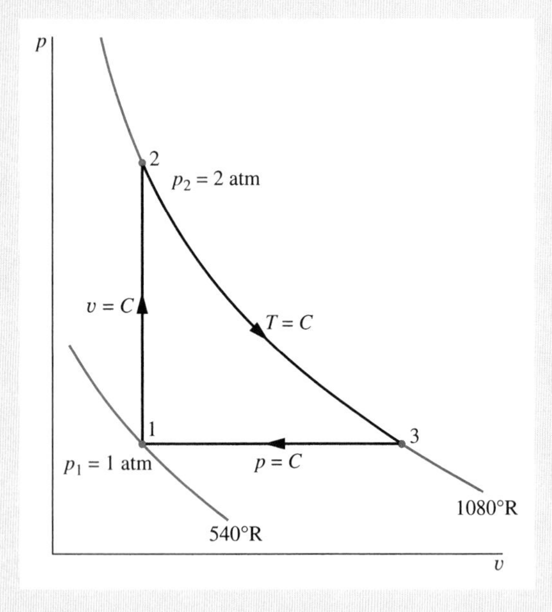


***Assumptions:***
1. The air is a closed system.
2. The air behaves as an ideal gas.

*Figure E4.7* ◁

***Analysis:*** (a) The cycle is shown on $p$–$v$ coordinates in the accompanying figure. Note that since $p = RT/v$ and temperature is constant, the variation of $p$ with $v$ for the process from 2 to 3 is nonlinear.
(b) Using $pv = RT$, the temperature at state 2 is

$$T_2 = p_2 v_2 / R$$

To obtain the specific volume $v_2$ required by this relationship, note that $v_2 = v_1$, so

$$v_2 = RT_1/p_1$$

Combining these two results gives

❶ $$T_2 = \frac{p_2}{p_1} T_1 = \left(\frac{2 \text{ atm}}{1 \text{ atm}}\right)(540°\text{R}) = 1080°\text{R} \triangleleft$$

(c) Since $pv = RT$, the specific volume at state 3 is

$$v_3 = RT_3/p_3$$

Noting that $T_3 = T_2$, $p_3 = p_1$, and $R = \overline{R}/M$

$$v_3 = \frac{\overline{R}T_2}{Mp_1}$$

$$= \left( \frac{1545 \dfrac{\text{ft} \cdot \text{lbf}}{\text{lbmol} \cdot {}^\circ\text{R}}}{28.97 \dfrac{\text{lb}}{\text{lbmol}}} \right) \frac{(1080^\circ\text{R})}{(14.7 \text{ lbf/in.}^2)|144 \text{ in.}^2/\text{ft}^2|}$$

$$= 27.2 \text{ ft}^3/\text{lb} \triangleleft$$

where the molecular weight of air is from Table T-1E.

❶ Carefully note that the equation of state $pv = RT$ requires the use of *absolute* temperature $T$ and *absolute* pressure $p$.

## 4.6 Internal Energy, Enthalpy, and Specific Heats of Ideal Gases

For a gas obeying the ideal gas model, specific internal energy depends only on temperature. Hence, the specific heat $c_v$, defined by Eq. 4.8, is also a function of temperature alone. That is,

$$c_v(T) = \frac{du}{dT} \quad \text{(ideal gas)} \tag{4.36}$$

This is expressed as an ordinary derivative because $u$ depends only on $T$.

By separating variables in Eq. 4.36

$$du = c_v(T)\, dT \tag{4.37}$$

On integration

$$u(T_2) - u(T_1) = \int_{T_1}^{T_2} c_v(T)\, dT \quad \text{(ideal gas)} \tag{4.38}$$

Similarly, for a gas obeying the ideal gas model, the specific enthalpy depends only on temperature, so the specific heat $c_p$, defined by Eq. 4.9, is also a function of temperature alone. That is

$$c_p(T) = \frac{dh}{dT} \quad \text{(ideal gas)} \tag{4.39}$$

Separating variables in Eq. 4.39

$$dh = c_p(T)\, dT \tag{4.40}$$

On integration

$$h(T_2) - h(T_1) = \int_{T_1}^{T_2} c_p(T)\, dT \quad \text{(ideal gas)} \tag{4.41}$$

An important relationship between the ideal gas specific heats can be developed by differentiating Eq. 4.35 with respect to temperature

$$\frac{dh}{dT} = \frac{du}{dT} + R$$

and introducing Eqs. 4.36 and 4.39 to obtain

$$c_p(T) = c_v(T) + R \qquad \text{(ideal gas)} \tag{4.42}$$

On a molar basis, this is written as

$$\bar{c}_p(T) = \bar{c}_v(T) + \bar{R} \qquad \text{(ideal gas)} \tag{4.43}$$

Although each of the two ideal gas specific heats is a function of temperature, Eqs. 4.42 and 4.43 show that the specific heats differ by just a constant: the gas constant. Knowledge of either specific heat for a particular gas allows the other to be calculated by using only the gas constant. The above equations also show that $c_p > c_v$ and $\bar{c}_p > \bar{c}_v$, respectively.

For an ideal gas, the specific heat ratio, $k$, is also a function of temperature only

$$k = \frac{c_p(T)}{c_v(T)} \qquad \text{(ideal gas)} \tag{4.44}$$

Since $c_p > c_v$, it follows that $k > 1$. Combining Eqs. 4.42 and 4.44 results in

$$c_p(T) = \frac{kR}{k-1} \tag{4.45}$$

$$c_v(T) = \frac{R}{k-1} \tag{4.46}$$

(ideal gas)

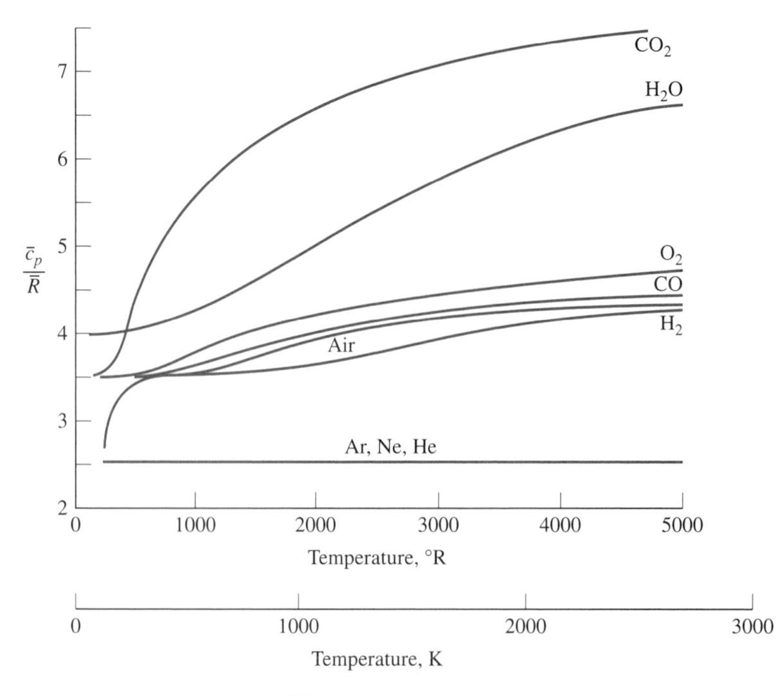


*Figure 4.10* Variation of $\bar{c}_p/\bar{R}$ with temperature for a number of gases modeled as ideal gases.

Similar expressions can be written for the specific heats on a molar basis, with $R$ being replaced by $\overline{R}$.

*Specific Heat Functions.* The foregoing expressions require the ideal gas specific heats as functions of temperature. These functions are available for gases of practical interest in various forms, including graphs, tables, and equations. Figure 4.10 illustrates the variation of $\overline{c}_p$ (molar basis) with temperature for a number of common gases. In the range of temperature shown, $\overline{c}_p$ increases with temperature for all gases, except for the monatonic gases Ar, Ne, and He. For these, $\overline{c}_p$ is closely constant at the value predicted by kinetic theory: $\overline{c}_p = \frac{5}{2}\overline{R}$. Tabular specific heat data for selected gases are presented versus temperature in Tables T-10.

## 4.7 Evaluating $\Delta u$ and $\Delta h$ of Ideal Gases

### Using Ideal Gas Tables

For a number of common gases, evaluations of specific internal energy and enthalpy changes are facilitated by the use of the *ideal gas tables,* Tables T-9 and T-11, which give $u$ and $h$ (or $\overline{u}$ and $\overline{h}$) versus temperature.

To obtain enthalpy versus temperature, write Eq. 4.41 as

$$h(T) = \int_{T_{\text{ref}}}^{T} c_p(T)\, dT + h(T_{\text{ref}})$$

where $T_{\text{ref}}$ is an arbitrary reference temperature and $h(T_{\text{ref}})$ is an arbitrary value for enthalpy at the reference temperature. Tables T-9 and T-11 are based on the selection $h = 0$ at $T_{\text{ref}} = 0$ K. Accordingly, a tabulation of enthalpy versus temperature is developed through the integral

$$h(T) = \int_{0}^{T} c_p(T)\, dT \tag{4.47}$$

Tabulations of internal energy versus temperature are obtained from the tabulated enthalpy values by using $u = h - RT$.

For air as an ideal gas, $h$ and $u$ are given in Table T-9 with units of kJ/kg and in Table T-9E in units of Btu/lb. Values of molar specific enthalpy $\overline{h}$ and internal energy $\overline{u}$ for several other common gases modeled as ideal gases are given in Tables T-11 with units of kJ/kmol or Btu/lbmol. Quantities other than specific internal energy and enthalpy appearing in these tables are introduced in Chap. 7 and should be ignored at present. Tables T-9 and T-11 are convenient for evaluations involving ideal gases, not only because the variation of the specific heats with temperature is accounted for automatically but also because the tables are easy to use.

***For Example...*** let us use Table T-9 to evaluate the change in specific enthalpy, in kJ/kg, for air from a state where $T_1 = 400$ K to a state where $T_2 = 900$ K. At the respective temperatures, the ideal gas table for air, Table T-9, gives

$$h_1 = 400.98 \frac{\text{kJ}}{\text{kg}}, \qquad h_2 = 932.93 \frac{\text{kJ}}{\text{kg}}$$

Then, $h_2 - h_1 = 531.95$ kJ/kg. ▲

### Assuming Constant Specific Heats

When the specific heats are taken as constants, Eqs. 4.38 and 4.41 reduce, respectively, to

$$u(T_2) - u(T_1) = c_v(T_2 - T_1) \tag{4.48}$$

$$h(T_2) - h(T_1) = c_p(T_2 - T_1) \tag{4.49}$$

Equations 4.48 and 4.49 are often used for thermodynamic analyses involving ideal gases because they enable simple closed-form equations to be developed for many processes.

The constant values of $c_v$ and $c_p$ in Eqs. 4.48 and 4.49 are, strictly speaking, mean values calculated as follows:

$$c_v = \frac{\int_{T_1}^{T_2} c_v(T)\, dT}{T_2 - T_1}, \qquad c_p = \frac{\int_{T_1}^{T_2} c_p(T)\, dT}{T_2 - T_1}$$

However, when the variation of $c_v$ or $c_p$ over a given temperature interval is slight, little error is normally introduced by taking the specific heat required by Eq. 4.48 or 4.49 as the arithmetic average of the specific heat values at the two end temperatures. Alternatively, the specific heat at the average temperature over the interval can be used. These methods are particularly convenient when tabular specific heat data are available, as in Tables T-10, for then the *constant* specific heat values often can be determined by inspection.

## Using Computer Software (CD-ROM)

The next example illustrates the use of the ideal gas tables, together with the closed system energy balance.

## *Example 4.8* Using the Energy Balance and Ideal Gas Tables

A piston–cylinder assembly contains 2 lb of air at a temperature of 540°R and a pressure of 1 atm. The air is compressed to a state where the temperature is 840°R and the pressure is 6 atm. During the compression, there is a heat transfer from the air to the surroundings equal to 20 Btu. Using the ideal gas model for air, determine the work during the process, in Btu.

### Solution

***Known:*** Two pounds of air are compressed between two specified states while there is heat transfer from the air of a known amount.

***Find:*** Determine the work, in Btu.

***Schematic and Given Data:***

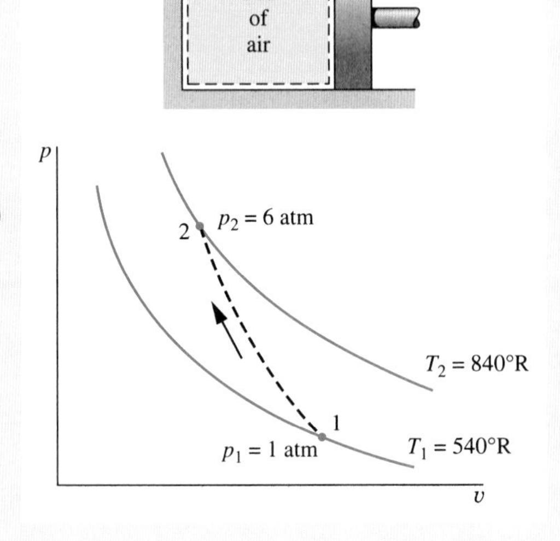


❶

***Assumptions:***

1. The air is a closed system.
2. The initial and final states are equilibrium states. There is no change in kinetic or potential energy.
3. The air is modeled as an ideal gas.

*Figure E4.8*

***Analysis:*** An energy balance for the closed system is

$$\Delta \cancel{KE}^{0} + \Delta \cancel{PE}^{0} + \Delta U = Q - W$$

where the kinetic and potential energy terms vanish by assumption 2. Solving for $W$

❷
$$W = Q - \Delta U = Q - m(u_2 - u_1)$$

From the problem statement, $Q = -20$ Btu. Also, from Table T-9E at $T_1 = 540°\text{R}$, $u_1 = 92.04$ Btu/lb, and at $T_2 = 840°\text{R}$, $u_2 = 143.98$ Btu/lb. Accordingly

$$W = -20 - (2)(143.98 - 92.04) = -123.9 \text{ Btu} \triangleleft$$

The minus sign indicates that work is done on the system in the process.

---

❶ Although the initial and final states are assumed to be equilibrium states, the intervening states are not necessarily equilibrium states, so the process has been indicated on the accompanying $p$–$v$ diagram by a dashed line. This dashed line does not define a "path" for the process.

❷ In principle, the work could be evaluated through $\int p\, dV$, but because the variation of pressure at the piston face with volume is not known, the integration cannot be performed without more information.

The next example illustrates the use of software for problem solving with the ideal gas model. The results obtained are compared with those determined assuming the specific heat $\bar{c}_v$ is constant.

## *Example 4.9* Using the Energy Balance and Software

One kmol of carbon dioxide gas ($CO_2$) in a piston–cylinder assembly undergoes a constant-pressure process at 1 bar from $T_1 = 300$ K to $T_2$. Plot the heat transfer to the gas, in kJ, versus $T_2$ ranging from 300 to 1500 K. Assume the ideal gas model, and determine the specific internal energy change of the gas using

(a) $\bar{u}$ data from *IT*.

(b) a constant $\bar{c}_v$ evaluated at $T_1$ from *IT*.

### Solution (CD-ROM)

The following example illustrates the use of the closed system energy balance, together with the ideal gas model and the assumption of constant specific heats.

## *Example 4.10* Using the Energy Balance and Constant Specific Heats

Two tanks are connected by a valve. One tank contains 2 kg of carbon monoxide gas at 77°C and 0.7 bar. The other tank holds 8 kg of the same gas at 27°C and 1.2 bar. The valve is opened and the gases are allowed to mix while receiving energy by heat transfer from the surroundings. The final equilibrium temperature is 42°C. Using the ideal gas model, determine (a) the final equilibrium pressure, in bar; (b) the heat transfer for the process, in kJ.

### Solution

***Known:*** Two tanks containing different amounts of carbon monoxide gas at initially different states are connected by a valve. The valve is opened and the gas allowed to mix while receiving a certain amount of energy by heat transfer. The final equilibrium temperature is known.

***Find:*** Determine the final pressure and the heat transfer for the process.

***Schematic and Given Data:***

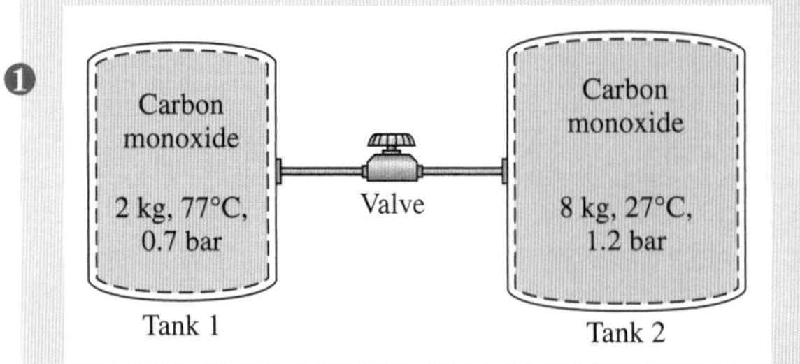


❶

***Assumptions:***

1. The total amount of carbon monoxide gas is a closed system.
2. The gas is modeled as an ideal gas with constant $c_v$.
3. The gas initially in each tank is in equilibrium. The final state is an equilibrium state.
4. No energy is transferred to, or from, the gas by work.
5. There is no change in kinetic or potential energy.

*Figure E4.10*

***Analysis:***

(a) The final equilibrium pressure $p_f$ can be determined from the ideal gas equation of state

$$p_f = \frac{mRT_f}{V}$$

where $m$ is the sum of the initial amounts of mass present in the two tanks, $V$ is the total volume of the two tanks, and $T_f$ is the final equilibrium temperature. Thus

$$p_f = \frac{(m_1 + m_2)RT_f}{V_1 + V_2}$$

Denoting the initial temperature and pressure in tank 1 as $T_1$ and $p_1$, respectively, $V_1 = m_1RT_1/p_1$. Similarly, if the initial temperature and pressure in tank 2 are $T_2$ and $p_2$, $V_2 = m_2RT_2/p_2$. Thus, the final pressure is

$$p_f = \frac{(m_1 + m_2)RT_f}{\left(\frac{m_1RT_1}{p_1}\right) + \left(\frac{m_2RT_2}{p_2}\right)} = \frac{(m_1 + m_2)T_f}{\left(\frac{m_1T_1}{p_1}\right) + \left(\frac{m_2T_2}{p_2}\right)}$$

Inserting values

$$p_f = \frac{(10\text{ kg})(315\text{ K})}{\frac{(2\text{ kg})(350\text{ K})}{0.7\text{ bar}} + \frac{(8\text{ kg})(300\text{ K})}{1.2\text{ bar}}} = 1.05\text{ bar} \triangleleft$$

(b) The heat transfer can be found from an energy balance, which reduces with assumptions 4 and 5 to give

$$\Delta U = Q - \cancelto{0}{W}$$

or

$$Q = U_f - U_i$$

$U_i$ is the initial internal energy, given by

$$U_i = m_1u(T_1) + m_2u(T_2)$$

where $T_1$ and $T_2$ are the initial temperatures of the CO in tanks 1 and 2, respectively. The final internal energy is $U_f$

$$U_f = (m_1 + m_2)u(T_f)$$

Introducing these expressions for internal energy, the energy balance becomes

$$Q = m_1[u(T_f) - u(T_1)] + m_2[u(T_f) - u(T_2)]$$

Since the specific heat $c_v$ is constant (assumption 2)

$$Q = m_1c_v(T_f - T_1) + m_2c_v(T_f - T_2)$$

Evaluating $c_v$ as the mean of the values listed in Table T-10 at 300 K and 350 K, $c_v = 0.745$ kJ/kg · K. Hence

❷

$$Q = (2\text{ kg})\left(0.745\frac{\text{kJ}}{\text{kg}\cdot\text{K}}\right)(315\text{ K} - 350\text{ K}) + (8\text{ kg})\left(0.745\frac{\text{kJ}}{\text{kg}\cdot\text{K}}\right)(315\text{ K} - 300\text{ K}) = +37.25\text{ kJ} \triangleleft$$

The plus sign indicates that the heat transfer is into the system.

---

❶ Since the specific heat $c_v$ of CO varies little over the temperature interval from 300 to 350 K (Table T-10), it can be treated as a constant.

❷ As an exercise, evaluate $Q$ using specific internal energy values from the ideal gas table for CO, Table T-11. Observe that specific internal energy is given in Table T-11 with units of kJ/kmol.

## 4.8 Polytropic Process of an Ideal Gas

Recall that a *polytropic* process of a closed system is described by a pressure–volume relationship of the form

$$pV^n = constant \tag{4.50}$$

where $n$ is a constant (Sec. 3.3). For a polytropic process between two states

$$p_1V_1^n = p_2V_2^n$$

or

$$\frac{p_2}{p_1} = \left(\frac{V_1}{V_2}\right)^n \tag{4.51}$$

The exponent $n$ may take on any value from $-\infty$ to $+\infty$, depending on the particular process. When $n = 0$, the process is an isobaric (constant-pressure) process, and when $n = \pm\infty$, the process is an isometric (constant-volume) process.

For a polytropic process

$$\int_1^2 p\,dV = \frac{p_2V_2 - p_1V_1}{1 - n} \qquad (n \neq 1) \tag{4.52}$$

for any exponent $n$ except $n = 1$. When $n = 1$,

$$\int_1^2 p\,dV = p_1V_1 \ln\frac{V_2}{V_1} \qquad (n = 1) \tag{4.53}$$

Example 3.1 provides the details of these integrations.

Equations 4.50 through 4.53 apply to *any* gas (or liquid) undergoing a polytropic process. When the *additional* idealization of ideal gas behavior is appropriate, further relations can be derived. Thus, when the ideal gas equation of state is introduced into Eqs. 4.51, 4.52, and 4.53, the following expressions are obtained, respectively:

$$\frac{T_2}{T_1} = \left(\frac{p_2}{p_1}\right)^{(n-1)/n} = \left(\frac{V_1}{V_2}\right)^{n-1} \qquad \text{(ideal gas)} \tag{4.54}$$

$$\int_1^2 p\,dV = \frac{mR(T_2 - T_1)}{1 - n} \qquad \text{(ideal gas, } n \neq 1\text{)} \tag{4.55}$$

$$\int_1^2 p\,dV = mRT \ln\frac{V_2}{V_1} \qquad \text{(ideal gas, } n = 1\text{)} \tag{4.56}$$

For an ideal gas, the case $n = 1$ corresponds to an isothermal (constant-temperature) process, as can readily be verified. In addition, when the specific heats are constant, the value of the exponent $n$ corresponding to an adiabatic polytropic process of an ideal gas is the specific heat ratio $k$ (see discussion of Eq. 7.36).

Example 4.11 illustrates the use of the closed system energy balance for a system consisting of an ideal gas undergoing a polytropic process.

## *Example 4.11* Polytropic Process of Air as an Ideal Gas

Air undergoes a polytropic compression in a piston–cylinder assembly from $p_1 = 1$ atm, $T_1 = 70°F$ to $p_2 = 5$ atm. Employing the ideal gas model, determine the work and heat transfer per unit mass, in Btu/lb, if $n = 1.3$.

### Solution

***Known:*** Air undergoes a polytropic compression process from a given initial state to a specified final pressure.

***Find:*** Determine the work and heat transfer, each in Btu/lb.

***Schematic and Given Data:***

❶

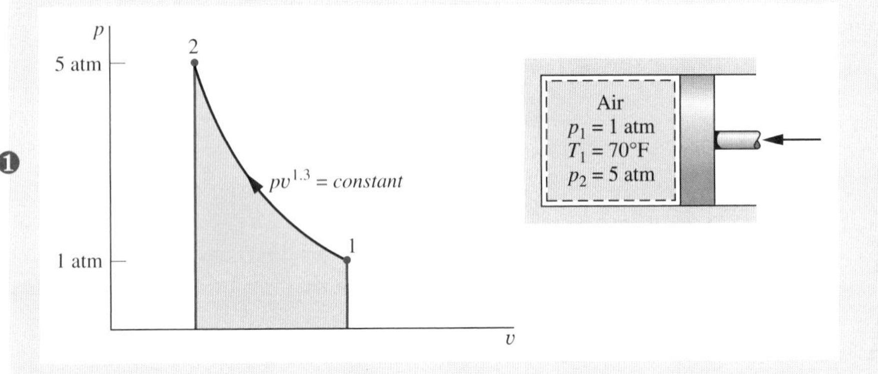


***Assumptions:***

1. The air is a closed system.
2. The air behaves as an ideal gas.
3. The compression is polytropic with $n = 1.3$.
4. There is no change in kinetic or potential energy.

*Figure E4.11*

***Analysis:*** The work can be evaluated in this case from the expression

$$W = \int_1^2 p\, dV$$

With Eq. 4.55

$$\frac{W}{m} = \frac{R(T_2 - T_1)}{1 - n}$$

The temperature at the final state, $T_2$, is required. This can be evaluated from Eq. 4.54

$$T_2 = T_1\left(\frac{p_2}{p_1}\right)^{(n-1)/n} = 530\left(\frac{5}{1}\right)^{(1.3-1)/1.3} = 768°R$$

The work is then

$$\frac{W}{m} = \frac{R(T_2 - T_1)}{1 - n} = \left(\frac{1.986}{28.97}\frac{\text{Btu}}{\text{lb}\cdot°\text{R}}\right)\left(\frac{768°\text{R} - 530°\text{R}}{1 - 1.3}\right)$$
$$= -54.39 \text{ Btu/lb} \triangleleft$$

The heat transfer can be evaluated from an energy balance. Thus

$$\frac{Q}{m} = \frac{W}{m} + (u_2 - u_1) = -54.39 + (131.88 - 90.33)$$
$$= -13.34 \text{ Btu/lb} \triangleleft$$

where the specific internal energy values are obtained from Table T-9E.

❶ The states visited in the polytropic compression process are shown by the curve on the accompanying $p$–$v$ diagram. The magnitude of the work per unit of mass is represented by the shaded area *below* the curve.

## 4.9 Chapter Summary and Study Guide

In this chapter, we have considered property relations for a broad range of substances in tabular, graphical, and equation form. Although computer retrieval of property data has been considered, primary emphasis has been placed on the use of tabular data.

A key aspect of thermodynamic analysis is fixing states. This is guided by the state principle for simple compressible systems, which indicates that the intensive state is fixed by the values of *two* independent, intensive properties. Another important aspect of thermodynamic analysis is locating principal states of processes on appropriate diagrams: $p$–$v$, $T$–$v$, and $p$–$T$ diagrams. The skills of fixing states and using property diagrams are particularly important when solving problems involving the energy balance.

The ideal gas model is introduced in the second part of this chapter, using the compressibility factor as a point of departure. This arrangement emphasizes the limitations of the ideal gas model. When it is appropriate to use the ideal gas model, we stress that specific heats generally vary with temperature, and feature the use of the ideal gas tables in problem solving.

The following checklist provides a study guide for this chapter. When your study of the text and end-of-chapter exercises has been completed you should be able to

- write out the meanings of the terms listed in the margins throughout the chapter and understand each of the related concepts. The subset of key terms listed here in the margin is particularly important in subsequent chapters.
- retrieve property data from Tables T-1 through T-11, using the state principle to fix states and linear interpolation when required.
- sketch $T$–$v$, $p$–$v$, and $p$–$T$ diagrams, and locate principal states on such diagrams.
- apply the closed system energy balance with property data.
- evaluate the properties of two-phase, liquid–vapor mixtures using Eqs. 4.1, 4.2, 4.6, and 4.7.
- estimate the properties of liquids using Eqs. 4.11, 4.12, and 4.14.
- apply the incompressible substance model.
- apply the ideal gas model for thermodynamic analysis appropriately, using ideal gas table data or constant specific heat data to determine $\Delta u$ and $\Delta h$.

***state principle***
***simple compressible system***
***p–v–T surface***
***p–v, T–v, p–T diagrams***
***saturation temperature***
***saturation pressure***
***two-phase, liquid–vapor mixture***
***quality***
***enthalpy***
***specific heats $c_p$, $c_v$***
***ideal gas model***

## Problems

### Using $p$–$v$–$T$ Data

**4.1** Determine the phase or phases in a system consisting of $H_2O$ at the following conditions and sketch $p$–$v$ and $T$–$v$ diagrams showing the location of each state.
(a) $p$ = 80 lbf/in.$^2$, $T$ = 312.07°F.
(b) $p$ = 80 lbf/in.$^2$, $T$ = 400°F.
(c) $T$ = 400°F, $p$ = 360 lbf/in.$^2$
(d) $T$ = 320°F, $p$ = 70 lbf/in.$^2$
(e) $T$ = 10°F, $p$ = 14.7 lbf/in.$^2$

**4.2** Determine the phase or phases in a system consisting of $H_2O$ at the following conditions and sketch $p$–$v$ and $T$–$v$ diagrams showing the location of each state.
(a) $p$ = 5 bar, $T$ = 151.9°C.
(b) $p$ = 5 bar, $T$ = 200°C.

(c) $T = 200°C$, $p = 2.5$ MPa.
(d) $T = 160°C$, $p = 4.8$ bar.
(e) $T = -12°C$, $p = 1$ bar.

**4.3** Data encountered in solving problems often do not fall exactly on the grid of values provided by property tables, and *linear interpolation* between adjacent table entries becomes necessary. The following table lists temperatures and specific volumes of water vapor at two pressures:

| $p = 1.0$ MPa | | $p = 1.5$ MPa | |
|---|---|---|---|
| $T$(°C) | $v$(m$^3$/kg) | $T$(°C) | $v$(m$^3$/kg) |
| 200 | 0.2060 | 200 | 0.1325 |
| 240 | 0.2275 | 240 | 0.1483 |
| 280 | 0.2480 | 280 | 0.1627 |

(a) Determine the specific volume at $T = 240°C$, $p = 1.25$ MPa, in m$^3$/kg.
(b) Determine the temperature at $p = 1.5$ MPa, $v = 0.1555$ m$^3$/kg, in °C.
(c) Determine the specific volume at $T = 220°C$, $p = 1.4$ MPa, in m$^3$/kg.

**4.4** Data encountered in solving problems often do not fall exactly on the grid of values provided by property tables, and *linear interpolation* between adjacent table entries becomes necessary. The following table lists temperatures and specific volumes of ammonia vapor at two pressures:

| $p = 50$ lbf/in.$^2$ | | $p = 60$ lbf/in.$^2$ | |
|---|---|---|---|
| $T$(°F) | $v$(ft$^3$/lb) | $T$(°F) | $v$(ft$^3$/lb) |
| 100 | 6.836 | 100 | 5.659 |
| 120 | 7.110 | 120 | 5.891 |
| 140 | 7.380 | 140 | 6.120 |

(a) Determine the specific volume at $T = 120°F$, $p = 54$ lbf/in.$^2$, in ft$^3$/lb.
(b) Determine the temperature at $p = 60$ lbf/in.$^2$, $v = 5.982$ ft$^3$/lb, in °F.
(c) Determine the specific volume at $T = 110°F$, $p = 58$ lbf/in.$^2$, in ft$^3$/lb.

**4.5** Determine the quality of a two-phase liquid–vapor mixture of
(a) $H_2O$ at 100°C with a specific volume of 0.8 m$^3$/kg.
(b) Refrigerant 134a at 0°C with a specific volume of 0.7721 cm$^3$/g.

**4.6** Determine the quality of a two-phase liquid–vapor mixture of
(a) $H_2O$ at 100 lbf/in.$^2$ with a specific volume of 3.0 ft$^3$/lb.
(b) Refrigerant 134a at −40°F with a specific volume of 5.7173 ft$^3$/lb.

**4.7** Ten kg of a two-phase, liquid–vapor mixture of methane ($CH_4$) exists at 160 K in a 0.3 m$^3$ tank. Determine the quality of the mixture, if the values of specific volume for saturated liquid and saturated vapor methane at 160 K are $v_f = 2.97 \times 10^{-3}$ m$^3$/kg and $v_g = 3.94 \times 10^{-2}$ m$^3$/kg, respectively.

**4.8** A two-phase liquid–vapor mixture of $H_2O$ at 200 lbf/in.$^2$ has a specific volume of 1.5 ft$^3$/lb. Determine the quality of a two-phase liquid–vapor mixture at 100 lbf/in.$^2$ with the same specific volume.

**4.9** Determine the volume, in ft$^3$, occupied by 2 lb of $H_2O$ at a pressure of 1000 lbf/in.$^2$ and
(a) a temperature of 600°F.
(b) a quality of 80%.
(c) a temperature of 200°F.

**4.10** Calculate the volume, in m$^3$, occupied by 2 kg of a two-phase liquid–vapor mixture of Refrigerant 134a at −10°C with a quality of 80%.

**4.11** A two-phase liquid–vapor mixture of $H_2O$ has a temperature of 300°C and occupies a volume of 0.05 m$^3$. The masses of saturated liquid and vapor present are 0.75 kg and 2.26 kg, respectively. Determine the specific volume of the mixture, in m$^3$/kg.

**4.12** (CD-ROM)

**4.13** Five kilograms of $H_2O$ are contained in a closed rigid tank at an initial pressure of 20 bar and a quality of 50%. Heat transfer occurs until the tank contains only saturated vapor. Determine the volume of the tank, in m$^3$, and the final pressure, in bar.

**4.14** (CD-ROM)

**4.15** Two thousand kg of water, initially a saturated liquid at 150°C, is heated in a closed, rigid tank to a final state where the pressure is 2.5 MPa. Determine the final temperature, in °C, the volume of the tank, in m$^3$, and sketch the process on $T$–$v$ and $p$–$v$ diagrams.

**4.16** Steam is contained in a closed rigid container. Initially, the pressure and temperature of the steam are 15 bar and 240°C, respectively. The temperature drops as a result of heat transfer to the surroundings. Determine the pressure at which condensation first occurs, in bar, and the fraction of the total mass that has condensed when the temperature reaches 100°C. What percentage of the volume is occupied by saturated liquid at the final state?

**4.17** Water vapor is heated in a closed, rigid tank from saturated vapor at 160°C to a final temperature of 400°C. Determine the initial and final pressures, in bar, and sketch the process on $T$–$v$ and $p$–$v$ diagrams.

**4.18** (CD-ROM)

**4.19** A two-phase liquid–vapor mixture of $H_2O$ is initially at a pressure of 30 bar. If on heating at fixed volume, the critical point is attained, determine the quality at the initial state.

**4.20** (CD-ROM)

**4.21** Three lb of saturated water vapor, contained in a closed rigid tank whose volume is 13.3 ft$^3$, is heated to a final temperature of 400°F. Sketch the process on a $T$–$v$ diagram. Determine the pressures at the initial and final states, each in lbf/in.$^2$

**4.22** Refrigerant 134a undergoes a constant-pressure process at 1.4 bar from $T_1 = 20°C$ to saturated vapor. Determine the work for the process, in kJ per kg of refrigerant.

4.23 (CD-ROM)

4.24 Two pounds mass of Refrigerant 134a, initially at $p_1$ = 180 lbf/in.$^2$ and $T_1$ = 120°F, undergo a constant-pressure process to a final state where the quality is 76.5%. Determine the work for the process, in Btu.

4.25 Water vapor initially at 3.0 MPa and 300°C is contained within a piston–cylinder assembly. The water is cooled at constant volume until its temperature is 200°C. The water is then condensed isothermally to saturated liquid. For the water as the system, evaluate the work, in kJ/kg.

4.26 A piston–cylinder assembly contains 0.04 lb of Refrigerant 134a. The refrigerant is compressed from an initial state where $p_1$ = 10 lbf/in.$^2$ and $T_1$ = 20°F to a final state where $p_2$ = 160 lbf/in.$^2$ During the process, the pressure and specific volume are related by $pv$ = *constant*. Determine the work, in Btu, for the refrigerant.

4.27 (CD-ROM)

4.28 (CD-ROM)

## Using *u–h* Data

4.29 Using the tables for water, determine the specified property data at the indicated states. In each case, locate the state on sketches of the $p$–$v$ and $T$–$v$ diagrams.

(a) At $p$ = 3 bar, $T$ = 240°C, find $v$ in m$^3$/kg and $u$ in kJ/kg.
(b) At $p$ = 3 bar, $v$ = 0.5 m$^3$/kg, find $T$ in °C and $u$ in kJ/kg.
(c) At $T$ = 400°C, $p$ = 10 bar, find $v$ in m$^3$/kg and $h$ in kJ/kg.
(d) At $T$ = 320°C, $v$ = 0.03 m$^3$/kg, find $p$ in MPa and $u$ in kJ/kg.
(e) At $p$ = 28 MPa, $T$ = 520°C, find $v$ in m$^3$/kg and $h$ in kJ/kg.
(f) At $T$ = 100°C, $x$ = 60%, find $p$ in bar and $v$ in m$^3$/kg.
(g) At $T$ = 10°C, $v$ = 100 m$^3$/kg, find $p$ in kPa and $h$ in kJ/kg.
(h) At $p$ = 4 MPa, $T$ = 160°C, find $v$ in m$^3$/kg and $u$ in kJ/kg.

4.30 Using the tables for water, determine the specified property data at the indicated states. In each case, locate the state on sketches of the $p$–$v$ and $T$–$v$ diagrams.

(a) At $p$ = 20 lbf/in.$^2$, $T$ = 400°F, find $v$ in ft$^3$/lb and $u$ in Btu/lb.
(b) At $p$ = 20 lbf/in.$^2$, $v$ = 16 ft$^3$/lb, find $T$ in °F and $u$ in Btu/lb.
(c) At $T$ = 900°F, $p$ = 170 lbf/in.$^2$, find $v$ in ft$^3$/lb and $h$ in Btu/lb.
(d) At $T$ = 600°F, $v$ = 0.6 ft$^3$/lb, find $p$ in lbf/in.$^2$ and $u$ in Btu/lb.
(e) At $p$ = 700 lbf/in.$^2$, $T$ = 650°F, find $v$ in ft$^3$/lb and $h$ in Btu/lb.
(f) At $T$ = 400°F, $x$ = 90%, find $p$ in lbf/in.$^2$ and $v$ in ft$^3$/lb.
(g) At $T$ = 40°F, $v$ = 1950 ft$^3$/lb, find $p$ in lbf/in.$^2$ and $h$ in Btu/lb.
(h) At $p$ = 600 lbf/in.$^2$, $T$ = 320°F, find $v$ in ft$^3$/lb and $u$ in Btu/lb.

4.31 (CD-ROM)

4.32 (CD-ROM)

4.33 A quantity of water is at 15 MPa and 100°C. Evaluate the specific volume, in m$^3$/kg, and the specific enthalpy, in kJ/kg, using

(a) data from Table T-5.
(b) saturated liquid data from Table T-2.

4.34 Evaluate the specific volume, in ft$^3$/lb, and the specific enthalpy, in Btu/lb, of water at 200°F and a pressure of 2000 lbf/in.$^2$

4.35 Evaluate the specific volume, in ft$^3$/lb, and the specific enthalpy, in Btu/lb, of Refrigerant 134a at 95°F and 150 lbf/in.$^2$

4.36 Evaluate the specific volume, in m$^3$/kg, and the specific enthalpy, in kJ/kg, of Refrigerant 134a at 41°C and 1.4 MPa.

4.37 (CD-ROM)

## Using the Energy Balance with Property Data

4.38 A closed, rigid tank contains 3 kg of saturated water vapor initially at 140°C. Heat transfer occurs, and the pressure drops to 200 kPa. Kinetic and potential energy effects are negligible. For the water as the system, determine the amount of energy transfer by heat, in kJ.

4.39 Refrigerant 134a is compressed with no heat transfer in a piston–cylinder assembly from 30 lbf/in.$^2$, 20°F to 160 lbf/in.$^2$ The mass of refrigerant is 0.04 lb. For the refrigerant as the system, $W$ = −0.56 Btu. Kinetic and potential energy effects are negligible. Determine the final temperature, in °F.

4.40 Saturated liquid water contained in a closed, rigid tank is cooled to a final state where the temperature is 50°C and the masses of saturated vapor and liquid present are 0.03 and 1999.97 kg, respectively. Determine the heat transfer for the process, in kJ.

4.41 Refrigerant 134a undergoes a process for which the pressure–volume relation is $pv^n$ = *constant*. The initial and final states of the refrigerant are fixed by $p_1$ = 200 kPa, $T_1$ = −10°C and $p_2$ = 1000 kPa, $T_2$ = 50°C, respectively. Calculate the work and heat transfer for the process, each in kJ per kg of refrigerant.

4.42 A rigid, well-insulated tank contains a two-phase mixture consisting of 0.07 lb of saturated liquid water and 0.07 lb of saturated water vapor, initially at 20 lbf/in.$^2$ A paddle wheel stirs the mixture until only saturated vapor remains in the tank. Kinetic and potential energy effects are negligible. For the water, determine the amount of energy transfer by work, in Btu.

4.43 (CD-ROM)

4.44 (CD-ROM)

4.45 (CD-ROM)

4.46 Five kilograms of water, initially a saturated vapor at 100 kPa, are cooled to saturated liquid while the pressure is maintained constant. Determine the work and heat transfer for the process, each in kJ. Show that the heat transfer equals the change in enthalpy of the water in this case.

4.47 A system consisting of 2 lb of water vapor, initially at 300°F and occupying a volume of 20 ft$^3$, is compressed isothermally to a volume of 9.05 ft$^3$. The system is then heated at constant volume to a final pressure of 120 lbf/in.$^2$ During the isothermal compression there is energy transfer by work of magnitude 90.8 Btu *into* the system. Kinetic and potential energy effects are negligible. Determine the heat transfer, in Btu, for each process.

**4.48** (CD-ROM)

**4.49** (CD-ROM)

**4.50** (CD-ROM)

**4.51** (CD-ROM)

**4.52** (CD-ROM)

**4.53** (CD-ROM)

**4.54** A system consisting of 1 kg of $H_2O$ undergoes a power cycle composed of the following processes:

*Process 1–2:* Constant-pressure heating at 10 bar from saturated vapor.

*Process 2–3:* Constant-volume cooling to $p_3 = 5$ bar, $T_3 = 160°C$.

*Process 3–4:* Isothermal compression with $Q_{34} = -815.8$ kJ.

*Process 4–1:* Constant-volume heating.

Sketch the cycle on $T$–$v$ and $p$–$v$ diagrams. Neglecting kinetic and potential energy effects, determine the thermal efficiency.

**4.55** A well-insulated copper tank of mass 13 kg contains 4 kg of liquid water. Initially, the temperature of the copper is 27°C and the temperature of the water is 50°C. An electrical resistor of neglible mass transfers 100 kJ of energy to the contents of the tank. The tank and its contents come to equilibrium. What is the final temperature, in °C?

**4.56** A steel bar (AISI 316) of mass 50 lb, initially at 200°F, is placed in an open tank together with 5 ft$^3$ of liquid water, initially at 70°F. For the water and the bar as the system, determine the final equilibrium temperature, in °F, ignoring heat transfer between the tank and its surroundings.

### Using Generalized Compressibility Data (CD-ROM)

**4.57** (CD-ROM)

**4.58** (CD-ROM)

**4.59** (CD-ROM)

**4.60** (CD-ROM)

**4.61** (CD-ROM)

**4.62** (CD-ROM)

**4.63** (CD-ROM)

### Using the Ideal Gas Model

**4.64** A tank contains 0.042 m$^3$ of oxygen at 21°C and 15 MPa. Determine the mass of oxygen, in kg, using the ideal gas model.

**4.65** Show that water vapor can be accurately modeled as an ideal gas at temperatures below about 60°C (140°F).

**4.66** Determine the percent error in using the ideal gas model to determine the specific volume of
(a) water vapor at 2000 lbf/in.$^2$, 700°F.
(b) water vapor at 1 lbf/in.$^2$, 200°F.

**4.67** Check the applicability of the ideal gas model for Refrigerant 134a at a temperature of 80°C and a pressure of
(a) 1.6 MPa.
(b) 0.10 MPa.

**4.68** Determine the temperature, in K, of 5 kg of air at a pressure of 0.3 MPa and a volume of 2.2 m$^3$. Ideal gas behavior can be assumed for air under these conditions.

**4.69** A 40-ft$^3$ tank contains air at 560°R with a pressure of 50 lbf/in.$^2$ Determine the mass of the air, in lb. Ideal gas behavior can be assumed for air under these conditions.

**4.70** Compare the densities, in kg/m$^3$, of helium and air, each at 300 K, 100 kPa. Assume ideal gas behavior.

**4.71** Assuming the ideal gas model, determine the volume, in ft$^3$, occupied by 1 lbmol of carbon dioxide ($CO_2$) gas at 200 lbf/in.$^2$ and 600°R.

### Using the Energy Balance with the Ideal Gas Model

**4.72** A rigid tank, with a volume of 2 ft$^3$, contains air initially at 20 lbf/in$^2$, 500°R. If the air receives a heat transfer of magnitude 6 Btu, determine the final temperature, in °R, and the final pressure, in lbf/in.$^2$ Assume ideal gas behavior, and use
(a) a constant specific heat value from Table T-10E evaluated at 500°R.
(b) data from Table T-9E.

**4.73** One kilogram of air, initially at 5 bar, 350 K, and 3 kg of carbon dioxide ($CO_2$), initially at 2 bar, 450 K, are confined to opposite sides of a rigid, well-insulated container, as illustrated in Fig. P4.73. The partition is free to move and allows conduction from one gas to the other without energy storage in the partition itself. The air and carbon dioxide each behave as ideal gases. Determine the final equilibrium temperature, in K, and the final pressure, in bar, assuming constant specific heats.

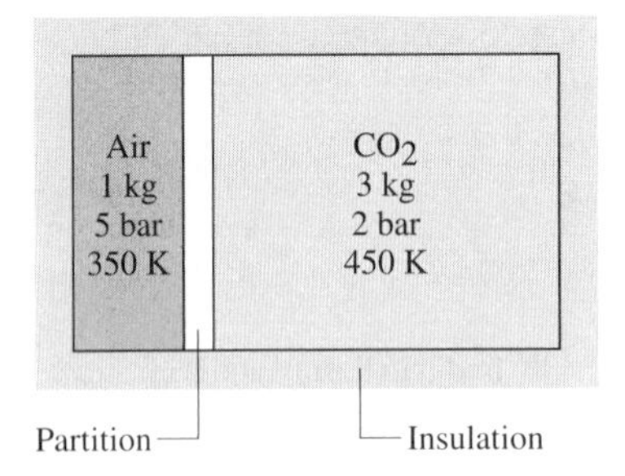


*Figure P4.73*

**4.74** Argon (Ar) gas initially at 1 bar, 100 K undergoes a polytropic process, with $n = k$, to a final pressure of 15.59 bar. Determine the work and heat transfer for the process, each in kJ per kg of argon. Assume ideal gas behavior with $\bar{c}_p = 2.5\,\bar{R}$.

**4.75** Carbon dioxide ($CO_2$) gas, initially at $T_1 = 530°R$, $p_1 = 15$ lbf/in.$^2$, and $V_1 = 1$ ft$^3$, is compressed in a piston–cylinder assembly. During the process, the pressure and specific volume are related by $pv^{1.2} = constant$. The amount of energy transfer *to the gas* by work is 45 Btu per lb of $CO_2$. Assuming ideal gas behavior, determine the final temperature, in °R, and the heat transfer, in Btu per lb of gas.

**4.76** A gas is confined to one side of a rigid, insulated container divided by a partition. The other side is initially evacuated. The following data are known for the initial state of the gas: $p_1 = 3$ bar, $T_1 = 380$ K, and $V_1 = 0.025$ m$^3$. When the partition is removed, the gas expands to fill the entire container and achieves a final equilibrium pressure of 1.5 bar. Assuming ideal gas behavior, determine the final volume, in m$^3$.

**4.77** (CD-ROM)

**4.78** (CD-ROM)

**4.79** (CD-ROM)

**4.80** A piston–cylinder assembly contains 1 kg of nitrogen gas ($N_2$). The gas expands from an initial state where $T_1 = 700$ K and $p_1 = 5$ bar to a final state where $p_2 = 2$ bar. During the process the pressure and specific volume are related by $pv^{1.3} = constant$. Assuming ideal gas behavior and neglecting kinetic and potential energy effects, determine the heat transfer during the process, in kJ, using

(a) a constant specific heat evaluated at 300 K.
(b) a constant specific heat evaluated at 700 K.
(c) data from Table T-11.

**4.81** Air is compressed adiabatically from $p_1 = 1$ bar, $T_1 = 300$ K to $p_2 = 15$ bar, $v_2 = 0.1227$ m$^3$/kg. The air is then cooled at constant volume to $T_3 = 300$ K. Assuming ideal gas behavior, and ignoring kinetic and potential energy effects, calculate the work for the first process and the heat transfer for the second process, each in kJ per kg of air. Solve the problem each of two ways:

(a) using data from Table T-9.
(b) using a constant specific heat evaluated at 300 K.

**4.82** A system consists of 2 kg of carbon dioxide gas initially at state 1, where $p_1 = 1$ bar, $T_1 = 300$ K. The system undergoes a power cycle consisting of the following processes:

***Process 1–2:*** constant volume to $p_2 = 4$ bar

***Process 2–3:*** expansion with $pv^{1.28} = constant$

***Process 3–1:*** constant-pressure compression

Assuming the ideal gas model and neglecting kinetic and potential energy effects,

(a) sketch the cycle on a $p$–$v$ diagram.
(b) determine the thermal efficiency.

**4.83** One lb of air undergoes a power cycle consisting of the following processes:

***Process 1–2:*** constant volume from $p_1 = 20$ lbf/in.$^2$, $T_1 = 500$°R to $T_2 = 820$°R

***Process 2–3:*** adiabatic expansion to $v_3 = 1.4v_2$

***Process 3–1:*** constant-pressure compression

Sketch the cycle on a $p$–$v$ diagram. Assuming ideal gas behavior, determine

(a) the pressure at state 2, in lbf/in.$^2$
(b) the temperature at state 3, in°R.
(c) the thermal efficiency of the cycle.

**4.84** (CD-ROM)

# CONTROL VOLUME ANALYSIS USING ENERGY

## *Introduction...*

*chapter objective*

The ***objective*** of this chapter is to develop and illustrate the use of the control volume forms of the conservation of mass and conservation of energy principles. Mass and energy balances for control volumes are introduced in Secs. 5.1 and 5.2, respectively. These balances are applied in Sec. 5.3 to control volumes at steady state.

Although devices such as turbines, pumps, and compressors through which mass flows can be analyzed in principle by studying a particular quantity of matter (a closed system) as it passes through the device, it is normally preferable to think of a region of space through which mass flows (a control volume). As in the case of a closed system, energy transfer across the boundary of a control volume can occur by means of work and heat. In addition, another type of energy transfer must be accounted for—the energy accompanying mass as it enters or exits.

## 5.1 Conservation of Mass for a Control Volume

In this section an expression of the conservation of mass principle for control volumes is developed and illustrated. As a part of the presentation, the one-dimensional flow model is introduced.

### Developing the Mass Rate Balance

The mass rate balance for control volumes is introduced by reference to Fig. 5.1, which shows a control volume with mass flowing in at $i$ and flowing out at $e$, respectively. When applied to such a control volume, the ***conservation of mass*** principle states

*conservation of mass*

$$\begin{bmatrix} \text{time } \textit{rate of change} \text{ of} \\ \text{mass contained within} \\ \text{the control volume } \textit{at time } t \end{bmatrix} = \begin{bmatrix} \text{time } \textit{rate} \text{ of flow} \\ \text{of mass } \textit{in} \text{ across} \\ \text{inlet } i \textit{ at time } t \end{bmatrix} - \begin{bmatrix} \text{time } \textit{rate} \text{ of flow} \\ \text{of mass } \textit{out} \text{ across} \\ \text{exit } e \textit{ at time } t \end{bmatrix}$$

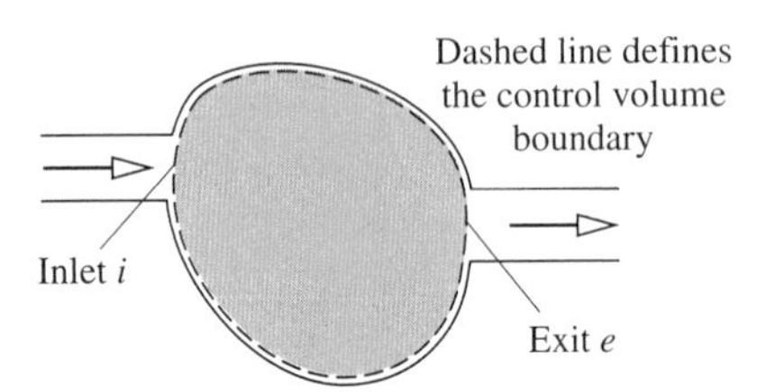


*Figure 5.1* One-inlet, one-exit control volume.

Denoting the mass contained within the control volume at time $t$ by $m_{cv}(t)$, this statement of the conservation of mass principle can be expressed in symbols as

$$\frac{dm_{cv}}{dt} = \dot{m}_i - \dot{m}_e \tag{5.1}$$

where $\dot{m}_i$ and $\dot{m}_e$ are the instantaneous ***mass flow rates*** at the inlet and exit, respectively. As for the symbols $\dot{W}$ and $\dot{Q}$, the dots in the quantities $\dot{m}_i$ and $\dot{m}_e$ denote time rates of transfer. In SI, all terms in Eq. 5.1 are expressed in kg/s. Other units employed in this text are lb/s and slug/s.

*mass flow rates*

In general, there may be several locations on the boundary through which mass enters or exits. This can be accounted for by summing, as follows:

$$\frac{dm_{cv}}{dt} = \sum_i \dot{m}_i - \sum_e \dot{m}_e \tag{5.2}$$

*mass rate balance*

Equation 5.2 is the ***mass rate balance*** for control volumes with several inlets and exits. It is a form of the conservation of mass principle commonly employed in engineering. Other forms of the mass rate balance are considered in discussions to follow.

## One-dimensional Flow

When a flowing stream of matter entering or exiting a control volume adheres to the following idealizations, the flow is said to be ***one-dimensional:*** (1) The flow is normal to the boundary at locations where mass enters or exits the control volume. (2) *All* intensive properties, including velocity and specific volume, are *uniform with position* (bulk average values) over each inlet or exit area through which matter flows. In subsequent control volume analyses in thermodynamics we routinely assume that the boundary of the control volume can be selected so that these idealizations are appropriate. Accordingly, the assumption of one-dimensional flow is not listed explicitly in the accompanying solved examples.

*one-dimensional flow*

METHODOLOGY UPDATE

Figure 5.2 illustrates the meaning of one-dimensional flow. The area through which mass flows is denoted by A. The symbol V denotes a single value that represents the velocity of the flowing air. Similarly $T$ and $v$ are single values that represent the temperature and specific volume, respectively, of the flowing air.

When the flow is one-dimensional, the mass flow rate can be evaluated using

$$\dot{m} = \frac{\mathrm{AV}}{v} \quad \text{(one-dimensional flow)} \tag{5.3a}$$

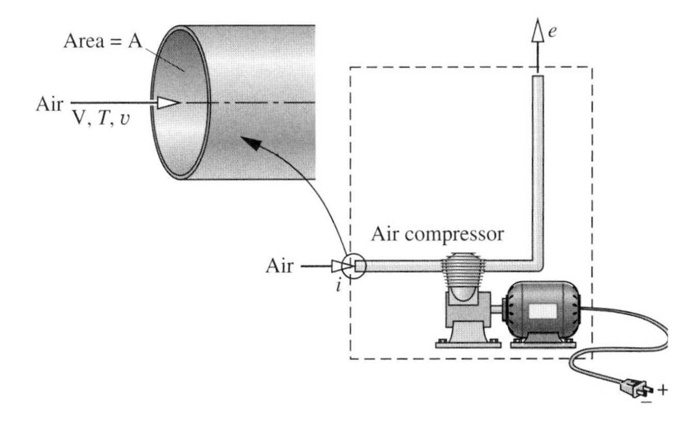


*Figure 5.2* Figure illustrating the one-dimensional flow model.

or in terms of density

$$\dot{m} = \rho AV \quad \text{(one-dimensional flow)} \tag{5.3b}$$

When area is in $m^2$, velocity is in m/s, and specific volume is in $m^3$/kg, the mass flow rate found from Eq. 5.3a is in kg/s, as can be verified.

*volumetric flow rate*

The product AV in Eqs. 5.3 is the ***volumetric flow rate.*** The volumetric flow rate is expressed in units of $m^3$/s or $ft^3$/s.

## Steady-state Form

*steady state*

Many engineering systems can be idealized as being at ***steady state,*** meaning that *all* properties are unchanging in time. For a control volume at steady state, the identity of the matter within the control volume changes continuously, but the total amount present at any instant remains constant, so $dm_{cv}/dt = 0$ and Eq. 5.2 reduces to

$$\sum_i \dot{m}_i = \sum_e \dot{m}_e \tag{5.4}$$

That is, the total incoming and outgoing rates of mass flow are equal.

Equality of total incoming and outgoing rates of mass flow does not necessarily mean that a control volume is at steady state. Although the total amount of mass within the control volume at any instant would be constant, other properties such as temperature and pressure might be varying with time. When a control volume is at steady state, *every* property is independent of time. Note that the steady-state assumption and the one-dimensional flow assumption are independent idealizations. One does not imply the other.

The following example illustrates an application of the rate form of the mass balance to a control volume *at steady state*. The control volume has two inlets and one exit.

## *Example 5.1* Feedwater Heater at Steady State

A feedwater heater operating at steady state has two inlets and one exit. At inlet 1, water vapor enters at $p_1 = 7$ bar, $T_1 = 200°C$ with a mass flow rate of 40 kg/s. At inlet 2, liquid water at $p_2 = 7$ bar, $T_2 = 40°C$ enters through an area $A_2 = 25$ cm$^2$. Saturated liquid at 7 bar exits at 3 with a volumetric flow rate of 0.06 $m^3$/s. Determine the mass flow rates at inlet 2 and at the exit, in kg/s, and the velocity at inlet 2, in m/s.

### Solution

***Known:*** A stream of water vapor mixes with a liquid water stream to produce a saturated liquid stream at the exit. The states at inlets and exit are specified. Mass flow rate and volumetric flow rate data are given at one inlet and at the exit, respectively.

***Find:*** Determine the mass flow rates at inlet 2 and at the exit, and the velocity $V_2$.

***Schematic and Given Data:***

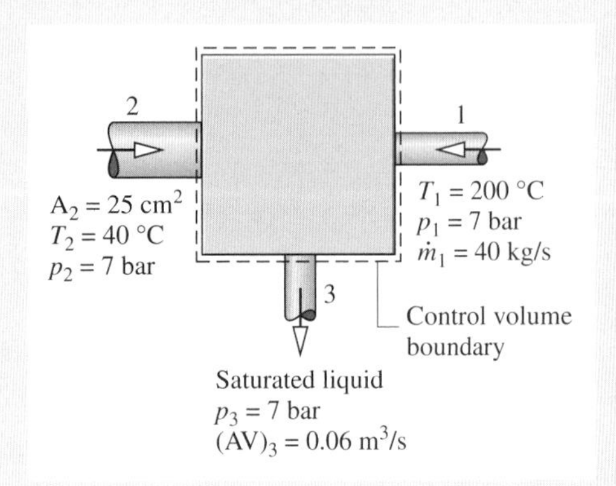


***Assumption:*** The control volume shown on the accompanying figure is at steady state.

*Figure E5.1*

***Analysis:*** The principal relations to be employed are the mass rate balance (Eq. 5.2) and the expression $\dot{m} = \mathrm{AV}/v$ (Eq. 5.3a). At steady state the mass rate balance becomes

❶
$$\frac{dm_{\text{cv}}}{dt}^{0} = \dot{m}_1 + \dot{m}_2 - \dot{m}_3$$

Solving for $\dot{m}_2$

$$\dot{m}_2 = \dot{m}_3 - \dot{m}_1$$

The mass flow rate $\dot{m}_1$ is given. The mass flow rate at the exit can be evaluated from the given volumetric flow rate

$$\dot{m}_3 = \frac{(\mathrm{AV})_3}{v_3}$$

where $v_3$ is the specific volume at the exit. In writing this expression, one-dimensional flow is assumed. From Table T-3, $v_3 = 1.108 \times 10^{-3}\ \mathrm{m^3/kg}$. Hence

$$\dot{m}_3 = \frac{0.06\ \mathrm{m^3/s}}{(1.108 \times 10^{-3}\ \mathrm{m^3/kg})} = 54.15\ \mathrm{kg/s}$$

The mass flow rate at inlet 2 is then

$$\dot{m}_2 = \dot{m}_3 - \dot{m}_1 = 54.15 - 40 = 14.15\ \mathrm{kg/s} \triangleleft$$

For one-dimensional flow at 2, $\dot{m}_2 = \mathrm{A_2V_2}/v_2$, so

$$\mathrm{V}_2 = \dot{m}_2 v_2/\mathrm{A}_2$$

State 2 is a compressed liquid. The specific volume at this state can be approximated by $v_2 \approx v_\mathrm{f}(T_2)$ (Eq. 4.11). From Table T-2 at 40°C, $v_2 = 1.0078 \times 10^{-3}\ \mathrm{m^3/kg}$. So

$$\mathrm{V}_2 = \frac{(14.15\ \mathrm{kg/s})(1.0078 \times 10^{-3}\ \mathrm{m^3/kg})}{25\ \mathrm{cm^2}} \left| \frac{10^4\ \mathrm{cm^2}}{1\ \mathrm{m^2}} \right| = 5.7\ \mathrm{m/s} \triangleleft$$

❶ At steady state the mass flow rate at the exit equals the sum of the mass flow rates at the inlets. It is left as an exercise to show that the volumetric flow rate at the exit does not equal the sum of the volumetric flow rates at the inlets.

Example 5.2 illustrates an unsteady, or *transient,* application of the mass rate balance. In this case, a barrel is filled with water.

### *Example 5.2* Filling a Barrel with Water

Water flows into the top of an open barrel at a constant mass flow rate of 30 lb/s. Water exits through a pipe near the base with a mass flow rate proportional to the height of liquid inside: $\dot{m}_e = 9L$, where $L$ is the instantaneous liquid height, in ft. The area of the base in 3 ft$^2$, and the density of water is 62.4 lb/ft$^3$. If the barrel is initially empty, plot the variation of liquid height with time and comment on the result.

**Solution (CD-ROM)**

## 5.2 Conservation of Energy for a Control Volume

In this section an expression of the conservation of energy principle for control volumes is developed and illustrated.

### 5.2.1 Developing the Energy Rate Balance for a Control Volume

The conservation of energy principle applied to a control volume states: *The time rate of change of energy stored within the control volume equals the difference between the total incoming and total outgoing rates of energy transfer.*

From our discussion of energy in Chap. 3 we know that energy can enter and exit a closed system by work and heat transfer. The same is true of a control volume. For a control volume, energy also enters and exits with flowing streams of matter. Accordingly, for the one-inlet one-exit control volume with one-dimensional flow shown in Fig. 5.3 the energy rate balance is

$$\frac{dE_{\text{cv}}}{dt} = \dot{Q} - \dot{W} + \underline{\dot{m}_i\left(u_i + \frac{\mathrm{V}_i^2}{2} + gz_i\right)} - \underline{\dot{m}_e\left(u_e + \frac{\mathrm{V}_e^2}{2} + gz_e\right)} \tag{5.5}$$

where $E_{\text{cv}}$ denotes the energy of the control volume at time $t$. The terms $\dot{Q}$ and $\dot{W}$ account, respectively, for the net rate of energy transfer by heat and work across the boundary of the control volume at $t$. The underlined terms account for the rates of transfer of internal, kinetic, and potential energy of the entering and exiting streams. If there is no mass flow in or out, the respective mass flow rates vanish and the underlined terms of Eq. 5.5 drop out. The equation then reduces to the rate form of the energy balance for closed systems: Eq. 3.13.

#### Evaluating Work for a Control Volume

Next, we will place Eq. 5.5 in an alternative form that is more convenient for subsequent applications. This will be accomplished primarily by recasting the work term $\dot{W}$, which represents the net rate of energy transfer by work across *all* portions of the boundary of the control volume.

Because work is always done on or by a control volume where matter flows across the boundary, it is convenient to separate the work term $\dot{W}$ into *two contributions*. One is the work associated with the fluid pressure as mass is introduced at inlets and removed at exits. The other contribution, denoted by $\dot{W}_{\text{cv}}$, includes *all other* work effects, such as those associated with rotating shafts, displacement of the boundary, and electrical effects.

Consider the work at an exit $e$ associated with the pressure of the flowing matter. Recall from Eq. 3.4 that the rate of energy transfer by work can be expressed as the product of a force and the velocity at the point of application of the force. Accordingly, the rate at which work is done at the exit by the normal force (normal to the exit area in the direction of flow) due to pressure is the product of the normal force, $p_e\mathrm{A}_e$, and the fluid velocity, $\mathrm{V}_e$. That is

$$\begin{bmatrix} \text{time rate of energy transfer} \\ \text{by work } \textit{from} \text{ the control} \\ \text{volume at exit } e \end{bmatrix} = (p_e\mathrm{A}_e)\mathrm{V}_e$$

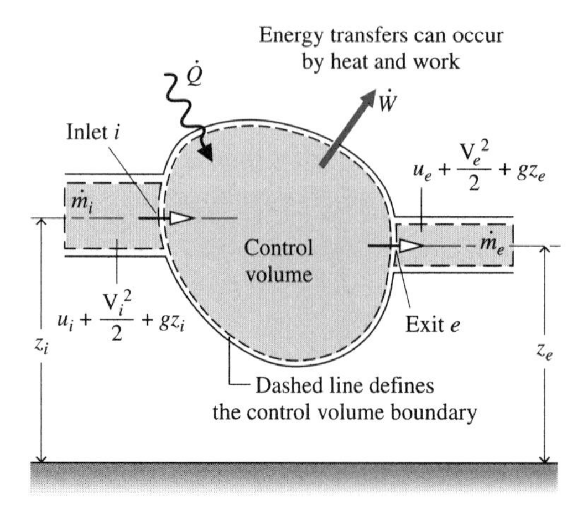


*Figure 5.3* Figure used to develop Eq. 5.5.

where $p_e$ is the pressure, $A_e$ is the area, and $V_e$ is the velocity at exit $e$, respectively. A similar expression can be written for the rate of energy transfer by work into the control volume at inlet $i$.

With these considerations, the work term $\dot{W}$ of the energy rate equation, Eq. 5.5, can be written as

$$\dot{W} = \dot{W}_{cv} + (p_e A_e)V_e - (p_i A_i)V_i \tag{5.6a}$$

where, in accordance with the sign convention for work, the term at the inlet has a negative sign because energy is transferred into the control volume there. A positive sign precedes the work term at the exit because energy is transferred out of the control volume there. With $AV = \dot{m}v$ from Eq. 5.3a, the above expression for work can be written as

$$\dot{W} = \dot{W}_{cv} + \dot{m}_e(p_e v_e) - \dot{m}_i(p_i v_i) \tag{5.6b}$$

where $\dot{m}_i$ and $\dot{m}_e$ are the mass flow rates and $v_i$ and $v_e$ are the specific volumes evaluated at the inlet and exit, respectively. In Eq. 5.6b, the terms $\dot{m}_i(p_i v_i)$ and $\dot{m}_e(p_e v_e)$ account for the work associated with the pressure at the inlet and exit, respectively. The term $\dot{W}_{cv}$ accounts for *all other* energy transfers by work across the boundary of the control volume.

The product $pv$ appearing in Eq. 5.6b is commonly referred to as ***flow work*** because it originates here in a work analysis. However, since $pv$ is a property, the term ***flow energy*** also is appropriate.

***flow work***

***flow energy***

## 5.2.2 Forms of the Control Volume Energy Rate Balance

Substituting Eq. 5.6b in Eq. 5.5 and collecting all terms referring to the inlet and the exit into separate expressions, the following form of the control volume energy rate balance results:

$$\frac{dE_{cv}}{dt} = \dot{Q}_{cv} - \dot{W}_{cv} + \dot{m}_i\left(u_i + p_i v_i + \frac{V_i^2}{2} + gz_i\right) - \dot{m}_e\left(u_e + p_e v_e + \frac{V_e^2}{2} + gz_e\right) \tag{5.7}$$

The subscript "cv" has been added to $\dot{Q}$ to emphasize that this is the heat transfer rate over the boundary (control surface) of the *control volume*.

The last two terms of Eq. 5.7 can be rewritten using the specific enthalpy $h$ introduced in Sec. 4.3.2. With $h = u + pv$, the energy rate balance becomes

$$\frac{dE_{cv}}{dt} = \dot{Q}_{cv} - \dot{W}_{cv} + \dot{m}_i\left(h_i + \frac{V_i^2}{2} + gz_i\right) - \dot{m}_e\left(h_e + \frac{V_e^2}{2} + gz_e\right) \tag{5.8}$$

The appearance of the sum $u + pv$ in the control volume energy equation is the principal reason for introducing enthalpy previously. It is brought in solely as a *convenience:* The algebraic form of the energy rate balance is simplified by the use of enthalpy and, as we have seen, enthalpy is normally tabulated along with other properties.

In practice there may be several locations on the boundary through which mass enters or exits. This can be accounted for by introducing summations as in the mass balance. Accordingly, the ***energy rate balance*** is

$$\frac{dE_{cv}}{dt} = \dot{Q}_{cv} - \dot{W}_{cv} + \sum_i \dot{m}_i\left(h_i + \frac{V_i^2}{2} + gz_i\right) - \sum_e \dot{m}_e\left(h_e + \frac{V_e^2}{2} + gz_e\right) \tag{5.9}$$

***energy rate balance***

Equation 5.9 is an *accounting* balance for the energy of the control volume. It states that the rate of energy increase or decrease within the control volume equals the difference between the rates of energy transfer in and out across the boundary. The mechanisms of energy transfer are heat and work, as for closed systems, and the energy that accompanies the mass entering and exiting.

Equation 5.9 provides a starting point for applying the conservation of energy principle to a wide range of problems of engineering importance, including *transient* control volumes in which the state changes with time. Transient examples include the startup or shutdown of turbines, compressors, and motors. Additional examples are provided by containers being filled or emptied, as considered in Example 5.2 and in the discussion of Fig. 2.3. Because property values, work and heat transfer rates, and mass flow rates may vary with time during transient operation, special care must be exercised when applying the mass and energy rate balances. Transient control volume applications are beyond the scope of this introductory presentation of engineering thermodynamics. Only steady-state control volumes are studied, as considered next.

## 5.3 Analyzing Control Volumes at Steady State

In this section steady-state forms of the mass and energy rate balances are developed and applied to a variety of cases of engineering interest. Steady-state cases are commonly encountered in engineering.

### 5.3.1 Steady-state Forms of the Mass and Energy Rate Balances

For a control volume at steady state, the conditions of the mass within the control volume and at the boundary do not vary with time. The mass flow rates and the rates of energy transfer by heat and work are also constant with time. There can be no accumulation of mass within the control volume, so $dm_{cv}/dt = 0$ and the mass rate balance, Eq. 5.2, takes the form

$$\underset{\text{(mass rate in)}}{\sum_i \dot{m}_i} = \underset{\text{(mass rate out)}}{\sum_e \dot{m}_e} \tag{5.4}$$

Furthermore, at steady state $dE_{cv}/dt = 0$, so Eq. 5.9 can be written as

$$0 = \dot{Q}_{cv} - \dot{W}_{cv} + \sum_i \dot{m}_i\left(h_i + \frac{V_i^2}{2} + gz_i\right) - \sum_e \dot{m}_e\left(h_e + \frac{V_e^2}{2} + gz_e\right) \tag{5.10a}$$

Alternatively

$$\underset{\text{(energy rate in)}}{\dot{Q}_{cv} + \sum_i \dot{m}_i\left(h_i + \frac{V_i^2}{2} + gz_i\right)} = \underset{\text{(energy rate out)}}{\dot{W}_{cv} + \sum_e \dot{m}_e\left(h_e + \frac{V_e^2}{2} + gz_e\right)} \tag{5.10b}$$

Equation 5.4 asserts that at steady state the total rate at which mass enters the control volume equals the total rate at which mass exits. Similarly, Eqs. 5.10 assert that the total rate at which energy is transferred into the control volume equals the total rate at which energy is transferred out.

Many important applications involve one-inlet, one-exit control volumes at steady state. It is instructive to apply the mass and energy rate balances to this special case. The mass rate balance reduces simply to $\dot{m}_1 = \dot{m}_2$. That is, the mass flow must be the same at the exit, 2, as it is at the inlet, 1. The common mass flow rate is designated simply by $\dot{m}$. Next, applying the energy rate balance and factoring the mass flow rate gives

$$0 = \dot{Q}_{cv} - \dot{W}_{cv} + \dot{m}\left[(h_1 - h_2) + \frac{(V_1^2 - V_2^2)}{2} + g(z_1 - z_2)\right] \tag{5.11a}$$

Or, dividing by the mass flow rate

$$0 = \frac{\dot{Q}_{cv}}{\dot{m}} - \frac{\dot{W}_{cv}}{\dot{m}} + (h_1 - h_2) + \frac{(V_1^2 - V_2^2)}{2} + g(z_1 - z_2) \tag{5.11b}$$

The enthalpy, kinetic energy, and potential energy terms all appear in Eqs. 5.11 as *differences* between their values at the inlet and exit. This illustrates that the datums used to assign values to specific enthalpy, velocity, and elevation cancel, provided the same ones are used at the inlet and exit. In Eq. 5.11b, the ratios $\dot{Q}_{cv}/\dot{m}$ and $\dot{W}_{cv}/\dot{m}$ are rates of energy transfer *per unit mass flowing through the control volume.*

The foregoing steady-state forms of the energy rate balance relate only energy transfer quantities evaluated at the *boundary* of the control volume. No details concerning properties *within* the control volume are required by, or can be determined with, these equations. When applying the energy rate balance in any of its forms, it is necessary to use the same units for all terms in the equation. For instance, *every* term in Eq. 5.11b must have a unit such as kJ/kg or Btu/lb. Appropriate unit conversions are emphasized in examples to follow.

## 5.3.2 Modeling Control Volumes at Steady State

In this section, we consider the modeling of control volumes *at steady state.* In particular, several examples are given in Sec. 5.3.3 showing the use of the principles of conservation of mass and energy, together with relationships among properties for the analysis of control volumes at steady state. The examples are drawn from applications of general interest to engineers and are chosen to illustrate points common to all such analyses. Before studying them, it is recommended that you review the methodology for problem solving outlined in Sec. 2.6. As problems become more complicated, the use of a systematic problem-solving approach becomes increasingly important.

When the mass and energy rate balances are applied to a control volume, simplifications are normally needed to make the analysis manageable. That is, the control volume of interest is *modeled* by making assumptions. The *careful* and *conscious* step of listing assumptions is necessary in every engineering analysis. Therefore, an important part of this section is devoted to considering various assumptions that are commonly made when applying the conservation principles to different types of devices. As you study the examples presented, it is important to recognize the role played by careful assumption-making in arriving at solutions. In each case considered, steady-state operation is assumed. The flow is regarded as one-dimensional at places where mass enters and exits the control volume. Also, at each of these locations equilibrium property relations are assumed to apply.

In several of the examples to follow, the heat transfer term $\dot{Q}_{cv}$ is set to zero in the energy rate balance because it is small relative to other energy transfers across the boundary. This may be the result of one or more of the following factors: (1) The outer surface of the control volume is well insulated. (2) The outer surface area is too small for there to be effective heat transfer. (3) The temperature difference between the control volume and its surroundings is so small that the heat transfer can be ignored. (4) The gas or liquid passes through the control volume so quickly that there is not enough time for significant heat transfer to occur. The work term $\dot{W}_{cv}$ drops out of the energy rate balance when there are no rotating shafts, displacements of the boundary, electrical effects, or other work mechanisms associated with the control volume being considered. The kinetic and potential energies of the matter entering and exiting the control volume are neglected when they are small relative to other energy transfers.

## 5.3.3 Illustrations

In this section, we present brief discussions and examples illustrating the analysis of several devices of interest in engineering, including nozzles and diffusers, turbines, compressors and

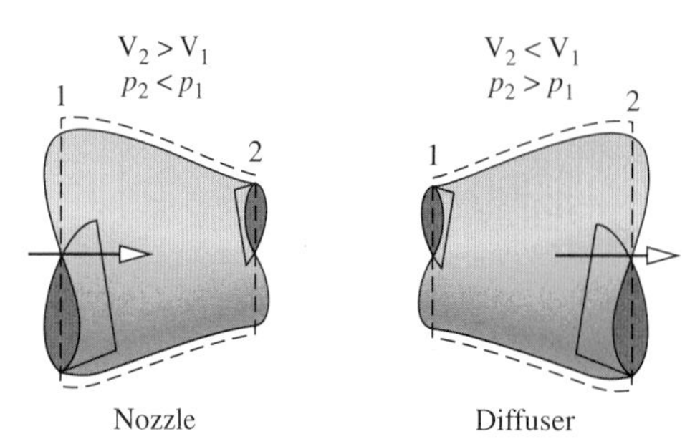


*Figure 5.4* Illustration of a nozzle and a diffuser.

pumps, heat exchangers, and throttling devices. The discussions highlight some common applications of each device and the important modeling assumptions used in thermodynamic analysis. The section also considers system integration, in which devices are combined to form an overall system serving a particular purpose.

## Nozzles and Diffusers

*nozzle*
*diffuser*

A ***nozzle*** is a flow passage of varying cross-sectional area in which the velocity of a gas or liquid increases in the direction of flow. In a ***diffuser,*** the gas or liquid decelerates in the direction of flow. Figure 5.4 shows a nozzle in which the cross-sectional area decreases in the direction of flow and a diffuser in which the walls of the flow passage diverge. In Fig. 5.5, a nozzle and diffuser are combined in a wind-tunnel test facility. Nozzles and diffusers for high-speed gas flows formed from a converging section followed by diverging section are encountered in engineering practice.

For nozzles and diffusers, the only work is *flow work* at locations where mass enters and exits the control volume, so the term $\dot{W}_{cv}$ drops out of the energy rate equation for these devices. The change in potential energy from inlet to exit is negligible under most conditions. At steady state the mass and energy rate balances reduce, respectively, to

$$\overset{0}{\cancel{\frac{dm_{cv}}{dt}}} = \dot{m}_1 - \dot{m}_2$$

$$\overset{0}{\cancel{\frac{dE_{cv}}{dt}}} = \dot{Q}_{cv} - \overset{0}{\cancel{\dot{W}_{cv}}} + \dot{m}_1\left(h_1 + \frac{V_1^2}{2} + gz_1\right) - \dot{m}_2\left(h_2 + \frac{V_2^2}{2} + gz_2\right)$$

where 1 denotes the inlet and 2 the exit. By combining these into a single expression and dropping the potential energy change from inlet to exit

$$0 = \frac{\dot{Q}_{cv}}{\dot{m}} + (h_1 - h_2) + \left(\frac{V_1^2 - V_2^2}{2}\right) \tag{5.12}$$

where $\dot{m}$ is the mass flow rate. The term $\dot{Q}_{cv}/\dot{m}$ representing heat transfer with the surroundings per unit of mass flowing through the nozzle or diffuser is often small enough relative to the enthalpy and kinetic energy changes that it can be dropped, as in the next example.

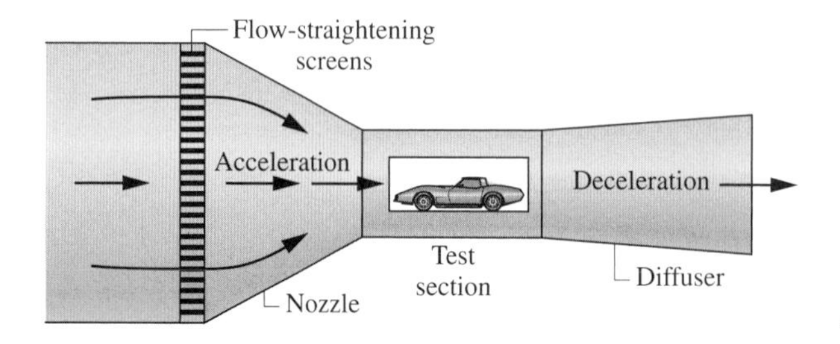


*Figure 5.5* Wind-tunnel test facility.

## *Example 5.3* Calculating Exit Area of a Steam Nozzle

Steam enters a converging–diverging nozzle operating at steady state with $p_1 = 40$ bar, $T_1 = 400°C$, and a velocity of 10 m/s. The steam flows through the nozzle with negligible heat transfer and no significant change in potential energy. At the exit, $p_2 = 15$ bar, and the velocity is 665 m/s. The mass flow rate is 2 kg/s. Determine the exit area of the nozzle, in m$^2$.

### Solution

***Known:*** Steam flows at steady state through a nozzle with known properties at the inlet and exit, a known mass flow rate, and negligible effects of heat transfer and potential energy.

***Find:*** Determine the exit area.

***Schematic and Given Data:***

❶

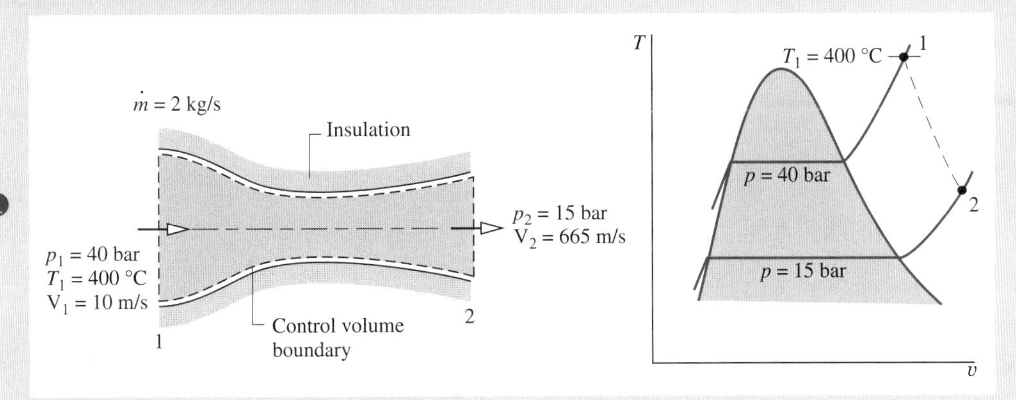


*Figure E5.3*

***Assumptions:***

1. The control volume shown on the accompanying figure is at steady state.
2. Heat transfer is negligible and $\dot{W}_{cv} = 0$.
3. The change in potential energy from inlet to exit can be neglected.

***Analysis:*** The exit area can be determined from the mass flow rate $\dot{m}$ and Eq. 5.3a, which can be arranged to read

$$A_2 = \frac{\dot{m}v_2}{V_2}$$

To evaluate $A_2$ from this equation requires the specific volume $v_2$ at the exit, and this requires that the exit state be fixed.

The state at the exit is fixed by the values of two independent intensive properties. One is the pressure $p_2$, which is known. The other is the specific enthalpy $h_2$, determined from the steady-state energy rate balance

$$0 = \overset{0}{\cancel{\dot{Q}_{cv}}} - \overset{0}{\cancel{\dot{W}_{cv}}} + \dot{m}\left(h_1 + \frac{V_1^2}{2} + gz_1\right) - \dot{m}\left(h_2 + \frac{V_2^2}{2} + gz_2\right)$$

where $\dot{Q}_{cv}$ and $\dot{W}_{cv}$ are deleted by assumption 2. The change in specific potential energy drops out in accordance with assumption 3 and $\dot{m}$ cancels, leaving

$$0 = (h_1 - h_2) + \left(\frac{V_1^2 - V_2^2}{2}\right)$$

Solving for $h_2$

$$h_2 = h_1 + \left(\frac{V_1^2 - V_2^2}{2}\right)$$

From Table T-4, $h_1 = 3213.6$ kJ/kg. The velocities $V_1$ and $V_2$ are given. Inserting values and converting the units of the kinetic energy terms to kJ/kg results in

❷

$$h_2 = 3213.6 \text{ kJ/kg} + \left[\frac{(10)^2 - (665)^2}{2}\right]\left(\frac{\text{m}^2}{\text{s}^2}\right)\left|\frac{1 \text{ N}}{1 \text{ kg} \cdot \text{m/s}^2}\right|\left|\frac{1 \text{ kJ}}{10^3 \text{ N} \cdot \text{m}}\right|$$

$$= 3213.6 - 221.1 = 2992.5 \text{ kJ/kg}$$

Finally, referring to Table T-4 at $p_2 = 15$ bar with $h_2 = 2992.5$ kJ/kg, the specific volume at the exit is $v_2 = 0.1627$ m³/kg. The exit area is then

❸
$$A_2 = \frac{(2 \text{ kg/s})(0.1627 \text{ m}^3/\text{kg})}{665 \text{ m/s}} = 4.89 \times 10^{-4} \text{ m}^2 \triangleleft$$

❶ Although equilibrium property relations apply at the inlet and exit of the control volume, the intervening states of the steam are not necessarily equilibrium states. Accordingly, the expansion through the nozzle is represented on the $T$–$v$ diagram as a dashed line.

❷ Care must be taken in converting the units for specific kinetic energy to kJ/kg.

❸ The area at the nozzle inlet can be found similarly, using $A_1 = \dot{m}v_1/V_1$.

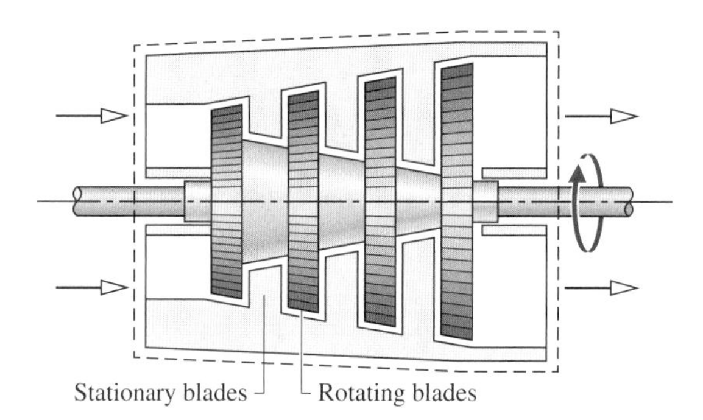


*Figure 5.6* Schematic of an axial-flow turbine.

## Turbines

*turbine*

A ***turbine*** is a device in which work is developed as a result of a gas or liquid passing through a set of blades attached to a shaft free to rotate. A schematic of an axial-flow steam or gas turbine is shown in Fig. 5.6. Turbines are widely used in vapor power plants, gas turbine power plants, and aircraft engines. In these applications, superheated steam or a gas enters the turbine and expands to a lower exit pressure as work is developed. A hydraulic turbine installed in a dam is shown in Fig. 5.7. In this application, water falling through the propeller causes the shaft to rotate and work is developed.

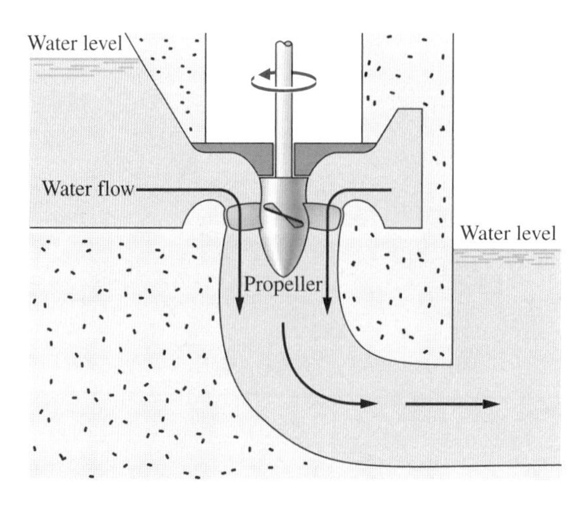


*Figure 5.7* Hydraulic turbine installed in a dam.

For a turbine at steady state the mass and energy rate balances reduce to give Eq. 5.11b. When gases are under consideration, the potential energy change is typically negligible. With a proper selection of the boundary of the control volume enclosing the turbine, the kinetic energy change is usually small enough to be neglected. The only heat transfer between the turbine and surroundings would be unavoidable heat transfer, and as illustrated in the next example, this is often small relative to the work and enthalpy terms.

## *Example 5.4* Calculating Heat Transfer from a Steam Turbine

Steam enters a turbine operating at steady state with a mass flow rate of 4600 kg/h. The turbine develops a power output of 1000 kW. At the inlet, the pressure is 60 bar, the temperature is 400°C, and the velocity is 10 m/s. At the exit, the pressure is 0.1 bar, the quality is 0.9 (90%), and the velocity is 50 m/s. Calculate the rate of heat transfer between the turbine and surroundings, in kW.

### Solution

***Known:*** A steam turbine operates at steady state. The mass flow rate, power output, and states of the steam at the inlet and exit are known.

***Find:*** Calculate the rate of heat transfer.

***Schematic and Given Data:***

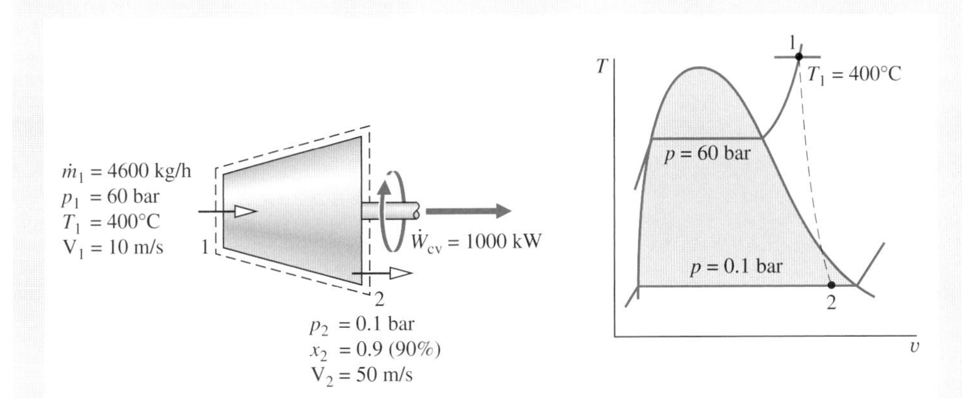


*Figure E5.4*

***Assumptions:***

1. The control volume shown on the accompanying figure is at steady state.
2. The change in potential energy from inlet to exit can be neglected.

***Analysis:*** To calculate the heat transfer rate, begin with the one-inlet, one-exit form of the energy rate balance for a control volume at steady state

$$0 = \dot{Q}_{cv} - \dot{W}_{cv} + \dot{m}\left(h_1 + \frac{V_1^2}{2} + gz_1\right) - \dot{m}\left(h_2 + \frac{V_2^2}{2} + gz_2\right)$$

where $\dot{m}$ is the mass flow rate. Solving for $\dot{Q}_{cv}$ and dropping the potential energy change from inlet to exit

$$\dot{Q}_{cv} = \dot{W}_{cv} + \dot{m}\left[(h_2 - h_1) + \left(\frac{V_2^2 - V_1^2}{2}\right)\right]$$

To compare the magnitudes of the enthalpy and kinetic energy terms, and stress the unit conversions needed, each of these terms is evaluated separately.

First, the specific *enthalpy difference* $h_2 - h_1$ is found. Using Table T-4, $h_1 = 3177.2$ kJ/kg. State 2 is a two-phase liquid–vapor mixture, so with data from Table T-3 and the given quality

$$h_2 = h_{f2} + x_2(h_{g2} - h_{f2})$$
$$= 191.83 + (0.9)(2392.8) = 2345.4 \text{ kJ/kg}$$

Hence

$$h_2 - h_1 = 2345.4 - 3177.2 = -831.8 \text{ kJ/kg}$$

Consider next the specific *kinetic energy difference*. Using the given values for the velocities

❶
$$\left(\frac{V_2^2 - V_1^2}{2}\right) = \left[\frac{(50)^2 - (10)^2}{2}\right]\left(\frac{m^2}{s^2}\right)\left|\frac{1 \text{ N}}{1 \text{ kg} \cdot \text{m/s}^2}\right|\left|\frac{1 \text{ kJ}}{10^3 \text{ N} \cdot \text{m}}\right|$$
$$= 1.2 \text{ kJ/kg}$$

Calculating $\dot{Q}_{cv}$ from the above expression

❷
$$\dot{Q}_{cv} = (1000 \text{ kW}) + \left(4600 \frac{\text{kg}}{\text{h}}\right)(-831.8 + 1.2)\left(\frac{\text{kJ}}{\text{kg}}\right)\left|\frac{1 \text{ h}}{3600 \text{ s}}\right|\left|\frac{1 \text{ kW}}{1 \text{ kJ/s}}\right|$$
$$= -61.3 \text{ kW} \quad \triangleleft$$

---

❶ The magnitude of the change in specific kinetic energy from inlet to exit is very much smaller than the specific enthalpy change.

❷ The negative value of $\dot{Q}_{cv}$ means that there is heat transfer from the turbine to its surroundings, as would be expected. The magnitude of $\dot{Q}_{cv}$ is small relative to the power developed.

## Compressors and Pumps

*compressor*
*pump*

***Compressors*** are devices in which work is done on a *gas* passing through them in order to raise the pressure. In ***pumps,*** the work input is used to change the state of a *liquid* passing through. A reciprocating compressor is shown in Fig. 5.8. Figure 5.9 gives schematic diagrams of three different rotating compressors: an axial-flow compressor, a centrifugal compressor, and a Roots type.

The mass and energy rate balances reduce for compressors and pumps at steady state, as for the case of turbines considered previously. For compressors, the changes in specific kinetic and potential energies from inlet to exit are often small relative to the work done per unit of mass passing through the device. Heat transfer with the surroundings is frequently a secondary effect in both compressors and pumps.

The next two examples illustrate, respectively, the analysis of an air compressor and a power washer. In each case the objective is to determine the power required to operate the device.

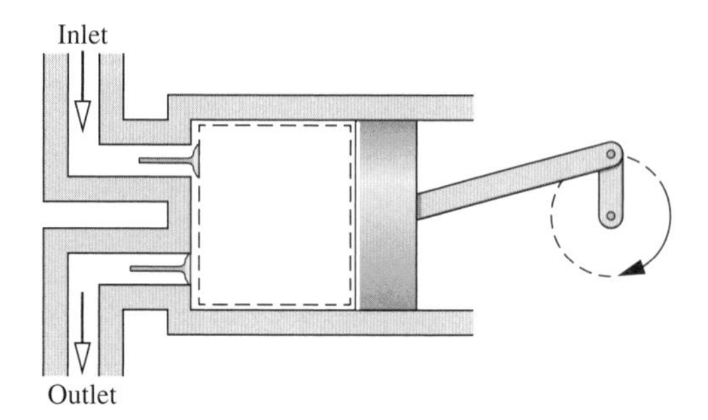


*Figure 5.8* Reciprocating compressor.

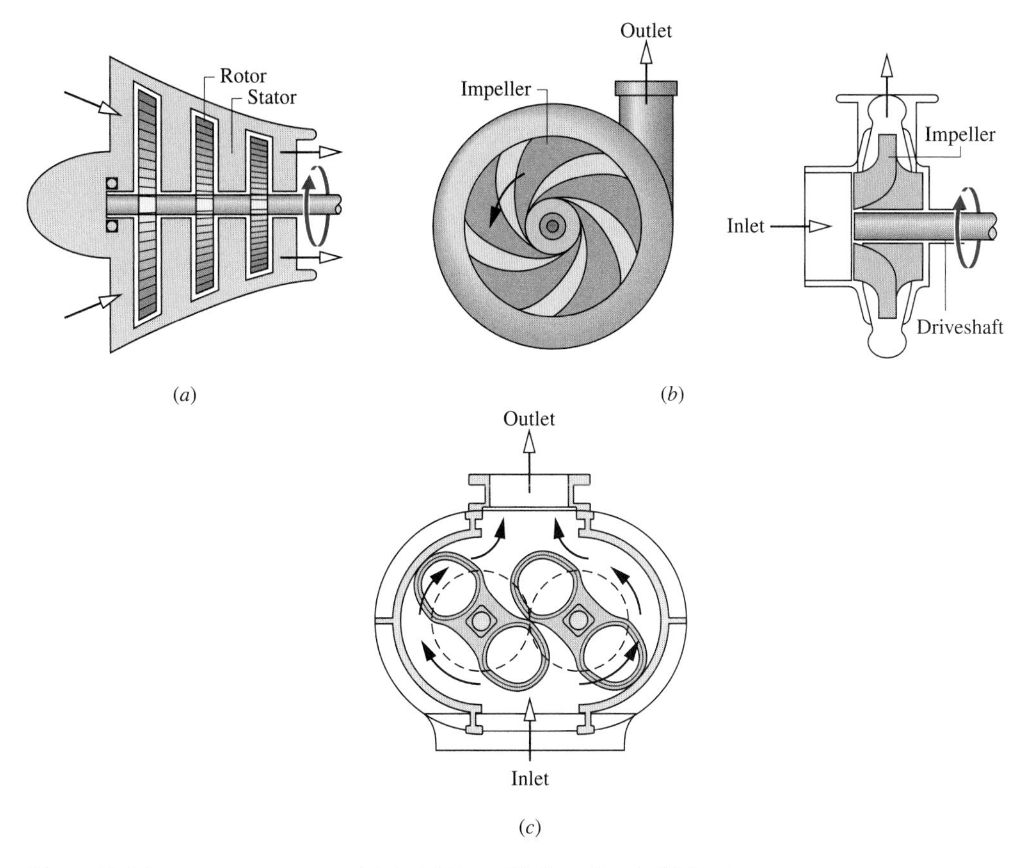


*Figure 5.9* Rotating compressors. (*a*) Axial flow. (*b*) Centrifugal. (*c*) Roots type.

## *Example 5.5* Calculating Compressor Power

Air enters a compressor operating at steady state at a pressure of 1 bar, a temperature of 290 K, and a velocity of 6 m/s through an inlet with an area of 0.1 m$^2$. At the exit, the pressure is 7 bar, the temperature is 450 K, and the velocity is 2 m/s. Heat transfer from the compressor to its surroundings occurs at a rate of 180 kJ/min. Employing the ideal gas model, calculate the power input to the compressor, in kW.

### Solution

***Known:*** An air compressor operates at steady state with known inlet and exit states and a known heat transfer rate.

***Find:*** Calculate the power required by the compressor.

***Schematic and Given Data:***

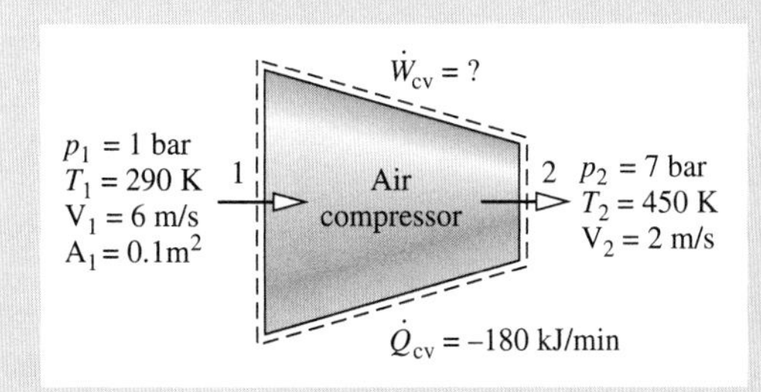


*Figure E5.5*

***Assumptions:***

1. The control volume shown on the accompanying figure is at steady state.
2. The change in potential energy from inlet to exit can be neglected.
3. The ideal gas model applies for the air.

***Analysis:*** To calculate the power input to the compressor, begin with the energy rate balance for the one-inlet, one-exit control volume at steady state:

$$0 = \dot{Q}_{cv} - \dot{W}_{cv} + \dot{m}\left(h_1 + \frac{V_1^2}{2} + gz_1\right) - \dot{m}\left(h_2 + \frac{V_2^2}{2} + gz_2\right)$$

Solving

$$\dot{W}_{cv} = \dot{Q}_{cv} + \dot{m}\left[(h_1 - h_2) + \left(\frac{V_1^2 - V_2^2}{2}\right)\right]$$

The change in potential energy from inlet to exit drops out by assumption 2.

The mass flow rate $\dot{m}$ can be evaluated with given data at the inlet and the ideal gas equation of state.

$$\dot{m} = \frac{A_1V_1}{v_1} = \frac{A_1V_1p_1}{(\bar{R}/M)T_1} = \frac{(0.1\text{ m}^2)(6\text{ m/s})(10^5\text{ N/m}^2)}{\left(\dfrac{8314}{28.97}\dfrac{\text{N}\cdot\text{m}}{\text{kg}\cdot\text{K}}\right)(290\text{ K})} = 0.72\text{ kg/s}$$

The specific enthalpies $h_1$ and $h_2$ can be found from Table T-9. At 290 K, $h_1$ = 290.16 kJ/kg. At 450 K, $h_2$ = 451.8 kJ/kg. Substituting values into the expression for $\dot{W}_{cv}$

$$\dot{W}_{cv} = \left(-180\frac{\text{kJ}}{\text{min}}\right)\left|\frac{1\text{ min}}{60\text{ s}}\right| + 0.72\frac{\text{kg}}{\text{s}}\left[(290.16 - 451.8)\frac{\text{kJ}}{\text{kg}} + \left(\frac{(6)^2 - (2)^2}{2}\right)\left(\frac{\text{m}^2}{\text{s}^2}\right)\left|\frac{1\text{ N}}{1\text{ kg}\cdot\text{m/s}^2}\right|\left|\frac{1\text{ kJ}}{10^3\text{ N}\cdot\text{m}}\right|\right]$$

❶
$$= -3\frac{\text{kJ}}{\text{s}} + 0.72\frac{\text{kg}}{\text{s}}(-161.64 + 0.02)\frac{\text{kJ}}{\text{kg}}$$

❷
$$= -119.4\frac{\text{kJ}}{\text{s}}\left|\frac{1\text{ kW}}{1\text{ kJ/s}}\right| = -119.4\text{ kW} \triangleleft$$

❶ The contribution of the kinetic energy is negligible in this case. Also, the heat transfer rate is seen to be small relative to the power input.

❷ In this example $\dot{Q}_{cv}$ and $\dot{W}_{cv}$ have negative values, indicating that the direction of the heat transfer is *from* the compressor and work is done *on* the air passing through the compressor. The magnitude of the power *input* to the compressor is 119.4 kW.

## *Example 5.6* Power Washer

A power washer is being used to clean the siding of a house. Water enters at 20°C, 1 atm, with a volumetric flow rate of 0.1 liter/s through a 2.5-cm-diameter hose. A jet of water exits at 23°C, 1 atm, with a velocity of 50 m/s at an elevation of 5 m. At steady state, the magnitude of the heat transfer rate *from* the power unit *to* the surroundings is 10% of the power input. The water can be considered incompressible with $c$ = 4.18 kJ/kg · K, and $g$ = 9.81 m/s². Determine the power input to the motor, in kW.

### Solution (CD-ROM)

### Heat Exchangers

*heat exchanger*

Devices that transfer energy between fluids at different temperatures by heat transfer modes such as discussed in Sec. 3.5.1 are called ***heat exchangers.*** One common type of heat exchanger is a vessel in which hot and cold streams are mixed directly as shown in Fig. 5.10*a*. An open feedwater heater is an example of this type of device. Another common type of heat exchanger is one in which a gas or liquid is *separated* from another gas or liquid by a wall through which energy is conducted. These heat exchangers, known as recuperators, take many different forms. Counterflow and parallel tube-within-a-tube configurations are shown in Figs. 5.10*b* and 5.10*c*, respectively.

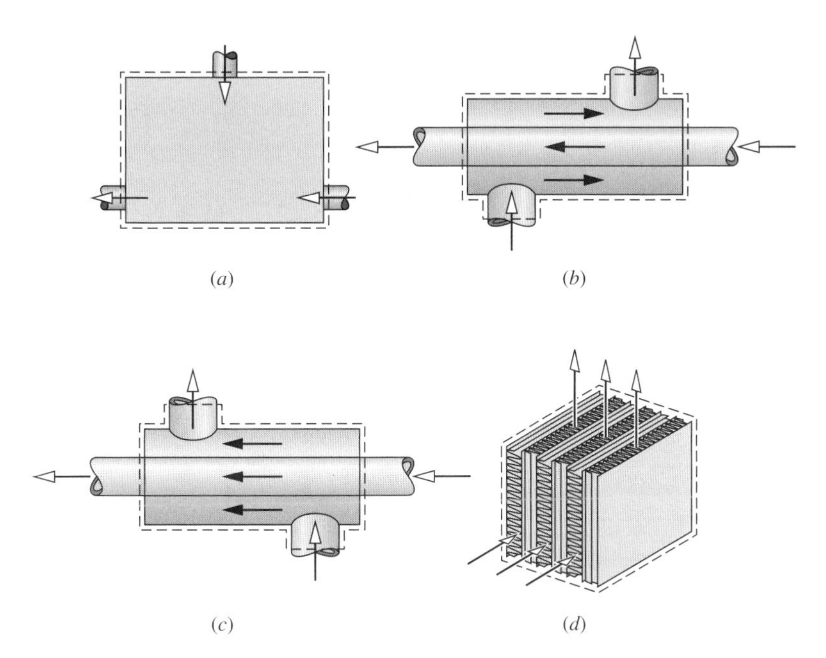


*Figure 5.10* Common heat exchanger types. (*a*) Direct contact heat exchanger. (*b*) Tube-within-a-tube counterflow heat exchanger. (*c*) Tube-within-a-tube parallel flow heat exchanger. (*d*) Cross-flow heat exchanger.

Other configurations include cross-flow, as in automobile radiators, and multiple-pass shell-and-tube condensers and evaporators. Figure 5.10*d* illustrates a cross-flow heat exchanger.

The only work interaction at the boundary of a control volume enclosing a heat exchanger is flow work at the places where matter enters and exits, so the term $\dot{W}_{cv}$ of the energy rate balance can be set to zero. Although high rates of energy transfer may be achieved from stream to stream, the heat transfer from the outer surface of the heat exchanger to the surroundings is often small enough to be neglected. In addition, the kinetic and potential energies of the flowing streams can often be ignored at the inlets and exits. See Sec. 17.5 for further discussion of heat exchangers.

The next example illustrates how the mass and energy rate balances are applied to a condenser at steady state. Condensers are commonly found in power plants and refrigeration systems.

## *Example 5.7* Power Plant Condenser

Steam enters the condenser of a vapor power plant at 0.1 bar with a quality of 0.95 and condensate exits at 0.1 bar and 45°C. Cooling water enters the condenser in a separate stream as a liquid at 20°C and exits as a liquid at 35°C with no change in pressure. Heat transfer from the outside of the condenser and changes in the kinetic and potential energies of the flowing streams can be ignored. For steady-state operation, determine

(a) the ratio of the mass flow rate of the cooling water to the mass flow rate of the condensing stream.

(b) the rate of energy transfer from the condensing steam to the cooling water, in kJ per kg of steam passing through the condenser.

### Solution

***Known:*** Steam is condensed at steady state by interacting with a separate liquid water stream.

***Find:*** Determine the ratio of the mass flow rate of the cooling water to the mass flow rate of the steam and the rate of energy transfer from the steam to the cooling water.

***Schematic and Given Data:***

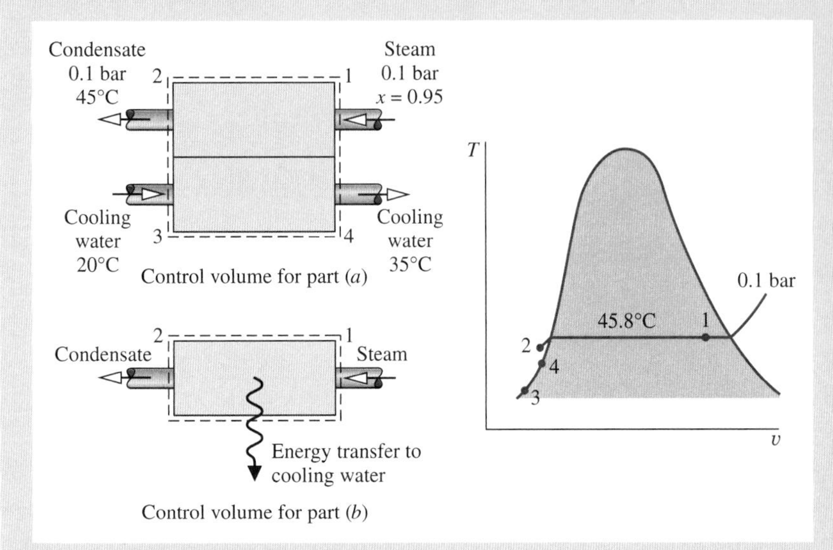


*Figure E5.7*

***Assumptions:***

1. Each of the two control volumes shown on the accompanying sketch is at steady state.
2. There is no significant heat transfer between the overall condenser and its surroundings, and $\dot{W}_{cv} = 0$.
3. Changes in the kinetic and potential energies of the flowing streams from inlet to exit can be ignored.
4. At states 2, 3, and 4, $h \approx h_f(T)$ (Eq. 4.14).

***Analysis:*** The steam and the cooling water streams do not mix. Thus, the mass rate balances for each of the two streams reduce at steady state to give

$$\dot{m}_1 = \dot{m}_2 \quad \text{and} \quad \dot{m}_3 = \dot{m}_4$$

**(a)** The ratio of the mass flow rate of the cooling water to the mass flow rate of the condensing steam, $\dot{m}_3/\dot{m}_1$, can be found from the steady-state form of the energy rate balance applied to the overall condenser as follows:

$$0 = \underline{\dot{Q}_{cv}} - \underline{\dot{W}_{cv}} + \dot{m}_1\left(h_1 + \underline{\frac{V_1^2}{2} + gz_1}\right) + \dot{m}_3\left(h_3 + \underline{\frac{V_3^2}{2} + gz_3}\right)$$
$$-\dot{m}_2\left(h_2 + \underline{\frac{V_2^2}{2} + gz_2}\right) - \dot{m}_4\left(h_4 + \underline{\frac{V_4^2}{2} + gz_4}\right)$$

The underlined terms drop out by assumptions 2 and 3. With these simplifications, together with the above mass flow rate relations, the energy rate balance becomes simply

$$0 = \dot{m}_1(h_1 - h_2) + \dot{m}_3(h_3 - h_4)$$

Solving, we get

$$\frac{\dot{m}_3}{\dot{m}_1} = \frac{h_1 - h_2}{h_4 - h_3}$$

The specific enthalpy $h_1$ can be determined using the given quality and data from Table T-3. From Table T-3 at 0.1 bar, $h_f$ = 191.83 kJ/kg and $h_g$ = 2584.7 kJ/kg, so

$$h_1 = 191.83 + 0.95(2584.7 - 191.83) = 2465.1 \text{ kJ/kg}$$

Using assumption 4, the specific enthalpy at 2 is given by $h_2 \approx h_f(T_2) = 188.45$ kJ/kg. Similarly, $h_3 \approx h_f(T_3)$ and $h_4 \approx h_f(T_4)$, giving $h_4 - h_3 = 62.7$ kJ/kg. Thus

$$\frac{\dot{m}_3}{\dot{m}_1} = \frac{2465.1 - 188.45}{62.7} = 36.3 \triangleleft$$

(b) For a control volume enclosing the steam side of the condenser only, the steady-state form of energy rate balance is

❶
$$0 = \dot{Q}_{cv} - \underline{\dot{W}_{cv}} + \dot{m}_1\left(h_1 + \underline{\frac{V_1^2}{2} + gz_1}\right) - \dot{m}_2\left(h_2 + \underline{\frac{V_2^2}{2} + gz_2}\right)$$

The underlined terms drop out by assumptions 2 and 3. Combining this equation with $\dot{m}_1 = \dot{m}_2$, the following expression for the rate of energy transfer between the condensing steam and the cooling water results:

$$\dot{Q}_{cv} = \dot{m}_1(h_2 - h_1)$$

Dividing by the mass flow rate of the steam, $\dot{m}_1$, and inserting values

$$\frac{\dot{Q}_{cv}}{\dot{m}_1} = h_2 - h_1 = 188.45 - 2465.1 = -2276.7 \text{ kJ/kg} \triangleleft$$

where the minus sign signifies that energy is transferred *from* the condensing steam *to* the cooling water.

❶ Depending on where the boundary of the control volume is located, two different formulations of the energy rate balance are obtained. In part (a), both streams are included in the control volume. Energy transfer between them occurs internally and not across the boundary of the control volume, so the term $\dot{Q}_{cv}$ drops out of the energy rate balance. With the control volume of part (b), however, the term $\dot{Q}_{cv}$ must be included.

Excessive temperatures in electronic components are avoided by providing appropriate cooling, as illustrated in the next example.

## *Example 5.8* Cooling Computer Components

The electronic components of a computer are cooled by air flowing through a fan mounted at the inlet of the electronics enclosure. At steady state, air enters at 20°C, 1 atm. For noise control, the velocity of the entering air cannot exceed 1.3 m/s. For temperature control, the temperature of the air at the exit cannot exceed 32°C. The electronic components and fan receive, respectively, 80 W and 18 W of electric power. Determine the smallest fan inlet diameter, in cm, for which the limits on the entering air velocity and exit air temperature are met.

### Solution

***Known:*** The electronic components of a computer are cooled by air flowing through a fan mounted at the inlet of the electronics enclosure. Conditions are specified for the air at the inlet and exit. The power required by the electronics and the fan are also specified.

***Find:*** Determine for these conditions the smallest fan inlet diameter.

***Schematic and Given Data:***

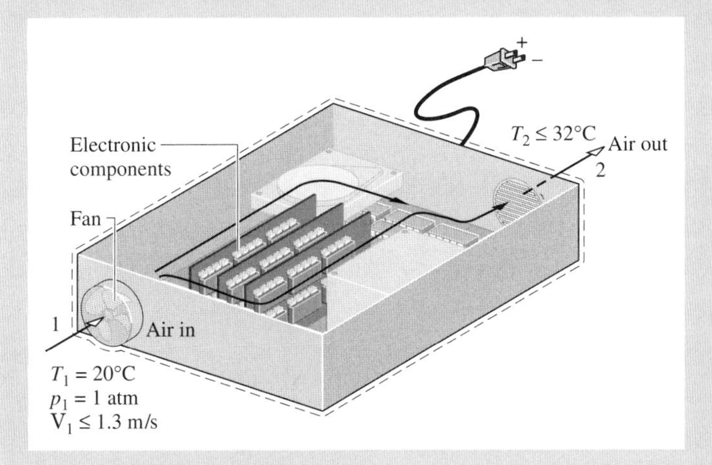


*Figure E5.8*

***Assumptions:***

1. The control volume shown on the accompanying figure is at steady state.
2. Heat transfer from the *outer* surface of the electronics enclosure to the surroundings is negligible. Thus, $\dot{Q}_{cv} = 0$.
3. ❶ Changes in kinetic and potential energies can be ignored.
4. ❷ Air is modeled as an ideal gas with $c_p = 1.005$ kJ/kg · K.

***Analysis:*** The inlet area $A_1$ can be determined from the mass flow rate $\dot{m}$ and Eq. 5.3a, which can be rearranged to read

$$A_1 = \frac{\dot{m} v_1}{V_1}$$

The mass flow rate can be evaluated, in turn, from the steady-state energy rate balance

$$0 = \underline{\dot{Q}_{cv}} - \dot{W}_{cv} + \dot{m}\left[(h_1 - h_2) + \underline{\left(\frac{V_1^2 - V_2^2}{2}\right)} + \underline{g(z_1 - z_2)}\right]$$

The underlined terms drop out by assumptions 2 and 3, leaving

$$0 = -\dot{W}_{cv} + \dot{m}(h_1 - h_2)$$

where $\dot{W}_{cv}$ accounts for the *total* electric power provided to the electronic components and the fan: $\dot{W}_{cv} = (-80 \text{ W}) + (-18 \text{ W}) = -98$ W. Solving for $\dot{m}$, and using assumption 4 with Eq. 4.49 to evaluate $(h_1 - h_2)$

$$\dot{m} = \frac{(-\dot{W}_{cv})}{c_p(T_2 - T_1)}$$

Introducing this into the expression for $A_1$ and using the ideal gas model to evaluate the specific volume $v_1$

$$A_1 = \frac{1}{V_1}\left[\frac{(-\dot{W}_{cv})}{c_p(T_2 - T_1)}\right]\left(\frac{RT_1}{p_1}\right)$$

From this expression we see that $A_1$ *increases* when $V_1$ and/or $T_2$ *decrease*. Accordingly, since $V_1 \leq 1.3$ m/s and $T_2 \leq 305$ K (32°C), the inlet area must satisfy

$$A_1 \geq \frac{1}{1.3 \text{ m/s}}\left[\frac{98 \text{ W}}{\left(1.005 \frac{\text{kJ}}{\text{kg} \cdot \text{K}}\right)(305 - 293)\text{K}}\left|\frac{1 \text{ kJ}}{10^3 \text{ J}}\right|\left|\frac{1 \text{ J/s}}{1 \text{ W}}\right|\right]\left(\frac{\left(\frac{8314 \text{ N} \cdot \text{m}}{28.97 \text{ kg} \cdot \text{K}}\right)293 \text{ K}}{1.01325 \times 10^5 \text{ N/m}^2}\right)$$

$$\geq 0.005 \text{ m}^2$$

Then, since $A_1 = \pi D_1^2/4$

$$D_1 \geq \sqrt{\frac{(4)(0.005 \text{ m}^2)}{\pi}} = 0.08 \text{ m}\left|\frac{10^2 \text{ cm}}{1 \text{ m}}\right|$$

$$D_1 \geq 8 \text{ cm}$$

For the specified conditions, the smallest fan inlet diameter is 8 cm. ◁

---

❶ Cooling air typically enters and exits electronic enclosures at low velocities, and thus kinetic energy effects are insignificant.

❷ Since the temperature of the air increases by no more than 12°C, the specific heat $c_p$ is nearly constant (Table T-10).

### Throttling Devices

A significant reduction in pressure can be achieved simply by introducing a restriction into a line through which a gas or liquid flows. This is commonly done by means of a partially opened valve or a porous plug, as illustrated in Fig. 5.11.

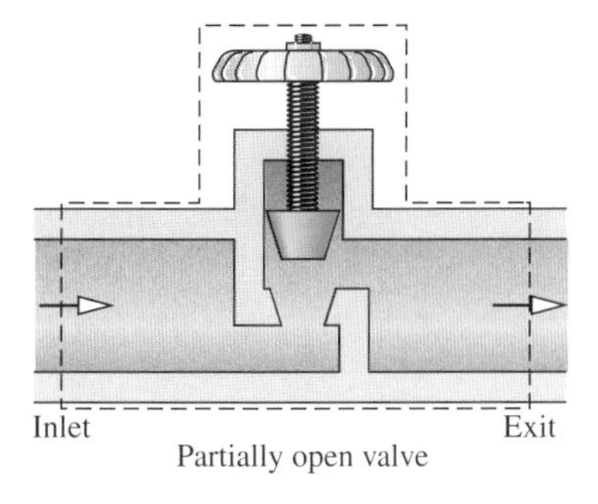


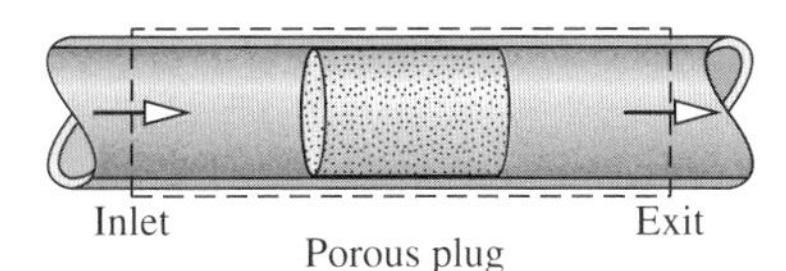


*Figure 5.11* Examples of throttling devices.

For a control volume enclosing such a device, the mass and energy rate balances reduce at steady state to

$$0 = \dot{m}_1 - \dot{m}_2$$

$$0 = \dot{Q}_{cv} - \cancelto{0}{\dot{W}_{cv}} + \dot{m}_1\left(h_1 + \frac{V_1^2}{2} + gz_1\right) - \dot{m}_2\left(h_2 + \frac{V_2^2}{2} + gz_2\right)$$

There is usually no significant heat transfer with the surroundings, and the change in potential energy from inlet to exit is negligible. With these idealizations, the mass and energy rate balances combine to give

$$h_1 + \frac{V_1^2}{2} = h_2 + \frac{V_2^2}{2}$$

Although velocities may be relatively high in the vicinity of the restriction, measurements made upstream and downstream of the reduced flow area show in most cases that the change in the specific kinetic energy of the gas or liquid between these locations can be neglected. With this further simplification, the last equation reduces to

$$h_1 = h_2 \tag{5.13}$$

*throttling process*

When the flow through a valve or other restriction is idealized in this way, the process is called a ***throttling process.***

An application of the throttling process occurs in vapor-compression refrigeration systems, where a valve is used to reduce the pressure of the refrigerant from the pressure at the exit of the *condenser* to the lower pressure existing in the *evaporator*. We consider this further in Chap. 8. Another application of the throttling process involves the ***throttling calorimeter,*** which is a device for determining the quality of a two-phase liquid–vapor mixture. The throttling calorimeter is considered in the next example.

*throttling calorimeter*

## *Example 5.9* Measuring Steam Quality

A supply line carries a two-phase liquid–vapor mixture of steam at 300 $\text{lbf/in.}^2$ A small fraction of the flow in the line is diverted through a throttling calorimeter and exhausted to the atmosphere at 14.7 $\text{lbf/in.}^2$ The temperature of the exhaust steam is measured as 250°F. Determine the quality of the steam in the supply line.

### Solution

***Known:*** Steam is diverted from a supply line through a throttling calorimeter and exhausted to the atmosphere.

***Find:*** Determine the quality of the steam in the supply line.

***Schematic and Given Data:***

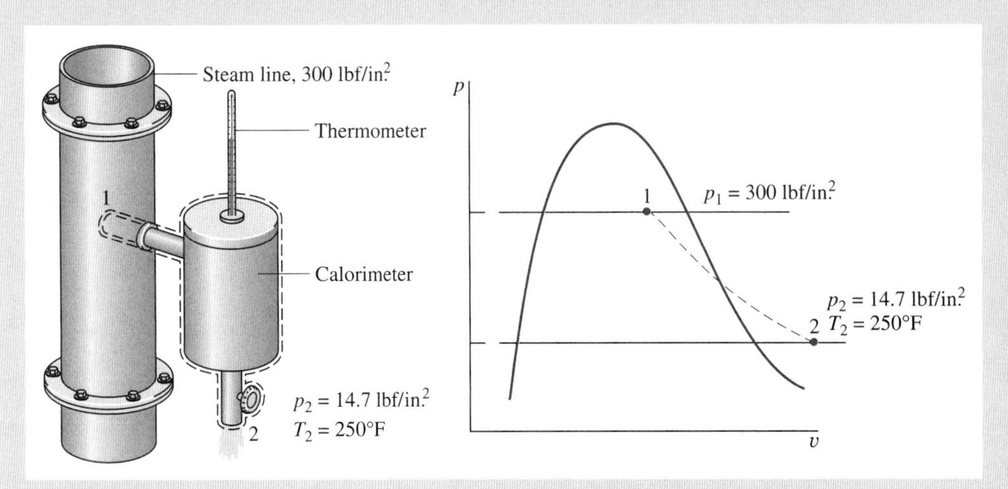


*Figure E5.9*

***Assumptions:***

1. The control volume shown on the accompanying figure is at steady state.
2. The diverted steam undergoes a throttling process.

***Analysis:*** For a throttling process, the energy and mass balances reduce to give $h_2 = h_1$, which agrees with Eq. 5.13. Thus, with state 2 fixed, the specific enthalpy in the supply line is known, and state 1 is fixed by the known values of $p_1$ and $h_1$.

❶ As shown on the accompanying $p$–$v$ diagram, state 1 is in the two-phase liquid–vapor region and state 2 is in the superheated vapor region. Thus

$$h_2 = h_1 = h_{f1} + x_1(h_{g1} - h_{f1})$$

Solving for $x_1$

$$x_1 = \frac{h_2 - h_{f1}}{h_{g1} - h_{f1}}$$

From Table T-3E at 300 lbf/in.$^2$, $h_{f1} = 394.1$ Btu/lb and $h_{g1} = 1203.9$ Btu/lb. At 14.7 lbf/in.$^2$ and 250°F, $h_2 = 1168.8$ Btu/lb from Table T-4E. Inserting values into the above expression, the quality in the line is $x_1 = 0.957$ (95.7%). ◁

❶ For throttling calorimeters exhausting to the atmosphere, the quality in the line must be greater than about 94% to ensure that the steam leaving the calorimeter is superheated.

## System Integration

Thus far, we have studied several types of components selected from those commonly seen in practice. These components are usually encountered in combination, rather than individually. Engineers often must creatively combine components to achieve some overall objective, subject to constraints such as minimum total cost. This important engineering activity is called *system integration.*

Many readers are already familiar with a particularly successful system integration: the simple power plant shown in Fig. 5.12. This system consists of four components in series: a turbine-generator, condenser, pump, and boiler. We consider such power plants in detail in subsequent sections of the book. The example to follow provides another illustration. Many more are considered in later sections and in end-of-chapter problems.

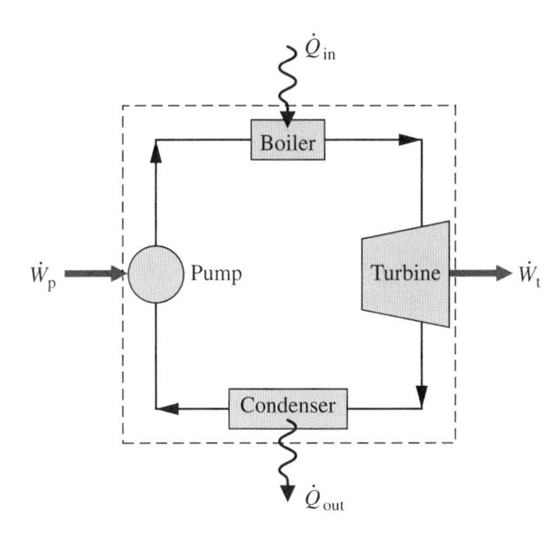


*Figure 5.12* Simple vapor power plant.

## *Example 5.10* Waste Heat Recovery System

An industrial process discharges $2 \times 10^5$ ft$^3$/min of gaseous combustion products at 400°F, 1 atm. As shown in Fig. E5.10, a proposed system for utilizing the combustion products combines a heat-recovery steam generator with a turbine. At steady state, combustion products exit the steam generator at 260°F, 1 atm and a separate stream of water enters at 40 lbf/in.$^2$, 102°F with a mass flow rate of 275 lb/min. At the exit of the turbine, the pressure is 1 lbf/in.$^2$ and the quality is 93%. Heat transfer from the outer surfaces of the steam generator and turbine can be ignored, as can the changes in kinetic and potential energies of the flowing streams. There is no significant pressure drop for the water flowing through the steam generator. The combustion products can be modeled as air as an ideal gas.

**(a)** Determine the power developed by the turbine, in Btu/min.

**(b)** Determine the turbine inlet temperature, in °F.

**(c)** Evaluating the power developed at \$0.08 per kW · h, which is a typical rate for electricity, determine the value of the power, in \$/year, for 8000 hours of operation annually.

**Solution (CD-ROM)**

## 5.4 Chapter Summary and Study Guide

The conservation of mass and energy principles for control volumes are embodied in the mass and energy rate balances developed in this chapter. The emphasis is on control volumes at steady-state for which one-dimensional flow is assumed.

The use of mass and energy balances for control volumes at steady state is illustrated for nozzles and diffusers, turbines, compressors and pumps, heat exchangers, throttling devices, and integrated systems. An essential aspect of all such applications is the careful and explicit listing of appropriate assumptions. Such model-building skills are stressed throughout the chapter.

The following checklist provides a study guide for this chapter. When your study of the text and end-of-chapter exercises has been completed you should be able to

- write out the meanings of the terms listed in the margins throughout the chapter and understand each of the related concepts. The subset of key terms listed here in the margin is particularly important in subsequent chapters.
- list the typical modeling assumptions for nozzles and diffusers, turbines, compressors and pumps, heat exchangers, and throttling devices.
- apply Eqs. 5.3a, 5.4, 5.10a, and 5.11b to control volumes at steady state, using appropriate assumptions and property data for the case at hand.

*mass flow rate*
*mass rate balance*
*one-dimensional flow*
*volumetric flow rate*
*steady state*
*energy rate balance*
*flow work*
*nozzle*
*diffuser*
*turbine*
*compressor*
*pump*
*heat exchanger*
*throttling process*

# Problems

## Conservation of Mass for Control Volumes at Steady State

**5.1** Air enters a one-inlet, one-exit control volume at 10 bar, 400 K, and 20 m/s through a flow area of 20 $cm^2$. At the exit, the pressure is 6 bar, the temperature is 345.7 K, and the velocity is 330.2 m/s. The air behaves as an ideal gas. For steady-state operation, determine
(a) the mass flow rate, in kg/s.
(b) the exit flow area, in $cm^2$.

**5.2** A substance flows through a 1-in.-diameter pipe with a velocity of 30 ft/s at a particular location. Determine the mass flow rate, in lb/s, if the substance is
(a) water at 30 lbf/in.$^2$, 60°F.
(b) air as an ideal gas at 100 lbf/in.$^2$, 100°F.
(c) Refrigerant 134a at 100 lbf/in.$^2$, 100°F.

**5.3** Air enters a 0.6-m-diameter fan at 16°C, 101 kPa, and is discharged at 18°C, 105 kPa, with a volumetric flow rate of 0.35 $m^3$/s. Assuming ideal gas behavior, determine for steady-state operation
(a) the mass flow rate of air, in kg/s.
(b) the volumetric flow rate of air at the inlet, in $m^3$/s.
(c) the inlet and exit velocities, in m/s.

**5.4** (CD-ROM)

**5.5** Steam at 120 bar, 520°C, enters a control volume operating at steady state with a volumetric flow rate of 460 $m^3$/min. Twenty-two percent of the entering mass flow exits at 10 bar, 220°C, with a velocity of 20 m/s. The rest exits at another location with a pressure of 0.06 bar, a quality of 86.2%, and a velocity of 500 m/s. Determine the diameters of each exit duct, in m.

**5.6** Air enters a household electric furnace at 75°F, 1 atm, with a volumetric flow rate of 800 $ft^3$/min. The furnace delivers air at 120°F, 1 atm to a duct system with three branches consisting of two 6-in.-diameter ducts and a 12-in. duct. The air behaves as an ideal gas. If the velocity in each 6-in. duct is 10 ft/s, determine for steady-state operation
(a) the mass flow rate of air entering the furnace, in lb/s.
(b) the volumetric flow rate in each 6-in. duct, in $ft^3$/min.
(c) the velocity in the 12-in. duct, in ft/s.

**5.7** Liquid water at 70°F enters a pump with a volumetric flow rate of 7.71 $ft^3$/min through an inlet pipe having a diameter of 6 in. The pump operates at steady state and supplies water to two exit pipes having diameters of 3 and 4 in., respectively. The mass flow rate of water in the smaller of the two exit pipes is 4 lb/s, and the temperature of the water exiting each pipe is 72°F. Determine the water velocity in each of the exit pipes, in ft/s.

**5.8** Air enters a compressor operating at steady state with a pressure of 14.7 lbf/in.$^2$, a temperature of 80°F, and a volumetric flow rate of 1000 $ft^3$/min. The diameter of the exit pipe is 1 in. and the exit pressure is 100 lbf/in.$^2$ The air behaves as an ideal gas. If each unit mass of air passing from inlet to exit undergoes a process described by $pv^{1.32} = constant$, determine the exit velocity, in ft/s, and the exit temperature, in °F.

**5.9** (CD-ROM)

**5.10** (CD-ROM)

**5.11** (CD-ROM)

## Energy Analysis of Control Volumes at Steady State

**5.12** Steam enters a nozzle operating at steady state at 30 bar, 320°C, with a velocity of 100 m/s. The exit pressure and temperature are 10 bar and 200°C, respectively. The mass flow rate is 2 kg/s. Neglecting heat transfer and potential energy, determine
(a) the exit velocity, in m/s.
(b) the inlet and exit flow areas, in $cm^2$.

**5.13** Steam enters a well-insulated nozzle at 200 lbf/in.$^2$, 500°F, with a velocity of 200 ft/s and exits at 60 lbf/in.$^2$ with a velocity of 1700 ft/s. For steady-state operation, and neglecting potential energy effects, determine the exit temperature, in °F.

**5.14** Air enters a nozzle operating at steady state at 800°R with negligible velocity and exits the nozzle at 570°R. Heat transfer occurs from the air to the surroundings at a rate of 10 Btu per lb of air flowing. Assuming ideal gas behavior and neglecting potential energy effects, determine the velocity at the exit, in ft/s.

**5.15** (CD-ROM)

**5.16** (CD-ROM)

**5.17** Steam enters a diffuser operating at steady state with a pressure of 14.7 lbf/in.$^2$, a temperature of 300°F, and a velocity of 500 ft/s. Steam exits the diffuser as a saturated vapor with negligible kinetic energy. Heat transfer occurs from the steam to its surroundings at a rate of 19.59 Btu per lb of steam flowing. Neglecting potential energy effects, determine the exit pressure, in lbf/in.$^2$

**5.18** Air enters an insulated diffuser operating at steady state with a pressure of 1 bar, a temperature of 57°C, and a velocity of 200 m/s. At the exit, the pressure is 1.13 bar and the temperature is 69°C. Potential energy effects can be neglected. Using the ideal gas model with a constant specific heat $c_p$ evaluated at the inlet temperature, determine
(a) the ratio of the exit flow area to the inlet flow area.
(b) the exit velocity, in m/s.

**5.19** The inlet ducting to a jet engine forms a diffuser that steadily decelerates the entering air to zero velocity relative to the engine before the air enters the compressor. Consider a jet airplane flying at 1000 km/h where the local atmospheric pressure is 0.6 bar and the air temperature is 8°C. Assuming ideal gas behavior and neglecting heat transfer and potential energy effects, determine the temperature, in °C, of the air entering the compressor.

**5.20** (CD-ROM)

**5.21** Carbon dioxide gas enters a well-insulated diffuser at 20 lbf/in.$^2$, 500°R, with a velocity of 800 ft/s through a flow area of 1.4 in.$^2$ At the exit, the flow area is 30 times the inlet area, and the velocity is 20 ft/s. The potential energy change from inlet to exit is negligible. For steady-state operation, determine the exit temperature, in °R, the exit pressure, in lbf/in.$^2$, and the mass flow rate, in lb/s.

**5.22** Air expands through a turbine from 10 bar, 900 K to 1 bar, 500 K. The inlet velocity is small compared to the exit velocity of 100 m/s. The turbine operates at steady state and develops a power output of 3200 kW. Heat transfer between the turbine and its surroundings and potential energy effects are negligible. Calculate the mass flow rate of air, in kg/s, and the exit area, in m$^2$.

**5.23** Air expands through a turbine operating at steady state on an instrumented test stand. At the inlet, $p_1$ = 150 lbf/in.$^2$, $T_1$ = 1500°R, and at the exit, $p_2$ = 14.5 lbf/in.$^2$ The volumetric flow rate of air entering the turbine is 2000 ft$^3$/min, and the power developed is measured as 2000 horsepower. Neglecting heat transfer and kinetic and potential energy effects, determine the exit temperature, $T_2$, in °R.

**5.24** Steam enters a turbine operating at steady state at 700°F and 600 lbf/in.$^2$ and leaves at 0.6 lbf/in.$^2$ with a quality of 90%. The turbine develops 12,000 hp, and heat transfer from the turbine to the surroundings occurs at a rate of 2.5 × 10$^6$ Btu/h. Neglecting kinetic and potential energy changes from inlet to exit, determine the mass flow rate of the steam, in lb/h.

**5.25** Nitrogen gas enters a turbine operating at steady state through a 2-in.-diameter duct with a velocity of 200 ft/s, a pressure of 50 lbf/in.$^2$, and a temperature of 1000°R. At the exit, the velocity is 2 ft/s, the pressure is 20 lbf/in.$^2$, and the temperature is 700°R. Heat transfer from the surface of the turbine to the surroundings occurs at a rate of 16 Btu per lb of nitrogen flowing. Neglecting potential energy effects and using the ideal gas model, determine the power developed by the turbine, in horsepower.

**5.26** (CD-ROM)

**5.27** (CD-ROM)

**5.28** The intake to a hydraulic turbine installed in a flood control dam is located at an elevation of 10 m above the turbine exit. Water enters at 20°C with negligible velocity and exits from the turbine at 10 m/s. The water passes through the turbine with no significant changes in temperature or pressure between the inlet and exit, and heat transfer is negligible. The acceleration of gravity is constant at $g$ = 9.81 m/s$^2$. If the power output at steady state is 500 kW, what is the mass flow rate of water, in kg/s?

**5.29** A well-insulated turbine operating at steady state is sketched in Fig. P5.29. Steam enters at 3 MPa, 400°C, with a volumetric flow rate of 85 m$^3$/min. Some steam is extracted from the turbine at a pressure of 0.5 MPa and a temperature of 180°C. The rest expands to a pressure of 6 kPa and a quality of 90%. The total power developed by the turbine is 11,400 kW. Kinetic and potential energy effects can be neglected. Determine

(a) the mass flow rate of the steam at each of the two exits, in kg/h.

(b) the diameter, in m, of the duct through which steam is extracted, if the velocity there is 20 m/s.

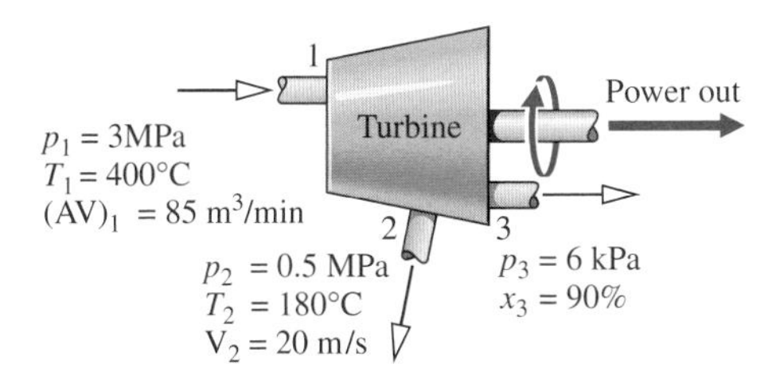


*Figure P5.29*

**5.30** (CD-ROM)

**5.31** (CD-ROM)

**5.32** At steady state, a well-insulated compressor takes in air at 60°F, 14.2 lbf/in.$^2$, with a volumetric flow rate of 1200 ft$^3$/min, and compresses it to 500°F, 120 lbf/in.$^2$ Kinetic and potential energy changes from inlet to exit can be neglected. Determine the compressor power, in hp, and the volumetric flow rate at the exit, in ft$^3$/min.

**5.33** Air enters a compressor with a pressure of 14.7 lbf/in.$^2$, a temperature of 70°F, and a volumetric flow rate of 40 ft$^3$/s. Air exits the compressor at 50 lbf/in.$^2$ and 190°F. Heat transfer from the compressor to its surroundings occurs at a rate of 20.5 Btu per lb of air flowing. Determine the compressor power, in hp, for steady-state operation.

**5.34** A compressor operates at steady state with Refrigerant 134a as the working fluid. The refrigerant enters at 0.2 MPa, 0°C, with a volumetric flow rate of 0.6 m$^3$/min. The diameters of the inlet and exit pipes are 3 and 1.5 cm, respectively. At the exit, the pressure is 1.0 MPa and the temperature is 50°C. If the magnitude of the heat transfer rate from the compressor to its surroundings is 5% of the compressor power input, determine the power input, in kW.

**5.35** Carbon dioxide gas is compressed at steady state from a pressure of 20 lbf/in.$^2$ and a temperature of 32°F to a pressure of 50 lbf/in.$^2$ and a temperature of 580°R. The gas enters the compressor through a 6-in.-diameter duct with a velocity of 30 ft/s and leaves with a velocity of 80 ft/s. The magnitude of the heat transfer rate from the compressor to its surroundings is 20% of the compressor power input. Using the ideal gas model and neglecting potential energy effects, determine the compressor power input, in horsepower.

**5.36** (CD-ROM)

**5.37** (CD-ROM)

**5.38** (CD-ROM)

**5.39** Refrigerant 134a is compressed at steady state from 2.4 bar, 0°C, to 12 bar, 50°C. Refrigerant enters the compressor with a volumetric flow rate of 0.38 m$^3$/min, and the power input to the compressor is 2.6 kW. Cooling water circulating through a water jacket enclosing the compressor experiences a temperature rise of 4°C from inlet to exit with a negligible change in pressure. Heat transfer from the outside of the water jacket and all kinetic and potential energy effects can be neglected. Deter-mine the mass flow rate of the cooling water, in kg/s.

**5.40** A pump steadily draws water from a pond at a mass flow rate of 20 lb/s through a pipe. At the pipe inlet, the pressure is 14.7 lbf/in.$^2$, the temperature is 68°F, and the velocity is 10 ft/s. At the pump exit, the pressure is 20 lbf/in.$^2$, the temperature is 68°F, and the velocity is 40 ft/s. The pump exit is located 50 ft above the pipe inlet. Determine the power required by the pump, in Btu/s and horsepower. The local acceleration of gravity is 32.0 ft/s$^2$. Neglect heat transfer.

**5.41** A pump steadily delivers water through a hose terminated by a nozzle. The exit of the nozzle has a diameter of 0.6 cm and is located 10 m above the pump inlet pipe, which has a diameter of 1.2 cm. The pressure is equal to 1 bar at both the inlet and the exit, and the temperature is constant at 20°C. The magnitude of the power input required by the pump is 1.5 kW, and the acceleration of gravity is $g = 9.81 \text{ m/s}^2$. Determine the mass flow rate delivered by the pump, in kg/s.

**5.42** An oil pump operating at steady state delivers oil at a rate of 12 lb/s through a 1-in.-diameter pipe. The oil, which can be modeled as incompressible, has a density of 100 lb/ft$^3$ and experiences a pressure rise from inlet to exit of 40 lbf/in.$^2$ There is no significant elevation difference between inlet and exit, and the inlet kinetic energy is negligible. Heat transfer between the pump and its surroundings is negligible, and there is no significant change in temperature as the oil passes through the pump. If pumps are available in 1/4-horsepower increments, determine the horsepower rating of the pump needed for this application.

**5.43** Refrigerant 134a enters a heat exchanger operating at steady state as a superheated vapor at 10 bar, 60°C, where it is cooled and condensed to saturated liquid at 10 bar. The mass flow rate of the refrigerant is 10 kg/min. A separate stream of air enters the heat exchanger at 22°C, 1 bar and exits at 45°C, 1 bar. Ignoring heat transfer from the outside of the heat exchanger and neglecting kinetic and potential energy effects, determine the mass flow rate of the air, in kg/min.

**5.44** (CD-ROM)

**5.45** (CD-ROM)

**5.46** Carbon dioxide gas is heated as it flows at steady state through a 2.5-cm-diameter pipe. At the inlet, the pressure is 2 bar, the temperature is 300 K, and the velocity is 100 m/s. At the exit, the pressure and velocity are 0.9413 bar and 400 m/s, respectively. The gas can be treated as an ideal gas with constant specific heat $c_p = 0.94 \text{ kJ/kg} \cdot \text{K}$. Neglecting potential energy effects, determine the rate of heat transfer to the carbon dioxide, in kW.

**5.47** A feedwater heater in a vapor power plant operates at steady state with liquid entering at inlet 1 with $T_1 = 40°\text{C}$ and $p_1 = 7.0$ bar. Water vapor at $T_2 = 200°\text{C}$ and $p_2 = 7.0$ bar enters at inlet 2. Saturated liquid water exits with a pressure of $p_3 = 7.0$ bar. Ignoring heat transfer with the surroundings and all kinetic and potential energy effects, determine the ratio of mass flow rates, $\dot{m}_1/\dot{m}_2$.

**5.48** Refrigerant 134a enters a heat exchanger in a refrigeration system operating at steady state as saturated vapor at 0°F and exits at 20°F with no change in pressure. A separate liquid stream of Refrigerant 134a passes in counterflow to the vapor stream, entering at 105°F, 160 lbf/in.$^2$, and exiting at a lower temperature while experiencing no pressure drop. The outside of the heat exchanger is well insulated, and the streams have equal mass flow rates. Neglecting kinetic and potential energy effects, determine the exit temperature of the liquid stream, in °F.

**5.49** (CD-ROM)

**5.50** Figure P5.50 shows a solar collector panel with a surface area of 32 ft$^2$. The panel receives energy from the sun at a rate of 150 Btu/h per ft$^2$ of collector surface. Thirty-six percent of the incoming energy is lost to the surroundings. The remainder is used to heat liquid water from 110 to 140°F. The water passes through the solar collector with a negligible pressure drop. Neglecting kinetic and potential energy effects, determine at steady state the mass flow rate of water, in lb/min. How many gallons of water at 140°F can eight collectors provide in a 30-min time period?

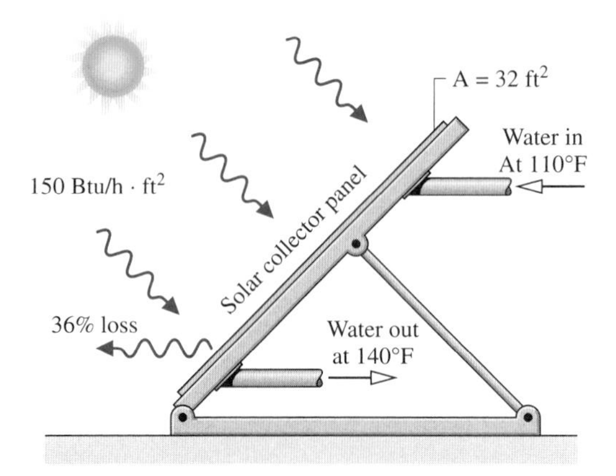


*Figure P5.50*

**5.51** (CD-ROM)

**5.52** A feedwater heater operates at steady state with liquid water entering at inlet 1 at 7 bar, 42°C, and a mass flow rate of 70 kg/s. A separate stream of water enters at inlet 2 as a two-phase liquid–vapor mixture at 7 bar with a quality of 98%. Saturated liquid at 7 bar exits the feedwater heater at 3. Ignoring heat transfer with the surroundings and neglecting kinetic and potential energy effects, determine the mass flow rate, in kg/s, at inlet 2.

**5.53** Figure P5.53 shows data for a portion of the ducting in a ventilation system operating at steady state. Air flows through the ducts with negligible heat transfer with the surroundings, and the pressure is very nearly 1 atm throughout. Determine the temperature of the air at the exit, in °F, and the exit diameter, in ft.

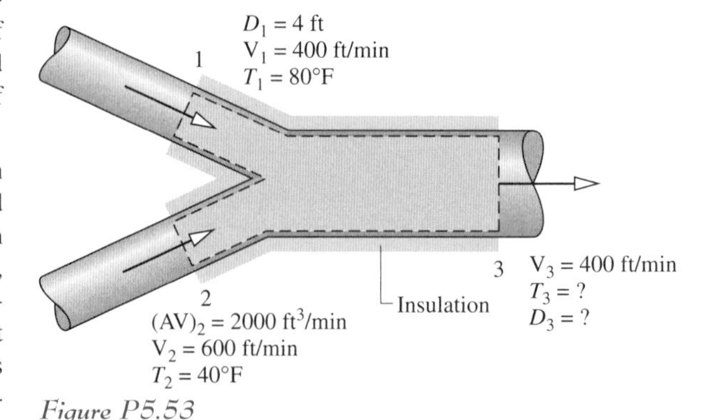


*Figure P5.53*

**5.54** The electronic components of a computer consume 0.1 kW of electrical power. To prevent overheating, cooling air is supplied by a 25-W fan mounted at the inlet of the electronics enclosure. At steady state, air enters the fan at 20°C, 1 bar and exits the electronics enclosure at 35°C. There is no significant energy transfer by heat from the outer surface of the enclosure to the surroundings and the effects of kinetic and potential energy can be ignored. Determine the volumetric flow rate of the entering air, in $m^3/s$.

**5.55** (CD-ROM)

**5.56** As shown in Fig. P5.56, electronic components mounted on a flat plate are cooled by air flowing over the top surface and by liquid water circulating through a U-tube bonded to the plate. At steady state, water enters the tube at 20°C and a velocity of 0.4 m/s and exits at 24°C with a negligible change in pressure. The electrical components receive 0.5 kW of electrical power. The rate of heat transfer from the top of the plate-mounted electronics is estimated to be 0.08 kW. Kinetic and potential energy effects can be ignored. Determine the tube diameter, in cm.

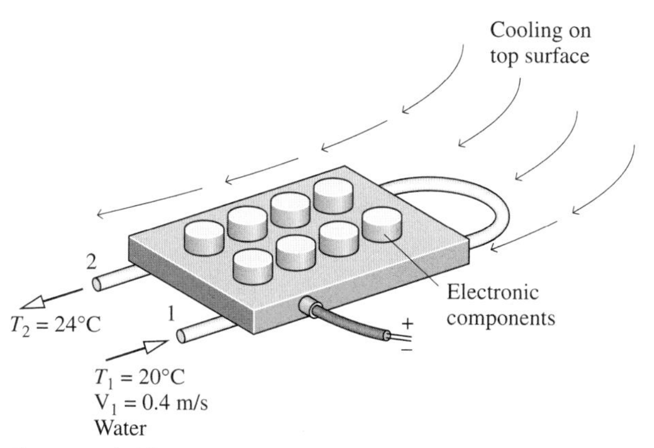


*Figure P5.56*

**5.57** (CD-ROM)

**5.58** Refrigerant 134a enters the expansion valve of a refrigeration system at a pressure of 1.2 MPa and a temperature of 38°C and exits at 0.24 MPa. If the refrigerant undergoes a throttling process, what is the quality of the refrigerant exiting the expansion valve?

**5.59** A large pipe carries steam as a two-phase liquid–vapor mixture at 1.0 MPa. A small quantity is withdrawn through a throttling calorimeter, where it undergoes a throttling process to an exit pressure of 0.1 MPa. For what range of exit temperatures, in °C, can the calorimeter be used to determine the quality of the steam in the pipe? What is the corresponding range of steam quality values?

**5.60** (CD-ROM)

**5.61** (CD-ROM

**5.62** As shown in Fig. P5.62, a steam turbine at steady state is operated at part load by throttling the steam to a lower pressure before it enters the turbine. Before throttling, the pressure and temperature are, respectively, 200 lbf/in.$^2$ and 600°F. After throttling, the pressure is 120 lbf/in.$^2$ At the turbine exit, the steam is at 1 lbf/in.$^2$ and a quality of 90%. Heat transfer with the surroundings and all kinetic and potential energy effects can be ignored. Determine

(a) the temperature at the turbine inlet, in °F.
(b) the power developed by the turbine, in Btu per lb of steam flowing.

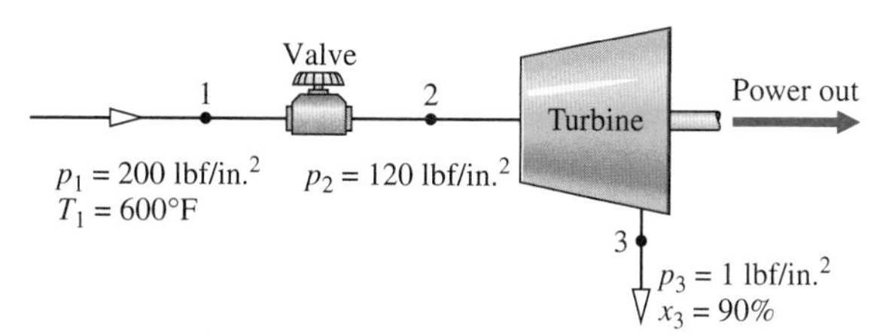


*Figure P5.62*

**5.63** Refrigerant 134a enters the flash chamber operating at steady state shown in Fig. P5.63 at 10 bar, 36°C, with a mass flow rate of 482 kg/h. Saturated liquid and saturated vapor exit as separate streams, each at 4 bar. Heat transfer to the surroundings and kinetic and potential energy effects can be ignored. Determine the mass flow rates of the exiting streams, each in kg/h.

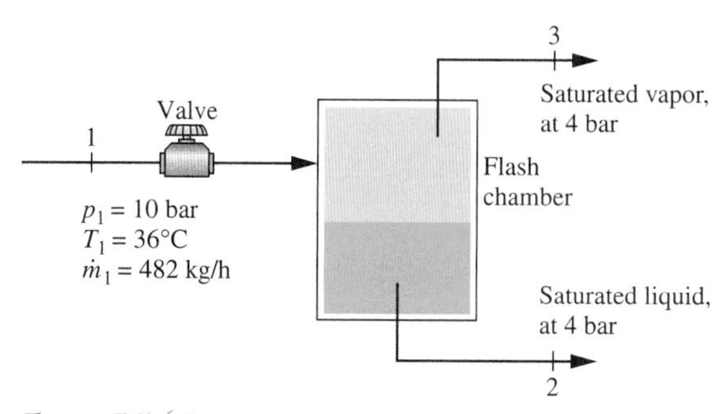


*Figure P5.63*

**5.64** At steady state, water enters the waste heat recovery steam generator shown in Fig. P5.64 at 42 lbf/in.$^2$, 220°F, and exits

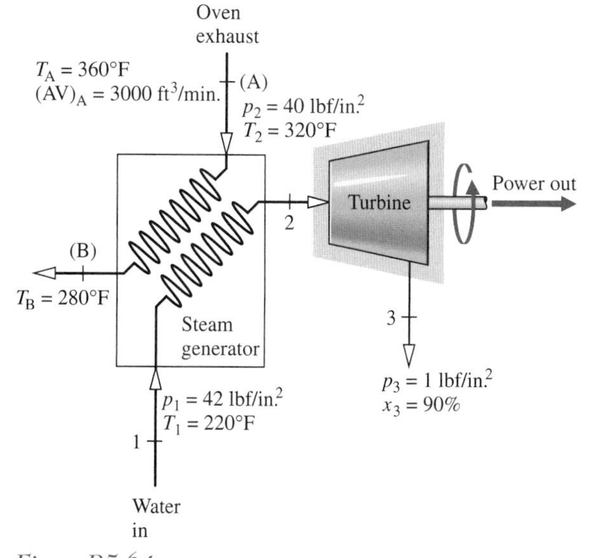


*Figure P5.64*

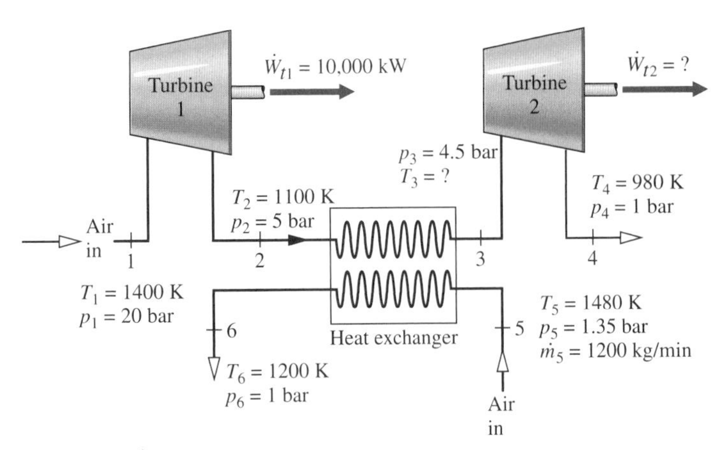


*Figure P5.65*

at 40 lbf/in.$^2$, 320°F. The steam is then fed into a turbine from which it exits at 1 lbf/in.$^2$ and a quality of 90%. Air from an oven exhaust enters the steam generator at 360°F, 1 atm, with a volumetric flow rate of 3000 ft$^3$/min, and exits at 280°F, 1 atm. Ignore all stray heat transfer with the surroundings and all kinetic and potential energy effects.

**(a)** Determine the power developed by the turbine, in horsepower.

**(b)** Evaluating the power developed at 8 cents per kW · h, determine its value, in $/year, for 8000 hours of operation annually, and comment.

**5.65** Air as an ideal gas flows through the turbine and heat exchanger arrangement shown in Fig. P5.65. Data for the two flow streams are shown on the figure. Heat transfer to the surroundings can be neglected, as can all kinetic and potential energy effects. Determine $T_3$, in K, and the power output of the second turbine, in kW, at steady state.

**5.66** (CD-ROM)

**5.67** (CD-ROM)

**5.68** A simple gas turbine power plant operating at steady state is illustrated schematically in Fig. P5.68. The power plant consists of an air compressor mounted on the same shaft as the turbine. Relevant data are given on the figure. Kinetic and potential energy effects are negligible, and the compressor and turbine operate adiabatically. Using the ideal gas model, determine the power required by the compressor and the net power developed, each in horsepower.

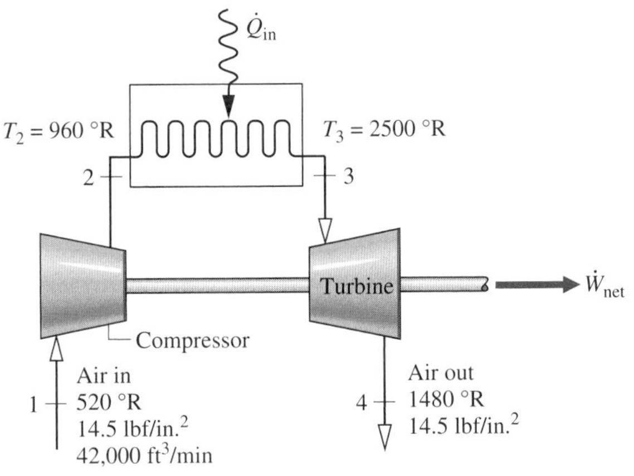


*Figure P5.68*

# THE SECOND LAW OF THERMODYNAMICS

## *Introduction...*

The presentation to this point has considered thermodynamic analysis using the conservation of mass and conservation of energy principles together with property relations. In Chaps. 3 through 5 these fundamentals are applied to increasingly complex situations. The conservation principles do not always suffice, however, and often the second law of thermodynamics is also required for thermodynamic analysis. The ***objective*** of this chapter is to introduce the second law of thermodynamics. A number of deductions that may be called corollaries of the second law are also considered, including performance limits for thermodynamic cycles. The current presentation provides the basis for subsequent developments involving the second law in Chap. 7.

*chapter objective*

## 6.1 Introducing the Second Law

The objectives of the present section are to (1) motivate the need for and the usefulness of the second law, and (2) to introduce statements of the second law that serve as the point of departure for its application.

### 6.1.1 Motivating the Second Law

It is a matter of everyday experience that there is a definite direction for *spontaneous* processes. This can be brought out by considering Fig. 6.1. Air held at a high pressure $p_i$ in a closed tank would flow spontaneously to the lower pressure surroundings at $p_0$ if the interconnecting valve were opened. Eventually fluid motions would cease and all of the air

*Figure 6.1* Illustrations of a spontaneous expansion and the eventual attainment of equilibrium with the surroundings.

would be at the same pressure as the surroundings. Drawing on experience, it should be clear that the *inverse* process would not take place *spontaneously*, even though energy could be conserved: Air would not flow spontaneously from the surroundings at $p_0$ into the tank, returning the pressure to its initial value. The initial condition can be restored, but not in a spontaneous process. An auxiliary device such as an air compressor would be required to return the air to the tank and restore the initial air pressure.

This illustration suggests that not every process consistent with the principle of energy conservation can occur. Generally, an energy balance alone neither enables the preferred direction to be predicted nor permits the processes that can occur to be distinguished from those that cannot. In elementary cases such as the one considered, experience can be drawn upon to deduce whether particular spontaneous processes occur and to deduce their directions. For more complex cases, where experience is lacking or uncertain, a guiding principle would be helpful. This is provided by the *second law.*

The foregoing discussion also indicates that when left to themselves, systems tend to undergo spontaneous changes until a condition of equilibrium is achieved, both internally and with their surroundings. In some cases equilibrium is reached quickly, in others it is achieved slowly. For example, some chemical reactions reach equilibrium in fractions of seconds; an ice cube requires a few minutes to melt; and it may take years for an iron bar to rust away. Whether the process is rapid or slow, it must of course satisfy conservation of energy. However, this alone would be insufficient for determining the final equilibrium state. Another general principle is required. This is provided by the *second law.*

By exploiting the spontaneous process shown in Fig. 6.1, it is possible for work to be developed as equilibrium is attained: Instead of permitting the air to expand aimlessly into the lower-pressure surroundings, the stream could be passed through a turbine and work could be developed. Accordingly, in this case there is a possibility for developing work that would not be exploited in an uncontrolled expansion. Recognizing this possibility for work, we can pose two questions:

- What is the theoretical maximum value for the work that could be obtained?
- What are the factors that would preclude the realization of the maximum value?

That there should be a maximum value is fully in accord with experience, for if it were possible to develop unlimited work, few concerns would be voiced over our dwindling fuel supplies. Also in accord with experience is the idea that even the best devices would be subject to factors such as friction that would preclude the attainment of the theoretical maximum work. The second law of thermodynamics provides the means for determining the theoretical maximum and evaluating quantitatively the factors that preclude attaining the maximum.

*Summary.* The preceding discussions can be summarized by noting that the second law and deductions from it are useful because they provide means for

1. predicting the direction of processes.
2. establishing conditions for equilibrium.
3. determining the best *theoretical* performance of cycles, engines, and other devices.
4. evaluating quantitatively the factors that preclude the attainment of the best theoretical performance level.

   Additional uses of the second law include its roles in

5. defining a temperature scale independent of the properties of any thermometric substance.
6. developing means for evaluating properties such as $u$ and $h$ in terms of properties that are more readily obtained experimentally.

Scientists and engineers have found many additional applications of the second law and deductions from it. It also has been used in economics, philosophy, and other areas far removed from engineering thermodynamics.

The six points listed can be thought of as aspects of the second law of thermodynamics and not as independent and unrelated ideas. Nonetheless, given the variety of these topic areas, it is easy to understand why there is no single statement of the second law that brings out each one clearly. There are several alternative, yet equivalent, formulations of the second law. In the next section, two equivalent statements of the second law are introduced as a *point of departure* for our study of the second law and its consequences. Although the exact relationship of these particular formulations to each of the second law aspects listed above may not be immediately apparent, all aspects listed can be obtained by deduction from these formulations or their corollaries. It is important to add that in every instance where a consequence of the second law has been tested directly or indirectly by experiment, it has been unfailingly verified. Accordingly, the basis of the second law of thermodynamics, like every other physical law, is experimental evidence.

### 6.1.2 Statements of the Second Law

Among many alternative statements of the second law, two are frequently used in engineering thermodynamics. They are the *Clausius* and *Kelvin–Planck* statements. The objective of this section is to introduce these two equivalent second law statements. The equivalence of the Clausius and Kelvin–Planck statements can be demonstrated by showing that the violation of each statement implies the violation of the other.

#### Clausius Statement of the Second Law

*Clausius statement*

The ***Clausius statement*** of the second law asserts that: *It is impossible for any system to operate in such a way that the sole result would be an energy transfer by heat from a cooler to a hotter body.*

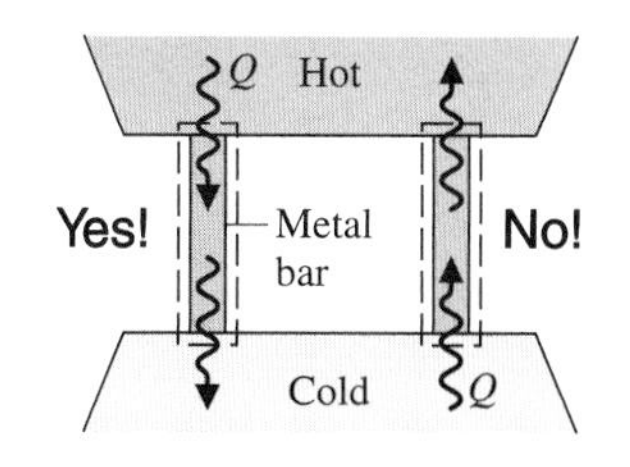


The Clausius statement does not rule out the possibility of transferring energy by heat from a cooler body to a hotter body, for this is exactly what refrigerators and heat pumps accomplish. However, as the words "sole result" in the statement suggest, when a heat transfer from a cooler body to a hotter body occurs, there must be some *other effect* within the system accomplishing the heat transfer, its surroundings, or both. If the system operates in a thermodynamic cycle, its initial state is restored after each cycle, so the only place that must be examined for such *other* effects is its surroundings. ***For Example...*** cooling of food is accomplished by refrigerators driven by electric motors requiring work from their surroundings to operate. The Clausius statement implies that it is impossible to construct a refrigeration cycle that operates without an input of work. ▲

#### Kelvin–Planck Statement of the Second Law

*thermal reservoir*

Before giving the Kelvin–Planck statement of the second law, the concept of a ***thermal reservoir*** is introduced. A thermal reservoir, or simply a reservoir, is a special kind of system that always remains at constant temperature even though energy is added or removed by heat transfer. A reservoir is an idealization, of course, but such a system can be approximated in a number of ways—by the earth's atmosphere, large bodies of water (lakes, oceans), a large block of copper, and so on. Extensive properties of a thermal reservoir such as internal energy can change in interactions with other systems even though the reservoir temperature remains constant.

*Kelvin–Planck statement*

Having introduced the thermal reservoir concept, we give the ***Kelvin–Planck statement*** of the second law: *It is impossible for any system to operate in a thermodynamic cycle and deliver a net amount of energy by work to its surroundings while receiving energy by heat* *transfer from a single thermal reservoir.* The Kelvin–Planck statement does not rule out the possibility of a system developing a net amount of work from a heat transfer drawn from a single reservoir. It only denies this possibility if the system undergoes a thermodynamic cycle.

The Kelvin–Planck statement can be expressed analytically. To develop this, let us study a system undergoing a cycle while exchanging energy by heat transfer with a *single* reservoir. The first and second laws each impose constraints:

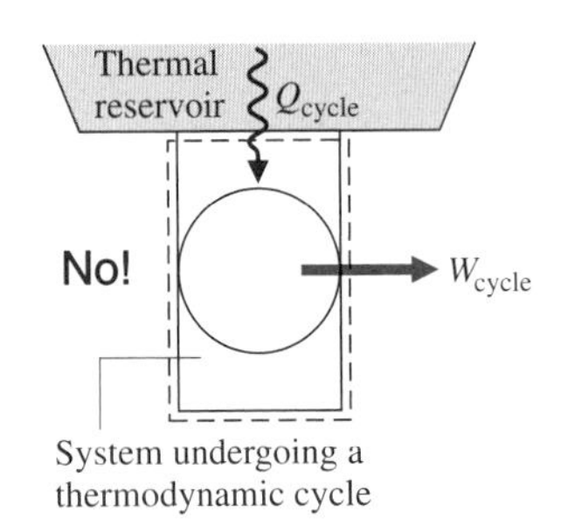


- A constraint is imposed by the first law on the net work and heat transfer between the system and its surroundings. According to the cycle energy balance, Eq. 3.15

  $$W_{cycle} = Q_{cycle}$$

  In words, the net work done by the system undergoing a cycle equals the net heat transfer to the system. Although the cycle energy balance allows the net work $W_{cycle}$ to be positive or negative, the second law imposes a constraint on its direction, as considered next.

- According to the Kelvin–Planck statement, a system undergoing a cycle while communicating thermally with a single reservoir *cannot* deliver a net amount of work to its surroundings. That is, the net work of the cycle *cannot be positive.* However, the Kelvin–Planck statement does not rule out the possibility that there is a net work transfer of energy *to* the system during the cycle or that the net work is zero. Thus, the ***analytical form of the Kelvin–Planck statement*** is

*analytical form: Kelvin–Planck statement*

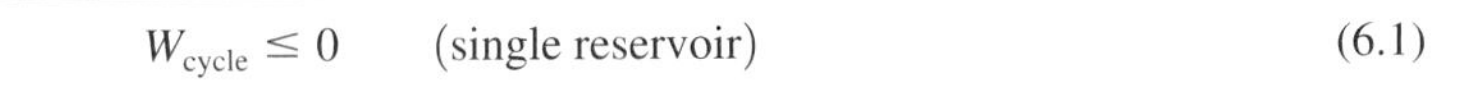

$$W_{cycle} \leq 0 \qquad \text{(single reservoir)} \tag{6.1}$$

The words *single reservoir* in Eq. 6.1 emphasize that the system communicates thermally only with a single reservoir as it executes the cycle. It can be shown that the "less than" and "equal to" signs of Eq. 6.1 correspond to the presence and absence of *internal irreversibilities*, respectfully. The concept of irreversibilities is considered next.

## 6.2 Identifying Irreversibilities

One of the important uses of the second law of thermodynamics in engineering is to determine the best theoretical performance of systems. By comparing actual performance with the best theoretical performance, insights often can be gained into the potential for improvement. As might be surmised, the best performance is evaluated in terms of idealized processes. In this section such idealized processes are introduced and distinguished from actual processes involving *irreversibilities.*

*irreversible and reversible processes*

A process is called ***irreversible*** if the system and all parts of its surroundings cannot be exactly restored to their respective initial states after the process has occurred. A process is ***reversible*** if both the system and surroundings can be returned to their initial states. Irreversible processes are the subject of the present discussion. The reversible process is considered again later in the section.

A system that has undergone an irreversible process is not necessarily precluded from being restored to its initial state. However, were the system restored to its initial state, it would not be possible also to return the surroundings to the state they were in initially. It might be apparent from the discussion of the Clausius statement of the second law that any process involving a spontaneous heat transfer from a hotter body to a cooler body is irreversible. Otherwise, it would be possible to return this energy from the cooler body to the hotter body with no other effects within the two bodies or their surroundings. However, this possibility is contrary to our experience and is denied by the Clausius statement. Processes involving other kinds of spontaneous events are irreversible, such as the unrestrained expansion of a gas considered in Fig. 6.1. Friction, electrical resistance, hysteresis, and inelastic deformation are examples of effects whose presence during a process renders it irreversible.

In summary, irreversible processes normally include one or more of the following ***irreversibilities:***

*irreversibilities*

1. Heat transfer through a finite temperature difference
2. Unrestrained expansion of a gas or liquid to a lower pressure
3. Spontaneous chemical reaction
4. Spontaneous mixing of matter at different compositions or states
5. Friction—sliding friction as well as friction in the flow of fluids
6. Electric current flow through a resistance
7. Magnetization or polarization with hysteresis
8. Inelastic deformation

Although the foregoing list is not exhaustive, it does suggest that *all actual processes are irreversible.* That is, every process involves effects such as those listed, whether it is a naturally occurring process or one involving a device of our construction, from the simplest mechanism to the largest industrial plant. The term "irreversibility" is used to identify any of these effects. The list given previously comprises a few of the irreversibilities that are commonly encountered.

As a system undergoes a process, irreversibilities may be found within the system as well as within its surroundings, although in certain instances they may be located predominately in one place or the other. For many analyses it is convenient to divide the irreversibilities present into two classes. ***Internal irreversibilities*** are those that occur within the system. ***External irreversibilities*** are those that occur within the surroundings, often the immediate surroundings. As this distinction depends solely on the location of the boundary, there is some arbitrariness in the classification, for by extending the boundary to take in a portion of the surroundings, all irreversibilities become "internal." Nonetheless, as shown by subsequent developments, this distinction between irreversibilities is often useful.

*internal and external irreversibilities*

Engineers should be able to recognize irreversibilities, evaluate their influence, and develop practical means for reducing them. However, certain systems, such as brakes, rely on the effect of friction or other irreversibilities in their operation. The need to achieve profitable rates of production, high heat transfer rates, rapid accelerations, and so on invariably dictates the presence of significant irreversibilities. Furthermore, irreversibilities are tolerated to some degree in every type of system because the changes in design and operation required to reduce them would be too costly. Accordingly, although improved thermodynamic performance can accompany the reduction of irreversibilities, steps taken in this direction are constrained by a number of practical factors often related to costs.

***For Example...*** consider two bodies at different temperatures that are able to communicate thermally. With a *finite* temperature difference between them, a spontaneous heat transfer would take place and, as discussed previously, this would be a source of irreversibility. It might be expected that the importance of this irreversibility would diminish as the temperature difference narrows, and this is the case. As the difference in temperature between the bodies approaches zero, the heat transfer would approach reversibility. From the study of heat transfer (Sec. 15.1), it is known that the transfer of a finite amount of energy by heat between bodies whose temperatures differ only slightly would require a considerable amount of time, a larger (more costly) heat transfer surface area, or both. To approach reversibility, therefore, a heat transfer would require an infinite amount of time and/or an infinite surface area. ▲

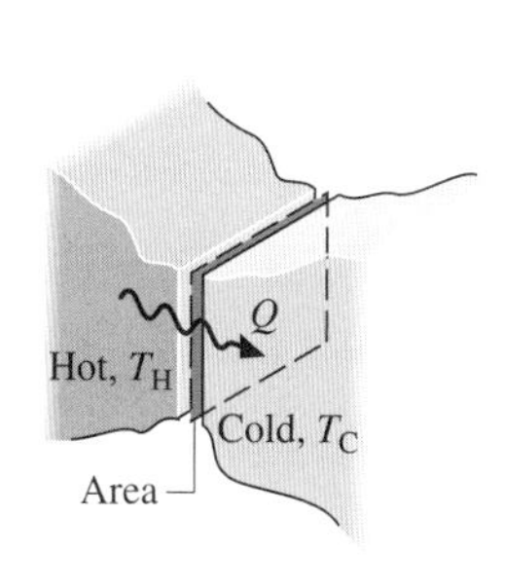

### Internally Reversible Processes

*internally reversible process*

In an irreversible process, irreversibilities are present within the system, its surroundings, or both. A reversible process is one in which there are no internal or external irreversibilities. An ***internally reversible process*** is one in which *there are no irreversibilities within the system.* Irreversibilities may be located within the surroundings, however, as when there is heat transfer between a portion of the boundary that is at one temperature and the surroundings at another.

At every intermediate state of an internally reversible process of a closed system, all intensive properties are uniform throughout each phase present. That is, the temperature, pressure, specific volume, and other intensive properties do not vary with position. If there were a spatial variation in temperature, say, there would be a tendency for a spontaneous energy transfer to occur *within* the system in the direction of decreasing temperature. For reversibility, however, no spontaneous processes can be present. From these considerations it can be concluded that the internally reversible process consists of a series of equilibrium states: It is a quasiequilibrium process. To avoid having two terms that refer to the same thing, in subsequent discussions we will refer to *any* such process as an internally reversible process.

The use of the internally reversible process concept in thermodynamics is comparable to the idealizations made in mechanics: point masses, frictionless pulleys, rigid beams, and so on. In much the same way as these are used in mechanics to simplify an analysis and arrive at a manageable model, simple thermodynamic models of complex situations can be obtained through the use of internally reversible processes. Initial calculations based on internally reversible processes would be adjusted with efficiencies or correction factors to obtain reasonable estimates of actual performance under various operating conditions. Internally reversible processes are also useful in determining the best thermodynamic performance of systems.

The internally reversible process concept can be employed to refine the definition of the thermal reservoir introduced in Sec. 6.1.2. In subsequent discussions we assume that no internal irreversibilities are present within a thermal reservoir. Accordingly, every process of a thermal reservoir is an internally reversible process.

## 6.3 Applying the Second Law to Thermodynamic Cycles

Several important applications of the second law related to power cycles and refrigeration and heat pump cycles are presented in this section. These applications further our understanding of the implications of the second law and provide the basis for important deductions from the second law introduced in subsequent sections. Familiarity with thermodynamic cycles is required, and we recommend that you review Sec. 3.7, where cycles are considered from an energy, or first law, perspective and the thermal efficiency of power cycles and coefficients of performance for refrigeration and heat pump cycles are introduced.

### 6.3.1 Power Cycles Interacting with Two Reservoirs

A significant limitation on the performance of systems undergoing power cycles can be brought out using the Kelvin–Planck statement of the second law. Consider Fig. 6.2, which shows a system that executes a cycle while communicating thermally with *two* thermal reservoirs, a hot reservoir and a cold reservoir, and developing net work $W_{\text{cycle}}$. The thermal efficiency of the cycle is

$$\eta = \frac{W_{\text{cycle}}}{Q_{\text{H}}} = 1 - \frac{Q_{\text{C}}}{Q_{\text{H}}} \tag{6.2}$$

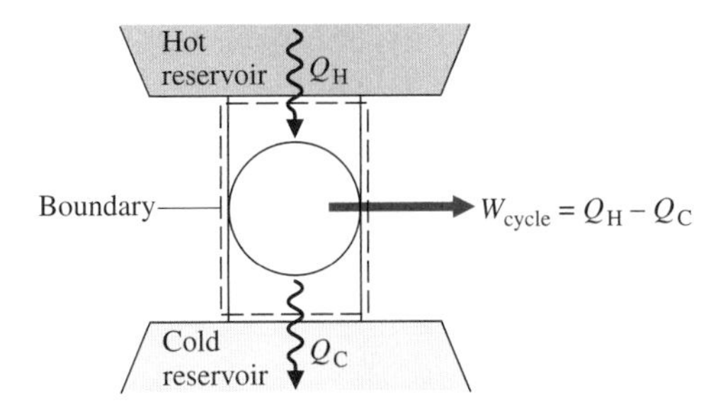


*Figure 6.2* System undergoing a power cycle while exchanging energy by heat transfer with two reservoirs.

where $Q_H$ is the amount of energy received by the system from the hot reservoir by heat transfer and $Q_C$ is the amount of energy discharged from the system to the cold reservoir by heat transfer. The energy transfers labeled on Fig. 6.2 are in the directions indicated by the arrows.

If the value of $Q_C$ were zero, the system of Fig. 6.2 would withdraw energy $Q_H$ from the hot reservoir and produce an equal amount of work, while undergoing a cycle. The thermal efficiency of such a cycle would have a value of unity (100%). However, this method of operation would violate the Kelvin–Planck statement and thus is not allowed. It follows that for any system executing a power cycle while operating between two reservoirs, only a portion of the heat transfer $Q_H$ can be obtained as work, and the remainder, $Q_C$, must be discharged by heat transfer to the cold reservoir. That is, the thermal efficiency must be less than 100%. In arriving at this conclusion it was *not* necessary to (1) identify the nature of the substance contained within the system, (2) specify the exact series of processes making up the cycle, or (3) indicate whether the processes are actual processes or somehow idealized. The conclusion that the thermal efficiency must be less than 100% applies to *all* power cycles whatever their details of operation. This may be regarded as a corollary of the second law. Other corollaries follow.

***Carnot Corollaries.*** Since no power cycle can have a thermal efficiency of 100%, it is of interest to investigate the maximum theoretical efficiency. The maximum theoretical efficiency for systems undergoing power cycles while communicating thermally with two thermal reservoirs at different temperatures is evaluated in Sec. 6.4 with reference to the following two corollaries of the second law, called the ***Carnot corollaries.***

*Carnot corollaries*

- *The thermal efficiency of an irreversible power cycle is always less than the thermal efficiency of a reversible power cycle when each operates between the same two thermal reservoirs.*
- *All reversible power cycles operating between the same two thermal reservoirs have the same thermal efficiency.*

A cycle is considered *reversible* when there are no irreversibilities within the system as it undergoes the cycle and heat transfers between the system and reservoirs occur reversibly.

The idea underlying the first Carnot corollary is in agreement with expectations stemming from the discussion of the second law thus far. Namely, the presence of irreversibilities during the execution of a cycle is expected to exact a penalty. If two systems operating between the same reservoirs each receive the same amount of energy $Q_H$ and one executes a reversible cycle while the other executes an irreversible cycle, it is in accord with intuition that the net work developed by the irreversible cycle will be less, and it will therefore have the smaller thermal efficiency.

The second Carnot corollary refers only to reversible cycles. All processes of a reversible cycle are perfectly executed. Accordingly, if two reversible cycles operating between the same reservoirs each receive the same amount of energy $Q_H$ but one could produce more work than the other, it could only be as a result of more advantageous selections for the substance making up the system (it is conceivable that, say, air might be better than water vapor) *or* the series of processes making up the cycle (nonflow processes might be preferable to flow processes). This corollary denies both possibilities and indicates that the cycles must have the same efficiency whatever the choices for the working substance or the series of processes.

The two Carnot corollaries can be demonstrated using the Kelvin–Planck statement of the second law. (CD-ROM)

### 6.3.2 Refrigeration and Heat Pump Cycles Interacting with Two Reservoirs

The second law of thermodynamics places limits on the performance of refrigeration and heat pump cycles as it does for power cycles. Consider Fig. 6.4, which shows a system undergoing a cycle while communicating thermally with two thermal reservoirs, a hot and a cold reservoir. The energy transfers labeled on the figure are in the directions indicated by the arrows. In accord with the conservation of energy principle, the cycle discharges energy $Q_H$ by heat transfer to the hot reservoir equal to the sum of the energy $Q_C$ received by heat transfer from the cold reservoir and the net work input. This cycle might be a refrigeration cycle or a heat pump cycle, depending on whether its function is to remove energy $Q_C$ from the cold reservoir or deliver energy $Q_H$ to the hot reservoir.

For a refrigeration cycle the coefficient of performance is

$$\beta = \frac{Q_C}{W_{cycle}} = \frac{Q_C}{Q_H - Q_C} \tag{6.3}$$

The coefficient of performance for a heat pump cycle is

$$\gamma = \frac{Q_H}{W_{cycle}} = \frac{Q_H}{Q_H - Q_C} \tag{6.4}$$

As the net work input to the cycle $W_{cycle}$ tends to zero, the coefficients of performance given by Eqs. 6.3 and 6.4 approach infinity. If $W_{cycle}$ were identically zero, the system of Fig. 6.4 would withdraw energy $Q_C$ from the cold reservoir and deliver energy $Q_C$ to the hot reservoir, while undergoing a cycle. However, this method of operation would violate the Clausius statement of the second law and thus is not allowed. It follows that these coefficients of performance must invariably be finite in value. This may be regarded as another corollary of the second law. Further corollaries follow.

***Corollaries for Refrigeration and Heat Pump Cycles.*** The maximum theoretical coefficients of performance for systems undergoing refrigeration and heat pump cycles while communicating thermally with two reservoirs at different temperatures are evaluated in Sec. 6.4 with reference to the following corollaries of the second law:

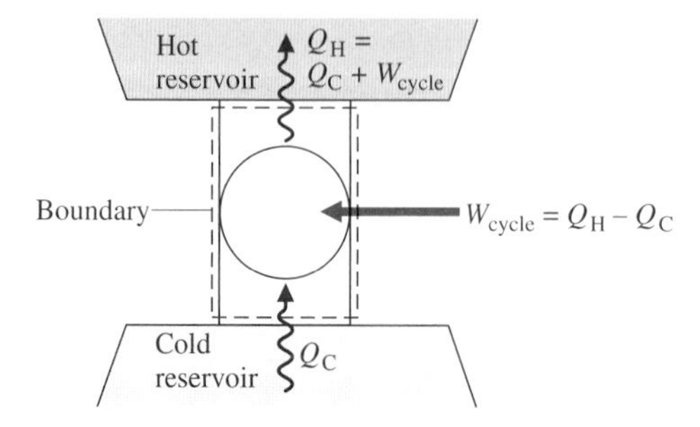


*Figure 6.4* System undergoing a refrigeration or heat pump cycle while exchanging energy by heat transfer with two reservoirs.

- *The coefficient of performance of an irreversible refrigeration cycle is always less than the coefficient of performance of a reversible refrigeration cycle when each operates between the same two thermal reservoirs.*
- *All reversible refrigeration cycles operating between the same two thermal reservoirs have the same coefficient of performance.*

By replacing the term *refrigeration* with *heat pump*, we obtain counterpart corollaries for heat pump cycles.

## 6.4 Maximum Performance Measures for Cycles Operating between Two Reservoirs

The results of Sec. 6.3 establish theoretical upper limits on the performance of power, refrigeration, and heat pump cycles communicating thermally with two reservoirs. Expressions for the *maximum* theoretical thermal efficiency of power cycles and the *maximum* theoretical coefficients of performance of refrigeration and heat pump cycles are developed in this section using the Kelvin temperature scale defined next.

### 6.4.1 Defining the Kelvin Temperature Scale

From the second Carnot corollary we know that all reversible power cycles operating between the same two reservoirs have the same thermal efficiency, regardless of the nature of the substance making up the system executing the cycle or the series of processes. Since the efficiency is independent of these factors, its value can depend only on the nature of the reservoirs themselves. Noting that it is the difference in *temperature* between the two reservoirs that provides the impetus for heat transfer between them, and thereby for the production of work during the cycle, we reason that the reversible power cycle efficiency depends *only* on the temperatures of the two reservoirs.

From Eq. 6.2 it also follows that for such reversible power cycles the ratio of the heat transfers $Q_C/Q_H$ depends only on the reservoir temperatures. This conclusion provides a basis for defining a thermodynamic temperature scale independent of the properties of any substance.

The thermodynamic temperature scale called the *Kelvin scale* is defined so that two temperatures are in the same ratio as the values of the heat transfers absorbed and rejected, respectively, by a system undergoing a reversible cycle while communicating thermally with reservoirs at these temperatures. That is, the ***Kelvin scale*** is based on

$$\left(\frac{Q_C}{Q_H}\right)_{\substack{\text{rev}\\\text{cycle}}} = \frac{T_C}{T_H} \tag{6.5}$$

*Kelvin scale*

where "rev cycle" emphasizes that the expression applies only to systems undergoing reversible cycles while operating between thermal reservoirs at $T_C$ and $T_H$.

If a reversible power cycle were operated in the opposite direction as a refrigeration or heat pump cycle, the magnitudes of the energy transfers $Q_C$ and $Q_H$ would remain the same, but the energy transfers would be oppositely directed. Accordingly, Eq. 6.5 applies to each type of cycle considered thus far, provided the system undergoing the cycle operates between two thermal reservoirs and the cycle is reversible.

### 6.4.2 Power Cycles

The use of Eq. 6.5 in Eq. 6.2 results in an expression for the thermal efficiency of a system undergoing a reversible *power cycle* while operating between thermal reservoirs at temperatures

$T_H$ and $T_C$. That is

$$\eta_{max} = 1 - \frac{T_C}{T_H} \tag{6.6}$$

*Carnot efficiency*

which is known as the ***Carnot efficiency.*** As temperatures on the Rankine scale differ from Kelvin temperatures only by the factor 1.8, the $T$'s in Eq. 6.6 may be on either scale of temperature.

Recalling the two Carnot corollaries, it should be evident that the efficiency given by Eq. 6.6 is the thermal efficiency of *all* reversible power cycles operating between two reservoirs at temperatures $T_H$ and $T_C$, and the *maximum* efficiency *any* power cycle can have while operating between the two reservoirs. By inspection, the value of the Carnot efficiency increases as $T_H$ increases and/or $T_C$ decreases.

Equation 6.6 is presented graphically in Fig. 6.5. The temperature $T_C$ used in constructing the figure is 298 K in recognition that actual power cycles ultimately discharge energy by heat transfer at about the temperature of the local atmosphere or cooling water drawn from a nearby river or lake. Note that the possibility of increasing the thermal efficiency by reducing $T_C$ below that of the environment is not practical, for maintaining $T_C$ lower than the ambient temperature would require a refrigerator that would have to be supplied work to operate.

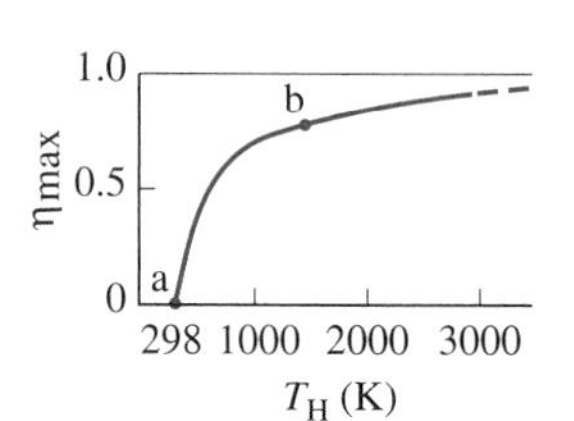


*Figure 6.5* Carnot efficiency versus $T_H$, for $T_C = 298$ K.

Figure 6.5 shows that the thermal efficiency increases with $T_H$. Referring to segment a–b of the curve, where $T_H$ and $\eta$ are relatively low, we can see that $\eta$ increases rapidly as $T_H$ increases, showing that in this range even a small increase in $T_H$ can have a large effect on efficiency. Though these conclusions, drawn as they are from Fig. 6.5, apply strictly only to systems undergoing reversible cycles, they are qualitatively correct for actual power cycles. The thermal efficiencies of actual cycles are observed to increase as the *average* temperature at which energy is added by heat transfer increases and/or the *average* temperature at which energy is discharged by heat transfer is reduced. However, maximizing the thermal efficiency of a power cycle may not be the only objective. In practice, other considerations such as cost may be overriding.

***Comment.*** Conventional power-producing cycles have thermal efficiencies ranging up to about 40%. This value may seem low, but the comparison should be made with an appropriate limiting value and not 100%. ***For Example...*** consider a system executing a power cycle for which the average temperature of heat addition is 745 K and the average temperature at which heat is discharged is 298 K. For a reversible cycle receiving and discharging energy by heat transfer at these temperatures, the thermal efficiency given by Eq. 6.6 is 60%. When compared to this value, an actual thermal efficiency of 40% does not appear to be so low. The cycle would be operating at two-thirds of the theoretical maximum. ▲

A more complete discussion of power cycles is provided in Chaps. 8 and 9.

### 6.4.3 Refrigeration and Heat Pump Cycles

Equation 6.5 is also applicable to reversible refrigeration and heat pump cycles operating between two thermal reservoirs, but for these $Q_C$ represents the heat added to the cycle from the cold reservoir at temperature $T_C$ on the Kelvin scale and $Q_H$ is the heat discharged to the hot reservoir at temperature $T_H$. Introducing Eq. 6.5 in Eq. 6.3 results in the following expression for the coefficient of performance of any system undergoing a reversible refrigeration cycle while operating between the two reservoirs

$$\beta_{max} = \frac{T_C}{T_H - T_C} \tag{6.7}$$

Similarly, substituting Eq. 6.5 into Eq. 6.4 gives the following expression for the coefficient of performance of any system undergoing a reversible heat pump cycle while operating between the two reservoirs

$$\gamma_{\max} = \frac{T_H}{T_H - T_C} \tag{6.8}$$

The development of Eqs. 6.7 and 6.8 is left as an exercise. Note that the temperatures used to evaluate $\beta_{\max}$ and $\gamma_{\max}$ must be absolute temperatures on the Kelvin or Rankine scale.

From the discussion of Sec. 6.3.2, it follows that Eqs. 6.7 and 6.8 are the maximum coefficients of performance that any refrigeration and heat pump cycles can have while operating between reservoirs at temperatures $T_H$ and $T_C$. As for the case of the Carnot efficiency, these expressions can be used as standards of comparison for actual refrigerators and heat pumps. A more complete discussion of refrigeration and heat pump cycles is provided in Chap. 8.

### 6.4.4 Applications

In this section, three examples are provided that illustrate the use of the second law corollaries of Secs. 6.3.1 and 6.3.2 together with Eqs. 6.6, 6.7, and 6.8, as appropriate.

The first example uses Eq. 6.6 to evaluate an inventor's claim.

## *Example 6.1* Evaluating a Power Cycle Performance Claim

An inventor claims to have developed a power cycle capable of delivering a net work output of 410 kJ for an energy input by heat transfer of 1000 kJ. The system undergoing the cycle receives the heat transfer from hot gases at a temperature of 500 K and discharges energy by heat transfer to the atmosphere at 300 K. Evaluate this claim.

### Solution

***Known:*** A system operates in a cycle and produces a net amount of work while receiving and discharging energy by heat transfer at fixed temperatures.

***Find:*** Evaluate the claim that the cycle can develop 410 kJ of work for an energy input by heat of 1000 kJ.

***Schematic and Given Data:***

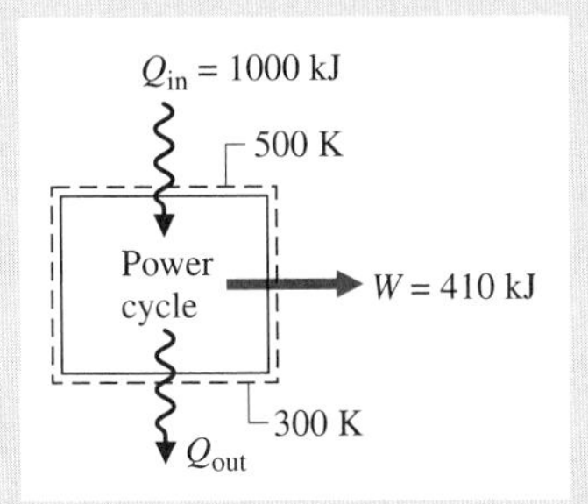


***Assumptions:***

1. The system is shown on the accompanying figure.
2. The hot gases and the atmosphere play the roles of hot and cold reservoirs, respectively.

*Figure E6.1*

***Analysis:*** Inserting the values supplied by the inventor in Eq. 6.2, the cycle thermal efficiency is

$$\eta = \frac{410 \text{ kJ}}{1000 \text{ kJ}} = 0.41(41\%)$$

The maximum thermal efficiency *any* power cycle can have while operating between reservoirs at $T_H = 500$ K and $T_C = 300$ K is given by Eq. 6.6. That is

❶
$$\eta_{max} = 1 - \frac{T_C}{T_H} = 1 - \frac{300 \text{ K}}{500 \text{ K}} = 0.40\ (40\%)$$

Since the thermal efficiency of the actual cycle exceeds the maximum theoretical value, the claim cannot be valid. ◁

❶ The temperatures used in evaluating $\eta_{max}$ *must* be in K or °R.

In the next example, we evaluate the coefficient of performance of a refrigerator and compare it with the maximum theoretical value.

## *Example 6.2* Evaluating Refrigerator Performance

By steadily circulating a refrigerant at low temperature through passages in the walls of the freezer compartment, a refrigerator maintains the freezer compartment at −5°C when the air surrounding the refrigerator is at 22°C. The rate of heat transfer from the freezer compartment to the refrigerant is 8000 kJ/h and the power input required to operate the refrigerator is 3200 kJ/h. Determine the coefficient of performance of the refrigerator and compare with the coefficient of performance of a reversible refrigeration cycle operating between reservoirs at the same two temperatures.

### Solution

***Known:*** A refrigerator maintains a freezer compartment at a specified temperature. The rate of heat transfer from the refrigerated space, the power input to operate the refrigerator, and the ambient temperature are known.

***Find:*** Determine the coefficient of performance and compare with that of a reversible refrigerator operating between reservoirs at the same two temperatures.

***Schematic and Given Data:***

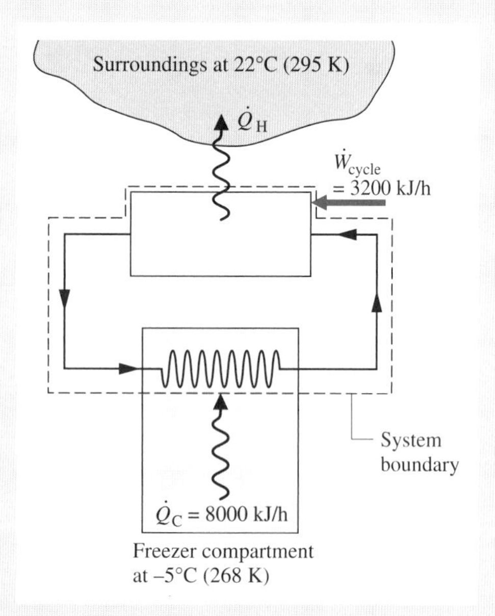


***Assumptions:***

1. The system shown on the accompanying figure is at steady state.
2. The freezer compartment and the surrounding air play the roles of cold and hot reservoirs, respectively.

*Figure E6.2*

***Analysis:*** Inserting the given operating data in Eq. 6.3, the coefficient of performance of the refrigerator is

$$\beta = \frac{\dot{Q}_C}{\dot{W}_{cycle}} = \frac{8000 \text{ kJ/h}}{3200 \text{ kJ/h}} = 2.5 \quad ◁$$

Substituting values into Eq. 6.7 gives the coefficient of performance of a reversible refrigeration cycle operating between reservoirs at $T_C = 268$ K and $T_H = 295$ K. That is

➊ $$\beta_{max} = \frac{T_C}{T_H - T_C} = \frac{268\text{ K}}{295\text{ K} - 268\text{ K}} = 9.9 \quad \triangleleft$$

➊ The difference between the actual and maximum coefficients of performance suggests that there may be some potential for improving the thermodynamic performance. This objective should be approached judiciously, however, for improved performance may require increases in size, complexity, and cost.

In Example 6.3, we determine the minimum theoretical work input and cost for one day of operation of an electric heat pump.

## *Example 6.3* Evaluating Heat Pump Performance

A dwelling requires $6 \times 10^5$ Btu per day to maintain its temperature at 70°F when the outside temperature is 32°F. **(a)** If an electric heat pump is used to supply this energy, determine the minimum theoretical work input for one day of operation, in Btu/day. **(b)** Evaluating electricity at 8 cents per kW · h, determine the minimum theoretical cost to operate the heat pump, in \$/day.

### Solution

***Known:*** A heat pump maintains a dwelling at a specified temperature. The energy supplied to the dwelling, the ambient temperature, and the unit cost of electricity are known.

***Find:*** Determine the *minimum* theoretical work required by the heat pump and the corresponding electricity cost.

***Schematic and Given Data:***

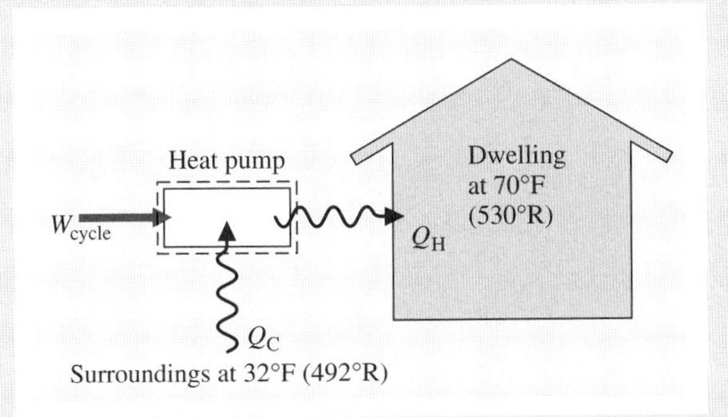


***Assumptions:***

1. The system is shown on the accompanying figure.
2. The dwelling and the outside air play the roles of hot and cold reservoirs, respectively.

*Figure E6.3*

***Analysis:*** (a) Using Eq. 6.4, the work for any heat pump cycle can be expressed as $W_{cycle} = Q_H/\gamma$. The coefficient of performance $\gamma$ of an actual heat pump is less than, or equal to, the coefficient of performance $\gamma_{max}$ of a reversible heat pump cycle when each operates between the same two thermal reservoirs: $\gamma \leq \gamma_{max}$. Accordingly, for a given value of $Q_H$, and using Eq. 6.8 to evaluate $\gamma_{max}$, we get

$$W_{cycle} \geq \frac{Q_H}{\gamma_{max}}$$

$$\geq \left(1 - \frac{T_C}{T_H}\right) Q_H$$

Inserting values

❶
$$W_{\text{cycle}} \geq \left(1 - \frac{492°\text{R}}{530°\text{R}}\right)\left(6 \times 10^5 \frac{\text{Btu}}{\text{day}}\right) = 4.3 \times 10^4 \frac{\text{Btu}}{\text{day}} \triangleleft$$

The *minimum* theoretical work input is $4.3 \times 10^4$ Btu/day.

**(b)** Using the result of part (a) together with the given cost data and an appropriate conversion factor

❷
$$\begin{bmatrix} \text{minimum} \\ \text{theoretical} \\ \text{cost per day} \end{bmatrix} = \left(4.3 \times 10^4 \frac{\text{Btu}}{\text{day}} \left| \frac{1 \text{ kW} \cdot \text{h}}{3413 \text{ Btu}} \right| \right)\left(0.08 \frac{\$}{\text{kW} \cdot \text{h}}\right) = 1.01 \frac{\$}{\text{day}} \triangleleft$$

❶ Note that the reservoir temperatures $T_C$ and $T_H$ must be expressed here in °R.

❷ Because of irreversibilities, an actual heat pump must be supplied more work than the minimum to provide the same heating effect. The actual daily cost could be substantially greater than the minimum theoretical cost.

## 6.5 Carnot Cycle

*Carnot cycle*

The Carnot cycle introduced in this section provides a specific example of a reversible power cycle operating between two thermal reservoirs. In a ***Carnot cycle,*** the system executing the cycle undergoes a series of four internally reversible processes: two adiabatic processes alternated with two isothermal processes.

Figure 6.6 shows the schematic and accompanying *p–v* diagram of a Carnot cycle executed by water steadily circulating through a series of four interconnected components that has features in common with the simple vapor power plant shown in Fig. 5.12. As the water flows through the boiler, a *change of phase* from liquid to vapor at constant temperature $T_H$ occurs as a result of heat transfer from the hot reservoir. Since temperature remains constant, pressure also remains constant during the phase change. The steam exiting the boiler expands adiabatically through the turbine and work is developed. In this process the temperature decreases to the temperature of the cold reservoir, $T_C$, and there is an accompanying decrease in pressure. As the steam passes through the condenser, a heat transfer to the cold reservoir occurs and some of the vapor condenses at constant temperature $T_C$. Since temperature remains constant, pressure also remains constant as the water passes through the condenser.

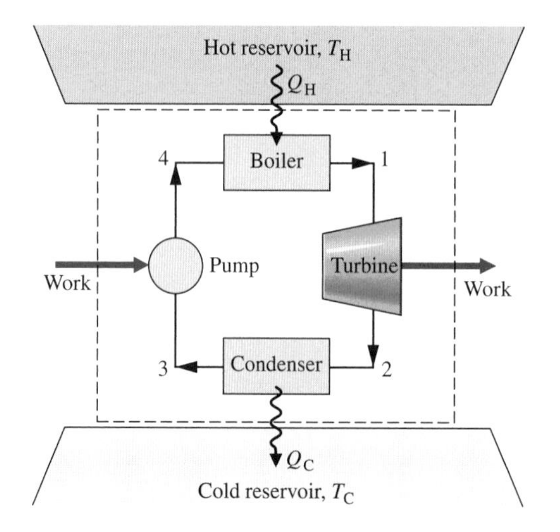


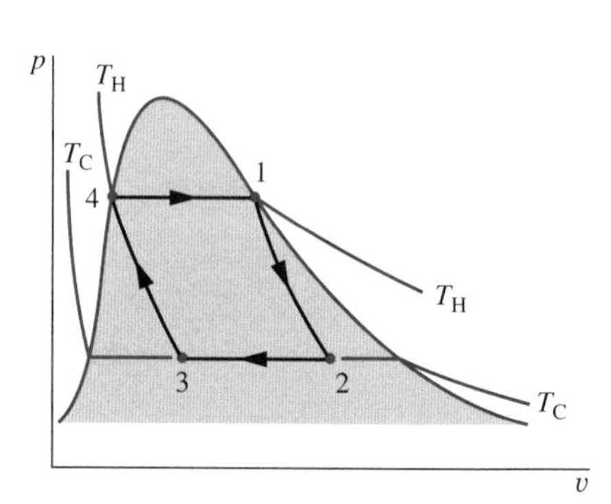


*Figure 6.6* Carnot vapor power cycle.

The fourth component is a pump, or compressor, that receives a two-phase liquid–vapor mixture from the condenser and returns it adiabatically to the state at the boiler entrance. During this process, which requires a work input to increase the pressure, the temperature increases from $T_C$ to $T_H$.

Carnot cycles also can be devised that are composed of processes in which a gas in a piston-cylinder is expanded and compressed, a capacitor is charged and discharged, a paramagnetic substance is magnetized and demagnetized, and so on. However, regardless of the type of device or the working substance used, the Carnot cycle always has the same four internally reversible processes: two adiabatic processes alternated with two isothermal processes. Moreover, the thermal efficiency is always given by Eq. 6.6 in terms of the temperatures of the two reservoirs evaluated on the Kelvin or Rankine scale.

If a Carnot power cycle is operated in the opposite direction, the magnitudes of all energy transfers remain the same but the energy transfers are oppositely directed. Such a cycle may be regarded as a reversible refrigeration or heat pump cycle, for which the coefficients of performance are given by Eqs. 6.7 and 6.8, respectively.

## 6.6 Chapter Summary and Study Guide

In this chapter, we motivate the need for and usefulness of the second law of thermodynamics, and provide the basis for subsequent applications involving the second law in Chap. 7. Two equivalent statements of the second law, the Clausius and Kelvin–Planck statements, are introduced together with several corollaries that establish the best theoretical performance for systems undergoing cycles while interacting with thermal reservoirs. The irreversibility concept is introduced and the related notions of irreversible, reversible, and internally reversible processes are discussed. The Kelvin temperature scale is defined and used to obtain expressions for the maximum performance measures of power, refrigeration, and heat pump cycles operating between two thermal reservoirs. Finally, the Carnot cycle is introduced to provide a specific example of a reversible cycle operating between two thermal reservoirs.

The following checklist provides a study guide for this chapter. When your study of the text and end-of-chapter exercises has been completed you should be able to

- write out the meanings of the terms listed in the margins throughout the chapter and understand each of the related concepts. The subset of key terms listed here in the margin is particularly important in subsequent chapters.
- give the Kelvin–Planck statement of the second law, correctly interpreting the "less than" and "equal to" signs in Eq. 6.1.
- list several important irreversibilities.
- apply the corollaries of Secs. 6.3.1 and 6.3.2 together with Eqs. 6.6, 6.7, and 6.8 to assess the performance of power cycles and refrigeration and heat pump cycles.
- describe the Carnot cycle.

*Kelvin–Planck statement*
*irreversible process*
*internal and external irreversibilities*
*internally reversible process*
*Carnot corollaries*
*Kelvin temperature scale*
*Carnot efficiency*

# *Problems*

**Exploring the Second Law**

**6.1** A heat pump receives energy by heat transfer from the outside air at 0°C and discharges energy by heat transfer to a dwelling at 20°C. Is this in violation of the Clausius statement of the second law of thermodynamics? Explain.

**6.2** Air as an ideal gas expands isothermally at 20°C from a volume of 1 $m^3$ to 2 $m^3$. During this process there is heat transfer to the air from the surrounding atmosphere, modeled as a thermal reservoir, and the air does work. Evaluate the work and heat transfer for the process, in kJ/kg. Is this process in violation of the second law of thermodynamics? Explain.

**6.3** Methane gas within a piston–cylinder assembly is compressed in a *quasiequilibrium* process. Is this process internally reversible? Is this process reversible?

**6.4** Water within a piston–cylinder assembly cools isothermally at 100°C from saturated vapor to saturated liquid while interacting thermally with its surroundings at 20°C. Is the process an internally reversible process? Is it reversible? Discuss.

**6.5** (CD-ROM)

**6.6** (CD-ROM)

**6.7** To increase the thermal efficiency of a reversible power cycle operating between reservoirs at $T_H$ and $T_C$, would you increase $T_H$ while keeping $T_C$ constant, or decrease $T_C$ while keeping $T_H$ constant? Are there any *natural* limits on the increase in thermal efficiency that might be achieved by such means?

**6.8** (CD-ROM)

**6.9** (CD-ROM)

**6.10** The data listed below are claimed for a power cycle operating between reservoirs at 727 and 127°C. For each case, determine if any principles of thermodynamics would be violated.
**(a)** $Q_H = 600$ kJ, $W_{cycle} = 200$ kJ, $Q_C = 400$ kJ.
**(b)** $Q_H = 400$ kJ, $W_{cycle} = 240$ kJ, $Q_C = 160$ kJ.
**(c)** $Q_H = 400$ kJ, $W_{cycle} = 210$ kJ, $Q_C = 180$ kJ.

**6.11** A power cycle operating between two reservoirs receives energy $Q_H$ by heat transfer from a hot reservoir at $T_H = 2000$ K and rejects energy $Q_C$ by heat transfer to a cold reservoir at $T_C = 400$ K. For each of the following cases determine whether the cycle operates reversibly, irreversibly, or is impossible:
**(a)** $Q_H = 1200$ kJ, $W_{cycle} = 1020$ kJ.
**(b)** $Q_H = 1200$ kJ, $Q_C = 240$ kJ.
**(c)** $W_{cycle} = 1400$ kJ, $Q_C = 600$ kJ.
**(d)** $\eta = 40\%$.

**6.12** A refrigeration cycle operating between two reservoirs receives energy $Q_C$ from a cold reservoir at $T_C = 250$ K and rejects energy $Q_H$ to a hot reservoir at $T_H = 300$ K. For each of the following cases determine whether the cycle operates reversibly, irreversibly, or is impossible:
**(a)** $Q_C = 1000$ kJ, $W_{cycle} = 400$ kJ.
**(b)** $Q_C = 1500$ kJ, $Q_H = 1800$ kJ.
**(c)** $Q_H = 1500$ kJ, $W_{cycle} = 200$ kJ.
**(d)** $\beta = 4$.

## Power Cycle Applications

**6.13** A reversible power cycle receives 1000 Btu of energy by heat transfer from a reservoir at 1540°F and discharges energy by heat transfer to a reservoir at 40°F. Determine the thermal efficiency and the net work developed, in Btu.

**6.14** A power cycle operates between a reservoir at temperature $T$ and a lower-temperature reservoir at 280 K. At steady state, the cycle develops 40 kW of power while rejecting 1000 kJ/min of energy by heat transfer to the cold reservoir. Determine the minimum theoretical value for $T$, in K.

**6.15** A certain reversible power cycle has the same thermal efficiency for hot and cold reservoirs at 1000 and 500 K, respectively, as for hot and cold reservoirs at temperature $T$ and 1000 K. Determine $T$, in K.

**6.16** A reversible power cycle whose thermal efficiency is 50% operates between a reservoir at 1800 K and a reservoir at a lower temperature $T$. Determine $T$, in K.

**6.17** An inventor claims to have developed a device that executes a power cycle while operating between reservoirs at 900 and 300 K that has a thermal efficiency of **(a)** 66%, **(b)** 50%. Evaluate the claim for each case.

**6.18** At steady state, a new power cycle is claimed by its inventor to develop 6 horsepower for a heat addition rate of 400 Btu/min. If the cycle operates between reservoirs at 2400 and 1000°R, evaluate this claim.

**6.19** (CD-ROM)

**6.20** A proposed power cycle is to have a thermal efficiency of 40% while receiving energy by heat transfer from steam condensing from saturated vapor to saturated liquid at temperature $T$ and discharging energy by heat transfer to a nearby lake at 70°F. Determine the *lowest* possible temperature $T$, in °F, and the corresponding steam pressure, in lbf/in.$^2$

**6.21** At steady state, a power cycle having a thermal efficiency of 38% generates 100 MW of electricity while discharging energy by heat transfer to cooling water at an average temperature of 70°F. The average temperature of the steam passing through the boiler is 900°F. Determine
**(a)** the rate at which energy is discharged to the cooling water, in Btu/h.
**(b)** the *minimum* theoretical rate at which energy could be discharged to the cooling water, in Btu/h. Compare with the actual rate and discuss.

**6.22** *Ocean temperature energy conversion* (*OTEC*) power plants generate power by utilizing the naturally occurring decrease with depth of the temperature of ocean water. Near Florida, the ocean surface temperature is 27°C, while at a depth of 700 m the temperature is 7°C.
**(a)** Determine the maximum thermal efficiency for any power cycle operating between these temperatures.
**(b)** The thermal efficiency of existing OTEC plants is approximately 2 percent. Compare this with the result of part (a) and comment.

**6.23** Geothermal power plants harness underground sources of hot water or steam for the production of electricity. One such plant receives a supply of hot water at 167°C and rejects energy by heat transfer to the atmosphere, which is at 13°C. Determine the maximum possible thermal efficiency for any power cycle operating between these temperatures.

**6.24** During January, at a location in Alaska winds at −23°F can be observed. Several meters below ground the temperature remains at 55°F, however. An inventor claims to have devised a power cycle exploiting this situation that has a thermal efficiency of 10%. Discuss this claim.

**6.25** Figure P6.25 shows a system for collecting solar radiation and utilizing it for the production of electricity by a power cycle. The solar collector receives solar radiation at the rate of 0.315 kW per m$^2$ of area and provides energy to a storage unit whose temperature remains constant at 220°C. The power cycle

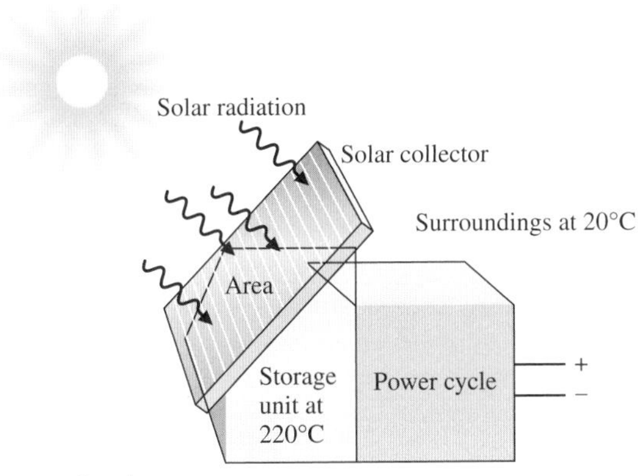


Figure P6.25

receives energy by heat transfer from the storage unit, generates electricity at the rate 0.5 MW, and discharges energy by heat transfer to the surroundings at 20°C. For operation at steady state, determine the minimum theoretical collector area required, in m$^2$.

### Refrigeration and Heat Pump Cycle Applications

**6.26** An inventor claims to have developed a refrigeration cycle that requires a net power input of 0.7 horsepower to remove 12,000 Btu/h of energy by heat transfer from a reservoir at 0°F and discharge energy by heat transfer to a reservoir at 70°F. There are no other energy transfers with the surroundings and operation is at steady state. Evaluate this claim.

**6.27** Determine if a tray of ice cubes could remain frozen when placed in a food freezer having a coefficient of performance of 9 operating in a room where the temperature is 32°C (90°F).

**6.28** The refrigerator shown in Fig. P6.28 operates at steady state with a coefficient of performance of 4.5 and a power input of 0.8 kW. Energy is rejected from the refrigerator to the surroundings at 20°C by heat transfer from metal coils attached to the back. Determine

(a) the rate energy is rejected, in kW.

(b) the lowest theoretical temperature *inside* the refrigerator, in K.

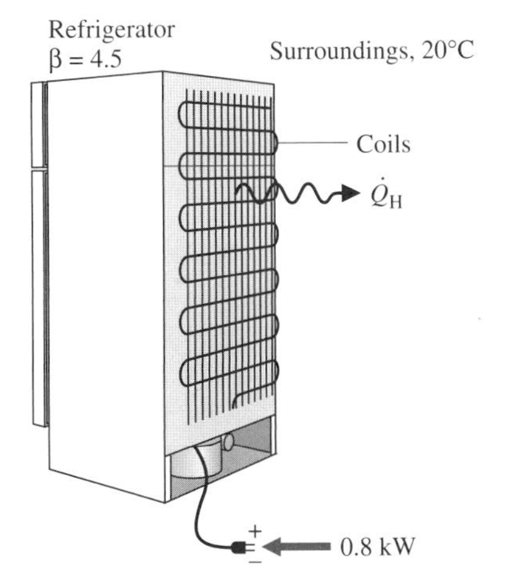


Figure P6.28

**6.29** Determine the minimum theoretical power, in Btu/s, required at steady state by a refrigeration system to maintain a cryogenic sample at −195°F in a laboratory at 70°F, if energy *leaks* by heat transfer to the sample from its surroundings at a rate of 0.085 Btu/s.

**6.30** (CD-ROM)

**6.31** At steady state, a refrigeration cycle driven by a 1-horsepower motor removes 200 Btu/min of energy by heat transfer from a space maintained at 20°F and discharges energy by heat transfer to surroundings at 75°F. Determine

(a) the coefficient of performance of the refrigerator and the rate at which energy is discharged to the surroundings, in Btu/min.

(b) the minimum theoretical net power input, in horsepower, for any refrigeration cycle operating between reservoirs at these two temperatures.

**6.32** At steady state, a refrigeration cycle removes 150 kJ/min of energy by heat transfer from a space maintained at −50°C and discharges energy by heat transfer to surroundings at 15°C. If the coefficient of performance of the cycle is 30 percent of that of a reversible refrigeration cycle operating between thermal reservoirs at these two temperatures, determine the power input to the cycle, in kW.

**6.33** A refrigeration cycle having a coefficient of performance of 3 maintains a computer laboratory at 18°C on a day when the outside temperature is 30°C. The *thermal load* at steady state consists of energy entering through the walls and windows at a rate of 30,000 kJ/h and from the occupants, computers, and lighting at a rate of 6000 kJ/h. Determine the power required by this cycle and compare with the *minimum* theoretical power required for any refrigeration cycle operating under these conditions, each in kW.

**6.34** If heat transfer through the walls and roof of a dwelling is $6.5 \times 10^5$ Btu per day, determine the *minimum* theoretical power, in hp, to drive a heat pump operating at steady state between the dwelling at 70°F and

(a) the outdoor air at 32°F.

(b) a pond at 40°F.

(c) the ground at 55°F.

**6.35** A heat pump operating at steady state is driven by a 1-kW electric motor and provides heating for a building whose interior is to be kept at 20°C. On a day when the outside temperature is 0°C and energy is lost through the walls and roof at a rate of 60,000 kJ/h, would the heat pump suffice?

**6.36** A heat pump operating at steady state maintains a dwelling at 70°F when the outside temperature is 40°F. The heat transfer rate through the walls and roof is 1300 Btu/h per degree temperature difference between the inside and outside. Determine the *minimum* theoretical power required to drive the heat pump, in horsepower.

**6.37** A building for which the heat transfer rate through the walls and roof is 1400 Btu/h per degree temperature difference between the inside and outside is to be maintained at 68°F. For a day when the outside temperature is 38°F, determine the power required at steady state, in hp, to heat the building using electrical resistance elements and compare with the *minimum* theoretical power that would be required by a heat pump.

**6.38** (CD-ROM)

**6.39** At steady state, a refrigerator whose coefficient of performance is 3 removes energy by heat transfer from a freezer compartment at 0°C at the rate of 6000 kJ/h and discharges energy by heat transfer to the surroundings, which are at 20°C.

**(a)** Determine the power input to the refrigerator and compare with the power input required by a reversible refrigeration cycle operating between reservoirs at these two temperatures.

**(b)** If electricity costs 8 cents per kW · h, determine the actual and minimum theoretical operating costs, each in $/day.

**6.40** At steady state, a heat pump provides 30,000 Btu/h to maintain a dwelling at 68°F on a day when the outside temperature is 35°F. The power input to the heat pump is 5 hp. If electricity costs 8 cents per kW · h, compare the actual operating cost with the minimum theoretical operating cost for each day of operation.

**6.41** By supplying energy to a dwelling at a rate of 8 kW, a heat pump maintains the temperature of the dwelling at 21°C when the outside air is at 0°C. If electricity costs 8 cents per kW · h, determine the minimum theoretical operating cost for each day of operation at steady state.

**6.42** At steady state, a refrigeration cycle maintains a food freezer at 0°F by removing energy by heat transfer from the inside at a rate of 2000 Btu/h. The cycle discharges energy by heat transfer to the surroundings at 72°F. If electricity costs 8 cents per kW · h, determine the minimum theoretical operating cost for each day of operation.

**6.43** By supplying energy at an average rate of 21,100 kJ/h, a heat pump maintains the temperature of a dwelling at 21°C. If electricity costs 8 cents per kW · h, determine the minimum theoretical operating cost for each day of operation if the heat pump receives energy by heat transfer from

**(a)** the outdoor air at −5°C.

**(b)** well water at 8°C.

**6.44** A heat pump with a coefficient of performance of 3.8 provides energy at an average rate of 75,000 kJ/h to maintain a building at 21°C on a day when the outside temperature is 0°C. If electricity costs 8 cents per kW · h

**(a)** Determine the actual operating cost and the minimum theoretical operating cost, each in $/day.

**(b)** Compare the results of part (a) with the cost of electrical-resistance heating.

**6.45** (CD-ROM)

**6.46** (CD-ROM)

# USING ENTROPY

*Introduction...*

Up to this point, our study of the second law has been concerned primarily with what it says about systems undergoing thermodynamic cycles. In this chapter means are introduced for analyzing systems from the second law perspective as they undergo processes that are not necessarily cycles. The property *entropy* plays a prominent part in these considerations. The ***objective*** of the present chapter is to introduce entropy and show its use for thermodynamic analysis.

*chapter objective*

The word *energy* is so much a part of the language that you were undoubtedly familiar with the term before encountering it in early science courses. This familiarity probably facilitated the study of energy in these courses and in the current course. In the present chapter you will see that the analysis of systems from a second law perspective is conveniently accomplished in terms of the property *entropy*. Energy and entropy are both abstract concepts. However, unlike energy, the word entropy is seldom heard in everyday conversation, and you may never have dealt with it quantitatively before. Energy and entropy play important roles in thermal systems engineering.

## 7.1 Introducing Entropy

Corollaries of the second law are developed in Chap. 6 for systems undergoing cycles while communicating thermally with *two* reservoirs, a hot reservoir and a cold reservoir. In the present section a corollary of the second law known as the Clausius inequality is introduced that is applicable to *any* cycle without regard for the body, or bodies, from which the cycle receives energy by heat transfer or to which the cycle rejects energy by heat transfer. The Clausius inequality provides the basis for introducing the property entropy and means for evaluating entropy change.

### 7.1.1 Clausius Inequality

The ***Clausius inequality*** states that for any thermodynamic cycle

$$\oint \left( \frac{\delta Q}{T} \right)_{\text{b}} \leq 0 \tag{7.1}$$

*Clausius inequality*

where $\delta Q$ represents the heat transfer at a part of the system boundary during a portion of the cycle, and $T$ is the absolute temperature at that part of the boundary. The subscript "b" serves as a reminder that the integrand is evaluated at the boundary of the system executing the cycle. The symbol $\oint$ indicates that the integral is to be performed over all parts of the boundary and over the entire cycle. The equality and inequality have the same interpretation as in the Kelvin–Planck statement: the equality applies when there are no internal irreversibilities as the system executes the cycle, and the inequality applies when internal irreversibilities are present. The Clausius inequality can be demonstrated using the Kelvin–Planck statement of the second law (CD-ROM).

Equation 7.1 can be expressed equivalently as

$$\oint \left(\frac{\delta Q}{T}\right)_{\mathrm{b}} = -\sigma_{\mathrm{cycle}} \tag{7.2}$$

where $\sigma_{\mathrm{cycle}}$ can be viewed as representing the "strength" of the inequality. The value of $\sigma_{\mathrm{cycle}}$ is positive when internal irreversibilities are present, zero when no internal irreversibilities are present, and can never be negative. In summary, the nature of a cycle executed by a system is indicated by the value for $\sigma_{\mathrm{cycle}}$ as follows:

| | |
|---|---|
| $\sigma_{\mathrm{cycle}} = 0$ | no irreversibilities present within the system |
| $\sigma_{\mathrm{cycle}} > 0$ | irreversibilities present within the system |
| $\sigma_{\mathrm{cycle}} < 0$ | impossible |

Accordingly, $\sigma_{\mathrm{cycle}}$ is a measure of the effect of the irreversibilities present within the system executing the cycle. This point is developed further in Sec. 7.4, where $\sigma_{\mathrm{cycle}}$ is identified as the *entropy produced* (or *generated*) by internal irreversibilities during the cycle.

### 7.1.2 Defining Entropy Change

A quantity is a property if, and only if, its change in value between two states is independent of the process (Sec. 2.2). This aspect of the property concept is used in the present section together with Eq. 7.2 to introduce entropy.

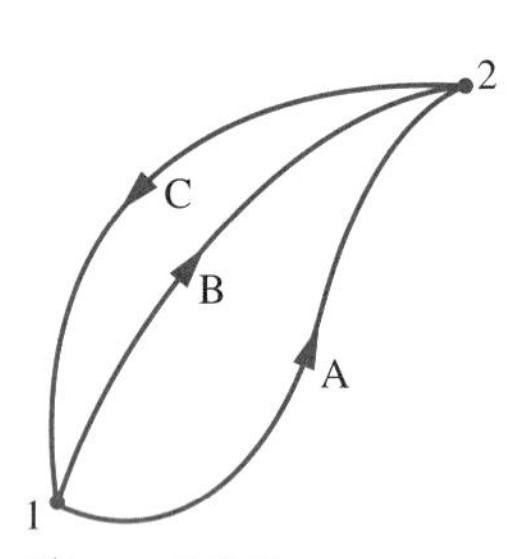

*Figure 7.2* Two internally reversible cycles.

Two cycles executed by a closed system are represented in Fig. 7.2. One cycle consists of an internally reversible process A from state 1 to state 2, followed by internally reversible process C from state 2 to state 1. The other cycle consists of an internally reversible process B from state 1 to state 2, followed by the same process C from state 2 to state 1 as in the first cycle. For the first cycle, Eq. 7.2 takes the form

$$\left(\int_1^2 \frac{\delta Q}{T}\right)_{\mathrm{A}} + \left(\int_2^1 \frac{\delta Q}{T}\right)_{\mathrm{C}} = -\overset{0}{\not\sigma}_{\mathrm{cycle}} \tag{7.3a}$$

and for the second cycle

$$\left(\int_1^2 \frac{\delta Q}{T}\right)_{\mathrm{B}} + \left(\int_2^1 \frac{\delta Q}{T}\right)_{\mathrm{C}} = -\overset{0}{\not\sigma}_{\mathrm{cycle}} \tag{7.3b}$$

In writing Eqs. 7.3, the term $\sigma_{\mathrm{cycle}}$ has been set to zero since the cycles are composed of internally reversible processes.

When Eq. 7.3b is subtracted from Eq. 7.3a

$$\left(\int_1^2 \frac{\delta Q}{T}\right)_{\mathrm{A}} = \left(\int_1^2 \frac{\delta Q}{T}\right)_{\mathrm{B}}$$

This shows that the integral of $\delta Q/T$ is the same for both processes. Since A and B are arbitrary, it follows that the integral of $\delta Q/T$ has the same value for *any* internally reversible process between the two states. In other words, the value of the integral depends on the end states only. It can be concluded, therefore, that the integral represents the change in some property of the system.

Selecting the symbol $S$ to denote this property, which is called *entropy,* its change is given by

$$S_2 - S_1 = \left( \int_1^2 \frac{\delta Q}{T} \right)_{\substack{\text{int} \\ \text{rev}}} \tag{7.4a}$$

*definition of entropy change*

where the subscript "int rev" is added as a reminder that the integration is carried out for any internally reversible process linking the two states. Equation 7.4a is the ***definition of entropy change.*** On a differential basis, the defining equation for entropy change takes the form

$$dS = \left( \frac{\delta Q}{T} \right)_{\substack{\text{int} \\ \text{rev}}} \tag{7.4b}$$

Entropy is an extensive property.

The ***SI unit for entropy*** is J/K. However, in this book it is convenient to work in terms of kJ/K. Another commonly employed ***unit for entropy*** is Btu/°R. Units in SI for *specific* entropy are kJ/kg · K for $s$ and kJ/kmol · K for $\bar{s}$. Other units for *specific* entropy are Btu/lb · °R and Btu/lbmol · °R.

*units for entropy*

Since entropy is a property, the change in entropy of a system in going from one state to another is the same for *all* processes, both internally reversible and irreversible, between these two states. Thus, Eq. 7.4a allows the determination of the change in entropy, and once it has been evaluated, this is the magnitude of the entropy change for all processes of the system between the two states. The evaluation of entropy change is discussed further in the next section.

It should be clear that entropy is defined and evaluated in terms of a particular integral for which *no accompanying physical picture is given.* We encountered this previously with the property enthalpy. Enthalpy is introduced without physical motivation in Sec. 4.3.2. Then, in Chap. 5, enthalpy is shown to be useful for thermodynamic analysis. As for the case of enthalpy, to gain an appreciation for entropy you need to understand *how* it is used and *what* it is used for.

## 7.2 Retrieving Entropy Data

In Chap. 4, we introduced means for retrieving property data, including tables, graphs, equations, and software available with this book. The emphasis there is on evaluating the properties $p$, $v$, $T$, $u$, and $h$ required for application of the conservation of mass and energy principles. For application of the second law, entropy values are usually required. In this section, means for retrieving entropy data are considered.

### 7.2.1 General Considerations

The defining equation for entropy change, Eq. 7.4a, serves as the basis for evaluating entropy relative to a reference value at a reference state. Both the reference value and the reference state can be selected arbitrarily. The value of the entropy at any state $y$ relative to the value at the reference state $x$ is obtained in principle from

$$S_y = S_x + \left( \int_x^y \frac{\delta Q}{T} \right)_{\substack{\text{int} \\ \text{rev}}} \tag{7.5}$$

where $S_x$ is the reference value for entropy at the specified reference state. The use of entropy values determined relative to an arbitrary reference state is unambiguous as long as they are used in calculations involving entropy differences, for then the reference value cancels.

## Entropy Data for Water and Refrigerants

Tables of thermodynamic data are introduced in Sec. 4.3 for water, Refrigerant 134a, and other substances. Specific entropy is tabulated in the same way as considered there for the properties $v$, $u$, and $h$, and entropy values are retrieved similarly.

*Vapor Data.* In the superheat regions of the tables for water and Refrigerant 134a, specific entropy is tabulated along with $v$, $u$, and $h$ versus temperature and pressure.

*For Example...* consider two states of water. At state 1, the pressure is 3 MPa and the temperature is 500°C. At state 2, the pressure is $p_2 = 0.3$ MPa and the specific entropy is the same as at state 1, $s_2 = s_1$. The object is to determine the temperature at state 2. Using $T_1$ and $p_1$, we find the specific entropy at state 1 from Table T-4 as $s_1 = 7.2338$ kJ/kg · K. State 2 is fixed by the pressure, $p_2 = 0.3$ MPa, and the specific entropy, $s_2 = 7.2338$ kJ/kg · K. Returning to Table T-4 at 0.3 MPa and interpolating with $s_2$ between 160 and 200°C results in $T_2 = 183$°C. ▲

*Saturation Data.* For saturation states, the values of $s_f$ and $s_g$ are tabulated as a function of either saturation pressure or saturation temperature. The specific entropy of a two-phase liquid–vapor mixture is calculated using the quality

$$s = (1 - x)s_f + xs_g = s_f + x(s_g - s_f) \tag{7.6}$$

These relations are identical in form to those for $v$, $u$, and $h$ (Sec. 4.3).

*For Example...* let us determine the specific entropy of Refrigerant 134a at a state where the temperature is 0°C and the specific internal energy is 138.43 kJ/kg. Referring to Table T-6, we see that the given value for $u$ falls between $u_f$ and $u_g$ at 0°C, so the system is a two-phase liquid–vapor mixture. The quality of the mixture can be determined from the known specific internal energy

$$x = \frac{u - u_f}{u_g - u_f} = \frac{138.43 - 49.79}{227.06 - 49.79} = 0.5$$

Then with values from Table T-6

$$\begin{aligned} s &= (1 - x)s_f + xs_g \\ &= (0.5)(0.1970) + (0.5)(0.9190) = 0.5580 \text{ kJ/kg} \cdot \text{K} \end{aligned}$$ ▲

*Liquid Data.* Compressed liquid data are presented for water in Tables T-5. In these tables $s$, $v$, $u$, and $h$ are tabulated versus temperature and pressure as in the superheat tables, and the tables are used similarly. In the absence of compressed liquid data, the value of the specific entropy can be estimated in the same way as estimates for $v$ and $u$ are obtained for liquid states (Sec. 4.3.6), by using the saturated liquid value at the given temperature

$$s(T, p) \approx s_f(T) \tag{7.7}$$

*For Example...* suppose the value of specific entropy is required for water at 25 bar, 200°C. The specific entropy is obtained directly from Table T-5 as $s = 2.3294$ kJ/kg · K. Using the saturated liquid value for specific entropy at 200°C from Table T-2, the specific entropy is approximated with Eq. 7.7 as $s = 2.3309$ kJ/kg · K, which agrees closely with the previous value. ▲

The specific entropy values for water and the refrigerants given in the tables accompanying this book are relative to the following *reference states and values*. For water, the entropy of saturated liquid at 0.01°C (32.02°F) is set to zero. For the refrigerants, the entropy of the saturated liquid at −40°C (−40°F) is assigned a value of zero.

***Computer Retrieval of Entropy Data.*** (CD-ROM)

## Using Graphical Entropy Data

The use of property diagrams as an adjunct to problem solving is emphasized throughout this book. When applying the second law, it is frequently helpful to locate states and plot processes on diagrams having entropy as a coordinate. Two commonly used figures having entropy as one of the coordinates are the temperature–entropy diagram and the enthalpy–entropy diagram.

***Temperature–Entropy Diagram.*** The main features of a temperature–entropy diagram are shown in Fig. 7.3. Observe that lines of constant enthalpy are shown on these figures. Also note that in the superheated vapor region constant specific volume lines have a steeper slope than constant-pressure lines. Lines of constant quality are shown in the two-phase liquid–vapor region. On some figures, lines of constant quality are marked as *percent moisture* lines. The percent moisture is defined as the ratio of the mass of liquid to the total mass.

In the superheated vapor region of the $T$–$s$ diagram, constant specific enthalpy lines become nearly horizontal as pressure is reduced. These states are shown as the shaded area on Fig. 7.3. For states in this region of the diagram, the enthalpy is determined primarily by the temperature: $h(T, p) \approx h(T)$. This is the region of the diagram where the ideal gas model provides a reasonable approximation. For superheated vapor states outside the shaded area, both temperature and pressure are required to evaluate enthalpy, and the ideal gas model is not suitable.

***Enthalpy–Entropy Diagram.*** The essential features of an enthalpy–entropy diagram, commonly known as a ***Mollier diagram,*** are shown in Fig. 7.4. Note the location of the critical point and the appearance of lines of constant temperature and constant pressure. Lines of

*Mollier diagram*

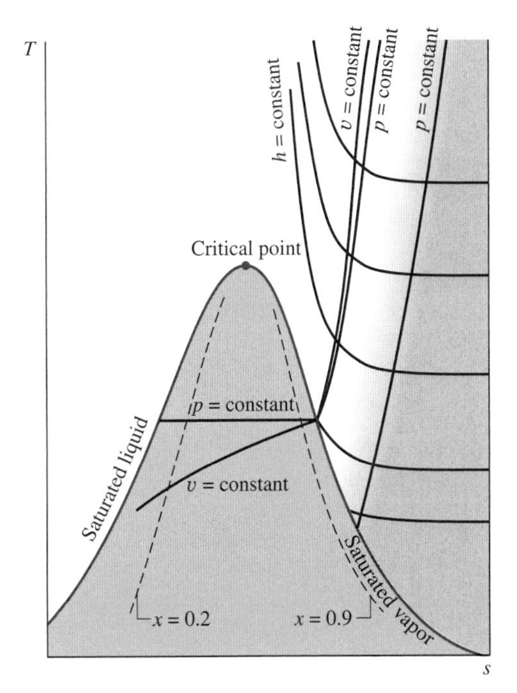


*Figure 7.3* Temperature–entropy diagram.

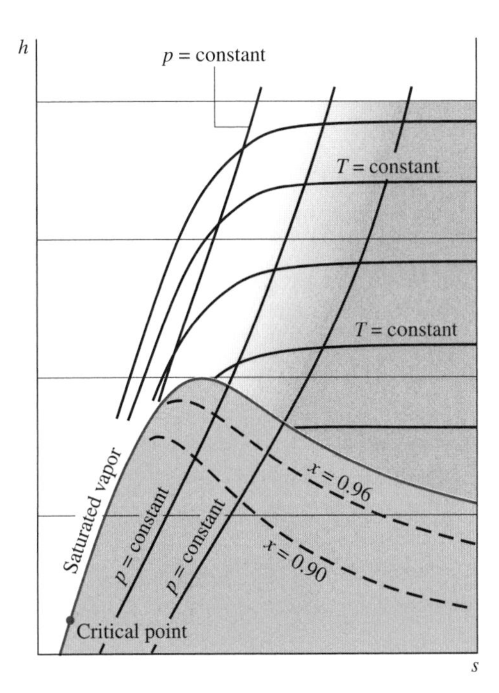


*Figure 7.4* Enthalpy–entropy diagram.

constant quality are shown in the two-phase liquid–vapor region (some figures give lines of constant percent moisture). The figure is intended for evaluating properties at superheated vapor states and for two-phase liquid–vapor mixtures. Liquid data are seldom shown. In the superheated vapor region, constant-temperature lines become nearly horizontal as pressure is reduced. These states are shown, approximately, as the shaded area on Fig. 7.4. This area corresponds to the shaded area on the temperature–entropy diagram of Fig. 7.3, where the ideal gas model provides a reasonable approximation.

## Using the *T dS* Equations

Although the change in entropy between two states can be determined in principle by using Eq. 7.4a, such evaluations are generally conducted using the *T dS* equations introduced in this section. The *T dS* equations allow entropy changes to be evaluated from other more readily determined property data. The use of the *T dS* equations to evaluate entropy changes for ideal gases is illustrated in Sec. 7.2.2 and for incompressible substances in Sec. 7.2.3.

The ***T dS* equations** can be written on a unit mass basis as

*T dS equations*

$$T\,ds = du + p\,dv \tag{7.8a}$$

$$T\,ds = dh - v\,dp \tag{7.8b}$$

or on a per mole basis as

$$T\,d\bar{s} = d\bar{u} + p\,d\bar{v} \tag{7.8c}$$

$$T\,d\bar{s} = d\bar{h} - \bar{v}\,dp \tag{7.8d}$$

To show the use of the *T dS* equations, consider a change in phase from saturated liquid to saturated vapor at constant temperature and pressure. Since pressure is constant, Eq. 7.8b reduces to give

$$ds = \frac{dh}{T}$$

Then, because temperature is also constant during the phase change

$$s_g - s_f = \frac{h_g - h_f}{T} \tag{7.9}$$

This relationship shows how $s_g - s_f$ is calculated for tabulation in property tables.

*For Example...* consider Refrigerant 134a at 0°C. From Table T-6, $h_g - h_f = 197.21$ kJ/kg, so with Eq. 7.9

$$s_g - s_f = \frac{197.21 \text{ kJ/kg}}{273.15 \text{ K}} = 0.7220 \frac{\text{kJ}}{\text{kg} \cdot \text{K}}$$

which is the value calculated using $s_f$ and $s_g$ from the table. To give another example, consider Refrigerant 134a at 0°F. From Table T-6E, $h_g - h_f = 90.12$ Btu/lb, so

$$s_g - s_f = \frac{90.12 \text{ Btu/lb}}{459.67 \text{°R}} = 0.1961 \frac{\text{Btu}}{\text{lb} \cdot \text{°R}}$$

which agrees with the value calculated using $s_f$ and $s_g$ from the table. ▲

***Developing the T dS Equations.*** (CD-ROM)

### 7.2.2 Entropy Change of an Ideal Gas

For an ideal gas, $du = c_v(T)dT$, $dh = c_p(T)dT$, and $pv = RT$. With these relations, the $T\,dS$ equations (Eqs. 7.8a and 7.8b) give, respectively,

$$ds = c_v(T)\frac{dT}{T} + R\frac{dv}{v} \quad \text{and} \quad ds = c_p(T)\frac{dT}{T} - R\frac{dp}{p}$$

On integration, we get the following expressions for entropy change of an ideal gas:

$$s(T_2, v_2) - s(T_1, v_1) = \int_{T_1}^{T_2} c_v(T)\frac{dT}{T} + R\ln\frac{v_2}{v_1} \tag{7.12}$$

$$s(T_2, p_2) - s(T_1, p_1) = \int_{T_1}^{T_2} c_p(T)\frac{dT}{T} - R\ln\frac{p_2}{p_1} \tag{7.13}$$

***Using Ideal Gas Tables.*** As for internal energy and enthalpy changes, the evaluation of entropy changes for ideal gases can be reduced to a convenient tabular approach. To introduce this, we begin by selecting a reference state and reference value: The value of the specific entropy is set to zero at the state where the temperature is 0 K and the pressure is 1 atmosphere. Then, using Eq. 7.13, the specific entropy at a state where the temperature is $T$ and the pressure is 1 atm is determined relative to this reference state and reference value as

$$s^\circ(T) = \int_0^T \frac{c_p(T)}{T}\,dT \tag{7.14}$$

The symbol $s^\circ(T)$ denotes the specific entropy at temperature $T$ and *a pressure of 1 atm.* Because $s^\circ$ depends only on temperature, it can be tabulated versus temperature, like $h$ and $u$. For air as an ideal gas, $s^\circ$ with units of kJ/kg · K or Btu/lb · °R is given in Tables T-9. Values of $\bar{s}^\circ$ for several other common gases are given in Tables T-11 with units of kJ/kmol · K or Btu/lbmol · °R. Since the integral of Eq. 7.13 can be expressed in terms of $s^\circ$

$$\int_{T_1}^{T_2} c_p\frac{dT}{T} = \int_0^{T_2} c_p\frac{dT}{T} - \int_0^{T_1} c_p\frac{dT}{T}$$
$$= s^\circ(T_2) - s^\circ(T_1)$$

it follows that Eq. 7.13 can be written as

$$s(T_2, p_2) - s(T_1, p_1) = s^\circ(T_2) - s^\circ(T_1) - R\ln\frac{p_2}{p_1} \tag{7.15a}$$

or on a per mole basis as

$$\bar{s}(T_2, p_2) - \bar{s}(T_1, p_1) = \bar{s}^\circ(T_2) - \bar{s}^\circ(T_1) - \bar{R}\ln\frac{p_2}{p_1} \tag{7.15b}$$

Using Eqs. 7.15 and the tabulated values for $s^\circ$ or $\bar{s}^\circ$, as appropriate, entropy changes can be determined that account explicitly for the variation of specific heat with temperature.

*For Example...* let us evaluate the change in specific entropy, in kJ/kg · K, of air modeled as an ideal gas from a state where $T_1 = 300$ K and $p_1 = 1$ bar to a state where $T_2 = 1000$ K and $p_2 = 3$ bar. Using Eq. 7.15a and data from Table T-9

$$\begin{aligned} s_2 - s_1 &= s°(T_2) - s°(T_1) - R \ln \frac{p_2}{p_1} \\ &= (2.96770 - 1.70203) \frac{\text{kJ}}{\text{kg} \cdot \text{K}} - \frac{8.314}{28.97} \frac{\text{kJ}}{\text{kg} \cdot \text{K}} \ln \frac{3 \text{ bar}}{1 \text{ bar}} \\ &= 0.9504 \text{ kJ/kg} \cdot \text{K} \quad \blacktriangle \end{aligned}$$

*Assuming Constant Specific Heats.* When the specific heats $c_v$ and $c_p$ are taken as constants, Eqs. 7.12 and 7.13 reduce, respectively, to

$$s(T_2, v_2) - s(T_1, v_1) = c_v \ln \frac{T_2}{T_1} + R \ln \frac{v_2}{v_1} \tag{7.16}$$

$$s(T_2, p_2) - s(T_1, p_1) = c_p \ln \frac{T_2}{T_1} - R \ln \frac{p_2}{p_1} \tag{7.17}$$

These equations, along with Eqs. 4.48 and 4.49 giving $\Delta u$ and $\Delta h$, respectively, are applicable when assuming the ideal gas model with constant specific heats.

*For Example...* let us determine the change in specific entropy, in kJ/kg · K, of air as an ideal gas undergoing a process from $T_1 = 300$ K, $p_1 = 1$ bar to $T_2 = 400$ K, $p_2 = 5$ bar. Because of the relatively small temperature range, we assume a constant value of $c_p$ evaluated at 350 K. Using Eq. 7.17 and $c_p = 1.008$ kJ/kg · K from Table T-10

$$\begin{aligned} \Delta s &= c_p \ln \frac{T_2}{T_1} - R \ln \frac{p_2}{p_1} \\ &= \left(1.008 \frac{\text{kJ}}{\text{kg} \cdot \text{K}}\right) \ln\left(\frac{400 \text{ K}}{300 \text{ K}}\right) - \left(\frac{8.314}{28.97} \frac{\text{kJ}}{\text{kg} \cdot \text{K}}\right) \ln\left(\frac{5 \text{ bar}}{1 \text{ bar}}\right) \\ &= -0.1719 \text{ kJ/kg} \cdot \text{K} \quad \blacktriangle \end{aligned}$$

*Using Computer Software to Evaluate Ideal Gas Entropy.* (CD-ROM)

### 7.2.3 Entropy Change of an Incompressible Substance

The incompressible substance model introduced in Sec. 4.3.6 assumes that the specific volume (density) is constant and the specific heat depends solely on temperature, $c_v = c(T)$. Accordingly, the differential change in specific internal energy is $du = c(T)\, dT$ and Eq. 7.8a reduces to

$$ds = \frac{c(T)\, dT}{T} + \frac{p \cancelto{0}{dv}}{T} = \frac{c(T)\, dT}{T}$$

On integration, the change in specific entropy is

$$s_2 - s_1 = \int_{T_1}^{T_2} \frac{c(T)}{T}\, dT \qquad \text{(incompressible)}$$

When the specific heat is assumed constant, this becomes

$$s_2 - s_1 = c \ln \frac{T_2}{T_1} \qquad \text{(incompressible, constant } c\text{)} \tag{7.18}$$

Equation 7.18, along with Eqs. 4.20 and 4.21 giving $\Delta u$ and $\Delta h$, respectively, are applicable to liquids and solids modeled as incompressible. Specific heats of some common liquids and solids are given in Tables HT-1, 2, 4, and 5.

## 7.3 Entropy Change in Internally Reversible Processes

In this section the relationship between entropy change and heat transfer for internally reversible processes is considered. The concepts introduced have important applications in subsequent sections of the book. The present discussion is limited to the case of closed systems. Similar considerations for control volumes are presented in Sec. 7.8.

As a closed system undergoes an internally reversible process, its entropy can increase, decrease, or remain constant. This can be brought out using Eq. 7.4b

$$dS = \left(\frac{\delta Q}{T}\right)_{\substack{\text{int}\\ \text{rev}}}$$

which indicates that when a closed system undergoing an internally reversible process receives energy by heat transfer, the system experiences an increase in entropy. Conversely, when energy is removed from the system by heat transfer, the entropy of the system decreases. This can be interpreted to mean that an entropy transfer *accompanies* heat transfer. The direction of the entropy transfer is the same as that of the heat transfer. In an *adiabatic* internally reversible process, the entropy would remain constant. A constant-entropy process is called an ***isentropic process.***

***isentropic process***

On rearrangement, the above expression gives

$$(\delta Q)_{\substack{\text{int}\\ \text{rev}}} = T\,dS$$

Integrating from an initial state 1 to a final state 2

$$Q_{\substack{\text{int}\\ \text{rev}}} = \int_1^2 T\,dS \qquad (7.19)$$

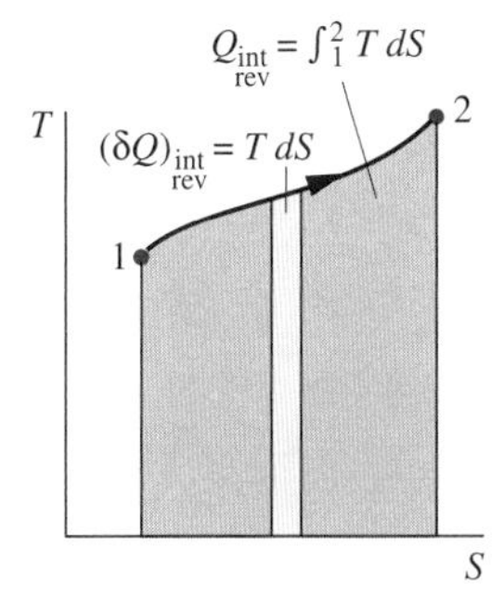


*Figure 7.5* Area representation of heat transfer for an internally reversible process of a closed system.

From Eq. 7.19 it can be concluded that an energy transfer by heat to a closed system during an internally reversible process can be represented as an area on a temperature–entropy diagram. Figure 7.5 illustrates the area interpretation of heat transfer for an arbitrary internally reversible process in which temperature varies. Carefully note that temperature must be in kelvins or degrees Rankine, and the area is the entire area under the curve (shown shaded). Also note that the area interpretation of heat transfer is not valid for irreversible processes, as discussed in Example 7.2.

To illustrate concepts introduced in this section, the next example considers water undergoing an internally reversible process while contained in a piston–cylinder assembly.

### *Example 7.1* Internally Reversible Process of Water

Water, initially a saturated liquid at 100°C, is contained in a piston–cylinder assembly. The water undergoes a process to the corresponding saturated vapor state, during which the piston moves freely in the cylinder. If the change of state is brought about by heating the water as it undergoes an internally reversible process at constant pressure and temperature, determine the work and heat transfer per unit of mass, each in kJ/kg.

## Solution

***Known:*** Water contained in a piston–cylinder assembly undergoes an internally reversible process at 100°C from saturated liquid to saturated vapor.

***Find:*** Determine the work and heat transfer per unit mass.

***Schematic and Given Data:***

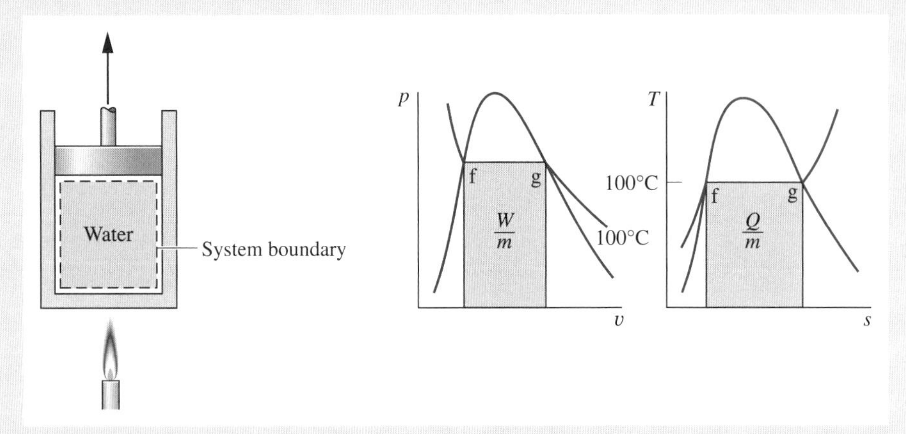


*Figure E7.1*

***Assumptions:***

1. The water in the piston–cylinder assembly is a closed system.
2. The process is internally reversible.
3. Temperature and pressure are constant during the process.
4. There is no change in kinetic or potential energy between the two end states.

***Analysis:*** At constant pressure the work is

$$\frac{W}{m} = \int_f^g p\,dv = p(v_g - v_f)$$

With values from Table T-2

$$\frac{W}{m} = (1.014 \text{ bar})(1.673 - 1.0435 \times 10^{-3})\left(\frac{\text{m}^3}{\text{kg}}\right)\left|\frac{10^5 \text{ N/m}^2}{1 \text{ bar}}\right|\left|\frac{1 \text{ kJ}}{10^3 \text{ N}\cdot\text{m}}\right|$$

$$= 170 \text{ kJ/kg} \triangleleft$$

Since the process is internally reversible and at constant temperature, Eq. 7.19 gives

$$Q = \int_f^g T\,dS = m\int_f^g T\,ds$$

or

$$\frac{Q}{m} = T(s_g - s_f)$$

With values from Table T-2

❶ $$\frac{Q}{m} = (373.15 \text{ K})(7.3549 - 1.3069) \text{ kJ/kg}\cdot\text{K} = 2257 \text{ kJ/kg} \triangleleft$$

As shown in the accompanying figure, the work and heat transfer can be represented as areas on $p$–$v$ and $T$–$s$ diagrams, respectively.

---

❶ The heat transfer can be evaluated alternatively from an energy balance written on a unit mass basis as

$$u_g - u_f = \frac{Q}{m} - \frac{W}{m}$$

Introducing $W/m = p(v_g - v_f)$ and solving

$$\frac{Q}{m} = (u_g - u_f) + p(v_g - v_f)$$
$$= (u_g + pv_g) - (u_f + pv_f)$$
$$= h_g - h_f$$

From Table T-2 at 100°C, $h_g - h_f = 2257$ kJ/kg, which gives the same value for $Q/m$ as obtained in the solution above.

## 7.4 Entropy Balance for Closed Systems

In this section, the Clausius inequality expressed by Eq. 7.2 and the defining equation for entropy change are used to develop the *entropy balance* for closed systems. The entropy balance is an expression of the second law that is particularly convenient for thermodynamic analysis. The current presentation is limited to closed systems. The entropy balance is extended to control volumes in Sec. 7.5.

### 7.4.1 Developing the Entropy Balance

Shown in Fig. 7.6 is a cycle executed by a closed system. The cycle consists of process I, during which internal irreversibilities are present, followed by internally reversible process R. For this cycle, Eq. 7.2 takes the form

$$\int_1^2 \left(\frac{\delta Q}{T}\right)_b + \int_2^1 \left(\frac{\delta Q}{T}\right)_{\substack{\text{int}\\\text{rev}}} = -\sigma \tag{7.20}$$

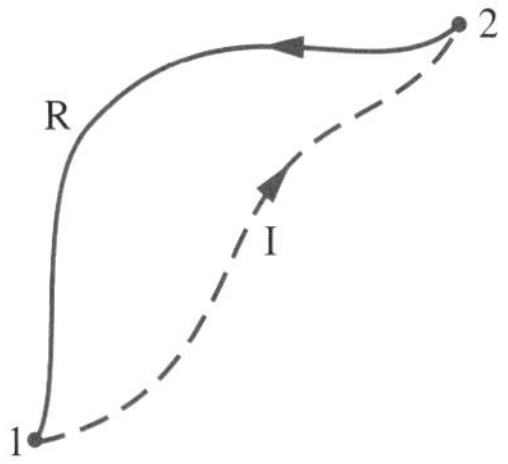

*Figure 7.6* Cycle used to develop the entropy balance.

where the first integral is for process I and the second is for process R. The subscript b in the first integral serves as a reminder that the integrand is evaluated at the system boundary. The subscript is not required in the second integral because temperature is uniform throughout the system at each intermediate state of an internally reversible process. Since no irreversibilities are associated with process R, the term $\sigma_{cycle}$ of Eq. 7.2, which accounts for the effect of irreversibilities during the cycle, refers only to process I and is shown in Eq. 7.20 simply as $\sigma$.

Applying the definition of entropy change, we can express the second integral of Eq. 7.20 as

$$S_1 - S_2 = \int_2^1 \left(\frac{\delta Q}{T}\right)_{\substack{\text{int}\\\text{rev}}}$$

With this, Eq. 7.20 becomes

$$\int_1^2 \left(\frac{\delta Q}{T}\right)_b + (S_1 - S_2) = -\sigma$$

Finally, on rearranging the last equation, the ***closed system entropy balance*** results

*closed system entropy balance*

$$\underbrace{S_2 - S_1}_{\substack{\text{entropy}\\\text{change}}} = \underbrace{\int_1^2 \left(\frac{\delta Q}{T}\right)_b}_{\substack{\text{entropy}\\\text{transfer}}} + \underbrace{\sigma}_{\substack{\text{entropy}\\\text{production}}} \tag{7.21}$$

If the end states are fixed, the entropy change on the left side of Eq. 7.21 can be evaluated independently of the details of the process. However, the two terms on the right side depend explicitly on the nature of the process and cannot be determined solely from knowledge of the end states. The first term on the right side of Eq. 7.21 is associated with heat transfer to or from the system during the process. This term can be interpreted as the ***entropy transfer accompanying heat transfer.*** The direction of entropy transfer is the same as the direction of the heat transfer, and the same sign convention applies as for heat transfer: A positive value means that entropy is transferred into the system, and a negative value means that entropy is transferred out. When there is no heat transfer, there is no entropy transfer.

*entropy transfer accompanying heat transfer*

The entropy change of a system is not accounted for solely by the entropy transfer, but is due in part to the second term on the right side of Eq. 7.21 denoted by $\sigma$. The term $\sigma$ is positive when internal irreversibilities are present during the process and vanishes when no internal irreversibilities are present. This can be described by saying that ***entropy is produced*** within the system by the action of irreversibilities. The second law of thermodynamics can be interpreted as requiring that entropy is produced by irreversibilities and conserved only in the limit as irreversibilities are reduced to zero. Since $\sigma$ measures the effect of irreversibilities present within the system during a process, its value depends on the nature of the process and not solely on the end states. It is *not* a property.

*entropy production*

When applying the entropy balance to a closed system, it is essential to remember the requirements imposed by the second law on entropy production: The second law requires that entropy *production* be positive, or zero, in value

$$\sigma: \begin{cases} > 0 & \text{irreversibilities present within the system} \\ = 0 & \text{no irreversibilities present within the system} \end{cases} \tag{7.22}$$

The value of the entropy production cannot be negative. By contrast, the *change* in entropy of the system may be positive, negative, or zero:

$$S_2 - S_1: \begin{cases} > 0 \\ = 0 \\ < 0 \end{cases} \tag{7.23}$$

Like other properties, entropy change can be determined without knowledge of the details of the process.

***For Example...*** to illustrate the entropy transfer and entropy production concepts, as well as the accounting nature of entropy balance, consider Fig. 7.7. The figure shows a system consisting of a gas or liquid in a rigid container stirred by a paddle wheel while receiving a heat transfer $Q$ from a reservoir. The temperature at the portion of the boundary where heat transfer occurs is the same as the constant temperature of the reservoir, $T_b$. By definition, the reservoir is free of irreversibilities; however, the system is not without irreversibilities, for fluid friction is evidently present, and there may be other irreversibilities within the system.

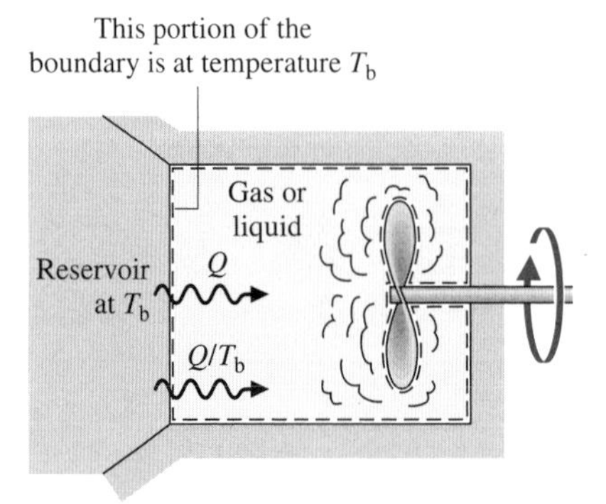


*Figure 7.7* Illustration of the entropy transfer and entropy production concepts.

Let us now apply the entropy balance to the system and to the reservoir. Since $T_b$ is constant, the integral in Eq. 7.21 is readily evaluated, and the entropy balance for the *system* reduces to

$$S_2 - S_1 = \frac{Q}{T_b} + \sigma \tag{7.24}$$

where $Q/T_b$ accounts for entropy transfer into the system accompanying heat transfer $Q$. The entropy balance for the *reservoir* takes the form

$$\Delta S]_{res} = \frac{Q_{res}}{T_b} + \cancelto{0}{\sigma_{res}}$$

where the entropy production term is set equal to zero because the reservoir is without irreversibilities. Since $Q_{res} = -Q$, the last equation becomes

$$\Delta S]_{res} = -\frac{Q}{T_b}$$

The minus sign signals that entropy is carried out of the reservoir accompanying heat transfer. Hence, the entropy of the reservoir decreases by an amount equal to the entropy transferred from it to the system. However, as shown by Eq. 7.24, the entropy change of the system *exceeds* the amount of entropy transferred to it because of entropy production within the system. ▲

If the heat transfer were oppositely directed in the above example, passing instead from the system to the reservoir, the magnitude of the entropy transfer would remain the same, but its direction would be reversed. In such a case, the entropy of the system would *decrease* if the amount of entropy transferred *from* the system to the reservoir *exceeded* the amount of entropy produced within the system due to irreversibilities. Finally, observe that there is no entropy transfer associated with work.

### 7.4.2 Other Forms of the Entropy Balance

The entropy balance can be expressed in various forms convenient for particular analyses. For example, if heat transfer takes place at several locations on the boundary of a system where the temperatures do not vary with position or time, the entropy transfer term can be expressed as a sum, so Eq. 7.21 takes the form

$$S_2 - S_1 = \sum_j \frac{Q_j}{T_j} + \sigma \tag{7.25}$$

where $Q_j/T_j$ is the amount of entropy transferred through the portion of the boundary at temperature $T_j$.

On a time rate basis, the ***closed system entropy rate balance*** is

$$\frac{dS}{dt} = \sum_j \frac{\dot{Q}_j}{T_j} + \dot{\sigma} \tag{7.26}$$

*closed system entropy rate balance*

where $dS/dt$ is the time rate of change of entropy of the system. The term $\dot{Q}_j/T_j$ represents the time rate of entropy transfer through the portion of the boundary whose instantaneous temperature is $T_j$. The term $\dot{\sigma}$ accounts for the time rate of entropy production due to irreversibilities within the system.

### 7.4.3 Evaluating Entropy Production and Transfer

Regardless of the form taken by the entropy balance, the objective in many applications is to evaluate the entropy production term. However, the value of the entropy production for a given process of a system often does not have much significance by itself. The significance is normally determined through comparison. For example, the entropy production within a given component might be compared to the entropy production values of the other components included in an overall system formed by these components. By comparing entropy production values, the components where appreciable irreversibilities occur can be identified and rank ordered. This allows attention to be focused on the components that contribute most to inefficient operation of the overall system.

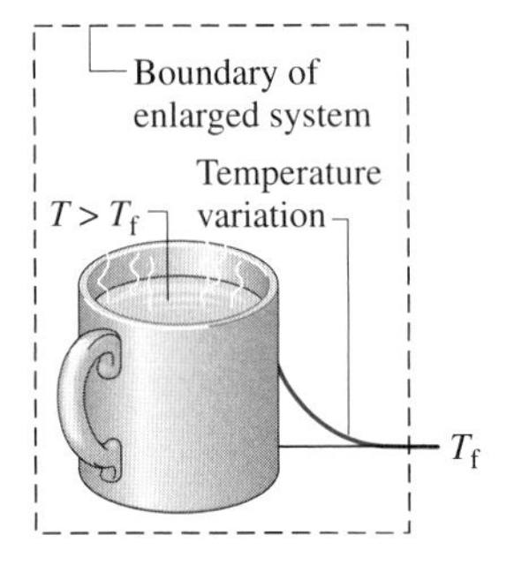


To evaluate the entropy transfer term of the entropy balance requires information regarding both the heat transfer and the temperature on the boundary where the heat transfer occurs. The entropy transfer term is not always subject to direct evaluation, however, because the required information is either unknown or not defined, such as when the system passes through states sufficiently far from equilibrium. In such applications, it may be convenient, therefore, to enlarge the system to include enough of the immediate surroundings that the temperature on the boundary of the *enlarged system* corresponds to the temperature of the surroundings away from the immediate vicinity of the system, $T_f$. The entropy transfer term is then simply $Q/T_f$. However, as the irreversibilities present would not be just for the system of interest but for the enlarged system, the entropy production term would account for the effects of internal irreversibilities within the original system and external irreversibilities present within that portion of the surroundings included within the enlarged system.

### 7.4.4 Illustrations

The following examples illustrate the use of the energy and entropy balances for the analysis of closed systems. Property relations and property diagrams also contribute significantly in developing solutions. The first example reconsiders the system and end states of Example 7.1 to demonstrate that entropy is produced when internal irreversibilities are present and that the amount of entropy production is not a property.

## *Example 7.2* Irreversible Process of Water

Water initially a saturated liquid at 100°C is contained within a piston–cylinder assembly. The water undergoes a process to the corresponding saturated vapor state, during which the piston moves freely in the cylinder. There is no heat transfer with the surroundings. If the change of state is brought about by the action of a paddle wheel, determine the net work per unit mass, in kJ/kg, and the amount of entropy produced per unit mass, in kJ/kg · K.

### Solution

***Known:*** Water contained in a piston–cylinder assembly undergoes an adiabatic process from saturated liquid to saturated vapor at 100°C. During the process, the piston moves freely, and the water is rapidly stirred by a paddle wheel.

***Find:*** Determine the net work per unit mass and the entropy produced per unit mass.

***Schematic and Given Data:***

❶

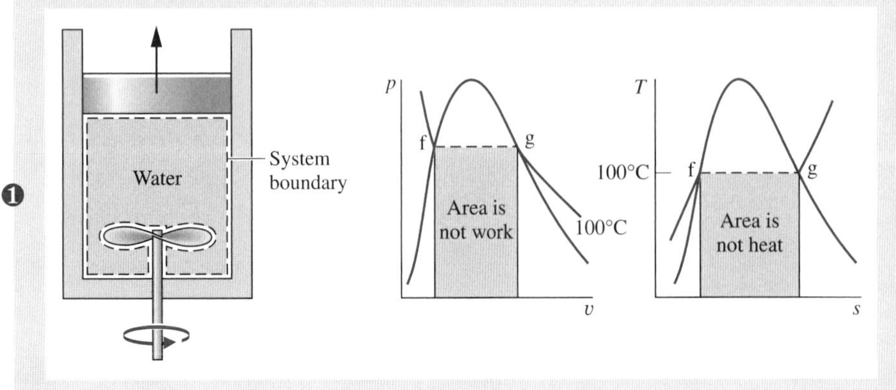


*Figure E7.2*

***Assumptions:***

1. The water in the piston–cylinder assembly is a closed system.
2. There is no heat transfer with the surroundings.
3. The system is at an equilibrium state initially and finally. There is no change in kinetic or potential energy between these two states.

***Analysis:*** As the volume of the system increases during the process, there is an energy transfer by work from the system during the expansion, as well as an energy transfer by work to the system via the paddle wheel. The *net* work can be evaluated from an energy balance, which reduces with assumptions 2 and 3 to

$$\Delta U + \cancelto{0}{\Delta \text{KE}} + \cancelto{0}{\Delta \text{PE}} = \cancelto{0}{Q} - W$$

On a unit mass basis, the energy balance reduces to

$$\frac{W}{m} = -(u_g - u_f)$$

With specific internal energy values from Table T-2 at 100°C

$$\frac{W}{m} = -2087.56 \, \frac{\text{kJ}}{\text{kg}} \quad \triangleleft$$

The minus sign indicates that the work input by stirring is greater in magnitude than the work done by the water as it expands.

The amount of entropy produced is evaluated by applying an entropy balance. Since there is no heat transfer, the term accounting for entropy transfer vanishes

$$\Delta S = \cancelto{0}{\int_1^2 \left(\frac{\delta Q}{T}\right)_b} + \sigma$$

On a unit mass basis, this becomes on rearrangement

$$\frac{\sigma}{m} = s_g - s_f$$

With specific entropy values from Table T-2 at 100°C

❷
$$\frac{\sigma}{m} = 6.048 \, \frac{\text{kJ}}{\text{kg} \cdot \text{K}} \quad \triangleleft$$

---

❶ Although each end state is an equilibrium state at the same pressure and temperature, the pressure and temperature are not necessarily uniform throughout the system at *intervening* states, nor are they necessarily constant in value during the process. Accordingly, there is no well-defined "path" for the process. This is emphasized by the use of dashed lines to represent the process on these $p-v$ and $T-s$ diagrams. The dashed lines indicate only that a process has taken place, and no "area" should be associated with them. In particular, note that the process is adiabatic, so the "area" below the dashed line on the $T-s$ diagram can have no significance as heat transfer. Similarly, the work cannot be associated with an area on the $p-v$ diagram.

❷ The change of state is the same in the present example as in Example 7.1. However, in Example 7.1 the change of state is brought about by heat transfer while the system undergoes an internally reversible process. Accordingly, the value of entropy production for the process of Example 7.1 is zero. Here, fluid friction is present during the process and the entropy production is positive in value. Accordingly, different values of entropy production are obtained for two processes between the *same* end states. This demonstrates that entropy production is not a property.

As an illustration of second law reasoning, the next example uses the fact that the entropy production term of the entropy balance cannot be negative.

## *Example 7.3* Evaluating Minimum Theoretical Compression Work

Refrigerant 134a is compressed adiabatically in a piston–cylinder assembly from saturated vapor at 10°F to a final pressure of 120 lbf/in.[2] Determine the minimum theoretical work input required per unit mass of refrigerant, in Btu/lb.

### Solution (CD-ROM)

To pinpoint the relative significance of the internal and external irreversibilities, the next example illustrates the application of the entropy rate balance to a system and to an enlarged system consisting of the system and a portion of its immediate surroundings.

## *Example 7.4* Pinpointing Irreversibilities

Referring to Example 3.4, evaluate the rate of entropy production $\dot{\sigma}$, in kW/K, for **(a)** the gearbox as the system and **(b)** an enlarged system consisting of the gearbox and enough of its surroundings that heat transfer occurs at the temperature of the surroundings away from the immediate vicinity of the gearbox, $T_f = 293$ K (20°C).

### Solution

***Known:*** A gearbox operates at steady state with known values for the power input through the high-speed shaft, power output through the low-speed shaft, and heat transfer rate. The temperature on the outer surface of the gearbox and the temperature of the surroundings away from the gearbox are also known.

***Find:*** Evaluate the entropy production rate $\dot{\sigma}$ for each of the two specified systems shown in the schematic.

***Schematic and Given Data:***

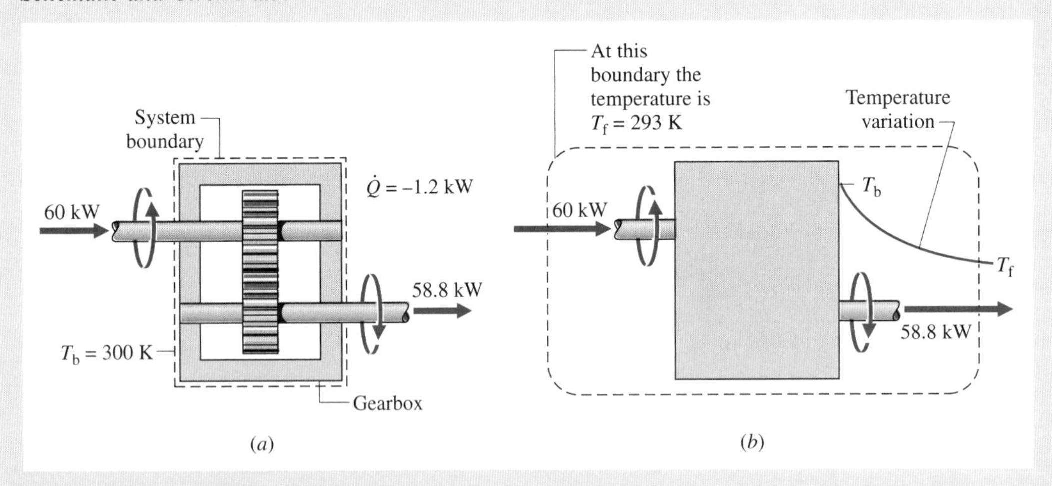


*Figure E7.4*

***Assumptions:***

**1.** In part (a), the gearbox is taken as a closed system operating at steady state, as shown on the accompanying sketch labeled with data from Example 3.4.

**2.** In part (b) the gearbox and a portion of its surroundings are taken as a closed system, as shown on the accompanying sketch labeled with data from Example 3.4.

**3.** The temperature of the outer surface of the gearbox and the temperature of the surroundings are each uniform.

***Analysis:*** **(a)** To obtain an expression for the entropy production rate, begin with the entropy balance for a closed system on a time rate basis: Eq. 7.26. Since heat transfer takes place only at temperature $T_b$, the entropy rate balance reduces at steady state to

$$\overset{0}{\cancel{\frac{dS}{dt}}} = \frac{\dot{Q}}{T_b} + \dot{\sigma}$$

Solving

$$\dot{\sigma} = -\frac{\dot{Q}}{T_b}$$

Introducing the known values for the heat transfer rate $\dot{Q}$ and the surface temperature $T_b$

$$\dot{\sigma} = -\frac{(-1.2 \text{ kW})}{(300 \text{ K})} = 4 \times 10^{-3} \text{ kW/K} \quad \triangleleft$$

(b) Since heat transfer takes place at temperature $T_f$ for the enlarged system, the entropy rate balance reduces at steady state to

$$\cancelto{0}{\frac{dS}{dt}} = \frac{\dot{Q}}{T_f} + \dot{\sigma}$$

Solving

$$\dot{\sigma} = -\frac{\dot{Q}}{T_f}$$

Introducing the known values for the heat transfer rate $\dot{Q}$ and the temperature $T_f$

❶
$$\dot{\sigma} = -\frac{(-1.2 \text{ kW})}{(293 \text{ K})} = 4.1 \times 10^{-3} \text{ kW/K} \quad \triangleleft$$

❶ The value of the entropy production rate calculated in part (a) gauges the significance of irreversibilities associated with friction and heat transfer *within* the gearbox. In part (b), an additional source of irreversibility is included in the enlarged system, namely the irreversibility associated with the heat transfer from the outer surface of the gearbox at $T_b$ to the surroundings at $T_f$. In this case, the irreversibilities within the gearbox are dominant, accounting for 97.6% of the total rate of entropy production.

## 7.5 Entropy Rate Balance for Control Volumes

Thus far the discussion of the entropy balance concept has been restricted to the case of closed systems. In the present section the entropy balance is extended to control volumes.

Like mass and energy, entropy is an extensive property, so it too can be transferred into or out of a control volume by streams of matter. Since this is the principal difference between the closed system and control volume forms, the ***control volume entropy rate balance*** can be obtained by modifying Eq. 7.26 to account for these entropy transfers. The result is

$$\underbrace{\frac{dS_{cv}}{dt}}_{\substack{\text{rate of} \\ \text{entropy} \\ \text{change}}} = \underbrace{\sum_j \frac{\dot{Q}_j}{T_j} + \sum_i \dot{m}_i s_i - \sum_e \dot{m}_e s_e}_{\substack{\text{rates of} \\ \text{entropy} \\ \text{transfer}}} + \underbrace{\dot{\sigma}_{cv}}_{\substack{\text{rate of} \\ \text{entropy} \\ \text{production}}} \tag{7.27}$$

*control volume entropy rate balance*

where $dS_{cv}/dt$ represents the time rate of change of entropy within the control volume. The terms $\dot{m}_i s_i$ and $\dot{m}_e s_e$ account, respectively, for rates of entropy *transfer* into and out of the control volume accompanying mass flow. In writing Eq. 7.27, one-dimensional flow is assumed at locations where mass enters and exits. The term $\dot{Q}_j$ represents the time rate of heat transfer at the location on the boundary where the instantaneous temperature is $T_j$. The ratio $\dot{Q}_j/T_j$ accounts for the accompanying rate of entropy *transfer*. The term $\dot{\sigma}_{cv}$ denotes the time rate of entropy *production* due to irreversibilities *within* the control volume.

### 7.5.1 Analyzing Control Volumes at Steady State

Since many engineering analyses involve control volumes at steady state, it is instructive to list steady-state forms of the balances developed for mass, energy, and entropy. At steady state, the conservation of mass principle takes the form

$$\sum_i \dot{m}_i = \sum_e \dot{m}_e \tag{5.4}$$

The energy rate balance at steady state is

$$0 = \dot{Q}_{cv} - \dot{W}_{cv} + \sum_i \dot{m}_i\left(h_i + \frac{V_i^2}{2} + gz_i\right) - \sum_e \dot{m}_e\left(h_e + \frac{V_e^2}{2} + gz_e\right) \tag{5.10a}$$

Finally, the ***steady-state form of the entropy rate balance*** is obtained by reducing Eq. 7.27 to give

*steady-state entropy rate balance*

$$0 = \sum_j \frac{\dot{Q}_j}{T_j} + \sum_i \dot{m}_i s_i - \sum_e \dot{m}_e s_e + \dot{\sigma}_{cv} \tag{7.28}$$

These equations often must be solved simultaneously, together with appropriate property relations.

Mass and energy are conserved quantities, but entropy is not conserved. Equation 5.4 indicates that at steady state the total rate of mass flow into the control volume equals the total rate of mass flow out of the control volume. Similarly, Eq. 5.10a indicates that the total rate of energy transfer into the control volume equals the total rate of energy transfer out of the control volume. However, Eq. 7.28 requires that the rate at which entropy is transferred out must *exceed* the rate at which entropy enters, the difference being the rate of entropy production within the control volume owing to irreversibilities.

#### One-inlet, One-exit Control Volumes

Since many applications involve one-inlet, one-exit control volumes at steady state, let us also list the form of the entropy rate balance for this important case:

$$0 = \sum_j \frac{\dot{Q}_j}{T_j} + \dot{m}(s_1 - s_2) + \dot{\sigma}_{cv}$$

Or, on dividing by the mass flow rate $\dot{m}$ and rearranging

$$s_2 - s_1 = \frac{1}{\dot{m}}\left(\sum_j \frac{\dot{Q}_j}{T_j}\right) + \frac{\dot{\sigma}_{cv}}{\dot{m}} \tag{7.29}$$

The two terms on the right side of Eq. 7.29 denote, respectively, the rate of entropy transfer accompanying heat transfer and the rate of entropy production within the control volume, each *per unit of mass flowing through the control volume*. From Eq. 7.29 it can be concluded that the entropy of a unit of mass passing from inlet to exit can increase, decrease, or remain the same. Furthermore, because the value of the second term on the right can never be negative, a decrease in the specific entropy from inlet to exit can be realized only when more entropy is transferred out of the control volume accompanying heat transfer than is produced by irreversibilities within the control volume. When the value of this entropy transfer term is positive, the specific entropy at the exit is greater than the specific entropy at the inlet, whether internal irreversibilities are present or not. In the special case where there is no entropy transfer accompanying heat transfer, Eq. 7.29 reduces to

$$s_2 - s_1 = \frac{\dot{\sigma}_{cv}}{\dot{m}} \tag{7.30}$$

Accordingly, when irreversibilities are present within the control volume, the entropy of a unit of mass increases as it passes from inlet to exit. In the limiting case in which no irreversibilities are present, the unit mass passes through the control volume with no change in its entropy—that is, isentropically.

### 7.5.2 Illustrations

The following examples illustrate the use of the mass, energy, and entropy balances for the analysis of control volumes at steady state. Carefully note that property relations and property diagrams also play important roles in arriving at solutions.

In the first example, we evaluate the rate of entropy production within a turbine operating at steady state when there is heat transfer from the turbine.

## *Example 7.5* Entropy Production in a Steam Turbine

Steam enters a turbine with a pressure of 30 bar, a temperature of 400°C, and a velocity of 160 m/s. Saturated vapor at 100°C exits with a velocity of 100 m/s. At steady state, the turbine develops work equal to 540 kJ per kg of steam flowing through the turbine. Heat transfer between the turbine and its surroundings occurs at an average outer surface temperature of 350 K. Determine the rate at which entropy is produced within the turbine per kg of steam flowing, in kJ/kg · K. Neglect the change in potential energy between inlet and exit.

### Solution

***Known:*** Steam expands through a turbine at steady state for which data are provided.

***Find:*** Determine the rate of entropy production per kg of steam flowing.

***Schematic and Given Data:***

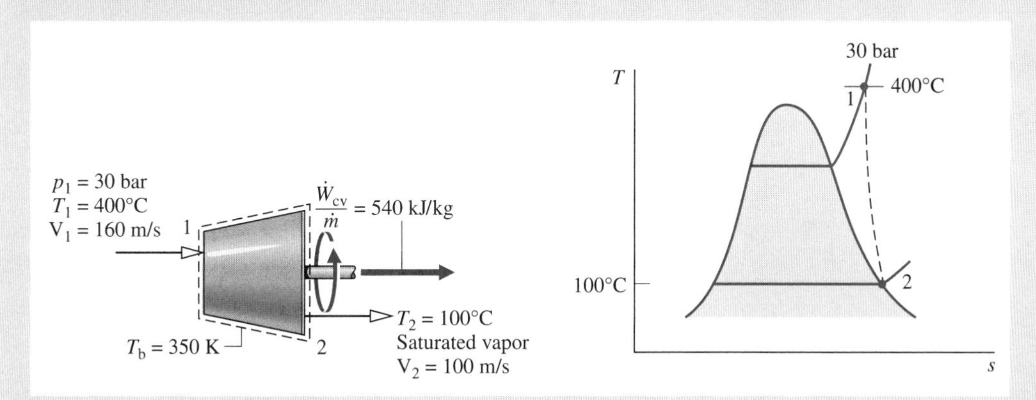


*Figure E7.5*

***Assumptions:***
1. The control volume shown on the accompanying sketch is at steady state.
2. Heat transfer from the turbine to the surroundings occurs at a specified average outer surface temperature.
3. The change in potential energy between inlet and exit can be neglected.

***Analysis:*** To determine the entropy production per unit mass flowing through the turbine, begin with mass and entropy rate balances for the one-inlet, one-exit control volume at steady state:

$$0 = \dot{m}_1 - \dot{m}_2$$

$$0 = \sum_j \frac{\dot{Q}_j}{T_j} + \dot{m}_1 s_1 - \dot{m}_2 s_2 + \dot{\sigma}_{cv}$$

Since heat transfer occurs only at $T_b = 350$ K, the first term on the right side of the entropy rate balance reduces to $\dot{Q}_{cv}/T_b$. Combining the mass and entropy rate balances

$$0 = \frac{\dot{Q}_{cv}}{T_b} + \dot{m}(s_1 - s_2) + \dot{\sigma}_{cv}$$

where $\dot{m}$ is the mass flow rate. Solving for $\dot{\sigma}_{cv}/\dot{m}$

$$\frac{\dot{\sigma}_{cv}}{\dot{m}} = -\frac{\dot{Q}_{cv}/\dot{m}}{T_b} + (s_2 - s_1)$$

The heat transfer rate, $\dot{Q}_{cv}/\dot{m}$, required by this expression is evaluated next.

Reduction of the mass and energy rate balances results in

$$\frac{\dot{Q}_{cv}}{\dot{m}} = \frac{\dot{W}_{cv}}{\dot{m}} + (h_2 - h_1) + \left(\frac{V_2^2 - V_1^2}{2}\right)$$

where the potential energy change from inlet to exit is dropped by assumption 3. From Table T-4 at 30 bar, 400°C, $h_1 = 3230.9$ kJ/kg, and from Table T-2, $h_2 = h_g(100°C) = 2676.1$ kJ/kg. Thus

$$\frac{\dot{Q}_{cv}}{\dot{m}} = 540\frac{\text{kJ}}{\text{kg}} + (2676.1 - 3230.9)\left(\frac{\text{kJ}}{\text{kg}}\right) + \left[\frac{(100)^2 - (160)^2}{2}\right]\left(\frac{\text{m}^2}{\text{s}^2}\right)\left|\frac{1\ \text{N}}{1\ \text{kg}\cdot\text{m/s}^2}\right|\left|\frac{1\ \text{kJ}}{10^3\ \text{N}\cdot\text{m}}\right|$$

$$= 540 - 554.8 - 7.8 = -22.6\ \text{kJ/kg}$$

From Table T-2, $s_2 = 7.3549$ kJ/kg · K, and from Table T-4, $s_1 = 6.9212$ kJ/kg · K. Inserting values into the expression for entropy production

$$\frac{\dot{\sigma}_{cv}}{\dot{m}} = -\frac{(-22.6\ \text{kJ/kg})}{350\ \text{K}} + (7.3549 - 6.9212)\left(\frac{\text{kJ}}{\text{kg}\cdot\text{K}}\right)$$

❶ $$= 0.0646 + 0.4337 = 0.4983\ \text{kJ/kg}\cdot\text{K} \triangleleft$$

❶ If the boundary were located to include a portion of the immediate surroundings so heat transfer would take place at the temperature of the surroundings, say $T_f = 293$ K, the entropy production for the enlarged control volume would be 0.511 kJ/kg · K. It is left as an exercise to verify this value and to explain why the entropy production for the enlarged control volume would be greater than for a control volume consisting of the turbine only.

In Example 7.6, the mass, energy, and entropy rate balances are used to test a performance claim for a device to produce hot and cold streams of air from a single stream of air at an intermediate temperature.

## *Example 7.6* Evaluating a Performance Claim

An inventor claims to have developed a device requiring no energy transfer by work or heat transfer, yet able to produce hot and cold streams of air from a single stream of air at an intermediate temperature. The inventor provides steady-state test data indicating that when air enters at a temperature of 70°F and a pressure of 5.1 atm, separate streams of air exit at temperatures of 0 and 175°F, respectively, and each at a pressure of 1 atm. Sixty percent of the mass entering the device exits at the lower temperature. Evaluate the inventor's claim, employing the ideal gas model for air and ignoring changes in the kinetic and potential energies of the streams from inlet to exit.

### Solution

***Known:*** Data are provided for a device that at steady state produces hot and cold streams of air from a single stream of air at an intermediate temperature without energy transfers by work or heat.

***Find:*** Evaluate whether the device can operate as claimed.

*Schematic and Given Data:*

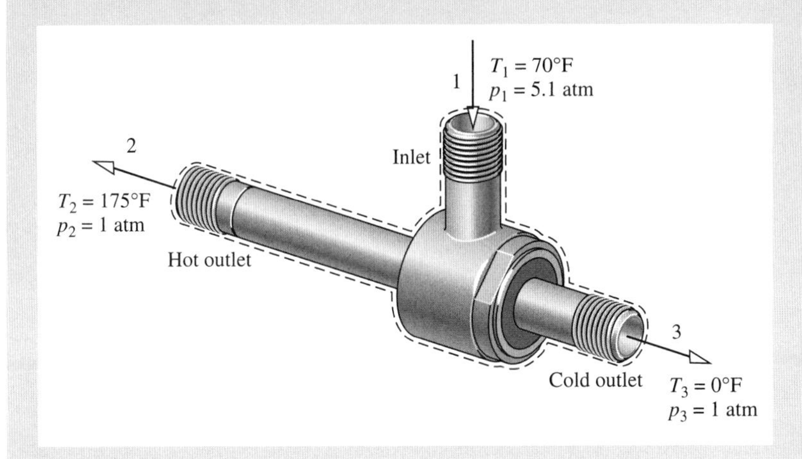


*Figure E7.6*

***Assumptions:***

1. The control volume shown on the accompanying sketch is a steady state.
2. For the control volume, $\dot{W}_{cv} = 0$ and $\dot{Q}_{cv} = 0$.
3. Changes in the kinetic and potential energies from inlet to exit can be ignored.
4. The air is modeled as an ideal gas with constant $c_p = 0.24$ Btu/lb · °R. ❶

***Analysis:*** For the device to operate as claimed, the conservation of mass and energy principles must be satisfied. The second law of thermodynamics also must be satisfied; and in particular the rate of entropy production cannot be negative. Accordingly, the mass, energy, and entropy rate balances are considered in turn.

With assumptions 1–3, the mass and energy rate balances reduce, respectively, to

$$\dot{m}_1 = \dot{m}_2 + \dot{m}_3$$

$$0 = \dot{m}_1 h_1 - \dot{m}_2 h_2 - \dot{m}_3 h_3$$

Since $\dot{m}_3 = 0.6\dot{m}_1$, it follows from the mass rate balance that $\dot{m}_2 = 0.4\dot{m}_1$. By combining the mass and energy rate balances and evaluating changes in specific enthalpy using constant $c_p$, the energy rate balance is also satisfied. That is

❷
$$\begin{aligned} 0 &= (\dot{m}_2 + \dot{m}_3)h_1 - \dot{m}_2 h_2 - \dot{m}_3 h_3 \\ &= \dot{m}_2(h_1 - h_2) + \dot{m}_3(h_1 - h_3) \\ &= 0.4\dot{m}_1[c_p(T_1 - T_2)] + 0.6\dot{m}_1[c_p(T_1 - T_3)] \\ &= 0.4(-105) + 0.6(70) \\ &= 0 \end{aligned}$$

Accordingly, with the given data the conservation of mass and energy principles are satisfied.

Since no significant heat transfer occurs, the entropy rate balance at steady state reads

$$0 = \cancel{\sum_j \frac{\dot{Q}_j}{T_j}}^{\,0} + \dot{m}_1 s_1 - \dot{m}_2 s_2 - \dot{m}_3 s_3 + \dot{\sigma}_{cv}$$

Combining the mass and entropy rate balances

$$\begin{aligned} 0 &= (\dot{m}_2 + \dot{m}_3)s_1 - \dot{m}_2 s_2 - \dot{m}_3 s_3 + \dot{\sigma}_{cv} \\ &= \dot{m}_2(s_1 - s_2) + \dot{m}_3(s_1 - s_3) + \dot{\sigma}_{cv} \\ &= 0.4\dot{m}_1(s_1 - s_2) + 0.6\dot{m}_1(s_1 - s_3) + \dot{\sigma}_{cv} \end{aligned}$$

Solving for $\dot{\sigma}_{cv}/\dot{m}_1$ and using Eq. 7.17 to evaluate changes in specific entropy

$$\frac{\dot{\sigma}_{cv}}{\dot{m}_1} = 0.4\left[c_p \ln\frac{T_2}{T_1} - R\ln\frac{p_2}{p_1}\right] + 0.6\left[c_p \ln\frac{T_3}{T_1} - R\ln\frac{p_3}{p_1}\right]$$

❸
$$= 0.4\left[\left(0.24\frac{\text{Btu}}{\text{lb}\cdot{}^\circ\text{R}}\right)\ln\frac{635}{530} - \left(\frac{1.986}{28.97}\frac{\text{Btu}}{\text{lb}\cdot{}^\circ\text{R}}\right)\ln\frac{1}{5.1}\right]$$

$$+ 0.6\left[\left(0.24\frac{\text{Btu}}{\text{lb}\cdot{}^\circ\text{R}}\right)\ln\frac{460}{530} - \left(\frac{1.986}{28.97}\frac{\text{Btu}}{\text{lb}\cdot{}^\circ\text{R}}\right)\ln\frac{1}{5.1}\right]$$

❹
$$= 0.1086\frac{\text{Btu}}{\text{lb}\cdot{}^\circ\text{R}}$$

Thus, the second law of thermodynamics is also satisfied.

❺ On the basis of this evaluation, the inventor's claim does not violate principles of thermodynamics. ◁

---

❶ Since the specific heat $c_p$ of air varies little over the temperature interval from 0 to 175°F, $c_p$ can be taken as constant. From Table T-10, $c_p = 0.24$ Btu/lb · °R.

❷ Since temperature *differences* are involved in this calculation, the temperatures can be either in °R or °F.

❸ In this calculation involving temperature *ratios*, the temperatures must be in °R.

❹ If the value of the rate of entropy production had been negative or zero, the claim would be rejected. A negative value is impossible by the second law and a zero value would indicate operation without irreversibilities.

❺ Such devices *do* exist. They are known as *vortex tubes* and are used in industry for *spot cooling*.

In Example 7.7, we evaluate and compare the rates of entropy production for three components of a heat pump system. Heat pumps are studied in Chap. 8.

## *Example 7.7* Entropy Production in Heat Pump Components

Components of a heat pump for supplying heated air to a dwelling are shown in the schematic below. At steady state, Refrigerant 22 enters the compressor at −5°C, 3.5 bar and is compressed adiabatically to 75°C, 14 bar. From the compressor, the refrigerant passes through the condenser, where it condenses to liquid at 28°C, 14 bar. The refrigerant then expands through a throttling valve to 3.5 bar. The states of the refrigerant are shown on the accompanying *T–s* diagram. Return air from the dwelling enters the condenser at 20°C, 1 bar with a volumetric flow rate of 0.42 $m^3/s$ and exits at 50°C with a negligible change in pressure. Using the ideal gas model for the air and neglecting kinetic and potential energy effects, **(a)** determine the rates of entropy production, in kW/K, for control volumes enclosing the condenser, compressor, and expansion valve, respectively. **(b)** Discuss the sources of irreversibility in the components considered in part (a).

**Solution (CD-ROM)**

## 7.6 Isentropic Processes

The term *isentropic* means constant entropy. Isentropic processes are encountered in many subsequent discussions. The object of the present section is to explain how properties are related at any two states of a process in which there is no change in specific entropy.

### 7.6.1 General Considerations

The properties at states having the same specific entropy can be related using the graphical and tabular property data discussed in Sec. 7.2. For example, as illustrated by Fig. 7.8, temperature–entropy and enthalpy–entropy diagrams are particularly convenient for determining

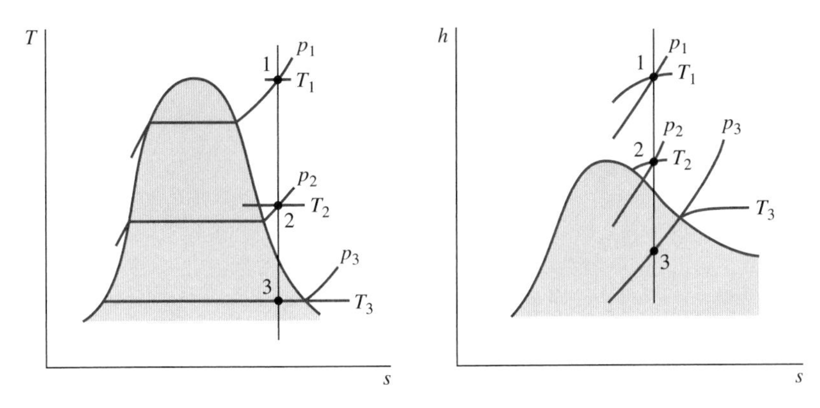


*Figure 7.8* $T$–$s$ and $h$–$s$ diagrams showing states having the same value of specific entropy.

properties at states having the same value of specific entropy. All states on a vertical line passing through a given state have the same entropy. If state 1 on Fig. 7.8 is fixed by pressure $p_1$ and temperature $T_1$, states 2 and 3 are readily located once one additional property, such as pressure or temperature, is specified. The values of several other properties at states 2 and 3 can then be read directly from the figures.

Tabular data also can be used to relate two states having the same specific entropy. For the case shown in Fig. 7.8, the specific entropy at state 1 could be determined from the superheated vapor table. Then, with $s_2 = s_1$ and one other property value, such as $p_2$ or $T_2$, state 2 could be located in the superheated vapor table. The values of the properties $v$, $u$, and $h$ at state 2 can then be read from the table. Note that state 3 falls in the two-phase liquid–vapor regions of Fig. 7.8. Since $s_3 = s_1$, the quality at state 3 could be determined using Eq. 7.6. With the quality known, other properties such as $v$, $u$, and $h$ could then be evaluated. Computer retrieval of entropy data provides an alternative to tabular data.

### 7.6.2 Using the Ideal Gas Model

Figure 7.9 shows two states of an ideal gas having the same value of specific entropy. Let us consider relations among pressure, specific volume, and temperature at these states, first using the ideal gas tables and then assuming specific heats are constant.

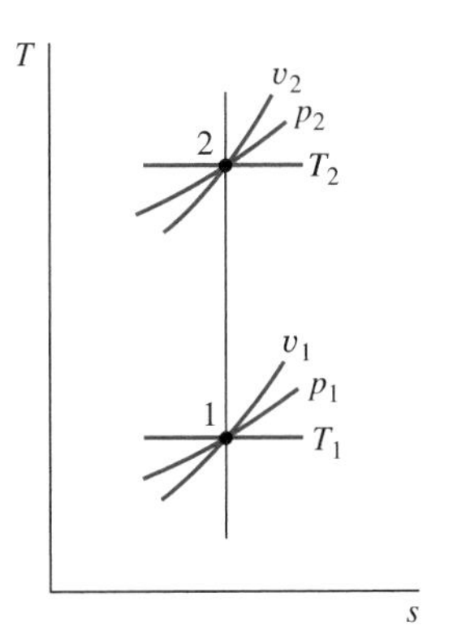


*Figure 7.9* Two states of an ideal gas where $s_2 = s_1$.

#### Ideal Gas Tables

For two states having the same specific entropy, Eq. 7.15a reduces to

$$0 = s°(T_2) - s°(T_1) - R \ln \frac{p_2}{p_1} \tag{7.31a}$$

Equation 7.31a involves four property values: $p_1$, $T_1$, $p_2$, and $T_2$. If any three are known, the fourth can be determined. If, for example, the temperature at state 1 and the pressure ratio $p_2/p_1$ are known, the temperature at state 2 can be determined from

$$s°(T_2) = s°(T_1) + R \ln \frac{p_2}{p_1} \tag{7.31b}$$

Since $T_1$ is known, $s°(T_1)$ would be obtained from the appropriate table, the value of $s°(T_2)$ would be calculated, and temperature $T_2$ would then be determined by interpolation. If $p_1$, $T_1$, and $T_2$ are specified and the pressure at state 2 is the unknown, Eq. 7.31a would be solved to obtain

$$p_2 = p_1 \exp\left[\frac{s°(T_2) - s°(T_1)}{R}\right] \tag{7.31c}$$

Equations 7.31 can be used when $s°$ (or $\bar{s}°$) data are known, as for the gases of Tables T-9 and T-11.

*Air.* For the special case of *air* modeled as an ideal gas, Eq. 7.31c provides the basis for an alternative tabular approach for relating the temperatures and pressures at two states having the same specific entropy. To introduce this, rewrite the equation as

$$\frac{p_2}{p_1} = \frac{\exp[s^\circ(T_2)/R]}{\exp[s^\circ(T_1)/R]}$$

The quantity $\exp[s^\circ(T)/R]$ appearing in this expression is solely a function of temperature, and is given the symbol $p_r(T)$. A tabulation of $p_r$ versus temperature for *air* is provided in Tables T-9. In terms of the function $p_r$, the last equation becomes

$$\frac{p_2}{p_1} = \frac{p_{r2}}{p_{r1}} \quad (s_1 = s_2, \text{ air only}) \tag{7.32}$$

where $p_{r1} = p_r(T_1)$ and $p_{r2} = p_r(T_2)$.

A relation between specific volumes and temperatures for two states of air having the same specific entropy also can be developed in the form

$$\frac{v_2}{v_1} = \frac{v_{r2}}{v_{r1}} \quad (s_1 = s_2, \text{ air only}) \tag{7.33}$$

where $v_{r1} = v_r(T_1)$ and $v_{r2} = v_r(T_2)$. Values of $v_r$ for *air* are tabulated versus temperature in Tables T-9. Finally, note that $p_r$ and $v_r$ have no physical significance.

### Assuming Constant Specific Heats

Let us consider next how properties are related for isentropic processes of an ideal gas when the specific heats are constants. For any such case, Eqs. 7.16 and 7.17 reduce to the equations

$$0 = c_p \ln\frac{T_2}{T_1} - R\ln\frac{p_2}{p_1}$$

$$0 = c_v \ln\frac{T_2}{T_1} + R\ln\frac{v_2}{v_1}$$

Introducing the ideal gas relations, Eqs. 4.45 and 4.46

$$c_p = \frac{kR}{k-1}, \qquad c_v = \frac{R}{k-1}$$

these equations can be solved, respectively, to give

$$\frac{T_2}{T_1} = \left(\frac{p_2}{p_1}\right)^{(k-1)/k} \quad (s_1 = s_2, \text{ constant } k) \tag{7.34}$$

$$\frac{T_2}{T_1} = \left(\frac{v_1}{v_2}\right)^{k-1} \quad (s_1 = s_2, \text{ constant } k) \tag{7.35}$$

The following relation can be obtained by eliminating the temperature ratio from Eqs. 7.34 and 7.35:

$$\frac{p_2}{p_1} = \left(\frac{v_1}{v_2}\right)^{k} \quad (s_1 = s_2, \text{ constant } k) \tag{7.36}$$

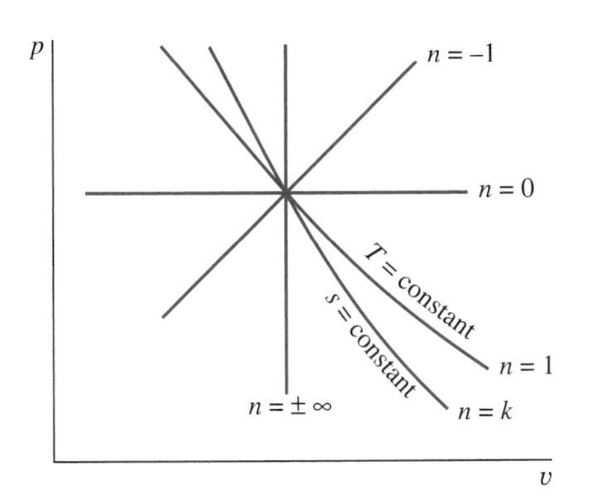


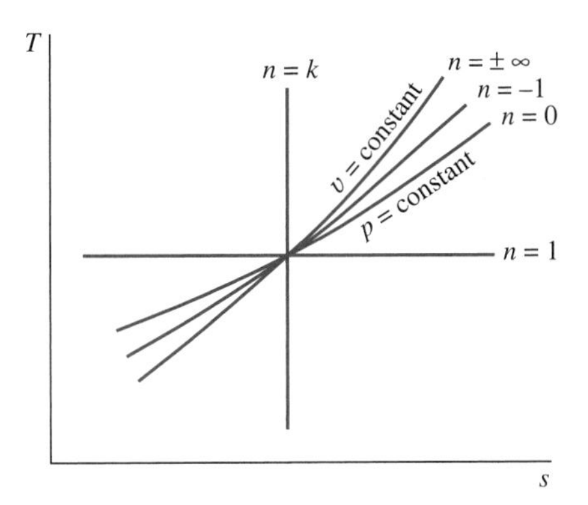


*Figure 7.10* Polytropic processes on $p$–$v$ and $T$–$s$ diagrams.

From the form of Eq. 7.36, it can be concluded that a polytropic process $pv^k = constant$ of an ideal gas with constant $k$ is an isentropic process. We noted in Sec. 4.8 that a polytropic process of an ideal gas for which $n = 1$ is an isothermal (constant-temperature) process. For *any* fluid, $n = 0$ corresponds to an isobaric (constant-pressure) process and $n = \pm\infty$ corresponds to an isometric (constant-volume) process. Polytropic processes corresponding to these values of $n$ are shown in Fig. 7.10 on $p$–$v$ and $T$–$s$ diagrams.

The foregoing means for evaluating data for an isentropic process of air modeled as an ideal gas are considered in the next example.

## *Example 7.8* Isentropic Process of Air

Air undergoes an isentropic process from $p_1 = 1$ atm, $T_1 = 540°$R to a final state where the temperature is $T_2 = 1160°$R. Employing the ideal gas model, determine the final pressure $p_2$, in atm. Solve using **(a)** $p_r$ data from Table T-9E, **(b)** a constant specific heat ratio $k$ evaluated at the mean temperature, 850°R, from Table T-10E, **(c)** *Interactive Thermodynamics: IT*.

### Solution

***Known:*** Air undergoes an isentropic process from a state where pressure and temperature are known to a state where the temperature is specified.

***Find:*** Determine the final pressure using (a) $p_r$ data, (b) a constant value for the specific heat ratio $k$, (c) *IT*.

***Schematic and Given Data:***

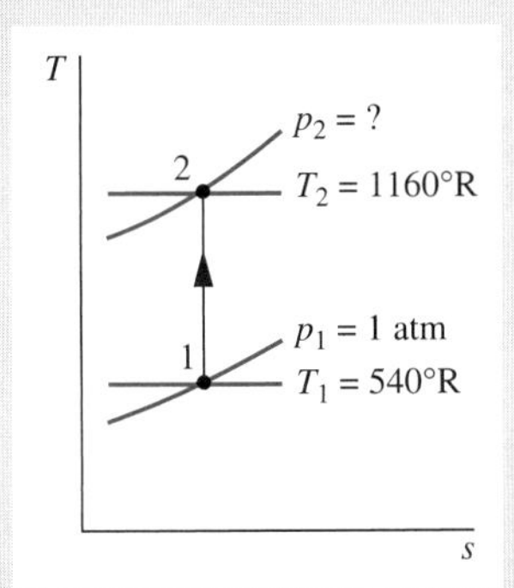


***Assumptions:***

1. A quantity of air as the system undergoes an isentropic process.
2. The air can be modeled as an ideal gas.
3. In part (b) the specific heat ratio is constant.

*Figure E7.8*

***Analysis:*** **(a)** The pressures and temperatures at two states of an ideal gas having the same specific entropy are related by Eq. 7.32

$$\frac{p_2}{p_1} = \frac{p_{r2}}{p_{r1}}$$

Solving

$$p_2 = p_1 \frac{p_{r2}}{p_{r1}}$$

With $p_r$ values from Table T-9E

$$p = (1\ \text{atm}) \frac{21.18}{1.3860} = 15.28\ \text{atm} \triangleleft$$

**(b)** When the specific heat ratio $k$ is assumed constant, the temperatures and pressures at two states of an ideal gas having the same specific entropy are related by Eq. 7.34. Thus

$$p_2 = p_1 \left(\frac{T_2}{T_1}\right)^{k/(k-1)}$$

From Table T-10E at 390°F (850°R), $k = 1.39$. Inserting values into the above expression

❶

$$p_2 = (1\ \text{atm}) \left(\frac{1160}{540}\right)^{1.39/0.39} = 15.26\ \text{atm} \triangleleft$$

(c) IT Solution. (CD-ROM)

---

❶ The close agreement between the answers obtained in parts (a) and (b) is attributable to the use of an appropriate value for the specific heat ratio $k$.

## 7.7 Isentropic Efficiencies of Turbines, Nozzles, Compressors, and Pumps

Engineers make frequent use of efficiencies and many different efficiency definitions are employed. In the present section, *isentropic* efficiencies for turbines, nozzles, compressors, and pumps are introduced. Isentropic efficiencies involve a comparison between the actual performance of a device and the performance that would be achieved under idealized circumstances for the same inlet state and the same exit pressure. These efficiencies are frequently used in subsequent sections of the book.

### Isentropic Turbine Efficiency

To introduce the isentropic turbine efficiency, refer to Fig. 7.11, which shows a turbine expansion on a Mollier diagram. The state of the matter entering the turbine and the exit pressure are fixed. Heat transfer between the turbine and its surroundings is ignored, as are kinetic and potential energy effects. With these assumptions, the mass and energy rate balances reduce, at steady state, to give the work developed per unit of mass flowing through the turbine

$$\frac{\dot{W}_{cv}}{\dot{m}} = h_1 - h_2$$

Since state 1 is fixed, the specific enthalpy $h_1$ is known. Accordingly, the value of the work depends on the specific enthalpy $h_2$ only, and increases as $h_2$ is reduced. The *maximum* value for the turbine work corresponds to the smallest *allowed* value for the specific enthalpy at the turbine exit. This can be determined using the second law.

Since there is no heat transfer, the allowed exit states are constrained by Eq. 7.30

$$\frac{\dot{\sigma}_{cv}}{\dot{m}} = s_2 - s_1 \geq 0$$

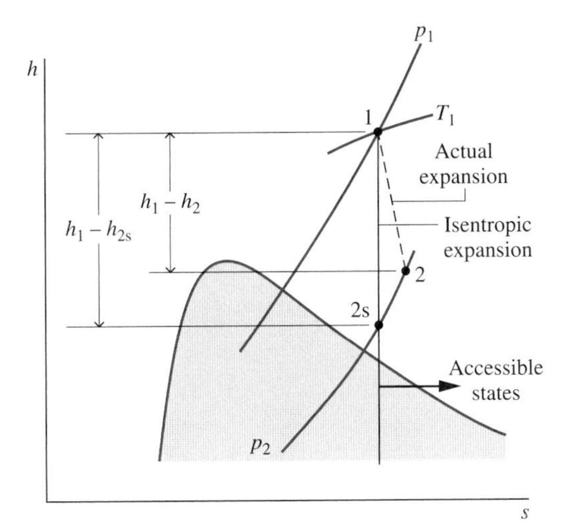


*Figure 7.11* Comparison of actual and isentropic expansions through a turbine.

Because the entropy production $\dot{\sigma}_{cv}/\dot{m}$ cannot be negative, states with $s_2 < s_1$ are not accessible in an adiabatic expansion. The only states that can be attained are those with $s_2 > s_1$. The state labeled "2s" on Fig. 7.11 would be attained only in the limit of no internal irreversibilities. This corresponds to an isentropic expansion through the turbine. For fixed exit pressure, the specific enthalpy $h_2$ decreases as the specific entropy $s_2$ decreases. Therefore, the *smallest allowed* value for $h_2$ corresponds to state 2s, and the *maximum* value for the turbine work is

$$\left(\frac{\dot{W}_{cv}}{\dot{m}}\right)_s = h_1 - h_{2s}$$

In an actual expansion through the turbine $h_2 > h_{2s}$, and thus less work than the maximum would be developed. This difference can be gauged by the ***isentropic turbine efficiency*** defined by

$$\eta_t = \frac{\dot{W}_{cv}/\dot{m}}{(\dot{W}_{cv}/\dot{m})_s} \tag{7.37}$$

***isentropic turbine efficiency***

Both the numerator and denominator of this expression are evaluated for the same inlet state and the same exit pressure. The value of $\eta_t$ is typically 0.7 to 0.9 (70–90%).

## Isentropic Nozzle Efficiency

A similar approach to that for turbines can be used to introduce the isentropic efficiency of nozzles operating at steady state. The ***isentropic nozzle efficiency*** is defined as the ratio of the actual specific kinetic energy of the gas leaving the nozzle, $V_2^2/2$, to the kinetic energy at the exit that would be achieved in an isentropic expansion between the same inlet state and the same exhaust pressure, $(V_2^2/2)_s$

$$\eta_{nozzle} = \frac{V_2^2/2}{(V_2^2/2)_s} \tag{7.38}$$

***isentropic nozzle efficiency***

Nozzle efficiencies of 95% or more are common, indicating that well-designed nozzles are nearly free of internal irreversibilities.

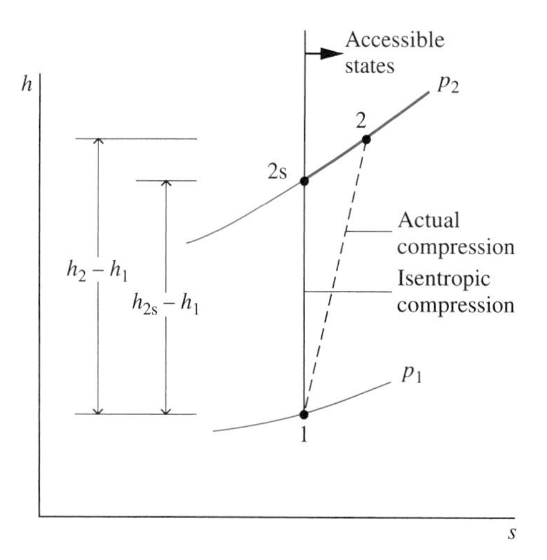


*Figure 7.12* Comparison of actual and isentropic compressions.

## Isentropic Compressor and Pump Efficiencies

The form of the isentropic efficiency for compressors and pumps is taken up next. Refer to Fig. 7.12, which shows a compression process on a Mollier diagram. The state of the matter entering the compressor and the exit pressure are fixed. For negligible heat transfer with the surroundings and no appreciable kinetic and potential energy effects, the work *input* per unit of mass flowing through the compressor is

$$\left(-\frac{\dot{W}_{cv}}{\dot{m}}\right) = h_2 - h_1$$

Since state 1 is fixed, the specific enthalpy $h_1$ is known. Accordingly, the value of the work input depends on the specific enthalpy at the exit, $h_2$. The above expression shows that the magnitude of the work input decreases as $h_2$ decreases. The *minimum* work input corresponds to the smallest *allowed* value for the specific enthalpy at the compressor exit. With similar reasoning as for the turbine, the smallest allowed enthalpy at the exit state would be achieved in an isentropic compression from the specified inlet state to the specified exit pressure. The minimum work *input* is given, therefore, by

$$\left(-\frac{\dot{W}_{cv}}{\dot{m}}\right)_s = h_{2s} - h_1$$

In an actual compression, $h_2 > h_{2s}$, and thus more work than the minimum would be required. This difference can be gauged by the ***isentropic compressor efficiency*** defined by

*isentropic compressor efficiency*

$$\eta_c = \frac{(-\dot{W}_{cv}/\dot{m})_s}{(-\dot{W}_{cv}/\dot{m})} \tag{7.39}$$

Both the numerator and denominator of this expression are evaluated for the same inlet state and the same exit pressure. The value of $\eta_c$ is typically 75 to 85% for compressors. An ***isentropic pump efficiency,*** $\eta_p$, is defined similarly.

*isentropic pump efficiency*

The series of four examples to follow illustrate various aspects of isentropic efficiencies of turbines, nozzles, and compressors. Example 7.9 is a direct application of the isentropic turbine efficiency $\eta_t$ to a steam turbine. Here, $\eta_t$ is known and the objective is to determine the turbine work.

## *Example 7.9* Evaluating Turbine Work Using the Isentropic Efficiency

A steam turbine operates at steady state with inlet conditions of $p_1 = 5$ bar, $T_1 = 320°C$. Steam leaves the turbine at a pressure of 1 bar. There is no significant heat transfer between the turbine and its surroundings, and kinetic and potential energy changes between inlet and exit are negligible. If the isentropic turbine efficiency is 75%, determine the work developed per unit mass of steam flowing through the turbine, in kJ/kg.

### Solution

***Known:*** Steam expands adiabatically through a turbine operating at steady state from a specified inlet state to a specified exit pressure. The turbine efficiency is known.

***Find:*** Determine the work developed per unit mass of steam flowing through the turbine.

***Schematic and Given Data:***

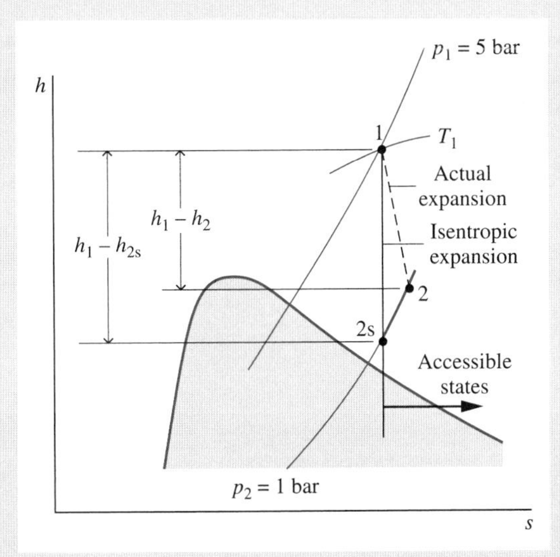


*Figure E7.9*

***Assumptions:***

1. A control volume enclosing the turbine is at steady state.
2. The expansion is adiabatic and changes in kinetic and potential energy between the inlet and exit can be neglected.

***Analysis:*** The work developed can be determined using the isentropic turbine efficiency, Eq. 7.37, which on rearrangement gives

$$\frac{\dot{W}_{cv}}{\dot{m}} = \eta_t \left(\frac{\dot{W}_{cv}}{\dot{m}}\right)_s = \eta_t(h_1 - h_{2s})$$

From Table T-4, $h_1 = 3105.6$ kJ/kg and $s_1 = 7.5308$ kJ/kg · K. The exit state for an isentropic expansion is fixed by $p_2 = 1$ bar and $s_{2s} = s_1$. Interpolating with specific entropy in Table T-4 at 1 bar gives $h_{2s} = 2743.0$ kJ/kg. Substituting values

❶ $$\frac{\dot{W}_{cv}}{\dot{m}} = 0.75(3105.6 - 2743.0) = 271.95 \text{ kJ/kg} \triangleleft$$

❶ The effect of irreversibilities is to exact a penalty on the work output of the turbine. The work is only 75% of what it would be for an isentropic expansion between the given inlet state and the turbine exhaust pressure. This is clearly illustrated in terms of enthalpy differences on the accompanying $h$–$s$ diagram.

The next example is similar to Example 7.9, but here the working substance is air as an ideal gas. Moreover, in this case the turbine work is known and the objective is to determine the isentropic turbine efficiency.

## *Example 7.10* Evaluating the Isentropic Turbine Efficiency

A turbine operating at steady state receives air at a pressure of $p_1 = 3.0$ bar and a temperature of $T_1 = 390$ K. Air exits the turbine at a pressure of $p_2 = 1.0$ bar. The work developed is measured as 74 kJ per kg of air flowing through the turbine. The turbine operates adiabatically, and changes in kinetic and potential energy between inlet and exit can be neglected. Using the ideal gas model for air, determine the turbine efficiency.

### Solution

***Known:*** Air expands adiabatically through a turbine at steady state from a specified inlet state to a specified exit pressure. The work developed per kg of air flowing through the turbine is known.

***Find:*** Determine the turbine efficiency.

***Schematic and Given Data:***

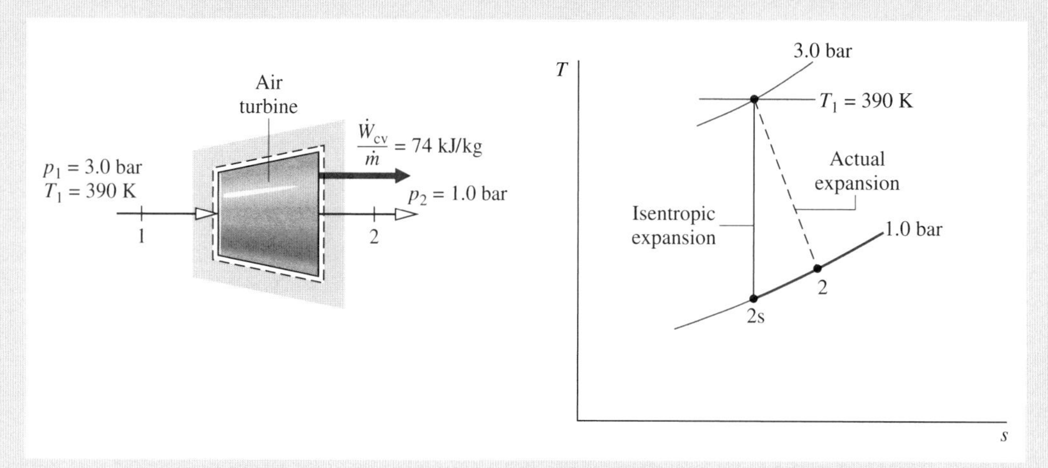


*Figure E7.10*

***Assumptions:***

1. The control volume shown on the accompanying sketch is at steady state.
2. The expansion is adiabatic and changes in kinetic and potential energy between inlet and exit can be neglected.
3. The air is modeled as an ideal gas.

***Analysis:*** The numerator of the isentropic turbine efficiency, Eq. 7.37, is known. The denominator is evaluated as follows.

The work developed in an isentropic expansion from the given inlet state to the specified exit pressure is

$$\left(\frac{\dot{W}_{cv}}{\dot{m}}\right)_s = h_1 - h_{2s}$$

From Table T-9 at 390 K, $h_1 = 390.88$ kJ/kg. To determine $h_{2s}$, use Eq. 7.32

$$p_r(T_{2s}) = \left(\frac{p_2}{p_1}\right)p_r(T_1)$$

With $p_1 = 3.0$ bar, $p_2 = 1.0$ bar, and $p_{r1} = 3.481$ from Table T-9 at 390 K

$$p_r(T_{2s}) = \left(\frac{1.0}{3.0}\right)(3.481) = 1.1603$$

Interpolation in Table T-9 gives $h_{2s} = 285.27$ kJ/kg. Thus

$$\left(\frac{\dot{W}_{cv}}{\dot{m}}\right)_s = 390.88 - 285.27 = 105.6 \text{ kJ/kg}$$

Substituting values into Eq. 7.37

$$\eta_t = \frac{\dot{W}_{cv}/\dot{m}}{(\dot{W}_{cv}/\dot{m})_s} = \frac{74 \text{ kJ/kg}}{105.6 \text{ kJ/kg}} = 0.70\ (70\%) \quad \triangleleft$$

In the next example, the objective is to determine the isentropic efficiency of a steam nozzle.

### *Example 7.11* Evaluating the Isentropic Nozzle Efficiency

Steam enters a nozzle operating at steady state at $p_1 = 140$ lbf/in.$^2$ and $T_1 = 600$°F with a velocity of 100 ft/s. The pressure and temperature at the exit are $p_2 = 40$ lbf/in.$^2$ and $T_2 = 350$°F. There is no significant heat transfer between the nozzle and its surroundings, and changes in potential energy between inlet and exit can be neglected. Determine the nozzle efficiency.

**Solution (CD-ROM)**

In Example 7.12, the isentropic efficiency of a refrigerant compressor is evaluated, first using data from property tables and then using *IT*.

### *Example 7.12* Evaluating the Isentropic Compressor Efficiency

For the compressor of the heat pump system in Example 7.7, determine the power, in kW, and the isentropic efficiency using **(a)** data from property tables, **(b)** *Interactive Thermodynamics: IT.*

**Solution (CD-ROM)**

## 7.8 Heat Transfer and Work in Internally Reversible, Steady-state Flow Processes

This section concerns one-inlet, one-exit control volumes at steady state. The objective is to derive expressions for the heat transfer and the work in the absence of internal irreversibilities. The resulting expressions have several important applications.

### Heat Transfer

For a control volume at steady state in which the flow is both *isothermal* and *internally reversible,* the appropriate form of the entropy rate balance Eq. 7.28 is

$$0 = \frac{\dot{Q}_{cv}}{T} + \dot{m}(s_1 - s_2) + \cancelto{0}{\dot{\sigma}_{cv}}$$

where 1 and 2 denote the inlet and exit, respectively, and $\dot{m}$ is the mass flow rate. Solving this equation, the heat transfer per unit of mass passing through the control volume is

$$\frac{\dot{Q}_{cv}}{\dot{m}} = T(s_2 - s_1)$$

More generally, the temperature would vary as the gas or liquid flows through the control volume. However, we can consider the temperature variation to consist of a series of infinitesimal steps. Then, the heat transfer per unit of mass would be given as

$$\left(\frac{\dot{Q}_{cv}}{\dot{m}}\right)_{\substack{\text{int}\\\text{rev}}} = \int_1^2 T\,ds \tag{7.40}$$

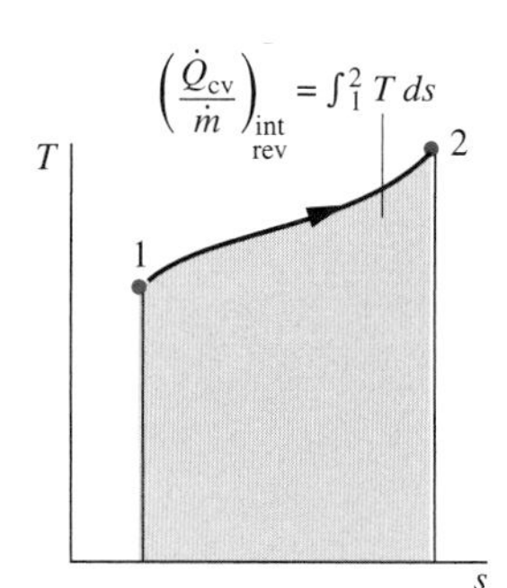


*Figure 7.13* Area representation of heat transfer for an internally reversible flow process.

The subscript "int rev" serves to remind us that the expression applies only to control volumes in which there are no internal irreversibilities. The integral of Eq. 7.40 is performed from inlet to exit. When the states visited by a unit mass as it passes reversibly from inlet to exit are described by a curve on a $T$–$s$ diagram, the magnitude of the heat transfer per unit of mass flowing can be represented as the area *under* the curve, as shown in Fig. 7.13.

## Work

The work per unit of mass passing through the control volume can be found from an energy rate balance, which reduces at steady state to give

$$\frac{\dot{W}_{cv}}{\dot{m}} = \frac{\dot{Q}_{cv}}{\dot{m}} + (h_1 - h_2) + \left(\frac{V_1^2 - V_2^2}{2}\right) + g(z_1 - z_2)$$

This equation is a statement of the conservation of energy principle that applies when irreversibilities are present within the control volume as well as when they are absent. However, if consideration is restricted to the internally reversible case, Eq. 7.40 can be introduced to obtain

$$\left(\frac{\dot{W}_{cv}}{\dot{m}}\right)_{\substack{int\\rev}} = \int_1^2 T\,ds + (h_1 - h_2) + \left(\frac{V_1^2 - V_2^2}{2}\right) + g(z_1 - z_2) \tag{7.41}$$

where the subscript "int rev" has the same significance as before. Since internal irreversibilities are absent, a unit of mass traverses a sequence of equilibrium states as it passes from inlet to exit. Entropy, enthalpy, and pressure changes are therefore related by Eq. 7.8b

$$T\,ds = dh - v\,dp$$

which on integration gives

$$\int_1^2 T\,ds = (h_2 - h_1) - \int_1^2 v\,dp$$

Introducing this relation, Eq. 7.41 becomes

$$\left(\frac{\dot{W}_{cv}}{\dot{m}}\right)_{\substack{int\\rev}} = -\int_1^2 v\,dp + \left(\frac{V_1^2 - V_2^2}{2}\right) + g(z_1 - z_2) \tag{7.42}$$

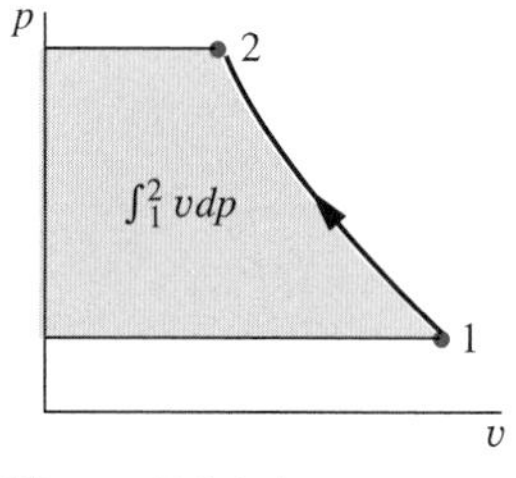


*Figure 7.14* Area representation of $\int_1^2 v\,dp$.

When the states visited by a unit of mass as it passes reversibly from inlet to exit are described by a curve on a $p$–$v$ diagram as shown in Fig. 7.14, the magnitude of the integral $\int v\,dp$ is represented by the shaded area *behind* the curve.

Equation 7.42 may be applied to devices such as turbines, compressors, and pumps. In many of these cases, there is no significant change in kinetic or potential energy from inlet to exit, so

$$\left(\frac{\dot{W}_{cv}}{\dot{m}}\right)_{\substack{int\\rev}} = -\int_1^2 v\,dp \qquad (\Delta ke = \Delta pe = 0) \tag{7.43a}$$

This expression shows that the work is related to the magnitude of the specific volume of the gas or liquid as it flows from inlet to exit. ***For Example...*** consider two devices: a pump through which liquid water passes and a compressor through which water vapor passes. For the *same pressure rise,* the pump would require a much smaller work *input* per unit of mass flowing than would the compressor because the liquid specific volume is much smaller than that of vapor. This conclusion is also qualitatively correct for actual pumps and compressors, where irreversibilities are present during operation. ▲

If the specific volume remains approximately constant, as in many applications with liquids, Eq. 7.43a becomes

$$\left(\frac{\dot{W}_{cv}}{\dot{m}}\right)_{\substack{int\\rev}} = -v(p_2 - p_1) \qquad (v = \text{constant}, \Delta\text{ke} = \Delta\text{pe} = 0) \tag{7.43b}$$

## Work in Polytropic Processes

When each unit of mass undergoes a *polytropic* process as it passes through the control volume, the relationship between pressure and specific volume is $pv^n = constant$. Introducing this into Eq. 7.43a and performing the integration

$$\left(\frac{\dot{W}_{cv}}{\dot{m}}\right)_{\substack{int\\rev}} = -\int_1^2 v\,dp = -(constant)^{1/n}\int_1^2 \frac{dp}{p^{1/n}}$$

$$= -\frac{n}{n-1}(p_2v_2 - p_1v_1) \qquad (\text{polytropic}, n \neq 1) \tag{7.44}$$

for any value of $n$ except $n = 1$. When $n = 1$, $pv = constant$, and the work is

$$\left(\frac{\dot{W}_{cv}}{\dot{m}}\right)_{\substack{int\\rev}} = -\int_1^2 v\,dp = -constant\int_1^2 \frac{dp}{p}$$

$$= -(p_1v_1)\ln(p_2/p_1) \qquad (\text{polytropic}, n = 1) \tag{7.45}$$

Equations 7.44 and 7.45 apply generally to polytropic processes of *any* gas (or liquid).

*Ideal Gas Case.* For the special case of an ideal gas, Eq. 7.44 becomes

$$\left(\frac{\dot{W}_{cv}}{\dot{m}}\right)_{\substack{int\\rev}} = -\frac{nR}{n-1}(T_2 - T_1) \qquad (\text{ideal gas}, n \neq 1) \tag{7.46a}$$

For a polytropic process of an ideal gas, Eq. 4.54 applies:

$$\frac{T_2}{T_1} = \left(\frac{p_2}{p_1}\right)^{(n-1)/n}$$

Thus, Eq. 7.46a can be expressed alternatively as

$$\left(\frac{\dot{W}_{cv}}{\dot{m}}\right)_{\substack{int\\rev}} = -\frac{nRT_1}{n-1}\left[\left(\frac{p_2}{p_1}\right)^{(n-1)/n} - 1\right] \qquad (\text{ideal gas}, n \neq 1) \tag{7.46b}$$

For the case of an ideal gas, Eq. 7.45 becomes

$$\left(\frac{\dot{W}_{cv}}{\dot{m}}\right)_{\substack{int\\rev}} = -RT\ln(p_2/p_1) \qquad (\text{ideal gas}, n = 1) \tag{7.47}$$

In the next example, we consider air modeled as an ideal gas undergoing a polytropic compression process at steady state.

## *Example 7.13* Polytropic Compression of Air

An air compressor operates at steady state with air entering at $p_1 = 1$ bar, $T_1 = 20°C$, and exiting at $p_2 = 5$ bar. Determine the work and heat transfer per unit of mass passing through the device, in kJ/kg, if the air undergoes a polytropic process with $n = 1.3$. Neglect changes in kinetic and potential energy between the inlet and the exit. Use the ideal gas model for air.

### Solution

***Known:*** Air is compressed in a polytropic process from a specified inlet state to a specified exit pressure.

***Find:*** Determine the work and heat transfer per unit of mass passing through the device.

***Schematic and Given Data:***

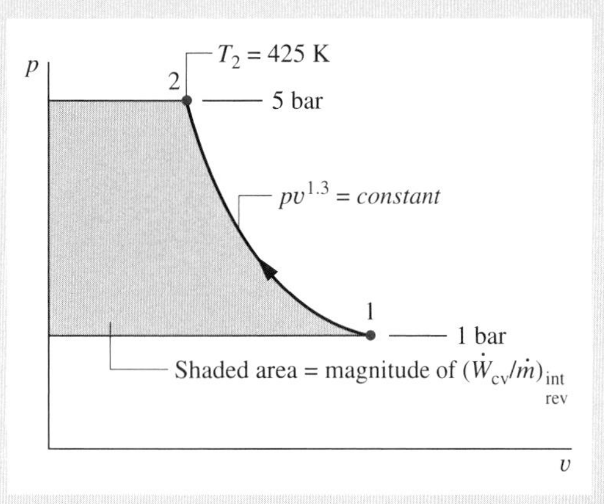


❶

***Assumptions:***

1. A control volume enclosing the compressor is at steady state.
2. The air undergoes a polytropic process with $n = 1.3$.
3. The air behaves as an ideal gas.
4. Changes in kinetic and potential energy from inlet to exit can be neglected.

*Figure E7.13*

***Analysis:*** The work is obtained using Eq. 7.46a, which requires the temperature at the exit, $T_2$. The temperature $T_2$ can be found using Eq. 4.54

$$T_2 = T_1\left(\frac{p_2}{p_1}\right)^{(n-1)/n} = 293\left(\frac{5}{1}\right)^{(1.3-1)/1.3} = 425 \text{ K}$$

Substituting known values into Eq. 7.46a then gives

$$\frac{\dot{W}_{cv}}{\dot{m}} = -\frac{nR}{n-1}(T_2 - T_1) = -\frac{1.3}{1.3-1}\left(\frac{8.314}{28.97}\frac{\text{kJ}}{\text{kg}\cdot\text{K}}\right)(425 - 293)\text{ K}$$
$$= -164.2 \text{ kJ/kg} \ \triangleleft$$

The heat transfer is evaluated by reducing the mass and energy rate balances with the appropriate assumptions to obtain

$$\frac{\dot{Q}_{cv}}{\dot{m}} = \frac{\dot{W}_{cv}}{\dot{m}} + h_2 - h_1$$

Using the temperatures $T_1$ and $T_2$, the required specific enthalpy values are obtained from Table T-9 as $h_1 = 293.17$ kJ/kg and $h_2 = 426.35$ kJ/kg. Thus

$$\frac{\dot{Q}_{cv}}{\dot{m}} = -164.15 + (426.35 - 293.17) = -31 \text{ kJ/kg} \ \triangleleft$$

❶ The states visited in the polytropic compression process are shown by the curve on the accompanying $p$–$v$ diagram. The magnitude of the work per unit of mass passing through the compressor is represented by the shaded area *behind* the curve.

## 7.9 Accounting for Mechanical Energy

The objective of this section is to introduce the *mechanical energy* and *Bernoulli equations*. These equations have several important applications in thermal systems engineering.

As in Sec. 7.8, we begin by considering a one-inlet, one-exit control volume at steady state in the absence of internal irreversibilities. When the flowing substance is modeled as *incompressible* ($v$ = *constant*), Eq. 7.42 becomes

$$\left(\frac{\dot{W}_{cv}}{\dot{m}}\right)_{\substack{int\\rev}} = -v(p_2 - p_1) + \left(\frac{V_1^2 - V_2^2}{2}\right) + g(z_1 - z_2)$$

where 1 and 2 denote the inlet and exit, respectively, and "int rev" indicates that no internal irreversibilities are present in the control volume. On rearrangement, we get

$$p_1 v + \frac{V_1^2}{2} + gz_1 = p_2 v + \frac{V_2^2}{2} + gz_2 + \left(\frac{\dot{W}_{cv}}{\dot{m}}\right)_{\substack{int\\rev}} \tag{7.48}$$

As discussed in Sec. 5.2.1, $V^2/2$ and $gz$ account for kinetic and potential energy, respectively, and $pv$ accounts for flow energy (flow work). Each of these quantities is a form of ***mechanical energy*** associated with the flowing substance. The term $\dot{W}_{cv}$ represents work due to devices such as rotating shafts that transfer mechanical energy across the control volume boundary. In Eq. 7.48, the usual sign convention for work applies: the work term would be positive if mechanical energy were transferred from the control volume, as by a turbine, and negative if mechanical energy were transferred into the control volume, as by a pump.

*mechanical energy*

Equation 7.48 states that *in the absence of friction and other internal irreversibilities,* the total mechanical energy entering the control volume equals the total mechanical energy exiting the control volume, each expressed per unit of mass flowing through the control volume. Equation 7.48 is the point of departure for introducing the mechanical energy and Bernoulli equations.

## Mechanical Energy Equation

We might expect that the presence of irreversibilities exacts a penalty on mechanical energy, and this is the case: an *irreversible conversion* of mechanical energy into internal energy occurs. Accordingly, for a one-inlet, one-exit control volume at steady state, the total mechanical energy entering *exceeds* the total mechanical energy exiting. That is

$$p_1 v + \frac{V_1^2}{2} + gz_1 > p_2 v + \frac{V_2^2}{2} + gz_2 + \left(\frac{\dot{W}_{cv}}{\dot{m}}\right) \tag{7.49}$$

It is convenient to express Eq. 7.49 as an equality rather than an inequality. That is

$$p_1 v + \frac{V_1^2}{2} + gz_1 = p_2 v + \frac{V_2^2}{2} + gz_2 + \left(\frac{\dot{W}_{cv}}{\dot{m}}\right) + loss \tag{7.50a}$$

where each term in this equation has units of energy per unit of mass flowing through the control volume (kJ/kg, Btu/lb, ft · lbf/slug). In Eq. 7.50a, the term denoted as *loss* accounts for the irreversible conversion of mechanical energy to internal energy due to effects such as friction. *Loss* is always a positive number when irreversibilities are present within the control volume, zero when the process within the control volume is internally reversible, and can never be negative. When irreversibilities are present, the internal energy gain in such a conversion is observed as heat transfer from the control volume to the surroundings, a temperature rise from inlet to exit, or both.

Equation 7.50a can be placed in an alternative form by dividing each term by $g$ to obtain the ***mechanical energy equation.*** That is

$$\frac{p_1}{\gamma} + \frac{V_1^2}{2g} + z_1 = \frac{p_2}{\gamma} + \frac{V_2^2}{2g} + z_2 + \frac{(\dot{W}_{cv}/\dot{m})}{g} + h_L \tag{7.50b}$$

*mechanical energy equation*

where $\gamma = \rho g$, called the ***specific weight,*** represents the weight per unit volume (lbf/ft$^3$, N/m$^3$), and $h_L = loss/g$. Each term in Eq. 7.50b has units of length. In this form, the terms

*specific weight*

*head loss*

are often referred to as *head*. That is, $p/\gamma$, $V^2/2g$, and $z$, are called the pressure head, velocity head, and elevation head, respectively. The work term denotes the turbine (or pump) head, and $h_L$ is called ***head loss.***

***Bernoulli Equation.*** Returning to consideration of Eq. 7.50a, in the absence of internal irreversibilities and when $\dot{W}_{cv} = 0$, the last two terms on the right side drop out and we get

$$p_1 v + \frac{V_1^2}{2} + gz_1 = p_2 v + \frac{V_2^2}{2} + gz_2 \tag{7.51}$$

Equation 7.51 shows that in such an idealized case the total mechanical energy values at states 1 and 2 are equal. Since any state downstream of state 1 can be regarded as state 2, the following must be satisfied at each state

$$pv + \frac{V^2}{2} + gz = constant \tag{7.52a}$$

Each term of this equation has units of energy per unit of mass flowing (kJ/kg, Btu/lb, ft · lbf/slug).

Equation 7.52a can be placed in an alternative form by dividing each term by the specific volume and introducing the specific weight to obtain the ***Bernoulli equation***

*Bernoulli equation*

$$p + \frac{\rho V^2}{2} + \gamma z = constant \tag{7.52b}$$

In this form, each term has units of pressure.

A second alternative form is obtained by dividing each term of Eq. 7.52b by the specific weight to get

$$\frac{p}{\gamma} + \frac{V^2}{2g} + z = constant \tag{7.53}$$

Each term in this equation represents head and has units of length.

Applications of the mechanical energy and Bernoulli equations are provided in Chapter 12.

## 7.10 Accounting for Internal Energy

The *general* concept of energy as used in thermal systems engineering is introduced in Chaps. 3 and 5. In those chapters, we present various forms of the energy balance to account for energy. In Sec. 7.9, we identify mechanical energy as an important *special* aspect of energy and introduce the mechanical energy equation. In the mechanical energy equation, we make an interpretation that is particularly important in fluid mechanics: The loss term accounts for the irreversible conversion of mechanical energy into internal energy when irreversibilities such as friction are present in the control volume. In the present section, we focus further on *internal energy* as another *special* aspect of energy and show how to account for internal energy in systems involving incompressible substances. This aspect of energy is important in later discussions of heat transfer.

***Introduction.*** In Sec. 7.9, the case of an incompressible substance flowing through a one-inlet, one-exit control volume at steady state is considered. The mechanical energy equation is given by Eq. 7.50b. With $\gamma = \rho g$, it can be written as

$$0 = -\frac{\dot{W}_{cv}}{\dot{m}} + \frac{p_1 - p_2}{\rho} + \frac{V_1^2 - V_2^2}{2} + g(z_1 - z_2) - gh_L \tag{7.54}$$

When ρ is constant, the energy balance, Eq. 5.11b, takes the form

$$0 = \frac{\dot{Q}_{cv}}{\dot{m}} - \frac{\dot{W}_{cv}}{\dot{m}} + \left[u_1 - u_2 + \frac{p_1 - p_2}{\rho}\right] + \frac{V_1^2 - V_2^2}{2} + g(z_1 - z_2) \tag{7.55}$$

Subtracting the mechanical energy equation, Eq. 7.54, from the energy balance, Eq. 7.55, we get

$$0 = \frac{\dot{Q}_{cv}}{\dot{m}} + (u_1 - u_2) + gh_L \tag{7.56}$$

Equation 7.56 accounts for internal energy per unit of mass flowing from inlet to exit. The first term on the right side represents internal energy transfer into (or out of) the control volume due to heat transfer across the boundary. The second term represents the difference in specific internal energy of a unit mass flowing between inlet and exit. The third term, which cannot be negative, represents the irreversible conversion of mechanical energy into internal energy. In subsequent discussions, this effect is referred to as ***internal energy generation.***

*internal energy generation*

***Internal Energy Equation.*** Using the discussion of Eq. 7.56 as a point of departure, we now generalize the idea of accounting for internal energy by presenting the ***internal energy equation,*** which applies to systems involving incompressible substances:

$$\begin{bmatrix}\text{time } \textit{rate of change} \\ \text{of the internal} \\ \text{energy contained} \\ \text{within the system} \\ \textit{at time } t\end{bmatrix} = \begin{bmatrix}\textit{rate} \text{ at which} \\ \text{internal energy} \\ \text{is being} \\ \text{transferred } \textit{in} \\ \textit{at time } t\end{bmatrix} - \begin{bmatrix}\textit{rate} \text{ at which} \\ \text{internal energy} \\ \text{is being} \\ \text{transferred } \textit{out} \\ \textit{at time } t\end{bmatrix} + \begin{bmatrix}\textit{rate} \text{ at which} \\ \text{internal energy is} \\ \text{being generated} \\ \text{within the system} \\ \textit{at time } t\end{bmatrix} \tag{7.57}$$

*internal energy equation*

Internal energy can be transferred in or out by heat transfer. For control volumes, internal energy also can be transferred in or out with streams of matter. The internal energy generation term accounts for irreversible conversion of mechanical energy into internal energy, as in the case of fluid friction. It also can account for other irreversible effects such as the passage of current through an electric resistance. Spontaneous chemical reactions and the absorption of neutrons liberated in nuclear fission also can be regarded as *sources* of internal energy. It is left as an exercise to show that Eq. 7.56 is a special case of 7.57.

The *internal energy equation,* which in the field of heat transfer is also referred to as the *thermal energy* equation, provides the basis for applying the conservation of energy principle in the heat transfer section of this book. See Sec. 15.2 for further discussion.

## 7.11 Chapter Summary and Study Guide

In this chapter, we have introduced the property of entropy and illustrated its use for thermodynamic analysis. Like mass and energy, entropy is an extensive property that can be transferred across system boundaries. Entropy transfer accompanies both heat transfer and mass flow. Unlike mass and energy, entropy is not conserved but is *produced* within systems whenever internal irreversibilities are present.

The use of entropy balances is featured in this chapter. Entropy balances are expressions of the second law that account for the entropy of systems in terms of entropy transfers and entropy production. For processes of closed systems, the entropy balance is Eq. 7.21, and a corresponding rate form is Eq. 7.26. For control volumes, rate forms include Eq. 7.27 and the companion steady-state expression given by Eq. 7.28. In this chapter, the mechanical energy, Bernoulli, and internal energy equations also are developed for later use in fluid mechanics and heat transfer.

The following checklist provides a study guide for this chapter. When your study of the text and end-of-chapter exercises has been completed you should be able to

- write out meanings of the terms listed in the margins throughout the chapter and understand each of the related concepts. The subset of key terms listed here in the margin is particularly important in subsequent chapters.
- apply entropy balances in each of several alternative forms, appropriately modeling the case at hand, correctly observing sign conventions, and carefully applying SI and other units.
- use entropy data appropriately, to include
  –retrieving data from Tables T-2 through T-8, using Eq. 7.6 to evaluate the specific entropy of two-phase liquid–vapor mixtures, sketching $T$–$s$ and $h$–$s$ diagrams and locating states on such diagrams, and appropriately using Eqs. 7.7 and 7.18 for liquids and solids.
  –determining $\Delta s$ of ideal gases using Eq. 7.15 for variable specific heats together with Tables T–9 and T–11, and using Eqs. 7.16 and 7.17 for constant specific heats.
  –evaluating isentropic efficiencies for turbines, nozzles, compressors, and pumps from Eqs. 7.37, 7.38, and 7.39, respectively, including for ideal gases the appropriate use of Eqs. 7.31–7.33 for variable specific heats and Eqs. 7.34–7.35 for constant specific heats.
- apply Eq. 7.19 for closed systems and Eqs. 7.40 and 7.42 for one-inlet, one-exit control volumes at steady state, correctly observing the restriction to internally reversible processes.

*entropy change*
*entropy transfer*
*entropy production*
*entropy balance*
*entropy rate balance*
*T ds equations*
*T–s, h-s diagrams*
*isentropic efficiencies*
*mechanical energy equation*
*Bernoulli equation*
*internal energy equation*

# Problems

## Exploring Entropy and the Second Law

**7.1** A system executes a power cycle while receiving 2000 Btu by heat transfer at a temperature of 1000°R and discharging energy by heat transfer at a temperature of 500°R. There are no other heat transfers. Applying Eq. 7.2, determine $\sigma_{\text{cycle}}$ if the thermal efficiency is **(a)** 75%, **(b)** 50%, **(c)** 25%. Identify the cases (if any) that are internally reversible or impossible.

**7.2** A system executes a power cycle while receiving 750 kJ by heat transfer at a temperature of 1500 K and discharging 100 kJ by heat transfer at 500 K. A heat transfer from the system also occurs at a temperature of 1000 K. There are no other heat transfers. If no internal irreversibilities are present, determine the thermal efficiency.

**7.3** (CD-ROM)

**7.4** (CD-ROM)

**7.5** Answer the following true or false. If false, explain why.

**(a)** The change of entropy of a closed system is the same for every process between two specified states.

**(b)** The entropy of a fixed amount of an ideal gas increases in every isothermal compression.

**(c)** The specific internal energy and enthalpy of an ideal gas are each functions of temperature alone but its specific entropy depends on two independent intensive properties.

**(d)** One of the $T\,ds$ equations has the form $T\,ds = du - p\,dv$.

**(e)** The entropy of a fixed amount of an incompressible substance increases in every process in which temperature decreases.

**7.6** Answer the following true or false. If false, explain why.

**(a)** A process that violates the second law of thermodynamics violates the first law of thermodynamics.

**(b)** When a net amount of work is done on a closed system undergoing an internally reversible process, a net heat transfer of energy from the system also occurs.

**(c)** One corollary of the second law of thermodynamics states that the change in entropy of a closed system must be greater than zero or equal to zero.

**(d)** A closed system can experience an increase in entropy only when irreversibilities are present within the system during the process.

**(e)** Entropy is produced in every internally reversible process of a closed system.

**(f)** In an adiabatic and internally reversible process of a closed system, the entropy remains constant.

**(g)** The energy of an isolated system must remain constant, but the entropy can only decrease.

**7.7** Taken together, a certain closed system and its surroundings make up an *isolated* system. Answer the following true or false. If false, explain why.

**(a)** No process is allowed in which the entropies of both the system and the surroundings increase.

(b) During a process, the entropy of the system might decrease, while the entropy of the surroundings increases, and conversely.
(c) No process is allowed in which the entropies of both the system and the surroundings remain unchanged.
(d) A process can occur in which the entropies of both the system and the surroundings decrease.

**7.8** (CD-ROM)

**7.9** A quantity of air is shown in Fig. 7.7. Consider a process in which the temperature of the air increases by some combination of stirring and heating. Assuming the ideal gas model for the air, suggest how this might be done with
(a) minimum entropy production.
(b) maximum entropy production.

### Using Entropy Data

**7.10** Using the tables for water, determine the specific entropy at the indicated states, in kJ/kg · K. In each case, locate the state on a sketch of the *T–s* diagram.
(a) $p = 5.0$ MPa, $T = 400°C$
(b) $p = 5.0$ MPa, $T = 100°C$
(c) $p = 5.0$ MPa, $u = 1872.5$ kJ/kg
(d) $p = 5.0$ MPa, saturated vapor

**7.11** Using the tables for water, determine the specific entropy at the indicated states, in Btu/lb · °R. In each case, locate the state on a sketch of the *T–s* diagram.
(a) $p = 1000$ lbf/in.$^2$, $T = 750°F$
(b) $p = 1000$ lbf/in.$^2$, $T = 300°C$
(c) $p = 1000$ lbf/in.$^2$, $h = 932.4$ Btu/lb
(d) $p = 1000$ lbf/in.$^2$, saturated vapor

**7.12** Using the appropriate table, determine the change in specific entropy between the specified states, in kJ/kg · K.
(a) water, $p_1 = 10$ MPa, $T_1 = 400°C$, $p_2 = 10$ MPa, $T_2 = 100°C$.
(b) Refrigerant 134a, $h_1 = 111.44$ kJ/kg, $T_1 = -40°C$, saturated vapor at $p_2 = 5$ bar.
(c) air as an ideal gas, $T_1 = 7°C$, $p_1 = 2$ bar, $T_2 = 327°C$, $p_2 = 1$ bar.

**7.13** Using the appropriate table, determine the change in specific entropy between the specified states, in Btu/lb · °R.
(a) water, $p_1 = 1000$ lbf/in.$^2$, $T_1 = 800°F$, $p_2 = 1000$ lbf/in.$^2$, $T_2 = 100°F$.
(b) Refrigerant 134a, $h_1 = 47.91$ Btu/lb, $T_1 = -40°F$, saturated vapor at $p_2 = 40$ lbf/in.$^2$
(c) air as an ideal gas, $T_1 = 40°F$, $p_1 = 2$ atm, $T_2 = 420°F$, $p_2 = 1$ atm.
(d) carbon dioxide as an ideal gas, $T_1 = 820°F$, $p_1 = 1$ atm, $T_2 = 77°F$, $p_2 = 3$ atm.

**7.14** (CD-ROM)

**7.15** One pound mass of water undergoes a process with no change in specific entropy from an initial state where $p_1 = 100$ lbf/in.$^2$, $T_1 = 650°F$ to a state where $p_2 = 5$ lbf/in.$^2$ Determine the temperature at the final state, if superheated, or the quality, if saturated, using steam table data.

**7.16** Employing the ideal gas model, determine the change in specific entropy between the indicated states, in kJ/kg · K. Solve two ways: Use the appropriate ideal gas table, and a constant specific heat value from Table T-10.
(a) air, $p_1 = 100$ kPa, $T_1 = 20°C$, $p_2 = 100$ kPa, $T_2 = 100°C$.
(b) air, $p_1 = 1$ bar, $T_1 = 27°C$, $p_2 = 3$ bar, $T_2 = 377°C$.
(c) carbon dioxide, $p_1 = 150$ kPa, $T_1 = 30°C$, $p_2 = 300$ kPa, $T_2 = 300°C$.
(d) carbon monoxide, $T_1 = 300$ K, $v_1 = 1.1$ m$^3$/kg, $T_2 = 500$ K, $v_2 = 0.75$ m$^3$/kg.
(e) nitrogen, $p_1 = 2$ MPa, $T_1 = 800$ K, $p_2 = 1$ MPa, $T_2 = 300$ K.

**7.17** (CD-ROM)

**7.18** Using the appropriate table, determine the indicated property for a process in which there is no change in specific entropy between state 1 and state 2.
(a) water, $p_1 = 14.7$ lbf/in.$^2$, $T_1 = 500°F$, $p_2 = 100$ lbf/in.$^2$ Find $T_2$ in °F.
(b) water, $T_1 = 10°C$, $x_1 = 0.75$, saturated vapor at state 2. Find $p_2$ in bar.
(c) air as an ideal gas, $T_1 = 27°C$, $p_1 = 1.5$ bar, $T_2 = 127°C$. Find $p_2$ in bar.
(d) air as an ideal gas, $T_1 = 100°F$, $p_1 = 3$ atm, $p_2 = 2$ atm. Find $T_2$ in °F.
(e) Refrigerant 134a, $T_1 = 20°C$, $p_1 = 5$ bar, $p_2 = 1$ bar. Find $v_2$ in m$^3$/kg.

**7.19** Two kilograms of water undergo a process from an initial state where the pressure is 2.5 MPa and the temperature is 400°C to a final state of 2.5 MPa, 100°C. Determine the entropy change of the water, in kJ/K, assuming the process is
(a) irreversible.
(b) internally reversible.

**7.20** A quantity of liquid water undergoes a process from 80°C, 5 MPa to saturated liquid at 40°C. Determine the change in specific entropy, in kJ/kg · K, using
(a) Tables T-2 and T-5.
(b) saturated liquid data only from Table T–2.
(c) the incompressible liquid model with a constant specific heat from Table HT-5.

**7.21** One-tenth kmol of carbon monoxide gas (CO) undergoes a process from $p_1 = 1.5$ bar, $T_1 = 300$ K to $p_2 = 5$ bar, $T_2 = 370$ K. For the process $W = -300$ kJ. Employing the ideal gas model, determine
(a) the heat transfer, in kJ.
(b) the change in entropy, in kJ/K.
(c) Show the initial and final states on a *T–s* diagram.

**7.22** (CD-ROM)

### Internally Reversible Processes

**7.23** A quantity of air amounting to $2.42 \times 10^{-2}$ kg undergoes a thermodynamic cycle consisting of three internally reversible processes in series.

***Process 1–2:*** constant-volume heating at $V = 0.02$ m$^3$ from $p_1 = 0.1$ MPa to $p_2 = 0.42$ MPa

***Process 2–3:*** constant-pressure cooling

***Process 3–1:*** isothermal heating to the initial state

Employing the ideal gas model with $c_p = 1$ kJ/kg · K, evaluate the change in entropy, in kJ/K, for each process. Sketch the cycle on $p$–$v$ and $T$–$s$ coordinates.

**7.24** One kilogram of water initially at 160°C, 1.5 bar undergoes an isothermal, internally reversible compression process to the saturated liquid state. Determine the work and heat transfer, each in kJ. Sketch the process on $p$–$v$ and $T$–$s$ coordinates. Associate the work and heat transfer with areas on these diagrams.

**7.25** (CD-ROM)

**7.26** A gas initially at 14 bar and 60°C expands to a final pressure of 2.8 bar in an isothermal, internally reversible process. Determine the heat transfer and the work, each in kJ per kg of gas, if the gas is **(a)** Refrigerant 134a, **(b)** air as an ideal gas. Sketch the processes on $p$–$v$ and $T$–$s$ coordinates.

**7.27** (CD-ROM)

**7.28** Air initially occupying 1 m$^3$ at 1.5 bar, 20°C undergoes an internally reversible compression during which $pV^{1.27} = constant$ to a final state where the temperature is 120°C. Determine

**(a)** the pressure at the final state, in bar.
**(b)** the work and heat transfer, each in kJ.
**(c)** the entropy change, in kJ/K.

**7.29** Air initially occupying a volume of 1 m$^3$ at 1 bar, 20°C undergoes two internally reversible processes in series

***Process 1–2:*** compression to 5 bar, 110°C during which $pV^n = constant$

***Process 2–3:*** adiabatic expansion to 1 bar

**(a)** Sketch the two processes on $p$–$v$ and $T$–$s$ coordinates.
**(b)** Determine $n$.
**(c)** Determine the temperature at state 3, in °C.
**(d)** Determine the net work, in kJ.

**7.30** (CD-ROM)

**7.31** (CD-ROM)

## Entropy Balance—Closed Systems

**7.32** A closed system undergoes a process in which work is done on the system and the heat transfer $Q$ occurs only at temperature $T_b$. For each case, determine whether the entropy change of the system is positive, negative, zero, or indeterminate.

**(a)** internally reversible process, $Q > 0$.
**(b)** internally reversible process, $Q = 0$.
**(c)** internally reversible process, $Q < 0$.
**(d)** internal irreversibilities present, $Q > 0$.
**(e)** internal irreversibilities present, $Q = 0$.
**(f)** internal irreversibilities present, $Q < 0$.

**7.33** For each of the following systems, specify whether the entropy change during the indicated process is positive, negative, zero, or indeterminate.

**(a)** One kilogram of water vapor undergoing an adiabatic compression process.
**(b)** Two pounds mass of nitrogen heated in an internally reversible process.
**(c)** One kilogram of Refrigerant 134a undergoing an adiabatic process during which it is stirred by a paddle wheel.
**(d)** One pound mass of carbon dioxide cooled isothermally.
**(e)** Two pounds mass of oxygen modeled as an ideal gas undergoing a constant-pressure process to a higher temperature.
**(f)** Two kilograms of argon modeled as an ideal gas undergoing an isothermal process to a lower pressure.

**7.34** An insulated piston–cylinder assembly contains Refrigerant 134a, initially occupying 0.6 ft$^3$ at 90 lbf/in.$^2$, 100°F. The refrigerant expands to a final state where the pressure is 50 lbf/in.$^2$ The work developed by the refrigerant is measured as 5.0 Btu. Can this value be correct?

**7.35** One pound mass of air is initially at 1 atm and 140°F. Can a final state at 2 atm and 60°F be attained in an adiabatic process?

**7.36** One kilogram of Refrigerant 134a contained within a piston–cylinder assembly undergoes a process from a state where the pressure is 7 bar and the quality is 50% to a state where the temperature is 16°C and the refrigerant is saturated liquid. Determine the change in specific entropy of the refrigerant, in kJ/kg · K. Can this process be accomplished adiabatically?

**7.37** Air as an ideal gas is compressed from a state where the pressure is 0.1 MPa and the temperature is 27°C to a state where the pressure is 0.5 MPa and the temperature is 207°C. Can this process occur adiabatically? If yes, determine the work per unit mass of air, in kJ/kg, for an adiabatic process between these states. If no, determine the direction of the heat transfer.

**7.38** Air as an ideal gas with $c_p = 0.241$ Btu/lb · °R is compressed from a state where the pressure is 3 atm and the temperature is 80°F to a state where the pressure is 10 atm and the temperature is 240°F. Can this process occur adiabatically? If yes, determine the work per unit mass of air, in Btu/lb, for an adiabatic process between these states. If no, determine the direction of the heat transfer.

**7.39** A piston–cylinder assembly contains 1 lb of Refrigerant 134a initially as saturated vapor at −10°F. The refrigerant is compressed adiabatically to a final volume of 0.8 ft$^3$. Determine if it is possible for the pressure of the refrigerant at the final state to be

**(a)** 60 lbf/in.$^2$
**(b)** 70 lbf/in.$^2$

**7.40** (CD-ROM)

**7.41** A gearbox operating at steady state receives 2 hp along the input shaft and delivers 1.89 hp along the output shaft. The outer surface of the gearbox is at 110°F and has an area of 1.4 ft.$^2$ The temperature of the surroundings away from the immediate vicinity of the gearbox is 70°F. For the gearbox, determine

**(a)** the rate of heat transfer, in Btu/s.
**(b)** the rate at which entropy is produced, in Btu/°R · s.

**7.42** For the silicon chip of Example 3.5, determine the rate of entropy production, in kW/K. What is the cause of entropy production in this case?

**7.43** At steady state, a 15-W curling iron has an outer surface temperature of 90°C. For the curling iron, determine the rate of heat transfer, in kW, and the rate of entropy production, in kW/K.

**7.44** (CD-ROM)

**7.45** (CD-ROM)

**7.46** (CD-ROM)

**7.47** Two insulated tanks are connected by a valve. One tank initially contains 0.5 kg of air at 80°C, 1 bar, and the other contains 1.0 kg of air at 50°C, 2 bar. The valve is opened and the two quantities of air are allowed to mix until equilibrium is attained. Employing the ideal gas model with $c_v = 0.72$ kJ/kg · K, determine
(a) the final temperature, in °C.
(b) the final pressure, in bar.
(c) the amount of entropy produced, in kJ/K.

**7.48** (CD-ROM)

## Entropy Balance—Control Volumes

**7.49** A gas flows through a one-inlet, one-exit control volume operating at steady state. Heat transfer at the rate $\dot{Q}_{cv}$ takes place only at a location on the boundary where the temperature is $T_b$. For each of the following cases, determine whether the specific entropy of the gas at the exit is greater than, equal to, or less than the specific entropy of the gas at the inlet:
(a) no internal irreversibilities, $\dot{Q}_{cv} = 0$.
(b) no internal irreversibilities, $\dot{Q}_{cv} < 0$.
(c) no internal irreversibilities, $\dot{Q}_{cv} > 0$.
(d) internal irreversibilities, $\dot{Q}_{cv} < 0$.
(e) internal irreversibilities, $\dot{Q}_{cv} \geq 0$.

**7.50** Steam at 3.0 MPa, 500°C, 70 m/s enters an insulated turbine operating at steady state and exits at 0.3 MPa, 140 m/s. The work developed per kg of steam flowing is claimed to be (a) 667 kJ/kg, (b) 619 kJ/kg. Can either claim be correct? Explain.

**7.51** Figure 7.51 provides steady-state test data for a steam turbine operating with negligible heat transfer with its surroundings and negligible changes in kinetic and potential energy. A faint photocopy of the data sheet indicates that the power developed is either 3080 or 3800 horsepower. Determine if either or both of these power values can be correct.

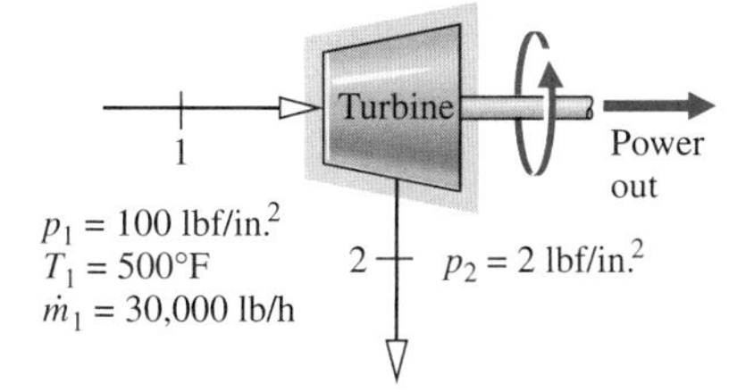


*Figure P7.51*

**7.52** Air enters an insulated turbine operating at steady state at 4.89 bar, 597°C and exits at 1 bar, 297°C. Neglecting kinetic and potential energy changes and assuming the ideal gas model, determine
(a) the work developed, in kJ per kg of air flowing through the turbine.
(b) whether the expansion is internally reversible, irreversible, or impossible.

**7.53** (CD-ROM)

**7.54** Figure P7.54 provides steady-state operating data for a well-insulated device with air entering at one location and exiting at another with a mass flow rate of 10 kg/s. Assuming ideal gas behavior and negligible potential energy effects, determine (a) the direction of flow and (b) the power, in kW.

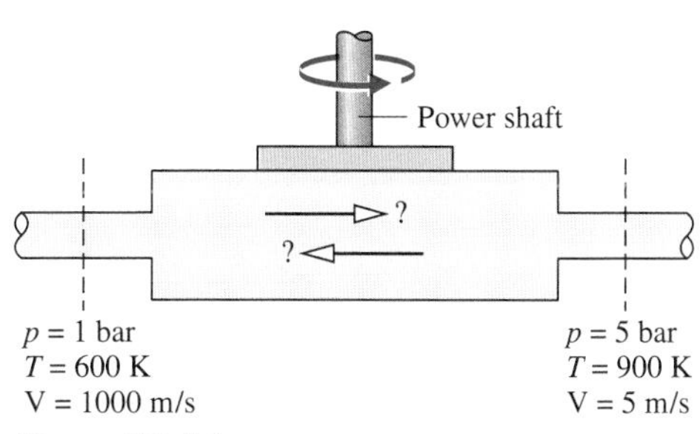


*Figure P7.54*

**7.55** An inventor claims to have developed a device requiring no work input or heat transfer, yet able to produce at steady state hot and cold air streams as shown in Fig. P7.55. Employing the ideal gas model for air and ignoring kinetic and potential energy effects, evaluate this claim.

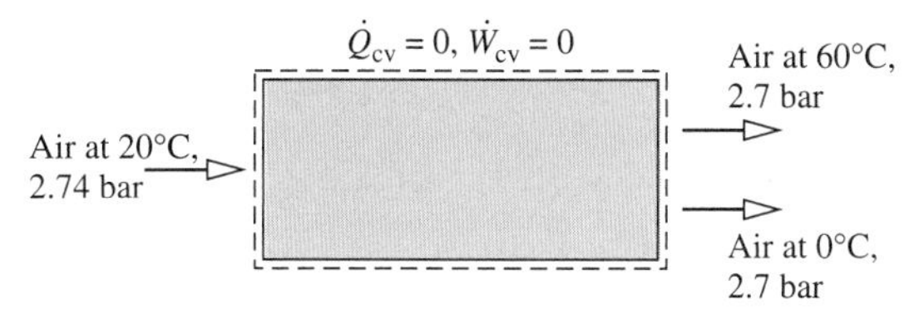


*Figure P7.55*

**7.56** (CD-ROM)

**7.57** According to test data, a new type of engine takes in streams of water at 400°F, 40 lbf/in.$^2$ and 200°F, 40 lbf/in.$^2$ The mass flow rate of the higher temperature stream is twice that of the other. A single stream exits at 40 lbf/in.$^2$ with a mass flow rate of 90 lb/min. There is no significant heat transfer between the engine and its surroundings, and kinetic and potential energy effects are negligible. For operation at steady state, determine the *maximum* theoretical rate that power can be developed, in horsepower.

**7.58** Figure P7.58 shows a proposed device to develop power using energy supplied to the device by heat transfer from a high-temperature industrial process together with a steam input. The figure provides data for steady-state operation. All surfaces are well insulated except for the one at 527°C, through which heat transfer occurs at a rate of 4.21 kW. Ignoring changes in kinetic and potential energy, evaluate the maximum theoretical power that can be developed, in kW.

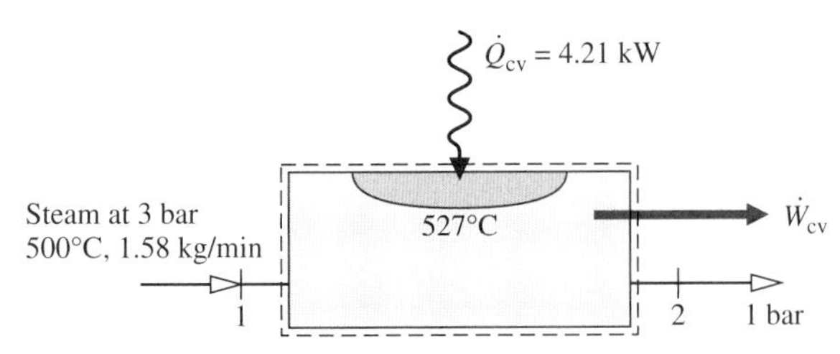


*Figure P7.58*

**7.59** Steam enters a turbine operating at steady state at a pressure of 3 MPa, a temperature of 400°C, and a velocity of 160 m/s. Saturated vapor exits at 100°C, with a velocity of 100 m/s. Heat transfer from the turbine to its surroundings takes place at the rate of 30 kJ per kg of steam at a location where the average surface temperature is 350 K.

**(a)** For a control volume including only the turbine and its contents, determine the work developed, in kJ, and the rate at which entropy is produced, in kJ/K, each per kg of steam flowing.

**(b)** The steam turbine of part (a) is located in a factory where the ambient temperature is 27°C. Determine the rate of entropy production, in kJ/K per kg of steam flowing, for an enlarged control volume that includes the turbine and enough of its immediate surroundings so that heat transfer takes place from the control volume at the ambient temperature.

Explain why the entropy production value of part (b) differs from that calculated in part (a).

**7.60** Air enters a turbine operating at steady state with a pressure of 75 lbf/in.$^2$, a temperature of 800°R, and a velocity of 400 ft/s. At the turbine exit, the conditions are 15 lbf/in.$^2$, 600°R, and 100 ft/s. Heat transfer from the turbine to its surroundings takes place at a location where the average surface temperature is 620°R. The rate of heat transfer is 10 Btu per lb of air passing through the turbine.

**(a)** For a control volume including only the turbine and its contents, determine the work developed, in Btu, and the rate at which entropy is produced, in Btu/°R, each per lb of air flowing.

**(b)** For a control volume including the turbine and a portion of its immediate surroundings so that the heat transfer occurs at the ambient temperature, 40°F, determine the rate of entropy production in Btu/°R per lb of air passing through the turbine.

Explain why the entropy production value of part (b) differs from that calculated in part (a).

**7.61** (CD-ROM)

**7.62** (CD-ROM)

**7.63** Air is compressed in an axial-flow compressor operating at steady state from 27°C, 1 bar to a pressure of 2.1 bar. The work input required is 94.6 kJ per kg of air flowing through the compressor. Heat transfer from the compressor occurs at the rate of 14 kJ per kg at a location on the compressor's surface where the temperature is 40°C. Kinetic and potential energy changes can be ignored. Determine

**(a)** the temperature of the air at the exit, in °C.

**(b)** the rate at which entropy is produced within the compressor, in kJ/K per kg of air flowing.

**7.64** Air enters a compressor operating at steady state at 1 bar, 20°C with a volumetric flow rate of 9 m$^3$/min and exits at 5 bar, 160°C. Cooling water is circulated through a water jacket enclosing the compressor at a rate of 8.6 kg/min, entering at 17°C, and exiting at 25°C with a negligible change in pressure. There is no significant heat transfer from the outer surface of the water jacket, and all kinetic and potential effects are negligible. For the water-jacketed compressor as the control volume, determine the power required, in kW, and the rate of entropy production, in kW/K.

**7.65** (CD-ROM)

**7.66** A counterflow heat exchanger operates at steady state with negligible kinetic and potential energy effects. In one stream, liquid water enters at 17°C and exits at 25°C with a negligible change in pressure. In the other stream, Refrigerant 134a enters at 14 bar, 80°C with a mass flow rate of 5 kg/min and exits as saturated liquid at 52°C. Heat transfer from the outer surface of the heat exchanger can be ignored. Determine

**(a)** the mass flow rate of the liquid water stream, in kg/min.

**(b)** the rate of entropy production within the heat exchanger, in kW/K.

**7.67** (CD-ROM)

**7.68** Air as an ideal gas flows through the compressor and heat exchanger shown in Fig. P7.68. A separate liquid water stream also flows through the heat exchanger. The data given are for operation at steady state. Stray heat transfer to the surroundings can be neglected, as can all kinetic and potential energy changes. Determine

**(a)** the compressor power, in kW, and the mass flow rate of the cooling water, in kg/s.

**(b)** the rates of entropy production, each in kW/K, for the compressor and heat exchanger.

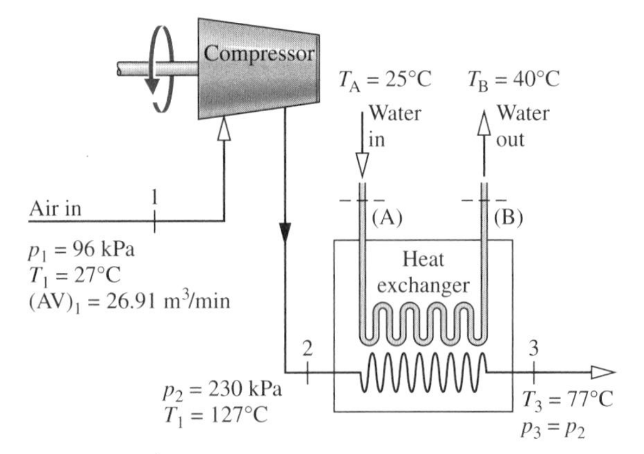


*Figure P7.68*

### Isentropic Processes/Efficiencies

**7.69** A piston–cylinder assembly initially contains 0.1 m$^3$ of carbon dioxide gas at 0.3 bar and 400 K. The gas is compressed isentropically to a state where the temperature is 560 K. Employing the ideal gas model and neglecting kinetic and potential energy effects, determine the final pressure, in bar, and the work in kJ, using

(a) data from Table T-11.
(b) a constant specific heat ratio from Table T-10 at the mean temperature, 480 K.
(c) a constant specific heat ratio from Table T-10 at 300 K.

**7.70** Air enters a turbine operating at steady state at 6 bar and 1100 K and expands isentropically to a state where the temperature is 700 K. Employing the ideal gas model and ignoring kinetic and potential energy changes, determine the pressure at the exit, in bar, and the work, in kJ per kg of air flowing, using

(a) data from Table T-9
(b) a constant specific heat ratio from Table T-10 at the mean temperature, 900 K.
(c) a constant specific heat ratio from Table T-10 at 300 K.

**7.71** (CD-ROM)

**7.72** Air enters a 3600-kW turbine operating at steady state with a mass flow rate of 18 kg/s at 800°C, 3 bar and a velocity of 100 m/s. The air expands adiabatically through the turbine and exits at a velocity of 150 m/s. The air then enters a diffuser where it is decelerated isentropically to a velocity of 10 m/s and a pressure of 1 bar. Employing the ideal gas model, determine

(a) the pressure and temperature of the air at the turbine exit, in bar and °C, respectively.
(b) the rate of entropy production in the turbine, in kW/K.
(c) Show the processes on a $T$–$s$ diagram.

**7.73** Steam at 140 lbf/in.$^2$, 1000°F enters an insulated turbine operating at steady state with a mass flow rate of 3.24 lb/s and exits at 2 lbf/in.$^2$ Kinetic and potential energy effects are negligible.

(a) Determine the maximum theoretical power that can be developed by the turbine, in hp, and the corresponding exit temperature, in °F.
(b) If the steam exits the turbine at 200°F, determine the isentropic turbine efficiency.

**7.74** Steam at 5 MPa and 600°C enters an insulated turbine operating at steady state and exits as saturated vapor at 50 kPa. Kinetic and potential energy effects are negligible. Determine

(a) the work developed by the turbine, in kJ per kg of steam flowing through the turbine.
(b) the isentropic turbine efficiency.

**7.75** Air at 4.5 bar, 550 K enters an insulated turbine operating at steady state and exits at 1.5 bar, 426 K. Kinetic and potential energy effects are negligible. Determine

(a) the work developed, in kJ per kg of air flowing.
(b) the isentropic turbine efficiency.

**7.76** Water vapor enters an insulated nozzle operating at steady state at 60 lbf/in.$^2$, 350°F, 10 ft/s and exits at 35 lbf/in.$^2$ If the isentropic nozzle efficiency is 94%, determine the exit velocity, in ft/s.

**7.77** Water vapor enters an insulated nozzle operating at steady state at 100 lbf/in.$^2$, 500°F, 100 ft/s and expands to 40 lbf/in.$^2$ If the isentropic nozzle efficiency is 95%, determine the velocity at the exit, in ft/s.

**7.78** Air enters an insulated nozzle operating at steady state at 80 lbf/in.$^2$, 120°F, 10 ft/s with a mass flow rate of 0.4 lb/s. At the exit, the velocity is 914 ft/s and the pressure is 50 lbf/in.$^2$ Determine

(a) the isentropic nozzle efficiency.
(b) the exit area, in ft$^2$.

**7.79** Refrigerant 134a enters a compressor operating at steady state as saturated vapor at −4°C and exits at a pressure of 8 bar. There is no significant heat transfer with the surroundings, and kinetic and potential energy effects can be ignored.

(a) Determine the minimum theoretical work input required, in kJ per kg of refrigerant flowing through the compressor, and the corresponding exit temperature, in °C.
(b) If the refrigerant exits at a temperature of 40°C, determine the isentropic compressor efficiency.

**7.80** Air enters an insulated compressor operating at steady state at 1.05 bar, 23°C with a mass flow rate of 1.8 kg/s and exits at 2.9 bar. Kinetic and potential energy effects are negligible.

(a) Determine the minimum theoretical power input required, in kW, and the corresponding exit temperature, in °C.
(b) If the exit temperature is 147°C, determine the power input, in kW, and the isentropic compressor efficiency.

**7.81** Refrigerant 134a enters a compressor operating at steady state as saturated vapor at −4°C and exits at a pressure of 14 bar. The isentropic compressor efficiency is 75%. Heat transfer between the compressor and its surroundings can be ignored. Kinetic and potential energy effects are also negligible. Determine

(a) the exit temperature, in °C.
(b) the work input, in kJ per kg of refrigerant flowing.

**7.82** (CD-ROM)

**7.83** (CD-ROM)

**7.84** Figure P7.84 shows liquid water at 80 lbf/in.$^2$, 300°F entering a flash chamber through a valve at the rate of 22 lb/s. At the valve exit, the pressure is 42 lbf/in.$^2$ Saturated liquid at 40 lbf/in.$^2$ exits from the bottom of the flash chamber and saturated vapor at 40 lbf/in.$^2$ exits from near the top. The vapor stream is fed to a steam turbine having an isentropic efficiency of 90% and an exit pressure of 2 lbf/in.$^2$ For steady-state

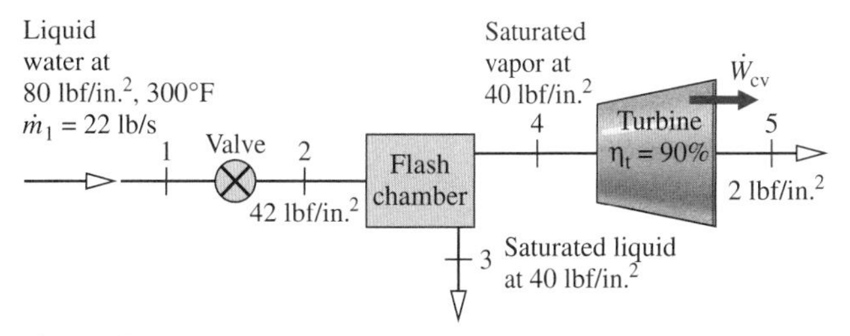


*Figure P7.84*

operation, negligible heat transfer with the surroundings, and no significant kinetic and potential energy effects, determine the
(a) power developed by the turbine, in Btu/s.
(b) rates of entropy production, each in Btu/s · °R, for the valve, the flash chamber, and the turbine. Compare.

### Internally Reversible Flow Processes and Related Applications

**7.85** Air enters a compressor operating at steady state at 17°C, 1 bar and exits at a pressure of 5 bar. Kinetic and potential energy changes can be ignored. If there are no internal irreversibilities, evaluate the work and heat transfer, each in kJ per kg of air flowing, for the following cases:
(a) isothermal compression.
(b) polytropic compression with $n = 1.3$.
(c) adiabatic compression.

Sketch the processes on $p$–$v$ and $T$–$s$ coordinates and associate areas on the diagrams with the work and heat transfer in each case. Referring to your sketches, compare for these cases the magnitudes of the work, heat transfer, and final temperatures, respectively.

**7.86** (CD-ROM)

**7.87** Refrigerant 134a enters a compressor operating at steady state as saturated vapor at 2 bar with a volumetric flow rate of $1.9 \times 10^{-2}$ m³/s. The refrigerant is compressed to a pressure of 8 bar in an internally reversible process according to $pv^{1.03} =$ *constant*. Neglecting kinetic and potential energy effects, determine
(a) the power required, in kW.
(b) the rate of heat transfer, in kW.

**7.88** Compare the work required at steady state to compress *water vapor* isentropically to 3 MPa from the saturated vapor state at 0.1 MPa to the work required to pump *liquid water* isentropically to 3 MPa from the saturated liquid state at 0.1 MPa, each in kJ per kg of water flowing through the device. Kinetic and potential energy effects can be ignored.

**7.89** (CD-ROM)

**7.90** As shown in Fig. P7.90, water flows from an elevated reservoir through a hydraulic turbine. The pipe diameter is constant, and operation is at steady state. Estimate the minimum mass flow rate, in kg/s, that would be required for a turbine power output of 1 MW. The local acceleration of gravity is 9.8 m/s².

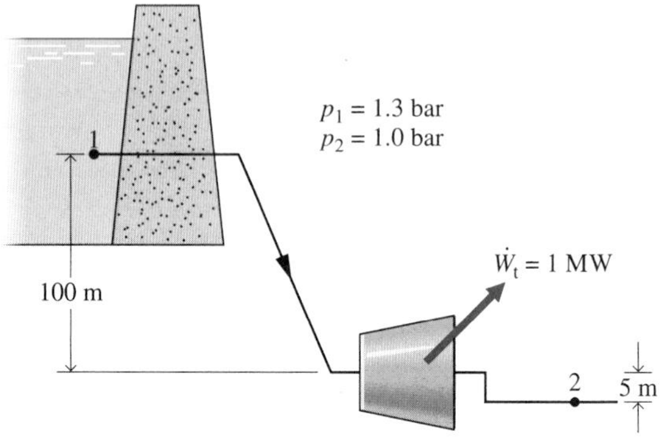


*Figure P7.90*

**7.91** Liquid water at 70°F, 1 ft/s enters a pipe and flows to a location where the pressure is 14.7 lbf/in.², the velocity is 20 ft/s, and the elevation is 30 ft above the inlet. The local acceleration of gravity is 32 ft/s². Ignoring internal irreversibilities, determine the pressure, in lbf/in.², required at the pipe inlet. Would the actual pressure required at the pipe inlet be greater or less than the calculated value? Explain.

**7.92** A pump operating at steady state draws water at 55°F from 10 ft underground where the pressure is 15 lbf/in.² and delivers it 12 ft above ground at a pressure of 45 lbf/in.² and a mass flow rate of 30 lb/s. In the absence of internal irreversibilities, determine the power required by the pump, in horsepower, ignoring kinetic energy effects. The local acceleration of gravity is 32.2 ft/s². Would the actual power required by the pump be greater or less than the calculated value? Explain.

**7.93** (CD-ROM)

**7.94** (CD-ROM)

# VAPOR POWER AND REFRIGERATION SYSTEMS

## *Introduction...*

An important engineering goal is to devise systems that accomplish desired types of energy conversion. The ***objective*** of the present chapter is to study *vapor* power and refrigeration systems in which the *working fluid* is alternatively vaporized and condensed. In the first part of the chapter vapor power systems are considered. Vapor refrigeration systems, including heat pump systems, are discussed in the second part of the chapter.

*chapter objective*

## Vapor Power Systems

This part of the chapter is concerned with vapor power-generating systems that produce a net power output from a fossil fuel, solar, or nuclear input. We describe some of the practical arrangements employed for power production and illustrate how such power plants can be modeled as thermal systems. In Chapter 9, we study internal combustion engines and gas turbines in which the working fluid remains a gas.

## 8.1 Modeling Vapor Power Systems

The processes taking place in power-generating systems are sufficiently complicated that idealizations are required to develop thermodynamic models. Such modeling is an important initial step in engineering design. Although the study of simplified models generally leads only to qualitative conclusions about the performance of the corresponding actual devices, models often allow deductions about how changes in major operating parameters affect actual performance. They also provide relatively simple settings in which to discuss the functions and benefits of features intended to improve overall performance.

The vast majority of electrical generating plants are variations of vapor power plants in which water is the working fluid. The basic components of a simplified fossil-fuel vapor power plant are shown schematically in Fig. 8.1. To facilitate thermodynamic analysis, the overall plant can be broken down into the four major subsystems identified by the letters A through D on the diagram. The focus of our considerations in this part of the chapter is subsystem A, where the important energy conversion from *heat to work* occurs. But first, let us briefly consider the other subsystems.

The function of subsystem B is to supply the energy required to vaporize the water passing through the boiler. In fossil-fuel plants, this is accomplished by heat transfer *to* the working fluid passing through tubes and drums in the boiler *from* the hot gases produced by the combustion of a fossil fuel. In nuclear plants, the origin of the energy is a controlled nuclear reaction taking place in an isolated reactor building. Pressurized water, a liquid metal,

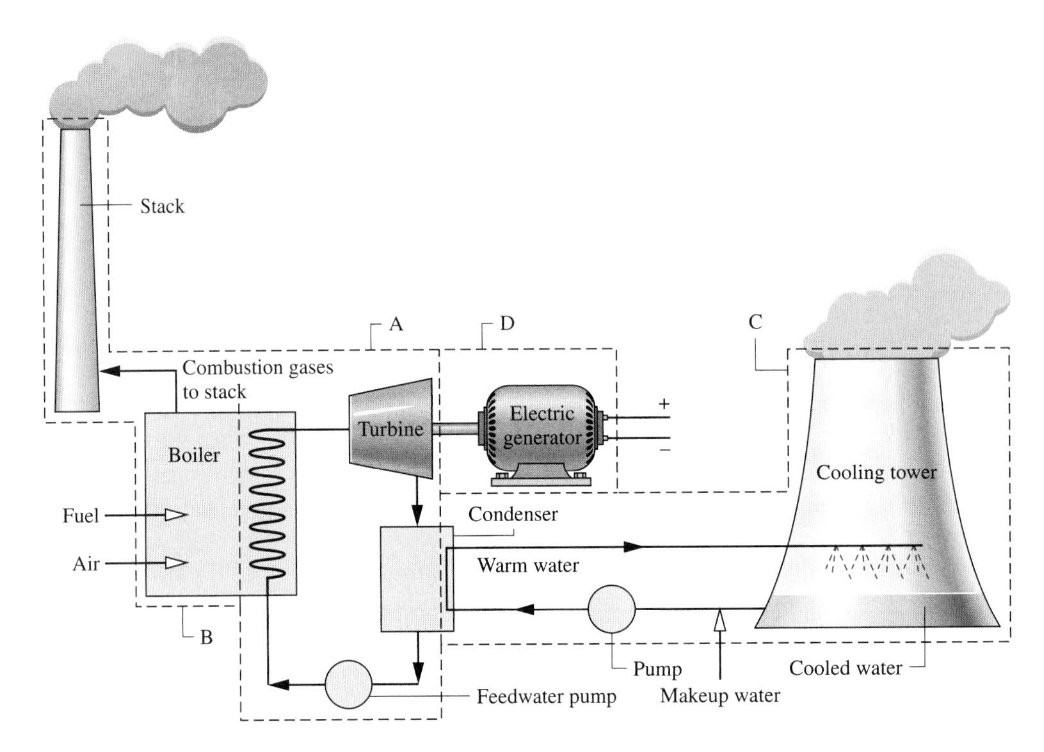


*Figure 8.1* Components of a simple vapor power plant.

or a gas such as helium can be used to transfer energy released in the nuclear reaction to the working fluid in specially designed heat exchangers. Solar power plants have receivers for concentrating and collecting solar radiation to vaporize the working fluid. Regardless of the energy source, the vapor produced in the boiler passes through a turbine, where it expands to a lower pressure. The shaft of the turbine is connected to an electric generator (subsystem D). The vapor leaving the turbine passes through the condenser, where it condenses on the outside of tubes carrying cooling water. The cooling water circuit comprises subsystem C. For the plant shown, the cooling water is sent to a cooling tower, where energy taken up in the condenser is rejected to the atmosphere. The cooling water is then recirculated through the condenser.

Concern for the environment and safety considerations govern what is allowable in the interactions between subsystems B and C and their surroundings. One of the major difficulties in finding a site for a vapor power plant is access to sufficient quantities of cooling water. For this reason and to minimize *thermal pollution* effects, most power plants now employ cooling towers. In addition to the question of cooling water, the safe processing and delivery of fuel, the control of pollutant discharges, and the disposal of wastes are issues that must be dealt with in both fossil-fueled and nuclear-fueled plants to ensure safety and operation with an acceptable level of environmental impact. Solar power plants are generally regarded as nonpolluting and safe but as yet are not widely used.

Returning now to subsystem A of Fig. 8.1, observe that each unit of mass periodically undergoes a thermodynamic cycle as the working fluid circulates through the series of four interconnected components. Accordingly, several concepts related to thermodynamic *power cycles* introduced in previous chapters are important for the present discussions. You will recall that the conservation of energy principle requires that the net work developed by a power cycle equals the net heat added. An important deduction from the second law is that the thermal efficiency, which indicates the extent to which the heat added is converted to a net work output, must be less than 100%. Previous discussions also have indicated that improved thermodynamic performance accompanies the reduction of irreversibilities. The extent to which irreversibilities can be reduced in power-generating systems depends on thermodynamic, economic, and other factors, however.

## 8.2 Analyzing Vapor Power Systems—Rankine Cycle

All of the fundamentals required for the thermodynamic analysis of power-generating systems already have been introduced. They include the conservation of mass and conservation of energy principles, the second law of thermodynamics, and thermodynamic data. These principles apply to individual plant components such as turbines, pumps, and heat exchangers as well as to the most complicated overall power plants. The object of this section is to introduce the ***Rankine cycle,*** which is a thermodynamic cycle that models the subsystem labeled A on Fig. 8.1. The presentation begins by considering the thermodynamic analysis of this subsystem.

*Rankine cycle*

### 8.2.1 Evaluating Principal Work and Heat Transfers

METHODOLOGY UPDATE

The principal work and heat transfers of subsystem A are illustrated in Fig. 8.2. In subsequent discussions, these energy transfers are taken to be *positive in the directions of the arrows.* The unavoidable stray heat transfer that takes place between the plant components and their surroundings is neglected here for simplicity. Kinetic and potential energy changes are also ignored. Each component is regarded as operating at steady state. Using the conservation of mass and conservation of energy principles together with these idealizations, we develop expressions for the energy transfers shown on Fig. 8.2 beginning at state 1 and proceeding through each component in turn.

***Turbine.*** Vapor from the boiler at state 1, having an elevated temperature and pressure, expands through the turbine to produce work and then is discharged to the condenser at state 2 with relatively low pressure. Neglecting heat transfer with the surroundings, the mass and energy rate balances for a control volume around the turbine reduce at steady state to give

$$0 = \dot{Q}_{cv}^{\,0} - \dot{W}_t + \dot{m}\left[h_1 - h_2 + \frac{V_1^2 - V_2^2}{2}^{\,0} + g(z_1 - z_2)^{\,0}\right]$$

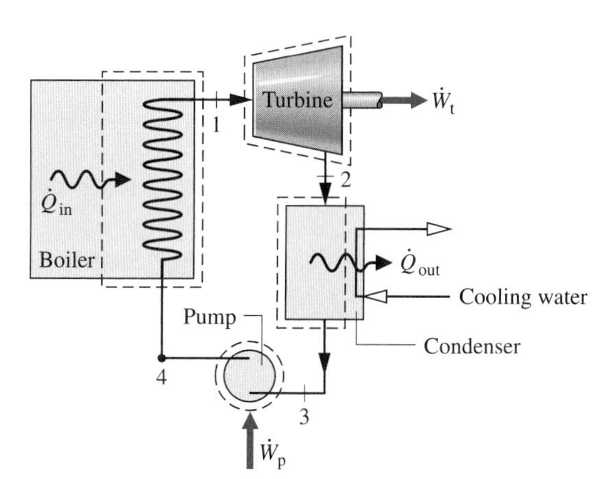


*Figure 8.2* Principal work and heat transfers of subsystem A.

which reduces to

$$\frac{\dot{W}_t}{\dot{m}} = h_1 - h_2 \tag{8.1}$$

where $\dot{m}$ denotes the mass flow rate of the working fluid, and $\dot{W}_t/\dot{m}$ is the rate at which work is developed per unit of mass of steam passing through the turbine. As noted above, kinetic and potential energy changes are ignored.

***Condenser.*** In the condenser there is heat transfer from the vapor to cooling water flowing in a separate stream. The vapor condenses and the temperature of the cooling water increases. At steady state, mass and energy rate balances for a control volume enclosing the condensing side of the heat exchanger give

$$\frac{\dot{Q}_{out}}{\dot{m}} = h_2 - h_3 \tag{8.2}$$

where $\dot{Q}_{out}/\dot{m}$ is the rate at which energy is transferred by heat *from* the working fluid to the cooling water per unit mass of working fluid passing through the condenser. This energy transfer is positive in the direction of the arrow on Fig. 8.2.

***Pump.*** The liquid condensate leaving the condenser at 3 is pumped from the condenser into the higher pressure boiler. Taking a control volume around the pump and assuming no heat transfer with the surroundings, mass and energy rate balances give

$$\frac{\dot{W}_p}{\dot{m}} = h_4 - h_3 \tag{8.3}$$

where $\dot{W}_p/\dot{m}$ is the rate of power *input* per unit of mass passing through the pump. This energy transfer is positive in the direction of the arrow on Fig. 8.2.

*feedwater*

***Boiler.*** The working fluid completes a cycle as the liquid leaving the pump at 4, called the boiler ***feedwater,*** is heated to saturation and evaporated in the boiler. Taking a control volume enclosing the boiler tubes and drums carrying the feedwater from state 4 to state 1, mass and energy rate balances give

$$\frac{\dot{Q}_{in}}{\dot{m}} = h_1 - h_4 \tag{8.4}$$

where $\dot{Q}_{in}/\dot{m}$ is the rate of heat transfer from the energy source into the working fluid per unit mass passing through the boiler.

*thermal efficiency*

***Performance Parameters.*** The thermal efficiency gauges the extent to which the energy input to the working fluid passing through the boiler is converted to the *net* work output. Using the quantities and expressions just introduced, the ***thermal efficiency*** of the power cycle of Fig. 8.2 is

$$\eta = \frac{\dot{W}_t/\dot{m} - \dot{W}_p/\dot{m}}{\dot{Q}_{in}/\dot{m}} = \frac{(h_1 - h_2) - (h_4 - h_3)}{h_1 - h_4} \tag{8.5a}$$

The net work output equals the net heat input. Thus, the thermal efficiency can be expressed alternatively as

$$\eta = \frac{\dot{Q}_{in}/\dot{m} - \dot{Q}_{out}/\dot{m}}{\dot{Q}_{in}/\dot{m}} = 1 - \frac{\dot{Q}_{out}/\dot{m}}{\dot{Q}_{in}/\dot{m}}$$

$$= 1 - \frac{(h_2 - h_3)}{(h_1 - h_4)} \tag{8.5b}$$

The *heat rate* is the amount of energy added by heat transfer to the cycle, usually in Btu, to produce a unit of net work output, usually in kW · h. Accordingly, the heat rate, which is inversely proportional to the thermal efficiency, has units of Btu/kW · h.

*heat rate*

Another parameter used to describe power plant performance is the *back work ratio,* or bwr, defined as the ratio of the pump work input to the work developed by the turbine. With Eqs. 8.1 and 8.3, the back work ratio for the power cycle of Fig. 8.2 is

*back work ratio*

$$\text{bwr} = \frac{\dot{W}_p/\dot{m}}{\dot{W}_t/\dot{m}} = \frac{(h_4 - h_3)}{(h_1 - h_2)} \tag{8.6}$$

Examples to follow illustrate that the change in specific enthalpy for the expansion of vapor through the turbine is normally many times greater than the increase in enthalpy for the liquid passing through the pump. Hence, the back work ratio is characteristically quite low for vapor power plants.

Provided states 1 through 4 are fixed, Eqs. 8.1 through 8.6 can be applied to determine the thermodynamic performance of a simple vapor power plant. Since these equations have been developed from mass and energy rate balances, they apply equally for actual performance when irreversibilities are present and for idealized performance in the absence of such effects. It might be surmised that the irreversibilities of the various power plant components can affect overall performance, and this is the case. Even so, it is instructive to consider an idealized cycle in which irreversibilities are assumed absent, for such a cycle establishes an *upper limit* on the performance of the Rankine cycle. The ideal cycle also provides a simple setting in which to study various aspects of vapor power plant performance.

## 8.2.2 Ideal Rankine Cycle

If the working fluid passes through the various components of the simple vapor power cycle without irreversibilities, frictional pressure drops would be absent from the boiler and condenser, and the working fluid would flow through these components at constant pressure. Also, in the absence of irreversibilities and heat transfer with the surroundings, the processes through the turbine and pump would be isentropic. A cycle adhering to these idealizations is the *ideal Rankine cycle* shown in Fig. 8.3.

*ideal Rankine cycle*

Referring to Fig. 8.3, we see that the working fluid undergoes the following series of internally reversible processes:

***Process 1–2:*** Isentropic expansion of the working fluid through the turbine from saturated vapor at state 1 to the condenser pressure.

***Process 2–3:*** Heat transfer *from* the working fluid as it flows at constant pressure through the condenser with saturated liquid at state 3.

***Process 3–4:*** Isentropic compression in the pump to state 4 in the compressed liquid region.

***Process 4–1:*** Heat transfer *to* the working fluid as it flows at constant pressure through the boiler to complete the cycle.

The ideal Rankine cycle also includes the possibility of superheating the vapor, as in cycle 1′–2′–3–4–1′. The importance of superheating is discussed in Sec. 8.3.

Since the ideal Rankine cycle consists of internally reversible processes, areas under the process lines of Fig. 8.3 can be interpreted as heat transfers per unit of mass flowing. Applying Eq. 7.40, area 1-b-c-4-a-1 represents the heat transfer to the working fluid passing through the boiler and area 2-b-c-3-2 is the heat transfer from the working fluid passing through the condenser, each per unit of mass flowing. The enclosed area 1-2-3-4-a-1 can be interpreted as the net heat input or, equivalently, the net work output, each per unit of mass flowing.

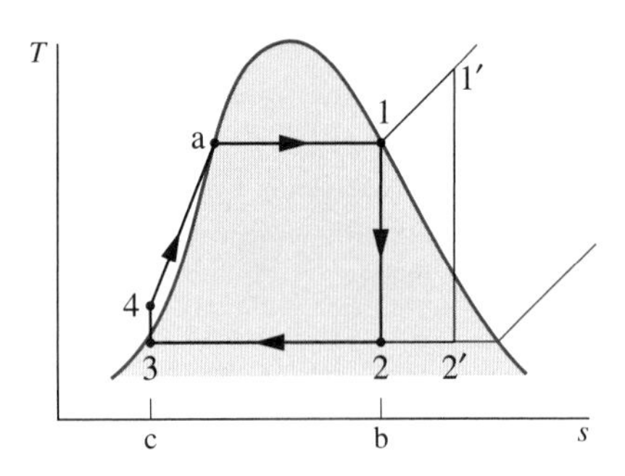


*Figure 8.3* Temperature–entropy diagram of the ideal Rankine cycle.

Because the pump is idealized as operating without irreversibilities, Eq. 7.43a can be invoked as an alternative to Eq. 8.3 for evaluating the pump work. That is,

$$\left(\frac{\dot{W}_p}{\dot{m}}\right)_{\substack{\text{int}\\\text{rev}}} = \int_3^4 v\, dp \tag{8.7a}$$

where the minus sign has been dropped for consistency with the positive value for pump work in Eq. 8.3. The subscript "int rev" has been retained as a reminder that this expression is restricted to an internally reversible process through the pump. No such designation is required by Eq. 8.3, however, because it expresses the conservation of mass and energy principles and thus is not restricted to processes that are internally reversible.

Evaluation of the integral of Eq. 8.7a requires a relationship between the specific volume and pressure for the process. Because the specific volume of the liquid normally varies only slightly as the liquid flows from the inlet to the exit of the pump, a plausible approximation to the value of the integral can be had by taking the specific volume at the pump inlet, $v_3$, as constant for the process. Then

$$\left(\frac{\dot{W}_p}{\dot{m}}\right)_{\substack{\text{int}\\\text{rev}}} \approx v_3(p_4 - p_3) \tag{8.7b}$$

The next example illustrates the analysis of an ideal Rankine cycle. Note that a minor departure from our usual problem-solving methodology is used in this example and examples to follow. In the ***Properties*** portion of the solution, attention is focused on the systematic evaluation of specific enthalpies and other required property values at each numbered state in the cycle. This eliminates the need to interrupt the solution repeatedly with property determinations and reinforces what is known about the processes in each component, since given information and assumptions are normally required to fix each of the numbered states.

## *Example 8.1* Ideal Rankine Cycle

Steam is the working fluid in an ideal Rankine cycle. Saturated vapor enters the turbine at 8.0 MPa and saturated liquid exits the condenser at a pressure of 0.008 MPa. The *net* power output of the cycle is 100 MW. Determine for the cycle **(a)** the thermal efficiency, **(b)** the back work ratio, **(c)** the mass flow rate of the steam, in kg/h, **(d)** the rate of heat transfer, $\dot{Q}_{in}$, into the working fluid as it passes through the boiler, in MW, **(e)** the rate of heat transfer, $\dot{Q}_{out}$, from the condensing steam as it passes through the condenser, in MW, **(f)** the mass flow rate of the condenser cooling water, in kg/h, if cooling water enters the condenser at 15°C and exits at 35°C.

### Solution

***Known:*** An ideal Rankine cycle operates with steam as the working fluid. The boiler and condenser pressures are specified, and the net power output is given.

***Find:*** Determine the thermal efficiency, the back work ratio, the mass flow rate of the steam, in kg/h, the rate of heat transfer to the working fluid as it passes through the boiler, in MW, the rate of heat transfer from the condensing steam as it passes through the condenser, in MW, the mass flow rate of the condenser cooling water, which enters at 15°C and exits at 35°C.

***Schematic and Given Data:***

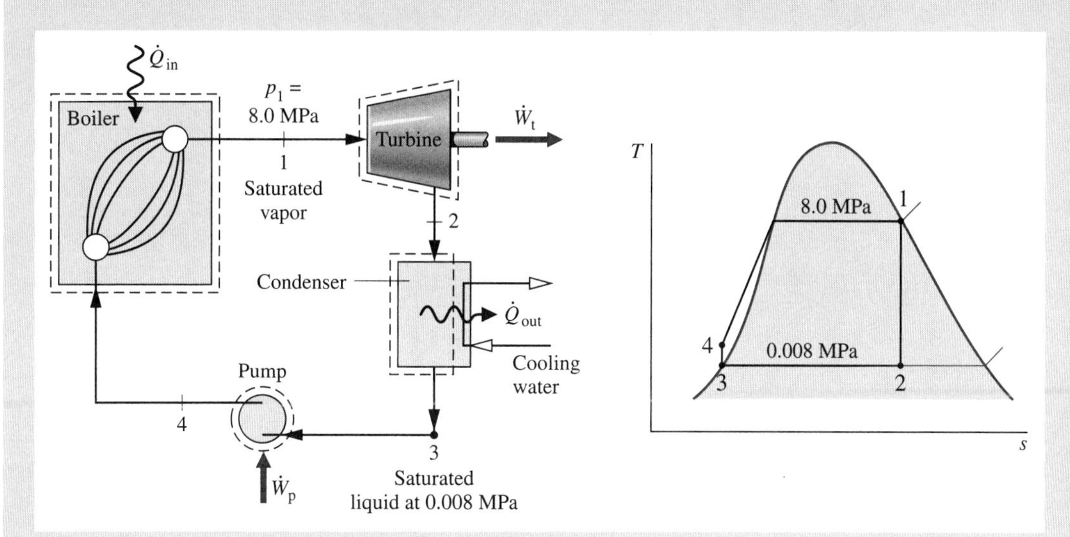


*Figure E8.1*

***Assumptions:***

1. Each component of the cycle is analyzed as a control volume at steady state. The control volumes are shown on the accompanying sketch by dashed lines.
2. All processes of the working fluid are internally reversible.
3. The turbine and pump operate adiabatically.
4. Kinetic and potential energy effects are negligible.
5. Saturated vapor enters the turbine. Condensate exits the condenser as saturated liquid.

***Properties:*** We begin by fixing each of the principal states located on the accompanying schematic and $T$–$s$ diagrams. Starting at the inlet to the turbine, the pressure is 8.0 MPa and the steam is a saturated vapor, so from Table T-3, $h_1 = 2758.0$ kJ/kg and $s_1 = 5.7432$ kJ/kg · K. ❶

State 2 is fixed by $p_2 = 0.008$ MPa and the fact that the specific entropy is constant for the adiabatic, internally reversible expansion through the turbine. Using saturated liquid and saturated vapor data from Table T-3, we find that the quality at state 2 is

$$x_2 = \frac{s_2 - s_f}{s_g - s_f} = \frac{5.7432 - 0.5926}{7.6361} = 0.6745$$

The enthalpy is then

$$h_2 = h_f + x_2 h_{fg} = 173.88 + (0.6745)2403.1$$
$$= 1794.8 \text{ kJ/kg}$$

State 3 is saturated liquid at 0.008 MPa, so $h_3 = 173.88$ kJ/kg.
State 4 is fixed by the boiler pressure $p_4$ and the specific entropy $s_4 = s_3$. The specific enthalpy $h_4$ can be found by interpolation in the compressed liquid tables. However, because compressed liquid data are relatively sparse, it is more convenient to solve Eq. 8.3 for $h_4$, using Eq. 8.7b to approximate the pump work. With this approach

$$h_4 = h_3 + \dot{W}_p/\dot{m} = h_3 + v_3(p_4 - p_3)$$

By inserting property values from Table T-3

$$h_4 = 173.88 \text{ kJ/kg} + (1.0084 \times 10^{-3} \text{ m}^3\text{/kg})(8.0 - 0.008)\text{MPa} \left| \frac{10^6 \text{ N/m}^2}{1 \text{ MPa}} \right| \left| \frac{1 \text{ kJ}}{10^3 \text{ N} \cdot \text{m}} \right|$$
$$= 173.88 + 8.06 = 181.94 \text{ kJ/kg}$$

***Analysis:*** **(a)** The *net* power developed by the cycle is

$$\dot{W}_{\text{cycle}} = \dot{W}_t - \dot{W}_p$$

Mass and energy rate balances for control volumes around the turbine and pump give, respectively

$$\frac{\dot{W}_t}{\dot{m}} = h_1 - h_2 \quad \text{and} \quad \frac{\dot{W}_p}{\dot{m}} = h_4 - h_3$$

where $\dot{m}$ is the mass flow rate of the steam. The rate of heat transfer to the working fluid as it passes through the boiler is determined using mass and energy rate balances as

$$\frac{\dot{Q}_{\text{in}}}{\dot{m}} = h_1 - h_4$$

The thermal efficiency is then

$$\eta = \frac{\dot{W}_t - \dot{W}_p}{\dot{Q}_{\text{in}}} = \frac{(h_1 - h_2) - (h_4 - h_3)}{h_1 - h_4}$$
$$= \frac{[(2758.0 - 1794.8) - (181.94 - 173.88)] \text{ kJ/kg}}{(2758.0 - 181.94) \text{ kJ/kg}}$$
$$= 0.371 \; (37.1\%) \; \triangleleft$$

**(b)** The back work ratio is

❷
$$\text{bwr} = \frac{\dot{W}_p}{\dot{W}_t} = \frac{h_4 - h_3}{h_1 - h_2} = \frac{(181.94 - 173.88) \text{ kJ/kg}}{(2758.0 - 1794.8) \text{ kJ/kg}}$$
$$= \frac{8.06}{963.2} = 8.37 \times 10^{-3} \; (0.84\%) \; \triangleleft$$

**(c)** The mass flow rate of the steam can be obtained from the expression for the net power given in part (a). Thus

$$\dot{m} = \frac{\dot{W}_{\text{cycle}}}{(h_1 - h_2) - (h_4 - h_3)}$$
$$= \frac{(100 \text{ MW})|10^3 \text{ kW/MW}||3600 \text{ s/h}|}{(963.2 - 8.06) \text{ kJ/kg}}$$
$$= 3.77 \times 10^5 \text{ kg/h} \; \triangleleft$$

**(d)** With the expression for $\dot{Q}_{\text{in}}$ from part (a) and previously determined specific enthalpy values

$$\dot{Q}_{\text{in}} = \dot{m}(h_1 - h_4)$$
$$= \frac{(3.77 \times 10^5 \text{ kg/h})(2758.0 - 181.94) \text{ kJ/kg}}{|3600 \text{ s/h}||10^3 \text{ kW/MW}|}$$
$$= 269.77 \text{ MW} \; \triangleleft$$

**(e)** Mass and energy rate balances applied to a control volume enclosing the steam side of the condenser give

$$\dot{Q}_{\text{out}} = \dot{m}(h_2 - h_3)$$
$$= \frac{(3.77 \times 10^5 \text{ kg/h})(1794.8 - 173.88) \text{ kJ/kg}}{|3600 \text{ s/h}||10^3 \text{ kW/MW}|}$$
$$= 169.75 \text{ MW} \; \triangleleft$$

❸ Note that the ratio of $\dot{Q}_{\text{out}}$ to $\dot{Q}_{\text{in}}$ is 0.629 (62.9%).

Alternatively, $\dot{Q}_{out}$ can be determined from an energy rate balance on the *overall* vapor power plant. At steady state, the net power developed equals the net rate of heat transfer to the plant

$$\dot{W}_{cycle} = \dot{Q}_{in} - \dot{Q}_{out}$$

Rearranging this expression and inserting values

$$\dot{Q}_{out} = \dot{Q}_{in} - \dot{W}_{cycle} = 269.77 \text{ MW} - 100 \text{ MW} = 169.77 \text{ MW}$$

The slight difference from the above value is due to round-off.

**(f)** Taking a control volume around the condenser, the mass and energy rate balances give at steady state

$$0 = \cancelto{0}{\dot{Q}_{cv}} - \cancelto{0}{\dot{W}_{cv}} + \dot{m}_{cw}(h_{cw,\,in} - h_{cw,\,out}) + \dot{m}(h_2 - h_3)$$

where $\dot{m}_{cw}$ is the mass flow rate of the cooling water. Solving for $\dot{m}_{cw}$

$$\dot{m}_{cw} = \frac{\dot{m}(h_2 - h_3)}{(h_{cw,\,out} - h_{cw,\,in})}$$

The numerator in this expression is evaluated in part (e). For the cooling water, $h \approx h_f(T)$, so with saturated liquid enthalpy values from Table T-2 at the entering and exiting temperatures of the cooling water

$$\dot{m}_{cw} = \frac{(169.75 \text{ MW})|10^3 \text{ kW/MW}||3600 \text{ s/h}|}{(146.68 - 62.99) \text{ kJ/kg}} = 7.3 \times 10^6 \text{ kg/h} \triangleleft$$

❶ Note that a slightly revised problem-solving methodology is used in this example problem: We begin with a systematic evaluation of the specific enthalpy at each numbered state.

❷ Note that the back work ratio is relatively low for the Rankine cycle. In the present case, the work required to operate the pump is less than 1% of the turbine output.

❸ In this example, 62.9% of the energy added to the working fluid by heat transfer is subsequently discharged to the cooling water. Although considerable energy is carried away by the cooling water, its usefulness is very limited because the water exits at a temperature only a few degrees greater than that of the surroundings.

## 8.2.3 Effects of Boiler and Condenser Pressures on the Rankine Cycle

In Sec. 6.4.2 we observed that the thermal efficiency of power cycles tends to increase as the average temperature at which energy is added by heat transfer increases and/or the average temperature at which energy is rejected decreases. Let us apply this idea to study the effects on performance of the ideal Rankine cycle of changes in the boiler and condenser pressures. Although these findings are obtained with reference to the ideal Rankine cycle, they also hold qualitatively for actual vapor power plants.

Figure 8.4*a* shows two ideal cycles having the same condenser pressure but different boiler pressures. By inspection, the average temperature of heat addition is seen to be greater for the higher-pressure cycle 1′–2′–3′–4′–1′ than for cycle 1–2–3–4–1. It follows that increasing the boiler pressure of the ideal Rankine cycle tends to increase the thermal efficiency.

Figure 8.4*b* shows two cycles with the same boiler pressure but two different condenser pressures. One condenser operates at atmospheric pressure and the other at *less than* atmospheric pressure. The temperature of heat rejection for cycle 1–2–3–4–1 condensing at atmospheric pressure is 100°C (212°F). The temperature of heat rejection for the lower-pressure cycle 1–2″–3″–4″–1 is corresponding lower, so this cycle has the greater thermal efficiency. It follows that decreasing the condenser pressure tends to increase the thermal efficiency.

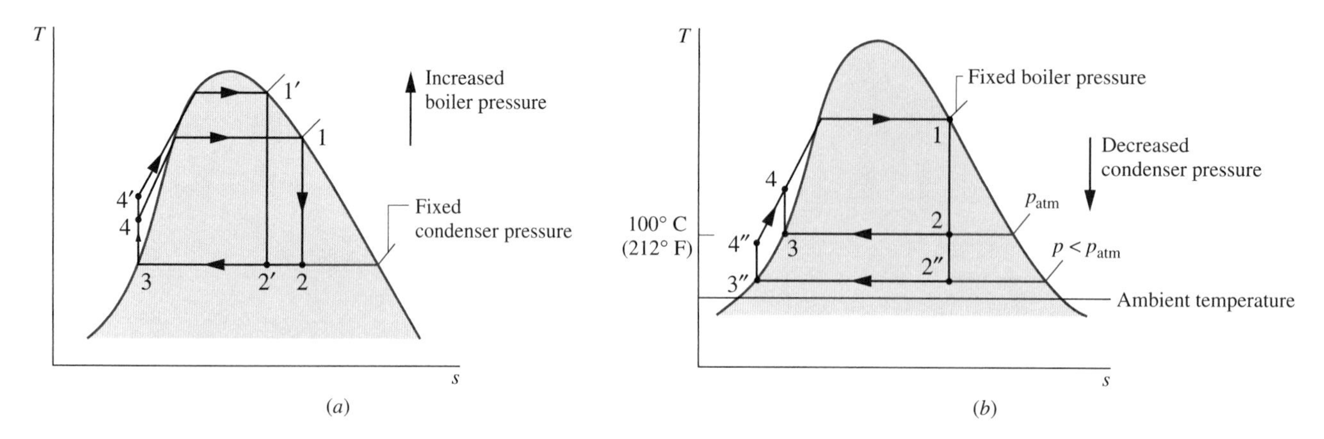


*Figure 8.4* Effects of varying operating pressures on the ideal Rankine cycle. (*a*) Effect of boiler pressure. (*b*) Effect of condenser pressure.

The lowest feasible condenser pressure is the saturation pressure corresponding to the ambient temperature, for this is the lowest possible temperature for heat rejection to the surroundings. The goal of maintaining the lowest practical turbine exhaust (condenser) pressure is a primary reason for including the condenser in a power plant. Liquid water at atmospheric pressure could be drawn into the boiler by a pump, and steam could be discharged directly to the atmosphere at the turbine exit. However, by including a condenser in which the steam side is operated at a pressure *below atmospheric,* the turbine has a lower-pressure region in which to discharge, resulting in a significant increase in net work and thermal efficiency. The addition of a condenser also allows the working fluid to flow in a closed loop. This arrangement permits continual circulation of the working fluid, so purified water that is less corrosive than tap water can be used.

## 8.2.4 Principal Irreversibilities and Losses

Irreversibilities and losses are associated with each of the four subsystems shown in Fig. 8.1. Some of these effects have a more pronounced influence on performance than others. Let us consider the irreversibilities and losses associated with the Rankine cycle.

***Turbine.*** The principal irreversibility experienced by the working fluid is associated with the expansion through the turbine. Heat transfer from the turbine to the surroundings represents a loss, but since it is usually of secondary importance, this loss is ignored in subsequent discussions. As illustrated by Process 1–2 of Fig. 8.5, an actual adiabatic expansion

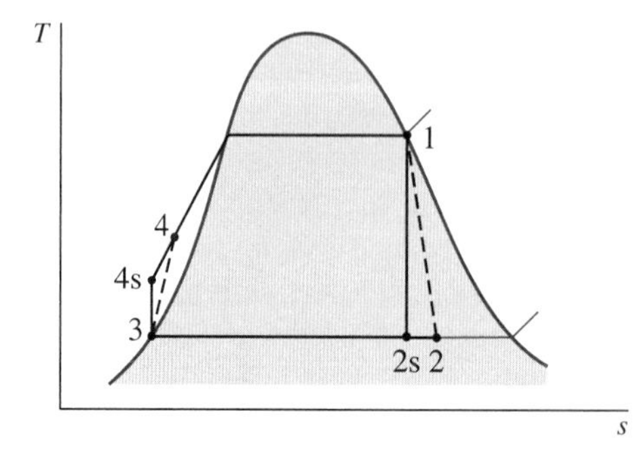


*Figure 8.5* Temperature–entropy diagram showing the effects of turbine and pump irreversibilities.

through the turbine is accompanied by an increase in entropy. The work developed per unit of mass in this process is less than for the corresponding isentropic expansion 1–2s. The isentropic turbine efficiency $\eta_t$ introduced in Sec. 7.7 allows the effect of irreversibilities within the turbine to be accounted for in terms of the actual and isentropic work amounts. Designating the states as in Fig. 8.5, the isentropic turbine efficiency is

$$\eta_t = \frac{(\dot{W}_t/\dot{m})}{(\dot{W}_t/\dot{m})_s} = \frac{h_1 - h_2}{h_1 - h_{2s}} \tag{8.8}$$

where the numerator is the actual work developed per unit of mass passing through the turbine and the denominator is the work for an isentropic expansion from the turbine inlet state to the turbine exhaust pressure. Irreversibilities within the turbine significantly reduce the net power output of the plant.

***Pump.*** The work input to the pump required to overcome frictional effects also reduces the net power output of the plant. In the absence of heat transfer to the surroundings, there would be an increase in entropy across the pump. Process 3–4 of Fig. 8.5 illustrates the actual pumping process. The work input for this process is *greater* than for the corresponding isentropic process 3–4s. The isentropic pump efficiency $\eta_p$ introduced in Sec. 7.7 allows the effect of irreversibilities within the pump to be accounted for in terms of the actual and isentropic work amounts. Designating the states as in Fig. 8.5, the isentropic pump efficiency is

$$\eta_p = \frac{(\dot{W}_p/\dot{m})_s}{(\dot{W}_p/\dot{m})} = \frac{h_{4s} - h_3}{h_4 - h_3} \tag{8.9}$$

In this expression, the pump work for the isentropic process appears in the numerator. The actual pump work, being the larger quantity, is the denominator. Because the pump work is so much less than the turbine work, irreversibilities in the pump have a much smaller impact on the net work of the cycle than do irreversibilities in the turbine.

***Other Nonidealities.*** The turbine and pump irreversibilities mentioned above are *internal* irreversibilities experienced by the working fluid as it flows around the closed loop of the Rankine cycle. In addition, there are other sources of nonideality. For example, frictional effects resulting in pressure drops are sources of internal irreversibility as the working fluid flows through the boiler, condenser, and piping connecting the various components. However, for simplicity such effects are ignored in the subsequent discussions. Thus, Fig. 8.5 shows no pressure drops for flow through the boiler and condenser or between plant components.

The most significant sources of irreversibility for a fossil-fueled vapor power plant are associated with the combustion of the fuel and the subsequent heat transfer from the hot combustion products to the cycle working fluid. These effects occur in the surroundings of the subsystem labeled A on Fig. 8.1 and thus are *external* irreversibilities for the Rankine cycle.

Another effect that occurs in the surroundings is the energy discharge to the cooling water as the working fluid condenses. Although considerable energy is carried away by the cooling water, its *usefulness* is severely limited. For condensers in which steam condenses near the ambient temperature, the cooling water experiences a temperature rise of *only a few degrees* over the temperature of the surroundings in passing through the condenser and thus has limited usefulness. Accordingly, the significance of the cooling water loss is *far less* than suggested by the magnitude of the energy transferred to the cooling water.

In the next example, the ideal Rankine cycle of Example 8.1 is modified to include the effects of irreversibilities in the turbine and pump.

## *Example 8.2* Rankine Cycle with Irreversibilities

Reconsider the vapor power cycle of Example 8.1, but include in the analysis that the turbine and the pump each have an isentropic efficiency of 85%. Determine for the modified cycle (a) the thermal efficiency, (b) the mass flow rate of steam, in kg/h, for a net power output of 100 MW, (c) the rate of heat transfer $\dot{Q}_{in}$ into the working fluid as it passes through the boiler, in MW, (d) the rate of heat transfer $\dot{Q}_{out}$ from the condensing steam as it passes through the condenser, in MW, (e) the mass flow rate of the condenser cooling water, in kg/h, if cooling water enters the condenser at 15°C and exits as 35°C. Discuss the effects on the vapor cycle of irreversibilities within the turbine and pump.

## Solution

***Known:*** A vapor power cycle operates with steam as the working fluid. The turbine and pump both have efficiencies of 85%.

***Find:*** Determine the thermal efficiency, the mass flow rate, in kg/h, the rate of heat transfer to the working fluid as it passes through the boiler, in MW, the heat transfer rate from the condensing steam as it passes through the condenser, in MW, and the mass flow rate of the condenser cooling water, in kg/h. Discuss.

***Schematic and Given Data:***

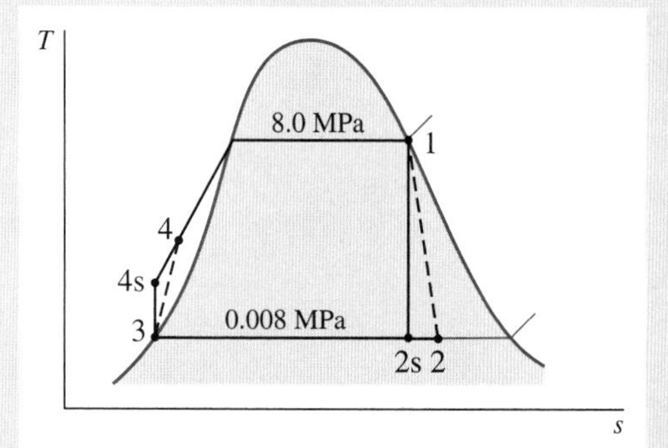


***Assumptions:***

1. Each component of the cycle is analyzed as a control volume at steady state.
2. The working fluid passes through the boiler and condenser at constant pressure. Saturated vapor enters the turbine. The condensate is saturated at the condenser exit.
3. The turbine and pump each operate adiabatically with an efficiency of 85%.
4. Kinetic and potential energy effects are negligible.

*Figure E8.2*

***Properties:*** Owing to the presence of irreversibilities during the expansion of the steam through the turbine, there is an increase in specific entropy from turbine inlet to exit, as shown on the accompanying $T$–$s$ diagram. Similarly, there is an increase in specific entropy from pump inlet to exit. Let us begin by fixing each of the principal states. State 1 is the same as in Example 8.1, so $h_1 = 2758.0$ kJ/kg and $s_1 = 5.7432$ kJ/kg · K.

The specific enthalpy at the turbine exit, state 2, can be determined using the turbine efficiency.

$$\eta_t = \frac{\dot{W}_t/\dot{m}}{(\dot{W}_t/\dot{m})_s} = \frac{h_1 - h_2}{h_1 - h_{2s}}$$

where $h_{2s}$ is the specific enthalpy at state 2s on the accompanying $T$–$s$ diagram. From the solution to Example 8.1, $h_{2s} = 1794.8$ kJ/kg. Solving for $h_2$ and inserting known values

$$\begin{aligned} h_2 &= h_1 - \eta_t(h_1 - h_{2s}) \\ &= 2758 - 0.85(2758 - 1794.8) = 1939.3 \text{ kJ/kg} \end{aligned}$$

State 3 is the same as in Example 8.1, so $h_3 = 173.88$ kJ/kg.

To determine the specific enthalpy at the pump exit, state 4, reduce mass and energy rate balances for a control volume around the pump to obtain $\dot{W}_p/\dot{m} = h_4 - h_3$. On rearrangement, the specific enthalpy at state 4 is

$$h_4 = h_3 + \dot{W}_p/\dot{m}$$

To determine $h_4$ from this expression requires the pump work, which can be evaluated using the pump efficiency $\eta_p$, as follows. By definition

$$\eta_p = \frac{(\dot{W}_p/\dot{m})_s}{(\dot{W}_p/\dot{m})}$$

The term $(\dot{W}_p/\dot{m})_s$ can be evaluated using Eq. 8.7b. Then solving for $\dot{W}_p/\dot{m}$ results in

$$\frac{\dot{W}_p}{\dot{m}} = \frac{v_3(p_4 - p_3)}{\eta_p}$$

The numerator of this expression was determined in the solution to Example 8.1. Accordingly,

$$\frac{\dot{W}_p}{\dot{m}} = \frac{8.06 \text{ kJ/kg}}{0.85} = 9.48 \text{ kJ/kg}$$

The specific enthalpy at the pump exit is then

$$h_4 = h_3 + \dot{W}_p/\dot{m} = 173.88 + 9.48 = 183.36 \text{ kJ/kg}$$

***Analysis:*** **(a)** The net power developed by the cycle is

$$\dot{W}_{cycle} = \dot{W}_t - \dot{W}_p = \dot{m}[(h_1 - h_2) - (h_4 - h_3)]$$

The rate of heat transfer to the working fluid as it passes through the boiler is

$$\dot{Q}_{in} = \dot{m}(h_1 - h_4)$$

Thus, the thermal efficiency is

$$\eta = \frac{(h_1 - h_2) - (h_4 - h_3)}{h_1 - h_4}$$

Inserting values

$$\eta = \frac{(2758 - 1939.3) - 9.48}{2758 - 183.36} = 0.314\ (31.4\%) \ \triangleleft$$

**(b)** With the net power expression of part (a), the mass flow rate of the steam is

$$\dot{m} = \frac{\dot{W}_{cycle}}{(h_1 - h_2) - (h_4 - h_3)}$$

$$= \frac{(100 \text{ MW})|3600 \text{ s/h}||10^3 \text{ kW/MW}|}{(818.7 - 9.48) \text{ kJ/kg}} = 4.449 \times 10^5 \text{ kg/h} \ \triangleleft$$

**(c)** With the expression for $\dot{Q}_{in}$ from part (a) and previously determined specific enthalpy values

$$\dot{Q}_{in} = \dot{m}(h_1 - h_4)$$

$$= \frac{(4.449 \times 10^5 \text{ kg/h})(2758 - 183.36) \text{ kJ/kg}}{|3600 \text{ s/h}||10^3 \text{ kW/MW}|} = 318.2 \text{ MW} \ \triangleleft$$

**(d)** The rate of heat transfer from the condensing steam to the cooling water is

$$\dot{Q}_{out} = \dot{m}(h_2 - h_3)$$

$$= \frac{(4.449 \times 10^5 \text{ kg/h})(1939.3 - 173.88) \text{ kJ/kg}}{|3600 \text{ s/h}||10^3 \text{ kW/MW}|} = 218.2 \text{ MW} \ \triangleleft$$

**(e)** The mass flow rate of the cooling water can be determined from

$$\dot{m}_{cw} = \frac{\dot{m}(h_2 - h_3)}{(h_{cw,out} - h_{cw,in})}$$

$$= \frac{(218.2 \text{ MW})|10^3 \text{ kW/MW}||3600 \text{ s/h}|}{(146.68 - 62.99) \text{ kJ/kg}} = 9.39 \times 10^6 \text{ kg/h} \ \triangleleft$$

The effect of irreversibilities within the turbine and pump can be gauged by comparing the present values with their counterparts in Example 8.1. In this example, the turbine work per unit of mass is less and the pump work per unit of mass is greater than in Example 8.1. The thermal efficiency in the present case is less than in the ideal case of the previous example. For a fixed net power output (100 MW), the smaller net work output per unit mass in the present case dictates a greater mass flow rate of steam. The magnitude of the heat transfer to the cooling water is greater in this example than in Example 8.1; consequently, a greater mass flow rate of cooling water would be required.

## 8.3 Improving Performance—Superheat and Reheat

The representations of the vapor power cycle considered thus far do not depict actual vapor power plants faithfully, for various modifications are usually incorporated to improve overall performance. In this section we consider two cycle modifications known as *superheat* and *reheat*. Both features are normally incorporated into vapor power plants.

Let us begin the discussion by noting that an increase in the boiler pressure or a decrease in the condenser pressure may result in a reduction of the steam quality at the exit of the turbine. This can be seen by comparing states 2′ and 2″ of Figs. 8.4*a* and 8.4*b* (p.194) to the corresponding state 2 of each diagram. If the quality of the mixture passing through the turbine becomes too low, the impact of liquid droplets in the flowing liquid–vapor mixture can erode the turbine blades, causing a decrease in the turbine efficiency and an increased need for maintenance. Accordingly, common practice is to maintain at least 90% quality ($x \geq 0.9$) at the turbine exit. The cycle modifications known as *superheat* and *reheat* permit advantageous operating pressures in the boiler and condenser and yet offset the problem of low quality of the turbine exhaust.

*superheat*

***Superheat.*** First, let us consider ***superheat.*** As we are not limited to having saturated vapor at the turbine inlet, further energy can be added by heat transfer to the steam, bringing it to a superheated vapor condition at the turbine inlet. This is accomplished in a separate heat exchanger called a superheater. The combination of boiler and superheater is referred to as a *steam generator*. Figure 8.3 (p. 190) shows an ideal Rankine cycle with superheated vapor at the turbine inlet: cycle 1′–2′–3–4–1′. The cycle with superheat has a higher average temperature of heat addition than the cycle without superheating (cycle 1–2–3–4–1), so the thermal efficiency is higher. Moreover, the quality at turbine exhaust state 2′ is greater than at state 2, which would be the turbine exhaust state without superheating. Accordingly, superheating also tends to alleviate the problem of low steam quality at the turbine exhaust. With sufficient superheating, the turbine exhaust state may even fall in the superheated vapor region.

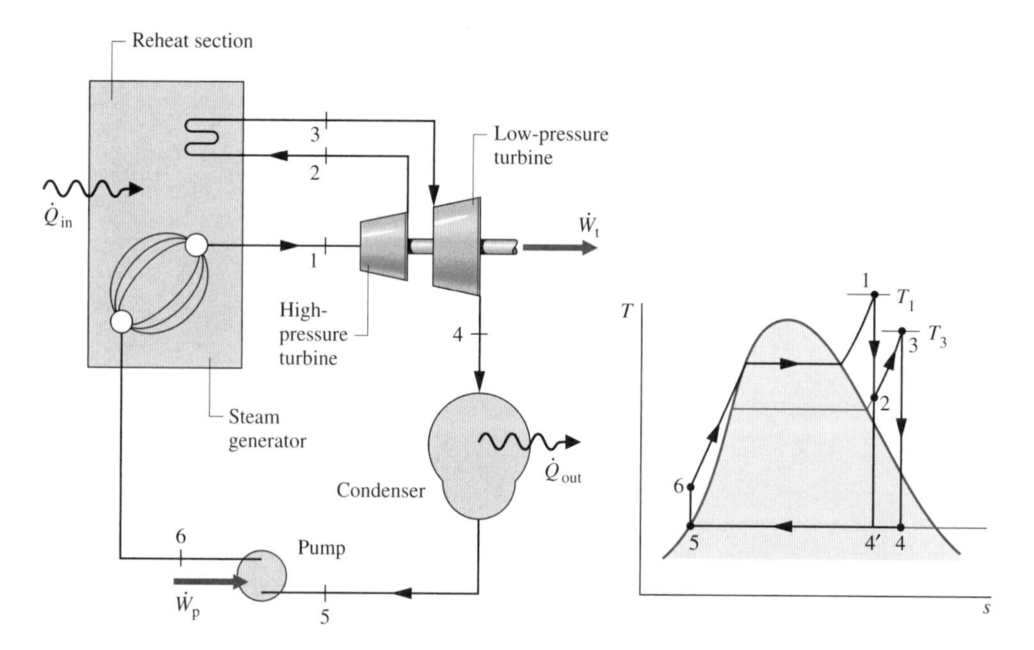


*Figure 8.6* Ideal reheat cycle.

***Reheat.*** A further modification normally employed in vapor power plants is ***reheat.*** With reheat, a power plant can take advantage of the increased efficiency that results with higher boiler pressures and yet avoid low-quality steam at the turbine exhaust. In the ideal reheat cycle shown in Fig. 8.6, the steam does not expand to the condenser pressure in a single stage. The steam expands through a first-stage turbine (Process 1–2) to some pressure between the steam generator and condenser pressures. The steam is then reheated in the steam generator (Process 2–3). Ideally, there would be no pressure drop as the steam is reheated. After reheating, the steam expands in a second-stage turbine to the condenser pressure (Process 3–4). The principal advantage of reheat is to increase the quality of the steam at the turbine exhaust. This can be seen from the $T$–$s$ diagram of Fig. 8.6 by comparing state 4 with state 4′, the turbine exhaust state without reheating. When computing the thermal efficiency of a reheat cycle, it is necessary to account for the work output of both turbine stages as well as the total heat addition occurring in the vaporization/superheating and reheating processes. This calculation is illustrated in Example 8.3, where the ideal Rankine of Example 8.1 is modified to include superheat, reheat, and the effect of turbine irreversibilities.

*reheat*

## *Example 8.3* Reheat Cycle

Steam is the working fluid in a Rankine cycle with superheat and reheat. Steam enters the first-stage turbine at 8.0 MPa, 480°C, and expands to 0.7 MPa. It is then reheated to 440°C before entering the second-stage turbine, where it expands to the condenser pressure of 0.008 MPa. The *net* power output is 100 MW. If the turbine stages and pump are isentropic, determine **(a)** the thermal efficiency of the cycle, **(b)** the mass flow rate of steam, in kg/h, **(c)** the rate of heat transfer $\dot{Q}_{out}$ from the condensing steam as it passes through the condenser, in MW. Discuss the effects of reheat on the vapor power cycle. **(d)** If each turbine stage has an isentropic efficiency of 85%, determine the thermal efficiency. **(e)** Plot the thermal efficiency versus the turbine stage efficiency ranging from 85 to 100%.

### Solution

***Known:*** A reheat cycle operates with steam as the working fluid. Operating pressures and temperatures are specified, and the net power output is given.

***Find:*** If the turbine stages and pump are isentropic, determine the thermal efficiency, the mass flow rate of the steam, in kg/h, and the heat transfer rate from the condensing steam as it passes through the condenser, in MW. Discuss. If each turbine stage has a specified isentropic efficiency, determine the thermal efficiency. Plot.

***Schematic and Given Data:***

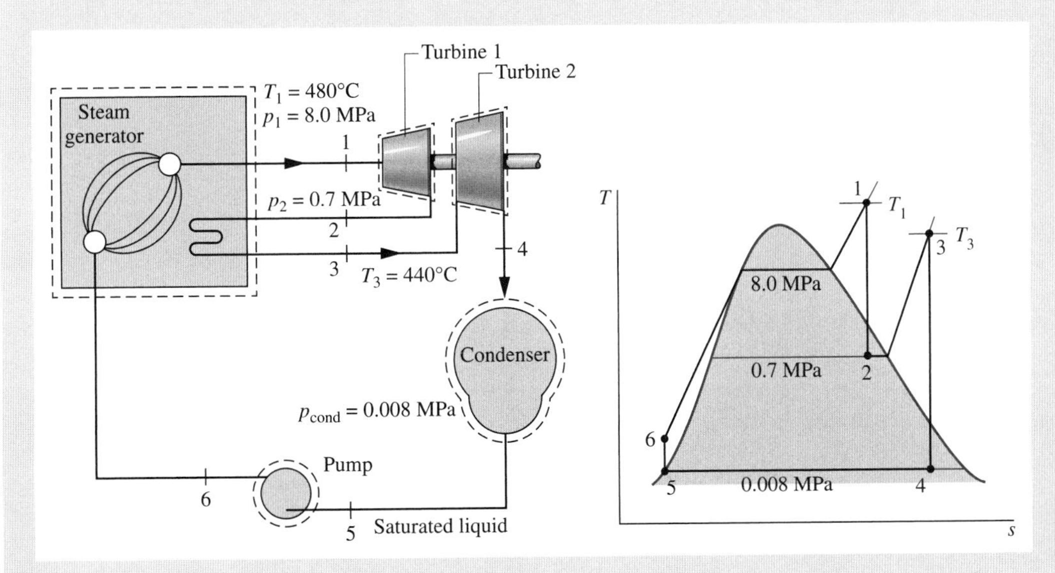


*Figure E8.3a*

***Assumptions:***

1. Each component in the cycle is analyzed as a control volume at steady state. The control volumes are shown on the accompanying sketch by dashed lines.
2. In parts (a)–(c), all processes of the working fluid are internally reversible. In parts (d) and (e), isentropic efficiencies are specified for the turbine stages.
3. The turbine and pump operate adiabatically.
4. Condensate exits the condenser as saturated liquid.
5. Kinetic and potential energy effects are negligible.

***Properties:*** To begin, let us fix each of the principal states of the ideal cycle shown in Fig. E8.3*a*. Starting at the inlet to the first turbine stage, the pressure is 8.0 MPa and the temperature is 480°C, so the steam is a superheated vapor. From Table T-4, $h_1 = 3348.4$ kJ/kg and $s_1 = 6.6586$ kJ/kg · K.

State 2 is fixed by $p_2 = 0.7$ MPa and $s_2 = s_1$ for the isentropic expansion through the first-stage turbine. Using saturated liquid and saturated vapor data from Table T-3, the quality at state 2 is

$$x_2 = \frac{s_2 - s_f}{s_g - s_f} = \frac{6.6586 - 1.9922}{6.708 - 1.9922} = 0.9895$$

The specific enthalpy is then

$$\begin{aligned} h_2 &= h_f + x_2 h_{fg} \\ &= 697.22 + (0.9895)2066.3 = 2741.8 \text{ kJ/kg} \end{aligned}$$

State 3 is superheated vapor with $p_3 = 0.7$ MPa and $T_3 = 440$°C, so from Table T-4, $h_3 = 3353.3$ kJ/kg and $s_3 = 7.7571$ kJ/kg · K.

To fix state 4, use $p_4 = 0.008$ MPa and $s_4 = s_3$ for the isentropic expansion through the second-stage turbine. With data from Table T-3, the quality at state 4 is

$$x_4 = \frac{s_4 - s_f}{s_g - s_f} = \frac{7.7571 - 0.5926}{8.2287 - 0.5926} = 0.9382$$

The specific enthalpy is

$$h_4 = 173.88 + (0.9382)2403.1 = 2428.5 \text{ kJ/kg}$$

State 5 is saturated liquid at 0.008 MPa, so $h_5 = 173.88$ kJ/kg. Finally, the state at the pump exit is the same as in Example 8.1, so $h_6 = 181.94$ kJ/kg.

***Analysis:*** (a) The *net* power developed by the cycle is

$$\dot{W}_{cycle} = \dot{W}_{t1} + \dot{W}_{t2} - \dot{W}_p$$

Mass and energy rate balances for the two turbine stages and the pump reduce to give, respectively

$$\begin{aligned} &\text{Turbine 1:} \quad \dot{W}_{t1}/\dot{m} = h_1 - h_2 \\ &\text{Turbine 2:} \quad \dot{W}_{t2}/\dot{m} = h_3 - h_4 \\ &\text{Pump:} \quad \dot{W}_p/\dot{m} = h_6 - h_5 \end{aligned}$$

where $\dot{m}$ is the mass flow rate of the steam.

The total rate of heat transfer to the working fluid as it passes through the boiler–superheater and reheater is

$$\frac{\dot{Q}_{in}}{\dot{m}} = (h_1 - h_6) + (h_3 - h_2)$$

Using these expressions, the thermal efficiency is

$$\eta = \frac{(h_1 - h_2) + (h_3 - h_4) - (h_6 - h_5)}{(h_1 - h_6) + (h_3 - h_2)}$$

$$= \frac{(3348.4 - 2741.8) + (3353.3 - 2428.5) - (181.94 - 173.88)}{(3348.4 - 181.94) + (3353.3 - 2741.8)}$$

$$= \frac{606.6 + 924.8 - 8.06}{3166.5 + 611.5} = \frac{1523.3 \text{ kJ/kg}}{3778 \text{ kJ/kg}} = 0.403 \, (40.3\%) \quad \triangleleft$$

**(b)** The mass flow rate of the steam can be obtained with the expression for net power given in part (a).

$$\dot{m} = \frac{\dot{W}_{\text{cycle}}}{(h_1 - h_2) + (h_3 - h_4) - (h_6 - h_5)}$$

$$= \frac{(100 \text{ MW})|3600 \text{ s/h}||10^3 \text{ kW/MW}|}{(606.6 + 924.8 - 8.06) \text{ kJ/kg}} = 2.363 \times 10^5 \text{ kg/h} \quad \triangleleft$$

**(c)** The rate of heat transfer from the condensing steam to the cooling water is

$$\dot{Q}_{\text{out}} = \dot{m}(h_4 - h_5)$$

$$= \frac{2.363 \times 10^5 \text{ kg/h} \, (2428.5 - 173.88) \text{ kJ/kg}}{|3600 \text{ s/h}||10^3 \text{ kW/MW}|} = 148 \text{ MW} \quad \triangleleft$$

To see the effects of reheat, we compare the present values with their counterparts in Example 8.1. With superheat and reheat, the thermal efficiency is increased over that of the cycle of Example 8.1. For a specified net power output (100 MW), a larger thermal efficiency means that a smaller mass flow rate of steam is required. Moreover, with a greater thermal efficiency the rate of heat transfer to the cooling water is also less, resulting in a reduced demand for cooling water. With reheating, the steam quality at the turbine exhaust is substantially increased over the value for the cycle of Example 8.1.

**(d)** The *T-s* diagram for the reheat cycle with irreversible expansions through the turbine stages is shown in Fig. E8.3*b*. The following specific enthalpy values are known from part (a), in kJ/kg: $h_1 = 3348.4$, $h_{2s} = 2741.8$, $h_3 = 3353.3$, $h_{4s} = 2428.5$, $h_5 = 173.88$, $h_6 = 181.94$.

The specific enthalpy at the exit of the first-stage turbine, $h_2$, can be determined by solving the expression for the turbine efficiency to obtain

$$h_2 = h_1 - \eta_t(h_1 - h_{2s})$$
$$= 3348.4 - 0.85(3348.4 - 2741.8) = 2832.8 \text{ kJ/kg}$$

The specific enthalpy at the exit of the second-stage turbine can be found similarly:

$$h_4 = h_3 - \eta_t(h_3 - h_{4s})$$
$$= 3353.3 - 0.85(3353.3 - 2428.5) = 2567.2 \text{ kJ/kg}$$

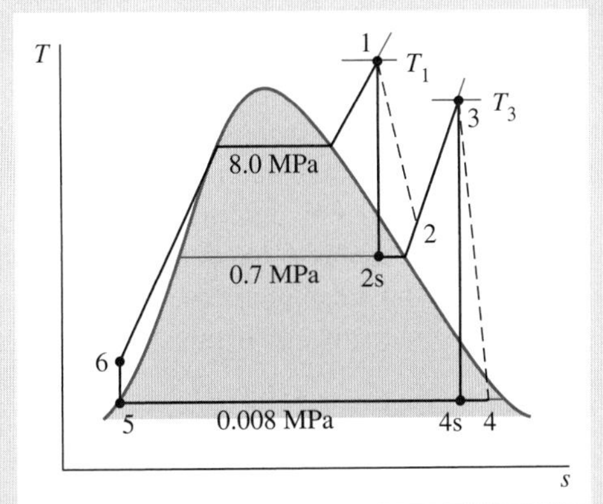


*Figure E8.3b*

The thermal efficiency is then

$$\eta = \frac{(h_1 - h_2) + (h_3 - h_4) - (h_6 - h_5)}{(h_1 - h_6) + (h_3 - h_2)}$$

❶
$$= \frac{(3348.4 - 2832.8) + (3353.3 - 2567.2) - (181.94 - 173.88)}{(3348.4 - 181.94) + (3353.3 - 2832.8)}$$

$$= \frac{1293.6 \text{ kJ/kg}}{3687.0 \text{ kJ/kg}} = 0.351 \ (35.1\%) \ \triangleleft$$

(e) (CD-ROM)

❶ Owing to the irreversibilities present in the turbine stages, the net work per unit of mass developed in the present case is significantly less than in part (a). The thermal efficiency is also considerably less.

## 8.4 Improving Performance—Regenerative Vapor Power Cycle

*regeneration*

Another commonly used method for increasing the thermal efficiency of vapor power plants is *regenerative feedwater heating,* or simply ***regeneration.*** This is the subject of the present section.

To introduce the principle underlying regenerative feedwater heating, consider Fig. 8.3 (p.190) once again. In cycle 1–2–3–4–a–1, the working fluid would enter the boiler as a compressed liquid at state 4 and be heated while in the liquid phase to state a. With regenerative feedwater heating, the working fluid would enter the boiler at a state *between* 4 and a. As a result, the average temperature of heat addition would be increased, thereby tending to increase the thermal efficiency.

### 8.4.1 Open Feedwater Heaters

*open feedwater heater*

Let us consider how regeneration can be accomplished using an ***open feedwater heater,*** a direct contact-type heat exchanger in which streams at different temperatures mix to form a stream at an intermediate temperature. Shown in Fig. 8.7 are the schematic diagram and the associated *T–s* diagram for a regenerative vapor power cycle having one open feedwater heater. For this cycle, the working fluid passes isentropically through the turbine stages and pumps, and the flow through the steam generator, condenser, and feedwater heater takes place with no pressure drop in any or these components. Steam enters the first-stage turbine at state 1 and expands to state 2, where a fraction of the total flow is *extracted,* or *bled,* into an open feedwater heater operating at the extraction pressure, $p_2$. The rest of the steam expands through the second-stage turbine to state 3. This portion of the total flow is condensed to saturated liquid, state 4, and then pumped to the extraction pressure and introduced into the feedwater heater at state 5. A single mixed stream exits the feedwater heater at state 6. For the case shown in Fig. 8.7, the mass flow rates of the streams entering the feedwater heater are chosen so that the stream exiting the feedwater heater is a saturated liquid at the extraction pressure. The liquid at state 6 is then pumped to the steam generator pressure and enters the steam generator at state 7. Finally, the working fluid is heated from state 7 to state 1 in the steam generator.

Referring to the *T–s* diagram of the cycle, note that the heat addition would take place from state 7 to state 1, rather than from state a to state 1, as would be the case without regeneration. Accordingly, the amount of energy that must be supplied from the combustion of a fossil fuel, or another source, to vaporize and superheat the steam would be reduced. This is the desired outcome. Only a portion of the total flow expands through the

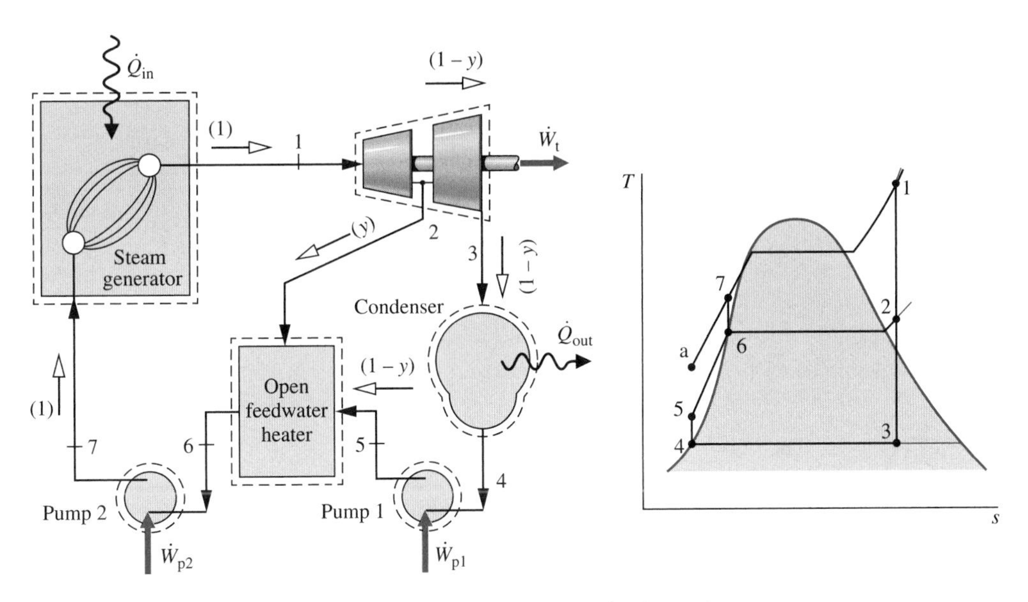


*Figure 8.7* Regenerative vapor power cycle with one open feedwater heater.

second-stage turbine (Process 2–3), however, so less work would be developed as well. In practice, operating conditions are chosen so that the reduction in heat added more than offsets the decrease in net work developed, resulting in an increased thermal efficiency in regenerative power plants.

*Cycle Analysis.* Consider next the thermodynamic analysis of the regenerative cycle illustrated in Fig. 8.7. An important initial step in analyzing any regenerative vapor cycle is the evaluation of the mass flow rates through each of the components. Taking a single control volume enclosing both turbine stages, the mass rate balance reduces at steady state to

$$\dot{m}_2 + \dot{m}_3 = \dot{m}_1 \tag{8.10a}$$

where $\dot{m}_1$ is the rate at which mass enters the first-stage turbine at state 1, $\dot{m}_2$ is the rate at which mass is extracted and exits at state 2, and $\dot{m}_3$ is the rate at which mass exits the second-stage turbine at state 3. Dividing by $\dot{m}_1$ places this on the basis of a *unit of mass* passing through the first-stage turbine

$$\frac{\dot{m}_2}{\dot{m}_1} + \frac{\dot{m}_3}{\dot{m}_1} = 1 \tag{8.10b}$$

Denoting the fraction of the total flow extracted at state 2 by $y$ ($y = \dot{m}_2/\dot{m}_1$), the fraction of the total flow passing through the second-stage turbine is

$$\frac{\dot{m}_3}{\dot{m}_1} = 1 - y \tag{8.11}$$

The fractions of the total flow at various locations are indicated on Fig. 8.7.

The fraction $y$ can be determined by applying the conservation of mass and conservation of energy principles to a control volume around the feedwater heater. Assuming no heat transfer between the feedwater heater and its surroundings and ignoring kinetic and potential energy effects, the mass and energy rate balances reduce at steady state to give

$$0 = yh_2 + (1 - y)h_5 - h_6$$

Solving for $y$

$$y = \frac{h_6 - h_5}{h_2 - h_5} \tag{8.12}$$

Equation 8.12 allows the fraction $y$ to be determined when states 2, 5, and 6 are fixed.

Expressions for the principal work and heat transfers of the regenerative cycle can be determined by applying mass and energy rate balances to control volumes around the individual components. Beginning with the turbine, the total work is the sum of the work developed by each turbine stage. Neglecting kinetic and potential energy effects and assuming no heat transfer with the surroundings, we can express the total turbine work on the basis of a unit of mass passing through the first-stage turbine as

$$\frac{\dot{W}_t}{\dot{m}_1} = (h_1 - h_2) + (1 - y)(h_2 - h_3) \tag{8.13}$$

The total pump work is the sum of the work required to operate each pump individually. On the basis of a unit of mass passing through the first-stage turbine, the total pump work is

$$\frac{\dot{W}_p}{\dot{m}_1} = (h_7 - h_6) + (1 - y)(h_5 - h_4) \tag{8.14}$$

The energy added by heat transfer to the working fluid passing through the steam generator, per unit of mass expanding through the first-stage turbine, is

$$\frac{\dot{Q}_{in}}{\dot{m}_1} = h_1 - h_7 \tag{8.15}$$

and the energy rejected by heat transfer to the cooling water is

$$\frac{\dot{Q}_{out}}{\dot{m}_1} = (1 - y)(h_3 - h_4) \tag{8.16}$$

The following example illustrates the analysis of a regenerative cycle with one open feedwater heater, including the evaluation of properties at state points around the cycle and the determination of the fractions of the total flow at various locations.

## *Example 8.4* Regenerative Cycle with Open Feedwater Heater

Consider a regenerative vapor power cycle with one open feedwater heater. Steam enters the turbine at 8.0 MPa, 480°C and expands to 0.7 MPa, where some of the steam is extracted and diverted to the open feedwater heater operating at 0.7 MPa. The remaining steam expands through the second-stage turbine to the condenser pressure of 0.008 MPa. Saturated liquid exits the open feedwater heater at 0.7 MPa. The isentropic efficiency of each turbine stage is 85% and each pump operates isentropically. If the net power output of the cycle is 100 MW, determine **(a)** the thermal efficiency and **(b)** the mass flow rate of steam entering the first turbine stage, in kg/h.

### Solution

***Known:*** A regenerative vapor power cycle operates with steam as the working fluid. Operating pressures and temperatures are specified; the efficiency of each turbine stage and the net power output are also given.

***Find:*** Determine the thermal efficiency and the mass flow rate into the turbine, in kg/h.

***Schematic and Given Data:***

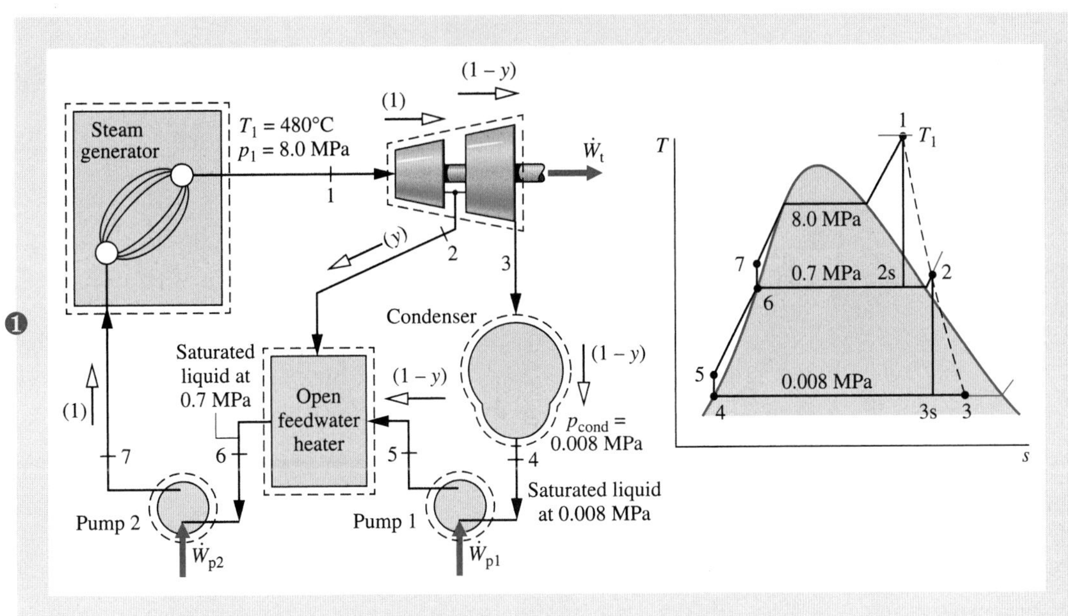


Figure E8.4

❶

***Assumptions:***

1. Each component in the cycle is analyzed as a steady-state control volume. The control volumes are shown in the accompanying sketch by dashed lines.
2. All processes of the working fluid are internally reversible, except for the expansions through the two turbine stages and mixing in the open feedwater heater.
3. The turbines, pumps, and feedwater heater operate adiabatically.
4. Kinetic and potential energy effects are negligible.
5. Saturated liquid exits the open feedwater heater, and saturated liquid exits the condenser.

***Properties:*** The specific enthalpy at states 1 and 4 can be read from the steam tables. The specific enthalpy at state 2 is evaluated in part (d) of the solution to Example 8.3. The specific entropy at state 2 can be obtained from the steam tables using the known values of enthalpy and pressure at this state. In summary, $h_1$ = 3348.4 kJ/kg, $h_2$ = 2832.8 kJ/kg, $s_2$ = 6.8606 kJ/kg · K, $h_4$ = 173.88 kJ/kg.

The specific enthalpy at state 3 can be determined using the efficiency of the second-stage turbine

$$h_3 = h_2 - \eta_t(h_2 - h_{3s})$$

With $s_{3s} = s_2$, the quality at state 3s is $x_{3s} = 0.8208$; using this, we get $h_{3s}$ = 2146.3 kJ/kg. Hence

$$h_3 = 2832.8 - 0.85(2832.8 - 2146.3) = 2249.3 \text{ kJ/kg}$$

State 6 is saturated liquid at 0.7 MPa. Thus, $h_6$ = 697.22 kJ/kg.

Since the pumps are assumed to operate with no irreversibilities, the specific enthalpy values at states 5 and 7 can be determined as

$$\begin{aligned} h_5 &= h_4 + v_4(p_5 - p_4) \\ &= 173.88 + (1.0084 \times 10^{-3})(\text{m}^3/\text{kg})(0.7 - 0.008)\,\text{MPa}\left|\frac{10^6\ \text{N/m}^2}{1\ \text{MPa}}\right|\left|\frac{1\ \text{kJ}}{10^3\ \text{N}\cdot\text{m}}\right| \\ &= 174.6 \text{ kJ/kg} \\ h_7 &= h_6 + v_6(p_7 - p_6) \\ &= 697.22 + (1.1080 \times 10^{-3})(8.0 - 0.7)|10^3| \\ &= 705.3 \text{ kJ/kg} \end{aligned}$$

***Analysis:*** Applying mass and energy rate balances to a control volume enclosing the open heater, we find the fraction $y$ of the flow extracted at state 2 from

$$y = \frac{h_6 - h_5}{h_2 - h_5} = \frac{697.22 - 174.6}{2832.8 - 174.6} = 0.1966$$

**(a)** On the basis of a unit of mass passing through the first-stage turbine, the total turbine work output is

$$\frac{\dot{W}_t}{\dot{m}_1} = (h_1 - h_2) + (1 - y)(h_2 - h_3)$$
$$= (3348.4 - 2832.8) + (0.8034)(2832.8 - 2249.3)$$
$$= 984.4 \text{ kJ/kg}$$

The total pump work per unit of mass passing through the first-stage turbine is

$$\frac{\dot{W}_p}{\dot{m}_1} = (h_7 - h_6) + (1 - y)(h_5 - h_4)$$
$$= (705.3 - 697.22) + (0.8034)(174.6 - 173.88)$$
$$= 8.7 \text{ kJ/kg}$$

The heat added in the steam generator per unit of mass passing through the first-stage turbine is

$$\frac{\dot{Q}_{in}}{\dot{m}_1} = h_1 - h_7 = 3348.4 - 705.3 = 2643.1 \text{ kJ/kg}$$

The thermal efficiency is then

$$\eta = \frac{\dot{W}_t/\dot{m}_1 - \dot{W}_p/\dot{m}_1}{\dot{Q}_{in}/\dot{m}_1} = \frac{984.4 - 8.7}{2643.1} = 0.369 \ (36.9\%) \ \triangleleft$$

**(b)** The mass flow rate of the steam entering the turbine, $\dot{m}_1$, can be determined using the given value for the net power output, 100 MW. Since

$$\dot{W}_{cycle} = \dot{W}_t - \dot{W}_p$$

and

$$\frac{\dot{W}_t}{\dot{m}_1} = 984.4 \text{ kJ/kg} \quad \text{and} \quad \frac{\dot{W}_p}{\dot{m}_1} = 8.7 \text{ kJ/kg}$$

it follows that

$$\dot{m}_1 = \frac{(100 \text{ MW})|3600 \text{ s/h}|}{(984.4 - 8.7) \text{ kJ/kg}} \left| \frac{10^3 \text{ kJ/s}}{1 \text{ MW}} \right| = 3.69 \times 10^5 \text{ kg/h} \ \triangleleft$$

❶ Note that the fractions of the total flow at various locations are labeled on the figure.

### 8.4.2 Closed Feedwater Heaters (CD-ROM)

## Vapor Refrigeration and Heat Pump Systems

In this part of the chapter, we consider vapor refrigeration and heat pump systems. Refrigeration systems for food preservation and air conditioning play prominent roles in our everyday lives. Heat pumps also are used for heating buildings and for producing industrial process heat. There are many other examples of commercial and industrial uses of refrigeration, including air separation to obtain liquid oxygen and liquid nitrogen, liquefaction of natural gas, and production of ice. In this part of the chapter we describe the most common type of vapor refrigeration and heat pump systems presently in use and illustrate how such systems can be modeled thermodynamically.

## 8.5 Vapor Refrigeration Systems

The purpose of a refrigeration system is to maintain a *cold* region at a temperature below the temperature of its surroundings. This is commonly achieved using the vapor refrigeration systems that are the subject of the present section.

### Carnot Refrigeration Cycle

To introduce some important aspects of vapor refrigeration, let us begin by considering a Carnot vapor refrigeration cycle. This cycle is obtained by reversing the Carnot vapor power cycle introduced in Sec. 6.5. Figure 8.10 shows the schematic and accompanying *T–s* diagram of a Carnot refrigeration cycle operating between a region at temperature $T_C$ and another region at a higher temperature $T_H$. The cycle is executed by a refrigerant circulating steadily through a series of components. All processes are internally reversible. Also, since heat transfers between the refrigerant and each region occur with no temperature differences, there are no external irreversibilities. The energy transfers shown on the diagram are positive in the directions indicated by the arrows.

Let us follow the refrigerant as it passes steadily through each of the components in the cycle, beginning at the inlet to the evaporator. The refrigerant enters the evaporator as a two-phase liquid–vapor mixture at state 4. In the evaporator, some of the refrigerant changes phase from liquid to vapor as a result of heat transfer from the region at temperature $T_C$ to the refrigerant. The temperature and pressure of the refrigerant remain constant during the process from state 4 to state 1. The refrigerant is then compressed adiabatically from state 1, where it is a two-phase liquid–vapor mixture, to state 2, where it is a saturated vapor. During this process, the temperature of the refrigerant increases from $T_C$ to $T_H$, and the pressure also increases. The refrigerant passes from the compressor into the condenser, where it changes phase from saturated vapor to saturated liquid as a result of heat transfer to the region at temperature $T_H$. The temperature and pressure remain constant in the process from state 2 to state 3. The refrigerant returns to the state at the inlet of the evaporator by expanding adiabatically through a turbine. In this process, from state 3 to state 4, the temperature decreases from $T_H$ to $T_C$, and there is a decrease in pressure.

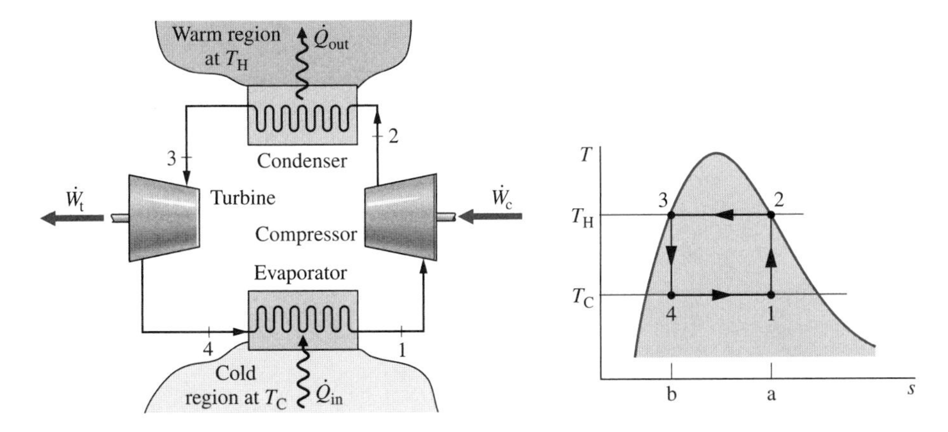


*Figure 8.10* Carnot vapor refrigeration cycle.

Since the Carnot vapor refrigeration cycle is made up of internally reversible processes, areas on the $T$–$s$ diagram can be interpreted as heat transfers. Applying Eq. 7.40, area 1–a–b–4–1 is the heat added to the refrigerant from the cold region per unit mass of refrigerant flowing. Area 2–a–b–3–2 is the heat rejected from the refrigerant to the warm region per unit mass of refrigerant flowing. The enclosed area 1–2–3–4–1 is the *net* heat transfer *from* the refrigerant. The net heat transfer *from* the refrigerant equals the net work done *on* the refrigerant. The net work is the difference between the compressor work input and the turbine work output.

The coefficient of performance $\beta$ of *any* refrigeration cycle is the ratio of the refrigeration effect to the net work input required to achieve that effect. For the Carnot vapor refrigeration cycle shown in Fig. 8.10, the coefficient of performance is

$$\begin{aligned}\beta_{\max} &= \frac{\dot{Q}_{\text{in}}/\dot{m}}{\dot{W}_{\text{c}}/\dot{m} - \dot{W}_{\text{t}}/\dot{m}} \\ &= \frac{\text{area 1–a–b–4–1}}{\text{area 1–2–3–4–1}} = \frac{T_{\text{C}}(s_{\text{a}} - s_{\text{b}})}{(T_{\text{H}} - T_{\text{C}})(s_{\text{a}} - s_{\text{b}})} \\ &= \frac{T_{\text{C}}}{T_{\text{H}} - T_{\text{C}}}\end{aligned} \tag{8.18}$$

This equation, which corresponds to Eq. 6.7, represents the *maximum* theoretical coefficient of performance of any refrigeration cycle operating between regions at $T_{\text{C}}$ and $T_{\text{H}}$.

## Departures from the Carnot Cycle

Actual vapor refrigeration systems depart significantly from the Carnot cycle considered above and have coefficients of performance lower than would be calculated from Eq. 8.18. Three ways actual systems depart from the Carnot cycle are considered next.

- One of the most significant departures is related to the heat transfers between the refrigerant and the two regions. In actual systems, these heat transfers are not accomplished reversibly as presumed above. In particular, to achieve a rate of heat transfer sufficient to maintain the temperature of the cold region at $T_{\text{C}}$ with a practical-sized evaporator requires the temperature of the refrigerant in the evaporator, $T'_{\text{C}}$, to be several degrees *below* $T_{\text{C}}$. This is illustrated by the placement of the temperature $T'_{\text{C}}$ on the $T$–$s$ diagram of Fig. 8.11. Similarly, to obtain a sufficient heat transfer rate from the refrigerant to the warm region requires that the refrigerant temperature in the condenser, $T'_{\text{H}}$, be several degrees *above* $T_{\text{H}}$. This is illustrated by the placement of the temperature $T'_{\text{H}}$ on the $T$–$s$ diagram of Fig. 8.11.

  Maintaining the refrigerant temperatures in the heat exchangers at $T'_{\text{C}}$ and $T'_{\text{H}}$ rather than at $T_{\text{C}}$ and $T_{\text{H}}$, respectively, has the effect of reducing the coefficient of

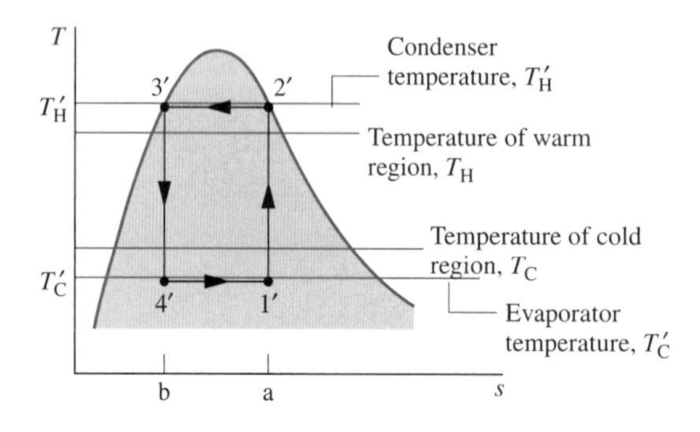


*Figure 8.11* Comparison of the condenser and evaporator temperatures with those of the warm and cold regions.

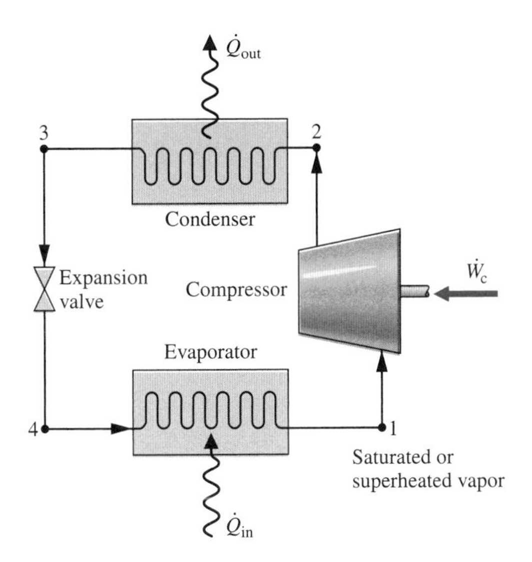


*Figure 8.12* Components of a vapor-compression refrigeration system.

performance. This can be seen by expressing the coefficient of performance of the refrigeration cycle designated by 1′–2′–3′–4′–1′ on Fig. 8.11 as

$$\beta' = \frac{\text{area } 1'\text{–a–b–}4'\text{–}1}{\text{area } 1'\text{–}2'\text{–}3'\text{–}4'\text{–}1'} = \frac{T'_C}{T'_H - T'_C} \tag{8.19}$$

Comparing the areas underlying the expressions for $\beta_{max}$ and $\beta'$ given above, we conclude that the value of $\beta'$ is less than $\beta_{max}$. This conclusion about the effect of refrigerant temperature on the coefficient of performance also applies to the vapor-compression systems considered in Sec. 8.6.

- Even when the temperature differences between the refrigerant and warm and cold regions are taken into consideration, there are other features that make the vapor refrigeration cycle of Fig. 8.11 impractical as a prototype. Referring again to the figure, note that the compression process from state 1′ to state 2′ occurs with the refrigerant as a two-phase liquid–vapor mixture. This is commonly referred to as *wet compression*. Wet compression is normally avoided because the presence of liquid droplets in the flowing liquid–vapor mixture can damage the compressor. In actual systems, the compressor handles vapor only. This is known as *dry compression.*
- Another feature that makes the cycle of Fig. 8.11 impractical is the expansion process from the saturated liquid state 3′ to the low-quality, two-phase liquid–vapor mixture state 4′. This expansion produces a relatively small amount of work compared to the work input in the compression process. The work output achieved by an actual turbine would be smaller yet because turbines operating under these conditions typically have low efficiencies. Accordingly, the work output of the turbine is normally sacrificed by substituting a simple throttling valve for the expansion turbine, with consequent savings in initial and maintenance costs. The components of the resulting cycle are illustrated in Fig. 8.12, where dry compression is presumed. This cycle, known as the *vapor-compression refrigeration cycle,* is the subject of the section to follow.

## 8.6 Analyzing Vapor-Compression Refrigeration Systems

***vapor-compression refrigeration***

***Vapor-compression refrigeration*** systems are the most common refrigeration systems in use today. The object of this section is to introduce some important features of systems of this type and to illustrate how they are modeled thermodynamically.

### 8.6.1 Evaluating Principal Work and Heat Transfers

Let us consider the steady-state operation of the vapor-compression system illustrated in Fig. 8.12. Shown on the figure are the principal work and heat transfers, which are positive in the directions of the arrows. Kinetic and potential energy changes are neglected in the following analyses of the components. We begin with the evaporator, where the desired refrigeration effect is achieved.

- As the refrigerant passes through the evaporator, heat transfer from the refrigerated space results in the vaporization of the refrigerant. For a control volume enclosing the refrigerant side of the evaporator, the mass and energy rate balances reduce to give the rate of heat transfer per unit mass of refrigerant flowing.

$$\frac{\dot{Q}_{\text{in}}}{\dot{m}} = h_1 - h_4 \tag{8.20}$$

where $\dot{m}$ is the mass flow rate of the refrigerant. The heat transfer rate $\dot{Q}_{\text{in}}$ is referred to as the ***refrigeration capacity.*** In the SI unit system, the capacity is normally expressed in kW. The refrigeration capacity also may be expressed in Btu/h. Another commonly used unit for the refrigeration capacity is the ***ton of refrigeration,*** which is equal to 200 Btu/min or about 211 kJ/min.

*refrigeration capacity*

*ton of refrigeration*

- The refrigerant leaving the evaporator is compressed to a relatively high pressure and temperature by the compressor. Assuming no heat transfer to or from the compressor, the mass and energy rate balances for a control volume enclosing the compressor give

$$\frac{\dot{W}_c}{\dot{m}} = h_2 - h_1 \tag{8.21}$$

where $\dot{W}_c/\dot{m}$ is the rate of power *input* per unit mass of refrigerant flowing.

- Next, the refrigerant passes through the condenser, where the refrigerant condenses and there is heat transfer from the refrigerant to the cooler surroundings. For a control volume enclosing the refrigerant side of the condenser, the rate of heat transfer from the refrigerant per unit mass of refrigerant flowing is

$$\frac{\dot{Q}_{\text{out}}}{\dot{m}} = h_2 - h_3 \tag{8.22}$$

- Finally, the refrigerant at state 3 enters the expansion valve and expands to the evaporator pressure. This process is usually modeled as a *throttling* process (p. 115) for which

$$h_4 = h_3 \tag{8.23}$$

The refrigerant pressure decreases in the irreversible adiabatic expansion, and there is an accompanying increase in specific entropy. The refrigerant exits the valve at state 4 as a two-phase liquid–vapor mixture.

In the vapor-compression system, the net power input is equal to the compressor power, since the expansion valve involves no power input or output. Using the quantities and expressions introduced above, the coefficient of performance of the vapor-compression refrigeration system of Fig. 8.12 is

$$\beta = \frac{\dot{Q}_{\text{in}}/\dot{m}}{\dot{W}_c/\dot{m}} = \frac{h_1 - h_4}{h_2 - h_1} \tag{8.24}$$

Provided states 1 through 4 are fixed, Eqs. 8.20 through 8.24 can be used to evaluate the principal work and heat transfers and the coefficient of performance of the vapor-compression system shown in Fig. 8.12. Since these equations have been developed by reducing mass and

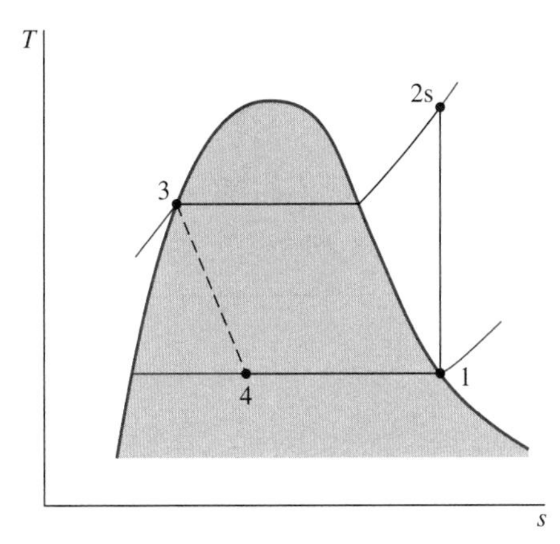


*Figure 8.13* $T$–$s$ diagram of an ideal vapor-compression cycle.

energy rate balances, they apply equally for actual performance when irreversibilities are present in the evaporator, compressor, and condenser and for idealized performance in the absence of such effects. Although irreversibilities in the evaporator, compressor, and condenser can have a pronounced effect on overall performance, it is instructive to consider an idealized cycle in which they are assumed absent. Such a cycle establishes an upper limit on the performance of the vapor-compression refrigeration cycle. It is considered next.

## 8.6.2 Performance of Vapor-Compression Systems

If irreversibilities within the evaporator and condenser are ignored, there are no frictional pressure drops, and the refrigerant flows at constant pressure through the two heat exchangers. If compression occurs without irreversibilities, and stray heat transfer to the surroundings is also ignored, the compression process is isentropic. With these considerations, the vapor-compression refrigeration cycle labeled 1–2s–3–4–1 on the $T$–$s$ diagram of Fig. 8.13 results. The cycle consists of the following series of processes:

***Process 1–2s:*** *Isentropic* compression of the refrigerant from state 1 to the condenser pressure at state 2s.

***Process 2s–3:*** Heat transfer *from* the refrigerant as it flows at constant pressure through the condenser. The refrigerant exits as a liquid at state 3.

***Process 3–4:*** *Throttling* process from state 3 to a two-phase liquid–vapor mixture at 4.

***Process 4–1:*** Heat transfer *to* the refrigerant as it flows at constant pressure through the evaporator to complete the cycle.

All of the processes in the above cycle are internally reversible except for the throttling process. Despite the inclusion of this irreversible process, the cycle is commonly referred to as the ***ideal vapor-compression cycle.***

*ideal vapor-compression cycle*

The following example illustrates the application of the first and second laws of thermodynamics along with property data to analyze an ideal vapor-compression cycle.

## *Example 8.5* Ideal Vapor-Compression Refrigeration Cycle

Refrigerant 134a is the working fluid in an ideal vapor-compression refrigeration cycle that communicates thermally with a cold region at 0°C and a warm region at 26°C. Saturated vapor enters the compressor at 0°C and saturated liquid leaves the condenser at 26°C. The mass flow rate of the refrigerant is 0.08 kg/s. Determine (a) the compressor power, in kW, (b) the refrigeration capacity, in tons, (c) the coefficient of performance, and (d) the coefficient of performance of a Carnot refrigeration cycle operating between warm and cold regions at 26 and 0°C, respectively.

## Solution

***Known:*** An ideal vapor-compression refrigeration cycle operates with Refrigerant 134a. The states of the refrigerant entering the compressor and leaving the condenser are specified, and the mass flow rate is given.

***Find:*** Determine the compressor power, in kW, the refrigeration capacity, in tons, the coefficient of performance, and the coefficient of performance of a Carnot vapor refrigeration cycle operating between warm and cold regions at the specified temperatures.

***Schematic and Given Data:***

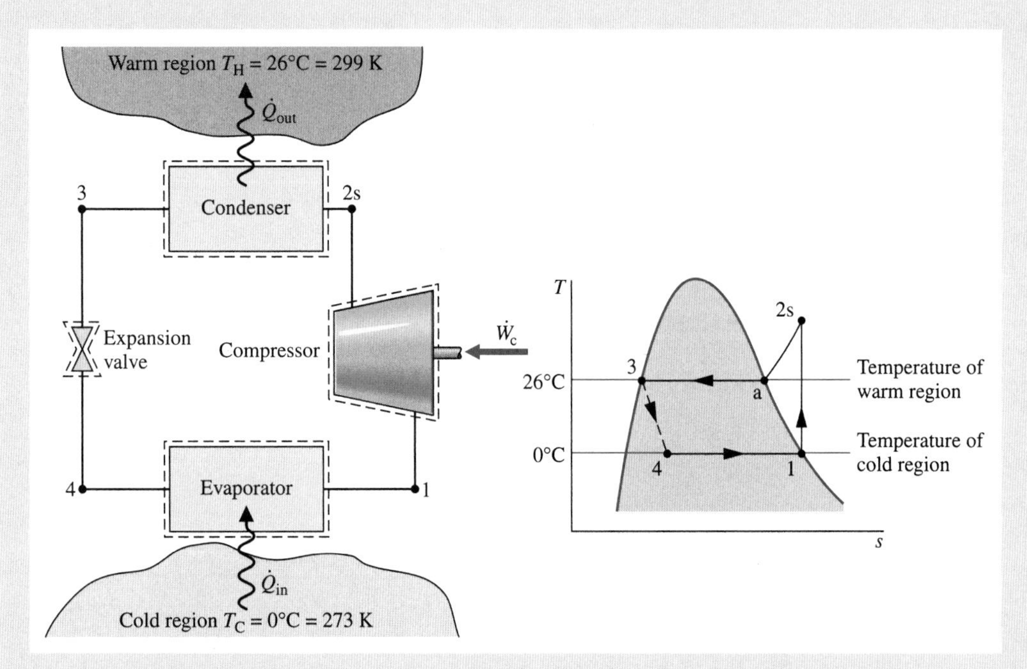


*Figure E8.5*

***Assumptions:***

1. Each component of the cycle is analyzed as a control volume at steady state. The control volumes are indicated by dashed lines on the accompanying sketch.
2. Except for the expansion through the valve, which is a throttling process, all processes of the refrigerant are internally reversible.
3. The compressor and expansion valve operate adiabatically.
4. Kinetic and potential energy effects are negligible.
5. Saturated vapor enters the compressor, and saturated liquid leaves the condenser.

***Properties:*** Let us begin by fixing each of the principal states located on the accompanying schematic and $T$–$s$ diagrams. At the inlet to the compressor, the refrigerant is a saturated vapor at 0°C, so from Table T-6, $h_1 = 247.23$ kJ/kg and $s_1 = 0.9190$ kJ/kg · K.

The pressure at state 2s is the saturation pressure corresponding to 26°C, or $p_2 = 6.853$ bar. State 2s is fixed by $p_2$ and the fact that the specific entropy is constant for the adiabatic, internally reversible compression process. The refrigerant at state 2s is a superheated vapor with $h_{2s} = 264.7$ kJ/kg. ❶

State 3 is saturated liquid at 26°C, so $h_3 = 85.75$ kJ/kg. The expansion through the valve is a throttling process (assumption 2), so $h_4 = h_3$.

***Analysis:*** (a) The compressor work input is

$$\dot{W}_c = \dot{m}(h_{2s} - h_1)$$

where $\dot{m}$ is the mass flow rate of refrigerant. Inserting values

$$\dot{W}_c = (0.08 \text{ kg/s})(264.7 - 247.23) \text{ kJ/kg} \left| \frac{1 \text{ kW}}{1 \text{ kJ/s}} \right|$$

$$= 1.4 \text{ kW} \triangleleft$$

(b) The refrigeration capacity is the heat transfer rate to the refrigerant passing through the evaporator. This is given by

$$\dot{Q}_{in} = \dot{m}(h_1 - h_4)$$

$$= (0.08 \text{ kg/s})|60 \text{ s/min}|(247.23 - 85.75) \text{ kJ/kg} \left| \frac{1 \text{ ton}}{211 \text{ kJ/min}} \right|$$

$$= 3.67 \text{ ton} \triangleleft$$

(c) The coefficient of performance β is

$$\beta = \frac{\dot{Q}_{in}}{\dot{W}_c} = \frac{h_1 - h_4}{h_{2s} - h_1} = \frac{247.23 - 85.75}{264.7 - 247.23} = 9.24 \triangleleft$$

(d) For a Carnot vapor refrigeration cycle operating at $T_H = 299$ K and $T_C = 273$ K, the coefficient of performance determined from Eq. 8.18 is

❷

$$\beta_{max} = \frac{T_C}{T_H - T_C} = 10.5 \triangleleft$$

❶ The value for $h_{2s}$ can be obtained by double interpolation in Table T-8 or by using the *Interactive Thermodynamics: IT* software that accompanies this book.

❷ As expected, the ideal vapor-compression cycle has a lower coefficient of performance than a Carnot cycle operating between the temperatures of the warm and cold regions. The smaller value can be attributed to the effects of the external irreversibility associated with desuperheating the refrigerant in the condenser (Process 2s–a on the *T–s* diagram) and the internal irreversibility of the throttling process.

Figure 8.14 illustrates several features exhibited by *actual* vapor-compression systems. As shown in the figure, the heat transfers between the refrigerant and the warm and cold regions are not accomplished reversibly: the refrigerant temperature in the evaporator is less than the cold region temperature, $T_C$, and the refrigerant temperature in the condenser is greater than the warm region temperature, $T_H$. Such irreversible heat transfers have a significant effect on performance. In particular, the coefficient of performance decreases as the average temperature of the refrigerant in the evaporator decreases and as the average temperature of the refrigerant in the condenser increases. Example 8.6 provides an illustration.

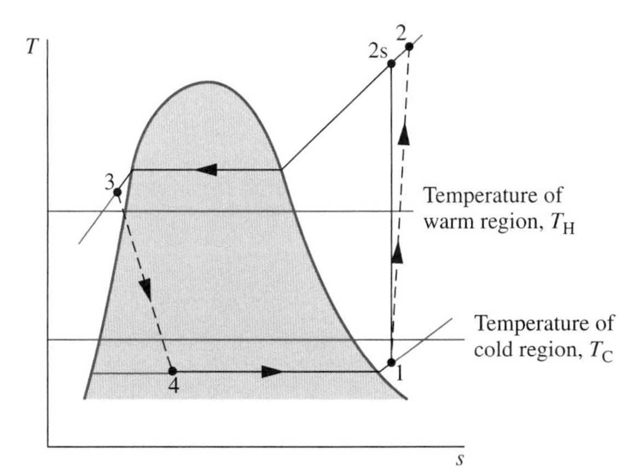


*Figure 8.14* *T–s* diagram of an actual vapor-compression cycle.

## *Example 8.6* Effect of Irreversible Heat Transfer on Performance

Modify Example 8.5 to allow for temperature differences between the refrigerant and the warm and cold regions as follows. Saturated vapor enters the compressor at −10°C. Saturated liquid leaves the condenser at a pressure of 9 bar. Determine for the modified vapor-compression refrigeration cycle (a) the compressor power, in kW, (b) the refrigeration capacity, in tons, (c) the coefficient of performance. Compare results with those of Example 8.5.

### Solution

***Known:*** An ideal vapor-compression refrigeration cycle operates with Refrigerant 134a as the working fluid. The evaporator temperature and condenser pressure are specified, and the mass flow rate is given.

***Find:*** Determine the compressor power, in kW, the refrigeration capacity, in tons, and the coefficient of performance. Compare results with those of Example 8.5.

***Schematic and Given Data:***

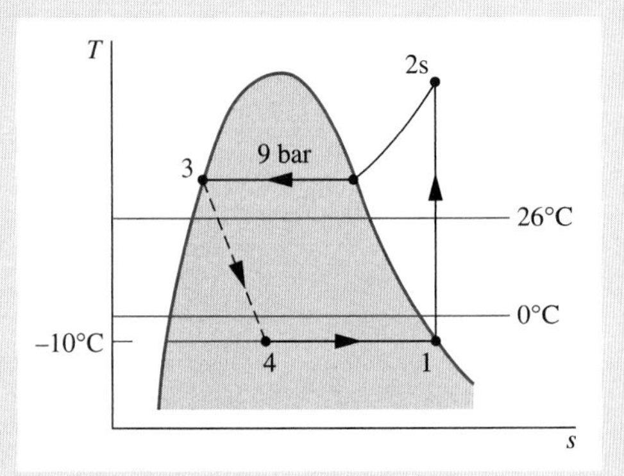


***Assumptions:***

1. Each component of the cycle is analyzed as a control volume at steady state. The control volumes are indicated by dashed lines on the sketch accompanying Example 8.5.
2. Except for the process through the expansion valve, which is a throttling process, all processes of the refrigerant are internally reversible.
3. The compressor and expansion valve operate adiabatically.
4. Kinetic and potential energy effects are negligible.
5. Saturated vapor enters the compressor, and saturated liquid exits the condenser.

*Figure E8.6*

***Properties:*** Let us begin by fixing each of the principal states located on the accompanying *T–s* diagram. Starting at the inlet to the compressor, the refrigerant is a saturated vapor at −10°C, so from Table T-6, $h_1 = 241.35$ kJ/kg and $s_1 = 0.9253$ kJ/kg · K.

The superheated vapor at state 2s is fixed by $p_2 = 9$ bar and the fact that the specific entropy is constant for the adiabatic, internally reversible compression process. Interpolating in Table T-8 gives $h_{2s} = 272.39$ kJ/kg.

State 3 is a saturated liquid at 9 bar, so $h_3 = 99.56$ kJ/kg. The expansion through the valve is a throttling process; thus, $h_4 = h_3$.

***Analysis:*** (a) The compressor power input is

$$\dot{W}_c = \dot{m}(h_{2s} - h_1)$$

where $\dot{m}$ is the mass flow rate of refrigerant. Inserting values

$$\dot{W}_c = (0.08\text{ kg/s})(272.39 - 241.35)\text{ kJ/kg}\left|\frac{1\text{ kW}}{1\text{ kJ/s}}\right|$$

$$= 2.48\text{ kW} \lhd$$

(b) The refrigeration capacity is

$$\dot{Q}_{in} = \dot{m}(h_1 - h_4)$$

$$= (0.08\text{ kg/s})|60\text{ s/min}|(241.35 - 99.56)\text{ kJ/kg}\left|\frac{1\text{ ton}}{211\text{ kJ/min}}\right|$$

$$= 3.23\text{ ton} \lhd$$

(c) The coefficient of performance $\beta$ is

$$\beta = \frac{\dot{Q}_{in}}{\dot{W}_c} = \frac{h_1 - h_4}{h_{2s} - h_1} = \frac{241.35 - 99.56}{272.39 - 241.35} = 4.57 \triangleleft$$

Comparing the results of the present example with those of Example 8.5, we see that the power input required by the compressor is greater in the present case. Furthermore, the refrigeration capacity and coefficient of performance are smaller in this example than in Example 8.5. This illustrates the considerable influence on performance of irreversible heat transfer between the refrigerant and the cold and warm regions.

Referring again to Fig. 8.14, we can identify another key feature of actual vapor-compression system performance. This is the effect of irreversibilities during compression, suggested by the use of a dashed line for the compression process from state 1 to state 2. The dashed line is drawn to show the increase in specific entropy that would accompany an *adiabatic* irreversible compression. Comparing cycle 1–2–3–4–1 with cycle 1–2s–3–4–1, the refrigeration capacity would be the same for each, but the work input would be greater in the case of irreversible compression than in the ideal cycle. Accordingly, the coefficient of performance of cycle 1–2–3–4–1 is less than that of cycle 1–2s–3–4–1. The effect of irreversible compression can be accounted for by using the isentropic compressor efficiency, which for states designated as in Fig. 8.14 is given by

$$\eta_c = \frac{(\dot{W}_c/\dot{m})_s}{(\dot{W}_c/\dot{m})} = \frac{h_{2s} - h_1}{h_2 - h_1} \tag{8.25}$$

Additional departures from ideality stem from frictional effects that result in pressure drops as the refrigerant flows through the evaporator, condenser, and piping connecting the various components. These pressure drops are not shown on the *T–s* diagram of Fig. 8.14 and are ignored in subsequent discussions for simplicity.

Finally, two additional features exhibited by actual vapor-compression systems are shown in Fig. 8.14. One is the superheated vapor condition at the evaporator exit (state 1), which differs from the saturated vapor condition shown in Fig. 8.13. Another is the subcooling of the condenser exit state (state 3), which differs from the saturated liquid condition shown in Fig. 8.13.

Example 8.7 illustrates the effects of irreversible compression and condenser exit subcooling on the performance of the vapor-compression refrigeration system.

## *Example 8.7* Actual Vapor-Compression Refrigeration Cycle

Reconsider the vapor-compression refrigeration cycle of Example 8.6, but include in the analysis that the compressor has an efficiency of 80%. Also, let the temperature of the liquid leaving the condenser be 30°C. Determine for the modified cycle (a) the compressor power, in kW, (b) the refrigeration capacity, in tons, and (c) the coefficient of performance.

### Solution

***Known:*** A vapor-compression refrigeration cycle has a compressor efficiency of 80%.

***Find:*** Determine the compressor power, in kW, the refrigeration capacity, in tons, and the coefficient of performance.

***Schematic and Given Data:***

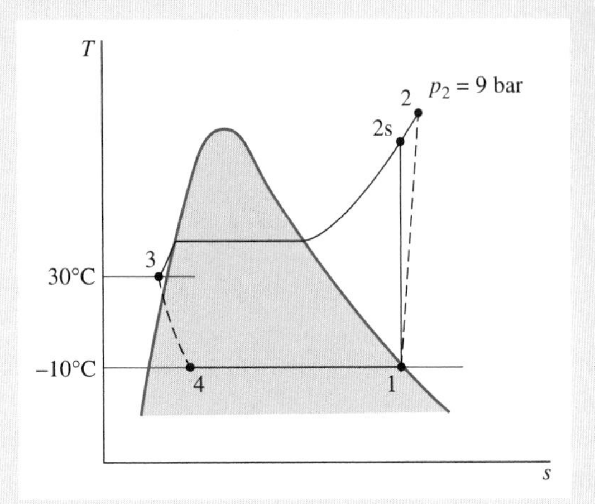


***Assumptions:***

1. Each component of the cycle is analyzed as a control volume at steady state.
2. There are no pressure drops through the evaporator and condenser.
3. The compressor operates adiabatically with an efficiency of 80%. The expansion through the valve is a throttling process.
4. Kinetic and potential energy effects are negligible.
5. Saturated vapor enters the compressor, and liquid at 30°C leaves the condenser.

*Figure E8.7*

***Properties:*** Let us begin by fixing the principal states. State 1 is the same as in Example 8.6, so $h_1 = 241.35$ kJ/kg. Owing to the presence of irreversibilities during the adiabatic compression process, there is an increase in specific entropy from compressor inlet to exit. The state at the compressor exit, state 2, can be fixed using the compressor efficiency

$$\eta_c = \frac{(\dot{W}_c/\dot{m})_s}{\dot{W}_c/\dot{m}} = \frac{(h_{2s} - h_1)}{(h_2 - h_1)}$$

where $h_{2s}$ is the specific enthalpy at state 2s, as indicated on the accompanying $T$–$s$ diagram. From the solution to Example 8.6, $h_{2s} = 272.39$ kJ/kg. Solving for $h_2$ and inserting known values

$$h_2 = \frac{h_{2s} - h_1}{\eta_c} + h_1 = \frac{(272.39 - 241.35)}{(0.80)} + 241.35 = 280.15 \text{ kJ/kg}$$

The state at the condenser exit, state 3, is in the liquid region. The specific enthalpy is approximated using Eq. 4.14, together with saturated liquid data at 30°C, as follows: $h_3 \approx h_f = 91.49$ kJ/kg.

***Analysis:*** (a) The compressor power is

$$\dot{W}_c = \dot{m}(h_2 - h_1)$$

$$= (0.08 \text{ kg/s})(280.15 - 241.35) \text{ kJ/kg} \left| \frac{1 \text{ kW}}{1 \text{ kJ/s}} \right| = 3.1 \text{ kW} \lhd$$

(b) The refrigeration capacity is

$$\dot{Q}_{in} = \dot{m}(h_1 - h_4)$$

$$= (0.08 \text{ kg/s})|60 \text{ s/min}|(241.35 - 91.49) \text{ kJ/kg} \left| \frac{1 \text{ ton}}{211 \text{ kJ/min}} \right|$$

$$= 3.41 \text{ ton} \lhd$$

(c) The coefficient of performance is

❶ $$\beta = \frac{(h_1 - h_4)}{(h_2 - h_1)} = \frac{(241.35 - 91.49)}{(280.15 - 241.35)} = 3.86 \lhd$$

❶ Irreversibilities in the compressor result in an increased compressor power requirement compared to the isentropic compression of Example 8.6. As a consequence, the coefficient of performance is lower.

## 8.7 Vapor-Compression Heat Pump Systems

The objective of a heat pump is to maintain the temperature within a dwelling or other building above the temperature of the surroundings or to provide a heat transfer for certain industrial processes that occur at elevated temperatures. Vapor-compression heat pump systems have many features in common with the refrigeration systems considered thus far.

In particular, the method of analysis of vapor-compression heat pumps is the same as that of vapor-compression refrigeration cycles considered previously. Also, the previous discussions concerning the departure of actual systems from ideality apply for vapor-compression heat pump systems as for vapor-compression refrigeration cycles.

*vapor-compression heat pump*

As illustrated by Fig. 8.15, a typical ***vapor-compression heat pump*** for space heating has the same basic components as the vapor-compression refrigeration system: compressor, condenser, expansion valve, and evaporator. The objective of the system is different, however. In a heat pump system, $\dot{Q}_{in}$ comes from the surroundings, and $\dot{Q}_{out}$ is directed to the dwelling as the desired effect. A net work input is required to accomplish this effect.

The coefficient of performance of a simple vapor-compression heat pump with states as designated on Fig. 8.15 is

$$\gamma = \frac{\dot{Q}_{out}/\dot{m}}{\dot{W}_c/\dot{m}} = \frac{h_2 - h_3}{h_2 - h_1} \tag{8.26}$$

The value of $\gamma$ can never be less than unity.

Many possible sources are available for heat transfer to the refrigerant passing through the evaporator. These include the outside air, the ground, and water in lakes, rivers, or wells. Liquid circulated through a solar collector and stored in an insulated tank also can be used as a source for a heat pump. Industrial heat pumps employ waste heat or warm liquid or gas streams as the low-temperature source and are capable of achieving relatively high condenser temperatures.

*air-source heat pump*

In the most common type of vapor-compression heat pump for space heating, the evaporator communicates thermally with the outside air. Such ***air-source heat pumps*** also can be used to provide cooling in the summer with the use of a reversing valve, as illustrated in

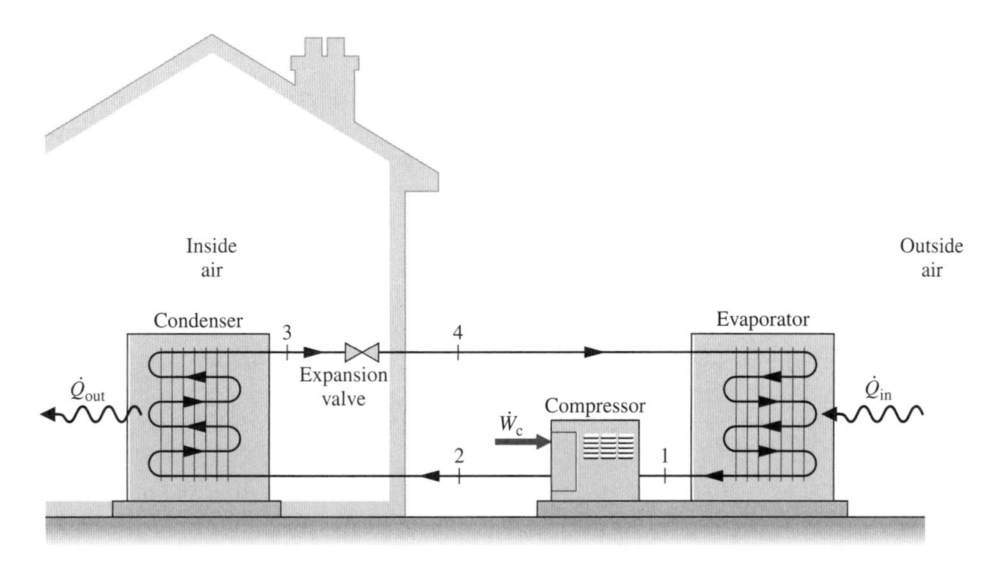


*Figure 8.15* Air-source vapor-compression heat pump system.

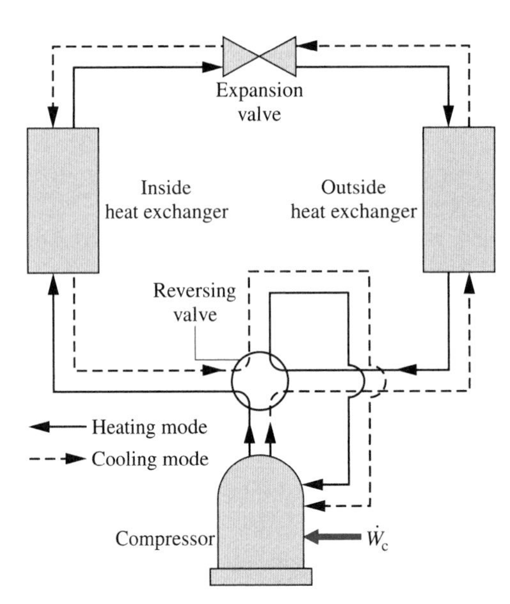


*Figure 8.16* Example of an air-to-air reversing heat pump.

Fig. 8.16. The solid lines show the flow path of the refrigerant in the heating mode, as described previously. To use the same components as an air conditioner, the valve is actuated, and the refrigerant follows the path indicated by the dashed line. In the cooling mode, the outside heat exchanger becomes the condenser, and the inside heat exchanger becomes the evaporator. Although heat pumps can be more costly to install and operate than other direct heating systems, they can be competitive when the potential for dual use is considered.

## 8.8 Working Fluids for Vapor Power and Refrigeration Systems

Water is used as the working fluid in the vast majority of vapor power systems because it is plentiful, low in cost, nontoxic, chemically stable, and relatively noncorrosive. Vapor power systems for special uses may employ other working fluids that have better characteristics than water for the particular applications. For example, vapor power systems for use in arctic regions might use propane, which at 1 atm condenses at about −40°C. Still, no other single working fluid has been found that is more satisfactory overall for large electrical generating plants than water.

For vapor refrigeration and heat pump applications, classes of chlorine-containing CFCs (chlorofluorocarbons), such as Refrigerant 12 ($CCL_2F_2$), commonly known as *Freon,* were believed to be suitable working fluids up to the early 1990's. However, owing to concern about the effects of such chlorine-containing refrigerants on the earth's protective ozone layer, international agreements now have been implemented that have phased out the use of CFCs. One class of refrigerants in which hydrogen atoms replace the chlorine atoms, called HFCs, contains no chlorine and is considered to be an environmentally acceptable substitute for CFCs. The HFC Refrigerant 134a ($CF_3CH_2F$) featured in this book has replaced Refrigerant 12 in many refrigeration and heat pump applications.

## 8.9 Chapter Summary and Study Guide

In this chapter we have studied vapor power systems and vapor refrigeration and heat pump systems. We have considered practical arrangements for such systems, illustrated how they are modeled, and discussed the principal irreversibilities and losses associated with their operation.

The main components of *simple* vapor power plants are modeled by the Rankine cycle. We also have introduced modifications to the simple vapor power cycle aimed at improving overall performance. These include superheat, reheat, and regeneration. We have evaluated the principal work and heat transfers along with the thermal efficiency. We also have considered the effects of irreversibilities on performance. The principal internal irreversibility is associated with turbine expansions, and is accounted for using the isentropic turbine efficiency.

The performance of simple vapor refrigeration and heat pump systems is described in terms of the vapor-compression cycle. For this cycle, we have evaluated the principal work and heat transfers along with two important performance parameters: the coefficient of performance and the refrigeration capacity. We have considered the effects on performance of irreversibilities during the compression process and in the expansion across the valve, as well as the effects of irreversible heat transfer between the refrigerant and the warm and cold regions.

The following checklist provides a study guide for this chapter. When your study of the text and end-of-chapter exercises has been completed you should be able to

- write out the meanings of the terms listed in the margin throughout the chapter and understand each of the related concepts. The subset of key terms listed here in the margin is particularly important.
- sketch schematic diagrams and accompanying $T$–$s$ diagrams of Rankine, reheat, and regenerative vapor power cycles.
- apply conservation of mass and energy, the second law, and property data to determine power cycle performance, including thermal efficiency, net power output, and mass flow rates.
- discuss the effects on Rankine cycle performance of varying steam generator pressure, condenser pressure, and turbine inlet temperature.
- sketch the $T$–$s$ diagrams of vapor-compression refrigeration and heat pump cycles, correctly showing the relationship of the refrigerant temperature to the temperatures of the warm and cold regions.
- apply conservation of mass and energy, the second law, and property data to determine the performance of vapor-compression refrigeration and heat pump cycles, including evaluation of the power required, the coefficient of performance, and the capacity.

*Rankine cycle*
*thermal efficiency*
*back work ratio*
*superheat*
*reheat*
*regeneration*
*vapor-compression refrigeration cycle*
*coefficient of performance*
*refrigeration capacity*
*ton of refrigeration*
*vapor-compression heat pump*

# Problems

**Rankine Cycle**

8.1 Water is the working fluid in an ideal Rankine cycle. The condenser pressure is 8 kPa, and saturated vapor enters the turbine at (a) 18 MPa and (b) 4 MPa. The net power output of the cycle is 100 MW. Determine for each case the mass flow rate of steam, in kg/h, the heat transfer rates for the working fluid passing through the boiler and condenser, each in kW, and the thermal efficiency.

8.2 Water is the working fluid in an ideal Rankine cycle. Superheated vapor enters the turbine at 8 MPa, 480°C. The condenser pressure is 8 kPa. The net power output of the cycle is 100 MW. Determine for the cycle

(a) the rate of heat transfer to the working fluid passing through the steam generator, in kW.
(b) the thermal efficiency.
(c) the mass flow rate of condenser cooling water, in kg/h, if the cooling water enters the condenser at 15°C and exits at 35°C with negligible pressure change.

8.3 (CD-ROM)

8.4 (CD-ROM)

8.5 Water is the working fluid in an ideal Rankine cycle. Saturated vapor enters the turbine at 18 MPa. The condenser pressure is 6 kPa. Determine

(a) the net work per unit mass of steam flow, in kJ/kg.
(b) the heat transfer to the steam passing through the boiler, in kJ per kg of steam flowing.
(c) the thermal efficiency.
(d) the heat transfer to cooling water passing through the condenser, in kJ per kg of steam condensed.

**8.6** (CD-ROM)

**8.7** Water is the working fluid in an ideal Rankine cycle. The pressure and temperature at the turbine inlet are 1200 lbf/in.$^2$ and 1000°F, respectively, and the condenser pressure is 1 lbf/in.$^2$ The mass flow rate of steam entering the turbine is $1.4 \times 10^6$ lb/h. The cooling water experiences a temperature increase from 60 to 80°F, with negligible pressure drop, as it passes through the condenser. Determine for the cycle
(a) the net power developed, in Btu/h.
(b) the thermal efficiency.
(c) the mass flow rate of cooling water, in lb/h.

**8.8** (CD-ROM)

**8.9** (CD-ROM)

**8.10** Refrigerant 134a is the working fluid in a solar power plant operating on an ideal Rankine cycle. Saturated vapor at 60°C enters the turbine, and the condenser operates at a pressure of 6 bar. The rate of energy input to the collectors from solar radiation is 0.4 kW per m$^2$ of collector surface area. Determine the solar collector surface area, in m$^2$, per kW of power developed by the plant.

**8.11** Reconsider the analysis of Problem 8.2, but include in the analysis that the turbine and pump have isentropic efficiencies of 85 and 70%, respectively. Determine for the modified cycle
(a) the thermal efficiency.
(b) the mass flow rate of steam, in kg/h, for a net power output of 100 MW.
(c) the mass flow rate of condenser cooling water, in kg/h, if the cooling water enters the condenser at 15°C and exits at 35°C with negligible pressure change.

**8.12** (CD-ROM)

**8.13** Reconsider the cycle of Problem 8.7, but include in the analysis that the turbine and pump have isentropic efficiencies of 88%. The mass flow rate is unchanged. Determine for the modified cycle
(a) the net power developed, in Btu/h.
(b) the rate of heat transfer to the working fluid passing through the steam generator, in Btu/h.
(c) the thermal efficiency.
(d) the volumetric flow rate of cooling water entering the condenser, in ft$^3$/min.

**8.14** (CD-ROM)

**8.15** (CD-ROM)

**8.16** Superheated steam at 8 MPa and 480°C leaves the steam generator of a vapor power plant. Heat transfer and frictional effects in the line connecting the steam generator and the turbine reduce the pressure and temperature at the turbine inlet to 7.6 MPa and 440°C, respectively. The pressure at the exit of the turbine is 10 kPa, and the turbine operates adiabatically. Liquid leaves the condenser at 8 kPa, 36°C. The pressure is increased to 8.6 MPa across the pump. The turbine and pump isentropic efficiencies are 88%. The mass flow rate of steam is 79.53 kg/s. Determine
(a) the net power output, in kW.
(b) the thermal efficiency.
(c) the rate of heat transfer from the line connecting the steam generator and the turbine, in kW.
(d) the mass flow rate of condenser cooling water, in kg/s, if the cooling water enters at 15°C and exits at 35°C with negligible pressure change.

**8.17** Modify Problem 8.7 as follows. Steam leaves the steam generator at 1200 lbf/in.$^2$, 1000°F, but due to heat transfer and frictional effects in the line connecting the steam generator and the turbine, the pressure and temperature at the turbine inlet are reduced to 1100 lbf/in.$^2$ and 900°F, respectively. Also, condensate leaves the condenser at 0.8 lbf/in.$^2$, 90°F and is pumped to 1250 lbf/in.$^2$ before entering the steam generator. Determine for the cycle
(a) the net power developed, in Btu/h.
(b) the thermal efficiency.
(c) the heat rate, in Btu/kW · h.
(d) the mass flow rate of cooling water, in lb/h.

**8.18** Superheated steam at 18 MPa, 560°C, enters the turbine of a vapor power plant. The pressure at the exit of the turbine is 0.06 bar, and liquid leaves the condenser at 0.045 bar, 26°C. The pressure is increased to 18.2 MPa across the pump. The turbine and pump have isentropic efficiencies of 82 and 77%, respectively. For the cycle, determine
(a) the net work per unit mass of steam flow, in kJ/kg.
(b) the heat transfer to steam passing through the boiler, in kJ per kg of steam flowing.
(c) the thermal efficiency.
(d) the heat transfer to cooling water passing through the condenser, in kJ per kg of steam condensed.

### Reheat Cycle

**8.19** Steam at 10 MPa, 600°C enters the first-stage turbine of an ideal Rankine cycle with reheat. The steam leaving the reheat section of the steam generator is at 500°C, and the condenser pressure is 6 kPa. If the quality at the exit of the second-stage turbine is 90%, determine the cycle thermal efficiency.

**8.20** The ideal Rankine cycle of Problem 8.7 is modified to include reheat. In the modified cycle, steam expands through the first-stage turbine to saturated vapor and then is reheated to 900°F. If the mass flow rate of steam in the modified cycle is the same as in Problem 8.7, determine for the modified cycle
(a) the net power developed, in Btu/h.
(b) the rate of heat transfer to the working fluid in the reheat process, in Btu/h.
(c) the thermal efficiency.

**8.21** The ideal Rankine cycle of Problem 8.2 is modified to include reheat. In the modified cycle, steam expands though the first-stage turbine to 0.7 MPa and then is reheated to 480°C. If the net power output of the modified cycle is 100 MW, determine for the modified cycle
(a) the rate of heat transfer to the working fluid passing through the steam generator, in MW.
(b) the thermal efficiency.
(c) the rate of heat transfer to cooling water passing through the condenser, in MW.

8.22 (CD-ROM)

8.23 (CD-ROM)

8.24 (CD-ROM)

## Regenerative Cycle

8.25 Modify the ideal Rankine cycle of Problem 8.2 to include one open feedwater heater operating at 0.7 MPa. Saturated liquid exits the feedwater heater at 0.7 MPa. Answer the same questions about the modified cycle as in Problem 8.2 and discuss the results.

8.26 A power plant operates on a regenerative vapor power cycle with one open feedwater heater. Steam enters the first turbine stage at 12 MPa, 520°C and expands to 1 MPa, where some of the steam is extracted and diverted to the open feedwater heater operating at 1 MPa. The remaining steam expands through the second turbine stage to the condenser pressure of 6 kPa. Saturated liquid exits the open feedwater heater at 1 MPa. For isentropic processes in the turbines and pumps, determine for the cycle (a) the thermal efficiency and (b) the mass flow rate into the first turbine stage, in kg/h, for a net power output of 330 MW.

8.27 Compare the results of Problem 8.26 with those for an ideal Rankine cycle having the same turbine inlet conditions and condenser pressure, but no regenerator.

8.28 Modify the ideal Rankine cycle of Problem 8.7 to include one open feedwater heater operating at 100 $lbf/in.^2$ Saturated liquid exits the open feedwater heater at 100 $lbf/in.^2$ The mass flow rate of steam into the first turbine stage is the same as the mass flow rate of steam in Problem 8.7. Answer the same questions about the modified cycle as in Problem 8.7 and discuss the results.

8.29 Reconsider the cycle of Problem 8.28, but include in the analysis that the isentropic efficiency of each turbine stage is 88% and of each pump is 80%.

8.30 Modify the ideal Rankine cycle of Problem 8.5 to include superheated vapor entering the first turbine stage at 18 MPa, 560°C, and one open feedwater heater operating at 1 MPa. Saturated liquid exits the open feedwater heater at 1 MPa. Determine for the modified cycle

(a) the net work, in kJ per kg of steam entering the first turbine stage.
(b) the thermal efficiency.
(c) the heat transfer to cooling water passing through the condenser, in kJ per kg of steam entering the first turbine stage.

8.31 Reconsider the cycle of Problem 8.30, but include in the analysis that each turbine stage and pump has an isentropic efficiency of 85%.

8.32 (CD-ROM)

8.33 (CD-ROM)

8.34 (CD-ROM)

8.35 (CD-ROM)

8.36 (CD-ROM)

## Vapor Refrigeration Systems

8.37 A Carnot vapor refrigeration cycle uses Refrigerant 134a as the working fluid. The refrigerant enters the condenser as saturated vapor at 28°C and leaves as saturated liquid. The evaporator operates at a temperature of $-10$°C. Determine, in kJ per kg of refrigerant flow,

(a) the work input to the compressor.
(b) the work developed by the turbine.
(c) the heat transfer to the refrigerant passing through the evaporator.

What is the coefficient of performance of the cycle?

8.38 A Carnot vapor refrigeration cycle is used to maintain a cold region at 0°F when the ambient temperature is 70°F. Refrigerant 134a enters the condenser as saturated vapor at 100 $lbf/in.^2$ and leaves as saturated liquid at the same pressure. The evaporator pressure is 20 $lbf/in.^2$ The mass flow rate of refrigerant is 12 lb/min. Calculate

(a) the compressor and turbine power, each in Btu/min.
(b) the coefficient of performance.

8.39 An ideal vapor-compression refrigeration cycle operates at steady state with Refrigerant 134a as the working fluid. Saturated vapor enters the compressor at $-10$°C, and saturated liquid leaves the condenser at 28°C. The mass flow rate of refrigerant is 5 kg/min. Determine

(a) the compressor power, in kW.
(b) the refrigerating capacity, in tons.
(c) the coefficient of performance.

8.40 Modify the cycle in Problem 8.39 to have saturated vapor entering the compressor at 1.6 bar and saturated liquid leaving the condenser at 9 bar. Answer the same questions for the modified cycle as in Problem 8.39.

8.41 (CD-ROM)

8.42 An ideal vapor-compression refrigeration system operates at steady state with Refrigerant 134a as the working fluid. Superheated vapor enters the compressor at 30 $lbf/in.^2$, 20°F, and saturated liquid leaves the condenser at 140 $lbf/in.^2$ The refrigeration capacity is 5 tons. Determine

(a) the compressor power, in horsepower.
(b) the rate of heat transfer from the working fluid passing through the condenser, in Btu/min.
(c) the coefficient of performance.

8.43 Refrigerant 134a enters the compressor of an ideal vapor-compression refrigeration system as saturated vapor at $-16$°C with a volumetric flow rate of 1 $m^3$/min. The refrigerant leaves the condenser at 36°C, 10 bar. Determine

(a) the compressor power, in kW.
(b) the refrigerating capacity, in tons.
(c) the coefficient of performance.

8.44 (CD-ROM)

8.45 (CD-ROM)

8.46 (CD-ROM)

8.47 (CD-ROM)

**8.48** Modify the cycle in Problem 8.40 to have an isentropic compressor efficiency of 80% and let the temperature of the liquid leaving the condenser be 32°C. Determine, for the modified cycle,
- **(a)** the compressor power, in kW.
- **(b)** the refrigerating capacity, in tons.
- **(c)** the coefficient of performance.

**8.49** Modify the cycle in Problem 8.42 to have an isentropic compressor efficiency of 85% and let the temperature of the liquid leaving the condenser be 95°F. Determine, for the modified cycle,
- **(a)** the compressor power, in horsepower.
- **(b)** the rate of heat transfer from the working fluid passing through the condenser, in Btu/min.
- **(c)** the coefficient of performance.

**8.50** A vapor-compression refrigeration system circulates Refrigerant 134a at a rate of 6 kg/min. The refrigerant enters the compressor at −10°C, 1.4 bar, and exits at 7 bar. The isentropic compressor efficiency is 67%. There are no appreciable pressure drops as the refrigerant flows through the condenser and evaporator. The refrigerant leaves the condenser at 7 bar, 24°C. Ignoring heat transfer between the compressor and its surroundings, determine
- **(a)** the coefficient of performance.
- **(b)** the refrigerating capacity, in tons.

**8.51** (CD-ROM)

**8.52** (CD-ROM)

**8.53** (CD-ROM)

**8.54** (CD-ROM)

### Vapor-Compression Heat Pump Systems

**8.55** An ideal vapor-compression heat pump cycle with Refrigerant 134a as the working fluid provides heating at a rate of 15 kW to maintain a building at 20°C when the outside temperature is 5°C. Saturated vapor at 2.4 bar leaves the evaporator, and saturated liquid at 8 bar leaves the condenser. Calculate
- **(a)** the power input to the compressor, in kW.
- **(b)** the coefficient of performance.
- **(c)** the coefficient of performance of a Carnot heat pump cycle operating between thermal reservoirs at 20 and 5°C.

**8.56** A vapor-compression heat pump system uses Refrigerant 134a as the working fluid. The refrigerant enters the compressor at 2.4 bar, 0°C, with a volumetric flow rate of 0.6 $m^3$/min. Compression is adiabatic to 9 bar, 60°C, and saturated liquid exits the condenser at 9 bar. Determine
- **(a)** the power input to the compressor, in kW.
- **(b)** the heating capacity of the system, in kW and tons.
- **(c)** the coefficient of performance.
- **(d)** the isentropic compressor efficiency.

**8.57** (CD-ROM)

**8.58** Refrigerant 134a enters the compressor of a vapor-compression heat pump at 30 lbf/in.$^2$, 20°F and is compressed adiabatically to 200 lbf/in.$^2$, 160°F. Liquid enters the expansion valve at 200 lbf/in.$^2$, 120°F. At the valve exit, the pressure is 30 lbf/in.$^2$ Determine
- **(a)** the isentropic compressor efficiency.
- **(b)** the coefficient of performance.

**8.59** (CD-ROM)

**8.60** (CD-ROM)

# GAS POWER SYSTEMS

## *Introduction...*

The vapor power systems studied in Chap. 8 use working fluids that are alternately vaporized and condensed. The ***objective*** of the present chapter is to study power systems utilizing working fluids that are always a gas. Included in this group are gas turbines and internal combustion engines of the spark-ignition and compression-ignition types. In the first part of the chapter, internal combustion engines are considered. Gas turbine power plants are discussed in the second part of the chapter.

*chapter objective*

## Internal Combustion Engines

This part of the chapter deals with *internal* combustion engines. Although most gas turbines are also internal combustion engines, the name is usually applied to *reciprocating* internal combustion engines of the type commonly used in automobiles, trucks, and buses. These engines also differ from the power plants considered thus far because the processes occur within reciprocating piston–cylinder arrangements and not in interconnected series of different components.

Two principal types of reciprocating internal combustion engines are the ***spark-ignition*** engine and the ***compression-ignition*** engine. In a spark-ignition engine, a mixture of fuel and air is ignited by a spark plug. In a compression-ignition engine, air is compressed to a high enough pressure and temperature that combustion occurs spontaneously when fuel is injected. Spark-ignition engines have advantages for applications requiring power up to about 225 kW (300 horsepower). Because they are relatively light and lower in cost, spark-ignition engines are particularly suited for use in automobiles. Compression-ignition engines are normally preferred for applications when fuel economy and relatively large amounts of power are required (heavy trucks and buses, locomotives and ships, auxiliary power units). In the middle range, spark-ignition and compression-ignition engines are used.

*spark-ignition*
*compression-ignition*

## 9.1 Engine Terminology

Figure 9.1 is a sketch of a reciprocating internal combustion engine consisting of a piston that moves within a cylinder fitted with two valves. The sketch is labeled with some special terms. The *bore* of the cylinder is its diameter. The *stroke* is the distance the piston moves in one direction. The piston is said to be at *top dead center* when it has moved to a position where the cylinder volume is a minimum. This minimum volume is known as the *clearance* volume. When the piston has moved to the position of maximum cylinder volume, the piston is at *bottom dead center*. The volume swept out by the piston as it moves from the top dead center to the bottom dead center position is called the *displacement volume*. The ***compression ratio*** $r$ is defined as the volume at bottom dead center divided by the volume at top dead center. The reciprocating motion of the piston is converted to rotary motion by a crank mechanism.

*compression ratio*

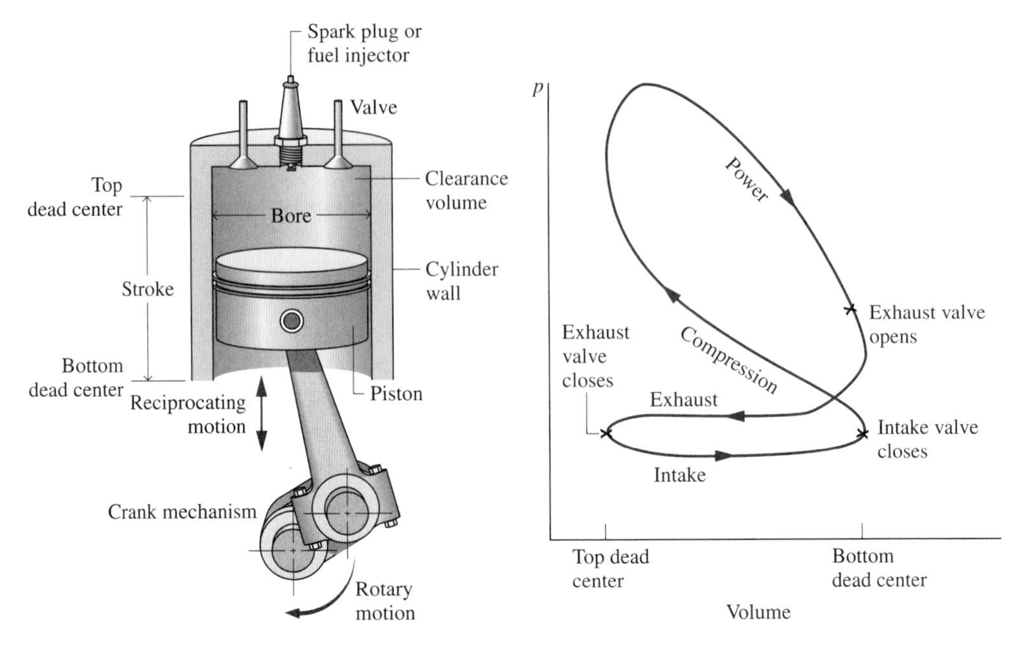


*Figure 9.1* Nomenclature for reciprocating piston–cylinder engines.

*Figure 9.2* Pressure–volume diagram for a reciprocating internal combustion engine.

In a *four-stroke* internal combustion engine, the piston executes four distinct strokes within the cylinder for every two revolutions of the crankshaft. Figure 9.2 gives a pressure–volume diagram such as might be displayed electronically. With the intake valve open, the piston makes an *intake stroke* to draw a fresh charge into the cylinder. For spark-ignition engines, the charge is a combustible mixture of fuel and air. Air alone is the charge in compression-ignition engines. Next, with both valves closed, the piston undergoes a *compression stroke,* raising the temperature and pressure of the charge. This requires work input from the piston to the cylinder contents. A combustion process is then initiated (both valves closed), resulting in a high-pressure, high-temperature gas mixture. Combustion is induced near the end of the compression stroke in spark-ignition engines by the spark plug. In compression-ignition engines, combustion is initiated by injecting fuel into the hot compressed air, beginning near the end of the compression stroke and continuing through the first part of the expansion. A *power* stroke follows the compression stroke, during which the gas mixture expands and work is done on the piston as it returns to bottom dead center. The piston then executes an *exhaust stroke* in which the burned gases are purged from the cylinder through the open exhaust valve. Smaller engines operate on two-stroke cycles. In two-stroke engines, the intake, compression, expansion, and exhaust operations are accomplished in one revolution of the crankshaft. Although internal combustion engines undergo *mechanical* cycles, the cylinder contents do not execute a *thermodynamic* cycle, for matter is introduced with one composition and is later discharged at a different composition.

A parameter used to describe the performance of reciprocating piston engines is the *mean effective pressure,* or mep. The ***mean effective pressure*** is the theoretical constant pressure that, if it acted on the piston during the power stroke, would produce the same *net* work as actually developed in one cycle. That is

***mean effective pressure***

$$\text{mep} = \frac{\text{net work for one cycle}}{\text{displacement volume}} \tag{9.1}$$

For two engines of equal displacement volume, the one with a higher mean effective pressure would produce the greater net work and, if the engines run at the same speed, greater power.

***Air-Standard Analysis.*** A detailed study of the performance of a reciprocating internal combustion engine would take into account many features. These would include the combustion process occurring within the cylinder and the effects of irreversibilities associated with friction and with pressure and temperature gradients. Heat transfer between the gases in the cylinder and the cylinder walls and the work required to charge the cylinder and exhaust the products of combustion also would be considered. Owing to these complexities, accurate modeling of reciprocating internal combustion engines normally involves computer simulation. To conduct *elementary* thermodynamic analyses of internal combustion engines, considerable simplification is required. One procedure is to employ an ***air-standard analysis*** having the following elements: (1) A fixed amount of air modeled as an ideal gas is the working fluid. (2) The combustion process is replaced by a heat transfer from an external source. (3) There are no exhaust and intake processes as in an actual engine. The cycle is completed by a constant-volume heat transfer process taking place while the piston is at the bottom dead center position. (4) All processes are internally reversible. In addition, in a ***cold air-standard analysis,*** the specific heats are assumed constant at their ambient temperature values. With an air-standard analysis, we avoid dealing with the complexities of the combustion process and the change of composition during combustion. A comprehensive analysis requires that such complexities be considered, however.

***air-standard analysis***

***cold air-standard analysis***

Although an air-standard analysis simplifies the study of internal combustion engines considerably, values for the mean effective pressure and operating temperatures and pressures calculated on this basis may depart significantly from those of actual engines. Accordingly, air-standard analysis allows internal combustion engines to be examined only qualitatively. Still, insights concerning actual performance can result with such an approach.

In the remainder of this part of the chapter, we consider two cycles that adhere to air-standard cycle idealizations: the Otto and Diesel cycles. These cycles differ from each other only in the way the heat addition process that replaces combustion in the actual cycle is modeled.

## 9.2 Air-Standard Otto Cycle

The air-standard Otto cycle is an ideal cycle that assumes the heat addition occurs instantaneously while the piston is at top dead center. The ***Otto cycle*** is shown on the $p$–$v$ and $T$–$s$ diagrams of Fig. 9.3. The cycle consists of four internally reversible processes in series. Process 1–2 is an isentropic compression of the air as the piston moves from bottom dead center to top dead center. Process 2–3 is a constant-volume heat transfer to the air from an

***Otto cycle***

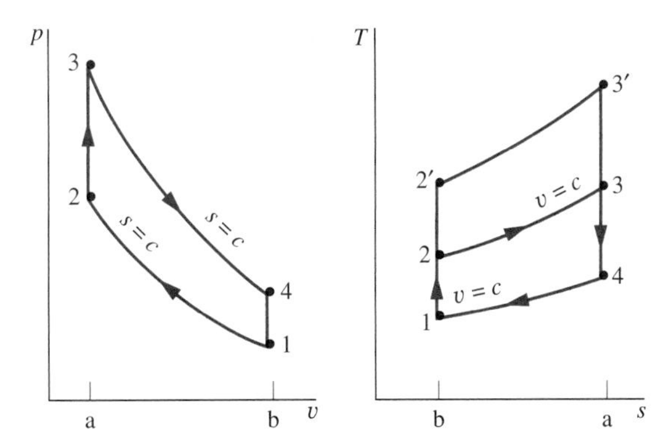


*Figure 9.3* $p$-$v$ and $T$–$s$ diagrams of the air-standard Otto cycle.

external source while the piston is at top dead center. This process is intended to represent the ignition of the fuel–air mixture and the subsequent rapid burning. Process 3–4 is an isentropic expansion (power stroke). The cycle is completed by the constant-volume Process 4–1 in which heat is rejected from the air while the piston is at bottom dead center.

Since the air-standard Otto cycle is composed of internally reversible processes, areas on the $T$–$s$ and $p$–$v$ diagrams of Fig. 9.3 can be interpreted as heat and work, respectively. On the $T$–$s$ diagram, area 2–3–a–b–2 represents the heat added per unit of mass and area 1–4–a–b–1 the heat rejected per unit of mass. On the $p$–$v$ diagram, area 1–2–a–b–1 represents the work input per unit of mass during the compression process and area 3–4–b–a–3 is the work done per unit of mass in the expansion process. The enclosed area of each figure can be interpreted as the net work output or, equivalently, the net heat added.

*Cycle Analysis.* The air-standard Otto cycle consists of two processes in which there is work but no heat transfer, Processes 1–2 and 3–4, and two processes in which there is heat transfer but no work, Processes 2–3 and 4–1. Expressions for these energy transfers are obtained by reducing the closed system energy balance assuming that changes in kinetic and potential energy can be ignored. The results are

$$\frac{W_{12}}{m} = u_2 - u_1, \qquad \frac{W_{34}}{m} = u_3 - u_4$$
$$\frac{Q_{23}}{m} = u_3 - u_2, \qquad \frac{Q_{41}}{m} = u_4 - u_1 \tag{9.2}$$

METHODOLOGY UPDATE

Carefully note that in writing Eqs. 9.2, we have departed from our usual sign convention for heat and work. When analyzing cycles, it is frequently convenient to regard all work and heat transfers as positive quantities. Thus, $W_{12}/m$ is a positive number representing the work *input* during compression and $Q_{41}/m$ is a positive number representing the heat *rejected* in Process 4–1. The net work of the cycle is expressed as

$$\frac{W_{\text{cycle}}}{m} = \frac{W_{34}}{m} - \frac{W_{12}}{m} = (u_3 - u_4) - (u_2 - u_1)$$

Alternatively, the net work can be evaluated as the net heat added

$$\frac{W_{\text{cycle}}}{m} = \frac{Q_{23}}{m} - \frac{Q_{41}}{m} = (u_3 - u_2) - (u_4 - u_1)$$

which, on rearrangement, can be placed in the same form as the previous expression for net work.

The thermal efficiency is the ratio of the net work of the cycle to the heat added.

$$\eta = \frac{(u_3 - u_2) - (u_4 - u_1)}{u_3 - u_2} = 1 - \frac{u_4 - u_1}{u_3 - u_2} \tag{9.3}$$

When air table data are used to conduct an analysis involving an air-standard Otto cycle, the specific internal energy values required by Eq. 9.3 can be obtained from Tables T-9 or T-9E as appropriate. The following relationships introduced in Sec. 7.6.2 apply for the isentropic processes 1–2 and 3–4

$$v_{r2} = v_{r1}\left(\frac{V_2}{V_1}\right) = \frac{v_{r1}}{r} \tag{9.4}$$

$$v_{r4} = v_{r3}\left(\frac{V_4}{V_3}\right) = r v_{r3} \tag{9.5}$$

where $r$ denotes the compression ratio. Note that since $V_3 = V_2$ and $V_4 = V_1$, $r = V_1/V_2 = V_4/V_3$. The parameter $v_r$ is tabulated versus temperature for air in Tables T-9.

When the Otto cycle is analyzed on a cold air-standard basis, the following expressions introduced in Sec. 7.6.2 would be used for the isentropic processes in place of Eqs. 9.4 and 9.5, respectively

$$\frac{T_2}{T_1} = \left(\frac{V_1}{V_2}\right)^{k-1} = r^{k-1} \qquad \text{(constant } k\text{)} \tag{9.6}$$

$$\frac{T_4}{T_3} = \left(\frac{V_3}{V_4}\right)^{k-1} = \frac{1}{r^{k-1}} \qquad \text{(constant } k\text{)} \tag{9.7}$$

where $k$ is the specific heat ratio, $k = c_p/c_v$.

*Effect of Compression Ratio on Performance.* By referring to the $T$–$s$ diagram of Fig. 9.3, we can conclude that the Otto cycle thermal efficiency increases as the compression ratio increases. An increase in the compression ratio changes the cycle from 1–2–3–4–1 to 1–2′–3′–4–1. Since the average temperature of heat addition is greater in the latter cycle and both cycles have the same heat rejection process, cycle 1–2′–3′–4–1 would have the greater thermal efficiency. The increase in thermal efficiency with compression ratio is also brought out simply by the following development on a cold air-standard basis. For constant $c_v$, Eq. 9.3 becomes

$$\eta = 1 - \frac{c_v(T_4 - T_1)}{c_v(T_3 - T_2)}$$

$$= 1 - \frac{T_1}{T_2}\left(\frac{T_4/T_1 - 1}{T_3/T_2 - 1}\right)$$

From Eqs. 9.6 and 9.7 above, $T_4/T_1 = T_3/T_2$, so

$$\eta = 1 - \frac{T_1}{T_2}$$

Finally, introducing Eq. 9.6

$$\eta = 1 - \frac{1}{r^{k-1}} \qquad \text{(constant } k\text{)} \tag{9.8}$$

Equation 9.8 indicates that the cold air-standard Otto cycle thermal efficiency is a function of compression ratio and specific heat ratio. This relationship is shown in Fig. 9.4 for $k = 1.4$.

The foregoing discussion suggests that it is advantageous for internal combustion engines to have high compression ratios, and this is the case. The possibility of autoignition, or "knock," places an upper limit on the compression ratio of spark-ignition engines, however. After the spark has ignited a portion of the fuel–air mixture, the rise in pressure accompanying combustion compresses the remaining charge. Autoignition can occur if the

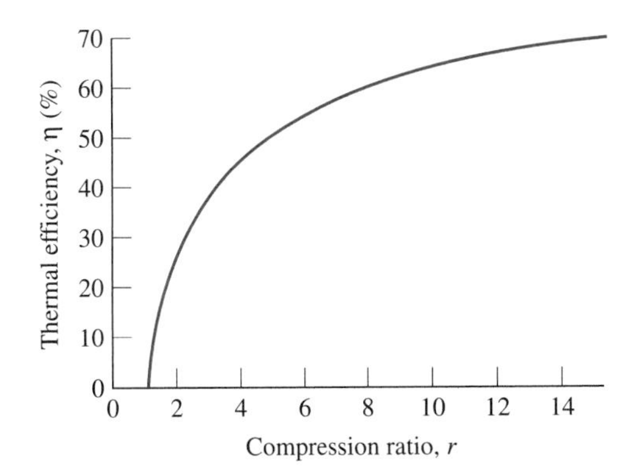


*Figure 9.4* Thermal efficiency of the cold air-standard Otto cycle, $k = 1.4$.

temperature of the unburned mixture becomes too high before the mixture is consumed by the flame front. Since the temperature attained by the air–fuel mixture during the compression stroke increases as the compression ratio increases, the likelihood of autoignition occurring increases with the compression ratio. Autoignition may result in high-pressure waves in the cylinder (manifested by a knocking or pinging sound) that can lead to loss of power as well as engine damage. Fuels formulated with tetraethyl lead are resistant to autoignition and thus allow relatively high compression ratios. The *unleaded* gasoline in common use today because of environmental concerns over air pollution limits the compression ratios of spark-ignition engines to approximately 9. Higher compression ratios can be achieved in compression-ignition engines because air alone is compressed. Compression ratios in the range of 12 to 20 are typical. Compression-ignition engines also can use less refined fuels having higher ignition temperatures than the volatile fuels required by spark-ignition engines.

In the next example, we illustrate the analysis of the air-standard Otto cycle. Results are compared with those obtained on a cold air-standard basis.

## *Example 9.1* Analyzing the Otto Cycle

The temperature at the beginning of the compression process of an air-standard Otto cycle with a compression ratio of 8 is 540°R, the pressure is 1 atm, and the cylinder volume is 0.02 ft$^3$. The maximum temperature during the cycle is 3600°R. Determine **(a)** the temperature and pressure at the end of each process of the cycle, **(b)** the thermal efficiency, and **(c)** the mean effective pressure, in atm.

### Solution

***Known:*** An air-standard Otto cycle with a given value of compression ratio is executed with specified conditions at the beginning of the compression stroke and a specified maximum temperature during the cycle.

***Find:*** Determine the temperature and pressure at the end of each process, the thermal efficiency, and mean effective pressure, in atm.

***Schematic and Given Data:***

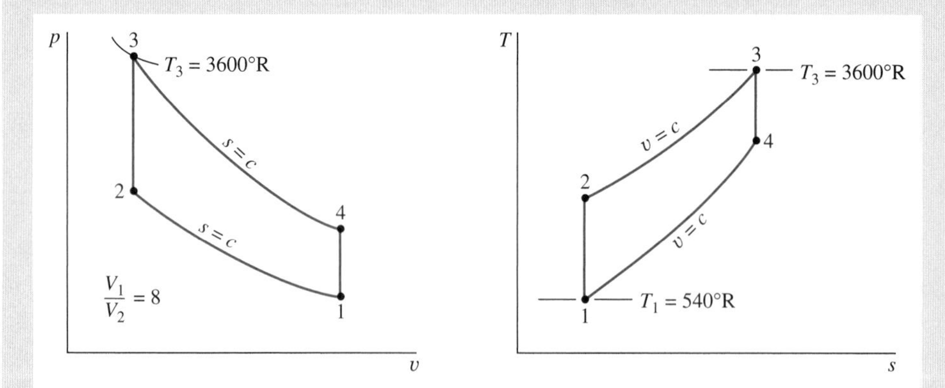


*Figure E9.1*

***Assumptions:***

1. The air in the piston–cylinder assembly is the closed system.
2. The compression and expansion processes are adiabatic.
3. All processes are internally reversible.
4. The air is modeled as an ideal gas.
5. Kinetic and potential energy effects are negligible.

***Analysis:*** **(a)** The analysis begins by determining the temperature, pressure, and specific internal energy at each principal state of the cycle. At $T_1 = 540°R$, Table T-9E gives $u_1 = 92.04$ Btu/lb and $v_{r1} = 144.32$.

For the isentropic compression Process 1–2

$$v_{r2} = \frac{V_2}{V_1} v_{r1} = \frac{v_{r1}}{r} = \frac{144.32}{8} = 18.04$$

Interpolating with $v_{r2}$ in Table T-9E, we get $T_2 = 1212°R$ and $u_2 = 211.3$ Btu/lb. With the ideal gas equation of state

$$p_2 = p_1 \frac{T_2}{T_1}\frac{V_1}{V_2} = (1 \text{ atm})\left(\frac{1212°R}{540°R}\right)8 = 17.96 \text{ atm} \triangleleft$$

The pressure at state 2 can be evaluated alternatively by using the isentropic relationship, $p_2 = p_1(p_{r2}/p_{r1})$.

Since Process 2–3 occurs at constant volume, the ideal gas equation of state gives

$$p_3 = p_2 \frac{T_3}{T_2} = (17.96 \text{ atm})\left(\frac{3600°R}{1212°R}\right) = 53.3 \text{ atm} \triangleleft$$

At $T_3 = 3600°R$, Table T-9E gives $u_3 = 721.44$ Btu/lb and $v_{r3} = 0.6449$.

For the isentropic expansion Process 3–4

$$v_{r4} = v_{r3}\frac{V_4}{V_3} = v_{r3}\frac{V_1}{V_2} = 0.6449(8) = 5.16$$

Interpolating in Table T-9E with $v_{r4}$ gives $T_4 = 1878°R$, $u_4 = 342.2$ Btu/lb. The pressure at state 4 can be found using the isentropic relationship $p_4 = p_3(p_{r4}/p_{r3})$ or the ideal gas equation of state applied at states 1 and 4. With $V_4 = V_1$, the ideal gas equation of state gives

$$p_4 = p_1 \frac{T_4}{T_1} = (1 \text{ atm})\left(\frac{1878°R}{540°R}\right) = 3.48 \text{ atm} \triangleleft$$

**(b)** The thermal efficiency is

$$\eta = 1 - \frac{Q_{41}/m}{Q_{23}/m} = 1 - \frac{u_4 - u_1}{u_3 - u_2}$$

$$= 1 - \frac{342.2 - 92.04}{721.44 - 211.3} = 0.51(51\%) \triangleleft$$

**(c)** To evaluate the mean effective pressure requires the net work per cycle. That is

$$W_{cycle} = m[(u_3 - u_4) - (u_2 - u_1)]$$

where $m$ is the mass of the air, evaluated from the ideal gas equation of state as follows:

$$m = \frac{p_1 V_1}{(\bar{R}/M)T_1}$$

$$= \frac{(14.696 \text{ lbf/in.}^2)|144 \text{ in.}^2/\text{ft}^2|(0.02 \text{ ft}^3)}{\left(\frac{1545}{28.97}\frac{\text{ft}\cdot\text{lbf}}{\text{lb}\cdot°R}\right)(540°R)}$$

$$= 1.47 \times 10^{-3} \text{ lb}$$

Inserting values into the expression for $W_{cycle}$

$$W_{cycle} = (1.47 \times 10^{-3} \text{ lb})[(721.44 - 342.2) - (211.3 - 92.04)] \text{ Btu/lb}$$

$$= 0.382 \text{ Btu}$$

The displacement volume is $V_1 - V_2$, so the mean effective pressure is given by

$$\text{mep} = \frac{W_{\text{cycle}}}{V_1 - V_2} = \frac{W_{\text{cycle}}}{V_1(1 - V_2/V_1)}$$

❶
$$= \frac{0.382 \text{ Btu}}{(0.02 \text{ ft}^3)(1 - 1/8)} \left| \frac{778 \text{ ft} \cdot \text{lbf}}{1 \text{ Btu}} \right| \left| \frac{1 \text{ ft}^2}{144 \text{ in.}^2} \right|$$

$$= 118 \text{ lbf/in.}^2 = 8.03 \text{ atm} \triangleleft$$

❶ This solution utilizes Table T-9E for air, which accounts explicitly for the variation of the specific heats with temperature. A solution also can be developed on a cold air-standard basis in which constant specific heats are assumed. This solution is left as an exercise, but for comparison the results are presented in the following table for the case $k = 1.4$, representing atmospheric air:

| Parameter | Air-Standard Analysis | Cold Air-Standard Analysis, $k = 1.4$ |
|---|---|---|
| $T_2$ | 1212°R | 1241°R |
| $T_3$ | 3600°R | 3600°R |
| $T_4$ | 1878°R | 1567°R |
| $\eta$ | 0.51 (51%) | 0.565 (56.5%) |
| mep | 8.03 atm | 7.05 atm |

## 9.3 Air-Standard Diesel Cycle

*Diesel cycle*

The air-standard Diesel cycle is an ideal cycle that assumes the heat addition occurs during a constant-pressure process that starts with the piston at top dead center. The ***Diesel cycle*** is shown on $p$–$v$ and $T$–$s$ diagrams in Fig. 9.5. The cycle consists of four internally reversible processes in series. The first process from state 1 to state 2 is the same as in the Otto cycle: an isentropic compression. Heat is not transferred to the working fluid at constant volume as in the Otto cycle, however. In the Diesel cycle, heat is transferred to the working fluid at *constant pressure.* Process 2–3 also makes up the first part of the power stroke. The isentropic expansion from state 3 to state 4 is the remainder of the power stroke. As in the Otto cycle, the cycle is completed by constant-volume Process 4–1 in which heat is rejected from the air while the piston is at bottom dead center. This process replaces the exhaust and intake processes of the actual engine.

Since the air-standard Diesel cycle is composed of internally reversible processes, areas on the $T$–$s$ and $p$–$v$ diagrams of Fig. 9.5 can be interpreted as heat and work, respectively.

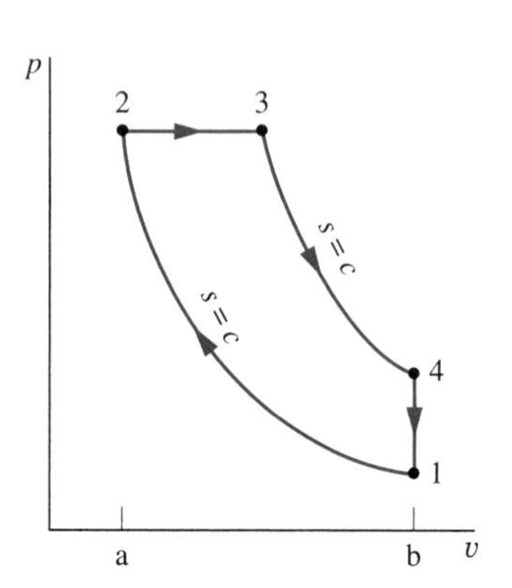


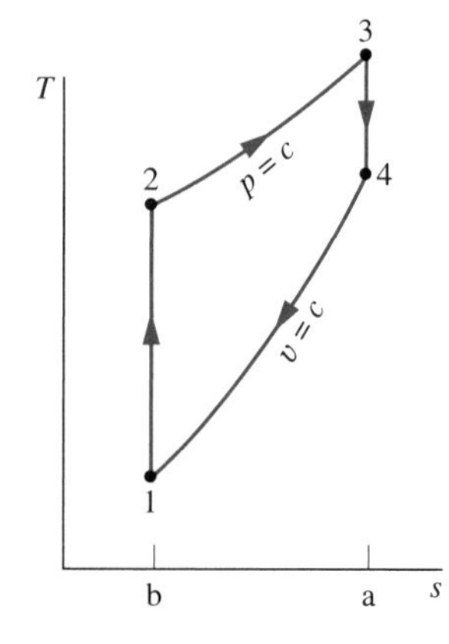


*Figure 9.5* $p$-$v$ and $T$–$s$ diagrams of the air-standard Diesel cycle.

On the $T$–$s$ diagram, area 2–3–a–b–2 represents the heat added per unit of mass and area 1–4–a–b–1 is the heat rejected per unit of mass. On the $p$–$v$ diagram, area 1–2–a–b–1 is the work input per unit of mass during the compression process. Area 2–3–4–b–a–2 is the work done per unit of mass as the piston moves from top dead center to bottom dead center. The enclosed area of each figure is the net work output, which equals the net heat added.

*Cycle Analysis.* In the Diesel cycle the heat addition takes place at constant pressure. Accordingly, Process 2–3 involves both work and heat. The work is given by

$$\frac{W_{23}}{m} = \int_2^3 p\,dv = p_2(v_3 - v_2) \tag{9.9}$$

The heat added in Process 2–3 can be found by applying the closed system energy balance

$$m(u_3 - u_2) = Q_{23} - W_{23}$$

Introducing Eq. 9.9 and solving for the heat transfer

$$\begin{aligned}\frac{Q_{23}}{m} &= (u_3 - u_2) + p(v_3 - v_2) = (u_3 + pv_3) - (u_2 + pv_2) \\ &= h_3 - h_2 \end{aligned} \tag{9.10}$$

where the specific enthalpy is introduced to simplify the expression. As in the Otto cycle, the heat rejected in Process 4–1 is given by

$$\frac{Q_{41}}{m} = u_4 - u_1$$

The thermal efficiency is the ratio of the net work of the cycle to the heat added

$$\eta = \frac{W_{\text{cycle}}/m}{Q_{23}/m} = 1 - \frac{Q_{41}/m}{Q_{23}/m} = 1 - \frac{u_4 - u_1}{h_3 - h_2} \tag{9.11}$$

As for the Otto cycle, the thermal efficiency of the Diesel cycle increases with the compression ratio.

To evaluate the thermal efficiency from Eq. 9.11 requires values for $u_1$, $u_4$, $h_2$, and $h_3$ or equivalently the temperatures at the principal states of the cycle. Let us consider next how these temperatures are evaluated. For a given initial temperature $T_1$ and compression ratio $r$, the temperature at state 2 can be found using the following isentropic relationship and $v_r$ data

$$v_{r2} = \frac{V_2}{V_1} v_{r1} = \frac{1}{r} v_{r1}$$

To find $T_3$, note that the ideal gas equation of state reduces with $p_3 = p_2$ to give

$$T_3 = \frac{V_3}{V_2} T_2 = r_c T_2$$

where $r_c = V_3/V_2$, called the ***cutoff ratio,*** has been introduced.

*cutoff ratio*

Since $V_4 = V_1$, the volume ratio for the isentropic process 3–4 can be expressed as

$$\frac{V_4}{V_3} = \frac{V_4}{V_2}\frac{V_2}{V_3} = \frac{V_1}{V_2}\frac{V_2}{V_3} = \frac{r}{r_c} \tag{9.12}$$

where the compression ratio $r$ and cutoff ratio $r_c$ have been introduced for conciseness.

Using Eq. 9.12 together with $v_{r3}$ at $T_3$, the temperature $T_4$ can be determined by interpolation once $v_{r4}$ is found from the isentropic relationship

$$v_{r4} = \frac{V_4}{V_3} v_{r3} = \frac{r}{r_c} v_{r3}$$

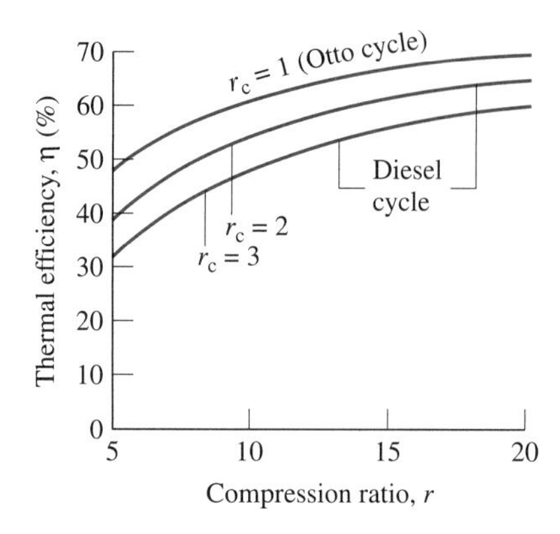


*Figure 9.6* Thermal efficiency of the cold air-standard Diesel cycle, $k = 1.4$.

In a *cold air-standard analysis,* the appropriate expression for evaluating $T_2$ is provided by

$$\frac{T_2}{T_1} = \left(\frac{V_1}{V_2}\right)^{k-1} = r^{k-1} \qquad \text{(constant } k\text{)}$$

The temperature $T_4$ is found similarly from

$$\frac{T_4}{T_3} = \left(\frac{V_3}{V_4}\right)^{k-1} = \left(\frac{r_c}{r}\right)^{k-1} \qquad \text{(constant } k\text{)}$$

where Eq. 9.12 has been used to replace the volume ratio.

***Effect of Compression Ratio on Performance.*** As for the Otto cycle, the thermal efficiency of the Diesel cycle increases with increasing compression ratio. This can be brought out simply using a *cold* air-standard analysis. On a cold air-standard basis, the thermal efficiency of the Diesel cycle can be expressed as

$$\eta = 1 - \frac{1}{r^{k-1}}\left[\frac{r_c^k - 1}{k(r_c - 1)}\right] \qquad \text{(constant } k\text{)} \qquad (9.13)$$

where $r$ is the compression ratio and $r_c$ the cutoff ratio. The derivation is left as an exercise. This relationship is shown in Fig. 9.6 for $k = 1.4$. Equation 9.13 for the Diesel cycle differs from Eq. 9.8 for the Otto cycle only by the term in brackets, which for $r_c > 1$ is greater than unity. Thus, when the compression ratio is the same, the thermal efficiency of the cold air-standard Diesel cycle would be less than that of the cold air-standard Otto cycle.

In the next example, we illustrate the analysis of the air-standard Diesel cycle.

## *Example 9.2* Analyzing the Diesel Cycle

At the beginning of the compression process of an air-standard Diesel cycle operating with a compression ratio of 18, the temperature is 300 K and the pressure is 0.1 MPa. The cutoff ratio for the cycle is 2. Determine **(a)** the temperature and pressure at the end of each process of the cycle, **(b)** the thermal efficiency, **(c)** the mean effective pressure, in MPa.

### Solution

***Known:*** An air-standard Diesel cycle is executed with specified conditions at the beginning of the compression stroke. The compression and cutoff ratios are given.

***Find:*** Determine the temperature and pressure at the end of each process, the thermal efficiency, and mean effective pressure.

***Schematic and Given Data:***

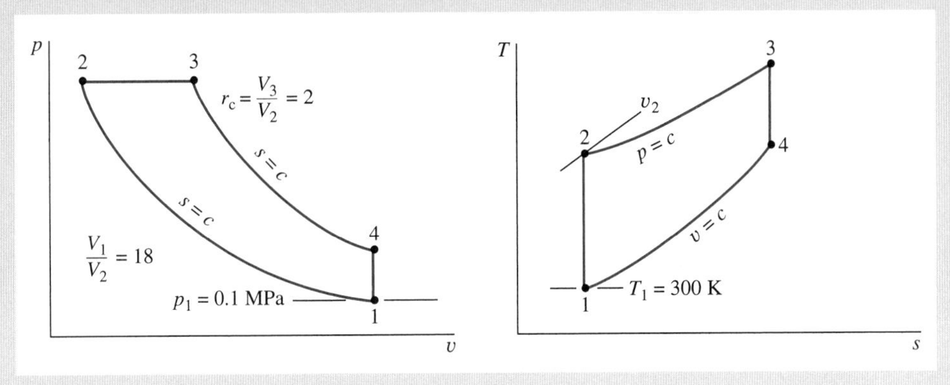


*Figure E9.2*

***Assumptions:***

1. The air in the piston–cylinder assembly is the closed system.
2. The compression and expansion processes are adiabatic.
3. All processes are internally reversible.
4. The air is modeled as an ideal gas.
5. Kinetic and potential energy effects are negligible.

***Analysis:*** **(a)** The analysis begins by determining properties at each principal state of the cycle. With $T_1 = 300$ K, Table T-9 gives $u_1 = 214.07$ kJ/kg and $v_{r1} = 621.2$. For the isentropic compression process 1–2

$$v_{r2} = \frac{V_2}{V_1} v_{r1} = \frac{v_{r1}}{r} = \frac{621.2}{18} = 34.51$$

Interpolating in Table T-9, we get $T_2 = 898.3$ K and $h_2 = 930.98$ kJ/kg. With the ideal gas equation of state

$$p_2 = p_1 \frac{T_2}{T_1} \frac{V_1}{V_2} = (0.1)\left(\frac{898.3}{300}\right)(18) = 5.39 \text{ MPa} \triangleleft$$

The pressure at state 2 can be evaluated alternatively using the isentropic relationship, $p_2 = p_1(p_{r2}/p_{r1})$.

Since Process 2–3 occurs at constant pressure, the ideal gas equation of state gives

$$T_3 = \frac{V_3}{V_2} T_2$$

Introducing the cutoff ratio, $r_c = V_3/V_2$

$$T_3 = r_c T_2 = 2(898.3) = 1796.6 \text{ K} \triangleleft$$

From Table T-9, $h_3 = 1999.1$ kJ/kg and $v_{r3} = 3.97$.

For the isentropic expansion process 3–4

$$v_{r4} = \frac{V_4}{V_3} v_{r3} = \frac{V_4}{V_2} \frac{V_2}{V_3} v_{r3}$$

Introducing $V_4 = V_1$, the compression ratio $r$, and the cutoff ratio $r_c$, we have

$$v_{r4} = \frac{r}{r_c} v_{r3} = \frac{18}{2}(3.97) = 35.73$$

By interpolating in Table T-9 with $v_{r4}$, $u_4 = 664.3$ kJ/kg and $T_4 = 887.7$ K. The pressure at state 4 can be found using the isentropic relationship $p_4 = p_3(p_{r4}/p_{r3})$ or the ideal gas equation of state applied at states 1 and 4. With $V_4 = V_1$, the ideal gas equation of state gives

$$p_4 = p_1 \frac{T_4}{T_1} = (0.1 \text{ MPa})\left(\frac{887.7 \text{ K}}{300 \text{ K}}\right) = 0.3 \text{ MPa} \triangleleft$$

**(b)** The thermal efficiency is found using

$$\eta = 1 - \frac{Q_{41}/m}{Q_{23}/m} = 1 - \frac{u_4 - u_1}{h_3 - h_2}$$

➊ $$= 1 - \frac{664.3 - 214.07}{1999.1 - 930.98} = 0.578(57.8\%) \triangleleft$$

**(c)** The mean effective pressure written in terms of specific volumes is

$$\text{mep} = \frac{W_{\text{cycle}}/m}{v_1 - v_2} = \frac{W_{\text{cycle}}/m}{v_1(1 - 1/r)}$$

The net work of the cycle equals the net heat added

$$\frac{W_{\text{cycle}}}{m} = \frac{Q_{23}}{m} - \frac{Q_{41}}{m} = (h_3 - h_2) - (u_4 - u_1)$$

$$= (1999.1 - 930.98) - (664.3 - 214.07)$$

$$= 617.9 \text{ kJ/kg}$$

The specific volume at state 1 is

$$v_1 = \frac{(\overline{R}/M)T_1}{p_1} = \frac{\left(\frac{8314}{28.97}\frac{\text{N}\cdot\text{m}}{\text{kg}\cdot\text{K}}\right)(300 \text{ K})}{10^5 \text{ N/m}^2} = 0.861 \text{ m}^3/\text{kg}$$

Inserting values

$$\text{mep} = \frac{617.9 \text{ kJ/kg}}{0.861(1 - 1/18)\text{m}^3/\text{kg}}\left|\frac{10^3 \text{ N}\cdot\text{m}}{1 \text{ kJ}}\right|\left|\frac{1 \text{ MPa}}{10^6 \text{ N/m}^2}\right|$$

$$= 0.76 \text{ MPa} \triangleleft$$

➊ This solution uses the air tables, which account explicitly for the variation of the specific heats with temperature. Note that Eq. 9.13 based on the assumption of *constant* specific heats has not been used to determine the thermal efficiency. The cold air-standard solution of this example is left as an exercise.

# Gas Turbine Power Plants

This part of the chapter deals with gas turbine power plants. Gas turbines tend to be lighter and more compact than the vapor power plants studied in Chap. 8. The favorable power-output-to-weight ratio of gas turbines makes them well suited for transportation applications (aircraft propulsion, marine power plants, and so on). Gas turbines are also commonly used for stationary power generation.

## 9.4 Modeling Gas Turbine Power Plants

Gas turbine power plants may operate on either an open or closed basis. The open mode pictured in Fig. 9.7*a* is more common. This is an engine in which atmospheric air is continuously drawn into the compressor, where it is compressed to a high pressure. The air then enters a combustion chamber, or combustor, where it is mixed with fuel and combustion occurs, resulting in combustion products at an elevated temperature. The combustion products expand through the turbine and are subsequently discharged to the surroundings. Part of the

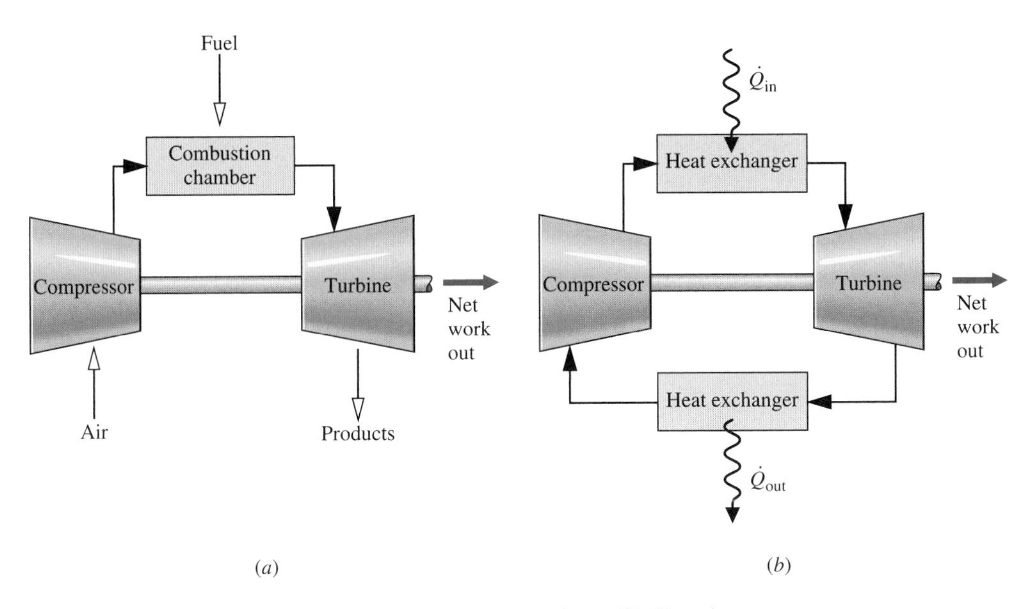


*Figure 9.7* Simple gas turbine. (*a*) Open to the atmosphere. (*b*) Closed.

turbine work developed is used to drive the compressor; the remainder is available to generate electricity, to propel a vehicle, or for other purposes. In the system pictured in Fig. 9.7*b*, the working fluid receives an energy input by heat transfer from an external source, for example a gas-cooled nuclear reactor. The gas exiting the turbine is passed through a heat exchanger, where it is cooled prior to reentering the compressor.

An idealization often used in the study of open gas turbine power plants is that of an ***air-standard analysis.*** In an air-standard analysis, two assumptions are always made: (1) The working fluid is air, which behaves as an ideal gas, and (2) the temperature rise that would be brought about by combustion is accomplished by a heat transfer from an external source. With an air-standard analysis, we avoid dealing with the complexities of the combustion process and the change of composition during combustion. An air-standard analysis simplifies the study of gas turbine power plants considerably. However, numerical values calculated on this basis may provide only qualitative indications of power plant performance. Sufficient information about combustion and the properties of products of combustion is known that the study of gas turbines can be conducted without the foregoing assumptions. Nevertheless, in the interest of simplicity the current presentation proceeds on the basis of an air-standard analysis.

*air-standard analysis*

## 9.5 Air-Standard Brayton Cycle

A schematic diagram of an air-standard gas turbine is shown in Fig. 9.8. The directions of the principal energy transfers are indicated on this figure by arrows. In accordance with the assumptions of an air-standard analysis, the temperature rise that would be achieved in the combustion process is brought about by a heat transfer to the working fluid from an external source and the working fluid is considered to be air as an ideal gas. With the air-standard idealizations, air would be drawn into the compressor at state 1 from the surroundings and later returned to the surroundings at state 4 with a temperature greater than the ambient temperature. After interacting with the surroundings, each unit mass of discharged air would eventually return to the same state as the air entering the compressor, so we may think of the air passing through the components of the gas turbine as undergoing a thermodynamic cycle. A simplified representation of the states visited by the air

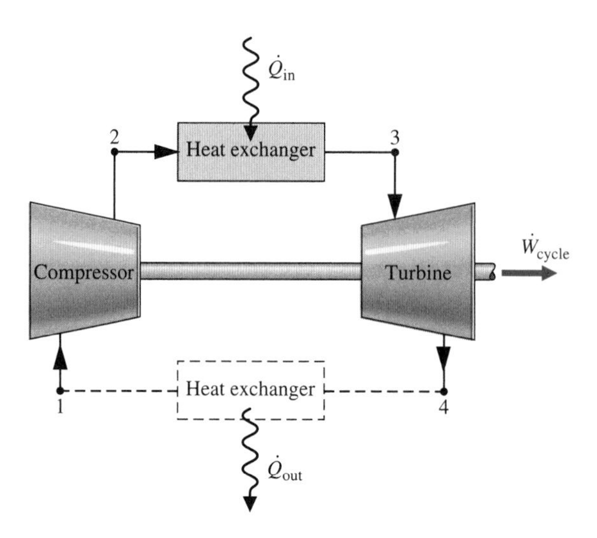


*Figure 9.8* Air-standard gas turbine cycle.

in such a cycle can be devised by regarding the turbine exhaust air as restored to the compressor inlet state by passing through a heat exchanger where heat rejection to the surroundings occurs. The cycle that results with this further idealization is called the air-standard ***Brayton cycle.***

*Brayton cycle*

## 9.5.1 Evaluating Principal Work and Heat Transfers

The following expressions for the work and heat transfers of energy that occur at steady state are readily derived by reduction of the control volume mass and energy rate balances. These energy transfers are positive in the directions of the arrows in Fig. 9.8. Assuming the turbine operates adiabatically and with negligible effects of kinetic and potential energy, the work developed per unit of mass is

$$\frac{\dot{W}_t}{\dot{m}} = h_3 - h_4 \tag{9.14}$$

where $\dot{m}$ denotes the mass flow rate. With the same assumptions, the compressor work per unit of mass is

$$\frac{\dot{W}_c}{\dot{m}} = h_2 - h_1 \tag{9.15}$$

The symbol $\dot{W}_c$ denotes work *input* and takes on a positive value. The heat added to the cycle per unit of mass is

$$\frac{\dot{Q}_{in}}{\dot{m}} = h_3 - h_2 \tag{9.16}$$

The heat rejected per unit of mass is

$$\frac{\dot{Q}_{out}}{\dot{m}} = h_4 - h_1 \tag{9.17}$$

where $\dot{Q}_{out}$ is positive in value.

The thermal efficiency of the cycle in Fig. 9.8 is

$$\eta = \frac{\dot{W}_t/\dot{m} - \dot{W}_c/\dot{m}}{\dot{Q}_{in}/\dot{m}} = \frac{(h_3 - h_4) - (h_2 - h_1)}{h_3 - h_2} \tag{9.18}$$

The back work ratio for the cycle is

$$\text{bwr} = \frac{\dot{W}_c/\dot{m}}{\dot{W}_t/\dot{m}} = \frac{h_2 - h_1}{h_3 - h_4} \tag{9.19}$$

For the same pressure rise, a gas turbine compressor would require a much greater work input per unit of mass flow than the pump of a vapor power plant because the average specific volume of the gas flowing through the compressor would be many times greater than that of the liquid passing through the pump (see discussion of Eq. 7.43a in Sec. 7.8). Hence, a relatively large portion of the work developed by the turbine is required to drive the compressor. Typical back work ratios of gas turbines range from 40 to 80%. In comparison, the back work ratios of vapor power plants are normally only 1 or 2%.

If the temperatures at the numbered states of the cycle are known, the specific enthalpies required by the foregoing equations are readily obtained from the ideal gas table for air, Table T-9 or Table T-9E. Alternatively, with the sacrifice of some accuracy, the variation of the specific heats with temperature can be ignored and the specific heats taken as constant. The air-standard analysis is then referred to as a *cold air-standard analysis.* As illustrated by the discussion of internal combustion engines given previously, the chief advantage of the assumption of constant specific heats is that simple expressions for quantities such as thermal efficiency can be derived, and these can be used to deduce qualitative indications of cycle performance without involving tabular data.

Since Eqs. 9.14 through 9.19 have been developed from mass and energy rate balances, they apply equally when irreversibilities are present and in the absence of irreversibilities. Although irreversibilities and losses associated with the various power plant components have a pronounced effect on overall performance, it is instructive to consider an idealized cycle in which they are assumed absent. Such a cycle establishes an upper limit on the performance of the air-standard Brayton cycle. This is considered next.

## 9.5.2 Ideal Air-Standard Brayton Cycle

Ignoring irreversibilities as the air circulates through the various components of the Brayton cycle, there are no frictional pressure drops, and the air flows at constant pressure through the heat exchangers. If stray heat transfers to the surroundings are also ignored, the processes through the turbine and compressor are isentropic. The ideal cycle shown on the $p$–$v$ and $T$–$s$ diagrams in Fig. 9.9 adheres to these idealizations.

Areas on the $T$–$s$ and $p$–$v$ diagrams of Fig. 9.9 can be interpreted as heat and work, respectively, per unit of mass flowing. On the $T$–$s$ diagram, area 2–3–a–b–2 represents the heat added per unit of mass and area 1–4–a–b–1 is the heat rejected per unit of mass. On the $p$–$v$ diagram, area 1–2–a–b–1 represents the compressor work input per unit of mass and area 3–4–b–a–3 is the turbine work output per unit of mass (Sec. 7.8). The enclosed area on each figure can be interpreted as the net work output or, equivalently, the net heat added.

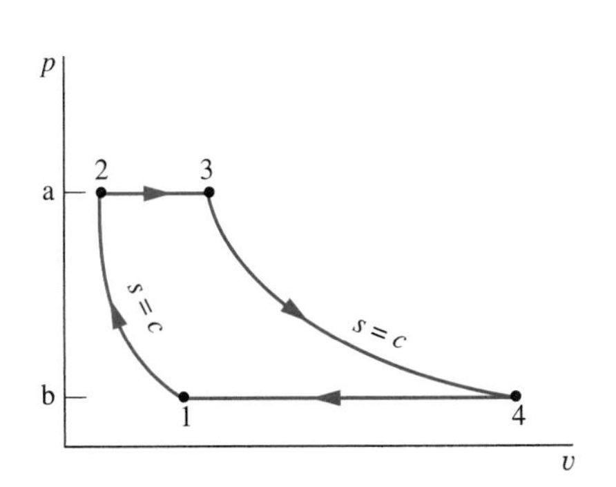


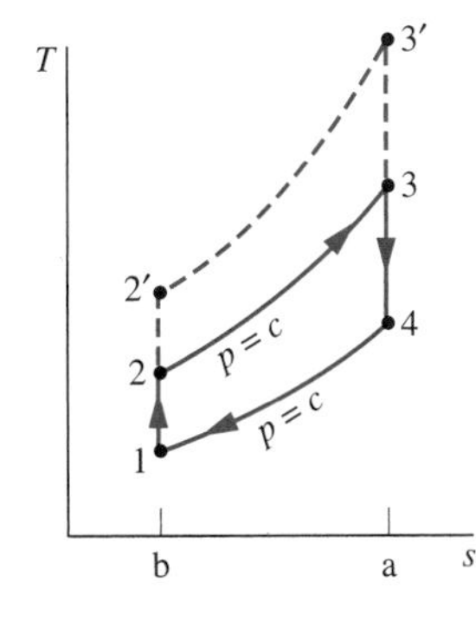


*Figure 9.9* Air-standard ideal Brayton cycle.

When air table data are used to conduct an analysis involving the ideal Brayton cycle, the following relationships, introduced in Sec. 7.6.2, apply for the isentropic processes 1–2 and 3–4

$$p_{r2} = p_{r1}\frac{p_2}{p_1} \tag{9.20}$$

$$p_{r4} = p_{r3}\frac{p_4}{p_3} = p_{r3}\frac{p_1}{p_2} \tag{9.21}$$

Recall that $p_r$ is tabulated versus temperature in Table T-9. Since the air flows through the heat exchangers of the ideal cycle at constant pressure, it follows that $p_4/p_3 = p_1/p_2$. This relationship has been used in writing Eq. 9.21.

When an ideal Brayton cycle is analyzed on a cold air-standard basis, the specific heats are taken as constant. Equations 9.20 and 9.21 are then replaced, respectively, by the following expressions, introduced in Sec. 7.6.2

$$T_2 = T_1\left(\frac{p_2}{p_1}\right)^{(k-1)/k} \tag{9.22}$$

$$T_4 = T_3\left(\frac{p_4}{p_3}\right)^{(k-1)/k} = T_3\left(\frac{p_1}{p_2}\right)^{(k-1)/k} \tag{9.23}$$

where $k$ is the specific heat ratio, $k = c_p/c_v$.

In the next example, we illustrate the analysis of an ideal air-standard Brayton cycle and compare results with those obtained on a cold air-standard basis.

## *Example 9.3* Analyzing the Ideal Brayton Cycle

Air enters the compressor of an ideal air-standard Brayton cycle at 100 kPa, 300 K, with a volumetric flow rate of 5 m³/s. The compressor pressure ratio is 10. The turbine inlet temperature is 1400 K. Determine **(a)** the thermal efficiency of the cycle, **(b)** the back work ratio, **(c)** the *net* power developed, in kW.

### Solution

***Known:*** An ideal air-standard Brayton cycle operates with given compressor inlet conditions, given turbine inlet temperature, and a known compressor pressure ratio.

***Find:*** Determine the thermal efficiency, the back work ratio, and the *net* power developed, in kW.

***Schematic and Given Data:***

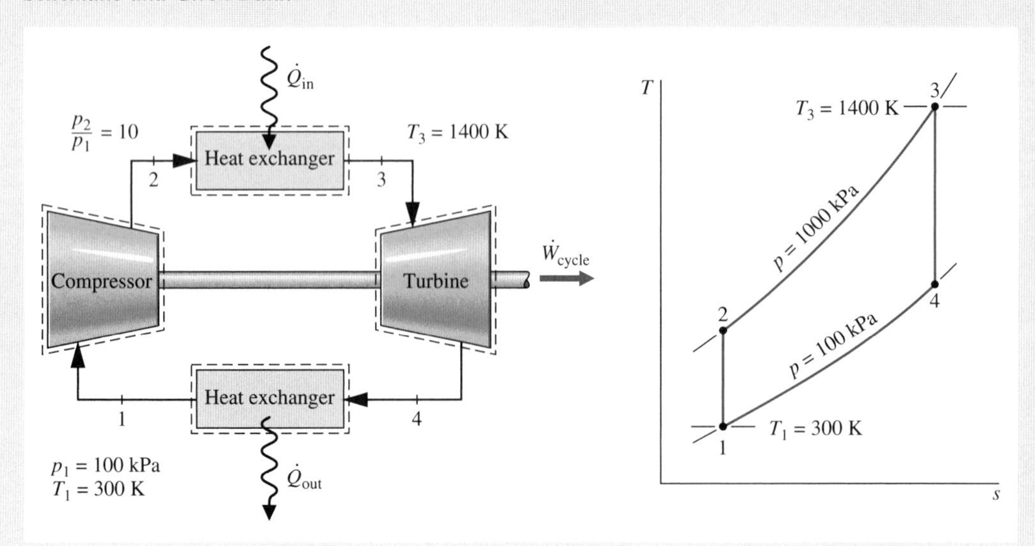


*Figure E9.3*

***Assumptions:***

1. Each component is analyzed as a control volume at steady state. The control volumes are shown on the accompanying sketch by dashed lines.
2. The turbine and compressor processes are isentropic.
3. There are no pressure drops for flow through the heat exchangers.
4. Kinetic and potential energy effects are negligible.
5. The working fluid is air modeled as an ideal gas.

***Properties:*** The analysis begins by determining the specific enthalpy at each numbered state of the cycle. At state 1, the ❶ temperature is 300 K. From Table T-9, $h_1 = 300.19$ kJ/kg and $p_{r1} = 1.386$.

Since the compressor process is isentropic, the following relationship can be used to determine $h_2$

$$p_{r2} = \frac{p_2}{p_1} p_{r1} = (10)(1.386) = 13.86$$

Then, interpolating in Table T-9, we obtain $h_2 = 579.9$ kJ/kg.

The temperature at state 3 is given as $T_3 = 1400$ K. With this temperature, the specific enthalpy at state 3 from Table T-9 is $h_3 = 1515.4$ kJ/kg. Also, $p_{r3} = 450.5$.

The specific enthalpy at state 4 is found by using the isentropic relation

$$p_{r4} = p_{r3} \frac{p_4}{p_3} = (450.5)(1/10) = 45.05$$

Interpolating in Table T-9, we get $h_4 = 808.5$ kJ/kg.

***Analysis:*** (a) The thermal efficiency is

$$\begin{aligned} \eta &= \frac{(\dot{W}_t/\dot{m}) - (\dot{W}_c/\dot{m})}{\dot{Q}_{in}/\dot{m}} \\ &= \frac{(h_3 - h_4) - (h_2 - h_1)}{h_3 - h_2} = \frac{(1515.4 - 808.5) - (579.9 - 300.19)}{1515.4 - 579.9} \\ &= \frac{706.9 - 279.7}{935.5} = 0.457\ (45.7\%) \ \triangleleft \end{aligned}$$

(b) The back work ratio is

❷
$$\text{bwr} = \frac{\dot{W}_c/\dot{m}}{\dot{W}_t/\dot{m}} = \frac{h_2 - h_1}{h_3 - h_4} = \frac{279.7}{706.9} = 0.396\ (39.6\%) \ \triangleleft$$

(c) The net power developed is

$$\dot{W}_{cycle} = \dot{m}[(h_3 - h_4) - (h_2 - h_1)]$$

To evaluate the net power requires the mass flow rate $\dot{m}$, which can be determined from the volumetric flow rate and specific volume at the compressor inlet as follows

$$\dot{m} = \frac{(\text{AV})_1}{v_1}$$

Since $v_1 = (\overline{R}/M)T_1/p_1$, this becomes

$$\dot{m} = \frac{(\text{AV})_1 p_1}{(\overline{R}/M)T_1} = \frac{(5\ \text{m}^3/\text{s})(100 \times 10^3\ \text{N/m}^2)}{\left(\dfrac{8314}{28.97}\dfrac{\text{N}\cdot\text{m}}{\text{kg}\cdot\text{K}}\right)(300\ \text{K})} = 5.807\ \text{kg/s}$$

Finally,

$$\dot{W}_{cycle} = (5.807 \text{ kg/s})(706.9 - 279.7)\left(\frac{\text{kJ}}{\text{kg}}\right)\left|\frac{1 \text{ kW}}{1 \text{ kJ/s}}\right| = 2481 \text{ kW} \triangleleft$$

❶ The use of the ideal gas table for air is featured in this solution. A solution also can be developed on a cold air-standard basis in which constant specific heats are assumed. The details are left as an exercise, but for comparison the results are presented in the following table for the case $k = 1.4$, representing atmospheric air:

| Parameter | Air-Standard Analysis | Cold Air-Standard Analysis, $k = 1.4$ |
|---|---|---|
| $T_2$ | 574.1 K | 579.2 K |
| $T_4$ | 787.7 K | 725.1 K |
| $\eta$ | 0.457 | 0.482 |
| bwr | 0.396 | 0.414 |
| $\dot{W}_{cycle}$ | 2481 kW | 2308 kW |

❷ The value of the back work ratio in the present gas turbine case is significantly greater than the back work ratio of the simple vapor power cycle of Example 8.1.

***Effect of Pressure Ratio on Performance.*** Conclusions that are qualitatively correct for actual gas turbines can be drawn from a study of the ideal Brayton cycle. The first of these conclusions is that the thermal efficiency increases with increasing pressure ratio across the compressor. ***For Example…*** referring again to the $T$–$s$ diagram of Fig. 9.9, we see that an increase in the pressure ratio changes the cycle from 1–2–3–4–1 to 1–2′–3′–4–1. Since the average temperature of heat addition is greater in the latter cycle and both cycles have the same heat rejection process, cycle 1–2′–3′–4–1 would have the greater thermal efficiency. ▲

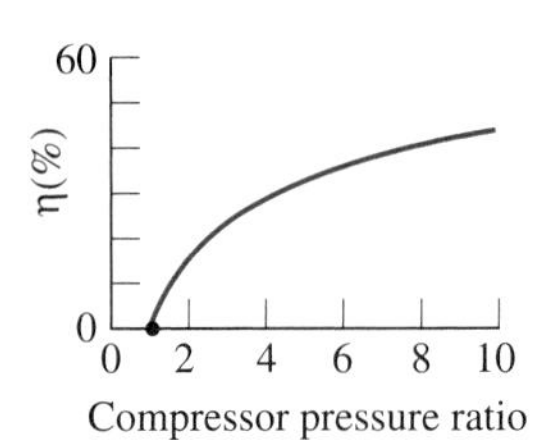


*Figure 9.10* Ideal Brayton cycle thermal efficiency versus compressor pressure ratio.

The increase in thermal efficiency with the compressor pressure ratio is shown in Fig. 9.10. There is a limit of about 1700 K (3060°R) imposed by metallurgical considerations on the maximum allowed temperature at the turbine inlet. It is instructive, therefore, to consider the effect of compressor pressure ratio on thermal efficiency when the turbine inlet temperature is restricted to the maximum allowable temperature. The $T$–$s$ diagrams of two ideal Brayton cycles having the same turbine inlet temperature but different compressor pressure ratios are shown in Fig. 9.11. Cycle A has a greater pressure ratio than cycle B and thus the greater

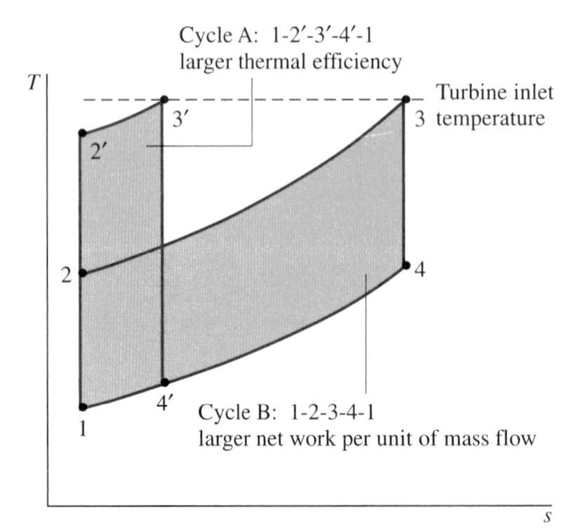


*Figure 9.11* Ideal Brayton cycles with different pressure ratios and the same turbine inlet temperature.

thermal efficiency. However, cycle B has a larger enclosed area and thus the greater net work developed per unit of mass flow. Accordingly, for cycle A to develop the same net *power* output as cycle B, a larger mass flow rate would be required, and this might dictate a larger system. These considerations are important for gas turbines intended for use in vehicles where engine weight must be kept small. For such applications, it is desirable to operate near the compressor pressure ratio that yields the most work per unit of mass flow and not the pressure ratio for the greatest thermal efficiency.

### 9.5.3 Gas Turbine Irreversibilities and Losses

The principal state points of an air-standard gas turbine might be shown more realistically as in Fig. 9.12*a*. Because of frictional effects within the compressor and turbine, the working fluid would experience increases in specific entropy across these components. Owing to friction, there also would be pressure drops as the working fluid passes through the heat exchangers. However, because frictional pressure drops are less significant sources of irreversibility, we ignore them in subsequent discussions and for simplicity show the flow through the heat exchangers as occurring at constant pressure. This is illustrated by Fig. 9.12*b*. Stray heat transfers from the power plant components to the surroundings represent losses, but these effects are usually of secondary importance and are also ignored in subsequent discussions.

As the effect of irreversibilities in the turbine and compressor becomes more pronounced, the work developed by the turbine decreases and the work input to the compressor increases, resulting in a marked decrease in the net work of the power plant. Accordingly, if an appreciable amount of net work is to be developed by the plant, relatively high turbine and compressor efficiencies are required. After decades of developmental effort, efficiencies of 80 to 90% can now be achieved for the turbines and compressors in gas turbine power plants. Designating the states as in Fig. 9.12*b*, the isentropic turbine and compressor efficiencies are given by

$$\eta_t = \frac{(\dot{W}_t/\dot{m})}{(\dot{W}_t/\dot{m})_s} = \frac{h_3 - h_4}{h_3 - h_{4s}} \tag{9.24}$$

$$\eta_c = \frac{(\dot{W}_c/\dot{m})_s}{(\dot{W}_c/\dot{m})} = \frac{h_{2s} - h_1}{h_2 - h_1} \tag{9.25}$$

The effect of irreversibilities in the turbine and compressor is important. Still, among the irreversibilities of actual gas turbine power plants, combustion irreversibility is the most significant by far. The simplified air-standard analysis does not allow this irreversibility to be evaluated, however.

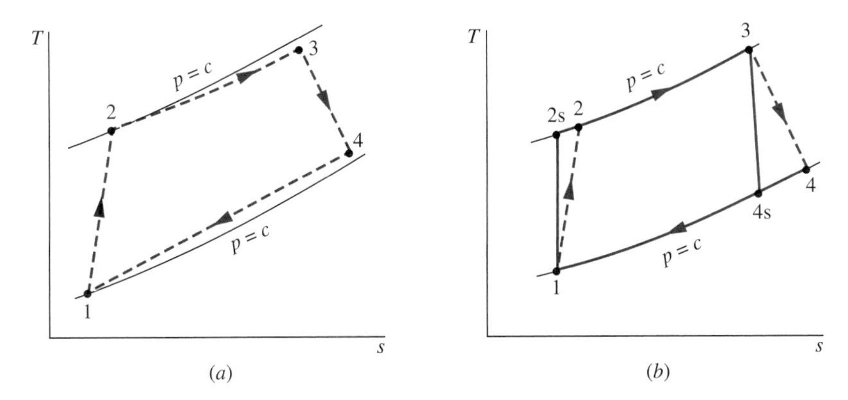


*Figure 9.12* Effects of irreversibilities on the air-standard gas turbine.

Example 9.4 brings out the effect of turbine and compressor irreversibilities on plant performance.

## *Example 9.4* Brayton Cycle with Irreversibilities

Reconsider Example 9.3, but include in the analysis that the turbine and compressor each have an isentropic efficiency of 80%. Determine for the modified cycle **(a)** the thermal efficiency of the cycle, **(b)** the back work ratio, **(c)** the *net* power developed, in kW.

### Solution

***Known:*** An air-standard Brayton cycle operates with given compressor inlet conditions, given turbine inlet temperature, and known compressor pressure ratio. The compressor and turbine each have an isentropic efficiency of 80%.

***Find:*** Determine the thermal efficiency, the back work ratio, and the net power developed, in kW.

***Schematic and Given Data:***

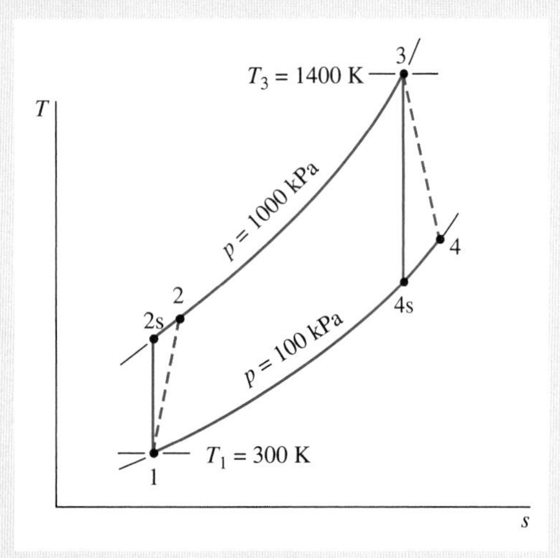


***Assumptions:***

1. Each component is analyzed as a control volume at steady state.
2. The compressor and turbine are adiabatic.
3. There are no pressure drops for flow through the heat exchangers.
4. Kinetic and potential energy effects are negligible.
5. The working fluid is air modeled as an ideal gas.

*Figure E9.4*

***Analysis:*** **(a)** The thermal efficiency is given by

$$\eta = \frac{(\dot{W}_t/\dot{m}) - (\dot{W}_c/\dot{m})}{\dot{Q}_{in}/\dot{m}}$$

The work terms in the numerator of this expression are evaluated using the given values of the compressor and turbine isentropic efficiencies as follows:

The turbine work per unit of mass is

$$\frac{\dot{W}_t}{\dot{m}} = \eta_t \left(\frac{\dot{W}_t}{\dot{m}}\right)_s$$

❶ where $\eta_t$ is the turbine efficiency. The value of $(\dot{W}_t/\dot{m})_s$ is determined in the solution to Example 9.3 as 706.9 kJ/kg. Thus

$$\frac{\dot{W}_t}{\dot{m}} = 0.8(706.9) = 565.5 \text{ kJ/kg}$$

For the compressor, the work per unit of mass is

$$\frac{\dot{W}_c}{\dot{m}} = \frac{(\dot{W}_c/\dot{m})_s}{\eta_c}$$

where $\eta_c$ is the compressor efficiency. The value of $(\dot{W}_c/\dot{m})_s$ is determined in the solution to Example 9.3 as 279.7 kJ/kg, so

$$\frac{\dot{W}_c}{\dot{m}} = \frac{279.7}{0.8} = 349.6 \text{ kJ/kg}$$

The specific enthalpy at the compressor exit, $h_2$, is required to evaluate the denominator of the thermal efficiency expression. This enthalpy can be determined by solving

$$\frac{\dot{W}_c}{\dot{m}} = h_2 - h_1$$

to obtain

$$h_2 = h_1 + \dot{W}_c/\dot{m}$$

Inserting known values

$$h_2 = 300.19 + 349.6 = 649.8 \text{ kJ/kg}$$

The heat transfer to the working fluid per unit of mass flow is then

$$\frac{\dot{Q}_{in}}{\dot{m}} = h_3 - h_2 = 1515.4 - 649.8 = 865.6 \text{ kJ/kg}$$

where $h_3$ is from the solution to Example 9.3.

Finally, the thermal efficiency is

$$\eta = \frac{565.5 - 349.6}{865.6} = 0.249\ (24.9\%) \triangleleft$$

**(b)** The back work ratio is

$$\text{bwr} = \frac{\dot{W}_c/\dot{m}}{\dot{W}_t/\dot{m}} = \frac{349.6}{565.5} = 0.618\ (61.8\%) \triangleleft$$

**(c)** The mass flow rate is the same as in Example 9.3. The net power developed by the cycle is then

❷ $$\dot{W}_{cycle} = \left(5.807 \frac{\text{kg}}{\text{s}}\right)(565.5 - 349.6)\frac{\text{kJ}}{\text{kg}}\left|\frac{1 \text{ kW}}{1 \text{ kJ/s}}\right| = 1254 \text{ kW} \triangleleft$$

❶ The solution to this example on a cold air-standard basis is left as an exercise.

❷ Irreversibilities within the turbine and compressor have a significant impact on the performance of gas turbines. This is brought out by comparing the results of the present example with those of Example 9.3. Irreversibilities result in an increase in the work of compression and a reduction in work output of the turbine. The back work ratio is greatly increased and the thermal efficiency significantly decreased.

## 9.6 Regenerative Gas Turbines

The turbine exhaust temperature of a gas turbine is normally well above the ambient temperature. Accordingly, the hot turbine exhaust gas has a potential for use that would be irrevocably lost were the gas discarded directly to the surroundings. One way of utilizing this potential is by means of a heat exchanger called a ***regenerator,*** which allows the air exiting the compressor to be *preheated* before entering the combustor, thereby reducing the amount of fuel that must be burned in the combustor.

*regenerator*

An air-standard Brayton cycle modified to include a regenerator is illustrated in Fig. 9.13. The regenerator shown is a counterflow heat exchanger through which the hot turbine exhaust gas and the cooler air leaving the compressor pass in opposite directions. Ideally, no frictional pressure drop occurs in either stream. The turbine exhaust gas is cooled from state 4 to state y, while the air exiting the compressor is heated from state 2 to state x. Hence, a

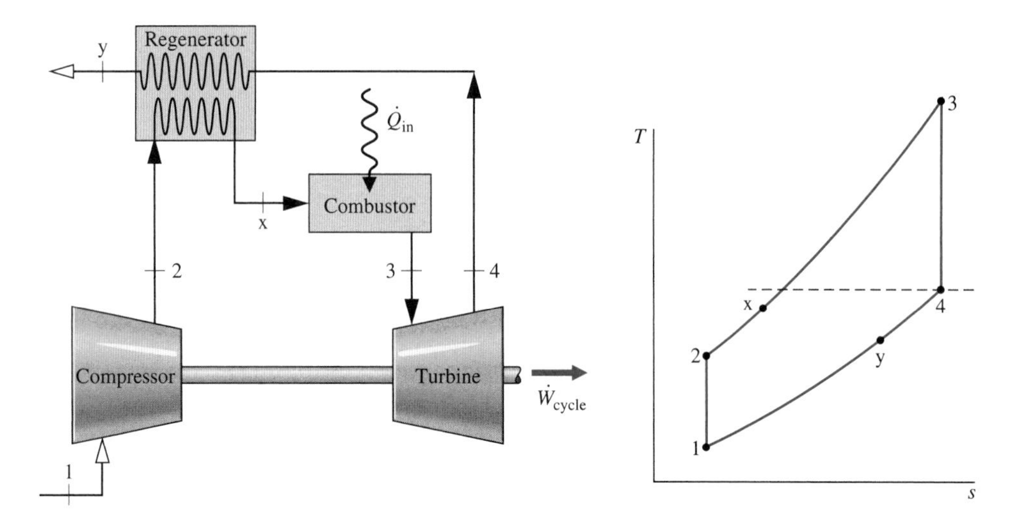


*Figure 9.13* Regenerative air-standard gas turbine cycle.

heat transfer from a source external to the cycle is required only to increase the air temperature from state x to state 3, rather than from state 2 to state 3, as would be the case without regeneration. The heat added per unit of mass is then given by

$$\frac{\dot{Q}_{in}}{\dot{m}} = h_3 - h_x \tag{9.26}$$

The net work developed per unit of mass flow is not altered by the addition of a regenerator. Thus, since the heat added is reduced, the thermal efficiency increases.

***Regenerator Effectiveness.*** From Eq. 9.26 it can be concluded that the external heat transfer required by a gas turbine power plant decreases as the specific enthalpy $h_x$ increases and thus as the temperature $T_x$ increases. Evidently, there is an incentive in terms of fuel saved for selecting a regenerator that provides the greatest practical value for this temperature. To consider the *maximum* theoretical value for $T_x$, refer to Fig. 9.14*a*, which shows typical temperature variations of the hot and cold streams of a counterflow heat exchanger. Since a finite temperature difference between the streams is required for heat transfer to occur, the temperature of the cold stream at each location, denoted by the coordinate $z$, is less than that of the hot stream. In particular, the temperature of the cold stream as it exits the heat exchanger is less than the temperature of the incoming hot stream. If the heat transfer area were increased, providing more opportunity for heat transfer between the two streams, there would be a smaller temperature difference at each location. In the limiting case of infinite heat transfer area, the temperature difference would approach zero at all locations, as illustrated in Fig. 9.14*b*, and the heat transfer would approach reversibility. In this limit, the exit temperature of the cold stream would approach the temperature of the incoming hot stream. Thus, the highest possible temperature that could be achieved by the cold stream is the temperature of the incoming hot gas.

Referring again to the regenerator of Fig. 9.13, we can conclude from the discussion of Fig. 9.14 that the maximum theoretical value for the temperature $T_x$ is the turbine exhaust temperature $T_4$, obtained if the regenerator were operating reversibly. The *regenerator effectiveness,* $\eta_{reg}$, is a parameter that gauges the departure of an actual regenerator from such an ideal regenerator. It is defined as the ratio of the actual enthalpy increase of the air flowing through the compressor side of the regenerator to the maximum theoretical enthalpy increase.

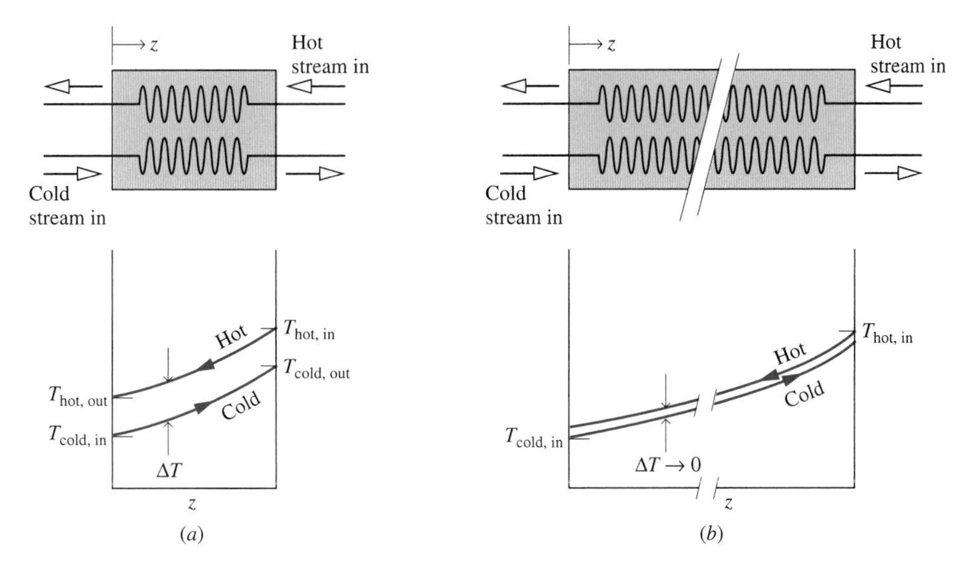


*Figure 9.14* Temperature distributions in counterflow heat exchangers. (*a*) Actual. (*b*) Reversible.

That is, the ***regenerator effectiveness*** is

$$\eta_{reg} = \frac{h_x - h_2}{h_4 - h_2} \tag{9.27}$$

*regenerator effectiveness*

As heat transfer approaches reversibility, $h_x$ approaches $h_4$ and $\eta_{reg}$ tends to unity (100%).

In practice, regenerator effectiveness values typically range from 60 to 80%, and thus the temperature $T_x$ of the air exiting on the compressor side of the regenerator is normally well below the turbine exhaust temperature. To increase the effectiveness above this range would require greater heat transfer area, resulting in equipment costs that cancel any advantage due to fuel savings. Moreover, the greater heat transfer area that would be required for a larger effectiveness can result in a significant frictional pressure drop for flow through the regenerator, thereby affecting overall performance. The decision to add a regenerator is influenced by considerations such as these, and the final decision is primarily an economic one. For further discussion of heat exchangers see Sec. 17.5.

In Example 9.5, we analyze an air-standard Brayton cycle with regeneration and explore the effect on thermal efficiency as the regenerator effectiveness varies.

## *Example 9.5* Brayton Cycle with Regeneration

A regenerator is incorporated in the cycle of Example 9.3. **(a)** Determine the thermal efficiency for a regenerator effectiveness of 80%. **(b)** Plot the thermal efficiency versus regenerator effectiveness ranging from 0 to 80%.

### Solution

***Known:*** A regenerative gas turbine operates with air as the working fluid. The compressor inlet state, turbine inlet temperature, and compressor pressure ratio are known.

***Find:*** For a regenerator effectiveness of 80%, determine the thermal efficiency. Also plot the thermal efficiency versus the regenerator effectiveness ranging from 0 to 80%.

***Schematic and Given Data:***

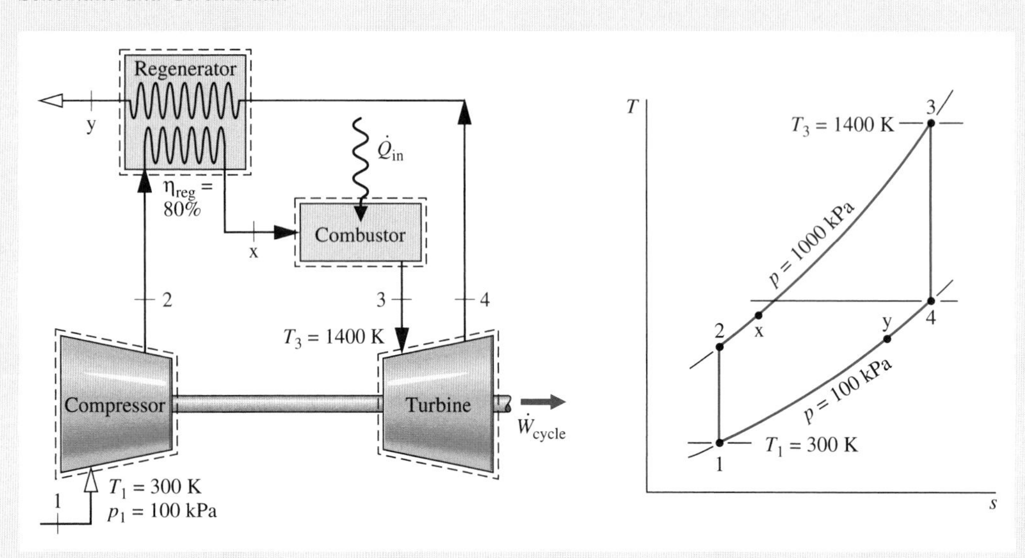


*Figure E9.5*

***Assumptions:***

1. Each component is analyzed as a control volume at steady state. The control volumes are shown on the accompanying sketch by dashed lines.
2. The compressor and turbine processes are isentropic.
3. There are no pressure drops for flow through the heat exchangers.
4. The regenerator effectiveness is 80% in part (a).
5. Kinetic and potential energy effects are negligible.
6. The working fluid is air modeled as an ideal gas.

***Properties:*** The specific enthalpy values at the numbered states on the $T$–$s$ diagram are the same as those in Example 9.3: $h_1 = 300.19$ kJ/kg, $h_2 = 579.9$ kJ/kg, $h_3 = 1515.4$ kJ/kg, $h_4 = 808.5$ kJ/kg.

To find the specific enthalpy $h_x$, the regenerator effectiveness is used as follows: By definition

$$\eta_{reg} = \frac{h_x - h_2}{h_4 - h_2}$$

Solving for $h_x$

$$\begin{aligned} h_x &= \eta_{reg}(h_4 - h_2) + h_2 \\ &= (0.8)(808.5 - 579.9) + 579.9 = 762.8 \text{ kJ/kg} \end{aligned}$$

***Analysis:*** **(a)** With the specific enthalpy values determined above, the thermal efficiency is

❶
$$\begin{aligned} \eta &= \frac{(\dot{W}_t/\dot{m}) - (\dot{W}_c/\dot{m})}{(\dot{Q}_{in}/\dot{m})} = \frac{(h_3 - h_4) - (h_2 - h_1)}{(h_3 - h_x)} \\ &= \frac{(1515.4 - 808.5) - (579.9 - 300.19)}{(1515.4 - 762.8)} \end{aligned}$$

❷
$$= 0.568 \ (56.8\%) \ \triangleleft$$

**(b)** Plot. (CD-ROM)

---

❶ The values for work per unit of mass flow of the compressor and turbine are unchanged by the addition of the regenerator. Thus, the back work ratio and net work output are not affected by this modification.

❷ Comparing the present thermal efficiency value with the one determined in Example 9.3, it should be evident that the thermal efficiency can be increased significantly by means of regeneration.

## 9.7 Gas Turbines for Aircraft Propulsion (CD ROM)

## 9.8 Chapter Summary and Study Guide

In this chapter, we have studied the thermodynamic modeling of internal combustion engines and gas turbine power plants. The modeling of cycles is based on the use of air-standard analysis, where the working fluid is considered to be air as an ideal gas.

The processes in internal combustion engines are described in terms of two air-standard cycles: the Otto and Diesel cycles, which differ from each other only in the way the heat addition process is modeled. For these cycles, we have evaluated the principal work and heat transfers along with two important performance parameters: the mean effective pressure and the thermal efficiency. The effect of varying compression ratio on cycle performance is also investigated.

The performance of simple gas turbine power plants is described in terms of the air-standard Brayton cycle. For this cycle, we evaluate the principal work and heat transfers along with two important performance parameters: the back-work ratio and the thermal efficiency. We also consider the effects on performance of irreversibilities and of varying compressor pressure ratio. The regenerative gas turbine also is discussed.

The following list provides a study guide for this chapter. When your study of the text and end-of-chapter exercises has been completed, you should be able to

- write out the meanings of the terms listed in the margin throughout the chapter and understand each of the related concepts. The subset of key terms listed here in the margin is particularly important.
- sketch $p$–$v$ and $T$–$s$ diagrams of the Otto and Diesel cycles. Apply the closed system energy balance and the second law along with property data to determine the performance of these cycles, including mean effective pressure, thermal efficiency, and the effects of varying compression ratio.
- sketch schematic diagrams and accompanying $T$–$s$ diagrams of the Brayton cycle and the regenerative gas turbine. In each case, be able to apply mass and energy balances, the second law, and property data to determine gas turbine power cycle performance, including thermal efficiency, back work ratio, net power output, and the effects of varying compressor pressure ratio.

*mean effective pressure*
*air-standard analysis*
*Otto cycle*
*Diesel cycle*
*Brayton cycle*
*regenerator effectiveness*

# *Problems*

### Otto Cycle

**9.1** An air-standard Otto cycle has a compression ratio of 8.5. At the beginning of compression, $p_1 = 100$ kPa and $T_1 = 300$ K. The heat addition per unit mass of air is 1400 kJ/kg. Determine
(a) the net work, in kJ per kg of air.
(b) the thermal efficiency of the cycle.
(c) the mean effective pressure, in kPa.
(d) the maximum temperature in the cycle, in K.

**9.2** Solve Problem 9.1 on a cold air-standard basis with specific heats evaluated at 300 K.

**9.3** At the beginning of the compression process of an air-standard Otto cycle, $p_1 = 1$ bar, $T_1 = 290$ K, $V_1 = 400$ cm$^3$. The maximum temperature in the cycle is 2200 K and the compression ratio is 8. Determine
(a) the heat addition, in kJ.
(b) the net work, in kJ.
(c) the thermal efficiency.
(d) the mean effective pressure, in bar.

**9.4** (CD-ROM)

**9.5** Solve Problem 9.3 on a cold air-standard basis with specific heats evaluated at 300 K.

**9.6** Consider the cycle in Problem 9.3 as a model of the processes in each cylinder of a spark-ignition engine. If the engine has four cylinders and the cycle is repeated 1200 times per min in each cylinder, determine the net power output, in kW.

**9.7** An air-standard Otto cycle has a compression ratio of 6 and the temperature and pressure at the beginning of the compression process are 520°R and 14.2 lbf/in.$^2$, respectively. The heat addition per unit mass of air is 600 Btu/lb. Determine
(a) the maximum temperature, in °R.
(b) the maximum pressure, in lbf/in.$^2$
(c) the thermal efficiency.

**9.8** Solve Problem 9.7 on a cold air-standard basis with specific heats evaluated at 520°R.

**9.9** (CD-ROM)

**9.10** (CD-ROM)

**9.11** An air-standard Otto cycle has a compression ratio of 9. At the beginning of compression, $p_1 = 95$ kPa and $T_1 = 37$°C. The mass of air is 3 g, and the maximum temperature in the cycle is 1020 K. Determine
**(a)** the heat rejection, in kJ.
**(b)** the net work, in kJ.
**(c)** the thermal efficiency.
**(d)** the mean effective pressure, in kPa.

**9.12** The compression ratio of a cold air-standard Otto cycle is 8. At the end of the expansion process, the pressure is 90 lbf/in.$^2$ and the temperature is 900°R. The heat rejection from the cycle is 70 Btu per lb of air. Assuming $k = 1.4$, determine
**(a)** the net work, in Btu per lb of air.
**(b)** the thermal efficiency.
**(c)** the mean effective pressure, in lbf/in.$^2$

**9.13** (CD-ROM)

**9.14** A four-cylinder, four-stroke internal combustion engine has a bore of 3.75 in. and a stroke of 3.45 in. The clearance volume is 17% of the cylinder volume at bottom dead center and the crankshaft rotates at 2600 RPM. The processes within each cylinder are modeled as an air-standard Otto cycle with a pressure of 14.6 lbf/in.$^2$ and a temperature of 60°F at the beginning of compression. The maximum temperature in the cycle is 5200°R. Based on this model, calculate the net work per cycle, in Btu, and the power developed by the engine, in horsepower.

**9.15** (CD-ROM)

**9.16** (CD-ROM)

### Diesel Cycle

**9.17** The pressure and temperature at the beginning of compression of an air-standard Diesel cycle are 95 kPa and 290 K, respectively. At the end of the heat addition, the pressure is 6.5 MPa and the temperature is 2000 K. Determine
**(a)** the compression ratio.
**(b)** the cutoff ratio.
**(c)** the thermal efficiency of the cycle.
**(d)** the mean effective pressure, in kPa.

**9.18** Solve Problem 9.17 on a cold air-standard basis with specific heats evaluated at 300 K.

**9.19** The compression ratio of an air-standard Diesel cycle is 17 and the conditions at the beginning of compression are $p_1 = 14.0$ lbf/in.$^2$, $V_1 = 2$ ft$^3$, and $T_1 = 520$°R. The maximum temperature in the cycle is 4000°R. Calculate
**(a)** the net work for the cycle, in Btu.
**(b)** the thermal efficiency.
**(c)** the mean effective pressure, in lbf/in.$^2$
**(d)** the cutoff ratio.

**9.20** Solve Problem 9.19 on a cold air-standard basis with specific heats evaluated at 520°R.

**9.21** The conditions at the beginning of compression in an air-standard Diesel cycle are fixed by $p_1 = 200$ kPa, $T_1 = 380$ K. The compression ratio is 20 and the heat addition per unit mass is 900 kJ/kg. Determine
**(a)** the maximum temperature, in K.
**(b)** the cutoff ratio.
**(c)** the net work per unit mass of air, in kJ/kg.
**(d)** the thermal efficiency.
**(e)** the mean effective pressure, in kPa.

**9.22** An air-standard Diesel cycle has a compression ratio of 16 and a cutoff ratio of 2. At the beginning of compression, $p_1 =$ 14.2 lbf/in.$^2$, $V_1 = 0.5$ ft$^3$, and $T_1 = 520$°R. Calculate
**(a)** the heat added, in Btu.
**(b)** the maximum temperature in the cycle, in °R.
**(c)** the thermal efficiency.
**(d)** the mean effective pressure, in lbf/in.$^2$

**9.23** The displacement volume of an internal combustion engine is 3 liters. The processes within each cylinder of the engine are modeled as an air-standard Diesel cycle with a cutoff ratio of 2.5. The state of the air at the beginning of compression is fixed by $p_1 = 95$ kPa, $T_1 = 22$°C, and $V_1 = 3.2$ liters. Determine the net work per cycle, in kJ, the power developed by the engine, in kW, and the thermal efficiency, if the cycle is executed 2000 times per min.

**9.24** The state at the beginning of compression of an air-standard Diesel cycle is fixed by $p_1 = 100$ kPa and $T_1 = 310$ K. The compression ratio is 15. For a cutoff ratio of 1.5 find
**(a)** the maximum temperature, in K.
**(b)** the pressure at the end of the expansion, in kPa.
**(c)** the net work per unit mass of air, in kJ/kg.
**(d)** the thermal efficiency.

**9.25** An air-standard Diesel cycle has a maximum temperature of 1800 K. At the beginning of compression, $p_1 = 95$ kPa and $T_1 = 300$ K. The mass of air is 12 g. For a compression ratio of 15, determine
**(a)** the net work of the cycle, in kJ.
**(b)** the thermal efficiency.
**(c)** the mean effective pressure, in kPa.

**9.26** At the beginning of compression in an air-standard Diesel cycle, $p_1 = 96$ kPa, $V_1 = 0.016$ m$^3$, and $T_1 = 290$ K. The compression ratio is 15 and the maximum cycle temperature is 1290 K. Determine
**(a)** the mass of air, in kg.
**(b)** the heat addition and heat rejection per cycle, each in kJ.
**(c)** the net work, in kJ, and the thermal efficiency.

**9.27** (CD-ROM)

### Brayton Cycle

**9.28** Air enters the compressor of an ideal air-standard Brayton cycle at 100 kPa, 300 K, with a volumetric flow rate of 5 m$^3$/s. The compressor pressure ratio is 10. For a turbine inlet temperature of 1000 K, find

(a) the thermal efficiency of the cycle.
(b) the back work ratio.
(c) the net power developed, in kW.

**9.29** Air enters the compressor of an ideal air-standard Brayton cycle at 100 kPa, 300 K, with a volumetric flow rate of 5 $m^3/s$. The turbine inlet temperature is 1400 K. For a compressor pressure ratio of 8, determine
(a) the thermal efficiency of the cycle.
(b) the back work ratio.
(c) the net power developed, in kW.

**9.30** The rate of heat addition to an air-standard Brayton cycle is $5.2 \times 10^8$ Btu/h. The pressure ratio for the cycle is 12 and the minimum and maximum temperatures are 520°R and 2800°R, respectively. Determine
(a) the thermal efficiency of the cycle.
(b) the mass flow rate of air, in lb/h.
(c) the net power developed by the cycle, in Btu/h.

**9.31** Solve Problem 9.30 on a cold air-standard basis with specific heats evaluated at 520°R.

**9.32** (CD-ROM)

**9.33** (CD-ROM)

**9.34** The compressor and turbine of a simple gas turbine each have isentropic efficiencies of 90%. The compressor pressure ratio is 12. The minimum and maximum temperatures are 290 K and 1400 K, respectively. On the basis of an air-standard analysis, compare the values of (a) the net work per unit mass of air flowing, in kJ/kg, (b) the heat rejected per unit mass of air flowing, in kJ/kg, and (c) the thermal efficiency to the same quantities evaluated for an ideal cycle.

**9.35** Air enters the compressor of a simple gas turbine at $p_1 =$ 14 lbf/in.$^2$, $T_1 = 520$°R. The isentropic efficiencies of the compressor and turbine are 83 and 87%, respectively. The compressor pressure ratio is 14 and the temperature at the turbine inlet is 2500°R. The net power developed is $5 \times 10^6$ Btu/h. On the basis of an air-standard analysis, calculate
(a) the volumetric flow rate of the air entering the compressor, in $ft^3$/min.
(b) the temperatures at the compressor and turbine exits, each in °R.
(c) the thermal efficiency of the cycle.

**9.36** Solve Problem 9.35 on a cold air-standard basis with specific heats evaluated at 520°R.

### Regenerative Gas Turbines

**9.37** Reconsider Problem 9.34, but include a regenerator in the cycle. For a regenerator effectiveness of 80%, determine
(a) the heat addition per unit mass of air flowing, in kJ/kg.
(b) the thermal efficiency.

**9.38** Reconsider Problem 9.35, but include a regenerator in the cycle. For a regenerator effectiveness of 78% determine
(a) the thermal efficiency.
(b) the percent decrease in heat addition to the air.

**9.39** An air-standard Brayton cycle has a compressor pressure ratio of 10. Air enters the compressor at $p_1 = 14.7$ lbf/in.$^2$, $T_1 = 70$°F, with a mass flow rate of 90,000 lb/h. The turbine inlet temperature is 2200°R. Calculate the thermal efficiency and the net power developed, in horsepower, if
(a) the turbine and compressor isentropic efficiencies are each 100%.
(b) the turbine and compressor isentropic efficiencies are 88 and 84%, respectively.
(c) the turbine and compressor isentropic efficiencies are 88 and 84%, respectively, and a regenerator with an effectiveness of 80% is incorporated.

**9.40** Air enters the compressor of a regenerative gas turbine with a volumetric flow rate of $1.4 \times 10^5$ $ft^3$/min at 14 lbf/in.$^2$, 540°R, and is compressed to 70 lbf/in.$^2$ The air then passes through the regenerator and exits at 1060°R. The temperature at the turbine inlet is 1540°R. The compressor and turbine each have an isentropic efficiency of 80%. Using an air-standard analysis, calculate
(a) the thermal efficiency of the cycle.
(b) the regenerator effectiveness.
(c) the net power output, in Btu/h.

**9.41** (CD-ROM)

### Gas Turbines for Aircraft Propulsion

**9.42** (CD-ROM)

**9.43** (CD-ROM)

**9.44** (CD-ROM)

**9.45** (CD-ROM)

**9.46** (CD-ROM)

**9.47** (CD-ROM)

# PSYCHROMETRIC APPLICATIONS (CD-ROM)

## *Introduction…*

*chapter objective*

The *objective* of this chapter is to study systems involving mixtures of dry air and water vapor. A liquid water phase also may be present. Knowledge of the behavior of such systems is essential for the analysis and design of air-conditioning devices, cooling towers, and industrial processes requiring close control of the vapor content in air. The study of systems involving dry air and water is known as *psychrometrics.*

*psychrometrics*

**The complete material for Chapter 10 is available on the CD-ROM only.**

# GETTING STARTED IN FLUID MECHANICS: FLUID STATICS

## *Introduction...*

Fluid mechanics is that discipline within the broad field of applied mechanics concerned with the behavior of fluids at rest or in motion. Both liquids and gases are fluids. (A more complete definition of a fluid is given in Section 12.1.) This field of mechanics obviously encompasses a vast array of problems that may vary from the study of blood flow in the capillaries (which are only a few microns in diameter) to the flow of crude oil across Alaska through an 800-mile-long, 4-ft-diameter pipe. Fluid mechanics principles explain why airplanes are made streamlined with smooth surfaces, whereas golf balls are made with rough surfaces (dimpled).

In addition, as discussed in Chap. 1, fluid mechanics principles and concepts are often involved in the study and analysis of thermal systems. Thus, it is very likely that during your career as an engineer you will be involved in the analysis and design of systems that require a good understanding of fluid mechanics. This introductory material will provide you with a sound foundation of the fundamental aspects of fluid mechanics.

In this chapter we will consider an important class of problems in which the fluid is at *rest.* In this case the only forces of interest will be due to the pressure acting on the surface of a fluid particle and the weight of the particle. Thus, the ***objective*** of this chapter is to investigate pressure and its variation throughout a fluid at rest, and the effect of pressure on submerged or partially submerged surfaces.

*chapter objective*

## 11.1 Pressure Variation in a Fluid at Rest

As is briefly discussed in Sec. 2.4.2, the term pressure is used to indicate the normal force per unit area at a given point acting on a given plane within a fluid mass of interest. The purpose of this section is to determine how the pressure in a fluid at rest varies from point to point.

Consider a small, stationary element of fluid removed from some arbitrary position within a mass of fluid as illustrated in Fig. 11.1. There are two types of forces acting on this element: ***surface forces*** due to the pressure, and a ***body force*** equal to the weight of the element. The weight, $\delta\mathcal{W}$, acts in the negative $z$-direction and can be written as

*surface force*
*body force*

$$\delta\mathcal{W} = \gamma\delta x\delta y\delta z \tag{11.1}$$

where the specific weight, $\gamma = \rho g$, is the fluid weight per unit volume. (Section 7.9)

The pressure forces on the sides, top, and bottom of the fluid element are shown in Fig. 11.1. The resultant forces in the $x$ and $y$ directions are

$$\delta F_x = (p_B - p_F)\delta y\delta z \qquad \text{and} \qquad \delta F_y = (p_L - p_R)\delta x\delta z \tag{11.2}$$

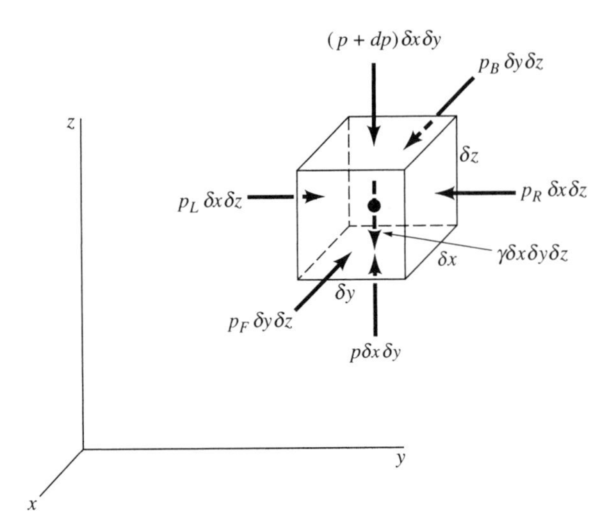


*Figure 11.1* Surface and body forces acting on small fluid element.

where the subscripts L, R, B, and F refer to the left, right, back, and front surfaces of the fluid element, respectively.

In the $z$ direction the resultant force is

$$\delta F_z = p\delta x\delta y - (p + dp)\delta x\delta y - \gamma\delta x\delta y\delta z = -dp\delta x\delta y - \gamma\delta x\delta y\delta z \tag{11.3}$$

where $dp$ is the pressure difference between the top and the bottom of the fluid element.

For equilibrium of the fluid element (since it is at rest)

$$\Sigma F_x = 0 \qquad \Sigma F_y = 0 \qquad \Sigma F_z = 0 \tag{11.4}$$

By combining the resultant forces (Eqs. 11.2 and 11.3) with the equilibrium conditions (Eq. 11.4) we obtain

$$(p_B - p_F)\delta y\delta z = 0 \qquad (p_L - p_R)\delta x\delta z = 0 \qquad -dp\delta x\delta y - \gamma\delta x\delta y\delta z = 0$$

Thus, in the $x$ and $y$ directions we obtain $p_B = p_F$, and $p_L = p_R$. These equations show that the pressure does not depend on $x$ or $y$. Accordingly, as we move from point to point in a horizontal plane (any plane parallel to the $x - y$ plane), the pressure does not change. In the $z$ direction the force balance becomes $dp = -\gamma\delta z$.

That is,

$$\frac{dp}{dz} = -\gamma \tag{11.5}$$

Equation 11.5 is the fundamental equation for fluids at rest and can be used to determine how pressure changes with elevation. This equation indicates that the pressure gradient in the vertical direction is negative; that is, the pressure decreases as we move upward in a fluid at rest. There is no requirement that $\gamma$ be a constant. Thus, Eq. 11.5 is valid for fluids with constant specific weight, such as liquids, as well as fluids whose specific weight may vary with elevation, such as air or other gases.

For an *incompressible fluid* ($\rho$ = constant) at a constant $g$, Eq. 11.5 can be directly integrated

$$\int_{p_1}^{p_2} dp = -\gamma \int_{z_1}^{z_2} dz$$

to yield

$$p_1 - p_2 = \gamma(z_2 - z_1) \qquad (11.6)$$

where $p_1$ and $p_2$ are pressures at the vertical elevations $z_1$ and $z_2$, as is illustrated in Fig. 11.2. Equation 11.6 can be written in the compact form

$$p_1 - p_2 = \gamma h \qquad (11.7)$$

where $h$ is the distance, $z_2 - z_1$, which is the depth of fluid measured downward from the location of $p_2$. This type of pressure distribution is commonly called a *hydrostatic pressure distribution,* and Eq. 11.7 shows that in an incompressible fluid at rest the pressure varies linearly with depth. The pressure must increase with depth to support the fluid above it.

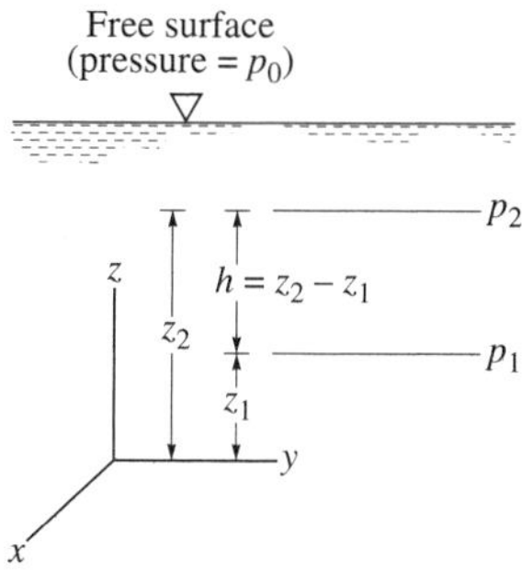

It can also be observed from Eq. 11.7 that the pressure difference between two points can be specified by the distance $h$ since

*Figure 11.2* Notation for pressure variation in a fluid at rest with a free surface.

$$h = \frac{p_1 - p_2}{\gamma}$$

In this case $h$ is called the ***pressure head*** and is interpreted as the height of a column of fluid of specific weight $\gamma$ required to give a pressure difference $p_1 - p_2$.

***pressure head***

*For Example...* for water with a specific weight of $\gamma = 62.4\ \text{lbf/ft}^3$, a pressure difference of $100\ \text{lbf/ft}^2$ is equal to a pressure head of $h = 100\ \text{lbf/ft}^2/62.4\ \text{lbf/ft}^3 = 1.60$ ft of water. ▲

For applications with liquids there is often a free surface, as is illustrated in Fig. 11.2, and it is convenient to use this surface as a reference plane. The reference pressure $p_0$ would correspond to the pressure acting on the free surface (which would frequently be atmospheric pressure), and thus if we let $p_2 = p_0$ in Eq. 11.7, it follows that the ***hydrostatic pressure distribution*** for the pressure $p$ at any depth $h$ below the free surface is given by the equation

$$p = \gamma h + p_0 \qquad (11.8)$$

***hydrostatic pressure distribution***

As is demonstrated by Eq. 11.8, the pressure in an incompressible fluid at rest depends on the depth of the fluid relative to some reference plane, and it is *not* influenced by the *size* or *shape* of the tank or container in which the fluid is held.

It should be emphasized that if the specific weight, $\gamma$, of the fluid is not constant, then Eq. 11.8 is not valid and the manner in which $\gamma$ varies must be specified before Eq. 11.5 can be integrated.

## *Example 11.1* Pressure Variation with Depth

Because of a leak in a buried storage tank, water has seeped in to the depth shown in Fig. E11.1. The pressures at the gasoline-water interface and at the bottom of the tank are greater than the atmospheric pressure at the top of the open standpipe connected to the tank. Express these pressures relative to atmospheric pressure in units of $\text{lbf/ft}^2$, $\text{lbf/in.}^2$, and as a pressure head in feet of water.

### Solution

***Known:*** Gasoline and water are contained in a storage tank. Both liquids are at rest.

***Find:*** The pressure and pressure head at the gasoline-water interface and at the bottom of the tank.

***Schematic and Given Data:***

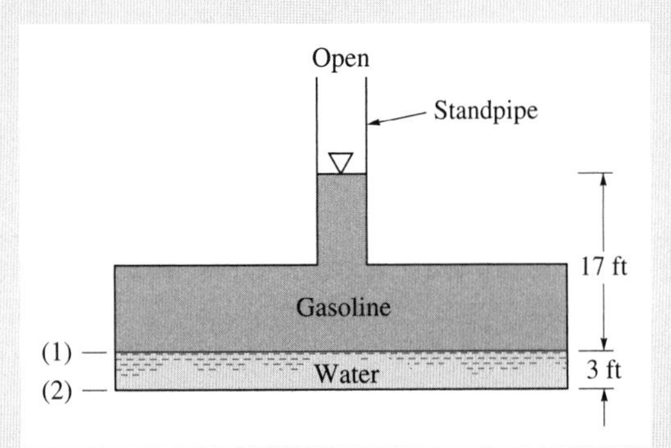


***Assumptions:***

1. The fluids are modeled as incompressible.
2. The fluids are at rest.
3. The specific weights of water and gasoline are $\gamma_{H_2O} = 62.4$ lbf/ft$^3$ and $\gamma_{gasoline} = 42.5$ lbf/ft$^3$. Note: These and other properties for common fluids can be found in the tables of Appendix FM-1.

*Figure E11.1*

***Analysis:*** Since we are dealing with liquids at rest, the pressure distribution will be hydrostatic, and therefore the pressure variation can be found from Eq. 11.8 as

$$p = \gamma h + p_0$$

With $p_0$ corresponding to the pressure at the free surface of the gasoline, the pressure at the interface is

$$\begin{aligned} p_1 &= \gamma_{gasoline} h_{gasoline} + p_0 \\ &= (42.5 \text{ lbf/ft}^3)(17 \text{ ft}) + p_0 \\ &= (722 + p_0) \text{ lbf/ft}^2 \end{aligned}$$

If we measure the pressure relative to atmospheric pressure

$$\begin{aligned} (p_1 - p_0) &= 722 \text{ lbf/ft}^2 \quad \triangleleft \\ &= 722 \text{ lbf/ft}^2 \left| \frac{1 \text{ ft}^2}{144 \text{ in.}^2} \right| = 5.02 \text{ lbf/in.}^2 \quad \triangleleft \end{aligned}$$

The corresponding pressure head in feet of water is therefore

❶
$$\frac{(p_1 - p_0)}{\gamma_{H_2O}} = \frac{722 \text{ lbf/ft}^2}{62.4 \text{ lbf/ft}^3} = 11.6 \text{ ft}$$

We can now apply the same relationship to determine the pressure (relative to atmospheric pressure) at the tank bottom; that is,

$$\begin{aligned} p_2 &= \gamma_{H_2O} h_{H_2O} + p_1 \\ &= (62.4 \text{ lbf/ft}^3)(3 \text{ ft}) + 722 \text{ lbf/ft}^2 + p_0 \end{aligned}$$

or

❷,❸
$$\begin{aligned} (p_2 - p_0) &= 909 \text{ lbf/ft}^2 \quad \triangleleft \\ &= 909 \text{ lbf/ft}^2 \left| \frac{1 \text{ ft}^2}{144 \text{ in.}^2} \right| = 6.31 \text{ lbf/in.}^2 \quad \triangleleft \end{aligned}$$

The corresponding pressure head in feet of water is therefore

$$\frac{(p_2 - p_0)}{\gamma_{H_2O}} = \frac{909 \text{ lbf/ft}^2}{62.4 \text{ lbf/ft}^3} = 14.6 \text{ ft} \quad \triangleleft$$

---

❶ It is noted that a rectangular column of water 11.6 ft tall and 1 ft$^2$ in cross section weighs 722 lbf. A similar column with a 1-in.$^2$ cross section weighs 5.02 lbf.

❷ The units of pressure lbf/in.$^2$ are often abbreviated as psi.

❸ If we wish to express these pressures in terms of *absolute* pressure, we would have to add the local atmospheric pressure (in appropriate units) to the above results. Thus, if the atmospheric pressure is 14.7 lbf/in.$^2$, the absolute pressure at the bottom of the tank would be $p_2 = (6.31 + 14.7)$ lbf/in.$^2$ = 21.01 lbf/in.$^2$.

## 11.2 Measurement of Pressure

Since pressure is a very important characteristic of a fluid, it is not surprising that numerous devices and techniques are used in its measurement.

The pressure at a point within a fluid mass can be designated as either an ***absolute pressure*** or a ***gage pressure.*** Absolute pressure is measured relative to absolute zero pressure, whereas gage pressure is measured relative to the local atmospheric pressure. ***For Example...*** referring to Fig. 11.3, a gage pressure of zero corresponds to a pressure that is equal to the local atmospheric pressure. ▲

*absolute pressure*
*gage pressure*

Absolute pressures are always positive, but gage pressure can be either positive or negative depending on whether the pressure is above or below atmospheric pressure. A negative gage pressure is also referred to as a *suction* or a ***vacuum pressure. For Example...*** an absolute pressure of 10 psi (i.e., 10 lbf/in.$^2$) could be expressed as $-4.7$ psi gage if the local atmospheric pressure is 14.7 psi, or alternatively as a 4.7 psi suction or a 4.7 psi vacuum. ▲

*vacuum pressure*

**METHODOLOGY UPDATE**

As indicated in Sec. 2.4.2, thermodynamic analyses use absolute pressure. On the other hand, for most fluid mechanics analyses it is convenient and customary practice to use gage pressure. Thus, in the fluid mechanics portion of this text, Chaps. 11 through 14, pressures typically will be gage pressures unless otherwise noted.

The measurement of atmospheric pressure is usually accomplished with a mercury ***barometer,*** which in its simplest form consists of a glass tube closed at one end with the open end immersed in a container of mercury as shown in Fig. 11.4. The tube is initially filled with mercury (inverted with its open end up) and then turned upside down (open end down) with the open end in the container of mercury. The column of mercury will come to an equilibrium position where its weight plus the force due to the vapor pressure (which develops in the space above the column) balances the force due to the atmospheric pressure. Thus,

*barometer*

$$p_{atm} = \gamma h + p_{vapor} \tag{11.9}$$

where $\gamma$ is the specific weight of mercury. For most practical purposes the contribution of the vapor pressure can be neglected since it is very small [for mercury, $p_{vapor} = 2.3 \times 10^{-5}$ lbf/in.$^2$ (absolute) at a temperature of 68°F] so that $p_{atm} \approx \gamma h$. It is convenient to specify atmospheric pressure in terms of the height, $h$, in millimeters or inches of mercury. ***For Example...*** since standard atmospheric pressure is 14.7 lbf/in.$^2$ (absolute) and mercury weighs 847 lbf/ft$^3$, it follows that $h = p_{atm}/\gamma_{mercury} = 14.7 \text{ lbf/in.}^2\ |144 \text{ in.}^2/\text{ft}^2| / 847 \text{ lbf/ft}^3 = 2.50 \text{ ft} = 30.0$ in. of mercury. ▲

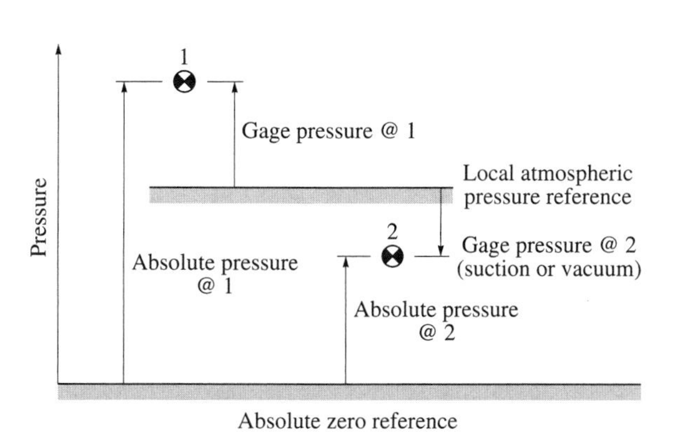


*Figure 11.3* Graphical representation of gage and absolute pressure.

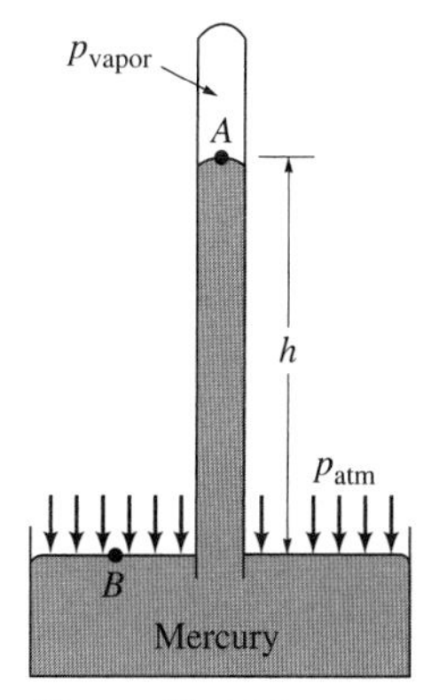


*Figure 11.4* Mercury barometer.

## 11.3 Manometry

*manometer*

A standard technique for measuring pressure involves the use of liquid columns in vertical or inclined tubes. Pressure measuring devices based on this technique are called ***manometers.*** The mercury barometer is an example of one type of manometer, but there are many other configurations possible, depending on the particular application. Two common types of manometers include the piezometer tube and the U-tube manometer.

### 11.3.1 Piezometer Tube

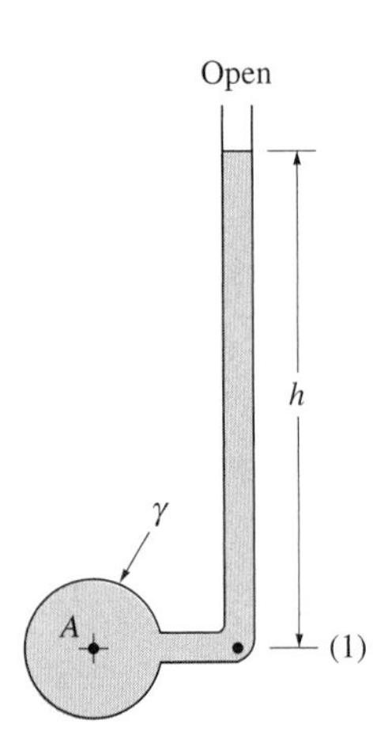

*Figure 11.5* Piezometer tube.

The simplest type of manometer consists of a vertical tube, open at the top, and attached to the container in which the pressure is desired, as illustrated in Fig. 11.5. Since manometers involve columns of fluids at rest, the fundamental equation describing their use is Eq. 11.8

$$p = \gamma h + p_0$$

which gives the pressure at any elevation within a homogeneous fluid in terms of a reference pressure $p_0$ and the vertical distance $h$ between $p$ and $p_0$. Remember that in a fluid at rest pressure will *increase* as we move *downward,* and will *decrease* as we move *upward.* Application of this equation to the piezometer tube of Fig. 11.5 indicates that the gage pressure $p_A$ can be determined by a measurement of $h$ through the relationship

$$p_A = \gamma h$$

where $\gamma$ is the specific weight of the liquid in the container. Note that since the tube is open at the top, the gage pressure $p_0$ is equal to zero. Since point (1) and point $A$ within the container are at the same elevation, $p_A = p_1$.

Although the piezometer tube is a very simple and accurate pressure measuring device, it has several disadvantages. It is only suitable if the pressure in the container is greater than atmospheric pressure (otherwise air would be sucked into the system), and the pressure to be measured must be relatively small so the required height of the column is reasonable. Also, the fluid in the container in which the pressure is to be measured must be a liquid rather than a gas.

### 11.3.2 U-Tube Manometer

*gage fluid*

*U-tube manometer*

To overcome the difficulties noted previously, another type of manometer that is widely used consists of a tube formed into the shape of a U as is shown in Fig. 11.6. The fluid in the manometer is called the ***gage fluid.*** To find the pressure $p_A$ in terms of the various column heights, we start at one end of the system and work our way around to the other end, simply utilizing Eq. 11.8. Thus, for the ***U-tube manometer*** shown in Fig. 11.6, we will start at point $A$ and work around to the open end. The pressure at points $A$ and (1) are the same, and as we move from point (1) to (2) the pressure will increase by $\gamma_1 h_1$. The pressure at point (2) is equal to the pressure at point (3), since the pressures at equal elevations in a continuous mass of fluid at rest must be the same. Note that we could not simply "jump across" from point (1) to a point at the same elevation in the right-hand tube since these would not be points within the same continuous mass of fluid. With the pressure at point (3) specified we now move to the open end where the gage pressure is zero. As we move vertically upward, the pressure decreases by an amount $\gamma_2 h_2$. In equation form these various steps can be expressed as

$$p_A + \gamma_1 h_1 - \gamma_2 h_2 = 0$$

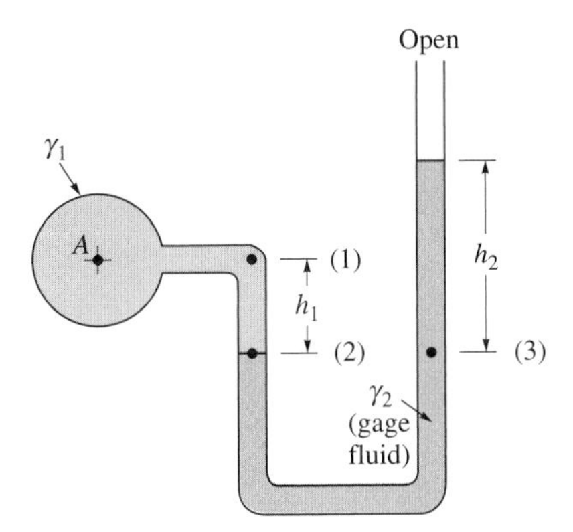


*Figure 11.6* Simple U-tube manometer.

and, therefore, the pressure $p_A$ can be written in terms of the column heights as

***V11.1 Blood pressure measurement***

$$p_A = \gamma_2 h_2 - \gamma_1 h_1 \tag{11.10}$$

A major advantage of the U-tube manometer lies in the fact that the gage fluid can be different from the fluid in the container in which the pressure is to be determined. For example, the fluid in $A$ in Fig. 11.6 can be either a liquid or a gas. If $A$ does contain a gas, the contribution of the gas column, $\gamma_1 h_1$, is almost always negligible so that $p_A \approx p_2$ and in this instance Eq. 11.10 becomes

$$p_A = \gamma_2 h_2$$

The specific weight, $\gamma$, of a liquid such as the gage fluid is often expressed in terms of the ***specific gravity,*** SG, by the following relationship

*specific gravity*

$$\gamma = SG\, \gamma_{\text{water}} = SG\, g\rho_{\text{water}}$$

with $\rho_{\text{water}} = 1000 \text{ kg/m}^3 = 1.94 \text{ slug/ft}^3$.

The U-tube manometer is also widely used to measure the *difference* in pressure between two containers or two points in a given system. Consider a manometer connected between containers $A$ and $B$ as is shown in Fig. 11.7. The difference in pressure between $A$ and $B$ can be found by again starting at one end of the system and working around to the other end.

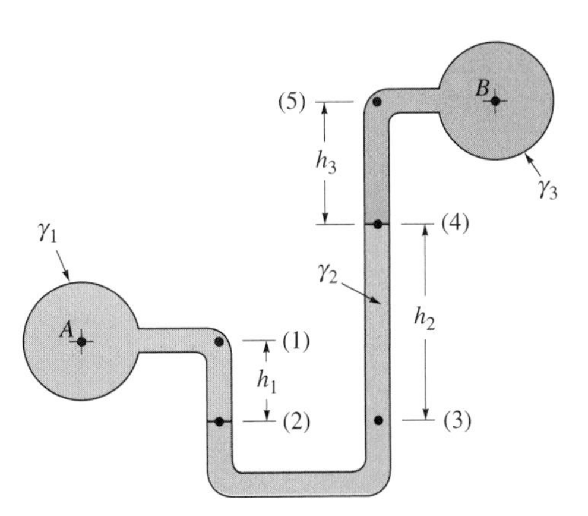


*Figure 11.7* Differential U-tube manometer.

*For Example...* at $A$ the pressure is $p_A$, which is equal to $p_1$, and as we move to point (2) the pressure increases by $\gamma_1 h_1$. The pressure at $p_2$ is equal to $p_3$, and as we move upward to point (4) the pressure decreases by $\gamma_2 h_2$. Similarly, as we continue to move upward from point (4) to (5) the pressure decreases by $\gamma_3 h_3$. Finally, $p_5 = p_B$, since they are at equal elevations. Thus,

$$p_A + \gamma_1 h_1 - \gamma_2 h_2 - \gamma_3 h_3 = p_B$$

and the pressure difference is

$$p_A - p_B = \gamma_2 h_2 + \gamma_3 h_3 - \gamma_1 h_1 \quad \blacktriangle$$

## *Example 11.2* U-tube Manometer

A closed tank contains compressed air and oil ($SG_{oil} = 0.90$) as is shown in Fig. E11.2. A U-tube manometer using mercury ($SG_{Hg} = 13.6$) is connected to the tank as shown. For column heights $h_1 = 36$ in., $h_2 = 6$ in., and $h_3 = 9$ in., determine the pressure reading of the gage.

### Solution

***Known:*** The various column heights and properties of the liquids in the U-tube manometer connected to the pressurized tank.

***Find:*** Determine the pressure reading of the gage at the top of the tank.

***Schematic and Given Data:***

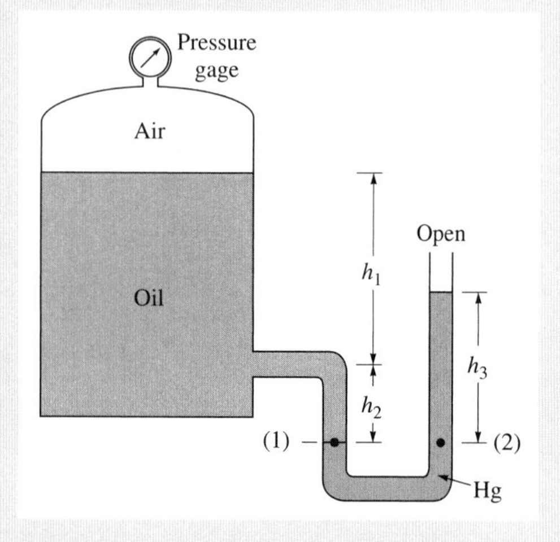


***Assumptions:***
1. The oil and mercury are modeled as incompressible liquids.
2. The variation in the pressure in the air between the oil surface and the gage is negligible.
3. All of the fluids in the system are at rest.
4. The specific weight of water is 62.4 lbf/ft$^3$.

*Figure E11.2*

***Analysis:*** Following the general procedure of starting at one end of the manometer system and working around to the other, we will start at the air–oil interface in the tank and proceed to the open end where the gage pressure is zero. The pressure at level (1) is

$$p_1 = p_{air} + \gamma_{oil}(h_1 + h_2)$$

This pressure is equal to the pressure at level (2), since these two points are at the same elevation in a homogeneous fluid at rest. As we move from level (2) to the open end, the pressure must decrease by $\gamma_{Hg} h_3$, and at the open end the gage pressure is zero. Thus, the manometer equation can be expressed as

$$p_{air} + \gamma_{oil}(h_1 + h_2) - \gamma_{Hg} h_3 = 0$$

or

$$p_{air} + (SG_{oil})(\gamma_{H_2O})(h_1 + h_2) - (SG_{Hg})(\gamma_{H_2O})h_3 = 0$$

For the values given

$$p_{\text{air}} = -(0.9)(62.4\ \text{lbf/ft}^3)\left(\frac{36 + 6}{12}\ \text{ft}\right) + (13.6)(62.4\ \text{lbf/ft}^3)\left(\frac{9}{12}\ \text{ft}\right)$$

so that

$$p_{\text{air}} = 440\ \text{lbf/ft}^2$$

Since the specific weight of the air above the oil is much smaller than the specific weight of the oil, the gage should read the pressure we have calculated; that is,

$$p_{\text{gage}} = 440\ \text{lbf/ft}^2 \left|\frac{1\ \text{ft}^2}{144\ \text{in.}^2}\right| = 3.06\ \text{lbf/in.}^2\,(\text{psi}) \quad \triangleleft$$

❶ Manometers can have a variety of configurations, but the method of analysis remains the same. Start at one end of the system and work around to the other simply making use of the equation for a hydrostatic pressure distribution (Eq. 11.8).

## 11.4 Mechanical and Electronic Pressure Measuring Devices

Although manometers are widely used, they are not well suited for measuring very high pressures, or pressures that are changing rapidly with time. In addition, they require the measurement of one or more column heights, which, although not particularly difficult, can be time consuming. To overcome some of these problems, numerous other types of pressure-measuring instruments have been developed. Most of these make use of the idea that when a pressure acts on an elastic structure the structure will deform, and this deformation can be related to the magnitude of the pressure. Probably the most commonly used device of this kind is the ***Bourdon pressure gage,*** which is shown in Fig. 11.8*a*. The essential mechanical element in this gage is the hollow, elastic curved tube (Bourdon tube), which is connected to the pressure source as shown in Fig. 11.8*b*. As the pressure within the tube increases the tube tends to straighten, and although the deformation is small, it can be translated into the motion of a pointer on a dial as illustrated. Since it is the difference in pressure between the outside of the tube (atmospheric pressure) and the inside of the tube that causes the

*Bourdon pressure gage*

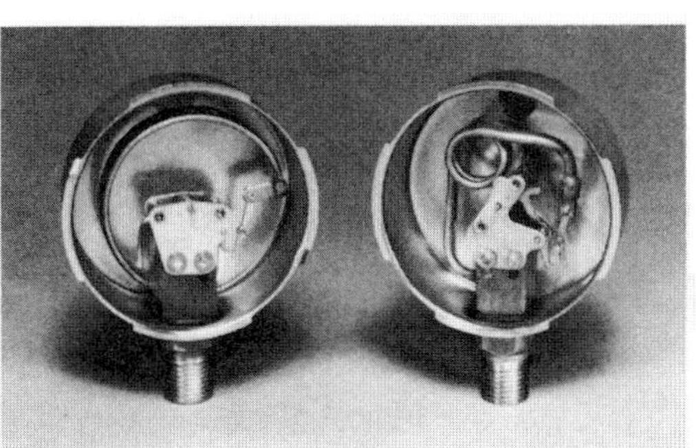

***V11.2 Bourdon gage***

*Figure 11.8* (*a*) Liquid-filled Bourdon pressure gages for various pressure ranges. (*b*) Internal elements of Bourdon gages. The "C-shaped" Bourdon tube is shown on the left, and the "coiled spring" Bourdon tube for high pressures of 1000 psi and above is shown on the right. (Photographs courtesy of Weiss Instruments, Inc.)

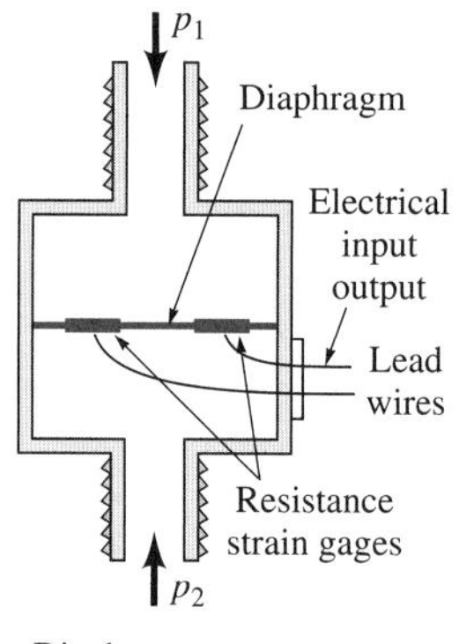


Diaphragm-type electrical pressure tranducer

*pressure transducer*

movement of the tube, the indicated pressure is gage pressure. The Bourdon gage must be calibrated so that the dial reading can directly indicate the pressure in suitable units such as psi or pascals. A zero reading on the gage indicates that the measured pressure is equal to the local atmospheric pressure. This type of gage can be used to measure a negative gage pressure (vacuum) as well as positive pressures.

For many applications in which pressure measurements are required, the pressure must be measured with a device that converts the pressure into an electrical output. For example, it may be desirable to continuously monitor a pressure that is changing with time. This type of pressure measuring device is called a ***pressure transducer,*** and many different designs are used.

## 11.5 Hydrostatic Force on a Plane Surface

***V11.3 Hoover dam***

When a surface is submerged in a fluid, forces develop on the surface due to the fluid. The determination of these forces is important in the design of storage tanks, ships, dams, and other hydraulic structures. For fluids at rest the force must be *perpendicular* to the surface. We also know that this pressure will vary linearly with depth if the fluid is incompressible. For a horizontal surface, such as the bottom of a liquid-filled tank (Fig. 11.9*a*), the magnitude of the resultant force is simply $F_R = pA$, where $p$ is the uniform pressure on the bottom and $A$ is the area of the bottom. For the open tank shown, $p = \gamma h$. Note that if atmospheric pressure acts on both sides, as is illustrated, the *resultant* force on the bottom is simply due to the liquid in the tank. Since the pressure is constant and uniformly distributed over the bottom, the resultant force acts through the centroid of the area as shown in Fig. 11.9*a*. Note that as indicated in Fig. 11.9*b* the pressure is not uniform on the vertical ends of the tank.

For the more general case in which a submerged plane surface is inclined, as is illustrated in Fig. 11.10, the determination of the resultant force acting on the surface is more involved. We assume that the free surface is open to the atmosphere. Let the plane in which the surface lies intersect the free surface at 0 and make an angle $\theta$ with this surface as in Fig. 11.10. The $x$-$y$ coordinate system is defined so that 0 is the origin and $y$ is directed along the surface as shown. The area can have an arbitrary shape as shown. We wish to determine the direction, location, and magnitude of the resultant force acting on one side of this area due to the liquid in contact with the area.

At any given depth, $h$, the force acting on $dA$ (the differential area of Fig. 11.10) is $dF = \gamma h\, dA$ and is perpendicular to the surface. Thus, the magnitude of the resultant force can be

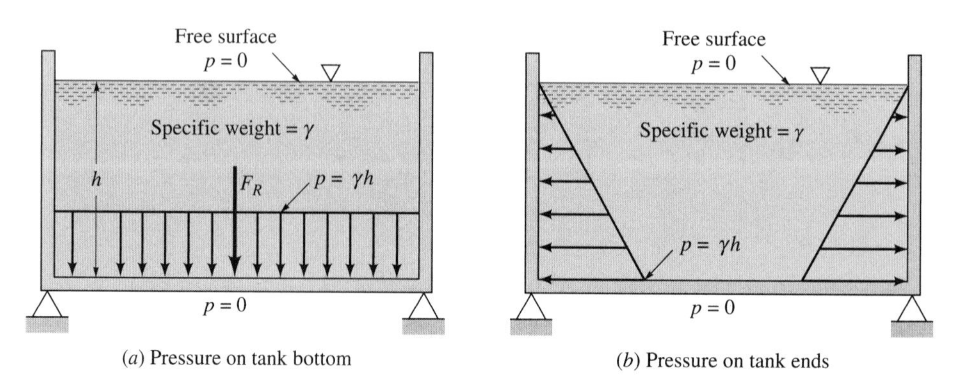


*Figure 11.9* Pressure and resultant hydrostatic force developed on the bottom of an open tank.

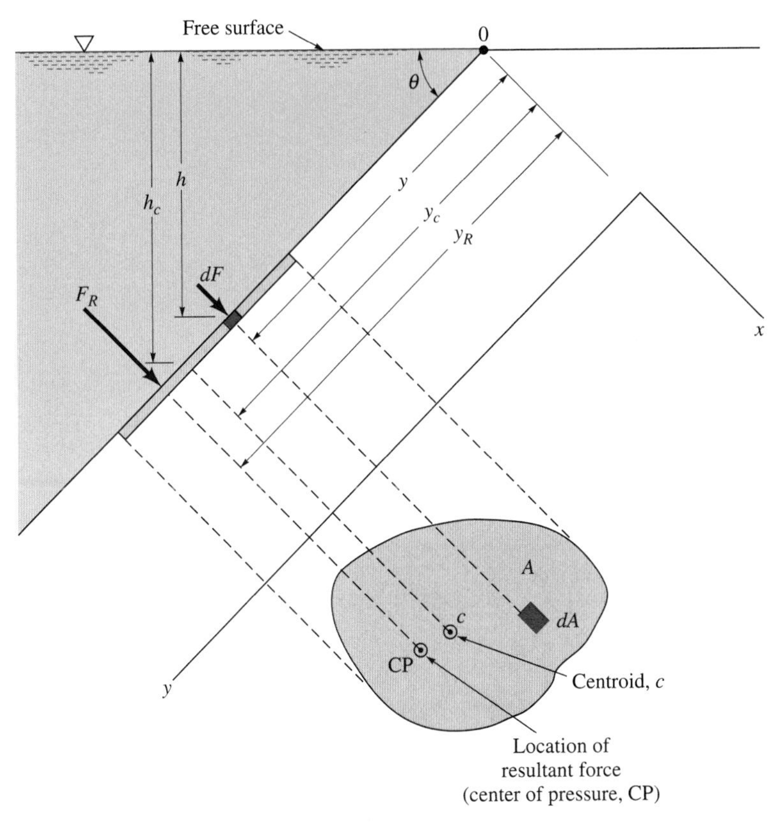


*Figure 11.10* Notation for hydrostatic force on an inclined plane surface of arbitrary shape.

found by summing these differential forces over the entire surface. In equation form

$$F_R = \int_A \gamma h \, dA = \int_A \gamma y \sin\theta \, dA \tag{11.11}$$

where $h = y \sin\theta$. For constant $\gamma$ and $\theta$

$$F_R = \gamma \sin\theta \int_A y \, dA \tag{11.12}$$

The integral appearing in Eq. 11.11 is the *first moment of the area* with respect to the $x$ axis, and can be expressed as

$$\int_A y \, dA = y_c A$$

where $y_c$ is the $y$ coordinate of the centroid measured from the $x$ axis, which passes through 0.

Equation 11.12 can thus be written as

$$F_R = \gamma A y_c \sin\theta$$

Then, with $h_c = y_c \sin\theta$, as shown in Fig. 11.10, we obtain

$$F_R = \gamma h_c A \tag{11.13}$$

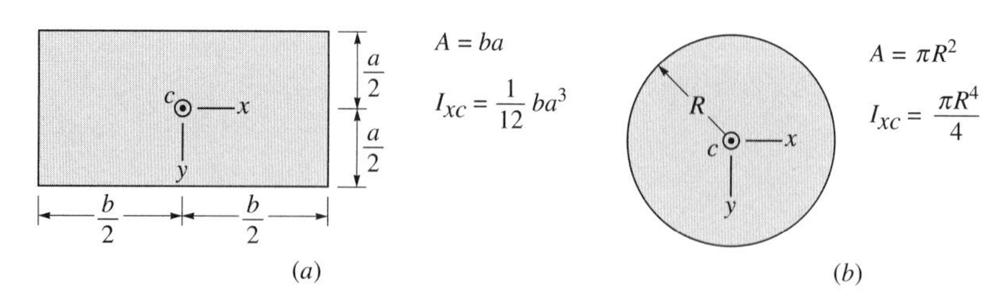


*Figure 11.12* Geometric properties of two common shapes.

where $h_c$ is the vertical distance from the fluid surface to the centroid of the area. Note that the magnitude of the force is independent of the angle $\theta$ and depends only on the specific weight of the fluid, the total area of the plane surface, and the depth of the centroid of the area below the surface. Equation 11.13 indicates that *the magnitude of the resultant force is equal to the pressure at the centroid of the area multiplied by the total area.* Since all the differential forces that were summed to obtain $F_R$ are perpendicular to the surface, the resultant $F_R$ must also be perpendicular to the surface.

Although our intuition might suggest that the resultant force should pass through the centroid of the area, this is not actually the case. The point through which the resultant force acts is called the ***center of pressure*** and its location relative to the centroid of the area A is indicated in Fig. 11.11. The $y$ coordinate, $y_R$, of the resultant force can be determined by summation of moments around the $x$ axis. That is, the moment of the resultant force must equal the moment of the distributed pressure force, or

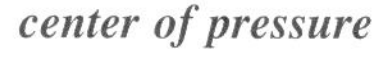


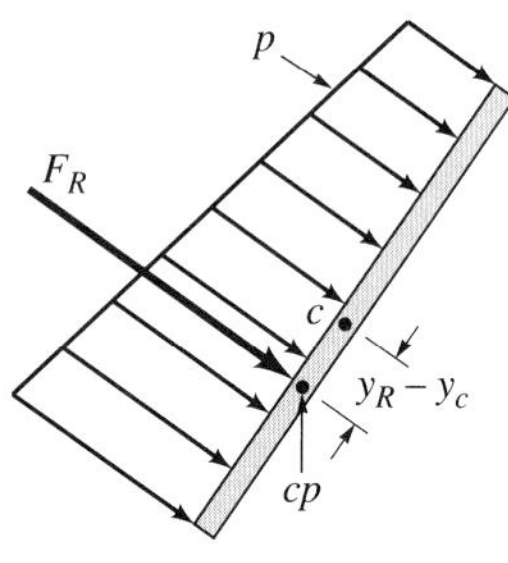


*Figure 11.11*

$$F_R y_R = \int_A y \, dF = \int_A \gamma \sin\theta \, y^2 \, dA \tag{11.14}$$

where we have used $dF = p\,dA = \gamma h\,dA$ together with $h = y \sin\theta$. It can be shown that this moment relationship leads to the following equation that gives the distance $y_R - y_c$ between the center of pressure and the centroid

$$y_R - y_c = \frac{I_{xc}}{y_c A} \tag{11.15}$$

The quantity $I_{xc}$, termed the second moment of the plane area $A$ with respect to an axis that passes through the centroid of $A$, is a geometric property of the area $A$. Values of $I_{xc}$ needed for applications in this book (rectangles and circles) are given in Fig. 11.12. Since $I_{xc}/y_c A > 0$, Eq. 11.15 clearly shows that the center of pressure is always *below* the centroid.

## *Example 11.3* Force on Plane Area

The 2-m-wide, 4-m-tall rectangular gate shown in Fig. E11.3*a* is hinged to pivot about point (1). For the water depth indicated, determine the magnitude and location of the resultant force exerted on the gate by the water.

### Solution

***Known:*** A rectangular gate is mounted on a hinge and located in the inclined wall of a tank containing water.

***Find:*** Determine the magnitude and location of the force of the water acting on the gate.

***Schematic and Given Data:***

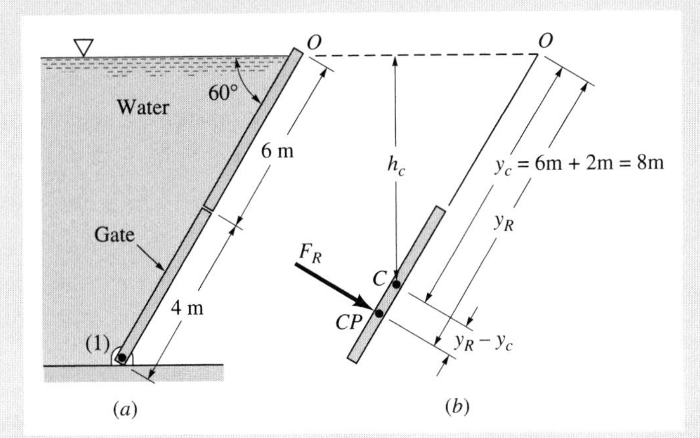


*Figure E11.3*

***Assumptions:***

1. The water is modeled as an incompressible fluid with a specific weight of $\gamma = 9.80 \times 10^3$ N/m$^3$.
2. The water is at rest.

***Analysis:*** One way to obtain the magnitude of the force of the water on the gate is to integrate the pressure distribution over the area of the gate as shown in Eq. 11.12. That is,

$$F_R = \gamma \sin\theta \int_A y\, dA = \gamma \sin\theta \int_A yb\, dy$$

where $b = 2$ m is the gate width. Thus,

❶ $$F_R = (9.80 \times 10^3 \text{ N/m}^3)(\sin 60°)(2 \text{ m}) \int_{6 \text{ m}}^{10 \text{ m}} y\, dy = 5.43 \times 10^5 \text{ N} \triangleleft$$

Alternatively, one could use the general formula given in Eq. 11.13 to obtain the same result more easily. That is, since $h_c = 8 \sin 60°$ m (see Fig. E11.3) it follows that

$$F_R = \gamma h_c A = (9.80 \times 10^3 \text{ N/m}^3)(8 \sin 60° \text{ m})(2 \text{ m} \times 4 \text{ m}) = 5.43 \times 10^5 \text{ N}$$

One way to determine the location of the resultant force is to use Eq. 11.14. That is,

$$F_R\, y_R = \gamma \sin\theta \int_A y^2\, dA = \gamma \sin\theta \int_A y^2 b\, dy$$

Thus, with $F_R = 5.43 \times 10^5$ N we obtain

$$(5.43 \times 10^5 \text{ N}) y_R = (9.80 \times 10^3 \text{ N/m}^3)(\sin 60°)(2 \text{ m}) \int_{y=6 \text{ m}}^{y=10 \text{ m}} y^2\, dy$$

or

$$y_R = 8.17 \text{ m} \triangleleft$$

Thus, the distance between the center of pressure and the centroid as measured along the inclined gate is $y_R - y_c = 8.17 \text{ m} - 8 \text{ m} = 0.17$ m.
Alternately, one could use the general formula given by Eq. 11.15 to obtain the same result more easily. That is,

$$y_R = \frac{I_{xc}}{y_c A} + y_c$$

where from Fig. 11.12, for the rectangular gate

$$I_{xc} = (ba^3)/12 = (2 \text{ m})(4 \text{ m})^3/12 = 10.67 \text{ m}^4$$

Hence,

$$y_R = (10.67 \text{ m}^4)/[(8 \text{ m})(2 \text{ m} \times 4 \text{ m})] + 8 \text{ m} = 8.17 \text{ m}$$

❶ Note that the $y$ coordinate is measured downward from the free surface in the direction *parallel to the area A*, whereas the depth to the centroid, $h_c$, is measured *vertically* downward from the free surface.

## 11.6 Buoyancy

*buoyant force*

***V11.4 Cartesian Driver***

When a body is completely submerged in a fluid, or floating so that it is only partially submerged, the resultant fluid force acting on the body is called the ***buoyant force.*** A net upward vertical force results because pressure increases with depth (see Eq. 11.8) and the pressure forces acting from below are larger than the pressure forces acting from above.

It is known from elementary physics that the buoyant force, $F_B$, is given by the equation

$$F_B = \gamma V \tag{11.16}$$

where $\gamma$ is the specific weight of the fluid and $V$ is the volume of the fluid displaced by the body. Thus, *the buoyant force has a magnitude equal to the weight of the fluid displaced by the body, and is directed vertically upward.* This result is commonly referred to as ***Archimedes' principle.*** The *buoyant force passes through the centroid of the displaced volume,* and the point through which the buoyant force acts is called the ***center of buoyancy.***

*Archimedes' principle*

*center of buoyancy*

These same results apply to floating bodies that are only partially submerged, if the specific weight of the fluid above the liquid surface is very small compared with the liquid in which the body floats. Since the fluid above the surface is usually air, for such applications this condition is satisfied.

Many important problems can be analyzed using the concept of the buoyant force.

***V11.5 Hydrometer***

***For Example...*** consider a spherical buoy having a diameter of 1.5 m and weighing 8.50 kN that is anchored to the sea floor with a cable as shown in Fig. 11.13*a*. Assume that the buoy is completely immersed as illustrated. In this case, what would be the tension in the cable? To solve this problem we first draw a free-body diagram of the buoy as shown in Fig. 11.13*b*, where $F_B$ is the buoyant force acting on the buoy, $\mathcal{W}$ is the weight of the buoy, and $T$ is the tension in the cable. For equilibrium it follows that

$$T = F_B - \mathcal{W}$$

From Eq. 11.16,

$$F_B = \gamma V$$

where for seawater Table FM-1 gives $\gamma = 10.1 \text{ kN/m}^3$. Thus, with $V = \pi d^3/6$ the buoyant force is

$$F_B = (10.1 \times 10^3 \text{ N/m}^3)[(\pi/6)(1.5 \text{ m})^3] = 1.785 \times 10^4 \text{ N}$$

The tension in the cable can now be calculated as

$$T = 1.785 \times 10^4 \text{ N} - 0.850 \times 10^4 \text{ N} = 9.35 \text{ kN}$$

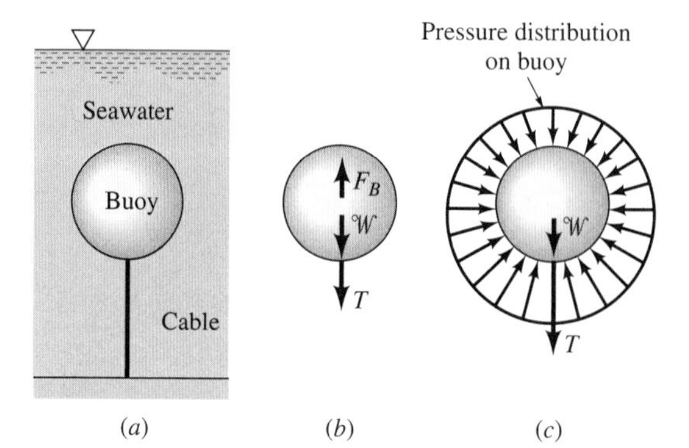


*Figure 11.13*

Note that we replaced the effect of the hydrostatic pressure force on the body by the buoyant force, $F_B$. Another correct free-body diagram of the buoy is shown in Fig. 11.13*c*. The net effect of the pressure forces on the surface of the buoy is equivalent to the upward force of magnitude $F_B$ (the buoyant force). Do not include both the buoyant force and the hydrostatic pressure effects in your calculations—use one or the other. ▲

## 11.7 Chapter Summary and Study Guide

In this chapter the pressure variation in a fluid at *rest* is considered, along with some important consequences of this type of pressure variation. It is shown that for incompressible fluids at rest, the pressure varies linearly with depth. This type of variation is commonly referred to as a *hydrostatic pressure distribution.* The distinction between absolute and gage pressure is discussed along with a consideration of barometers for the measurement of atmospheric pressure.

Pressure measuring devices called manometers, which utilize static liquid columns, are analyzed in detail. A brief discussion of mechanical and electronic pressure gages is also included. Equations for determining the magnitude and location of the resultant fluid force acting on a plane surface in contact with a static fluid are developed. For submerged or floating bodies the concept of the buoyant force and the use of Archimedes' principle are reviewed.

The following check list provides a study guide for this chapter. When your study of the text and end-of-chapter exercises has been completed you should be able to

- write out the meanings of the terms listed in the margins throughout the chapter and understand each of the related concepts. The subset of key terms listed here in the margin is particularly important.
- calculate the pressure at various locations within an incompressible fluid at rest.
- use the concept of a hydrostatic pressure distribution to determine pressures from measurements on various types of manometers.
- determine the magnitude of the resultant hydrostatic force acting on a plane surface using Eq. 11.13, and the location of this force using Eq. 11.15.
- use Archimedes' principle to calculate the resultant fluid force acting on floating or submerged bodies.

*hydrostatic pressure distribution*
*pressure head*
*absolute pressure*
*gage pressure*
*barometer*
*manometer*
*U-tube manometer*
*Bourdon pressure gage*
*center of pressure*
*buoyant force*
*Archimedes' principle*

## *Problems*

**Note:** Unless otherwise indicated in the problem statement, use values of fluid properties given in the tables of Appendix FM-1 when solving these problems.

**Pressure**

**11.1** The water level in an open tank is 90 ft above the ground. What is the static pressure at a fire hydrant that is connected to the tank and located at ground level? Express your answer in psi.

**11.2** Bathyscaphes are capable of submerging to great depths in the ocean. What is the pressure at a depth of 6 km, assuming that seawater has a constant specific weight of 10.1 $kN/m^3$? Express your answer in pascals and psi.

**11.3** A barometric pressure of 29.4 in. Hg corresponds to what value of atmospheric pressure in psi, and in pascals?

**11.4** A pressure of 7 psi absolute corresponds to what gage pressure for standard atmospheric pressure of 14.7 psi absolute?

**11.5** Blood pressure is usually given as a ratio of the maximum pressure (systolic pressure) to the minimum pressure (diastolic pressure). Such pressures are commonly measured with a mercury manometer. A typical value for this ratio for a human would be 120/70, where the pressures are in mm Hg. **(a)** What would these pressures be in pascals? **(b)** If your car tire was inflated to 120 mm Hg, would it be sufficient for normal driving?

**11.6** On the suction side of a pump a Bourdon pressure gage reads 40 kPa vacuum. What is the corresponding absolute pressure if the local atmospheric pressure is 100 kPa absolute?

**11.7** (CD-ROM)

**11.8** The closed tank of Fig. P11.8 is filled with water. The pressure gage on the tank reads 7 psi. Determine: (**a**) the height, *h*, in the open water column, (**b**) the gage pressure acting on the bottom tank surface *AB*, and (**c**) the absolute pressure of the air in the top of the tank if the local atmospheric pressure is 14.7 psi absolute.

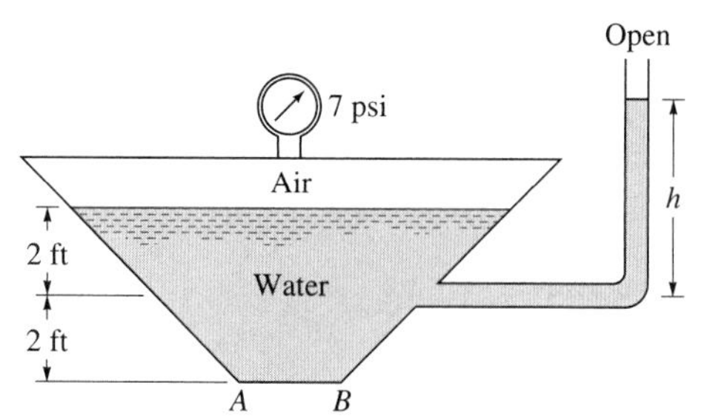


*Figure P11.8*

## Manometers

**11.9** In Fig. P11.9 pipe *A* contains carbon tetrachloride ($SG$ = 1.60) and the closed storage tank *B* contains a salt brine ($SG$ = 1.15). Determine the air pressure in tank *B* if the gage pressure in pipe *A* is 25 psi.

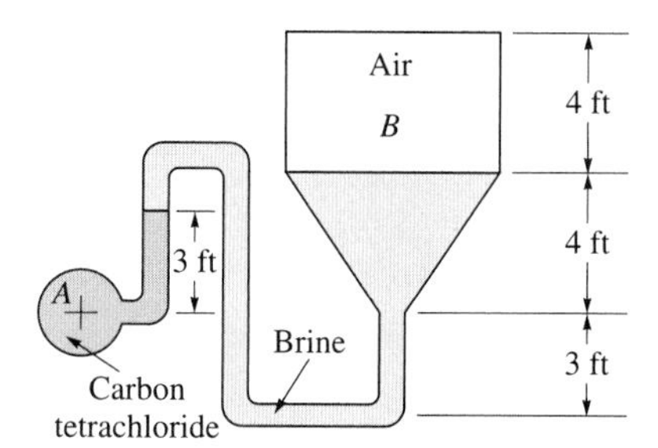


*Figure P11.9*

**11.10** A U-tube mercury manometer is connected to a closed pressurized tank as illustrated in Fig. P11.10. If the air pressure is 2 psi, determine the differential reading, *h*. The specific weight of the air is negligible.

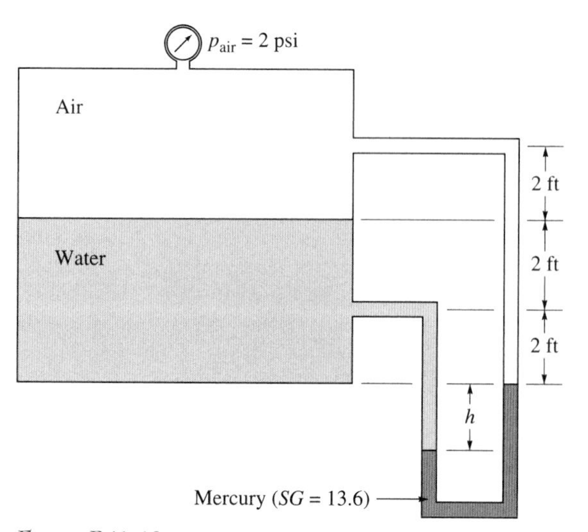


*Figure P11.10*

**11.11** (CD-ROM)

**11.12** Water, oil, and an unknown fluid are contained in the open vertical tubes shown in Fig. P11.12. Determine the density of the unknown fluid.

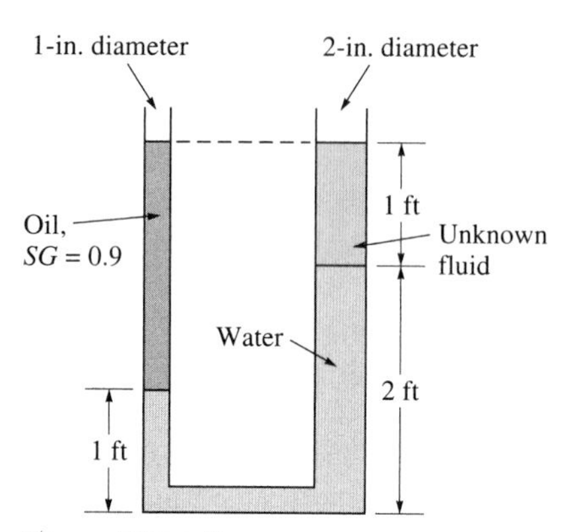


*Figure P11.12*

**11.13** The mercury manometer of Fig. P11.13 indicates a differential reading of 0.30 m when the pressure in pipe *A* is 30-mm Hg vacuum. Determine the pressure in pipe *B*.

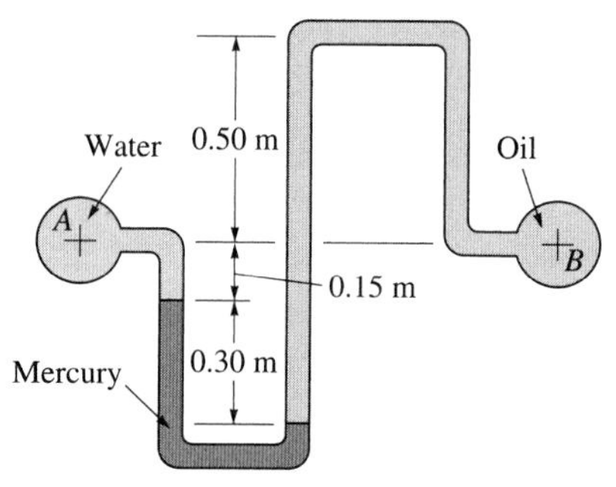


*Figure P11.13*

**11.14** Determine the angle θ of the inclined tube shown in Fig. P11.14 if the pressure at *A* is 2 psi greater than that at *B*.

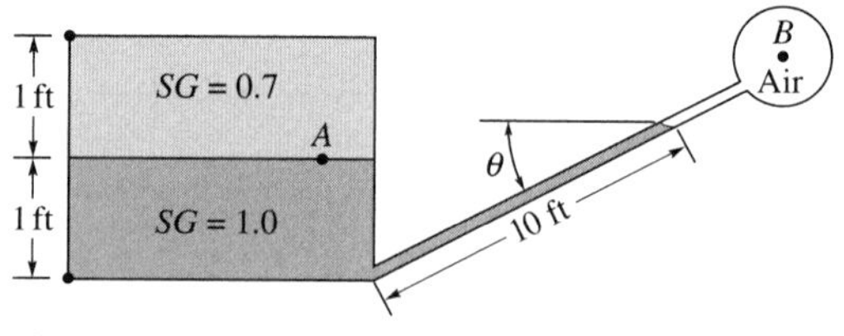


*Figure P11.14*

**11.15** (CD-ROM)

## Forces on plane areas

**11.16** A rectangular gate having a width of 5 ft is located in the sloping side of a tank as shown in Fig. P11.16. The gate is hinged along its top edge and is held in position by the force *P*. Friction at the hinge and the weight of the gate can be neglected. Determine the required value of *P*.

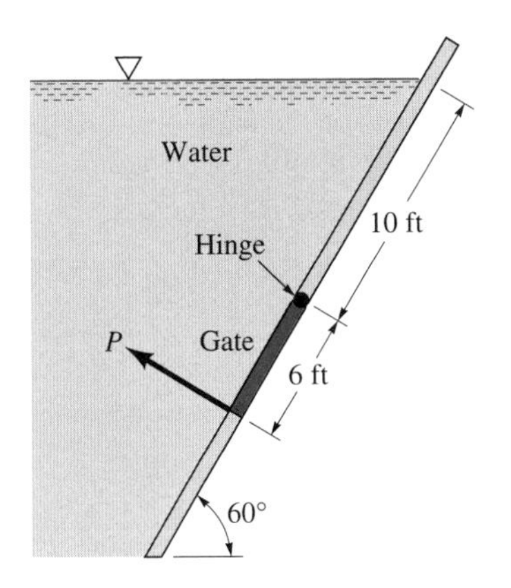

Figure P11.16

**11.17** (CD-ROM)

**11.18** A square gate (4 m by 4 m) is located on the 45° face of a dam. The top edge of the gate lies a vertical distance 8 m below the water surface. Determine the force of the water on the gate and the point through which it acts.

**11.19** A large, open tank contains water and is connected to a 6-ft-diameter conduit as shown in Fig. P11.19. A circular plug is used to seal the conduit. Determine the magnitude, direction, and location of the force of the water on the plug.

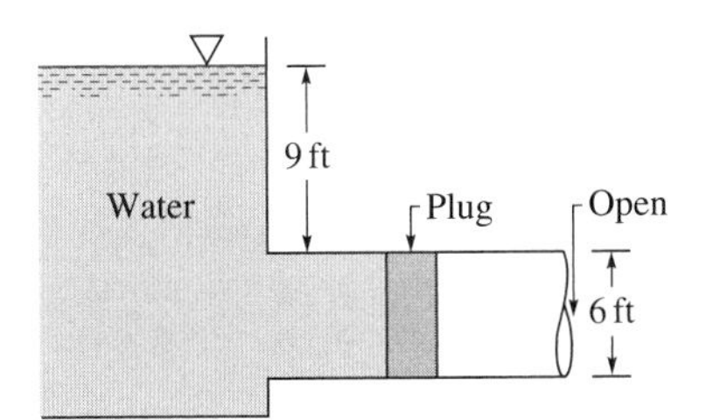

Figure P11.19

**11.20** A homogeneous, 4-ft-wide, 8-ft-long rectangular gate weighing 800 lbf is held in place by a horizontal flexible cable as shown in Fig. P11.20. Water acts against the gate, which is hinged at point *A*. Friction in the hinge is negligible. Determine the tension in the cable.

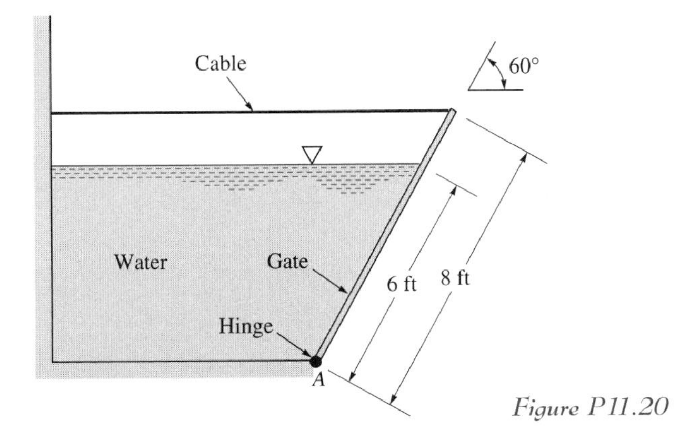

Figure P11.20

**11.21** (CD-ROM)

**11.22** Two square gates close two openings in a conduit connected to an open tank of water as shown in Fig. P11.22. When the water depth, *h*, reaches 5 m it is desired that both gates open at the same time. Determine the weight of the homogeneous horizontal gate and the horizontal force, *R*, acting on the vertical gate that is required to keep the gates closed until this depth is reached. The weight of the vertical gate is negligible, and both gates are hinged at one end as shown. Friction in the hinges is negligible.

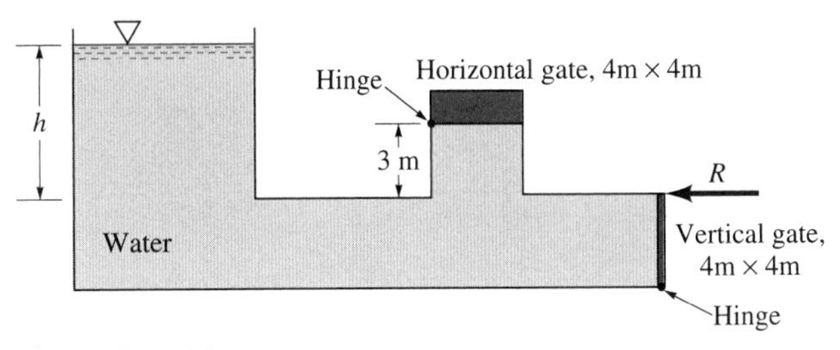

Figure P11.22

**11.23** The rigid gate, *OAB*, of Fig. P11.23 is hinged at *O* and rests against a rigid support at *B*. What minimum horizontal force, *P*, is required to hold the gate closed if its width is 3 m? Neglect the weight of the gate and friction in the hinge. The back of the gate is exposed to the atmosphere.

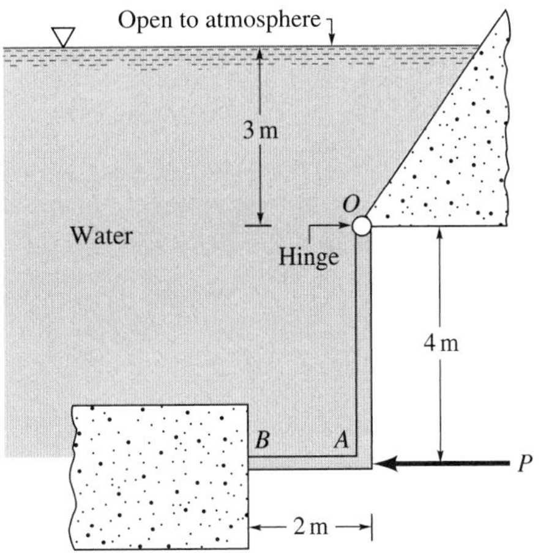

Figure P11.23

**11.24** A gate having the cross section shown in Fig. P11.24 is 4 ft wide and is hinged at *C*. The gate weighs 18,000 lbf, and

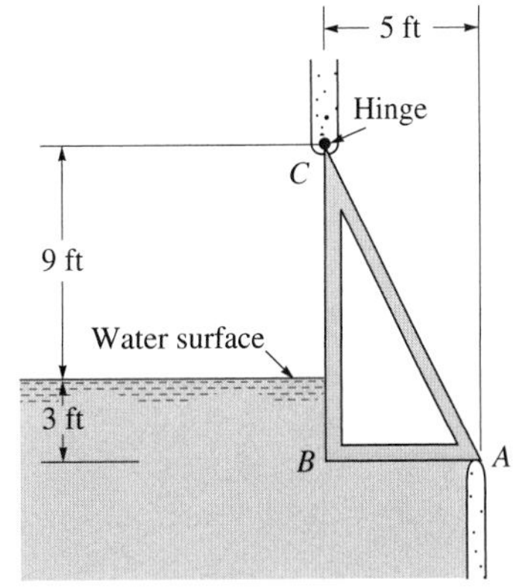

Figure P11.24

its mass center is 1.67 ft to the right of the plane $BC$. Determine the vertical reaction at $A$ on the gate when the water level is 3 ft above the base. All contact surfaces are smooth.

**11.25** (CD-ROM)

### Buoyancy

**11.26** A solid cube floats in water with a 0.5-ft-thick oil layer on top as shown in Fig. P11.26. Determine the weight of the cube.

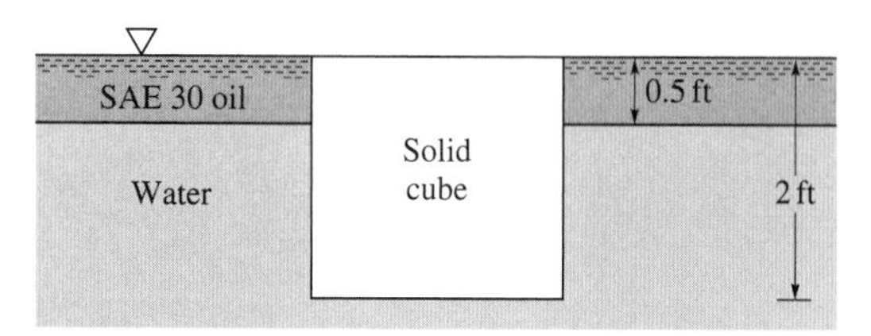


*Figure P11.26*

**11.27** The homogeneous timber $AB$ of Fig. P11.27 is 0.15 m by 0.35 m in cross section. Determine the specific weight of the timber and the tension in the rope.

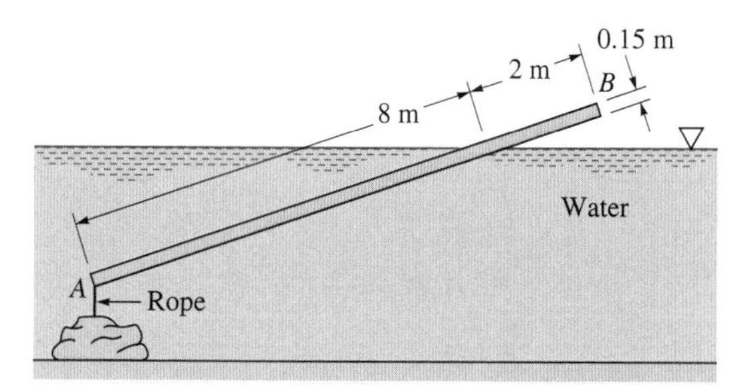


*Figure P11.27*

**11.28** (CD-ROM)

**11.29** (CD-ROM)

### Fluid Statics—general

**11.30** A plate of negligible weight closes a 1-ft-diameter hole in a tank containing air and water as shown in Fig. P11.30. A block of concrete (specific weight = 150 lbf/ft$^3$), having a volume of 1.5 ft$^3$, is suspended from the plate and is completely immersed in the water. As the air pressure is increased, the differential reading, $\Delta h$, on the inclined-tube mercury manometer increases. Determine $\Delta h$ just before the plate starts to lift off the hole. The weight of the air has a negligible effect on the manometer reading.

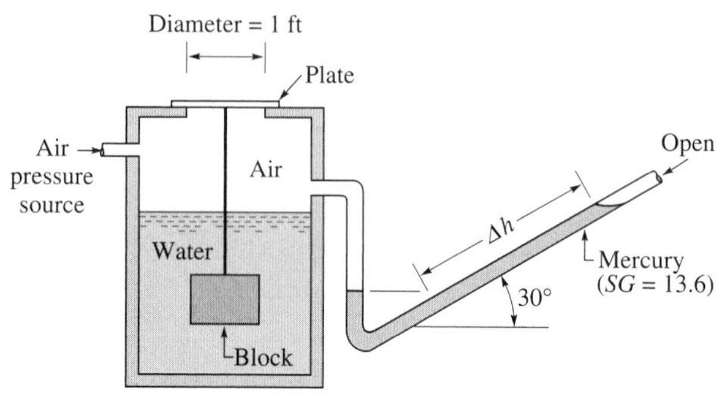


*Figure P11.30*

**11.31** A 1-ft-diameter, 2-ft-long cylinder floats in an open tank containing a liquid having a specific weight $\gamma$. A U-tube manometer is connected to the tank as shown in Fig. P11.31. When the pressure in pipe $A$ is 0.1 psi below atmospheric pressure, the various fluid levels are as shown. Determine the weight of the cylinder. Note that the top of the cylinder is flush with the fluid surface.

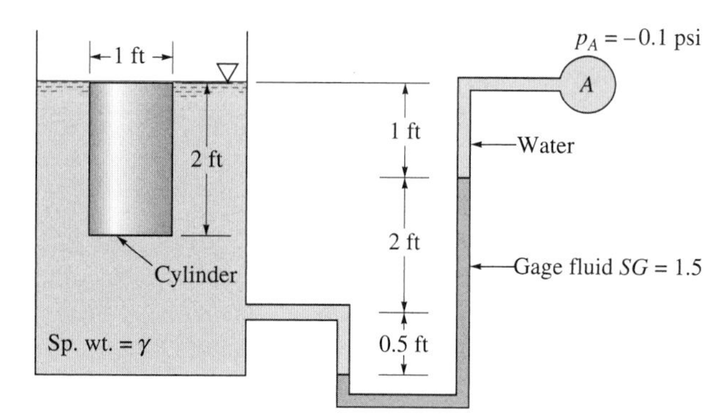


*Figure P11.31*

**11.32** (CD-ROM)

# THE MOMENTUM AND MECHANICAL ENERGY EQUATIONS

*Introduction...*

A fluid's behavior is governed by a set of fundamental physical laws expressed by an appropriate set of equations. The application of laws such as the conservation of mass, Newton's laws of motion, and the laws of thermodynamics form the foundation of fluid mechanics analyses. In this chapter we consider three equations that are mathematical representations of these laws—the momentum equation, the Bernoulli equation, and the mechanical energy equation. These equations deal with a flowing fluid, unlike the equations in the previous chapter that involved stationary fluids. Thus, the ***objective*** of this chapter is to show the use of these equations in thermal systems engineering. A discussion of compressible flow is also provided on the accompanying CD-ROM (Secs. 12.8–12.10).

*chapter objective*

## 12.1 Fluid Flow Preliminaries

In this section we introduce some important concepts relating to fluid flow that support not only the discussions of the present chapter but also those of Chaps. 13 and 14. These concepts include body and surface forces, viscosity, and the incompressible flow model.

### 12.1.1 Body and Surface Forces

In fluid mechanics, the resultant force on the contents of a control volume is commonly represented as the sum of all the surface and body forces acting. The only ***body force*** we consider is that associated with the action of gravity. We experience this body force as weight, $\mathcal{W}$. ***For Example...*** when a ball is thrown, its weight is one of the forces that clearly alters the motion of the ball and affects its trajectory. Similarly, the weight of a fluid may affect its motion. ▲

*body force*

***Surface forces*** are exerted on the contents of the control volume by material just outside the control volume in contact with material just inside the control volume. Surface forces can be written in terms of components normal and tangential to the surface. As introduced in Sec. 2.4.2 and used extensively in Chap. 11, the normal component of force is $F_{\text{normal}} = pA$, where $p$ is the pressure. Similarly, the tangential component of force is $F_{\text{tangential}} = \tau A$, where $\tau$ is the shear stress, the tangential force per unit area. Although it is possible to generate a shear stress in a stationary solid (e.g., apply a horizontal force to a table top), it is impossible to generate a shear stress in a fluid without the fluid being in motion. In fact, the ***definition of a fluid*** is that it is a material in which the application of any shear stress (no matter how small) will cause motion.

*surface force*

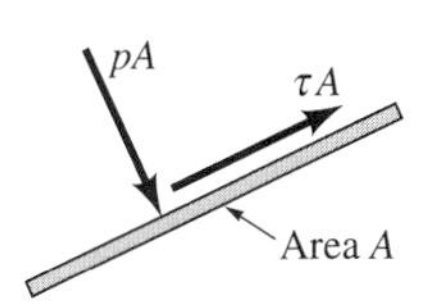


*fluid definition*

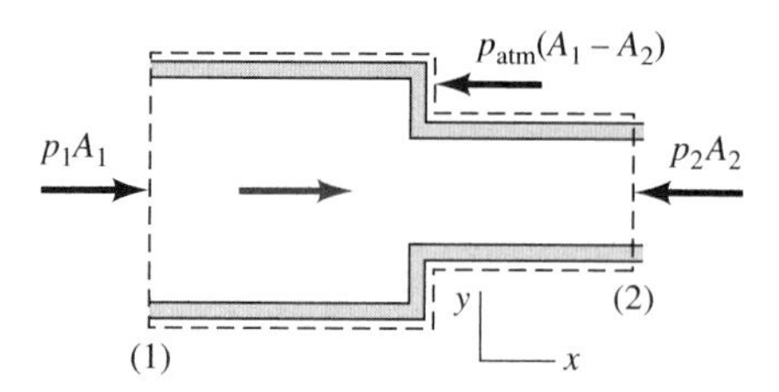


*Figure 12.1* Pressure forces acting on a sudden contraction.

The shear force concept is studied in the next section together with the property viscosity. We conclude the present discussion of surface forces by considering the role that gage pressure can play when evaluating the forces acting on a control volume. Figure 12.1 shows a control volume enclosing a horizontal contraction consisting of a smaller diameter pipe following one having a larger diameter. Surface forces acting on the control volume include the force $p_1A_1$ acting in the direction of flow and the oppositely directed force $p_2A_2$. Also, as shown in the figure, the atmospheric pressure $p_{atm}$ acts on the washer-shaped area $(A_1 - A_2)$, producing a surface force $p_{atm}(A_1 - A_2)$ acting opposite to the flow direction. The net pressure force acting axially is

$$\Sigma F_x = p_1A_1 - p_2A_2 - p_{atm}(A_1 - A_2) \tag{12.1a}$$

where $p_1$, $p_2$, and $p_{atm}$ are absolute pressures. By rearranging terms, Eq. 12.1a can be written in terms of gage pressures as

$$\Sigma F_x = (p_1 - p_{atm})A_1 - (p_2 - p_{atm})A_2$$

where $(p_1 - p_{atm})$ and $(p_2 - p_{atm})$ are recognized as the gage pressures at (1) and (2), respectively. That is,

$$\Sigma F_x = p_1(\text{gage})A_1 - p_2(\text{gage})A_2 \tag{12.1b}$$

Thus, the net pressure force can be expressed in terms of absolute pressure, as in Eq. 12.1a, or gage pressure, as in Eq. 12.1b.

### 12.1.2 Viscosity

Shear stresses play an important role in subsequent developments involving fluid flow. Accordingly, in this section we provide background material required by these discussions.

***V12.1 No-slip condition***

The character of the shear stress developed depends on the specific flow situation. Consider a flat plate of area $A$ located a distance $b$ above a fixed parallel plate and the gap between the plates filled with a viscous fluid. As shown in Fig. 12.2, a force, $F$, applied to the top plate causes it to move at a constant speed, $U$. Experimental observations show that the fluid sticks to both plates so that the fluid velocity is zero on the bottom plate and $U$ on the top plate. In the gap between the two plates the fluid velocity profile is linear and given by $u = u(y) = Uy/b$. The moving plate exerts a shear stress on the fluid layer at $y = b$. Similarly,

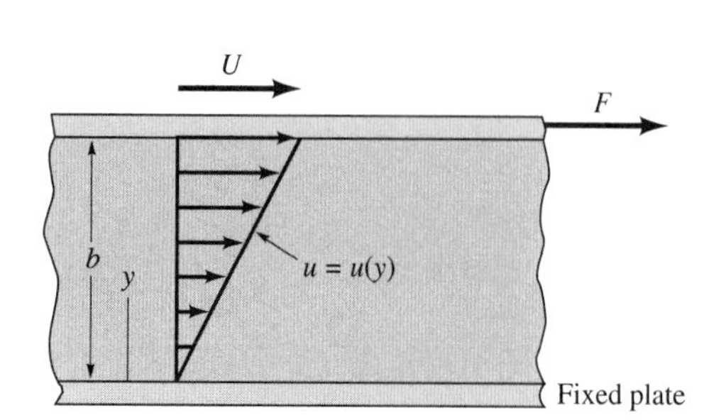


*Figure 12.2* Behavior of a fluid placed between two parallel plates.

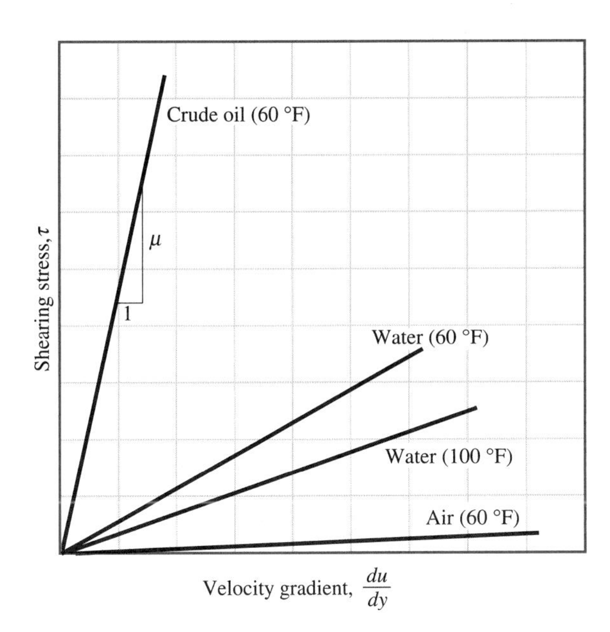


*Figure 12.3* Linear variation of shearing stress with velocity gradient for common fluids.

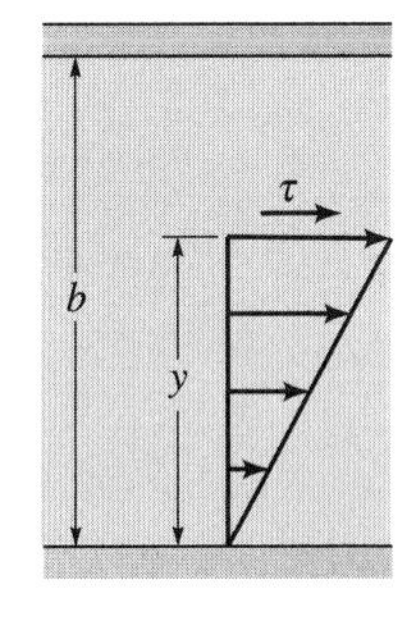


the more rapidly moving fluid above each fluid layer in the gap exerts a shear stress, $\tau$, on that layer.

For most common fluids such as air, water, and motor oil, the shear stress is found to be directly proportional to the velocity gradient $du/dy$. That is,

$$\tau = \mu \frac{du}{dy} \tag{12.2}$$

where the proportionality factor designated by $\mu$ is called the ***viscosity.*** Such fluids are called ***Newtonian fluids.*** For the flow shown in Fig. 12.2, since $u = Uy/b$, the velocity gradient is $du/dy = U/b$. Thus, from Eq. 12.2, the shear stress exerted on each fluid layer by the fluid above it is $\tau = \mu U/b$.

*viscosity*

*Newtonian fluids*

Viscosity is a property. From Eq. 12.2, it can be deduced that alternative units for viscosity are $\text{N} \cdot \text{s/m}^2$ or $\text{lbf} \cdot \text{s/ft}^2$. In accordance with Eq. 12.2, plots of $\tau$ versus $du/dy$ should be linear with the slope equal to the viscosity as illustrated in Fig. 12.3. The actual value of the viscosity depends on the particular fluid, and for a particular fluid the viscosity is also highly dependent on temperature, as illustrated in Fig. 12.3 with the two curves for water. Values of viscosity for several common gases and liquids are listed in the tables of Appendix FM-1.

***V12.2 Viscous fluids***

Quite often viscosity appears in fluid flow problems combined with the density in the form

$$\nu = \frac{\mu}{\rho} \tag{12.3}$$

This ratio is called the ***kinematic viscosity*** and is denoted by the Greek symbol $\nu$ (nu). Alternative units for kinematic viscosity include $\text{m}^2/\text{s}$ and $\text{ft}^2/\text{s}$. Values of kinematic viscosity for some common liquids and gases are given in Appendix FM-1.

*kinematic viscosity*

In some applications, fluids are considered to be ***inviscid.*** That is, the fluid is considered to have zero viscosity. Shear stresses cannot play a role in such applications.

*inviscid*

## 12.1.3 Incompressible Flow

As discussed in Sec. 4.3.6, the density (and specific volume) of liquids varies little with pressure at fixed temperature. Accordingly, to simplify evaluations involving the flow of liquids, the density is often taken as constant. When this assumption is made, the flow is called an ***incompressible flow.***

*incompressible flow*

As we have seen in previous sections, the density of air, and other gases, can vary significantly. Still, flowing air can often be modeled as incompressible provided the velocity of the air is not too great and the temperature is nearly constant. As a *rule of thumb,* air flows having velocities less than about 100 m/s (330 ft/s or 225 mi/h) can be modeled as incompressible. At higher velocities, density change becomes important, and then the compressible flow principles of Secs. 12.8–12.10 apply.

*steady flow*

In some discussions that follow, a flow may be modeled as both incompressible and steady. A ***steady flow*** is a flow in which nothing changes with time at a given location in the flow. This usage is consistent with the *steady state* concept discussed in Sec. 5.1.

## 12.2 Momentum Equation

*Newton's second law of motion*

***Newton's second law of motion*** for a single particle of mass $m$ involves the familiar form $\boldsymbol{F} = m\boldsymbol{a}$, where $\boldsymbol{F}$ is the resultant force acting on the particle and $\boldsymbol{a}$ is the acceleration. Since the mass of a particle is constant and $\boldsymbol{a} = d\mathbf{V}/dt$, an alternate form of this equation is $\boldsymbol{F} = d(m\mathbf{V})/dt$, where $m\mathbf{V}$ denotes momentum. That is, the resultant force on the particle is equal to the time rate of change of the particle's momentum. The object of the present section is to introduce Newton's second law of motion in a form appropriate for application to a fluid flowing through a control volume. Consider flow through the control volume shown in Fig. 12.4. For simplicity, we assume that the control volume has one inlet, (1), one outlet, (2) and that the flow is one-dimensional (Sec. 5.1). As discussed in Chaps. 5 and 7, the fluid flowing across the control surface carries mass, energy, and entropy across the surface, into or out of the control volume. Similarly, the flow also transfers momentum into or out of the control volume. Such transfers can be accounted for as

$$\left[\begin{array}{c}\text{time rate of momentum}\\ \text{transfer into or}\\ \text{out of a control volume}\\ \text{accompanying mass flow}\end{array}\right] = \dot{m}\mathbf{V}$$

where $\dot{m}$ is the mass flow rate (kg/s, lb/s, or slug/s) across the inlet or outlet of the control volume. In this expression, the momentum per unit of mass flowing across the boundary of the control volume is given by the velocity vector $\mathbf{V}$. In accordance with the one-dimensional flow model, the vector is normal to the inlet or exit and oriented in the direction of flow.

In words, Newton's second law of motion for control volumes is

$$\left[\begin{array}{c}\text{time rate of change}\\ \text{of momentum contained}\\ \text{within the control volume}\end{array}\right] = \left[\begin{array}{c}\text{resultant force}\\ \text{acting } on \text{ the}\\ \text{control volume}\end{array}\right] + \left[\begin{array}{c}\text{net rate at which momentum is}\\ \text{transferred into the control}\\ \text{volume accompanying mass flow}\end{array}\right]$$

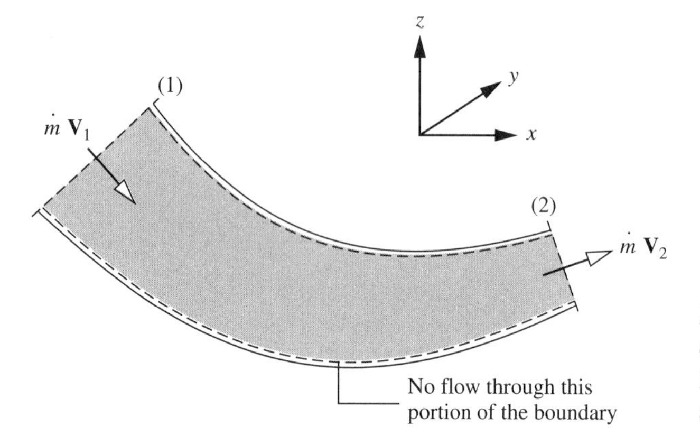


*Figure 12.4* One-inlet, one-exit control volume at steady state labeled with momentum transfers accompanying mass flow.

At steady state, the total amount of momentum contained in the control volume is constant with time. Accordingly, when applying Newton's second law of motion to control volumes at steady state, it is necessary to consider only the momentum accompanying the incoming and outgoing streams of matter and the forces acting on the control volume. Newton's law then states that the resultant force $\mathbf{F}$ acting *on* the control volume equals the difference between the rates of momentum exiting and entering the control volume accompanying mass flow. This is expressed by the following ***momentum equation:***

$$\mathbf{F} = \dot{m}_2 \mathbf{V}_2 - \dot{m}_1 \mathbf{V}_1 = \dot{m}(\mathbf{V}_2 - \mathbf{V}_1) \quad (12.4)$$

***momentum equation***

***V12.3 Sink flow***

Since $\dot{m}_1 = \dot{m}_2$ at steady state, the common mass flow is designated in this expression simply as $\dot{m}$.

The momentum equation for a control volume, Eq. 12.4, is a *vector equation.* In this text, components of vectors are resolved along rectangular coordinates. Thus, the $x$-, $y$-, and $z$-components of $\mathbf{F}$ are denoted $F_x$, $F_y$, and $F_z$, respectively. The components of the velocity vector $\mathbf{V}$ are denoted $u$, $v$, and $w$, respectively. The mass flow rate is evaluated using $\dot{m} = \rho A V$, where V is the *magnitude* of the velocity at the inlet or exit of the control volume where the flow rate is determined.

## 12.3 Applying the Momentum Equation

In this section we consider three applications of the momentum equation, Eq. 12.4. The applications have been selected to bring out important aspects of the momentum concept. The first case involves the deflection of a fluid jet by a fixed vane.

***V12.4 Force due to a water jet***

### *Example 12.1* Deflection of a Fluid Jet

As shown in Fig. E12.1*a*, a jet of water exits a nozzle with uniform velocity V = 10 ft/s, strikes a vane, and is turned through an angle θ. **(a)** Determine the anchoring force needed to hold the vane stationary as a function of θ. **(b)** Discuss the results.

### Solution

***Known:*** The direction of a jet of water is changed by a vane.

***Find:*** Determine the force needed to hold the vane stationary as a function of θ.

***Schematic and Given Data:***

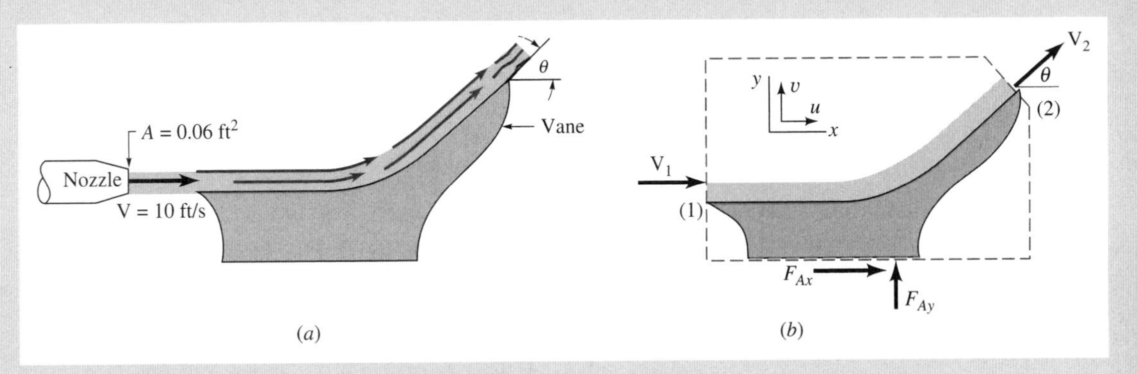


*Figure E12.1a*

***Assumptions:***

1. The control volume shown on the accompanying figure is at steady state.
2. Water is incompressible with $\rho = 1.94$ slug/ft$^3$.
3. At the inlet and outlet of the control volume, sections (1) and (2), the flow is one-dimensional and each cross-sectional area is 0.06 ft$^2$.
4. The pressure is atmospheric on the entire control surface.
5. The flow occurs in the horizontal $x$-$y$ plane.

***Analysis:*** (a) We select a control volume that includes the vane and a portion of the water (see Fig. E12.1*a*) and apply the momentum equation to this fixed control volume. The $x$- and $y$-components of Eq. 12.4 become

$$\Sigma F_x = \dot{m}(u_2 - u_1) \tag{1}$$

and

$$\Sigma F_y = \dot{m}(v_2 - v_1) \tag{2}$$

where $u$ and $v$ are the $x$- and $y$-components of velocity, and $\Sigma F_x$ and $\Sigma F_y$ are the $x$- and $y$-components of force acting on the contents of the control volume. Since the pressure is atmospheric on the entire control volume surface, the net pressure force on the control volume surface is zero. Thus, the only forces applied to the control volume contents are the $x$- and $y$-components of the anchoring force, $F_{Ax}$ and $F_{Ay}$, respectively. Although $F_{Ax}$ and $F_{Ay}$ are shown on the schematic as acting in the positive $x$- and $y$-directions, their magnitudes and directions will be determined as a part of the analysis.

We begin by evaluating the velocity components required in Eqs. (1) and (2). With assumptions 1, 2, and 3, a mass rate balance for the control volume shows that the *magnitude* of the velocities at (1) and (2) are equal: $V_1 = V_2 = V = 10$ ft/s. Accordingly, at inlet (1) we have $u_1 = V$ and $v_1 = 0$; at exit (2), $u_2 = V\cos\theta$ and $v_2 = V\sin\theta$.

Thus, Eqs. (1) and (2) can be written as

$$F_{Ax} = \dot{m}[V\cos\theta - V] = -\dot{m}V[1 - \cos\theta] \tag{3}$$

$$F_{Ay} = \dot{m}[V\sin\theta - 0] = \dot{m}V\sin\theta \tag{4}$$

where $\dot{m} = \rho AV$. Thus,

$$F_{Ax} = -\rho AV^2[1 - \cos\theta] \tag{5}$$

$$F_{Ay} = \rho AV^2 \sin\theta \tag{6}$$

By introducing known data, the components of the anchoring force are

$$F_{Ax} = -(1.94\ \text{slug/ft}^3)(0.06\ \text{ft}^2)(10\ \text{ft/s})^2(1 - \cos\theta)\left|\frac{1\ \text{lbf}}{1\ \text{slug}\cdot\text{ft/s}^2}\right|$$

$$= -11.64(1 - \cos\theta)\ \text{lbf} \ \triangleleft$$

and similarly

$$F_{Ay} = +11.64\sin\theta\ \text{lbf} \ \triangleleft$$

The minus sign in the expression for $F_{Ax}$ indicates that this component of the anchoring force is exerted to the left, in the negative $x$-direction. The plus sign in the expression for $F_{Ay}$ indicates that this component is exerted in the positive $y$-direction.

(b) The product $\dot{m}V$ in Eqs. (3) and (4) accounts for the rate at which momentum enters and exits the control volume. Although this product has the same magnitude at locations (1) and (2), namely 11.64 lbf, the direction is different. A change of direction requires a force, the components of which are $F_{Ax}$ and $F_{Ay}$.

For example, if $\theta = 90°$, the forces are $F_{Ax} = -11.64$ lbf and $F_{Ay} = +11.64$ lbf. Thus, as shown in Fig. E12.1*b*, the anchoring force must oppose the entering fluid momentum and supply the exiting momentum. If $\theta = 180°$, the jet is turned back on itself and the forces are $F_{Ax} = -23.28$ lbf and $F_{Ay} = 0$. As shown in Fig. E12.1*b*, the force must oppose the entering fluid momentum and supply the exiting momentum, neither of which has a *y* component.

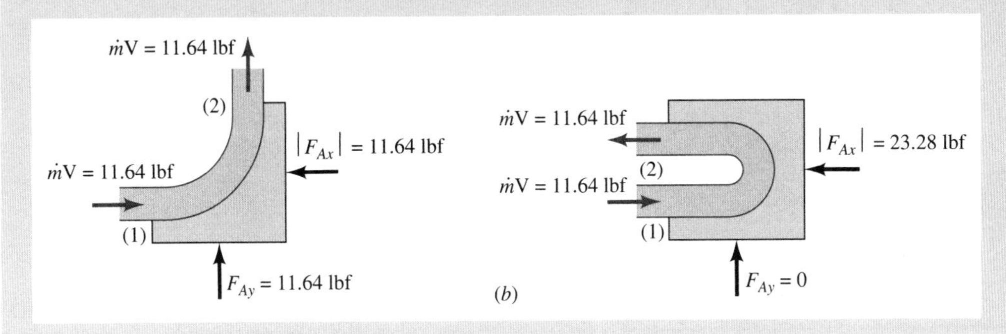


*Figure E12.1b*

In the previous example the anchoring force needed to hold the vane in place is a result of the change in direction of the fluid momentum. The pressure is uniform around the entire control surface and, therefore, provides no contribution to the force. In the next example the pressure is not uniform and is a factor in determining the anchoring force.

## *Example 12.2* Force Generated by Flow in a Pipe Bend

Water flows through a horizontal, 180° pipe bend as illustrated in Fig. E12.2*a*. The flow cross-sectional area is constant at a value of 0.1 ft² through the bend. The flow velocity at the entrance and exit of the bend is axial and 50 ft/s. The gage pressures at the entrance and exit of the bend are 30 psi and 24 psi, respectively. Calculate the horizontal (*x* and *y*) components of the anchoring force required to hold the bend in place.

### Solution

***Known:*** Water flows under given conditions in a horizontal, 180° pipe bend.

***Find:*** Determine the *x*- and *y*-components of the force needed to hold the bend in place.

***Schematic and Given Data:***

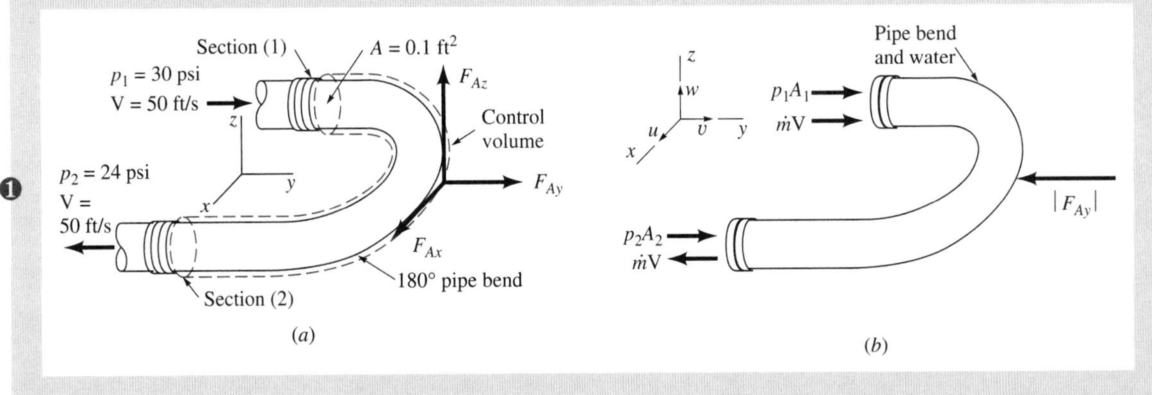


*Figure E12.2*

***Assumptions:***

1. The control volume shown on the accompanying figure is at steady state.
2. Water is incompressible with $\rho = 1.94$ slug/ft$^3$.
3. At (1) and (2) the flow is one-dimensional and each cross-sectional area is 0.1 ft$^2$.
4. The pressure is atmospheric on the outside of the pipe bend.

***Analysis:*** Since we want to evaluate components of the anchoring force needed to hold the pipe bend in place, an appropriate control volume (see dashed line in Fig. E12.2*a*) contains the bend and the water in the bend at an instant. The components of the anchoring force are $F_{Ax}$, $F_{Ay}$, and $F_{Az}$. Note that the weight of the water is vertical (in the negative $z$ direction) and does not contribute to the $x$ and $y$ components of the anchoring force. Although $F_{Ax}$ and $F_{Ay}$ are shown as acting in the positive $x$- and $y$-directions, respectively, their magnitudes *and* directions will be determined as a part of the analysis.

Since atmospheric pressure acts uniformly over the outside of the pipe bend, the effect of atmospheric pressure *in the x direction* cancels. Also, at locations (1) and (2) there are no $x$-components of the fluid velocity: $u_1 = u_2 = 0$. Accordingly, Eq. 12.4 applied in the $x$-direction reduces to

$$F_{Ax} = 0 \qquad (1) \triangleleft$$

❶ Since atmospheric pressure acts on the outside of the pipe bend, the *net* pressure force exerted on the control volume in the $y$ direction is $(p_1A + p_2A)$, where $p_1$ and $p_2$ are *gage pressures* at locations (1) and (2), respectively. At section (1) the flow is in the positive $y$ direction, so $v_1 = \text{V}$. At section (2) the flow is in the negative $y$ direction, so $v_2 = -\text{V}$. Accordingly, when Eq. 12.4 is applied in the $y$ direction we obtain

$$\begin{aligned} F_{Ay} + p_1A + p_2A &= \dot{m}[v_2 - v_1] \\ &= \dot{m}[(-\text{V}) - \text{V}] \end{aligned} \qquad (2) \triangleleft$$

Hence,

❷
$$F_{Ay} = -2\dot{m}\text{V} - (p_1 + p_2)A \qquad (3) \triangleleft$$

The mass flow rate is

$$\dot{m} = \rho A\text{V} = (1.94 \text{ slug/ft}^3)(0.1 \text{ ft}^2)(50 \text{ ft/s}) = 9.70 \text{ slug/s}$$

Inserting this into Eq. (3) together with the given data gives

$$\begin{aligned} F_{Ay} &= -2(9.70 \text{ slug/s})(50 \text{ ft/s})\left|\frac{1 \text{ lbf}}{1 \text{ slug} \cdot \text{ft/s}^2}\right| \\ &\quad -(30 + 24) \text{ lbf/in.}^2\left|\frac{144 \text{ in.}^2}{1 \text{ ft}^2}\right|(0.1 \text{ ft}^2) \\ &= -970 \text{ lbf} - 778 \text{ lbf} = -1748 \text{ lbf} \triangleleft \end{aligned}$$

The minus sign indicates that the force is exerted to the left, in the negative $y$ direction.

---

❶ Note that the pressure force is directed into the control volume at both the inlet and exit sections, independent of the direction of flow, which is in at the inlet and out at the exit. Pressure is a compressive stress.

❷ The product $\dot{m}\text{V}$ in Eq. (3) accounts for the rate at which momentum enters and exits the control volume. Although this product has the same magnitude at locations (1) and (2), the direction is different. Accordingly, as shown by the factor 2 in Eq. (3), the anchoring force must oppose the entering fluid momentum *and* supply the exiting momentum. The anchoring force must also oppose the net pressure force acting on the control volume. These forces and rates of momentum transfer are shown on Fig. 12.2*b*

In each of the previous two examples, the magnitude of the velocity of the fluid stream remains constant, but its direction changes. The change in direction requires an anchoring force whether or not pressure plays an explicit role. In the next example, an anchoring force is evaluated for a case where the direction of the flow is unchanged, but the magnitude of the velocity increases in the flow direction. Pressure and weight are also important.

## *Example 12.3* Force on a Nozzle

Determine the anchoring force required to hold in place a conical nozzle attached to the end of a laboratory sink faucet (see Fig. E12.3*a*) when the water volumetric flow rate is 0.6 liter/s. The nozzle weight is 1 N and the weight of the water in the nozzle at any instant is 0.03 N. The nozzle inlet and exit diameters are 16 mm and 5 mm, respectively. The nozzle axis is vertical and the gage pressures at sections (1) and (2) are 464 kPa and 0, respectively.

### Solution

***Known:*** Water flows vertically at a known volumetric flow rate through a given nozzle.

***Find:*** Determine the force needed to hold the nozzle in place.

***Schematic and Given Data:***

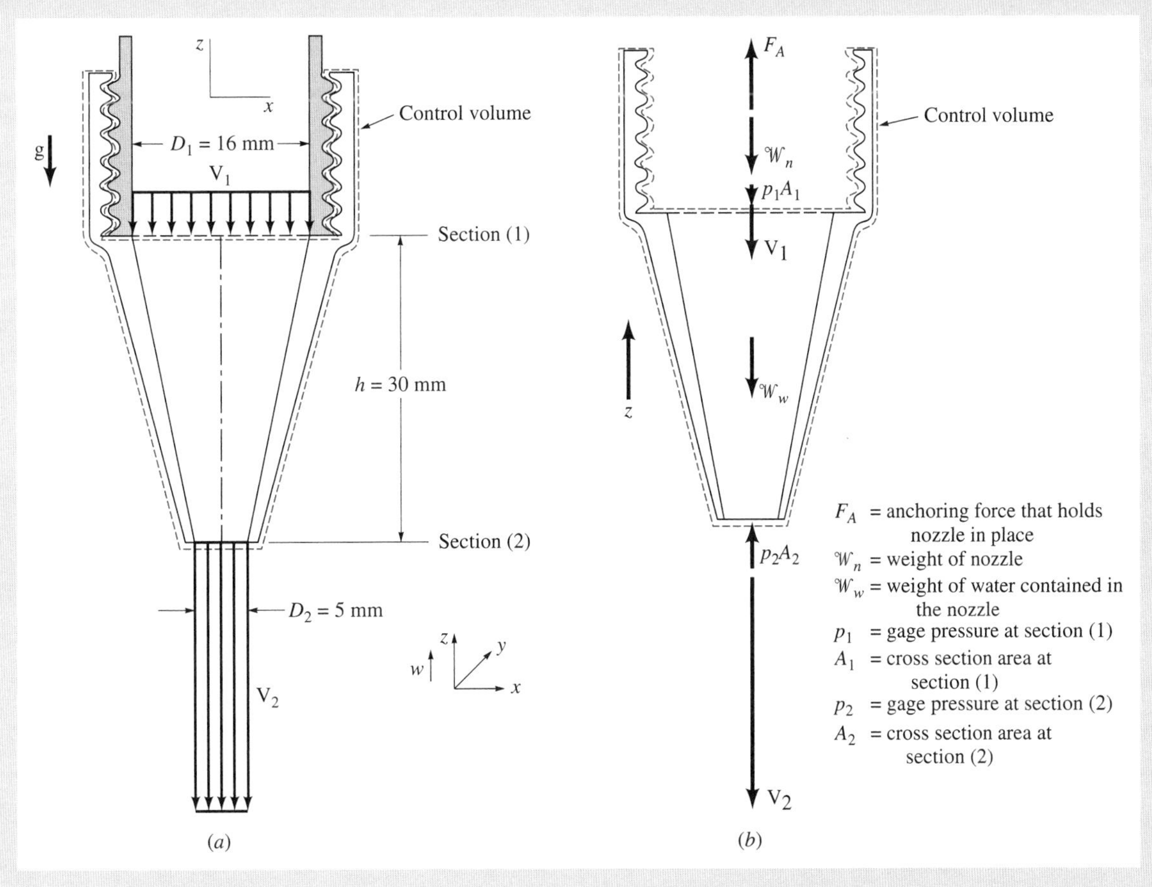


*Figure E12.3*

***Assumptions:***

1. The control volume shown in the accompanying figure is at steady state.
2. Water is incompressible with $\rho = 999\ \text{kg/m}^3$.
3. At sections (1) and (2) the flow is one-dimensional.
4. The water leaves the nozzle at atmospheric pressure (zero gage pressure).

***Analysis:*** The anchoring force sought, $F_A$, is the reaction force between the faucet and nozzle threads. To evaluate this force we select a control volume that includes the entire nozzle and the water contained in the nozzle at an instant, as is indicated in Figs. E12.3*a* and E12.3*b*. All of the vertical forces acting on the contents of this control volume are identified in Fig. E12.3*b*. Since atmospheric pressure acts on the outside of the nozzle, the *net* pressure force in the *z* direction can be evaluated using gage pressures.

Application of Eq. 12.4 to the $z$ direction gives

$$F_A - \mathcal{W}_n - \mathcal{W}_w - p_1A_1 + p_2A_2 = \dot{m}[w_2 - w_1] \tag{1}$$

where $w_1$ and $w_2$ are the $z$-components of velocity at (1) and (2). Since the flow is in the negative $z$ direction, $w_1 = -V_1$ and $w_2 = -V_2$. Thus, solving Eq.(1) for the anchoring force gives

$$F_A = \dot{m}[V_1 - V_2] + \mathcal{W}_n + \mathcal{W}_w + p_1A_1 - p_2A_2 \tag{2}$$

To complete this example, we use values given in the problem statement to quantify terms on the right-hand side of Eq. (2). The mass flow rate is

$$\dot{m} = \rho V_1A_1 = \rho Q = (999 \text{ kg/m}^3)(0.6 \text{ liter/s})|10^{-3} \text{ m}^3\text{/liter}| = 0.599 \text{ kg/s}$$

where $Q = V_1A_1 = V_2A_2$ is the volumetric flow rate.

Thus,

$$V_1 = \frac{Q}{A_1} = \frac{Q}{\pi(D_1^2/4)} = \frac{(0.6 \text{ liter/s})|10^{-3} \text{ m}^3\text{/liter}|}{\pi(16 \text{ mm})^2/4|1000^2 \text{ mm}^2\text{/m}^2|} = 2.98 \text{ m/s}$$

and

❶

$$V_2 = \frac{Q}{A_2} = \frac{Q}{\pi(D_2^2/4)} = \frac{(0.6 \text{ liter/s})|10^{-3} \text{ m}^3\text{/liter}|}{\pi(5 \text{ mm})^2/4|1000^2 \text{ mm}^2\text{/m}^2|} = 30.6 \text{ m/s}$$

Also, we have $\mathcal{W}_n = 1$ N, $\mathcal{W}_w = 0.03$ N, $p_1 = 464$ kPa, and $p_2 = 0$. Thus, from Eq. (2)

❷

$$\begin{aligned} F_A &= (0.599 \text{ kg/s})(2.98 - 30.6)\text{m/s}\left|\frac{1 \text{ N}}{1 \text{ kg}\cdot\text{m/s}^2}\right| + 1 \text{ N} + 0.03 \text{ N} \\ &\quad + (464 \text{ kPa})\left|\frac{10^3 \text{ N/m}^2}{1 \text{ kPa}}\right|\left[\pi\left(16 \text{ mm}\left|\frac{1 \text{ m}}{10^3 \text{ mm}}\right|\right)^2/4\right] - 0 \\ &= (-16.54 + 1 + 0.03 + 93.29)\text{N} = 77.8 \text{ N} \triangleleft \end{aligned}$$

Since the anchoring force, $F_A$, is positive, it acts upward in the $z$ direction—the nozzle would be pushed off the pipe if it were not fastened securely.

❶ In accord with the discussion of nozzles in Sec. 5.3.3, note that $V_2 > V_1$. The water accelerates as it flows through the nozzle.

❷ It is instructive to note how the anchoring force is affected by the different actions involved. As expected, the nozzle weight, $\mathcal{W}_n$, the water weight, $\mathcal{W}_w$, and the pressure force at section (1), $p_1A_1$, all increase the anchoring force. Of these, the effect of the pressure at section (1) is far more important than the total weight. Since $V_2 > V_1$, the contribution to the anchoring force from the momentum effect, $\dot{m}(V_1 - V_2)$, is negative, and thus decreases the anchoring force.

## 12.4 The Bernoulli Equation

The Bernoulli equation is introduced in Sec. 7.9 as an application of thermodynamic principles to a special case involving a one-inlet, one-outlet control volume at steady state. Equations 7.52b and 7.52c are the results of this development. In fluid mechanics it is customary to obtain the Bernoulli equation as an application of Newton's second law to a fluid particle moving along a streamline. As indicated in Fig. 12.5, ***streamlines*** are lines that are tangent to the velocity vector at any location in the flow. For steady flow a streamline can be thought of as the path along which a fluid particle moves when traveling from one location in the flow, point (1), to another location, point (2).

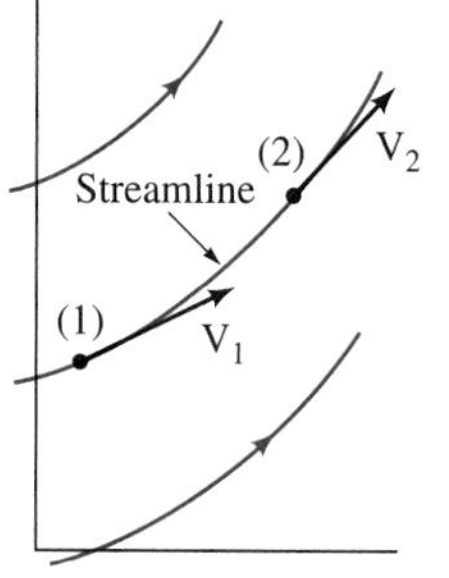

*Figure 12.5* Streamlines.

***streamlines***

When shear forces due to viscosity (friction) are negligible, Newton's second law takes the form:

(Net pressure force on a particle) + (net gravity force on a particle)
= (particle mass) × (particle acceleration)

In addition, we assume that the flow is steady and the fluid is modeled as incompressible. Under such conditions Newton's second law can be integrated along a streamline to give the ***Bernoulli equation*** in the form

$$p + \tfrac{1}{2}\rho V^2 + \gamma z = \text{constant along a streamline} \qquad (12.5)$$

*Bernoulli equation*

where the $z$ coordinate is positive vertically upward. Details of this development are provided in Sec. 12.4.1. Equation 12.5 corresponds to Eq. 7.52b.

### 12.4.1 Derivation of the Bernoulli Equation from Newton's Second Law (CD-ROM)

### 12.4.2 Static, Stagnation, Dynamic, and Total Pressure

Each term of the Bernoulli equation, Eq. 12.5, has the units of pressure: psi, $lbf/ft^2$, $N/m^2$. The first term, $p$, is the actual thermodynamic pressure of the fluid as it flows. To measure its value, one could move along with the fluid, thus being "static" relative to the moving fluid. Hence, $p$ is normally termed the ***static pressure.*** Another way to measure the static pressure would be to drill a hole in a flat surface and fasten a piezometer tube as indicated by the location of point (3) in Fig. 12.7.

*static pressure*

The third term in Eq. 12.5, $\gamma z$, is termed the ***hydrostatic pressure,*** in obvious regard to the hydrostatic pressure variation discussed in Chapter 11. It is not actually a pressure, but does represent the change in pressure possible due to potential energy variations of the fluid as a result of elevation changes.

*hydrostatic pressure*

The second term in the Bernoulli equation, $\rho V^2/2$, is termed the ***dynamic pressure.*** Its interpretation can be seen in Fig. 12.7 by considering the pressure at the end of a small tube inserted into the flow and pointing upstream. This type of tube is termed a *Pitot tube.* After the initial transient motion has died out, the liquid will fill the tube to a height of $H$ as shown. The fluid in the tube, including that at its tip, (2), will be stationary. That is, $V_2 = 0$, or point (2) is a ***stagnation point.***

*dynamic pressure*

*stagnation point*

If we apply the Bernoulli equation between points (1) and (2), using $V_2 = 0$ and assuming that $z_1 = z_2$, we find that

$$p_2 = p_1 + \tfrac{1}{2}\rho V_1^2$$

Hence, the pressure, $p_2$, at the stagnation point is greater than the static pressure, $p_1$, by an amount $\rho V_1^2/2$, the dynamic pressure. It can be shown that there is a stagnation point on any stationary body that is placed into a flowing fluid.

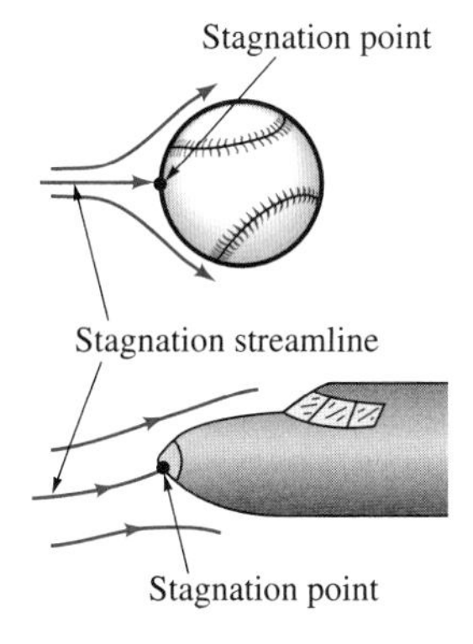


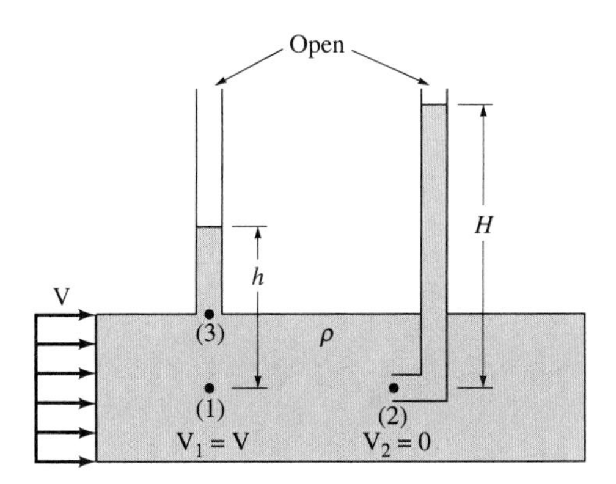


*Figure 12.7* Measurement of static and stagnation pressures.

***V12.5 Stagnation point flow***

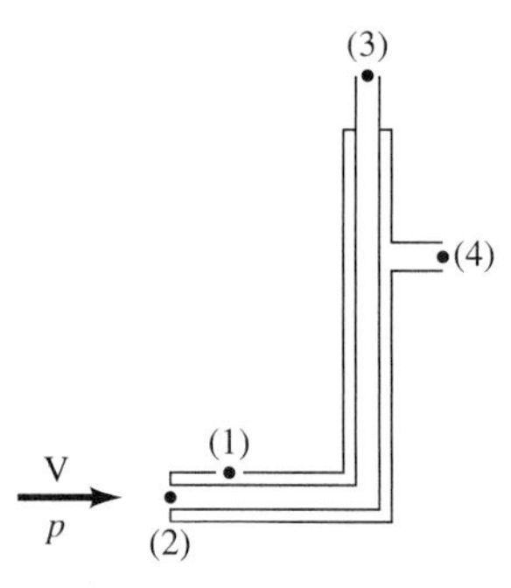


*Figure 12.8* Pitot-static tube.

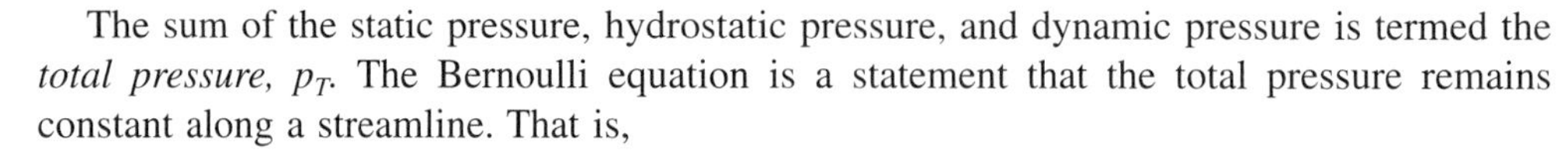

The sum of the static pressure, hydrostatic pressure, and dynamic pressure is termed the *total pressure*, $p_T$. The Bernoulli equation is a statement that the total pressure remains constant along a streamline. That is,

$$p + \tfrac{1}{2}\rho V^2 + \gamma z = p_T = \text{constant along a streamline} \qquad (12.10)$$

If the values of the static and stagnation pressures in a fluid are known, the fluid velocity can be calculated. This is the principle on which the *Pitot-static tube* is based. As shown in Fig. 12.8, two concentric tubes are attached to two pressure gages. The center tube measures the stagnation pressure at its open tip. If the elevation difference between (2) and (3) is negligible, then $p_3 = p_2$. The relationship between the stagnation, static, and dynamic pressures is then

$$p_3 = p + \tfrac{1}{2}\rho V^2$$

where $p$ and V are the pressure and velocity of the fluid flowing upstream of point (2). The outer tube is made with several small holes at an appropriate distance from the tip so that they measure the static pressure. If the elevation difference between (1) and (4) is negligible, then

$$p_4 = p_1 = p$$

***V12.6 Airspeed indicator***

These two equations can be rearranged to give

$$V = \sqrt{2(p_3 - p_4)/\rho} \qquad (12.11)$$

Accordingly, the velocity of the fluid can be determined by measuring the pressure difference ($p_3 - p_4$), which is readily accomplished.

An alternate but equivalent form of the Bernoulli equation is obtained by dividing each term of Eq. 12.5 by the specific weight, $\gamma$, to obtain

$$\frac{p}{\gamma} + \frac{V^2}{2g} + z = \text{constant along a streamline} \qquad (12.12)$$

*head*

This equation corresponds to Eq. 7.53. Each of the terms in this equation has the units of length and represents a certain type of ***head.***

*elevation, pressure, and velocity head*

The elevation term, $z$, is related to the potential energy of the particle and is called the ***elevation head.*** The pressure term, $p/\gamma$, is called the ***pressure head*** and represents the height of a column of the fluid that is needed to produce the pressure $p$. The velocity term, $V^2/2g$, is the ***velocity head*** and represents the vertical distance needed for the fluid to fall freely (neglecting friction) if it is to reach velocity V from rest. The Bernoulli equation states that the sum of the pressure head, the velocity head, and the elevation head is constant along a streamline.

## 12.5 Further Examples of Use of the Bernoulli Equation

In this section we consider applications of the Bernoulli equation for free jets and confined flows.

***Free Jets.*** Consider flow of a liquid from a large reservoir as is shown in Fig. 12.9. A jet of liquid of diameter $d$ flows from the nozzle with velocity V. Application of Eq. 12.5 between points (1) and (2) on the streamline shown gives

$$p_1 + \tfrac{1}{2}\rho V_1^2 + \gamma z_1 = p_2 + \tfrac{1}{2}\rho V_2^2 + \gamma z_2 \qquad (12.13)$$

***V12.7 Flow from a tank***

We use the facts that $z_1 = h$, $z_2 = 0$, the reservoir is large ($V_1 \cong 0$), open to the atmosphere ($p_1 = 0$ gage), and the fluid leaves as a "free jet" at atmospheric pressure ($p_2 = 0$ gage). Thus, we obtain

$$\gamma h = \tfrac{1}{2}\rho V^2$$

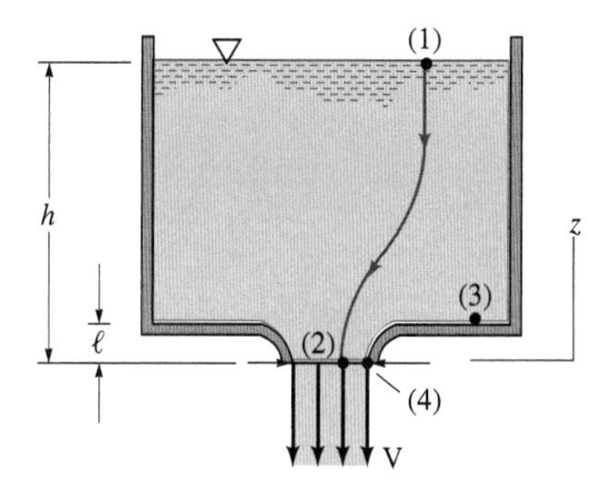


*Figure 12.9* Vertical flow from a tank.

or since $\gamma = \rho g$,

$$V = \sqrt{2\frac{\gamma h}{\rho}} = \sqrt{2gh} \tag{12.14}$$

Equation 12.14 could also be obtained by writing the Bernoulli equation between points (3) and (4) using the fact that $z_4 = 0$, $z_3 = \ell$. Also, $V_3 = 0$ since it is far from the nozzle, and from hydrostatics, $p_3 = \gamma(h - \ell)$.

*Confined Flows.* In many cases the fluid is physically constrained within a device so that pressure cannot be prescribed on the boundary as was done for the free jet example above. For many such situations it is necessary to use the mass balance together with the Bernoulli equation as illustrated in the following example.

***V12.8 Confined flow***

## *Example 12.4* Confined Flow

Water flows through a pipe reducer with volumetric flow rate Q as shown in Fig. E12.4. The difference in the static pressures at (1) and (2) is measured by the inverted U-tube manometer containing oil of specific gravity, *SG*, less than one. Determine the manometer reading, h, in terms of the volumetric flow rate and other pertinent quantities.

### Solution

***Known:*** Water flows through a variable area pipe that has a manometer attached.
***Find:*** Determine the manometer reading in terms of the volumetric flow rate.

***Schematic and Given Data:***

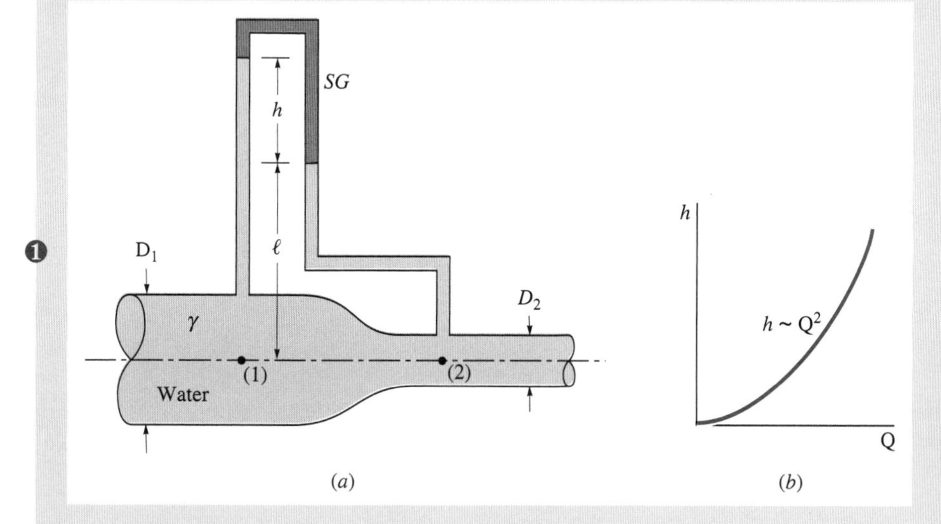


*Figure E12.4*

***Assumptions:***
1. The flow is steady, inviscid, and incompressible.
2. The pressure taps to which the manometer is fastened measure the difference in static pressure of the water between (1) and (2).
3. The water velocity is uniform (one-dimensional flow) across sections (1) and (2), and $z_1 = z_2$.

***Analysis:*** With the assumptions of steady, inviscid, incompressible flow, the Bernoulli equation can be written along the streamline between (1) and (2) as

$$p_1 + \tfrac{1}{2}\rho V_1^2 + \gamma z_1 = p_2 + \tfrac{1}{2}\rho V_2^2 + \gamma z_2 \tag{1}$$

With assumptions 1 and 3, the mass balance, $\rho A_1 V_1 = \rho A_2 V_2$, gives

$$V_1 = (A_2/A_1)V_2 \tag{2}$$

and

$$Q = A_1 V_1 = A_2 V_2$$

By combining Eqs. (1) and (2), and noting that $z_1 = z_2$, we obtain

$$p_1 - p_2 = \tfrac{1}{2}\rho V_2^2[1 - (A_2/A_1)^2] \tag{3}$$

This pressure difference is measured by the manometer and can be determined by using the pressure-depth ideas developed in Chapter 11. Thus,

$$p_1 - \gamma \ell - \gamma h + SG\gamma h + \gamma \ell = p_2$$

or

$$p_1 - p_2 = (1 - SG)\gamma h \tag{4}$$

Equations 3 and 4 can be combined to give the desired result as follows:

$$(1 - SG)\gamma h = \frac{1}{2}\rho V_2^2\left[1 - \left(\frac{A_2}{A_1}\right)^2\right]$$

or since $V_2 = Q/A_2$ and $\gamma = \rho g$,

$$h = (Q/A_2)^2 \frac{1 - (A_2/A_1)^2}{2g(1 - SG)} \quad \triangleleft$$

❶ Various types of flow meters are based on a variable area pipe geometry similar to that shown in this example. That is, the flow rate, Q, in a pipe can be determined if the manometer reading is known. Note that the manometer reading is proportional to the square of the flow rate.

## 12.6 The Mechanical Energy Equation

The Bernoulli equation considered in Secs. 12.4 and 12.5 is a form of the mechanical energy equation introduced in Sec. 7.9. In terms of heads, the Bernoulli equation states that the sum of the velocity head, the elevation head, and the pressure head is constant along a streamline. Key underlying assumptions include: (1) the flow is inviscid (frictionless), and (2) there are no mechanical devices such as pumps or turbines within the control volume. To account for such effects the full mechanical energy equation must be applied. That is,

$$\frac{p_1}{\gamma} + \frac{V_1^2}{2g} + z_1 = \frac{p_2}{\gamma} + \frac{V_2^2}{2g} + z_2 + \frac{(\dot{W}_{cv}/\dot{m})}{g} + h_L \tag{7.50b}$$

where $h_L$, the ***head loss,*** accounts for the irreversible conversion of mechanical energy into internal energy due to friction. The term $\dot{W}_{cv}$, which represents power due to devices that transfer mechanical energy across the control volume boundary, can be expressed as

*head loss*

$$\dot{W}_{cv} = \dot{W}_t - \dot{W}_p$$

where $\dot{W}_t$ is the power (kW, horsepower) removed from the control volume by a turbine, and $\dot{W}_p$ is the power added by a pump. Thus, Eq. 7.50b can be written as

$$\frac{p_1}{\gamma} + \frac{V_1^2}{2g} + z_1 + h_p - h_L - h_t = \frac{p_2}{\gamma} + \frac{V_2^2}{2g} + z_2 \tag{12.15}$$

where $(p_1/\gamma + V_1^2/2g + z_1)$ is the sum of the pressure head, velocity head, and elevation head at the inlet to the control volume [section (1)] and $(p_2/\gamma + V_2^2/2g + z_2)$ is the sum of these quantities at the exit [section (2)]. In Eq. 12.15, $h_p$ is the ***pump head*** and $h_t$ is the ***turbine head*** defined by

$$h_p = \frac{\dot{W}_p/\dot{m}}{g} = \frac{\dot{W}_p}{\gamma Q} \tag{12.16}$$

*pump head*

and

$$h_t = \frac{\dot{W}_t/\dot{m}}{g} = \frac{\dot{W}_t}{\gamma Q} \tag{12.17}$$

*turbine head*

where we have used $\dot{m} = \rho Q$ and $\gamma = \rho g$.

The head loss, pump head, and turbine head must satisfy the following constraints:

$$h_L \geq 0, h_p \geq 0, h_t \geq 0 \tag{12.18}$$

The head loss is zero if there are no frictional effects within the control volume. The pump and turbine heads are zero if there is no pump and turbine within the control volume. Otherwise, these quantities must be positive. As can be seen from Eq. 12.15, a pump adds head (or mechanical energy) to what was available at the inlet, whereas both a turbine and friction reduce the amount of head (or mechanical energy) available at the outlet.

***V12.9 Water plant aerator***

## 12.7 Applying the Mechanical Energy Equation

The mechanical energy equation plays an important role in fluid mechanics. In Chap. 14 it is used in the study of pipe systems. In the present section three introductory examples are considered. The first of these involves the evaluation of head loss in a pipe.

## *Example 12.5* Head Loss in a Pipe

An incompressible liquid flows steadily along the pipe shown in Fig. E12.5. Determine the direction of flow and the head loss over the 6-m length of pipe.

### Solution

***Known:*** Two pressure taps along a pipe indicate the pressure head in the constant diameter pipe.
***Find:*** Determine the direction of flow and the head loss for the flow.

***Schematic and Given Data:***

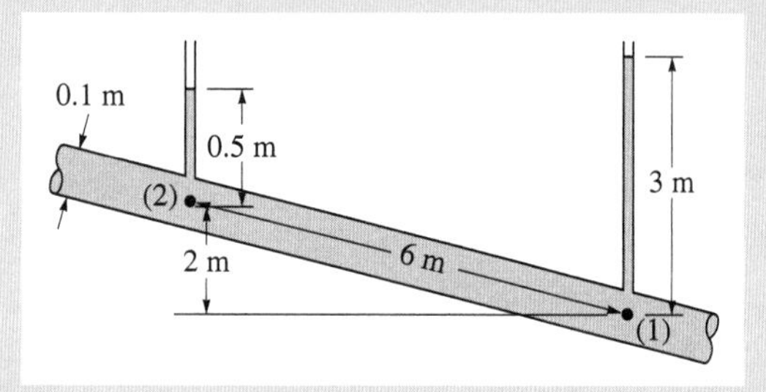


***Assumptions:***
1. The flow is steady and incompressible.
2. The pipe diameter is constant.
3. The two vertical liquid filled tubes measure the pressures $p_1$ and $p_2$.
4. There are no pumps or turbines within the section of pipe of interest.

*Figure E12.5*

***Analysis:*** The direction of flow can be obtained by determining which flow direction (uphill, or downhill) will give a positive head loss. Negative head losses cannot occur. Thus, we assume the flow is uphill and use the mechanical energy equation to determine $h_L$. From Eq. 12.15

$$\frac{p_1}{\gamma} + \frac{V_1^2}{2g} + z_1 + h_p - h_L - h_t = \frac{p_2}{\gamma} + \frac{V_2^2}{2g} + z_2 \qquad (1)$$

where $h_p = h_t = 0$ because there are no pumps or turbines. Since the flow areas $A_1$ and $A_2$ are equal and the liquid is incompressible, the mass rate balance gives $V_1 = V_2$. Thus, Eq. (1) reduces to

$$\frac{p_1}{\gamma} + z_1 - h_L = \frac{p_2}{\gamma} + z_2$$

where $p_1/\gamma = 3$ m, $p_2/\gamma = 0.5$ m, $z_1 = 0$, and $z_2 = 2$ m so that

❶ $$h_L = 0.5 \text{ m} \triangleleft$$

Since the head loss is positive, the flow is uphill as assumed.

---

❶ If we assume the flow is downhill, we would obtain $h_L = -0.5 \text{ m} < 0$, which is impossible since it would violate the Second Law of Thermodynamics.

The two following examples involve flows for which the turbine and pump heads play a significant role.

## *Example 12.6* Hydroelectric Turbine

Determine the maximum possible power output of the hydroelectric turbine shown in Fig. E12.6

### Solution

***Known:*** Water flows from a lake and through a turbine under known conditions.
***Find:*** Determine the maximum power that the turbine can extract from the water.

*Schematic and Given Data:*

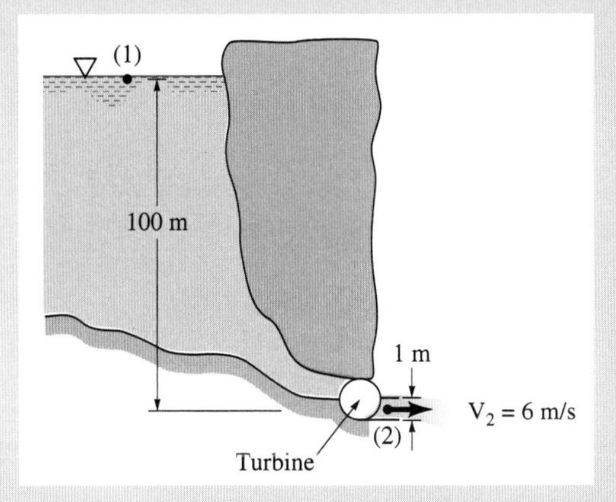


*Assumptions:*
1. The flow is steady and incompressible and $g = 9.81$ m/s$^2$.
2. At (1) the velocity is essentially zero because the surface area is large; also the pressure is atmospheric.
3. At (2) the water exits at a specified velocity and as a free jet at atmospheric pressure.

*Figure E12.6*

***Analysis:*** The power output from the turbine, $\dot{W}_t$, can be found from Eq. 12.17 as

$$\dot{W}_t = \gamma Q h_t \tag{1}$$

where from the given data the volumetric flow rate is

$$Q = A_2 V_2 = \pi(1 \text{ m})^2(6 \text{ m/s})/4 = 4.72 \text{ m}^3/\text{s} \tag{2}$$

The turbine head, $h_t$, can be obtained from Eq 12.15 as

$$h_t = \frac{p_1}{\gamma} + \frac{V_1^2}{2g} + z_1 - \frac{p_2}{\gamma} - \frac{V_2^2}{2g} - z_2 - h_L \tag{3}$$

From assumptions 2 and 3, it follows that $p_1 = p_2$ and $V_1 = 0$. Thus, with $z_2 = 0$ (arbitrary datum) and $z_1 = 100$ m, Eq. (3) becomes

$$h_t = z_1 - \frac{V_2^2}{2g} - h_L = 100 \text{ m} - \frac{(6 \text{ m/s})^2}{2(9.81 \text{ m/s}^2)} - h_L$$

$$= 98.2 \text{ m} - h_L \tag{4}$$

For the given flow rate, the maximum power output corresponds to the maximum turbine head. From Eq. 4 this clearly occurs when there is no head loss, $h_L = 0$.
Thus, $h_t = 98.2$ m and the maximum power output is

$$\dot{W}_t = \gamma Q h_t = 9.80 \times 10^3 \text{ N/m}^3(4.72 \text{ m}^3/\text{s})(98.2 \text{ m})$$

$$= 4.54 \times 10^6 \text{ N} \cdot \text{m/s} \left| \frac{1 \text{ kW}}{10^3 \text{ N} \cdot \text{m/s}} \right|$$

or

$$\dot{W}_t = 4.54 \times 10^3 \text{ kW} \triangleleft$$

## *Example 12.7* Pump System Head Loss

The pump shown in Fig. E12.7 adds 10 horsepower to the water as it pumps 2 ft$^3$/s from the lower lake to the upper lake. The elevation difference between the lake surfaces is 30 ft. Determine the head loss, in ft and in horsepower.

### Solution

***Known:*** The pump power, the elevation difference, and the volumetric flow rate are known.
***Find:*** Determine the head loss.

***Schematic and Given Data:***

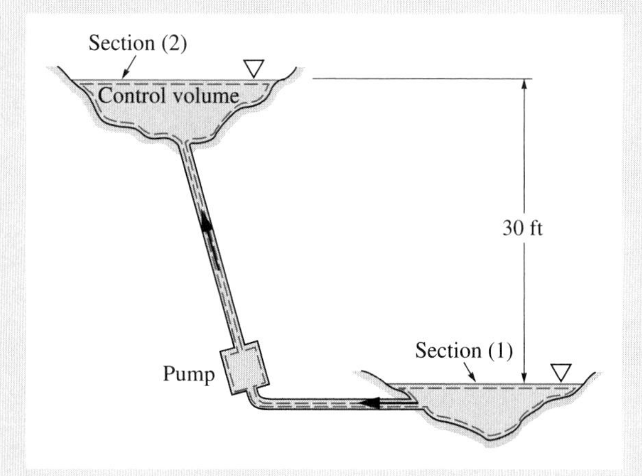


***Assumptions:***
1. The flow is steady and incompressible.
2. At each surface the pressure is atmospheric. Also, the water velocities on each surface are essentially zero because each surface area is large.

*Figure E12.7*

***Analysis:*** The head loss can be obtained from Eq. 12.15 as

$$h_L = \frac{p_1 - p_2}{\gamma} + \frac{V_1^2 - V_2^2}{2g} + z_1 - z_2 + h_p \tag{1}$$

where from assumption 2, the first two terms on the right drop out. Also, if we set $z_1 = 0$ (arbitrary datum), then $z_2 = 30$ ft. The pump head is found from Eq. 12.16 to be

$$h_p = \frac{\dot{W}_p}{\gamma Q} = \frac{10 \text{ hp}}{(62.4 \text{ lbf/ft}^3)(2 \text{ ft}^3/\text{s})}\left|\frac{550 \text{ ft} \cdot \text{lbf}}{1 \text{ hp}}\right| = 44.1 \text{ ft}$$

Thus, Eq. 1 gives

$$h_L = 0 + 0 + 0 - 30 \text{ ft} + 44.1 \text{ ft} = 14.1 \text{ ft} \triangleleft$$

In this case, a portion of the power input (a 44.1 ft head) is required to lift the water (30 ft head) and a portion is required to overcome the head loss (a 14.1 ft head). When expressed on a power basis, the head loss is

$$\gamma Q h_L = \left(62.4 \frac{\text{lbf}}{\text{ft}^3}\right)\left(2 \frac{\text{ft}^3}{\text{s}}\right)(14.1 \text{ ft})\left|\frac{1 \text{ hp}}{550 \text{ ft} \cdot \text{lbf/s}}\right|$$

❶ $$= 3.20 \text{ hp} \triangleleft$$

❶ The 3.20 hp portion of the power input is irreversibly converted into internal energy. The remaining 10 hp − 3.20 hp = 6.80 hp that the pump adds to the water is used to lift the water from the lower to the upper lake. This mechanical energy is stored as potential energy.

## 12.8 Compressible Flow (CD-ROM)

## 12.9 One-dimensional Steady Flow in Nozzles and Diffusers (CD-ROM)

## 12.10 Flow in Nozzles and Diffusers of Ideal Gases with Constant Specific Heats (CD-ROM)

## 12.11 Chapter Summary and Study Guide

In this chapter we have considered several preliminary concepts that are essential in the analysis of fluid motion. These concepts include surface and body forces, viscosity, and the steady, incompressible flow model.

In addition we have considered the application of Newton's second law to obtain the momentum equation for fluids flowing through control volumes. For steady flow, the sum of all the forces acting on the contents of the control volume equals the difference between the outflow and inflow rates of momentum across the control volume surface.

We have also considered two forms of the mechanical energy balance. The Bernoulli equation is valid for steady, inviscid, incompressible flows and provides the relationship between pressure, elevation, and velocity for such flows. The more general mechanical energy equation can be used in situations where viscous effects are important and pumps or turbines add or remove mechanical energy to or from the flowing fluid.

The following checklist provides a study guide for this chapter. When your study of the text and end-of-chapter exercises has been completed you should be able to

- write out the meanings of the terms listed in the margin throughout the chapter and understand each of the related concepts. The subset of key terms listed here in the margin is particularly important.
- explain the concepts of surface and body forces.
- explain the concepts of viscosity and shear stress.
- select a control volume and apply the momentum equation to analyze the flow through the control volume.
- apply the Bernoulli equation appropriately to analyze flow situations applicable to the use of this equation.
- use the concepts of head loss, pump head, and turbine head in the analysis of various flow situations.
- use the mechanical energy equation appropriately to analyze various flows.

*body and surface forces*
*viscosity*
*momentum equation*
*Bernoulli equation*
*static pressure*
*dynamic pressure*
*mechanical energy equation*
*head*
*head loss*
*pump and turbine heads*

## Problems

**Note:** Unless otherwise indicated in the problem statement, use values of fluid properties given in the tables of Appendix FM-1 when solving these problems. Also, except for the problems under the Compressible Flow heading, all problems are for steady, *incompressible* flow.

### Viscosity

**12.1** Crude oil having a viscosity of $9.52 \times 10^{-4}$ lbf · s/ft$^2$ is contained between parallel plates. (see Fig. 12.2). The bottom plate is fixed and the upper plate moves when a force $F$ is applied. If the distance between the two plates is 0.1 in., what value of $F$ is required to translate the plate with a velocity of 3 ft/s? The effective area of the upper plate is 200 in.$^2$.

**12.2** A layer of water flows down an inclined fixed surface with the velocity profile shown in Fig. P12.2. Determine the magnitude and direction of the shearing stress that the water exerts on the fixed surface for $U = 3$ m/s and $h = 0.1$ m.

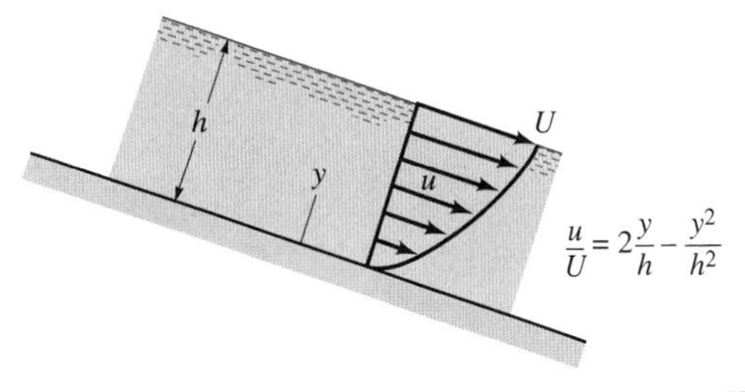


*Figure P12.2*

**12.3** (CD-ROM)

**12.4** (CD-ROM)

### Mass Balance

**12.5** A hydroelectric turbine passes 2 million gal/min through its blades. If the average velocity of the flow in the circular cross-section conduit leading to the turbine is not to exceed

30 ft/s, determine the minimum allowable diameter of the conduit.

**12.6** (CD-ROM)

### Momentum Equation

**12.7** Air flows into the atmosphere from a nozzle and strikes a vertical plate as shown in Fig. P12.7. A horizontal force of 9 N is required to hold the plate in place. Determine the velocity at the exit, $V_1$, and the velocity within the pipe, $V_2$.

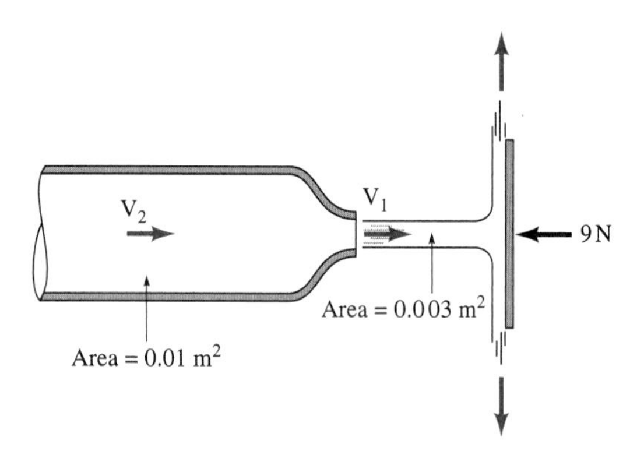


*Figure P12.7*

**12.8** Water, contained in a large open tank, discharges steadily into the atmosphere from a curved pipe as shown in Fig. P12.8. The tank rests on a smooth surface, and to prevent it from sliding, a horizontal flexible cable is to be connected to hooks on either the right or left side of the tank. Assuming the cable can only support a tensile force, would you connect it to the right or left side? What tensile force does the cable have to support? Assume the flow to be frictionless.

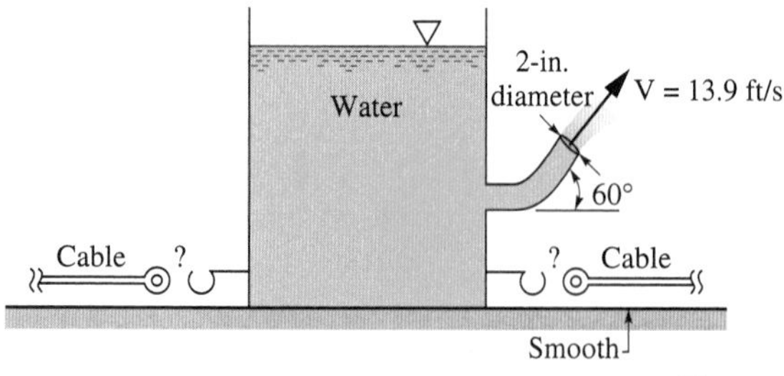


*Figure P12.8*

**12.9** A circular plate having a diameter of 300 mm is held perpendicular to an axisymmetric horizontal jet of air having a velocity of 40 m/s and a diameter of 80 mm as shown in Fig. P12.9. A hole at the center of the plate results in a discharge jet of air having a velocity of 40 m/s and a diameter of 20 mm. Determine the horizontal component of force required to hold the plate stationary.

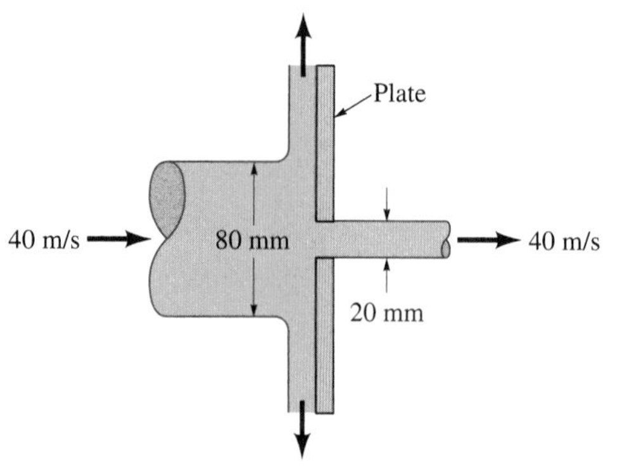


*Figure P12.9*

**12.10** (CD-ROM)

**12.11** A 10-mm-diameter jet of water is deflected by a homogeneous rectangular block (15 mm by 200 mm by 100 mm) that weighs 6 N as shown in Fig. P12.11. Determine the minimum volumetric flow rate needed to tip the block about point 0.

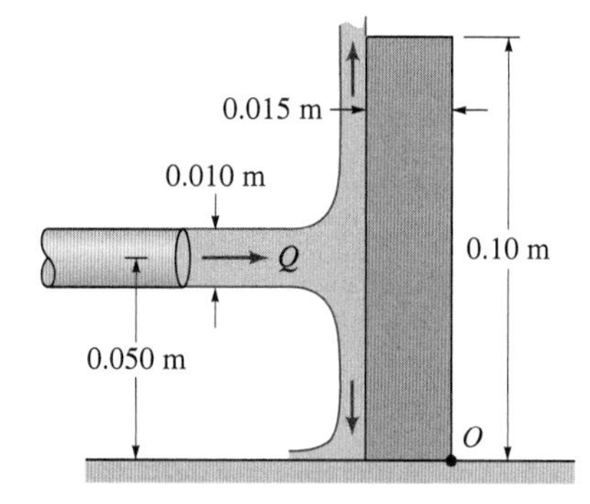


*Figure P12.11*

**12.12** (CD-ROM)

**12.13** (CD-ROM)

**12.14** Thrust vector control is a new technique that can be used to greatly improve the maneuverability of military fighter aircraft. It consists of using a set of vanes in the exit of a jet engine to deflect the exhaust gases as shown in Fig. P12.14. By how much is the thrust (force along the centerline of the aircraft) reduced for the case indicated compared to normal flight when the exhaust is parallel to the centerline?

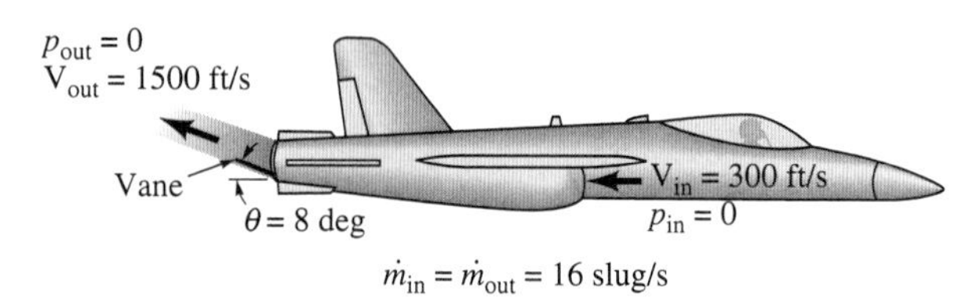


*Figure P12.14*

**12.15** (CD-ROM)

**12.16** Water flows through a right angle valve at the rate of 1000 lb/s as is shown in Fig. P12.16. The pressure just upstream of the valve is 90 psi and the pressure drop across the valve is 5 psi. The inside diameters of the valve inlet and exit pipes are 12 and 24 in. If the flow through the valve occurs in a horizontal plane, determine the $x$ and $y$ components of the anchoring force required to hold the valve stationary.

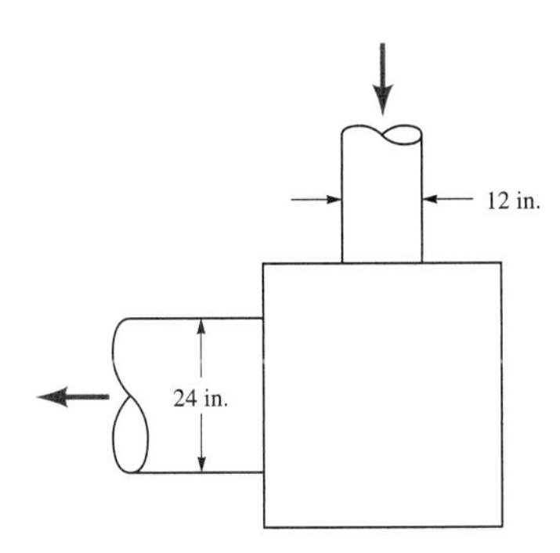


*Figure P12.16*

**12.17** Water flows through a 2-ft-diameter pipe arranged horizontally in a circular arc as shown in Fig. P12.17. If the pipe discharges to the atmosphere ($p_2 = 0$ gage), determine the $x$ and $y$ components of the anchoring force needed to hold the piping between sections (1) and (2) stationary. The steady flow rate is 3000 ft$^3$/min. The loss in pressure due to fluid friction between sections (1) and (2) is 25 psi.

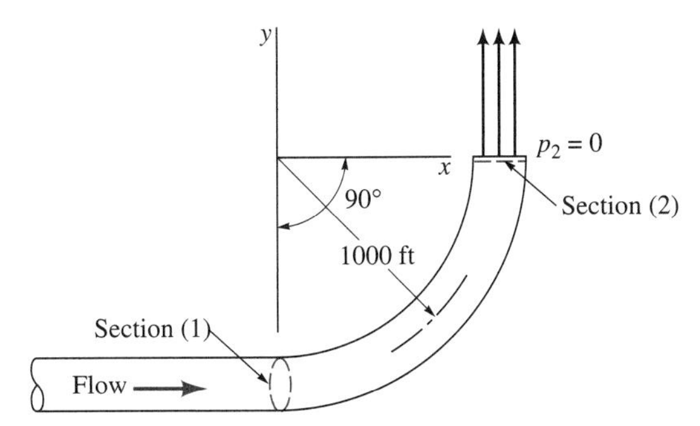


*Figure P12.17*

**12.18** Water enters the horizontal, circular cross-sectional, sudden contraction nozzle sketched in Fig. P12.18 at section (1) with a velocity of 25 ft/s and a pressure of 75 psi. The water exits from the nozzle into the atmosphere at section (2) where the velocity is 100 ft/s. Determine the axial component of the anchoring force required to hold the contraction in place.

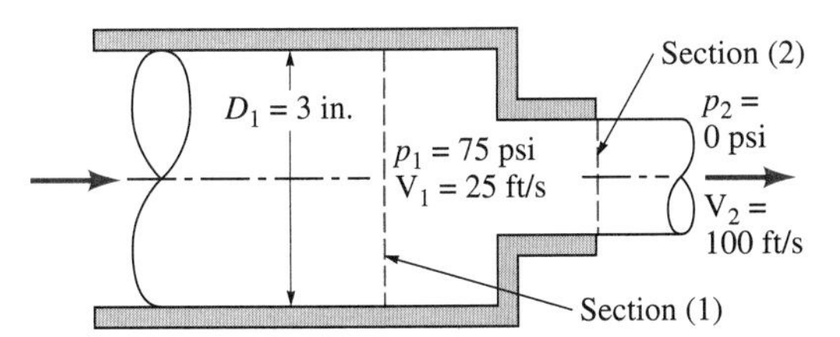


*Figure P12.18*

**12.19** Determine the magnitude and direction of the $x$ and $y$ components of the anchoring force required to hold in place the horizontal 180° elbow and nozzle combination shown in Fig. P12.19. Neglect gravity.

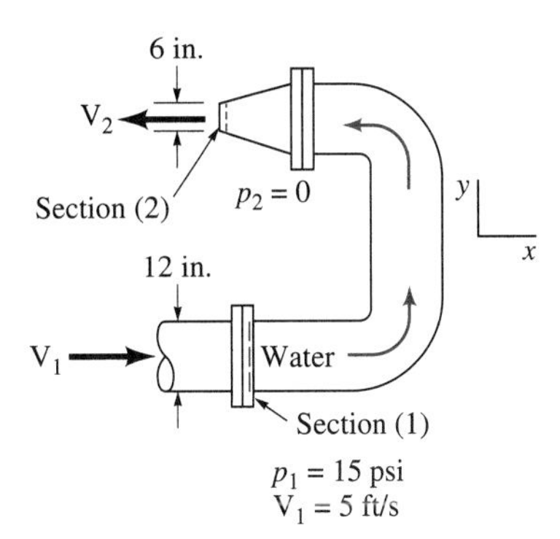


*Figure P12.19*

**12.20** A converging elbow (see Fig. P12.20) turns water through an angle of 135° in a vertical plane. The flow cross-sectional diameter is 400 mm at the elbow inlet, section (1), and 200 mm at the elbow outlet, section (2). The elbow flow passage volume is 0.2 m$^3$ between sections (1) and (2). The water volumetric flow rate is 0.4 m$^3$/s and the elbow inlet and outlet pressures are 150 kPa and 90 kPa. The elbow mass is 12 kg. Calculate the horizontal ($x$ direction) and vertical ($z$ direction) anchoring forces required to hold the elbow in place.

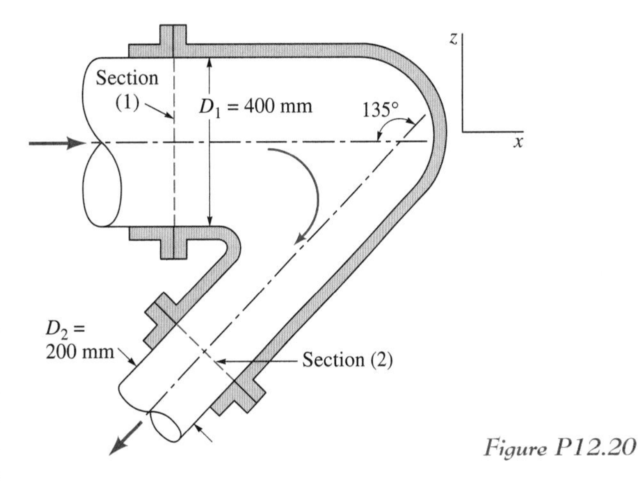


*Figure P12.20*

**12.21** (CD-ROM)

### Bernoulli Equation

**12.22** A hang glider soars through the air with an airspeed of 10 m/s. **(a)** What is the gage pressure at a stagnation point on the structure if it is at sea level where the air density is 1.23 kg/m$^3$? **(b)** Repeat the problem if the hang glider is at an altitude of 3000 m where the density is 0.909 kg/m$^3$.

**12.23** A person holds her hand out of an open car window while the car drives through still air at 65 mph. Under standard atmospheric conditions with $\rho = 0.00238$ slug/ft$^3$, what is the maximum pressure on her hand? What would be the maximum pressure if the "car" were an Indy 500 racer traveling 200 mph?

**12.24** A 4-in.-diameter pipe carries 300 gal/min of water at a pressure of 60 psi. Determine **(a)** the pressure head in feet of water, **(b)** the velocity head.

**12.25** A fire hose nozzle has a diameter of $1\frac{1}{8}$ in. According to some fire codes, the nozzle must be capable of delivering at least 300 gal/min. If the nozzle is attached to a 3-in.-diameter hose, what pressure must be maintained just upstream of the nozzle to deliver this volumetric flow rate?

**12.26** The pressure in domestic water pipes is typically 70 psi above atmospheric. If viscous effects are neglected, determine the height reached by a jet of water through a small hole in the top of the pipe.

**12.27** The circular stream of water from a faucet is observed to taper from a diameter of 20 mm to 10 mm in a distance of 40 cm. Determine the volumetric flow rate.

**12.28** (CD-ROM)

**12.29** A plastic tube of 50-mm diameter is used to siphon water from the large tank shown in Fig. P12.29. If the pressure on the outside of the tube is more than 30 kPa greater than the pressure within the tube, the tube will collapse and the siphon will stop. If viscous effects are negligible, determine the minimum value of $h$ allowed without the siphon stopping.

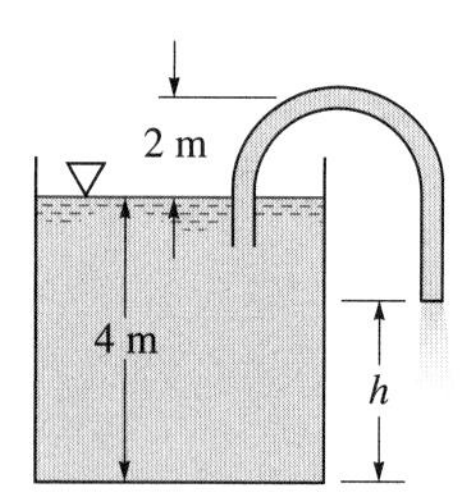


*Figure P12.29*

**12.30** Water flows steadily from the pipe shown in Fig. P12.30 with negligible viscous effects. Determine the maximum volumetric flow rate if the water is not to flow from the open vertical tube at $A$.

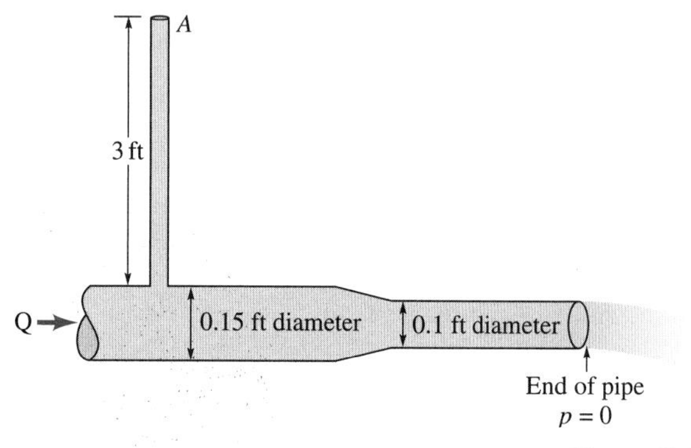


*Figure P12.30*

**12.31** (CD-ROM)

**12.32** Determine the volumetric flow rate through the Venturi meter shown in Fig. P12.32 if viscous effects are negligible and the fluid is water.

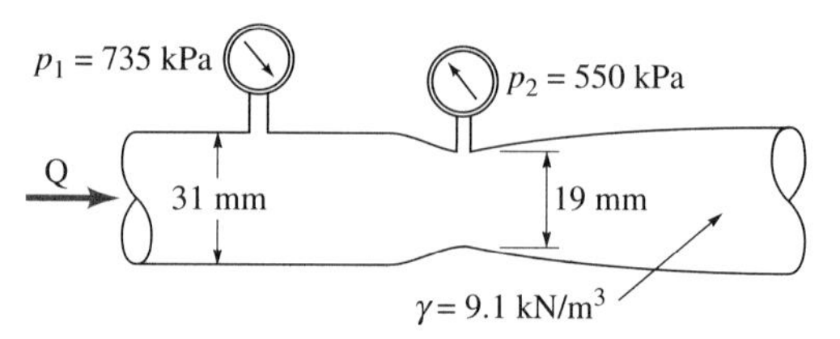


*Figure P12.32*

**12.33** (CD-ROM)

## Mechanical Energy Equation

**12.34** Water flows steadily from one location to another in the inclined pipe shown in Fig. P12.34. At one section, the static pressure is 8 psi. At the other section, the static pressure is 5 psi. Which way is the water flowing? Explain.

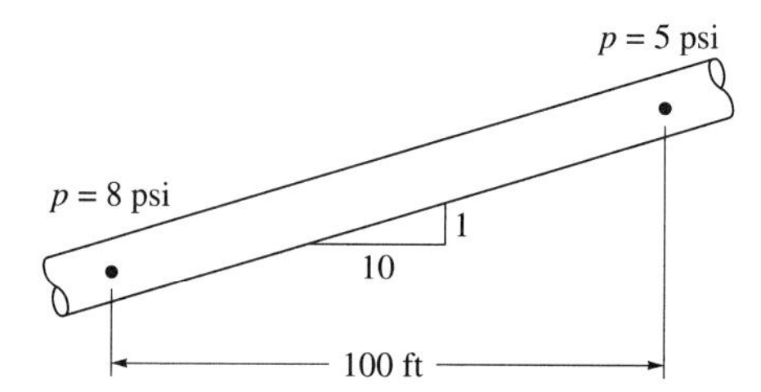


*Figure P12.34*

**12.35** Oil ($SG = 0.9$) flows downward through a vertical pipe contraction as shown in Fig. P12.35. If the mercury manometer reading, $h$, is 120 mm, determine the volumetric flow rate for frictionless flow. Is the actual flow rate more or less than the frictionless value? Explain.

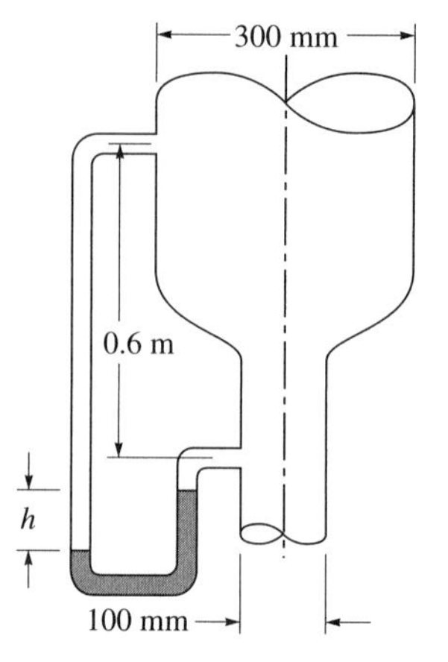


*Figure P12.35*

**12.36** (CD-ROM)

**12.37** A fire hose nozzle is designed to deliver water that will rise 30 m vertically. Calculate the stagnation pressure required at the nozzle inlet if (a) no loss is assumed, (b) a head loss of 10 m is assumed.

**12.38** (CD-ROM)

**12.39** (CD-ROM)

## Pumps and Turbines

**12.40** Water is to be moved from one large reservoir to another at a higher elevation, as indicated in Fig. P12.40. The head loss associated with 2.5 $ft^3/s$ being pumped from section (1) to (2) is $61V^2/2g$ ft, where V is the average velocity of water in the 8-in. inside diameter piping involved. Determine the amount of pumping power required.

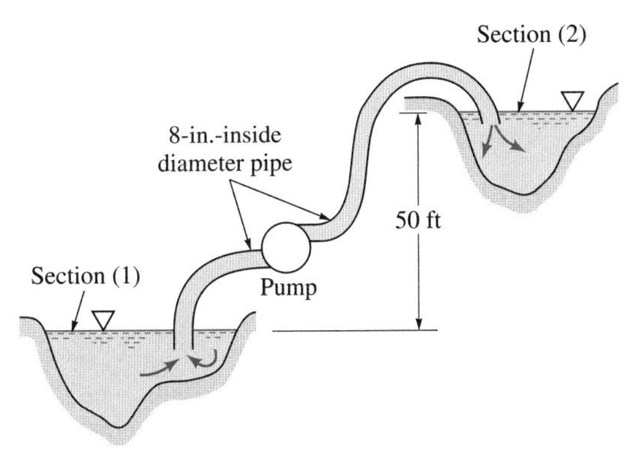


*Figure P12.40*

**12.41** Water flows by gravity from one lake to another as sketched in Fig. P12.41 at the steady rate of 100 gallons per minute. What is the head loss associated with this flow? If this same amount of head loss is associated with pumping the fluid from the lower lake to the higher one at the same flow rate, estimate the amount of pumping power required.

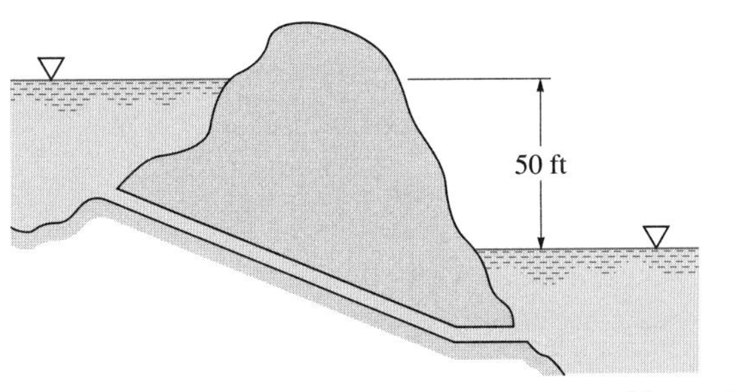


*Figure P12.41*

**12.42** (CD-ROM)

**12.43** Water is pumped from the tank shown in Fig. P12.43*a*. The head loss is known to be 1.2 $V^2/2g$, where V is the average velocity in the pipe. According to the pump manufacturer, the relationship between the pump head and the flow rate is as shown in Fig. P12.43*b*: $h_p = 20 - 2000\ Q^2$, where $h_p$ is in meters and Q is in $m^3/s$. Determine the volumetric flow rate, Q.

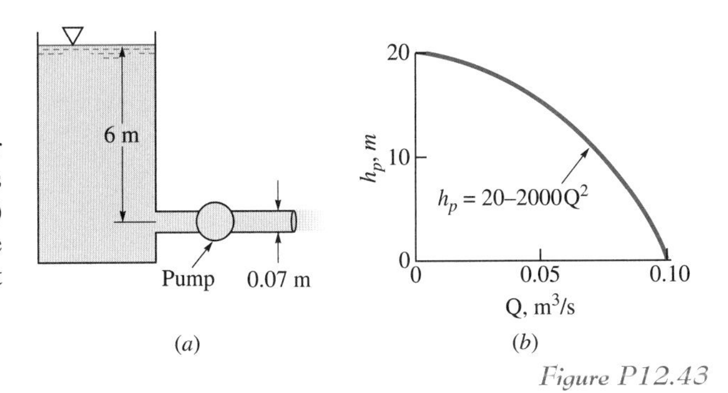


*Figure P12.43*

**12.44** (CD-ROM)

**12.45** Water flows through a hydroelectric turbine at a rate of 4 million gal/min. The elevation difference between the reservoir surface and the turbine outlet is 100 ft. What is the maximum amount of power output possible? Why will the actual amount be less?

**12.46** The turbine shown in Fig. P12.46 develops 100 hp when the volumetric flow rate of water is 20 $ft^3/s$. If all losses are negligible, determine (a) the elevation $h$, (b) the pressure difference across the turbine, and (c) if the turbine were removed, the volumetric flow rate expected.

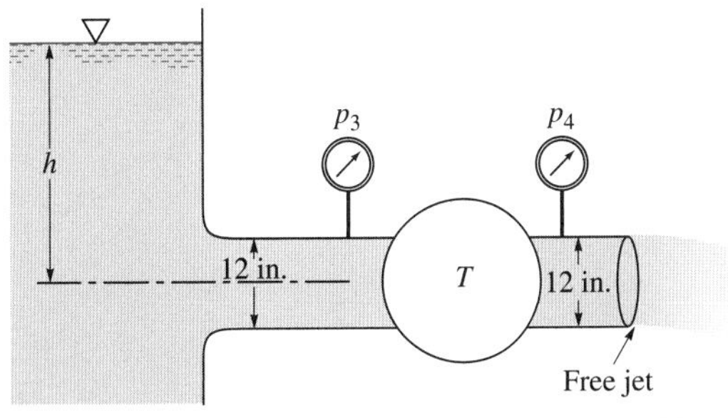


*Figure P12.46*

**12.47** A hydraulic turbine is provided with 4.25 $m^3/s$ of water at 415 kPa. A vacuum gage in the turbine discharge 3 m below the turbine inlet centerline reads 250 mm Hg vacuum. If the turbine shaft output power is 1100 kW, calculate the frictional power loss through the turbine. The supply and discharge pipe inside diameters are identically 800 mm.

**12.48** (CD-ROM)

## Compressible Flow

**12.49** (CD-ROM)

**12.50** (CD-ROM)

**12.51** (CD-ROM)

**12.52** (CD-ROM)

**12.53** (CD-ROM)

**12.54** (CD-ROM)

**12.55** (CD-ROM)

**12.56** (CD-ROM)

**12.57** (CD-ROM)

**12.58** (CD-ROM)

**12.59** (CD-ROM)

**12.60** (CD-ROM)

**12.61** (CD-ROM)

**12.62** (CD-ROM)

**12.63** (CD-ROM)

**12.64** (CD-ROM)

**12.65** (CD-ROM)

# SIMILITUDE, DIMENSIONAL ANALYSIS, AND MODELING

## *Introduction...*

There are a large number of practical engineering problems involving fluid mechanics that rely on experimentally obtained data for their solution. An obvious goal of any experiment is to make the results as widely applicable as possible. To achieve this end, the concept of *similitude* is often used so that measurements made on one system (for example, in the laboratory) can be used to describe the behavior of other similar systems (outside the laboratory). The laboratory systems are usually thought of as *models* and are used to study the phenomenon of interest under carefully controlled conditions. From these model studies, empirical formulations can be developed, or specific predictions of one or more characteristics of some other similar system can be made. To do this, it is necessary to establish the relationship between the laboratory model and the "other" system. The ***objective*** of this chapter is to determine how to use similitude, dimensional analysis, and modeling to simplify the experimental investigation of fluid mechanics problems.

*chapter objective*

## 13.1 Dimensional Analysis

To illustrate a typical fluid mechanics problem in which experimentation is required, consider the steady flow of an incompressible, viscous fluid through a long, smooth-walled, horizontal, circular pipe. An important characteristic of this system, which would be of interest to an engineer designing a pipeline, is the pressure drop per unit length that develops along the pipe as a result of friction. Although this would appear to be a relatively simple flow problem, it cannot generally be solved analytically (even using large computers) without the use of experimental data.

The first step in the planning of an experiment to study this problem would be to decide on the factors, or variables, that will have an effect on the pressure drop per unit length, $\Delta p_\ell$, which has units of $(\text{N/m}^2)/\text{m}$ or psi/ft, for example. On the basis of a prior analysis, including experimental observation, we expect the list of variables to include the pipe diameter, $D$, the fluid density, $\rho$, and fluid viscosity, $\mu$, and the mean velocity, V, at which the fluid is flowing through the pipe. (See Sec. 12.1.2 for a discussion about the fluid property viscosity, $\mu$.) Thus, we can express this relationship as

$$\Delta p_\ell = f(D, \rho, \mu, V) \tag{13.1}$$

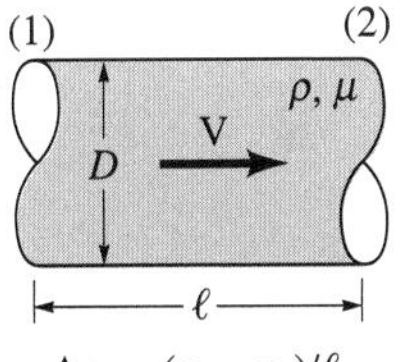

$\Delta p_\ell = (p_1 - p_2)/\ell$

which simply indicates mathematically that we expect the pressure drop per unit length to be some function of the factors contained within the parentheses. At this point the nature of the function is unknown. The objective of the experiments to be performed is to determine this function.

To perform the experiments in a meaningful and systematic manner, it would be necessary to change one of the variables, such as the velocity, while holding all others constant, and measure the corresponding pressure drop. This approach to determining the functional relationship between the pressure drop and the various factors that influence it, although logical in concept, is difficult. Some of the experiments would be hard to carry out—for example, it would be necessary to vary fluid density while holding viscosity constant. How would you do this? Finally, once we obtained the various curves, how could we combine these data to obtain the desired general functional relationship between $\Delta p_\ell$, $D$, $\rho$, $\mu$, and V that would be valid for any similar pipe system?

Fortunately, there is a much simpler approach to this problem that will eliminate the difficulties described above. In the following sections we will show that rather than working with the original list of variables, as described in Eq. 13.1, we can collect these into two combinations of variables called ***dimensionless products*** (or *dimensionless groups*) so that

*dimensionless products*

$$\frac{D\,\Delta p_\ell}{\rho V^2} = \phi\left(\frac{\rho VD}{\mu}\right) \tag{13.2}$$

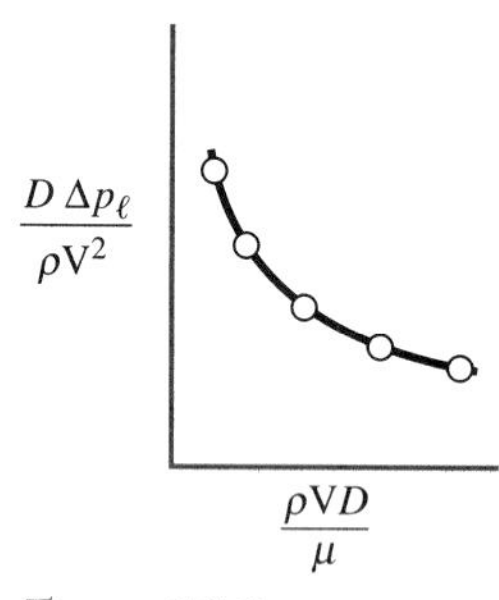

*Figure 13.1*

Thus, instead of having to work with five variables, we now have only two. The necessary experiment would simply consist of varying the dimensionless product $\rho VD/\mu$ and determining the corresponding value of $D\,\Delta p_\ell/\rho V^2$. The results of the experiment could then be represented by a single, universal curve (see Fig. 13.1). As described in the following section, the basis for this simplification lies in a consideration of the so-called *dimensions* of the variables involved in a given problem.

## 13.2 Dimensions, Dimensional Homogeneity, and Dimensional Analysis

### 13.2.1 Dimensions

Since in our study of fluid mechanics we will be dealing with a variety of fluid characteristics, it is necessary to develop a system for describing these characteristics both *qualitatively* and *quantitatively.* The qualitative aspect serves to identify the nature, or type, of the characteristics (such as length, time, stress, and velocity), whereas the quantitative aspect provides a numerical measure of the characteristics. The quantitative description requires both a number and a standard by which various quantities can be compared. A standard for length might be a meter or foot, for time an hour or second, and for mass a slug or kilogram. Such standards are called ***units,*** and systems of units in common use are as described in Chapter 2. The qualitative description is conveniently given in terms of certain *primary quantities,* such as length, L, time, t, mass, M, and temperature, T. These primary quantities can then be used to provide a qualitative description of any other *secondary quantity,* for example, area $\doteq L^2$, velocity $\doteq Lt^{-1}$, density $\doteq ML^{-3}$, and so on, where the symbol $\doteq$ is used to indicate the ***dimensions*** of the secondary quantity in terms of the primary quantities. Thus, to describe qualitatively a velocity, V, we would write

*units*

*dimensions*

$$V \doteq Lt^{-1}$$

and say that "the dimensions of a velocity equal length divided by time." The primary quantities are also referred to as ***basic dimensions.***

*basic dimensions*

For a wide variety of problems involving fluid mechanics, only the three basic dimensions, L, t, and M, are required. Alternatively, L, t, and F could be used, where F is the basic dimension of force. Since Newton's law states that force is equal to mass times acceleration, it follows that $F \doteq MLt^{-2}$ or $M \doteq FL^{-1}t^{2}$. Thus, secondary quantities expressed in terms of M can be expressed in terms of F through the relationship above. For example, pressure, $p$, is a force per unit area, so that $p \doteq FL^{-2}$, but an equivalent dimensional equation is $p \doteq ML^{-1}t^{-2}$. Table 13.1 provides a list of dimensions for a number of common physical quantities that occur frequently in fluid mechanics. It should be noted that in thermodynamic and heat transfer analyses, an additional basic dimension, temperature, T, is often involved. Therefore, in such cases there are often four basic dimensions: L, t, M, and T rather than the three basic dimensions L, t, and M considered in this fluid mechanics chapter.

The dimensions of the variables in the pipe flow example are $\Delta p_\ell \doteq FL^{-3}$, $D \doteq L$, $\rho \doteq FL^{-4}t^{2}$, $\mu \doteq FL^{-2}t$, and $V \doteq Lt^{-1}$. A quick check of the dimensions of the two groups that appear in Eq. 13.2 shows that they are in fact *dimensionless* products; that is,

$$\frac{D\,\Delta p_\ell}{\rho V^2} \doteq \frac{L(FL^{-3})}{(FL^{-4}t^{2})(Lt^{-1})^{2}} \doteq F^0L^0t^0$$

and

$$\frac{\rho V D}{\mu} \doteq \frac{(FL^{-4}t^{2})(LT^{-1})(L)}{(FL^{-2}t)} \doteq F^0L^0t^0$$

Not only have we reduced the numbers of variables from five to two, but the new groups are dimensionless combinations of variables, which means that the results will be independent of the system of units we choose to use.

**Table 13.1** Dimensions Associated with Common Fluid Mechanics Physical Quantities

| | FLt System | MLt System |
|---|---|---|
| Acceleration | $Lt^{-2}$ | $Lt^{-2}$ |
| Angle | $F^0L^0t^0$ | $M^0L^0t^0$ |
| Angular velocity | $t^{-1}$ | $t^{-1}$ |
| Area | $L^2$ | $L^2$ |
| Density | $FL^{-4}t^2$ | $ML^{-3}$ |
| Energy | FL | $ML^2t^{-2}$ |
| Force | F | $MLt^{-2}$ |
| Frequency | $t^{-1}$ | $t^{-1}$ |
| Length | L | L |
| Mass | $FL^{-1}t^2$ | M |
| Moment of a force | FL | $ML^2t^{-2}$ |
| Moment of inertia (area) | $L^4$ | $L^4$ |
| Momentum | Ft | $MLt^{-1}$ |
| Power | $FLt^{-1}$ | $ML^2t^{-3}$ |
| Pressure | $FL^{-2}$ | $ML^{-1}t^{-2}$ |
| Specific weight | $FL^{-3}$ | $ML^{-2}t^{-2}$ |
| Stress | $FL^{-2}$ | $ML^{-1}t^{-2}$ |
| Time | t | t |
| Torque | FL | $ML^2t^{-2}$ |
| Velocity | $Lt^{-1}$ | $Lt^{-1}$ |
| Viscosity (dynamic) | $FL^{-2}t$ | $ML^{-1}t^{-1}$ |
| Viscosity (kinematic) | $L^2t^{-1}$ | $L^2t^{-1}$ |
| Volume | $L^3$ | $L^3$ |
| Work | FL | $ML^2t^{-2}$ |

### 13.2.2 Dimensional Homogeneity

*dimensionally homogeneous*

We accept as a fundamental premise that all equations describing physical phenomena are ***dimensionally homogeneous.*** That is, the dimensions of the left side of the equation must be the same as those on the right side, and all additive separate terms must have the same dimensions. ***For Example...*** the equation for the velocity, V, of a uniformly accelerated body is

$$V = V_0 + at \tag{13.3}$$

where $V_0$ is the initial velocity, $a$ the acceleration, and $t$ the time. In terms of dimensions the equation is

$$Lt^{-1} \doteq Lt^{-1} + Lt^{-2}t$$

and thus Eq. 13.3 is dimensionally homogeneous. ▲

### 13.2.3 Dimensional Analysis

*dimensional analysis*

The use of dimensions, along with the concept of dimensional homogeneity, forms the foundation of a very useful approach for investigating a wide variety of engineering problems. This approach, generally referred to as ***dimensional analysis,*** is based on the fact that the number of dimensionless variables needed to describe a physical phenomenon is fewer than the number of physical (dimensional) variables needed to describe it. Any reduction in the number of required variables represents a considerable simplification in the analysis of the problem.

*For Example...* in the problem involving the pressure drop per unit length of smooth pipe discussed in Sec. 13.1, there are 5 physical variables ($\Delta p_\ell$, $D$, $\rho$, $\mu$, and V) and 2 dimensionless products ($\rho VD/\mu$ and $D\Delta p_\ell/\rho V^2$). As shown at the beginning of this section, it takes three basic dimensions (either F, L, t or M, L, t) to describe the physical variables of this problem. Note that the difference between the number of physical variables (5) and the number of basic dimensions needed to describe them (3) is equal to the number of dimensionless products needed to describe the physical phenomenon (2). ▲

As noted above, the number of dimensionless groups needed to describe a phenomenon is fewer than the number of the original physical variables. That is, if it takes $k$ physical variables to describe the problem and these variables involve $r$ basic dimensions, then the problem can be described in terms of $k - r$ dimensionless products. These dimensionless products are called *pi terms.* In the above example,

$$\Pi_1 = D\Delta p_\ell/\rho V^2 \quad \text{and} \quad \Pi_2 = \rho VD/\mu$$

where $\Pi_1$ is some function of $\Pi_2$. That is, $\Pi_1 = \varphi(\Pi_2)$. The functional relationship between $\Pi_1$ and $\Pi_2$ (i.e., the shape of the curve shown in Fig. 13.1) is unknown until a detailed analysis or experiment is carried out.

The number of pi terms needed to describe a flow depends on the particular situation involved. *For Example...* consider the aerodynamic drag, $\mathscr{D}$, on a high-speed airplane. If we assume that the drag for a given shaped airplane is a function of its length, $\ell$, the speed, V, at which it flies, the density, $\rho$, and viscosity, $\mu$, of the air, and the speed of sound in the air, $c$, then $k = 6$. That is,

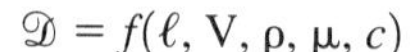

$$\mathscr{D} = f(\ell, V, \rho, \mu, c)$$

It takes six physical variables to describe this situation. Also, $r = 3$. That is, the physical variables can be described in terms of 3 basic dimensions (either F, L, t or M, L, t). Thus, $k - r = 6 - 3 = 3$. The flow can be described in terms of 3 pi terms as

$$\Pi_1 = \varphi(\Pi_2, \Pi_3)$$

where, for example,

$$\Pi_1 = \frac{\mathscr{D}}{\frac{1}{2}\rho V^2 \ell^2}, \; \Pi_2 = \frac{\rho V \ell}{\mu}, \text{ and } \Pi_3 = \frac{V}{c}$$

These three pi terms occur quite often in fluid mechanics and are named the *drag coefficient,* $C_D$, the *Reynolds number, Re*, and the *Mach number, M*, respectively. More information about these and other commonly used pi terms is given in Sec. 13.5. ▲

## 13.3 Buckingham Pi Theorem and Pi Terms

A fundamental question in dimensional analysis is how many dimensionless products are required to replace the original list of variables? The answer to this question is supplied by the basic theorem of dimensional analysis that states the following:

> If an equation involving $k$ variables is dimensionally homogeneous, it can be reduced to a relationship among $k - r$ independent dimensionless products, where $r$ is the minimum number of basic dimensions required to describe the variables.

*Buckingham pi theorem*

The dimensionless products are frequently referred to as "pi terms," and the theorem is called the ***Buckingham pi theorem.*** Buckingham used the symbol $\Pi$ to represent a dimensionless product, and this notation is commonly used. Although the pi theorem is a simple one, its proof is not so simple and we will not include it here.

The pi theorem is based on the idea of dimensional homogeneity. Essentially we assume that for any physically meaningful equation involving $k$ variables, such as

$$u_1 = f(u_2, u_3, \ldots, u_k)$$

the dimensions of the variable on the left side of the equal sign must be equal to the dimensions of any term that stands by itself on the right side of the equal sign. It then follows that we can rearrange the equation into a set of dimensionless products (***pi terms***) so that

*pi terms*

$$\Pi_1 = \phi(\Pi_2, \Pi_3, \ldots, \Pi_{k-r})$$

The required number of pi terms is fewer than the number of original variables by $r$, where $r$ is the minimum number of basic dimensions required to describe the original list of variables. Usually in fluid mechanics the basic dimensions required to describe the variables will be the basic dimensions M, L, and t or F, L, and t. However, in some instances perhaps only two dimensions, such as L and t, are required, or maybe just one, such as L.

Since the only restriction placed on the pi terms are that they be (1) correct in number, (2) dimensionless, and (3) independent, it is usually possible simply to form the pi terms by inspection. ***For Example...*** to illustrate this approach, we again consider the pressure drop per unit length along a smooth pipe (see Sec. 13.1). Regardless of the technique to be used, the starting point remains the same—determine the physical variables, which in this case are

$$\Delta p_\ell = f(D, \rho, \mu, V)$$

Next, by inspection of Table 13.1, the dimensions of the variables are listed:

$$\Delta p_\ell \doteq FL^{-3}$$
$$D \doteq L$$
$$\rho \doteq FL^{-4}t^2$$
$$\mu \doteq FL^{-2}t$$
$$V \doteq Lt^{-1}$$

Since there are $k = 5$ variables involving $r = 3$ basic dimensions, this relationship can be written in terms of $k - r = 2$ dimensionless pi terms.

Once the number of pi terms is known, we can form each pi term by inspection, simply making use of the fact that each pi term must be dimensionless. *We will always let $\Pi_1$ contain the dependent variable,* which in this example is $\Delta p_\ell$. Since this variable has the dimensions $FL^{-3}$, we need to combine it with other variables so that a dimensionless product will result. For example, divide $\Delta p_\ell$ by $\rho$ (to eliminate F), then by $V^2$ (to eliminate t), and finally multiply by $D$ (to eliminate L) to obtain

$$\Pi_1 = \frac{\Delta p_\ell D}{\rho V^2}$$

Next, we will form the second pi term by selecting the variable that was not used in $\Pi_1$, which in this case is $\mu$. We simply combine $\mu$ with the other variables to make the combination dimensionless (but do not use $\Delta p_\ell$ in $\Pi_2$, since we want the dependent variable to appear only in $\Pi_1$). For example, divide $\mu$ by $\rho$ (to eliminate F), then by V (to eliminate t), and finally by $D$ (to eliminate L). Thus,

$$\Pi_2 = \frac{\mu}{\rho V D}$$

and, therefore,

$$\frac{\Delta p_\ell D}{\rho V^2} = \tilde{\phi}\left(\frac{\mu}{\rho V D}\right)$$

Since the $\Pi_2$ term given above is dimensionless, it follows that its inverse, $\rho V D/\mu$, is also dimensionless. Hence, both the representation given above and that in Eq. 13.2 are equally valid dimensionless relationships. ▲

## 13.4 Method of Repeating Variables

Several methods can be used to form the dimensionless products, or pi terms, that arise in a dimensional analysis. Essentially we are looking for a method that will allow us to systematically form the pi terms so that we are sure that they are dimensionless and independent, and that we have the right number. The method we will describe in detail in this section is called the ***method of repeating variables.***

*method of repeating variables*

It will be helpful to break the repeating variable method down into a series of distinct steps that can be followed for any given problem. With a little practice you will be able to readily complete a dimensional analysis for your problem.

**Step 1.** **List all the variables that are involved in the problem.** This step is the most difficult one and it is, of course, vitally important that all pertinent variables be included.

**Step 2.** **Express each of the variables in terms of basic dimensions.** For the typical fluid mechanics problem the basic dimensions will be either M, L, t or F, L, t.

**Step 3.** **Determine the required number of pi terms.** This can be accomplished by means of the Buckingham pi theorem, which indicates that the number of pi terms is equal to $k - r$, where $k$ is the number of variables in the problem (which is determined from Step 1) and $r$ is the number of basic dimensions required to describe these variables (which is determined from Step 2).

**Step 4.** **Select the repeating variables, where the number required is equal to the number of basic dimensions.** What we are doing here is selecting from the original list of variables several that can be combined with each of the remaining variables to form a pi term.

**Step 5.** **Form a pi term by multiplying one of the nonrepeating variables by the product of the repeating variables, each raised to an exponent that will make the combination dimensionless.** Each pi term will be of the form $u_i u_1^{a_i} u_2^{b_i} u_3^{c_i}$, where $u_i$ is one of the nonrepeating variables; $u_1$, $u_2$, and $u_3$ are the repeating variables; and the exponents $a_i$, $b_i$, and $c_i$ are determined so that the combination is dimensionless.

**Step 6.** **Repeat Step 5 for each of the remaining nonrepeating variables.** The resulting set of pi terms will correspond to the required number obtained from Step 3.

**Step 7.** **Check all the resulting pi terms carefully to make sure they are dimensionless.**

**Step 8.** **Express the final form as a relationship among the pi terms, and think about what it means.** Typically the final form can be written as

$$\Pi_1 = \phi(\Pi_2, \Pi_3, \ldots, \Pi_{k-r})$$

where $\Pi_1$ would contain the dependent variable in the numerator.

***For Example...*** to illustrate these various steps we will again consider the problem discussed earlier in this chapter involving the steady flow of an incompressible, viscous fluid through a long, smooth-walled-horizontal, circular pipe. We are interested in the pressure drop per unit length, $\Delta p_\ell$, along the pipe. According to Step 1 we must list all of the pertinent variables that are involved based on the experimenter's knowledge of the problem. In this problem we assume that

$$\Delta p_\ell = f(D, \rho, \mu, \mathrm{V})$$

where $D$ is the pipe diameter, $\rho$ and $\mu$ are the fluid density and viscosity, respectively, and V is the mean velocity.

Next (Step 2), using Table 13.1 we express all the variables in terms of basic dimensions. For F, L, and t as basic dimensions it follows that

$$\begin{aligned} \Delta p_\ell &\doteq (\mathrm{FL}^{-2})/\mathrm{L} = \mathrm{FL}^{-3} \\ D &\doteq \mathrm{L} \\ \rho &\doteq \mathrm{FL}^{-4}\mathrm{t}^2 \\ \mu &\doteq \mathrm{FL}^{-2}\mathrm{t} \\ \mathrm{V} &\doteq \mathrm{Lt}^{-1} \end{aligned}$$

We could also use M, L, and t as basic dimensions if desired—the final result will be the same. Do not mix the basic dimensions; that is, use either F, L, and t or M, L, and t.

We can now apply the pi theorem to determine the required number of pi terms (Step 3). An inspection of the dimensions of the variables from Step 2 reveals that three basic dimensions are required to describe the variables. Since there are five ($k = 5$) variables (do not forget to count the dependent variable, $\Delta p_\ell$) and three required dimensions ($r = 3$), then according to the pi theorem there will be $(5 - 3)$, or two pi terms required.

The repeating variables to be used to form the pi terms (Step 4) need to be selected from the list $D$, $\rho$, $\mu$, and V. We do not want to use the dependent variable as one of the repeating variables. Since three dimensions are required, we will need to select three repeating variables. Generally, we would try to select for repeating variables those that are the simplest, dimensionally. For example, if one of the variables has the dimension of a length, choose it as one of the repeating variables. In this example we will use $D$, V, and $\rho$ as repeating variables. Note that these are dimensionally independent, since $D$ is a length, V involves both length and time, and $\rho$ involves force, length, and time.

We are now ready to form the two pi terms (Step 5). Typically, we would start with the dependent variable and combine it with the repeating variables to form the first pi term. That is,

$$\Pi_1 = \Delta p_\ell D^a V^b \rho^c$$

Since this combination is to be dimensionless, it follows that

$$(FL^{-3})(L)^a(Lt^{-1})^b(FL^{-4}t^2)^c \doteq F^0L^0t^0$$

The exponents $a$, $b$, and $c$ must be determined such that the resulting exponent for each of the basic dimensions—F, L, and t—must be zero (so that the resulting combination is dimensionless). Thus, we can write

$$\begin{aligned} 1 + c &= 0 \qquad \text{(for F)} \\ -3 + a + b - 4c &= 0 \qquad \text{(for L)} \\ -b + 2c &= 0 \qquad \text{(for t)} \end{aligned}$$

The solution of this system of equations gives the desired values for $a$, $b$, and $c$. It follows that $a = 1$, $b = -2$, $c = -1$, and, therefore,

$$\Pi_1 = \frac{\Delta p_\ell D}{\rho V^2}$$

The process is now repeated for the remaining nonrepeating variables (Step 6). In this example there is only one additional variable ($\mu$) so that

$$\Pi_2 = \mu D^a V^b \rho^c$$

or

$$(FL^{-2}t)(L)^a(Lt^{-1})^b(FL^{-4}t^2)^c \doteq F^0L^0t^0$$

and, therefore,

$$\begin{aligned} 1 + c &= 0 \qquad \text{(for F)} \\ -2 + a + b - 4c &= 0 \qquad \text{(for L)} \\ 1 - b + 2c &= 0 \qquad \text{(for t)} \end{aligned}$$

Solving these equations simultaneously it follows that $a = -1$, $b = -1$, $c = -1$ so that

$$\Pi_2 = \frac{\mu}{DV\rho}$$

Note that we end up with the correct number of pi terms as determined from Step 3.

At this point stop and check to make sure the pi terms are actually dimensionless (Step 7). Finally (Step 8), we can express the result of the dimensional analysis as

$$\frac{\Delta p_\ell D}{\rho V^2} = \tilde{\phi}\left(\frac{\mu}{DV\rho}\right)$$

This result indicates that this problem can be studied in terms of these two pi terms, rather than the original five variables we started with. However, dimensional analysis will *not* provide the form of the function $\tilde{\phi}$. This can only be obtained from a suitable set of experiments. If desired, the pi terms can be rearranged; that is, the reciprocal of $\mu/DV\rho$ could be used, and of course the order in which we write the variables can be changed. Thus, for example, $\Pi_2$ could be expressed as

$$\Pi_2 = \frac{\rho V D}{\mu}$$

and the relationship between $\Pi_1$ and $\Pi_2$ as

$$\frac{D\,\Delta p_\ell}{\rho V^2} = \phi\left(\frac{\rho V D}{\mu}\right)$$

This is the form we previously used in our initial discussion of this problem (Eq. 13.2). The dimensionless product, $\rho VD/\mu$, is an important one in fluid mechanics—the Reynolds number. ▲

## 13.5 Common Dimensionless Groups in Fluid Mechanics

The heading of Table 13.2 lists variables that commonly arise in fluid mechanics problems. The list is not exhaustive, but does indicate a broad range of variables likely to be found in typical problems. Fortunately, not all of these variables would be encountered in each problem. However, when these variables are present, it is standard practice to combine them into some of the common dimensionless groups (pi terms) given in Table 13.2. These combinations appear so frequently that special names are associated with them as indicated in the table.

It is often possible to provide a physical interpretation to the dimensionless groups. This can be helpful in assessing their influence in a particular application. The interpretation is based on the ratio of typical forces such as weight, viscous (friction) force, and inertial force (mass times acceleration) that may be important in a particular flow. Thus, as indicated in Table 13.2, the Reynolds number, *Re*, one of the most important dimensionless parameters in fluid mechanics, represents a measure of the ratio of inertial to viscous effects. If the

***V13.1 Reynolds number***

**Table 13.2** Some Common Variables and Dimensionless Groups in Fluid Mechanics

Variables: Acceleration of gravity, $g$; Characteristic length, $\ell$; Density, $\rho$; Pressure, $p$ (or $\Delta p$); Speed of sound, $c$; Velocity, V; Viscosity, $\mu$

| Dimensionless Groups | Name | Interpretation | Types of Applications |
|---|---|---|---|
| $\frac{\rho V \ell}{\mu}$ | Reynolds number, *Re* | $\frac{\text{inertia force}}{\text{viscous force}}$ | Generally important in all types of fluid mechanics problems |
| $\frac{V}{\sqrt{g\ell}}$ | Froude number, *Fr* | $\frac{\text{inertia force}}{\text{gravitational force}}$ | Flow with a free surface |
| $\frac{p}{\rho V^2}$ | Euler number, *Eu* | $\frac{\text{pressure force}}{\text{inertia force}}$ | Problems in which pressure or pressure differences are important |
| $\frac{V}{c}$ | Mach number, *M* | $\frac{\text{inertia force}}{\text{compressibility force}}$ | Flows in which variation of fluid density is important |

Reynolds number for a given flow is large, the inertia of the fluid is dominant over the viscous effects. For flows with small *Re*, viscous effects are dominant. ***For Example...*** when one stirs cream into a mug of coffee, the Reynolds number is on the order of $Re = \rho V\ell/\mu \approx 10{,}000$. Inertia is dominant and the coffee continues to swirl in the mug after the spoon is removed. On the other hand, the Reynolds number associated with stirring a mug of highly viscous molasses would be on the order of $Re \approx 1$ and the motion would cease almost immediately upon removal of the spoon. ▲

Other common dimensionless parameters and their corresponding force ratios are indicated in Table 13.2.

## 13.6 Correlation of Experimental Data

One of the most important uses of dimensional analysis is as an aid in the efficient handling, interpretation, and correlation of experimental data. Since the field of fluid mechanics relies heavily on empirical data, it is not surprising that dimensional analysis is such an important tool in this field. As noted previously, a dimensional analysis cannot give a complete answer to any given problem, since the analysis only provides the dimensionless groups describing the phenomenon, and not the specific relationship among the groups. To determine this relationship, suitable experimental data must be obtained. The degree of difficulty involved in this process depends on the number of pi terms, and the nature of the experiments. The simplest problems are obviously those involving the fewest pi terms. For example, if a given phenomenon can be described with two pi terms such that

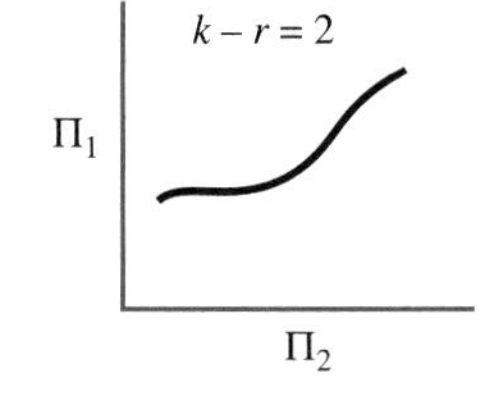


$$\Pi_1 = \phi(\Pi_2)$$

the functional relationship among the variables can then be determined by varying $\Pi_2$ and measuring the corresponding values of $\Pi_1$. For this case the results can be conveniently presented in graphical form by plotting $\Pi_1$ versus $\Pi_2$. It should be emphasized that the resulting curve would be a "universal" one for the particular phenomenon studied. This means that if the variables and the resulting dimensional analysis are correct, then there is only a single relationship between $\Pi_1$ and $\Pi_2$.

In addition to presenting the data graphically, it may be possible (and desirable) to obtain an empirical equation relating $\Pi_1$ and $\Pi_2$ by using a standard curve fitting technique. These considerations are illustrated in Example 13.1.

### *Example 13.1* Correlation of Pipe Flow Data

The relationship between the pressure drop per unit length along a smooth walled, horizontal pipe and the variables that affect the pressure drop is to be determined experimentally. In the laboratory the pressure drop was measured over a 5-ft length of smooth walled pipe having an inside diameter of 0.496 in. The fluid used was water at 60°F ($\mu = 2.34 \times 10^{-5}$ lbf · s/ft$^2$, $\rho = 1.94$ slug/ft$^3$). A total of eight tests was run in which the velocity was varied and the corresponding pressure drop measured. The results of these tests are shown below:

| Test | 1 | 2 | 3 | 4 | 5 | 6 | 7 | 8 |
|---|---|---|---|---|---|---|---|---|
| Velocity (ft/s) | 1.17 | 1.95 | 2.91 | 5.84 | 11.13 | 16.92 | 23.34 | 28.73 |
| Pressure drop (lbf/ft$^2$) (for 5-ft length) | 6.26 | 15.6 | 30.9 | 106 | 329 | 681 | 1200 | 1730 |

Use these data to obtain a general relationship between the pressure drop per unit length and the other variables.

## Solution

***Known:*** Experimental data relating the pressure drop and velocity for the flow of water through a smooth walled pipe.
***Find:*** Based on the given experimental data, determine an empirical relationship between the pressure drop per unit length and the other variables that affect the pressure drop.

***Assumptions:***
1. The variables used in the analysis are correct, i.e., we have not included any extraneous variables or omitted any important ones.
2. The experimental data are accurate.

***Analysis:*** The first step is to perform a dimensional analysis during the planning stage *before* the experiments are actually run. As was discussed in Sec. 13.1, we will assume that the pressure drop per unit length, $\Delta p_\ell$, is a function of the pipe diameter, $D$, fluid density, $\rho$, fluid viscosity, $\mu$, and the velocity, V. Thus,

$$\Delta p_\ell = f(D, \rho, \mu, V)$$

which, as shown previously, can be written in dimensionless form as

$$\frac{D\,\Delta p_\ell}{\rho V^2} = \phi\left(\frac{\rho V D}{\mu}\right)$$

To determine the form of the relationship, we need to vary the Reynolds number, $\rho VD/\mu$, and to measure the corresponding values of $D\,\Delta p_\ell/\rho V^2$. The Reynolds number could be varied by changing any one of the variables, $\rho$, V, $D$, or $\mu$, or any combination of them. However, the simplest way to do this is to vary the velocity, since this will allow us to use the same fluid and pipe. Based on the data given, values for the two pi terms can be computed with the result:

| Test | 1 | 2 | 3 | 4 | 5 | 6 | 7 | 8 |
|---|---|---|---|---|---|---|---|---|
| $D\,\Delta p_\ell/\rho V^2$ | 0.0195 | 0.0175 | 0.0155 | 0.0132 | 0.0113 | 0.0101 | 0.00939 | 0.00893 |
| $\rho VD/\mu$ | $4.01 \times 10^3$ | $6.68 \times 10^3$ | $9.97 \times 10^3$ | $2.00 \times 10^4$ | $3.81 \times 10^4$ | $5.80 \times 10^4$ | $8.00 \times 10^4$ | $9.85 \times 10^4$ |

These are dimensionless groups so their values are independent of the system of units used as long as a consistent system is used. For example, if the velocity is in ft/s, then the diameter should be in feet, not inches or meters.

A plot of these two pi terms can now be made with the results shown in Fig. E13.1*a*. The correlation appears to be quite good, and if it was not, this would suggest that either we had large experimental measurement errors, or that we had perhaps omitted an important variable. The curve shown in Fig. E13.1*a* represents the general relationship between the pressure drop and the other factors in the range of Reynolds numbers between $4.01 \times 10^3$ and $9.85 \times 10^4$. Thus, for this range of Reynolds numbers it is *not* necessary to repeat the tests for other pipe sizes or other fluids provided the assumed independent variables ($D$, $\rho$, $\mu$, V) are the only important ones.

Since the relationship between $\Pi_1$ and $\Pi_2$ is nonlinear, it is not immediately obvious what form of empirical equation might be used to describe the relationship. If, however, the same data are plotted logarithmically, as shown in Fig. E13.1*b*, the data form a straight line, suggesting that a suitable equation is of the form $\Pi_1 = A\Pi_2^n$ where $A$ and $n$ are empirical constants to be determined from the data by using a suitable curve fitting technique, such as a nonlinear regression program. For the data given in this example, a good fit of the data is obtained with the equation.

$$\Pi_1 = 0.150\,\Pi_2^{-0.25}$$

giving

❶
$$\frac{D\,\Delta p_\ell}{\rho V^2} = 0.150\left(\frac{\rho V D}{\mu}\right)^{-1/4} \quad \triangleleft$$

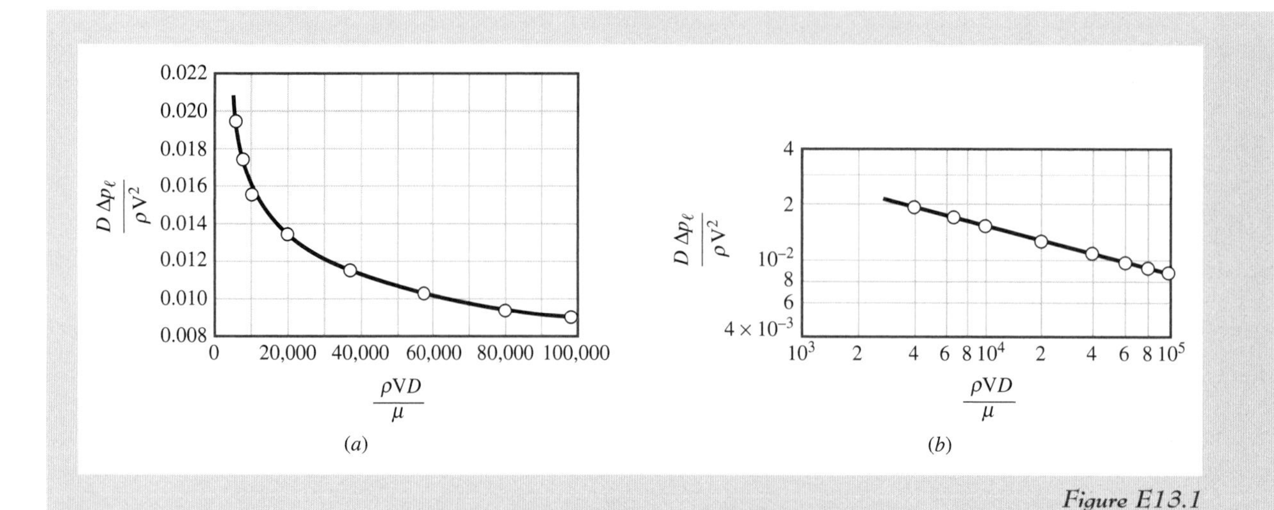


*Figure E13.1*

❶ In 1911, H. Blasius, a German fluid mechanician, established a similar empirical equation that is used widely for predicting the pressure drop in smooth pipes in the range $4 \times 10^3 < Re < 10^5$. This equation can be expressed in the form

$$\frac{D\,\Delta p_\ell}{\rho V^2} = 0.1582 \left(\frac{\rho VD}{\mu}\right)^{-1/4}$$

This so-called Blasius formula is based on numerous experimental results of the type used in this example. Flow in pipes is discussed in more detail in the next chapter, where it is shown how pipe roughness (which introduces another variable) may affect the results given in this example, which is for smooth walled pipes.

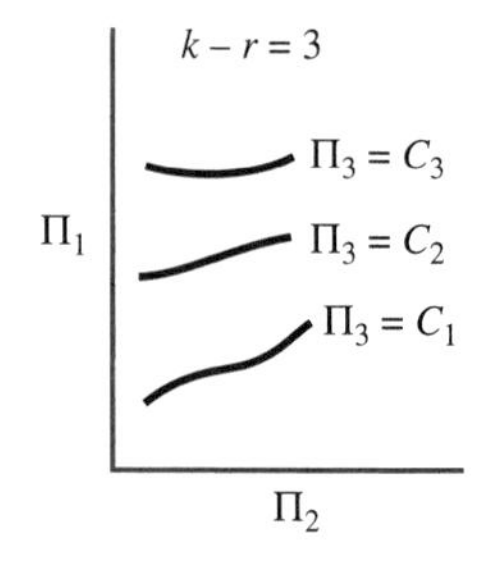


As the number of required pi terms increases, it becomes more difficult to display the results in a convenient graphical form and to determine a specific empirical equation that describes the phenomenon. For problems involving three pi terms

$$\Pi_1 = \phi(\Pi_2, \Pi_3)$$

it is still possible to show data correlations on simple graphs by plotting families of curves. This is an informative and useful way of representing the data in a general way. (The generalized compressibility diagram of Fig. 4.9 is an example.) It may also be possible to determine a suitable empirical equation relating the three pi terms. However, as the number of pi terms continues to increase, corresponding to an increase in the general complexity of the problem of interest, both the graphical presentation and the determination of a suitable empirical equation become unmanageable. For these more complicated problems, it is often more feasible to use *models* to predict specific characteristics of the system rather than to try to develop general correlations. The concept of modeling is discussed in the next section.

## 13.7 Modeling and Similitude

*model*

Models are widely used in fluid mechanics. Major engineering projects involving structures, aircraft, ships, rivers, harbors, dams, air and water pollution, and so on, frequently involve the use of models. Although the term "model" is used in many different contexts, the "engineering model" generally conforms to the following definition. *A **model** is a representation of a physical system that can be used to predict the behavior of the system in some desired respect.* The physical system for which the predictions are to be made is called the

***prototype.*** Although *mathematical* or *computer* models may also conform to this definition, our interest will be in physical models, that is, models that resemble the prototype but are generally of a different size, may involve different fluids, and often operate under different conditions (pressures, velocities, etc.). Usually a model is smaller than the prototype. Therefore, it is more easily handled in the laboratory and less expensive to construct and operate than the larger prototype. With the successful development of a valid model, it is possible to predict the behavior of the prototype under a certain set of conditions.

*prototype*

***V13.2 Environmental models***

In the following paragraphs we develop procedures for designing models so that the model and prototype will behave in a similar fashion. The theory of models can be readily developed by using the principles of dimensional analysis. Thus, as discussed in Sec. 13.3, consider a problem that can be described in terms of a set of pi terms as

$$\Pi_1 = \phi(\Pi_2, \Pi_3, \ldots, \Pi_n) \tag{13.4}$$

***V13.3 Wind tunnel train model***

In formulating this relationship, only a knowledge of the general nature of the physical phenomenon, and the variables involved, is required. Specific values for variables (size of components, fluid properties, and so on) are not needed to perform the dimensional analysis. Accordingly, Eq. 13.4 applies to any system that is governed by the same variables. If Eq. 13.4 describes the behavior of a particular prototype, a similar relationship can be written for a model of this prototype; that is,

$$\Pi_{1m} = \phi(\Pi_{2m}, \Pi_{3m}, \ldots, \Pi_{nm}) \tag{13.5}$$

where the form of the function will be the same as long as the same phenomenon is involved in both the prototype and the model. Variables, or pi terms, without a subscript $m$ will refer to the prototype, whereas the subscript $m$ will be used to designate the model variables or pi terms.

The pi terms can be developed so that $\Pi_1$ contains the variable that is to be predicted from observations made on the model. Therefore, if the model is designed and operated under the following *model design conditions,* also called *similarity requirements* or ***modeling laws,***

$$\begin{aligned} \Pi_{2m} &= \Pi_2 \\ \Pi_{3m} &= \Pi_3 \\ &\vdots \\ \Pi_{nm} &= \Pi_n \end{aligned} \tag{13.6}$$

*modeling laws*

and since the form of $\phi$ is the same for model and prototype, it follows that

$$\Pi_1 = \Pi_{1m} \tag{13.7}$$

*prediction equation*

Equation 13.7 is the desired ***prediction equation*** and indicates that the measured value of $\Pi_{1m}$ obtained with the model will be equal to the corresponding $\Pi_1$ for the prototype as long as the other pi terms are equal.

To illustrate the procedure, consider the problem of determining the drag, $\mathscr{D}$, on a thin rectangular plate ($w \times h$ in size) placed normal to a fluid with velocity, V. Assume that the drag is a function of $w$, $h$, V, the fluid viscosity, $\mu$, and fluid density, $\rho$, so that

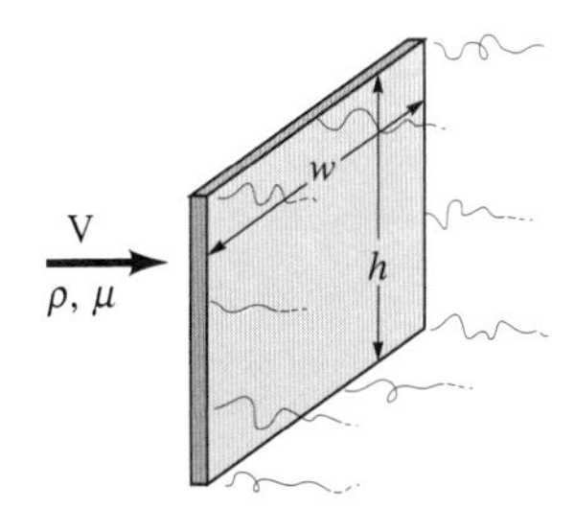


$$\mathscr{D} = f(w, h, \mu, \rho, V)$$

Since this problem involves 6 physical variables that can be described in terms of 3 basic dimensions (M, L, t *or* F, L, t), it can be written in terms of 3 pi terms as

$$\frac{\mathscr{D}}{w^2 \rho V^2} = \phi\left(\frac{w}{h}, \frac{\rho V w}{\mu}\right) \tag{13.8}$$

where $\rho V w / \mu$ is recognized as the Reynolds number.

*V13.4 Wind engineering models*

We are now concerned with designing a model that could be used to predict the drag on a certain prototype (which presumably has a different size than the model). Since the relationship expressed by Eq. 13.8 applies to both prototype and model, Eq. 13.8 is assumed to govern the prototype, with a similar relationship

$$\frac{\mathcal{D}_m}{w_m^2 \rho_m V_m^2} = \phi\left(\frac{w_m}{h_m}, \frac{\rho_m V_m w_m}{\mu_m}\right) \tag{13.9}$$

for the model. The modeling laws (similarity requirements) are therefore

$$\frac{w_m}{h_m} = \frac{w}{h} \qquad \frac{\rho_m V_m w_m}{\mu_m} = \frac{\rho V w}{\mu}$$

The size of the model is obtained from the first requirement, which indicates that

$$w_m = \frac{h_m}{h} w \tag{13.10}$$

We are free to select the height ratio, $h_m/h$, desired for the model test (i.e., a ½ sized model or a 1/10 sized model, for example), but then the model plate width, $w_m$, is fixed in accordance with Eq. 13.10. This assures that the model is a *geometrically scaled* model.

The second similarity requirement indicates that the model and prototype must be operated at the same Reynolds number. Thus, the required velocity for the model is obtained from the relationship

$$V_m = \frac{\mu_m}{\mu} \frac{\rho}{\rho_m} \frac{w}{w_m} V \tag{13.11}$$

Note that this model design requires not only geometric scaling, as specified by Eq. 13.10, but also the correct scaling of the velocity in accordance with Eq. 13.11. This result is typical of most model designs—there is more to the design than simply scaling the geometry.

With the foregoing similarity requirements satisfied, the prediction equation for the drag is

$$\frac{\mathcal{D}}{w^2 \rho V^2} = \frac{\mathcal{D}_m}{w_m^2 \rho_m V_m^2}$$

or

$$\mathcal{D} = \left(\frac{w}{w_m}\right)^2 \left(\frac{\rho}{\rho_m}\right) \left(\frac{V}{V_m}\right)^2 \mathcal{D}_m \tag{13.12}$$

Thus, a measured drag on the model, $\mathcal{D}_m$, must be multiplied by the ratio of the square of the plate widths, the ratio of the fluid densities, and the ratio of the square of the velocities to obtain the predicted value of the prototype drag, $\mathcal{D}$.

Generally, to achieve similarity between model and prototype behavior, *all the corresponding pi terms must be equated between model and prototype.* ***For Example...*** assume that air flowing with a velocity of 20 m/s normal to a 2-m-tall by 1-m-wide prototype plate is to be modeled by a 0.2-m-tall model plate in water. The model and prototype parameters are indicated in the table below.

| | $w$, m | $h$, m | $\mu$, N · s/m$^2$ | $\rho$, kg/m$^3$ | V, m/s | $\mathcal{D}$, N |
|---|---|---|---|---|---|---|
| Prototype | 1 | 2 | $1.79 \times 10^{-5}$ | 1.23 | 20 | ? |
| Model | ? | 0.2 | $1.12 \times 10^{-3}$ | 999 | ? | $\mathcal{D}_m$ |

There are three question marks in the table, one for each of the three pi terms. To achieve similarity, the width of the model plate, $w_m$, is determined from Eq. 13.10 as

$$w_m = (0.2/2)(1 \text{ m}) = 0.1 \text{ m}.$$

In addition, for the flow past the model plate to be similar to the flow past the prototype plate, the water model speed is obtained from Eq. 13.11 as

$$V_m = (1.12 \times 10^{-3}/1.79 \times 10^{-5})(1.23/999)(1/0.1)(20 \text{ m/s}) = 15.4 \text{ m/s}$$

Finally, when operating under the similar conditions given above, the predicted drag on the prototype plate can be determined from Eq. 13.12 to be

$$\mathcal{D} = (1/0.1)^2(1.23/999)(20/15.4)^2 \mathcal{D}_m = 0.208 \mathcal{D}_m$$

where $\mathcal{D}_m$ is determined from the model experiment in water. ▲

## *Example 13.2* Pump Modeling

The power input, $\dot{W}_p$, required to run a centrifugal pump is a function of the diameter, $D$, and angular velocity, $\omega$, of the pump impeller, the volumetric flow rate, Q, and the density, $\rho$, of the fluid being pumped. Data for a particular test of an 8-in.-diameter model pump are shown in the table below.

| $D_m$, in. | $\omega_m$, rev/min | $Q_m$, ft$^3$/s | $\rho_m$, slug/ft$^3$ | $\dot{W}_{pm}$, horsepower (hp) |
|---|---|---|---|---|
| 8.0 | 1200 | 2.33 | 1.94 | 12.0 |

It is desired to make a geometrically similar, larger pump with $D$ = 12 in. Based on the above experimental data for the smaller pump, predict the power required to run the 12-in.-diameter pump at 1000 rev/min with flow conditions similar to those of the small pump. In each case the working fluid is water.

### Solution

***Known:*** Variables affecting the pump power. Some prototype (12-in.-diameter pump) and model (8-in.-diameter pump) data.

***Find:*** The power required to run the larger pump at conditions similar to those of the smaller pump.

***Schematic and Given Data:***

❶

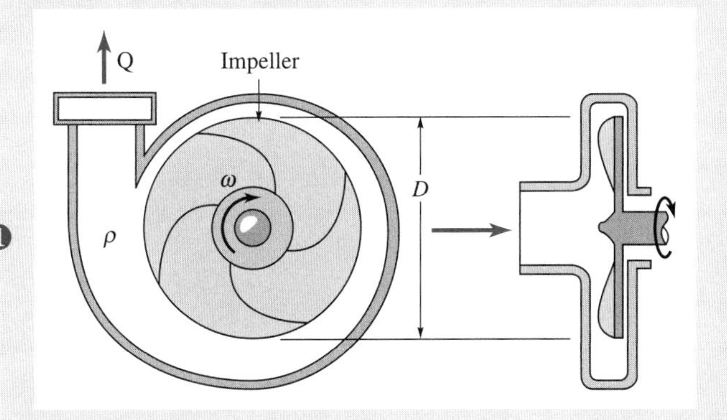


***Assumptions:***
1. The specified physical variables that the pump power is a function of are correct.
2. The model and prototype pumps are geometrically similar.

*Figure E13.2*

***Analysis:*** From the statement of the problem we can write

$$\dot{W}_p = f(D, \omega, Q, \rho)$$

We see that there are five physical variables that can be written in terms of three basic dimensions (either M, L, t *or* F, L, t). Thus, this flow can be described in terms of two dimensionless pi terms as

$$C_{\mathcal{P}} = \phi(C_Q) \tag{1}$$

where $C_{\mathcal{P}} = \dot{W}_p/(\rho\omega^3D^5)$ is the *power coefficient* and $C_Q = Q/(\omega D^3)$ is the *flow coefficient*. A simple check of the dimensions involved will show that these terms are indeed dimensionless.

For similar flow conditions the flow coefficient for the prototype pump must be the same as that for the model pump. That is,

$$\frac{Q}{\omega D^3} = \left(\frac{Q}{\omega D^3}\right)_m$$

where the subscript $m$ refers to the model. Thus, the prototype volumetric flow rate must be

$$Q = (\omega/\omega_m)(D/D_m)^3 Q_m = (1000\text{ rpm}/1200\text{ rpm})(12\text{ in.}/8\text{ in.})^3\,(2.33\text{ ft}^3/\text{s}) = 6.55\text{ ft}^3/\text{s}$$

From Eq. 1, if the model and prototype flow coefficients are the same, then the model and prototype power coefficients are also the same. That is

$$\frac{\dot{W}_p}{\rho\omega^3 D^5} = \left(\frac{\dot{W}_p}{\rho\omega^3 D^5}\right)_m$$

Thus, the power required to run the 12-in.-diameter prototype is

$$\dot{W}_p = (\rho/\rho_m)(\omega/\omega_m)^3(D/D_m)^5 \dot{W}_{pm}$$

Since $\rho = \rho_m$ (both pump water), we obtain the prototype power as

❷ $$\dot{W}_p = (1000\text{ rpm}/1200\text{ rpm})^3(12\text{ in.}/8\text{ in.})^5\,(12.0\text{ hp}) = 52.7\text{ hp} \quad \triangleleft$$

Among other parameters of interest in the design of pumps is the head rise across the pump, $h_p$. This parameter can be put into dimensionless form as a *head rise coefficient*, $C_H$, where

$$C_H = \frac{gh_p}{\omega^2 D^2}$$

As with the power coefficient, the head rise coefficient is also a function of the flow coefficient, $C_Q$. Thus, if the model and prototype flow coefficients are equal, then $C_H = C_{Hm}$, or

$$\frac{gh_p}{\omega^2 D^2} = \left(\frac{gh_p}{\omega^2 D^2}\right)_m$$

Since $g_m = g$ it follows that $h_p = (\omega/\omega_m)^2(D/D_m)^2\, h_{pm}$. Thus, for the model and prototype pumps of this example

$$h_p = (1000\text{ rpm}/1200\text{ rpm})^2(12\text{ in.}/8\text{ in.})^2\, h_{pm} = 1.56\, h_{pm}$$

That is, the 12-in.-diameter pump operating under conditions similar to those given in the above table for the 8-in.-diameter pump will produce a head rise 1.56 times greater than that for the smaller pump.

❶ If the 12-in.-diameter prototype pump were not geometrically similar to the 8-in.-diameter model, it would be inappropriate to use the model data to predict the prototype performance.

❷ The power calculated above is that for only one set of operating parameters for the prototype pump. By doing a series of model tests at various operating conditions, one could obtain the corresponding predicted performance for the prototype over a range of operating conditions.

## 13.8 Chapter Summary and Study Guide

Many practical engineering problems involving fluid mechanics require experimental data for their solution. Thus, laboratory studies and experimentation play a vital role in this field. It is important to develop good procedures for the design of experiments so they can be efficiently completed with as broad applicability as possible. To achieve this end the concept of *similitude* is often used in which measurements made in the laboratory can be utilized for predicting the behavior of other similar systems. In this chapter *dimensional analysis* is used for designing such experiments, as an aid for correlating experimental data, and as the basis for the design of physical models.

Dimensional analysis simplifies a given problem described by a certain set of variables by reducing the number of variables that need to be considered. In addition to being fewer in number, the new variables are dimensionless products of the original variables. Typically these new dimensionless variables are much simpler to work with in performing the desired experiments. It is shown how the use of dimensionless variables can be of assistance in planning experiments and as an aid in correlating experimental data.

For problems in which there are a large number of variables, the use of physical models is described. Models are used to make specific predictions from laboratory tests rather than formulating a general relationship for the phenomenon of interest. The correct design of a model is obviously imperative so that accurate predictions of other similar, but usually larger, systems can be made. It is shown how dimensional analysis can be used to establish a valid model design.

The following check list provides a study guide for this chapter. When your study of the text and end-of-chapter exercises has been completed you should be able to

- write out the meanings of the terms listed in the margins throughout the chapter and understand each of the related concepts. The subset of key terms listed here in the margin is particularly important.
- form a set of dimensionless variables from a set of physical variables.
- use dimensionless variables as an aid in interpreting and correlating experimental data.
- establish a set of modeling laws and the prediction equation for a model to be used to predict the behavior of another similar system (the prototype).

*dimensionless products*
*basic dimensions*
*dimensionally homogeneous*
*dimensional analysis*
*Buckingham pi theorem*
*pi term*
*model*
*prototype*
*prediction equation*
*modeling laws*

## Problems

**Note:** Unless otherwise indicated in the problem statement, use values of fluid properties given in the tables of Appendix FM-1 when solving these problems.

### Dimensionless Variables

**13.1** The Reynolds number, $\rho VD/\mu$, is a very important parameter in fluid mechanics. Verify that the Reynolds number is dimensionless, using both the FLt system and the MLt system for basic dimensions, and determine its value for water flowing at a velocity of 2 m/s through a 1-in.-diameter pipe.

**13.2** Some common variables in fluid mechanics include: volumetric flow rate, Q, acceleration of gravity, $g$, viscosity, $\mu$, density, $\rho$, and a length, $\ell$. Which of the following combinations of these variables are dimensionless? **(a)** $Q^2/g\ell^2$. **(b)** $\rho Q/\mu\ell$. **(c)** $g\ell^2/Q$. **(d)** $\rho Q\ell/\mu$.

### Forming Dimensionless Parameters

**13.3** The pressure rise, $\Delta p$, across a pump can be expressed as

$$\Delta p = f(D, \rho, \omega, Q)$$

where $D$ is the impeller diameter, $\rho$ the fluid density, $\omega$ the rotational speed, and Q the volumetric flow rate. Determine a suitable set of dimensionless parameters.

**13.4** The drag, $\mathcal{D}$, on a washer-shaped plate placed normal to a stream of fluid can be expressed as

$$\mathcal{D} = f(d_1, d_2, V, \mu, \rho)$$

where $d_1$ is the outer diameter, $d_2$ the inner diameter, V the fluid velocity, $\mu$ the fluid viscosity, and $\rho$ the fluid density. Some experiments are to be performed in a wind tunnel to determine the drag. What dimensionless parameters would you use to organize these data?

**13.5** The velocity, V, of a spherical particle falling slowly in a very viscous liquid can be expressed as

$$V = f(d, \mu, \gamma, \gamma_s)$$

where $d$ is the particle diameter, $\mu$ is the liquid viscosity, and $\gamma$ and $\gamma_s$ are the specific weight of the liquid and particle, respectively. Develop a set of dimensionless parameters that can be used to investigate this problem.

**13.6** Assume that the drag, $\mathcal{D}$, on an aircraft flying at supersonic speeds is a function of its velocity, V, fluid density, $\rho$, speed of sound, $c$, and a series of lengths, $\ell_1, \ldots, \ell_i$, which describe the geometry of the aircraft. Develop a set of pi terms that could be used to investigate experimentally how the drag is affected by the various factors listed.

### Repeating Variable Method

**13.7** At a sudden contraction in a pipe the diameter changes from $D_1$ to $D_2$. The pressure drop, $\Delta p$, which develops across the contraction is a function of $D_1$ and $D_2$, as well as the velocity, V, in the larger pipe, the fluid density, ρ, and viscosity, μ. Use $D_1$, V, and μ as repeating variables to determine a suitable set of dimensionless parameters. Why would it be incorrect to include the velocity in the smaller pipe as an additional variable?

**13.8** Assume that the power, $\dot{W}$, required to drive a fan is a function of the fan diameter, $D$, the air density, ρ, the rotational speed, ω, and the flow rate, Q. Use $D$, ω, and ρ as repeating variables to determine a suitable set of pi terms.

**13.9** It is desired to determine the wave height when wind blows across a lake. The wave height, $H$, is assumed to be a function of the wind speed, V, the water density, ρ, the air density, $\rho_a$, the water depth, $d$, the distance from the shore, ℓ, and the acceleration of gravity, $g$, as shown in Fig. P13.9. Use $d$, V, and ρ as repeating variables to determine a suitable set of pi terms that could be used to describe this problem.

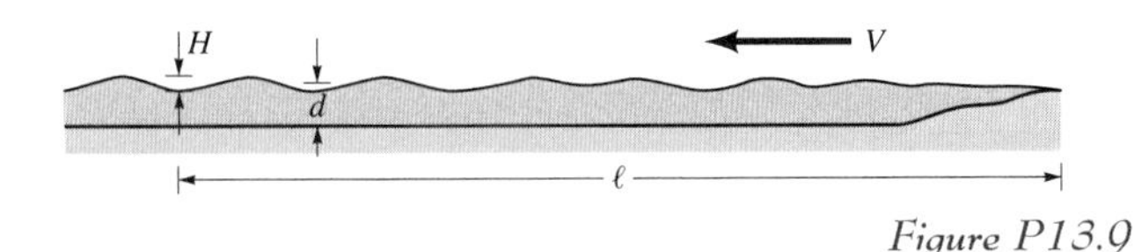


*Figure P13.9*

## Using Dimensional Analysis—General

**13.10** The pressure drop across a short hollowed plug placed in a circular tube through which a liquid is flowing (see Fig. P13.10) can be expressed as

$$\Delta p = f(\rho, V, D, d)$$

where ρ is the fluid density, and V is the velocity in the tube. Some experimental data obtained with $D = 0.2$ ft, $\rho = 2.0$ slug/ft$^3$, and $V = 2$ ft/s are given in the following table:

| Test | 1 | 2 | 3 | 4 |
|---|---|---|---|---|
| $d$ (ft) | 0.06 | 0.08 | 0.10 | 0.15 |
| $\Delta p$ (lbf/ft$^2$) | 493.8 | 156.2 | 64.0 | 12.6 |

Using suitable dimensionless parameters, plot the results of these tests logarithmically. Use a standard curve-fitting technique to determine a general equation for $\Delta p$. What are the limits of applicability of the equation?

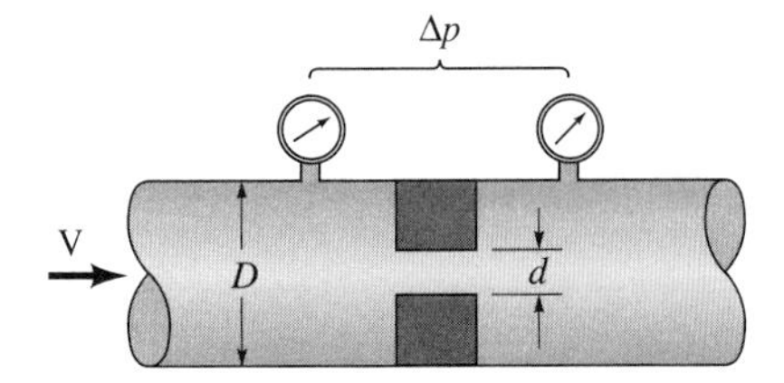


*Figure P13.10*

**13.11** The pressure drop per unit length, $\Delta p_\ell = \Delta p/\ell$, (N/m$^2$)/m, for the flow of blood through a horizontal small diameter tube is a function of the volumetric rate of flow, Q, the diameter, $D$, and the blood viscosity, μ. For a series of tests in which $D = 2$ mm and $\mu = 0.004$ N · s/m$^2$, the following data were obtained, where the $\Delta p$ listed was measured over the length, $\ell = 300$ mm.

| Q (m$^3$/s) | $\Delta p$ (N/m$^2$) |
|---|---|
| $3.6 \times 10^{-6}$ | $1.1 \times 10^4$ |
| $4.9 \times 10^{-6}$ | $1.5 \times 10^4$ |
| $6.3 \times 10^{-6}$ | $1.9 \times 10^4$ |
| $7.9 \times 10^{-6}$ | $2.4 \times 10^4$ |
| $9.8 \times 10^{-6}$ | $3.0 \times 10^4$ |

Perform a dimensional analysis for this problem, and make use of the data given to determine a general relationship between $\Delta p_\ell$ and Q that is valid for other values of $D$, ℓ, and μ.

**13.12** When a very viscous fluid flows slowly past a vertical plate of height $h$ and width $b$ (see Fig. P13.12), pressure develops on the face of the plate. Assume that the pressure, $p$, at the midpoint of the plate is a function of plate height and width, the approach velocity, V, and the fluid viscosity, μ. Make use of dimensional analysis to determine how the pressure, $p$, will change when the fluid velocity, V, is doubled.

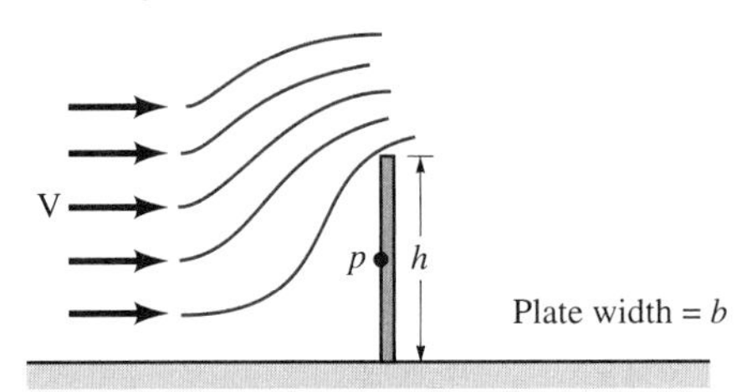


*Figure P13.12*

**13.13** The viscosity, μ of a liquid can be measured by determining the time, $t$, it takes for a sphere of diameter, $d$, to settle slowly through a distance, ℓ, in a vertical cylinder of diameter, $D$, containing the liquid (see Fig. P13.13). Assume that

$$t = f(\ell, d, D, \mu, \Delta\gamma)$$

where $\Delta\gamma$ is the difference in specific weights between the sphere and the liquid. Use dimensional analysis to show how $t$ is related to μ, and describe how such an apparatus might be used to measure viscosity.

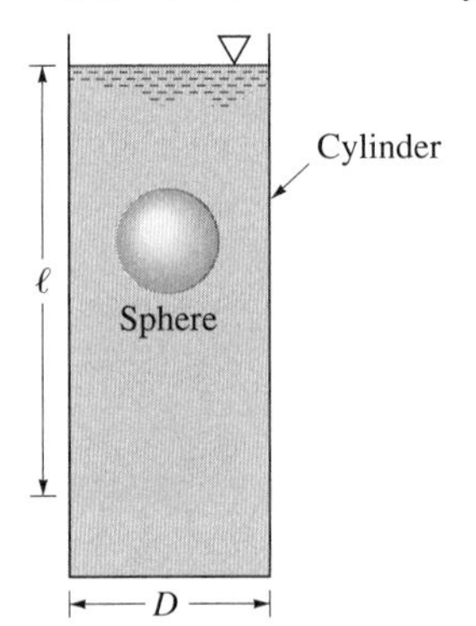


*Figure P13.13*

**13.14** (CD-ROM)

**13.15** (CD-ROM)

## Using Dimensional Analysis—Models

**13.16** SAE 30 oil at 60°F is pumped through a 3-ft-diameter pipeline at a rate of 5700 gal/min. A model of this pipeline is to be designed using a 2-in.-diameter pipe and water at 60°F as the working fluid. To maintain Reynolds number similarity between these two systems, what fluid velocity will be required in the model?

**13.17** The design of a river model is to be based on Froude number similarity, where the Froude number, $Fr = V/(gy)^{1/2}$ is a function of the water velocity, V, the water depth, $y$, and the acceleration of gravity, $g$. If the river depth is 3 m and the model depth is 100 mm, what prototype velocity corresponds to a model velocity of 2 m/s?

**13.18** (CD-ROM)

**13.19** The lift and drag developed on a hydrofoil are to be determined through wind tunnel tests using standard air. If full scale tests are to be run, what is the required wind tunnel velocity corresponding to a hydrofoil velocity in seawater at 20 mph? Assume Reynolds number similarity is required.

**13.20** The drag on a 2-m-diameter satellite dish due to an 80 km/hr wind is to be determined through a wind tunnel test using a geometrically similar 0.40-m-diameter model dish. Standard air is used for both the model and the prototype. **(a)** Assuming Reynolds number similarity, at what air speed should the model test be run? **(b)** With all similarity conditions satisfied, the measured drag on the model was determined to be 179 N. What is the predicted drag on the prototype dish?

**13.21** The pressure rise, $\Delta p$, across a centrifugal pump of a given shape (see Fig. P13.21$a$) can be expressed as

$$\Delta p = f(D, \omega, \rho, Q)$$

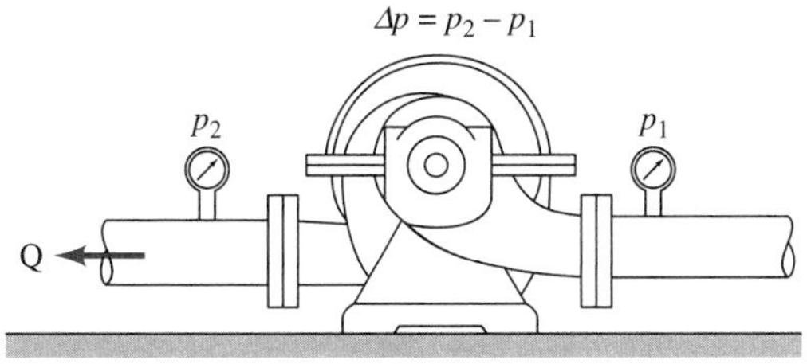

$\Delta p = p_2 - p_1$

Centrifugal pump

($a$)

Model data
$\omega_m = 40\pi$ rad/s
$D_m = 8$ in.

$\Delta p_m$ (psi)

$Q_m$ (ft$^3$/s)

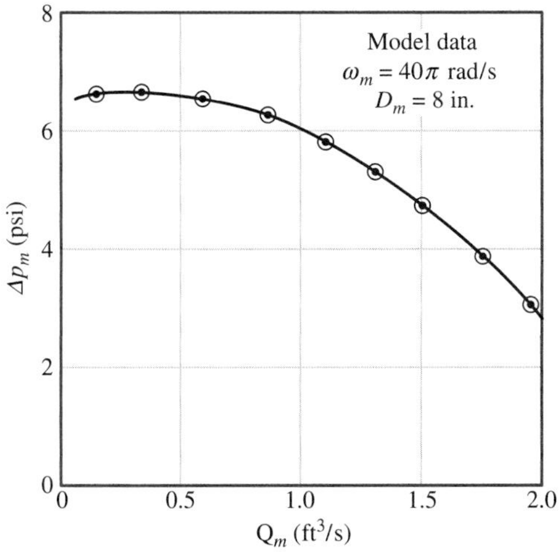

($b$)

*Figure P13.21*

where $D$ is the impeller diameter, $\omega$ the angular velocity of the impeller, $\rho$ the fluid density, and Q the volumetric rate of flow through the pump. A model pump having a diameter of 8 in. is tested in the laboratory using water. When operated at an angular velocity of 40 $\pi$ rad/s the model pressure rise as a function of Q is shown in Fig. P13.21$b$. Use this curve to predict the pressure rise across a geometrically similar pump (prototype) for a prototype flow rate of 6 ft$^3$/s. The prototype has a diameter of 12 in. and operates at an angular velocity of 60 $\pi$ rad/s. The prototype fluid is also water.

**13.22** When small particles of diameter $d$ are transported by a moving fluid having a velocity V, they settle to the ground at some distance $\ell$ after starting from a height $h$ as shown in Fig. P13.22. The variation in $\ell$ with various factors is to be studied with a $1/10$ size scale model. Assume that

$$\ell = f(h, d, V, \gamma, \mu)$$

where $\gamma$ is the particle specific weight and $\mu$ is the fluid viscosity. The same fluid is to be used in both the model and the prototype, but $\gamma$ (model) $= 9 \times \gamma$ (prototype). **(a)** If V = 50 mph, at what velocity should the model tests be run? **(b)** During a certain model test it was found that $\ell$ (model) = 0.8 ft. What would be the predicted $\ell$ for the prototype?

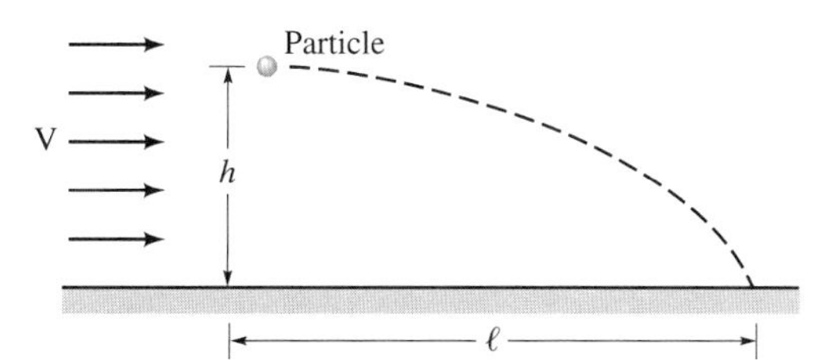


*Figure P13.22*

**13.23** A square parking lot of width $w$ is bounded on all sides by a curb of height $d$ with only one opening of width $b$ as shown in Fig. P13.23. During a heavy rain the lot fills with water and it is of interest to determine the time, $t$, it takes for the water to completely drain from the lot after the rain stops. A scale model is to be used to study this problem, and it is assumed that

$$t = f(w, b, d, g, \mu, \rho)$$

where $g$ is the acceleration of gravity, $\mu$ is the fluid viscosity, and $\rho$ is the fluid density. **(a)** A dimensional analysis indicates that two important dimensionless parameters are $b/w$ and $d/w$. What additional dimensionless parameters are required?

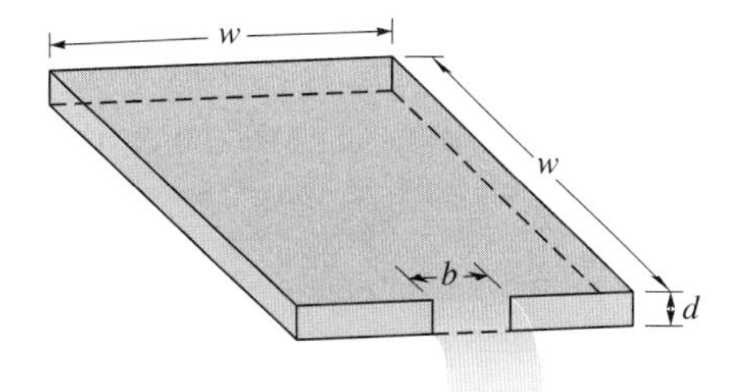


*Figure P13.23*

(b) For a geometrically similar 1/10 size model, what is the relationship between the drain time for the model and the corresponding drain time for the actual parking lot? Assume all similarity requirements are satisfied. Can water be used as the model fluid? Explain and justify your answer.

13.24 (CD-ROM)

13.25 (CD-ROM)

13.26 (CD-ROM)

13.27 The drag on a sphere moving in a fluid is known to be a function of the sphere diameter, the velocity, and the fluid viscosity and density. Laboratory tests on a 4-in.-diameter sphere were performed in a water tunnel and some model data are plotted in Fig. P13.27. For these tests the viscosity of the water was $2.3 \times 10^{-5}$ lbf · s/ft² and the water density was 1.94 slug/ft³. Estimate the drag on an 8-ft-diameter balloon moving in air at a velocity of 3 ft/s. Assume the air to have a viscosity of $3.7 \times 10^{-7}$ lbf · s/ft² and a density of $2.38 \times 10^{-3}$ slug/ft³. Assume Reynolds number similarity.

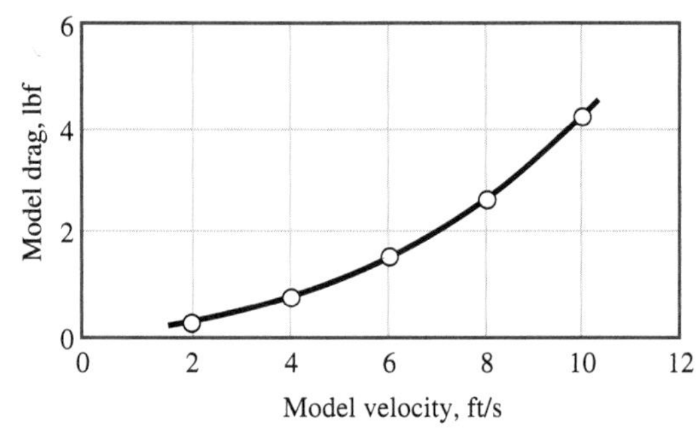


*Figure P13.27*

13.28 A circular cylinder of diameter $d$ is placed in a uniform stream of fluid as shown in Fig. P13.28*a*. Far from the cylinder the velocity is V and the pressure is atmospheric. The gage pressure, $p$, at point A on the cylinder surface is to be determined from a model study for an 18-in.-diameter prototype placed in an air stream having a speed of 8 ft/s. A 1/12 scale model is to be used with water as the working fluid. Some experimental data obtained from the model are shown in Fig. P13.28*b*. Predict the prototype pressure. Assume Reynolds number similarity.

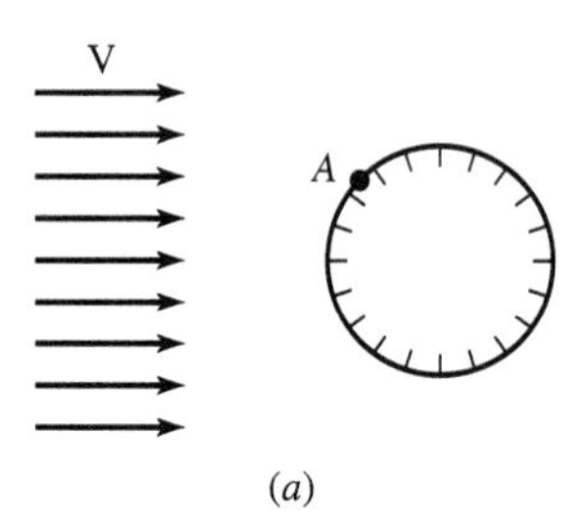


(*a*)

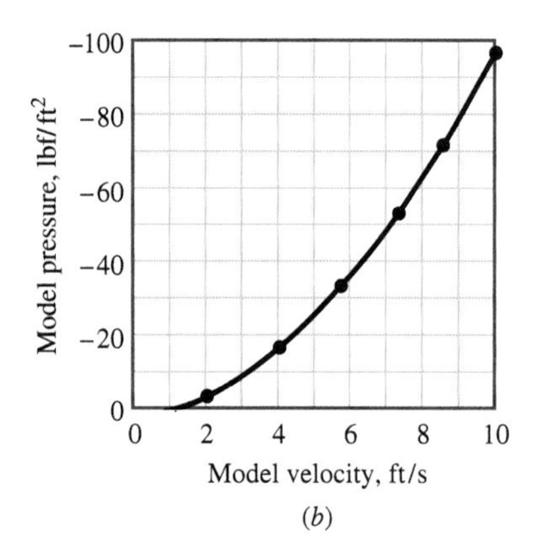


(*b*)

*Figure P13.28*

# INTERNAL AND EXTERNAL FLOW

## *Introduction...*

Fluid mechanics problems concerned with flowing fluids can be broadly classified as ***internal*** or ***external flow*** problems. Flows enclosed by boundaries are considered to be internal flows. Examples of internal flows include flow though pipes, ducts, valves, and various pipe fittings. Flow around bodies completely surrounded by a fluid are considered to be external flows. Examples of external flows include flow around airplanes, automobiles, buildings, and submarines. The ***objective*** of this chapter is to study the internal flow of a viscous fluid through pipe systems and the external flow around familiar geometric shapes.

*internal and external flow*

*chapter objective*

## Internal Flow

This part of the chapter deals with the internal flow of a viscous fluid in a ***pipe system.*** Some of the basic components of a typical pipe system are shown in Fig. 14.1. They include the pipes themselves (perhaps of more than one diameter), the various fittings used to connect the individual pipes to form the desired system, the flow control devices (valves), and the pumps or turbines that add mechanical energy to or remove mechanical energy from the fluid.

*pipe system*

Before we apply the various governing equations to pipe flow examples, we will discuss some of the basic concepts of pipe flow. Unless otherwise specified, we will assume that the pipe is round and that it is completely filled with the fluid being transported.

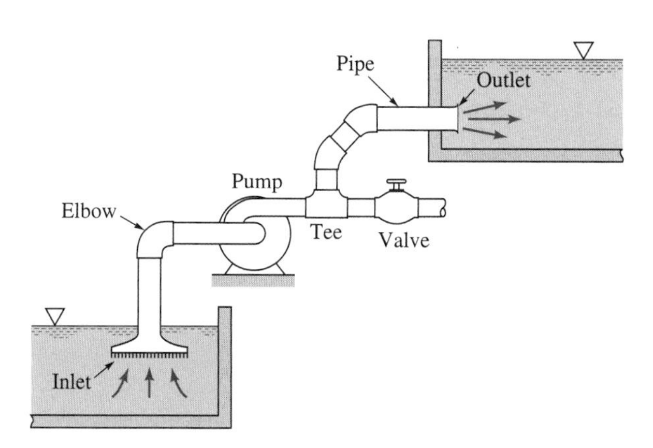


*Figure 14.1* Typical pipe system components.

## 14.1 General Characteristics of Pipe Flow

### 14.1.1 Laminar or Turbulent Flow

The flow of a fluid in a pipe may be laminar flow or it may be turbulent flow. Osborne Reynolds, a British scientist and mathematician, was the first to distinguish the difference between these two classifications of flow by using a simple apparatus as shown in Fig. 14.2*a*. For "small enough flow rates" the dye streak will remain as a well-defined line as it flows along, with only slight blurring due to molecular diffusion of the dye into the surrounding water. For a somewhat larger "intermediate flow rate" the dye streak fluctuates in time and space, and intermittent bursts of irregular behavior appear along the streak. On the other hand, for "large enough flow rates" the dye streak almost immediately becomes blurred and spreads across the entire pipe in a random fashion. These three characteristics, denoted as *laminar, transitional,* and *turbulent* flow, respectively, are illustrated in Fig. 14.2*b*.

***V14.1 Laminar/ turbulent pipe flow***

In the previous paragraph the term flow rate should be replaced by Reynolds number, $Re = \rho VD/\mu$, where V is the average velocity in the pipe. That is, the flow in a pipe is laminar, transitional, or turbulent provided the Reynolds number is "small enough," "intermediate," or "large enough." It is not only the fluid velocity that determines the character of the flow—its density, viscosity, and the pipe size are of equal importance. These parameters combine to produce the Reynolds number. Recall from Sec. 13.5 that the Reynolds number is a measure of the relative importance of inertial and viscous effects in the flow.

*laminar flow*
*turbulent flow*

For most engineering applications of flow in a round pipe, the following values are appropriate: The flow is ***laminar*** if the Reynolds number is less than approximately 2100 to 2300. The flow is ***turbulent*** if the Reynolds number is greater than approximately 4000. For Reynolds numbers between these two limits, the flow may switch between laminar and turbulent conditions. Such flow, which represents the onset of turbulence, is called ***transitional.***

*transitional flow*

### 14.1.2 Entrance Region and Fully Developed Flow

*entrance region*

Any fluid flowing in a pipe had to enter the pipe at some location. The region of flow near where the fluid enters the pipe is termed the ***entrance region*** and is illustrated in Fig. 14.3. As shown, the fluid typically enters the pipe with a nearly uniform velocity profile at section (1). As the fluid moves through the pipe, viscous effects cause it to stick to the pipe wall. That is, whether the fluid is air or a very viscous oil, at the motionless pipe wall the fluid velocity has a zero value.

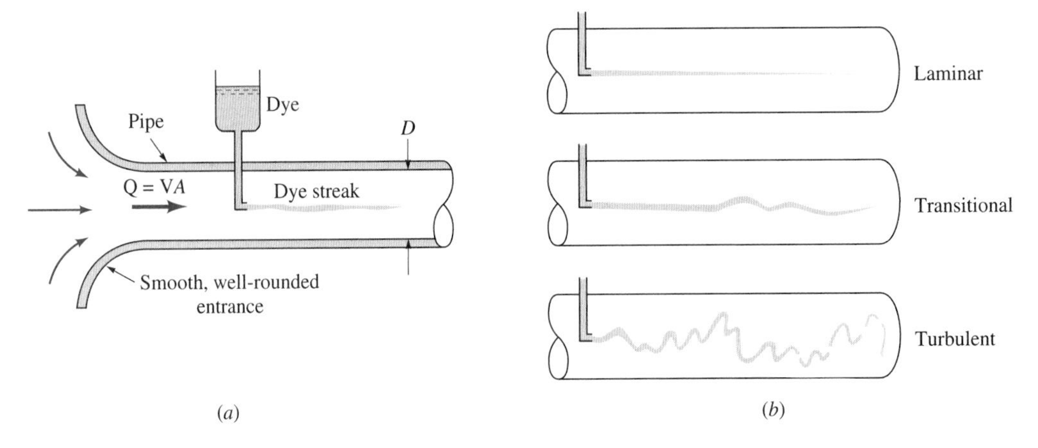


*Figure 14.2* (*a*) Experiment to illustrate type of flow. (*b*) Typical dye streaks.

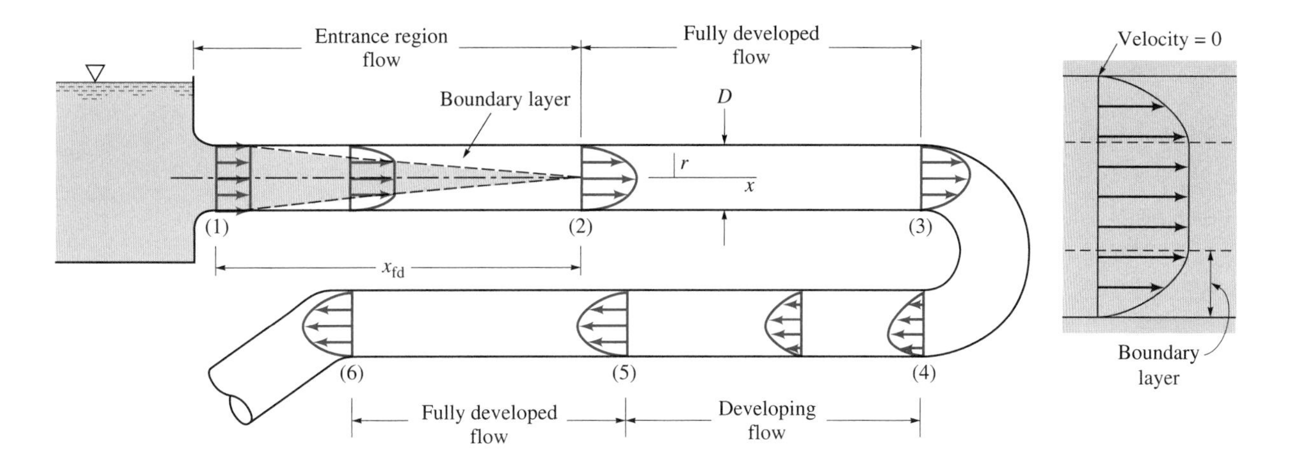


*Figure 14.3* Entrance region, developing flow, and fully developed flow in a pipe system.

As shown in Fig. 14.3, a ***boundary layer*** in which viscous effects are important is produced along the pipe wall such that the initial velocity profile changes with distance along the pipe, $x$, until the fluid reaches the end of the entrance length, section (2), beyond which the velocity profile does not vary with $x$. The boundary layer has grown in thickness to completely fill the pipe, and the flow is termed ***fully developed.***

*boundary layer*

*fully developed flow*

The shape of the velocity profile in the pipe depends on whether the flow is laminar or turbulent, as does the ***entrance length,*** $x_{fd}$. Typical entrance lengths are given by

*entrance length*

$$\frac{x_{fd}}{D} = 0.05\, Re \qquad \text{(laminar flow)} \tag{14.1}$$

and

$$10 \leq \frac{x_{fd}}{D} \leq 60 \qquad \text{(turbulent flow)} \tag{14.2}$$

Once the fluid reaches the end of the entrance region, section (2) of Fig. 14.3, the flow is simpler to describe because the velocity is a function of only the distance from the pipe centerline, $r$, and independent of the axial distance $x$. This is true until the character of the pipe changes in some way, such as a change in diameter, or the fluid flows through a bend, valve, or some other component at section (3). The flow between (2) and (3) is fully developed. Beyond the interruption of the fully developed flow [at section (4)], the flow gradually begins its return to its fully developed character [section (5)] and continues with this profile until the next pipe system component is reached [section (6)].

## 14.2 Fully Developed Laminar Flow

Flow in straight sections of pipe is a common occurrence. If the flow is fully developed, steady, and laminar, and the fluid is Newtonian, a detailed analysis reveals that the velocity distribution in the pipe is given by the equation

$$u(r) = V_c\left[1 - \left(\frac{2r}{D}\right)^2\right] \qquad \text{(laminar flow)} \tag{14.3}$$

where $u(r)$ is the velocity at a distance $r$ from the pipe centerline, $D$ is the pipe diameter, and $V_c$ is the centerline velocity. This important result indicates that the velocity distribution is *parabolic* for laminar pipe flow.

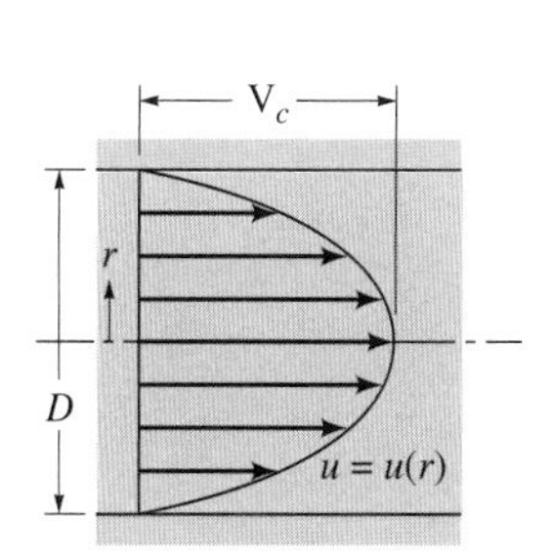

A further analysis reveals that the relationship between the volumetric flow rate, Q, through the pipe and the pressure drop, $\Delta p = p_1 - p_2$, along the pipe is given by the equation

*Poiseuille's law*

$$Q = \frac{\pi D^4 \Delta p}{128\mu\ell} \quad \text{(laminar flow)} \tag{14.4}$$

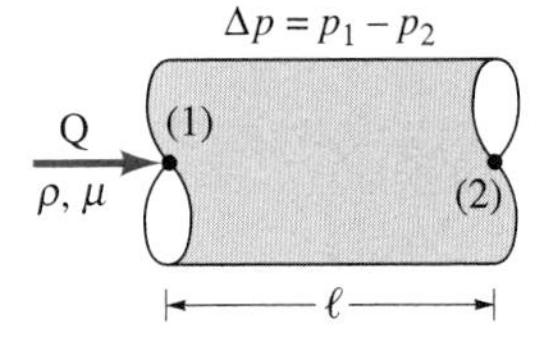


where $\mu$ is the fluid viscosity and $\ell$ is the length of pipe over which the pressure drop occurs. Equation 14.4 is commonly referred to as ***Poiseuille's law.*** It should be emphasized that these results are restricted to fully developed laminar flow (pipe flow in which the Reynolds number is less than approximately 2100 to 2300) in a horizontal pipe.

***For Example...*** to illustrate the use of Eq. 14.4, consider the following problem. An oil with a viscosity of $\mu = 0.40 \text{ N}\cdot\text{s/m}^2$ and density $\rho = 900 \text{ kg/m}^3$ is flowing through a pipe of diameter $D = 0.020$ m. What pressure drop, $\Delta p$, over a length of $\ell = 10$ m is needed to produce a volumetric flow rate of $Q = 2.0 \times 10^{-5} \text{ m}^3\text{/s}$? If the Reynolds number is less than 2100, the flow is laminar and Eq. 14.4 is valid. The average velocity is $V = Q/A = (2.0 \times 10^{-5} \text{ m}^3\text{/s})/[\pi(0.020 \text{ m})^2/4] = 0.0637$ m/s. Using this velocity along with the given density, viscosity, and diameter, the Reynolds number is calculated to be $Re = \rho VD/\mu = 2.87 < 2100$. Hence, the flow is laminar and from Eq. 14.4 with $\ell = 10$ m, the pressure drop is

$$\Delta p = \frac{128\mu\ell Q}{\pi D^4} = \frac{128(0.40 \text{ N}\cdot\text{s/m}^2)(10.0 \text{ m})(2.0 \times 10^{-5} \text{ m}^3\text{/s})}{\pi(0.020 \text{ m})^4}$$

or

$$\Delta p = 20{,}400 \text{ N/m}^2 \left| \frac{1 \text{ kPa}}{10^3 \text{ N/m}^2} \right| = 20.4 \text{ kPa} \quad \blacktriangle$$

## 14.3 Laminar Pipe Flow Characteristics (CD-ROM)

## 14.4 Fully Developed Turbulent Flow

In the previous sections various characteristics of fully developed laminar pipe flow were discussed. In a majority of practical situations, the combination of fluid properties (density and viscosity), pipe diameter, and flow rate are such that the flow is turbulent rather than laminar. Thus, it is necessary to obtain relevant information about turbulent pipe flow.

### 14.4.1 Transition from Laminar to Turbulent Pipe Flow

Flows are classified as laminar or turbulent. For any flow geometry, there is one (or more) dimensionless parameter such that with this parameter value below a particular value the flow is laminar, whereas with the parameter value larger than a certain value it is turbulent. For pipe flow this parameter is the Reynolds number. The value of the Reynolds number must be less than approximately 2100 to 2300 for laminar flow and greater than approximately 4000 for turbulent flow.

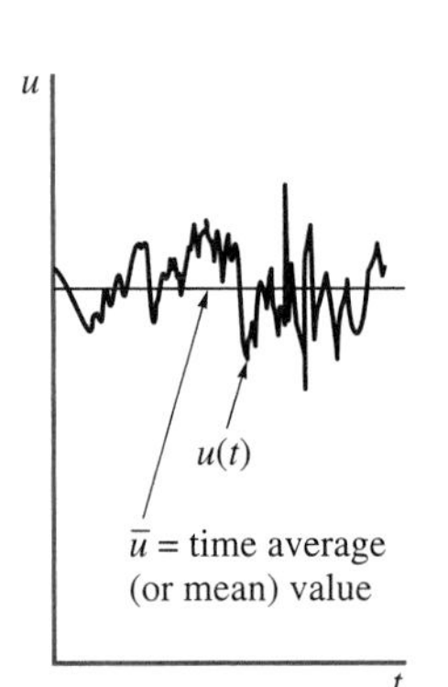


*Figure 14.5* Turbulent fluctuations and time average velocity.

A typical trace of the axial component of velocity, $u = u(t)$, measured at a given location in turbulent pipe flow is shown in Fig. 14.5. Its irregular, random nature is the distinguishing feature of turbulent flows. The character of many of the important features of the flow (pressure drop, heat transfer, etc.) depends strongly on the existence and nature of the turbulent fluctuations or randomness indicated.

For example, mixing processes and heat and mass transfer processes are considerably enhanced in turbulent flow compared to laminar flow. We are all familiar with the "rolling," vigorous eddy type motion of the water in a pan being heated on the stove (even if it is not

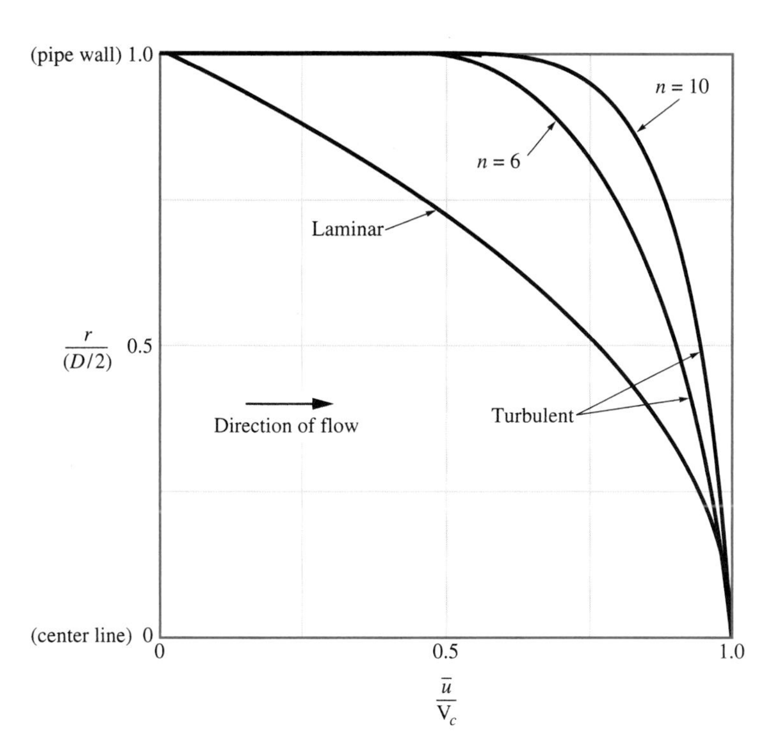


*Figure 14.6* Typical laminar flow and turbulent flow velocity profiles for flow in pipes.

heated to boiling). Such finite-sized random mixing is very effective in transporting energy and mass throughout the fluid, thereby increasing the various rate processes involved. Laminar flow, on the other hand, can be thought of as very small but finite-sized fluid particles flowing smoothly in layers, one over another. Randomness and mixing take place only on the molecular scale and result in relatively small heat, mass, and momentum transfer rates.

***V14.2 Turbulence in a bowl***

### 14.4.2 Turbulent Velocity Profile

Although considerable information concerning turbulent velocity profiles has been obtained through the use of dimensional analysis, experimentation, and semiempirical theoretical efforts, there is still no generally applicable expression for turbulent velocity profiles.

An often-used (and relatively easy-to-use) correlation is the empirical *power-law velocity profile*

*power-law velocity profile*

$$\frac{\bar{u}}{V_c} = \left(1 - \frac{r}{(D/2)}\right)^{1/n} \qquad \text{(turbulent flow)} \qquad (14.10)$$

where $\bar{u}$ is the time averaged velocity at a distance $r$ from the pipe centerline (see Fig. 14.5), and $V_c$ is the centerline velocity. In this representation, the value of $n$ is a function of the Reynolds number, with typical values between $n = 6$ and $n = 10$. Typical turbulent velocity profiles based on this power-law representation are shown in Fig. 14.6. Note that the turbulent profiles are much "flatter" than the laminar profile.

***V14.3 Laminar/turbulent velocity profiles***

Also note that the velocity gradient, $du/dy$, at the wall is much larger than in laminar flow. Hence, the shear stress is much greater in turbulent flow than in laminar flow.

## 14.5 Pipe Flow Head Loss

Most turbulent pipe flow analyses are based on experimental and semiempirical formulas, with the data conveniently expressed in dimensionless form. It is often necessary to determine the head loss, $h_L$, that occurs in a pipe flow so that the mechanical energy equation, Eq. 12.15, can be used in the analysis of pipe flow problems. As shown in Fig. 14.1, a typical pipe system usually consists of various lengths of straight pipe interspersed with various types of components (valves, elbows, etc.). The overall head loss for the pipe system consists of the head loss

*major loss*
*minor loss*

due to viscous effects in the straight pipes, termed the ***major loss*** and denoted $h_{L\text{ major}}$, and the head loss in the various pipe components, termed the ***minor loss*** and denoted $h_{L\text{ minor}}$. That is,

$$h_L = h_{L\text{ major}} + h_{L\text{ minor}}$$

The head loss designations of "major" and "minor" do not necessarily reflect the relative importance of each type of loss. For a pipe system that contains many components and a relatively short length of pipe, the minor loss may actually be larger than the major loss.

### 14.5.1 Major Losses

The major loss is associated with friction (viscous) effects as the fluid flows through the straight pipe and can be expressed in functional form as

$$h_{L\text{ major}} = F(\mathrm{V}, D, \ell, \varepsilon, \mu, \rho)$$

where V is the average velocity, $\ell$ is the pipe length, $D$ the pipe diameter, and $\varepsilon$ is a length characterizing the roughness of the pipe wall. Although the head loss or pressure drop for laminar pipe flow is found to be independent of the roughness of the pipe (e.g., the pipe roughness does not appear in Eq. 14.4), it is necessary to include this parameter when considering turbulent flow. The above relationship between the head loss and the other physical variables can be expressed as

$$h_{L\text{ major}} = f\frac{\ell}{D}\frac{\mathrm{V}^2}{2g} \tag{14.11}$$

*friction factor*

*relative roughness*

where $f$ is termed the ***friction factor.*** Equation 14.11 is called the *Darcy-Weisbach equation.* The dimensionless friction factor, $f$, is a function of two other dimensionless terms—the Reynolds number based on the pipe diameter, $Re = \rho \mathrm{V} D/\mu$, and the ***relative roughness,*** $\varepsilon/D$. That is, $f = f(Re, \varepsilon/D)$. As seen by Eq. 14.11, the head loss in a straight pipe is proportional to the friction factor, $f$, the length-to-diameter ratio, $\ell/D$, and the velocity head, $\mathrm{V}^2/2g$.

*Moody chart*

Figure 14.7 shows the experimentally determined functional dependence of $f$ on $Re$ and $\varepsilon/D$. This is called the ***Moody chart.*** Typical roughness values, $\varepsilon$, for various new, clean pipe surfaces are given in Table 14.1.

The following characteristics are observed from the data of Fig. 14.7. For laminar flow, the friction factor is independent of the relative roughness and is a function of the Reynolds number only:

$$f = 64/Re \qquad (\text{laminar, } Re < 2100) \tag{14.12}$$

*wholly turbulent*

For ***wholly turbulent flow,*** where the Reynolds number is relatively large, the friction factor is independent of the Reynolds number and is a function of the relative roughness only: $f = f(\varepsilon/D)$.

**Table 14.1** Equivalent Roughness for New Pipes.

| | Equivalent Roughness, $\varepsilon$ | |
|---|---|---|
| Pipe | Feet | Millimeters |
| Riveted steel | 0.003–0.03 | 0.9–9.0 |
| Concrete | 0.001–0.01 | 0.3–3.0 |
| Wood stave | 0.0006–0.003 | 0.18–0.9 |
| Cast iron | 0.00085 | 0.26 |
| Galvanized iron | 0.0005 | 0.15 |
| Commercial steel or wrought iron | 0.00015 | 0.045 |
| Drawn tubing | 0.000005 | 0.0015 |
| Plastic, glass | 0.0 (smooth) | 0.0 (smooth) |

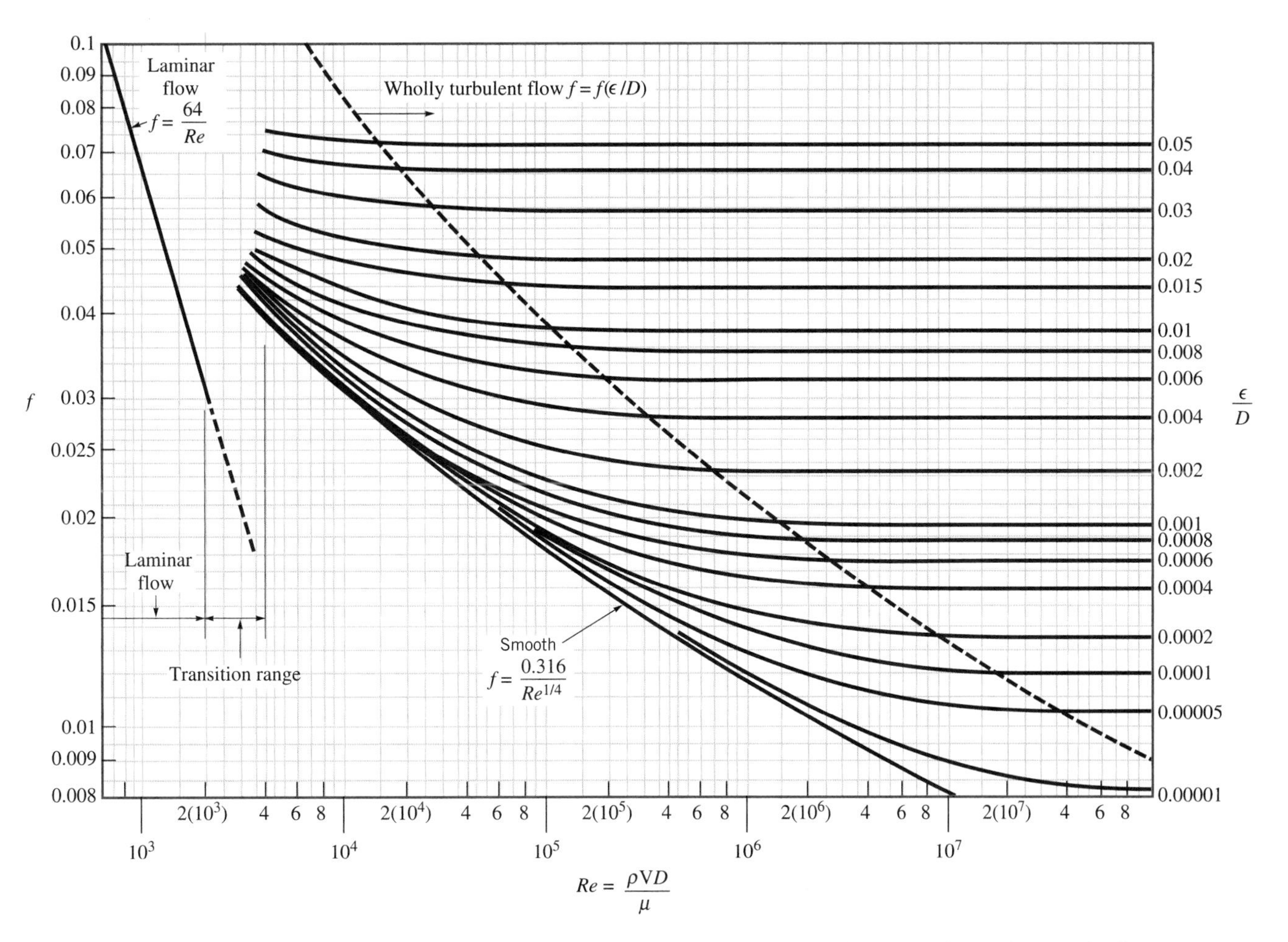


*Figure 14.7* Friction factor as a function of Reynolds number and relative roughness for round pipes—the Moody chart

Inspection of Fig. 14.7 also indicates that between the laminar flow and wholly turbulent flow regimes the friction factor depends on both the Reynolds number and the relative roughness.

For the entire turbulent flow range, friction factors can be read from the Moody chart or evaluated using the ***Colebrook formula***

$$\frac{1}{\sqrt{f}} = -2.0 \log\left(\frac{\varepsilon/D}{3.7} + \frac{2.51}{Re\sqrt{f}}\right) \quad \text{(turbulent)} \tag{14.13}$$

***Colebrook formula***

which is an empirical fit of the pipe flow data. For ***hydraulically smooth*** ($\varepsilon = 0$) pipes the friction factor is given by the ***Blasius formula***

$$f = 0.316/Re^{1/4} \quad (\text{turbulent, } \varepsilon = 0) \tag{14.14}$$

***Blasius formula***

## *Example 14.1* Turbulent Pipe Flow—Friction Factors

Air under standard conditions flows through a horizontal section of 4.0-mm-diameter drawn tubing with an average velocity of V = 50 m/s. Determine the pressure drop in a 0.1-m length of the tube.

### Solution

***Known:*** Air at standard conditions flows through a horizontal section of drawn tubing with a specified velocity.

***Find:*** Determine the pressure drop.

***Assumptions:***
1. The air is modeled as an incompressible fluid with a density of $\rho = 1.23\ \text{kg/m}^3$ and a viscosity of $\mu = 1.79 \times 10^{-5}\ \text{N} \cdot \text{s/m}^2$ (see Appendix FM-1).
2. The flow is fully developed and steady.
3. Minor losses are zero since we are considering only a straight portion of pipe.

***Analysis:*** The mechanical energy equation, Eq. 12.15, for this flow can be written as

$$\frac{p_1}{\gamma} + \frac{V_1^2}{2g} + z_1 = \frac{p_2}{\gamma} + \frac{V_2^2}{2g} + z_2 + h_L \tag{1}$$

where points (1) and (2) are located within the tube a distance 0.1 m apart.

Since the density and tube area are constant, the mass balance gives $V_1 = V_2$. Also, the tube is horizontal so $z_1 = z_2$. From Eq. 14.11, $h_L = f(\ell/D)(V^2/2g)$. Thus, with $\Delta p = p_1 - p_2$, Eq. 1 becomes

$$\Delta p = \gamma h_L = \rho g\, h_L = f\frac{\ell}{D}\frac{1}{2}\rho V^2 \tag{2}$$

Using known data, the Reynolds number is

$$Re = \frac{\rho V D}{\mu} = \frac{(1.23\ \text{kg/m}^3)(50\ \text{m/s})(0.004\ \text{m})}{1.79 \times 10^{-5}\ \text{N} \cdot \text{s/m}^2}\left|\frac{1\ \text{N}}{1\ \text{kg} \cdot \text{m/s}^2}\right| = 13{,}700$$

which indicates turbulent flow.

❶ For turbulent flow $f = f(Re, \varepsilon/D)$, where from Table 14.1, $\varepsilon = 0.0015$ mm so that $\varepsilon/D = 0.0015\ \text{mm}/4.0\ \text{mm} = 0.000375$. From the Moody chart (Fig. 14.7) with $Re = 1.37 \times 10^4$ and $\varepsilon/D = 0.000375$ we obtain $f = 0.028$. Thus, from Eq. 2

$$\Delta p = f\frac{\ell}{D}\frac{1}{2}\rho V^2 = (0.028)\frac{(0.1\ \text{m})}{(0.004\ \text{m})}\frac{1}{2}(1.23\ \text{kg/m}^3)(50\ \text{m/s})^2\left|\frac{1\ \text{N}}{1\ \text{kg} \cdot \text{m/s}^2}\right|\left|\frac{1\ \text{kPa}}{10^3\ \text{N/m}^2}\right|$$

or

$$\Delta p = 1.076\ \text{kPa} \quad \triangleleft$$

---

❶ An alternate method to determine the friction factor for the turbulent flow would be to use the Colebrook formula, Eq. 14.13. Thus,

$$\frac{1}{\sqrt{f}} = -2.0 \log\left(\frac{\varepsilon/D}{3.7} + \frac{2.51}{Re\sqrt{f}}\right) = -2.0 \log\left(\frac{0.000375}{3.7} + \frac{2.51}{1.37 \times 10^4\sqrt{f}}\right)$$

or

$$\frac{1}{\sqrt{f}} = -2.0 \log\left(1.01 \times 10^{-4} + \frac{1.83 \times 10^{-4}}{\sqrt{f}}\right)$$

A simple iterative solution of this equation gives $f = 0.0291$, which is in agreement (within the accuracy of reading the graph) with the Moody chart value of $f = 0.028$.

## 14.5.2 Minor Losses

Losses due to the components of pipe systems (other than the straight pipe itself) are termed minor losses and are given in terms of the dimensionless ***loss coefficient,*** $K_L$, as

*loss coefficient*

$$h_{L\,\text{minor}} = K_L \frac{V^2}{2g} \tag{14.15}$$

Numerical values of the loss coefficients for various components (elbows, valves, entrances, etc.) are determined experimentally.

Many pipe systems contain various transition sections in which the pipe diameter changes from one size to another. Any change in flow area contributes losses that are not accounted for by the friction factor. The extreme cases involve flow into a pipe from a reservoir (an entrance) or out of a pipe into a reservoir (an exit). Some loss coefficients for entrance and exit flows are shown in Fig. 14.8.

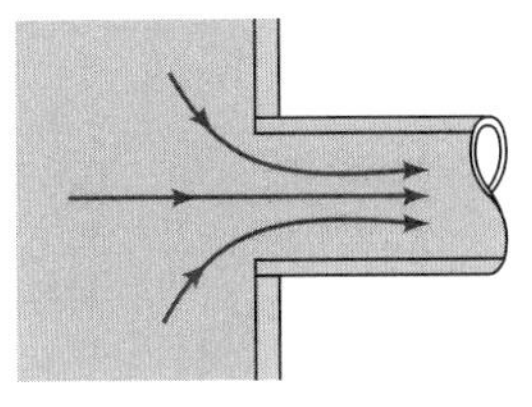

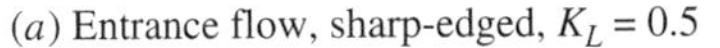

(*a*) Entrance flow, sharp-edged, $K_L = 0.5$

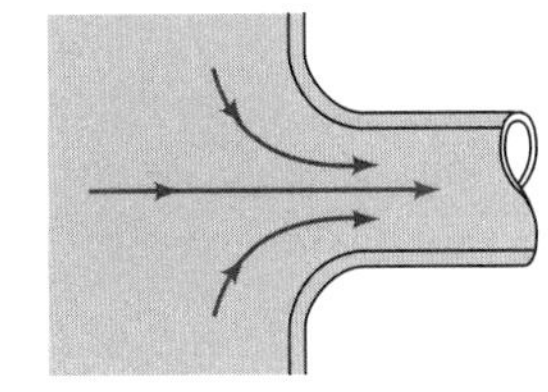

(*b*) Entrance flow, well-rounded, $K_L = 0.04$

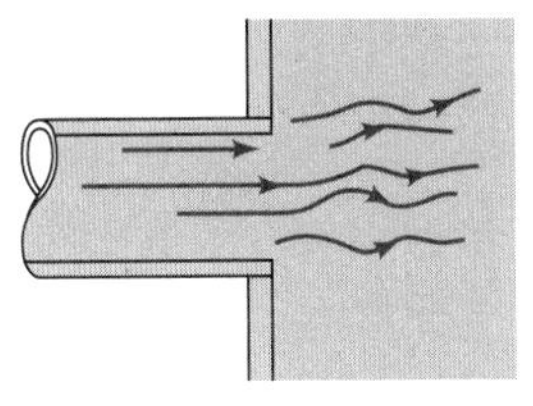

(*c*) Exit flow, sharp-edged, $K_L = 1.0$

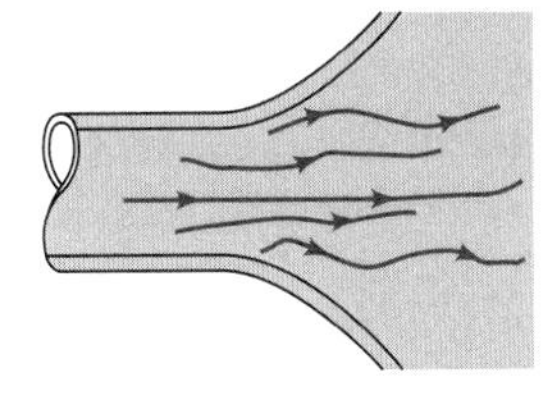

(*d*) Exit flow, well-rounded, $K_L = 1.0$

*Figure 14.8* Loss coefficient values for typical entrance and exit flows.

Another important category of pipe system components is that of commercially available pipe fittings such as elbows, tees, reducers, valves, and filters. The values of $K_L$ for such components depend strongly on the shape of the component and only very weakly on the Reynolds number for typical large *Re* flows. Thus, the loss coefficient for a 90° elbow depends on whether the pipe joints are threaded or flanged, but is, within the accuracy of the data, fairly independent of the pipe diameter, flow rate, or fluid properties—that is, independent of the Reynolds number. Typical values of $K_L$ for such components are given in Table 14.2.

**Table 14.2** Loss Coefficients for Pipe Components $\left(h_L = K_L \dfrac{V^2}{2g}\right)$

| Component | $K_L$ |
|---|---|
| a. Elbows | |
| Regular 90°, flanged | 0.3 |
| Regular 90°, threaded | 1.5 |
| Long radius 90°, flanged | 0.2 |
| Long radius 90°, threaded | 0.7 |
| Long radius 45°, flanged | 0.2 |
| Regular 45°, threaded | 0.4 |
| b. 180° return bends | |
| 180° return bend, flanged | 0.2 |
| 180° return bend, threaded | 1.5 |
| c. Tees | |
| Line flow, flanged | 0.2 |
| Line flow, threaded | 0.9 |
| Branch flow, flanged | 1.0 |
| Branch flow, threaded | 2.0 |
| d. Union, threaded | 0.08 |
| e. Valves | |
| Globe, fully open | 10 |
| Angle, fully open | 2 |
| Gate, fully open | 0.15 |
| Ball valve, fully open | 0.05 |

V → elbows V → return bend V → tees V → union

## 14.6 Pipe Flow Examples

In the previous sections of this chapter we discussed concepts concerning flow in pipes. The purpose of this section is to apply these ideas to the solutions of various practical problems. The nature of the solution procedure for pipe flow problems can depend strongly on which of the various parameters are independent parameters (the "known") and which is the dependent parameter (the "find"). The two most common types of problems are discussed below.

In a Type I problem we specify the desired volumetric flow rate or average velocity and determine the necessary pressure difference, head added by a pump, head removed by a turbine, or head loss. For example, if a volumetric flow rate of 2.0 gal/min is required for a dishwasher that is connected to the water heater by a given pipe system, what pressure is needed in the water heater?

In a Type II problem we specify the applied driving pressure (or, alternatively, the head loss) and determine the volumetric flow rate. For example, how many gal/min of hot water are supplied to the dishwasher if the pressure within the water heater is 60 psi and the pipe system details (length, diameter, roughness of the pipe; number of elbows; etc.) are specified?

### *Example 14.2* Pressure Drop with Major/Minor Losses (Type I)

Water at 60 °F ($\rho = 1.94$ slug/ft$^3$ and $\mu = 2.34 \times 10^{-5}$ lbf · s/ft$^2$) flows from the basement to the second floor through the 0.75-in. (0.0625-ft)-diameter copper pipe (a drawn tubing) at a rate of $Q = 12.0$ gal/min $= 0.0267$ ft$^3$/s and exits through a faucet of diameter 0.50 in. as shown in Fig. E14.2. Determine the pressure at point (1) if both major and minor losses are included.

#### Solution

***Known:*** Water, with specified properties, flows at a given flow rate through a piping system containing straight sections of pipe and various pipe fittings.

***Find:*** Determine the pressure at the upstream end of the piping system.

***Schematic and Given Data:***

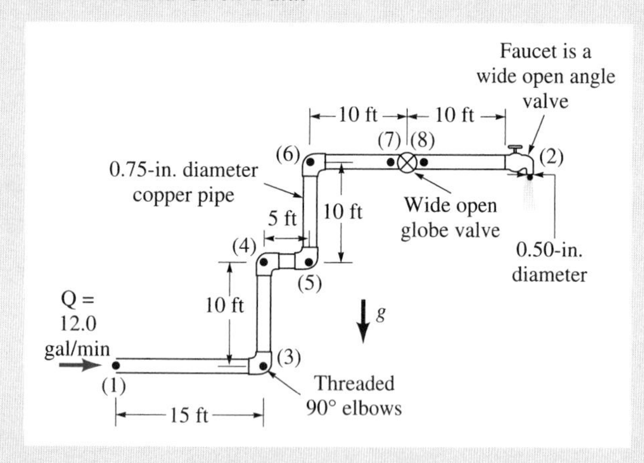


*Figure E14.2*

***Assumptions:***
1. The water is modeled as an incompressible fluid.
2. The flow is steady and fully developed in the straight sections of pipe.
3. The water flows from the faucet as a free jet at atmospheric pressure.

***Analysis:*** Since the fluid velocity in the pipe is given by $V_1 = Q/A_1 = Q/(\pi D^2/4) = (0.0267\text{ ft}^3/\text{s})/[\pi(0.0625\text{ ft})^2/4] =$ 8.70 ft/s, it follows that

$$Re = \frac{\rho V D}{\mu} = \frac{(1.94\text{ slug/ft}^3)(8.70\text{ ft/s})(0.0625\text{ ft})}{(2.34 \times 10^{-5}\text{ lbf}\cdot\text{s/ft}^2)}\left|\frac{1\text{ lbf}}{1\text{ slug}\cdot\text{ft/s}^2}\right|$$

$$= 45{,}000$$

Thus, the flow is turbulent.

The governing equation is the following form of the mechanical energy equation (Eq. 12.15).

$$\frac{p_1}{\gamma} + \frac{V_1^2}{2g} + z_1 = \frac{p_2}{\gamma} + \frac{V_2^2}{2g} + z_2 + h_L$$

where $z_1 = 0$, $z_2 = 20$ ft, $p_2 = 0$ (gage), $\gamma = \rho g = 62.4\ \text{lbf/ft}^3$. The velocity of the stream exiting the faucet is $V_2 = Q/A_2 = (0.0267\ \text{ft}^3/\text{s})/[\pi(0.50/12)^2\text{ft}^2/4] = 19.6$ ft/s. Solving for $p_1$

❶
$$p_1 = \gamma z_2 + \frac{1}{2}\rho(V_2^2 - V_1^2) + \gamma h_L \tag{1}$$

If the only losses were the major losses, the head loss would be

$$h_L = f\frac{\ell}{D}\frac{V_1^2}{2g}$$

From Table 14.1 the roughness for a 0.75-in.-diameter copper pipe (drawn tubing) is $\varepsilon = 0.000005$ ft so that $\varepsilon/D = 8 \times 10^{-5}$. With this $\varepsilon/D$ and the calculated Reynolds number ($Re = 45{,}000$), the value of $f$ is obtained from the Moody chart as $f = 0.0215$. Note that the Colebrook equation (Eq. 14.13) would give the same value of $f$. The total length of the pipe is $\ell = (15 + 10 + 5 + 10 + 20)$ ft $= 60$ ft and Eq. 1 gives

$$
\begin{aligned}
p_1 &= \gamma z_2 + \frac{1}{2}\rho(V_2^2 - V_1^2) + \rho f\frac{\ell}{D}\frac{V_1^2}{2} \\
&= (62.4\ \text{lbf/ft}^3)(20\ \text{ft}) + \frac{1.94\ \text{slug/ft}^3}{2}\left[\left(19.6\frac{\text{ft}}{\text{s}}\right)^2 - \left(8.70\frac{\text{ft}}{\text{s}}\right)^2\right]\left|\frac{1\ \text{lbf}}{1\ \text{slug}\cdot\text{ft/s}^2}\right| \\
&\quad + (1.94\ \text{slug/ft}^3)(0.0215)\left(\frac{60\ \text{ft}}{0.0625\ \text{ft}}\right)\frac{(8.70\ \text{ft/s})^2}{2}\left|\frac{1\ \text{lbf}}{1\ \text{slug}\cdot\text{ft/s}^2}\right| \\
&= (1248 + 299 + 1515)\ \text{lbf/ft}^2 = 3062\ \text{lbf/ft}^2 \quad \text{(gage)}
\end{aligned}
$$
❷

or

$$p_1 = 21.3\ \text{psi}$$

If major *and* minor losses are included, Eq. 1 becomes

$$p_1 = \gamma z_2 + \frac{1}{2}\rho(V_2^2 - V_1^2) + f\gamma\frac{\ell}{D}\frac{V_1^2}{2g} + \sum \rho K_L\frac{V_1^2}{2}$$

The sum of first three terms, which account for the elevation change, the kinetic energy change, and the major losses, has been evaluated above as 21.3 psi. The last term accounts for the minor losses. Accordingly,

$$p_1 = 21.3\ \text{psi} + \sum \rho K_L\frac{V_1^2}{2} \tag{2}$$

The loss coefficients of the components ($K_L = 1.5$ for each elbow, $K_L = 10$ for the wide-open globe valve, $K_L = 2$ for the wide open angle valve) are obtained from Table 14.2. Thus,

$$
\begin{aligned}
\sum \rho K_L\frac{V_1^2}{2} &= (1.94\ \text{slug/ft}^3)\frac{(8.70\ \text{ft/s})^2}{2}[10 + 4(1.5) + 2]\left|\frac{1\ \text{lbf}}{1\ \text{slug}\cdot\text{ft/s}^2}\right| \\
&= 1321\ \text{lbf/ft}^2\left|\frac{1\ \text{ft}^2}{144\ \text{in.}^2}\right| = 9.17\ \text{psi}
\end{aligned} \tag{3}
$$

By combining Eqs. 2 and 3 we obtain the entire pressure drop as

$$p_1 = (21.3 + 9.17)\ \text{psi} = 30.5\ \text{psi} \quad \text{(gage)} \ \triangleleft$$

---

❶ Losses typically play an important role in the analysis of flow through a pipe system. In this example a simple calculation reveals that if all losses are neglected (i.e., $f = 0$ and $\sum K_L = 0$) the pressure would be $p_1 = \gamma z_2 + \rho(V_2^2 - V_1^2)/2 = 10.7$ psi. This compares with the value of $p_1 = 30.5$ psi which was calculated including both major and minor losses. Thus, it is apparent that the neglect of losses for this type of problem would lead to very significant errors.

❷ Since we used gage pressure at point (2) (i.e., $p_2 = 0$), the result for $p_1$ is also gage pressure.

## *Example 14.3* Pipe Flow with Pumps (Type I)

Crude oil at 140°F with $\gamma = 53.7$ lbf/ft$^3$, $\rho = 1.67$ slug/ft$^3$, and $\mu = 8 \times 10^{-5}$ lbf · s/ft$^2$ (about four times the viscosity of water) is pumped across Alaska through the Alaskan pipeline, a 799-mile-long, 4-ft-diameter steel pipe, at a maximum rate of Q = 2.4 million barrels/day = 117 ft$^3$/s, or V = Q/A = 9.31 ft/s. Determine the power added to the fluid by the pumps that drive this large system.

### Solution

***Known:*** Oil, with specified properties, is pumped through a long, constant diameter pipeline at a given volumetric flow rate.
***Find:*** Determine the power added to the fluid by the pumps that drive this system.

***Schematic and Given Data:***

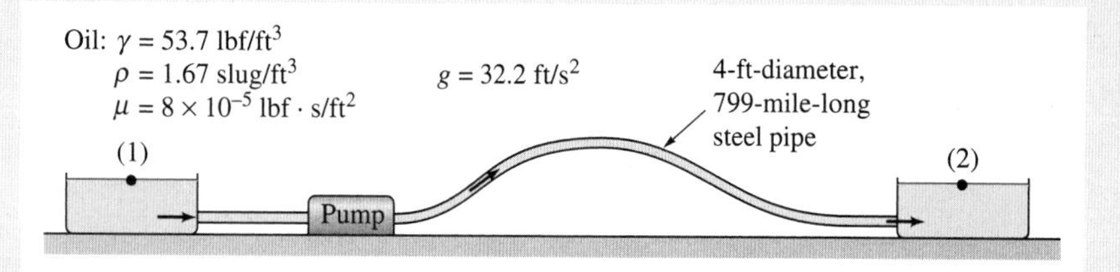


*Figure E14.3*

***Assumptions:***
1. The oil is modeled as an incompressible fluid with properties given in the problem statement.
2. The flow is fully developed and steady.
3. The oil is pumped from a large, open tank at sea level at the beginning of the pipe line to another large, open tank at sea level at the end.
4. Since the tank diameters are very large relative to the pipe diameters, the velocities at (1) and (2), $V_1$ and $V_2$, are much smaller than the velocity V within the pipe and can be neglected.
5. ❶ Because of the extremely large length-to-diameter ratio of the pipe, minor losses are assumed negligible compared to major (frictional) losses in the pipe.

***Analysis:*** From the mechanical energy equation (Eq. 12.15) we obtain

$$\frac{p_1}{\gamma} + \frac{V_1^2}{2g} + z_1 + h_p = \frac{p_2}{\gamma} + \frac{V_2^2}{2g} + z_2 + h_L \tag{1}$$

where points (1) and (2) represent locations within the large holding tanks at either end of the line. The pump head, $h_p = \dot{W}_p/\dot{m}g$, is the head provided to the oil by the pumps; $\dot{W}_p$ is the power provided to the oil by the pumps. Note that $h_p > 0$ and $\dot{W}_p > 0$. The pump adds head (mechanical energy) to the flow. With $z_1 = z_2$ (pumped from sea level to sea level), $p_1 = p_2 = V_1 = V_2 = 0$ (large, open tanks), and $h_L = h_{L\,\text{major}} = (f\ell/D)(V^2/2g)$, Eq. 1 becomes

$$h_p = h_L = f\frac{\ell}{D}\frac{V^2}{2g} \tag{2}$$

From Table 14.1, $\varepsilon = 0.00015$ ft so that $\varepsilon/D = (0.00015 \text{ ft}/4 \text{ ft}) = 3.75 \times 10^{-5}$. Using given data

$$Re = \frac{\rho V D}{\mu} = \frac{(1.67 \text{ slug/ft}^3)(9.31 \text{ ft/s})(4 \text{ ft})}{8 \times 10^{-5} \text{ lbf} \cdot \text{s/ft}^2}\left|\frac{1 \text{ lbf}}{1 \text{ slug} \cdot \text{ft/s}^2}\right|$$
$$= 7.76 \times 10^5$$

Thus, by inspection of Fig. 14.7, the friction factor is $f = 0.0125$ so that Eq. 2 gives

$$h_p = 0.0125\left(\frac{799 \text{ miles}}{4 \text{ ft}}\right)\left|\frac{5280 \text{ ft}}{\text{mile}}\right|\frac{(9.31 \text{ ft/s})^2}{2(32.2 \text{ ft/s}^2)}$$
$$= 17{,}700 \text{ ft}$$

With Eq. 12.16, the power required is $\dot{W}_p = \gamma Q h_p$, or

❷
$$\dot{W}_p = \left(53.7\,\frac{\text{lbf}}{\text{ft}^3}\right)\left(117\,\frac{\text{ft}^3}{\text{s}}\right)(17{,}700\text{ ft})\left|\frac{1\text{ hp}}{550\text{ ft}\cdot\text{lb/s}}\right|$$
$$= 202{,}000\text{ hp} \quad \triangleleft$$

❶ An indication of the relative importance of major and minor head losses can be seen by considering the ratio $h_{L\,\text{major}}/h_{L\,\text{minor}} = [(f\ell/D)V^2/2g]/[K_L V^2/2g] = (f\ell/D)/K_L$, which is directly proportional to the length-to-diameter ratio. For the Alaskan pipe line, $\ell/D = (799\text{ mi})|5280\text{ ft/mi}|/(4\text{ ft}) = 1.05 \times 10^6 \gg 1$, so that it is reasonable to neglect minor losses.

❷ There are many reasons why it is not practical to move the oil with a single pump of this size. First, there are no pumps this large. Second, if the pump were located near the tank at the beginning of the pipeline, application of the mechanical energy from the outlet of the pump to the end of the pipeline would show that the pressure at the pump outlet would need to be $p = \gamma h_L = (53.7\text{ lbf/ft}^3)(17{,}700\text{ ft})\,|1\text{ ft}^2/144\text{ in.}^2| = 6600$ psi. No practical 4-ft-diameter pipe would withstand this pressure.

To produce the desired flow, the actual system contains 12 pumping stations positioned at strategic locations along the pipeline. Each station contains four pumps, three of which operate at any one time (the fourth is in reserve in case of emergency).

Pipe flow problems in which it is desired to determine the volumetric flow rate for a given set of conditions (Type II problems) often require trial-and-error solution techniques. This is because it is necessary to know the value of the friction factor to carry out the calculations, but the friction factor is a function of the unknown velocity (flow rate) in terms of the Reynolds number. The solution procedure is indicated in Example 14.4.

### *Example 14.4* Pipe Flow Rate (Type II)

According to an appliance manufacturer, the 4-in.-diameter vent on a clothes dryer is not to contain more than 20 ft of pipe and four 90° elbows. Under these conditions determine the air volumetric flow rate if the gage pressure within the dryer is 0.20 inches of water.

Assume both the specific weight and the kinematic viscosity of the heated air to be constant and equal to $\gamma = 0.0709$ lbf/ft$^3$, and $\nu = \mu/\rho = 1.79 \times 10^{-4}$ ft$^2$/s, respectively, and that the roughness of the vent pipe surface is equivalent to that of galvanized iron.

**Solution (CD-ROM)**

## 14.7 Pipe Volumetric Flow Rate Measurement (CD-ROM)

## External Flow

In this part of the chapter we consider various aspects of the flow over bodies that are immersed in a fluid. Examples include the flow of air around airplanes, automobiles, and falling snowflakes, or the flow of water around submarines and fish. In these situations the object is completely surrounded by the fluid and the flows are termed *external flows.*

A body immersed in a moving fluid experiences a resultant force due to the interaction between the body and the fluid surrounding it. We can fix the coordinate system in the body and treat the situation as fluid flowing past a stationary body with velocity $U$, the ***upstream velocity.***

*upstream velocity*

The resultant force in the direction of the upstream velocity is termed the *drag*, $\mathscr{D}$, and the resultant force normal to the upstream velocity is termed the *lift*, $\mathscr{L}$. The lift and drag are often obtained from dimensionless ***lift coefficients***, $C_L$, and ***drag coefficients***, $C_D$, which are defined as follows

*lift coefficient*
*drag coefficient*

$$C_L = \frac{\mathscr{L}}{\frac{1}{2}\rho U^2 A} \quad \text{and} \quad C_D = \frac{\mathscr{D}}{\frac{1}{2}\rho U^2 A} \tag{14.18}$$

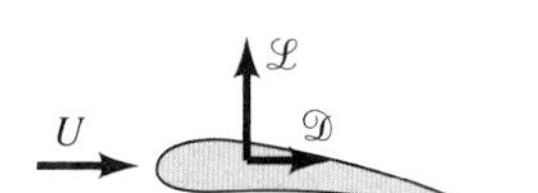


where $A$ is a characteristic area of the object. Typically, $A$ is taken to be the *frontal area*—the projected area seen by a person looking toward the object from a direction parallel to the upstream velocity, $U$. Values of the lift and drag coefficients are determined by an appropriate analysis, a numerical technique, or, most frequently, from experimental data.

External flows past objects encompass an extremely wide variety of fluid mechanics phenomena. For a given-shaped object, the flow characteristics may depend very strongly on various parameters such as size, orientation, speed, and fluid properties. As discussed in Chapter 13, to simplify the presentation and organization of the data and to more easily characterize the flow properties, the various physical data are normally presented in terms of dimensionless parameters. ***For Example...*** we normally use the dimensionless lift and drag coefficients of Eqs. 14.18 rather than lift and drag. ▲

## 14.8 Boundary Layer on a Flat Plate

Perhaps the simplest example of an external flow is the steady, incompressible flow past a flat plate parallel to the flow as illustrated in Fig. 14.12. The fluid approaches the plate with a uniform upstream velocity $U$. Since the fluid viscosity is not zero, it follows that the fluid must stick to the solid surface of the plate—the no-slip boundary condition. This experimentally observed condition is, perhaps, obvious for viscous fluids such as honey. It is equally valid for all fluids, even those like water and air which do not appear to be so viscous.

As indicated in Fig. 14.12, the fact that the fluid adheres to the surface requires the existence of a region in the flow in which the velocity of the fluid changes from 0 on the surface to $U$ some distance away from the surface. This relatively thin layer next to the surface is termed the hydrodynamic ***boundary layer.*** Outside of the boundary layer the fluid flows with velocity $U$ parallel to the plate as if the plate were not there.

*boundary layer*

*boundary layer thickness*

At each location $x$ along the plate, we define the ***boundary layer thickness,*** $\delta(x)$, as that distance from the plate at which the fluid velocity is within some arbitrary value of the upstream velocity $U$. As shown in Fig. 14.13, for simplicity in this introductory discussion, we take this arbitrary value as 0.99 (i.e., 99%). Thus, $u = 0$ at $y = 0$ and $u = 0.99U$ at $y = \delta$, with the velocity profile $u = u(x, y)$ bridging the boundary layer thickness.

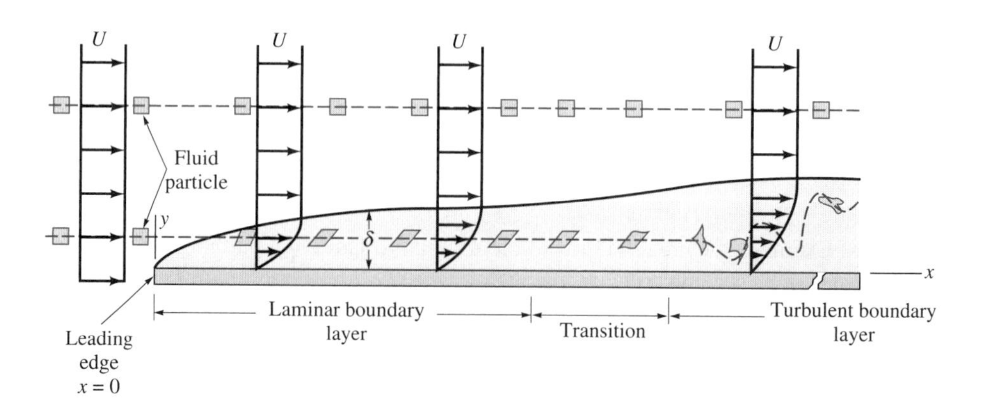


*Figure 14.12* Distortion of a fluid particle as it flows within the boundary layer.

An appreciation of the structure of the boundary layer flow can be obtained by considering what happens to a fluid particle that flows into the boundary layer. As is indicated in Fig. 14.12, a small rectangular particle retains its original shape as it flows in the uniform flow outside of the boundary layer. Once it enters the boundary layer, the particle begins to distort because of the velocity gradient within the boundary layer—the top of the particle has a larger velocity than its bottom. At some distance downstream from the leading edge, the boundary layer flow makes the transition to turbulent flow and the fluid particles become greatly distorted because of the random, irregular nature of the turbulence.

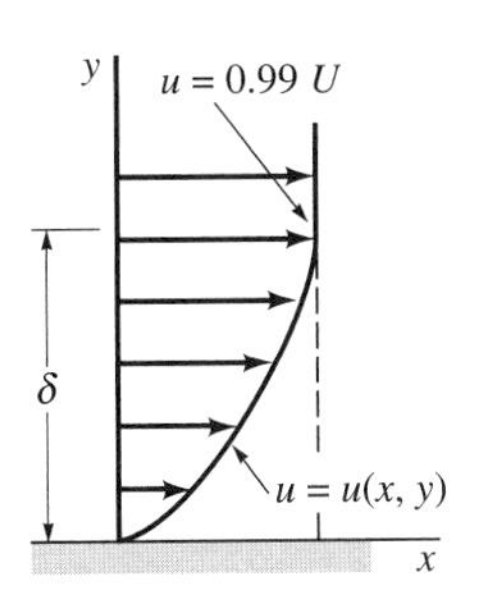


*Figure 14.13* Boundary layer thickness.

### 14.8.1 Laminar Boundary Layer Characteristics

Various characteristics of boundary layer flow can be calculated. For example, it can be shown that the boundary layer thickness for laminar boundary layer flow along a flat plate is given by

$$\delta(x) = 5\sqrt{\frac{\nu x}{U}} \quad \text{(laminar)} \tag{14.19}$$

where $\nu = \mu/\rho$ is the kinematic viscosity. Note that the boundary layer thickness increases in the downstream direction (increasing $x$) and decreases as the velocity, $U$, increases or the kinematic viscosity, $\nu$, decreases.

In addition, for a flat plate of length $\ell$ and width $b$, the drag, $\mathscr{D}$, on one side can be expressed in terms of the drag coefficient, $C_D$, as

$$C_D = \frac{\mathscr{D}}{\frac{1}{2}\rho U^2 b\ell} = \frac{1.328}{\sqrt{Re_\ell}} \quad \text{(laminar)} \tag{14.20}$$

where $Re_\ell = U\ell/\nu$ is the Reynolds number based on the plate length. Note that Eqs. 14.19 and 14.20 are valid only for laminar boundary layers.

### 14.8.2 Transition from Laminar to Turbulent Flat Plate Flow

*V14.7 Laminar/turbulent transition*

The analytical results given in Eqs. 14.19 and 14.20 are restricted to laminar boundary layer flows along a flat plate. They agree quite well with experimental results up to the point where the boundary layer flow becomes turbulent. A transition to turbulent flow will occur for any free stream velocity and any fluid provided the plate is long enough. The parameter that governs the transition to turbulent flow is the Reynolds number—in this case the Reynolds number based on the distance from the leading edge of the plate, $Re_x = Ux/\nu$.

The value of the Reynolds number at the transition location is a rather complex function of various parameters involved, including the roughness of the surface, the curvature of the surface (e.g., a flat plate or a sphere), and some measure of the disturbances in the flow outside the boundary layer. On a flat plate with a sharp leading edge in a typical fluid stream, the ***transition*** takes place at a distance $x_c$ from the leading edge and is given in terms of the ***critical Reynolds number,*** $Re_{x,c} = Ux_c/\nu = 5 \times 10^5$.

*transition*
*critical Reynolds number*

## *Example 14.6* Boundary Layer Thickness and Transition

A fluid flows steadily past a flat plate with a velocity of $U = 10$ ft/s. At approximately what location will the boundary layer become turbulent, and how thick is the boundary layer at that point if the fluid is (a) water at 60°F, (b) air at standard conditions, or (c) glycerin at 68°F?

### Solution

***Known:*** The fluids stick to the plate to form a boundary layer.
***Find:*** Determine the location of the transition point and the boundary layer thickness at that location.

***Assumptions:***

1. The boundary layer flow is laminar up to the transition point.
2. Transition to turbulent flow occurs at $Re_{x,c} = 5 \times 10^5$.
3. Values for the fluid viscosities obtained from Appendix FM-1 are given in the table below.

***Analysis:*** For any fluid, the laminar boundary layer thickness is found from Eq. 14.19 as

$$\delta(x) = 5\sqrt{\frac{\nu x}{U}} \tag{1}$$

The boundary layer remains laminar up to

$$x_c = \frac{\nu Re_{x,c}}{U} = \frac{5 \times 10^5}{10 \text{ ft/s}}\nu = (5 \times 10^4 \text{ s/ft})\,\nu \tag{2}$$

Combining Eqs. 1 and 2, the boundary layer thickness at the transition point is

$$\delta(x_c) = 5\left[\frac{\nu}{10}(5 \times 10^4\,\nu)\right]^{1/2} = (354 \text{ s/ft})\,\nu$$

The resulting $x_c$ and $\delta(x_c)$ values are listed in the table below along with the corresponding values of $\nu$ obtained from Appendix FM-1.

| Fluid | $\nu$ (ft²/s) | $x_c$ (ft) | $\delta(x_c)$ (ft) |
|---|---|---|---|
| ❶ a. Water | $1.21 \times 10^{-5}$ | 0.605 | 0.00428 ◁ |
| b. Air | $1.57 \times 10^{-4}$ | 7.85 | 0.0556 |
| c. Glycerin | $1.28 \times 10^{-2}$ | 640.0 | 4.53 |

❶ As shown by the data, laminar flow can be maintained on a longer portion of the plate if the viscosity is increased. However, the boundary layer flow eventually becomes turbulent, provided the plate is long enough. Similarly, the boundary layer thickness is greater if the viscosity is increased.

### 14.8.3 Turbulent Boundary Layer Flow

The structure of turbulent boundary layer flow is very complex, random, and irregular. It shares many of the characteristics described for turbulent pipe flow in Section 14.4. In particular, the velocity at any given location in the flow is unsteady in a random fashion. The flow can be thought of as a jumbled mix of intertwined eddies (or swirls) of different sizes (diameters and angular velocities). The various fluid quantities involved (i.e., mass, momentum, energy) are transported downstream as in a laminar boundary layer. For turbulent flow they are also transported across the boundary layer (in the direction perpendicular to the plate) by the random transport of finite-sized fluid particles associated with the turbulent eddies. There is considerable mixing involved with these finite-sized eddies—considerably more than is associated with the mixing found in laminar flow where it is confined to the molecular scale. Consequently, the drag for turbulent boundary layer flow along a flat plate is considerably greater than it is for laminar boundary layer flow.

Owing to the complexity of turbulent boundary layer flow, it is necessary to use an empirical relationship for the drag coefficient. In general, the drag coefficient, $C_D = \mathfrak{D}/\frac{1}{2}\rho U^2 A$, for a flat plate of length $\ell$, width b, and area $A = \ell b$ is a function of the Reynolds number, $Re_\ell$, and the relative roughness, $\varepsilon/\ell$, where $\varepsilon$ is the surface roughness. The results of numerous experiments covering a wide range of the parameters of interest are shown in Fig. 14.14. For laminar boundary layer flow the drag coefficient is a function of only the

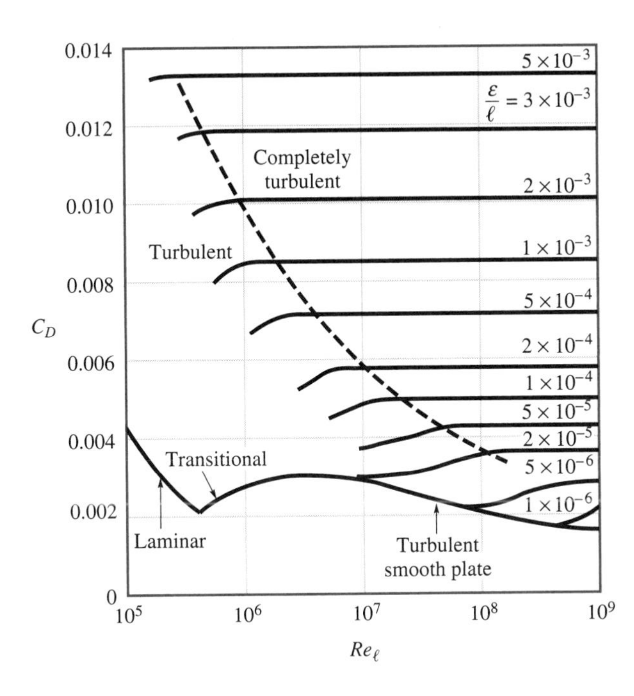


*Figure 14.14* Drag coefficient for a flat plate parallel to the upstream flow.

Reynolds number—surface roughness is not important. This is similar to laminar flow in a pipe. However, for turbulent flow, the surface roughness does affect the shear stress and, hence, the drag coefficient. This, also, is similar to turbulent pipe flow. Values of the roughness, $\varepsilon$, for different materials can be obtained from Table 14.1.

## *Example 14.7* Drag on a Flat Plate

A 4-ft by 8-ft piece of plywood is fastened to the roof rack of a car that is driven through still air at 55 mph = 80.7 ft/s. Estimate the drag caused by shear stress on the top of the plywood.

### Solution

***Known:*** Air flows past a 4-ft by 8-ft piece of plywood with a velocity of 80.7 ft/s.
***Find:*** Determine the drag on the top surface of the plywood.

***Assumptions:***
1. The top surface of the plywood is parallel to the upstream air flow, which has a velocity of 80.7 ft/s.
2. The straps used to tie the plywood to the roof rack do not significantly interfere with the boundary layer flow along the plywood surface.
3. The surface roughness of the plywood is $\varepsilon = 0.003$ ft (see Table 14.1).
4. The air is at standard conditions with $\rho = 0.00238$ slug/ft$^3$ and $\mu = 3.74 \times 10^{-7}$ lbf · s/ft$^2$ (see Appendix FM-1).
5. The plywood is aligned with its 8 ft edge parallel to the road.

❶ ***Analysis:*** If, as assumed, the plywood is aligned exactly parallel to the upstream flow, then the friction drag, $\mathfrak{D}$, caused by the shear stress on the top of the plywood can be determined from

$$\mathfrak{D} = \tfrac{1}{2}\rho U^2 \ell b C_D \tag{1}$$

where the drag coefficient, $C_D$, is a function of the relative roughness, $\varepsilon/\ell$, and Reynolds number, $Re_\ell = \rho U \ell/\mu$, as given in Fig. 14.14. Also, $A = \ell b = 4 \text{ ft} \times 8 \text{ ft} = 32 \text{ ft}^2$.

With the given data we obtain

❷
$$Re_\ell = \frac{\rho U \ell}{\mu} = \frac{(0.00238 \text{ slug/ft}^3)(80.7 \text{ ft/s})(8 \text{ ft})}{3.74 \times 10^{-7} \text{ lbf} \cdot \text{s/ft}^2} \left| \frac{1 \text{ lbf}}{1 \text{ slug} \cdot \text{ft/s}^2} \right| = 4.11 \times 10^6$$

and

$$\frac{\varepsilon}{D} = \frac{0.003 \text{ ft}}{8 \text{ ft}} = 3.75 \times 10^{-4}$$

❸ Thus, as seen in Fig. 14.14, the boundary layer flow is in the turbulent flow regime and the drag coefficient is read from the figure to be approximately $C_D = 0.0066$. Hence, from Eq. 1,

$$\begin{aligned}\mathfrak{D} &= (1/2)(0.00238 \text{ slug/ft}^3)(80.7 \text{ ft/s})^2(8 \text{ ft})(4 \text{ ft})(0.0066) \left| \frac{1 \text{ lbf}}{1 \text{ slug} \cdot \text{ft/s}^2} \right| \\ &= 1.64 \text{ lbf} \quad \triangleleft\end{aligned}$$

---

❶ In reality, the air velocity past the plywood is not equal to the upstream velocity of the air approaching the car because the air accelerates as it passes over the car. In addition, the air after passing over the car's hood and windshield is probably not aligned directly parallel to the plywood. This fact can dramatically affect the net force (lift and drag) on the plywood.

❷ If the same 4-ft by 8-ft plywood were aligned with its 4-ft edge parallel to the road, the Reynolds number, which is based on the length $\ell = 4$ ft, would be smaller by a factor of two ($Re_\ell = 2.05 \times 10^6$ rather than $4.11 \times 10^6$), and the corresponding drag coefficient would be somewhat smaller ($C_D = 0.0062$ rather than 0.0066, see Fig. 14.14). Thus, the drag is a function of the orientation of the plate (short or long edge parallel to the flow). This results from the fact that the shear stress is a function of the distance from the leading edge of the plate.

❸ If the plywood were made very smooth, $\varepsilon/D = 0$, the drag coefficient would be reduced to approximately $C_D = 0.0030$ and the drag would be reduced to approximately $\mathfrak{D} = 0.745$ lbf.

## 14.9 General External Flow Characteristics

*friction drag*
*pressure drag*

The characteristics of flow past a zero-thickness flat plate are discussed in Sec. 14.8. As discussed below, additional phenomena occur for flow past bodies with non-zero thickness. In particular, for such bodies, there are two contributions to the drag: ***friction drag*** and ***pressure drag.***

As a fluid flows past a body, friction exerts its effect in two ways. One is the direct application of a friction (viscous) force caused by the shear stress acting on the body. This is the friction drag. The other is related to the fact that frictional effects within the flowing fluid can drastically alter the path that the fluid takes when flowing around the body. Such frictional effects produce an irreversible pressure drop in the direction of flow, resulting in lower pressure on the back of the object than that on the front. This produces the pressure drag.

The total drag is found from

$$\mathfrak{D} = \tfrac{1}{2}\rho U^2 A C_D \tag{14.21}$$

where $\rho$ is the density of the fluid, $U$ is the upstream velocity, $A$ is the frontal area, and $C_D$ is the drag coefficient. As for the case of the flat plate discussed previously, the drag coefficient is a function of dimensionless parameters such as the Reynolds number and the relative roughness of the surface.

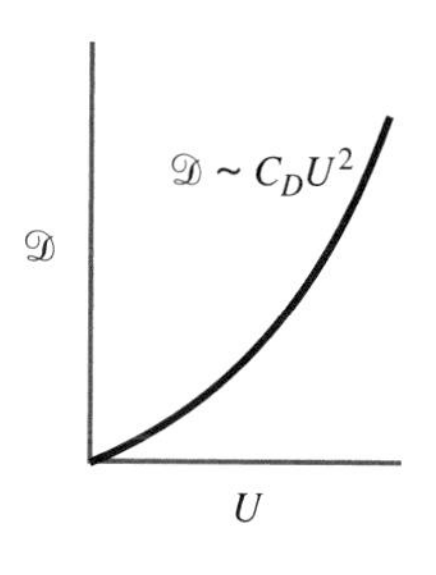


Consider the flow of an incompressible fluid past a smooth circular cylinder. For this case the drag coefficient is a function of the Reynolds number only, as shown in Fig. 14.15. (The figure also provides drag coefficient data for flow past a smooth sphere.) As noted from Eq. 14.21 the drag force is proportional to $C_D$ times $U^2$. Thus, although the value of $C_D$ may decrease with increasing Reynolds number, the drag tends to increase as the upstream velocity increases (i.e., as the Reynolds number increases).

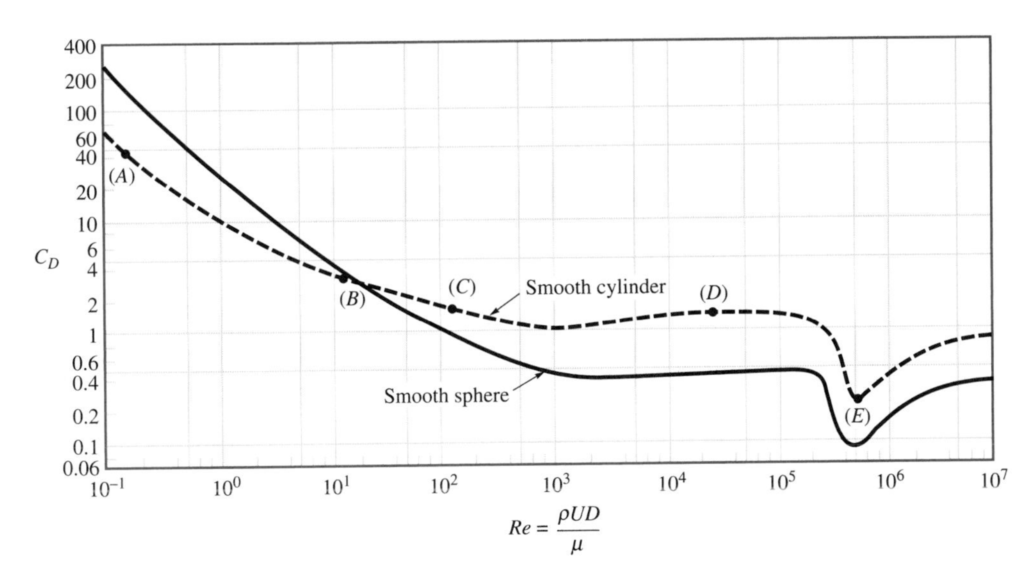


*Figure 14.15* Drag coefficient as a function of Reynolds number for a smooth circular cylinder and a smooth sphere.

Five cases labeled (A) through (E) are indicated in Fig. 14.15; typical flow patterns for these cases are shown in Fig. 14.16. Case (A) corresponds to a small Reynolds number flow ($Re < 1$). For this case the flow is essentially symmetric about the sphere. For somewhat larger Reynolds number [$Re \approx 10$, case (B)] the symmetry is lost and a stationary

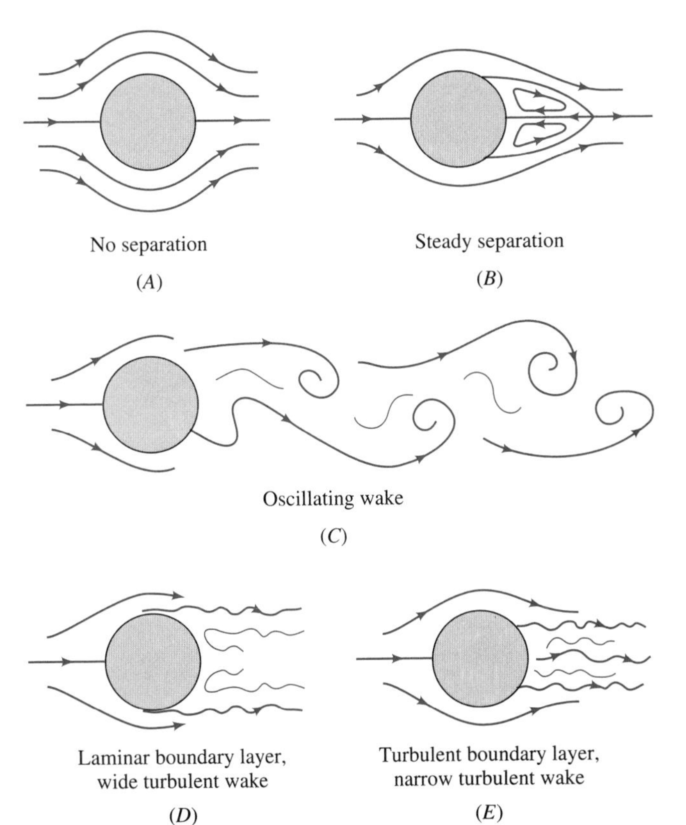


*Figure 14.16* Typical flow patterns for flow past a circular cylinder.

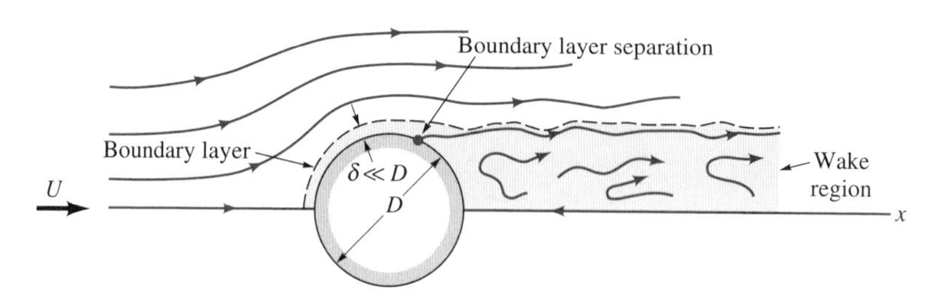


*Figure 14.17* Flow past a circular cylinder, $Re = 10^5$.

separation region occurs at the rear of the cylinder. At still larger Reynolds numbers [$Re \approx 100$, case (C)] the separation region grows, becomes unsteady, and forms an oscillating ***wake*** behind the cylinder, with swirls being shed alternately from the top and bottom of the cylinder. If the Reynolds number is large enough [cases (D) and (E)], a turbulent wake forms behind the cylinder.

*wake*

*boundary layer separation*

The wake is a result of ***boundary layer separation,*** a phenomenon in which at some location (the separation location) the fluid flowing around the object does not continue in the direction parallel to the surface, but veers away from the surface. The result is a low-pressure wake region behind the cylinder. If the boundary layer is laminar [case (D) in Fig. 14.16], the wake is wide; if the boundary layer is turbulent [case(E) in Fig. 14.16], the wake is relatively narrow. This narrowing of the wake as the boundary layer becomes turbulent is the cause of the dip in the $C_D$ vs $Re$ curve of Fig. 14.15 that occurs in the range of $10^5 < Re < 10^6$.

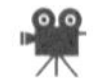

***V14.9 Snow drifts***

Figure 14.17 shows the flow past a circular cylinder for $Re = 10^5$, corresponding to case (D) considered above. The figure shows a relatively thin ($\delta \ll D$) boundary layer, the point of boundary layer separation, and the low-pressure wake region behind the cylinder.

A review of Fig. 14.16 would correctly suggest that the pressure drag is significant in cases such as (D) and (E) for which a low-pressure wake region occurs. On the other hand, in the low-Reynolds-number case (A) there is no separation and the pressure drag contribution to the total drag is much less than the friction drag. As the Reynolds number increases and a wake becomes established behind the cylinder, the pressure drag assumes a greater role in determining the total drag. Eventually, at large enough Reynolds number, pressure drag dominates.

## 14.10 Drag Coefficient Data

Most of the information pertaining to drag is a result of numerous experiments with wind tunnels, water tunnels, towing tanks, and other ingenious devices that are used to measure the drag on scale models. Typically, the result for a given shape is given as a drag coefficient, $C_D$.

Figure 14.14 gives drag coefficient data for a flat plate parallel to the upstream flow. Figure 14.15 gives data for flow past smooth cylinders and spheres. Figure 14.18 gives data for various objects of interest. It might be recalled that the racing bicycle data of this table are used in the discussion of power in Sec. 3.2.2.

Clearly the drag coefficient for an object depends on its shape, with shapes ranging from those that are streamlined to those that are blunt. The amount of streamlining can have a considerable effect on the drag. The goal in streamlining is to eliminate boundary layer separation. In such cases the pressure drag is minimal and the drag is mainly friction drag ***For Example...*** the total drag force on the two-dimensional streamlined strut of length $b$ shown in Fig. 14.19 is the same as for the circular cylinder (also of length $b$) shown to scale. Although the strut is 10 times larger than the cylinder ($A_{\text{strut}} = 10Db$ and $A_{\text{cylinder}} = Db$), its drag coefficient is 10 times smaller. There is no boundary layer separation for the

***V14.10 Streamlined and blunt bodies***

| Shape | Reference area $A$ ($b$ = length) | Drag coefficient $C_D = \dfrac{\mathscr{D}}{\frac{1}{2}\rho U^2 A}$ | |
| --- | --- | --- | --- |
| Square rod with rounded corners | $A = bD$ | $R/D$ | $C_D$ |
| | | 0 | 2.2 |
| | | 0.02 | 2.0 |
| | | 0.17 | 1.2 |
| | | 0.33 | 1.0 |
| Semicircular shell | $A = bD$ | → | 2.3 |
| | | ← | 1.1 |
| Rectangle | $A = bD$ | $\ell/D$ | $C_D$ |
| | | ≤0.1 | 1.9 |
| | | 0.5 | 2.5 |
| | | 0.65 | 2.9 |
| | | 1.0 | 2.2 |
| | | 2.0 | 1.6 |
| | | 3.0 | 1.3 |
| Streamlined strut | $A = bD$ | 0.12 | |
| Cube | $A = D^2$ | 1.05 | |
| Hollow hemisphere | $A = \frac{\pi}{4}D^2$ | → | 1.42 |
| | | ← | 0.38 |
| Circular rod parallel to flow | $A = \frac{\pi}{4}D^2$ | $\ell/D$ | $C_D$ |
| | | 0.5 | 1.1 |
| | | 1.0 | 0.93 |
| | | 2.0 | 0.83 |
| | | 4.0 | 0.85 |

| Shape | Reference area | Drag coefficient $C_D$ | | | |
| --- | --- | --- | --- | --- | --- |
| Parachute | Frontal area $A = \frac{\pi}{4}D^2$ | 1.4 | | | |
| Porous parabolic dish | Frontal area $A = \frac{\pi}{4}D^2$ | Porosity | 0 | 0.2 | 0.5 |
| | | → | 1.42 | 1.20 | 0.82 |
| | | ← | 0.95 | 0.90 | 0.80 |
| | | Porosity = open area/total area | | | |
| Average person | Standing | $C_D A = 9$ ft$^2$ | | | |
| | Sitting | $C_D A = 6$ ft$^2$ | | | |
| | Crouching | $C_D A = 2.5$ ft$^2$ | | | |
| Bikes | | | | | |
| Upright commuter | $A = 5.5$ ft$^2$ | 1.1 | | | |
| Racing | $A = 3.9$ ft$^2$ | 0.88 | | | |
| Drafting | $A = 3.9$ ft$^2$ | 0.50 | | | |
| Streamlined | $A = 5.0$ ft$^2$ | 0.12 | | | |
| Tractor-trailor trucks | | | | | |
| Standard | Frontal area | 0.96 | | | |
| Fairing: With fairing | Frontal area | 0.76 | | | |
| Gap seal: With fairing and gap seal | Frontal area | 0.70 | | | |
| Tree, $U = 10$ m/s | Frontal area | 0.43 | | | |
| $U = 20$ m/s | | 0.26 | | | |
| $U = 30$ m/s | | 0.20 | | | |

*Figure 14.18* Typical drag coefficients for objects of interest; $Re \geq 10^4$.

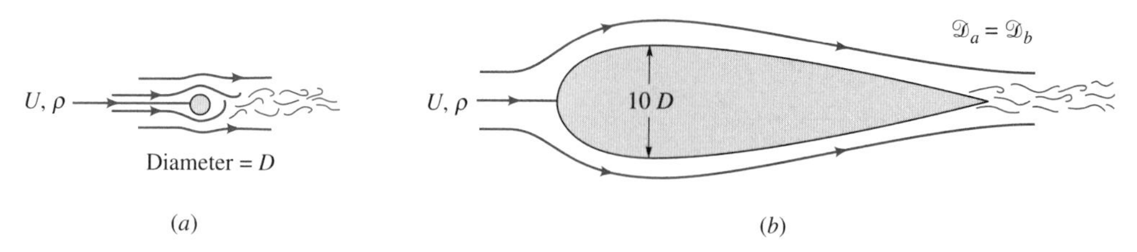


*Figure 14.19* Two objects of considerably different size that have the same drag force: (*a*) circular cylinder $C_D = 1.2$, (*b*) streamlined strut $C_D = 0.12$. The length of each object is *b*.

***V14.11 Skydiving practice***

streamlined strut, so its wake is very narrow, on the order of that for the much smaller circular cylinder whose low pressure wake is approximately the width of the cylinder. For the strut, friction drag is the main contributor to the total drag, whereas for the cylinder, pressure drag is most important. ▲

## *Example 14.8* Drag on an Automobile

As discussed in Sec. 1.2.2, the fuel economy of an automobile can be increased by decreasing the aerodynamic drag on the vehicle, especially at highway speeds. As indicated in Fig. E14.8, by appropriate consideration of numerous design aspects (from the overall shape of the vehicle to the use of recessed door handles and radio antennas), it has been possible to lower the drag coefficient from a typical value of 0.55 for a 1940 model car to 0.30 for a 2003 model. For each of the models shown in the figure, determine the aerodynamic drag and the power needed to overcome this aerodynamic drag at a highway speed of 65 mph = 95.3 ft/s.

### Solution

***Known:*** Air flows past two cars with known drag coefficients with a velocity of 95.3 ft/s.

***Find:*** Determine the drag on each of the cars and the power required to overcome the drag.

***Schematic and Given Data:***

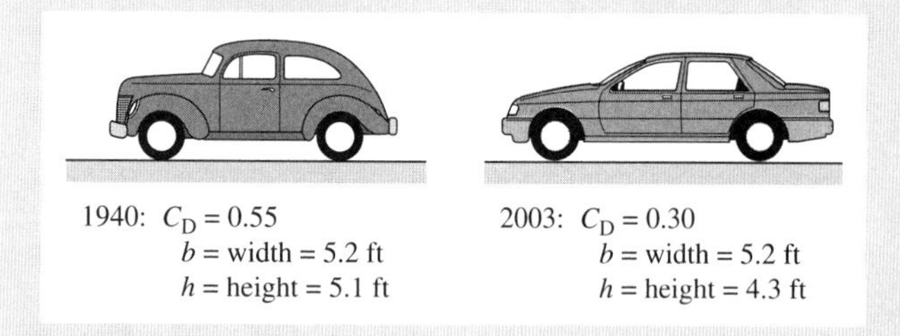


*Figure E14.8*

***Assumptions:***

1. The cars are driven steadily on a calm day so that the upstream velocity is 95.3 ft/s.
2. The air is at standard conditions with $\rho = 0.00238$ slug/ft$^3$ and $\mu = 3.74 \times 10^{-7}$ lbf · s/ft$^2$ (see Appendix FM-1).
3. The power, $\dot{W}$, needed to overcome the aerodynamic drag, $\mathfrak{D}$, is equal to the upstream velocity times the drag; $\dot{W} = U\mathfrak{D}$.

***Analysis:*** The drag on the cars can be calculated from

❶
$$\mathfrak{D} = \tfrac{1}{2}\rho U^2 A C_D \tag{1}$$

where the drag coefficients are given in Fig. E14.8. The frontal area is $A = bh$, where $b$ is the width of the car and $h$ is its height.

Thus, with the given data for the 1940 model, from Eq. 1 we obtain

$$\mathfrak{D}_{1940} = (1/2)(0.00238 \text{ slug/ft}^3)(95.3 \text{ ft/s})^2(5.20 \text{ ft} \times 5.10 \text{ ft})(0.55)\left|\frac{1 \text{ lbf}}{1 \text{ slug} \cdot \text{ft/s}^2}\right|$$

$$= 158 \text{ lbf} \quad \triangleleft$$

Similarly, for the 2003 model,

$$\mathfrak{D}_{2003} = (1/2)(0.00238 \text{ slug/ft}^3)(95.3 \text{ ft/s})^2(5.20 \text{ ft} \times 4.30 \text{ ft})(0.30)\left|\frac{1 \text{ lbf}}{1 \text{ slug} \cdot \text{ft/s}^2}\right|$$

$$= 72.5 \text{ lbf} \quad \triangleleft$$

❷ The power required to overcome the drag is obtained from the equation $\dot{W} = U\mathfrak{D}$. Hence for the 1940 model we obtain

$$\dot{W}_{1940} = (95.3 \text{ ft/s})(158 \text{ lbf})\left|\frac{1 \text{ hp}}{550 \text{ ft} \cdot \text{lb/s}}\right|$$

$$= 27.4 \text{ hp} \quad \triangleleft$$

whereas, for the 2003 model

$$\dot{W}_{2003} = (95.3 \text{ ft/s})(72.5 \text{ lbf})\left|\frac{1 \text{ hp}}{550 \text{ ft} \cdot \text{lb/s}}\right|$$

$$= 12.6 \text{ hp} \quad \triangleleft$$

❶ Since $\mathfrak{D} \sim A\,C_D$, the reduction in drag and power for the 2003 model relative to the 1940 model is due to two factors: (1) the smaller frontal area ($A_{2003} = 5.20 \text{ ft} \times 4.30 \text{ ft} = 22.4 \text{ ft}^2$ compared to $A_{1940} = 5.20 \text{ ft} \times 5.1 \text{ ft} = 26.5 \text{ ft}^2$) and (2) the more streamlined shape with a lower drag coefficient ($C_{D2003} = 0.30$ compared to $C_{D1940} = 0.55$).

❷ Note that the power required to overcome the aerodynamic drag is $\dot{W} = U\mathfrak{D} = \frac{1}{2}\rho U^3 A C_D$. That is, the power is proportional to the speed cubed. Hence, the power required to overcome the aerodynamic drag at 65 mph is $(65 \text{ mph}/55 \text{ mph})^3 = 1.65$ times greater than it is at 55 mph. High speed driving is not as energy efficient as driving at a lower speed.

## 14.11 Lift

Any object moving through a fluid will experience a net force of the fluid on the object. For symmetrical objects, this force will be in the direction of the upstream flow—a drag, $\mathfrak{D}$. If the object is not symmetrical (or if it does not produce a symmetrical flow, such as the flow around a rotating sphere), there may also be a force normal to the upstream flow—a lift, $\mathfrak{L}$.

The most important parameter that affects the lift is the shape of the object, and considerable effort has gone into designing optimally shaped lift-producing devices. Typically, the lift is given in terms of a lift coefficient, $C_L$ (see Eq. 14.18).

Most common lift-generating devices (e.g., airfoils, fans and spoilers on cars) operate in the large Reynolds number range in which the flow has a boundary layer character, with viscous effects confined to the boundary layers and wake regions. Most of the lift comes from the pressure acting on the surface.

Since most airfoils are thin, it is customary to use the planform area, $A = bc$, in the definition of the lift coefficient. Here $b$ is the length of the airfoil (wing tip to wing tip) and $c$ is the *chord length* (the distance from the leading edge to the trailing edge). Typical lift coefficients so defined are on the order of unity. That is, the lift force is on the order of the dynamic pressure times the planform area of the wing, $\mathfrak{L} \approx (\rho U^2/2)A$. The ***wing loading,*** defined as the average lift per unit area of the wing, $\mathfrak{L}/A$, therefore, increases with velocity. ***For Example...*** the wing loading of the 1903 Wright Flyer aircraft was 1.5 lbf/ft$^2$, while for the present-day Boeing 747 aircraft it is 150 lbf/ft$^2$. The wing loading for a bumble bee is approximately 1 lbf/ft$^2$. ▲

*wing loading*

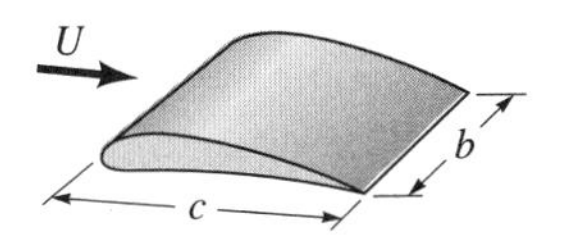

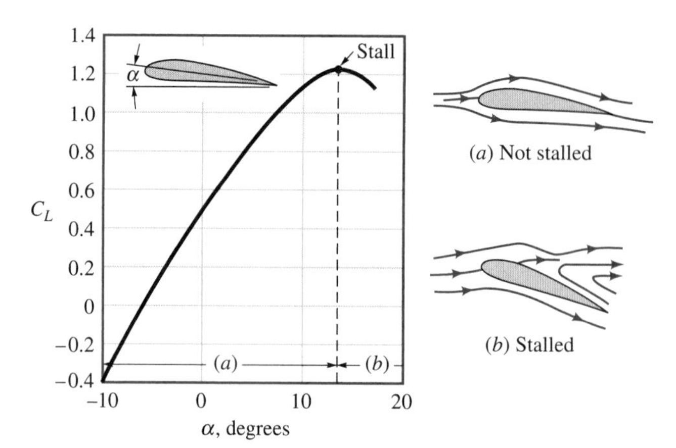


*Figure 14.20* Typical lift coefficient data as a function of angle of attack.

As indicated in Fig. 14.20, the lift coefficient for a wing is a function of the *angle of attack,* α. For small angles of attack, the lift coefficient increases with α, the air flows smoothly over the wing, and there is no boundary layer separation. The wing is a streamlined object. However, for sufficiently large angles of attack the airfoil behaves as a blunt object, there is boundary layer separation on the upper surface, and the lift coefficient suddenly decreases. This condition, termed ***stall,*** is a potentially dangerous situation, especially at low altitudes where there is not sufficient altitude for the plane to recover from the sudden loss of lift.

***stall***

## *Example 14.9* Human-Powered Flight

In 1977 the *Gossamer Condor* (see Fig. E14.9) was recognized for being the first human-powered aircraft to complete a prescribed figure-of-eight course around two turning points 0.5 mi apart. The following data pertain to this aircraft: flight velocity = $U$ = 15 ft/s, airfoil length = $b$ = 96 ft, chord length = $c$ = 7.5 ft (average), weight (including pilot) = $\mathcal{W}$ = 210 lbf. Determine the lift coefficient, $C_L$.

### Solution

***Known:*** Data for a human-powered aircraft.
***Find:*** Determine the lift coefficient needed.

***Schematic and Given Data:***

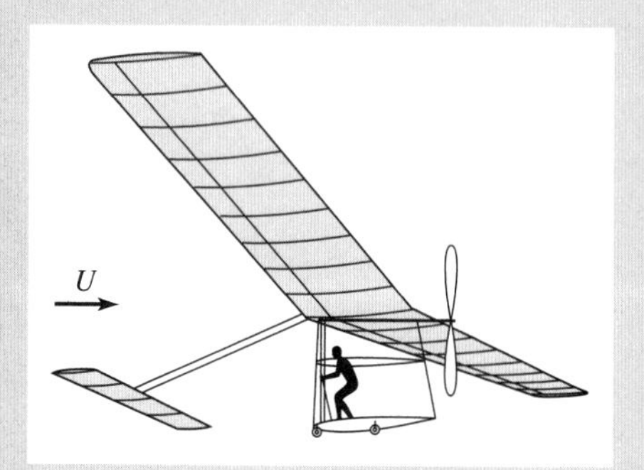


***Assumptions:***
1. The aircraft is flown at a steady speed and a constant altitude.
2. Portions of the aircraft other than the wing (i.e., the tail or fuselage) do not contribute to the lift.
3. The aircraft is flown through standard sea level air with $\rho = 0.00238$ slug/ft$^3$ (see Appendix FM-1).

*Figure E14.9*

***Analysis:*** For steady flight conditions the lift must be exactly balanced by the weight, or

$$\mathcal{W} = \mathcal{L} = \tfrac{1}{2}\rho U^2 A C_L$$

Thus,

$$C_L = \frac{2\mathscr{W}}{\rho U^2 A}$$

❶ where $A = bc = 96 \text{ ft} \times 7.5 \text{ ft} = 720 \text{ ft}^2$, $\mathscr{W} = 210 \text{ lbf}$, and $\rho = 2.38 \times 10^{-3} \text{ slug/ft}^3$ for standard air. This gives

$$C_L = \frac{2(210 \text{ lbf})}{(2.38 \times 10^{-3} \text{ slug/ft}^3)(15 \text{ ft/s})^2(720 \text{ ft}^2)} \left| \frac{1 \text{ slug} \cdot \text{ft/s}^2}{1 \text{ lbf}} \right|$$

or

❷ $$C_L = 1.09 \quad \triangleleft$$

❶ The wing loading for this aircraft is only $\mathscr{W}/A = 210 \text{ lbf}/720 \text{ ft}^2 = 0.292 \text{ lbf/ft}^2$, well below that of even the 1903 Wright Flyer ($1.5 \text{ lbf/ft}^2$) or a bumble bee ($1.0 \text{ lbf/ft}^2$).

❷ This calculated lift coefficient value is consistent with the data for the airfoil of Fig. 14.20.

## 14.12 Chapter Summary and Study Guide

In this chapter we have considered the application of fluid mechanics principles to internal flows through pipes and to external flows past various objects.

We have investigated how the flow in a pipe depends on system parameters such as the pipe diameter, length and material from which it is made; fluid properties such as viscosity and density; the pressure drop or head loss along the pipe; elevation differences along the pipe; and the mechanical energy that devices such as pumps and turbines add to or remove from the fluid. We have considered characteristics of both laminar and turbulent flows and have seen how the analysis of a pipe system can be quite different depending on which type of flow occurs.

We have described the general interaction between an object and the surrounding fluid flowing past it in terms of pressure and viscous forces and the boundary layer that develops along the surface of the object. We have indicated how the characteristics of the boundary layer flow (i.e., laminar or turbulent, separated or not separated) influence the drag on the object. We have discussed how to determine the lift and drag on objects by use of the lift and drag coefficients.

The following checklist provides a study guide for this chapter. When your study of the text and end-of-chapter exercises has been completed you should be able to

- write out the meanings of the terms listed in the margin throughout the chapter and understand each of the related concepts. The subset of key terms listed here in the margin is particularly important.
- determine whether the flow in a pipe is laminar or turbulent.
- determine friction factors using the Colebrook formula or the Moody chart.
- use the concept of friction factors and minor loss coefficients to determine flow rate and head loss for fully developed pipe flow situations.
- discuss the nature of boundary layer flow past a flat plate.
- determine the friction drag on a flat plate for either laminar or turbulent boundary layer flow.
- discuss how various factors affect the drag coefficient.
- determine the lift and drag on an object in terms of its lift and drag coefficients.

*laminar flow*
*turbulent flow*
*Poiseuille's Law*
*relative roughness*
*friction factor*
*Moody chart*
*Colebrook formula*
*minor losses*
*major losses*
*lift and drag*
*lift coefficient*
*drag coefficient*
*boundary layer*
*boundary layer separation*

## Problems

**Note:** Unless otherwise indicated in the problem statement, use values of fluid properties given in the tables of Appendix FM-1 when solving these problems.

### Reynolds Number and Entrance Length

**14.1** Rainwater runoff from a parking lot flows through a 3-ft-diameter pipe, completely filling it. Whether flow in a pipe is laminar or turbulent depends on the value of the Reynolds number. Would you expect the flow to be laminar or turbulent? Support your answer with appropriate calculations.

**14.2** Carbon dioxide at 20°C and a pressure of 550 kPa (abs) flows in a pipe at a rate of 0.04 N/s. Determine the maximum diameter allowed if the flow is to be turbulent.

**14.3** (CD-ROM)

**14.4** To cool a given room it is necessary to supply 4 $ft^3/s$ of air through an 8-in.-diameter pipe. Approximately how long is the entrance length in this pipe?

### Laminar Pipe Flow

**14.5** Water flows through a horizontal 1-mm-diameter tube to which are attached two pressure taps a distance 1 m apart. What is the maximum pressure drop allowed if the flow is to be laminar?

**14.6** (CD-ROM)

**14.7** (CD-ROM)

**14.8** Oil (specific weight = 8900 $N/m^3$, viscosity = 0.10 $N \cdot s/m^2$) flows through a horizontal 23-mm-diameter tube as shown in Fig. P14.8. A differential U-tube manometer is used to measure the pressure drop along the tube. Determine the range of values for $h$ for laminar flow.

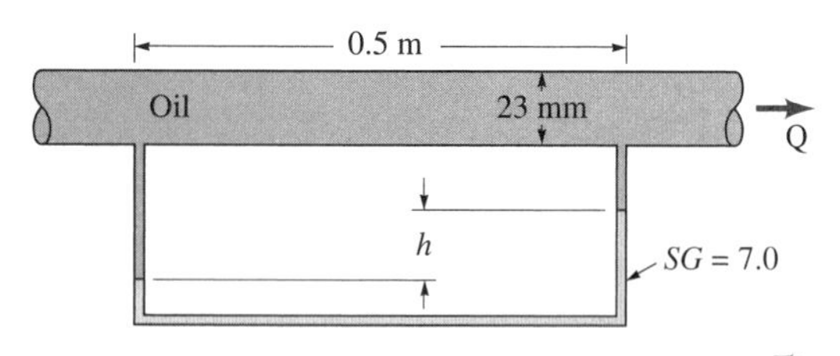


*Figure P14.8*

**14.9** A viscous fluid flows in a 0.10-m-diameter pipe such that its velocity measured 0.012 m away from the pipe wall is 0.8 m/s. If the flow is laminar, determine the centerline velocity and the volumetric flow rate.

### Turbulent Pipe Flow—Friction Factors

**14.10** Water flows through a 6-in.-diameter horizontal pipe at a rate of 2.0 $ft^3/s$ and a pressure drop of 4.2 psi per 100 ft of pipe. Determine the friction factor.

**14.11** A 70-ft-long, 0.5-in.-diameter hose with a roughness of $\varepsilon = 0.0009$ ft is fastened to a water faucet where the pressure is $p_1$. Determine $p_1$ if there is no nozzle attached to the hose and the average velocity in the hose is 6 ft/s. Neglect minor losses and elevation changes.

**14.12** Determine the pressure drop per 100-m length of horizontal new 0.20-m-diameter cast iron water pipe when the average velocity is 1.7 m/s.

**14.13** Natural gas ($\rho = 0.0044$ $slug/ft^3$ and $\nu = 5.2 \times 10^{-5}$ $ft^2/s$) is pumped through a horizontal 6-in.-diameter cast-iron pipe at a rate of 800 lb/hr. If the pressure at section (1) is 50 psi (abs), determine the pressure at section (2) 8 mi downstream if the flow is assumed incompressible.

**14.14** Water flows from one large tank to another at a rate of 0.50 $ft^3/s$ through a horizontal 3-in.-diameter cast-iron pipe of length 200 ft. If minor losses are neglected, determine the difference in elevation of the free surfaces of the tanks.

**14.15** A 3-ft-diameter duct is used to carry ventilating air into a vehicular tunnel at a rate of 9000 $ft^3/min$. Tests show that the pressure drop is 1.5 in. of water per 1500 ft of duct. What is the value of the friction factor for this duct and the approximate size of the equivalent roughness of the surface of the duct?

**14.16** Air flows through the 0.108-in.-diameter, 24-in.-long tube shown in Fig. P14.16. Determine the friction factor if the volumetric flow rate is $Q = 0.00191$ $ft^3/s$ when $h = 1.70$ in. Compare your results with the expression $f = 64/Re$. Is the flow laminar or turbulent?

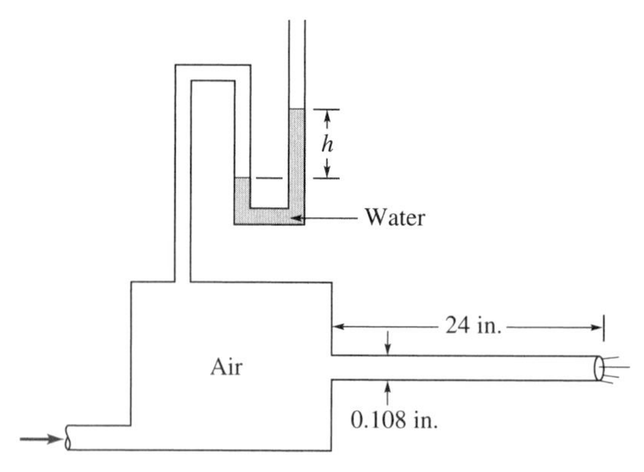


*Figure P14.16*

**14.17** Gasoline flows in a smooth pipe of 40-mm diameter at a rate of 0.001 $m^3/s$. Show that the flow is turbulent. What would be the ratio of the head loss for the actual turbulent flow compared to that if it were laminar flow?

### Minor Losses

**14.18** Air flows through the fine mesh gauze shown in Fig. P14.18 with an average velocity of 1.50 m/s in the pipe. Determine the loss coefficient for the gauze.

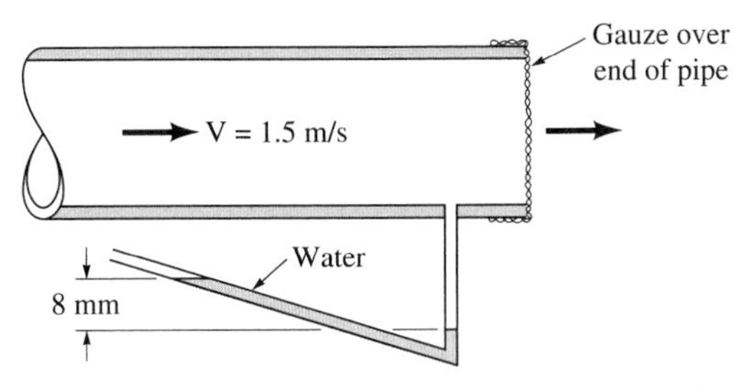


*Figure P14.18*

**14.19** Water flows through the screen in the pipe shown in Fig. P14.19 as indicated. Determine the loss coefficient for the screen.

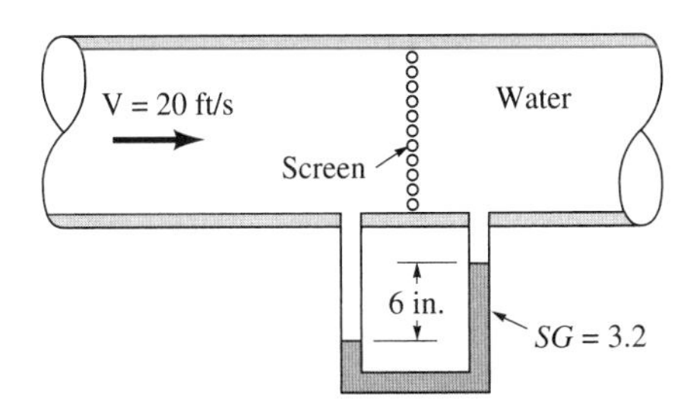


*Figure P14.19*

**14.20** (CD-ROM)

### Pipe Flow—Type I

**14.21** The $\frac{1}{2}$-in.-diameter hose shown in Fig. P14.21 can withstand a maximum pressure within it of 200 psi without rupturing. Determine the maximum length, $\ell$, allowed if the friction factor is 0.022 and the volumetric flow rate is 0.010 ft$^3$/s. Neglect minor losses. The fluid is water.

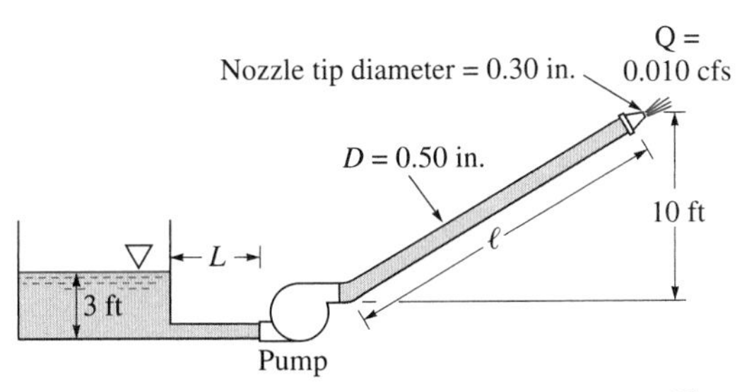


*Figure P14.21*

**14.22** The hose shown in Fig. P14.21 will collapse if the pressure within it is lower than 10 psi below atmospheric pressure. Determine the maximum length, $L$, allowed if the friction factor is 0.015 and the volumetric flow rate is 0.010 ft$^3$/s. Neglect minor losses. The fluid is water.

**14.23** Water flows through the coils of the heat exchanger as shown in Fig. P14.23 at a rate of 0.9 gal/min. Determine the pressure drop between the inlet and outlet of the horizontal device.

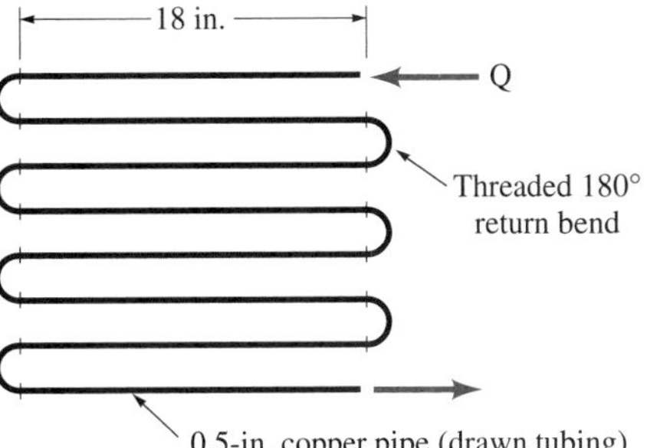


*Figure P14.23*

**14.24** As shown in Fig. P14.24, a water jet rises 3 in. above the exit of the vertical pipe attached to three horizontal pipe segments. The total length of the 0.75-in.-diameter galvanized iron pipe between point (1) and the exit is 21 inches. Determine the pressure needed at point (1) to produce this flow. Note that the velocity of the water exiting the pipe can be determined from the fact that the water rises 3 in. above the exit.

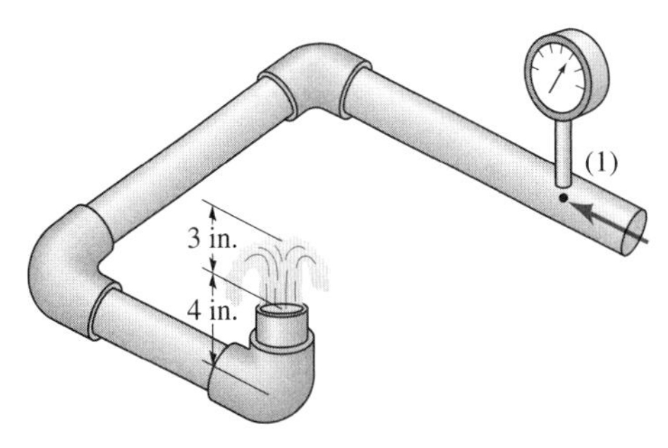


*Figure P14.24*

**14.25** (CD-ROM)

### Pipe Flow—Pumps/Turbines

**14.26** Water flows from a lake as is shown in Fig. P14.26 at a rate of 4.0 ft$^3$/s. Is the device inside the building a pump or a turbine? Explain. Determine the horsepower of the device. Neglect all minor losses and assume the friction factor is 0.025.

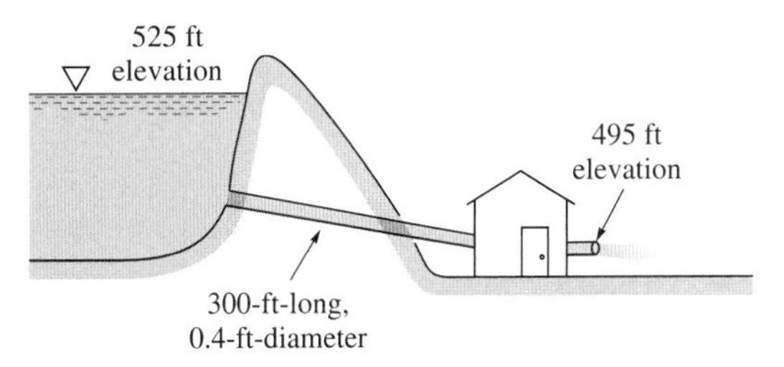


*Figure P14.26*

**14.27** The pump shown in Fig. P14.27 adds power equal to 25 kW to the water and causes a volumetric flow rate of 0.04 m$^3$/s.
(a) Determine the water depth, $h$, in the tank.

(b) If the pump is removed from the system, determine the flow rate expected. Assume $f = 0.016$ for either case and neglect minor losses.

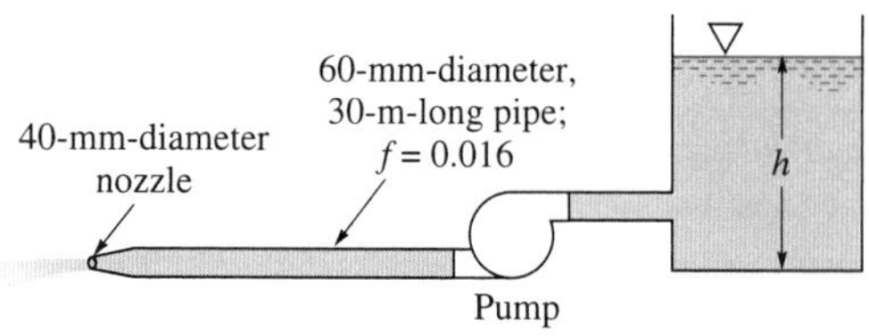


*Figure P14.27*

**Pipe Flow—Type II**

**14.28** A smooth plastic, 10-m-long garden hose with an inside diameter of 15 mm is used to drain a wading pool as is shown in Fig. P14.28. What is the volumetric flow rate from the pool? Assume $K_L = 0.8$ for the minor loss coefficient at the hose entrance.

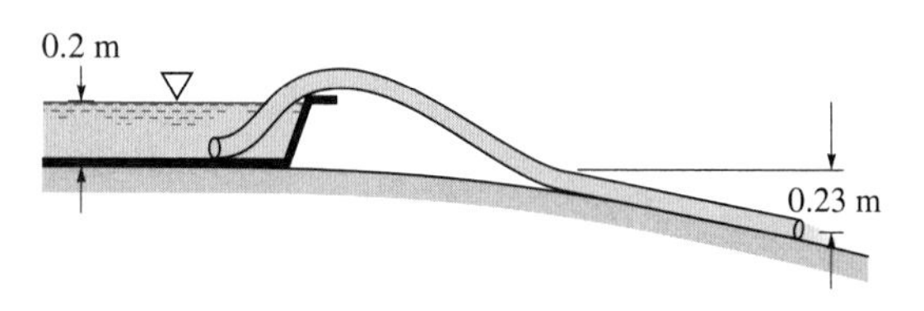


*Figure P14.28*

**14.29** Determine the diameter of a steel pipe that is to carry 2,000 gal/min of gasoline with a pressure drop of 5 psi per 100 ft of horizontal pipe.

**14.30** Water is circulated from a large tank through a filter, and back to the tank as shown in Fig. P14.30. The power added to the water by the pump is 200 ft · lbf/s. Determine the volumetric flow rate through the filter.

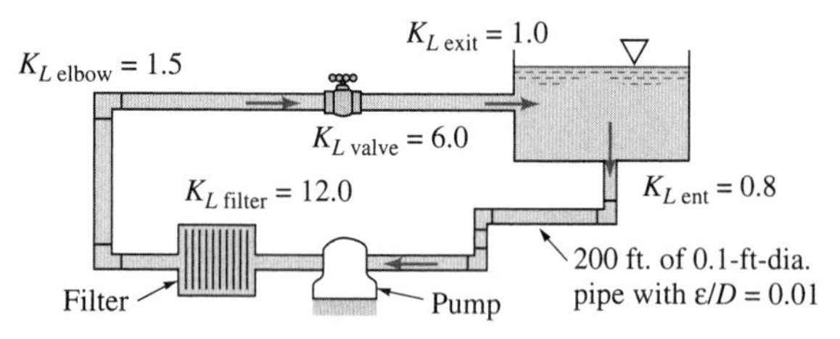


*Figure P14.30*

**14.31** (CD-ROM)

**14.32** (CD-ROM)

**Flowmeters**

**14.33** (CD-ROM)

**14.34** (CD-ROM)

**14.35** (CD-ROM)

**Boundary Layers–External Flow**

**14.36** A smooth flat plate of length $\ell = 6$ m and width $b = 4$ m is placed in water with an upstream velocity of $U = 0.5$ m/s. Determine the boundary layer thickness at the center and the trailing edge of the plate. Assume a laminar boundary layer.

**14.37** A viscous fluid flows past a flat plate such that the boundary layer thickness at a distance 1.3 m from the leading edge is 12 mm. Determine the boundary layer thickness at distances of 0.20, 2.0, and 20 m from the leading edge. Assume laminar flow.

**14.38** The net drag on one side of the two plates (each of size $\ell$ by $\ell/2$) parallel to the free stream shown in Fig. P14.38*a* is $\mathfrak{D}$. Determine the drag (in terms of $\mathfrak{D}$) on the same two plates when they are connected together as indicated in Fig. P14.38*b*. Assume laminar boundary flow. Explain your answer physically.

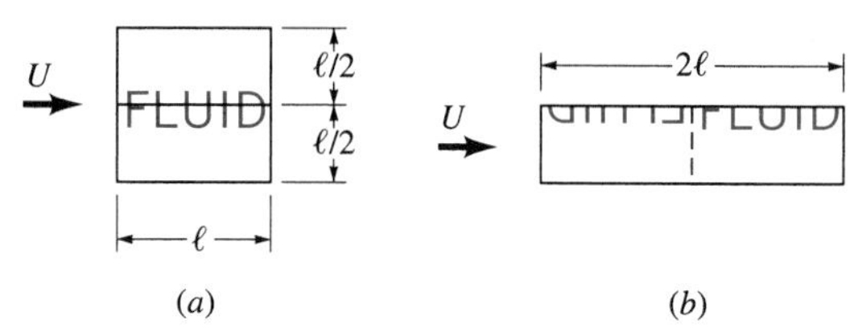


*Figure P14.38*

**14.39** If the drag on one side of a flat plate parallel to the upstream flow is $\mathfrak{D}$ when the upstream velocity is $U$, what will the drag be when the upstream velocity is $2U$; or $U/2$? Assume laminar flow.

**Drag**

**14.40** A 60-mph wind flows against an outdoor movie screen that is 70 ft wide and 20 ft tall. Estimate the wind force on the screen. (See Fig. 14.18 for drag coefficient data.)

**14.41** (CD-ROM)

**14.42** Determine the moment needed at the base of 30-m-tall, 0.12-m-diameter flag pole to keep it in place in a 20 m/s wind.

**14.43** Two bicycle racers ride 30 km/hr through still air. By what percentage is the power required to overcome aerodynamic drag for the second cyclist reduced if she drafts closely behind the first cyclist rather than riding alongside her? Neglect any forces other than aerodynamic drag. (See Fig. 14.18 for drag coefficient data.)

**14.44** A 12-mm-diameter cable is tautly strung between a series of poles that are 60 m apart. Determine the horizontal force this cable puts on each pole if the wind velocity is 30 m/s perpendicular to the cable.

**14.45** (CD-ROM)

**14.46** A 22 in. by 34 in. speed limit sign is supported on a 3-in. wide, 5-ft-long pole. Estimate the bending moment in the pole at ground level when a 30-mph wind blows against the sign. List any assumptions used in your calculations. (See Fig. 14.18 for drag coefficient data.)

**14.47** A 25-ton (50,000-lb) truck coasts down a steep 7% mountain grade without brakes, as shown in Fig. P14.47. The truck's ultimate steady-state speed, V, is determined by a balance between weight, rolling resistance, and aerodynamic drag. Determine V if the rolling resistance for a truck on concrete is

1.2% of the weight and the drag coefficient based on frontal area is 0.76.

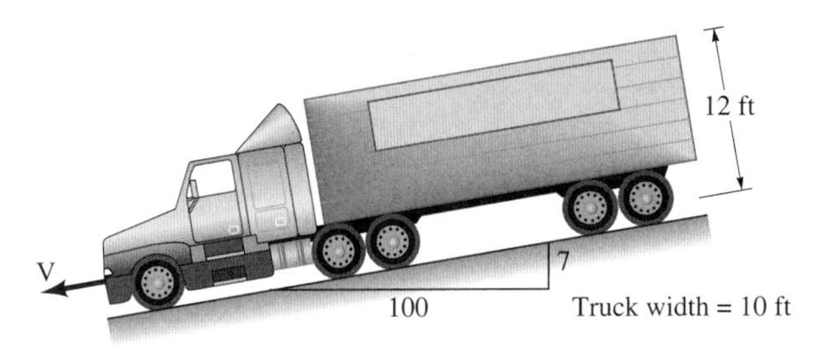


*Figure P14.47*

**14.48** A 38.1-mm-diameter, 0.0245-N table tennis ball is released from the bottom of a swimming pool. With what steady velocity does it rise to the surface?

**14.49** A regulation football is 6.78 in. in diameter and weighs 0.91 lbf. If its drag coefficient is $C_D = 0.2$, determine its deceleration if it has a speed of 20 ft/s at the top of its trajectory.

**14.50** (CD-ROM)

**14.51** A 1.2-lbf kite with an area of 6 ft$^2$ flies in a 20 ft/s wind such that the weightless string makes an angle of 55° relative to the horizontal. If the pull on the string is 1.5 lbf, determine the lift and drag coefficients based on the kite area.

**14.52** A vertical wind tunnel can be used for skydiving practice. Estimate the vertical wind speed needed if a 150-lbf person is to be able to "float" motionless when the person (**a**) curls up as in a crouching position or (**b**) lies flat as shown in Fig.P14.52. (See Fig. 14.18 for drag coefficient data.)

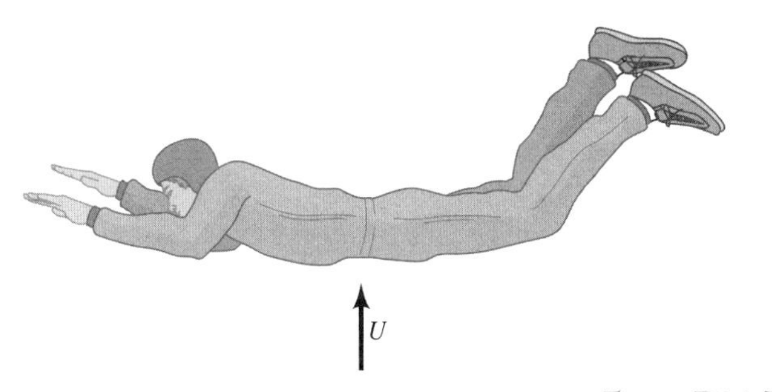


*Figure P14.52*

**Power**

**14.53** The power, $\dot{W}$, needed to overcome the aerodynamic drag on a vehicle traveling at a speed $U$ varies as $\dot{W} \sim U^n$. What is an appropriate value for the constant $n$? Explain.

**14.54** Estimate the power needed to overcome the aerodynamic drag of a person who runs at a rate of 100 yds in 10 s in still air. Repeat the calculations if the race is run into a 20-mph headwind. (See Fig. 14.18 for drag coefficient data.)

**14.55** If for a given vehicle it takes 20 hp to overcome aerodynamic drag while being driven at 55 mph, estimate the horsepower required at 65 mph.

**Lift**

**14.56** A Piper Cub airplane has a gross weight of 1750 lbf, a cruising speed of 115 mph, and a wing area of 179 ft$^2$. Determine the lift coefficient of this airplane for these conditions.

**14.57** (CD-ROM)

**14.58** A Boeing 747 aircraft weighing 580,000 lbf, when loaded with fuel and 100 passengers, takes off with an air speed of 140 mph. With the same configuration (i.e., angle of attack, flap settings, etc.), what is the takeoff speed if it is loaded with 372 passengers? Assume each passenger with luggage weighs 200 lbf.

**14.59** The landing speed of an airplane is dependent on the air density. By what percent must the landing speed be increased on a day when the temperature is 110 deg F compared to a day when it is 50 deg F? Assume the atmospheric pressure remains constant.

**14.60** (CD-ROM)

# GETTING STARTED IN HEAT TRANSFER: MODES, RATE EQUATIONS AND ENERGY BALANCES

## *Introduction...*

From the study of thermodynamics, you learned that energy can be transferred by interactions between a system and its surroundings. These interactions include energy transfer by heat and work, as well as energy transfer associated with mass flow. Thermodynamics deals with the end states of processes during which interactions occur, and also with the *net* amounts of energy transfer by heat and work for the processes. Fluid mechanics deals with the nature of fluid flow and forces that exist within fluids and at the boundaries between fluids and solids. In the following chapters, we extend thermodynamic and fluid mechanics analysis through the study of the *modes of heat transfer* and the development of relations to calculate heat transfer *rates*.

*chapter objective*

The ***objective*** of this chapter is to lay the foundation common to the modes of conduction (Chap. 16), convection (Chap. 17), and radiation (Chap. 18). We begin by addressing the questions of *What is heat transfer*? and *How is energy transferred by heat*? First, we want to help you develop an appreciation for the fundamental concepts and principles that underlie heat transfer processes. Second, we will illustrate the manner in which knowledge of heat transfer processes is used in conjunction with the first law of thermodynamics to solve problems in thermal systems engineering.

## 15.1 Heat Transfer Modes: Physical Origins and Rate Equations

A simple, yet general, definition provides sufficient response to the question: *What* is heat transfer?

*Heat transfer is energy in transit due to a temperature difference.*

*modes of heat transfer*

Whenever there exists a temperature difference in a medium or between media, heat transfer can occur. We refer to the different types of heat transfer processes as ***modes,*** which we subsequently term *conduction, convection,* and *radiation.*

### 15.1.1 Conduction

*conduction*

When a temperature gradient exists in a stationary medium, which may be a solid or a fluid, we use the term ***conduction*** to refer to the heat transfer that will occur across the medium. The *physical mechanism* of conduction involves concepts of atomic and molecular activity, which sustains the transfer of energy from the more energetic to the less energetic particles of a substance due to interactions between the particles.

Consider a *gas* occupying the space between two surfaces maintained at different temperatures and assume that there is *no bulk motion*. We associate the temperature at any point with the energy of the gas molecule. This energy is related to the random translational motion, as well as to the internal rotational and vibrational motions, of the molecules.

Higher temperatures are associated with higher molecular energies, and when neighboring molecules collide, as they are constantly doing, a transfer of energy from the more energetic to the less energetic molecules must occur. In the presence of a temperature gradient, energy transfer by conduction must then occur in the direction of *decreasing* temperature. We may speak of the net transfer of energy by this molecular motion as a *diffusion* of energy. The situation is much the same in *liquids,* although the molecules are more closely spaced and the molecular interactions are stronger and more frequent. In a *solid,* conduction is attributed to atomic activity in the form of lattice vibrations and electron migration. We treat the important properties associated with conduction phenomena in Chap. 16.

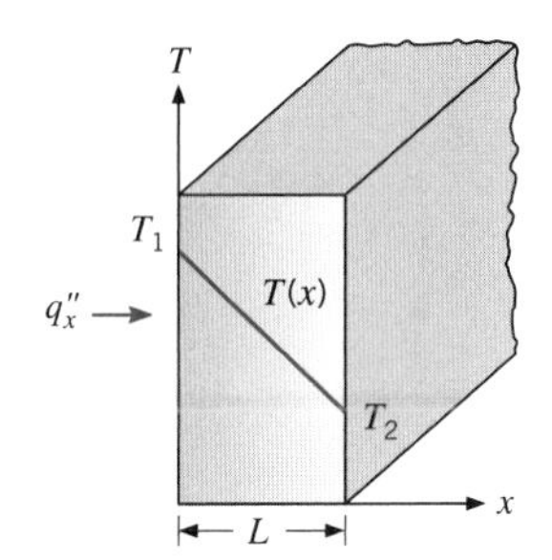


*Figure 15.1* One-dimensional heat transfer by conduction.

Occurrences of conduction heat transfer are legion. ***For Example...*** the exposed end of a metal spoon suddenly immersed in a cup of hot coffee will eventually be warmed due to the conduction of energy through the spoon. On a winter day there is significant energy transfer from a heated room to the outside air. This transfer is principally due to conduction heat transfer through the wall that separates the room air from the outside air. ▲

It is possible to quantify heat transfer processes in terms of appropriate *rate equations.* These equations can be used to compute the amount of energy being transferred per unit time. For heat conduction, the rate equation is known as ***Fourier's law.*** For the one-dimensional plane wall shown in Fig. 15.1, having a temperature distribution $T(x)$, the rate equation is expressed as

$$q''_x = -k\frac{dT}{dx} \tag{15.1}$$

*Fourier's law*

The *heat flux* $q''_x$ (W/m²) is the heat transfer rate in the $x$ direction *per* unit area *perpendicular* to the direction of transfer, and it is proportional to the *temperature gradient, $dT/dx$,* in this direction. The proportionality constant $k$ is a *transport* property known as the ***thermal conductivity*** (W/m · K), and is a characteristic of the wall material. The minus sign is a consequence of the fact that heat is transferred in the direction of decreasing temperature.

*thermal conductivity*

Under the *steady-state conditions* shown in Figure 15.1, where the temperature distribution is *linear,* the temperature gradient and heat flux, respectively, may be expressed as

$$\frac{dT}{dx} = \frac{T_2 - T_1}{L} \qquad q''_x = -k\frac{T_2 - T_1}{L}$$

We can also write this *rate equation* in the form

$$q''_x = k\frac{T_1 - T_2}{L} = k\frac{\Delta T}{L} \tag{15.2}$$

Note that this equation provides a ***heat flux,*** that is, the rate of heat transfer per *unit area.* The ***heat rate*** by conduction, $q_x$ (W), through a plane wall of area $A$, is then the product of the flux and the area, $q_x = q''_x \cdot A$.

*heat flux*
*heat rate*

## *Example 15.1* The Conduction Rate Equation, Fourier's Law

The wall of an industrial furnace is constructed from 0.15-m-thick fireclay brick having a thermal conductivity of 1.7 W/m · K. Measurements made during steady-state operation reveal temperatures of 1400 and 1150 K at the inner and outer surfaces, respectively. What is the rate of heat transfer through a wall that is 0.5 m by 1.2 m on a side?

## Solution

***Known:*** Steady-state conditions with prescribed wall thickness, area, thermal conductivity, and surface temperatures.
***Find:*** Heat transfer rate through the wall.

***Schematic and Given Data:***

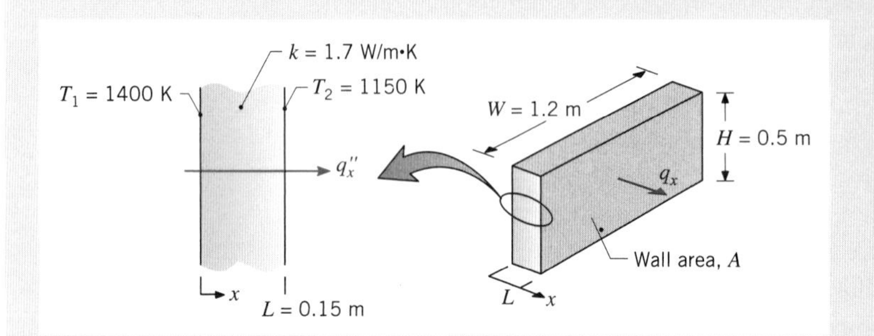


***Assumptions:***
1. Steady-state conditions.
2. One-dimensional conduction through the wall.
3. Constant thermal conductivity.

*Figure E15.1*

***Analysis:*** Since heat transfer through the wall is by conduction, the heat flux may be determined from Fourier's law. Using Eq. 15.2, we have

$$q''_x = k\frac{\Delta T}{L} = 1.7 \text{ W/m}\cdot\text{K} \times \frac{250 \text{ K}}{0.15 \text{ m}} = 2833 \text{ W/m}^2$$

The heat *flux* represents the rate of heat transfer through a section of unit area, and it is uniform across the surface of the wall. The heat *rate* through the wall of area $A = H \times W$ is then

$$q_x = (HW)\, q''_x = (0.5 \text{ m} \times 1.2 \text{ m})\, 2833 \text{ W/m}^2 = 1700 \text{ W} \triangleleft$$

***Comments:*** Note the direction of heat flow and the distinction between heat *flux* and heat *rate*.

### 15.1.2 Convection

*convection*

The term *convection* refers to heat transfer that will occur between a surface and a moving or stationary fluid when they are at different temperatures.

The convection heat transfer *mode* is comprised of *two mechanisms*. In addition to energy transfer due to *random molecular motion* (*conduction*), energy is also transferred by the *bulk*, or *macroscopic*, *motion* of the fluid. This fluid motion is associated with the fact that, at any instant, large numbers of molecules are moving collectively or as aggregates. Such motion, in the presence of a temperature gradient, contributes to heat transfer. Because the molecules in the aggregate retain their random motion, the total heat transfer is then due to a superposition of energy transport by the random motion of the molecules and by the bulk motion of the fluid. It is customary to use the term *convection* when referring to this cumulative transport, and the term *advection* when referring to transport due to bulk fluid motion.

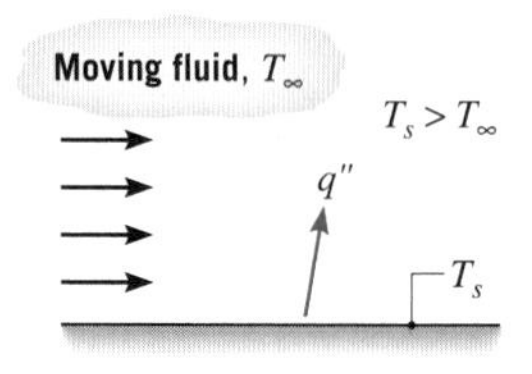


Convection from a surface to a moving fluid

You learned in Sec. 14.8 that, with fluid flow over a surface, viscous effects are important in the hydrodynamic (velocity) boundary layer and, for a Newtonian fluid, the frictional shear stresses are proportional to the velocity gradient. In the treatment of convection in Chap. 17, we will study the *thermal boundary layer,* the region that experiences a temperature distribution from that of the freestream $T_\infty$ to the surface $T_s$ (Fig. 15.2). Appreciation of boundary layer phenomena is essential to understanding convection heat transfer. It is for this reason that the discipline of fluid mechanics will play a vital role in our later analysis of convection.

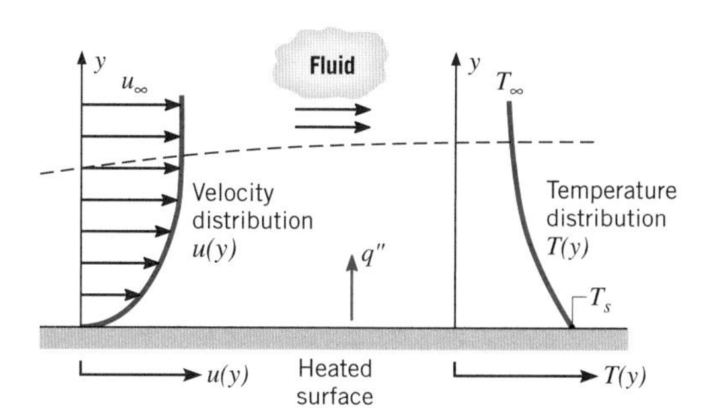


*Figure 15.2* Hydrodynamic and thermal boundary layer development in convection heat transfer.

Convection heat transfer may be classified according to the nature of the flow. We speak of ***forced convection*** when the flow is caused by external means, such as a fan, a pump, or atmospheric winds. In contrast, for ***free*** (or *natural*) ***convection,*** the flow is induced by buoyancy forces, which arise from density differences caused by temperature variations in the fluid. We speak also of ***external*** and ***internal*** flow. As you learned in Chap. 14, external flow is associated with immersed bodies for situations such as flow over plates, cylinders and foils. In internal flow, the flow is constrained by the tube or duct surface. You saw that the corresponding hydrodynamic boundary layer phenomena are quite different, so it is reasonable to expect that the convection processes for the two types of flow are distinctive.

*forced convection*
*free convection*

*external flow*
*internal flow*

Regardless of the particular nature of the convection heat transfer process, the appropriate *rate equation,* known as ***Newton's law of cooling,*** is of the form

$$q'' = h(T_s - T_\infty) \tag{15.3a}$$

*Newton's law of cooling*

where $q''$, the convective *heat flux* (W/m$^2$), is proportional to the difference between the surface and fluid temperatures, $T_s$ and $T_\infty$, respectively, and the proportionality constant $h$ (W/m$^2 \cdot$ K) is termed the ***convection heat transfer coefficient.*** When using Eq. 15.3a, the convection heat flux is presumed to be *positive* if the heat transfer is from the surface ($T_s > T_\infty$) and *negative* if the heat transfer is to the surface ($T_\infty > T_s$). However, if $T_\infty > T_s$, there is nothing to preclude us from expressing Newton's law of cooling as

*convection heat transfer coefficient*

$$q'' = h(T_\infty - T_s) \tag{15.3b}$$

in which case heat transfer is positive to the surface. The choice of Eq. 15.3a or 15.3b is made in the context of a particular problem as appropriate.

The convection coefficient depends on conditions in the boundary layer, which is influenced by surface geometry, the nature of fluid motion, and an assortment of fluid thermodynamic and transport properties. Any study of convection ultimately reduces to a study of the means by which $h$ may be determined. Although consideration of these means is deferred to Chap. 17, convection heat transfer will frequently appear as a boundary condition in the solution of conduction problems (Chap. 16). In the solution of such problems, we presume $h$ to be known, using typical values given in Table 15.1.

**Table 15.1** Typical values of the convection heat transfer coefficient

| Process | $h$ (W/m$^2 \cdot$ K) |
|---|---|
| Free convection | |
| Gases | 2–25 |
| Liquids | 50–1000 |
| Forced convection | |
| Gases | 25–250 |
| Liquids | 100–20,000 |

## 15.1.3 Radiation

The third mode of heat transfer is termed thermal *radiation.* All surfaces of finite temperature emit energy in the form of electromagnetic waves. Hence, in the absence of an intervening medium, there is net heat transfer by radiation between two surfaces at different temperatures.

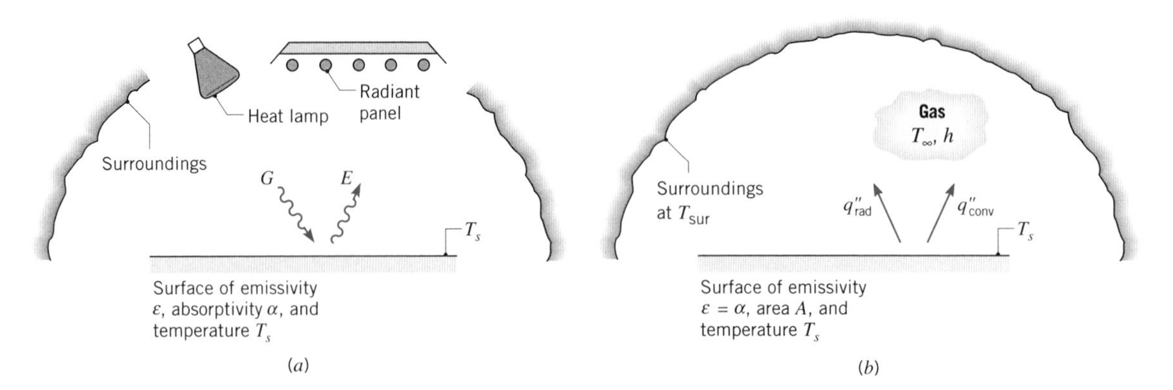


*Figure 15.3* Radiation exchange: (*a*) at a surface in terms of the irradiation *G* provided by different radiation sources and the surface emissive power *E*; and (*b*) between a small, gray surface and its large, isothermal surroundings.

*thermal radiation*

***Thermal radiation*** is energy *emitted* by matter that is at a finite temperature. Although we will focus on radiation from solid surfaces, emission may also occur from liquids and gases. Regardless of the form of matter, the emission may be attributed to changes in the electron configurations of the constituent atoms or molecules. The energy of the radiation field is transported by electromagnetic waves (or alternatively, photons). While the transfer of energy by conduction or convection requires the presence of a material medium, radiation does not. In fact, radiation transfer occurs most efficiently in a vacuum.

Consider radiation transfer processes for the surface of Fig. 15.3*a*. Radiation that is *emitted* by the surface originates from the internal energy of matter bounded by the surface, and the rate at which energy is released per unit area (W/m$^2$) is termed the surface *emissive power E*. There is an upper limit to the emissive power, which is prescribed by the ***Stefan–Boltzmann law***

*Stefan–Boltzmann law*

$$E_b = \sigma T_s^4 \tag{15.4}$$

where $T_s$ is the *absolute temperature* (*K*) of the surface and σ is the *Stefan–Boltzmann constant* ($\sigma = 5.67 \times 10^{-8}$ W/m$^2 \cdot$ K$^4$). Such a surface is called an ideal radiator or ***blackbody.***

*blackbody*

The radiant heat flux emitted by a *real surface* is less than that of a blackbody at the same temperature and is given by

$$E = \varepsilon \sigma T_s^4 \tag{15.5}$$

*emissivity*

where ε is a *radiative property* of the surface termed the ***emissivity.*** With values in the range $0 \leq \varepsilon \leq 1$, this property provides a measure of how efficiently a surface emits energy relative to a blackbody. It depends strongly on the surface material and finish, and representative values are provided in Chap. 18.

Radiation can also be *incident* on a surface. The radiation can originate from a special source, such as the sun, or from other surfaces to which the surface of interest is exposed. Irrespective of the source(s), we designate the rate at which all such radiation is incident on a unit area (W/m$^2$) of the surface as the ***irradiation*** *G* (Fig. 15.3*a*).

*irradiation*

A portion, or all, of the irradiation may be *absorbed* by the surface, thereby increasing the internal energy of the material. The rate at which radiant energy is absorbed per unit surface area may be evaluated from knowledge of a surface radiative property termed the ***absorptivity*** α. That is

*absorptivity*

$$G_{abs} = \alpha G \tag{15.6}$$

where $0 \leq \alpha \leq 1$. If $\alpha < 1$, a portion of the irradiation is not absorbed and may be *reflected* or *transmitted*.

Note that the value of $\alpha$ depends on the nature of the irradiation, as well as on the surface itself. For example, the absorptivity of a surface to solar radiation may differ from its absorptivity to radiation emitted by the walls of a furnace or a heat lamp.

A special case that occurs frequently involves radiation exchange between a *small surface* at $T_s$ and a much *larger, isothermal surface* that completely surrounds the smaller one (Fig. 15.3*b*). The ***surroundings*** could, for example, be the walls of a room or a furnace whose temperature $T_{sur}$ differs from that of an enclosed surface ($T_{sur} \neq T_s$). We will show in Chap. 18 that, for such a condition, the irradiation may be approximated by emission from a blackbody at $T_{sur}$, in which case $G = \sigma T^4_{sur}$. If the surface is assumed to be one for which $\alpha = \varepsilon$ (called a ***diffuse-gray surface***), the *net* rate of ***radiation exchange*** *leaving* the surface, expressed *per* unit area of the surface, is

*surroundings*

$$q''_{rad} = \frac{q}{A} = \varepsilon E_b(T_s) - \alpha G = \varepsilon\sigma(T_s^4 - T_{sur}^4) \tag{15.7}$$

*radiation exchange: diffuse-gray surface —large surroundings*

This expression provides the difference between internal energy that is released due to radiation emission and that which is gained due to radiation absorption.

There are many applications for which it is convenient to express the net radiation exchange in the form

$$q_{rad} = h_{rad} A(T_s - T_{sur}) \tag{15.8}$$

where, with Eq. 15.7, the ***radiation heat transfer coefficient*** $h_{rad}$ is

$$h_{rad} \equiv \varepsilon\sigma(T_s + T_{sur})(T_s^2 + T_{sur}^2) \tag{15.9}$$

*radiation heat transfer coefficient*

Here we have modeled the radiation mode in a manner similar to convection. In this sense we have *linearized* the radiation rate equation, making the heat transfer rate proportional to a temperature difference rather than to the difference between two temperatures to the fourth power. Note, however, that $h_{rad}$ *depends strongly on temperature,* while the temperature dependence of the convection heat transfer coefficient $h$ is generally weak.

The surfaces of Fig. 15.3 may also simultaneously experience convection heat transfer to an adjoining gas. For the conditions of Fig. 15.3*b*, the total rate of heat transfer *leaving* the surface is then

$$q = q_{conv} + q_{rad} = hA(T_s - T_\infty) + \varepsilon A\sigma(T_s^4 - T_{sur}^4) \tag{15.10}$$

## *Example 15.2* Rate Equations for Convection and Radiation Exchange

An uninsulated steam pipe passes through a large room in which the air and walls are at 25°C. The outside diameter of the pipe is 70 mm, and its surface temperature and emissivity are 200°C and 0.8, respectively. What are the surface emissive power and irradiation? If the coefficient associated with free convection heat transfer from the surface to the air is 15 W/m$^2 \cdot$ K and the surface is gray, what is the rate of heat transfer from the surface per unit length of pipe?

### Solution

***Known:*** Uninsulated pipe of prescribed diameter, emissivity, and surface temperature in a large room with fixed wall and air temperatures.

***Find:*** Surface emissive power, $E$, and irradiation, $G$. Pipe heat transfer per unit length, $q'$.

***Schematic and Given Data:***

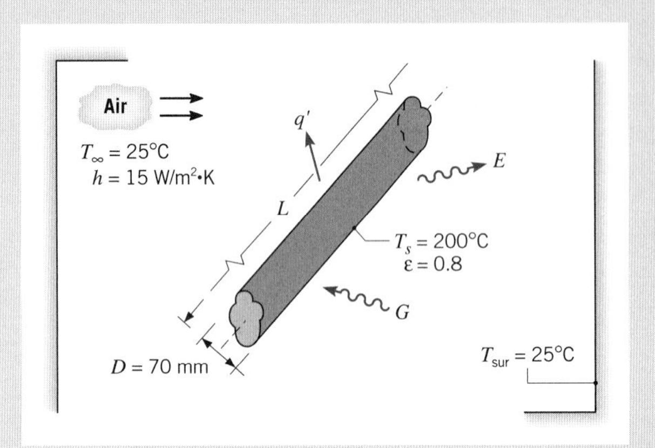


*Figure E15.2*

***Assumptions:***

1. Steady-state conditions.
2. Radiation exchange between the pipe and the room is between a small surface and large, isothermal surroundings.
3. The surface is diffuse-gray; that is, the emissivity and absorptivity are equal.

***Analysis:***

The surface emissive power may be evaluated from Equation 15.5, while the irradiation corresponds to $G = \sigma T^4_{sur}$. Hence

$$E = \varepsilon\sigma T_s^4 = 0.8(5.67 \times 10^{-8}\ \text{W/m}^2 \cdot \text{K}^4)(473\ \text{K})^4 = 2270\ \text{W/m}^2$$

$$G = \sigma T^4_{sur} = 5.67 \times 10^{-8}\ \text{W/m}^2 \cdot \text{K}^4\ (298\ \text{K})^4 = 447\ \text{W/m}^2$$

Heat transfer from the pipe is by convection to the room air and by radiation exchange with the walls. Hence, $q = q_{conv} + q_{rad}$ and from Equation 15.10, with $A = \pi DL$

$$q = h(\pi DL)(T_s - T_\infty) + \varepsilon(\pi DL)\sigma(T_s^4 - T^4_{sur})$$

The heat transfer per unit length of pipe is then

$$q' = \frac{q}{L} = 15\ \text{W/m}^2 \cdot \text{K}(\pi \times 0.07\ \text{m})(200 - 25)°\text{C} + 0.8(\pi \times 0.07\ \text{m})\ 5.67 \times 10^{-8}\ \text{W/m}^2 \cdot \text{K}^4\ (473^4 - 298^4)\ \text{K}^4$$

$$q' = 577\ \text{W/m} + 421\ \text{W/m} = 998\ \text{W/m} \ \triangleleft$$

***Comments:***

1. Note that temperature may be expressed in units of °C or K when evaluating the *temperature difference* for a convection (or conduction) heat transfer rate. However, temperature must be expressed in kelvins (K) when evaluating a radiation transfer rate.
2. In this situation the radiation and convection heat transfer rates are comparable because $T_s$ is large compared to $T_{sur}$, and the coefficient associated with free convection is small. For more moderate values of $T_s$ and the larger values of $h$ associated with forced convection, the effect of radiation may often be neglected. The radiation heat transfer coefficient may be computed from Equation 15.9, and for the conditions of this problem its value is $h_{rad} = 11\ \text{W/m}^2 \cdot \text{K}$.

## 15.2 Applying the First Law in Heat Transfer

The subjects of thermodynamics, fluid mechanics, and heat transfer are highly complementary. ***For Example…*** because it deals with the details of the *rate* at which energy is transferred by heat, the subject of heat transfer may be viewed as an extension of thermodynamics. Still, for many heat transfer problems, the principle of conservation of energy introduced in Chap. 3 is an essential tool. ▲

We have used conservation of energy throughout this text in the form of energy balances commonly encountered in thermodynamics (Sec. 3.6 and 5.2), and the mechanical energy equation used in fluid mechanics (Sec. 12.6). In this section, the conservation of energy principle will be applied, but in the form of the *internal energy equation* (Sec. 7.10) commonly used in heat transfer.

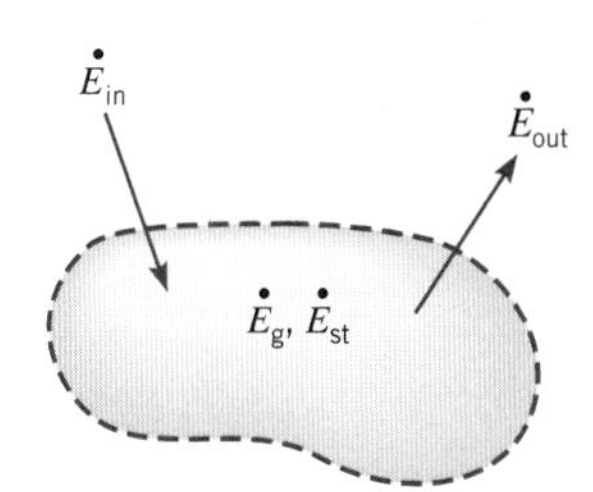


*Figure 15.4* Conservation of energy for a system. Application at an instant.

Consider applying the internal energy equation to the system identified by the dashed line in Fig. 15.4. Identified on the figure are relevant ***internal energy terms*** in the notation used in heat transfer:

***internal energy terms***

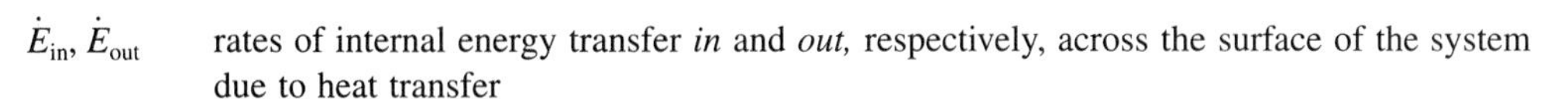

$\dot{E}_{in}, \dot{E}_{out}$ rates of internal energy transfer *in* and *out,* respectively, across the surface of the system due to heat transfer

$\dot{E}_g$ rate of internal energy *generation* within the system

$\dot{E}_{st}$ rate of internal energy *storage* within the system

Accordingly, the internal ***energy balance*** on a ***rate*** basis (Eq. 7.57) is written as

$$\dot{E}_{in} + \dot{E}_g - \dot{E}_{out} = \dot{E}_{st} \tag{15.11a}$$

***energy rate balance***

Equation 15.11a may be applied at any *instant of time.* An alternative form that applies for a process is obtained by integrating Eq. 15.11a over the *time interval* $\Delta t$. That is

$$E_{in} + E_g - E_{out} = \Delta E_{st} \tag{15.11b}$$

Equations 15.11a and 15.11b indicate that the internal energy *inflow* and generation act to increase the amount of internal energy stored in the system, whereas the *outflow* acts to decrease the stored internal energy.

The *inflow* and *outflow* terms are ***surface phenomena.*** That is, they are associated exclusively with processes occurring at the boundary or surface of the system. A common situation involves internal energy inflow and outflow due to conduction, convection, and/or radiation. In situations involving fluid flow across the surface of a control volume, we will apply the control volume energy balance, Eqs. 5.10 and 5.11.

***surface phenomena***

As noted in Sec. 7.10, the ***internal energy generation*** term accounts for the *conversion* of mechanical energy into internal energy, including the passage of current through an electrical resistance, as well as other effects such as electromagnetic absorption and chemical and nuclear reactions. ***For Example...*** if an exothermic chemical reaction occurs within a system, the temperature might increase spontaneously throughout the volume due to the local generation of internal energy by the reaction. ▲

***internal energy generation***

Each of the phenomena that lead to internal energy generation can be modeled as occurring in a distributed way throughout the volume, and the total rate of internal generation is proportional to the volume. Thus they are referred to as ***volumetric phenomena.*** If the internal energy generation $\dot{E}_g$ (W) occurs uniformly throughout the medium of volume $V$ (m$^3$), we can define the ***volumetric generation rate*** $\dot{q}$ (W/m$^3$)

***volumetric phenomena***

$$\dot{q} = \frac{\dot{E}_g}{V} \tag{15.12}$$

***volumetric energy generation rate***

In the case of electric current flow through a resistor, the energy generation rate, also known as *electrical power dissipation,* can be expressed as

$$\dot{E}_g = I^2 R_e \tag{15.13a}$$

where $I$ is the current, in amperes (A), and $R_e$ is the electrical resistance, in ohms (Ω), and the internal energy generation rate is in watts (W). When the resistance is expressed per unit length of the electrical conductor, or $R'_e = R_e/L$, then Eq. 15.13a takes the form

$$\dot{E}_g = I^2 R'_e L \tag{15.13b}$$

*internal energy storage*

The ***internal energy storage*** term represents the rate of accumulation (or reduction) of internal energy in the system. In the applications considered here, the change of internal energy is often indicated by increases (or decreases) in temperature at different locations within the system. In some cases, we consider phase changes from saturated liquid (or solid) to saturated vapor at constant pressure. In those instances, there is no change in temperature. *For systems at steady state, the internal energy storage term reduces to zero.*

Equations 15.11a,b are used to develop more specific forms of the conservation of energy for particular heat transfer applications as illustrated in the following examples.

## *Example 15.3* Applying the First Law on a Rate Basis

A long conducting rod of diameter $D$ and electrical resistance per unit length $R'_e$ is initially in thermal equilibrium with the ambient air and its surroundings. This equilibrium condition is disturbed when an electrical current $I$ is passed through the rod. Develop an equation that could be used to compute the variation of the rod temperature with time during passage of the current.

### Solution

***Known:*** The temperature of a rod changes with time due to passage of an electrical current.
***Find:*** Equation that governs temperature change with time for the rod.

***Schematic and Given Data:***

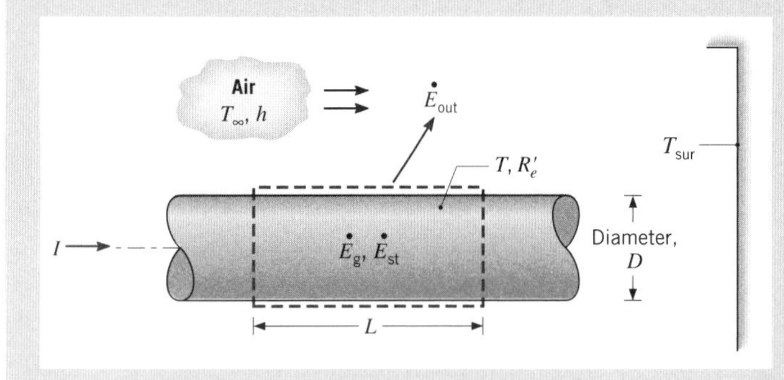


***Assumptions:***
1. At any time $t$ the temperature of the rod is uniform.
2. Constant properties.
3. Radiation exchange between the outer surface of the rod and the surroundings is between a small, diffuse-gray surface and large, isothermal surroundings.

*Figure E15.3a*

***Analysis:*** The first law of thermodynamics may often be used to determine an unknown temperature. In this case, relevant terms include heat transfer by convection and radiation from the surface, internal energy generation due to electrical current passage through the conductor, and a change in internal energy storage. Since we wish to determine the rate of change of the temperature, the first law should be applied at an instant of time. Hence, applying Eq. 15.11a to a system of length $L$ about the rod, it follows that

$$\dot{E}_g - \dot{E}_{out} = \dot{E}_{st}$$

where the energy generation due to the electric resistance heating is given by Eq. 15.13b

$$\dot{E}_g = I^2 R'_e L$$

Energy outflow due to convection and net radiation leaving the surface is given by Eq. 15.10

$$\dot{E}_{out} = h(\pi DL)(T - T_\infty) + \varepsilon\sigma(\pi DL)(T^4 - T_{sur}^4)$$

With Eq. 4.20, the change in energy storage due to the temperature change is

$$\dot{E}_{st} = \frac{dU}{dt} = \rho c V \frac{dT}{dt}$$

where $\rho$ and $c$ are the density and the specific heat, respectively, of the rod material, and $V$ is the volume of the rod $V = (\pi D^2/4)L$. Substituting the rate equations into the energy balance, it follows that

$$I^2 R'_e L - h(\pi DL)(T - T_\infty) - \varepsilon\sigma(\pi DL)(T^4 - T^4_{sur}) = \rho c\left(\frac{\pi D^2}{4}\right)L\frac{dT}{dt}$$

Hence, the time rate of change of the rod temperature is

$$\frac{dT}{dt} = \frac{I^2 R'_e - \pi Dh(T - T_\infty) - \pi D\varepsilon\sigma(T^4 - T^4_{sur})}{\rho c(\pi D^2/4)} \quad \triangleleft$$

***Comments:***

1. Internal energy generation occurs uniformly within the system and could also be expressed in terms of a volumetric generation rate $\dot{q}$ (W/m$^3$). The generation rate for the entire system is then $\dot{E}_g = \dot{q}V$, where $\dot{q} = I^2 R'_e/(\pi D^2/4)$.
2. The differential equation for $dT/dt$ could be solved for the time dependence of the rod temperature by integrating numerically. A *steady-state condition* would eventually be reached for which $dT/dt = 0$. The rod temperature is then determined by an algebraic equation of the form

$$0 = I^2 R'_e - \pi Dh(T - T_\infty) - \pi D\varepsilon\sigma(T^4 - T^4_{sur})$$

3. Parameter study: effect of current on temperature. (CD-ROM)
4. Parameter study: effect of convection coefficient on allowable current. (CD-ROM)
5. Using the *Interactive Heat Transfer* (*IHT*) software. (CD-ROM)

## *Example 15.4* Applying the First Law for an Interval of Time

Ice of mass $m$ at the fusion temperature ($T_f = 0°C$) is enclosed in a cubical cavity of width $W$ on a side. The cavity wall is of thickness $L$ and thermal conductivity $k$. If the outer surface of the wall is at a temperature $T_s > T_f$, obtain an expression for the time required to completely melt the ice.

**Solution (CD-ROM)**

## 15.3 The Surface Energy Balance

We will frequently have occasion to apply the conservation of energy requirement at the surface of a medium. In this special case, the control surface includes no mass or volume, and appears as shown in Fig. 15.5. Accordingly, the generation and storage terms of the conservation expression, Eq. 15.11a, are no longer relevant, and it is only necessary to deal with surface phenomena. For this case, the conservation of energy requirement becomes

$$\dot{E}_{in} - \dot{E}_{out} = 0 \qquad (15.14)$$

*surface energy balance*

which is called the ***surface energy balance.*** Eq. 15.14 indicates that the rate at which energy is transfered to the surface is equal to the rate at which energy is transfered from the surface. Even though energy generation may be occurring in the medium, the process would not affect the energy balance at the surface. Moreover, the surface energy balance holds for both *steady-state* and *transient* conditions.

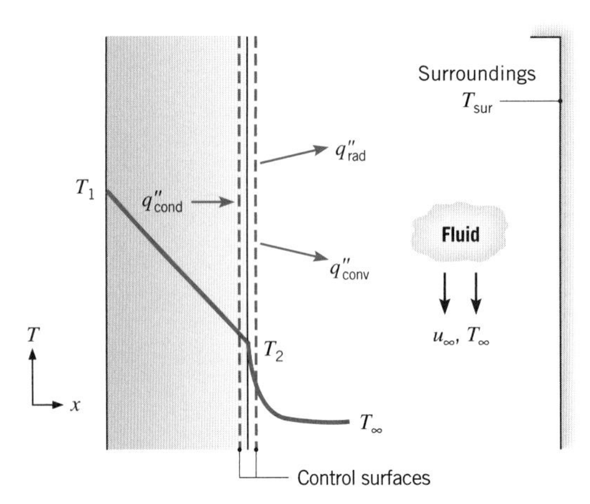


*Figure 15.5* The energy balance for conservation of energy at the surface of a medium.

In Fig. 15.5 three heat transfer processes are shown for the control surface. On a unit area basis, they are conduction from the medium *to* the control surface ($q''_{cond}$), convection *from* the surface to a fluid ($q''_{conv}$), and net radiation exchange from the surface to the surroundings ($q''_{rad}$). The surface energy balance then takes the form

$$q''_{cond} - q''_{conv} - q''_{rad} = 0 \tag{15.15}$$

and we can express each of the terms using the appropriate rate equations, Eqs. 15.2, 15.3, and 15.7.

## *Example 15.5* Applying the Surface Energy Balance with Multiple Heat Transfer Modes

The hot combustion gases of a furnace are separated from the ambient air and its surroundings, which are at 25°C, by a brick wall 0.15 m thick. The brick has a thermal conductivity of 1.2 W/m · K and a surface emissivity of 0.8. Under steady-state conditions an outer surface temperature of 100°C is measured. Free convection heat transfer to the air adjoining the surface is characterized by a convection coefficient of $h = 20\ \text{W/m}^2 \cdot \text{K}$. What is the brick inner surface temperature?

### Solution

***Known:*** Outer surface temperature of a furnace wall of prescribed thickness, thermal conductivity, and emissivity. Ambient conditions.

***Find:*** Wall inner surface temperature, $T_1$.

***Schematic and Given Data:***

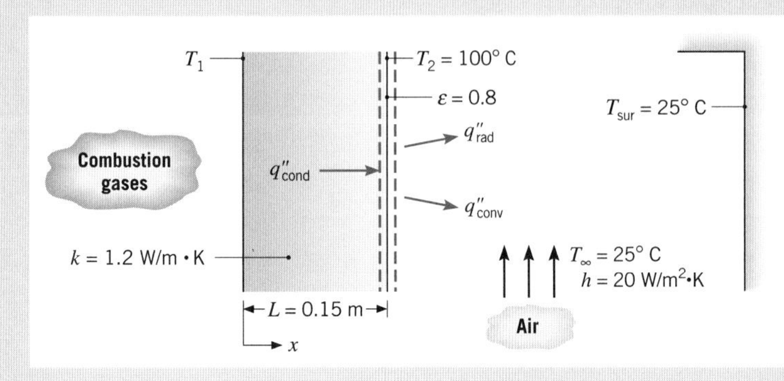


***Assumptions:***

1. Steady-state conditions.
2. One-dimensional heat transfer by conduction across the wall.
3. Radiation exchange between the outer surface of the wall and the surroundings is between a small, diffuse-gray surface and large, isothermal surroundings.

*Figure E15.5*

***Analysis:*** The inner surface temperature $T_1$ may be obtained by performing an energy balance at the outer surface. From Eq. 15.14

$$\dot{E}_{\text{in}} - \dot{E}_{\text{out}} = 0$$

it follows that, on a unit area basis

$$q''_{\text{cond}} - q''_{\text{conv}} - q''_{\text{rad}} = 0$$

or, rearranging and substituting from Eqs. 15.2, 15.3a, and 15.7

$$k\frac{T_1 - T_2}{L} = h(T_2 - T_\infty) + \varepsilon\sigma(T_2^4 - T_{\text{sur}}^4)$$

Therefore, substituting the appropriate numerical values, we find

$$\begin{aligned} 1.2\ \text{W/m}\cdot\text{K}\frac{(T_1 - 373)\ \text{K}}{0.15\ \text{m}} &= 20\ \text{W/m}^2\cdot\text{K}\,(373 - 298)\ \text{K} \\ &\quad + 0.8(5.67 \times 10^{-8}\ \text{W/m}^2\cdot\text{K}^4)(373^4 - 298^4)\ \text{K}^4 \\ &= 1500\ \text{W/m}^2 + 520\ \text{W/m}^2 = 2020\ \text{W/m}^2 \end{aligned}$$

Solving for $T_1$, find the inner wall temperature as

$$T_1 = 373\ \text{K} + \frac{0.15\ \text{m}}{1.2\ \text{W/m}\cdot\text{K}}(2020\ \text{W/m}^2) = 625\ \text{K} = 352^\circ\text{C} \quad \triangleleft$$

***Comments:***

1. Note that the contribution of radiation exchange to the total heat transfer rate from the outer surface is significant. The relative contribution would diminish, however, with increasing $h$ and/or decreasing $T_2$.
2. When using energy balances involving radiation exchange and other modes of heat transfer, it is good practice to express all temperatures in kelvin units. This practice is *necessary* when the unknown temperature appears in the radiation term and in one or more of the other terms.

## *Example 15.6* Curing a Coating with a Radiant Source

The coating on a plate is cured by exposure to an infrared lamp providing an irradiation of 2000 W/m$^2$. It absorbs 80% of the irradiation from the lamp and has an emissivity of 0.50. It is also exposed to an air flow and large surroundings for which temperatures are 20°C and 30°C, respectively. The convection coefficient between the coating and the ambient air is 15 W/m$^2\cdot$K, and the back side of the plate is insulated. What is the temperature of the coated plate?

### Solution

***Known:*** Coating with prescribed radiation properties is cured by irradiation from an infrared lamp. Heat transfer from the coating is by convection to ambient air and radiation exchange with the surroundings.

***Find:*** Temperature of the coated plate, $T_s$.

***Schematic and Given Data:***

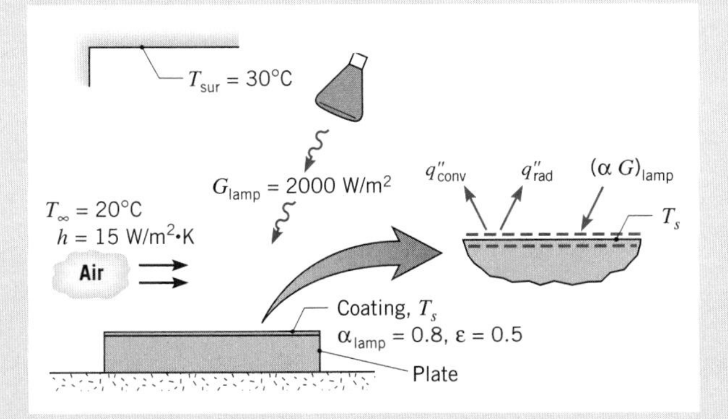


***Assumptions:***

1. Steady-state conditions.
2. Negligible heat loss from back side of the plate.
3. Plate is small object in large surroundings. Coating is diffuse-gray, having an absorptivity of $\alpha = \varepsilon = 0.5$ with respect to irradiation from the surroundings.
4. Absorptivity of lamp irradiation is $\alpha_{\text{lamp}} = 0.80$.

*Figure E15.6*

***Analysis:*** Since the process corresponds to steady-state conditions and there is no heat transfer at the plate back side, the plate and the coating are both at $T_s$. Hence the coating temperature may be determined by placing a control surface about the exposed surface and applying Eq. 15.14

$$\dot{E}_{\text{in}} - \dot{E}_{\text{out}} = 0$$

With energy inflow due to absorption of the lamp irradiation and outflow due to convection and net radiation transfer to the surroundings, it follows that

$$(\alpha G)_{\text{lamp}} - q''_{\text{conv}} - q''_{\text{rad}} = 0$$

Substituting the rate equations from Eqs. 15.3a and 15.7, we obtain

$$(\alpha G)_{\text{lamp}} - h(T_s - T_\infty) - \varepsilon\sigma(T_s^4 - T_{\text{sur}}^4) = 0$$

Substituting numerical values

$$0.8 \times 2000 \text{ W/m}^2 - 15 \text{ W/m}^2 \cdot \text{K}(T_s - 293) \text{ K} - 0.5 \times 5.67 \times 10^{-8} \text{ W/m}^2 \cdot \text{K}^4(T_s^4 - 303^4) \text{ K}^4 = 0$$

and solving iteratively, we obtain the coating temperature

$$T_s = 377 \text{ K} = 104°\text{C} \quad \triangleleft$$

***Comments:*** The coating (plate) temperature may be elevated by increasing $T_\infty$ and $T_{\text{sur}}$, as well as by decreasing the air velocity and hence the convection coefficient.

## *Example 15.7* Identifying Relevant Heat Transfer Modes

A closed container filled with hot coffee is in a room whose air and walls are at a fixed temperature. Identify all heat transfer processes that contribute to cooling of the coffee. Comment on features that would contribute to an improved container design.

### Solution

***Known:*** Hot coffee is separated from its cooler surroundings by a plastic flask, an air space, and a plastic cover.

***Find:*** Relevant heat transfer processes.

***Schematic and Given Data:***

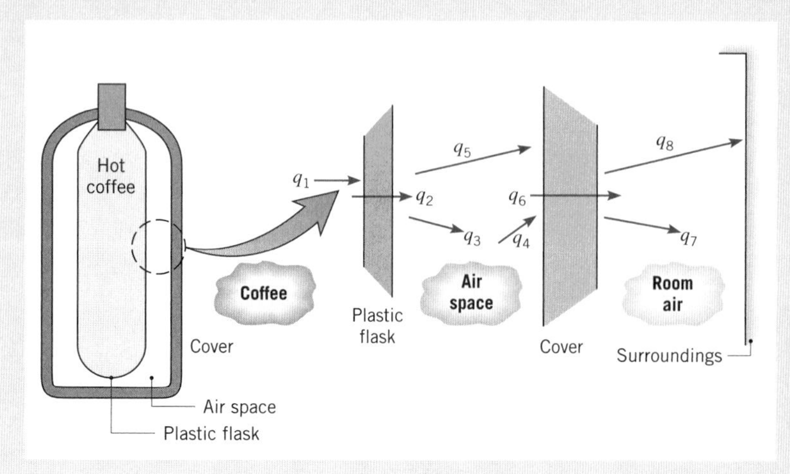


*Figure E15.7*

***Analysis:*** Pathways for energy transfer from the coffee are as follows:

$q_1$: free convection from the coffee to the flask
$q_2$: conduction through the flask
$q_3$: free convection from the flask to the air space
$q_4$: free convection from the air space to the cover
$q_5$: net radiation exchange between the outer surface of the flask and the inner surface of the cover
$q_6$: conduction through the cover
$q_7$: free convection from the cover to the room air
$q_8$: net radiation exchange between the outer surface of the cover and the surroundings

***Comments:*** Design improvements are associated with (1) use of aluminized (low emissivity) surfaces for the flask and cover to reduce net radiation exchange, and (2) evacuating the air space or using a filler material to retard free convection.

## 15.4 Summary and Study Guide

Although much of the material in this chapter will be treated in greater detail in later ones, you should be developing a reasonable overview. In particular, you should know the several *modes of heat transfer* and their *physical origins.* Moreover, given a physical situation, you should be able to identify the relevant transport phenomena (see Example 15.7). You should be able to use the *rate equations* in Table 15.2 (see following page) to compute transfer rates. The conservation of energy principle plays an important role in heat transfer, and as in thermodynamics and fluid mechanics, careful identification of *systems, control volumes,* and *control surfaces* is very important. The conservation of energy principle may be used with the rate equations to solve numerous heat transfer problems.

The following checklist provides a study guide for this chapter. When your study of the text and end-of-chapter problems has been completed you should be able to

- write out the meanings of the terms listed in the margins throughout the chapter and understand each of the related concepts. The subset of key terms listed below in the margin is particularly important.
- identify the physical mechanisms associated with heat transfer by conduction, convection, and radiation.
- explain the difference between a heat *flux* and a heat *rate,* and specify the appropriate units.
- write Fourier's law and explain its role in heat transfer.
- explain thermal conductivity and specify its units.
- write Newton's law of cooling and explain the role played by the convection heat transfer coefficient.
- write the Stefan-Boltzmann law, and identify what unit of temperature must be used with the law.
- explain emissivity and absorptivity and the role they play in characterizing radiation transfer at a surface.
- write the conservation of internal energy requirement for a system on a rate basis. Identify the terms representing surface and volumetric phenomena.
- write the surface energy balance and identify the terms.

***modes of heat transfer***
***heat flux, heat rate***
***Fourier's law***
***thermal conductivity***
***Newton's law of cooling***
***convection coefficient***
***Stefan-Boltzmann law***
***emissivity, absorptivity***
***energy rate balance***
***surface energy balance***

**Table 15.2** Summary of heat transfer processes

| Mode | Mechanism(s) | Rate Equation | Equation Number | Transport Property or Coefficient |
|---|---|---|---|---|
| Conduction | Energy transfer due to molecular/atomic activity | $q''_x(\text{W/m}^2) = -k\dfrac{dT}{dx}$ | (15.1) | $k$ (W/m · K) |
| Convection | Energy transfer due to molecular motion (conduction) plus energy transfer due to bulk motion (advection) | $q''(\text{W/m}^2) = h(T_s - T_\infty)$ | (15.3a) | $h$ (W/m$^2$ · K) |
| Radiation | Energy transfer by electromagnetic waves; radiation exchange, diffuse-gray surface-large surroundings | $q''(\text{W/m}^2) = \varepsilon\sigma(T_s^4 - T_{sur}^4)$<br>or $q''(W) = h_{rad}(T_s - T_{sur})$ | (15.7)<br>(15.8) | $\varepsilon$<br>$h_{rad}$ (W/m$^2$ · K) |

# Problems

## Conduction

**15.1** The horizontal concrete slab of a basement is 11 m long, 8 m wide, and 0.20 m thick. During the winter, temperatures are nominally 17°C and 10°C at the top and bottom surfaces, respectively. If the concrete has a thermal conductivity of 1.4 W/m · K, what is the heat transfer rate through the slab?

**15.2** A heat transfer rate of 3 kW is conducted through a section of an insulating material of cross-sectional area 10 m$^2$ and thickness 2.5 cm. If the inner (hot) surface temperature is 415°C and the thermal conductivity of the material is 0.2 W/m · K, what is the outer surface temperature?

**15.3** A concrete wall, which has a surface area of 20 m$^2$ and is 0.30 m thick, separates conditioned room air from ambient air. The temperature of the inner surface of the wall is maintained at 25°C, and the thermal conductivity of the concrete is 1 W/m · K.

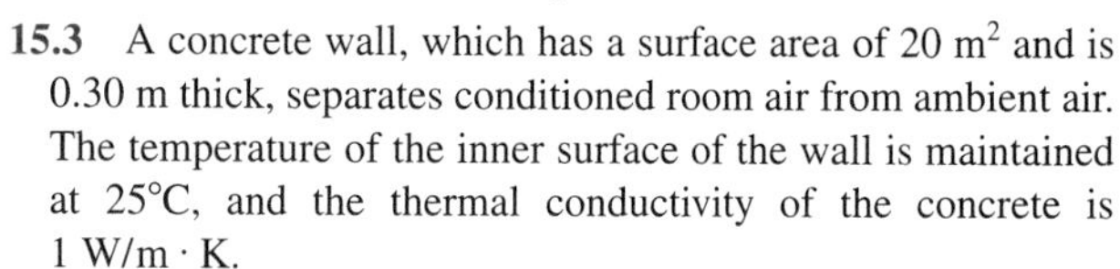

(a) Determine the heat transfer rate through the wall for outer surface temperatures ranging from −15°C to 38°C, which correspond to winter and summer extremes, respectively. Display your results graphically.

(b) On your graph, also plot the heat transfer rate as a function of the outer surface temperature for wall materials having thermal conductivities of 0.75 and 1.25 W/m · K. Explain the family of curves you have obtained.

**15.4** (CD-ROM)

**15.5** (CD-ROM)

## Convection

**15.6** The case of a power transistor, which is of length $L$ = 10 mm and diameter $D$ = 12 mm, is cooled by an air stream of temperature $T_\infty$ = 25°C as shown in Fig. P15.6.

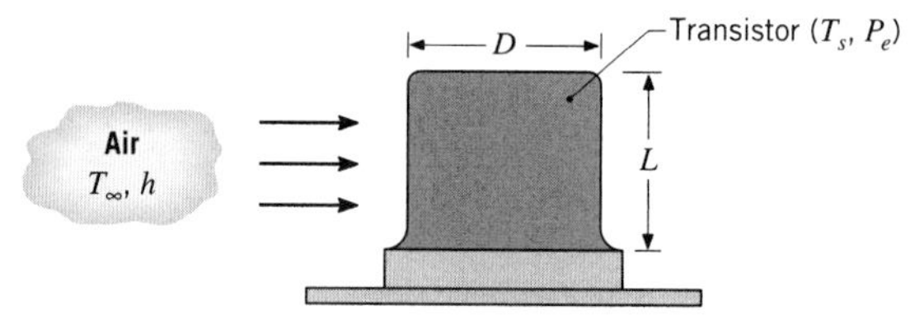


*Figure P15.6*

Under conditions for which the air maintains an average convection coefficient of $h$ = 100 W/m$^2$ · K on the surface of the case, what is the maximum allowable power dissipation $P_e$ if the surface temperature $T_s$ is not to exceed 85°C?

**15.7** A cartridge electrical heater is shaped as a cylinder of length $L$ = 200 mm and outer diameter $D$ = 20 mm. Under normal operating conditions the heater dissipates 2 kW, while submerged in a water flow that is at 20°C and provides a convection heat transfer coefficient of $h$ = 5000 W/m$^2$ · K. Neglecting heat transfer from the ends of the heater, determine its surface temperature $T_s$. If the water flow is inadvertently terminated while the heater continues to operate, the heater surface is exposed to air that is also at 20°C but for which $h$ = 50 W/m$^2$ · K. What is the corresponding surface temperature? What are the consequences of such an event?

**15.8** The temperature controller for a clothes dryer consists of a bimetallic switch mounted on an electrical heater attached to a wall-mounted insulation pad (Fig. P15.8). The switch is set to open at 70°C, the maximum dryer air temperature. In order to operate the dryer at a lower air temperature, sufficient power is supplied to the heater such that the switch reaches 70°C ($T_{set}$) when the air temperature $T_\infty$ is less than $T_{set}$. If the convection heat transfer coefficient between the air and the exposed switch

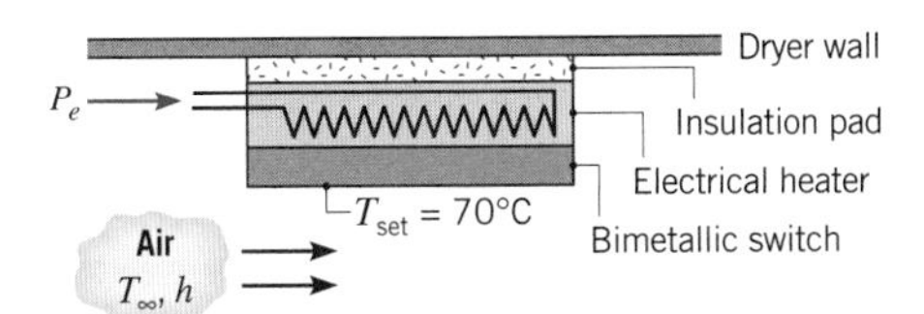


*Figure P15.8*

surface of 30 mm$^2$ is 25 W/m$^2\cdot$K, how much heater power $P_e$ is required when the desired dryer air temperature is $T_\infty = 50°C$?

**15.9** (CD-ROM)

**15.10** (CD-ROM)

## Radiation

**15.11** A spherical interplanetary probe of 0.5-m diameter contains electronics that dissipate 150 W. If the probe surface has an emissivity of 0.8 and the probe does not receive radiation from the sun or deep space, what is its surface temperature?

**15.12** A surface of area 0.5 m$^2$, emissivity 0.8, and temperature 150°C is placed in a large, evacuated chamber whose walls are maintained at 25°C. What is the rate at which radiation is *emitted* by the surface? What is the net rate at which radiation is *exchanged* between the surface and the chamber walls?

**15.13** A vacuum chamber, as used in sputtering electrically conducting thin films on microcircuits, is comprised of a baseplate maintained by an electrical heater at $T_s = 300$ K and a shroud within the enclosure maintained at 77 K by a liquid-nitrogen ($LN_2$) coolant loop. $LN_2$ enters as a saturated liquid, experiences evaporation, and exits the loop as saturated vapor. The baseplate, insulated on the lower side, is 0.3 m in diameter and has an emissivity of $\varepsilon = 0.25$.

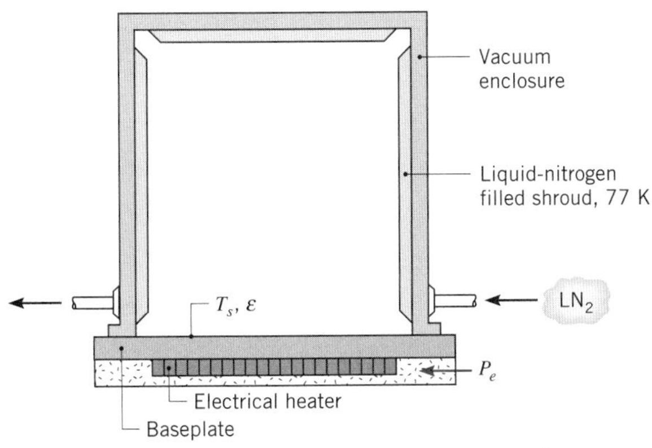


*Figure P15.13*

(a) How much electrical power $P_e$ must be provided to the baseplate heater?

(b) At what rate must liquid nitrogen be supplied to the shroud if its heat of vaporization ($h_{fg}$) is 125 kJ/kg?

(c) To reduce the liquid-nitrogen consumption, it is proposed to bond a thin sheet of aluminum foil ($\varepsilon = 0.09$) to the baseplate. Will this have the desired effect?

**15.14** (CD-ROM)

**15.15** (CD-ROM)

## Energy Balance and Multimode Effects

**15.16** Consider the steam pipe of Example 15.2. The facilities manager wants you to recommend methods for reducing the heat transfer to the room, and two options are proposed. The first option would restrict air movement around the outer surface of the pipe and thereby reduce the convection coefficient by a factor of two. The second option would coat the outer surface of the pipe with a low emissivity ($\varepsilon = 0.4$) paint. Which of the foregoing options would you recommend?

**15.17** The curing process of Example 15.6 involves exposure of the plate to irradiation from an infrared lamp and attendant cooling by convection and radiation exchange with the surroundings. Alternatively, in lieu of the lamp, heating may be achieved by inserting the plate in an oven whose walls (the surroundings) are maintained at an elevated temperature. Consider conditions for which the oven walls are at 200°C, air flow over the plate is characterized by $T_\infty = 20°C$ and $h = 15$ W/m$^2\cdot$K, and the coating has an emissivity of $\varepsilon = 0.5$. What is the temperature of the coating?

**15.18** (CD-ROM)

**15.19** (CD-ROM)

**15.20** In an orbiting space station, an electronic package is housed in a compartment having a surface area, $A_s = 1$ m$^2$, which is exposed to space. Under normal operating conditions, the electronics dissipate 1 kW, all of which must be transferred from the exposed surface to deep space (0 K). If the surface emissivity is 1.0 and the surface is not exposed to the sun, what is its steady-state temperature? If the surface is exposed to a solar flux of 750 W/m$^2$, and its absorptivity to solar radiation is 0.25, what is its steady-state temperature?

**15.21** Electronic power devices are mounted to a *heat sink* having an exposed surface area of 0.045 m$^2$ and an emissivity of 0.80 (Fig. P15.21). When the devices dissipate a total power of 20 W and the air and surroundings are at 27°C, the average sink temperature is 42°C. What average temperature will the heat sink reach when the devices dissipate 30 W for the same environmental conditions?

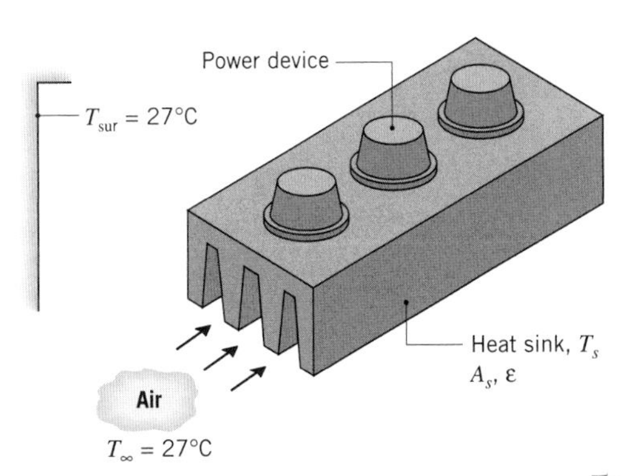


*Figure P15.21*

15.22 (CD-ROM)

15.23 (CD-ROM)

## Process Identification

15.24 In analyzing the performance of a thermal system, the engineer must be able to identify the relevant heat transfer processes. Only then can the system behavior be properly quantified. For the following systems, identify the pertinent processes, designating them by appropriately labeled arrows on a sketch of the system. Answer additional questions that appear in the problem statement.

(a) Identify the heat transfer processes that determine the temperature of an asphalt pavement on a summer day. Write an energy balance for the surface of the pavement.

(b) Consider an exposed portion of your body (e.g., your forearm with a short-sleeved shirt) while you are sitting in a room. Identify all heat transfer processes that occur at the surface of your skin. In the interest of conserving fuel and funds, you keep the thermostat of your home at 15°C (59°F) throughout the winter months. You are able to tolerate this condition if the outside (ambient) air temperature exceeds −10°C (14°F) but feel cold if the ambient temperature falls much below this value. Are you imagining things?

(c) A thermocouple junction is used to measure the temperature of a hot gas stream flowing through a channel by inserting the junction into the mainstream of the gas. The surface of the channel is cooled such that its temperature is well below that of the gas. Identify the heat transfer processes associated with the junction surface. Will the junction sense a temperature that is less than, equal to, or greater than the gas temperature? A radiation shield is a small, open-ended tube that encloses the thermocouple junction, yet allows for passage of the gas through the tube. How does use of such a shield improve the accuracy of the temperature measurement?

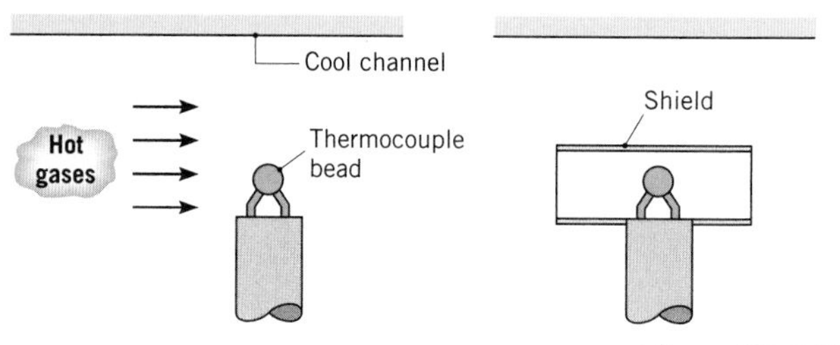


*Figure P15.24*

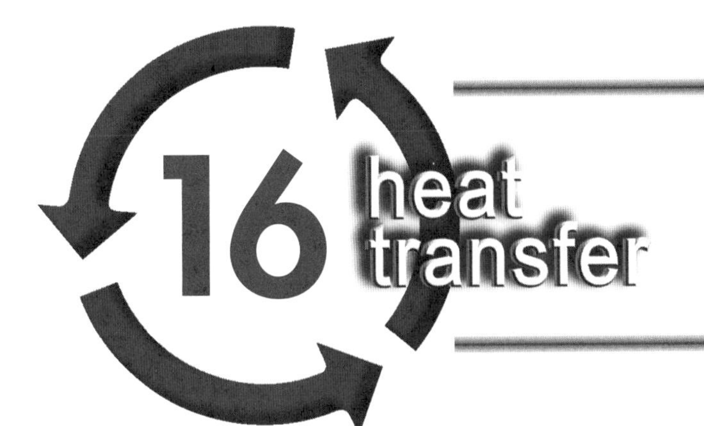


# HEAT TRANSFER BY CONDUCTION

## *Introduction...*

In Chap. 15, we learned that conduction heat transfer is governed by *Fourier's law.* We also learned that from knowledge of the manner in which the temperature varies within the medium, referred to as the *temperature distribution,* the law can be used to determine the *heat flux.* The aim in a conduction analysis is to determine the temperature distribution in a medium resulting from conditions imposed on its boundaries.

*chapter objectives*

The ***first objective*** of this chapter is to understand how the *heat equation,* based upon Fourier's law and the conservation of energy requirement, can be used to obtain the temperature distribution within a medium for *steady-state* and *transient* conditions. The ***second objective*** is to show how *thermal circuits* can be used to model steady-state heat flow in common geometries such as the plane wall, cylinder, sphere, and extended surface (fin). The ***third objective*** is to solve *transient conduction* problems using the *lumped capacitance method,* which is appropriate when a single temperature can be used to characterize the time response of the medium to the boundary change. When *spatial effects* must be considered, we will use analytical solutions to the heat equation.

## 16.1 Introduction to Conduction Analysis

Conduction analysis is about determining the temperature distribution within a medium resulting from conditions at its boundaries. With knowledge of the temperature distribution, the heat flux distribution can be determined using Fourier's law.

### 16.1.1 More About Fourier's Law

In Sec. 15.1.1, we introduced *Fourier's law,* Eq. 15.1, which relates the *heat flux* ($W/m^2$) in the $x$-direction, per unit area perpendicular to the direction of transfer, to the product of the *thermal conductivity* ($W/m \cdot K$) and the *temperature gradient,* $(dT/dx)$, in the $x$-direction

$$q''_x = -k\frac{dT}{dx} \tag{16.1}$$

Fourier's law, as written above, implies that the heat flux is a directional quantity. The relationship between the coordinate system, heat flow direction, and temperature gradient in *one dimension* is illustrated in Fig. 16.1*a*. If the temperature distribution is linear, the gradient is constant, and therefore, the heat flux is a constant: $q''_x$ is independent of $x$. When the *temperature distribution* is nonlinear with the $x$-coordinate as shown in Fig. 16.1*b*, the gradient

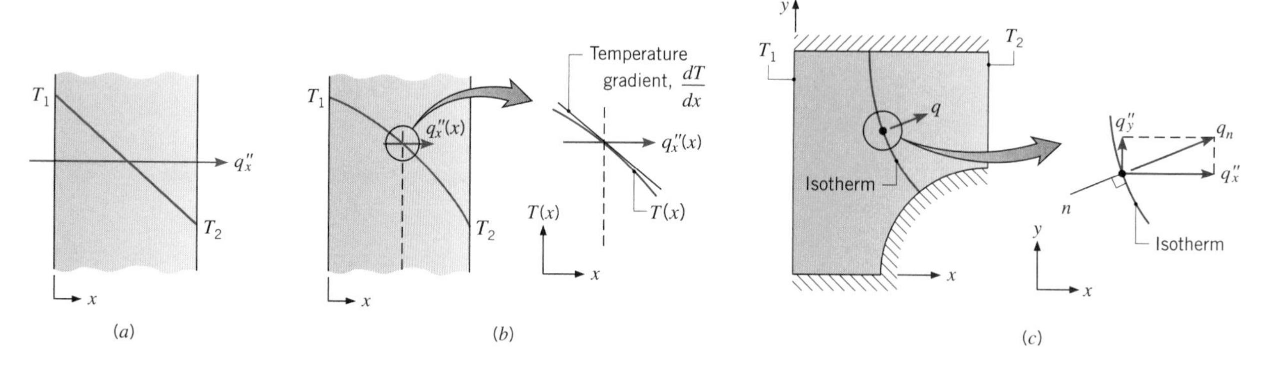


*Figure 16.1* Relationships between heat flux, temperature gradient, and coordinate system. One-dimensional temperature distributions: (*a*) Linear with constant heat flux, and (*b*) Nonlinear with variable heat flux. (*c*) Heat flux vector normal to an isotherm in a two-dimensional coordinate system.

is no longer a constant, and consequently, the heat flux will be a function of the $x$-coordinate, $q''_x(x)$. Later we'll explore what conditions give rise to nonlinear temperature and nonconstant ***heat flux distributions***.

*heat flux distribution*

Consider the object of Fig. 16.1*c* experiencing *two-dimensional* conduction. Notice the line of constant temperature, referred to as an *isotherm,* near the midpoint of the object. The heat flux, $\mathbf{q}''_n$, a vector quantity, is in the direction normal to the isotherm. The heat flux is sustained by a temperature gradient in the $n$-direction and can be expressed in terms of its $x$- and $y$-direction components

$$\mathbf{q}''_n = \mathbf{q}''_x + \mathbf{q}''_y$$

Each of the heat flux components could be expressed in terms of their respective gradients. While we treat only one-dimensional conduction in this text, recognize that the concepts you will learn can be extended to two- and three-dimensional conduction.

The origin of Fourier's law is phenomenological. That is, it is developed from observed phenomena—the generalization of extensive experimental evidence—rather than being derived from first principles. The expression defines the important material property, *thermal conductivity* (see Fig. 16.2), one of several *thermophysical* (*transport*) *properties* you will encounter in performing conduction analyses. Tabulated values of the thermophysical properties required for solution of heat transfer problems are provided in Appendix HT for selected technical materials (HT-1), common materials (HT-2), gases (HT-3), saturated liquids (HT-4), and saturated water (HT-5). Many of the example problems will demonstrate how to use these tables effectively.

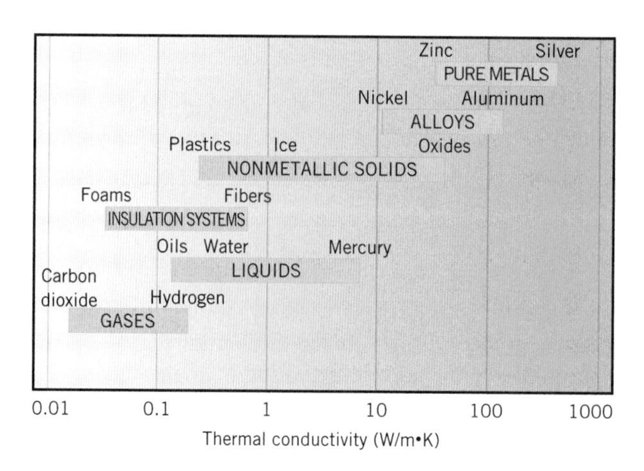


*Figure 16.2* Range of thermal conductivity for various phases of matter at normal temperatures and pressure.

### 16.1.2 The Heat Equation

We consider now the manner in which the temperature distribution in a medium resulting from conditions imposed on the boundaries can be determined.

We will determine the temperature distribution $T(x, t)$ associated with one-dimensional (Cartesian coordinate) heat transfer in a stationary, homogenous medium experiencing uniform volumetric energy generation $\dot{q}$ (W/m$^3$, see Eq. 15.12). We will define a differential system (element), identify relevant energy processes, introduce the appropriate rate equations, and apply the conservation of energy requirement. The result is a differential equation whose solution for the prescribed boundary and initial conditions provides the temperature distribution in the medium.

#### The Heat Equation: Derivation (CD-ROM)

#### The Heat Equation: Results

For the foregoing case of one-dimensional, transient conduction with volumetric energy generation, the ***heat equation*** is

$$\frac{\partial}{\partial x}\left(k\frac{\partial T}{\partial x}\right) + \dot{q} = \rho c \frac{\partial T}{\partial t} \tag{16.2}$$

*heat equation*

where the temperature is a function of the $x$ coordinate and time, $T(x, t)$.

In this text, we will show only the solutions for transient cases. However, we will derive solutions for the steady-state form of the heat diffusion equations for these cases:
*Steady-state conditions, with volumetric energy generation*

$$\frac{d}{dx}\left(k\frac{dT}{dx}\right) + \dot{q} = 0 \tag{16.3}$$

*Steady-state conditions, without volumetric energy generation*

$$\frac{d}{dx}\left(k\frac{dT}{dx}\right) = 0 \tag{16.4}$$

where the temperature depends only on the $x$ coordinate, $T(x)$.

Note the appearance of the properties $\rho$, $c$, and $k$ in the heat equation of Eq. 16.2. The product $\rho c$ (J/m$^3 \cdot$ K) is commonly termed the ***volumetric heat capacity,*** and measures the ability of a material to store energy. The thermal conductivity is a *transport property* since it is the rate coefficient associated with Fourier's law. In heat transfer analysis, for transient conduction and convection, the ratio of the thermal conductivity $k$ to the volumetric heat capacity is an important transport property termed the ***thermal diffusivity*** (m$^2$/s)

*volumetric heat capacity*

$$\alpha = \frac{k}{\rho c} \tag{16.5}$$

*thermal diffusivity*

It measures the ability of a material to conduct energy relative to its ability to store energy. Materials with large $\alpha$ will respond more quickly to changes in their thermal environment, while materials with a small $\alpha$ will respond more slowly, taking longer to reach a new equilibrium condition.

### 16.1.3 Boundary and Initial Conditions

To determine the temperature distribution in a medium, it is necessary to solve the appropriate form of the heat equation. However, the solution depends on the physical conditions existing at the *boundaries* of the medium and, if the situation is time dependent, on conditions existing in the medium at some *initial time*. With regard to the *boundary conditions,* there are several common possibilities that are simply expressed in mathematical form. Because

**Table 16.1** Boundary Conditions for the Heat Equation at the Surface ($x = 0$)

1. Constant surface temperature

$$T(0, t) = T_s \qquad (16.6)$$

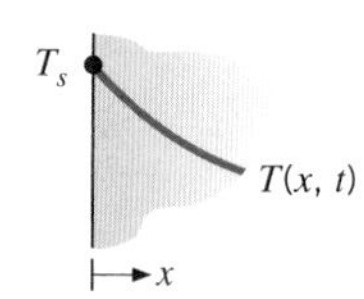


2. Constant surface heat flux

(a) Finite heat flux

$$-k\frac{\partial T}{\partial x}\bigg|_{x=0} = q_s'' \qquad (16.7)$$

$q_s''$
$T(x, t)$
$x$

(b) Adiabatic or insulated surface

$$\frac{\partial T}{\partial x}\bigg|_{x=0} = 0 \qquad (16.8)$$

$T(x, t)$
$x$

3. Convection surface condition

$$-k\frac{\partial T}{\partial x}\bigg|_{x=0} = h[T_\infty - T(0, t)] \qquad (16.9)$$

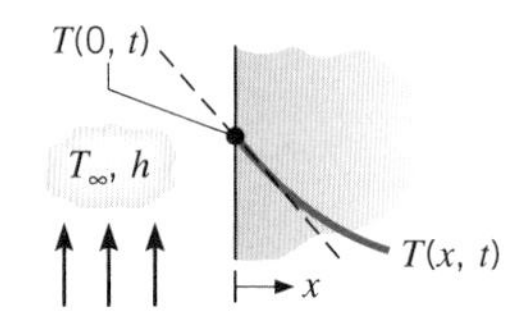


*initial condition*
*boundary conditions*

the heat equation, Eq. 16.2, is second order in the spatial coordinate, two boundary conditions must be expressed to describe the system. Because the equation is first order in time, however, only one condition, termed the ***initial condition,*** must be specified.

The three kinds of ***boundary conditions*** commonly encountered in heat transfer are summarized in Table 16.1. The conditions are specified at the surface $x = 0$ for a one-dimensional system. Heat transfer is in the positive $x$-direction with the temperature distribution, which may be time dependent, designated as $T(x, t)$.

*First kind. Constant surface temperature.* This condition is closely approximated, for example, when the surface experiences convection with an extremely high convection coefficient. Such conditions occur with boiling or condensation, and in both instances the surface remains at the temperature of the phase change process.

*Second kind. Constant surface heat flux.* The heat flux is related to the temperature gradient at the surface by Fourier's law. This condition could be realized by bonding a thin-film or patch-electric heater to the surface or by irradiating the surface with a heat lamp. A special case of this condition corresponds to the *perfectly insulated,* or *adiabatic,* surface for which the gradient is zero. If there is a *symmetry* in the temperature distribution, a surface corresponding to the maximum or minimum temperature could also represent an adiabatic surface.

*Third kind. Convection surface condition.* This condition corresponds to the existence of convection heating (or cooling) at the surface and is obtained from the surface energy balance discussed in Sec. 15.3.

## 16.2 Steady-State Conduction

Using the heat equation for steady-state conditions, we will determine the temperature distributions for one-dimensional plane walls and radial systems. We will introduce the concept of thermal resistance useful for representing systems and their boundary conditions by an equivalent thermal circuit.

### 16.2.1 The Plane Wall

For one-dimensional conduction in a plane wall under steady-state conditions, temperature is a function of the $x$ coordinate only, and heat transfer occurs exclusively in this direction. In Fig. 16.4*a*, a plane wall separates two fluids of different temperatures. Heat transfer occurs by convection from the hot fluid at $T_{\infty,1}$ to one surface of the wall at $T_{s,1}$, by conduction through the wall and by convection from the other surface of the wall at $T_{s,2}$ to the cold fluid at $T_{\infty,2}$. We begin by considering conditions within the wall. We first determine the temperature distribution, from which we can then obtain the conduction heat transfer rate.

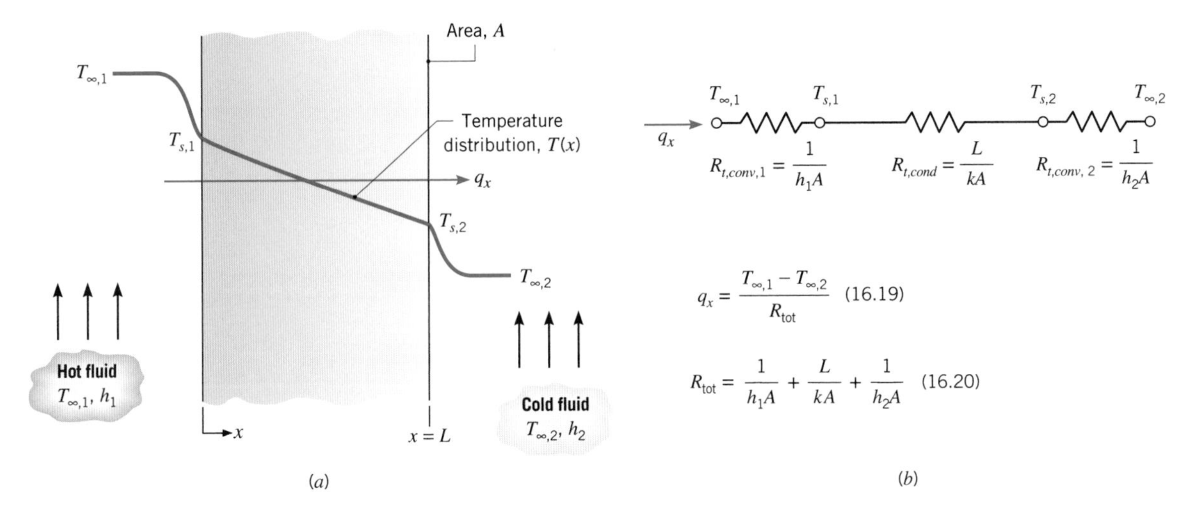


*Figure 16.4* Heat transfer through a plane wall. (*a*) Temperature distribution. (*b*) Equivalent thermal circuit.

## Temperature Distribution

The temperature distribution in the wall can be determined by solving the heat equation with the proper boundary conditions. For steady-state conditions with no energy generation within the wall, the appropriate form of the heat equation is Eq. 16.4

$$\frac{d}{dx}\left(k\frac{dT}{dx}\right) = 0$$

Note that the term in parenthesis represents the heat flux. It follows that for *one-dimensional conduction in a plane wall with no generation, the heat flux is a constant, independent of x.* If the thermal conductivity of the wall material is constant, the equation may be integrated twice to obtain the *general solution*

$$T(x) = C_1 x + C_2 \tag{16.10}$$

To obtain the constants of integration, $C_1$ and $C_2$, *boundary conditions* must be introduced. We choose to apply conditions of the *first kind* at $x = 0$ and $x = L$, in which case

$$T(0) = T_{s,1} \qquad \text{and} \qquad T(L) = T_{s,2}$$

Applying the condition at $x = 0$ to the general solution, it follows that

$$C_2 = T_{s,1}$$

Similarly, at $x = L$

$$T_{s,2} = C_1 L + C_2 = C_1 L + T_{s,1}$$

in which case

$$C_1 = \frac{T_{s,2} - T_{s,1}}{L}$$

Substituting into the general solution, the temperature distribution is then

$$T(x) = (T_{s,2} - T_{s,1})\frac{x}{L} + T_{s,1} \tag{16.11}$$

From this result it is evident that, *for one-dimensional, steady-state conduction in a plane wall with no energy generation and constant thermal conductivity, the temperature varies linearly with x.*

Now that we have the temperature distribution, we can use Fourier's law, Eq. 16.1, to determine the conduction *heat flux* ($W/m^2$). That is

$$q''_x = -k\frac{dT}{dx} = \frac{k}{L}(T_{s,1} - T_{s,2}) \tag{16.12}$$

For the plane wall, $A$ is the area of the wall normal to the direction of heat transfer and is a constant independent of $x$, so that the conduction *heat rate* (W) is

$$q_x = q''_x A = \frac{k}{L}A(T_{s,1} - T_{s,2}) \tag{16.13}$$

Equations 16.12 and 16.13 indicate that both the heat flux and the heat rate are constants, independent of $x$.

Note that we have opted to prescribe surface temperatures at $x = 0$ and $x = L$ as boundary conditions, even though it is the fluid temperatures, and not the surface temperatures, that are typically known. However, since adjoining fluid and surface temperatures are easily related through a surface energy balance (see Sec. 15.3), it is a simple matter to express Eqs. 16.11 to 16.13 in terms of fluid, rather than surface, temperatures. Alternatively, equivalent results could be obtained directly by using the surface energy balances as boundary conditions of the *third kind* in evaluating the constants of Eq. 16.10 (see Problem 16.7).

### Thermal Resistance and Thermal Circuits

At this point we note that a very important concept is suggested by Eq. 16.13. In particular, there exists an analogy between the conduction of heat and electrical current. Just as an electrical resistance is associated with electrical conduction, a thermal resistance may be associated with heat conduction. Defining resistance as the ratio of a driving potential to the corresponding transfer rate, it follows from Eq. 16.13 that the ***thermal resistance for conduction*** in a ***plane wall*** is

*thermal resistance: plane wall conduction*

$$R_{t,\,\text{cond}} \equiv \frac{T_{s,1} - T_{s,2}}{q_x} = \frac{L}{kA} \tag{16.14}$$

Similarly, for electrical conduction in the same system, Ohm's law provides an electrical resistance of the form

$$R_e = \frac{\mathcal{E}_{s,1} - \mathcal{E}_{s,2}}{I} = \frac{L}{\sigma_e A} \tag{16.15}$$

where the driving potential is $\mathcal{E}_{s,1} - \mathcal{E}_{s,2}$ (electrical potential difference), the transfer rate is $I$ (electrical current), and $\sigma_e$ is the electrical conductivity. The analogy between heat and electrical current flow is seen by comparing Eqs. 16.14 and 16.15.

A thermal resistance can also be associated with heat transfer by convection at a surface. From Newton's law of cooling

$$q = hA(T_s - T_\infty) \tag{16.16}$$

the ***thermal resistance for convection*** from a surface is then

*thermal resistance: convection*

$$R_{t,\,\text{conv}} \equiv \frac{T_s - T_\infty}{q} = \frac{1}{hA} \tag{16.17}$$

*thermal circuit*

Circuit representations provide a useful tool for both conceptualizing and quantifying heat transfer problems. The equivalent ***thermal circuit*** for the plane wall with convection surface conditions is shown in Fig. 16.4*b*. The circuit is comprised of resistance elements and ***nodes*** that represent surface or fluid temperatures. The heat transfer rate can be determined from separate or combined considerations of the elements and nodes in the network. Since $q_x$ is constant throughout the network, it follows that

$$q_x = \frac{T_{\infty,1} - T_{s,1}}{1/h_1A} = \frac{T_{s,1} - T_{s,2}}{L/kA} = \frac{T_{s,2} - T_{\infty,2}}{1/h_2A} \tag{16.18}$$

In terms of the *overall temperature difference*, $T_{\infty,1} - T_{\infty,2}$, and the *total thermal resistance*, $R_{\text{tot}}$, the heat transfer rate can also be expressed as

$$q_x = \frac{T_{\infty,1} - T_{\infty,2}}{R_{\text{tot}}} \tag{16.19}$$

Because the conduction and convection resistances are in series and may be summed, it follows that the ***total thermal resistance*** is

$$R_{\text{tot}} = R_{t,\text{conv},1} + R_{t,\text{cond}} + R_{t,\text{conv},2} = \frac{1}{h_1A} + \frac{L}{kA} + \frac{1}{h_2A} \tag{16.20}$$

*total thermal resistance*

Yet another resistance may be pertinent if a surface is exposed to *large, isothermal surroundings* (Sec. 15.1.3). In particular, radiation exchange between the surface and its surroundings may be important, and the rate can be determined from Eq. 15.8. It follows that a ***thermal resistance for radiation*** may be defined as

$$R_{t,\text{rad}} = \frac{T_s - T_{\text{sur}}}{q_{\text{rad}}} = \frac{1}{h_{\text{rad}}A} \tag{16.21}$$

*thermal resistance: radiation coefficient*

where $h_{\text{rad}}$, the *linearized radiation coefficient,* is determined from Eq. 15.9. Surface radiation and convection resistances act in parallel, and if $T_\infty = T_{\text{sur}}$, they can be combined to obtain a single, effective surface resistance.

## The Composite Wall

Equivalent thermal circuits may also be used for more complex systems, such as *composite walls*. Such walls may involve any number of series and parallel thermal resistances due to layers of different materials. Consider the *series composite wall* of Fig. 16.5. The one-dimensional heat transfer rate for this system may be expressed as

$$q_x = \frac{T_{\infty,1} - T_{\infty,3}}{R_{\text{tot}}} \tag{16.22}$$

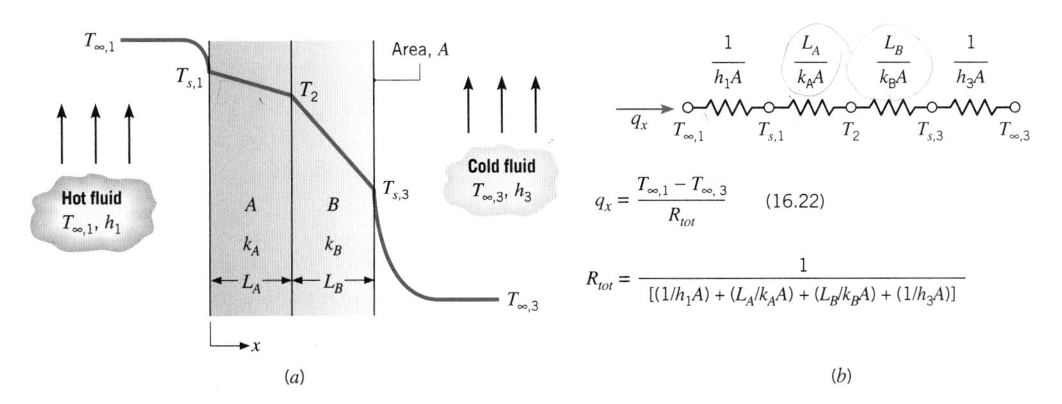


*Figure 16.5* Series-composite wall (*a*) with convection on both surfaces and (*b*) equivalent thermal circuit.

where $T_{\infty,1} - T_{\infty,3}$ is the *overall* temperature difference and $R_{\text{tot}}$ includes all thermal resistances. Hence

$$q_x = \frac{T_{\infty,1} - T_{\infty,3}}{[(1/h_1A) + (L_A/k_AA) + (L_B/k_BA) + (1/h_3A)]} \tag{16.23}$$

Alternatively, the heat transfer rate can be related to the temperature difference and resistance associated with each element. For example

$$q_x = \frac{T_{\infty,1} - T_{s,1}}{(1/h_1A)} = \frac{T_{s,1} - T_2}{(L_A/k_AA)} = \frac{T_2 - T_{s,3}}{(L_B/k_BA)} \tag{16.24}$$

With composite systems it is often convenient to work with an ***overall heat transfer coefficient,*** $U$, which is defined by an expression analogous to Newton's law of cooling. Accordingly

***overall heat transfer coefficient***

$$q_x \equiv UA\,\Delta T \tag{16.25}$$

where $\Delta T$ is the overall temperature difference. The overall heat transfer coefficient is related to the *total thermal resistance,* and from Eqs. 16.22 and 16.25 we see that $UA = 1/R_{\text{tot}}$. Hence, for the composite wall of Fig. 16.5

$$U = \frac{1}{R_{\text{tot}}A} = \frac{1}{[(1/h_1) + (L_A/k_A) + (L_B/k_B) + (1/h_3)]} \tag{16.26}$$

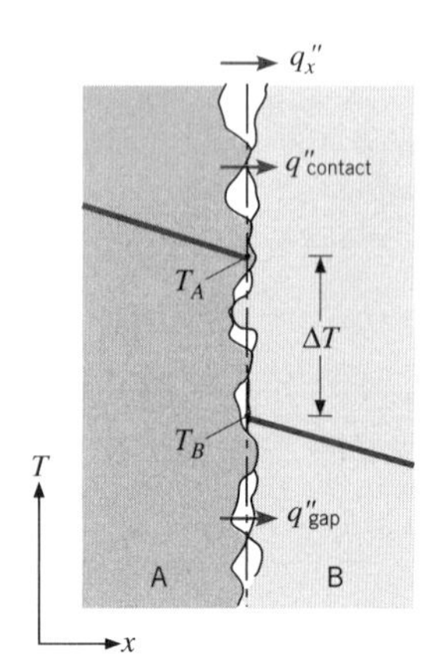


*Figure 16.7* Temperature drop due to thermal contact resistance.

In general, we may write

$$R_{\text{tot}} = \frac{\Delta T}{q} = \frac{1}{UA} \tag{16.27}$$

Composite walls may also be characterized by *series-parallel configurations* and the heat rate determined by a network comprised of thermal resistances in series and parallel arrangements. (CD-ROM)

## Contact Resistance

Although neglected until now, it is important to recognize that, in composite systems, the temperature drop across the *interface* between materials may be appreciable. This temperature change is attributed to what is known as the ***thermal contact resistance,*** $R_{t,c}$. The effect is shown in Fig. 16.7, and for a unit area of the interface, the resistance is defined as

***thermal contact resistance***

$$R''_{t,c} = \frac{T_A - T_B}{q''_x} \tag{16.28}$$

The existence of a finite contact resistance is due principally to surface roughness effects. Contact spots are interspersed with gaps that may be evacuated or filled with an interfacial medium such as a gas, grease, or paste. Heat transfer is therefore due to conduction across the actual contact area and to conduction and/or radiation across the gaps. The contact resistance may be viewed as two parallel resistances: that due to the contact spots and that due to the gaps. The contact area is typically small, and especially for rough surfaces, the major contribution to the resistance is made by the gaps. For problems in this text, we will neglect contact resistance unless otherwise noted.

Comprehensive reviews of thermal contact resistance results are available in the literature, especially for space thermal control and electronics applications. Thermal resistances of representative solid/solid interfaces are shown in Table 16.2. (CD-ROM)

## *Example 16.1* Thermal Circuit Analysis–the Plane Wall

A manufacturer of household appliances is proposing a self-cleaning oven design that involves use of a composite window separating the oven cavity from the room air. The composite is to consist of two high-temperature plastics (A and B) of thicknesses $L_A = 2L_B$ and thermal conductivities $k_A = 0.15$ W/m · K and $k_B = 0.08$ W/m · K. During the self-cleaning process, the inside window temperature $T_{s,i}$ is 385°C, while the room air temperature $T_\infty$ is 25°C and the outside convection coefficient is 25 W/m² · K. What is the minimum window thickness, $L = L_A + L_B$, needed to ensure a temperature that is 50°C or less at the outer surface of the window during steady-state operation? This temperature must not be exceeded for safety reasons.

### Solution

***Known:*** The properties and relative dimensions of plastic materials used for a composite oven window, and conditions associated with self-cleaning operation.

***Find:*** Composite thickness $L$ needed to ensure safe operation.

***Schematic and Given Data:***

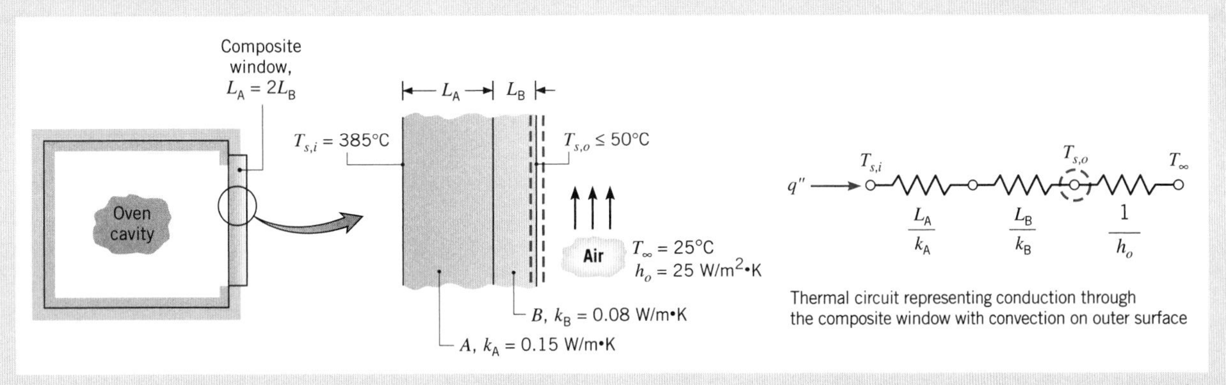


*Figure E16.1*

***Assumptions:***

1. Steady-state conditions exist.
2. Conduction through the window is one-dimensional.
3. Contact resistance between the plastics is negligible.
4. Radiation transfer through the window is negligible.
5. Plastics are homogenous with constant properties.

***Analysis:*** The thermal circuit can be constructed by recognizing that resistance to heat transfer through the composite window to the ambient air is associated with conduction in the plastics and convection at the outer surface. Since the outer surface temperature of the window, $T_{s,o}$, is prescribed, the required window thickness is obtained by applying an energy balance at this surface (see control surfaces on the Schematic). Referring to the circuit, recognize that an energy balance on the $T_{s,o}$ *node* is equivalent to the *surface energy balance*. Hence, the heat flux into the node (surface) is equal to the heat flux out of the node (surface). As such, the heat rate can be expressed as

$$q'' = \frac{T_{s,i} - T_{s,o}}{L_A/k_A + L_B/k_B} = \frac{T_{s,o} - T_\infty}{1/h_o}$$

With $L_B = L_A/2$, and substituting numerical values, find $L_A$

$$\frac{(385 - 50)°\text{C}}{(L_A/0.15 + 0.5L_A/0.08)\,\text{m} \cdot \text{K/W}} = \frac{(50 - 25)°\text{C}}{(1/25)\,\text{m}^2 \cdot \text{K/W}}$$

$$L_A = 0.0415 \text{ m}$$

Hence, the required thickness for the composite window is

$$L = L_A + L_B = (0.0415 + 0.5 \times 0.0415)\,\text{m} = 0.0622 \text{ m} = 62.2 \text{ mm} \quad \triangleleft$$

## *Example 16.2* Silicon Chip on a Substrate

A thin silicon chip and an 8-mm-thick aluminum substrate are separated by a thin epoxy joint with a thermal resistance of $R''_{t,c} = 0.9 \times 10^{-4}\ \text{m}^2 \cdot \text{K/W}$. The chip and substrate are each 10 mm on a side, and their exposed surfaces are cooled by air, which is at a temperature of 25°C and provides a convection coefficient of $100\ \text{W/m}^2 \cdot \text{K}$. If the chip dissipates electrical power $P''_e = 10^4\ \text{W/m}^2$ under normal conditions, will it operate below a maximum allowable temperature of 85°C?

### Solution

***Known:*** Dimensions, power dissipation, and maximum allowable temperature of a silicon chip. Thickness of aluminum substrate and thermal resistance of epoxy joint. Convection conditions at exposed chip and substrate surfaces.

***Find:*** Whether the temperature of the chip, $T_c$, exceeds the maximum allowed.

***Schematic and Given Data:***

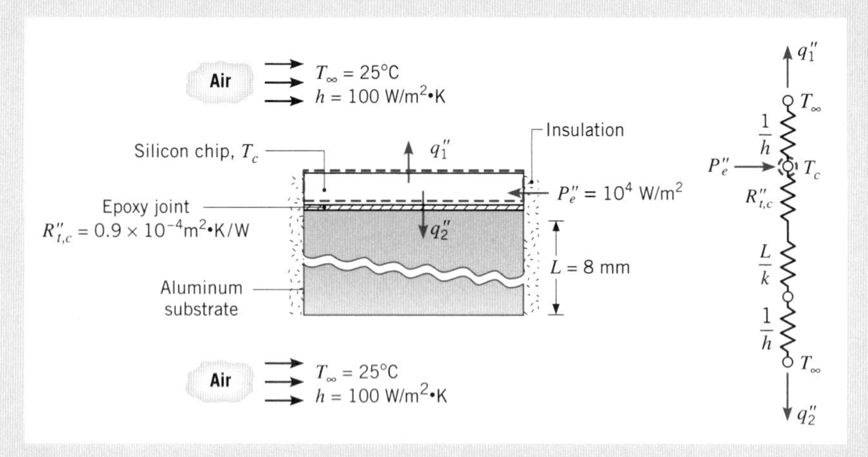


***Assumptions:***
1. Steady-state conditions.
2. One-dimensional conduction.
3. Isothermal chip with negligible thermal resistance.
4. Negligible radiation exchange with surroundings.
5. Constant properties.

*Figure E16.2*

***Properties:*** Table HT-1, pure aluminum ($T \approx 350$ K): $k = 238\ \text{W/m} \cdot \text{K}$.

***Analysis:*** Electrical power dissipated in the chip is transferred to the air directly from the exposed surface and indirectly through the joint and substrate. Performing an energy balance on a control surface about the chip (see dashed lines on schematic and thermal circuit), it follows that, on the basis of a unit surface area

$$P''_e = q''_1 + q''_2 = \frac{T_c - T_\infty}{(1/h)} + \frac{T_c - T_\infty}{R''_{t,c} + (L/k) + (1/h)}$$

Rearranging the above equation and substituting numerical values, the chip temperature is

$$T_c = T_\infty + P''_e \left[ \frac{1}{(1/h)} + \frac{1}{R''_{t,c} + (L/k) + (1/h)} \right]^{-1}$$

$$T_c = 25°\text{C} + 10^4\ \text{W/m}^2 \times \left[ \frac{1}{(1/100)} + \frac{1}{0.9 \times 10^{-4} + (0.008/238) + (1/100)} \right]^{-1} \text{m}^2 \cdot \text{K/W}$$

$$T_c = 25°\text{C} + \left[ \frac{1}{100 \times 10^{-4}} + \frac{1}{(0.9 + 3.4 + 100) \times 10^{-4}} \right]^{-1} \times 10^4 °\text{C}$$

$$T_c = 25°\text{C} + 50.3°\text{C} = 75.3°\text{C} \quad \triangleleft$$

Hence the chip will operate below its maximum allowable temperature.

***Comments:***

1. Note that we have used Appendix Table HT-1 to obtain the value for the thermal conductivity of pure aluminum, which was evaluated at the estimated average temperature of the substrate.
2. The joint and substrate conduction resistances are much less than the convection resistance. The joint resistance would have to increase to the unrealistically large value of $50 \times 10^{-4}\ \text{m}^2 \cdot \text{K/W}$ before the maximum allowable chip temperature would be exceeded.

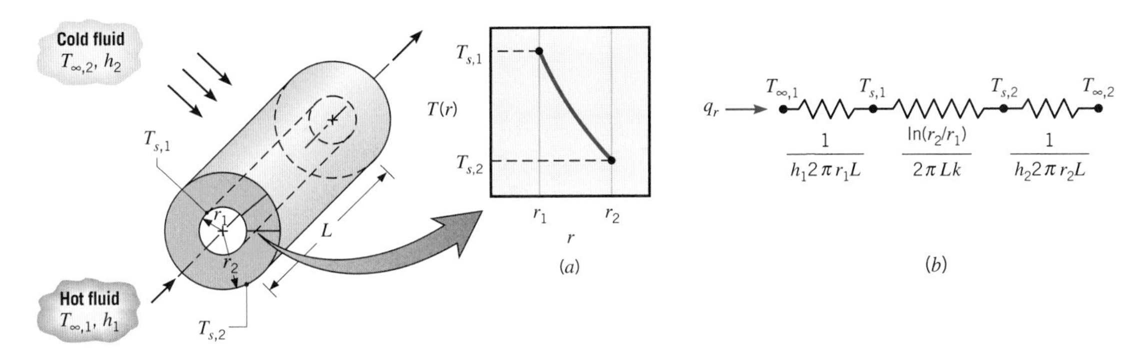


*Figure 16.8* Hollow cylinder with convection surface conditions. (*a*) Logarithmic temperature distribution. (*b*) Equivalent thermal circuit.

## 16.2.2 One-Dimensional Radial Systems

Cylindrical and spherical systems often experience temperature gradients in the radial direction only, and may therefore be treated as one-dimensional. As was shown for the plane wall, such systems may be analyzed using the heat equation to obtain the temperature distribution and heat rate. In this section, we have avoided the attendant derivations, and present the results that are used for building equivalent thermal circuits of radial systems.

### The Cylinder

A common configuration is the hollow cylinder whose inner and outer surfaces are exposed to fluids at different temperatures (Fig. 16.8). For steady-state conditions with no energy generation, the temperature distribution in the radial (cylindrical) coordinate system is

$$T(r) = \frac{T_{s,1} - T_{s,2}}{\ln(r_1/r_2)} \ln\left(\frac{r}{r_2}\right) + T_{s,2} \tag{16.29}$$

Note that the temperature distribution associated with radial conduction through a cylindrical wall is *logarithmic,* not linear as it is for the plane wall under the same conditions. The logarithmic distribution is shown in Fig. 16.8*a*.

The appropriate form of Fourier's law for the radial (cylindrical) coordinate system is

$$q_r = -kA_r \frac{dT}{dr} = -k(2\pi rL)\frac{dT}{dr} \tag{16.30}$$

where $A_r = 2\pi rL$ is the area normal to the direction of heat transfer. By applying an energy balance on a cylindrical control surface at any radius, we conclude that the conduction *heat transfer rate* (*not* the heat flux) is a *constant in the radial direction.*

If the temperature distribution, Eq. 16.29, is now used with Fourier's law, Eq. 16.30, we obtain the expression for the *heat transfer rate*

$$q_r = \frac{2\pi Lk(T_{s,1} - T_{s,2})}{\ln(r_2/r_1)} \tag{16.31}$$

From this result it is evident that for radial conduction, ***thermal resistance*** in a ***cylindrical wall*** is of the form

$$R_{t,\text{cond}} = \frac{\ln(r_2/r_1)}{2\pi Lk} \tag{16.32}$$

***thermal resistance: cylindrical wall***

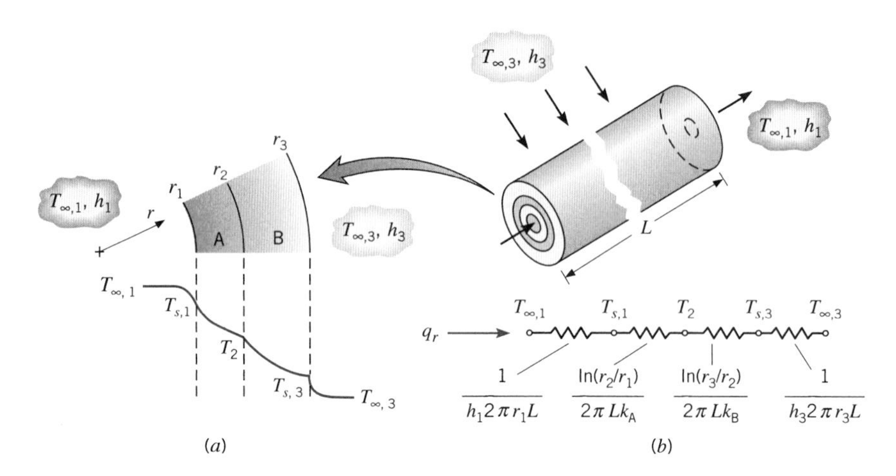


*Figure 16.9* Composite hollow cylinder with convection both surfaces: (*a*) Temperature distribution and (*b*) Equivalent thermal circuit.

This resistance is shown in the series circuit of Fig. 16.8*b* along with the convection resistances at the inner and outer surfaces.

Consider now the *composite cylindrical system* of Fig. 16.9. Recalling how we treated the composite plane wall and neglecting any interfacial contact resistances, the heat transfer rate may be expressed as

$$q_r = \frac{T_{\infty,1} - T_{\infty,3}}{\dfrac{1}{2\pi r_1 L h_1} + \dfrac{\ln(r_2/r_1)}{2\pi L k_A} + \dfrac{\ln(r_3/r_2)}{2\pi L k_B} + \dfrac{1}{2\pi r_3 L h_3}} \tag{16.33}$$

The foregoing result may also be expressed in terms of an overall heat transfer coefficient. That is

$$q_r = \frac{T_{\infty,1} - T_{\infty,3}}{R_{\text{tot}}} = UA(T_{\infty,1} - T_{\infty,3}) \tag{16.34}$$

If $U$ is defined in terms of the inside area, $A_1 = 2\pi r_1 L$, Eqs. 16.33 and 16.34 may be equated to yield

$$U_1 = \frac{1}{\dfrac{1}{h_1} + \dfrac{r_1}{k_A}\ln\dfrac{r_2}{r_1} + \dfrac{r_1}{k_B}\ln\dfrac{r_3}{r_2} + \dfrac{r_1}{r_3}\dfrac{1}{h_3}} \tag{16.35}$$

This definition is *arbitrary,* and the overall coefficient may also be defined in terms of $A_3$ or any of the intermediate areas. Note that

$$U_1 A_1 = U_2 A_2 = U_3 A_3 = R_{\text{tot}}^{-1} \tag{16.36}$$

and the specific forms of $U_2$ and $U_3$ can be inferred from Eqs. 16.33 and 16.34.

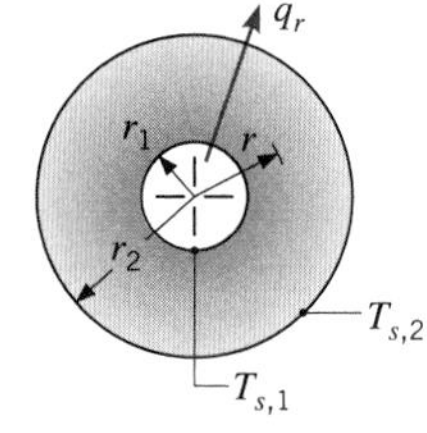


*Figure 16.10*

## The Sphere

Consider the spherical shell of Fig. 16.10, whose inner and outer surfaces are maintained at $T_{s,1}$ and $T_{s,2}$, respectively. For steady-state conditions with no energy generation, the temperature distribution is

$$T(r) = \frac{T_{s,2} - T_{s,1}}{1 - (r_1/r_2)}[1 - (r_1/r)] + T_{s,1} \tag{16.37}$$

**Table 16.3** One-Dimensional, Steady-State Solutions to the Heat Equation ($\dot{q} = 0$ and $k$ constant) and Associated Thermal Resistances

| | Plane Wall | | Cylindrical Wall | | Spherical Wall | |
|---|---|---|---|---|---|---|
| Heat equation | $\frac{d}{dx}\left(\frac{dT}{dx}\right) = 0$ | (16.4) | $\frac{1}{r}\frac{d}{dr}\left(r\frac{dT}{dr}\right) = 0$ | | $\frac{1}{r^2}\frac{d}{dr}\left(r^2\frac{dT}{dr}\right) = 0$ | |
| Temperature distribution[a] | $T_{s,1} - \Delta T\frac{x}{L}$ | (16.11) | $T_{s,2} + \Delta T\frac{\ln(r/r_2)}{\ln(r_1/r_2)}$ | (16.29) | $T_{s,1} - \Delta T\left[\frac{1-(r_1/r)}{1-(r_1/r_2)}\right]$ | (16.37) |
| Heat flux, $q''$ | $k\frac{\Delta T}{L}$ | (16.12) | $\frac{k\Delta T}{r\ln(r_2/r_1)}$ | | $\frac{k\Delta T}{r^2[(1/r_1)-(1/r_2)]}$ | |
| Heat rate, $q$ | $kA\frac{\Delta T}{L}$ | (16.13) | $\frac{2\pi Lk\Delta T}{\ln(r_2/r_1)}$ | (16.31) | $\frac{4\pi k\Delta T}{(1/r_1)-(1/r_2)}$ | (16.39) |
| Conduction thermal resistance, $R_{t,\text{cond}}$ | $\frac{L}{kA}$ | (16.14) | $\frac{\ln(r_2/r_1)}{2\pi Lk}$ | (16.32) | $\frac{(1/r_1)-(1/r_2)}{4\pi k}$ | (16.40) |
| Convection thermal resistance, $R_{t,\text{conv}}$ | $\frac{1}{hA}$ | (16.17) | $\frac{1}{h(2\pi r_2 L)}$ | (16.41) | $\frac{1}{h(4\pi r_2^2)}$ | (16.42) |

[a]The temperature difference, $\Delta T$, is defined as $\Delta T \equiv T_{s,1} - T_{s,2}$. See Figs. 16.4, 16.8, and 16.10 for geometrical representations of the walls.

The appropriate form of Fourier's law for the radial (spherical) coordinate system may be expressed as

$$q_r = -kA_r\frac{dT}{dr} = -k(4\pi r^2)\frac{dT}{dr} \qquad (16.38)$$

where $A_r = 4\pi r^2$ is the area normal to the direction of heat transfer. Applying an energy balance on a spherical control surface at any radius, we find that the conduction *heat transfer rate* (*not* the heat flux) is a *constant in the radial direction.*

If the temperature distribution, Eq. 16.37, is used with Fourier's law, Eq. 16.38, we obtain the expression for the *heat transfer rate*

$$q_r = \frac{4\pi k(T_{s,1} - T_{s,2})}{(1/r_1) - (1/r_2)} \qquad (16.39)$$

From this result it is evident that for radial conduction, the ***thermal resistance*** in a ***spherical wall*** is of the form

$$R_{t,\text{cond}} = \frac{1}{4\pi k}\left(\frac{1}{r_1} - \frac{1}{r_2}\right) \qquad (16.40)$$

*thermal resistance: spherical wall*

Spherical composites may be treated in much the same way as composite walls and cylinders, where the appropriate forms of the total resistance and overall heat transfer coefficient can be determined.

### 16.2.3 Summary of One-Dimensional Conduction Results

Many important problems are characterized by one-dimensional, steady-state conduction in plane, cylindrical, or spherical walls *without* energy generation. Key results for these three geometries are summarized in Table 16.3, where $\Delta T$ refers to the temperature difference, $T_{s,1} - T_{s,2}$, between the inner and outer surfaces identified in Figs. 16.4, 16.8, and 16.10.

## *Example 16.3* Thermal Circuit Analysis–Spherical System

A spherical, thin-walled metallic container is used to store liquid nitrogen at 77 K. The container has a diameter of 0.5 m and is covered with an evacuated, reflective insulation composed of silica powder. The insulation is 25 mm thick, and its outer surface is exposed to ambient air at 300 K. The convection coefficient is known to be 20 W/m$^2 \cdot$ K. The heat of vaporization and the density of liquid nitrogen are $2 \times 10^5$ J/kg and 804 kg/m$^3$, respectively.

**(a)** What is the rate of heat transfer to the liquid nitrogen?
**(b)** What is the rate of liquid boil-off (liters/day)?

## Solution

***Known:*** Liquid nitrogen is stored in a spherical container that is insulated and exposed to ambient air.

***Find:***

**(a)** The rate of heat transfer to the nitrogen.
**(b)** The rate of nitrogen boil-off.

***Schematic and Given Data:***

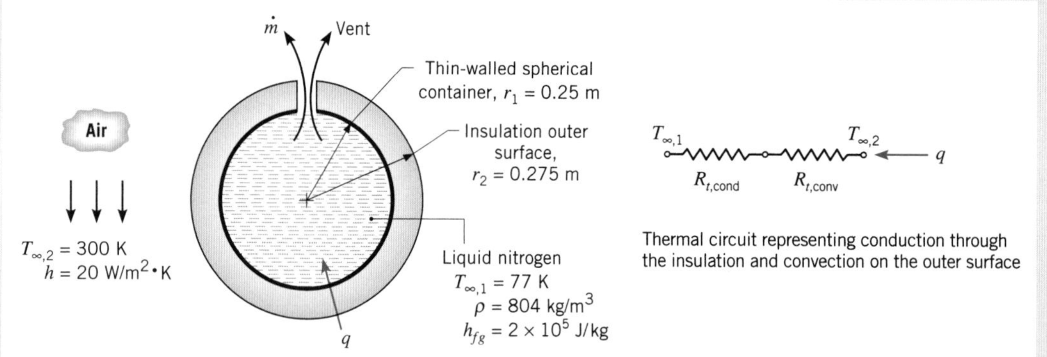


*Figure E16.3*

***Assumptions:***

1. Steady-state heat transfer.
2. One-dimensional transfer in the radial direction.
3. Negligible resistance to heat transfer through the container wall and from the container to the nitrogen.
4. Negligible radiation exchange between outer surface of insulation and surroundings.
5. Constant properties.

***Properties:*** Table HT-2, evacuated silica powder (300 K): $k = 0.0017$ W/m $\cdot$ K.

***Analysis:***

**(a)** By assumption 3, the only elements in the thermal circuit as shown above are the resistances due to conduction through the insulation and convection from the outer surface where, from Table 16.3

$$R_{t,\text{cond}} = \frac{1}{4\pi k}\left(\frac{1}{r_1} - \frac{1}{r_2}\right) \qquad R_{t,\text{conv}} = \frac{1}{h4\pi r_2^2}$$

The rate of heat transfer to the liquid nitrogen is then

$$q = \frac{T_{\infty,2} - T_{\infty,1}}{(1/4\pi k)[(1/r_1) - (1/r_2)] + (1/h4\pi r_2^2)}$$

$$q = \frac{(300 - 77)\text{ K}}{\left[\dfrac{1}{4\pi(0.0017 \text{ W/m}\cdot\text{K})}\left(\dfrac{1}{0.25\text{ m}} - \dfrac{1}{0.275\text{ m}}\right) + \dfrac{1}{(20\text{ W/m}^2\cdot\text{K})4\pi(0.275\text{ m})^2}\right]}$$

$$q = \frac{223}{17.02 + 0.05}\text{ W} = 13.06\text{ W} \quad \triangleleft$$

**(b)** The heat transfer to the liquid nitrogen provides energy to vaporize the liquid nitrogen by boiling

$$q = \dot{m}h_{fg}$$

and the mass rate of nitrogen boil-off is

$$\dot{m} = \frac{q}{h_{fg}} = \frac{13.06 \text{ J/s}}{2 \times 10^5 \text{ J/kg}} = 6.53 \times 10^{-5} \text{ kg/s}$$

The mass rate per day is

$$\dot{m} = 6.53 \times 10^{-5} \text{ kg/s} \left|\frac{3600 \text{ s}}{\text{h}}\right| \left|\frac{24 \text{ h}}{\text{day}}\right| = 5.64 \text{ kg/day}$$

or on a volumetric flow rate basis

$$\frac{\dot{m}}{\rho} = \frac{5.64 \text{ kg/day}}{804 \text{ kg/m}^3} = 0.007 \text{ m}^3\text{/day} \left|\frac{10^3 \text{ liters}}{\text{m}^3}\right| = 7 \text{ liters/day} \quad \triangleleft$$

***Comments:***

**1.** Since $R_{t,\text{conv}} \ll R_{t,\text{cond}}$, the dominant contribution to the total thermal resistance is that due to conduction in the insulation. Even if the convection coefficient were reduced by a factor of 10, thereby increasing the convection resistance by the same proportion, the effect on the boil-off rate would be small.

**2.** With a container volume of $(4/3)(\pi r_1^3) = 0.065 \text{ m}^3 = 65$ liters, the daily evaporation rate amounts to (7 liters/65 liters) 100% = 10.8% of capacity.

## 16.3 Conduction with Energy Generation

In the preceding section we considered conduction problems for which the temperature distribution in a medium was determined solely by conditions at the boundaries of the medium. For this situation we were able to represent conduction within the medium and the boundary heat transfer processes by resistances that comprise a thermal circuit.

We consider now the additional effect on the temperature distribution of processes that may be occurring *within* the medium. In particular, we will treat common geometries experiencing a uniform volumetric rate of *energy generation* $\dot{q}$ (W/m$^3$) arising from energy conversion processes as described in Sec. 15.2. For this situation we *cannot* represent the medium by a thermal circuit, but must solve the heat equation to obtain the temperature distribution, and hence the heat flux.

### 16.3.1 The Plane Wall

Consider the plane wall of Fig. 16.11*a*, in which there is *uniform* energy generation per unit volume ($\dot{q}$ is constant) and the surfaces are maintained at $T_{s,1}$ and $T_{s,2}$. For constant thermal conductivity $k$, the appropriate form of *the heat equation,* Eq. 16.3, is

$$\frac{d^2T}{dx^2} + \frac{\dot{q}}{k} = 0$$

The general solution for the *temperature distribution* is

$$T = -\frac{\dot{q}}{2k}x^2 + C_1x + C_2 \tag{16.43}$$

where $C_1$ and $C_2$ are the constants of integration. By substitution, it may be verified that Eq. 16.43 is indeed a solution to the heat equation. For the prescribed boundary conditions shown in Fig. 16.11*a*

$$T(-L) = T_{s,1} \qquad \text{and} \qquad T(L) = T_{s,2}$$

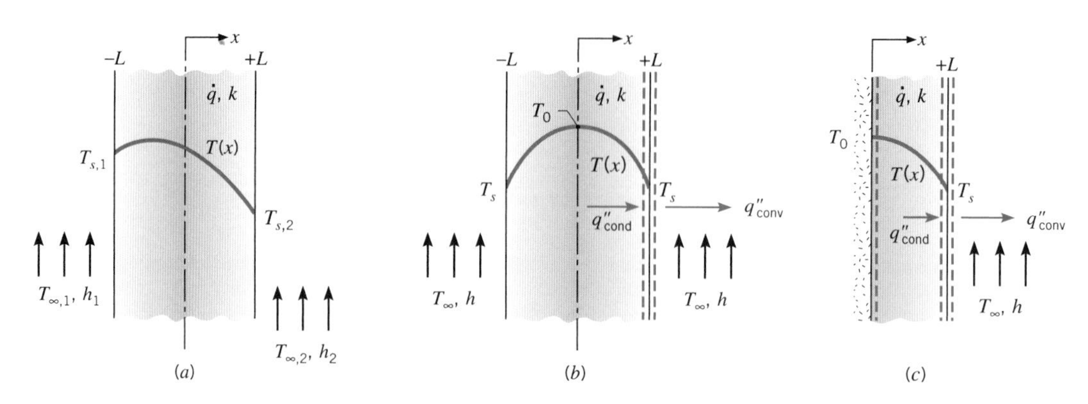


*Figure 16.11* Conduction in a plane wall with uniform heat generation. Temperature distributions for (*a*) Asymmetrical boundary conditions, Eq. 16.44, (*b*) Symmetrical boundary conditions, Eqs. 16.45–16.47 and (*c*) Adiabatic surface at midplane, Eqs. 16.45–16.47.

The constants can be evaluated and are of the form

$$C_1 = \frac{T_{s,2} - T_{s,1}}{2L} \quad \text{and} \quad C_2 = \frac{\dot{q}}{2k}L^2 + \frac{T_{s,1} + T_{s,2}}{2}$$

Then the temperature distribution for the *asymmetrical boundary conditions* case, Fig. 16.11*a*, is

$$T(x) = \frac{\dot{q}L^2}{2k}\left(1 - \frac{x^2}{L^2}\right) + \frac{T_{s,2} - T_{s,1}}{2}\frac{x}{L} + \frac{T_{s,1} + T_{s,2}}{2} \tag{16.44}$$

The heat flux at any point in the wall may, of course, be determined by using Eq. 16.44 with Fourier's law. Note, however, that *with generation the heat flux is no longer independent of* $x$. Furthermore, note that the temperature distribution is not linear, but is a quadratic function of $x$.

The preceding result simplifies when both surfaces are maintained at a common temperature, $T_{s,1} = T_{s,2} \equiv T_s$ as shown in Fig. 16.11*b*. The temperature distribution is then *symmetrical* about the midplane, and is given by

$$T(x) = \frac{\dot{q}L^2}{2k}\left(1 - \frac{x^2}{L^2}\right) + T_s \tag{16.45}$$

The maximum temperature exists at the midplane, $x = 0$

$$T(0) \equiv T_0 = \frac{\dot{q}L^2}{2k} + T_s \tag{16.46}$$

in which case the temperature distribution, Eq. 16.45, may be expressed alternatively as

$$\frac{T(x) - T_0}{T_s - T_0} = \left(\frac{x}{L}\right)^2 \tag{16.47}$$

It is important to note that at the *plane of symmetry* in Fig. 16.11*b*, the temperature gradient is zero, $(dT/dx)_{x=0} = 0$. Accordingly, there is no heat transfer across this plane, and it may be represented by the *adiabatic* surface shown in Fig. 16.11*c*. One implication of this result is that Eq. 16.45 also applies to plane walls that are perfectly insulated on one side ($x = 0$) and maintained at a fixed temperature $T_s$ on the other side ($x = L$).

To use the foregoing results, the surface temperature(s) $T_s$ must be known. However, a common situation is one for which it is the temperature of an adjoining fluid, $T_\infty$, and not $T_s$, which is known. It then becomes necessary to relate $T_s$ to $T_\infty$. This relation may be developed by applying a *surface energy balance.* Consider the surface at $x = L$ for the symmetrical plane wall (Fig. 16.11*b*) or the insulated plane wall (Fig. 16.11*c*). The surface energy balance, Eq. 15.14, has the form $q''_{\text{cond}} = q''_{\text{conv}}$, and substituting the appropriate rate equations

$$-k\frac{dT}{dx}\bigg|_{x=L} = h(T_s - T_\infty) \tag{16.48}$$

Substituting from Eq. 16.45 to obtain the temperature gradient at $x = L$, it follows that

$$T_s = T_\infty + \frac{\dot{q}L}{h} \tag{16.49}$$

Hence $T_s$ may be computed from knowledge of $T_\infty$, $\dot{q}$, $L$, and $h$.

Equation 16.49 may also be obtained by applying an *overall* energy balance to the plane wall of Fig. 16.11*b* or 16.11*c*. ***For Example…*** relative to a control surface about the wall of Fig. 16.11*c*, the rate at which energy is generated within the wall must be balanced by the rate at which energy leaves via convection at the boundary. Equation 15.11a reduces to

$$\dot{E}_g = \dot{E}_{\text{out}} \tag{16.50}$$

or, for a unit surface area

$$\dot{q}L = h(T_s - T_\infty) \tag{16.51}$$

Solving for $T_s$, Eq. 16.49 is obtained. ▲

## *Example 16.4* Energy Generation in a Plane Wall

A plane wall is a composite of two materials, A and B. The wall of material A has uniform energy generation $\dot{q} = 1.5 \times 10^6$ W/m$^3$, $k_A = 75$ W/m · K, and thickness $L_A = 50$ mm. The wall of material B has no generation, with $k_B = 150$ W/m · K and thickness $L_B = 20$ mm. The inner surface of material A is well insulated, while the outer surface of material B is cooled by a water stream with $T_\infty = 30°$C and $h = 1000$ W/m$^2$ · K.

(a) Determine the temperature $T_0$ of the insulated surface and the temperature $T_2$ of the cooled surface.

(b) Sketch the temperature distribution that exists in the composite under steady-state conditions.

### Solution

***Known:*** Plane wall of material A with internal energy generation is insulated on one side and bounded by a second wall of material B, which is without energy generation and is subjected to convection cooling.

***Find:***

(a) Inner and outer surface temperatures of the composite.

(b) Sketch of steady-state temperature distribution in the composite.

***Schematic and Given Data:***

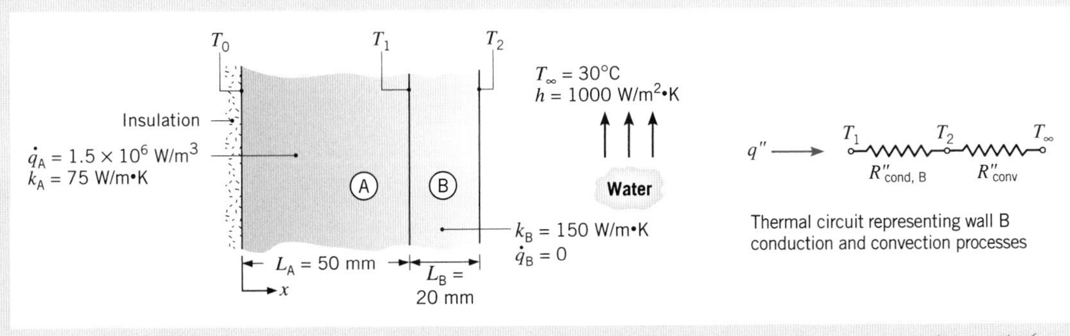


*Figure E16.4a*

***Assumptions:***

1. Steady-state conditions.
2. One-dimensional conduction in $x$-direction.
3. Negligible contact resistance between walls.
4. Inner surface of A is adiabatic.
5. Constant properties for materials A and B.

***Analysis:***

**(a)** The outer surface temperature $T_2$ can be obtained by performing an energy balance on a system about material B (Fig. E16.4*b*). Since there is no generation in this material, it follows that, for steady-state conditions and a unit surface area, the heat flux into the material at $x = L_A$ must equal the heat flux from the material due to convection at $x = L_A + L_B$. Hence

$$q'' = h(T_2 - T_\infty) \tag{1}$$

The heat flux $q''$ can be determined by performing a second energy balance about material A. In particular, since the surface at $x = 0$ is adiabatic, there is no inflow and the rate at which energy is generated must equal the outflow. Accordingly, for a unit surface area

$$\dot{q}L_A = q'' \tag{2}$$

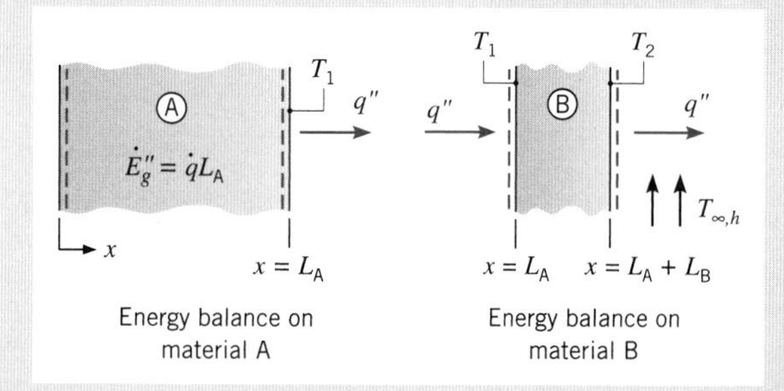


*Figure E16.4b*

Combining Eqs. 1 and 2, the outer surface temperature is

$$T_2 = T_\infty + \frac{\dot{q}L_A}{h} = 30°\text{C} + \frac{1.5 \times 10^6 \text{ W/m}^3 \times 0.05 \text{ m}}{1000 \text{ W/m}^2 \cdot \text{K}} = 105°\text{C} \ \triangleleft$$

From Eq. 16.46 the temperature at the insulated surface is

$$T_0 = \frac{\dot{q}L_A^2}{2k_A} + T_1 \tag{3}$$

where $T_1$ may be obtained from the thermal circuit shown in Fig. E16.4*a* representing the wall B conduction and convection processes. That is,

$$T_1 = T_\infty + (R''_{\text{cond,B}} + R''_{\text{conv}})\, q''$$

where the resistances for a unit surface area are

$$R''_{\text{cond, B}} = \frac{L_B}{k_B} \qquad R''_{\text{conv}} = \frac{1}{h}$$

Hence, the temperature at the composite interface is

$$T_1 = 30°\text{C} + \left(\frac{0.02 \text{ m}}{150 \text{ W/m} \cdot \text{K}} + \frac{1}{1000 \text{ W/m}^2 \cdot \text{K}}\right)(1.5 \times 10^6 \text{ W/m}^3)\, 0.05 \text{ m}$$

$$T_1 = 30°\text{C} + 85°\text{C} = 115°\text{C}$$

Substituting into Eq. 3, the inner surface temperature of the composite is

$$T_0 = \frac{1.5 \times 10^6 \text{ W/m}^3 (0.05 \text{ m})^2}{2 \times 75 \text{ W/m} \cdot \text{K}} + 115°\text{C} = 25°\text{C} + 115°\text{C} = 140°\text{C} \ \triangleleft$$

**(b)** From the prescribed physical conditions, the temperature distribution in the *composite* has the following features, as shown:

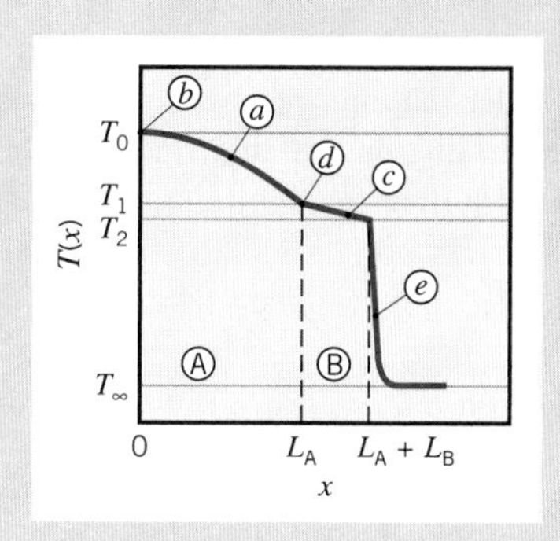


(**a**) Parabolic in material A.
(**b**) Zero slope at insulated boundary.
(**c**) Linear in material B.
(**d**) Slope change $= k_B/k_A$ at interface.

The temperature distribution in the *water* is characterized by large gradients near the surface (e).

*Figure E16.4c*

***Comments:***

1. Material A, having energy generation, cannot be represented by a thermal circuit element.
2. Since the resistance to heat transfer by convection is significantly larger than that due to conduction in material B, $R''_{conv}/R''_{cond} = 7.5$, the surface-to-fluid temperature difference is much larger than the temperature drop across material B, $(T_2 - T_\infty)/(T_1 - T_2) = 7.5$. This result is consistent with the temperature distribution plotted in Fig. E16.4*c*.

### 16.3.2 Radial Systems with Energy Generation (CD-ROM)

### 16.3.3 Application of Resistance Concepts

We conclude our discussion of energy generation effects with a word of caution. In particular, when such effects are present, the heat transfer rate is not a constant, independent of the spatial coordinate. Accordingly, it would be *incorrect* to use the conduction resistance concepts and the related heat rate equations developed in Sec. 16.2.

## 16.4 Heat Transfer from Extended Surfaces: Fins

In many industrial applications it is desirable to enhance the rate of heat transfer from a solid surface to an adjoining fluid. Consider the plane wall of Fig. 16.13*a*. If $T_s$ is fixed, the heat transfer rate may be increased by increasing the fluid velocity, which has the effect of increasing the convection coefficient $h$, and/or by increasing the difference between the surface and fluid temperatures $(T_s - T_\infty)$. However, there are many situations for which these changes might be insufficient, expensive, and/or impractical.

Accordingly, another option is shown in Fig. 16.13*b*. That is, the heat transfer rate may be increased by increasing the surface area across which the convection occurs. This may be accomplished by employing fins that *extend* from the wall into the surrounding fluid. Heat transfer occurs by *conduction within the fin,* and by *convection from the surfaces of the fin.*

The thermal conductivity of the fin material has a strong effect on the temperature distribution along the fin and therefore influences the degree to which the heat transfer rate is enhanced. Ideally, the fin material should have a large thermal conductivity to minimize temperature variations from its base to its tip. In the limit of infinite thermal conductivity, the entire fin would be at the temperature of the base surface, thereby providing the maximum possible heat transfer enhancement.

You are already familiar with several fin applications, including arrangements for cooling engine heads on motorcycles and lawn mowers, or for cooling electrical power transformers. Consider also the tubes with attached fins used to promote heat exchange between air and the working fluid of an air conditioner or heat pump. Two common finned-tube arrangements are shown in Fig. 16.14.

While there are numerous configurations of fins with different methods of attachments to surfaces, in our introductory treatment we will consider two common types widely used in thermal systems. First, we'll consider the *straight fin,* which is an extended surface that is attached to a

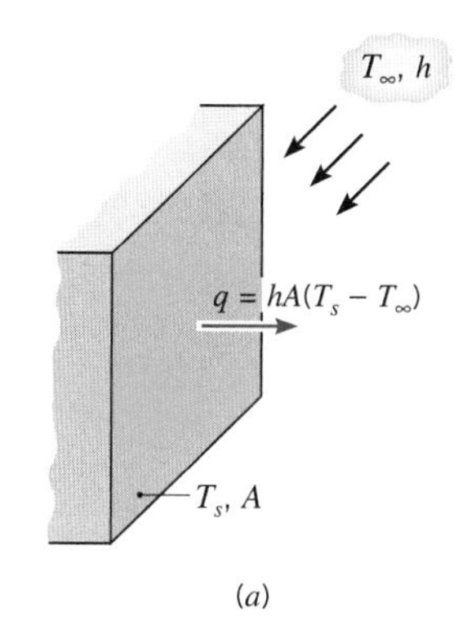


(*a*)

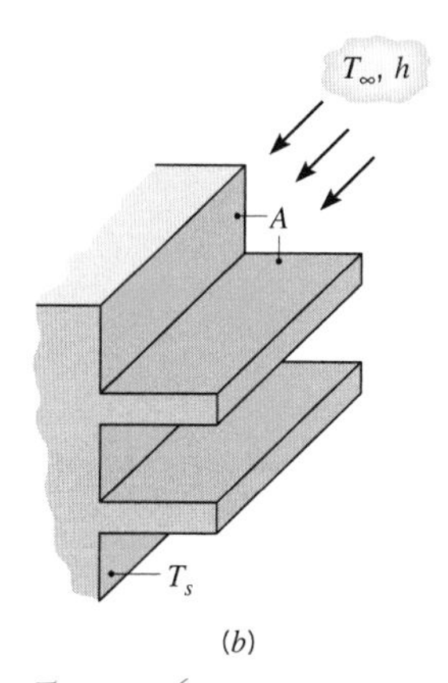


(*b*)

*Figure 16.13* Use of fins to enhance heat transfer from a plane wall.
(*a*) Bare surface.
(*b*) Finned surface.

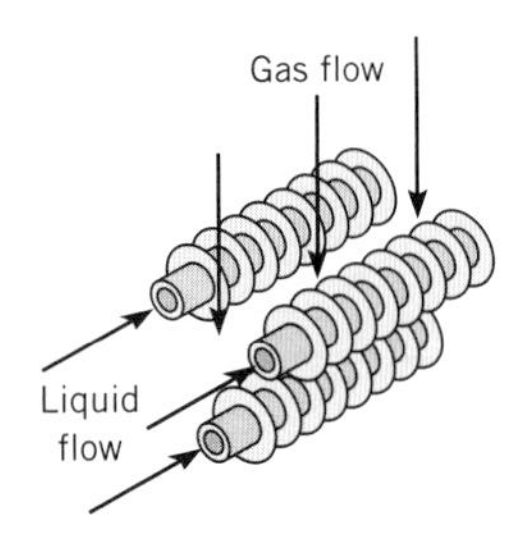


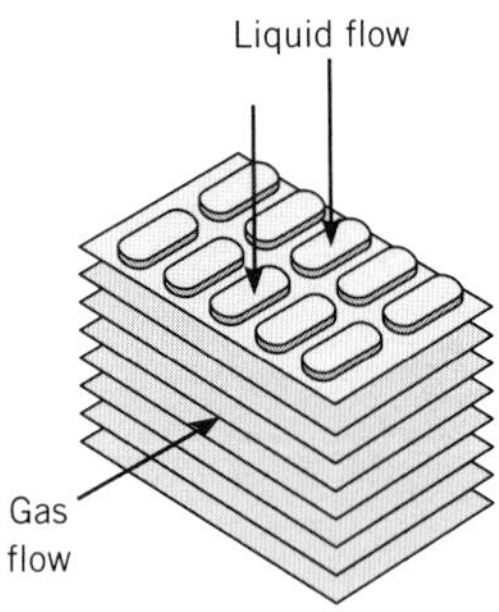


*Figure 16.14* Schematic of typical finned-tube heat exchangers.

plane wall. Examples of this type include the *straight fin with rectangular cross section,* Fig. 16.15*a*, and the *pin fin with a circular cross section,* Fig. 16.15*b*. Because of their uniform cross-sectional geometry, a simple analysis provides an understanding of the conduction-convection processes, as well as expressions for the temperature distribution and fin heat rate. Second, we will consider the *annular fin,* Fig. 16.15*c*, an extended surface that is circumferentially attached to a cylinder (tube or pipe). Because the analysis is much more complicated, we will introduce design charts that are widely used in engineering practice to determine fin heat transfer rates.

In the next section, we will perform a *conduction-convection analysis* on the fins of uniform cross section to obtain the temperature distribution. Using the results of this analysis in Sec. 16.4.2, we will obtain the fin heat rate, and in Sec. 16.4.3 we will identify key fin parameters useful for evaluating their performance in practical applications.

### 16.4.1 Conduction–Convection Analysis (CD-ROM)

### 16.4.2 Fin Temperature Distribution and Heat Rate

In the preceding section, we applied the conservation of energy requirement to a differential element (system) in fins of *uniform cross section,* Fig. 16.17*a,b*, experiencing conduction and convection processes. We defined the ***excess temperature*** as

*excess temperature, θ*

$$\theta = T(x) - T_\infty \tag{16.63}$$

and identified the ***fin parameter***

*fin parameter, m*

$$m = \sqrt{\frac{hP}{kA_c}} \tag{16.65}$$

where $P$ and $A_c$ are the fin perimeter and cross-sectional area, respectively. The resulting linear, second-order differential equation with constant coefficients has the general solution of the form

$$\theta(x) = C_1 e^{mx} + C_2 e^{-mx} \tag{16.66}$$

where $C_1$ and $C_2$ are arbitrary constants. To obtain the temperature distribution we need to evaluate the arbitrary constants from *two* boundary conditions representative of the fin physical situation.

To demonstrate the approach for obtaining the *fin temperature distribution,* consider case *A*, Fig. 16.17*c*, the *infinite fin.* At $x = \infty$, the tip temperature must equal that of the fluid, hence this boundary condition has the form

$$\theta(\infty) = T(\infty) - T_\infty = 0$$

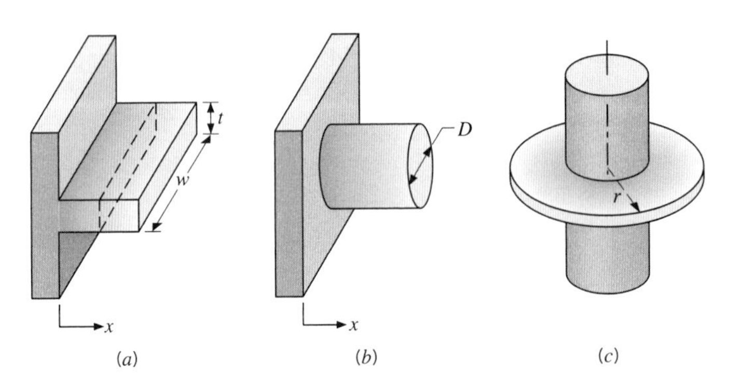


*Figure 16.15* Fin configurations. Straight fins with uniform cross-sectional area: (*a*) straight rectangular fin and (*b*) straight pin fin. (*c*) Annular fin with rectangular cross-sectional area.

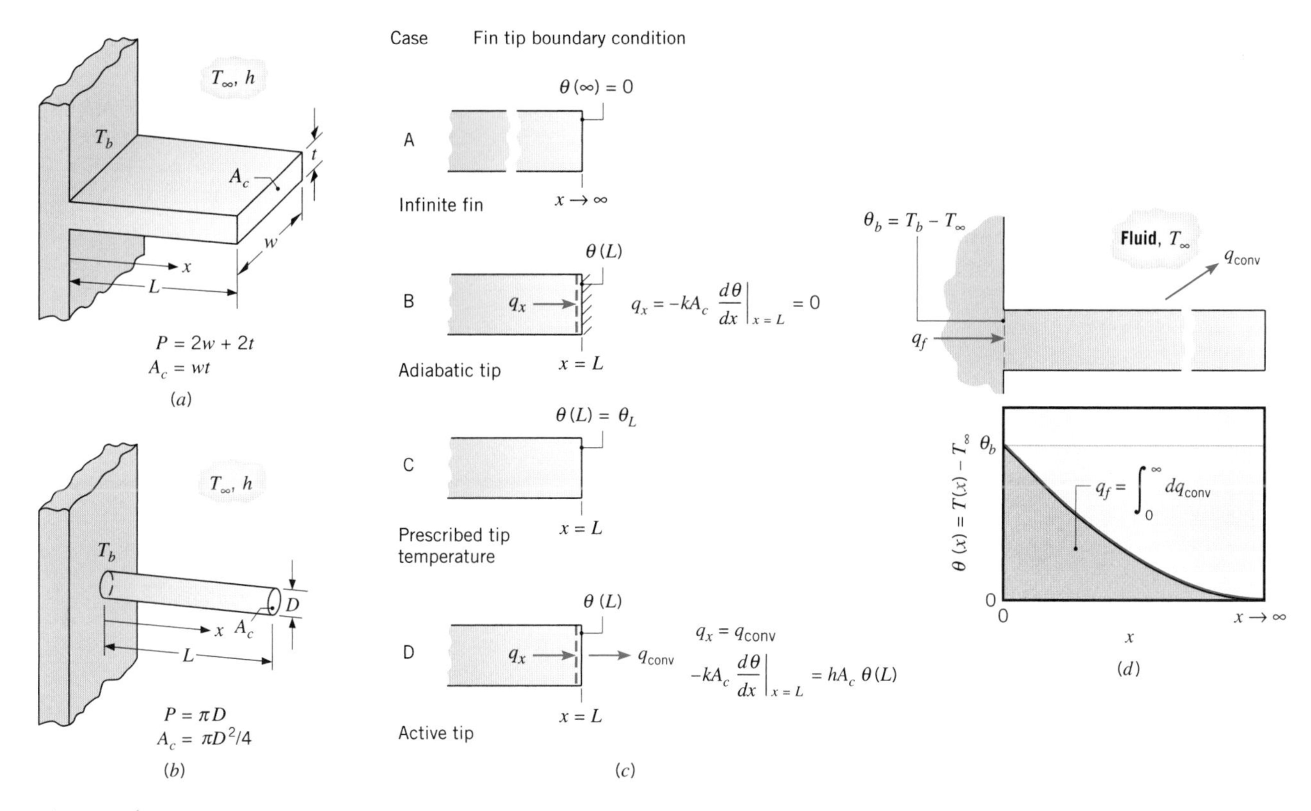


*Figure 16.17* Conduction and convection in a straight fin of uniform cross-sectional area. (*a*) Rectangular fin. (*b*) Pin fin. (*c*) Four common tip boundary conditions; see also Table 16.4. (*d*) Temperature distribution for the infinite fin ($x \to \infty$).

For the second boundary condition, we prescribe the temperature at the base of the fin, $T(0) = T_b$

$$\theta(0) = T_b - T_\infty \equiv \theta_b$$

Substituting the boundary condition at the fin tip into the general solution, Eq. 16.66

$$0 = C_1 e^{\infty} + C_2 e^{-\infty}$$

it follows that $C_1 = 0$. Substituting the boundary condition at the fin base, $x = 0$

$$\theta_b = C_1 e^{0} + C_2 e^{-0}$$

and with $C_1 = 0$, it follows that $C_2 = \theta_b$. Accordingly, the *temperature distribution* for the *infinite fin* is

$$\theta(x) = \theta_b e^{-mx} \qquad (16.67)$$

*infinite fin*

which is shown schematically in Fig. 16.17*d*. Note that the magnitude of the temperature gradient decreases with increasing $x$. This trend is a consequence of the reduction in the conduction heat transfer $q_x(x)$ with increasing $x$ due to continuous convection heat transfer from the surface.

The *fin heat rate* may be evaluated in two alternative ways, both of which involve use of the temperature distribution. The first involves applying *Fourier's law* at the fin base, $x = 0$, as shown in Fig. 16.18. That is

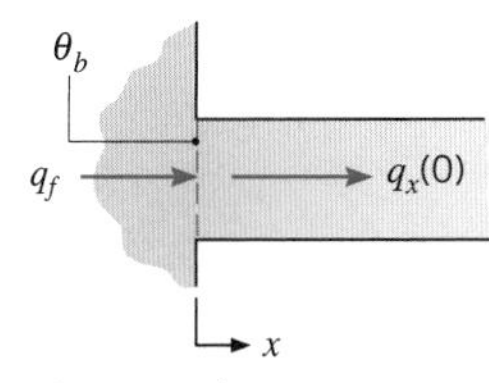


*Figure 16.18*

$$q_f = -kA_c \frac{dT}{dx}\Big|_{x=0} = -kA_c \frac{d\theta}{dx}\Big|_{x=0} = -kA_c(-m\theta_b e^{-m(0)})$$

with Eq. 16.65, the ***fin heat rate*** becomes

*fin heat rate*

$$q_f = \sqrt{hPkA_c}\,\theta_b \tag{16.68}$$

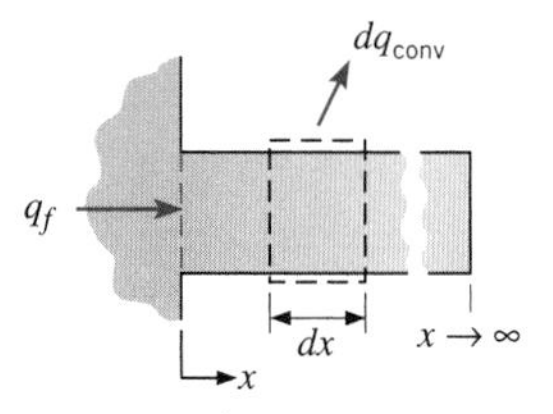


*Figure 16.19*

Conservation of energy dictates that the heat transfer rate by convection, $q_{conv}(x)$, must equal the heat transfer rate by conduction through the base of the fin, $q_f$. Accordingly, the alternative formulation for the fin heat rate, $q_f$, using *Newton's law of cooling* as shown in Fig. 16.19 is

$$q_f = q_{conv} = \int_0^\infty dq_{conv}$$

$$q_f = \int_0^\infty h[T(x) - T_\infty]P dx = \int_0^\infty h\theta(x) P dx = hP\theta_b \int_0^\infty e^{-mx}\, dx = hPm^{-1}\theta_b$$

$$q_f = \sqrt{hPkA_c}\,\theta_b$$

which agrees with Eq. 16.68. Recognize that the foregoing integral for the convective heat rate represents the area under the curve for the temperature distribution, $\theta$ vs. $x$, as shown in Fig. 16.17*d*.

In the same manner, but with greater mathematical complexity, we can obtain the fin temperature distribution and heat transfer rate for the other cases represented in Fig. 16.17*c*. These results are summarized in Table 16.4. Note also how these results can be used to determine the thermal resistance for a fin.

In lieu of the somewhat cumbersome expression for heat transfer from a straight fin with an active tip, Eq. 16.74 (Table 16.4), it has been shown that accurate predictions can be obtained by using the adiabatic tip result, Eq. 16.70, with the corrected length of the form $L_c = L + (t/2)$ for a *rectangular fin* and $L_c = L + (D/4)$ for a *pin fin*. The correction is based upon assuming equivalence between heat transfer from the actual fin with tip convection and heat transfer from a longer, hypothetical fin with an adiabatic tip.

**Table 16.4** Temperature Distribution and Heat Rate for Fins of Uniform Cross Section

| Case | Tip Condition[a] $(x = L)$ | Temperature Distribution $\theta/\theta_b$[b] | | Fin Heat Transfer Rate $q_f$[c] | |
|---|---|---|---|---|---|
| A | Infinite fin $(L \to \infty)$: $\theta(L) = 0$ | $e^{-mx}$ | (16.67) | $M = \sqrt{hPkA_c}\,\theta_b$ | (16.68) |
| | | $m = \sqrt{\dfrac{hP}{kA_c}}$ | (16.65) | | |
| B | Adiabatic: $d\theta/dx\vert_{x=L} = 0$ | $\dfrac{\cosh m(L-x)}{\cosh mL}$ | (16.69)[d] | $M \tanh mL$ | (16.70)[d] |
| C | Prescribed temperature: $\theta(L) = \theta_L$ | $\dfrac{(\theta_L/\theta_b)\sinh mx + \sinh m(L-x)}{\sinh mL}$ | (16.71) | $M\dfrac{(\cosh mL - \theta_L/\theta_b)}{\sinh mL}$ | (16.72) |
| D | Active, convection heat transfer: $h\theta(L) = -kd\theta/dx\vert_{x=L}$ | $\dfrac{\cosh m(L-x) + (h/mk)\sinh m(L-x)}{\cosh mL + (h/mk)\sinh mL}$ | (16.73) | $M\dfrac{\sinh mL + (h/mk)\cosh mL}{\cosh mL + (h/mk)\sinh mL}$ | (16.74)[e] |

[a]See Fig. 16.17*b* for relevant surface energy balances.

[b]Temperature excess definitions: $\theta \equiv T - T_\infty$ and $\theta_b = \theta(0) = T_b - T_\infty$.

[c]Fin thermal resistance is defined as $R_{t,f} = \theta_b/q_f$; see Eq. 16.76. ($q_f \longrightarrow \theta_b$ o–$R_{t,f}$–o $0$)

[d]A table of hyperbolic functions is provided in Appendix HT-6.

[e]Alternatively, use adiabatic tip result, Eq. 16.70, with corrected length $L_c$: $L_c = L + (t/2)$ for *rectangular fin* and $L_c = L + (D/4)$ for a *pin fin*.

## *Example 16.6* The Infinite Fin

A very long rod 5 mm in diameter has one end maintained at 100°C. The cylindrical (lateral) surface of the rod is exposed to ambient air at 25°C with a convection heat transfer coefficient of 100 W/m$^2\cdot$K.

**(a)** Assuming an infinite length, determine the steady-state temperature distributions along rods constructed from pure copper, 2024 aluminum alloy, and type AISI 316 stainless steel. What are the corresponding fin heat rates from the rods?

**(b)** How long must the rods be for the assumption of *infinite length* to yield a reasonable estimate of the heat loss?

### Solution

***Known:*** A long, circular rod exposed to ambient air.

***Find:***

**(a)** Temperature distribution and fin heat rate when rod is fabricated from copper, an aluminum alloy, or stainless steel.

**(b)** How long rods must be to assume infinite length.

***Schematic and Given Data:***

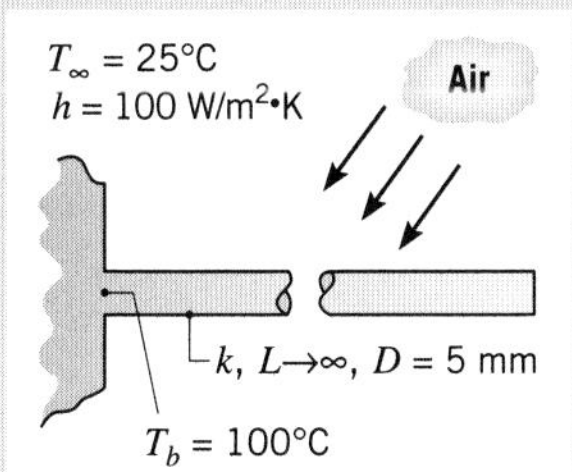


***Assumptions:***
1. Steady-state conditions.
2. One-dimensional conduction along the rod.
3. Infinitely long rod.
4. Negligible radiation exchange with surroundings.
5. Uniform heat transfer coefficient.
6. Constant properties.

*Figure E16.6a*

***Properties:*** Table HT-1, copper [$T = (T_b + T_\infty)/2 = 62.5°C \approx 335$ K]: $k = 398$ W/m$\cdot$K. Table HT-1, 2024 aluminum alloy (335 K): $k = 180$ W/m$\cdot$K. Table HT-1, stainless steel, AISI 316 (335 K): $k = 14$ W/m$\cdot$K.

***Analysis:*** **(a)** Subject to the assumption of an infinitely long fin, the temperature distributions are determined from Eq. 16.67, which may be expressed as

$$T = T_\infty + (T_b - T_\infty)e^{-mx}$$

where $m = (hP/kA_c)^{1/2} = (4h/kD)^{1/2}$. Substituting for $h$ and $D$, as well as for the thermal conductivities of copper, the aluminum alloy, and the stainless steel, respectively, the values of $m$ are 14.2, 21.2, and 75.6 m$^{-1}$. The temperature distributions may then be computed and plotted as shown in Fig. E16.6*b*

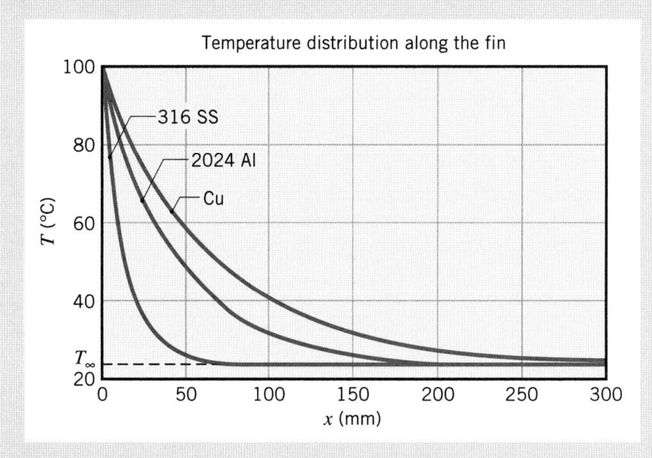


*Figure E16.6b*

From Eq. 16.68, the fin heat rate is

$$q_f = \sqrt{hPkA_c}\,\theta_b$$

Hence for the copper rod,

$$q_f = [100\text{ W/m}^2\cdot\text{K}(\pi \times 0.005\text{ m})(398\text{ W/m}\cdot\text{K})(\pi/4(0.005\text{ m})^2)]^{1/2}(100 - 25)°\text{C}$$

$$q_f = 8.3\text{ W} \quad \triangleleft$$

Similarly, for the aluminum alloy and stainless steel rods, respectively, the fin heat rates are $q_f = 5.6$ W and 1.6 W.

**(b)** From the temperature distributions in Fig. E16.6*b*, it is evident that there is little additional heat transfer associated with extending the length of the rod much beyond 250, 150, and 50 mm, respectively, for the copper, aluminum alloy, and stainless steel. Note also that the areas under the temperature distributions are in proportion to the fin heat rates for the three materials. (See also Fig. 16.17*d*.)

***Comments:*** Since there is no heat loss from the tip of an infinitely long rod, an estimate of the validity of this approximation may be made by comparing Eqs. 16.70 and 16.68 (Table 16.4). To a satisfactory approximation, the expressions provide equivalent results if $\tanh mL \geq 0.99$ or $mL \geq 2.65$. Hence a rod may be assumed to be infinitely long if

$$L \geq \frac{2.65}{m} = 2.65\left(\frac{kA_c}{hP}\right)^{1/2}$$

For copper,

$$L \geq 2.65\left[\frac{398 \text{ W/m}\cdot\text{K} \times (\pi/4)(0.005 \text{ m})^2}{100 \text{ W/m}^2\cdot\text{K} \times \pi(0.005 \text{ m})}\right]^{1/2} = 187 \text{ mm}$$

Results for the aluminum alloy and stainless steel are $L \geq 126$ mm and $L \geq 35$ mm, respectively. The estimates for the infinite length, based upon inspection of the temperature distributions of Fig. E16.6*b* and summarized in part (b), are in reasonable agreement with the quantitative approach based upon the fin heat rate considered here.

### 16.4.3 Fin Performance Parameters

Recall that fins are used to increase the heat transfer rate from a surface by increasing the effective surface area. However, the fin itself represents a conduction resistance to heat transfer from the original surface. For this reason, there is no assurance that the heat transfer rate will be increased through the use of fins. An assessment of this matter may be made by evaluating the ***fin effectiveness*** $\varepsilon_f$, which is defined as the *ratio of the fin heat transfer rate to the heat transfer rate that would exist without the fin*. That is

*fin effectiveness*

$$\varepsilon_f = \frac{q_f}{hA_c\theta_b} \tag{16.75}$$

where $A_c$ is the fin cross-sectional area. Subject to any one of the four tip conditions, the effectiveness for a fin of uniform cross section may be obtained by dividing the appropriate expression for $q_f$ in Table 16.4 by $hA_c\theta_b$. In any rational design the value of $\varepsilon_f$ should be as large as possible, and in general, the use of fins may rarely be justified unless $\varepsilon_f \gtrsim 2$. ***For Example…*** the fin effectiveness of the infinitely long copper, aluminum alloy, and stainless steel rods of Example 16.4 are 56.4, 38.0, and 10.9, respectively. ▲

Fin performance may also be quantified in terms of a thermal resistance. Treating the difference between the base and fluid temperatures as the driving potential, a ***fin resistance*** may be defined as

*fin resistance*

$$R_{t,f} = \frac{\theta_b}{q_f} \tag{16.76}$$

This result is extremely useful, particularly when representing a finned surface as a thermal circuit element. Note that, according to the fin tip condition, the appropriate expression for $q_f$ is obtained from Table 16.4.

Dividing Eq. 16.76 into the expression for the *thermal resistance* due to convection at the exposed base,

$$R_{t,b} = \frac{1}{hA_c} \tag{16.77}$$

and substituting from Eq. 16.75, it follows that

$$\varepsilon_f = \frac{R_{t,b}}{R_{t,f}} \tag{16.78}$$

Hence the fin effectiveness may be interpreted as a ratio of thermal resistances, and to increase $\varepsilon_f$ it is necessary to reduce the conduction–convection resistance of the fin. If the fin is to enhance heat transfer, its resistance must not exceed that of the exposed base.

Another measure of fin thermal performance is provided by the *fin efficiency* $\eta_f$. The maximum driving potential for convection is the temperature difference between the base ($x = 0$) and the fluid, $\theta_b = T_b - T_\infty$. Hence the maximum fin heat rate is the rate that would exist *if* the entire fin surface were at the base temperature: $q_{\text{max}} = hA_f\theta_b$ where $A_f$ is the total surface area of the fin. However, since any fin is characterized by a finite conduction resistance, a temperature gradient must exist along the fin and the above condition is an idealization. A logical definition of ***fin efficiency*** is therefore

$$\eta_f \equiv \frac{q_f}{q_{\text{max}}} = \frac{q_f}{hA_f\theta_b} \tag{16.79}$$

*fin efficiency*

***For Example...*** the fin efficiency of a 250-mm-long copper rod of Example 16.4, with $A_f = PL$, is

$$\eta_f = \frac{8.3\ \text{W}}{100\ \text{W/m}^2 \cdot \text{K}(\pi \times 0.005\ \text{m} \times 0.250\ \text{m})(100 - 25)°\text{C}} = 0.28\ \blacktriangle$$

For fins of uniform cross section, the expressions of Table 16.4 for the heat rate can be used to calculate the fin efficiency, $\eta_f$. For fins of *nonuniform cross section,* such expressions are very cumbersome, so that practitioners use graphs created by analytical or empirical treatment to obtain estimates for fin efficiency as a function of geometric parameters and the convection coefficient. An example of such a *design aid* is Fig. 16.20 for the *annular fin* of rectangular cross section (see also Fig. 16.15*c*). With knowledge of $\eta_f$, Eq. 16.79 is convenient to use for calculating the fin heat rate.

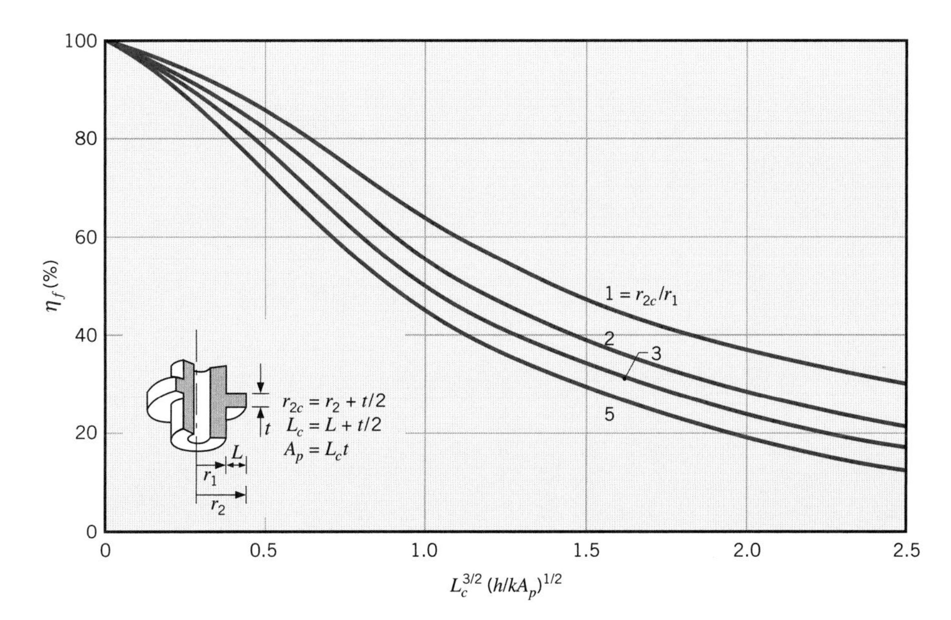


*Figure 16.20* Efficiency of annular fins of rectangular cross section.

## *Example 16.7* Cooling Fins for a Transistor Package

Heat transfer from a transistor may be enhanced by inserting it in an aluminum sleeve ($k = 200$ W/m · K) having 12 integrally machined longitudinal fins on its outer surface. The transistor radius and height are $r_1 = 2$ mm and $H = 6$ mm, respectively, while the fins are of length $L = r_3 - r_2 = 10$ mm and uniform thickness $t = 0.7$ mm. The thickness of the sleeve base is $r_2 - r_1 = 1$ mm, and the contact resistance of the sleeve–transistor interface is $R''_{t,c} = 10^{-3}$ m$^2$ · K/W. Air at $T_\infty = 20$°C flows over the fin surface, providing an approximately uniform convection coefficient of $h = 25$ W/m$^2$ · K.

**(a)** Assuming one-dimensional transfer in the radial direction, sketch the equivalent thermal circuit for heat transfer from the transistor case ($r = r_1$) to the air. Clearly label each resistance.

**(b)** Evaluate each of the resistances in the foregoing circuit. If the temperature of the transistor case is $T_1 = 80$°C, what is the rate of heat transfer from the sleeve?

## Solution

***Known:*** Dimensions of finned aluminum sleeve inserted over a transistor. Contact resistance between sleeve and transistor. Surface convection conditions and temperature of transistor case.

***Find:***

**(a)** Equivalent thermal circuit.

**(b)** Rate of heat transfer from sleeve.

***Schematic and Given Data:***

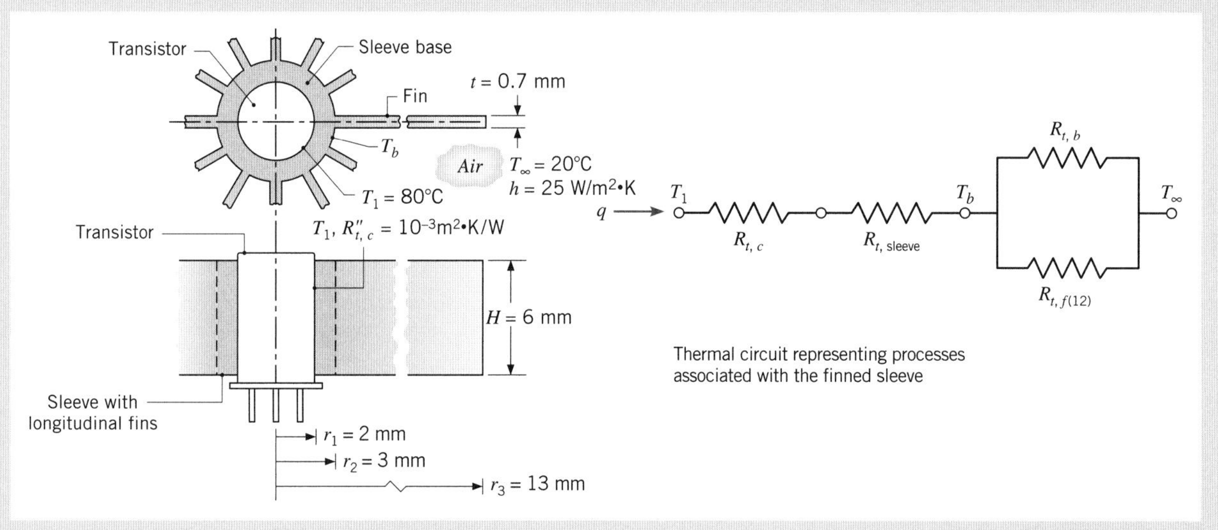


*Figure E16.7*

***Assumptions:***

1. Steady-state conditions.
2. Negligible heat transfer from the top and bottom surfaces of the transistor.
3. One-dimensional radial conduction through sleeve base.
4. Negligible radiation exchange between surfaces and surroundings.
5. Constant properties.

***Analysis:***

**(a)** The thermal circuit shown above accounts for the contact resistance between the transistor case and the sleeve, $R''_{t,c}$, conduction through the sleeve, $R_{t,\text{sleeve}}$, convection from the exposed sleeve base, $R''_{t,b}$, and conduction–convection from the twelve fins, $R''_{t,f(12)}$. Note that $R''_{t,f(12)}$ represents 12 fin-resistance elements in a parallel-circuit arrangement. The elements $R_{t,b}$ and $R_{t,f(12)}$ represent parallel heat flow paths from the sleeve outer surface ($r_2$) by convection and through the fins.

**(b)** Thermal resistances associated with the contact joint and sleeve (Eq. 16.32) are

$$R_{t,c} = \frac{R''_{t,c}}{2\pi r_1 H} = \frac{10^{-3}\ \text{m}^2 \cdot \text{K/W}}{2\pi(0.002\ \text{m})(0.006\ \text{m})} = 13.3\ \text{K/W}$$

$$R_{t,\text{sleeve}} = \frac{\ln(r_2/r_1)}{2\pi H k} = \frac{\ln(3/2)}{2\pi(0.006\ \text{m})(200\ \text{W/m} \cdot \text{K})} = 0.054\ \text{K/W}$$

For a *single* fin, the thermal resistance follows from Eq. 16.76 with Eq. 16.74 (Table 16.4, active tip condition) for the fin heat rate

$$R_{t,f} = \frac{\theta_b}{q_f} = \left[ (hPkA_c)^{1/2} \frac{\sinh mL + (h/mk) \cosh mL}{\cosh mL + (h/mk) \sinh mL} \right]^{-1}$$

With $P = 2(H + t) = 13.4$ mm $= 0.0134$ m and $A_c = t \times H = 4.2 \times 10^{-6}$ m$^2$, evaluate the parameters

$$m = \sqrt{\frac{hP}{kA_c}} = \left( \frac{25 \text{ W/m}^2 \cdot \text{K} \times 0.0134 \text{ m}}{200 \text{ W/m} \cdot \text{K} \times 4.2 \times 10^{-6} \text{ m}^2} \right)^{1/2} = 20.0 \text{ m}^{-1}$$

$$mL = 20 \text{ m}^{-1} \times 0.01 \text{ m} = 0.20$$

$$\frac{h}{mk} = \frac{25 \text{ W/m}^2 \cdot \text{K}}{20 \text{ m}^{-1} \times 200 \text{ W/m} \cdot \text{K}} = 0.00625$$

$$(hPkA_c)^{1/2} = (25 \text{ W/m}^2 \cdot \text{K} \times 0.0134 \text{ m} \times 200 \text{ W/m} \cdot \text{K} \times 4.2 \times 10^{-6} \text{ m}^2)^{1/2} = 0.0168 \text{ W/K}$$

Substituting numerical values, using Table HT-6 to evaluate the hyperbolic functions, the *thermal resistance for a single fin* is

$$R_{t,f} = \left[ \frac{0.0168 \text{ W/K} (0.201 + 0.00625 \times 1.020)}{1.020 + 0.00625 \times 0.201} \right]^{-1} = 293 \text{ K/W}$$

Hence, the *thermal resistance of 12 fins* in a parallel-circuit arrangement is

$$R_{t,f(12)} = \frac{R_{t,f}}{12} = 24.4 \text{ K/W}$$

For the *exposed base,* the thermal resistance due to convection is

$$R_{t,b} = \frac{1}{h(2\pi r_2 - 12t)H} = \frac{1}{25 \text{ W/m}^2 \cdot \text{K} (2\pi \times 0.003 - 12 \times 0.0007) \text{ m} \times 0.006 \text{ m}}$$

$$R_{t,b} = 638 \text{ K/W}$$

For the *parallel* resistances of the 12 fins, $R_{t,f(12)}$, and convection from the base, $R_{t,b}$, as shown in the thermal circuit of Fig. E16.7 the equivalent resistance is

$$R_{\text{equiv}} = [1/R_{t,f(12)} + 1/R_{t,b}]^{-1} = [(24.4)^{-1} + (638)^{-1}]^{-1} = 23.5 \text{ K/W}$$

so that the *total resistance of the finned sleeve* is

$$R_{\text{tot}} = R_{t,c} + R_{t,\text{ sleeve}} + R_{\text{equiv}} = (13.3 + 0.054 + 23.5) \text{ K/W} = 36.9 \text{ K/W}$$

and the heat transfer rate from the sleeve is

$$qt = \frac{T_1 - T_\infty}{R_{\text{tot}}} = \frac{(80 - 20)°\text{C}}{36.9 \text{ K/W}} = 1.63 \text{ W} \quad \triangleleft$$

***Comments:*** Without the finned sleeve, the convection resistance of the transistor case is $R_{\text{tran}} = (2\pi r_1 Hh)^{-1} = 531$ K/W. Hence there is considerable advantage to using the fins.

## 16.5 Transient Conduction

Many heat transfer applications involve *transient* or *unsteady* conduction resulting from a change with time of conditions within the system, and/or of the thermal environment of the system. In this section, we will consider transient conduction resulting from a change in convective boundary conditions. ***For Example...*** a hot metal billet, suddenly removed from a furnace and quenched in a large liquid bath, experiences conduction while the surface is cooled by convection heat transfer (Fig. 16.21). The billet eventually reaches a steady-state condition with a uniform temperature equal to that of the bath. ▲

Our objective in this section is to develop methods for determining the time dependence of the temperature distribution within an object during a transient process, as well as determining the heat transfer between the object and its environment.

*lumped capacitance method*

The method of analysis depends upon the nature of the temperature gradients within the object during the transient process. If the temperature of the object is *approximately uniform, a single temperature* can be used to characterize the time response of the object to the convective boundary change. Termed the ***lumped capacitance method,*** an overall energy balance is used to determine the variation of temperature with time (Sec. 16.5.1). Because conditions for the *validity* of the method can be clearly established, the method can be applied with confidence to appropriate applications.

If there are appreciable temperature differences within the object during the transient process, *spatial effects* must be considered and the temperature distribution must be determined by solving the *heat equation.* We will consider solutions for *finite solids* (plane walls and radial systems—long cylinders and spheres) in Secs. 16.5.2 and 16.5.3, and for *semi-infinite solids* in Sec. 16.5.4.

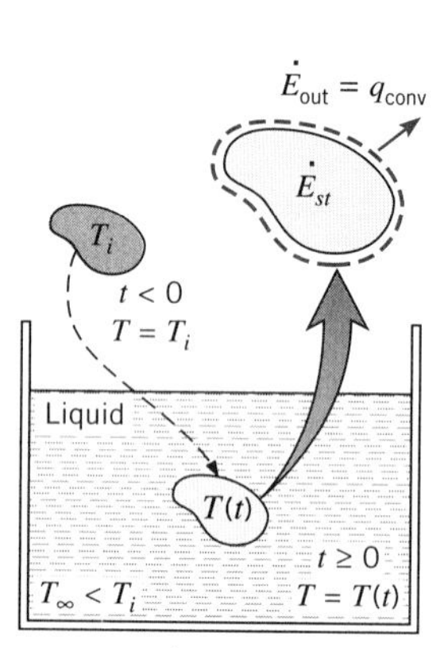


*Figure 16.21* Cooling of a hot metal forging.

## 16.5.1 The Lumped Capacitance Method

Consider again the hot metal billet undergoing the quenching process. The billet is initially at a uniform temperature $T_i$ and is suddenly immersed in a large liquid bath of lower temperature $T_\infty < T_i$ (Fig. 16.21). If the quenching is said to begin at time $t = 0$, the temperature of the billet, $T(t)$ will decrease for time $t > 0$, until eventually it reaches $T_\infty$. The temperature of the billet decreases as its internal energy is diminished due to convection heat transfer from the billet to the liquid bath. The essence of the lumped capacitance method is the assumption that the *temperature* of the solid is *approximately uniform* at any instant during the transient process. For now we assume that this is, in fact, the case; we will then determine under what conditions this *uniform temperature assumption* is valid.

### Temperature–Time History: Energy Balance

Because we have assumed a *single temperature* can be used to *characterize* the solid during the transient process of Fig. 16.21, we can determine the temperature response by formulating an *overall energy balance* on the solid. This balance will relate the convective heat transfer rate, at the surface to the rate of change of internal energy. Applying Eq. 15.11a to the system of Fig. 16.21, this requirement takes the form

$$-\dot{E}_{out} = \dot{E}_{st} \tag{16.80}$$

or

$$-hA_s(T - T_\infty) = \rho V c \frac{dT}{dt} \tag{16.81}$$

Introducing the temperature difference

$$\theta \equiv T - T_\infty \tag{16.82}$$

and recognizing that $(d\theta/dt) = (dT/dt)$, it follows that

$$\frac{\rho V c}{h A_s} \frac{d\theta}{dt} = -\theta$$

Separating variables and integrating from the *initial condition,* for which $t = 0$ and $T(0) = T_i$, we then obtain

$$\frac{\rho V c}{h A_s} \int_{\theta_i}^{\theta} \frac{d\theta}{\theta} = -\int_0^t dt$$

where

$$\theta_i \equiv T_i - T_\infty \tag{16.83}$$

Evaluating the integrals, the ***temperature-time history*** has the form

*temperature-time history*

$$\frac{\rho V c}{h A_s} \ln \frac{\theta_i}{\theta} = t \tag{16.84}$$

or rearranging to obtain the temperature explicitly

$$\frac{\theta}{\theta_i} = \frac{T - T_\infty}{T_i - T_\infty} = \exp\left[-\left(\frac{h A_s}{\rho V c}\right)t\right] \tag{16.85}$$

Equation 16.84 may be used to determine the time required for the solid to reach some temperature $T$, or, conversely, Eq. 16.85 may be used to compute the temperature reached by the solid at some time $t$.

The foregoing results indicate that the difference between the solid and fluid temperatures must decrease exponentially to zero as $t$ approaches infinity. This behavior is shown in Fig. 16.22. From Eq. 16.85 it is also evident that the quantity $(\rho Vc/hA_s)$ may be interpreted as a ***thermal time constant,*** which has the form

$$\tau_t = \left(\frac{1}{h A_s}\right)(\rho V c) = R_t C_t \tag{16.86}$$

*thermal time constant*

where $R_t$ is the *resistance* to *convection heat transfer* and $C_t$ is the ***lumped thermal capacitance*** of the solid. Any increase in $R_t$ or $C_t$ will cause a solid to respond more slowly to changes in its thermal environment and will increase the time required to reach thermal equilibrium ($\theta = 0$).

*lumped thermal capacitance*

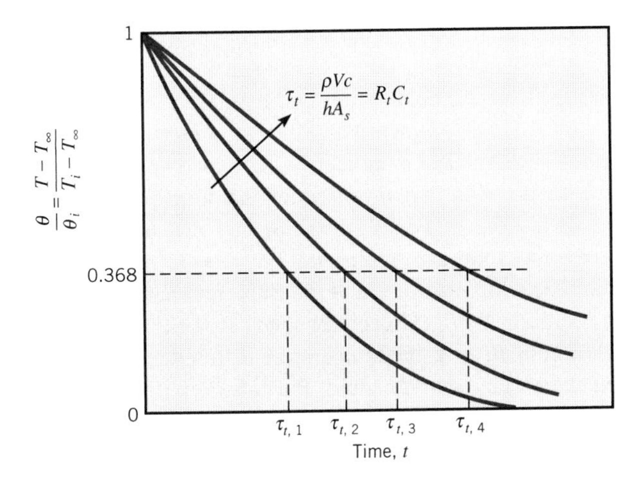


*Figure 16.22* Transient temperature response of lumped capacitance solids for different thermal time constants $\tau_t$.

To determine the *total energy transfer* $Q$ (J) occurring up to some time $t$, we simply write

$$Q = \int_0^t q\, dt = hA_s \int_0^t \theta\, dt$$

Substituting for $\theta$ from Eq. 16.85 and integrating, we obtain

$$Q = (\rho V c)\theta_i \left[1 - \exp\left(-\frac{t}{\tau_t}\right)\right] \tag{16.87}$$

## Validity of the Lumped Capacitance Method

It is important to determine under what conditions the lumped capacitance method may be used with reasonable accuracy. To develop a suitable criterion, consider steady-state conduction through a plane wall of area $A$ (Fig. 16.23). Although we are assuming *steady-state conditions,* we'll later see how this criterion is readily extended to transient processes. One surface is maintained at a temperature $T_{s,1}$ and the other surface is exposed to a fluid of temperature $T_\infty < T_{s,1}$. The temperature of this surface will be some intermediate value, $T_{s,2}$, for which $T_\infty < T_{s,2} < T_{s,1}$. Hence, under steady-state conditions the surface energy balance, Eq. 15.14 expressed as $q_{\text{cond}} = q_{\text{conv}}$, reduces to

$$\frac{kA}{L}(T_{s,1} - T_{s,2}) = hA(T_{s,2} - T_\infty)$$

where $k$ is the thermal conductivity of the solid. Rearranging, we then obtain

$$\frac{T_{s,1} - T_{s,2}}{T_{s,2} - T_\infty} = \frac{(L/kA)}{(1/hA)} = \frac{R_{\text{cond}}}{R_{\text{conv}}} = \frac{hL}{k} \equiv Bi \tag{16.88}$$

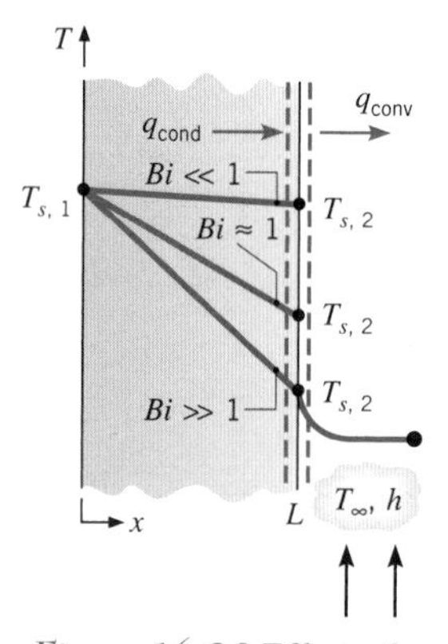


*Figure 16.23* Effect of Biot number on steady-state temperature distribution in a plane wall with surface convection.

The quantity $(hL/k)$ appearing in Eq. 16.88 is a *dimensionless parameter.* It is termed the ***Biot number,*** and it plays a fundamental role in conduction problems that involve surface convection effects. According to Eq. 16.88 and as illustrated in Fig. 16.23, the *Biot number provides a measure of the temperature drop in the solid relative to the temperature difference between the surface and the fluid.* Note especially the conditions corresponding to $Bi \ll 1$. The results suggest that, for these conditions, it is reasonable to *assume* a uniform temperature distribution across a solid at any time during a transient process. This result may also be associated with interpretation of the Biot number as a ratio of thermal resistances, Eq. 16.88. *If $Bi \ll 1$, the resistance to conduction within the solid is much less than the resistance to convection across the fluid boundary layer.* Hence, the assumption of a uniform temperature distribution is reasonable.

The significance of the Biot number to *transient* conduction is illustrated in the following situation. Consider the plane wall of Fig. 16.24, which is initially at a uniform temperature $T_i$, and experiences convection cooling when it is suddenly immersed in a fluid of $T_\infty < T_i$. The temperature variations with position and time within the solid, denoted as $T(x, t)$, are plotted for the two extreme conditions of Biot number, $Bi \ll 1$ and $Bi \gg 1$. For $Bi \ll 1$, the temperature gradients within the solid are small, and $T(x, t) \approx T(t)$. Virtually all the temperature difference is between the solid's surface and the fluid, and *the solid temperature remains nearly uniform* as it decreases to $T_\infty$. This corresponds to the condition required of the lumped capacitance method. For $Bi \gg 1$, the temperature difference across the solid is much larger than that between the surface and the fluid. For this condition, *spatial effects are important,* and we must use the heat equation to obtain the temperature distribution.

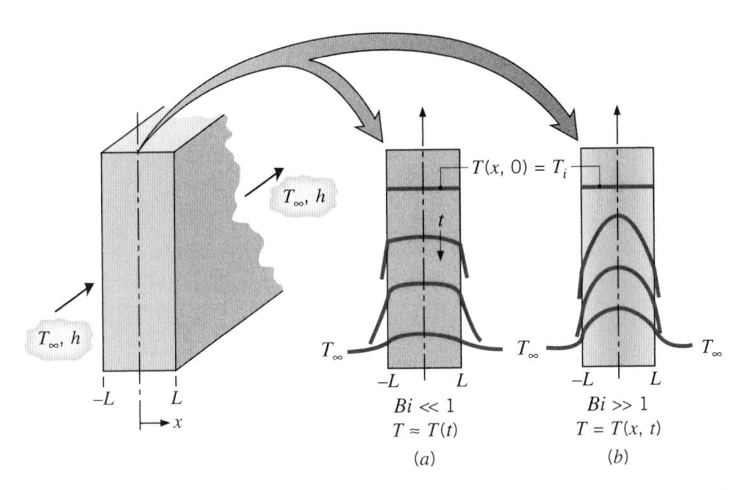


*Figure 16.24* Transient temperature distributions for extreme Biot numbers in a plane wall symmetrically cooled by convection. (*a*) $Bi \ll 1$, corresponding to the condition required of the lumped capacitance method, and (*b*) $Bi \gg 1$, conditions for which spatial effects are important.

When confronted with a transient conduction problem involving a sudden change in the thermal environment, you should calculate the Biot number. If the following ***validity criterion*** is satisfied

$$Bi = \frac{hL_c}{k} < 0.1 \tag{16.89}$$

*validity criterion*

the error associated with using the lumped capacitance method is small. For convenience it is customary to define the ***characteristic length,*** $L_c$, of Eq. 16.89 as the ratio of the solid's volume to surface area

$$L_c = \frac{V}{A_s} \tag{16.90}$$

*characteristic length*

Such a definition facilitates calculation of $L_c$ for solids of complicated shapes and reduces to the half-thickness for a plane wall of thickness $2L$ (Fig. 16.24), to $r_o/2$ for a long cylinder and to $r_o/3$ for a sphere.

Finally, we note that, with $L_c \equiv V/A_s$ and the thermal diffusivity $\alpha = k/\rho c$ from Eq. 16.5, the exponential term within brackets of Eq. 16.85 can be expressed as

$$\frac{hA_s t}{\rho V c} = \frac{ht}{\rho c L_c} = \frac{hL_c}{k}\frac{k}{\rho c}\frac{t}{L_c^2} = \frac{hL_c}{k}\frac{\alpha t}{L_c^2}$$

or

$$\frac{hA_s t}{\rho V c} = Bi \cdot Fo \tag{16.91}$$

where $Bi = hL/k$ and $Fo$ is the ***Fourier number***

$$Fo \equiv \frac{\alpha t}{L_c^2} \tag{16.92}$$

*Fourier number*

The Fourier number is a *dimensionless time,* which, with the Biot number, characterizes transient conduction problems. Substituting Eq. 16.91 into 16.85, we obtain

$$\frac{\theta}{\theta_i} = \frac{T - T_\infty}{T_i - T_\infty} = \exp(-Bi \cdot Fo) \tag{16.93}$$

## *Example 16.8* Lumped Capacitance Method: Cooling Process

In a materials evaluation program, dielectric-coated glass beads of 12.5 mm diameter are removed from a process oven with a uniform temperature of 225°C. The beads are cooled in an air stream for which $T_\infty$ = 20°C and the convection coefficient is 25 W/m$^2$ · K. What is the temperature of a bead after 6 min?

### Solution

***Known:*** A glass bead, initially at a uniform temperature, is suddenly subjected to a convection cooling process.
***Find:*** Temperature of the glass bead after 6 min.

***Schematic and Given Data:***

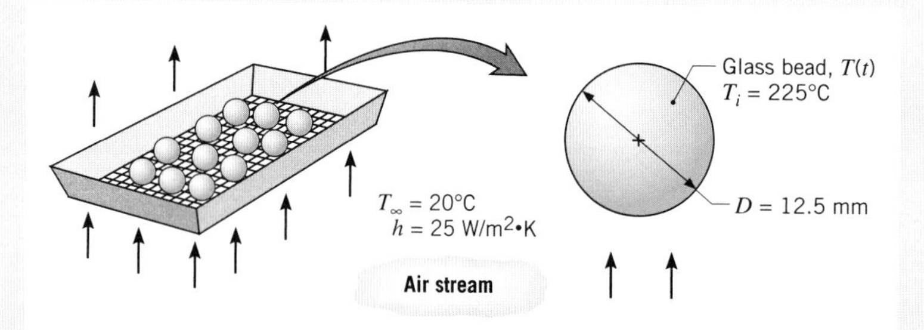


***Assumptions:***
1. Temperature of the bead is uniform at any instant.
2. The coating has negligible thermal resistance and capacitance.
3. Radiation exchange with the surroundings is negligible.
4. Constant properties.

*Figure E16.8*

***Properties:*** Table HT-2, glass, Pyrex (300 K): $\rho$ = 2225 kg/m$^3$, $c$ = 835 J/kg · K, $k$ = 1.4 W/m · K.

***Analysis:*** To establish the validity of the lumped capacitance method, calculate the Biot number. From Eq. 16.90, the characteristic length of the spherical bead is

$$L_c = \frac{V}{A_s} = \frac{\pi D^3/6}{\pi D^2} = \frac{D}{6}$$

and using Eq. 16.89, determine the Biot number,

$$Bi = \frac{hL_c}{k} = \frac{h(D/6)}{1.4} = \frac{25 \text{ W/m}^2 \cdot \text{K} (.0125 \text{ mm}/6)}{1.4 \text{ W/m} \cdot \text{K}} = 0.037$$

Accordingly, $Bi < 0.1$ so that the bead has a nearly uniform temperature during the cooling process. Using Eq. 16.85, with $L_c = D/6$ the temperature $T(t)$ after 6 min is

$$\frac{T(t) - T_\infty}{T_i - T_\infty} = \exp\left[-\left(\frac{h}{\rho L_c c}\right)t\right]$$

$$\frac{T(t) - 20°\text{C}}{(225 - 20)°\text{C}} = \exp\left[-\left(\frac{25 \text{ W/m}^2 \cdot \text{K}}{2225 \text{ kg/m}^3 (0.0125 \text{ m}/6)\, 835 \text{ J/kg}}\right) 360 \text{ s}\right]$$

$$T(t) = 20°\text{C} + (225 - 20)°\text{C} \times 0.0978 = 40.0°\text{C} \triangleleft$$

## *Example 16.9* Workpiece Temperature-Time History: Curing Operation

A 3-mm-thick panel of aluminum alloy ($\rho$ = 2770 kg/m$^3$, $c$ = 875 J/kg · K, and $k$ = 177 W/m · K) is finished on both sides with an epoxy coating that must be cured at or above $T_c$ = 150°C for at least 5 min. The curing operation is performed in a large oven with air at 175°C and convection coefficient of $h$ = 40 W/m$^2$ · K. The coating has an emissivity of $\varepsilon$ = 0.8, and the temperature of the oven walls is 175°C, providing an *effective* radiation coefficient of $h_{rad}$ = 12 W/m$^2$ · K. If the panel is placed in the oven at an initial temperature of 25°C, at what total elapsed time, $t_c$, will the cure process be completed?

### Solution

***Known:*** Operating conditions for a heating process in which a coated aluminum panel is maintained at or above the cure temperature $T_c$ = 150°C for at least 5 min.
***Find:*** Elapsed time for completion of the cure process, $t_c$.

*Schematic and Given Data:*

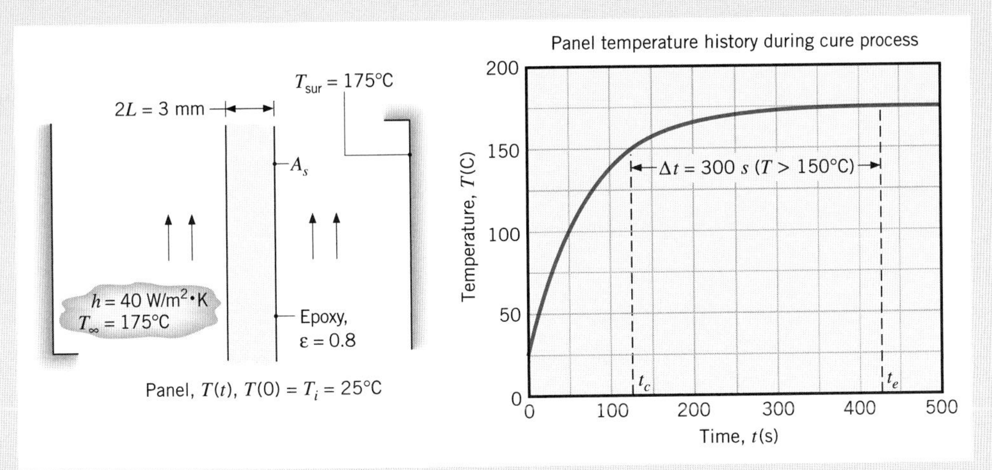


*Assumptions:*
1. Panel temperature is uniform at any instant.
2. Thermal resistance of epoxy is negligible.
3. Radiation exchange with the surroundings can be characterized by *an effective* linearized radiation coefficient, $h_{rad}$.
4. Constant properties.

*Figure E16.9*

***Analysis:*** During the cure operation, the panel surface experiences convection with the fluid at $T_\infty$, $q_{conv}$, and radiation exchange with the surrounding at $T_{sur} = T_\infty$, $q_{rad}$. The total heat rate from the panel surface is

$$q = q_{conv} + q_{rad} = (h + h_{rad})A_s(T - T_\infty)$$

where $h$ and $h_{rad}$ are the convection and *effective* radiation coefficients, respectively, and $(h + h_{rad})$ represents the *combined* convection-radiation coefficient. To assess the validity of the lumped capacitance method, we begin by calculating the Biot number, Eq. 16.89, using the combined convection-radiation coefficient

$$Bi = \frac{(h + h_{rad})L}{k} = \frac{(40 + 12)\ \text{W/m}^2 \times \text{K}\ (0.0015\ \text{m})}{177\ \text{W/m} \times \text{K}} = 4.68 \times 10^{-4}$$

Since $Bi < 0.1$, the lumped capacitance approximation is excellent. From Eq. 16.84, the time required for the panel to reach the cure temperature is

$$t_c = \frac{\rho V c}{hA_s} \ln\frac{\theta_i}{\theta} = \frac{\rho L c}{(h + h_{rad})} \ln\frac{T_i - T_\infty}{T_c - T_\infty}$$

$$t_c = \frac{2770\ \text{kg/m}^3 \times 0.0015\ \text{mm} \times 875\ \text{J/kg}\cdot\text{K}}{(40 + 12)\ \text{W/m}^2\cdot\text{K}} \ln\frac{25 - 175}{150 - 175} = 125\ \text{s}$$

The panel reaches the cure temperature of 150°C in 125 s, hence the total time to complete the 5-min duration cure is

$$t_e = t_c + 5 \times 60\ \text{s} = (125 + 300)\ \text{s} = 425\ \text{s} \quad \lhd$$

***Comments:***
1. Note that the characteristic length $L_c = V/A_s$ used in the analysis is the half-width of the plate, $L$.
2. The temperature-time history for the panel is shown in the graph of Fig. E16.9. Note that the panel has reached the oven air and wall temperature upon completion of the cure process ($t = t_e$). If you wanted to reduce the time-to-cure, what parameters would you change?
3. The *effective* linearized radiation heat transfer coefficient associated with radiation exchange between the panel and its surroundings was determined from Eq. 15.9. The estimate for $h_{rad}$ is based upon the average panel temperature during the heating process, $T_{avg} = (T_c + T_i)/2 = 87.5°\text{C} = 360.5$ K,

$$h_{rad} = \varepsilon\sigma(T_{avg} + T_{sur})(T_{avg}^2 + T_{sur}^2)$$
$$h_{rad} = 0.8(5.67 \times 10^{-8}\ \text{W/m}^2\cdot\text{K}^4)(360.5 + 448)(360.5^2 + 448^2)\text{K}^3 = 12.1\ \text{W/m}^2\cdot\text{K}$$

Remember to use absolute temperatures when evaluating the linearized radiation coefficient.
4. Use *Interactive Heat Transfer (IHT)* for analysis of the lumped capacitance method. (CD-ROM)

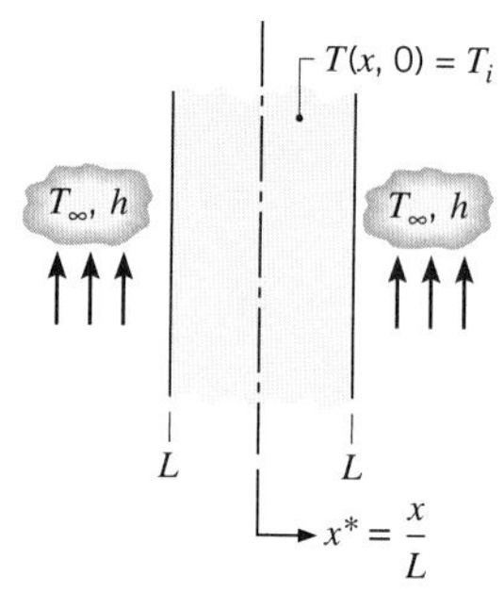


*Figure 16.25* Plane wall with an initial uniform temperature subjected to sudden convection conditions.

### 16.5.2 The Plane Wall with Convection

With $Bi > 0.1$, the lumped capacitance method is no longer appropriate, and the solid cannot be approximated by a single temperature during the transient process. *Spatial effects* must be considered as the temperature differences within the solid are appreciable.

Consider the plane wall of thickness $2L$ of Fig. 16.25 experiencing one-dimensional conduction in the $x$-direction. The wall is initially at a uniform temperature, $T(x, 0) = T_i$, and is suddenly immersed in a fluid of $T_\infty \neq T_i$. The resulting temperature distribution as a function of location and time, $T(x, t)$, may be obtained by solving the appropriate form of the heat equation with the relevant initial and boundary conditions.

#### The Heat Equation: Derivation (CD-ROM)

#### Solution for the Temperature Distribution

From a detailed treatment of the heat equation in the preceding section, we showed that the ***dimensionless solution*** for the transient temperature distribution is a universal function of $x^*$, $Fo$, and $Bi$. That is

*dimensionless solution*

$$\theta^* = f(x^*, Fo, Bi) \qquad (16.103)$$

$$\begin{cases} x^* = x/L & (16.100) \\ Fo = \alpha t/L^2 & (16.101) \\ Bi = hL/k & (16.102) \end{cases}$$

where $x^*$ is the dimensionless spatial coordinate, $Fo$ is the dimensionless time (Fourier number, Eq. 16.92), and $Bi$ is the ratio of thermal resistances (Biot number, Eq. 16.88).

Several mathematical techniques can be used to obtain exact solutions for such conduction problems, and typically the solutions are in the form of an infinite series. However, except for small values of the Fourier number, $Fo < 0.2$, the infinite series can be approximated by the first term of the series. Using the *one-term approximation,* the dimensionless form of the temperature distribution is

$$\theta^* = C \exp(-\zeta^2 Fo) \cos(\zeta x^*) \qquad (16.104a)$$

or alternatively as

$$\theta^* = \theta_o^* \cos(\zeta x^*) \qquad (16.104b)$$

where $\theta_o^* \equiv (T_o - T_\infty)/(T_i - T_\infty)$ represents the midplane ($x^* = 0$) dimensionless temperature, namely

$$\theta_o^* = C \exp(-\zeta^2 Fo) \qquad (16.105)$$

An important implication of Eq. 16.104b is that *the time dependence of the temperature at any location within the wall is the same as that of the midplane temperature.* The coefficients $C$ and $\zeta$ are given in Table 16.5 for a range of Biot numbers.

#### Total Energy Transfer

In many situations it is useful to know the total energy that has left (or entered) the wall up to any time $t$ in the transient process. The conservation of energy requirement, Eq. 15.11b, may be applied for the *time interval* bounded by the initial condition ($t = 0$) and any time $t > 0$

$$E_{in} - E_{out} = \Delta E_{st} \qquad (16.106)$$

Identifying the *energy transferred from the wall* $Q$ with $E_{out}$, and setting $E_{in} = 0$ and $\Delta E_{st} = E(t) - E(0)$, it follows that

$$Q = -[E(t) - E(0)] \qquad (16.107a)$$

or

$$Q = -\int \rho c[T(x, t) - T_i]dV \qquad (16.107b)$$

**Table 16.5** Coefficients Used in the One-Term Approximation to the Series Solution for Transient, One-Dimensional Conduction in the Plane Wall

| Plane Wall | | | | | | | | |
|---|---|---|---|---|---|---|---|---|
| $Bi^a$ | $\zeta$ | $C$ | $Bi$ | $\zeta$ | $C$ | $Bi$ | $\zeta$ | $C$ |
| 0.01 | 0.0998 | 1.0017 | 0.25 | 0.4801 | 1.0382 | 5.0 | 1.3138 | 1.2402 |
| 0.02 | 0.1410 | 1.0033 | 0.30 | 0.5218 | 1.0450 | 6.0 | 1.3496 | 1.2479 |
| 0.03 | 0.1732 | 1.0049 | 0.4 | 0.5932 | 1.0580 | 7.0 | 1.3766 | 1.2532 |
| 0.04 | 0.1987 | 1.0066 | 0.5 | 0.6533 | 1.0701 | 8.0 | 1.3978 | 1.2570 |
| 0.05 | 0.2217 | 1.0082 | 0.6 | 0.7051 | 1.0814 | 9.0 | 1.4149 | 1.2598 |
| 0.06 | 0.2425 | 1.0098 | 0.7 | 0.7506 | 1.0919 | 10.0 | 1.4289 | 1.2620 |
| 0.07 | 0.2615 | 1.0114 | 0.8 | 0.7910 | 1.1016 | 20.0 | 1.4961 | 1.2699 |
| 0.08 | 0.2791 | 1.0130 | 0.9 | 0.8274 | 1.1107 | 30.0 | 1.5202 | 1.2717 |
| 0.09 | 0.2956 | 1.0145 | 1.0 | 0.8603 | 1.1191 | 40.0 | 1.5325 | 1.2723 |
| 0.10 | 0.3111 | 1.0160 | 2.0 | 1.0769 | 1.1795 | 50.0 | 1.5400 | 1.2727 |
| 0.15 | 0.3779 | 1.0237 | 3.0 | 1.1925 | 1.2102 | 100.0 | 1.5552 | 1.2731 |
| 0.20 | 0.4328 | 1.0311 | 4.0 | 1.2646 | 1.2287 | $\infty$ | 1.5707 | 1.2733 |

[a] $Bi = hL/k$ for the plane wall. See Fig. 16.25.

where the integration is performed over the volume of the wall. It is convenient to nondimensionalize this result by introducing the quantity

$$Q_o = \rho c V(T_i - T_\infty) \tag{16.108}$$

which is the initial internal energy of the wall relative to the fluid temperature. It is also the *maximum* amount of energy transfer that could occur if the process were continued to time $t = \infty$. Hence, assuming constant properties, the ratio of the total energy transferred from the wall over the time interval $t$ to the maximum possible transfer is

$$\frac{Q}{Q_o} = \int \frac{-[T(x, t) - T_i]}{T_i - T_\infty}\frac{dV}{V} = \frac{1}{V}\int (1 - \theta^*)dV \tag{16.109}$$

Employing the approximate form of the temperature distribution for the plane wall, Eq. 16.104a, the integration prescribed by Eq. 16.109 can be performed to obtain the *energy transfer* relation

$$\frac{Q}{Q_o} = 1 - \frac{\sin \zeta}{\zeta}\theta_o^* \tag{16.110}$$

where $\theta_o^*$ can be determined from Eq. 16.105, using Table 16.5 for values of the coefficients $C$ and $\zeta$.

## Additional Considerations

Because the mathematical problem is precisely the same, the foregoing results can also be applied to a plane wall of thickness $L$, which is insulated on one side ($x^* = 0$) and experiences convective transport on the other side ($x^* = +1$). This equivalence is a consequence of the fact that, regardless of whether a symmetrical or an adiabatic requirement is prescribed at $x^* = 0$, the boundary condition is of the form $\partial\theta^*/\partial x^* = 0$.

It should also be noted that the foregoing results can be used to determine the transient response of a plane wall to a sudden change in *surface* temperature. The process is equivalent to having an infinite convection coefficient, in which case the Biot number is infinite ($Bi = \infty$) and the fluid temperature $T_\infty$ is replaced by the prescribed surface temperature $T_s$.

Finally, we note that *graphical representations* of the one-term approximation, referred to as the ***Heisler*** and ***Gröber charts,*** have been developed and are presented in Appendix HT-7. Although the associated charts provide a convenient means for solving transient conduction problems for $Fo > 0.2$, better accuracy can be obtained by using Eqs. 16.104 and 16.110.

*Heisler, Gröber charts*

## *Example 16.10* Plane Wall Experiencing Sudden Convective Heating Process

A large polymer slab of 50-mm thickness is suspended from a conveyor system that transports the slab through a heat treatment oven. The slab is at a uniform temperature of 25°C before it enters the oven, and experiences convection with the hot oven air at 175°C and a convection coefficient of 100 W/m$^2$ · K. The thermophysical properties of the polymer are $\rho$ = 2325 kg/m$^3$, $c$ = 800 J/kg · K, and $k$ = 1.0 W/m · K.

**(a)** What are the appropriate Biot and Fourier numbers 10 min after the slab enters the oven?
**(b)** At $t$ = 10 min, what is the temperature at the midplane of the slab, and at its surface?
**(c)** What is the heat flux $q''$ (W/m$^2$) to the slab from the oven air at $t$ = 10 min?
**(d)** How much energy per unit area has been transferred from the oven air to the slab at $t$ = 10 min?

### Solution

***Known:*** Polymer slab subjected to sudden change in convective surface conditions.

***Find:***

**(a)** Biot and Fourier numbers after 10 min.
**(b)** Slab midplane and surface temperatures after 10 min.
**(c)** Heat flux to the slab at 10 min.
**(d)** Energy transferred to the slab per unit surface area after 10 min.

***Schematic and Given Data:***

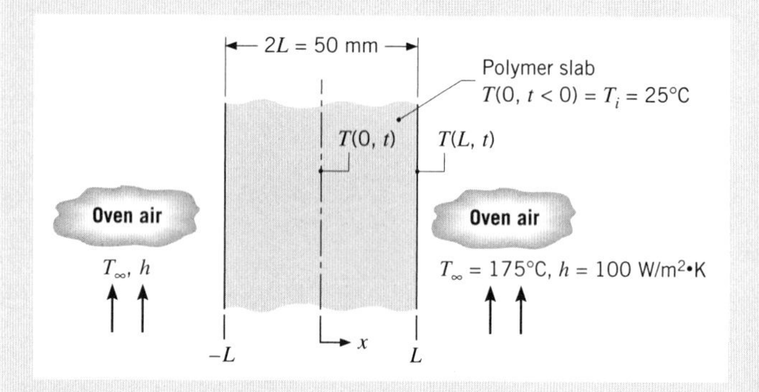


*Figure E16.10*

***Assumptions:***

1. Polymer slab can be approximated as a plane wall, with one-dimensional conduction, experiencing symmetrical convective heating.
2. Negligible radiation exchange with the surroundings.
3. Constant properties.

***Properties:*** Polymer (given) $\rho$ = 2325 kg/m$^3$, $c$ = 800 J/kg · K, $k$ = 1.0 W/m · K; from Eqs. 16.15, $\alpha = k/\rho c$ = 1.0 W/m · K/(2325 kg/m$^3$ × 800 J/kg · K) = 5.38 × 10$^{-7}$ m$^2$/s.

***Analysis:***

**(a)** At $t$ = 10 min, the Biot and Fourier numbers are computed from Eqs. 16.102 and 16.101, respectively. Hence

$$Bi = \frac{hL}{k} = \frac{100\ \text{W/m}^2 \cdot \text{K} \times 0.025\ \text{m}}{1.0\ \text{W/m} \cdot \text{K}} = 2.50 \quad \triangleleft$$

$$Fo = \frac{\alpha t}{L^2} = \frac{5.38 \times 10^{-7}\ \text{m}^2/\text{s} \times 10\ \text{min}}{(0.025\ \text{m})^2}\left|\frac{60\ \text{s}}{\text{min}}\right| = 0.516 \quad \triangleleft$$

**(b)** With $Bi$ = 2.50, use of the lumped capacitance method is inappropriate. However, since $Fo > 0.2$, and transient conditions in the slab correspond to those in a plane wall of thickness 2$L$, the desired results may be obtained from the one-term approximation. Thus, the *midplane temperature*, $T(0, t)$, can be determined from Eq. 16.105

$$\theta_o^* = \frac{T_o - T_\infty}{T_i - T_\infty} = C\exp(-\zeta^2 Fo)$$

where, with $Bi$ = 2.50, the coefficients are $\zeta$ = 1.1347 and $C$ = 1.1949 from Table 16.5. With $Fo$ = 0.516

$$\theta_o^* = 1.1949\exp[-(1.1347)^2 \times 0.516] = 0.615$$

Hence, after 10 min, the midplane temperature is

$$T(0, 10\ \text{min}) = T_\infty + \theta_o^*(T_i - T_\infty) = 175°\text{C} + 0.615(25 - 175)°\text{C} = 83°\text{C} \quad \triangleleft$$

The *surface temperature*, $T(L, t)$, can be determined from Eq. 16.104b, with $x^* = x/L = 1$

$$\theta^* = \frac{T(x^*, t) - T_\infty}{T_i - T_\infty} = \theta_o^* \cos(\zeta x^*) = 0.615 \cos(1.1347) = 0.257$$

$$T(L, 10 \text{ min}) = 175°\text{C} + 0.257(25 - 175)°\text{C} = 136°\text{C} \triangleleft$$

(c) Heat transfer to the slab outer surface at $x = L$ is by convection, and at any time $t$, the heat flux may be obtained from Newton's law of cooling. Hence, at $t = 10$ min

$$q_x''(L, 10 \text{ min}) = q_L'' = h[T(L, 10 \text{ min}) - T_\infty]$$

$$q_L'' = 100 \text{ W/m}^2 \cdot \text{K}[136 - 175]°\text{C} = -3860 \text{ W/m}^2 \triangleleft$$

(d) The energy transfer to the slab over the 10-min interval may be obtained from Eqs. 16.108 and 16.110. With

$$\frac{Q}{Q_o} = 1 - \frac{\sin(\zeta)}{\zeta}\theta_o^* = 1 - \frac{\sin(1.1347)}{1.1347} \times 0.615 = 0.509$$

and with the maximum possible energy transfer

$$Q_o = \rho c V(T_i - T_\infty)$$

where $V = AL$, the *energy per unit surface area* is

$$Q'' = 0.509 \times 2325 \text{ kg/m}^3 \times 800 \text{ J/kg} \cdot \text{K} \times 0.025 \text{ m}(25 - 175)°\text{C}$$

$$Q'' = -3.55 \times 10^6 \text{ J/m}^2 \triangleleft$$

***Comments:***

1. The minus sign associated with $q_L''$ and $Q''$ implies that the direction of the heat transfer is from the hot air to the slab.
2. The temperature distributions in the graphs below are for these conditions: (a) $T(x, 10 \text{ min})$ as a function of the $x$-coordinate and (b) $T(0, t)$ and $T(L, t)$ as a function of time $t$.

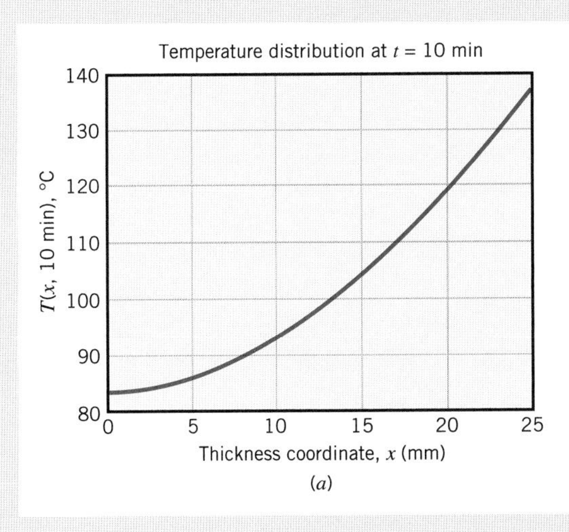


(a)

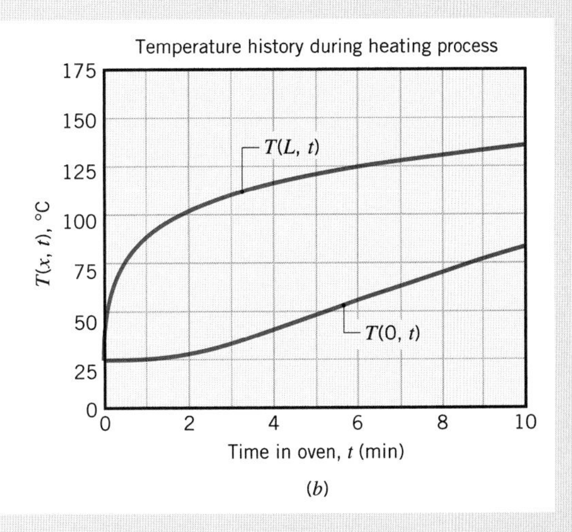


(b)

3. Applying the Heisler and Gröber charts, graphical representations of the one-term approximations. (CD-ROM)

### 16.5.3 Radial Systems with Convection (CD-ROM)

### 16.5.4 The Semi-Infinite Solid (CD-ROM)

## 16.7 Chapter Summary and Study Guide

One-dimensional, *steady-state heat transfer* occurs in numerous engineering applications. The method of analysis using thermal circuits is a powerful approach for practical thermal systems. You should be comfortable using equivalent *thermal circuits* and the expressions for the *conduction resistances* that pertain to the common geometries of the plane wall, hollow cylinder, and sphere, as well as thermal resistances for other processes including *surface convection, radiation exchange* between the surface and its surroundings, interfacial *thermal contact resistance,* and *fin heat rate*. Using these resistance elements, you should be able to construct *thermal circuits* representative of a system and its boundary conditions or surroundings, which could be used to solve for temperatures and heat rates.

You should also be familiar with how to use the *heat equation* and *Fourier's law* to obtain *temperature distributions* and the corresponding *heat fluxes*. You should recognize effects that *volumetric energy generation* have on the temperature and heat flux distributions.

When confronted with a transient conduction application, it is important to first calculate the *Biot number, Bi,* the measure of internal conduction thermal resistance to external convection thermal resistance. If $Bi < 0.1$, the temperature distribution in the object can be characterized by a single temperature, and you may use the *lumped capacitance method* to obtain the temperature-time history. However, if the Biot number does not meet this criterion, spatial effects must be considered. One-term *analytical results* were introduced for the *plane wall, the infinite cylinder, and the sphere*. Analytical solutions were provided for the temperature distribution in the *semi-infinite solid,* initially at a uniform temperature, suddenly subjected to three types of boundary conditions. (CD-ROM)

The following checklist provides a study guide for this chapter. When your study of the text and end-of-chapter problems has been completed you should be able to

- write out the meanings of the terms listed in the margins throughout the chapter and understand each of the related concepts. The subset of key terms listed here is particularly important.
- explain why the *temperature distribution* is linear for one-dimensional, steady-state conduction in a *plane wall* with no volumetric energy generation. You should also be able to explain whether the heat flux is constant, independent of the direction coordinate in a *plane wall, hollow cylinder,* and *hollow sphere.*
- define *thermal resistance* and identify the thermal resistances for these processes and represent them in a *thermal circuit:* conduction in a plane wall; convection from a surface to a fluid; radiation exchange between a surface and its surroundings; interfacial contact between surfaces. Is it proper to include a solid experiencing *volumetric energy generation* in a thermal circuit analysis?
- write an expression for the steady-state temperature difference between the exposed surface and the fluid for the case of a plane wall of thickness $L$ experiencing uniform volumetric energy generation $\dot{q}$ having one surface perfectly insulated and the other exposed to a convection process ($T_\infty$, $h$). You also should be able to determine the temperature difference if both surfaces are exposed to the convection process, and to sketch the temperature distributions for both cases.
- sketch the *temperature distributions* in a straight fin ($T_b > T_\infty$) for two cases: 100% efficiency, low efficiency.
- explain the physical interpretation of the *Biot number* and the *Fourier number.*
- list dimensionless parameters that are used to represent the temperature distribution for one-dimensional, transient conduction in a plane wall, a long cylinder or a sphere, with surface convection.
- sketch the *temperature distribution* ($T$-$x$ coordinates) for a plane wall initially at a uniform temperature, which experiences a sudden change in convection conditions. Show the distributions for the initial condition, the final condition, and two intermediate times.

***Fourier's law***
***thermal conductivity***
***heat equation***
***temperature distribution***
***heat flux, heat rate***
***thermal resistances***
***thermal circuits***
***energy generation***
***transient conduction***
***thermal diffusivity***
***lumped capacitance method***
***Biot number***
***Fourier number***

## Problems

*Note:* Unless otherwise indicated in the problem statement, use values of the required thermophysical properties given in the appropriate tables of Appendix HT when solving these problems.

### Fourier's Law and the Heat Equation

**16.1** Consider steady-state conditions for one-dimensional conduction in a plane wall having a thermal conductivity $k = 50\ \text{W/m}\cdot\text{K}$ and a thickness $L = 0.25$ m, with no energy generation.

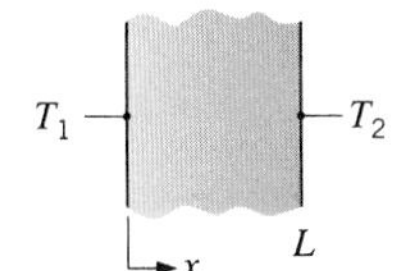


*Figure P16.1*

Determine the heat flux and the unknown quantity for each case and sketch the temperature distribution, indicating the direction of the heat flux.

| Case | $T_1$(°C) | $T_2$(°C) | $dT/dx$ (K/m) |
|---|---|---|---|
| 1 | 50 | −20 | |
| 2 | −30 | −10 | |
| 3 | 70 | | 160 |
| 4 | | 40 | −80 |
| 5 | | 30 | 200 |

**16.2** (CD-ROM)

**16.3** A one-dimensional system without energy generation has a thickness of 20 mm with surfaces maintained at temperatures of 275 and 325 K. Determine the heat flux through the system if it is constructed from **(a)** pure aluminum, **(b)** plain carbon steel, **(c)** AISI 316 stainless steel, **(d)** pyroceram, **(e)** Teflon, and **(f)** concrete.

**16.4** The steady-state temperature distribution in a one-dimensional wall of thermal conductivity 50 W/m · K and thickness 50 mm is observed to be $T(°C) = a + bx^2$, where $a = 200°C$, $b = -2000°C/m^2$, and $x$ is in meters.

**(a)** What is the volumetric energy generation rate $\dot{q}$ in the wall?

**(b)** Determine the heat fluxes at the two wall faces. In what manner are these heat fluxes related to the volumetric energy generation rate?

**16.5** (CD-ROM)

**16.6** One-dimensional, steady-state conduction with uniform energy generation occurs in a plane wall with a thickness of 50 mm and a constant thermal conductivity of 5 W/m · K (Fig. P16.6). For these conditions, the temperature distribution has the form $T(x) = a + bx + cx^2$. The surface at $x = 0$ has a temperature of $T(0) \equiv T_0 = 120°C$ and experiences convection with a fluid for which $T_\infty = 20°C$ and $h = 500\ \text{W/m}^2\cdot\text{K}$. The surface at $x = L$ is well insulated.

**(a)** Applying an overall energy balance to the wall, calculate the generation rate, $\dot{q}$.

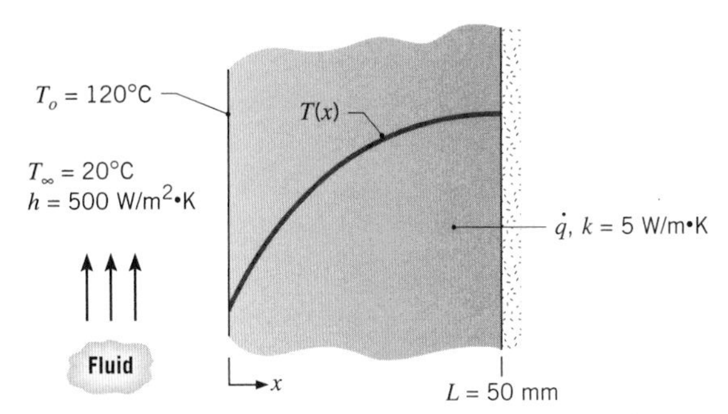


*Figure P16.6*

**(b)** Determine the coefficients $a$, $b$, and $c$ by applying the boundary conditions to the prescribed temperature distribution. Use the results to calculate and plot the temperature distribution.

### The Plane Wall

**16.7** Consider the plane wall of Figure 16.4, separating hot and cold fluids at temperatures $T_{\infty,1}$ and $T_{\infty,2}$, respectively. Using surface energy balances as boundary conditions at $x = 0$ and $x = L$ (see Eq. 16.9), obtain the temperature distribution within the wall and the heat flux in terms of $T_{\infty,1}$, $T_{\infty,2}$, $h_1$, $h_2$, $k$, and $L$.

**16.8** Consider the composite window of Example 16.1 with a thickness $L = 62.2$ mm. The outside convection coefficient is increased to 35 W/m² · K, while all other conditions remain the same.

**(a)** What is the temperature of the outer surface, $T_{s,o}$?

**(b)** Calculate the temperature at the interface between the two plastics (A and B).

**16.9** Consider the chip cooling arrangement of Example 16.2. We found that the chip operating temperature is $T_c = 75.3°C$ for a chip power dissipation of $10^4\ \text{W/m}^2$. Calculate the allowable power dissipation for the same prescribed cooling conditions when the chip temperature is 85°C.

**16.10** The walls of a refrigerator are typically constructed by sandwiching a layer of insulation between sheet metal panels. Consider a wall made from fiberglass insulation of thermal conductivity $k_i = 0.046\ \text{W/m}\cdot\text{K}$ and thickness $L_i = 50$ mm and steel panels, each of thermal conductivity $k_p = 60\ \text{W/m}\cdot\text{K}$ and thickness $L_p = 3$ mm. If the wall separates refrigerated air at $T_{\infty,i} = 4°C$ from ambient air at $T_{\infty,o} = 25°C$, what is the heat transfer rate per unit surface area? Coefficients associated with natural convection at the inner and outer surfaces may be approximated as $h_i = h_o = 5\ \text{W/m}^2\cdot\text{K}$.

**16.11** A house has a composite wall of wood, fiberglass insulation, and plaster board, as indicated in Fig. P16.11. On a cold winter day the convection heat transfer coefficients are $h_o = 60\ \text{W/m}^2\cdot\text{K}$ and $h_i = 30\ \text{W/m}^2\cdot\text{K}$. The total wall surface area is 350 m².

**(a)** Determine an expression in symbol form for the total thermal resistance of the wall, including inside and outside convection effects for the prescribed conditions.

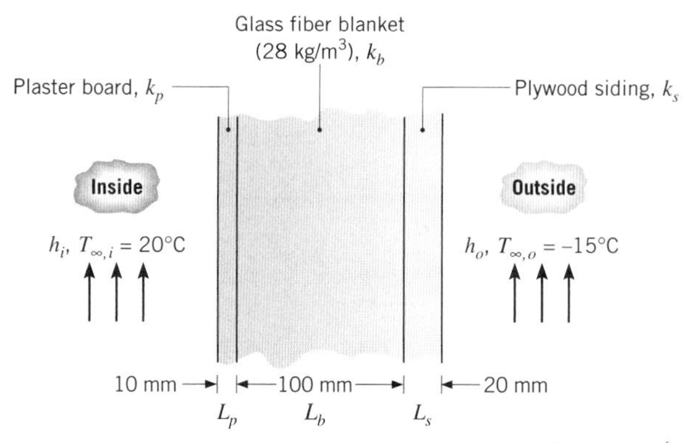


*Figure P16.11*

**(b)** Determine the total heat rate through the wall.

**(c)** If the wind were blowing violently, raising $h_o$ to 300 $W/m^2 \cdot K$, determine the percentage increase in the heat rate.

**(d)** What is the controlling resistance that determines the heat rate through the wall?

**16.12** (CD-ROM)

**16.13** (CD-ROM)

**16.14** The wall of a drying oven is constructed by sandwiching an insulation material of thermal conductivity $k = 0.05$ W/m · K between thin metal sheets. The oven air is at $T_{\infty,i} = 300°C$, and the corresponding convection coefficient is $h_i = 30$ $W/m^2 \cdot K$. The inner wall surface absorbs a radiant flux of $q''_{rad} = 100$ $W/m^2$ from hotter objects within the oven. The room air is at $T_{\infty,o} = 25°C$, and the overall coefficient for convection and radiation from the outer surface is $h_o = 10$ $W/m^2 \cdot K$.

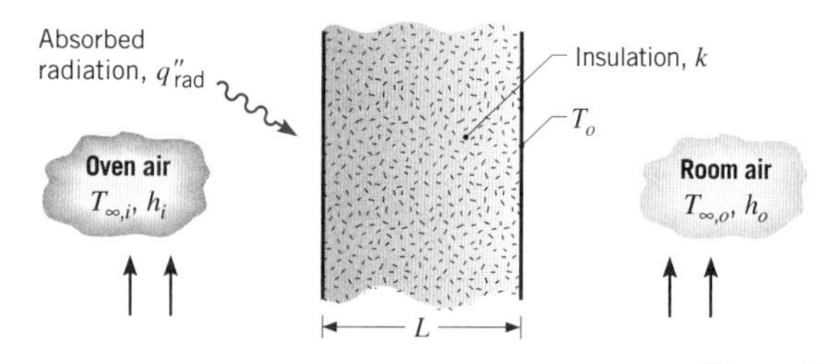


*Figure P16.14*

**(a)** Show the thermal circuit for the wall and label all temperatures, heat rates, and thermal resistances.

**(b)** What insulation thickness $L$ is required to maintain the outer wall surface at a *safe-to-touch* temperature of $T_o = 40°C$?

**16.15** The rear window of an automobile is defogged by passing warm air over its inner surface. If the warm air is at $T_{\infty,i} = 40°C$ and the corresponding convection coefficient is $h_i = 30$ $W/m^2 \cdot K$, what are the inner and outer surface temperatures of 4-mm-thick window glass, if the outside ambient air temperature is $T_{\infty,o} = -10°C$ and the associated convection coefficient is $h_o = 65$ $W/m^2 \cdot K$?

**16.16** (CD-ROM)

**16.17** In a manufacturing process, a transparent film is being bonded to a substrate as shown in Fig. P16.17. To cure the bond at a temperature $T_0$, a radiant source is used to provide a heat flux $q''_0$ ($W/m^2$), all of which is absorbed at the bonded surface. The back of the substrate is maintained at $T_1$ while the free surface of the film is exposed to air at $T_\infty$ and a convection heat transfer coefficient $h$.

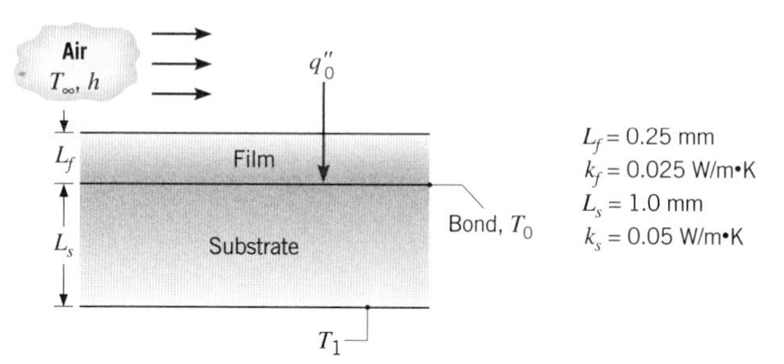


*Figure P16.17*

**(a)** Show the thermal circuit representing the steady-state heat transfer situation. Be sure to label *all* elements, nodes, and heat rates. Leave in symbol form.

**(b)** Assume the following conditions: $T_\infty = 20°C$, $h = 50$ $W/m^2 \cdot K$, and $T_1 = 30°C$. Calculate the heat flux $q''_0$ that is required to maintain the bonded surface at $T_0 = 60°C$.

**16.18** Consider a plane composite wall that is composed of two materials of thermal conductivities $k_A = 0.1$ W/m · K and $k_B = 0.04$ W/m · K and thicknesses $L_A = 10$ mm and $L_B = 20$ mm. The contact resistance at the interface between the two materials is known to be 0.30 $m^2 \cdot K/W$. Material A adjoins a fluid at 200°C for which $h = 10$ $W/m^2 \cdot K$, and material B adjoins a fluid at 40°C for which $h = 20$ $W/m^2 \cdot K$.

**(a)** What is the rate of heat transfer through a wall that is 2 m high by 2.5 m wide?

**(b)** Sketch the temperature distribution.

**16.19** A silicon chip is encapsulated such that, under steady-state conditions, all of the power it dissipates is transferred by convection to a fluid stream for which $h = 1000$ $W/m^2 \cdot K$ and $T_\infty = 25°C$. The chip is separated from the fluid by a 2-mm-thick aluminum cover plate, and the contact resistance of the chip/aluminum interface is $0.5 \times 10^{-4}$ $m^2 \cdot K/W$.

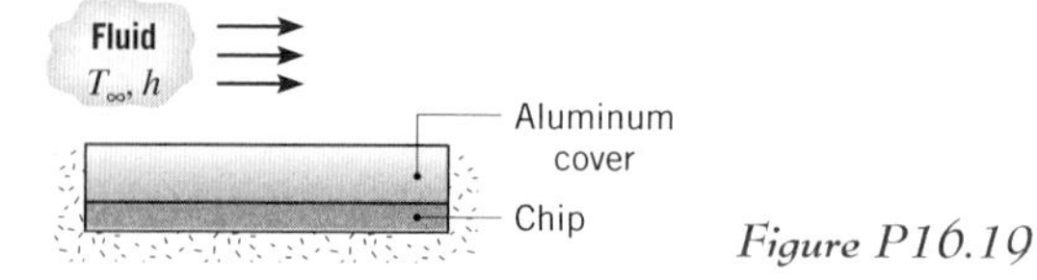


*Figure P16.19*

If the chip surface area is 100 $mm^2$ and its maximum allowable temperature is 85°C, what is the maximum allowable power dissipation in the chip?

**16.20** Approximately $10^6$ discrete electrical components can be placed on a single integrated circuit (chip), with electrical power dissipation as high as 30,000 $W/m^2$. The chip, which is very thin, is exposed to a dielectric liquid at its outer surface, with $h_o = 1000$ $W/m^2 \cdot K$ and $T_{\infty,o} = 20°C$, and is joined to a circuit board at its inner surface (see Fig. P16.20). The thermal contact resistance between the chip and the board is $10^{-4}$ m$^2 \cdot$ K/W, and the board thickness and thermal conductivity are $L_b$ = 5 mm and $k_b$ = 1 W/m · K, respectively. The other surface of the board is exposed to ambient air for which $h_i$ = 40 W/m$^2 \cdot$ K and $T_{\infty,i}$ = 20°C.

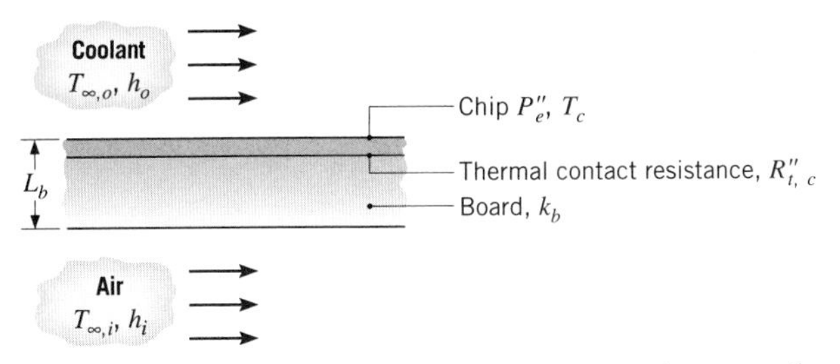


*Figure P16.20*

(a) Show the thermal circuit corresponding to steady-state conditions. Label appropriate resistances, temperatures, and heat fluxes.

(b) Under steady-state conditions for which the chip power dissipation is $P''_e$ = 30,000 W/m$^2$, what is the chip temperature?

**16.21** (CD-ROM)

## The Cylindrical Wall

**16.22** A steam pipe of 120-mm outside diameter is covered with a 20-mm-thick layer of calcium silicate insulation ($k$ = 0.089 W/m · K). The pipe surface temperature is 800 K, and the ambient air and surroundings temperatures are 300 K. The convection and radiation coefficients for the outer surface of the insulation are estimated as 5.5 and 10 W/m$^2 \cdot$ K, respectively. Determine the heat rate per unit length from the pipe (W/m) and the insulation outer surface temperature.

**16.23** A stainless steel ($k$ = 14 W/m · K) tube used to transport a chilled pharmaceutical has an inner diameter of 36 mm and a wall thickness of 2 mm. The pharmaceutical and ambient air are at temperatures of 6°C and 23°C, respectively, while the corresponding inner and outer convection coefficients are 400 W/m$^2 \cdot$ K and 6 W/m$^2 \cdot$ K, respectively.

(a) What is the heat transfer rate per unit tube length?

(b) What is the heat transfer rate per unit length if a 10-mm-thick layer of calcium silicate insulation ($k_{ins}$ = 0.050 W/m · K) is applied to the outer surface of the tube?

**16.24** A thin electrical heater is wrapped around the outer surface of a long cylindrical tube whose inner surface is maintained at a temperature of 5°C. The tube wall has inner and outer radii of 25 and 75 mm, respectively, and a thermal conductivity of 10 W/m · K. The thermal contact resistance between the heater and the outer surface of the tube (per unit length of the tube) is $R'_{t,c}$ = 0.01 m · K/W. The outer surface of the heater is exposed to a fluid with $T_\infty$ = −10°C and a convection coefficient of $h$ = 100 W/m$^2 \cdot$ K. Determine the heater power per unit length of tube required to maintain the heater at $T_o$ = 25°C.

**16.25** (CD-ROM)

**16.26** (CD-ROM)

## The Spherical Wall

**16.27** Consider the spherical liquid-nitrogen storage container of Example 16.3. The designer of the container has been asked to reduce the daily boil-off rate from 7 to 4 liters/day. What is the required thickness of the silica powder insulation?

**16.28** The wall of a spherical tank of 1-m diameter contains an exothermic chemical reaction and is at 200°C when the ambient air temperature is 25°C. What thickness of urethane foam is required to reduce the exterior temperature to 40°C, assuming the convection coefficient is 20 W/m$^2 \cdot$ K for both situations? What is the percentage reduction in heat rate achieved by using the insulation?

**16.29** A spherical, cryosurgical probe may be imbedded in diseased tissue for the purpose of freezing, and thereby destroying, the tissue. Consider a probe of 3-mm diameter whose surface is maintained at −30°C when imbedded in tissue that is at 37°C. A spherical layer of frozen tissue forms around the probe, with a temperature of 0°C existing at the phase front (interface) between the frozen and normal tissue. If the thermal conductivity of frozen tissue is approximately 1.5 W/m · K and heat transfer at the phase front may be characterized by a convection coefficient of 50 W/m$^2 \cdot$ K, what is the thickness of the layer of frozen tissue?

**16.30** (CD-ROM)

**16.31** (CD-ROM)

## Conduction with Energy Generation: The Plane Wall

**16.32** Consider the composite wall of Example 16.4. Calculate the temperature $T_0$ of the insulated surface if the energy generation rate is doubled ($\dot{q}$ = 3.0 × 10$^6$ W/m$^3$), while all other conditions remain unchanged. Sketch the temperature distributions for this case and for that represented in Example 16.4. Identify key differences between the distributions for the two cases.

**16.33** Consider the composite wall of Example 16.4. The analysis was performed assuming negligible contact resistance between materials A and B. Calculate the temperature $T_0$ of the insulated surface if the thermal contact resistance is $R''_{t,c}$ = $10^{-4}$ m$^2 \cdot$ K/W, while all other conditions remain unchanged. Sketch the temperature distributions for this case and for that represented in Example 16.4. Identify key similarities and differences between the distributions for the two cases.

**16.34** A plane wall of thickness 0.1 m and thermal conductivity 25 W/m · K having uniform volumetric energy generation of 0.3 MW/m$^3$ is insulated on one side, while the other side is exposed to a fluid at 92°C. The convection heat transfer coefficient between the wall and the fluid is 500 W/m$^2 \cdot$ K. Determine the maximum temperature in the wall.

**16.35** Bus bars proposed for use in a power transmission station have a rectangular cross section of height $H$ = 600 mm and width $W$ = 200 mm. The thermal conductivity of the bar material is $k$ = 165 W/m · K and the electrical resistance per unit length is $R'_e$ = 1.044 μΩ/m. The convection coefficient between the bar and the ambient air at 30°C is 19 W/m$^2 \cdot$ K.

(a) Assuming the bar has a uniform temperature $T$, calculate the steady-state temperature when a current of 60,000 A passes through the bar.

(b) Assuming the bar can be approximated as a one-dimensional plane wall of thickness $2L = W$ with uniform energy generation, estimate the temperature difference between the midplane and the surface of the bus bar. Is the uniform-temperature assumption of part (a) reasonable? Comment on the validity of the plane-wall assumption made for estimating the temperature difference.

**16.36** When passing an electrical current $I$ (A), a copper bus bar of rectangular cross section (6 mm × 150 mm) experiences uniform energy generation at a rate $\dot{q}$ (W/m$^3$) given by $\dot{q} = aI^2$ where $a = 0.015$ W/m$^3 \cdot$ A$^2$. If the bar is in ambient air with $h = 5$ W/m$^2 \cdot$ K and its maximum temperature must not exceed that of the air by more than 30°C, what is the allowable current capacity for the bus bar?

**16.37** The steady-state temperature distribution in a composite plane wall of three different materials, each of constant thermal conductivity, is shown as follows.

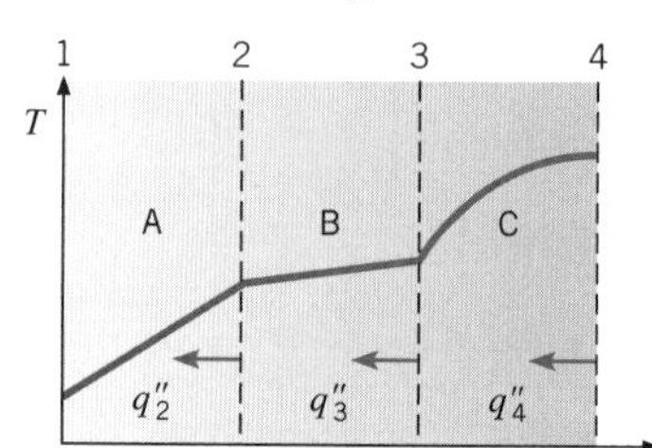


*Figure P16.37*

(a) Comment on the relative magnitudes of $q''_2$ and $q''_3$ and of $q''_3$ and $q''_4$.

(b) Comment on the relative magnitudes of $k_A$ and $k_B$ and of $k_B$ and $k_C$.

(c) Sketch the heat flux as a function of $x$.

**16.38** A nuclear fuel element of thickness $2L$ is covered with a steel cladding of thickness $b$. Energy generated within the nuclear fuel at a rate $\dot{q}$ is removed by a fluid at $T_\infty$, which adjoins one surface and is characterized by a convection coefficient $h$. The other surface is well insulated, and the fuel and steel have thermal conductivities of $k_f$ and $k_s$, respectively.

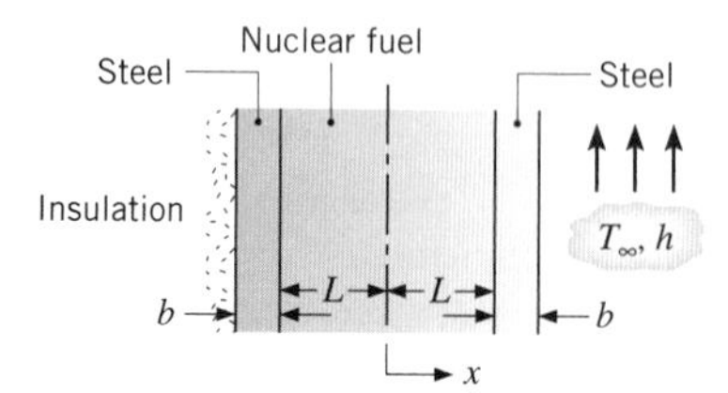


*Figure P16.38*

(a) Sketch the temperature distribution $T(x)$ for the entire system and describe key features of the distribution.

(b) For $k_f = 60$ W/m · K, $L = 15$ mm, $b = 3$ mm, $k_s = 15$ W/m · K, $h = 10{,}000$ W/m$^2 \cdot$ K, and $T_\infty = 200$°C, what are the highest and lowest temperatures in the fuel element if energy is generated uniformly at a volumetric rate of $\dot{q} = 2 \times 10^7$ W/m$^3$? What are the corresponding locations?

(c) If the insulation is removed and equivalent convection conditions are maintained at each surface, what is the corresponding form of the temperature distribution in the fuel element? For the conditions of part (a), what are the highest and lowest temperatures in the fuel? What are the corresponding locations?

**16.39** (CD-ROM)

### Conduction with Energy Generation: Radial Systems (CD-ROM)

**16.40** (CD-ROM)

**16.41** (CD-ROM)

**16.42** (CD-ROM)

**16.43** (CD-ROM)

**16.44** (CD-ROM)

### Extended Surfaces and Fins

**16.45** A long, circular aluminum rod is attached at one end to a heated wall and transfers heat by convection to a cold fluid.

(a) If the diameter of the rod is tripled, by how much would the heat rate change?

(b) If a copper rod of the same diameter is used in place of the aluminum, by how much would the heat rate change?

**16.46** A long rod passes through the opening in an oven having an air temperature of 400°C and is pressed firmly onto the surface of a billet (Fig. P16.46). Thermocouples imbedded in the rod at locations 25 and 120 mm from the billet register temperatures of 325 and 375°C, respectively. What is the temperature of the billet?

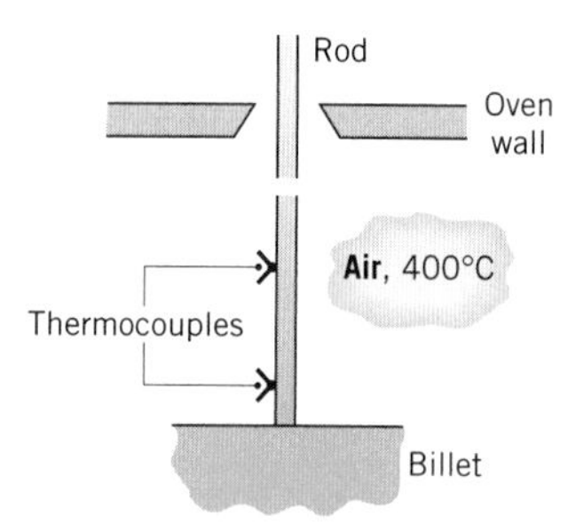


*Figure P16.46*

**16.47** Consider two long, slender rods of the same diameter but different materials. One end of each rod is attached to a base surface maintained at 100°C, while the surface of the rods are exposed to ambient air at 20°C. By traversing the length of each rod with a thermocouple, it was observed that the temperatures of the rods were equal at the positions $x_A = 0.15$ m and $x_B = 0.075$ m, where $x$ is measured from the base surface. If the thermal conductivity of rod A is known to be $k_A = 70$ W/m · K, determine the value of $k_B$ for rod B.

**16.48** The extent to which the tip condition affects the thermal performance of a fin depends on the fin geometry and thermal conductivity, as well as the convection coefficient. Consider an alloyed aluminum ($k = 180$ W/m · K) rectangular fin of length

$L = 10$ mm, thickness $t = 1$ mm, and width $w \gg t$. The base temperature of the fin is $T_b = 100°C$, and the fin is exposed to a fluid of temperature $T_\infty = 25°C$. Assuming a uniform convection coefficient of $h = 100\ W/m^2 \cdot K$ over the entire fin surface, determine the fin heat transfer rate per unit width $q'_f$, efficiency $\eta_f$, effectiveness $\varepsilon_f$, thermal resistance per unit width $R'_{t,f}$, and the tip temperature $T(L)$ for Cases B and D of Table 16.4. Contrast your results with those based on an *infinite fin* approximation.

**16.49** A straight fin of rectangular cross section fabricated from an aluminum alloy ($k = 185\ W/m \cdot K$) has a base thickness of $t = 3$ mm and a length of $L = 15$ mm. Its base temperature is $T_b = 100°C$, and it is exposed to a fluid for which $T_\infty = 20°C$ and $h = 50\ W/m^2 \cdot K$.

(a) For the foregoing conditions and a fin unit width, calculate the fin heat rate, efficiency, and effectiveness.

(b) Compare the foregoing results with those for a fin fabricated from pure copper ($k = 400\ W/m \cdot K$).

**16.50** (CD-ROM)

**16.51** Turbine blades mounted to a rotating disc in a gas turbine engine are exposed to a gas stream that is at $T_\infty = 1200°C$ and maintains a convection coefficient of $h = 250\ W/m^2 \cdot K$ over the blade.

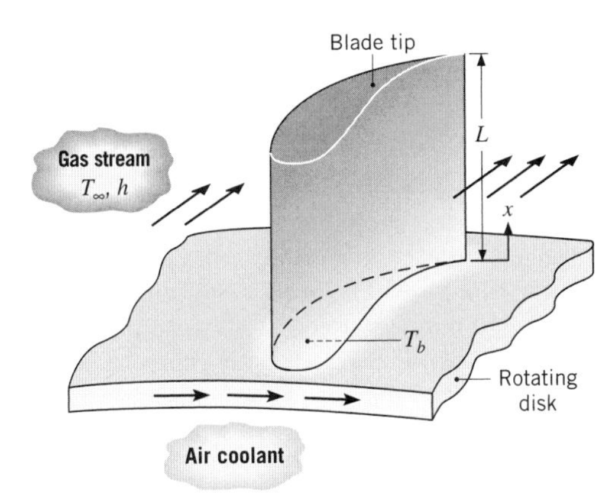


*Figure P16.51*

The blades, which are fabricated from Inconel, $k \approx 20\ W/m \cdot K$, have a length of $L = 50$ mm. The blade profile has a uniform cross-sectional area of $A_c = 6 \times 10^{-4}\ m^2$ and a perimeter of $P = 110$ mm. A proposed blade-cooling scheme, which involves routing air through the supporting disk, is able to maintain the base of each blade at a temperature of $T_b = 300°C$.

(a) If the maximum allowable blade temperature is 1050°C and the blade tip may be assumed to be adiabatic, is the proposed cooling scheme satisfactory?

(b) For the proposed cooling scheme, what is the heat transfer rate from each blade to the coolant?

**16.52** Pin fins are widely used in electronic systems to provide cooling as well as to support devices (Fig. P16.52). Consider the pin fin of uniform diameter $D$, length $L$, and thermal conductivity $k$ connecting two identical devices of length $L_g$ and surface area $A_g$. The devices are characterized by a uniform volumetric energy generation $\dot{q}$ and a thermal conductivity $k_g$. Assume that the exposed surfaces of the devices are at a uniform temperature corresponding to that of the pin base, $T_b$, and that convection heat transfer occurs from the exposed surfaces to an adjoining fluid. The back and sides of the devices are perfectly insulated. Derive an expression for the base temperature $T_b$ in terms of the device parameters ($k_g$, $\dot{q}$, $L_g$, $A_g$), the convection parameters ($T_\infty$, $h$), and the fin parameters ($k$, $D$, $L$).

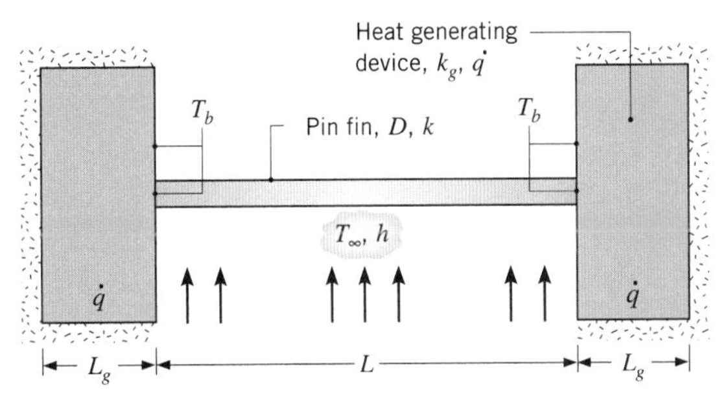


*Figure P16.52*

**16.53** A very thin disk-shaped electronic device of thickness $L_d$, diameter $D$, and thermal conductivity $k_d$ dissipates electrical power at a steady rate $P_e$. The device is bonded to a cooled base at $T_o$ using a thermal pad of thickness $L_p$ and thermal conductivity $k_p$. A long fin of diameter $D$ and thermal conductivity $k_f$ is bonded to the energy-generating surface of the device using an identical thermal pad. The fin is cooled by an air stream, which is at a temperature $T_\infty$ and provides a convection coefficient $h$.

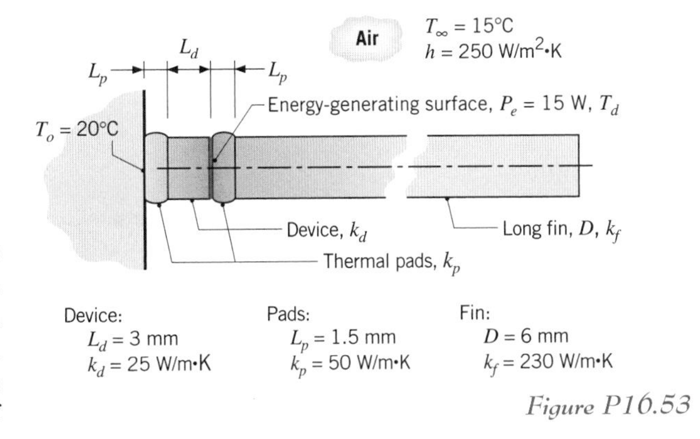


*Figure P16.53*

(a) Construct a thermal circuit of the system.

(b) Derive an expression for the temperature $T_d$ of the energy-generating surface of the device in terms of the circuit thermal resistances, $T_o$ and $T_\infty$. Express the thermal resistances in terms of appropriate parameters.

(c) Calculate $T_d$ for the prescribed conditions.

**16.54** (CD-ROM)

**16.55** Consider the finned aluminum sleeve of Example 16.7. We want to explore what measures could be taken to increase the heat transfer rate, while keeping the base temperature at 80°C.

(a) One option is to increase the velocity of air flowing over the finned surfaces in order to increase the convection coefficient. Use the model developed in the example to determine the increase in the heat rate by *doubling* the convection coefficient ($h = 50\ \text{W/m}^2 \cdot \text{K}$) while all other conditions remain unchanged.

(b) What other parameters would you consider changing in order to effect an improvement in the system's performance?

**16.56** Determine the percentage increase in heat transfer associated with attaching alloyed aluminum ($k = 180\ \text{W/m} \cdot \text{K}$) fins of rectangular profile to a plane wall. The fins are 50 mm long, 0.5 mm thick, and are equally spaced at a distance of 4 mm (250 fins/m). The convection coefficient associated with the bare wall is $40\ \text{W/m}^2 \cdot \text{K}$, and with the fin surfaces is $30\ \text{W/m}^2 \cdot \text{K}$.

**16.57** (CD-ROM)

**16.58** (CD-ROM)

**16.59** (CD-ROM)

### Lumped Capacitance Method

**16.60** Steel balls 12 mm in diameter are annealed by heating to 1150 K and then slowly cooling to 400 K in an air environment for which $T_\infty = 325$ K and $h = 20\ \text{W/m}^2 \cdot \text{K}$. Assuming the properties of the steel to be $k = 40\ \text{W/m} \cdot \text{K}$, $\rho = 7800\ \text{kg/m}^3$, and $c = 600\ \text{J/kg} \cdot \text{K}$, estimate the time required for the cooling process.

**16.61** The heat transfer coefficient for air flowing over a sphere is to be determined by observing the temperature-time history of a sphere fabricated from pure copper. The sphere, which is 12.7 mm in diameter, is at 66°C before it is inserted into an air stream having a temperature of 27°C. After the sphere has been inserted in the air stream for 69 s, the thermocouple on the outer surface indicates 55°C. Assume, and then justify, that the sphere behaves as a spacewise isothermal object and calculate the heat transfer coefficient.

**16.62** (CD-ROM)

**16.63** Carbon steel (AISI 1010) shafts of 0.1-m diameter are heat treated in a gas-fired furnace whose gases are at 1200 K and provide a convection coefficient of $100\ \text{W/m}^2 \cdot \text{K}$. If the shafts enter the furnace at 300 K, how long must they remain in the furnace to achieve a centerline temperature of 800 K?

**16.64** An energy storage unit consists of a large rectangular channel, which is well insulated on its outer surface and encloses alternating layers of the storage material and the flow passage (Fig. P16.64). Each layer of the storage material is an aluminum slab of width $W = 0.05$ m, which is at an initial temperature of 25°C. Consider conditions for which the storage unit is charged by passing a hot gas through the passages, with the gas temperature and the convection coefficient assumed to have constant values of $T_\infty = 600$°C and $h = 100\ \text{W/m}^2 \cdot \text{K}$ throughout the channel. How long will it take to achieve 75% of the maximum possible energy storage? What is the temperature of the aluminum at this time?

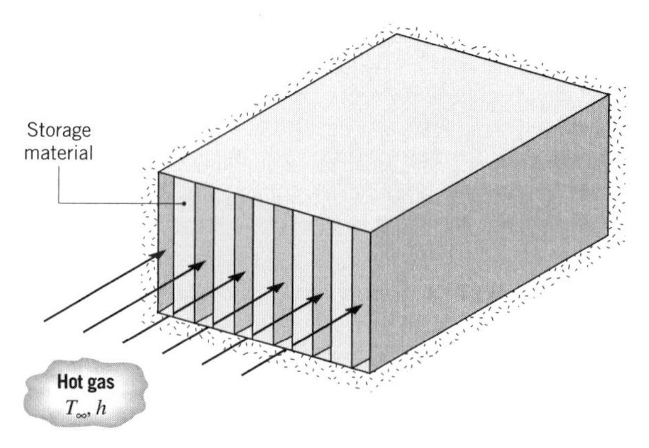


*Figure P16.64*

**16.65** (CD-ROM)

**16.66** A leaf spring of dimensions 32 mm × 10 mm × 1.1 m is sprayed with a thin anticorrosion coating, which is heat-treated by suspending the spring vertically in the lengthwise direction and passing it through a conveyor oven maintained at an air temperature of 175°C. Satisfactory coatings have been obtained on springs, initially at 25°C, with an oven residence time of 35 min. The coating supplier has specified that the coating should be treated for 10 min above a temperature of 140°C. How long should a spring of dimensions 76 mm × 35 mm × 1.6 m remain in the oven in order to properly heat treat the coating? Assume that both springs experience the same convection coefficient. The thermophysical properties of the spring material are $\rho = 8131\ \text{kg/m}^3$, $c = 473\ \text{J/kg} \cdot \text{K}$, and $k = 42\ \text{W/m} \cdot \text{K}$.

**16.67** A plane wall of a furnace is fabricated from plain carbon steel ($k = 60\ \text{W/m} \cdot \text{K}$, $\rho = 7850\ \text{kg/m}^3$, $c = 430\ \text{J/kg} \cdot \text{K}$) and is of thickness $L = 10$ mm. To protect it from the corrosive effects of the furnace combustion gases, one surface of the wall is coated with a thin ceramic film that, for a unit surface area, has a thermal resistance of $R''_{t,f} = 0.01\ \text{m}^2 \cdot \text{K/W}$. The opposite surface is well insulated from the surroundings.

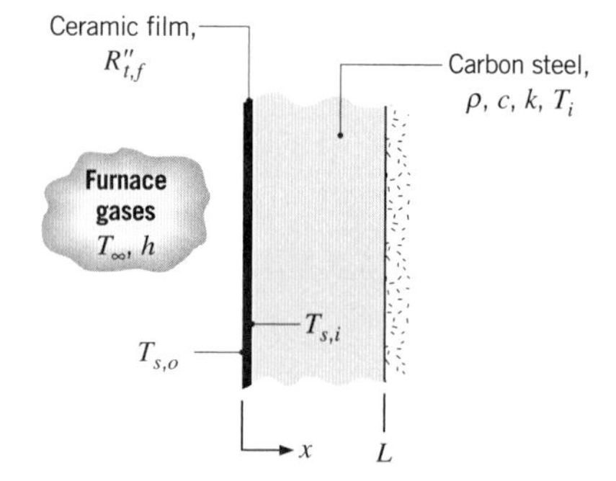


*Figure P16.67*

At furnace start-up the wall is at an initial temperature of $T_i = 300$ K, and combustion gases at $T_\infty = 1300$ K enter the furnace, providing a convection coefficient of $h = 25\ \text{W/m}^2 \cdot \text{K}$ at the ceramic film. *Hint:* In Eq. 16.85 for the temperature history, replace the convection coefficient by an overall coefficient $U\ (1/R_t)$ representing the external resistances due to convection and the conduction resistance of the film.

(a) Assuming the film to have negligible thermal capacitance, how long will it take for the inner surface of the steel to achieve a temperature of $T_{s,i} = 1200$ K?

(b) What is the temperature $T_{s,o}$ of the exposed surface of the ceramic film at this time?

**16.68** A tool used for fabricating semiconductor devices consists of a *chuck* (a thick metallic, cylindrical disk) onto which a very thin silicon wafer ($\rho = 2700$ kg/m$^3$, $c = 875$ J/kg · K, $k = 177$ W/m · K) is placed by a robotic arm (Fig. P16.68). Once in position, an electric field in the chuck is energized, creating an electrostatic force that holds the wafer firmly to the chuck. To ensure a reproducible thermal contact resistance between the chuck and the wafer from cycle-to-cycle, pressurized helium gas is introduced at the center of the chuck and flows (very slowly) radially outward between the asperites of the interface region.

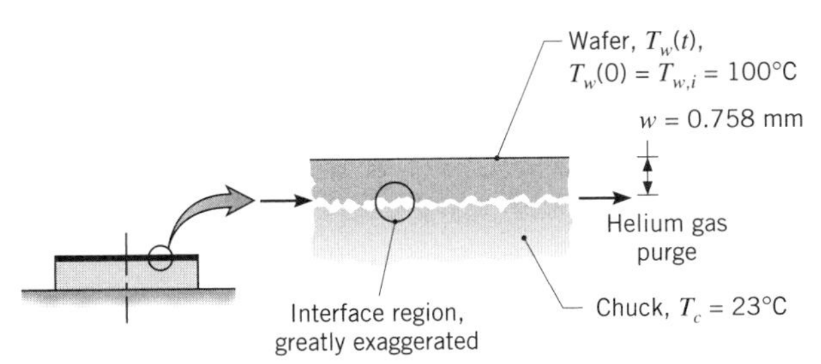


*Figure P16.68*

An experiment has been performed under conditions for which the wafer, initially at a uniform temperature $T_{w,i} = 100°C$, is suddenly placed on the chuck, which is at a uniform and constant temperature $T_c = 23°C$. With the wafer in place, the electrostatic force and the helium gas flow are applied. After 15 seconds, the temperature of the wafer is determined to be 33°C. What is the thermal contact resistance $R''_{t,c}$ (m$^2$ · K/W) between the wafer and chuck? Will the value of $R''_{t,c}$ increase, decrease, or remain the same if air, instead of helium, is used as the purge gas?

**16.69** An electronic device, such as a power transistor mounted on a finned *heat sink*, can be modeled as a spatially isothermal object with energy generation $\dot{E}_g$ and an external convection resistance $R_t$. Consider such a system of mass $m$, specific heat $c$, and surface area $A_s$, which is initially in equilibrium with the environment at $T_\infty$. Suddenly the electronic device is energized such that a constant energy generation $\dot{E}_g$ (W) occurs.

(a) Following the analysis of Sec. 16.5.1 and beginning with a properly defined system, show that the overall energy balance on the system has the form

$$-hA_s(T - T_\infty) + \dot{E}_g = mc\frac{dT}{dt}$$

(b) After the device has been energized, it will eventually ($t \to \infty$) reach a steady-state uniform temperature $T(\infty)$. Using the foregoing energy balance for the *steady-state condition,* show that

$$\dot{E}_g = hA_s(T(\infty) - T_\infty)$$

(c) Using the energy balance from part (a) and the expression for $\dot{E}_g$ from part (b), show that the transient temperature response of the device is

$$\frac{\theta}{\theta_i} = \exp\left(-\frac{t}{R_tC_t}\right)$$

where $\theta \equiv T - T(\infty)$ and $T(\infty)$ is the steady-state temperature corresponding to $t \to \infty$; $\theta_i \equiv T_i - T(\infty)$; $T_i$ is the initial temperature of the device; $R_t$ is the thermal resistance $1/hA_s$; and $C_t$ is the thermal capacitance $mc$.

**16.70** An electronic device, which dissipates 60 W, is mounted on an aluminum ($c = 875$ J/kg · K) *heat sink* with a mass of 0.31 kg and reaches a temperature of 100°C in ambient air at 20°C under steady-state conditions. If the device is initially at 20°C, what temperature will it reach 5 min after the power is switched on? *Hint:* See Problem 16.69 for the temperature response of this system experiencing energy generation and external convection resistance.

**16.71** (CD-ROM)

**16.72** (CD-ROM)

### One-Dimensional Conduction: The Plane Wall

**16.73** Consider the polymer slab of Example 16.10, which is suddenly subjected to the hot oven air. At what time will the slab surface temperature, $T(L, t)$, reach 125°C? What is the midplane temperature, $T(0, t)$, at this elapsed time?

**16.74** Annealing is a process by which steel is reheated and then cooled to make it less brittle. Consider reheat of a 100-mm-thick steel plate ($\rho = 7830$ kg/m$^3$, $c = 550$ J/kg · K, $k = 48$ W/m · K), which is initially at a uniform temperature of $T_i = 200°C$ and is to be heated to a minimum temperature of 550°C. Heating is effected in a gas-fired furnace, where products of combustion at $T_\infty = 800°C$ maintain a convection coefficient of $h = 250$ W/m$^2$ · K on both surfaces of the plate. How long should the plate be left in the furnace?

**16.75** (CD-ROM)

**16.76** A technique being evaluated for eliminating biochemical contamination of mail in the postal service processing centers uses an electron beam source that serves to chemically alter agents, but has the adverse effect of substantially heating the mail. After being exposed to an *e-beam* source for a prescribed period of time, tests indicate that the mail within the process container (150 mm × 300 mm × 600 mm) reaches a temperature of 50°C. The effective thermophysical properties of the mail packed within the container are $k = 0.15$ W/m · K and $\rho c = 2.0 \times 10^6$ J/m$^3$ · K. Estimate the time required for the contents of the container to reach a safe-to-touch temperature of 43°C when the container is subjected to convection cooling with ambient air at 25°C and a convection coefficient of 25 W/m$^2$ · K. *Hint:* Model the container as a plane wall with a thickness $2L = 150$ mm; this condition represents the limiting one-dimensional approximation to the container.

**16.77** Referring to the semiconductor processing tool of Problem 16.68, it is desired at some point in the manufacturing cycle to cool the chuck, which is made of aluminum alloy 2024.

The proposed cooling scheme passes air at 15°C between the air-supply head and the chuck surface as shown in Fig. P16.77. If the chuck is initially at a uniform temperature of 100°C, calculate the time required for its lower surface to reach 25°C, assuming a uniform convection coefficient of 50 $W/m^2 \cdot K$ at the head–chuck interface.

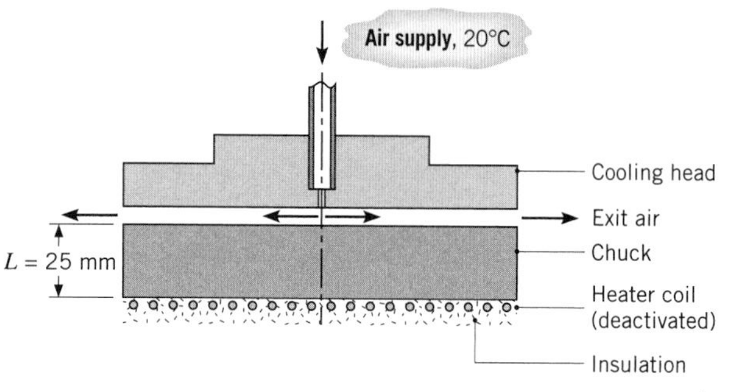


*Figure P16.77*

**16.78** (CD-ROM)

**16.79** An ice layer forms on a 5-mm-thick windshield of a car while parked during a cold night for which the ambient temperature is −20°C. Upon start-up using a new defrost system, the interior surface is suddenly exposed to an air stream at 30°C. Assuming that the ice behaves as an insulating layer on the exterior surface, what interior convection coefficient would allow the exterior surface to reach 0°C in 60 s? The windshield thermophysical properties are $\rho$ = 2200 $kg/m^3$, $c$ = 830 $J/kg \cdot K$, and $k$ = 1.2 $W/m \cdot K$.

**16.80** (CD-ROM)

**16.81** (CD-ROM)

**Transient Conduction: Radial Systems (CD-ROM)**

**16.82** (CD-ROM)

**16.83** (CD-ROM)

**16.84** (CD-ROM)

**16.85** (CD-ROM)

**16.86** (CD-ROM)

**16.87** (CD-ROM)

**Transient Conduction: The Semi-infinite Solid (CD-ROM)**

**16.88** (CD-ROM)

**16.89** (CD-ROM)

**16.90** (CD-ROM)

**16.91** (CD-ROM)

**16.92** (CD-ROM)

**16.93** (CD-ROM)

# HEAT TRANSFER BY CONVECTION

## *Introduction...*

Thus far we have focused on heat transfer by conduction and have considered convection only as a possible boundary condition for conduction problems. In Sec. 15.1.2, we used the term *convection* to describe heat transfer between a surface and an adjacent fluid when they are at different temperatures. Although molecular motion (conduction) contributes to this transfer, the dominant contribution is generally made by the bulk or gross motion of fluid particles. We learned also that knowledge of the convection coefficient is required to use Newton's law of cooling to determine the convective heat flux. In addition to depending upon *fluid properties,* the convection coefficient depends upon the *surface geometry* and the *flow conditions.* The multiplicity of independent variables results from the fact that convection transfer is determined by the boundary layers that develop on the surface. Determination of the convection coefficient by treating these effects is viewed as the *problem of convection.*

*chapter objectives*

In this chapter, our ***first objective*** is to develop an understanding of *boundary layer phenomena* and the features that control the convection coefficient. Our ***second objective*** is to learn how to *estimate convection coefficients* in order to perform analyses on thermal systems experiencing different types of flow and heat transfer situations.

We begin by addressing the *problem of convection.* We will build upon your understanding of the hydrodynamic (velocity) boundary layer concepts from Chap. 14 and introduce the *thermal boundary layer,* the region of the fluid next to the surface in which energy exchange is occurring, and discuss its influence on the convection coefficient. The chapter is then partitioned in three parts, each involving means to estimate the convection coefficient. In the first part, we will consider *forced convection* and introduce methods for estimating convection coefficients associated with *external* and *internal flows.* In the second part, we will consider *free convection* and present methods for estimating the convection coefficients for common geometries. The third part concludes the chapter with a discussion of *heat exchangers,* an extremely important thermal systems application involving convection heat transfer between two fluids separated by a solid surface.

## 17.1 The Problem of Convection

The *problem of convection* is to determine the effects of surface geometry and flow conditions on the convection coefficient resulting from boundary layers that develop on the surface. To introduce these effects, consider *forced convection flow* of a fluid with a

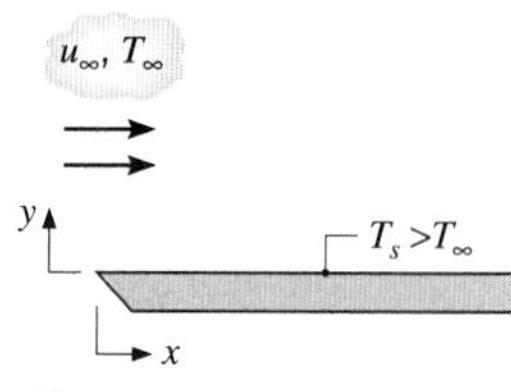


Figure 17.1

free stream velocity $u_\infty$ and temperature $T_\infty$ in parallel, steady, incompressible flow over a *flat plate* with a uniform temperature $T_s$ ($T_s > T_\infty$) as shown in Fig. 17.1. We use this situation to develop an understanding of geometry and flow effects and, in subsequent parts of the chapter, extend the concepts for other flow conditions.

### 17.1.1 The Thermal Boundary Layer

As you learned in Sec. 14.8, the *hydrodynamic boundary layer* is the thin region next to the surface in which the velocity of the fluid changes from zero at the surface (no slip condition) to the free stream velocity some distance from the surface. At each location $x$ along the plate, the boundary layer thickness, $\delta(x)$, was defined as the distance from the surface ($y = 0$) to that distance at which $u = 0.99u_\infty$ (see Fig. 14.13). The fluid flow is characterized by two distinct regions: a thin fluid layer (the *boundary layer*) in which velocity gradients and shear stresses are large, and a region outside the boundary layer (the *free stream*) in which velocity gradients and shear stresses are negligible. We use the subscript $\infty$ to designate conditions in the free stream outside the boundary layer.

Just as a hydrodynamic boundary layer develops when there is fluid flow over a surface, a *thermal boundary layer* develops if the free stream and surface temperatures differ. Consider *laminar* flow over the flat plate shown in Fig. 17.2*a*. At the leading edge, the temperature profile is uniform, with $T(0,y) = T_\infty$. Fluid particles that come into contact with the plate achieve the plate's surface temperature, $T_s$. In turn, these particles exchange energy with those in the adjoining layer, and the temperature gradients develop in the fluid. The region of the fluid in which these temperature gradients exist is the ***thermal boundary layer,*** and its ***thickness*** $\delta_t$ is typically defined as the value of $y$ for which the ratio $[(T_s - T)/(T_s - T_\infty)] = 0.99$. With increasing distance from the leading edge, the effects of heat transfer penetrate further into the free stream, and the thermal boundary layer grows in a similar manner as does the hydrodynamic boundary layer, Fig. 17.2*b*.

*thermal boundary layer thickness*

The relation between conditions in the thermal boundary layer and the convection coefficient can readily be demonstrated. As shown in Fig. 17.2*c*, at any distance $x$ from the leading edge, the *local* heat flux may be obtained by applying Fourier's law to the fluid at $y = 0$ in terms of the thermal conductivity of the fluid, $k$, and the *temperature gradient at the surface*. That is

$$q_s'' = -k \left. \frac{\partial T}{\partial y} \right|_{y=0} \tag{17.1}$$

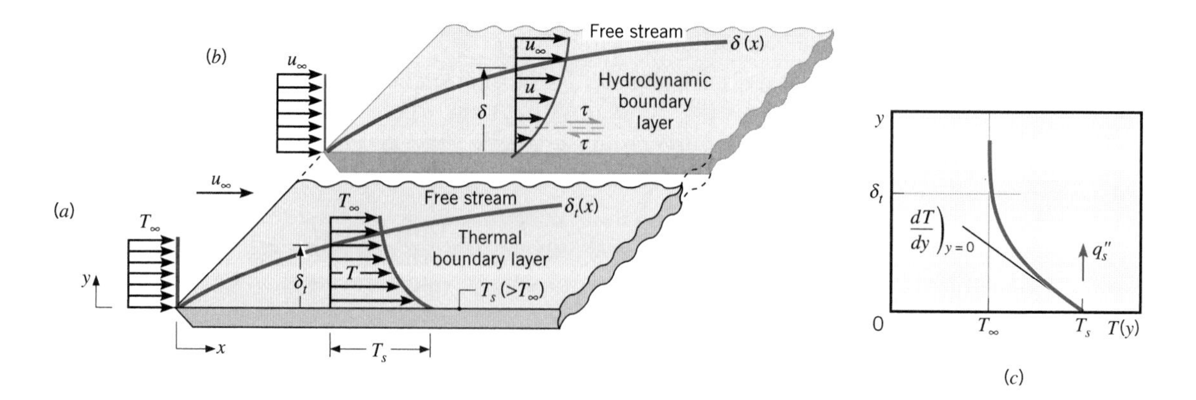


Figure 17.2 Fluid with uniform free stream velocity $u_\infty$ and temperature $T_\infty$ in laminar flow over a flat plate with uniform temperature $T_s$ ($T_s > T_\infty$). (*a*) Thermal boundary layer, (*b*) Hydrodynamic boundary layer, and (*c*) Local heat flux determined from the temperature gradient at the surface, Eq. 17.1.

This expression is appropriate because *at the surface, the fluid velocity is zero (no-slip condition) and energy transfer occurs by conduction*. Recognize that the surface heat flux is equal to the *convective* flux, which is expressed by Newton's law of cooling

$$q''_s = q''_{\text{conv}} = h_x(T_s - T_\infty) \tag{17.2}$$

By combining the foregoing equations, we obtain an expression for the ***local convection coefficient***

$$h_x = \frac{-k\,\partial T/\partial y|_{y=0}}{T_s - T_\infty} \tag{17.3}$$

*local convection coefficient*

Conditions in the thermal boundary layer strongly influence the temperature gradient at the surface, which from Eq. 17.1 determines the rate of heat transfer across the boundary layer, and from Eq. 17.2 determines the local convection coefficient.

Referring to the thermal boundary representation of Fig. 17.2*a*, note that as $\delta_t$ increases with $x$, the temperature gradients in the boundary layer must decrease with $x$. Accordingly, the magnitude of $\partial T/\partial y|_{y=0}$ decreases with $x$, and it follows that $q''_s$ and $h_x$ decrease with $x$, as shown in Fig. 17.3.

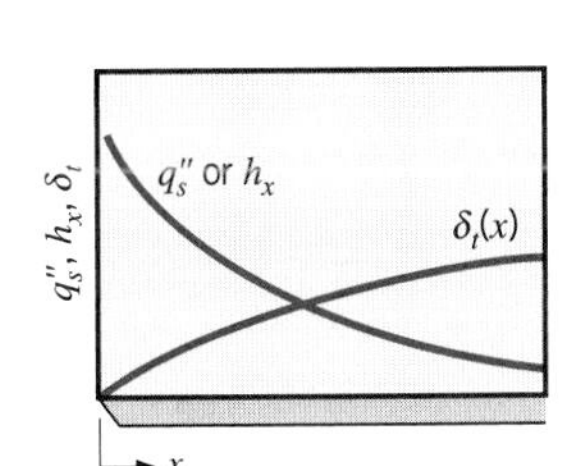


*Figure 17.3*

As you learned in Sec. 14.8 (see Fig. 14.12), the structure of the flow in the *hydrodynamic* boundary layer can undergo a transition from laminar flow near the leading edge to turbulent flow. As shown in Fig. 17.4, the *thermal* boundary has flow characteristics and temperature profiles that are a consequence of the hydrodynamic boundary layer behavior. In the *laminar* region, fluid motion is highly ordered and characterized by velocity components in both the $x$- and $y$-directions. The velocity component $v$ in the $y$-direction (normal to the surface) contributes to the transfer of energy (and momentum) through the boundary layer. The resulting ***temperature profile*** (Fig. 17.4) changes in a gradual manner over the thickness of the boundary layer.

*temperature profile*

At some distance from the leading edge, small disturbances in the flow are amplified, and transition to turbulent flow begins to occur. Fluid motion in the *turbulent* region is highly irregular and characterized by velocity fluctuations that enhance the transfer of energy. Due to fluid mixing resulting from the fluctuations, turbulent boundary layers are thicker. Accordingly, the temperature profiles are flatter, but the temperature gradients at the surface are steeper than for laminar flow. Consequently, from Eq. 17.3, we expect the local convection coefficients to be larger than for laminar flow, but to decrease with $x$ as shown in Fig. 17.4.

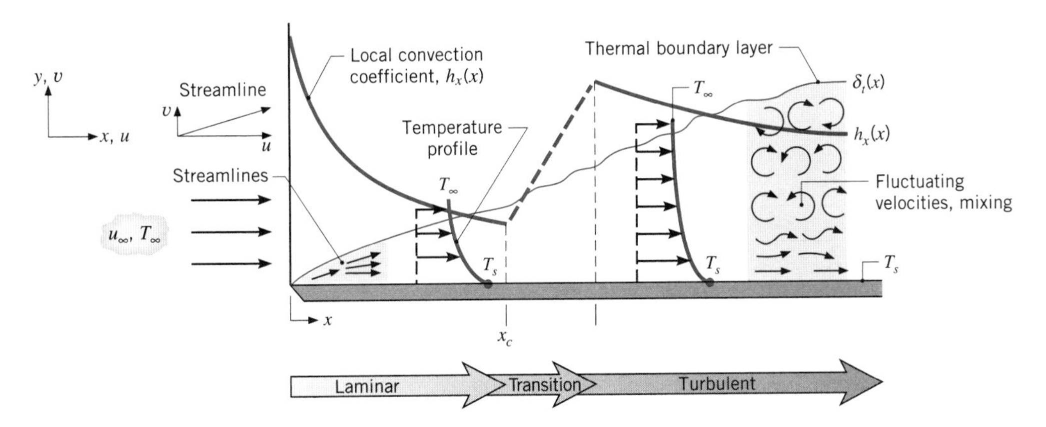


*Figure 17.4* Thermal boundary layer development on a flat plate showing changes in fluid temperature profiles and local convection coefficient in the laminar and turbulent flow regions.

In analyzing boundary layer behavior for the flat plate, we identify *transition* as occurring at the location $x_c$. The ***critical Reynolds number,*** $\mathrm{Re}_{x,c}$, corresponding to the onset of transition, is known to vary from $10^5$ to $3 \times 10^6$, depending upon surface roughness and turbulence level of the free stream. A representative value of

*critical Reynolds number*

$$\mathrm{Re}_{x,c} = \frac{u_\infty x_c}{\nu} = 5 \times 10^5 \tag{17.4}$$

is often assumed for heat transfer calculations and, unless otherwise noted, is used for the calculations of this text.

### 17.1.2 Local and Average Convection Coefficients

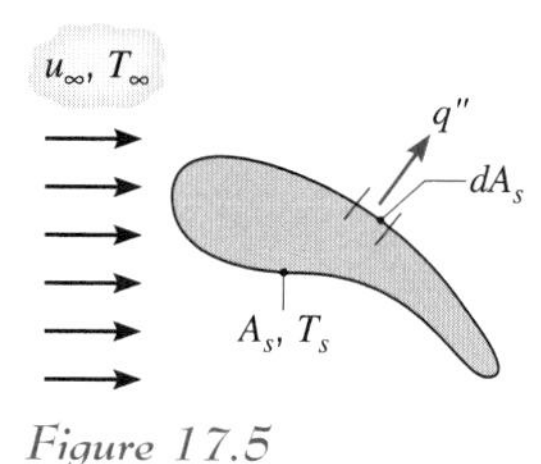

*Figure 17.5*

As we have seen for the case of parallel flow over a flat plate in Fig. 17.4, the local convection coefficient $h_x$ varies over the surface of the plate. Consider a surface of arbitrary shape and of area $A_s$, with a uniform surface temperature $T_s$ ($>T_\infty$) exposed to fluid flow with a free stream velocity $u_\infty$ and temperature $T_\infty$ ($<T_s$) (Fig. 17.5). We'd expect the convection coefficient to vary over the surface of the arbitrary shape, as well as for the flat plate. For both situations the *local heat flux, $q''$*, can be expressed as

$$q'' = h_x(T_s - T_\infty) \qquad \text{[local]} \tag{17.5}$$

The *total* heat transfer rate can be obtained by integrating the local heat flux over the entire surface $A_s$. That is

$$q = \bar{h} A_s(T_s - T_\infty) \qquad \text{[total]} \tag{17.6}$$

where $\bar{h}$ denotes the ***average convection coefficient*** obtained from

*average convection coefficient*

$$\bar{h} = \left(\frac{1}{A_s}\right) \int_{A_s} h_x dA_s \tag{17.7}$$

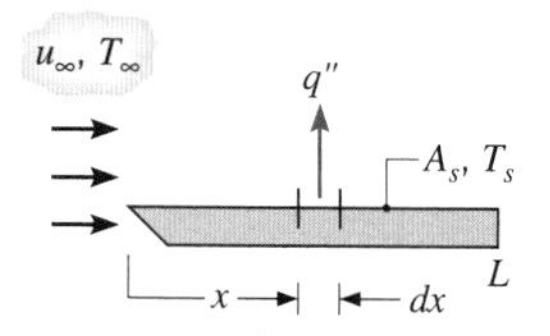

*Figure 17.6*

Note that for the special case of flow over a flat plate (Fig. 17.6), $h_x$ varies with distance $x$ from the leading edge and Eq. 17.7 reduces to

$$\bar{h}_x = \frac{1}{x} \int_0^x h_x \, dx \tag{17.8}$$

#### *Example 17.1* Average Coefficient from the Local Coefficient Variation

Experimental results for the *local* heat transfer coefficient $h_x$ for flow over a flat plate with an extremely rough surface were found to fit the relation

$$h_x(x) = ax^{-0.1}$$

where $a$ is a coefficient (W/m$^{1.9}\cdot$K) and $x$ (m) is the distance from the leading edge of the plate.

**(a)** Develop an expression for the ratio of the *average* heat transfer coefficient $\bar{h}_x$ for a plate of length $x$ to the *local* heat transfer coefficient $h_x$ at $x$.

**(b)** Show qualitatively the variation of $h_x$ and $\bar{h}_x$ as a function of $x$.

#### Solution

***Known:*** Variation of the *local* heat transfer coefficient, $h_x(x)$.

***Find:***

**(a)** The ratio of the *average* heat transfer coefficient $\bar{h}(x)$ to the *local* value $h_x(x)$.

**(b)** Sketch of the variation of $h_x$ and $\bar{h}_x$ with $x$.

*Schematic and Given Data:*

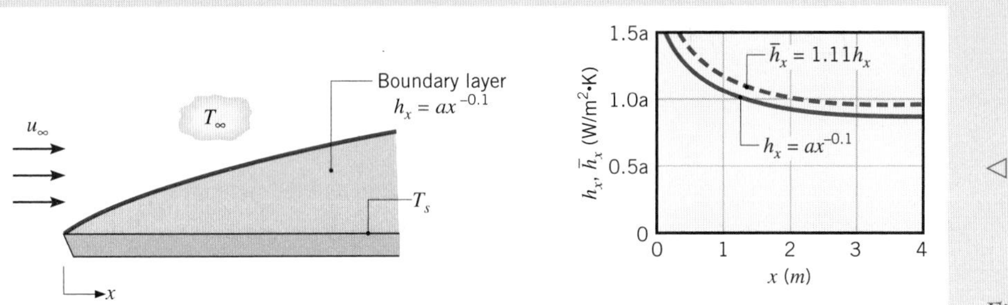


◁

*Figure E17.1*

***Analysis:***

**(a)** From Eq. 17.8 the average value of the convection heat transfer coefficient over the region from 0 to $x$ is

$$\bar{h}_x = \bar{h}_x(x) = \frac{1}{x}\int_0^x h_x(x)\,dx$$

Substituting the expression for the local heat transfer coefficient

$$h_x(x) = ax^{-0.1}$$

and integrating, we obtain

$$\bar{h}_x = \frac{1}{x}\int_0^x ax^{-0.1}\,dx = \frac{a}{x}\int_0^x x^{-0.1}\,dx = \frac{a}{x}\left(\frac{x^{+0.9}}{0.9}\right) = 1.11ax^{-0.1}$$

Rearranging, find the *ratio* of the *average* convection coefficient over the region 0 to $x$ to the *local* value at $x$

$$\frac{\bar{h}_x}{h_x} = 1.11 \quad \triangleleft$$

**(b)** The variation of $\bar{h}_x$ and $h_x$ with $x$ is shown in the graph above. Boundary layer development causes both the local and average coefficients to decrease as $x^{-0.1}$ in the flow direction. The *average* convection coefficient from the leading edge to a point $x$ on the plate is 1.11 times the *local* coefficient at that point.

## 17.1.3 Correlations: Estimating Convection Coefficients

Our primary aim in the problem of convection is to determine the convection coefficient for different flow conditions and geometries with different fluids. Recognizing that there are numerous variables associated with any flow situation, our interest is in identifying *universal functions* in terms of dimensionless parameters or groups that have *physical significance* for convective flow situations. The approach is the same one you followed in Sec. 13.5 in forming and using dimensionless groups that commonly arise in fluid mechanics. The dimensionless groups important to convective heat transfer are introduced in the following paragraphs and summarized in Table 17.1.

The ***Nusselt number,*** which represents the dimensionless temperature gradient at the surface (Eq. 17.3) and provides a measure of the convection coefficient, is defined as

$$\mathrm{Nu}_L = \frac{hL}{k} \tag{17.9}$$

*Nusselt number*

where $L$ is the characteristic length of the surface of interest. Based upon analytical solutions and experimental observations, it has been shown that for *forced convection,* the local and average convection coefficients can be correlated, respectively, by equations of the form

$$\mathrm{Nu}_x = f(x^*, \mathrm{Re}_x, \mathrm{Pr}) \qquad \overline{\mathrm{Nu}}_x = f(\mathrm{Re}_x, \mathrm{Pr}) \tag{17.10, 11}$$

**Table 17.1** Important Dimensionless Groups in Convection Heat Transfer

| Group[a] | Definition[a] | | Interpretation/Application |
|---|---|---|---|
| Nusselt number, $Nu_L$ | $\frac{hL}{k}$ | (17.9) | Dimensionless temperature gradient at the surface. Measure of the convection heat transfer coefficient. |
| Reynolds number, $Re_L$ | $\frac{VL}{\nu}$ | (17.12) | Ratio of the inertia and viscous forces. Characterizes forced convection flows. |
| Prandtl number, Pr | $\frac{c_p\mu}{k} = \frac{\nu}{\alpha}$ | (17.13) | Ratio of the momentum and thermal diffusivities. Property of the fluid. |
| Grashof number, $Gr_L$ | $\frac{g\beta(T_s - T_\infty)L^3}{\nu^2}$ | (17.16) | Ratio of buoyancy to viscous forces. Characterizes free convection flows. |
| Rayleigh number, $Ra_L$ | $\frac{g\beta(T_s - T_\infty)L^3}{\nu\alpha}$ | (17.19) | Product of Grashof and Prandtl numbers, Gr · Pr. Characterizes free convection flows. |

[a]The subscript $L$ represents the characteristic length on the surface of interest.

where the subscript $x$ has been added to emphasize our interest in conditions at a particular location on the surface identified by the dimensionless distance $x^*$. The overbar indicates an average over the surface from $x^* = 0$ to the location of interest.

The ***Reynolds number,*** $Re_L$, is the ratio of the inertia to viscous forces, and is used to characterize boundary layer flows (Sec. 13.5)

*Reynolds number*

$$Re_L = \frac{VL}{\nu} \tag{17.12}$$

where V represents the reference velocity of the fluid, $L$ is the characteristic length of the surface, and $\nu$ is the kinematic viscosity of the fluid.

The ***Prandtl number,*** Pr, is a transport property of the fluid and provides a measure of the relative effectiveness of momentum and energy transport in the hydrodynamic and thermal boundary layers, respectively

*Prandtl number*

$$Pr = \frac{c_p\mu}{k} = \frac{\nu}{\alpha} \tag{17.13}$$

where $\mu$ is the dynamic viscosity and $\alpha$ is the thermal diffusivity of the fluid (Eq. 16.5).

From Table HT-3, we see that the Prandtl number for gases is near unity, in which case momentum and energy transport are comparable. In contrast, for oils and some liquids with $Pr \gg 1$ (Tables HT-4, 5), momentum transport is more significant, and the effects extend further into the free stream. From this interpretation, it follows that the value of Pr strongly influences the relative growth of the velocity and thermal boundary layers. In fact, for a *laminar* boundary layer, it has been shown that

$$\frac{\delta}{\delta_t} = Pr^n \tag{17.14}$$

where $n$ is a positive constant, typically $n = 1/3$. Hence for a gas, $\delta_t \approx \delta$; for an oil $\delta_t \ll \delta$. However, for all fluids in the *turbulent* region, because of extensive mixing, we expect $\delta_t \approx \delta$.

The forms of the functions associated with Eqs. 17.10 and 17.11 are most commonly determined from extensive sets of experimental measurements performed on specific surface geometries and types of flows. Such functions are termed ***empirical correlations*** and are always accompanied by specifications regarding surface geometry and flow conditions. ***For Example...*** the most general correlation for forced convection *external flow* over flat plates and other immersed geometries has the form

*empirical correlations*

$$\overline{Nu}_x = C\,Re_x^m\,Pr^n \tag{17.15}$$

where $C$, $m$ and $n$ are independent of the fluid, but dependent upon the surface geometry and flow condition (laminar vs. turbulent). For forced convection *internal flow,* the same general correlation form applies, although the boundary layer flow regions have different characteristics than we've seen for external flow. ▲

In *free convection,* the boundary layer flow is induced by thermally driven buoyancy forces arising from a difference between the surface temperature $T_s$ and the adjoining fluid temperature $T_\infty$. The flow is characterized by the ***Grashof number,*** which is the ratio of the buoyancy to viscous forces

$$\mathrm{Gr}_L = \frac{g\beta(T_s - T_\infty)L^3}{\nu^2} \qquad (17.16)$$

*Grashof number*

where $g$ is the gravitational acceleration and $\beta$ is the volumetric thermal expansion coefficient. The local and average convection coefficients are correlated, respectively, by equations having the form

$$\mathrm{Nu}_x = f(x^*, \mathrm{Gr}_x, \mathrm{Pr}) \qquad \overline{\mathrm{Nu}}_x = f(\mathrm{Gr}_x, \mathrm{Pr}) \qquad (17.17, 17.18)$$

Note these forms are the same as for forced convection, Eqs. 17.10 and 17.11, where the Grashof number replaces the Reynolds number as the parameter to characterize the flow. Since the product of the Grashof and Prandtl number appears frequently in free convection correlations, it is convenient to represent the product as the ***Rayleigh number***

$$\mathrm{Ra}_L = \frac{g\beta(T_s - T_\infty)L^3}{\nu\alpha} \qquad (17.19)$$

*Rayleigh number*

which has the same physical interpretation as the Grashof number.

Table 17.1 lists the dimensionless groups that appear frequently in heat transfer practice. You should become familiar with the definitions and application of these important convection parameters.

*Correlation Selection Rules.* Thus far we have discussed forced convection correlations for flow over the flat plate, and described only major features for correlations associated with other flow situations. The selection and application of convection correlations *for any flow situation* are facilitated by following a few simple *rules:*

- *Identify the flow surface geometry.* Does the problem involve flow over a flat plate, a cylinder, or a sphere? Or flow through a tube of circular or non-circular cross-sectional area?
- *Specify the appropriate reference temperature and evaluate the pertinent fluid properties at that temperature.* For moderate boundary layer temperature differences, the ***film temperature,*** $T_f$, defined as the average of the surface and free stream temperatures

$$T_f = \frac{T_s + T_\infty}{2} \qquad (17.20)$$

*film temperature*

may be used for this purpose. However, we will consider correlations that require property evaluation at the free stream temperature, and include a property ratio accounting for the nonconstant property effect.

- *Calculate the Reynolds number.* Using the appropriate characteristic length, calculate the Reynolds number to determine the boundary layer flow conditions. If the geometry is the flat plate in parallel flow, determine whether the flow is laminar, turbulent, or mixed.
- *Decide whether a local or surface average coefficient is required.* The local coefficient is used to determine the heat flux at a point on the surface; the average coefficient is used to determine the heat transfer rate for the entire surface.
- *Select the appropriate correlation.*

At the end of the sections dealing with *forced convection* external flow (Sec. 17.2) and internal flow (Sec. 17.3), and with *free convection* (Sec. 17.4), the recommended correlations are summarized along with guidelines that will facilitate their selection for your problem.

# Forced Convection

## 17.2 External Flow

In the previous section, we learned that correlations for estimating convection coefficients for external forced convection flows provide the Nusselt number as a function of the Reynolds number and the Prandtl number, where the function depends upon the geometry of the surface, flow conditions, and fluid properties. We will introduce correlations useful for estimating coefficients over a flat plate and curved surfaces of a cylinder and sphere, and illustrate how they can be used to compute convection heat rates.

### 17.2.1 The Flat Plate in Parallel Flow

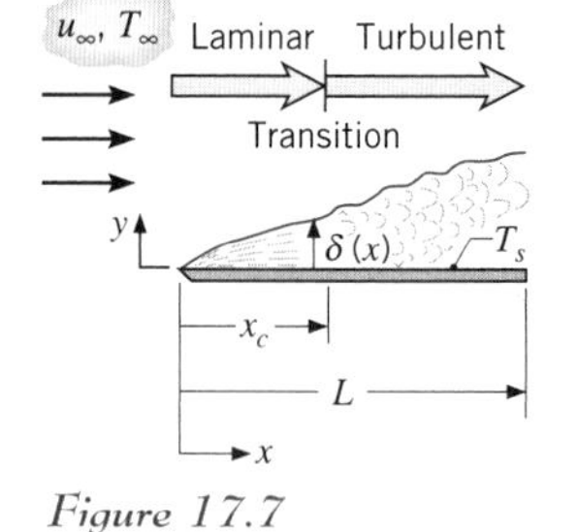

*Figure 17.7*

Despite its simplicity, parallel flow over a flat plate (Fig. 17.7) occurs in numerous engineering applications. As discussed initially in the previous section, boundary layer flow conditions are characterized by the Reynolds number, Eq. 17.12. In the absence of upstream disturbances, laminar boundary layer development begins at the leading edge ($x = 0$), and transition to turbulence may occur at a downstream location ($x_c$) for which the critical Reynolds number is $Re_{x,c} = 5 \times 10^5$ (Eq. 17.4). We will present correlations that are appropriate for calculating the boundary layer thickness as well as the convection coefficient. We begin by considering conditions in the laminar boundary layer.

#### Laminar Flow

As illustrated in Fig. 17.2, the hydrodynamic boundary layer thickness $\delta$ is defined as that value of $\delta(x)$ for which $u/u_\infty = 0.99$, and from Eq. 14.19 can be expressed as

$$\delta = 5x\,Re_x^{-1/2} \tag{17.21}$$

where the characteristic length in the Reynolds number is $x$, the distance from the leading edge

$$Re_x = \frac{u_\infty x}{\nu} \tag{17.22}$$

It is clear that $\delta$ increases with $x$, and decreases with increasing $u_\infty$. That is, the higher the free stream velocity, the thinner the boundary layer. The *local* Nusselt number is of the form

$$\mathrm{Nu}_x = \frac{h_x x}{k} = 0.332\,\mathrm{Re}_x^{1/2}\,\mathrm{Pr}^{1/3} \qquad [0.6 \leq \mathrm{Pr} \leq 50] \tag{17.23}$$

Note how we have designated a restriction on the range of applicability of the correlation, in this instance for the Prandtl number. The ratio of the hydrodynamic to thermal boundary layer thickness is

$$\frac{\delta}{\delta_t} \approx \mathrm{Pr}^{1/3} \tag{17.24}$$

where $\delta$ is given by Eq. 17.21.

The foregoing results may be used to compute the *local* boundary layer parameters for any $0 < x < x_c$, where $x_c$ is the distance from the leading edge at which transition begins. Equation 17.23 implies $h_x$ is, in principle, infinite at the leading edge and decreases as $x^{-1/2}$ in the flow direction, whereas from Equation 17.21 and 17.24, the thickness of the boundary layers increases as $x^{1/2}$ in the flow direction (see Fig. 17.8). Equation 17.24 also implies that, for values of Pr close to unity, which is the case for most gases, the hydrodynamic and thermal boundary layers experience nearly identical growth.

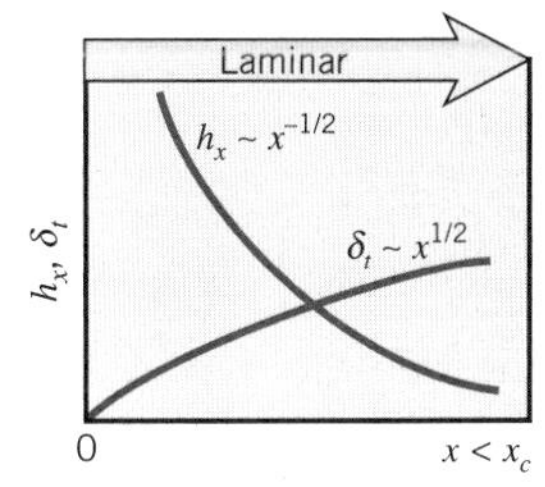

*Figure 17.8*

The expression for the *average* convection coefficient for any surface shorter than $x_c$ may be determined by performing the integration prescribed by Eq. 17.8 using Eq. 17.23 for the *local* coefficient. That is

$$\bar{h}_x = \frac{1}{x}\int_0^x h_x\,dx = 0.332\left(\frac{k}{x}\right)\mathrm{Pr}^{1/3}\left(\frac{u_\infty}{\nu}\right)^{1/2}\int_0^x \frac{dx}{x^{1/2}} \tag{17.25}$$

and since the definite integral has the value $2x^{1/2}$, it follows after some rearrangement that $\bar{h}_x = 2h_x$. Hence

$$\overline{\mathrm{Nu}}_x = \frac{\bar{h}_x x}{k} = 0.664\,\mathrm{Re}_x^{1/2}\,\mathrm{Pr}^{1/3} \qquad [0.6 \leq \mathrm{Pr} \leq 50] \tag{17.26}$$

If the flow is laminar over the entire surface, the subscript $x$ may be replaced by $L$, and Eq. 17.26 may be used to predict the average coefficient for the entire surface.

From the foregoing expressions for the local and average coefficients, Eqs. 17.23 and 17.26, respectively, we see that, for *laminar* flow over a flat plate, the *average* convection coefficient from the leading edge to a point $x$ on the surface is *twice* the *local* coefficient at that point. In using these expressions, the effect of variable properties is treated by evaluating all the properties at the film temperature defined as the average of the surface and free stream temperatures, Eq. 17.20.

## Turbulent Flow

For turbulent flows, to a reasonable approximation, the hydrodynamic boundary layer thickness can be expressed as

$$\delta = 0.37x\,\mathrm{Re}_x^{-1/5} \qquad [\mathrm{Re}_x \leq 10^8] \tag{17.27}$$

and the local Nusselt number is given as

$$\mathrm{Nu}_x = \frac{h_x x}{k} = 0.0296\,\mathrm{Re}_x^{4/5}\,\mathrm{Pr}^{1/3} \qquad \begin{bmatrix}\mathrm{Re}_x \leq 10^8 \\ 0.6 < \mathrm{Pr} < 60\end{bmatrix} \tag{17.28}$$

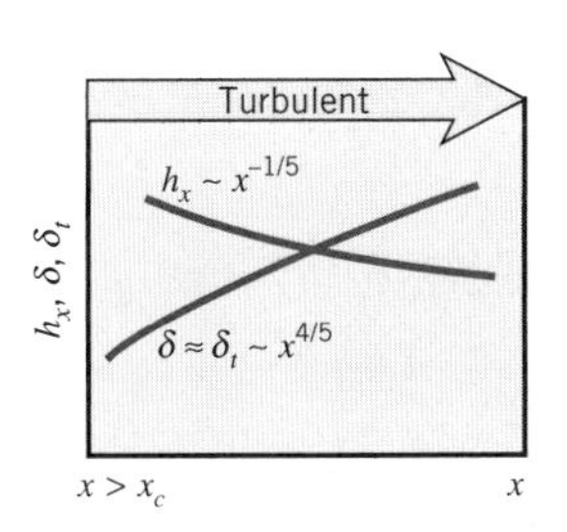

*Figure 17.9*

where all properties are evaluated at the film temperature, $T_f$, Eq. 17.20.

*Summary.* Comparing the results for the turbulent and laminar boundary layers, important differences should be noted:

- For *laminar flow,* the hydrodynamic and thermal boundary layer thicknesses depend on the Prandtl number, the dimensionless number representing the ratio of the momentum, and thermal diffusivities (Table 17.1). From Eq. 17.24

$$\text{Laminar} \qquad \delta \approx \delta_t\,\text{Pr}^{-1/3}$$

- For *turbulent flow,* the boundary layer development is strongly influenced by random velocity and less so by molecular motion. Hence, relative boundary layer growth does not depend on the Prandtl number, Eq. 17.27. That is, *the hydrodynamic and thermal boundary thicknesses are nearly equal*

$$\text{Turbulent} \qquad \delta \approx \delta_t$$

- *Thermal boundary layer growth* is more rapid in the flow direction for *turbulent* flow (Fig. 17.9) than for *laminar* flow (Fig. 17.8)

$$\text{Turbulent} \qquad \delta_t \sim x^{4/5}$$

$$\text{Laminar} \qquad \delta_t \sim x^{1/2}$$

- The *convection coefficient* for *turbulent* flow is larger than for laminar flow due to enhanced mixing in the boundary layer. For *turbulent* flow (Fig. 17.9), the decrease in the convection coefficient in the flow direction is more gradual than for *laminar* flow (Fig. 17.8)

$$\text{Turbulent} \qquad h_x \sim x^{-1/5}$$

$$\text{Laminar} \qquad h_x \sim x^{-1/2}$$

### Mixed Boundary Layer Conditions

An expression for the *average* coefficient can now be determined. However, since the turbulent boundary layer is generally preceded by a laminar boundary layer, we first consider *mixed* flow conditions.

For laminar flow over the *entire* plate, Eq. 17.26 can be used to compute the average coefficient for the plate. Moreover, if transition occurs toward the trailing edge of the plate, for example, in the range $0.95 \le (x_c/L) \le 1$, this equation will also provide a reasonable approximation. However, when transition occurs sufficiently upstream of the trailing edge, $(x_c/L) \le 0.95$, the surface average coefficient will be influenced by conditions in both the laminar and turbulent boundary layers.

In the *mixed boundary layer* situation, shown in Fig. 17.7, Eq. 17.8 can be used to obtain the average convection coefficient for the entire plate. Integrating over the laminar region $(0 \le x \le x_c)$ and then over the turbulent region $(x_c < x \le L)$, this equation may be expressed as

$$\bar{h}_L = \frac{1}{L}\left(\int_0^{x_c} h_{\text{lam}}\,dx + \int_{x_c}^{L} h_{\text{turb}}\,dx\right)$$

where it is assumed that transition occurs abruptly at $x = x_c$. Substituting from Eqs. 7.23 and 7.28, for $h_{\text{lam}}$ and $h_{\text{turb}}$, respectively, we obtain

$$\bar{h}_L = \left(\frac{k}{L}\right)\left[0.332\left(\frac{u_\infty}{\nu}\right)^{1/2}\int_0^{x_c}\frac{dx}{x^{1/2}} + 0.0296\left(\frac{u_\infty}{\nu}\right)^{4/5}\int_{x_c}^{L}\frac{dx}{x^{1/5}}\right]\text{Pr}^{1/3}$$

Integrating, we then obtain

$$\overline{\text{Nu}}_L = [0.664\,\text{Re}_{x,c}^{1/2} + 0.037(\text{Re}_L^{4/5} - \text{Re}_{x,c}^{4/5})]\text{Pr}^{1/3}$$

or

$$\overline{\text{Nu}}_L = (0.037\,\text{Re}_L^{4/5} - A)\text{Pr}^{1/3} \tag{17.29}$$

where the constant $A$ is determined by the value of the critical Reynolds number $\mathrm{Re}_{x,c}$. That is

$$A = 0.037\,\mathrm{Re}_{x,c}^{4/5} - 0.664\,\mathrm{Re}_{x,c}^{1/2} \tag{17.30}$$

If a representative *transition Reynolds number* of $\mathrm{Re}_{x,c} = 5 \times 10^5$ is assumed, Eq. 17.29 reduces to

$$\overline{\mathrm{Nu}}_L = (0.037\,\mathrm{Re}_L^{4/5} - 871)\mathrm{Pr}^{1/3} \qquad \begin{bmatrix} 0.6 < \mathrm{Pr} < 60 \\ 5 \times 10^5 < \mathrm{Re}_L \leq 10^8 \\ \mathrm{Re}_{x,c} = 5 \times 10^5 \end{bmatrix} \tag{17.31}$$

where the bracketed relations indicate the range of applicability.

It is important to recognize that the transition Reynolds number can be influenced by the roughness of the surface, and by disturbances upstream of the boundary layer caused by fluid machines such as fans, compressors, and pumps. There are many practical applications where it is desirable to use *turbulence promoters (turbulators),* such as a fine wire or screens, to *trip the boundary layer* at the leading edge. For such a condition, from Eq. 17.29 with $A = 0$ (corresponding to $\mathrm{Re}_{x,c} = 0$)

$$\overline{\mathrm{Nu}}_L = 0.037\,\mathrm{Re}_L^{4/5}\,\mathrm{Pr}^{1/3} \qquad \begin{bmatrix} \mathrm{Re}_{x,c} = 0 \\ 0.6 \leq \mathrm{Pr} \leq 50 \end{bmatrix} \tag{17.32}$$

where the boundary layer is assumed to be *fully turbulent* from the leading edge over the entire plate.

To facilitate the selection of correlations appropriate for your application, Table 17.3 (page 422) provides a summary of the correlations along with their limits of applicability.

## *Example 17.2* Laminar Flow over a Flat Plate

Air at atmospheric pressure and a temperature of 300°C flows steadily with a velocity of 10 m/s over a flat plate of length 0.5 m. Estimate the cooling rate per unit width of the plate needed to maintain a surface temperature of 27°C.

### Solution

***Known:*** Airflow over an isothermal flat plate.

***Find:*** Cooling rate per unit width of the plate, $q'$ (W/m).

***Schematic and Given Data:***

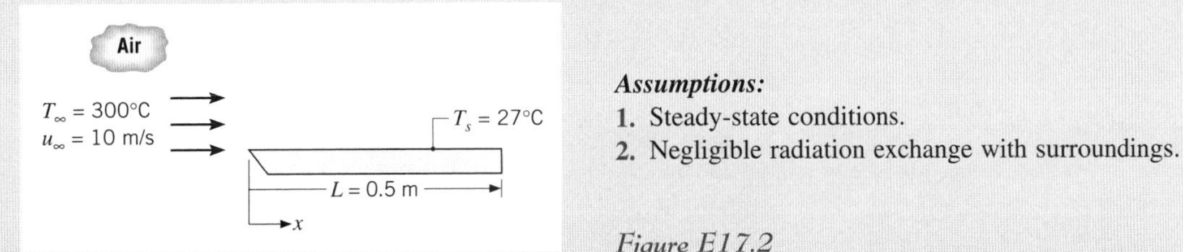


***Assumptions:***
1. Steady-state conditions.
2. Negligible radiation exchange with surroundings.

*Figure E17.2*

***Properties:*** Table HT-3, air ($T_f = 437$ K, $p = 1$ atm): $\nu = 30.84 \times 10^{-6}$ m²/s, $k = 36.4 \times 10^{-3}$ W/m · K, Pr = 0.687.

***Analysis:*** For a plate of unit width, it follows from Newton's law of cooling that the rate of convection heat transfer *to* the plate is

$$q' = \bar{h}L(T_\infty - T_s)$$

To select the appropriate convection correlation for estimating $\bar{h}$, the Reynolds number must be determined to characterize the flow

$$\mathrm{Re}_L = \frac{u_\infty L}{\nu} = \frac{10 \text{ m/s} \times 0.5 \text{ m}}{30.84 \times 10^{-6} \text{ m}^2\text{/s}} = 1.62 \times 10^5$$

Since $\mathrm{Re}_L < \mathrm{Re}_{x,c} = 5 \times 10^5$, the flow is laminar over the entire plate, and the appropriate correlation is given by Eq. 17.26 (see also Table 17.3, page 423)

$$\overline{\mathrm{Nu}}_L = 0.664\,\mathrm{Re}_L^{1/2}\,\mathrm{Pr}^{1/3} = 0.664(1.62 \times 10^5)^{1/2}(0.687)^{1/3} = 236$$

The average convection coefficient is then

$$\bar{h} = \frac{\overline{\mathrm{Nu}}_L\, k}{L} = \frac{236 \times 0.0364 \text{ W/m} \cdot \text{K}}{0.5 \text{ m}} = 17.2 \text{ W/m}^2 \cdot \text{K}$$

and the required cooling rate per unit width of plate is

$$q' = 17.2 \text{ W/m}^2 \cdot \text{K} \times 0.5 \text{ m}(300 - 27)^\circ\text{C} = 2348 \text{ W/m} \quad \triangleleft$$

***Comments:***

1. Note that the thermophysical properties are evaluated at the film temperature, $T_f = (T_s + T_\infty)/2$, Eq. 17.20.
2. Using Eq. 7.21, the hydrodynamic boundary layer thickness at the trailing edge of the plate ($x = L = 0.5$ m) is

$$\delta = 5L\,\mathrm{Re}_L^{-1/2} = 5 \times 0.5 \text{ m}(1.62 \times 10^5)^{-1/2} = 0.0062 \text{ m} = 6.2 \text{ mm}$$

The thermal boundary layer at the same location from Eq. 17.24 is

$$\delta_t = \delta\,\mathrm{Pr}^{-1/3} = 6.2 \text{ mm}(0.687)^{-1/3} = 7.0 \text{ mm}$$

Since $\mathrm{Pr} \approx 0.7 < 1$, we find that $\delta < \delta_t$. Still, note that the magnitudes of the boundary layer thicknesses, $\delta$ and $\delta_t$, are quite similar as expected for gases.

3. If upstream turbulence is promoted by a fan or grill, or a trip wire were placed at the leading edge, a *turbulent boundary condition could exist over the entire plate.* For such a condition, Eq. 17.32 is the appropriate correlation to estimate the convection coefficient

$$\overline{\mathrm{Nu}}_L = 0.037\,\mathrm{Re}_L^{4/5}\,\mathrm{Pr}^{1/3} = 0.037(1.62 \times 10^5)^{4/5}\,(0.687)^{1/3} = 480$$

$$\bar{h}_L = 480(36.4 \times 10^{-3} \text{ W/m} \cdot \text{K})/0.5 \text{ m} = 35.0 \text{ W/m}^2 \cdot \text{K}$$

The cooling rate per unit plate width is

$$q' = 35 \text{ W/m}^2 \cdot \text{K} \times 0.5 \text{ m}\,(300 - 27)^\circ\text{C} = 4778 \text{ W/m}$$

The effect of inducing turbulence over the entire plate is to double the convection coefficient, and hence, double the required cooling rate.

## *Example 17.3* Mixed Boundary Layer Flow, Segmented Flat Plate

A flat plate of width $w = 1$ m is maintained at a uniform surface temperature, $T_s = 230^\circ$C, by using independently controlled, electrical strip heaters, each of which is 50 mm long. If atmospheric air at 25°C flows over the plates at a velocity of 60 m/s, what is the electrical power requirement for the fifth heater?

### Solution

***Known:*** Air flow over a flat plate with segmented heaters.
***Find:*** Electrical power required for the fifth heater.

*Schematic and Given Data:*

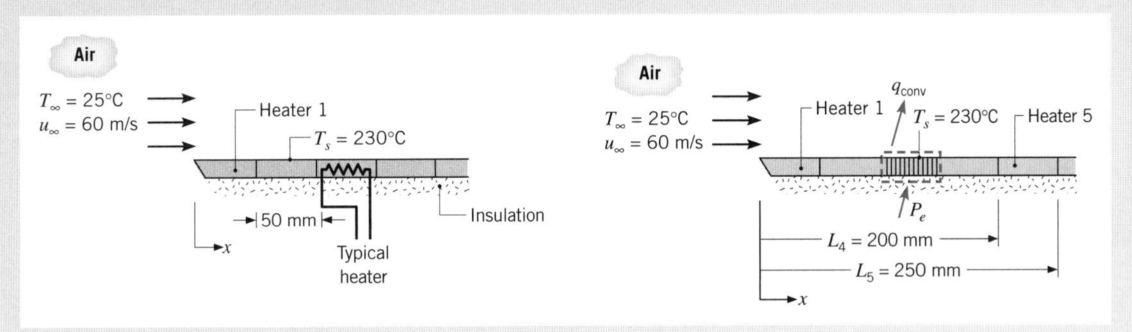


*Figure E17.3a*

***Assumptions:***

1. Steady-state conditions.
2. Negligible radiation effects.
3. Bottom surface of plate is adiabatic.

***Properties:*** Table HT-3, air ($T_f$ = 400 K, $p$ = 1 atm): $\nu = 26.41 \times 10^{-6}$ m²/s, $k$ = 0.0338 W/m · K, Pr = 0.690.

***Analysis:*** For each of the heaters, conservation of energy requires that

$$P_e = q_{\text{conv}}$$

The power requirement for the fifth heater may be obtained by subtracting the total convection heat transfer associated with the first four heaters from that associated with the first five heaters. Accordingly

$$q_{\text{conv},5} = \bar{h}_{1\text{-}5}L_5w(T_s - T_\infty) - \bar{h}_{1\text{-}4}L_4w(T_s - T_\infty)$$
$$q_{\text{conv},5} = (\bar{h}_{1\text{-}5}L_5 - \bar{h}_{1\text{-}4}L_4)w(T_s - T_\infty)$$

where $\bar{h}_{1\text{-}4}$ and $\bar{h}_{1\text{-}5}$ represent the *average* coefficients over heaters 1 to 4 ($0 \le x \le L_4$) and heaters 1 to 5 ($0 \le x \le L_5$), respectively. To characterize the flow, calculate the Reynolds numbers at $x = L_4$ and $x = L_5$

$$\text{Re}_4 = \frac{u_\infty L_4}{\nu} = \frac{60 \text{ m/s} \times 0.200 \text{ m}}{26.41 \times 10^{-6} \text{ m}^2/\text{s}} = 4.56 \times 10^5$$

$$\text{Re}_5 = \frac{u_\infty L_5}{\nu} = \frac{60 \text{ m/s} \times 0.250 \text{ m}}{26.41 \times 10^{-6} \text{ m}^2/\text{s}} = 5.70 \times 10^5$$

Since $\text{Re}_4 < \text{Re}_{x,c} = 5 \times 10^5$, the flow is *laminar* over the first four heaters, and $\bar{h}_{1\text{-}4}$ may be estimated from Eq. 17.26 where

$$\overline{\text{Nu}}_4 = \frac{\bar{h}_{1\text{-}4}L_4}{k} = 0.664\,\text{Re}_4^{1/2}\,\text{Pr}^{1/3} = 0.664(4.56 \times 10^5)^{1/2}(0.69)^{1/3} = 396$$

$$\bar{h}_{1\text{-}4} = \frac{396 \times 0.0338 \text{ W/m} \cdot \text{K}}{0.200 \text{ m}} = 67 \text{ W/m}^2 \cdot \text{K}$$

In contrast, since $\text{Re}_5 > \text{Re}_{x,c}$, the fifth heater is characterized by *mixed boundary layer conditions* and $\bar{h}_{1\text{-}5}$ must be obtained from Eq. 17.31 where

$$\overline{\text{Nu}}_5 = \frac{\bar{h}_{1\text{-}5}L_5}{k} = (0.037\,\text{Re}_5^{4/5} - 871)\text{Pr}^{1/3} = [0.037(5.70 \times 10^5)^{4/5} - 871](0.69)^{1/3} = 546$$

$$\bar{h}_{1\text{-}5} = \frac{546 \times 0.0338 \text{ W/m} \cdot \text{K}}{0.250 \text{ m}} = 74 \text{ W/m}^2 \cdot \text{K}$$

The rate of heat transfer from the fifth heater, and thus the electrical power required, is then

$$q_{\text{conv},5} = (74 \text{ W/m}^2 \cdot \text{K} \times 0.250 \text{ m} - 67 \text{ W/m}^2 \cdot \text{K} \times 0.200 \text{ m})1 \text{ m}(230 - 25)^\circ\text{C} = 1050 \text{ W} \quad \triangleleft$$

***Comments:*** The variation of the local convection coefficient along the flat plate may be determined from Eqs. 17.23 and 17.28 for laminar and turbulent flow, respectively, and the results are represented by the solid curves in the graph below.

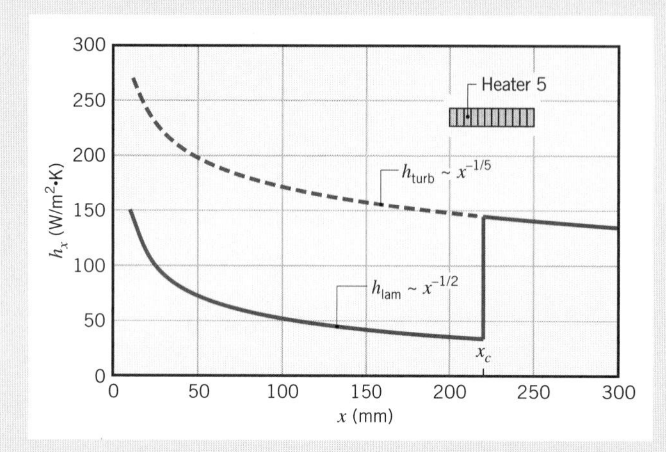


*Figure E17.3b*

The $x^{-1/2}$ decrease of the laminar convection coefficient is presumed to conclude abruptly at $x_c = 220$ mm, where transition yields more than a fourfold increase in the local convection coefficient. For $x > x_c$, the decrease in the convection coefficient is more gradual ($x^{-1/5}$). The dashed line for $h_{turb}$ would apply if fully turbulent conditions existed over the plate.

### 17.2.2 The Cylinder in Cross Flow

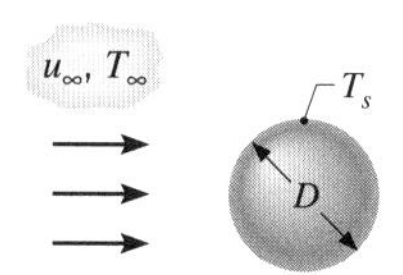


*Figure 17.10*

Another common external flow involves fluid motion normal to the axis of a circular cylinder. Here we consider the long cylinder of diameter $D$ with a uniform surface temperature $T_s$ experiencing cross flow by a free stream fluid of temperature $T_\infty$ with a uniform *upstream* velocity $u_\infty$ (Fig. 17.10).

In Sec. 14.9 you were introduced to the structure of the flow field and the hydrodynamic boundary layer characteristics that influenced the drag coefficient, which was shown to be a function of the Reynolds number based upon the cylinder diameter as the characteristic length

$$\mathrm{Re}_D \equiv \frac{\rho u_\infty D}{\mu} = \frac{u_\infty D}{\nu} \tag{17.33}$$

You learned that the free stream fluid is brought to rest at the *forward stagnation point,* and the thin hydrodynamic boundary layer begins to grow as the flow moves toward the rear of the cylinder. Depending upon the Reynolds number, a transition from *laminar* to *turbulent* conditions can occur. This transition influences the location of *separation* and the formation of the *wake* in the downstream region of the flow (see Figs. 14.16 and 14.17).

As you would expect from our understanding of boundary layer flow for the flat plate, the nature of the hydrodynamic boundary layer strongly influences the formation of the *thermal* boundary layer, and hence the variation of the *local* convection coefficient over the cylinder. The effects of transition, separation, and the formation of the wake control the temperature profile in a complicated manner, and the convection coefficient shows marked changes over the cylinder surface.

Correlations are available for the local Nusselt number. However, from the standpoint of engineering calculations, we are more interested in the overall average condition. From our discussion in Sec. 17.1.3, we expect to see correlations for the *average* convection coefficient

**Table 17.2** Constants for the Hilpert Correlation, Eq. 17.34, for Circular ($Pr \geq 0.7$) and Noncircular (Gases only) Cylinders in Cross Flow

| Geometry | $Re_D$ | $C$ | $m$ |
|---|---|---|---|
| Circular | 0.4–4 | 0.989 | 0.330 |
| | 4–40 | 0.911 | 0.385 |
| | 40–4000 | 0.683 | 0.466 |
| | 4000–40,000 | 0.193 | 0.618 |
| | 40,000–400,000 | 0.027 | 0.805 |

| Geometry | $Re_D$ | $C$ | $m$ |
|---|---|---|---|
| Square (diamond orientation) | $5 \times 10^3$–$10^5$ | 0.246 | 0.588 |
| Square (flat face to flow) | $5 \times 10^3$–$10^5$ | 0.102 | 0.675 |
| Hexagon (flat face to flow) | $5 \times 10^3$–$1.95 \times 10^4$ | 0.160 | 0.638 |
| | $1.95 \times 10^4$–$10^5$ | 0.0385 | 0.782 |
| Hexagon (corner to flow) | $5 \times 10^3$–$10^5$ | 0.153 | 0.638 |
| Vertical plate | $4 \times 10^3$–$1.5 \times 10^4$ | 0.228 | 0.731 |

with the Nusselt number as a function of the Reynolds and Prandtl numbers. The *Hilpert correlation* is one of the most widely used and has the form

$$\overline{Nu}_D = \frac{\bar{h}D}{k} = C\,Re_D^m\,Pr^{1/3} \qquad [Pr \geq 0.7] \qquad (17.34)$$

where the cylinder diameter $D$ is the characteristic length for the Nusselt number. The constants $C$ and $m$, which are dependent upon the Reynolds number range, are listed in Table 17.2. All properties are evaluated at the film temperature, $T_f$, Eq. 17.20.

The Hilpert correlation, Eq. 17.34, may also be used for *gas flow* over cylinders of *noncircular cross section,* with the characteristic length $D$ and the constants obtained from Table 17.2.

The *Churchill-Bernstein correlation* is a single comprehensive equation that covers a wide range of Reynolds and Prandtl numbers. The equation is recommended for all $Re_D\,Pr > 0.2$ and has the form

$$\overline{Nu}_D = 0.3 + \frac{0.62\,Re_D^{1/2}\,Pr^{1/3}}{[1 + (0.4/Pr)^{2/3}]^{1/4}}\left[1 + \left(\frac{Re_D}{282{,}000}\right)^{5/8}\right]^{4/5} \qquad [Re_D\,Pr > 0.2] \qquad (17.35)$$

where all properties are evaluated at the film temperature. This correlation is normally preferred, unless the simplicity of the Hilpert equation is advantageous.

## *Example 17.4* Cylindrical Test Section: Measurement of the Convection Coefficient

Experiments have been conducted to measure the convection coefficient on a polished metallic cylinder 12.7 mm in diameter and 94 mm long (Fig. E17.4*a*). The cylinder is heated internally by an electrical resistance heater and is subjected to a cross flow of air in a low-speed wind tunnel. Under a specific set of operating conditions for which the free stream air velocity and temperature were maintained at $u_\infty$ = 10 m/s and 26.2°C, respectively, the heater power dissipation was measured to be $P_e$ = 46 W, while the average cylinder surface temperature was determined to be $T_s$ = 128.4°C. It is estimated that 15% of the power dissipation is lost by conduction through the endpieces.

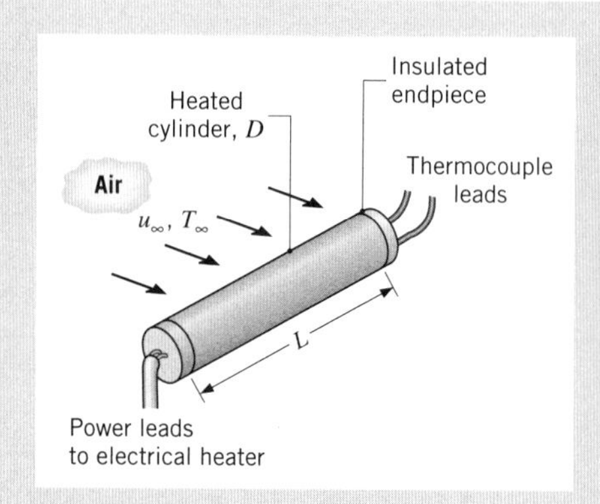


**(a)** Determine the convection heat transfer coefficient from the experimental observations.
**(b)** Compare the experimental result with the convection coefficient computed from an appropriate correlation.

*Figure E17.4a*

## Solution

***Known:*** Operating conditions for a heated cylinder.

***Find:***

**(a)** Convection coefficient associated with the operating conditions.
**(b)** Convection coefficient from an appropriate correlation.

***Schematic and Given Data:***

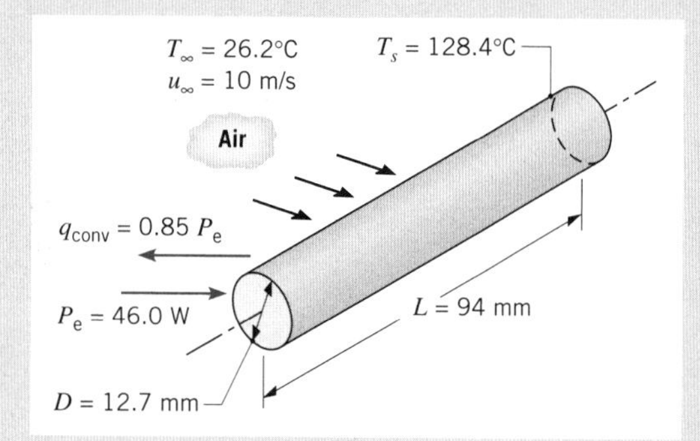


***Assumptions:***
1. Steady-state conditions.
2. Uniform cylinder surface temperature.
3. Negligible radiation exchange with surroundings.

*Figure E17.4b*

***Properties:*** Table HT-3, air ($T_f \approx 350$ K): $\nu = 20.92 \times 10^{-6}$ m$^2$/s, $k = 30 \times 10^{-3}$ W/m · K, Pr = 0.700.

***Analysis:***

**(a)** The convection heat transfer coefficient may be determined from the *experimental observations* by using Newton's law of cooling. That is

$$\bar{h} = \frac{q_{\text{conv}}}{A(T_s - T_\infty)}$$

Since 15% of the electrical power is transfered by conduction from the test section, it follows that $q_{\text{conv}} = 0.85P_e$, and with $A = \pi DL$

$$\bar{h} = \frac{0.85 \times 46 \text{ W}}{\pi \times 0.0127 \text{ m} \times 0.094 \text{ m} (128.4 - 26.2)°\text{C}} = 102 \text{ W/m}^2 \cdot \text{K} \triangleleft$$

**(b)** Using the *Churchill-Bernstein correlation,* Eq. 17.35

$$\overline{\text{Nu}}_D = 0.3 + \frac{0.62\,\text{Re}_D^{1/2}\,\text{Pr}^{1/3}}{[1 + (0.4/\text{Pr})^{2/3}]^{1/4}} \left[1 + \left(\frac{\text{Re}_D}{282{,}000}\right)^{5/8}\right]^{4/5}$$

With all properties evaluated at $T_f$, Pr = 0.70 and

$$\text{Re}_D = \frac{u_\infty D}{\nu} = \frac{10 \text{ m/s} \times 0.0127 \text{ m}}{20.92 \times 10^{-6} \text{ m}^2\text{/s}} = 6071$$

Note that $Re_D\, Pr = 6071 \times 0.700 = 4250 > 0.2$, so that the correlation is within the recommended range. Hence, the Nusselt number and the convection coefficient are

$$\overline{Nu}_D = 0.3 + \frac{0.62(6071)^{1/2}(0.70)^{1/3}}{[1 + (0.4/0.70)^{2/3}]^{1/4}}\left[1 + \left(\frac{6071}{282{,}000}\right)^{5/8}\right]^{4/5} = 40.6$$

$$\bar{h} = \overline{Nu}_D \frac{k}{D} = 40.6 \frac{0.30\ \mathrm{W/m \cdot K}}{0.0127\ \mathrm{m}} = 96\ \mathrm{W/m^2 \cdot K} \quad \triangleleft$$

***Comments:***

1. The Hilpert correlation, Eq. 17.34, is also appropriate for estimating the convection coefficient

$$\overline{Nu}_D = C\, Re_D^m\, Pr^{1/3}$$

With all properties evaluated at the film temperature, $Re_D = 6071$ and $Pr = 0.70$. Hence, from Table 17.2, find for the given Reynolds number that $C = 0.193$ and $m = 0.618$. The Nusselt number and the convection coefficient are then

$$\overline{Nu}_D = 0.193(6071)^{0.618}(0.700)^{0.333} = 37.3$$

$$\bar{h} = \overline{Nu}_D \frac{k}{D} = 37.3 \frac{0.030\ \mathrm{W/m \cdot K}}{0.0127\ \mathrm{m}} = 88\ \mathrm{W/m^2 \cdot K}$$

2. Uncertainties associated with measuring the air velocity, estimating the heat transfer from cylinder ends, and averaging the cylinder surface temperature, which varies axially and circumferentially, render the experimental result accurate to no better than 15%. Accordingly, calculations based on the two correlations used here are within the experimental uncertainty of the measured result.

### 17.2.3 The Sphere

Boundary layer effects associated with flow over a sphere are much like those for the circular cylinder, with transition and separation both playing prominent roles in influencing the variation of the local convection coefficient. From the standpoint of engineering calculations, our interest is in the average condition for the spherical surface. For this purpose, the *Whitaker correlation* is recommended and has the form

$$\overline{Nu}_D = 2 + (0.4\, Re_D^{1/2} + 0.06\, Re_D^{2/3})Pr^{0.4}\left(\frac{\mu}{\mu_s}\right)^{1/4} \quad \left[\begin{matrix} 0.71 < Pr < 380 \\ 3.5 < Re_D < 7.6 \times 10^4 \end{matrix}\right] \tag{17.36}$$

Note that for this correlation, $\mu_s$ is evaluated at the surface temperature $T_s$, and the remaining properties are evaluated at the free stream temperature $T_\infty$. The property ratio $(\mu/\mu_s)$ accounts for the nonconstant property effects in the boundary layer.

### *Example 17.5* Time to Cool a Sphere in an Air Stream

The decorative plastic film on a copper sphere of 10 mm in diameter is cured in an oven at 75°C. Upon removal from the oven, the sphere is subjected to an air stream at 1 atm and 23°C having a velocity of 10 m/s. Estimate how long it will take to cool the sphere to 35°C.

**Solution**

***Known:*** Sphere cooling in an air stream.

***Find:*** Time $t$ required to cool from $T_i = 75°C$ to $T(t) = 35°C$.

***Schematic and Given Data:***

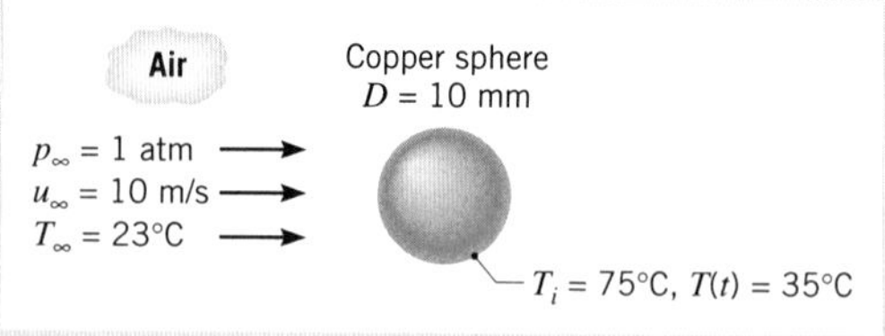

Air

$p_\infty = 1$ atm

$u_\infty = 10$ m/s

$T_\infty = 23°C$

Copper sphere $D = 10$ mm

$T_i = 75°C$, $T(t) = 35°C$

***Assumptions:***

1. Negligible thermal resistance and capacitance for the plastic film.
2. Spatially isothermal sphere with $Bi \ll 1$.
3. Negligible radiation effects.

*Figure E17.5*

***Properties:*** Table HT-1, copper ($\overline{T}_s = 328$ K): $\rho = 8933$ kg/m$^3$, $k = 399$ W/m · K, $c = 387$ J/kg · K. Table HT-3, air ($T_\infty = 296$ K): $\mu = 181.6 \times 10^{-7}$ N · s/m$^2$, $\nu = 15.36 \times 10^{-6}$ m$^2$/s, $k = 0.0258$ W/m · K, Pr = 0.709. Table HT-3, air ($T_s \approx 328$ K): $\mu = 197.8 \times 10^{-7}$ N · s/m$^2$.

***Analysis:*** The time required to complete the cooling process may be obtained from results for a lumped capacitance (see Comment 1). In particular, from Eq. 16.84

$$t = \frac{\rho V c}{\bar{h} A_s} \ln \frac{T_i - T_\infty}{T - T_\infty}$$

or, with $V = \pi D^3/6$ and $A_s = \pi D^2$

$$t = \frac{\rho c D}{6\bar{h}} \ln \frac{T_i - T_\infty}{T - T_\infty}$$

To estimate the average convection coefficient, use the *Whitaker correlation*, Eq. 17.36

$$\overline{\text{Nu}}_D = 2 + (0.4\,\text{Re}_D^{1/2} + 0.06\,\text{Re}_D^{2/3})\text{Pr}^{0.4}\left(\frac{\mu}{\mu_s}\right)^{1/4}$$

where the Reynolds number is

$$\text{Re}_D = \frac{u_\infty D}{\nu} = \frac{10 \text{ m/s} \times 0.01 \text{ m}}{15.36 \times 10^{-6} \text{ m}^2/\text{s}} = 6510$$

Hence the Nusselt number and the convection coefficient are

$$\overline{\text{Nu}}_D = 2 + [0.4(6510)^{1/2} + 0.06(6510)^{2/3}](0.709)^{0.4} \times \left(\frac{181.6 \times 10^{-7} \text{ N} \cdot \text{s/m}^2}{197.8 \times 10^{-7} \text{ N} \cdot \text{s/m}^2}\right)^{1/4} = 47.4$$

$$\bar{h} = \overline{\text{Nu}}_D \frac{k}{D} = 47.4 \frac{0.0258 \text{ W/m} \cdot \text{K}}{0.01 \text{ m}} = 122 \text{ W/m}^2 \cdot \text{K}$$

The time required for cooling is then

$$t = \frac{8933 \text{ kg/m}^3 \times 387 \text{ J/kg} \cdot \text{K} \times 0.01 \text{ m}}{6 \times 122 \text{ W/m}^2 \cdot \text{K}} \ln\left(\frac{75 - 23}{35 - 23}\right) = 69.2 \text{ s} \quad \triangleleft$$

***Comments:***

1. The validity of the lumped capacitance method may be determined by calculating the Biot number. With Eqs. 16.89 and 16.90

$$Bi = \frac{\bar{h} L_c}{k_s} = \frac{\bar{h}(r_o/3)}{k_s} = \frac{122 \text{ W/m}^2 \cdot \text{K} \times 0.005 \text{ m}/3}{399 \text{ W/m} \cdot \text{K}} = 5.1 \times 10^{-4}$$

and since $Bi < 0.1$, the criterion is satisfied.

2. Note that the thermophysical properties of copper and air corresponding to the average surface temperature were evaluated at $\overline{T}_s = (T_i + T(t))/2 = (75 + 35)°\text{C}/2 = 328$ K.
3. Although their definitions are similar, the Nusselt number is defined in terms of the thermal conductivity of the fluid, whereas the Biot number is defined in terms of the thermal conductivity of the solid.

**Table 17.3** Summary of Convection Heat Transfer Correlations for External Flow

| Flow | Coefficient | Correlation[a] | | Range of Applicability |
|---|---|---|---|---|
| ***Flat plate*** | | | | |
| *Laminar* | — | $\delta = 5x\,\mathrm{Re}_x^{-1/2}$ | (17.21) | |
| | Local | $\mathrm{Nu}_x = 0.332\,\mathrm{Re}_x^{1/2}\,\mathrm{Pr}^{1/3}$ | (17.23) | $0.6 \le \mathrm{Pr} \le 50$ |
| | — | $\delta_t = \delta\,\mathrm{Pr}^{-1/3}$ | (17.24) | |
| | Average | $\overline{\mathrm{Nu}}_L = 0.664\,\mathrm{Re}_L^{1/2}\,\mathrm{Pr}^{1/3}$ | (17.26) | $0.6 \le \mathrm{Pr} \le 50$ |
| *Turbulent* | Local | $\delta = 0.37x\,\mathrm{Re}_x^{-1/5}$ | (17.27) | $\mathrm{Re}_x \le 10^8$ |
| | Local | $\mathrm{Nu}_x = 0.0296\,\mathrm{Re}_x^{4/5}\,\mathrm{Pr}^{1/3}$ | (17.28) | $\mathrm{Re}_x \le 10^8$, $0.6 \le \mathrm{Pr} \le 60$ |
| | Average | $\overline{\mathrm{Nu}}_L = 0.037\,\mathrm{Re}_L^{4/5}\,\mathrm{Pr}^{1/3}$ | (17.32) | $\mathrm{Re}_{x,c} = 0$, $0.6 \le \mathrm{Pr} \le 60$ |
| *Mixed* | Average | $\overline{\mathrm{Nu}}_L = (0.037\,\mathrm{Re}_L^{4/5} - 871)\mathrm{Pr}^{1/3}$ | (17.31) | $\mathrm{Re}_{x,c} = 5 \times 10^5$, $10^5 \le \mathrm{Re}_L \le 10^8$<br>$0.6 \le \mathrm{Pr} \le 60$ |
| **Cylinders**[b] | Average | $\overline{\mathrm{Nu}}_D = C\,\mathrm{Re}_D^m\,\mathrm{Pr}^{1/3}$ (Table 7.2) | (17.34) | $\mathrm{Pr} \ge 0.70$ |
| | Average | $\overline{\mathrm{Nu}}_D = 0.3 + \{0.62\,\mathrm{Re}_D^{1/2}\,\mathrm{Pr}^{1/3} \times [1 + (0.4/\mathrm{Pr})^{2/3}]^{-1/4}\} \times [1 + (\mathrm{Re}_D/282{,}000)^{5/8}]^{4/5}$ | (17.35) | $\mathrm{Re}_D\,\mathrm{Pr} > 0.2$ |
| **Sphere** | Average | $\overline{\mathrm{Nu}}_D = 2 + (0.4\,\mathrm{Re}_D^{1/2} + 0.06\,\mathrm{Re}_D^{2/3})\mathrm{Pr}^{0.4}(\mu/\mu_s)^{1/4}$ | (17.36) | $3.5 < \mathrm{Re}_D < 7.6 \times 10^4$<br>$0.71 < \mathrm{Pr} < 380$ |

[a]Thermophysical properties are evaluated at the film temperature, $T_f = (T_\infty + T_s)/2$, for all the correlations except Eq. 17.36. For that correlation, properties are evaluated at the free stream temperature $T_\infty$ or at the surface temperature $T_s$ if designated with the subscript $s$.

[b]For the cylinder with noncircular cross section, use Eq. 17.34 with the constants listed in Table 17.2.

### 17.2.4 Guide for Selection of External Flow Correlations

In this section you have been introduced to empirical correlations to estimate the convection coefficients for forced convection flow over flat plates, cylinders, and spheres. For your convenience in selecting appropriate correlations for your problems, the recommended correlations have been summarized in Table 17.3. While specific conditions are associated with each of the correlations, you are reminded to follow the rules for performing convection calculations outlined in Sec. 17.1.3.

## 17.3 Internal Flow

In the previous section you saw that an external flow, such as for the flat plate, is one for which boundary layer development on a surface continues without external constraints. In contrast, for internal flow in a pipe or tube, the fluid is constrained by a surface, and hence eventually the boundary layer development will be constrained. In Chap. 14 you learned that when flow enters a tube, a hydrodynamic boundary layer forms in the *entrance region,* growing in thickness to eventually fill the tube. Beyond this location, referred to as the *fully developed region,* the velocity profile no longer changes in the flow direction.

We begin by considering thermal boundary layer formation in the entrance and fully developed regions, and how the convection coefficient is determined from the resulting temperature profile. We will introduce empirical correlations to estimate convection coefficients for laminar and turbulent flows in the fully developed region, deferring consideration of correlations for the entrance region to a more advanced course in heat transfer.

### 17.3.1 Hydrodynamic and Thermal Considerations

The development of the boundary layer for *laminar* flow in a circular tube is represented in Fig. 17.11*a* (see also Fig. 14.3). In Sec. 14.1.2 you learned that because of viscous effects, the uniform velocity profile at the entrance will gradually change to a parabolic distribution as the boundary layer $\delta$ begins to fill the tube in the entrance region. Beyond the *hydrodynamic entrance length,* $x_{fd,h}$, the velocity profile no longer changes, and we speak of the flow as *hydrodynamically fully developed*. The extent of the entrance region, as well as the shape of the velocity profile, depends upon Reynolds number, which for internal flow has the form

$$\mathrm{Re}_D = \frac{\rho u_m D}{\mu} = \frac{u_m D}{\nu} = \frac{4\dot{m}}{\pi D \mu} \tag{17.37}$$

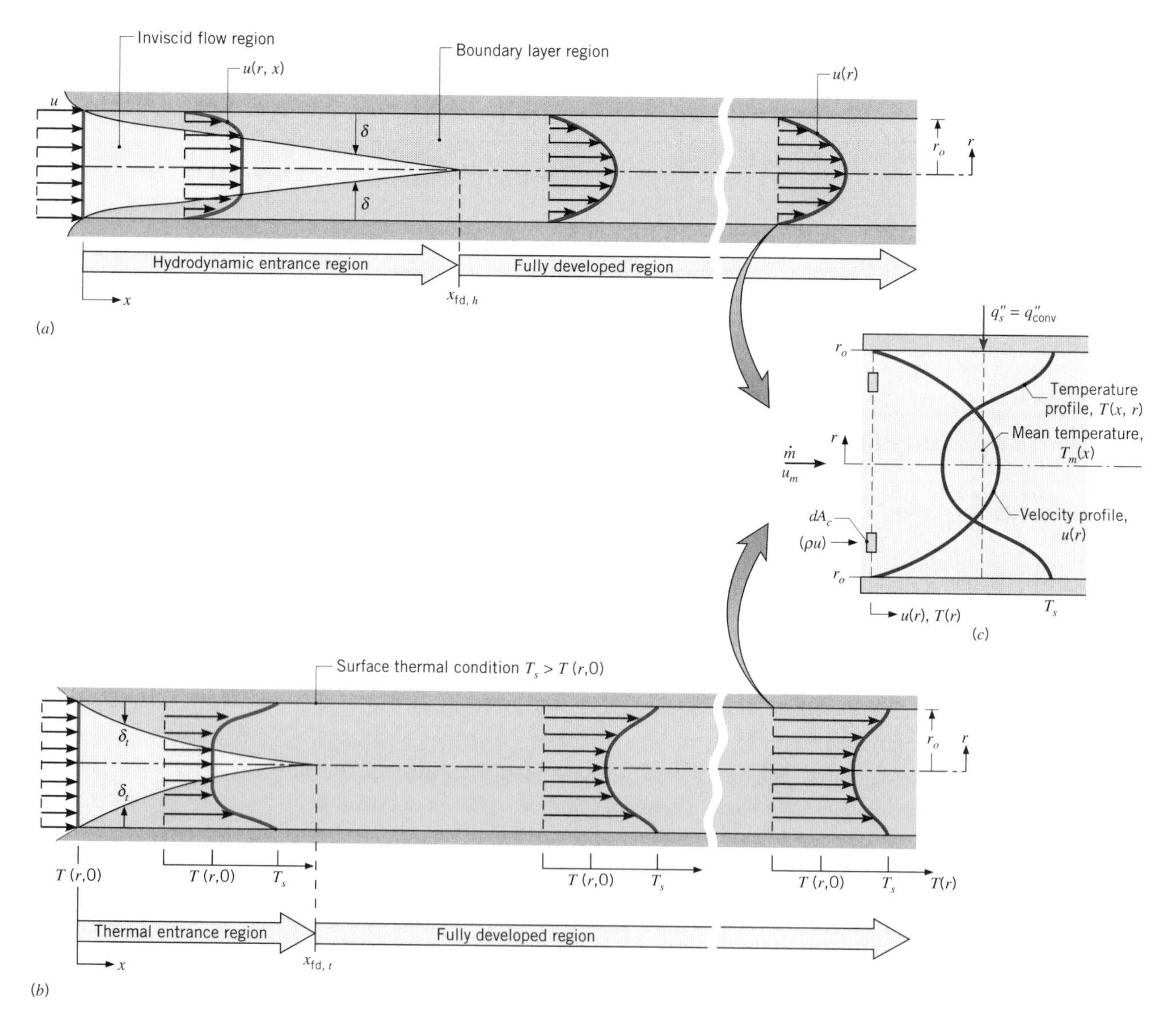


*Figure 17.11* Boundary layer development for laminar flow in a circular tube: (*a*) The hydrodynamic boundary layer and velocity profiles. (*b*) The thermal boundary layer and temperature profiles for *surface thermal condition: constant temperature,* $T_s$. (*c*) Velocity and temperature profiles for determining the mean (average) temperature at a location $x$.

where $u_m$ is the *mean* (average) *velocity; D*, the tube diameter, is the *characteristic length;* and $\dot{m}$ is the mass flow rate. In a fully developed flow, the *critical Reynolds number* corresponding to the onset of turbulence is

$$\text{Re}_{D,c} \approx 2300 \tag{17.38}$$

although much larger Reynolds numbers ($\text{Re}_{D,c} \approx 10{,}000$) are needed to achieve fully turbulent conditions. For *laminar flow* ($\text{Re}_D \leq 2300$), the *hydrodynamic entry length* has the form

$$\left(\frac{x_{fd,h}}{D}\right)_{\text{lam}} \leq 0.05\,\text{Re}_D \tag{17.39}$$

while for *turbulent flow,* the entry length is approximately independent of Reynolds number and that, as a first approximation

$$10 \leq \left(\frac{x_{fd,h}}{D}\right)_{\text{turb}} \leq 60 \tag{17.40}$$

METHODOLOGY UPDATE

For the purposes of this text, we shall assume fully developed turbulent flow for $(x/D) > 10$.

If fluid enters the tube of Fig. 17.11*b* at $x = 0$ with a uniform temperature $T(r, 0)$ that is less than the *constant* tube surface temperature, $T_s$, convection heat transfer occurs, and a thermal boundary layer begins to develop. In the ***thermal entrance region,*** the temperature of the central portion of the flow outside the *thermal boundary layer,* $\delta_t$, remains unchanged, but in the boundary layer, the temperature increases sharply to that of the tube surface. At the ***thermal entrance length,*** $x_{fd,t}$, the thermal boundary layer has filled the tube, the fluid at the centerline begins to experience heating, and the *thermally fully developed flow* condition has been reached.

*thermal entrance region*

*thermal entrance length*

For *laminar flow,* the thermal entry length may be expressed as

$$\left(\frac{x_{fd,t}}{D}\right)_{\text{lam}} \leq 0.05\,\text{Re}_D\,\text{Pr} \qquad [\text{Re}_D < 2300] \tag{17.41}$$

From this relation and by comparison of the hydrodynamic and thermal boundary layers of Fig. 17.11*a* and 17.11*b*, it is evident that we have represented a fluid with a $\text{Pr} < 1$ (gas), as the hydrodynamic boundary layer has developed more slowly than the thermal boundary layer ($x_{fd,h} > x_{fd,t}$). For liquids having $\text{Pr} > 1$, the inverse situation would occur.

For *turbulent flow,* conditions are nearly independent of Prandtl number, and to a first approximation the *thermal entrance length* is

$$\left(\frac{x_{fd,t}}{D}\right)_{\text{turb}} = 10 \qquad [\text{Re}_D \geq 10{,}000] \tag{17.42}$$

which is the same criterion as for the hydrodynamic entry length, Eq. 17.40.

In Fig. 17.11*b* we have shown temperature profiles for laminar flow experiencing heating with a *uniform surface temperature condition* ($T_s$ is constant). Note that the temperature gradient at the surface is steepest in the *entrance region,* implying that the convection coefficient in the entrance region is likely to be higher than in the fully developed region.

Since the fluid is being heated within the tube, we know that the mean (average) temperature of the fluid is increasing in the flow direction, and consequently *the temperature profile is changing shape.* Thermal conditions in the ***thermally fully developed region*** shown in Fig. 17.11*b* are characterized by three key features: the *mean temperature* is increasing (or decreasing if a cooling condition), the *relative shape* of the temperature profiles is constant, and the *convection coefficient* is constant. We'll now explain these features, and then introduce appropriate forms of the convection rate and energy balance equations for fully developed region flow analyses.

*thermal fully developed region*

*The Mean Temperature.* As shown in Fig. 17.11*c*, the temperature and velocity profiles at a *particular* location in the flow direction $x$ each depend on radius, $r$. The *mean* temperature of the fluid, also referred to as the average or bulk temperature, shown on the figure as $T_m(x)$, is defined in terms of the energy transported by the fluid as it moves past location $x$. For incompressible flow, with constant specific heat $c_p$, the *mean temperature* is found from

$$T_m = \frac{\int_{A_c} uTdA_c}{u_m A_c} \tag{17.43}$$

where $u_m$ is the mean velocity. For a circular tube, $dA_c = 2\pi r dr$, and it follows that

*mean temperature*

$$T_m = \frac{2}{u_m r_o^2}\int_0^{r_o} uTrdr \tag{17.44}$$

The *mean temperature* is the fluid reference temperature used for determining the convection heat rate with Newton's law of cooling and the overall energy balance.

*Newton's Law of Cooling.* To determine the convective heat flux at the tube surface, Fig. 17.11*c*, Newton's law of cooling, also referred to as the *convection rate equation,* is expressed as

*convection rate equation*

$$q_s'' = q_{\text{conv}}'' = h(T_s - T_m) \tag{17.45}$$

where $h$ is the *local* convection coefficient. Depending upon the method of surface heating (cooling), $T_s$ can be a constant or can vary, but the mean temperature will always change in the flow direction. Still, the *convection coefficient is a constant for the fully developed conditions* we examine next.

*Fully Developed Conditions.* The temperature profile can be conveniently represented as the dimensionless ratio $(T_s - T)/(T_s - T_m)$. While the temperature profile $T(r)$ continues to change with $x$, the *relative shape* of the profile given by this *temperature ratio* is independent of $x$ for fully developed conditions. The requirement for such a condition is mathematically stated as

$$\frac{\partial}{\partial x}\left[\frac{T_s(x) - T(r, x)}{T_s(x) - T_m(x)}\right]_{fd,t} = 0 \tag{17.46}$$

where $T_s$ is the tube surface temperature, $T$ is the local fluid temperature, and $T_m$ is the mean temperature as shown in Fig. 17.12*a*. Since the temperature ratio is independent of $x$, the

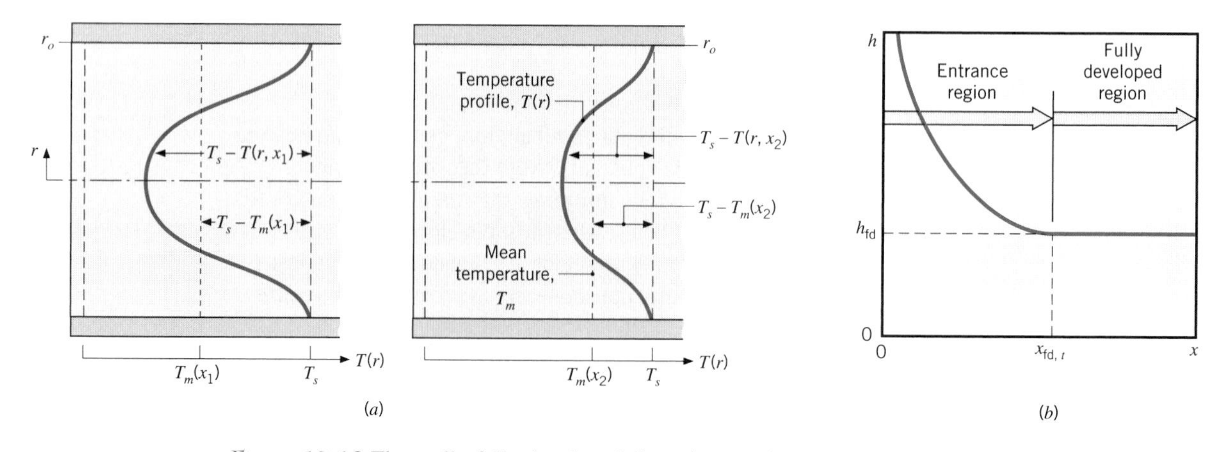


*Figure 17.12* Thermally fully developed flow characteristics for constant surface temperature heating. (*a*) Relative shape of the temperature profile remains unchanged in the flow direction ($x_2 > x_1$). (*b*) Convection coefficient is constant for $x > x_{fd,t}$.

derivative of this ratio with respect to $r$ must also be independent of $x$. Evaluating this derivative at the tube surface (note that $T_s$ and $T_m$ are constants insofar as differentiation with respect to $r$ is concerned), we obtain

$$\frac{\partial}{\partial r}\left(\frac{T_s - T}{T_s - T_m}\right)\bigg|_{r=r_o} = \frac{-\partial T/\partial r|_{r=r_o}}{T_s - T_m} \neq f(x)$$

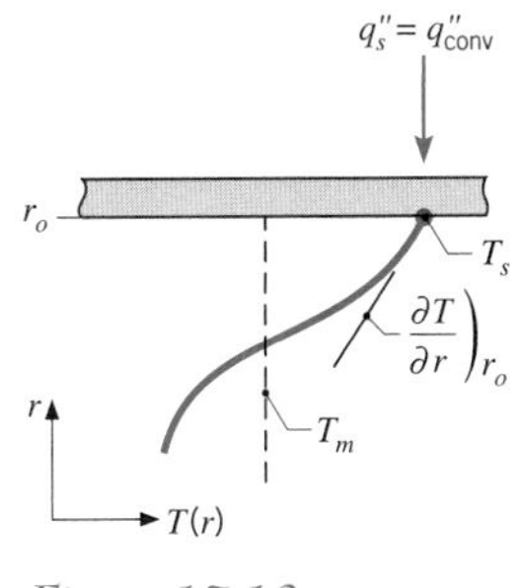

Figure 17.13

Substituting for $\partial T/\partial r$ from Fourier's law, which, from Fig. 17.1, is of the form

$$q_s'' = k\frac{\partial T}{\partial r}\bigg|_{r=r_o}$$

and for $q_s''$ from Newton's law of cooling, Eq. 17.45, we obtain

$$\frac{h}{k} \neq f(x) = \text{constant} \tag{17.47}$$

Hence, *in the thermally fully developed flow of a fluid with constant properties, the local convection coefficient is a constant, independent of x.* Equation 17.47 is not satisfied in the entrance region where $h$ varies with $x$.

Because the thermal boundary layer thickness is zero at the tube entrance, the coefficient is extremely large near $x = 0$, and decreases markedly as the boundary layer develops, until the constant value associated with the fully developed conditions is reached as shown in Fig. 17.12*b*.

## 17.3.2 Energy Balances and Methods of Heating

Because the flow in a tube is completely enclosed, an energy balance may be applied to determine the convection heat transfer rate, $q_{\text{conv}}$, in terms of the difference in temperatures at the tube inlet and outlet. From an energy balance applied to a differential control volume in the tube, we will determine how the mean temperature $T_m(x)$ varies in the flow direction with position along the tube for two *surface thermal conditions* (methods of heating/cooling).

*Overall Tube Energy Balance.* Consider the tube flow of Fig. 17.14*a*. Fluid moves at a constant flow rate $\dot{m}$, and convection heat transfer occurs along the wall surface. Assuming that fluid kinetic and potential energy changes are negligible, there is no shaft work, and regarding $c_p$ as constant, the energy rate balance, Eq. 5.11b, reduces to give

$$q_{\text{conv}} = \dot{m}c_p(T_{m,o} - T_{m,i}) \tag{17.48}$$

where $T_m$ denotes the mean fluid temperature and the subscripts $i$ and $o$ denote inlet and outlet conditions, respectively. It is important to recognize that this *overall energy balance*

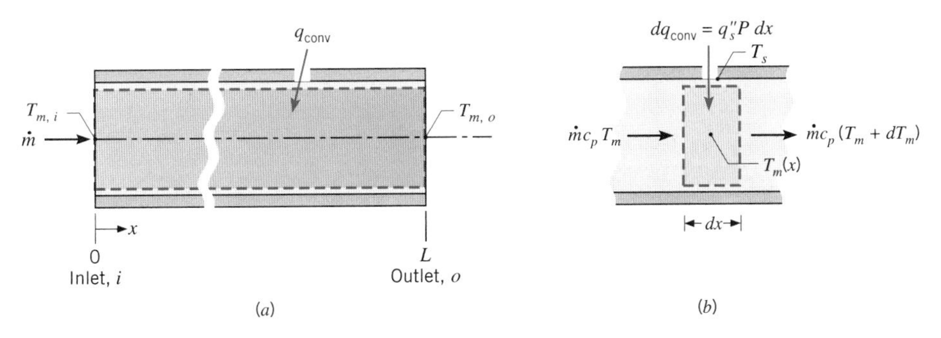


*Figure 17.14* Energy balances for steady flow in a tube. (*a*) Overall tube balance for the convection heat rate, Eq. 17.48. (*b*) Balance on a differential control volume for determining $T_m(x)$, Eq. 17.50.

*is a general expression that applies irrespective of the nature of the surface thermal or tube flow conditions.*

***Energy Balance on a Differential Control Volume.*** We can apply the same analysis to a differential control volume within the tube as shown in Fig. 17.14*b* by writing Eq. 17.48 in differential form

$$dq_{\text{conv}} = \dot{m}c_p dT_m \tag{17.49}$$

We can express the rate of convection heat transfer to the differential element in terms of the surface heat flux as

$$dq_{\text{conv}} = q_s'' P dx \tag{17.50}$$

where $P$ is the surface perimeter. Combining Eqs. 17.49 and 17.50, it follows that

$$q_s'' P dx = \dot{m}c_p dT_m$$

By rearranging this result, we obtain an expression for the axial variation of $T_m$ in terms of the *surface heat flux*

$$\frac{dT_m}{dx} = \frac{q_s'' P}{\dot{m}c_p} \qquad [\text{surface heat flux, } q_s''] \tag{17.51}$$

or, using Newton's law of cooling, Eq. 17.45, with $q_s'' = h(T_s - T_m)$, in terms of the tube wall *surface temperature*

$$\frac{dT_m}{dx} = \frac{P}{\dot{m}c_p} h(T_s - T_m) \qquad [\text{surface temperature, } T_s] \tag{17.52}$$

The manner in which the quantities on the right-hand side of Eqs. 17.51 and 17.52 vary with $x$ should be noted. For a circular tube of uniform diameter ($P = \pi D$), the quantity ($P/\dot{m}c_p$) is a constant. In the fully developed region, the convection coefficient $h$ is also constant, although it varies with $x$ in the entrance region (see Fig. 17.12*b*). Finally, although $T_s$ can be a constant, $T_m$ must always vary with $x$. The solutions to Eqs. 17.51 and 17.52 for $T_m(x)$ depend upon the ***surface thermal condition.*** We will now consider two special cases of interest: *constant surface heat flux* ($q_s''$) and *constant surface temperature* ($T_s$). It is common to find one of these conditions existing in practical applications to a reasonable approximation.

*surface thermal condition*

## Thermal Condition: Constant Surface Heat Flux, $q_s''$

For *constant surface heat flux* thermal condition (Fig. 17.15), we first note that it is a simple matter to determine the total heat transfer rate, $q_{\text{conv}}$. Since $q_s''$ is independent of $x$, it follows that

$$q_{\text{conv}} = q_s''(P \cdot L) \tag{17.53}$$

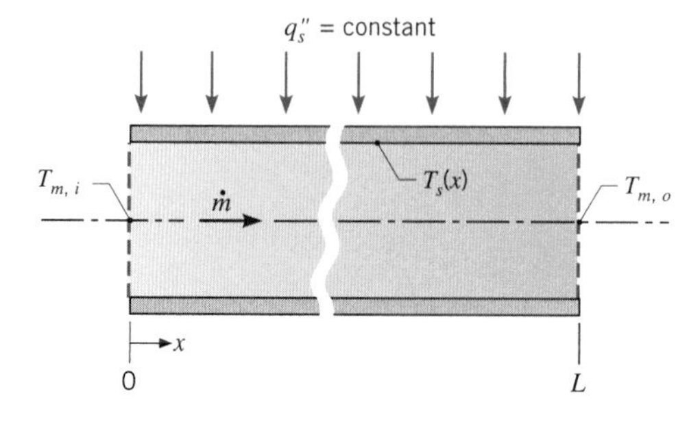


*Figure 17.15* Internal flow through a circular tube with the surface thermal condition corresponding to *constant surface heat flux, $q_s''$.*

This expression can be used with the overall energy balance, Eq. 17.48, to determine the fluid temperature change, $T_{m,o} - T_{m,i}$.

For constant $q_s''$ it also follows that the right-hand side of Eq. 17.51 is a constant independent of $x$. Hence

$$\frac{dT_m}{dx} = \frac{q_s''P}{\dot{m}c_p} = \text{constant}$$

Integrating from $x = 0$ to some axial position $x$, we obtain the *mean temperature distribution, $T_m(x)$*

$$T_m(x) = T_{m,i} + \frac{q_s''P}{\dot{m}c_p}x \qquad [q_s'' = \text{constant}] \tag{17.54}$$

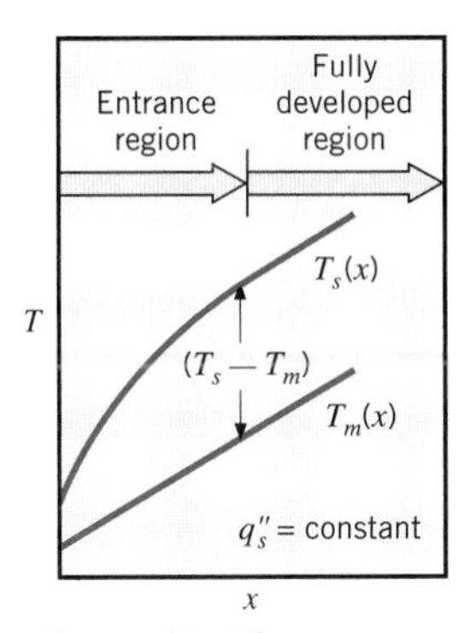

*Figure 17.16*

Accordingly, the mean temperature varies linearly with $x$ along the tube (Fig. 17.16). Moreover, from Newton's law of cooling, Eq. 17.45, we also expect the temperature difference $(T_s - T_m)$ to vary with $x$ as shown in Fig. 17.16. This difference is initially small (due to the large value of $h$ at the entrance) but increases with increasing $x$ due to the decrease in $h$ that occurs as the boundary layer develops. However, in the fully developed region we know that *$h$ is independent of $x$*. Hence from Eq. 17.45 it follows that $(T_s - T_m)$ must also be independent of $x$ in this region.

## *Example 17.6* Thermal Condition: Constant Surface Heat Flux $q_s''$

A system for heating water from an inlet temperature of $T_{m,i} = 20°\text{C}$ to an outlet temperature of $T_{m,o} = 60°\text{C}$ involves passing the water through a tube having inner and outer diameters of 20 and 40 mm. The outer surface of the tube is well insulated, and electrical power dissipation within the wall provides for a uniform volumetric generation rate of $\dot{q} = 10^6\ \text{W/m}^3$.

(a) For a water mass flow rate of $\dot{m} = 0.1$ kg/s, how long must the tube be to achieve the desired outlet temperature?
(b) Do fully developed hydrodynamic and thermal conditions exist in the flow?
(c) If the inner surface temperature of the tube is $T_s = 70°\text{C}$ at the outlet ($x = L$), what is the local convection heat transfer coefficient at the outlet?

### Solution

***Known:*** Internal flow through thick-walled tube having uniform volumetric energy generation.

***Find:***
(a) Length of tube needed to achieve the desired outlet temperature.
(b) Whether fully developed hydrodynamic and thermal conditions exist.
(c) Local convection coefficient at the outlet.

***Schematic and Given Data:***

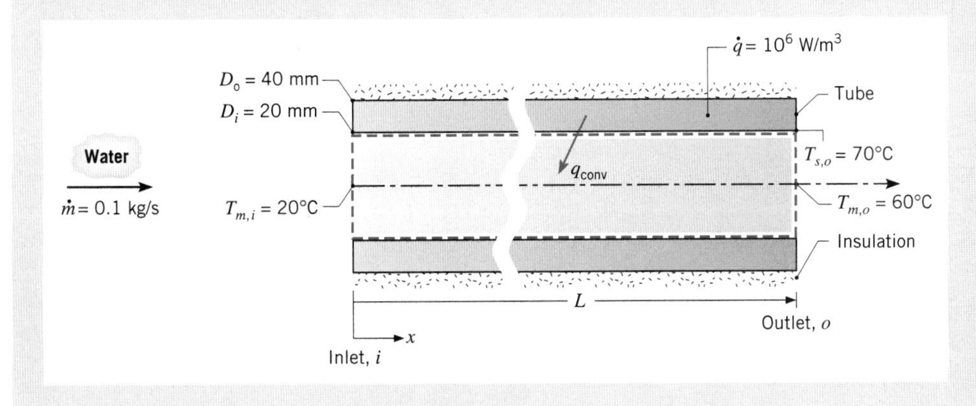


***Assumptions:***
1. Steady-state conditions.
2. Uniform heat flux.
3. Negligible potential energy and kinetic energy effects. No shaft work.
4. Constant properties.
5. Adiabatic outer tube surface.

*Figure E17.6a*

***Properties:*** Table HT-5, water ($\bar{T}_m = (T_{m,i} + T_{m,o})/2 = 313$ K): $c_p = 4.179$ kJ/kg · K, $\mu = 5.56 \times 10^{-4}$ N · s/m$^2$.

***Analysis:***

**(a)** Since the outer surface of the tube is adiabatic, the rate at which energy is generated within the tube wall must equal the rate at which it is convected to the water ($\dot{E}_g = q_{\text{conv}}$)

$$\dot{q}\frac{\pi}{4}(D_o^2 - D_i^2)L = q_{\text{conv}}$$

From the overall tube energy balance, Eq. 17.48, it follows that

$$\dot{q}\frac{\pi}{4}(D_o^2 - D_i^2)L = \dot{m}c_p(T_{m,o} - T_{m,i})$$

Solving for $L$ and substituting numerical values with $c_p$ evaluated at $\bar{T}_m = (T_{m,i} + T_{m,o})/2$, the required tube length is

$$L = \frac{4\dot{m}c_p}{\pi(D_o^2 - D_i^2)\dot{q}}(T_{m,o} - T_{m,i}) = \frac{4 \times 0.1 \text{ kg/s} \times 4179 \text{ J/kg} \cdot \text{K}}{\pi(0.04^2 - 0.02^2)\text{ m}^2 \times 10^6 \text{ W/m}^3}(60 - 20)°\text{C} = 17.7 \text{ m} \quad \triangleleft$$

**(b)** To determine whether fully developed conditions exist, calculate the Reynolds number to characterize the flow. From Eq. 17.37

$$\text{Re}_D = \frac{4\dot{m}}{\pi D\mu} = \frac{4 \times 0.1 \text{ kg/s}}{\pi(0.020 \text{ m})(6.57 \times 10^{-4}\text{N} \cdot \text{s/m}^2)} = 9696$$

Since $\text{Re}_D$ is nearly 10,000, the flow is fully turbulent. The hydrodynamic and thermal entry length is given as $x_{fd}/D = 10$ so that $x_{fd} = 10D = 10 \times 0.020$ m $= 0.2$ m. We conclude that, to a good approximation, fully developed conditions exist over the entire tube since $L \gg x_{fd}$ (17.7 m vs. 0.2 m).

**(c)** From Newton's law of cooling, Eq. 17.45, the local convection coefficient at the tube exit is

$$h_o = \frac{q_s''}{T_{s,o} - T_{m,o}}$$

Assuming that uniform generation in the wall provides a constant surface heat flux, with

$$q_s'' = \frac{q_{\text{conv}}}{\pi D_i L} = \dot{q}\frac{D_o^2 - D_i^2}{4D_i} = 10^6 \text{ W/m}^3\frac{(0.04^2 - 0.02^2)\text{ m}^2}{4 \times 0.02 \text{ m}} = 1.5 \times 10^4 \text{ W/m}^2$$

it follows that the local coefficient at the outlet is

$$h_o = \frac{1.5 \times 10^4 \text{ W/m}^2}{(70 - 60)°\text{C}} = 1500 \text{ W/m}^2 \cdot \text{K} \quad \triangleleft$$

***Comments:***

1. Since conditions are *fully developed over the entire tube,* the local convection coefficient and the temperature difference ($T_s - T_m$) are independent of $x$ for this *constant heat flux condition.* Hence, $h = 1500$ W/m$^2$ · K and ($T_s - T_m$) = 10°C over the entire tube. The inner surface temperature at the tube inlet is then $T_{s,i} = 30$°C. The fluid and tube surface temperature distributions are shown in Fig. E17.6*b*.
2. For the constant surface heat flux condition, the *exact shape* of the temperature profile in the *fully developed region* does not change in the flow direction ($x_2 > x_1$) as illustrated in Fig. E17.6*c*. Compare this behavior to that for *constant surface temperature condition,* Fig. 17.12*a*, where it is the *relative* shape that remains unchanged in the fully developed region.

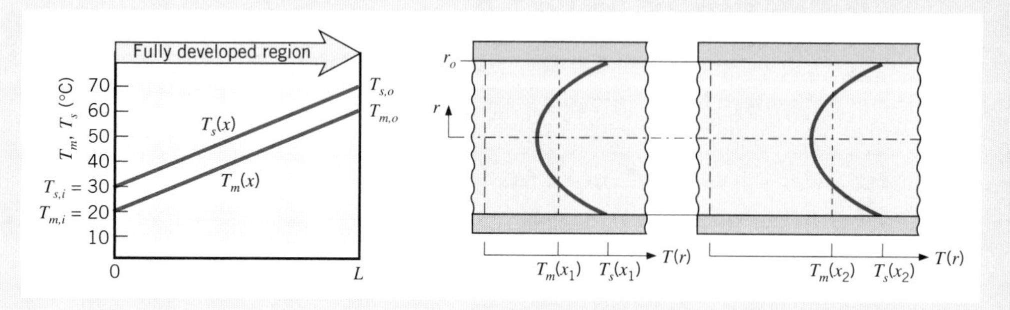


Figure E17.6b,c

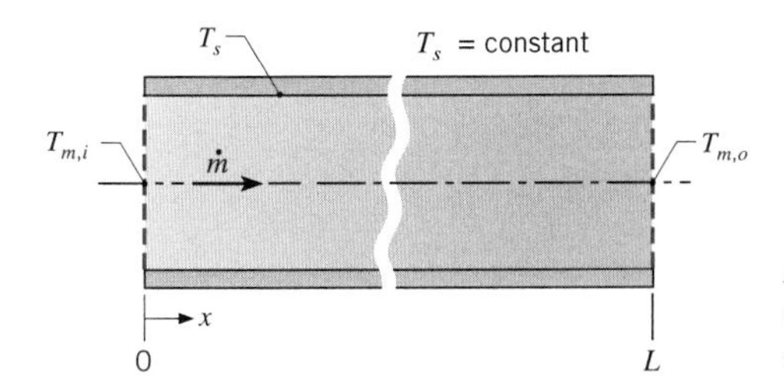


*Figure 17.17* Internal flow through a circular tube with the surface thermal condition corresponding to *constant surface temperature*, $T_s$.

### Thermal Condition: Constant Surface Temperature, $T_s$

Results for the total heat transfer rate and the axial distribution of the mean temperature are entirely different for the *constant surface temperature* condition (Fig. 17.17). Defining $\Delta T$ as $(T_s - T_m)$, Eq. 17.52 may be expressed as

$$\frac{dT_m}{dx} = -\frac{d(\Delta T)}{dx} = \frac{P}{\dot{m}c_p} h\,\Delta T$$

With $P/\dot{m}c_p$ constant, separate variables and integrate from the tube inlet to the outlet

$$\int_{\Delta T_i}^{\Delta T_o} \frac{d(\Delta T)}{\Delta T} = -\frac{P}{\dot{m}c_p}\int_0^L h\,dx$$

or

$$\ln\frac{\Delta T_o}{\Delta T_i} = -\frac{PL}{\dot{m}c_p}\left(\frac{1}{L}\int_0^L h\,dx\right)$$

From the definition of the average convection heat transfer coefficient, Eq. 17.8, it follows that

$$\ln\frac{\Delta T_o}{\Delta T_i} = -\frac{PL}{\dot{m}c_p}\bar{h}_L \qquad [T_s = \text{constant}] \tag{17.55a}$$

where $\bar{h}_L$, or simply $\bar{h}$, is the average value of $h$ for the entire tube. Alternatively, taking the exponent of both sides of the equation

$$\frac{\Delta T_o}{\Delta T_i} = \frac{T_s - T_{m,o}}{T_s - T_{m,i}} = \exp\left(-\frac{PL}{\dot{m}c_p}\bar{h}\right) \qquad [T_s = \text{constant}] \tag{17.55b}$$

If we had integrated from $x = 0$ to some axial position, we obtain the *mean temperature distribution*, $T_m(x)$

$$\frac{T_s - T_m(x)}{T_s - T_{m,i}} = \exp\left(-\frac{Px}{\dot{m}c_p}\bar{h}\right) \qquad [T_s = \text{constant}] \tag{17.56}$$

where $\bar{h}$ is now the average value of $h$ from the tube inlet to $x$. This result tells us that the temperature difference $(T_s - T_m)$ *decreases exponentially* with distance along the tube axis. The axial surface and mean temperature distributions are therefore as shown in Fig. 17.18.

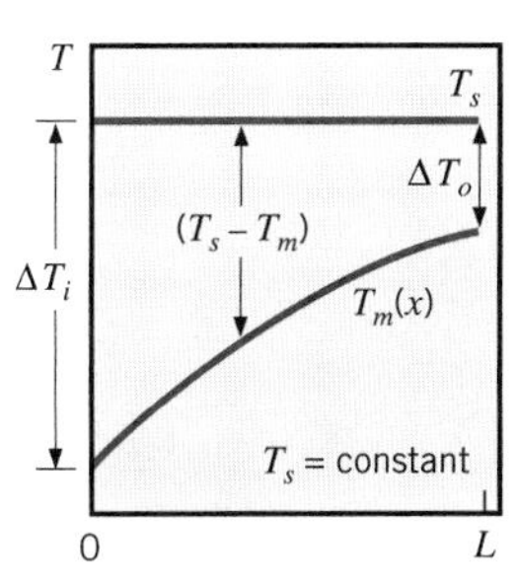


*Figure 17.18*

Determination of an expression for the total heat transfer rate $q_{\text{conv}}$ is complicated by the exponential nature of the temperature decrease. Expressing Eq. 17.48 in the form

$$q_{\text{conv}} = \dot{m}c_p[(T_s - T_{m,i}) - (T_s - T_{m,o})] = \dot{m}c_p(\Delta T_i - \Delta T_o)$$

and substituting for $\dot{m}c_p$ from Eq. 17.55a, we obtain the *convection rate equation*

$$q_{\text{conv}} = \bar{h}A_s\Delta T_{\text{lm}} \qquad [T_s = \text{constant}] \tag{17.57}$$

where $A_s$ is the tube surface area ($A_s = P \cdot L$) and $\Delta T_{lm}$ is the ***log mean temperature difference (LMTD)***

*log mean temperature difference (LMTD)*

$$\Delta T_{lm} \equiv \frac{\Delta T_o - \Delta T_i}{\ln(\Delta T_o / \Delta T_i)} \tag{17.58}$$

Equation 17.57 is a form of Newton's law of cooling for the entire tube, and $\Delta T_{lm}$ is the appropriate *average* of the temperature difference over the tube length. The logarithmic nature of this average temperature difference is due to the exponential nature of the temperature decrease.

A common variation of the foregoing constant surface temperature condition is one for which the outer tube surface is exposed to the freestream temperature of an external fluid, $T_\infty$. This case is treated in the following section.

## *Example 17.7* Thermal Condition: Constant Surface Temperature, $T_s$

Steam condensing on the outer surface of a thin-walled circular tube of 50-mm diameter and 6-m length maintains a uniform surface temperature of 100°C. Water flows through the tube at a rate of $\dot{m} = 0.25$ kg/s, and its inlet and outlet temperatures are $T_{m,i} = 15$°C and $T_{m,o} = 57$°C. What is the average convection coefficient associated with the water flow?

### Solution

***Known:*** Flow rate and inlet and outlet temperatures of water flowing through a tube of prescribed dimensions and surface temperature.

***Find:*** Average convection heat transfer coefficient.

***Schematic and Given Data:***

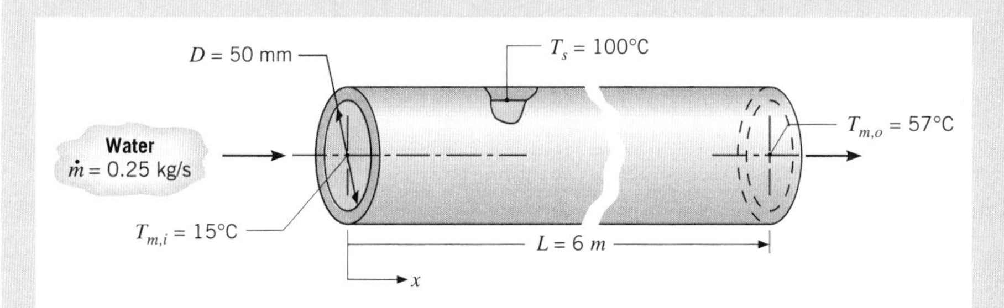


*Figure E17.7*

***Assumptions:***
1. Negligible outer surface convection resistance and tube wall conduction resistance; hence, tube inner surface is at $T_s = 100$°C.
2. Negligible kinetic and potential energy effects. No shaft work.
3. Constant properties.

***Properties:*** Table HT-5, water ($\overline{T}_m = (T_{m,i} + T_{m,o})/2 = 36$°C $= 309$ K): $c_p = 4178$ J/kg · K.

***Analysis:*** Combining the overall tube energy balance, Eq. 17.48, with the convection rate equation, Eq. 17.57, the average convection coefficient is given by

$$\overline{h} = \frac{\dot{m} c_p}{\pi D L} \frac{(T_{m,o} - T_{m,i})}{\Delta T_{lm}}$$

From Eq. 17.58, the log mean temperature difference is

$$\Delta T_{lm} = \frac{(T_s - T_{m,o}) - (T_s - T_{m,i})}{\ln[(T_s - T_{m,o})/(T_s - T_{m,i})]} = \frac{(100 - 57) - (100 - 15)}{\ln[(100 - 57)/(100 - 15)]} = 61.6\text{°C}$$

Hence, the average convection coefficient is

$$\overline{h} = \frac{0.25 \text{ kg/s} \times 4178 \text{ J/kg} \cdot \text{K}}{\pi \times 0.05 \text{ m} \times 6 \text{ m}} \frac{(57 - 15)\text{°C}}{61.6\text{°C}} = 756 \text{ W/m}^2 \cdot \text{K} \triangleleft$$

***Comments:*** Note that the properties for use in the energy balance and convection correlations are evaluated at the *average* mean temperature defined as $\overline{T}_m = (T_{m,i} + T_{m,o})/2$.

### Surface Thermal Condition: External Fluid (CD-ROM)

## 17.3.3 Convection Correlations for Tubes: Fully Developed Region

To use many of the foregoing results for internal flow, the convection coefficients must be known. In this section we present correlations for estimating the coefficients for *fully developed laminar* and *turbulent* flows in *circular* and *noncircular tubes*. The correlations for internal flow are summarized in Table 17.5 (page 438) along with guidelines to facilitate their selection for your application.

### Laminar Flow

The problem of laminar flow ($Re_D < 2300$) in tubes has been treated theoretically, and the results can be used to determine the convection coefficients. For flow in a *circular* tube characterized by *uniform surface heat flux* and *laminar, fully developed conditions,* the *Nusselt number is a constant,* independent of $Re_D$, Pr, and axial location

$$Nu_D = \frac{hD}{k} = 4.36 \qquad [q_s'' = \text{constant}] \tag{17.61}$$

When the thermal surface condition is characterized by a *constant surface temperature,* the results are of similar form, but with a smaller value for the Nusselt number

$$Nu_D = \frac{hD}{k} = 3.66 \qquad [T_s = \text{constant}] \tag{17.62}$$

In using these equations to determine $h$, the thermal conductivity should be evaluated at $T_m$.

**Table 17.4** Nusselt Numbers for Fully Developed Laminar Flow in Noncircular Tubes for Constant $T_s$ and $q_s''$ Surface Thermal Conditions[a]

| | | $Nu_D \equiv \frac{hD_h}{k}$ | |
|---|---|---|---|
| Cross Section | $\frac{b}{a}$ | Constant $q_s''$ | Constant $T_s$ |
| Circle | — | 4.36 | 3.66 |
| Rectangle (a × b) | 1.0 | 3.61 | 2.98 |
| Rectangle (a × b) | 1.43 | 3.73 | 3.08 |
| Rectangle (a × b) | 2.0 | 4.12 | 3.39 |
| Rectangle (a × b) | 3.0 | 4.79 | 3.96 |
| Rectangle (a × b) | 4.0 | 5.33 | 4.44 |
| Rectangle (a × b) | 8.0 | 6.49 | 5.60 |
| Parallel plates | $\infty$ | 8.23 | 7.54 |
| Parallel plates (one Heated, one Insulated) | $\infty$ | 5.39 | 4.86 |
| Triangle | — | 3.11 | 2.47 |

[a]The characteristic length is the hydraulic diameter, $D_h$, Eq. 17.63.

For applications involving convection transport in *noncircular tubes,* to at least a first approximation, the foregoing correlations can be applied by using the ***hydraulic diameter*** as the characteristic length

*hydraulic diameter*

$$D_h \equiv \frac{4A_c}{P} \tag{17.63}$$

where $A_c$ and $P$ are the flow cross-sectional area and the *wetted perimeter,* respectively. It is this diameter that should be used in calculating the Reynolds and Nusselt numbers. This approach is less accurate for noncircular tubes with cross sections characterized by sharp corners. Table 17.4 (previous page) presents correlations covering many of such cases for the same surface thermal conditions associated with the circular tube.

## *Example 17.9* Laminar Flow Application: Solar Collector

One concept used for solar energy collection involves placing a tube at the focal point of a parabolic reflector (concentrator) and passing a fluid through the tube.

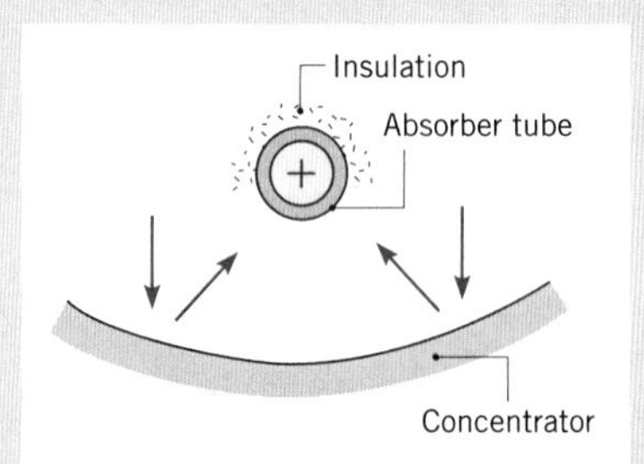


*Figure E17.9a*

The net effect of this arrangement *can be approximated* as one of creating a condition of uniform heating at the surface of the tube. That is, the resulting heat flux to the fluid $q''_s$ can be assumed to be a constant along the circumference and axis of the tube. Consider operation with a tube of diameter $D$ = 60 mm on a sunny day for which $q''_s$ = 2000 W/m$^2$.

**(a)** If pressurized water enters the tube at $\dot{m}$ = 0.01 kg/s and $T_{m,i}$ = 20°C, what tube length $L$ is required to obtain an exit temperature of 80°C?

**(b)** What is the surface temperature at the outlet of the tube, where fully developed conditions can be assumed to exist?

## Solution

***Known:*** Internal flow with uniform surface heat flux.

***Find:***

**(a)** Length of tube $L$ to achieve required heating.

**(b)** Surface temperature $T_s(L)$ at the outlet section, $x = L$.

***Schematic and Given Data:***

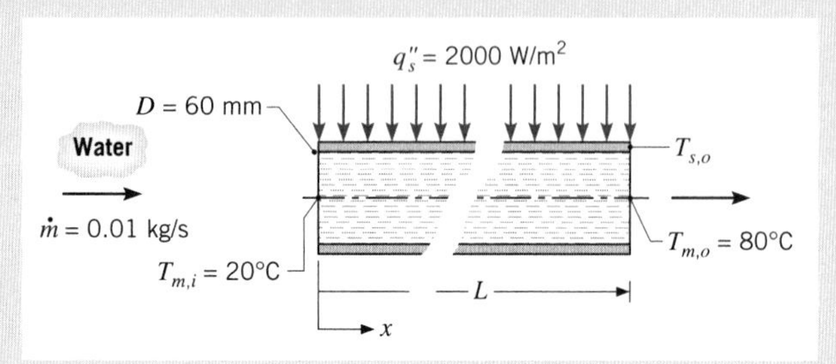


***Assumptions:***

1. Steady-state conditions.
2. Incompressible flow with constant properties.
3. Negligible kinetic and potential energy effects. No shaft work.
4. Constant properties.
5. Fully developed conditions at tube outlet.

*Figure E17.9b*

***Properties:*** Table HT-5, water ($\overline{T}_m = (T_{m,i} + T_{m,o})/2$ = 323 K): $c_p$ = 4181 J/kg · K. Table HT-5, water ($T_{m,o}$ = 353 K): $k$ = 0.670 W/m · K, $\mu = 352 \times 10^{-6}$ N · s/m$^2$, Pr = 2.2.

***Analysis:***

(a) For constant surface heat flux, Eq. 17.53 can be used with the overall tube energy balance, Eq. 17.48, to obtain

$$A_s = \pi DL = \frac{\dot{m}c_p(T_{m,o} - T_{m,i})}{q_s''} \quad \text{or} \quad L = \frac{\dot{m}c_p}{\pi D q_s''}(T_{m,o} - T_{m,i})$$

Substituting numerical values, the required tube length is

$$L = \frac{0.01\text{ kg/s} \times 4181\text{ J/kg}\cdot\text{K}}{\pi \times 0.060\text{ m} \times 2000\text{ W/m}^2}(80 - 20)^\circ\text{C} = 6.65\text{ m} \quad \triangleleft$$

(b) The surface temperature at the outlet can be obtained from Newton's law of cooling, Eq. 17.45, where

$$T_{s,o} = \frac{q_s''}{h} + T_{m,o}$$

To find the local convection coefficient at the tube outlet, the nature of the flow condition must first be established. From Equation 17.37

$$\text{Re}_D = \frac{4\dot{m}}{\pi D\mu} = \frac{4 \times 0.01\text{ kg/s}}{\pi \times 0.060\text{ m}(352 \times 10^{-6}\text{ N}\cdot\text{s/m}^2)} = 603$$

Hence the flow is laminar. With the assumption of fully developed conditions, the appropriate heat transfer correlation is Eq. 17.61

$$\text{Nu}_D = \frac{hD}{k} = 4.36$$

and the local coefficient is

$$h = 4.36\frac{k}{D} = 4.36\frac{0.670\text{ W/m}\cdot\text{K}}{0.06\text{ m}} = 48.7\text{ W/m}^2\cdot\text{K}$$

The surface temperature at the tube outlet is then

$$T_{s,o} = \frac{2000\text{ W/m}^2}{48.7\text{ W/m}^2\cdot\text{K}} + 80^\circ\text{C} = 121^\circ\text{C} \quad \triangleleft$$

***Comments:*** For this laminar flow condition, from Eq. 17.41, we find the thermal entry length, $(x_{fd}/D) = 0.05\,\text{Re}_D\,\text{Pr} = 66.3$, while $L/D = 110$. Hence the assumption of fully developed conditions is reasonable. Because the water is pressurized, we assume that local boiling does not occur even though $T_{s,o} > 100^\circ\text{C}$.

## Turbulent Flow

A commonly used expression for computing the *local* Nusselt number for *fully developed* (hydrodynamically and thermally) *turbulent* flow in a smooth *circular tube* is the *Dittus-Boelter correlation* of the form

$$\text{Nu}_D = 0.023\,\text{Re}_D^{4/5}\,\text{Pr}^n \qquad \begin{bmatrix} 0.6 \le \text{Pr} \le 160 \\ \text{Re}_D \gtrsim 10{,}000 \\ \dfrac{L}{D} \gtrsim 10 \end{bmatrix} \tag{17.64}$$

where $n = 0.4$ for heating $(T_s > T_m)$ and 0.3 for cooling $(T_s < T_m)$. These correlations have been confirmed experimentally for the range of conditions shown in the brackets. The correlations can be used for small to moderate temperature differences, $(T_s - T_m)$ with all properties evaluated at $\overline{T}_m$. For flows characterized by large property variations, the *Sieder-Tate correlation* is recommended

$$\text{Nu}_D = 0.027\,\text{Re}_D^{4/5}\,\text{Pr}^{1/3}\left(\frac{\mu}{\mu_s}\right)^{0.14} \qquad \begin{bmatrix} 0.7 \le \text{Pr} \le 16{,}700 \\ \text{Re}_D \gtrsim 10{,}000 \\ \dfrac{L}{D} \gtrsim 10 \end{bmatrix} \tag{17.65}$$

where all properties except $\mu_s$ are evaluated at $\overline{T}_m$. The foregoing correlations can be applied to *noncircular tubes* by using the hydraulic diameter, Eq. 17.63, as the characteristic length for the Reynolds and Nusselt numbers. To a good approximation, the foregoing correlations can be applied for *both constant heat flux* and *constant temperature surface thermal conditions*.

Although Eqs. 17.64 and 17.65 are easily applied and are certainly satisfactory for many purposes, errors as large as 25% can result from their use. Such errors can be reduced to less than 10% through the use of more comprehensive or application-specific correlations. Correlations that account for highly variable properties, laminar-turbulent transition regime effects, and surface roughness effects are available in the literature.

In many applications the tube length will exceed the thermal entry length, $10 \le (x_{fd}/D) \le 60$. Hence, it is often reasonable to assume that the average Nusselt number for the entire tube is equal to the value associated with the fully developed region, $\overline{\text{Nu}}_D \approx \text{Nu}_{D,fd}$. For short tubes, $(x_{fd}/D) \le 10$, $\overline{\text{Nu}}_D$ will exceed $\text{Nu}_{D,fd}$, requiring that entrance region effects must be considered.

## *Example 17.10* Turbulent Flow Application: Hot Water Supply

Water flows steadily at 2 kg/s through a 40-mm-diameter tube that is 4 m long. The water enters at 25°C, and the tube temperature is maintained at 95°C by steam condensing on the exterior surface. Determine the outlet temperature of the water and the rate of heat transfer to the water.

### Solution

***Known:*** Flow rate and inlet temperature of water passing through a tube of prescribed length, diameter, and surface temperature.

***Find:*** Outlet water temperature, $T_{m,o}$, and rate of heat transfer to the water, $q$, for the prescribed conditions.

***Schematic and Given Data:***

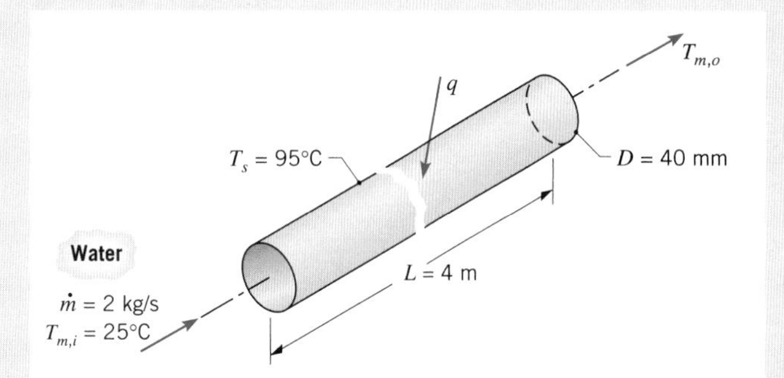


***Assumptions:***
1. Steady-state conditions.
2. Negligible kinetic and potential energy effects. No shaft work.
3. Constant properties.
4. Fully developed flow conditions since $L/D = 100$.

*Figure E17.10*

***Properties:*** Table HT-5, water (assume $T_{m,o} = 50°\text{C}$; hence $\overline{T}_m = (T_{m,o} + T_{m,i})/2 = 37.5°\text{C} \approx 310\text{K}$): $c_p = 4178$ J/kg · K, $\mu = 695 \times 10^{-6}$ N · s/m², $k = 0.628$ W/m · K, Pr = 4.62. Table HT-5, water ($T_s = 95°\text{C} = 368$ K): $\mu_s = 296 \times 10^{-6}$ N · s/m².

***Analysis:*** Since the tube surface temperature is constant, the water outlet temperature $T_{m,o}$ can be calculated from the energy rate expression of Eq. 17.55b

$$\frac{T_s - T_{m,o}}{T_s - T_{m,i}} = \exp\left(-\frac{PL}{\dot{m}c_p}\overline{h}\right) \tag{1}$$

Knowing $T_{m,o}$, the heat rate to the water follows from the overall energy balance, Eq. 17.48

$$q = \dot{m}c_p(T_{m,o} - T_{m,i}) \tag{2}$$

To select an appropriate correlation for estimating the average convection coefficient $\bar{h}$, calculate the Reynolds number, Eq. 17.37, to characterize the flow

$$\mathrm{Re}_D = \frac{4\dot{m}}{\pi D \mu} = \frac{4 \times 2\ \mathrm{kg/s}}{\pi(0.040\ \mathrm{m})695 \times 10^{-6}\ \mathrm{N \cdot s/m^2}} = 9.16 \times 10^4$$

Hence the flow is turbulent, and with the assumption of fully developed conditions, we select the *Dittus-Boelter correlation*, Eq. 17.64, with $n = 0.4$ since $T_s > T_m$

$$\overline{\mathrm{Nu}}_D = \frac{\bar{h}D}{k} = 0.023\,\mathrm{Re}_D^{4/5}\,\mathrm{Pr}^{0.4} = 0.023\,(9.16 \times 10^4)^{4/5}\,(4.62)^{0.4} = 396$$

$$\bar{h} = \frac{\overline{\mathrm{Nu}}_D k}{D} = \frac{396 \times 0.628\ \mathrm{W/m \cdot K}}{0.040\ \mathrm{m}} = 6210\ \mathrm{W/m^2 \cdot K}$$

Using the energy rate expression, Eq. (1) with $P = \pi D$, find $T_{m,o}$

$$\frac{95^\circ\mathrm{C} - T_{m,o}}{95^\circ\mathrm{C} - 25^\circ\mathrm{C}} = \exp\left(\frac{-\pi(0.040\ \mathrm{m})4\ \mathrm{m}}{2\ \mathrm{kg/s} \times 4178\ \mathrm{J/kg \cdot K}} 6210\ \mathrm{W/m^2 \cdot K}\right)$$

$$T_{m,o} = 46.8^\circ\mathrm{C} \quad \triangleleft$$

From the overall energy balance, Eq. (2), the heat rate *to* the water is

$$q = 2\ \mathrm{kg/s} \times 4176\ \mathrm{J/kg \cdot K}\,(46.8 - 25)^\circ\mathrm{C} = 182\ \mathrm{kW} \quad \triangleleft$$

***Comments:***

**1.** Since the flow is turbulent and $L/D = 100$, the assumption of fully developed conditions is justified according to Eq. 17.42.

**2.** In using the energy relations for the entire tube, properties are evaluated at $\bar{T}_m$. Not knowing $T_{m,o}$ at the outset, we *guessed* $T_{m,o} = 50^\circ\mathrm{C}$ and used $\bar{T}_m = 310$ K. This was a good guess since the analysis shows $\bar{T}_m = (T_{m,i} + T_{m,o})/2 = (25 + 46.8)^\circ\mathrm{C}/2 = 309$ K. For such a situation, recognize that you may have to iterate your analysis until the *guessed* and *calculated* temperatures are in satisfactory agreement.

**3.** The *Sieder-Tate* correlation, Eq. 17.65, would also be appropriate for this situation. Substituting numerical values, find

$$\overline{\mathrm{Nu}}_D = 0.027\,\mathrm{Re}_D^{4/5}\,\mathrm{Pr}^{1/3}\left(\frac{\mu}{\mu_s}\right)^{0.14} = 0.027(9.16 \times 10^4)^{4/5}\,4.62^{1/3}\left(\frac{695 \times 10^{-6}}{695 \times 10^{-6}}\right)^{0.14} = 523$$

$$\bar{h} = \overline{\mathrm{Nu}}_D \frac{k}{D} = 523\,\frac{0.628\ \mathrm{W/m \cdot K}}{0.040\ \mathrm{m}} = 8214\ \mathrm{W/m \cdot K}$$

Using Eqs. (1) and (2), find $T_{m,o} = 50.3^\circ\mathrm{C}$ and $q = 212$ kW. The results of the two correlations differ by approximately 15%, which is within the uncertainty normally associated with such correlations. Note that all properties are evaluated at $\bar{T}_m$, except for $\mu_s$, which is evaluated at $T_s$.

### 17.3.4 Guide for Selection of Internal Flow Correlations

In this section you have been introduced to empirical correlations to estimate the convection coefficients for *fully developed* laminar and turbulent flows in circular and noncircular tubes. For your convenience in selecting appropriate correlations for your problems, the recommended correlations have been summarized in Table 17.5 (next page). While specific conditions are associated with each of the correlations, you are reminded to follow the rules for performing convection calculations outlined in Sec. 17.1.3.

**Table 17.5** Summary of Forced Convection Heat Transfer Correlations for Internal Flow in Smooth Circular Tubes[c]

| Flow/Surface Thermal Conditions | Correlation[a,b] | | Restrictions on Applicability |
|---|---|---|---|
| Laminar, fully developed, $(x_{fd}/D) > 0.05\,Re_D Pr$ | | | |
| Constant $q_s''$ | $Nu_D = 4.36$ | (17.61) | $Pr \geq 0.6$, $Re_D \leq 2300$ |
| Constant $T_s$ | $Nu_D = 3.66$ | (17.62) | $Pr \geq 0.6$, $Re_D \leq 2300$ |
| Turbulent, fully developed, $(x_{fd}/D) > 10$ | | | |
| Constant $q_s''$ or $T_s$ (*Dittus-Boelter*) | $Nu_D = 0.023\,Re_D^{4/5}\,Pr^n$ | (17.64) | $0.6 \leq Pr \leq 160$, $Re_D \geq 10{,}000$, $n = 0.4$ for $T_s > T_m$ and $n = 0.3$ for $T_s < T_m$ |
| Constant $q_s''$ or $T_s$ (*Sieder-Tate*) | $Nu_D = 0.027\,Re_D^{4/5}\,Pr^{1/3}\left(\frac{\mu}{\mu_s}\right)^{0.14}$ | (17.65) | $0.7 \leq Pr \leq 16{,}700$, $Re_D \gtrsim 10{,}000$ |

[a]Thermophysical properties in Eqs. 17.61, 17.62, and 17.64 are based upon the mean temperature, $T_m$. If the correlations are used to estimate the *average* Nusselt number over the entire tube length, the properties should be based upon the average of the mean temperatures, $\overline{T}_m = (T_{m,i} + T_{m,o})/2$.
[b]Thermophysical properties in Eq. 17.65 should be evaluated at $T_m$ or $\overline{T}_m$, except for $\mu_s$, which is evaluated at the tube wall temperature $T_s$ or $\overline{T}_s$.
[c]For tubes of *noncircular cross section,* use the hydraulic diameter, $D_h$, Eq. 17.63, as the characteristic length for the Reynolds and Nusselt numbers. Results for fully developed *laminar* flow are provided in Table 17.4. For *turbulent* flow, Eq. 17.64 may be used as a first approximation.

# Free Convection

## 17.4 Free Convection

In the preceding sections of this chapter, we considered convection heat transfer in fluid flows that originate from an *external forcing* condition. Now we consider situations for which there is no forced motion, but heat transfer occurs because of *convection currents* that are induced by *buoyancy forces,* which arise from density differences caused by temperature variations in the fluid. Heat transfer by this means is referred to as *free* (or *natural*) convection.

Since free convection flow velocities are generally much smaller than those associated with forced convection, the corresponding heat transfer rates are also smaller. However, in many thermal systems, free convection may provide the largest resistance to heat transfer and therefore plays an important role in the design or performance of the system. Free convection is often the preferred mode of convection heat transfer, especially in electronic systems, for reasons of space limitations, maintenance-free operation, and reduced operating costs. Free convection strongly influences heat transfer from pipes, transmission lines, transformers, baseboard heaters, as well as appliances such as your stereo, television and laptop computer. It is also relevant to the environmental sciences, where it is responsible for oceanic and atmospheric motions.

We begin by considering the physical origins and nature of buoyancy-driven flows, and introduce empirical correlations to estimate convection coefficients for common geometries.

### 17.4.1 Flow and Thermal Considerations

To illustrate the nature of the boundary layer development in free convection flows, consider the heated vertical plate (Fig. 17.20*a*) that is immersed in a cooler *extensive, quiescent fluid.* An extensive medium is, in principle, an infinite one; a quiescent fluid is one that is otherwise at rest, except in the vicinity of the surface.

Since the plate is hotter than the fluid, $T_s > T_\infty$, the fluid close to the plate is less dense than fluid in the quiescent region. The fluid density gradient and the gravitational field create the buoyancy force that induces the free convection *boundary layer flow* in which the heated fluid rises. The boundary layer grows as more fluid from the quiescent region is

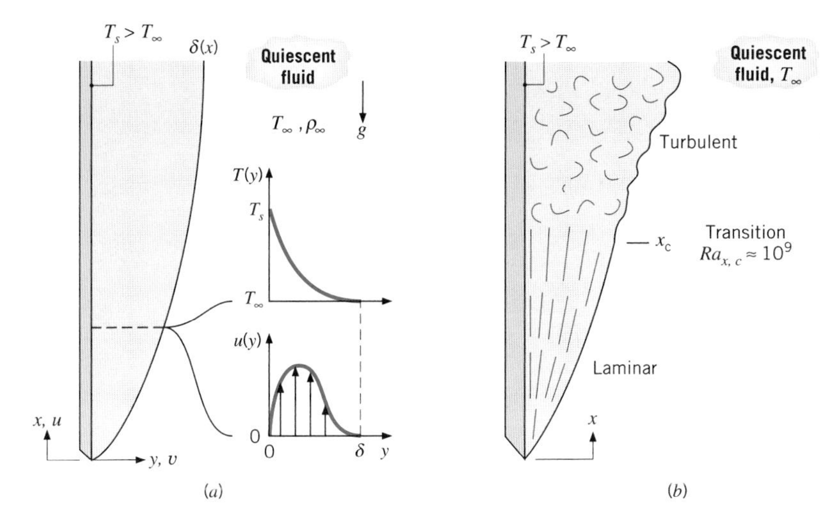


*Figure 17.20* Boundary layer development on a heated vertical plate. (*a*) Velocity and temperature profiles in the boundary layer at the location $x$. (*b*) Boundary layer transitional flow conditions.

involved (entrained). The resulting *velocity distribution* is illustrated in Fig. 17.20*a*. The velocity at the surface is zero (the no-slip condition), as was the case for forced convection. With increasing distance $y$ from the plate, the velocity increases to a maximum value, then decreases to zero as $y \rightarrow \infty$ in the quiescent region. Note also that beyond the maximum velocity value the velocity gradient decreases and becomes zero (no-shear condition). These conditions define the boundary layer thickness $\delta(x)$.

In free convection, the hydrodynamic and thermal boundary layer flows are *coupled*: thermal effects induce flow, which in turn affects the temperature distribution. This situation is in contrast with forced convection flows where the hydrodynamic conditions control the energy transport. The *temperature distribution* associated with the velocity distribution is shown in Fig. 17.20*a*. At $y = 0$, the fluid is at the surface temperature, $T_s$, and the profile has a steep gradient at the surface ($y = 0$), which decreases in the $y$ direction as the temperature eventually decreases to that of the quiescent fluid, $T_\infty$. Note also that the temperature gradient eventually becomes zero, corresponding to no heat transfer into the quiescent region.

The *convection coefficient* is related to the temperature gradient in the fluid at the surface in the same manner as we found for forced convection. That is, steeper gradients are associated with thinner boundary layers and larger heat fluxes. For the vertical plate of Fig. 17.20*a*, we expect the convection coefficient to be largest near the leading edge ($x = 0$) and to decrease with $x$.

As with forced convection flows, free convection flows can experience instabilities that cause disturbances in the flow to be amplified, leading to *transition* from *laminar* to *turbulent* flow (Fig. 17.20*b*). Transition in a free convection boundary layer depends upon the relative magnitude of the buoyancy and viscous forces in the fluid. It is customary to correlate the occurence of transition in terms of the Rayleigh number, which is the product of the Grashof and Prandtl numbers. For vertical plates, the *critical Rayleigh number* is

$$\mathrm{Ra}_{x,c} = \mathrm{Gr}_{x,c}\,\mathrm{Pr} = \frac{g\beta(T_s - T_\infty)x_c^3}{\nu\alpha} \approx 10^9 \tag{17.66}$$

The dimensionless parameter that plays the role of characterizing free convection flows is the *Grashof number* (see also Table 17.1)

$$\mathrm{Gr}_L = \frac{g\beta(T_s - T_\infty)L^3}{\nu^2} \tag{17.67}$$

which indicates the *ratio of the buoyancy force to the viscous force*. The key buoyancy-related parameters are the temperature difference, $(T_s - T_\infty)$, or if a heating process, $(T_\infty - T_s)$, and the ***volumetric thermal expansion coefficient***

*volumetric thermal expansion coefficient*

$$\beta = -\frac{1}{\rho}\left(\frac{\partial \rho}{\partial T}\right)_p \tag{17.68}$$

which is a thermodynamic property relating the variation of density with temperature. For an ideal gas, $\rho = p/RT$, and it follows that

$$\beta = -\frac{1}{\rho}\left(\frac{\partial \rho}{\partial T}\right)_p = \frac{1}{\rho}\frac{p}{RT^2} = \frac{1}{T} \tag{17.69}$$

where $T$ is the *absolute* temperature. For liquids and nonideal gases, $\beta$ must be obtained from appropriate tables (see Appendixes HT-4 and HT-5).

For free convection flows, we expect that the convection coefficient can be functionally expressed by equations of the form

$$\overline{\mathrm{Nu}}_L = f(\mathrm{Gr}_L, \mathrm{Pr})$$

The overbar indicates an average over the surface of the immersed geometry of characteristic length $L$. The most common empirical correlations suitable for engineering calculations have the form

$$\overline{\mathrm{Nu}}_L = \frac{\bar{h}L}{k} = C\,\mathrm{Ra}_L^n \tag{17.70}$$

where the *Rayleigh number*

$$\mathrm{Ra}_L = \mathrm{Gr}_L\,\mathrm{Pr} = \frac{g\beta(T_s - T_\infty)L^3}{\nu\alpha} \tag{17.71}$$

is based on the characteristic length $L$ of the geometry. Typically, $n = 1/4$ and $1/3$ for laminar and turbulent flows, respectively. For turbulent flow it then follows from Eqs. 17.70 and 17.71 that $\bar{h}_L$ is independent of $L$. Note that all properties are evaluated at the *film temperature*, $T_f \equiv (T_s + T_\infty)/2$.

We'll consider now specific forms of correlations for the immersed geometries of the vertical and horizontal plate, the long horizontal cylinder, and the sphere. The recommended correlations are summarized at the end of these sections in Table 17.6 (page 446).

### 17.4.2 Correlations: The Vertical Plate

Expressions of the form given by Eq. 17.70 have been developed for the vertical plate

$$\overline{\mathrm{Nu}}_L = 0.59\,\mathrm{Ra}_L^{1/4} \qquad [10^4 \le \mathrm{Ra}_L \le 10^9] \tag{17.72}$$

$$\overline{\mathrm{Nu}}_L = 0.10\,\mathrm{Ra}_L^{1/3} \qquad [10^9 \le \mathrm{Ra}_L \le 10^{13}] \tag{17.73}$$

The *Churchill-Chu correlation* may be applied over the entire range of $\mathrm{Ra}_L$ and has the form

$$\overline{\mathrm{Nu}}_L = \left\{0.825 + \frac{0.387\,\mathrm{Ra}_L^{1/6}}{[1 + (0.492/\mathrm{Pr})^{9/16}]^{8/27}}\right\}^2 \tag{17.74}$$

Although Eq. 17.74 is suitable for most engineering calculations, slightly better accuracy can be obtained for *laminar* flow by using

$$\overline{\mathrm{Nu}}_L = 0.68 + \frac{0.670\,\mathrm{Ra}_L^{1/4}}{[1 + (0.492/\mathrm{Pr})^{9/16}]^{4/9}} \qquad [\mathrm{Ra}_L \lesssim 10^9] \tag{17.75}$$

The foregoing results can also be applied to *vertical* cylinders of height $L$, if the boundary layer thickness $\delta$ is much less than the cylinder diameter $D$, a condition that is generally satisfied when $(D/L) \geq (35/\mathrm{Gr}_L^{1/4})$.

For *laminar* flow of gases ($\mathrm{Pr} = 0.7$), the boundary layer thickness ($\delta \approx \delta_t$) can be estimated using the expression

$$\frac{\delta}{x} = 6(\mathrm{Gr}_x/4)^{-1/4} \qquad [\mathrm{Pr} = 0.7, \mathrm{Ra}_L \leq 10^9] \tag{17.76}$$

## *Example 17.11* Vertical Plate: Glass-Door Firescreen

A glass-door firescreen, used to reduce loss of room air through a chimney, has a height of 0.71 m and a width of 1.02 m and reaches a temperature of 232°C. If the room temperature is 23°C, estimate the convection heat rate from the fireplace to the room.

### Solution

***Known:*** Glass screen situated in fireplace opening.

***Find:*** Heat transfer by free convection between firescreen and room air.

***Schematic and Given Data:***

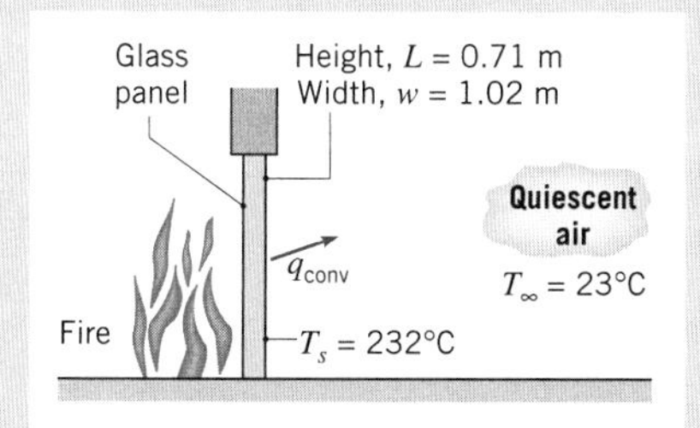


***Assumptions:***

1. Firescreen is at a uniform temperature $T_s$.
2. Room air is quiescent.
3. Costant properties.

*Figure E17.11*

***Properties:*** Table HT-3, air ($T_f = (T_s + T_\infty)/2 = 400$ K): $k = 33.8 \times 10^{-3}$ W/m · K, $\nu = 26.4 \times 10^{-6}$ m²/s, $\alpha = 38.3 \times 10^{-6}$ m²/s, $\mathrm{Pr} = 0.690$, $\beta = (1/T_f) = 0.0025\ \mathrm{K}^{-1}$.

***Analysis:*** The rate of heat transfer by free convection from the firescreen to the room is given by Newton's law of cooling

$$q = \bar{h}A_s(T_s - T_\infty)$$

where $\bar{h}$ may be obtained from knowledge of the Rayleigh number. Using Eq. 17.71

$$\mathrm{Ra}_L = \frac{g\beta(T_s - T_\infty)L^3}{\alpha\nu}$$

$$\mathrm{Ra}_L = \frac{9.8\ \mathrm{m/s^2}\,(1/400\ \mathrm{K})(232 - 23)°\mathrm{C} \times (0.71\ \mathrm{m})^3}{(38.3 \times 10^{-6}\ \mathrm{m^2/s})(26.4 \times 10^{-6}\ \mathrm{m^2/s})} = 1.813 \times 10^9$$

and from Eq. 17.66 it follows that transition to turbulence occurs on the panel. Using the *Churchill-Chu correlation,* Eq. 17.74, and substituting for the Rayleigh number, find

$$\overline{\mathrm{Nu}}_L = \left\{0.825 + \frac{0.387\,\mathrm{Ra}_L^{1/6}}{[1 + (0.492/\mathrm{Pr})^{9/16}]^{8/27}}\right\}^2$$

$$\overline{\mathrm{Nu}}_L = \left\{0.825 + \frac{0.387(1.813 \times 10^9)^{1/6}}{[1 + (0.492/0.690)^{9/16}]^{8/27}}\right\}^2 = 147$$

Hence, the average convection coefficient is

$$\bar{h} = \frac{\overline{\mathrm{Nu}}_L k}{L} = \frac{147(33.8 \times 10^{-3}\ \mathrm{W/m \cdot K})}{0.71\ \mathrm{m}} = 7.0\ \mathrm{W/m^2 \cdot K}$$

and the heat transfer by free convection between the firescreen and room air is

$$q = 7.0\ \mathrm{W/m^2 \cdot K}(1.02 \times 0.71)\mathrm{m^2}\,(232 - 23)°\mathrm{C} = 1060\ \mathrm{W} \quad \triangleleft$$

***Comments:***

1. If $\bar{h}$ were computed using the simpler correlation of Eq. 17.73, we would obtain $\bar{h} = 5.8\ \text{W/m}^2 \cdot \text{K}$, and the heat transfer prediction would be approximately 20% lower than the foregoing result. This difference is within the uncertainty normally associated with using such correlations.
2. Radiation heat transfer effects are often significant relative to free convection. Using the radiative exchange rate equation, Eq. 15.7, and assuming $\varepsilon = 1.0$ for the glass surface and $T_{\text{sur}} = 23°\text{C}$, the net rate of radiation heat transfer between the firescreen and the surroundings is

$$q_{\text{rad}} = \varepsilon A_s \sigma (T_s^4 - T_{\text{sur}}^4) = 1(1.02 \times 0.71)\text{m}^2\,(5.67 \times 10^{-8}\ \text{W/m}^2 \cdot \text{K}^4)\,(505^4 - 296^4)\text{K}^4$$

$$q_{\text{rad}} = 2355\ \text{W}$$

The linearized radiation coefficient is given by Eq. 15.9

$$h_{\text{rad}} = \varepsilon\sigma(T_s + T_{\text{sur}})(T_s^2 + T_{\text{sur}}^2) = 1(5.67 \times 10^5\ \text{W/m}^2 \cdot \text{K}^4)(505 + 296)(505^2 + 296^2)\text{K}^3$$

$$h_{\text{rad}} = 15.6\ \text{W/m}^2 \cdot \text{K}$$

Note that the radiation coefficient (radiation heat rate) is more than twice the convection coefficient (convection heat rate) for this application.

### 17.4.2 Correlations: The Horizontal Plate

For a vertical plate, heated (or cooled) relative to an ambient fluid, the plate is aligned with the gravitational field, and the buoyant force induces fluid motion in the upward (or downward) direction. If the plate is *horizontal,* the buoyancy force is normal to the surface. The flow patterns and heat transfer rate depend strongly on whether the surface is hot *or* cold *and* on whether it is facing upward *or* downward. These four combinations and the general features of their convection currents are represented in Fig. 17.21.

For a *hot surface facing downward* (Case A) and a *cold surface facing upward* (Case B), the tendency of the fluid to ascend and descend, respectively, is impeded by the plate. The flow must move horizontally before it can ascend or descend from the edges of the plate, and convection heat transfer is somewhat ineffective.

For a *hot surface facing upward* (Case C) and a *cold surface facing downward* (Case D), flow is driven by ascending and descending parcels of fluids, respectively. Conservation of mass dictates that warm fluid ascending (cold fluid descending) from a surface be replaced by descending cooler fluid (ascending warmer fluid) from the ambient, and heat transfer is much more effective than for cases A and B.

The correlations widely used for horizontal plates corresponding to these arrangements use the *characteristic length L* defined as

$$L \equiv \frac{A_s}{P} \tag{17.77}$$

where $A_s$ and $P$ are the plate surface area and perimeter, respectively. The recommended correlations for the average Nusselt number are

***Hot Surface Facing Downward or Cold Surface Facing Upward (Cases A and B)***

$$\overline{\text{Nu}}_L = 0.27\,\text{Ra}_L^{1/4} \qquad [10^5 \lesssim \text{Ra}_L \lesssim 10^{10}] \tag{17.78}$$

***Hot Surface Facing Upward or Cold Surface Facing Downward (Cases C and D)***

$$\overline{\text{Nu}}_L = 0.54\,\text{Ra}_L^{1/4} \qquad [10^4 \lesssim \text{Ra}_L \lesssim 10^7] \tag{17.79}$$

$$\overline{\text{Nu}}_L = 0.15\,\text{Ra}_L^{1/3} \qquad [10^7 \lesssim \text{Ra}_L \lesssim 10^{11}] \tag{17.80}$$

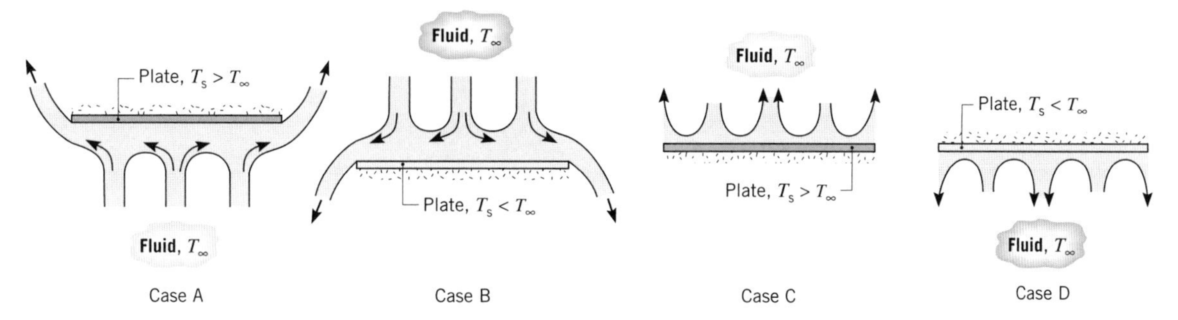


*Figure 17.21* Free convection buoyancy-driven flows for hot ($T_s > T_\infty$) and cold ($T_s < T_\infty$) horizontal plates: Case A — hot surface facing downwards, Case B — cold surface facing upwards, Case C — hot surface facing upwards, and Case D — cold surface facing downwards.

## *Example 17.12* Horizontal Plate: Cooling an Electronic Equipment Enclosure

An array of power-dissipating electrical components is mounted on the bottom side of a 1.2 m by 1.2 m horizontal aluminum alloy plate ($\varepsilon = 0.25$), while the top side is cooled by free convection with ambient, quiescent air at $T_\infty = 300$ K and by radiation exchange with the surroundings at $T_{sur} = 300$ K. The plate is sufficiently thick to ensure a nearly uniform upper surface temperature and is attached to a well-insulated enclosure.

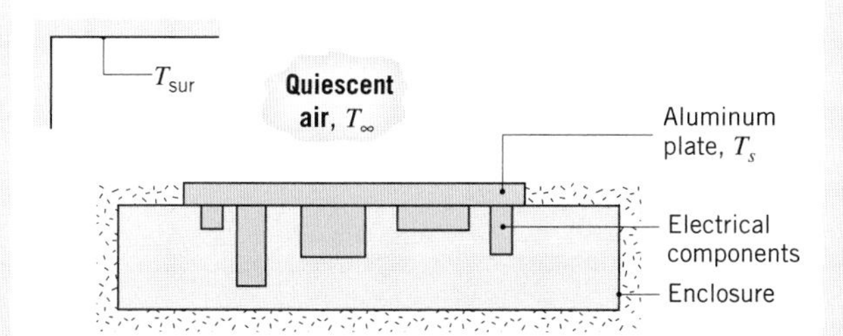


*Figure E17.12a*

If the temperature of the plate is not to exceed 57°C, what is the maximum allowable power dissipation in the electrical components?

### Solution

***Known:*** Horizontal plate and maximum allowable temperature experiencing free convection and radiation exchange.

***Find:*** Maximum allowable electrical power dissipation, $P_{elec}$.

***Schematic and Given Data:***

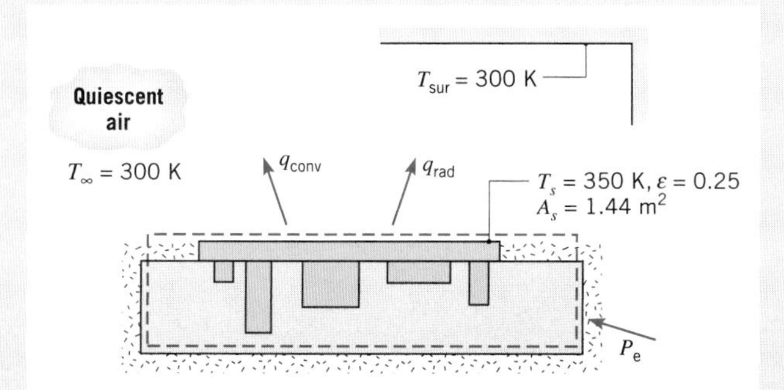


***Assumptions:***

1. Steady-state conditions.
2. Plate is isothermal.
3. Negligible heat transfer from other surfaces of the enclosure.
4. Radiation exchange is between a small, gray object (plate) and large isothermal surroundings.
5. Constant properties.

*Figure E17.12b*

***Properties:*** Table HT-3, air ($T_f = 325$ K, 1 atm): $\nu = 18.4 \times 10^{-6}$ m$^2$/s, $k = 0.028$ W/m · K, $\alpha = 26.2 \times 10^{-6}$ m$^2$/s.

***Analysis:*** From an overall energy balance on the enclosure and plate, the electrical power dissipation is the sum of the heat transfer rates by free convection and radiation exchange (Eq. 15.7)

$$P_e = q_{conv} + q_{rad}$$
$$P_e = \bar{h}A_s(T_s - T_\infty) + \varepsilon A_s\sigma(T_s^4 - T_{sur}^4)$$

For free convection from the horizontal plate, the characteristic length from Eq. 17.77 is

$$L = A_s/P = (1.2 \times 1.2\ \text{m}^2)/(4 \times 1.2\ \text{m}) = 0.3\ \text{m}$$

and from Eq. 17.71, the Rayleigh number with $\beta = 1/T_f$ (Eq. 17.69) is

$$\text{Ra}_L = \frac{g\beta(T_s - T_\infty)L^3}{\nu\alpha} = \frac{9.8\ \text{m/s}^2(325\ \text{K})^{-1}(50\ \text{K})(0.3\ \text{m})^3}{(18.4 \times 10^{-6}\ \text{m}^2/\text{s})(26.2 \times 10^{-6}\ \text{m}^2/\text{s})} = 8.44 \times 10^7$$

Using the correlation of Eq. 17.80 for a *hot surface facing upward* (Case C), find the average convection coefficient

$$\overline{\text{Nu}}_L = \frac{\bar{h}L}{k} = 0.15\,\text{Ra}_L^{1/3} = 0.15(8.44 \times 10^7)^{1/3} = 65.8$$

$$\bar{h}_L = 65.8\,\frac{0.028\ \text{W/m}\cdot\text{K}}{0.3\ \text{m}} = 6.2\ \text{W/m}^2\cdot\text{K}$$

The allowable electrical power is

$$P_e = [6.1\ \text{W/m}^2\cdot\text{K}(350 - 300)\text{K} + 0.25(5.67 \times 10^{-8}\ \text{W/m}^2\cdot\text{K}^4)(350^4 - 300^4)\text{K}^4](1.44\ \text{m}^2)$$
$$P_e = 446\ \text{W} + 141\ \text{W} = 587\ \text{W}$$

***Comments:*** Note that heat transfer by free convection and radiation exchange comprise 76% and 24%, respectively, of the total heat rate. It would be beneficial to apply a high emissivity coating to the plate as a means to enhance radiative heat transfer, and hence, the allowable electrical power.

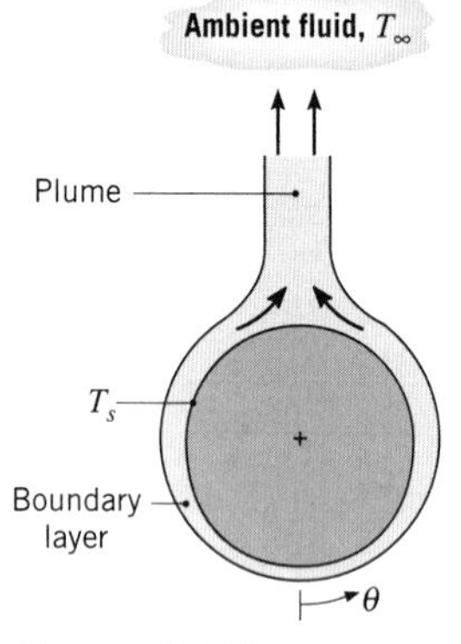

*Figure 17.22*

## 17.5.3 Correlations: The Horizontal Cylinder and Sphere

As shown in Fig. 17.22 for a heated *cylinder,* the boundary layer development begins at $\theta = 0°$ and concludes at $\theta < 180°$ with the formation of a *plume* ascending from the cylinder. If the flow remains laminar over the entire surface, the distribution of the local convection coefficient is characterized by a maximum at $\theta = 0°$, and a decrease with increasing $\theta$. This steady decrease could be disrupted at Rayleigh numbers sufficiently large ($\text{Ra}_D \gtrsim 10^9$) by the occurrence of transition to turbulence within the boundary layer. If the cylinder is cooled relative to the ambient fluid, the plume descends from the cylinder.

Expressions of the form given by Eq. 17.70 for prescribed Rayleigh number ranges have been developed by *Morgan* for the *long, horizontal cylinder:*

$$\overline{\text{Nu}}_D = 0.850\,\text{Ra}_D^{0.188} \quad [10^2 \le \text{Ra}_D \le 10^4] \tag{17.81}$$
$$\overline{\text{Nu}}_D = 0.480\,\text{Ra}_D^{0.250} \quad [10^4 \le \text{Ra}_D \le 10^7] \tag{17.82}$$
$$\overline{\text{Nu}}_D = 0.125\,\text{Ra}_D^{0.333} \quad [10^7 \le \text{Ra}_D \le 10^{12}] \tag{17.83}$$

In contrast, the *Churchill-Chu correlation* is recommended for a wide Rayleigh number range

$$\overline{\text{Nu}}_D = \left\{0.60 + \frac{0.387\,\text{Ra}_D^{1/6}}{[1 + (0.559/\text{Pr})^{9/16}]^{8/27}}\right\}^2 \quad [\text{Ra}_D \lesssim 10^{12}] \tag{17.84}$$

Boundary layer development for the isothermal *sphere* is similar to that for the cylinder with the formation of a plume. The *Churchill correlation* is recommended for estimating the average convection coefficient

$$\overline{\text{Nu}}_D = 2 + \frac{0.589\,\text{Ra}_D^{1/4}}{[1 + (0.469\,\text{Pr})^{9/16}]^{4/9}} \quad [\text{Pr} \ge 0.7,\ \text{Ra}_D \lesssim 10^{11}] \tag{17.85}$$

## *Example 17.13* Horizontal Cylinder: High Pressure Steam Line

A horizontal, high-pressure steam pipe of 0.1-m outside diameter passes through a large room whose wall and air temperatures are 23°C. The pipe has an outside surface temperature of 165°C and an emissivity of $\varepsilon = 0.85$. Estimate the heat transfer from the pipe per unit length.

## Solution

***Known:*** Surface temperature of a horizontal steam pipe.
***Find:*** Heat transfer from the pipe per unit length $q'$ (W/m).

***Schematic and Given Data:***

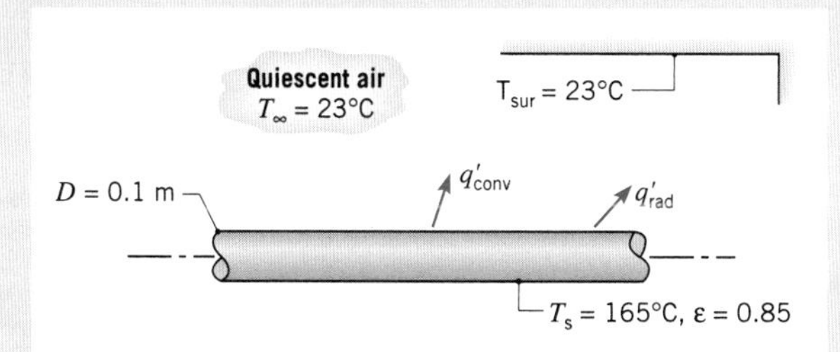


***Assumptions:***
1. Pipe surface area is small compared to surroundings.
2. Room air is quiescent.
3. Radiation exchange is between a small, gray surface (pipe) and large isothermal surroundings.
4. Constant properties.

*Figure E17.13*

***Properties:*** Table HT-3, air ($T_f = 367$ K): $k = 0.0313$ W/m · K, $\nu = 22.8 \times 10^{-6}$ m$^2$/s, $\alpha = 32.8 \times 10^{-6}$ m$^2$/s, Pr = 0.697, $\beta = 2.725 \times 10^{-3}$ K$^{-1}$.

***Analysis:*** The total heat transfer per unit length of pipe due to convection and radiation exchange (Eq. 15.7) is

$$q' = q'_{conv} + q'_{rad} = \bar{h}\pi D(T_s - T_\infty) + \varepsilon\pi D\sigma(T_s^4 - T_{sur}^4)$$

The free convection coefficient may be estimated with the *Churchill-Chu correlation,* Eq. 17.84

$$\overline{\mathrm{Nu}}_D = \left\{0.60 + \frac{0.387\,\mathrm{Ra}_D^{1/6}}{[1 + (0.559/\mathrm{Pr})^{9/16}]^{8/27}}\right\}^2$$

where the Rayleigh number from Eq. 17.71 is

$$\mathrm{Ra}_D = \frac{g\beta(T_s - T_\infty)D^3}{\nu\alpha}$$

$$\mathrm{Ra}_D = \frac{9.8\ \mathrm{m/s^2}(2.725 \times 10^{-3}\ \mathrm{K^{-1}})(165 - 23)°\mathrm{C}\ (0.1\ \mathrm{m})^3}{(22.8 \times 10^{-6}\ \mathrm{m^2/s})(32.8 \times 10^{-6}\ \mathrm{m^2/s})} = 5.073 \times 10^6$$

Substituting for the Rayleigh number into the correlation, find

$$\overline{\mathrm{Nu}}_D = \left\{0.60 + \frac{0.387(5.073 \times 10^6)^{1/6}}{[1 + (0.559/0.697)^{9/16}]^{8/27}}\right\}^2 = 23.3$$

and the average convection coefficient for the cylinder is

$$\bar{h} = \frac{k}{D}\overline{\mathrm{Nu}}_D = \frac{0.0313\ \mathrm{W/m \cdot K}}{0.1\ \mathrm{m}} \times 23.3 = 7.29\ \mathrm{W/m^2 \cdot K}$$

The total heat transfer rate from the pipe is

$$q' = 7.29\ \mathrm{W/m^2 \cdot K}\ (\pi \times 0.1\ \mathrm{m})(165 - 23)°\mathrm{C}$$
$$+\ 0.85\ (\pi \times 0.1\ \mathrm{m})(5.67 \times 10^{-8}\ \mathrm{W/m^2 \cdot K^4})(438^4 - 296^4)\ \mathrm{K^4}$$
$$q' = (325 + 441)\ \mathrm{W/m} = 766\ \mathrm{W/m} \triangleleft$$

***Comments:*** 1. Note that the heat transfer by free convection and radiation exchange comprise 42 and 58%, respectively, of the total heat rate. It would be beneficial to apply a low emissivity coating to the pipe as a means to reduce the radiation exchange, and hence the heat transfer from the pipe to the room.

2. Equation 17.82 could also be used to estimate the Nusselt number and the convection coefficient, with the result that $\overline{\mathrm{Nu}}_D = 22.8$ and $\bar{h} = 7.14$ W/m$^2$ · K. These results are about 2% lower than the foregoing ones. Generally we expect differences between correlation results of 10–15%, rather than the excellent agreement found here.

**Table 17.6** Summary of Free Convection Correlations for Immersed Geometries

| Geometry | Recommended Correlation | | Restrictions |
|---|---|---|---|
| Vertical plates[a] | $\overline{\mathrm{Nu}}_L = \left\{0.825 + \dfrac{0.387\,\mathrm{Ra}_L^{1/6}}{[1 + (0.492/\mathrm{Pr})^{9/16}]^{8/27}}\right\}^2$ | (17.74) | $\mathrm{Ra}_L \lesssim 10^{13}$ |
| Horizontal plates[b] Case A or B: Hot surface down or cold surface up | $\overline{\mathrm{Nu}}_L = 0.27\,\mathrm{Ra}_L^{1/4}$ | (17.78) | $10^5 \lesssim \mathrm{Ra}_L \lesssim 10^{10}$ |
| Case C or D: Hot surface up or cold surface down | $\overline{\mathrm{Nu}}_L = 0.54\,\mathrm{Ra}_L^{1/4}$ | (17.79) | $10^4 \lesssim \mathrm{Ra}_L \lesssim 10^7$ |
| | $\overline{\mathrm{Nu}}_L = 0.15\,\mathrm{Ra}_L^{1/3}$ | (17.80) | $10^7 \lesssim \mathrm{Ra}_L \lesssim 10^{11}$ |
| Horizontal cylinder | $\overline{\mathrm{Nu}}_D = \left\{0.60 + \dfrac{0.387\,\mathrm{Ra}_D^{1/6}}{[1 + (0.559/\mathrm{Pr})^{9/16}]^{8/27}}\right\}^2$ | (17.84) | $\mathrm{Ra}_D \lesssim 10^{12}$ |
| Sphere | $\overline{\mathrm{Nu}}_D = 2 + \dfrac{0.589\,\mathrm{Ra}_D^{1/4}}{[1 + (0.469\,\mathrm{Pr})^{9/16}]^{4/9}}$ | (17.85) | $\mathrm{Ra}_D \lesssim 10^{11}$, $\mathrm{Pr} \gtrsim 0.7$ |

[a]The correlation may be applied to a vertical cylinder if $(D/L) \gtrsim (35/\mathrm{Gr}_L^{1/4})$.

[b]The characteristic length is defined as $L \equiv A_s/P$, Eq. 17.77.

### 17.4.5 Guide for Selection of Free Convection Correlations

In this section you have been introduced to empirical correlations to estimate the convection coefficients for free convection heat transfer for vertical and horizontal plates, the horizontal cylinder, and the sphere. For your convenience in selecting appropriate correlations for your problems, the recommended correlations have been summarized in Table 17.6. Specific conditions are associated with each of the correlations, and you are reminded to follow the rules for peforming convection calculations outlined in Sec. 17.1.3.

## Convection Application: Heat Exchangers

## 17.5 Heat Exchangers

The process of heat exchange between two fluids that are at different temperatures and separated by a solid wall occurs in many engineering applications. The device used to implement this exchange is termed a *heat exchanger,* and specific applications can be found in space heating and air-conditioning, power production, waste heat recovery, and chemical processing.

In Sec. 5.3 you considered the form of the control volume energy balance and its application to a heat exchanger (Example 5.7). In this section we will extend heat exchanger analysis to include the convection rate equation, and demonstrate the methodology for predicting exchanger performance.

## 17.5.1 Heat Exchanger Types

Heat exchangers are typically classified according to *flow arrangement* and *type of construction.* In this introductory treatment, we will consider three types that are representative of a wide variety of exchangers used in industrial practice.

The simplest heat exchanger is one for which the hot and cold fluids flow in the same or opposite directions in a ***concentric-tube*** (or *double-pipe*) construction. In the *parallel-flow* arrangement of Fig. 17.23*a*, the hot and cold fluids enter at the same end, flow in the same direction, and leave at the same end. In the *counterflow* arrangement, Fig. 17.23*b*, the fluids enter at opposite ends, flow in opposite directions, and leave at opposite ends.

*concentric-tube heat exchanger*

A common configuration for power plant and large industrial applications is the ***shell-and-tube heat exchanger,*** shown in Fig. 17.23*c*. This exchanger has one shell with multiple tubes, but the flow makes one pass through the shell. Baffles are usually installed to increase the convection coefficient of the shell side by inducing turbulence and a cross-flow velocity component.

*shell-and-tube heat exchanger*

The ***cross-flow heat exchanger,*** Fig. 17.23*d*, is constructed with a stack of thin plates bonded to a series of parallel tubes. The plates function as fins to enhance convection heat transfer and to ensure cross-flow over the tubes. Usually it is a gas that flows over the fin surfaces and the tubes, while a liquid flows in the tube. Such exchangers are used for air-conditioner and refrigeration heat rejection applications.

*cross-flow heat exchanger*

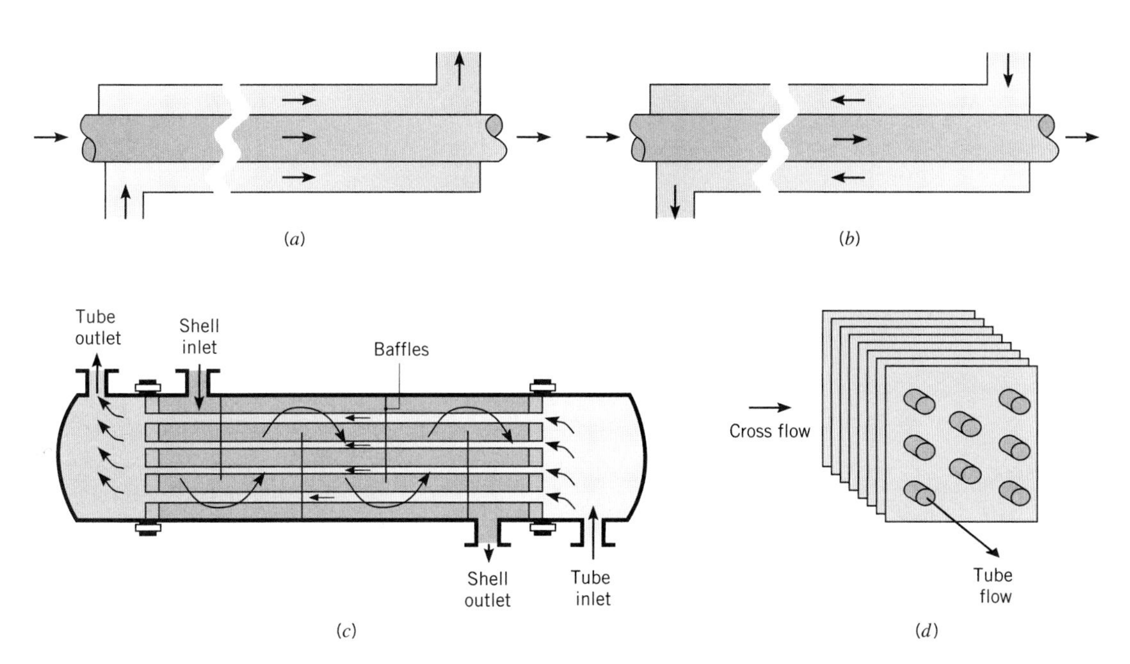


*Figure 17.23* Types of heat exchangers. Concentric tube heat exchangers: (*a*) Parallel flow and (*b*) Counterflow. (*c*) Shell-and-tube exchanger with one shell pass and one tube pass (showing 4 tubes, cross-counterflow mode of operation). (*d*) Cross-flow heat exchanger.

### 17.5.2 Heat Exchanger Analysis: Energy Balances, Rate Equation, Overall Coefficient

To predict the performance of a heat exchanger, it is necessary to relate the total heat transfer rate to parameters such as the fluid flow rates, inlet and outlet fluid temperatures, the overall heat transfer coefficient, and the total surface area for heat transfer.

***The Fluid Energy Balances.*** Consider the schematic representation of the heat exchanger shown in Fig. 17.24*a*. Assuming steady state, negligible kinetic and potential energy changes, no shaft work, and no stray heat transfer to the surroundings, and regarding $c_p$ as a constant, the energy rate balance, Equation 5.11b, reduces to give

$$q = \dot{m}_h c_{p,h} (T_{h,i} - T_{h,o}) \tag{17.86a}$$

$$q = \dot{m}_c c_{p,c} (T_{c,o} - T_{c,i}) \tag{17.87a}$$

where the temperatures are the *mean* fluid temperatures and the subscripts $h$ and $c$ refer to the hot and cold fluids, respectively. As before, $i$ and $o$ designate the fluid inlet and outlet conditions. Note that these equations have been written so that the heat rate $q$ is a positive value for both the hot and the cold fluids.

Equations 17.86a and 17.87a representing the *fluid energy balances* can be expressed as

$$q = C_h(T_{h,i} - T_{h,o}) \tag{17.86b}$$

$$q = C_c(T_{c,o} - T_{c,i}) \tag{17.87b}$$

where $C_h$ and $C_c$ are the hot and cold ***capacity rates*** (W/K), respectively

*capacity rates*

$$C_h = \dot{m}_h c_{p,h} \qquad C_c = \dot{m}_c c_{p,c} \tag{17.86c, 17.87c}$$

*Note that these equations are independent of the flow arrangement, heat exchanger type, as well as physical dimensions (surface area).*

***The Convection Rate Equation.*** We seek another expression for relating the heat rate to an appropriate temperature difference between the hot and cold fluids, where

$$\Delta T \equiv T_h - T_c \tag{17.88}$$

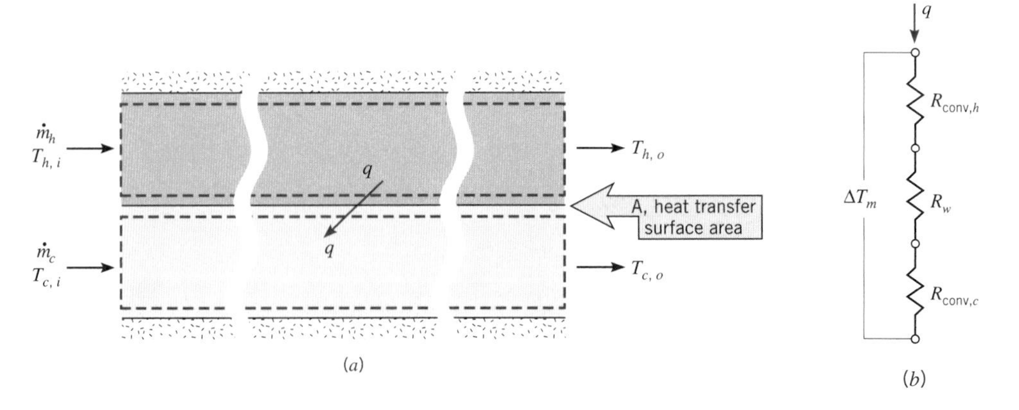


*Figure 17.24* Heat exchanger analysis. (*a*) Energy balances for the hot and cold fluids. (*b*) The convection rate equation in terms of the thermal resistances for convection and wall conduction and the mean fluid temperature difference.

Such an expression would be an extension of Newton's law of cooling, with the *overall heat transfer coefficient U* used in place of the single convection coefficient.

However, since $\Delta T$ varies with position in the heat exchanger, it is necessary to work with a *convection rate equation* of the form

$$q = UA\Delta T_m \tag{17.89}$$

where $\Delta T_m$ is an appropriate *mean temperature difference* and $A$ is the heat exchange surface area. As we'll see, this expression depends upon the heat exchanger configuration, and flow conditions, as well as physical dimensions.

***The Overall Coefficient.*** The convection rate equation, Eq. 17.89, can be represented by the thermal circuit shown in Fig. 17.24*b* in terms of the *convection thermal resistances* on the hot- and cold-fluid sides and the *wall conduction resistance*. It follows that the *overall heat transfer coefficient* may be expressed as

$$\frac{1}{UA} = R_{\text{conv},h} + R_w + R_{\text{conv},c} \tag{17.90a}$$

$$\frac{1}{UA} = \left(\frac{1}{hA}\right)_h + R_w + \left(\frac{1}{hA}\right)_c \tag{17.90b}$$

Note that the calculation of the $UA$ product can be based on the *hot* or *cold side* since

$$\frac{1}{UA} = \frac{1}{U_h A_h} = \frac{1}{U_c A_c} \tag{17.91}$$

However, a choice of the *hot-* or *cold-side* surface area must be specified because $U_h \neq U_c$ if $A_h \neq A_c$.

The convection coefficients for the hot and cold side can be estimated using empirical correlations appropriate for the flow geometry and conditions. The conduction resistance $R_w$ is obtained from Eq. 16.14 for a plane wall or Eq. 16.32 for a cylindrical wall. During normal heat exchanger operation, surfaces are subjected to fouling by fluid impurities, rust formation, and scale depositions, which can markedly increase the resistance to heat transfer between the fluids. For such situations, you would add the *fouling resistance* (cold and/or hot-side) to Eq. 17.90.

The *fluid energy balances,* Eqs. 17.86 and 17.87, and the *convection rate equation,* Eq. 17.89, provide the means to perform the *heat exchanger analysis.* Before this can be done, however, the specific form of $\Delta T_m$ must be established. The appropriate forms of $\Delta T_m$ for parallel and counterflow heat exchangers are presented in Secs. 17.5.3 and 17.5.4, respectively.

### 17.5.3 The Parallel Flow Heat Exchanger

The hot and cold fluid temperature distributions associated with a *parallel flow exchanger* are shown in Fig. 17.25. The temperature difference $\Delta T$ is initially very large, but decreases rapidly with increasing $x$, approaching zero asymptotically. It is important to note that, for such an exchanger, *the outlet temperature of the cold fluid never exceeds that of the hot fluid.* In Fig. 17.25, the subscripts 1 and 2 designate opposite ends of the heat exchanger. This *convention* is also used for the counterflow heat exchanger considered in Sec. 17.5.4.

The form of the appropriate *mean temperature difference,* $\Delta T_m$, for the parallel flow exchanger may be determined by applying an energy balance to differential control volumes (elements) in the hot and cold fluids as shown in the derivation that follows.

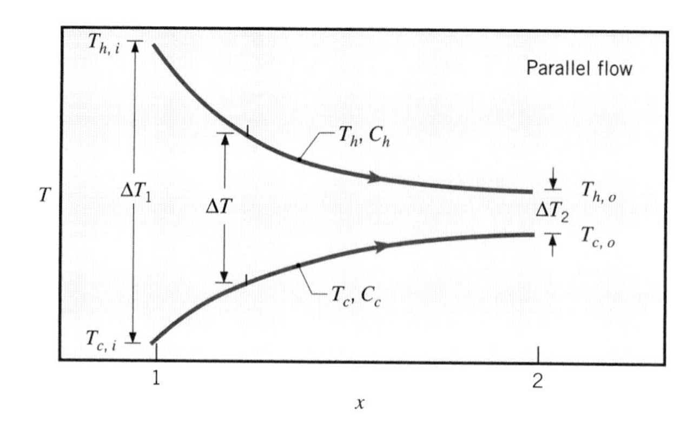


*Figure 17.25* Temperature distributions for a parallel-flow heat exchanger.

## Log Mean Temperature Difference: Derivation (CD-ROM)

## Log Mean Temperature Difference

From the derivation in the previous section, we found that the appropriate mean temperature difference required in the convection rate equation, Eq. 17.89

$$q = UA\Delta T_m$$

is the ***log mean temperature difference*** (LMTD) and has the form

*log mean temperature difference*

$$\Delta T_{\text{lm}} = \frac{\Delta T_2 - \Delta T_1}{\ln \Delta T_2/\Delta T_1} = \frac{\Delta T_1 - \Delta T_2}{\ln \Delta T_1/\Delta T_2} \tag{17.96}$$

where, from Fig. 17.25, the ***endpoint temperatures,*** $\Delta T_1$ and $\Delta T_2$, for the ***parallel flow exchanger*** are

*endpoint temperatures: parallel flow exchanger*

$$\Delta T_1 = T_{h,i} - T_{c,i} \qquad \Delta T_2 = T_{h,o} - T_{c,o} \tag{17.97}$$

### 17.5.4 The Counterflow Heat Exchanger

The hot and cold fluid temperature distributions associated with a *counterflow exchanger* are shown in Fig. 17.27. In contrast to the parallel-flow exchanger, this configuration provides for heat transfer between the hotter portions of the two fluids at one end, as well as between the colder portions at the other. For this reason, the change in the temperature difference, $\Delta T = T_h - T_c$, with respect to $x$ is nowhere as large as it is for the inlet region of the parallel-flow exchanger. Note that *the outlet temperature of the cold fluid may now exceed the outlet temperature of the hot fluid.*

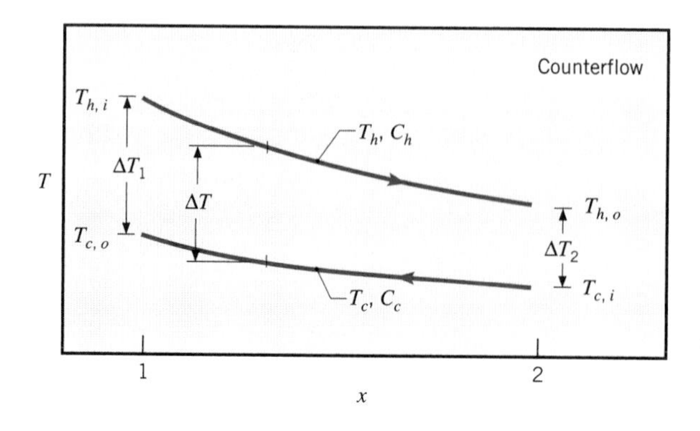


*Figure 17.27* Temperature distributions for a counterflow heat exchanger.

The form of the appropriate *mean temperature difference,* $\Delta T_m$, for the counterflow exchanger may be obtained from a derivation such as was performed for the parallel flow exchanger. The outcome is the same, except for the manner in which the endpoint temperatures, $\Delta T_1$ and $\Delta T_2$, are defined.

The appropriate mean temperature difference required in the convection rate equation, Eq. 17.89

$$q = UA\Delta T_m$$

is the *log mean temperature difference* and has the form

$$\Delta T_{\text{lm}} = \frac{\Delta T_2 - \Delta T_1}{\ln \Delta T_2/\Delta T_1} = \frac{\Delta T_1 - \Delta T_2}{\ln \Delta T_1/\Delta T_2} \tag{17.96}$$

where, from Fig. 17.27, the ***endpoint temperatures,*** $\Delta T_1$ and $\Delta T_2$, for the ***counterflow exchanger*** are

$$\Delta T_1 = T_{h,i} - T_{c,o} \qquad \Delta T_2 = T_{h,o} - T_{c,i} \tag{17.98}$$

*endpoint temperatures: counterflow exchanger*

Important differences in the operation of parallel flow and counterflow heat exchangers should be noted. For the *same* inlet and outlet fluid temperatures:

- The log mean temperature difference for counterflow exceeds that for parallel flow, $\Delta T_{\text{lm},CF} > \Delta T_{\text{lm},PF}$, and, hence,
- The surface area required to effect a prescribed heat transfer rate $q$ is smaller for counterflow than for the parallel flow arrangement, for the same value of $U$.
- Note also that $T_{c,o}$ can exceed $T_{h,o}$ for the counterflow arrangement, but not for parallel flow.

### 17.5.5 Special Heat Exchanger Operating Conditions

In Fig. 17.28, we've shown the temperature distributions associated with three special conditions under which heat exchangers may be operated.

- $C_h \gg C_c$. For this case, the hot fluid capacity rate $C_h$ is much larger than the cold fluid capacity rate $C_c$. As shown in Fig. 17.28*a*, the hot fluid temperature remains approximately constant throughout the exchanger, while the temperature of the cold fluid increases. The same condition could be achieved if the hot fluid is a *condensing vapor*. Condensation occurs at a constant temperature, and for all practical purposes, $C_h \rightarrow \infty$.

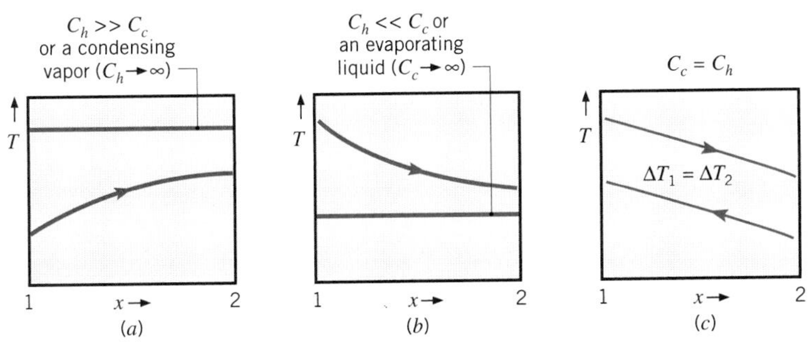


*Figure 17.28* Special heat exchanger conditions. (*a*) $C_h \gg C_c$ or a condensing vapor. (*b*) $C_h \ll C_c$ or an evaporating liquid. (*c*) A counterflow heat exchanger with equivalent fluid heat capacities ($C_h = C_c$).

- $C_h \ll C_c$. For this case, as shown in Fig. 17.28*b*, the cold fluid temperature remains approximately constant throughout the exchanger, while the temperature of the hot fluid decreases. The same effect is achieved if the cold fluid experiences *evaporation* for which $C_c \to \infty$. Note that *with evaporation and condensation, the fluid energy balances would be written in terms of the phase change enthalpies.*
- $C_h = C_c$. The third case, Fig. 17.28*c*, involves a *counterflow exchanger* for which the heat capacity rates are equal. The temperature difference $\Delta T$ must be constant throughout the exchanger, in which case, $\Delta T_1 = \Delta T_2 = \Delta T_{lm}$.

## *Example 17.14* Counterflow, Concentric Tube Heat Exchanger Analysis

A counterflow, concentric tube heat exchanger is used to cool the lubricating oil for a large industrial gas turbine engine. The flow rate of cooling water through the inner tube ($D_i$ = 25 mm) is 0.2 kg/s. The flow rate of hot oil through the outer annulus ($D_o$ = 45 mm) is 0.1 kg/s. The convection coefficient associated with the oil flow is $h_o = 40\ \text{W/m}^2 \cdot \text{K}$. The oil and water enter at temperatures of 100 and 30°C, respectively. What is the required tube length for an oil outlet temperature of 60°C?

### Solution

***Known:*** Fluid flow rates and inlet temperatures for a counterflow, concentric tube heat exchanger of prescribed inner and outer diameter.

***Find:*** Tube length to achieve a desired hot fluid outlet temperature, $T_{h,o}$ = 60°C.

***Schematic and Given Data:***

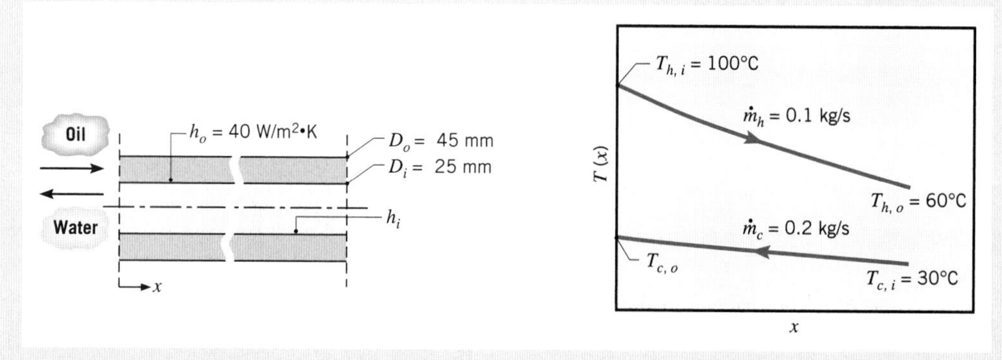


*Figure E17.14*

***Assumptions:***

1. Negligible heat loss to the surroundings.
2. Negligible kinetic and potential energy effects. No shaft work.
3. Constant properties.
4. Negligible tube wall thermal resistance and fouling factors.
5. Fully developed conditions for water flow.

***Properties:*** Table HT-5, water (assume $\overline{T}_c$ = 35°C = 308 K): $c_p = 4178$ J/kg · K, $\mu = 725 \times 10^{-6}$ N · s/m², $k$ = 0.625 W/m · K, Pr = 4.85. Table HT-4, oil ($\overline{T}_h$ = 80°C = 353 K): $c_p$ = 2131 J/kg · K.

***Analysis:*** The heat transfer rate can be obtained from the hot (oil) fluid energy balance, Eq. 17.86a

$$q = \dot{m}_h c_{p,h}(T_{h,i} - T_{h,o}) = 0.1\ \text{kg/s} \times 2131\ \text{J/kg} \cdot \text{K}\ (100 - 60)\text{°C} = 8524\ \text{W}$$

Applying the cold fluid energy balance, Eq. 17.87a, the water outlet temperature is

$$T_{c,o} = \frac{q}{\dot{m}_c c_{p,c}} + T_{c,i} = \frac{8524\ \text{W}}{0.2\ \text{kg/s} \times 4178\ \text{J/kg} \cdot \text{K}} + 30\text{°C} = 40.2\text{°C}$$

Accordingly, the use of $\overline{T}_c$ = 35°C, the average temperature of the cold fluid, to evaluate the water properties was a good choice. The required heat exchanger length may now be obtained from the convection rate equation, Eq. 17.89

$$q = UA\,\Delta T_{lm}$$

where $A = \pi D_i L$, and from Eqs. 17.96 and 17.98, the log mean temperature difference is

$$\Delta T_{\text{lm}} = \frac{(T_{h,i} - T_{c,o}) - (T_{h,o} - T_{c,i})}{\ln\left[(T_{h,i} - T_{c,o})/(T_{h,o} - T_{c,i})\right]} = \frac{59.8 - 30}{\ln(59.8/30)} = 43.2°\text{C}$$

From Eq. 17.90b, the overall heat transfer coefficient in terms of the water-side ($i$) and oil-side ($o$) convection coefficients is

$$U = \frac{1}{(1/h_i) + (1/h_o)}$$

To estimate $h_i$ for the *water-side* (cold fluid), calculate the Reynolds number from Eq. 17.37 to characterize the flow and select a correlation

$$\text{Re}_D = \frac{4\dot{m}_c}{\pi D_i \mu} = \frac{4 \times 0.2\ \text{kg/s}}{\pi(0.025\ \text{m})(725 \times 10^6\ \text{N}\cdot\text{s/m}^2)} = 14{,}050$$

Accordingly, the flow is turbulent, and the convection coefficient may be estimated using the *Dittus-Boelter correlation*, Eq. 17.64, with $n = 0.4$ since $T_s > T_m$

$$\text{Nu}_D = 0.023\,\text{Re}_D^{4/5}\,\text{Pr}^{0.4} = 0.023(14{,}050)^{4/5}(4.85)^{0.4} = 90$$

$$h_i = \text{Nu}_D \frac{k}{D_i} = \frac{90 \times 0.625\ \text{W/m}\cdot\text{K}}{0.025\ \text{m}} = 2250\ \text{W/m}^2\cdot\text{K}$$

Since the convection coefficient for the *oil-side* (hot fluid) is $h_o = 40\ \text{W/m}^2\cdot\text{K}$, the overall coefficient is then

$$U = \frac{1}{(1/2250\ \text{W/m}^2\cdot\text{K}) + (1/40\ \text{W/m}^2\cdot\text{K})} = 39.3\ \text{W/m}^2\cdot\text{K}$$

and from the convection rate equation it follows that the required length of the exchanger is

$$L = \frac{q}{U \pi D_i \Delta T_{\text{lm}}} = \frac{8524\ \text{W}}{39.3\ \text{W/m}^2\cdot\text{K}\ \pi(0.025\ \text{m})(43.2°\text{C})} = 63.9\ \text{m} \quad \triangleleft$$

***Comments:*** **1.** The oil-side convection coefficient controls the rate of heat transfer between the two fluids, and the low value of $h_o$ is responsible for the large value of $L$. In practice, multiple-pass construction would be required for a concentric tube exchanger with such a large tube length. Alternately, another exchanger type should be considered for this application.
**2.** Since the water flow is turbulent and $L/D = 2556$, the assumption of fully developed flow is justified according to Eq. 17.42.

### 17.5.6 The Shell-and-Tube and the Cross-Flow Heat Exchangers

The flow conditions in the shell-and-tube and the cross-flow tube heat exchangers shown in Fig. 17.23*c* and 17.23*d* are more complicated than for the concentric tube exchangers. However, the *fluid energy balances,* Eqs. 17.86 and 17.87, and the *convection rate equation,* Eq. 17.89, can still be used if the following modification is made to the *log mean temperature difference*

$$\Delta T_{\text{lm}} = F \Delta T_{\text{lm},CF} \tag{17.99}$$

*correction factor*

That is, the appropriate form of $\Delta T_m$ is obtained by applying a ***correction factor*** to the value of $\Delta T_{\text{lm}}$ that would be computed *under the assumption of counterflow conditions,* Eqs. 17.96 and 17.98.

Algebraic expressions for the *correction factor F* have been developed for various common heat exchanger configurations. The results for the exchangers of interest are shown in

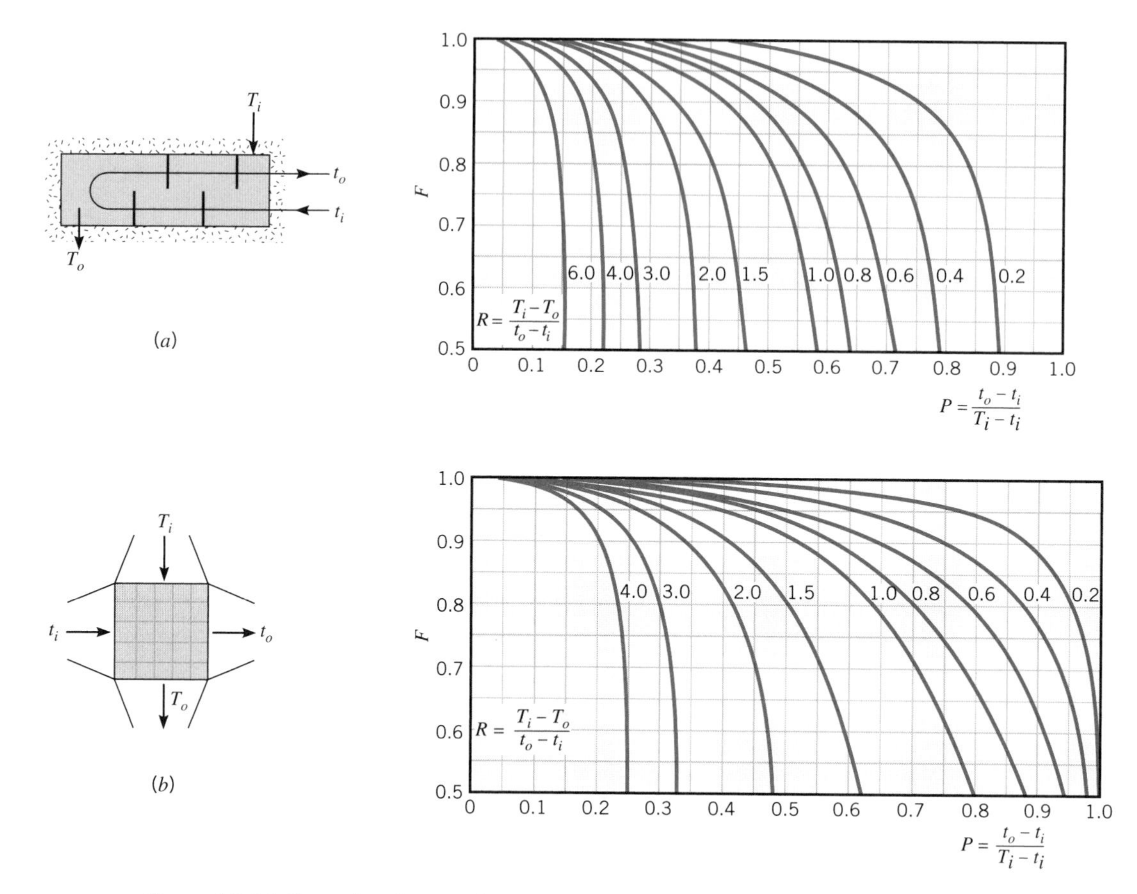


*Figure 17.29* Correction factor $F$ for heat exchangers: (*a*) Shell-and-tube configuration with one shell and any multiple of two tube passes (two, four, etc., tube passes) as shown in Fig. 17.23*c* and (*b*) cross-flow configuration as shown in Fig. 17.23*d*.

Fig. 17.29. The notation ($T$, $t$) is used on the figures to specify the fluid temperatures, with the variable $t$ always assigned to the tube-side fluid.

An important implication of Fig. 17.29 is that, *if the temperature change of one fluid is negligible,* either $P$ or $R$ is zero and $F$ is 1. Hence, *the exchanger behavior is independent of the specific configuration.* Such would be the case if one of the fluids underwent a phase change (see Fig. 17.28*a*,*b*).

*LMTD method*

The method of heat exchanger analysis that has been described here is referred to as the *LMTD method.* The use of the method is clearly facilitated by knowledge of the hot and cold fluid inlet and outlet temperatures. Such applications may be classified as *heat exchanger design problems;* that is, problems in which the temperatures and capacity rates are known, and it is desired to *size the exchanger*. Alternatively, if the exchanger type and size are known, and the fluid outlet temperatures need to be determined, the application is referred to as a *performance calculation problem*. Such problems are best analyzed by the *NTU-effectiveness method,* which is widely used in engineering practice and treated in more advanced courses in thermal system engineering.

*NTU-effectiveness method*

## *Example 17.15* Shell-and-Tube Heat Exchanger Analysis

A shell-and-tube heat exchanger must be designed to heat 2.5 kg/s of water from 15 to 85°C. The heating is to be accomplished by passing hot engine oil, which is available at 160°C, through the shell side of the exchanger. The oil is known to provide an average convection coefficient of $h_o = 400\ \text{W/m}^2 \cdot \text{K}$ on the outside of the tubes. Ten tubes pass the water through the shell. Each tube is thin walled, of diameter $D = 25$ mm, and makes eight passes through the shell. If the oil leaves the exchanger at 100°C, what is its flow rate? How long must each tube be to accomplish the desired heating?

## Solution

***Known:*** Fluid inlet and outlet temperatures for a shell-and-tube heat exchanger (one shell, eight tube passes; see also Fig. 17.23*c*) with $N = 10$ tubes.

***Find:***

(a) Oil flow rate required to achieve specified outlet temperature.

(b) Tube length required to achieve specified water heating.

***Schematic and Given Data:***

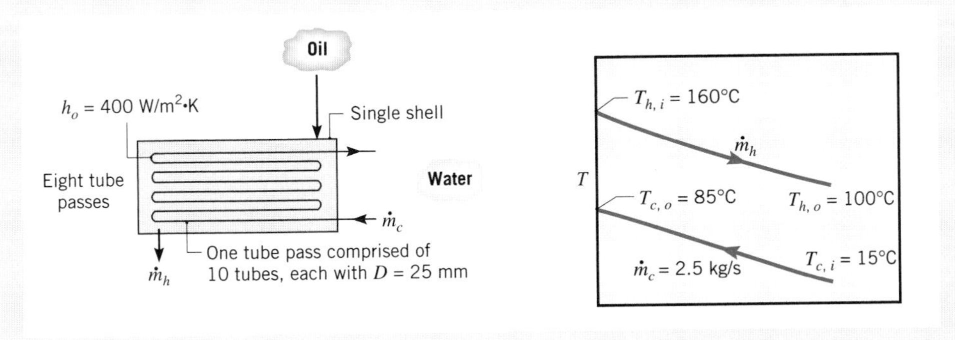


*Figure E17.15*

***Assumptions:***

1. Negligible heat loss to the surroundings.
2. Negligible kinetic and potential energy effects. No shaft work.
3. Constant properties.
4. Negligible tube wall thermal resistance and fouling effects.
5. Fully developed water flow in tubes.

***Properties:*** Table HT-4, unused engine oil ($\overline{T}_h = 130°\text{C} = 403$ K): $c_p = 2350$ J/kg · K. Table HT-5, water ($\overline{Tc} = 50°\text{C} = 323\ K$): $c_p = 4181$ J/kg · K, $\mu = 548 \times 10^{-6}\text{N} \cdot \text{s/m}^2$, $k = 0.643$ W/m · K, Pr = 3.56.

***Analysis:***

(a) From an energy balance on the cold fluid (water), Eq. 17.87a, the heat transfer required of the exchanger is

$$q = \dot{m}_c c_{p,c}(T_{c,o} - T_{c,i}) = 2.5\ \text{kg/s} \times 4181\ \text{J/kg} \cdot \text{K}\ (85 - 15)°\text{C} = 7.317 \times 10^5\ \text{W}$$

Hence, from an energy balance on the hot fluid, Eq. 17.86a, the required oil flow rate is

$$\dot{m}_h = \frac{q}{c_{p,h}(T_{h,i} - T_{h,o})} = \frac{7.317 \times 10^5\ \text{W}}{2350\ \text{J/kg} \cdot \text{K} \times (160 - 100)°\text{C}} = 5.19\ \text{kg/s} \triangleleft$$

(b) The required tube length can be obtained from the convection rate equation, Eq. 17.89, using the mean temperature difference from Eq. 17.99, where

$$q = UAF\ \Delta T_{\text{lm,CF}}$$

From Eq. 17.90b, the overall coefficient can be expressed in terms of the convection coefficients on the inside (water-side), $h_i$, and outside (oil-side), $h_o$, of the tube

$$U = \frac{1}{(1/h_i) + (1/h_o)}$$

where $h_i$ may be obtained by first calculating $Re_D$. With $\dot{m}_1 \equiv \dot{m}_c/N = 0.25$ kg/s defined as the water flow rate per tube, Eq. 17.37 yields

$$Re_D = \frac{4\dot{m}_1}{\pi D \mu} = \frac{4 \times 0.25 \text{ kg/s}}{\pi(0.025 \text{ m})548 \times 10^{-6} \text{ kg/s} \cdot \text{m}} = 23{,}234$$

Since $Re_D > 2300$, the water flow is turbulent, and an appropriate correlation is Eq. 17.64 (*Dittus-Boelter*) with $n = 0.4$ since $T_s > T_m$

$$Nu_D = 0.023\, Re_D^{4/5}\, Pr^{0.4} = 0.023(23{,}234)^{4/5}(3.56)^{0.4} = 119$$

$$h_i = \frac{k}{D} Nu_D = \frac{0.643 \text{ W/m} \cdot \text{K}}{0.025 \text{ m}} 119 = 3061 \text{ W/m}^2 \cdot \text{K}$$

Hence, the overall coefficient is

$$U = \frac{1}{(1/400) + (1/3061)} = 354 \text{ W/m}^2 \cdot \text{K}$$

Associating $T$ with the oil and $t$ with the water, the correction factor $F$ may be obtained from Fig. 17.29*a*, where

$$R = \frac{160 - 100}{85 - 15} = 0.86 \qquad P = \frac{85 - 15}{160 - 15} = 0.48$$

Hence, $F = 0.87$. From Eqs. 17.96 and 17.98, the log mean temperature difference for counterflow conditions is

$$\Delta T_{\text{lm,CF}} = \frac{(T_{h,i} - T_{c,o}) - (T_{h,o} - T_{c,i})}{\ln\left[(T_{h,i} - T_{c,o})/(T_{h,o} - T_{c,i})\right]} = \frac{75 - 85}{\ln(75/85)} = 79.9°\text{C}$$

Solving the convection rate equation for $L$, with $A = N\pi DL$, where $N = 10$ is the number of tubes, and substituting numerical values, find the required tube length

$$L = \frac{q}{UN(\pi D)F\, \Delta T_{\text{lm,CF}}} = \frac{7.317 \times 10^5 \text{ W}}{354 \text{ W/m}^2 \cdot \text{K} \times 10(\pi 0.025 \text{ m}) \times 0.87(79.9°\text{C})}$$

$$L = 37.9 \text{ m} \quad \triangleleft$$

***Comments:***

1. With $(L/D) = 37.9 \text{ m}/0.025 \text{ m} = 1516$, the assumption of fully developed conditions throughout the tube for the water flow is justified.
2. With eight passes, the shell length is approximately $L/8 = 4.7$ m.

## 17.6 Chapter Summary and Study Guide

The objectives of this chapter were to develop an understanding of the physical mechanisms that underlie convection heat transfer, and develop the means to estimate convection coefficients required for convection calculations. We found that *boundary layer phenomena* control the convection coefficient, and that empirical *correlations* to estimate the Nusselt number involving key dimensionless numbers (see Table 17.1) are available for common geometries and flow conditions. For forced and free convection flows, respectively, the Reynolds and Grashof (or Rayleigh) numbers characterize the flow conditions, while the Prandtl number incorporates the fluid properties into the analysis. Summaries of the correlations and guidelines for their selection are provided in Tables 17.3, 17.5, and 17.6 for forced convection *external* and *internal flows* and *free convection,* respectively.

We began our treatment by considering *external flow* over a flat plate and identified the characteristics of the *hydrodynamic* and *thermal boundary layers*. With negligible upstream disturbances, the boundary layer flow is *laminar* at the leading edge and experiences a *transition* to *turbulent* flow. We learned that the *convection coefficient* depends upon the temperature gradient at the surface, which is controlled by the thickness of the boundary layer, as well as by the nature of the flow condition. The convective heat flux is given by Newton's law of cooling in terms of the local coefficient and the difference in surface and free stream temperatures.

In external flow, the boundary layers grow unconstrained, while in *internal flow,* the boundary layers eventually fill the tube. We identified the *entrance* and *fully developed flow regions,* and recognized the distinctive nature of the velocity and temperature profiles in each region. In the fully developed flow region, the convection coefficient does not change in the flow direction, but remains constant. The concept of a *mean fluid temperature* was introduced for use in Newton's law of cooling to calculate the surface heat flux, $q'' = h(T_s - T_m)$. Correlations were presented for two types of *surface thermal conditions,* constant heat flux, $q_s''$, and constant surface temperature, $T_s$.

For heat transfer by *forced* convection, the flow originates because of forcing conditions by a pump or fan. Heat transfer by *free* convection occurs because of convection currents that are induced by *buoyancy forces,* due to fluid density differences arising from temperature gradients in the fluid near the immersed geometry surface.

The *heat exchanger* is a very common, important thermal system that requires application of key convection heat transfer concepts for analyzing performance. Considering the common concentric-tube, parallel and counterflow arrangements, two energy relations were developed. The *overall fluid energy balances* provide relations between the heat rate, capacity rate, and fluid inlet and outlet temperatures that are independent of the exchanger type and flow conditions. The *convection rate equation,* Newton's law of cooling, involved the overall heat transfer coefficient, surface area, and average temperature difference between the two fluids. The overall coefficient is determined by the convection coefficients associated with the fluids, and the average temperature difference is the *log mean temperature difference,* which depends upon the exchanger configuration.

The following checklist provides a study guide for this chapter. When your study of the text and end-of-chapter problems has been completed you should be able to

- write out the meanings of the terms listed in the margins throughout the chapter and understand each of the related concepts. The subset of the key terms listed here in the margin is particularly important.
- define the *Nusselt number* and discuss its physical interpretation.
- list the general forms of the *empirical correlations* to estimate convection coefficients for forced convection external and internal flow and free convection. Know the rules you should follow in selecting correlations for any flow situation.
- describe the major features of the *hydrodynamic* and *thermal boundary layers* for parallel flow over a *flat plate.* Explain the physical features that distinguish a *turbulent* flow from a *laminar* one. Define the *Reynolds number* and indicate its physical interpretation. Show how the convection coefficient varies over the plate.
- explain how the convection coefficient in internal flow varies with distance from the inlet for the *entry region* and the *fully developed region.* List the key *hydrodynamic* and *thermal* features of fully developed flow.
- explain under what conditions the Nusselt number associated with *internal flow* is equal to a constant value, independent of Reynolds number and Prandtl number.
- know the conditions required for free convection and provide the physical interpretation of the *Grashof number* and the *Rayleigh number.*
- explain the two possible flow arrangements for a *concentric tube heat exchanger:* parallel and counterflow. For each arrangement, list the restrictions on the fluid outlet temperatures and discuss the role of the *log mean temperature difference* in the convection rate equation.

*thermal boundary layer*
*convection coefficient*
*forced convection*
*external, internal flow*
*laminar, turbulent flow*
*fully developed conditions*
*free convection*
*convection correlations*
*Nusselt number*
*Reynolds number*
*Prandtl number*
*Grashof, Rayleigh numbers*

# Problems

*Note:* Unless otherwise indicated in the problem statement, use values of the required thermophysical properties given in the appropriate tables of Appendix HT when solving these problems.

## The Problem of Convection

17.1 In flow over a surface, the temperature profile has the form

$$T(y) = A + By + Cy^2 - Dy^3$$

where the coefficients $A$ through $D$ are constants. Obtain an expression for the convection coefficient $h$ in terms of $u_\infty$, $T_\infty$, and appropriate profile coefficients and fluid properties.

17.2 Consider conditions for which a fluid with a free stream velocity of $u_\infty = 1$ m/s flows over a surface with a characteristic length of $L = 1$ m, providing an average convection heat transfer coefficient of $\bar{h} = 100$ W/m$^2 \cdot$ K. Calculate the dimensionless parameters $\overline{Nu}_L$, $Re_L$, and Pr for the following fluids: air, engine oil, and water. Assume the fluids to be at 300 K.

17.3 To a good approximation, the dynamic viscosity $\mu$, the thermal conductivity $k$, and the specific heat $c_p$ are independent of pressure. In what manner do the kinematic viscosity $\nu$ and thermal diffusivity $\alpha$ vary with pressure for an incompressible liquid and for an ideal gas? Determine $\nu$ and $\alpha$ of air at 350 K for pressures of 1 and 10 atm.

17.4 Parallel flow of atmospheric air over a flat plate of length $L = 3$ m is disrupted by an array of stationary rods placed in the flow path over the plate.

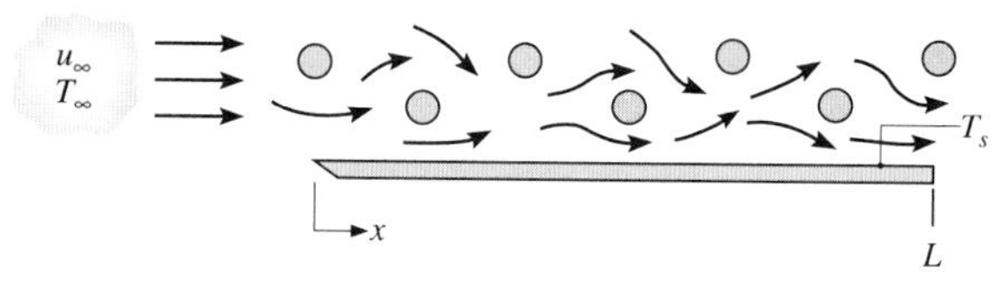


*Figure P17.4*

Laboratory measurements of the local convection coefficient at the surface of the plate are made for a prescribed value of $u_\infty$ and $T_s > T_\infty$. The results are correlated by an expression of the form $h_x = 0.7 + 13.6x - 3.4x^2$, where $h_x$ has units of W/m$^2 \cdot$ K and $x$ is in meters. Evaluate the average convection coefficient $\bar{h}_L$ for the entire plate and the ratio $\bar{h}_L/h_L$ at the trailing edge.

17.5 For laminar flow over a flat plate, the local heat transfer coefficient $h_x$ is known to vary as $x^{-1/2}$, where $x$ is the distance from the leading edge ($x = 0$) of the plate. What is the ratio of the average coefficient between the leading edge and some location $x$ on the plate to the local coefficient at $x$?

17.6 For laminar free convection from a heated vertical surface, the local convection coefficient may be expressed as $h_x = Cx^{-1/4}$, where $h_x$ is the coefficient at a distance $x$ from the leading edge of the surface, and the quantity $C$, which depends on the fluid properties, is independent of $x$. Obtain an expression for the ratio $\bar{h}_x/h_x$, where $\bar{h}_x$ is the average coefficient between the leading edge ($x = 0$) and the $x$ location. Sketch the variation of $h_x$ and $\bar{h}_x$ with $x$.

17.7 Experimental results for heat transfer over a flat plate with an extremely rough surface were found to be correlated by an expression of the form

$$Nu_x = 0.04\,Re_x^{0.9}\,Pr^{1/3}$$

where $Nu_x$ is the local value of the Nusselt number at a position $x$ measured from the leading edge of the plate. Obtain an expression for the ratio of the average heat transfer coefficient $\bar{h}_x$ to the local coefficient $h_x$.

17.8 (CD-ROM)

17.9 (CD-ROM)

## External Flow: Flat Plate Laminar and Turbulent Flows

17.10 Consider flow of air over the flat plate shown in Example 17.2. Because of the application requirements, it is important to maintain a *laminar* boundary layer flow over the plate. What is the maximum allowable air velocity that will satisfy this flow condition if all other parameters remain unchanged? What is the required cooling rate for this condition?

17.11 Consider the flat plate with segmented heaters of Example 17.3. If a wire were placed near the leading edge of the plate to induce turbulence over its entire length, what is the total electrical power required for the first five heaters?

17.12 Consider the following fluids at a film temperature of 300 K in parallel flow over a flat plate with velocity of 1 m/s: atmospheric air, water, and engine oil.

(a) For each fluid, determine the hydrodynamic and thermal boundary layer thickness at a distance of $x = 40$ mm from the leading edge.

(b) For each fluid, determine the *local* convection coefficient at $x = 40$ mm, and the *average* value over the distance from $x = 0$ to $x = 40$ mm.

17.13 Engine oil at 100°C and a velocity of 0.1 m/s flows over both surfaces of a 1-m-long flat plate maintained at 20°C. Determine the following:

(a) The hydrodynamic and thermal boundary layer thicknesses at the trailing edge.

(b) The local heat flux at the trailing edge.

(c) The total heat transfer per unit width of the plate.

17.14 Steel plates of length $L = 1$ m on a side are conveyed from a heat treatment process and are concurrently cooled by atmospheric air of velocity $u_\infty = 10$ m/s and $T_\infty = 20$°C in parallel flow over the plates (Fig. P17.4). For a plate temperature 300°C, what is the rate of heat transfer from the plate? The velocity of the air is much larger than that of the plate.

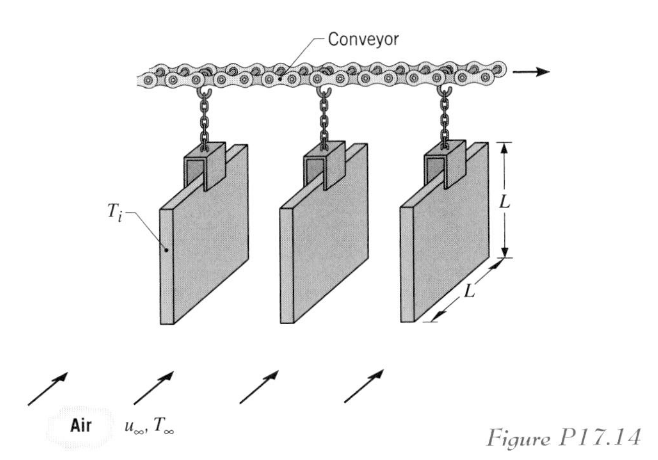


*Figure P17.14*

17.15 (CD-ROM)

17.16 (CD-ROM)

## External Flow: Flat Plate Mixed Flow Conditions

17.17 Consider flow of air over the plate with segmented electrical strip heaters as shown in Example 17.3. Calculate the power requirement for the *fourth* plate when the air velocity is 78 m/s, all other conditions remaining the same. Sketch the variation of the local convection coefficient with distance along the plate, and comment on key features.

17.18 An array of power-dissipating electrical components is mounted on the bottom side of a 1.2 m by 1.2 m horizontal aluminum plate, while the top side is cooled by an air stream for which $u_\infty = 15$ m/s and $T_\infty = 300$ K. The plate is attached to a well-insulated enclosure such that all the dissipated power must be transferred to the air. Also, the aluminum is sufficiently thick to ensure a nearly uniform plate temperature.

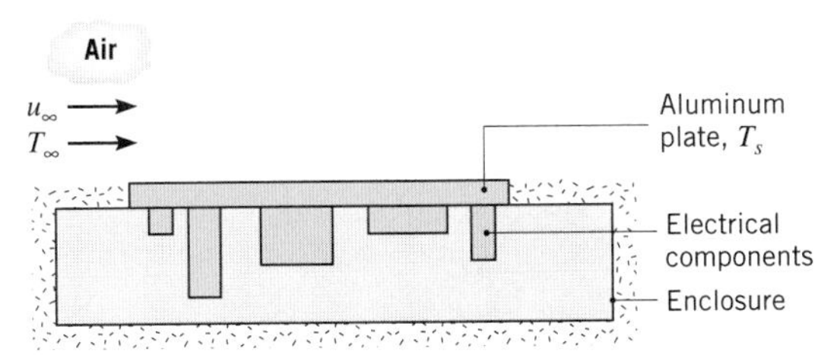


*Figure P17.18*

If the temperature of the plate is not to exceed 350 K, what is the maximum allowable power dissipation?

17.19 Air at a pressure and a temperature of 1 atm and 50°C, respectively, is in parallel flow over the top surface of a flat plate that is heated to a uniform temperature of 100°C. The plate has a length of 0.20 m (in the flow direction) and a width of 0.10 m. The Reynolds number based on the plate length is 40,000.

(a) What is the rate of heat transfer from the plate to the air?

(b) If the free stream velocity of the air is doubled and the pressure is increased to 10 atm, what is the rate of heat transfer? *Hint:* See Problem 17.3 for comments on the pressure dependence of the relevant thermophysical properties.

17.20 Consider atmospheric air at 25°C and a velocity of 25 m/s flowing over both surfaces of a 1-m-long flat plate that is maintained at 125°C. Determine the rate of heat transfer per unit width from the plate for values of the critical Reynolds number corresponding to $10^5$, $5 \times 10^5$, and $10^6$.

17.21 (CD-ROM)

## External Flow: Flat Plate Segmented Sections

17.22 Consider flow of air over the plate with segmented electrical strip heaters as shown in Example 17.3. Calculate the rate of heat transfer from the first and the sixth heater. Compare these results with that from Example 17.3 for the fifth heater. Relate their relative values to the plot shown in the example for the variation of the convection coefficient, $h(x)$.

17.23 An electric air heater consists of a horizontal array of thin metal strips that are each 10 mm long in the direction of an air stream that is in parallel flow over the top of the strips. Each strip is 0.2 m wide, and 25 strips are arranged side by side, forming a continuous and smooth surface over which the air flows at 2 m/s. During operation each strip is maintained at 500°C and the air is at 25°C.

(a) What is the rate of convection heat transfer from the first strip? The fifth strip? The tenth strip? All the strips?

(b) Repeat part (a), but under conditions for which the flow is fully turbulent over the entire array of strips.

17.24 Consider weather conditions for which the prevailing wind blows past the penthouse tower on a tall building. The tower length in the wind direction is 10 m, and there are 10 window panels.

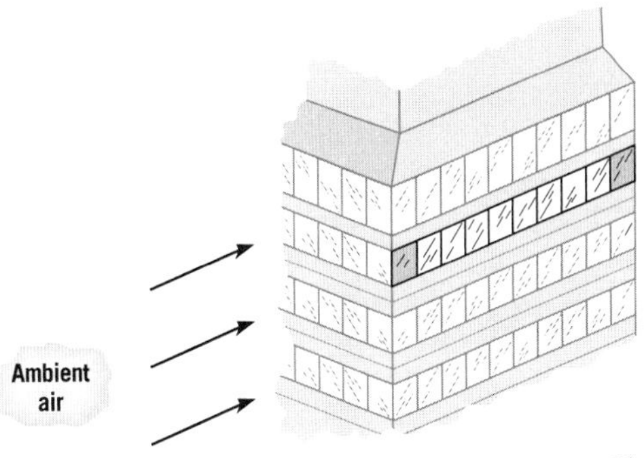


*Figure P17.24*

Calculate the average convection coefficient for the first, third, and tenth window panels when the wind speed is 5 m/s. Use a film temperature of 300 K to evaluate the thermophysical properties required of the correlation. Would this be a suitable value of the film temperature for ambient air temperatures in the range $-15 \leq T_\infty \leq 38$°C?

**17.25** Air at 27°C with a free stream velocity of 10 m/s is used to cool electronic devices mounted on a printed circuit board as shown in Fig. P17.25. Each device, 4 mm by 4 mm, dissipates 40 mW, which is transferred by convection from the top surface. A turbulator is located at the leading edge of the board, causing the boundary layer to be turbulent.

**(a)** Estimate the surface temperature of the fourth device located 15 mm from the leading edge of the board.

**(b)** What is the minimum free stream velocity if the surface temperature of this device is not to exceed 80°C?

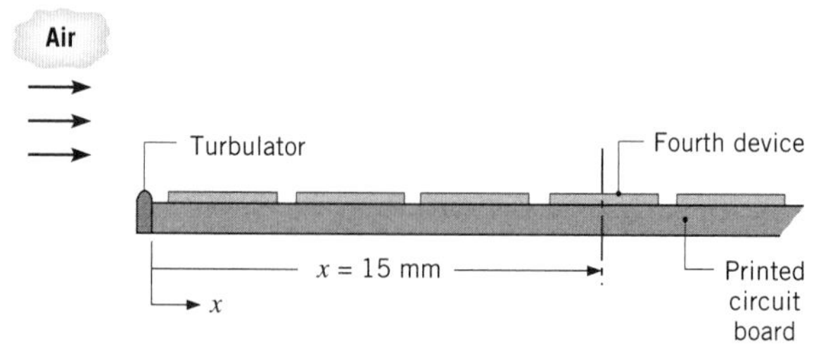


*Figure P17.25*

**17.26** (CD-ROM)

### External Flow: Flat Plate Energy Balance Applications

**17.27** The Weather Channel reports that it is a hot, muggy day with an air temperature of 90°F, a 10 mph breeze out of the southwest, and bright sunshine with a solar insolation of 400 $W/m^2$. Consider the wall of a metal building over which the prevailing wind blows. The length of the wall in the wind direction is 10 m, and the emissivity is 0.93. Assume that all the solar irradiation is absorbed, that irradiation from the sky is negligible, and that flow is fully turbulent over the wall. Estimate the average wall temperature.

**17.28** Consider the wing of an aircraft as a flat plate of 2.5-m length in the flow direction. The plane is moving at 100 m/s in air that is at a pressure of 0.7 bar and a temperature of −10°C. The top surface of the wing absorbs solar radiation at a rate of 800 $W/m^2$. Assume the wing to be of solid construction and to have a single, uniform temperature. Estimate the steady-state temperature of the wing.

**17.29** Initially the top surface of an oven measuring 0.5 m by 0.5 m is at a uniform temperature of 47°C under quiescent room air conditions (Fig. P17.29). The inside air temperature of the oven is 150°C, the room air temperature is 17°C, and the heat transfer from the surface is 40 W. In order to reduce the surface temperature and meet safety requirements, room air is blown across the top surface with a velocity of 20 m/s in a direction parallel to an edge.

**(a)** Calculate the thermal resistance due to the oven wall and internal convection associated with the quiescent room air condition (when the surface is at $T_s$ = 47°C). Represent this condition (case A) by a thermal circuit and label all elements.

**(b)** Assuming internal convection conditions to remain unchanged, determine the heat transfer from the top surface under forced convection conditions. Represent this condition (case B) by a thermal circuit and label all elements.

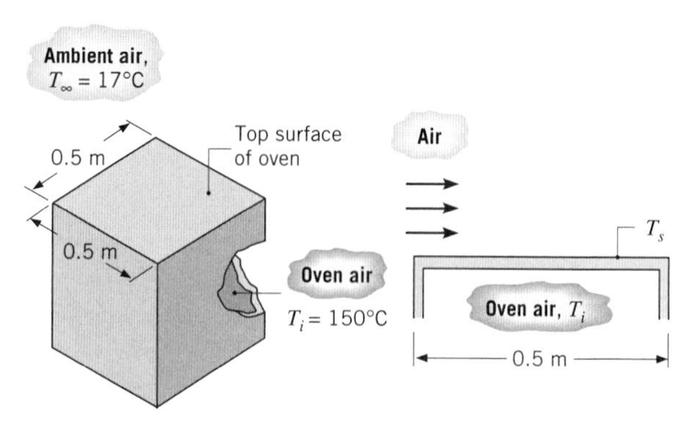


*Figure P17.29*

**(c)** Estimate the surface temperature achieved with the forced convection condition (case B).

**17.30** One-hundred electrical components, each dissipating 25 W, are attached to one surface of a square (0.2 m × 0.2 m) copper plate, and all the dissipated power is transferred to water in parallel flow over the opposite surface. A turbulator at the leading edge of the plate acts to *trip* the boundary layer, and the plate itself may be assumed to be isothermal. The water velocity and temperature are $u_\infty$ = 2 m/s and $T_\infty$ = 17°C, and its thermophysical properties may be approximated as $\nu = 0.96 \times 10^{-6}$ $m^2/s$, $k$ = 0.620 W/m · K, and Pr = 5.2.

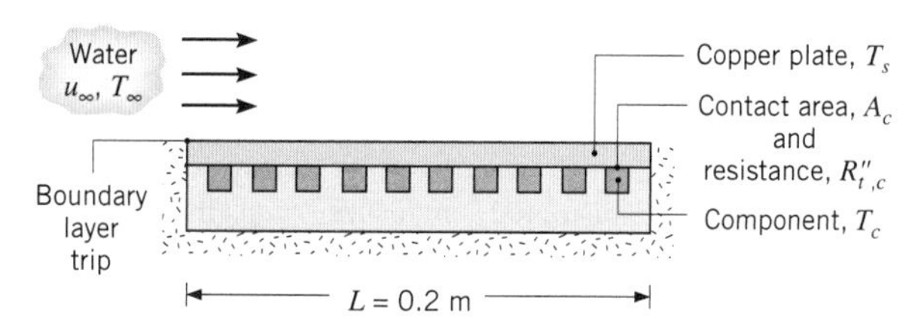


*Figure P17.30*

**(a)** What is the temperature of the copper plate?

**(b)** If each component has a plate contact surface area of 100 $mm^2$ and the corresponding contact resistance is $2 \times 10^{-4}$ $m^2 \cdot K/W$, what is the component temperature? Neglect the temperature variation across the thickness of the copper plate.

**17.31** (CD-ROM)

### External Flow: Cylinder in Cross Flow

**17.32** Consider the following fluids, each with a velocity of $u_\infty$ = 5 m/s and a temperature of $T_\infty$ = 20°C, in cross flow over a 10-mm-diameter cylinder maintained at 50°C: atmospheric air, saturated water, and engine oil. Calculate the rate of heat transfer per unit length, $q'$.

**17.33** Assume that a person can be approximated as a cylinder of 0.3-m diameter and 1.8-m height with a surface temperature of 24°C. Calculate the body energy loss while this person is subjected to a 15-m/s wind whose temperature is −5°C.

17.34 To enhance heat transfer from a silicon chip of width $W = 4$ mm on a side, a copper pin fin is brazed to the surface of the chip as shown in Fig. P17.34. The pin length and diameter are $L = 12$ mm and $D = 2$ mm, respectively, and atmospheric air at $u_\infty = 10$ m/s and $T_\infty = 300$ K is in cross flow over the pin. The surface of the chip, and hence the base of the pin, are maintained at a temperature of $T_b = 350$ K.

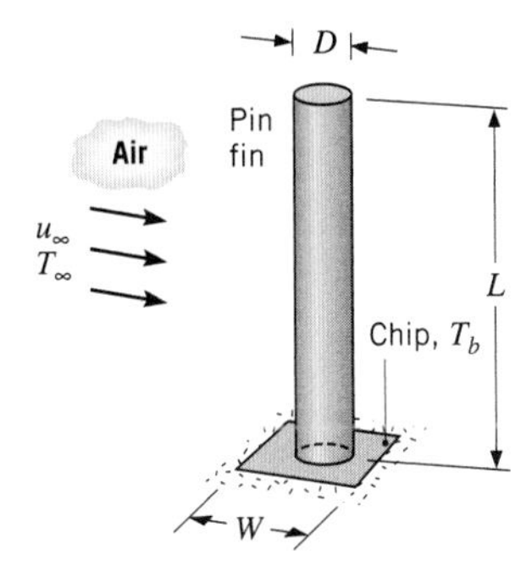


*Figure P17.34*

(a) Assuming the chip to have a negligible effect on flow over the pin, what is the average convection coefficient for the surface of the pin?
(b) Neglecting radiation and assuming the convection coefficient at the pin tip to equal that calculated in part (a), determine the pin heat transfer rate.
(c) Neglecting radiation and assuming the convection coefficient at the exposed chip surface to equal that calculated in part (a), determine the total rate of heat transfer from the chip.

17.35 A horizontal copper rod 10 mm in diameter and 100 mm long is inserted in the air space between surfaces of an electronic device to enhance heat dissipation. The ends of the rod are at 90°C, while air at 25°C is in cross flow over the cylinder with a velocity of 25 m/s. What is the temperature at the midplane of the rod? What is the rate of heat transfer from the rod?

17.36 A 25-mm-diameter, high-tension line has an electrical resistance of $10^{-4}$ Ω/m and is transmitting a current of 1000 A.
(a) If ambient air at 10°C and 10 m/s is in cross flow over the line, what is its surface temperature?
(b) If the line may be approximated as a solid copper rod, what is its centerline temperature?

17.37 Hot water at 50°C is routed from one building in which it is generated to an adjoining building in which it is used for space heating. Transfer between the buildings occurs in a steel pipe ($k = 60$ W/m · K) of 100-mm outside diameter and 8-mm wall thickness. During the winter, representative environmental conditions involve air at $T_\infty = -5$°C and $u_\infty = 3$ m/s in cross flow over the pipe.
(a) If the cost of producing the hot water is \$0.05 per kW · h, what is the representative daily cost of energy loss from an uninsulated pipe to the air per meter of pipe length? The convection resistance associated with water flow in the pipe may be neglected.
(b) Determine the savings associated with application of a 10-mm-thick coating of urethane insulation ($k = 0.026$ W/m · K) to the outer surface of the pipe.

17.38 (CD-ROM)

17.39 (CD-ROM)

17.40 (CD-ROM)

**External Flow: Spheres**

17.41 Water at 20°C flows over a 20-mm-diameter sphere with a velocity of 5 m/s. The surface of the sphere is at 60°C. What is the rate of heat transfer from the sphere?

17.42 Air at 25°C flows over a 10-mm-diameter sphere with a velocity of 25 m/s, while the surface of the sphere is maintained at 75°C.
(a) What is the rate of heat transfer from the sphere?
(b) Generate a plot of the heat transfer rate as a function of the air velocity for the range 1 to 25 m/s.

17.43 Atmospheric air at 25°C and a velocity of 0.5 m/s flows over a 50-W incandescent bulb whose surface temperature is at 140°C. The bulb may be approximated as a sphere of 50-mm diameter. What is the rate of heat transfer by convection to the air?

17.44 A spherical, underwater instrument pod used to make soundings and to measure conditions in the water has a diameter of 85 mm and dissipates electrical power of 300 W.
(a) Estimate the surface temperature of the pod when suspended in a bay where the current is 1 m/s and the water temperature is 15°C.
(b) Inadvertently, the pod is hauled out of the water and suspended in ambient air without deactivating the power. Estimate the surface temperature of the pod if the air temperature is 15°C and the wind speed is 3 m/s.

17.45 A spherical workpiece of pure copper with a diameter of 15 mm and an emissivity of 0.5 is suspended in a large furnace with walls at a uniform temperature of 600°C. Air flows over the workpiece at a temperature of 900°C and a velocity of 7.5 m/s.
(a) Determine the steady-state temperature of the workpiece.
(b) Estimate the time required for the workpiece to come within 5°C of the steady-state temperature if it is at an initial, uniform temperature of 25°C.

17.46 A thermocouple junction is inserted in a large duct to measure the temperature of hot gases flowing through the duct.

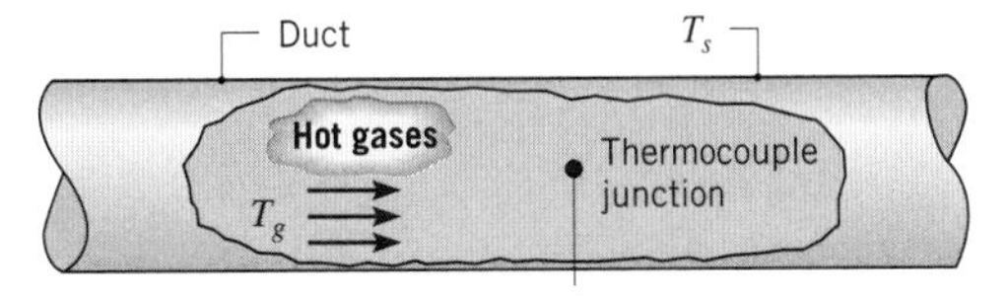


*Figure P17.46*

(a) If the duct surface temperature $T_s$ is less than the gas temperature $T_g$, will the thermocouple sense a temperature that is less than, equal to, or greater than $T_g$? Justify your answer on the basis of a simple analysis.
(b) A thermocouple junction in the shape of a 2-mm-diameter sphere with a surface emissivity of 0.60 is placed in a gas stream moving at 3 m/s. If the thermocouple senses a

temperature of 320°C when the duct surface temperature is 175°C, what is the actual gas temperature? The gas may be assumed to have the properties of air at atmospheric pressure.

**17.47** (CD-ROM)

### Internal Flow: Boundary Layer and Energy Balance Considerations

**17.48** Compare the hydrodynamic and thermal entry lengths for oil, water, and ethylene glycol flowing through a 25-mm-diameter tube with a mean velocity and temperature of $u_m = 50$ mm/s and $T_m = 27°C$, respectively.

**17.49** Velocity and temperature profiles for laminar flow in a tube of radius $r_o = 10$ mm have the form

$$u(r) = 0.1[1 - (r/r_o)^2]$$

$$T(r) = 344.8 + 75.0(r/r_o)^2 - 18.8(r/r_o)^4$$

with units of m/s and K, respectively. Determine the corresponding value of the mean (or bulk) temperature, $T_m$, at this axial position.

**17.50** Atmospheric air enters the heated section of a circular tube at a flow rate of 0.005 kg/s and a temperature of 20°C. The tube is of diameter $D = 50$ mm, and fully developed conditions with $h = 25$ W/m$^2 \cdot$ K exist over the entire length of $L = 3$ m. Within the heated section length, a uniform heat flux of $q_s'' = 1000$ W/m$^2$ is maintained.

(a) Determine the total heat transfer rate $q$ and the mean temperature of the air leaving the tube $T_{m,o}$.

(b) What is the value of the surface temperature at the tube inlet $T_{s,i}$ and outlet $T_{s,o}$?

(c) Sketch the axial variation of $T_s$ and $T_m$ with distance from the inlet $x$. On the same figure, also sketch (qualitatively) the axial variation of $T_s$ and $T_m$ for the more realistic case in which the local convection coefficient varies with $x$.

**17.51** Atmospheric air enters a 10-m-long, 150-mm-diameter uninsulated heating duct at 60°C and 0.04 kg/s. The air outlet temperature is 30°C and the duct surface temperature is approximately constant at $T_s = 15°C$.

(a) Determine the heat transfer rate.

(b) Calculate the *log mean temperature difference,* $\Delta T_{\text{lm}}$.

(c) What is the average convection coefficient for air flow $\bar{h}$?

(d) Sketch the axial variation of $T_s$ and $T_m$ with distance from the inlet $x$. Comment on the key features of the distributions.

### Internal Flow Applications: Fully Developed, Laminar Flow

**17.52** Ethylene glycol flows at 0.01 kg/s through a 3-mm-diameter, thin-walled tube. The tube is coiled and submerged in a well-stirred water bath maintained at 25°C. If the fluid enters the tube at 85°C, what heat rate and tube length are required for the fluid to leave at 35°C? Neglect heat transfer enhancement associated with the coiling.

**17.53** In the final stages of production, a pharmaceutical is sterilized by heating it from 25 to 75°C as it moves at 0.2 m/s through a straight thin-walled stainless steel tube of 12.7-mm diameter. A uniform heat flux is maintained by an electric resistance heater wrapped around the outer surface of the tube. If the tube is 10 m long, what is the required heat flux? Neglecting entrance effects, what is the surface temperature at the tube exit? Fluid properties may be approximated at $\rho = 1000$ kg/m$^3$, $c_p = 4000$ J/kg $\cdot$ K, $\mu = 2 \times 10^{-3}$ kg/s $\cdot$ m, $k = 0.48$ W/m $\cdot$ K, and Pr = 10.

**17.54** An electrical power transformer of diameter 300 mm and height 500 mm dissipates 1000 W. It is desired to maintain its surface temperature at 47°C by supplying glycerin at 24°C through thin-walled tubing of 20-mm diameter welded to the lateral surface of the transformer. All the power dissipated by the transformer is assumed to be transferred to the glycerin.

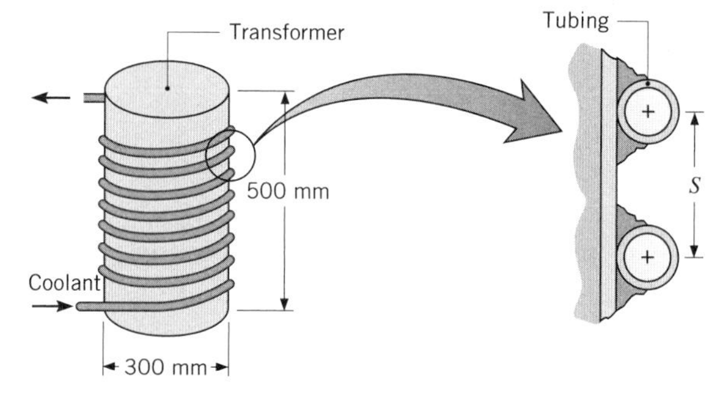


*Figure P17.54*

Assuming the maximum allowable temperature rise of the coolant to be 6°C and fully developed flow throughout the tube, determine the required coolant flow rate, the total length of tubing, and the lateral spacing $S$ between turns of the tubing.

**17.55** You are designing an operating room heat exchange device to cool blood (bypassed from a patient) from 40 to 30°C by passing the fluid through a coiled tube immersed in a vat of water–ice mixture. The volumetric flow rate is $10^{-4}$ m$^3$/min; the tube diameter ($D$) is 2.5 mm; and $T_{m,i}$ and $T_{m,o}$ represent the inlet and outlet temperatures of the blood.

(a) At what temperature would you evaluate the fluid properties in determining $\bar{h}$ for the entire tube length?

(b) If the properties of blood evaluated at the temperature for part (a) are $\rho = 1000$ kg/m$^3$, $\nu = 7 \times 10^{-7}$ m$^2$/s, $k = 0.5$ W/m $\cdot$ K, and $c_p = 4000$ J/kg $\cdot$ K, what is the Prandtl number for the blood?

(c) Is the blood flow laminar or turbulent?

(d) Neglecting all entrance effects and assuming fully developed conditions, calculate the value of $\bar{h}$ for heat transfer from the blood.

(e) What is the total heat rate from the blood as it passes through the tube?

(f) When free convection effects on the outside of the tube are included, the average overall heat transfer coefficient $\bar{U}$ between the blood and the ice–water mixture can be approximated as 300 W/m$^2 \cdot$ K. Determine the tube length $L$ required to obtain the outlet temperature $T_{m,o}$.

**17.56** Air flowing at $3 \times 10^{-4}$ kg/s and 27°C enters a rectangular duct that is 1 m long and 4 mm by 16 mm on a side. A uniform heat flux of 600 W/m$^2$ is imposed on the duct surface. What is the temperature of the air and of the duct surface at the outlet?

**17.57** Air flowing at $4 \times 10^{-4}$ kg/s and 27°C enters a triangular duct that is 20 mm on a side and 2 m long. The duct surface is maintained at 100°C. Assuming fully developed flow throughout the duct, determine the air outlet temperature.

**17.58** (CD ROM)

**17.59** (CD ROM)

### Internal Flow Applications: Fully Developed, Turbulent Flow

**17.60** Water flowing at 2 kg/s through a 40-mm-diameter tube is to be heated from 25 to 75°C by maintaining the tube surface temperature at 100°C. What is the required tube length for these conditions?

**17.61** Atmospheric air enters a 10-m-long, 150-mm-diameter uninsulated heating duct at 60°C and 0.04 kg/s. The duct surface temperature is approximately constant at $T_s = 15$°C. What are the outlet air temperature and the heat rate $q$ for these conditions?

**17.62** Water flows at 2 kg/s through a 40-mm-diameter tube 4 m long. The water enters the tube at 25°C, and the surface temperature is 90°C. What is the outlet temperature of the water? What is the rate of heat transfer to the water?

**17.63** Consider a thin-walled tube of 10-mm diameter and 2-m length. Water enters the tube from a large reservoir at $\dot{m} = 0.2$ kg/s and $T_{m,i} = 47$°C. If the tube surface is maintained at a uniform temperature of 27°C, what is the outlet temperature of the water, $T_{m,o}$? What is the rate of heat transfer from the water? To obtain the properties of water, assume an average mean temperature of $\overline{T}_m = 300$ K.

**17.64** The evaporator section of a heat pump is installed in a large tank of water, which is used as an energy source during the winter. As energy is extracted from the water, it begins to freeze, creating an ice/water bath at 0°C, which may be used for air conditioning during the summer. Consider summer cooling conditions for which air is passed through an array of copper tubes, each of inside diameter $D = 50$ mm, submerged in the bath.

(a) If air enters each tube at a mean temperature of $T_{m,i} = 24$°C and a flow rate of $\dot{m} = 0.01$ kg/s, what tube length $L$ is needed to provide an exit temperature of $T_{m,o} = 14$°C?

(b) With 10 tubes passing through a tank of total volume of 10 m$^3$, which initially contains 80% ice by volume, how long would it take to completely melt the ice? The density and heat of fusion of ice are 920 kg/m$^3$ and $3.34 \times 10^5$ J/kg, respectively.

**17.65** Cooling water flows through the 25.4-mm-diameter thin-walled tubes of a steam condenser at 1 m/s, and a surface temperature of 350 K is maintained by the condensing steam. The water inlet temperature is 290 K, and the tubes are 5 m long. What is the water outlet temperature? Evaluate water properties at an assumed average mean temperature, $\overline{T}_m = 300$ K.

**17.66** The core of a high-temperature, gas-cooled nuclear reactor has coolant tubes of 20-mm diameter and 780-mm length. *Helium* enters at 600 K and exits at 1000 K when the flow rate is $8 \times 10^{-3}$ kg/s per tube.

(a) Determine the uniform tube wall surface temperature for these conditions.

(b) If the coolant gas is *air,* determine the required flow rate if the heat transfer rate and tube wall surface temperature remain the same. What is the outlet temperature of the air?

**17.67** Heated air required for a food-drying process is generated by passing ambient air at 20°C through long, circular tubes ($D = 50$ mm, $L = 5$ m) housed in a steam condenser. Saturated steam at atmospheric pressure condenses on the outer surface of the tubes, maintaining a uniform surface temperature of 100°C. If an air flow rate of 0.01 kg/s is maintained in each tube, determine the air outlet temperature $T_{m,o}$ and the total heat rate $q$ for the tube.

**17.68** (CD-ROM)

**17.69** Fluid enters a thin-walled tube of 5 mm diameter and 2 m length with a flow rate of 0.04 kg/s and temperature of $T_{m,i} = 85$°C. The tube surface is maintained at a temperature of $T_s = 25$°C, and for this operating condition, the outlet temperature is $T_{m,o} = 31.1$°C. What is the outlet temperature if the flow rate is doubled? Fully developed, turbulent flow may be assumed to exist in both cases, and the fluid properties may be assumed to be independent of temperature.

**17.70** Air at 1 atm and 285 K enters a 2-m-long rectangular duct with cross section 75 mm by 150 mm. The duct is maintained at a constant surface temperature of 400 K, and the air mass flow rate is 0.10 kg/s. Determine the heat transfer rate from the duct to the air and the air outlet temperature.

**17.71** (CD-ROM)

**17.72** (CD-ROM)

**17.73** (CD-ROM)

**17.74** (CD-ROM)

### Internal Flow Applications: External Fluid Effects

**17.75** (CD-ROM)

**17.76** (CD-ROM)

**17.77** (CD-ROM)

**17.78** (CD-ROM)

**17.79** (CD-ROM)

**17.80** (CD-ROM)

### Free Convection: Vertical Plates

**17.81** A vertically mounted, square metallic plate 200 mm on a side is maintained at a uniform temperature of 15°C while exposed to quiescent air at 40°C. Calculate the average heat transfer coefficient for the plate using *all* of the appropriate correlations. Calculate the boundary layer thickness at the trailing edge.

**17.82** Consider a 0.25-m-long plate that is maintained at a uniform surface temperature of 70°C and is vertically suspended in quiescent air at 25°C and one atmosphere.

(a) Calculate the heat transfer rate from the plate by free convection.

(b) Estimate the boundary layer thickness at the trailing edge of the plate.

(c) How do the heat transfer rates and boundary layer thickness compare with those which would exist if the air were flowing over the plate with a free stream velocity of 5 m/s?

**17.83** The components of a vertical circuit board, 150 mm on a side, dissipate 5 W. The back surface is well insulated and the front surface is exposed to quiescent air at 27°C.

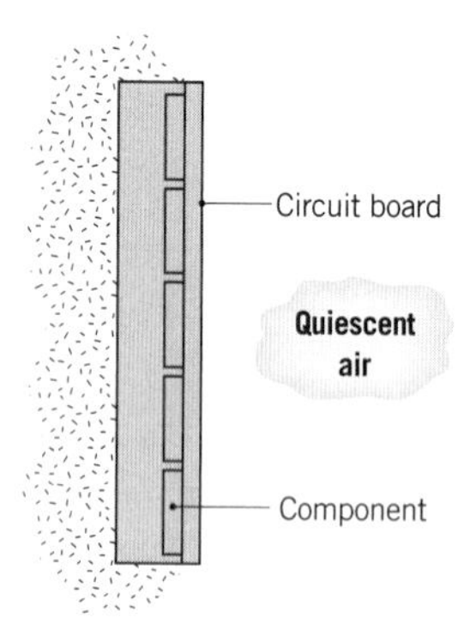


*Figure P17.83*

What is the temperature of the board for an isothermal surface condition?

**17.84** Consider an array of vertical rectangular fins which is to be used to cool an electronic device mounted in quiescent, atmospheric air at $T_\infty = 27°C$. Each fin has $L = 20$ mm and $H = 150$ mm and operates at an approximately uniform temperature of $T_s = 77°C$.

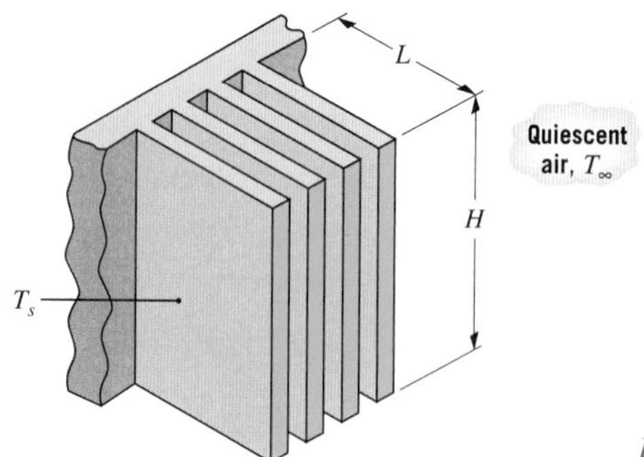


*Figure P17.84*

Viewing each fin surface as a vertical plate in an infinite, quiescent medium, estimate the rate of heat transfer from a fin by free convection. Comment on the effect of boundary layer formation on specifying the spacing between fins.

**17.85** During a winter day, the window of a patio door with a height of 1.8 m and width of 1.0 m shows a frost line near its base (Fig. P17.85). The room wall and air temperatures are 15°C.

*Figure P17.85*

(a) Estimate the heat transfer through the window due to free convection and radiation. Assume the window has a uniform temperature of 0°C and the emissivity of the glass surface is 0.94. If the room has electric baseboard heating, estimate the corresponding daily cost of the window energy loss for a utility rate of 0.08 \$/kW · h.

(b) Explain why the window would show a frost layer at the base rather than at the top.

**17.86** A thin-walled container with a hot process fluid at 50°C is placed in a quiescent, cold water bath at 10°C. Heat transfer at the inner and outer surfaces of the container may be approximated by free convection from a vertical plate.

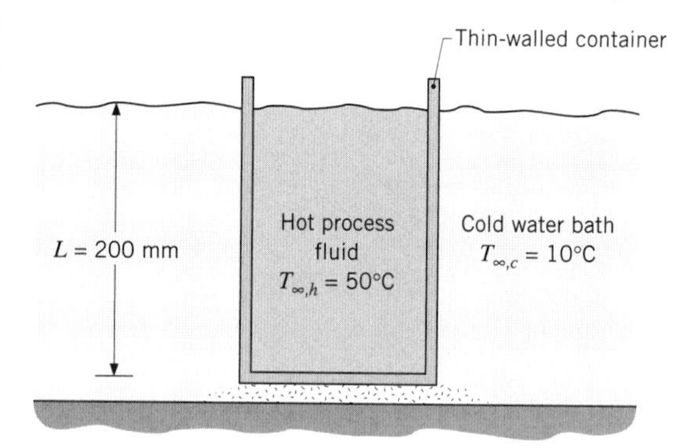


*Figure P17.86*

Determine the overall heat transfer coefficient between the hot process fluid and the cold water bath. Assume the properties of the hot process fluid are those of water. *Hint:* Assume the container surface temperature is 30°C for estimating the convection coefficients.

**17.87** (CD-ROM)

### Free Convection: Horizontal Plates

**17.88** A horizontal circular grill of diameter 0.25 m and emissivity 0.9 is maintained at a constant surface temperature of 130°C. What electrical power is required when the room air and surroundings are at 24°C?

**17.89** An electrical heater in the form of a horizontal disk of 400-mm diameter is used to heat the bottom of a tank filled with engine oil at a temperature of 5°C. Calculate the power required to maintain the heater surface temperature at 70°C.

**17.90** A horizontal opaque, black plate (300 mm × 300 mm) is exposed to a solar flux of 700 W/m$^2$ under still air and clear-sky conditions. The ambient air temperature is 20°C and the sky temperature is −10°C (treat the sky as large isothermal surroundings). Assuming that the backside of the plate is insulated, determine the steady-state temperature of the plate. *Hint:* Assume a value for the film temperature required for use of the convection correlation; after calculating the surface temperature, check to see if your assumption was reasonable.

**17.91** A 200 mm × 200 mm chill plate ($\varepsilon$ = 0.2) is being designed to maintain biological test samples at 12°C. The horizontal chill plate is located in a large glove box where the dry, quiescent ambient air and surroundings are at 25°C. The bottom of the chill plate is attached to a thermoelectric cooler, which operates with an efficiency of 12%. The efficiency is defined as the ratio of the heat rate into the system to the electrical power consumed by the system. Estimate the electrical power required to operate the cooler under these conditions.

**17.92** A horizontal plate 1 m by 1 m is exposed to a net radiation heat flux of 300 W/m$^2$ at its bottom surface. If the top surface of the plate is well insulated, estimate the temperature the plate reaches when the ambient air is quiescent and at a temperature of 0°C.

**17.93** Consider a horizontal 6-mm-thick, 100-mm-long straight fin fabricated from plain carbon steel ($k$ = 57 W/m · K, $\varepsilon$ = 0.5). The base of the fin is maintained at 150°C, while the quiescent ambient air and the surroundings are at 25°C. Assume the fin tip is adiabatic.

(a) Estimate the free convection coefficient for the upper and lower surfaces of the fin. *Hint:* Use an average fin surface temperature of 125°C for your analysis.

(b) Estimate the linearized radiation coefficient based upon the assumed average fin surface temperature.

(c) Using the foregoing results to represent an average combined convection-radiation coefficient, estimate the fin heat rate per unit width, $q'$ (W/m).

**17.94** (CD-ROM)

### Free Convection: Horizontal Cylinder and Sphere

**17.95** A horizontal electrical cable of 25-mm diameter has a power dissipation rate of 30 W/m. If the ambient air temperature is 27°C, estimate the surface temperature of the cable. Assume negligible radiation exchange.

**17.96** An electric immersion heater, 10 mm in diameter and 300 mm long, is rated at 550 W. If the heater is horizontally positioned in a large tank of water at 20°C, estimate its surface temperature. Estimate the surface temperature if the heater is accidentally operated in air at 20°C.

**17.97** Under steady-state operation, the surface temperature of a small 20-W incandescent light bulb is 125°C when the temperature of the room air and walls is 25°C. Approximating the bulb as a sphere 40 mm in diameter with a surface emissivity of 0.8, what is the rate of heat transfer from the surface of the bulb to the surroundings?

**17.98** A sphere of 25-mm diameter contains an embedded electrical heater. Calculate the power required to maintain the surface temperature at 94°C when the sphere is exposed to a quiescent medium at 20°C for (a) air at atmospheric pressure, (b) water, and (c) ethylene glycol.

**17.99** A 25-mm-diameter copper sphere with a low emissivity coating is removed from an oven at a uniform temperature of 85°C and allowed to cool in a quiescent fluid maintained at 25°C.

(a) Calculate the convection coefficient associated with the initial condition of the sphere if the quiescent fluid is air.

(b) Using the lumped-capacitance method with the convection coefficient estimated in part (a), estimate the time for the sphere to reach 30°C.

(c) Repeat your analysis to estimate the cooling time if the quiescent fluid is water.

**17.100** Consider a horizontal pin fin of 6-mm diameter and 60-mm length fabricated from plain carbon steel ($k$ = 57 W/m · K, $\varepsilon$ = 0.5). The base of the fin is maintained at 150°C, while the quiescent ambient air and the surroundings are at 25°C. Assume the fin tip is adiabatic. Estimate the fin heat rate, $q_f$. Use an average fin surface temperature of 125°C in estimating the free convection coefficient and the linearized radiation coefficient. How sensitive is this estimate to your choice of the average fin surface temperature?

**17.101** (CD-ROM)

**17.102** (CD-ROM)

### Heat Exchanger: Overall Heat Transfer Coefficient

**17.103** In a fire-tube boiler, hot products of combustion flowing through an array of thin-walled tubes are used to heat water flowing over the tubes. At the time of installation, the overall heat transfer coefficient was 400 W/m$^2$ · K. After 1 year of use, the inner and outer tube surfaces are fouled, with corresponding fouling factors of $R''_{f,i}$ = 0.0015 and $R''_{f,o}$ = 0.0005 m$^2$ · K/W, respectively. Should the boiler be scheduled for cleaning of the tube surfaces?

**17.104** Steel tubes ($k$ = 15 W/m · K) of inner and outer diameter $D_i$ = 10 mm and $D_o$ = 20 mm, respectively, are used in a condenser. Under normal operating conditions, a convection coefficient of $h_i$ = 7000 W/m$^2$ · K is associated with condensation on the inner surface of the tubes, while a coefficient of $h_o$ = 100 W/m$^2$ · K is maintained by air flow over the tubes. What is the hot-side overall convection coefficient $U_h$? Is the thermal resistance of the tube wall significant?

**17.105** A steel tube ($k$ = 50 W/m · K) of inner and outer diameters $D_i$ = 20 mm and $D_o$ = 26 mm, respectively, is used for heat transfer from hot gases flowing over the tube ($h_h$ = 200 W/m$^2$ · K) to cold water flowing through the tube ($h_c$ = 8000 W/m$^2$ · K). What is the cold-side overall heat transfer coefficient $U_c$?

**17.106** A copper tube of inner and outer diameters $D_i = 13$ mm and $D_o = 18$ mm, respectively, is used in a shell-and-tube heat exchanger (Fig. P17.06). The convection coefficient associated with the condensation process is 11,000 W/m$^2 \cdot$ K.

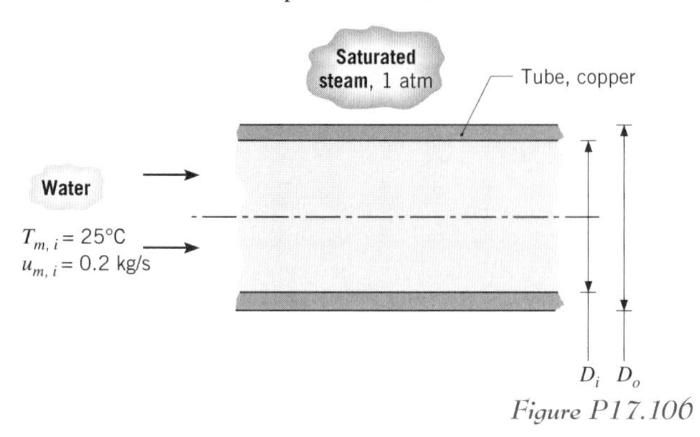


*Figure P17.106*

(a) Estimate the convection coefficient for the internal flow assuming fully developed conditions.
(b) Determine the overall heat transfer coefficient based upon the outside area of the tube $U_o$. Compare the thermal resistances due to internal flow convection, tube wall conduction, and condensation. Represent these resistances in a thermal circuit and label all elements.

### Heat Exchangers: Concentric Tube Type

**17.107** Consider the counterflow heat exchanger of Example 17.14. From the analysis, you saw that the overall coefficient is dominated by the hot-side convection coefficient $h_o$. The operations manager discovers that a spiral insert for the annulus that should increase $h_o$ by a factor of 10 is commercially available. If this enhancement could be achieved, what is the required tube length while all other conditions remain the same?

**17.108** Consider the counterflow heat exchanger of Example 17.14. The designer wishes to consider the effect of the cooling water flow rate on the tube length. All other conditions, including the outlet oil temperature of 60°C, remain the same. Calculate the required exchanger tube length $L$ and the water outlet temperature $T_{c,o}$ if the cooling water flow rate is doubled.

**17.109** Consider a concentric tube heat exchanger with an area of 50 m$^2$ operating under the following conditions:

| | Hot fluid | Cold fluid |
|---|---|---|
| Heat capacity rate, kW/K | 6 | 3 |
| Inlet temperature, °C | 60 | 30 |
| Outlet temperature, °C | — | 54 |

(a) Determine the outlet temperature of the hot fluid.
(b) Is the heat exchanger operating in counterflow or parallel flow, or, can't you tell from the available information?
(c) Calculate the overall heat transfer coefficient.

**17.110** Consider a *very long,* concentric tube heat exchanger having hot and cold water inlet temperatures of 85 and 15°C. The flow rate of the hot water is twice that of the cold water. Assuming equivalent hot and cold water specific heats, determine the hot water outlet temperature for the following modes of operation: (a) Counterflow and (b) Parallel flow.

**17.111** A counterflow, concentric tube heat exchanger used for engine cooling has been in service for an extended period of time. The heat transfer surface area of the exchanger is 5 m$^2$, and the *design value* of the overall convection coefficient is 38 W/m$^2 \cdot$ K. During a test run, engine oil flowing at 0.1 kg/s is cooled from 110°C to 66°C by water supplied at a temperature of 25°C and a flow rate of 0.2 kg/s. Determine whether fouling has occurred during the service period. If so, calculate the fouling factor, $R_f''$ (m$^2 \cdot$ K/W).

**17.112** A process fluid having a specific heat of 3500 J/kg · K and flowing at 2 kg/s is to be cooled from 80°C to 50°C with chilled water, which is supplied at a temperature of 15°C and a flow rate of 2.5 kg/s. Assuming an overall heat transfer coefficient of 2000 W/m$^2 \cdot$ K, calculate the required heat transfer areas for the following exchanger configurations: (a) Parallel flow, (b) Counterflow.

**17.113** Water at 225 kg/h is to be heated from 35 to 95°C by means of a concentric tube heat exchanger. Oil at 225 kg/h and 210°C, with a specific heat of 2095 J/kg · K, is to be used as the hot fluid. If the overall heat transfer coefficient based on the outer diameter of the inner tube is 550 W/m$^2 \cdot$ K, determine the length of the exchanger if the outer diameter is 100 mm.

**17.114** A concentric tube heat exchanger uses water, which is available at 15°C, to cool ethylene glycol from 100 to 60°C. The water and glycol flow rates are each 0.5 kg/s. Which is preferred, a parallel-flow or counterflow mode of operation?

**17.115** (CD-ROM)

**17.116** (CD-ROM)

**17.117** In a dairy operation, milk at a flow rate of 250 liter/hour and a *cow-body* temperature of 38.6°C must be chilled to a safe-to-store temperature of 13°C or less. Ground water at 10°C is available at a flow rate of 0.72 m$^3$/h. The density and specific heat of milk are 1030 kg/m$^3$ and 3860 J/kg · K, respectively.
(a) Determine the outlet temperature of the water.
(b) Determine the *UA* product of a counterflow heat exchanger required for the chilling process. Determine the length of the exchanger if the inner pipe has a 50-mm diameter and the overall heat transfer coefficient is $U = 1000$ W/m$^2 \cdot$ K.
(c) Using the value of *UA* found in part (a), determine the milk outlet temperature if the water flow rate is doubled. What is the outlet temperature if the flow rate is halved?

**17.118** In open heart surgery, the patient's blood is cooled before the surgery and rewarmed afterward. It is proposed that a concentric tube, counterflow heat exchanger of length 0.5 m be used for this purpose, with the thin-walled inner tube having a diameter of 55 mm. The specific heat of the blood is 3500 J/kg · K. The overall heat transfer coefficient is 500 W/m$^2 \cdot$ K. If water at $T_{h,i} = 60$°C and $\dot{m}_h = 0.10$ kg/s is used to heat blood entering the exchanger at $T_{c,i} = 18$°C and $\dot{m}_h = 0.05$ kg/s, what is the temperature of the blood leaving the exchanger?

**17.119** (CD-ROM)

### Heat Exchangers: Shell-and-Tube and Cross-Flow

**17.120** Hot exhaust gases are used in a shell-and-tube exchanger to heat 2.5 kg/s of water from 35 to 85°C. The gases, assumed to have the properties of air, enter at 200°C and leave at 93°C. The overall heat transfer coefficient is 180 $W/m^2 \cdot K$. Calculate the surface area of the heat exchanger.

**17.121** Saturated steam at 100°C condenses in a shell-and-tube exchanger (single shell, two tube passes) with an overall heat transfer coefficient of 2000 $W/m^2 \cdot K$. Water enters at 0.5 kg/s and 15°C and exits the exchanger at 48°C. Determine the required surface area and the rate of steam condensation.

**17.122** An automobile radiator may be viewed as a cross-flow heat exchanger. Water, which has a flow rate of 0.05 kg/s, enters the radiator at 400 K and is to leave at 330 K. The water is cooled by air that enters at 0.75 kg/s and 300 K. If the overall heat transfer coefficient is 200 $W/m^2 \cdot K$, what is the required heat transfer surface area?

**17.123** A cross-flow heat exchanger used in a cardiopulmonary bypass procedure cools blood flowing at 5 liter/min from a body temperature of 37°C to 25°C in order to induce body hypothermia, which reduces metabolic and oxygen requirements. The coolant is ice water at 0°C, and its flow rate is adjusted to provide an outlet temperature of 15°C. The heat exchanger overall heat transfer coefficient is 750 $W/m^2 \cdot K$. The density and specific heat of the blood are 1050 $kg/m^3$ and 3740 $J/kg \cdot K$, respectively.

(a) Determine the heat transfer rate for the exchanger.
(b) Calculate the water flow rate.
(c) What is the surface area of the heat exchanger?

**17.124** A single-pass, cross-flow heat exchanger uses hot exhaust gases to heat water from 30 to 80°C at a rate of 3 kg/s. The exhaust gases, having thermophysical properties similar to air, enter and exit the exchanger at 225 and 100°C, respectively. If the overall heat transfer coefficient is 200 $W/m^2 \cdot K$, estimate the required surface area.

# HEAT TRANSFER BY RADIATION

## *Introduction…*

When a temperature difference exists between two surfaces, a net heat transfer by radiation can occur even in the absence of any intervening medium. In Section 15.1.3, you were introduced to key *radiation processes* (emission, irradiation, absorption and reflection), the *radiative properties* (emissivity, absorptivity, and reflectivity), and the *rate equation* (Stefan-Boltzmann law).

In this chapter we will expand on those introductory concepts and present new methods to deal with radiation transfer from *spectrally-selective surfaces* and radiation exchange between surfaces in *enclosures*. Spectrally selective refers to preferential emission and absorption properties associated with different wavelengths of radiation. Examples of such behavior include absorption and reflection characteristics of *snow* and *solar collector* surfaces. Examples of enclosures are furnaces and ovens having important features such as radiation shields and insulated walls. We will treat enclosure surfaces as *gray* (without preferential spectral properties), an assumption that facilitates the analysis and is representative of many practical applications.

*chapter objectives*

The ***objectives*** of this chapter are threefold. The *first objectives* is to develop a deeper understanding of thermal radiation fundamentals. The *second objective* is to develop the methodology for performing energy balances on *single surfaces* having *spectrally-selective* radiative properties. The *third objective* is to develop the relationships required for calculating the radiative exchange between surfaces that comprise an *enclosure*.

We begin by treating the radiation field quantities, methods for performing surface energy balances, and characteristics of the blackbody. Thereafter the chapter is separated into two parts. The first part deals with *spectrally-selective* surfaces and detailed treatment is given to the radiative properties required to perform surface *energy balances*. The second part is about the *gray-surface enclosure,* which involves using the *view factor* to describe the geometrical features of the enclosure and the development of a network representation to facilitate radiative exchange analysis.

## 18.1 Fundamental Concepts

Consider an object that is initially at a higher temperature $T_s$ than that of its surroundings $T_{sur}$, but around which there exists a vacuum (Fig. 18.1). The presence of the vacuum precludes energy loss from the surface of the object by conduction or convection. However, our intuition tells us that the object will cool and eventually achieve thermal equilibrium with its surroundings. This cooling is associated with a reduction in the internal energy stored by the object and is a direct consequence of the *emission* of thermal radiation from the surface. In

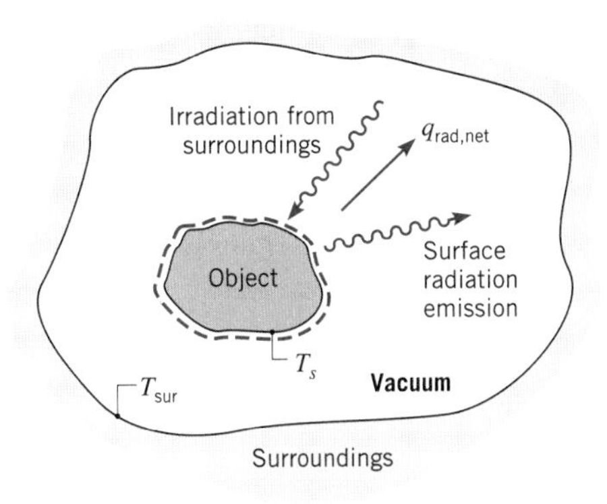


*Figure 18.1* A hot object experiences cooling by radiation transfer and eventually achieves thermal equilibrium with its surroundings.

turn, the surface will intercept and absorb radiation originating from the surroundings. However, if $T_s > T_{sur}$ the *net* heat transfer rate by radiation $q_{rad,net}$ is *from* the surface, and the surface will cool until $T_s$ reaches $T_{sur}$.

We associate ***thermal radiation*** with the rate at which energy is emitted by matter as a result of its temperature. At this moment, thermal radiation is being emitted by all the matter that surrounds you: by the furniture and walls of the room if you are indoors, or by the ground, the buildings, the atmosphere and the sun if you are outdoors. The mechanism of ***emission*** is related to energy released as a result of oscillations or transitions of the many electrons that constitute matter. These oscillations are, in turn, sustained by the internal energy, and therefore the temperature, of the matter. Hence, we associate the emission of thermal radiation with thermally excited conditions within the matter.

*thermal radiation*

*emission*

We know that radiation originates due to emission by matter and that its subsequent transport does not require the presence of any matter. But what is the nature of this transport? One theory views radiation as the *propagation* of a collection of particles termed *photons* or *quanta*. Alternatively, radiation may be viewed as the propagation of *electromagnetic waves*. In any case we attribute to radiation the standard wave properties of frequency $\nu$ and wavelength $\lambda$. For radiation propagating in a particular medium, the two properties are related by $\lambda = c/\nu$, where $c$ is the speed of light in the medium. For propagation in a vacuum, $c_o = 2.998 \times 10^8$ m/s. The unit of wavelength is commonly the *micrometer* ($\mu$m), where 1 $\mu$m $= 10^{-6}$ m.

The complete ***electromagnetic spectrum*** is delineated in Fig. 18.2. The short wavelength gamma rays, X rays, and ultraviolet (UV) radiation are primarily of interest to the

*electromagnetic spectrum*

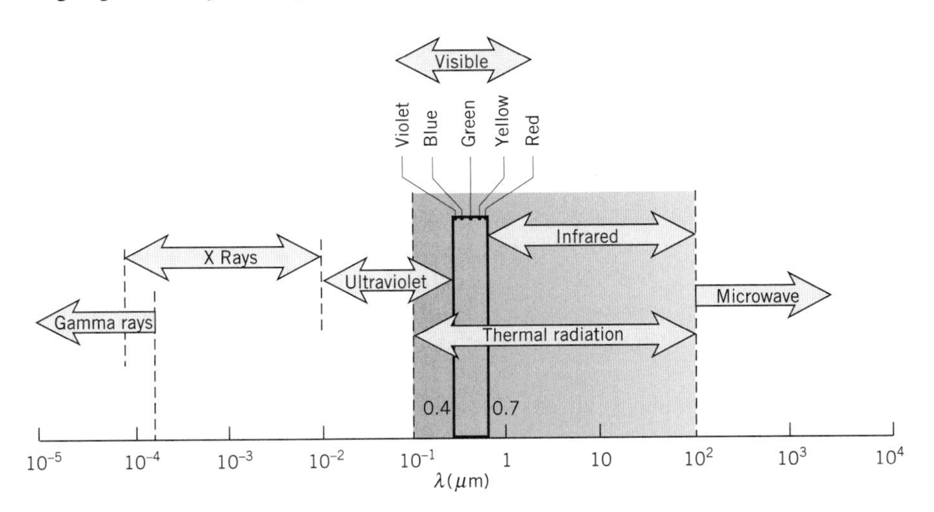


*Figure 18.2* Spectrum of electromagnetic radiation identifying the spectral region *thermal radiation* pertinent to heat transfer.

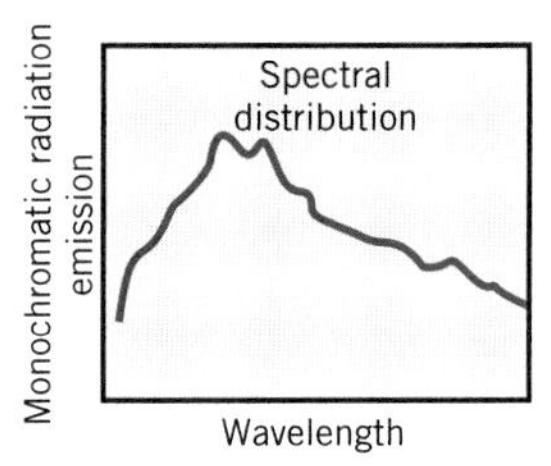


*Figure 18.3*

high-energy physicist and the nuclear engineer, while the long wavelength microwaves and radio waves are of concern to the electrical engineer. It is the intermediate portion of the spectrum, which extends from approximately 0.1 to 100 μm and includes a portion of the UV and all of the visible and infrared (IR) regions, that is termed *thermal radiation* and is pertinent to heat transfer.

*spectral distribution*

Thermal radiation emitted by a surface encompasses a range of wavelengths. As shown in Fig. 18.3, the magnitude of the radiation varies with wavelength, and the term *spectral* is used to refer to the nature of this dependence. Emitted radiation consists of a continuous, nonuniform distribution of *monochromatic* (single-wavelength) components. As we will find, both the magnitude of the radiation at any wavelength and the ***spectral distribution*** vary with the nature and temperature of the emitting surface.

*directional distribution*
*diffuse*

The spectral nature of thermal radiation is one of two features required for its description. The second feature relates to its *directionality*. While some surfaces may emit preferentially in certain directions, we show in Fig. 18.4 the ***directional distribution*** of emitted radiation that is uniform in all directions. We refer to this distribution as ***diffuse*** and to the surface as a *diffuse emitter*. The diffuse assumption greatly simplifies analysis, and provides a good approximation for many engineering applications.

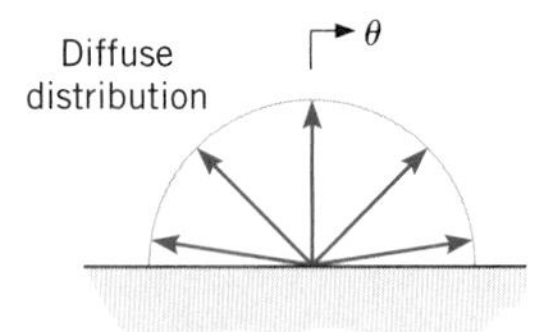


*Figure 18.4*

It is important to recognize that we have imposed restrictions on our treatment of radiation analysis. In addition to simplifying the directionality distribution for *diffuse* conditions, we will consider emission as a *surface phenomena,* deferring to more advanced study volumetric phenomena that are present with gases and semitransparent solids such as glasses and salts. Further, we will consider the medium that separates the surfaces experiencing radiation exchange to be *nonparticipating;* that is, it neither absorbs nor scatters surface radiation, and it emits no radiation. These restrictions will still allow you to perform engineering analysis on many practical thermal systems.

## 18.2 Radiation Quantities and Processes

We will consider three quantities that describe the thermal radiation field undergoing interactions (processes) with a surface. The *emissive power* and *irradiation* relate to the processes of emission from a surface and to radiation incident on a surface, respectively. The *radiosity* relates to radiation leaving a surface by emission *and* reflected irradiation. We assume that the radiation is *diffuse,* but distinguish between *spectral* and *total* wavelength conditions. Using these concepts, we will write surface energy balances in two different forms that will be useful for problem solving and developing analysis methods.

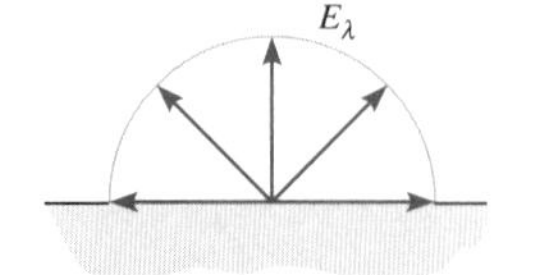


*Figure 18.5*

### 18.2.1 Emissive Power

*emissive power*

Recall that emission occurs from any surface that is at a finite temperature. The concept of ***emissive power*** is introduced to quantify the rate of radiation emitted per unit surface area. The *spectral emissive power,* $E_\lambda$ (W/m$^2 \cdot$ μm), is defined as the rate at which radiation of wavelength λ is emitted *in all directions into the hemispheric space* from a surface, *per* unit surface area and *per* unit wavelength interval $d\lambda$ about λ (Fig. 18.5).

The *total emissive power,* $E$ (W/m$^2$), is the rate at which radiation is emitted *per* unit area in all possible directions and at all possible wavelengths. Accordingly, as illustrated in Fig. 18.6

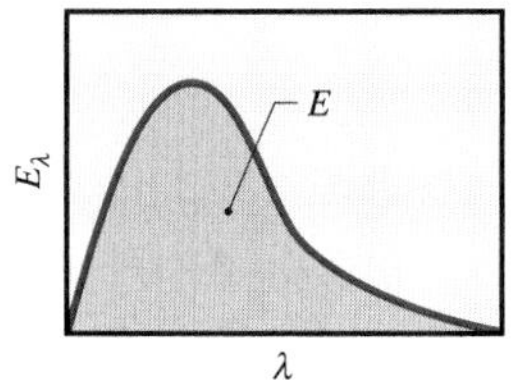

$E_\lambda$ — $E$ — $\lambda$

*Figure 18.6*

$$E = \int_0^\infty E_\lambda(\lambda)\, d\lambda \tag{18.1}$$

### 18.2.2 Irradiation

The foregoing approach may be applied to *incident* radiation (Figs. 18.7 and 18.8). Such radiation may originate from emission and reflection occurring at other surfaces or from the surroundings and radiation sources such as lamps. The incident radiation represents a radiative flux, termed the *irradiation,* which encompasses radiation incident from *all directions.*

*irradiation*

The *spectral irradiation,* $G_\lambda$ (W/m$^2 \cdot \mu$m), is defined as the rate at which radiation of wavelength $\lambda$ is incident on a surface, *per* unit area of the surface and *per* unit wavelength interval $d\lambda$ about $\lambda$.

If the *total irradiation,* $G$ (W/m$^2$), represents the rate at which radiation is incident *per* unit area from all directions and at all wavelengths. That is

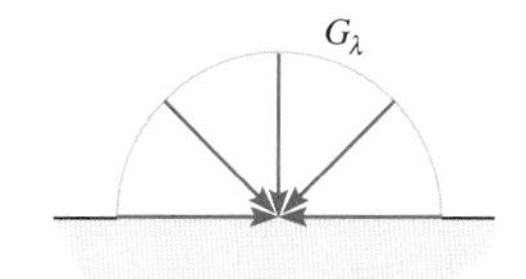


Figure 18.7

$$G = \int_0^\infty G_\lambda(\lambda)\, d\lambda \tag{18.2}$$

where $G_\lambda(\lambda)$ is given by the *spectral distribution* such as illustrated in Fig. 18.8.

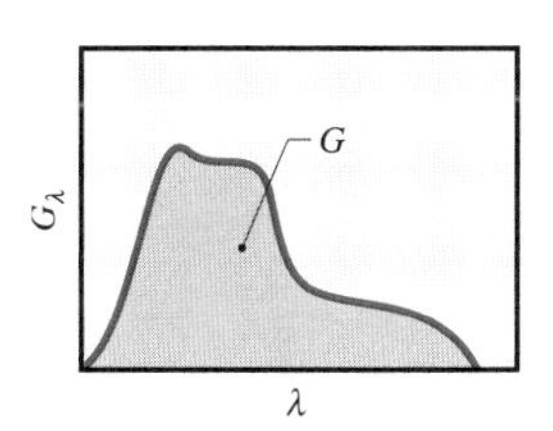


Figure 18.8

### 18.2.3 Radiosity

The third radiative flux of interest, termed *radiosity,* accounts for the radiant energy *leaving* a surface. Since this radiation includes the *reflected* portion of the irradiation, as well as direct emission (Fig. 18.9), *the radiosity is generally different from the emissive power.*

The *spectral radiosity,* $J_\lambda$ (W/m$^2 \cdot \mu$m), is defined as the rate at which radiation of wavelength $\lambda$ leaves a unit area of the surface, *per* unit wavelength interval $d\lambda$ about $\lambda$

$$J_\lambda \equiv E_\lambda + G_{\lambda,\text{ref}} \tag{18.3}$$

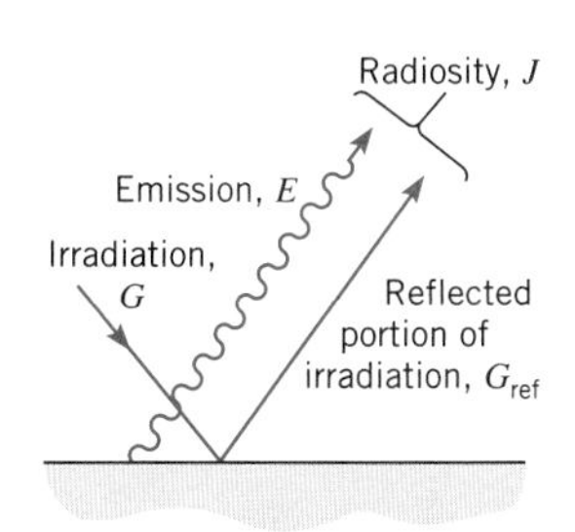


Figure 18.9

where $E_\lambda$ is the spectral emissive power representing the *direct emission* component, and $G_{\lambda,\text{ref}}$ is the *reflected portion* of the spectral irradiation $G_\lambda$. The *total radiosity,* $J$ (W/m$^2$), associated with the entire spectrum can be expressed as the integral of the spectral quantities

$$J = \int_0^\infty J_\lambda d\lambda = \int_0^\infty (E_\lambda + G_{\lambda,\text{ref}})\, d\lambda \tag{18.4a}$$

or in terms of the *total* emissive power and the reflected portion of the *total* irradiation

$$J = E + G_{\text{ref}} \tag{18.4b}$$

*total radiosity*

### 18.2.4 Surface Energy Balances with Radiation Processes

Following the methodology of Sec. 15.2, it is important to recognize two forms of surface energy balances that will be useful with radiation processes.

We are interested in performing energy balances on *spectrally-selective, single surfaces* experiencing emission and irradiation. Quite frequently, the properties that control the emission and absorption of irradiation are different, as we'll soon see. For this situation, from Eq. 15.14, the *surface energy balance* represented in Fig. 18.10 has the form

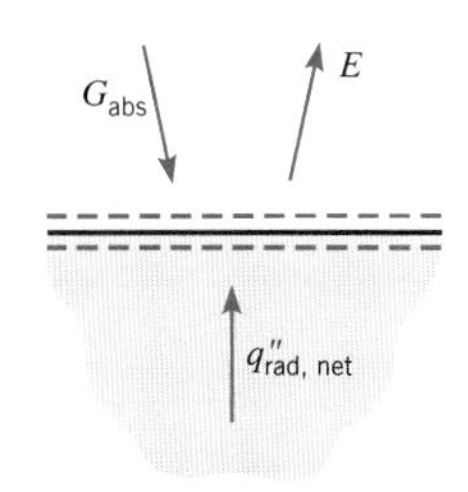


Figure 18.10

$$q''_{\text{rad,net}} = E - G_{\text{abs}} \tag{18.5a}$$

*surface energy balances*

where $q''_{\text{rad,net}}$ is the *net* radiative flux *leaving* the surface, $E$ is the total emissive power of the surface, and $G_{\text{abs}}$ is the *absorbed portion* of the total irradiation $G$. Since $q''_{\text{rad,net}}$ is the *net* radiative flux, it differs from the radiosity $J$, which represents *only* the radiant flux leaving the surface.

**Table 18.1** Glossary of Thermal Radiation Quantities

| Quantities | Definition |
|---|---|
| Emissive power | Rate of radiant energy *emitted by* a surface in all directions *per* unit area of the surface, $E_\lambda$ (W/m$^2 \cdot \mu$m) or $E$ (W/m$^2$); Eq. 18.1. Modifiers: *spectral or total.* |
| Irradiation | Rate at which radiation is *incident on* a surface from all directions *per* unit area of the surface, $G_\lambda$ (W/m$^2 \cdot \mu$m) or $G$ (W/m$^2$); Eq. 18.2. Modifiers: *spectral or total.* |
| Radiosity | Rate at which radiation leaves a surface due to *emission and reflection* (reflected irradiation) in all directions *per* unit area of the surface, $J_\lambda$ (W/m$^2 \cdot \mu$m) or $J$ (W/m$^2$); Eqs. 18.3 and 18.4. Modifiers: *spectral or total.* |
| **Modifiers** | **Definition** |
| Diffuse | Refers to directional uniformity of radiation field associated with emission, irradiation, and reflection. |
| Spectral | Refers to a single-wavelength (monochromatic) or narrow spectral band; denoted by the subscript $\lambda$. |
| Total | Refers to all wavelengths; integrated over all wavelengths ($0 < \lambda < \infty$) |

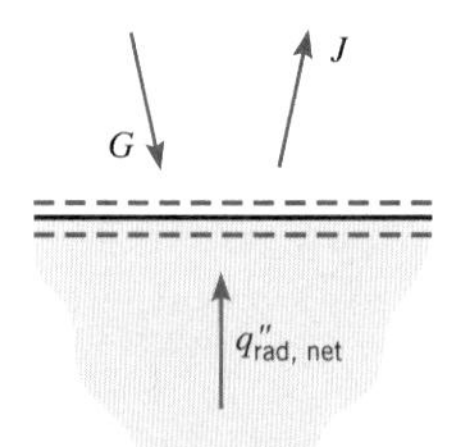


*Figure 18.11*

The surface energy balance can be written in an alternative form in terms of the *total radiosity* and *total irradiation*. For the surface of Fig. 18.11

$$q''_{\text{rad,net}} = J - G \tag{18.5b}$$

where $q''_{\text{rad,net}}$ is the net radiation *leaving* the surface by radiation, $J$ is the total radiosity of the surface, and $G$ is the total irradiation. Recall from Eq. 18.4b that the radiosity is the sum of the emitted and reflected irradiation leaving the surface. It is not always convenient to calculate the radiosity, and hence Eq. 18.5a may be more useful. However, when dealing with radiative exchange between surfaces in enclosures (Sec. 18.7), Eq. 18.5b will be required.

### 18.2.5 Summary: Radiation Quantities

The *emissive power*, *E*, *irradiation*, *G*, and *radiosity*, *J*, are the quantities that describe the radiation processes experienced by a surface. You should know their definitions and understand how they are employed in the surface energy balances of Eqs. 18.5a and 18.5b to calculate the net radiation leaving a surface. Table 18.1 summarizes the definition of these quantities and the other related terms.

## *Example 18.1* Radiation Processes and Surface Energy Balances

The total emissive power for the surface of a solar collector plate is 525 W/m$^2$. The spectral distribution of the surface irradiation is shown below, and 85% of the irradiation is absorbed, while 15% is reflected.

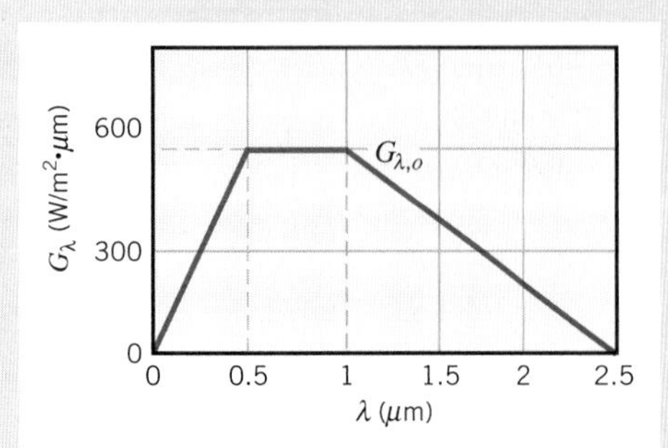


*Figure E18.1*

**(a)** What is the total irradiation on the plate, $G$? What is the *absorbed* total irradiation, $G_{\text{abs}}$? **(b)** What is the total radiosity, $J$, of the plate for these conditions? **(c)** What is the *net* radiative flux *leaving* the surface, $q''_{\text{rad,net}}$?

## Solution

***Known:*** Total emissive power, spectral distribution of irradiation, fraction of irradiation absorbed and reflected for a surface.

***Find:*** (a) The total irradiation and the *absorbed* total irradiation, (b) the total radiosity, and (c) *net* radiative flux *leaving* the surface.

***Analysis:*** (a) The total irradiation may be obtained from Eq. 18.2, where the integral is readily evaluated by breaking it into parts. The units for $G_\lambda$ and $\lambda$ are $W/m^2 \cdot \mu m$ and $\mu m$, respectively.

$$G = \int_0^\infty G_\lambda d\lambda = \int_0^{0.5\ \mu m} G_\lambda d\lambda + \int_{0.5}^{1\ \mu m} G_\lambda d\lambda + \int_1^{2.5\ \mu m} G_\lambda d\lambda + \int_{2.5}^\infty G_\lambda d\lambda$$

$$G = 0.5 G_{\lambda,o}(\lambda_1 - 0) + G_{\lambda,o}(\lambda_2 - \lambda_1) + 0.5 G_{\lambda,o}(\lambda_3 - \lambda_2) + 0$$

$$G = 0.5 \times 600(0.5 - 0) + 600(1.0 - 0.5) + 0.5 \times 600(2.5 - 1.0) + 0$$

$$G = 150 + 300 + 450 = 900\ W/m^2 \ \triangleleft$$

With $\alpha = 0.85$ representing the fraction of irradiation that is absorbed, the *absorbed* irradiation is

$$G_{abs} = \alpha G = 0.85 \times 900\ W/m^2 = 765\ W/m^2 \ \triangleleft$$

(b) The total radiosity is the sum of the direct emission and reflected irradiation. From Eq. 18.4b, with $\rho = 0.15$ representing the fraction of irradiation that is reflected, find

$$J = E + G_{ref} = E + \rho G$$

$$J = 525\ W/m^2 + 0.15 \times 900\ W/m^2 = 660\ W/m^2 \ \triangleleft$$

(c) From Eq. 18.5a, the surface energy balance can be written in terms of the emissive power and the *absorbed* total irradiation to obtain the *net* radiative flux *leaving* the surface

$$q''_{rad,net} = E - G_{abs} = 525\ W/m^2 - 765\ W/m^2 = -240\ W/m^2$$

Alternatively, from Equation 18.5b, the surface energy balance can be written in terms of the total radiosity and total irradiation

$$q''_{rad,net} = J - G = 660\ W/m^2 - 900\ W/m^2 = -240\ W/m^2 \ \triangleleft$$

Since the sign is negative, it follows that the net radiative flux is *into* the solar collector surface.

***Comments:***

1. Generally, radiation sources do not provide such a regular spectral distribution as prescribed above. However, the procedure for computing the total irradiation from knowledge of the spectral distribution remains the same, although evaluation of the integral is likely to involve more detail.
2. Be sure you recognize the equivalence of performing radiative surface energy balances using the two forms shown above, as we'll use both in subsequent sections.

## 18.3 Blackbody Radiation

Before attempting to describe the radiation characteristics of real surfaces, it is useful to introduce the concept of a blackbody. An *ideal surface* having the following ***properties*** is called a ***blackbody:***

- *A blackbody absorbs all incident radiation, regardless of wavelength and direction.*
- *For a prescribed temperature and wavelength, no surface can emit more energy than a blackbody.*
- *Although the radiation emitted by a blackbody is a function of wavelength and temperature, it is independent of direction. That is, the blackbody is a diffuse emitter.*

*blackbody properties*

As the *perfect absorber* and *diffuse emitter,* the blackbody serves as a *standard* against which the radiative properties of actual surfaces are compared.

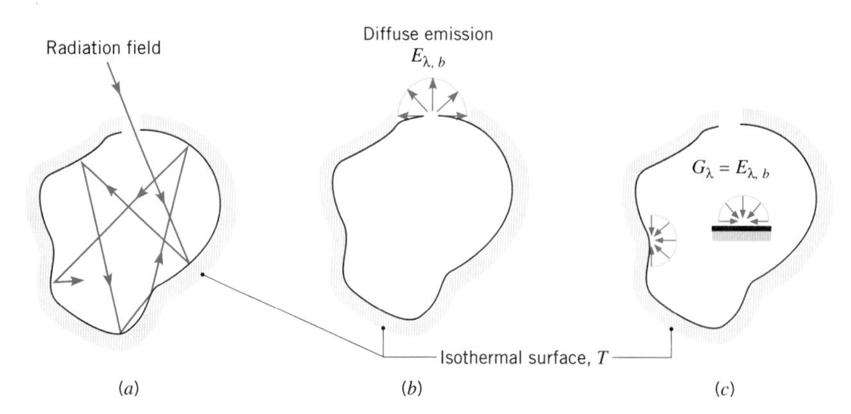


*Figure 18.12* Characteristics of an isothermal blackbody cavity. (*a*) Complete absorption. (*b*) Diffuse emission from an aperture. (*c*) Diffuse irradiation on interior surfaces.

*isothermal cavity*

Although closely approximated by some special surfaces, no surface has precisely the properties of a blackbody. The closest approximation is achieved by a cavity with a small aperture whose inner surface is at a uniform temperature (Fig. 18.12). This *isothermal cavity* has the following characteristics:

*complete absorption*

- *Complete absorption.* If radiation enters through the small aperture (Fig. 18.12*a*), it is likely to experience multiple reflections before reemergence. Hence, the radiation is almost entirely *absorbed* by the cavity, and blackbody behavior is approximated.

*blackbody emission*

- *Blackbody emission.* From thermodynamic principles it may be argued that the radiation leaving the aperture depends only on the surface temperature (Fig. 18.12*b*). The blackbody emission is diffuse and with the *blackbody spectral emissive power,* $E_{\lambda,b}$. Note the use of the subscript $b$ to denote *blackbody conditions.*

*blackbody irradiation on interior surfaces*

- *Blackbody irradiation on interior surfaces.* The radiation field within the cavity, which is the cumulative effect of emission and reflection from the cavity surface, must be of the same form as the radiation emerging from the aperture. It follows that a blackbody radiation field exists within the cavity. Accordingly, *any small surface in the cavity* (Fig. 18.12*c*) *experiences blackbody irradiation* for which

$$G_\lambda = E_{\lambda,b}(\lambda, T) \qquad \text{[isothermal cavity]} \tag{18.6}$$

Note that blackbody radiation exists within the cavity irrespective of whether the surface is highly reflecting or absorbing.

### 18.3.1 The Planck Distribution

The *spectral distribution* of blackbody emission is well known, having first been determined by *Planck.* It is of the form

*Planck spectral distribution*

$$E_{\lambda,b}(\lambda, T) = \frac{C_1}{\lambda^5[\exp(C_2/\lambda T) - 1]} \tag{18.7}$$

where the *first* and *second radiation constants* are $C_1 = 2\pi h c_o^2 = 3.742 \times 10^8\ \text{W} \cdot \mu\text{m}^4/\text{m}^2$ and $C_2 = (hc_o/k) = 1.439 \times 10^4\ \mu\text{m} \cdot \text{K}$, and $T$ is the *absolute temperature* of the blackbody.

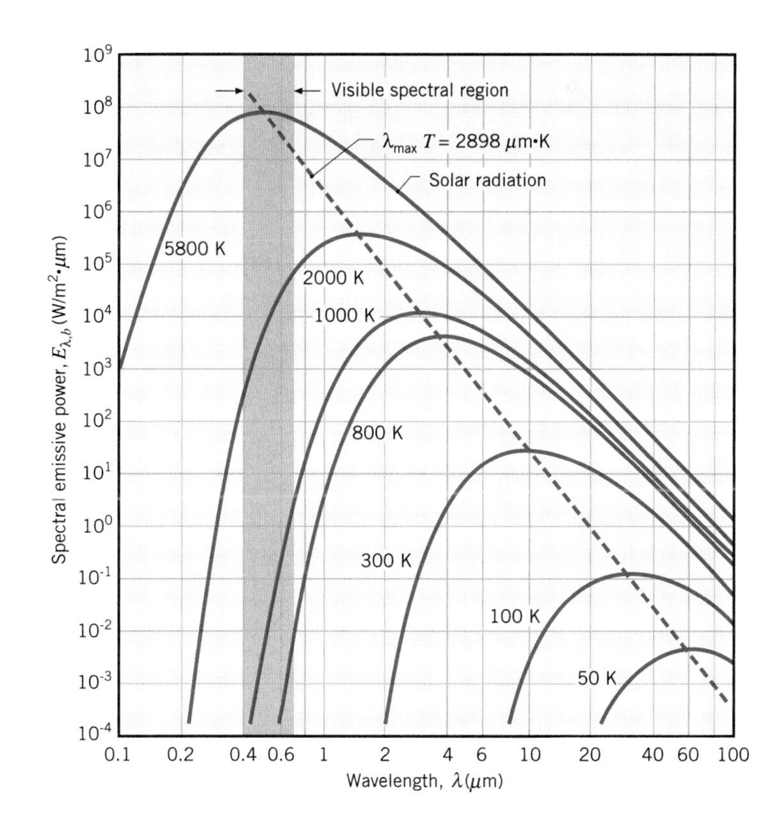


*Figure 18.13* Spectral blackbody emissive power (Planck spectral distribution).

Note that $C_1$ and $C_2$ are calculated from the universal constants, $h$, $k$, and $c_o$, which are the Planck constant, the Boltzmann constant and the speed of light in a vacuum, respectively. (See inside front cover for values of the physical constants.)

Equation 18.7, known as the *Planck spectral distribution,* is plotted in Fig. 18.13 for selected temperatures. Several important features should be noted:

- The emitted radiation varies *continuously* with wavelength.
- At any wavelength the magnitude of the emitted radiation increases with increasing temperature.
- The spectral region in which the radiation is concentrated depends on temperature, with *comparatively* more radiation appearing at shorter wavelengths as the temperature increases.

## 18.3.2 Wien's Displacement Law

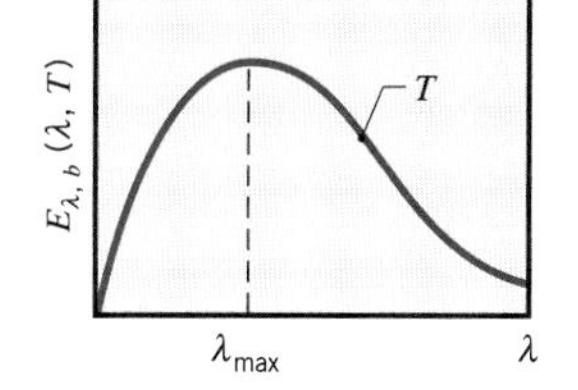


From Fig. 18.13 we see that the blackbody spectral distribution has a maximum and that the corresponding wavelength $\lambda_{max}$ depends on temperature. The nature of this dependence is obtained by differentiating Eq. 18.7 with respect to $\lambda$ and setting the result equal to zero. In so doing, we obtain

$$\lambda_{max} T = C_3 \tag{18.8}$$

*Wien's displacement law*

where the *third radiation constant* is $C_3 = 2897.8\ \mu m \cdot K$.

Equation 18.8 is known as ***Wien's displacement law,*** and the locus of points described by the law is plotted as the dashed line of Fig. 18.13. According to this result, the maximum spectral emissive power is displaced to shorter wavelengths with increasing temperature. For solar radiation, the maximum emission is in the middle of the visible spectrum ($\lambda \approx 0.50\ \mu m$) since the sun emits approximately as a blackbody at 5800 K. For a blackbody at 1000 K, peak emission occurs at 2.90 μm, with some of the emitted radiation appearing in the visible region as red light. With increasing temperature, shorter wavelengths become more prominent, until eventually significant emission occurs over the entire visible spectrum. ***For Example...*** a tungsten filament lamp operating at 2900 K ($\lambda_{max} = 1$ μm) emits white light, although most of the emission remains in the IR region. ▲

### 18.3.3 The Stefan–Boltzmann Law

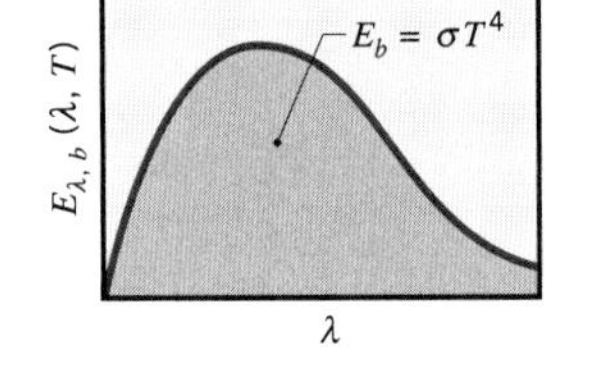


Substituting the Planck distribution, Eq. 18.7, into Eq. 18.1, the *total emissive power* of a blackbody, $E_b$, may be expressed as

$$E_b = \int_0^\infty E_{\lambda,b}\, d\lambda = \int_0^\infty \frac{C_1}{\lambda^5[\exp(C_2/\lambda T) - 1]}\, d\lambda$$

The result obtained from performing the integration is termed the ***Stefan–Boltzmann law*** having the form

*Stefan–Boltzmann law*

$$E_b = \sigma T^4 \tag{18.9}$$

where the *Stefan–Boltzmann constant,* which depends on $C_1$ and $C_2$, has the numerical value

$$\sigma = 5.670 \times 10^{-8} \text{ W/m}^2 \cdot \text{K}^4$$

This simple but important law enables the calculation of the amount of radiation emitted in all directions and over all wavelengths from knowledge of the blackbody temperature.

### 18.3.4 Blackbody Band Emission

It is often necessary to know the fraction of the total emission from a blackbody that is in a certain wavelength interval or spectral *band.* Such information is useful to determine the extent of spectral regions that influence radiative exchange, as well as for evaluation of the radiative properties as we'll see in the next section.

For a prescribed temperature and in the wavelength interval from 0 to λ, the ***band emission fraction*** is determined by the ratio of the shaded section to the total area under the blackbody curve of Fig. 18.14*a*. Accordingly, the band fraction has the form

*band emission fraction*

$$F_{(0\to\lambda)} \equiv \frac{\int_0^\lambda E_{\lambda,b}\, d\lambda}{\int_0^\infty E_{\lambda,b}\, d\lambda} = \frac{\int_0^\lambda E_{\lambda,b}\, d\lambda}{\sigma T^4} = \int_0^{\lambda T} \frac{E_{\lambda,b}}{\sigma T^5}\, d(\lambda T) = f(\lambda T) \tag{18.10a}$$

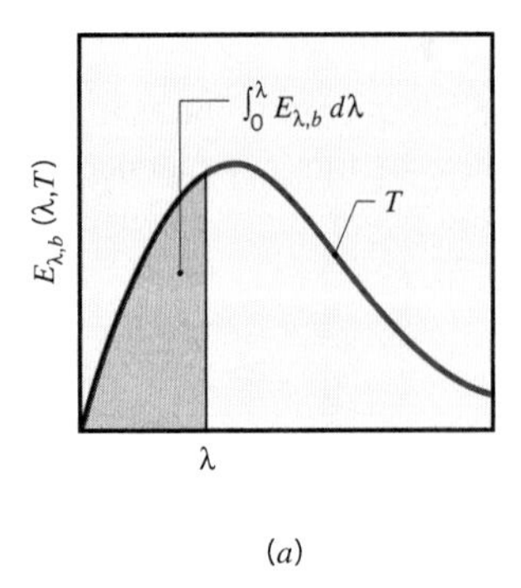


(*a*)

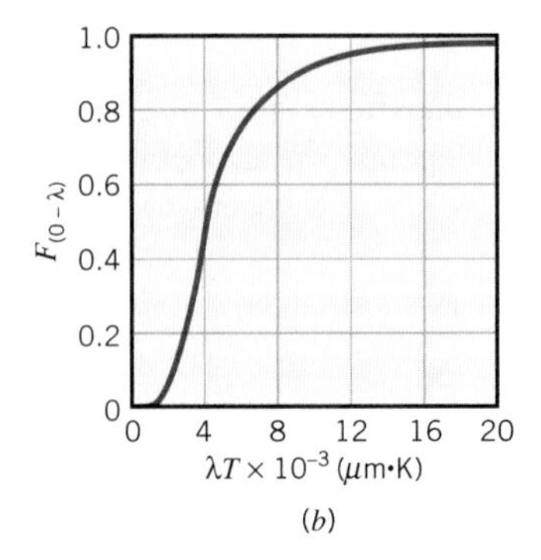


(*b*)

*Figure 18.14* The band emission fraction. (*a*) Radiation emission from a blackbody in the spectral band 0 to λ. (*b*) Fraction of the total blackbody emission in the spectral band from 0 to λ*T* as a function of λ*T*.

**Table 18.2** Blackbody Radiation Band Emission Fractions

| $\lambda T$ (μm · K) | $F_{(0\rightarrow\lambda)}$ | $\lambda T$ (μm · K) | $F_{(0\rightarrow\lambda)}$ | $\lambda T$ (μm · K) | $F_{(0\rightarrow\lambda)}$ |
|---|---|---|---|---|---|
| 200 | 0.000000 | 4,000 | 0.480877 | 8,000 | 0.856288 |
| 400 | 0.000000 | 4,200 | 0.516014 | 8,500 | 0.874608 |
| 600 | 0.000000 | 4,400 | 0.548796 | 9,000 | 0.890029 |
| 800 | 0.000016 | 4,600 | 0.579280 | 9,500 | 0.903085 |
| 1,000 | 0.000321 | 4,800 | 0.607559 | 10,000 | 0.914199 |
| 1,200 | 0.002134 | 5,000 | 0.633747 | 10,500 | 0.923710 |
| 1,400 | 0.007790 | 5,200 | 0.658970 | 11,000 | 0.931890 |
| 1,600 | 0.019718 | 5,400 | 0.680360 | 11,500 | 0.939959 |
| 1,800 | 0.039341 | 5,600 | 0.701046 | 12,000 | 0.945098 |
| 2,000 | 0.066728 | 5,800 | 0.720158 | 13,000 | 0.955139 |
| 2,200 | 0.100888 | 6,000 | 0.737818 | 14,000 | 0.962898 |
| 2,400 | 0.140256 | 6,200 | 0.754140 | 15,000 | 0.969981 |
| 2,600 | 0.183120 | 6,400 | 0.769234 | 18,000 | 0.980860 |
| 2,800 | 0.227897 | 6,600 | 0.783199 | 20,000 | 0.985602 |
| 2,898 | 0.250108 | 6,800 | 0.796129 | 25,000 | 0.992215 |
| 3,000 | 0.273232 | 7,000 | 0.808109 | 30,000 | 0.995340 |
| 3,200 | 0.318102 | 7,200 | 0.819217 | 40,000 | 0.997967 |
| 3,400 | 0.361735 | 7,400 | 0.829527 | 50,000 | 0.998953 |
| 3,600 | 0.403607 | 7,600 | 0.839102 | 75,000 | 0.999713 |
| 3,800 | 0.443382 | 7,800 | 0.848005 | 100,000 | 0.999905 |

*Note:* the shaded entry corresponds to the blackbody maximum, $\lambda_{\max}T = 2898$ μm · K, shown in Fig. 18.13.

Since the integrand $(E_{\lambda,b}/\sigma T^5)$ is exclusively a function of the wavelength–temperature product $\lambda T$, the integral of Eq. 18.10a can be evaluated to obtain $F_{(0\rightarrow\lambda)}$ as a function of only $\lambda T$. The results are presented in Table 18.2 and Fig. 18.14*b*.

The band emission fraction may also be used to obtain the fraction of the blackbody radiation in the spectral region between any two wavelengths $\lambda_1$ and $\lambda_2$, using

$$F_{(\lambda_1\rightarrow\lambda_2)} = \frac{\int_0^{\lambda_2} E_{\lambda,b}\,d\lambda - \int_0^{\lambda_1} E_{\lambda,b}\,d\lambda}{\sigma T^4} = F_{(0\rightarrow\lambda_2)} - F_{(0\rightarrow\lambda_1)} \tag{18.10b}$$

## *Example 18.2* Characteristics of Blackbody Radiation

Consider a large isothermal enclosure that is maintained at 2000 K. **(a)** Calculate the emissive power of the radiation that emerges from a small aperture on the enclosure surface. **(b)** What is the wavelength $\lambda_1$ below which 10% of the emission is concentrated? What is the wavelength $\lambda_2$ above which 10% of the emission is concentrated? **(c)** Determine the maximum spectral emissive power and the wavelength at which this emission occurs. **(d)** What is the irradiation incident on a small object placed inside the enclosure?

### Solution

***Known:*** Large isothermal enclosure at 2000 K.

***Find:***

**(a)** Emissive power of a small aperture on the enclosure.
**(b)** Wavelengths below which and above which 10% of the radiation is concentrated.
**(c)** Maximum spectral emissive power and wavelength at which it occurs.
**(d)** Irradiation on a small object inside the enclosure.

***Schematic and Given Data:***

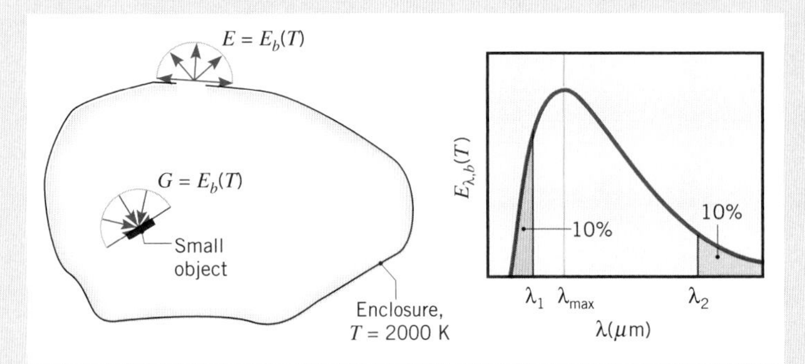


***Assumptions:*** Areas of aperture and object are very small relative to enclosure surface.

*Figure E18.2*

***Analysis:***

(a) Emission from the aperture of any isothermal enclosure will have the characteristics of blackbody radiation. Hence, from Eq. 18.9, the blackbody spectral emissive power is

$$E = E_b(T) = \sigma T^4 = 5.670 \times 10^{-8}\ \text{W/m}^2 \cdot \text{K}^4 (2000\ \text{K})^4$$

$$E = 9.07 \times 10^5\ \text{W/m}^2 \quad \triangleleft$$

(b) The wavelength $\lambda_1$ corresponds to the upper limit of the spectral band $(0 \rightarrow \lambda_1)$ containing 10% of the emitted radiation. With $F_{(0\rightarrow\lambda_1)} = 0.10$ it follows from Table 18.2 that $\lambda_1 T \approx 2200\ \mu\text{m} \cdot \text{K}$. Hence

$$\lambda_1 = \frac{(\lambda_1 T)}{T} = \frac{2200\ \mu\text{m} \cdot \text{K}}{2000\ \text{K}} = 1.1\ \mu\text{m} \quad \triangleleft$$

The wavelength $\lambda_2$ corresponds to the lower limit of the spectral band $(\lambda_2 \rightarrow \infty)$ containing 10% of the emitted radiation. With Eq. 18.10b

$$F_{(\lambda_2\rightarrow\infty)} = F_{(0\rightarrow\infty)} - F_{(0\rightarrow\lambda_2)} = 1 - F_{(0\rightarrow\lambda_2)} = 0.1$$

Recognizing that $F_{(0\rightarrow\lambda_2)} = 0.9$, it follows from Table 18.2 that $\lambda_2 T = 9382\ \mu\text{m} \cdot \text{K}$. Hence

$$\lambda_2 = \frac{(\lambda_2 T)}{T} = \frac{9382\ \mu\text{m} \cdot \text{K}}{1200\ \text{K}} = 4.69\ \mu\text{m} \quad \triangleleft$$

(c) From Wein's law, Eq. 18.8, when $T = 2000$ K, find $\lambda_{max} = 2898\ \mu\text{m} \cdot \text{K}/2000\ \text{K} = 1.45\ \mu\text{m}$. Hence, from Equation 18.7, the spectral emissive power corresponding to the peak of the blackbody curve is

$$E_{\lambda,b}(\lambda_{max}, T) = \frac{C_1}{\lambda_{max}^5[\exp(C_2/\lambda_{max}T) - 1]}$$

$$E_{\lambda,b}(\lambda_{max}, T) = \frac{3.742 \times 10^8\ \text{W} \cdot \mu\text{m}^4/\text{m}^2}{(1.45\ \mu\text{m})^5[\exp(1.439 \times 10^4\ \mu\text{m} \cdot \text{K}/(1.45\ \mu\text{m} \times 2000\ \text{K})) - 1]}$$

$$E_{\lambda,b}(1.45\ \mu\text{m}, 2000\ \text{K}) = 4.11 \times 10^5\ \text{W/m}^2 \cdot \mu\text{m} \quad \triangleleft$$

An approximate value for the blackbody spectral emissive power can be read from the Planck distribution plotted in Fig. 18.13.

(d) Irradiation of any small object inside the enclosure can be approximated as being equal to emission from a blackbody at the enclosure surface temperature. Hence, from Eq. 18.6, $G = E_b(T)$, in which case from part (a)

$$G = 9.07 \times 10^5\ \text{W/m}^2 \quad \triangleleft$$

## *Example 18.3* Blackbody Band Emission Fraction

A surface emits as a blackbody at 1500 K. What is the rate per unit area (W/m$^2$) at which it emits radiation in the wavelength interval $2\ \mu\text{m} \leq \lambda \leq 4\ \mu\text{m}$?

**Solution (CD-ROM)**

## Spectrally Selective Surfaces

Having developed the notion of a blackbody to describe ideal surface behavior, we consider now the behavior of *real* surfaces. In this part of the chapter we introduce the radiative properties and identify the characteristics of *spectrally selective surfaces*. The examples illustrate the methodology for performing energy balances on such surfaces.

## 18.4 Radiative Properties of Real Surfaces

We begin by defining the spectral and total emissivity, the *radiative properties* that describe the *emission* process. We will also introduce the radiative properties of *absorptivity, reflectivity*, and *transmissivity* that characterize the interception of *irradiation* with real surfaces. The interrelationships between the properties are developed. Representative radiative properties are provided for different classes of materials, and examples relevant to spectrally selective surface applications are presented.

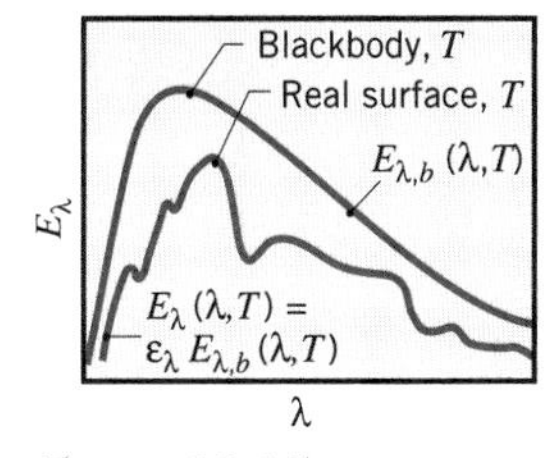

*Figure 18.15*

### 18.4.1 Surface Emission: Emissivity

It is important to acknowledge that, in general, the spectral radiation emitted by a *real* surface differs from the Planck spectral distribution (Fig. 18.15). Hence the emissivity may assume different values according to whether one is interested in emission at a given wavelength (*spectral*) or for all wavelengths (*total*).

The ***spectral emissivity***, $\varepsilon_\lambda$, is defined as the ratio of the *spectral* emissive power of a surface to that of a blackbody at the same temperature and for the same wavelength

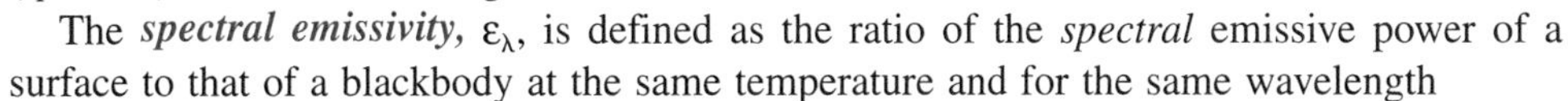

$$\varepsilon_\lambda(\lambda, T) \equiv \frac{E_\lambda(\lambda, T)}{E_{\lambda,b}(\lambda, T)} \tag{18.11}$$

*spectral emissivity*

The ***total emissivity***, $\varepsilon_\lambda$, is defined as the ratio of the *total* emissive power of a surface to that of a blackbody at the same temperature

$$\varepsilon(T) \equiv \frac{E(T)}{E_b(T)} \tag{18.12}$$

*total emissivity*

Substituting from Eqs. 18.1 and 18.11, it follows that

$$\varepsilon(T) = \frac{\displaystyle\int_0^\infty \varepsilon_\lambda(\lambda, T) E_{\lambda,b}(\lambda, T)\, d\lambda}{E_b(T)} \tag{18.13}$$

If the emissivities of the surfaces are known, it is a simple matter to compute their emissive powers. Specifically, if $\varepsilon_\lambda(\lambda,T)$ is known, it may be used with Eqs. 18.7 and 18.11 to compute the *spectral* emissive power at any wavelength and temperature. Similarly, if $\varepsilon(T)$ is known, it may be used with Eqs. 18.9 and 18.12 to compute the *total* emissive power of the surface at any temperature.

Typical values of the total emissivity for selected classes of materials are shown in Fig. 18.16.

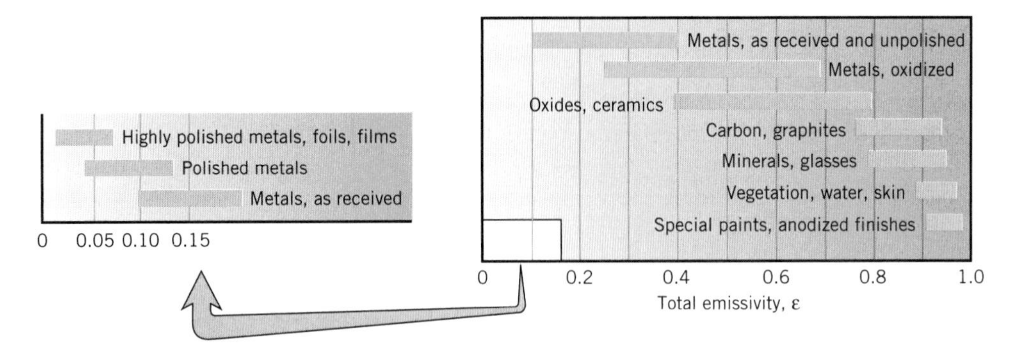


*Figure 18.16* Representative values of the total emissivity, ε.

## *Example 18.4* Total Emissivity from the Spectral Emissivity Distribution

A diffuse surface at 1600 K has the spectral emissivity shown as follows. Determine **(a)** the total emissivity and **(b)** the total emissive power.

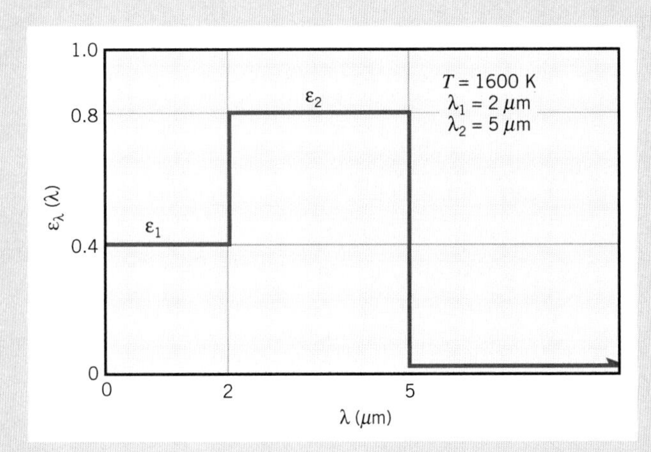


*Figure E18.4a*

### Solution

***Known:*** Spectral emissivity of a diffuse surface at 1600 K.

***Find:***

**(a)** Total emissivity, ε.

**(b)** Total emissive power, *E*.

***Assumptions:*** Surface is a diffuse emitter.

***Analysis:***

**(a)** The total emissivity is given by Eq. 18.13, where the integration can be performed in parts

$$\varepsilon = \frac{\int_0^\infty \varepsilon_\lambda E_{\lambda,b}\, d\lambda}{E_b} = \frac{\varepsilon_1 \int_0^{2\,\mu\text{m}} E_{\lambda,b}\, d\lambda}{E_b} + \frac{\varepsilon_2 \int_{2\,\mu\text{m}}^{5\,\mu\text{m}} E_{\lambda,b}\, d\lambda}{E_b}$$

and using Eq. 18.10a with the band emission fractions to represent the integrals

$$\varepsilon = \varepsilon_1 F_{(0\to 2\,\mu\text{m})} + \varepsilon_2 [F_{(0\to 5\,\mu\text{m})} - F_{(0\to 2\,\mu\text{m})}]$$

From Table 18.2 we obtain the bond emission fractions

$$\lambda_1 T = 2\ \mu\text{m} \times 1600\ \text{K} = 3200\ \mu\text{m}\cdot\text{K}: \qquad F_{(0\rightarrow 2\ \mu\text{m})} = 0.318$$

$$\lambda_2 T = 5\ \mu\text{m} \times 1600\ \text{K} = 8000\ \mu\text{m}\cdot\text{K}: \qquad F_{(0\rightarrow 5\ \mu\text{m})} = 0.856$$

Hence the total emissivity for this spectrally selective material at 1600 K is

$$\varepsilon = 0.4 \times 0.318 + 0.8[0.856 - 0.318] = 0.558 \ \triangleleft$$

(b) From Eq. 18.12, the total emissive power is

$$E = \varepsilon E_b = \varepsilon \sigma T^4$$

$$E = 0.558(5.67 \times 10^{-8}\ \text{W/m}^2\cdot\text{K}^4)(1600\ \text{K})^4 = 207\ \text{kW/m}^2 \ \triangleleft$$

***Comments:***

1. The *spectral* emissivity, $\varepsilon_\lambda$, is an intrinsic property of the material, which, to a first approximation, is independent of the material temperature. However, the *total emissivity can be strongly dependent upon temperature* because of the behavior of the product $\varepsilon_\lambda(\lambda,T)\cdot E_{\lambda,b}(\lambda,T)$ in the integration of Eq. 18.13. For a more detailed explanation and use of *Interactive Heat Transfer* (*IHT*) to evaluate $F_{(0\text{-}\lambda)}$, see Comment 2 **(CD-ROM).**

## 18.4.2 Irradiation: Absorptivity, Reflectivity, Transmissivity

In Sec. 18.2 we defined the *spectral* irradiation $G_\lambda$ (W/m$^2\cdot\mu$m) as the rate at which radiation of wavelength $\lambda$ is incident on a surface *per* unit area of the surface and *per* unit wavelength interval $d\lambda$ about $\lambda$. The *total* irradiation $G$ (W/m$^2$) encompasses all spectral components and may be evaluated from the spectral distribution, Eq. 18.2.

Consider the processes resulting from the interception of this radiation by a *semitransparent* medium such as a glass plate (Fig. 18.17). The term semitransparent refers to a medium in which radiation not reflected is absorbed or transmitted. For this general situation, portions of the spectral irradiation may be reflected (*ref*), absorbed (*abs*), and transmitted (*tr*). From a *radiation balance* on the medium, it follows that

$$G_\lambda = G_{\lambda,\text{ref}} + G_{\lambda,\text{abs}} + G_{\lambda,\text{tr}} \qquad (18.14)$$

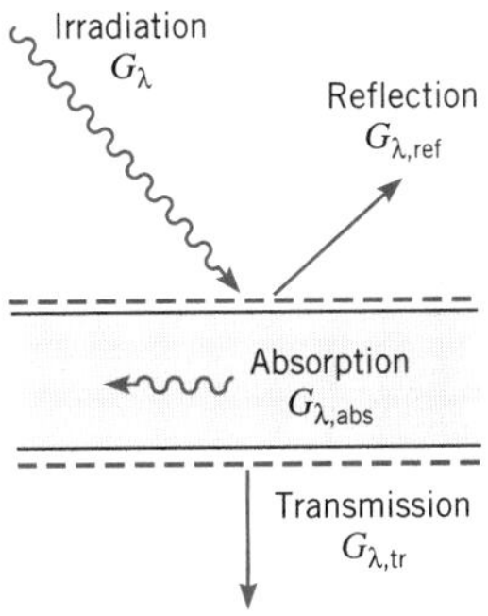

*Figure 18.17*

Recognize that there is no net effect of the reflected or transmitted radiation on the medium, while the absorbed radiation has the effect of increasing the internal energy of the medium. For an *opaque* medium, $G_{\lambda,\text{tr}} = 0$, and the spectral irradiation is either absorbed or reflected at the surface of the medium. In the subsections that follow, we will introduce the radiative properties to characterize the absorption, reflection, and transmission processes.

### Absorptivity

The absorptivity is the fraction of the irradiation absorbed by a medium. Recognizing that surfaces may exhibit selective absorption with respect to the wavelength of the incident radiation, we define the ***spectral absorptivity,*** $\alpha_\lambda(\lambda)$, as

$$\alpha_\lambda(\lambda) \equiv \frac{G_{\lambda,\text{abs}}(\lambda)}{G_\lambda(\lambda)} \qquad (18.15)$$

*spectral absorptivity*

The ***total absorptivity,*** $\alpha$, represents an integrated value over all wavelengths and is defined as the fraction of total irradiation absorbed by a surface

$$\alpha \equiv \frac{G_{\text{abs}}}{G} \qquad (18.16)$$

*total absorptivity*

Using Eqs. 18.2 and 18.15, the total absorptivity can be calculated from the *spectral* absorptivity, $\alpha_\lambda$, and the *spectral* irradiation, $G_\lambda$, as

$$\alpha = \frac{\int_0^\infty \alpha_\lambda(\lambda) G_\lambda(\lambda)\, d\lambda}{\int_0^\infty G_\lambda(\lambda)\, d\lambda} \tag{18.17}$$

Accordingly, $\alpha$ depends on the *spectral distribution of the incident radiation* ($G_\lambda$) as well as on the nature of the absorbing surface ($\alpha_\lambda$).

In general, $\alpha_\lambda$ is only weakly dependent upon the surface (medium) temperature; hence, the total absorptivity $\alpha$ *is nearly independent of the surface temperature.*

In contrast, the total emissivity $\varepsilon$ *is strongly dependent upon the surface temperature.* From Eq. 18.13, note that $\varepsilon$ depends upon the spectral distributions of the emission ($E_{\lambda,b}$) and the spectral emissivity ($\varepsilon_\lambda$). Although $\varepsilon_\lambda$ is also weakly dependent upon the surface temperature, $E_{\lambda,b}(\lambda,T)$ is strongly dependent upon temperature. For an illustration of this behavior, see Comment 1 of Ex. 18.4.

*Solar Absorptivity.* Because $\alpha$ depends on the spectral distribution of the irradiation, its value for a surface exposed to solar radiation may differ appreciably from its value for the same surface exposed to longer wavelength radiation originating from a source of lower temperature. Since the spectral distribution of solar radiation is nearly proportional to that of emission from a blackbody at $T_S = 5800$ K (see Fig. 18.13)

$$G_{\lambda,S} \sim E_{\lambda,b}(\lambda, T_S)$$

it follows from Eq. 18.17 that the total absorptivity to solar radiation, termed the ***solar absorptivity,*** $\alpha_S$, can be approximated as

*solar absorptivity*

$$\alpha_S \approx \frac{\int_0^\infty \alpha_\lambda(\lambda) E_{\lambda,b}(\lambda, 5800\text{ K})\, d\lambda}{\int_0^\infty E_{\lambda,b}(\lambda, 5800\text{ K})\, d\lambda} \tag{18.18}$$

The integrals of Eqs. 18.17 and 18.18 can be evaluated in the same manner as for the total emissivity, Eq. 18.13, using the band emission fractions, $F_{(0\to\lambda)}$, of Table 18.2.

## Reflectivity

The reflectivity is the fraction of the incident radiation reflected by a surface. The ***spectral reflectivity,*** $\rho(\lambda)$, is defined as the fraction of the *spectral* irradiation that is reflected from the surface

*spectral reflectivity*

$$\rho_\lambda(\lambda) \equiv \frac{G_{\lambda,\text{ref}}(\lambda)}{G_\lambda(\lambda)} \tag{18.19}$$

The ***total reflectivity,*** $\rho$, is then defined as

*total reflectivity*

$$\rho \equiv \frac{G_{\text{ref}}}{G} \tag{18.20}$$

in which case, the *total* reflectivity is related to the *spectral* reflectivity and *spectral* irradiation by

$$\rho = \frac{\int_0^\infty \rho_\lambda(\lambda) G_\lambda(\lambda)\, d\lambda}{\int_0^\infty G_\lambda(\lambda)\, d\lambda} \tag{18.21}$$

The reflection from surfaces can be idealized as *diffuse* (uniformly in all directions, rough surfaces) or *specular* (mirror-like, polished surfaces). In our treatment of the irradiation process and its associated radiative properties, we are assuming diffuse conditions, a reasonable assumption for most engineering applications.

### Transmissivity

The transmissivity is the fraction of the incident radiation transmitted through a *semitransparent* material. The ***spectral transmissivity*** is defined as the fraction of the *spectral* irradiation that is transmitted through the medium

$$\tau_\lambda = \frac{G_{\lambda,\mathrm{tr}}(\lambda)}{G_\lambda(\lambda)} \tag{18.22}$$

*spectral transmissivity*

The ***total transmissivity,*** $\tau$, is then defined as

$$\tau = \frac{G_{\mathrm{tr}}}{G} \tag{18.23}$$

*total transmissivity*

in which case, the *total* transmissivity is related to the *spectral* transmissivity and *spectral* irradiation by

$$\tau = \frac{\int_0^\infty G_{\lambda,\,\mathrm{tr}}(\lambda)\, d\lambda}{\int_0^\infty G_\lambda(\lambda)\, d\lambda} = \frac{\int_0^\infty \tau_\lambda(\lambda) G_\lambda(\lambda)\, d\lambda}{\int_0^\infty G_\lambda(\lambda)\, d\lambda} \tag{18.24}$$

## 18.4.3 Radiation Property Interrelationships

***Surface Radiation Balances.*** In Sec. 18.4.2, we considered the general situation of irradiation interacting with a *semitransparent* medium. From the radiation balance on the medium, including the processes of reflection, absorption, and transmission, Eq. 18.14, and the foregoing definitions of their respective *spectral* properties, it follows that

$$\rho_\lambda + a_\lambda + \tau_\lambda = 1 \tag{18.25}$$

and for the *total* properties over the entire spectrum,

$$\rho + a + \tau = 1 \tag{18.26}$$

If the medium is *opaque,* there is no transmission. Accordingly, absorption and reflection are the surface processes for which the properties on a *spectral* and *total* basis are related as

$$\rho_\lambda + a_\lambda = 1 \qquad \rho + a = 1 \qquad [\text{opaque medium}] \tag{18.27, 18.28}$$

Hence knowledge of one property implies knowledge of the other.

***Emission and Absorption Properties.*** In the foregoing sections we separately considered the *spectral* and *total* radiation properties associated with the processes of emission and absorption. For a surface that emits and reflects uniformly, called a ***diffuse surface,*** the spectral *emissivity* and *absorptivity* are *equal*

*diffuse surface*

$$\varepsilon_\lambda = a_\lambda \qquad [\text{diffuse surface}] \tag{18.29}$$

The proof for this equality involving the directionality characteristics of the radiation fields, referred to as *Kirchhoff's law,* is given in more advanced treatments of the subject.

Assuming the existence of a diffuse surface, we now consider what *additional* conditions must be satisfied for an equality between the *total* properties. From Eqs. 18.13 and 18.17, the equality between the *total emissivity* and the *total absorptivity* applies if

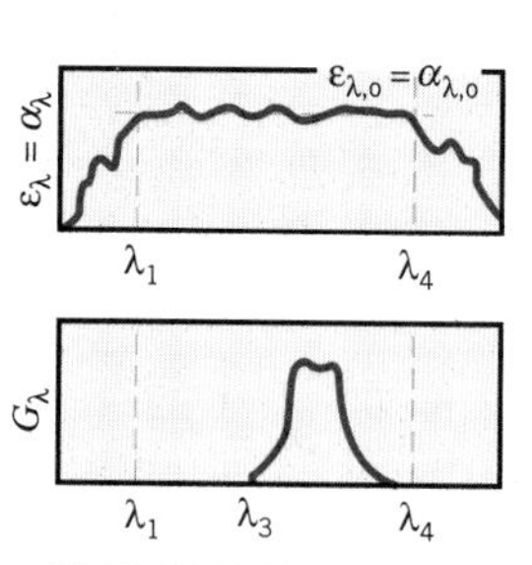


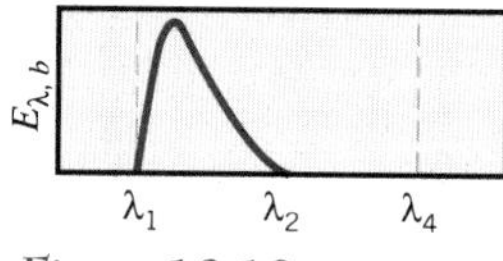


*Figure 18.18*

$$\varepsilon = \frac{\int_0^\infty \varepsilon_\lambda E_{\lambda,b}(\lambda, T)\, d\lambda}{E_b(T)} \overset{?}{=} \frac{\int_0^\infty \alpha_\lambda G_\lambda(\lambda)\, d\lambda}{G} = \alpha \tag{18.30}$$

With $\varepsilon_\lambda = \alpha_\lambda$, it follows by inspection that

$$\varepsilon = a \tag{18.31}$$

if either of the following conditions is satisfied:

- The irradiation corresponds to emission from a blackbody at the surface temperature $T$, in which case $G_\lambda(\lambda) = E_{\lambda,b}(\lambda, T)$ and $G = E_b(T)$, *or*
- The *surface* is *gray,* that is, $\alpha_\lambda$ and $\varepsilon_\lambda$ are independent of $\lambda$.

*gray surface*

The most likely situation corresponding to the first condition is when an object is in *thermal equilibrium* with its surroundings. The ***gray surface*** is one for which $\varepsilon_\lambda$ and $\alpha_\lambda$ are independent over the spectral regions of the irradiation and the surface emission.

A set of conditions for which *gray surface behavior* may be assumed is illustrated in Fig. 18.18. Note that the spectral distributions for the irradiation, $G_\lambda$, and surface emission, $E_{\lambda,b}$, are concentrated in a region for which the spectral properties of the surface are constant, $\lambda_1 < \lambda < \lambda_4$. Accordingly, from Eq. 18.30, $\varepsilon = \varepsilon_{\lambda,o}$ and $\alpha = \alpha_{\lambda,o}$, in which case $\alpha = \varepsilon$. However, if the irradiation or emission were in a spectral region corresponding to $\lambda < \lambda_1$ or $\lambda > \lambda_4$, gray surface behavior could not be assumed.

### 18.4.4 Summary: Spectrally Selective and Gray Surfaces

Considerable detail has been required to define the spectral and total radiative properties, as well as to introduce their interrelationships. Recall that the objectives of this chapter involve analyses with *spectrally selective* and *gray surfaces.* To make clear the distinction between them, the key *property concepts* are summarized.

- $\varepsilon_\lambda = \alpha_\lambda$: this equality holds for *diffuse conditions* associated with the surface and/or radiation processes. In this text we deal exclusively with diffuse conditions. Eq. 18.29.
- $\varepsilon \neq \alpha$: while $\varepsilon_\lambda = \alpha_\lambda$, values for $\varepsilon$ and $\alpha$ are separately determined from the spectral distributions for emission and irradiation, Eqs. 18.13 and 18.17 or 18.18, respectively. This surface is termed as *spectrally selective.*

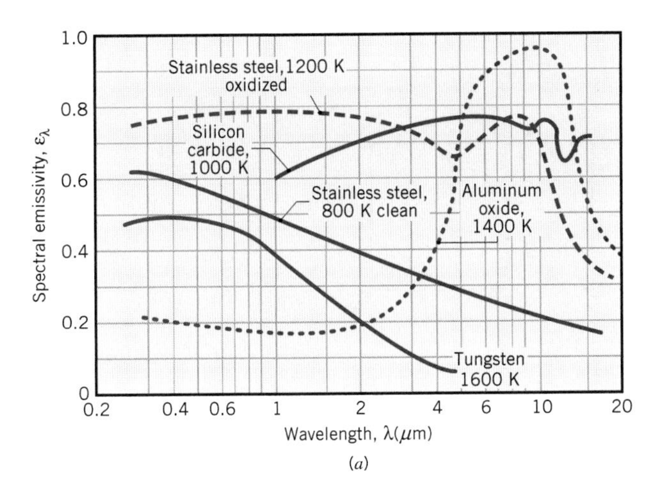


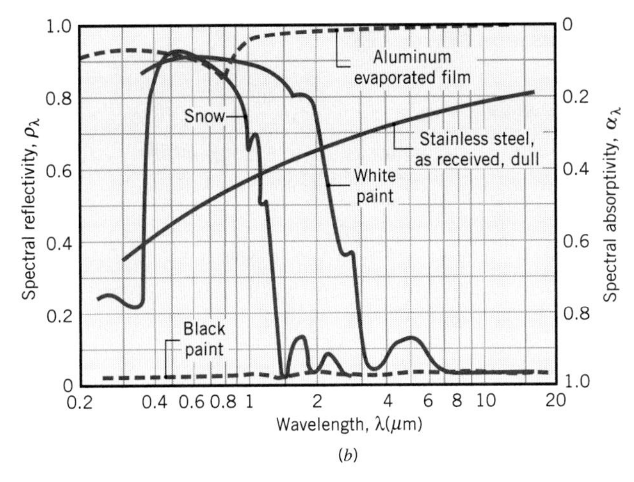


*Figure 18.19* Spectral dependence of the (*a*) spectral emissivity at elevated temperatures and (*b*) spectral reflectivity or absorptivity near room temperature of selected opaque materials.

- $\varepsilon = \alpha$: applies when any type of surface is in *thermal equilibrium* with its surroundings. Eq. 18.31.
- $\varepsilon = \alpha$: applies when the surface has no spectral character in the wavelength region of importance (Fig. 18.18) so that $\varepsilon_\lambda = \alpha_\lambda = \varepsilon = \alpha =$ constant. Such a surface is called a ***diffuse-gray surface.*** Eq. 18.31.

*diffuse-gray surface*

In engineering practice, when deciding on which type of surface is appropriate for an application, it is necessary to examine the spectral dependence of the material's properties. Spectral emissivity and reflectivity properties for selected opaque materials are shown in Fig. 18.19. We illustrate in Ex. 18.5 (Comment 2) and Ex. 18.6 (Comment 3) how to approximate the *spectral* behavior of real *spectrally selective* materials so that *total* properties can be readily evaluated.

In the next part of this chapter on enclosure analysis, we will treat the surfaces as *diffuse-gray*. In the following application-type examples, we illustrate how to perform energy balances on *spectrally selective surfaces*.

## *Example 18.5* Heating Application: Spectrally Selective Workpiece

A *small*, solid metallic sphere has an opaque, diffuse coating for which $\alpha_\lambda = 0.8$ for $\lambda \le 5$ μm and $\alpha_\lambda = 0.1$ for $\lambda > 5$ μm. The sphere, which is initially at a uniform temperature of 300 K, is inserted into a *large* furnace whose walls are at 1200 K. Eventually, the sphere reaches the furnace wall temperature. **(a)** For the *initial* condition, determine the total absorptivity and emissivity of the coating, and the net heat flux by radiation *leaving* the sphere. **(b)** For the *final* condition, determine the total absorptivity and emissivity of the coating.

### Solution

***Known:*** Small metallic sphere with spectrally selective absorptivity, *initially* at $T_{s,i} = 300$ K, is inserted into a large furnace at $T_f = 1200$ K. *Finally*, the sphere reaches $T_{s,f} = 1200$ K.

***Find:***

**(a)** Total absorptivity and emissivity of the coating, and net radiative heat flux leaving the sphere for the *initial* condition.
**(b)** Total absorptivity and emissivity of the coating for the *final* condition.

***Schematic and Given Data:***

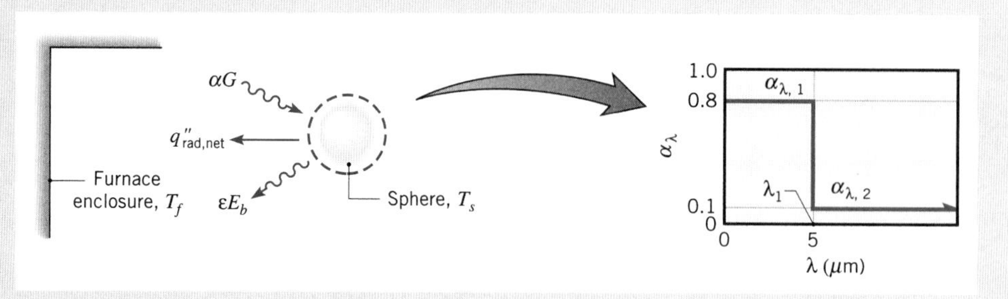


*Figure E18.5*

***Assumptions:***

1. Coating is opaque and diffuse.
2. Irradiation approximates emission from a blackbody at $T_f$ since the furnace surface is much larger than that of the sphere.

***Analysis:*** (a) From Eq. 18.17, the *total absorptivity* is expressed as

$$\alpha = \frac{\int_0^\infty \alpha_\lambda(\lambda) G_\lambda(\lambda)\, d\lambda}{\int_0^\infty G_\lambda(\lambda)\, d\lambda} = \frac{\int_0^\infty \alpha_\lambda(\lambda) E_{\lambda,b}(\lambda, 1200\text{ K})\, d\lambda}{E_b(1200\text{ K})}$$

where the spectral distribution of the irradiation is that of a blackbody (Eq. 16.7)

$$G_\lambda = E_{\lambda,b}(\lambda, T_f) = E_{\lambda,b}(\lambda, 1200\text{ K})$$

The integration can be performed in parts represented by band emission fractions, Eq. 18.10

$$\alpha = \alpha_{\lambda,1} \frac{\int_0^{\lambda_1} E_{\lambda,b}(\lambda, 1200\text{ K})\, d\lambda}{E_b(1200\text{ K})} + \alpha_{\lambda,2} \frac{\int_{\lambda_1}^\infty E_{\lambda,b}(\lambda, 1200\text{ K})\, d\lambda}{E_b(1200\text{ K})}$$

$$\alpha = \alpha_{\lambda,1} F_{(0\to\lambda_1)} + \alpha_{\lambda,2}[1 - F_{(0\to\lambda_1)}]$$

From Table 18.2, for the value $\lambda_1 T_f = 5\ \mu\text{m} \times 1200\text{ K} = 6000\ \mu\text{m}\cdot\text{K}$, find $F_{(0\to\lambda_1)} = 0.738$. Hence, the total absorptivity for the *initial condition* is

$$\alpha = 0.8 \times 0.738 + 0.1(1 - 0.738) = 0.62 \triangleleft$$

From Eq. 18.13, the *total emissivity* for the coating at the sphere *initial* temperature $T_{s,i}$ is expressed as

$$\varepsilon(T_{s,i}) = \frac{\int_0^\infty \varepsilon_\lambda E_{\lambda,b}(\lambda, T_{s,i})\, d\lambda}{E_b(T_{s,i})}$$

where $\varepsilon_\lambda = \alpha_\lambda$ since the coating is diffuse. The integration can be performed in parts represented by the band emission fractions, Eq. 18.10

$$\varepsilon = \alpha_{\lambda,1} \frac{\int_0^{\lambda_1} E_{\lambda,b}(\lambda, 300\text{ K})\, d\lambda}{E_b(300\text{ K})} + \alpha_{\lambda,2} \frac{\int_{\lambda_1}^\infty E_{\lambda,b}(\lambda, 300\text{ K})\, d\lambda}{E_b(300\text{ K})}$$

$$\varepsilon = \alpha_{\lambda,1} F_{(0\to\lambda_1)} + \alpha_{\lambda,2}[1 - F_{(0\to\lambda_1)}]$$

From Table 18.2, for the value $\lambda_1 T_{s,i} = 5\ \mu\text{m} \times 300\text{ K} = 1500\ \mu\text{m}\cdot\text{K}$, find $F_{(0\to\lambda_1)} = 0.014$. Hence, the total emissivity for the *initial* condition is

$$\varepsilon = 0.8 \times 0.014 + 0.1(1 - 0.014) = 0.11 \triangleleft$$

The sphere experiences emission and absorbed irradiation originating from the furnace wall. From an *energy balance* on the sphere as represented in Fig. E18.5 (see also Fig. 18.10 and Eq. 18.5a), the *net* radiation heat flux *leaving* the sphere is

$$q''_{\text{rad,net}} = E - \alpha G = \varepsilon E_b(T_{s,i}) - \alpha E_b(T_f)$$

Expressing the blackbody total emissive powers using the Stefan–Boltzmann law, Eq. 18.9, and substituting numerical values, find

$$q''_{\text{rad,net}} = \varepsilon\sigma T_{s,i}^4 - \alpha\sigma T_f^4$$

$$q''_{\text{rad,net}} = 0.11(5.67 \times 10^{-8}\,\text{W/m}^2\cdot\text{K}^4)(300\,\text{K})^4 - 0.62(5.67 \times 10^{-8}\,\text{W/m}^2\cdot\text{K}^4)(1200\,\text{K})^4$$

$$q''_{\text{rad,net}} = (0.11 \times 459)\text{W/m}^2 - 0.62(1.177 \times 10^5) = (50.5 - 7.29 \times 10^4)\text{W/m}^2 = -72.8\,\text{kW/m}^2 \triangleleft$$

The minus sign implies that the net radiant flux is *into* the sphere.

**(b)** Because the spectral characteristics of the coating and the furnace temperature remain fixed, there is no change in the value of $\alpha$ with increasing time. However, as $T_s$ increases with time, the value of $\varepsilon$ will change. After a sufficiently long time, $T_s = T_f$, which corresponds to the thermal equilibrium condition, so that $\varepsilon = \alpha$. That is, for the *final* condition,

$$\varepsilon = \alpha = 0.62 \triangleleft$$

***Comments:***

1. The equilibrium condition that eventually exists ($T_{s,f} = T_f$) satisfies the condition required for the equality of Eq. 18.31. Hence, $\alpha$ must equal $\varepsilon$ for the final condition.
2. The spectral emissivity distribution of the diffuse coating as shown in Fig. E18.5 corresponds to that for a heavily oxidized metallic or a nonmetallic material. This idealized distribution is representative of the spectra for *oxidized stainless steel* or *silicon carbide* as shown in Fig. 18.19*a*.
3. Approximating the sphere (mass $m$ with specific heat $c$) as a lumped capacitance and neglecting convection, an energy balance for the system, $\dot{E}_{\text{in}} - \dot{E}_{\text{out}} = \dot{E}_{\text{st}}$, can be expressed as

$$(\alpha G)A_s - (\varepsilon\sigma T_s^4)A_s = mc\frac{dT_s}{dt}$$

This differential equation could be solved to determine $T(t)$ for $t > 0$. However, the variation of $\varepsilon$ that occurs with increasing time would have to be included in the solution, or a suitable average value could be used as a first estimate.

## *Example 18.6* Solar Application: Spectrally Selective Spacecraft Panel

A spacecraft panel maintained at 300 K is coated with an opaque, diffuse white paint having the spectral reflectivity distribution shown below. The spacecraft is in a near-earth orbit and is exposed to solar irradiation of 1353 W/m$^2$ as well as to deep space at 0 K. What is the net radiative heat flux *leaving* the panel surface?

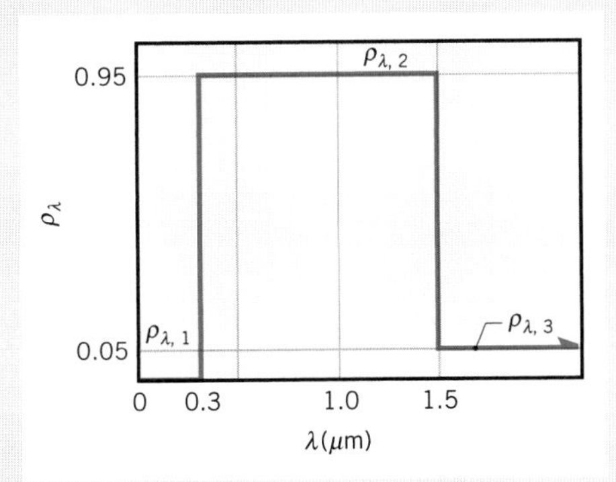


*Figure E18.6a*

### Solution

***Known:*** Panel with spectrally selective radiative properties operating at prescribed surface temperature in near-earth orbit.

***Find:*** Net radiative heat flux *leaving* the panel surface.

***Schematic and Given Data:***

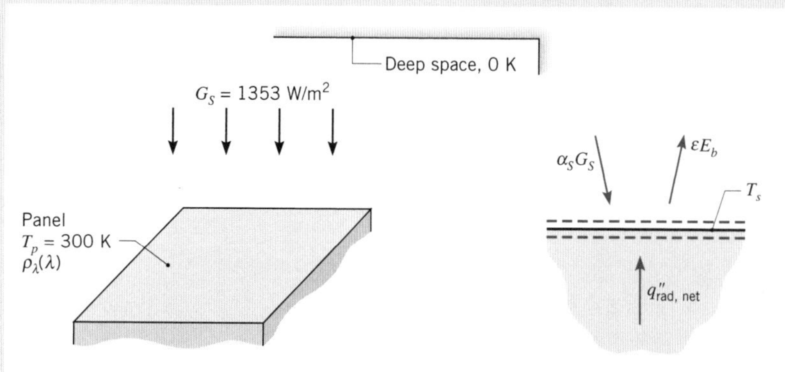


***Assumptions:***
1. Steady-state conditions.
2. Panel surface is opaque and diffuse.
3. Spectral distribution of solar irradiation is proportional to that for blackbody radiation at 5800 K.
4. Panel radiates to deep space at 0 K.

*Figure E18.6b*

***Analysis:*** As represented in Fig. E18.6b, from Equation 18.5a, the *surface energy balance* on the panel has the form

$$q''_{\text{rad,net}} = E - G_{\text{abs}} = \varepsilon E_b - \alpha_S G_S$$

The *solar absorptivity* can be obtained using Eq. 18.18, where the integration is performed in parts, which in turn are represented by band emission fractions

$$\alpha_S = \frac{\int_0^\infty \alpha_\lambda E_{\lambda,b}(\lambda, T_S)}{E_b(T_S)} = \frac{\int_0^{\lambda_1} \alpha_{\lambda,1} E_{\lambda,b}}{E_b} + \frac{\int_{\lambda_1}^{\lambda_2} \alpha_{\lambda,2} E_{\lambda,b}}{E_b} + \frac{\int_{\lambda_2}^{\infty} \alpha_{\lambda,3} E_{\lambda,b}}{E_b}$$

$$\alpha_S = \alpha_{\lambda,1} F_{(0\to\lambda_1)} + \alpha_{\lambda,2}[F_{(0\to\lambda_2)} - F_{(0\to\lambda_1)}] + \alpha_{\lambda,3}[1 - F_{(0\to\lambda_2)}]$$

$$\alpha_S = (1 - 0) \times 0.03345 + (1 - 0.95)[0.8808 - 0.03345] + (1 - 0.05)[1 - 0.8808]$$

$$\alpha_S = 0.226$$

Since the surface is opaque $\alpha_\lambda = 1 - \rho_\lambda$. From Fig. E18.6a, $\lambda_1 = 0.3\ \mu\text{m}$, $\lambda_2 = 1.5\ \mu\text{m}$, and $T_S = 5800$ K. The band emission fractions from Table 18.2 are

$$\lambda_1 T_S = 0.3 \times 5800 = 1740\ \mu\text{m}\cdot\text{K}: \quad F_{(0\to\lambda_1)} = 0.03345$$

$$\lambda_2 T_S = 1.5 \times 5800 = 8700\ \mu\text{m}\cdot\text{K}: \quad F_{(0\to\lambda_2)} = 0.8808$$

The *total emissivity* is calculated from Eq. 18.13, and the integration can be performed using appropriate band emission fractions. However, recognize that with $\lambda_2 T_p = 1.5\ \mu\text{m} \times 300\ \text{K} = 450\ \mu\text{m}\cdot\text{K}$ from Table 18.2, $F_{(0\to\lambda_2)} = 0.000$. Accordingly, the spectral region of importance is $\lambda > \lambda_2$, so that

$$\varepsilon = \alpha_{\lambda,3} = 1 - \rho_{\lambda,3} = 1 - 0.05 = 0.95$$

Substituting numerical values for the radiative properties into the energy balance, and with $E_b = \sigma T_p^4$, the net radiative heat flux *leaving* the panel surface is

$$q''_{\text{rad,net}} = 0.95(5.67 \times 10^{-8}\ \text{W/m}^2\cdot\text{K}^4)(300\ \text{K})^4 - 0.226 \times 1353\ \text{W/m}^2$$

$$q''_{\text{rad,net}} = 413\ \text{W/m}^2 - 305\ \text{W/m}^2 = 108\ \text{W/m}^2 \quad \triangleleft$$

***Comments:***

1. Since the net radiative flux leaving the panel is positive, the panel is behaving as a *radiator,* rejecting energy dissipated within the spacecraft to deep space.
2. Recognize that the absorption of solar irradiation and emission, respectively, are controlled by the short- and long-wavelength characteristics of the spectral absorptivity. The ratio of $\alpha_S/\varepsilon$ is an important parameter for spacecraft thermal control and solar collectors. The coating of this example has $\alpha_S/\varepsilon = 0.226/0.95 = 0.23$, and functions as a radiator. For a *collector* panel, a coating with a ratio greater than unity would be required.
3. The spectral reflectivity distribution of the diffuse white coating as shown in Fig. E18.6*a* is an idealized representation for a *white paint* such as that shown in Fig. 18.19*b*.

## Radiative Exchange Between Surfaces in Enclosures

Thus far we have restricted our attention to radiation processes that occur at a single surface. Now we will consider the problem of *radiative exchange between two or more surfaces.* In general, radiation may leave a surface due to both direct emission and reflection (radiosity), and upon reaching a second surface, experience absorption as well as reflection. The radiative exchange depends upon the surface geometries and their orientations, as well as on their radiative properties and temperatures.

We begin by establishing the *geometrical features* of the radiation exchange problem by developing the concept of the *view factor.* Using the view factor, we then treat *black surface exchange,* which does not have the complications of multiple-surface reflections present with non-black surfaces. Analyzing radiation exchange between *non-black,* opaque surfaces in an enclosure is greatly simplified through two major assumptions: the surfaces are *diffuse-gray* ($\varepsilon = \alpha$), and are characterized by a *uniform radiosity and irradiation.* The importance of these assumptions will become evident as we develop the means to calculate radiative exchange.

## 18.5 The View Factor

The view factor (also called a configuration or shape factor) accounts for the geometrical features for the radiation exchange between two surfaces. The ***view factor*** $F_{ij}$ is defined as the *fraction of the radiation leaving surface i that is intercepted by surface j.* For the arbitrarily oriented surfaces $A_i$ and $A_j$ of Fig. 18.20

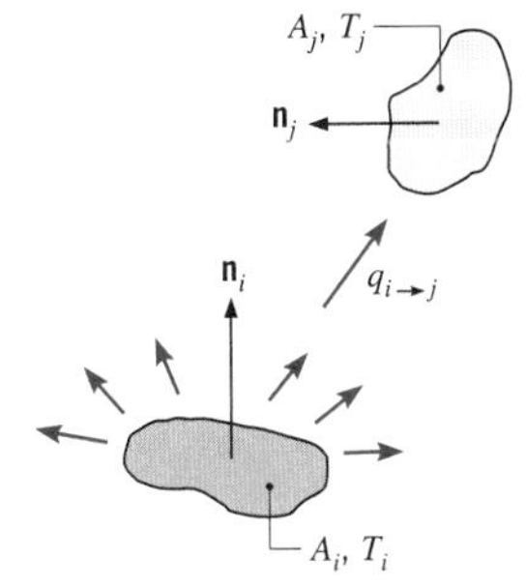

Figure 18.20

$$F_{ij} = \frac{q_{i\to j}}{A_i J_i} \quad (18.32)$$

*view factor*

where $q_{i\to j}$ is the radiative flux *leaving* $A_i$ that is *intercepted by* $A_j$; $J_i$ is the radiosity of surface $A_i$, which represents the radiative flux leaving $A_i$ in all directions. It is assumed that *the surfaces are isothermal, diffuse, and have a uniform radiosity.*

Two important relationships involving the view factors should be recognized. For the arbitrarily oriented surfaces (Fig. 18.20), we can write

$$A_i F_{ij} = A_j F_{ji} \quad (18.33)$$

*reciprocity relation*

This expression, termed the ***reciprocity relation,*** is useful in determining one view factor from knowledge of the other. This relation is a consequence of the diffuse nature of the radiation from the surfaces.

For surfaces forming an enclosure (Figure 18.21), the ***summation rule***

$$\sum_{j=1}^{N} F_{ij} = 1 \quad (18.34)$$

*summation rule*

can be applied to each of the $N$ surfaces in the enclosure. This rule follows from the requirement that all radiation leaving surface $i$ must be intercepted by the enclosure surfaces. The term $F_{ii}$ appearing in the summation represents the fraction of the radiation that leaves surface $i$ and is directly intercepted by $i$. If the surface is concave, it "*sees itself*" and $F_{ii}$ is nonzero. However, for a plane or convex surface, $F_{ii} = 0$.

There are several approaches for evaluating the view factors. For some situations, it may be possible to determine $F_{ij}$ by *inspection.* That is, by intuition stemming from the physical interpretation of $F_{ij}$, with consideration to the surface arrangement, you can sometimes recognize the fraction of radiation leaving $A_i$ that is intercepted by $A_j$.

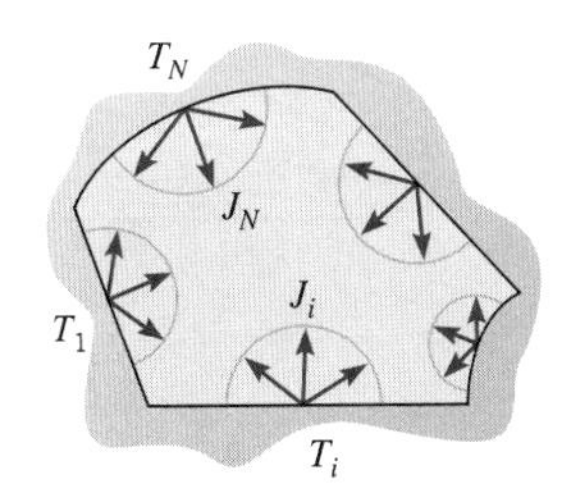

Figure 18.21

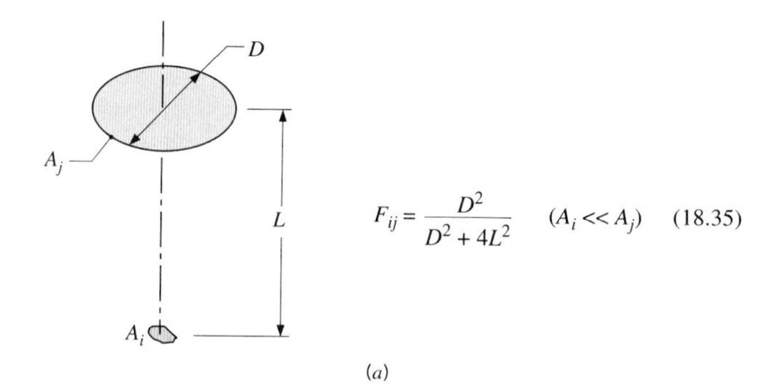


$$F_{ij} = \frac{D^2}{D^2 + 4L^2} \qquad (A_i << A_j) \qquad (18.35)$$

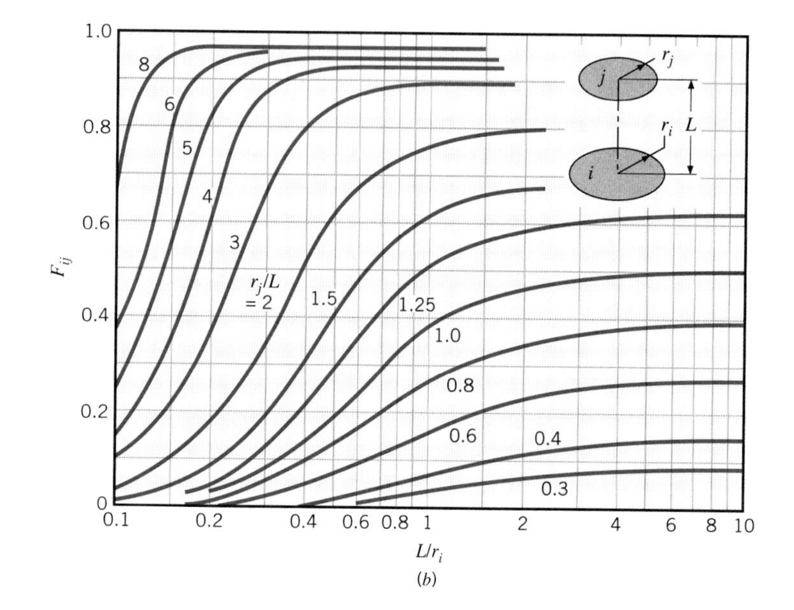


$$F_{ij} = \frac{1}{2}\left\{S - [S^2 - 4(r_j/r_i)^2]^{1/2}\right\} \qquad (18.36)$$

$$S = 1 + \frac{1 + R_j^2}{R_i^2} \qquad (18.37)$$

$$R_i = r_i/L, \; R_j = r_j/L \qquad (18.38)$$

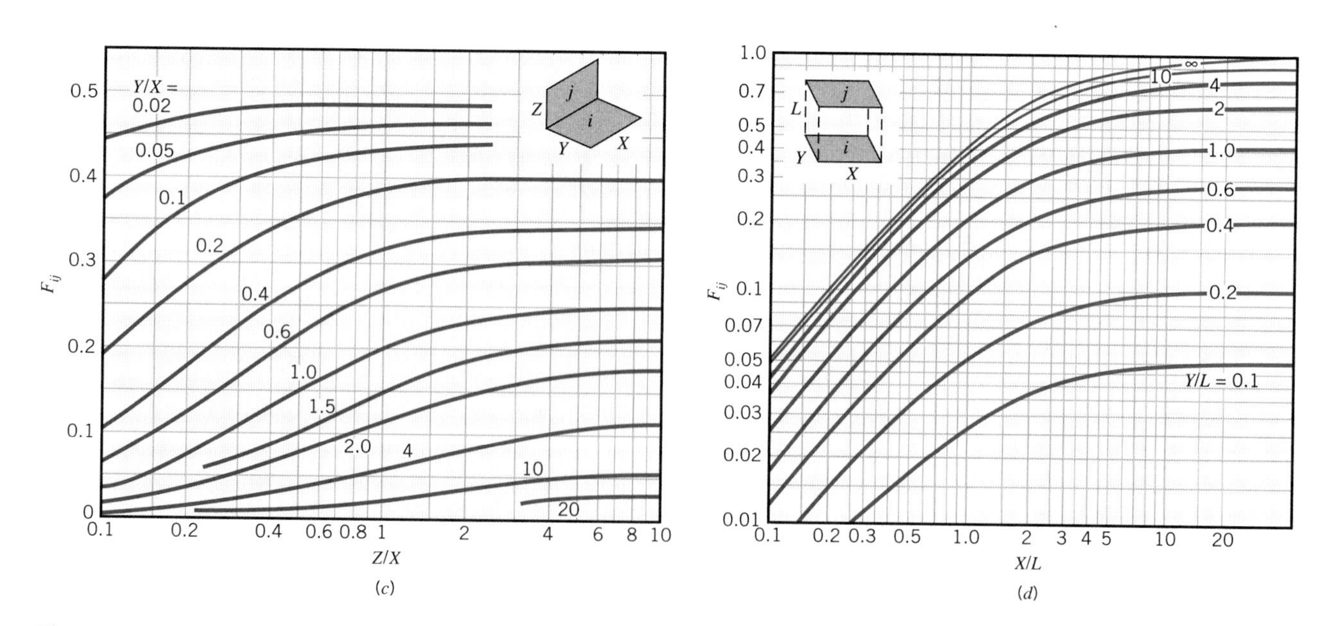


*Figure 18.22* View factors for three-dimensional geometries: (*a*) small surface coaxial, parallel to disk, (*b*) coaxial parallel disks (*c*) aligned parallel rectangles, and (*d*) perpendicular rectangles with a common edge.

**Table 18.3** View Factors for Two-Dimensional Geometries

Geometry/Relation

| Parallel Plates with Midlines Connected by Perpendicular | Inclined Parallel Plates of Equal Width and a Common Edge | Perpendicular Plates with a Common Edge |
|---|---|---|
| 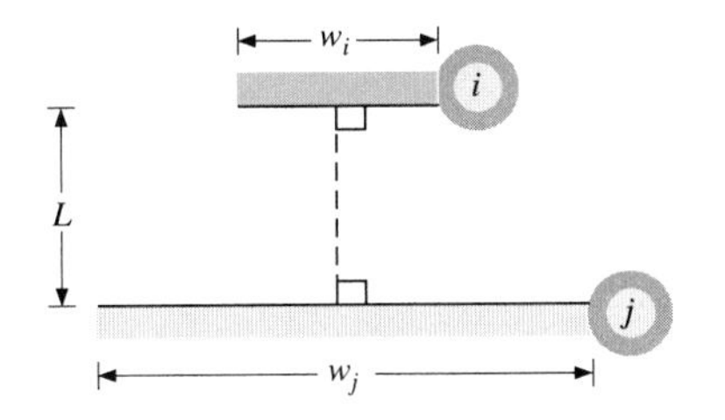  | 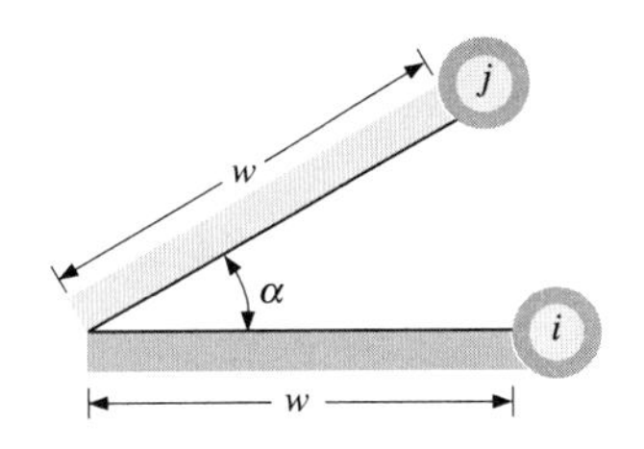  | 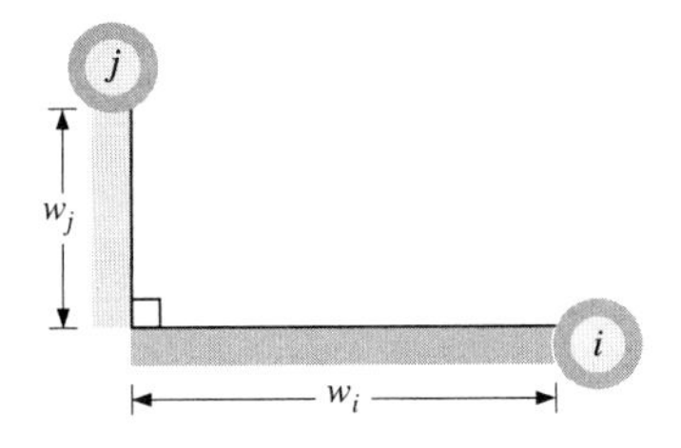  |
| $F_{ij} = \dfrac{[(W_i + W_j)^2 + 4]^{1/2} - [(W_j - W_i)^2 + 4]^{1/2}}{2W_i}$ <br> $W_i = w_i/L,\ W_j = w_j/L$ (18.39) | $F_{ij} = 1 - \sin\left(\dfrac{\alpha}{2}\right)$ (18.40) | $F_{ij} = \dfrac{1 + (w_j/w_i) - [1 + (w_j/w_i)^2]^{1/2}}{2}$ (18.41) |

Analytical solutions for $F_{ij}$ have been obtained for many common surface arrangements and are available in *equation, graphical,* and *tabular forms*. View factors for selected *two-dimensional* configurations (infinitely long in the direction perpendicular to the page) are presented in Table 18.3 (Eqs. 18.39–41). View factors for *three-dimensional* geometries are presented in Fig. 18.22 (Eqs. 18.35–38). In conjunction with these results, you may also use the *reciprocity relation* and *summation rule* to determine the required view factors.

## *Example 18.7* Calculating View Factors for Diffuse Surfaces

Determine the view factors $F_{12}$ and $F_{21}$ for the following geometries:

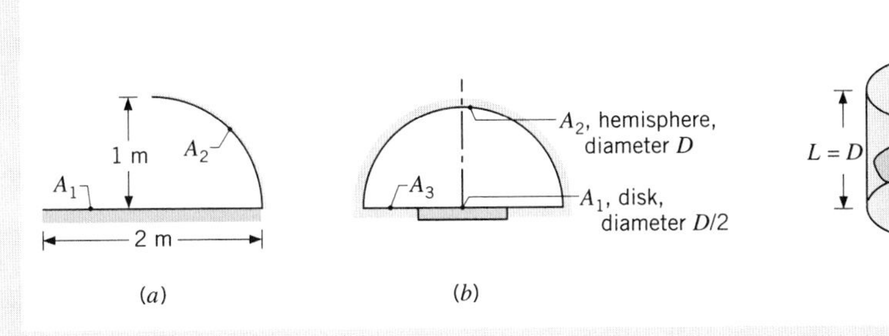


*Figure E18.7a,b,c*

(a) Long open channel.
(b) Hemispherical-disk arrangement; find also $F_{22}$ and $F_{23}$.
(c) End and side of a circular tube of equal length and diameter.

### Solution

***Known:*** Surface geometries.
***Find:*** View factors.

***Assumptions:*** Diffuse surfaces with uniform radiosities.

***Analysis:*** The desired view factors are obtained from inspection, the reciprocity relation, the summation rule and/or use of a chart.

(a) *Long open channel of length L.* Complete the enclosure by defining the third surface $A_3$, which is symmetrical in form to $A_2$. Applying the *summation rule* for surface $A_1$,

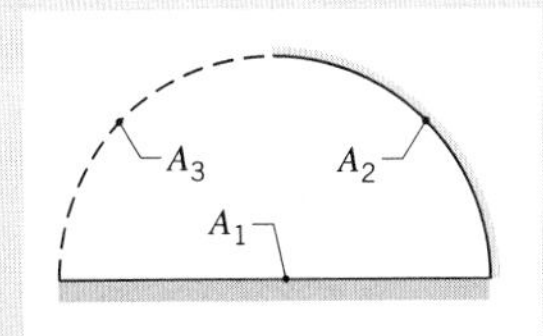


*Figure E18.7d*

$$F_{11} + F_{12} + F_{13} = 1$$

By *inspection:* since $A_1$ *does not see itself,* it follows that $F_{11} = 0$. Also, since $A_2$ and $A_3$ are symmetrically positioned with respect to $A_1$, *by symmetry* we recognize that $F_{12} = F_{13}$, hence

$$F_{12} = 0.5 \quad \triangleleft$$

From the *reciprocity relation* between surfaces $A_1$ and $A_2$

$$F_{21} = \frac{A_1}{A_2}F_{12} = \frac{2\text{ m} \times L}{(2\pi 1\text{ m}/4) \times L}0.5 = \frac{4}{\pi}0.5 = 0.637 \quad \triangleleft$$

**(b)** *Hemisphere of diameter D over a disk of diameter D/2.* By *inspection,* recognize that $A_1$ *sees* only $A_2$, hence, it follows that

$$F_{12} = 1.0 \quad \triangleleft$$

From the *reciprocity relation* between surfaces $A_1$ and $A_2$

$$F_{21} = \frac{A_1}{A_2}F_{12} = \frac{\pi(D/2)^2/4}{(\pi D^2)/2}1.0 = \frac{1}{8}1.0 = 0.125 \quad \triangleleft$$

By *inspection,* based upon a *symmetry* argument that $A_2$ *sees* as much of itself as it does of $A_1$ and $A_3$ combined, it follows that

$$F_{22} = 0.50 \quad \triangleleft$$

Applying the *summation rule* to surface $A_2$

$$F_{21} + F_{22} + F_{23} = 1$$

$$F_{23} = 1 - F_{21} + F_{22} = 1 - 0.125 - 0.50 = 0.375 \quad \triangleleft$$

**(c)** *Circular tube.* Apply the *summation rule* to surface $A_1$

$$F_{11} + F_{12} + F_{13} = 1$$

By *inspection,* $F_{11} = 0$, and from the *chart* for the coaxial, parallel disks, Fig. 18.22*b*, with $(r_3/L) = 0.5$ and $(L/r_1) = 2$, find

$$F_{13} = 0.17$$

Substituting numerical values into the summation rule

$$F_{12} = 1 - F_{11} - F_{13} = 1 - 0 - 0.17 = 0.83 \quad \triangleleft$$

From the *reciprocity relation* between surfaces $A_1$ and $A_2$, find

$$F_{21} = \frac{A_1}{A_2}F_{12} = \frac{\pi D^2/4}{\pi DL}0.83 = 0.21 \quad \triangleleft$$

***Comments:***

1. Note that the summation rule must be applied to an enclosure. To complete the enclosure in part (a), it was necessary to define a third hypothetical surface $A_3$ (shown by dashed lines), which we made symmetrical in form to $A_2$.
2. Recognize that the solutions follow a systematic procedure by applying the reciprocity relation and summation rule. Always look for instances to deduce the shape factor by *inspection* as has been illustrated in this example.

## 18.6 Blackbody Radiation Exchange

For surfaces that may be approximated as blackbodies, radiation leaves only as a result of emission, none of the incident radiation is reflected. We develop first the relation for the *net exchange between two black surfaces,* and then extend the treatment for determining the *net radiation from a black surface in an enclosure.*

Consider radiation exchange between two black surfaces of arbitrary shape (Fig. 18.23*a*). Recalling from our discussion in Sec. 18.6, $q_{i \to j}$ is the rate at which radiation *leaves* the surface *i* and is *intercepted* by surface *j*. From Eq. 18.32, it follows that

$$q_{i \to j} = (A_i J_i) F_{ij} \tag{18.42}$$

or, since *the radiosity equals the emissive power for a black surface,* $J_i = E_{bi}$

$$q_{i \to j} = A_i F_{ij} E_{bi}$$

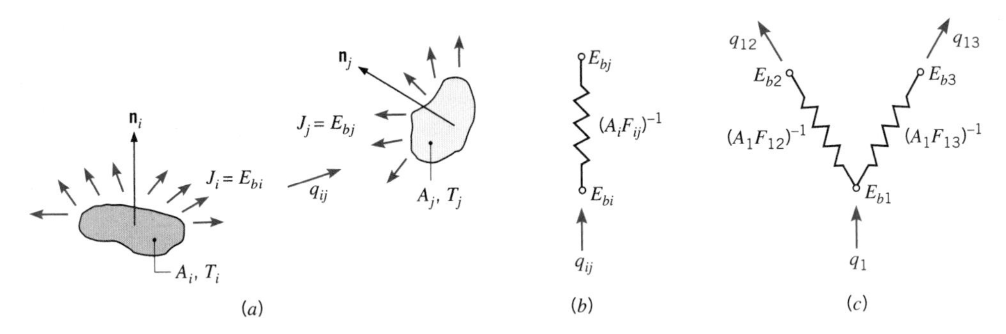


*Figure 18.23* Radiation transfer between black surfaces. (*a*) Net exchange between two surfaces, $q_{ij}$. Network elements representing (*b*) net exchange between two surfaces, $q_{ij}$, in terms of the *space radiative resistance* and blackbody emissive powers, and (*c*) net rate from surface $A_1$ due to exchange with the remaining surfaces ($A_2$, $A_3$) in a *three-surface enclosure,* $q_1$.

Similarly

$$q_{j\to i} = A_jF_{ji}E_{bj}$$

The *net radiative exchange* between the two surfaces can then be defined as

$$q_{ij} = q_{i\to j} - q_{j\to i} \tag{18.43}$$

from which it follows that

$$q_{ij} = A_iF_{ij}E_{bi} - A_jF_{ji}E_{bj}$$

Using the view factor reciprocity relation, Eq. 18.33, and rearranging, we find

$$q_{ij} = \frac{E_{bi} - E_{bj}}{(A_iF_{ij})^{-1}} \tag{18.44}$$

*network element*

or in terms of the surface temperatures using Eq. 18.9

$$q_{ij} = A_iF_{ij}\sigma(T_i^4 - T_j^4) \tag{18.45}$$

Note that the expression for the net exchange of Eq. 18.44 can be represented by the ***network element*** of Fig. 18.23*b* associated with the *driving potential* ($E_{bi} - E_{bj}$) and a ***space or geometrical radiative resistance*** of the form $(A_iF_{ij})^{-1}$.

*space resistance*

The foregoing results can also be used to evaluate the net radiation transfer from any surface in an *enclosure of black surfaces.* For an enclosure with *three* surfaces maintained at different temperatures, the *net rate* of radiation from surface $A_1$ is due to exchange with the remaining surfaces ($A_2$, $A_3$) and can be expressed as

$$q_1 = q_{12} + q_{13} \tag{18.46}$$

$$q_1 = \frac{E_{b1} - E_{b2}}{(A_1F_{12})^{-1}} + \frac{E_{b1} - E_{b3}}{(A_1F_{13})^{-1}} \tag{18.47}$$

These relations for the black surface can also be represented by *network* elements as shown in Fig. 18.23*c*. Recognize that we can write a similar relation for each of the surfaces in the enclosure, and can, of course, extend the treatment to more than three surfaces.

## *Example 18.8* Enclosure Analysis: Black Surface Exchange

A furnace cavity, which is in the form of a cylinder of 75-mm diameter and 150-mm length, is open at one end to large surroundings that are at 27°C. The sides and bottom may be approximated as blackbodies, are heated electrically, are well insulated, and are maintained at temperatures of 1350 and 1650°C, respectively.

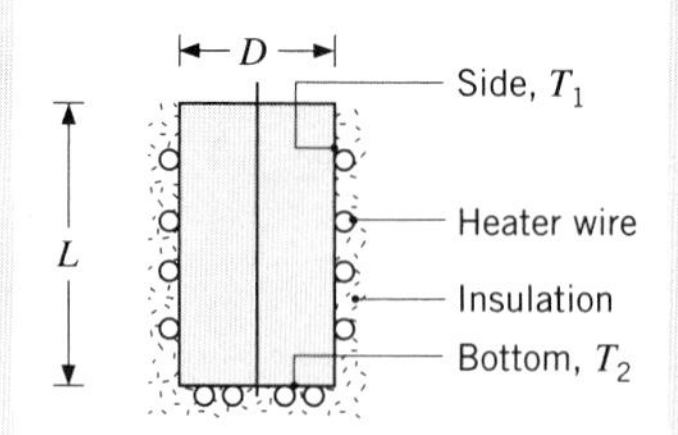


*Figure E18.8a*

How much electrical power is required to maintain the furnace under steady-state conditions?

## Solution

***Known:*** Surface temperatures of cylindrical furnace.

***Find:*** Electrical power required to maintain prescribed temperatures.

***Schematic and Given Data:***

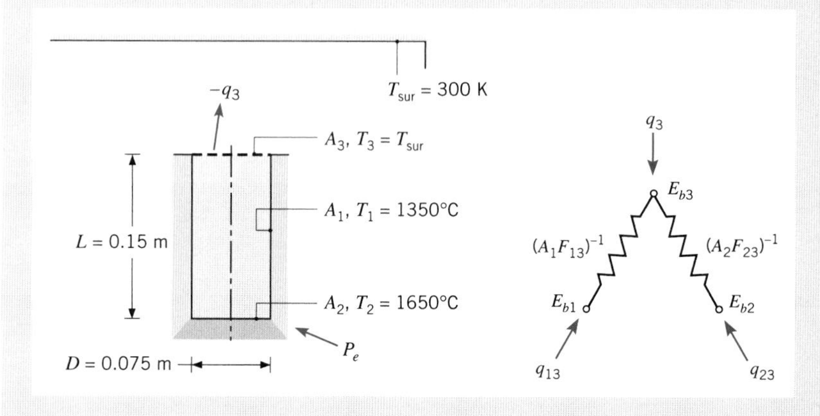


***Assumptions:***

1. Steady-state conditions.
2. Interior surfaces behave as blackbodies.
3. Heat transfer by convection is negligible.
4. Outer surface of furnace is adiabatic.

*Figure E18.8b,c*

***Analysis:*** The electrical power required to operate the furnace under steady-state conditions, $P_e$, must balance the heat transfer from the furnace opening. Subject to the foregoing assumptions, the only heat transfer is by radiation through the opening, which may be treated as the *hypothetical surface* of area $A_3$, which completes the furnace interior enclosure ($A_1$, $A_2$, $A_3$). Because the surroundings are large, radiation exchange between the furnace opening and the surroundings may be treated by approximating the surface $A_3$ as a blackbody at $T_3 = T_{sur}$. The processes associated with $A_3$ are represented by the network in Fig. E18.8*c* from which it follows that

$$P_e = -q_3 = q_{13} + q_{23}$$

$$P_e = A_1F_{13}\sigma(T_1^4 - T_3^4) + A_2F_{23}\sigma(T_2^4 - T_3^4)$$

where $q_3$ is the net radiative heat rate *leaving* $A_3$ (on the enclosure side of the surface). From Fig. 18.22*b* for the view factors between parellel, coaxial disks with $(r_j/L) = (0.0375 \text{ m}/0.15 \text{ m}) = 0.25$ and $(L/r_i) = (0.15 \text{ m}/0.0375 \text{ m}) = 4$, find that

$$F_{23} = 0.06$$

From the *summation rule* for surface $A_2$, with $F_{22} = 0$

$$F_{21} = 1 - F_{22} - F_{23} = 1 - 0 - 0.06 = 0.94$$

and using the *reciprocity relation*

$$F_{12} = \frac{A_2}{A_1}F_{21} = \frac{\pi(0.075 \text{ m})^2/4}{\pi(0.075 \text{ m})(0.15 \text{ m})} \times 0.94 = 0.118$$

From symmetry considerations, find that $F_{13} = F_{12} = 0.118$. With $A_1 = \pi DL$ and $A_2 = \pi D^2/4$, and substituting numerical values, the electrical power required is

$$P_e = (\pi \times 0.075 \text{ m} \times 0.15 \text{ m})0.118(5.67 \times 10^{-8} \text{ W/m}^2 \cdot \text{K}^4)[(1623 \text{ K})^4 - (300 \text{ K})^4]$$
$$+ (\pi(0.075 \text{ m})^2/4) \times 0.06(5.67 \times 10^{-8} \text{ W/m}^2 \cdot \text{K}^4)[(1923 \text{ K})^4 - (300 \text{ K})^4]$$
$$P_e = 1639 \text{ W} + 205 \text{ W} = 1844 \text{ W} \triangleleft$$

## 18.7 Radiation Exchange Between Diffuse-Gray Surfaces in an Enclosure

For an enclosure comprised of opaque, *nonblack* surfaces, radiation may leave a surface by emission *and* by reflection of irradiation that originates from other surfaces in the *enclosure* as shown in Fig. 18.24*a*. We begin the enclosure analysis problem by formulating surface energy balances to obtain relations for the *net radiation leaving a surface,* and representing the results with network elements. We will apply the network to the *two-surface enclosure,* considering also the special case of the *radiation shield,* and to the *three-surface enclosure* having *one reradiating* (insulated) *surface*.

### 18.7.1 Radiation Exchange Relations: Network Representation

The term $q_i$, which is the *net rate at which radiation leaves surface i*, represents the net effect of radiative interactions at the surface. As shown in Fig. 18.24*b* (see also Fig. 18.11 and Eq. 18.5b), it is equal to the difference between the surface radiosity and irradiation, and may be expressed as

$$q_i = A_i(J_i - G_i) \tag{18.48}$$

From the definition of the radiosity, Eq. 18.4b, with Eq. 18.12 for the emissive power and Eq. 18.20 for the reflected irradiation, find

$$J_i \equiv E_i + G_{\text{ref},i} = \varepsilon_i E_{bi} + \rho_i G_i \tag{18.49}$$

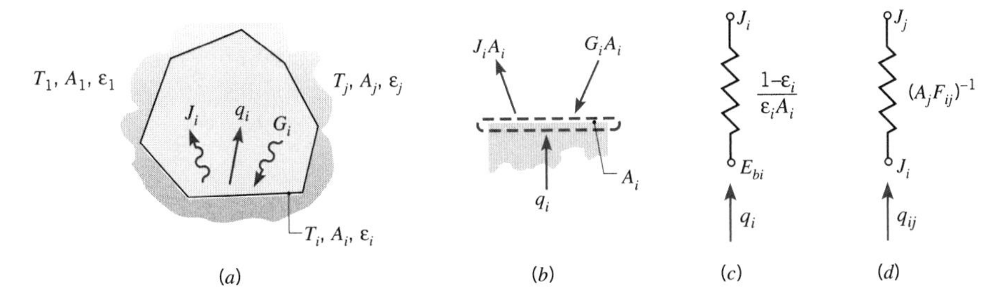


*Figure 18.24* Net radiation exchange in an enclosure of diffuse-gray surfaces. (*a*) Schematic of the enclosure. (*b*) Radiative balance according to Eq. 18.48. (*c*) Network element representing the net radiation transfer from the surface, $q_i$, in terms of the *surface radiative resistance*. (*d*) Network element representing the exchange between two surfaces, $q_{ij}$, in terms of the *space* or *geometrical radiative resistance*.

and solving for the irradiation $G$, find

$$G_i = \frac{J_i - \varepsilon_i E_{bi}}{\rho_i} = \frac{J_i - \varepsilon_i E_{bi}}{1 - \varepsilon_i} \tag{18.50}$$

where, from Eq. 18.28, $\rho_i = 1 - \alpha_i = 1 - \varepsilon_i$, since $\varepsilon_i = \alpha_i$ for an opaque, diffuse-gray surface (Eq. 18.31). Substituting Eq. 18.50 into 18.48, it follows that the ***net radiative heat rate leaving surface*** $A_i$ has the form

*net radiative heat rate leaving surface*

$$q_i = \frac{E_{bi} - J_i}{(1 - \varepsilon_i)/\varepsilon_i A_i} \tag{18.51}$$

*surface radiative resistance*

This relation may be represented by the network element of Fig. 18.24*c* where $(E_{bi} - J_i)$ is the *driving potential* and $(1 - \varepsilon_i)/\varepsilon_i A_i$ is viewed as the ***surface radiative resistance.*** This element represents a diffuse-gray surface; if the surface were black, this element would not appear in the network we are attempting to build.

***Space Radiative Resistance.*** Consider now radiation *exchange* between two of the surfaces in the enclosure of Fig. 18.24*a*. Following the treatment for blackbody exchange (see Eqs. 18.42–43), recall that the term $q_{i \to j}$ was defined as the rate at which radiation leaves surface $i$ and is intercepted by surface $j$. Hence, the *net radiation exchange between the two surfaces* can be expressed as

$$q_{ij} = q_{i \to j} - q_{j \to i} \tag{18.52}$$

From Eq. 18.42, in terms of the surface radiosities (not the emissive powers), it follows that

$$q_{ij} = (A_i J_i) F_{ij} - (A_j J_j) F_{ji} \tag{18.53}$$

and using the view factor reciprocity relation

*radiation exchange between surfaces*

$$q_{ij} = \frac{J_i - J_j}{(A_i F_{ij})^{-1}} \tag{18.54}$$

*space radiative resistance*

The *component* $q_{ij}$ may be represented by a network element for which $(J_i - J_j)$ is the *driving potential* and $(A_i F_{ij})^{-1}$ is a ***space or geometrical radiative resistance*** (Fig. 18.24*d*).

***Energy Balance on Node $J_i$.*** From Eq. 18.51, we see that the net radiation transfer (current flow) to surface $i$ through its *surface resistance*, $q_i$, must equal the net rate of radiation transfer (current flows) from $i$ to all the other surfaces through the corresponding *space resistances*, $q_{ij}$, as given by Eq. 18.54. These equations are represented by the *network* in Fig. 18.25 for the surface $A_1$ in a three-surface enclosure. The network corresponds to an *energy balance on the node representing the radiosity* (potential). From the network we can see that the *net rate* of radiation transfer from surface $A_1$, $q_1$, is equal to the sum of the components related to radiative exchange with the other surfaces in the enclosure and has the form

$$q_1 = q_{12} + q_{13} + \cdots = \frac{J_1 - J_2}{(A_1 F_{12})^{-1}} + \frac{J_1 - J_3}{(A_1 F_{13})^{-1}} + \cdots \tag{18.55}$$

The *network representation* of the diffuse-gray surface (Fig. 18.25) serves as a useful tool for visualizing and calculating radiation exchange. In the next section, we apply this network to a two-surface enclosure.

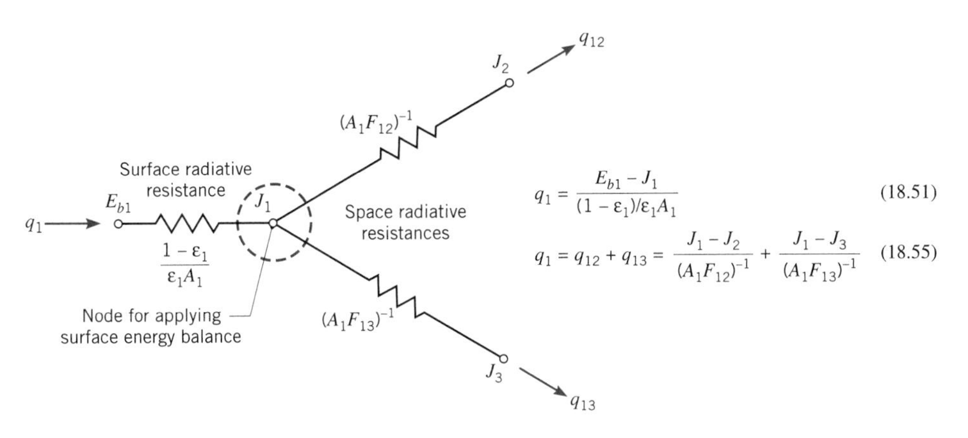


*Figure 18.25* Surface and space radiative resistances comprising the network representing surface $A_1$ in a *three*-surface enclosure. Equations 18.51 and 18.55 are energy balances on the $J_1$ surface node.

## 18.7.2 The Two-Surface Enclosure

The simplest example of an enclosure is one involving two surfaces that exchange radiation only with each other. Such a two-surface enclosure is shown schematically in Fig. 18.26*a*. From an *overall energy balance on the enclosure,* we recognize that the net rate of radiation transfer from surface 1, $q_1$, must equal the net rate of radiation transfer to surface 2, $q_2$. Since there are only two surfaces in the enclosure, it follows that both quantities must equal the net rate at which radiation is exchanged between 1 and 2, $q_{12}$. Accordingly

$$q_1 = -q_2 = q_{12} \tag{18.56}$$

We can use the network representation of Fig. 18.25 for each of the surfaces to construct the network corresponding to the *two-surface enclosure* shown in Fig. 18.26*b*. The total resistance to radiation exchange between surface 1 and 2 is comprised of two surface resistances and the space resistance. Hence the *net radiation exchange* between the two surfaces can be expressed as

$$q_1 = \frac{\sigma(T_1^4 - T_2^4)}{\dfrac{1-\varepsilon_1}{\varepsilon_1 A_1} + \dfrac{1}{A_1F_{12}} + \dfrac{1-\varepsilon_2}{\varepsilon_2 A_2}} \tag{18.57}$$

where $E_b = \sigma T^4$ from Eq. 18.9.

The forgoing result may be used for any two diffuse-gray surfaces that form an enclosure. The application of Eq. 18.57 to important common geometries is summarized in Table 18.4. Note that the net radiative heat rate, Eq. 18.61, for the *small convex object in large isothermal surroundings* corresponds to the exchange equation you first encountered in Chap. 15 (Eq. 15.15).

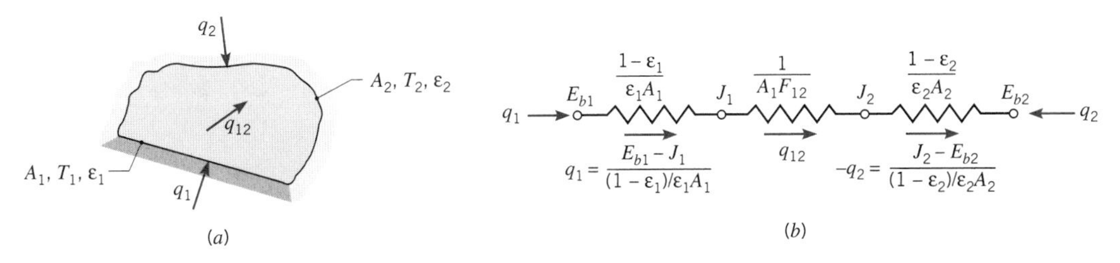


*Figure 18.26* The two-surface enclosure. (*a*) Schematic. (*b*) Network representation of the enclosure with two *surface*- and one *space*-radiative resistances, Eq. 18.57.

*Figure 18.27* Radiation exchange between large parallel planes with a radiation shield. (*a*) Schematic. (*b*) Network representation with four *surface-* and two *space-radiative resistances.*

*radiation shields*

***Radiation Shields.*** ***Radiation shields*** constructed from low emissivity (high reflectivity) materials, can be used to reduce the net radiation transfer between two surfaces. Consider placing a shield, surface 3, between the two, large parallel planes of Fig. 18.27*a*. *Without* the radiation shield, the net rate of radiation transfer between surfaces 1 and 2 is given by Eq. 18.57. However, *with* the radiation shield, additional resistances are present, as shown in Fig. 18.27*b*, and hence, the heat rate is reduced. Note that the emissivity associated with one side of the shield ($\varepsilon_{3,1}$) may differ from that associated with the opposite side ($\varepsilon_{3,2}$) and the radiosities will

**Table 18.4** Net Radiative Exchange Equations for Common Diffuse-Gray, Two-Surface Enclosures from Application of Eq. 18.57

**Large (Infinite) Parallel Planes**

$A_1, T_1, \varepsilon_1$; $A_2, T_2, \varepsilon_2$

$$A_1 = A_2 = A \qquad F_{12} = 1 \qquad q_{12} = \frac{A\sigma(T_1^4 - T_2^4)}{\dfrac{1}{\varepsilon_1} + \dfrac{1}{\varepsilon_2} - 1} \tag{18.58}$$

**Long (Infinite) Concentric Cylinders**

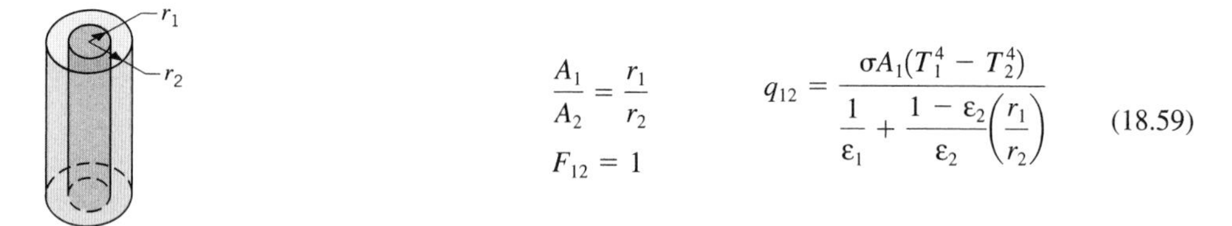

$$\frac{A_1}{A_2} = \frac{r_1}{r_2} \qquad F_{12} = 1 \qquad q_{12} = \frac{\sigma A_1(T_1^4 - T_2^4)}{\dfrac{1}{\varepsilon_1} + \dfrac{1 - \varepsilon_2}{\varepsilon_2}\left(\dfrac{r_1}{r_2}\right)} \tag{18.59}$$

**Concentric Spheres**

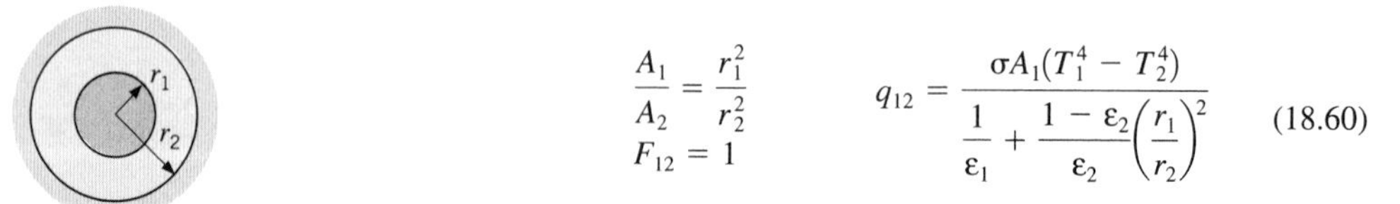

$$\frac{A_1}{A_2} = \frac{r_1^2}{r_2^2} \qquad F_{12} = 1 \qquad q_{12} = \frac{\sigma A_1(T_1^4 - T_2^4)}{\dfrac{1}{\varepsilon_1} + \dfrac{1 - \varepsilon_2}{\varepsilon_2}\left(\dfrac{r_1}{r_2}\right)^2} \tag{18.60}$$

**Small Convex Object in Large Surroundings**

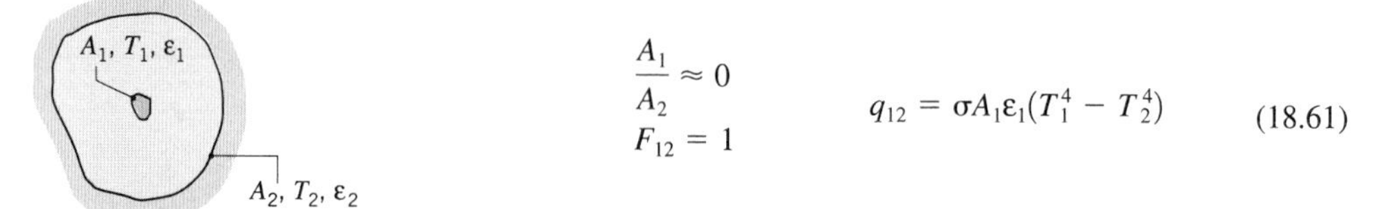

$$\frac{A_1}{A_2} \approx 0 \qquad F_{12} = 1 \qquad q_{12} = \sigma A_1 \varepsilon_1 (T_1^4 - T_2^4) \tag{18.61}$$

always differ. Summing the resistances and recognizing that $F_{13} = F_{32} = 1$, it follows that

$$q_{12} = \frac{A_1\sigma(T_1^4 - T_2^4)}{\dfrac{1}{\varepsilon_1} + \dfrac{1-\varepsilon_{3,1}}{\varepsilon_{3,1}} + \dfrac{1-\varepsilon_{3,2}}{\varepsilon_{3,2}} + \dfrac{1}{\varepsilon_2}} \tag{18.62}$$

Recognize that the resistances associated with the radiation shield become very large when the emissivities $\varepsilon_{3,1}$ and $\varepsilon_{3,2}$ are very small.

## *Example 18.9* Radiation Shield for a Cryogenic Fluid Transfer Line

A cryogenic fluid flows through a long tube of 20-mm diameter, the outer surface of which is diffuse and gray with $\varepsilon_1 = 0.02$ and $T_1 = 77$ K. This tube is concentric with a larger tube of 50-mm diameter, the inner surface of which is diffuse and gray with $\varepsilon_2 = 0.05$ and $T_2 = 300$ K. The space between the surfaces is evacuated. (a) Calculate the heat transfer to the cryogenic fluid per unit length of tubes. (b) If a thin radiation shield of 35-mm diameter and $\varepsilon_3 = 0.02$ (both sides) is inserted midway between the inner and outer surfaces, calculate the change (percentage) in heat transfer per unit length of the tubes.

### Solution

***Known:*** Concentric tube arrangement with diffuse-gray surfaces of different emissivities and temperatures.

***Find:***

(a) Heat transfer by the cryogenic fluid passing through the inner tube *without* the radiation shield.
(b) Percentage change in heat transfer *with* the radiation shield inserted midway between inner and outer tubes.

***Schematic and Given Data:***

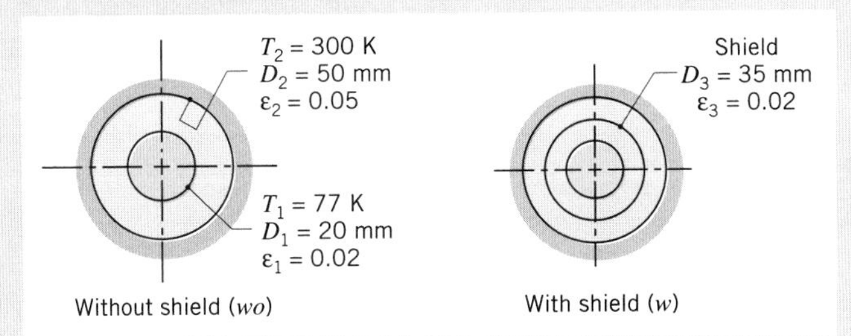


***Assumptions:***
1. Surfaces are diffuse and gray.
2. Space between tubes is evacuated.
3. Conduction resistance for radiation shield is negligible.
4. Concentric tubes form a two-surface enclosure (end effects are negligible).

*Figure E18.9*

***Analysis:***

(a) The network representation of the system *without* the shield ($wo$) is shown in Fig. 18.26, and the desired heat rate may be obtained from Eq. 18.59, where

$$q_{wo} = \frac{\sigma(\pi D_1 L)(T_1^4 - T_2^4)}{\dfrac{1}{\varepsilon_1} + \dfrac{1-\varepsilon_2}{\varepsilon_2}\left(\dfrac{D_1}{D_2}\right)}$$

Hence, the heat rate *from* the cryogenic fluid per unit tube length is

$$q'_{wo} = \frac{q_{wo}}{L} = \frac{5.67 \times 10^{-8}\ \text{W/m}^2\cdot\text{K}^4(\pi \times 0.02\ \text{m})[(77\ \text{K})^4 - (300\ \text{K})^4]}{\dfrac{1}{0.02} + \dfrac{1-0.05}{0.05}\left(\dfrac{0.02\ \text{m}}{0.05\ \text{m}}\right)}$$

$$q'_{wo} = -0.50\ \text{W/m} \quad \triangleleft$$

The negative sign implies that the radiative heat transfer is *into* the cryogenic fluid.

(b) The network representation of the system with the shield ($w$) is shown in Fig. 18.27, and the desired heat rate is now

$$q_w = \frac{E_{b1} - E_{b2}}{R_{\text{tot}}} = \frac{\sigma(T_1^4 - T_2^4)}{R_{\text{tot}}}$$

where the total radiation resistance is the sum of *four surface* and *two space* radiative resistances

$$R_{\text{tot}} = \frac{1-\varepsilon_1}{\varepsilon_1(\pi D_1 L)} + \frac{1}{(\pi D_1 L)F_{13}} + 2\left[\frac{1-\varepsilon_3}{\varepsilon_3(\pi D_3 L)}\right] + \frac{1}{(\pi D_3 L)F_{32}} + \frac{1-\varepsilon_2}{\varepsilon_2(\pi D_2 L)}$$

Substituting numerical values, find

$$R_{tot} = \frac{1}{L}\left\{\frac{1 - 0.02}{0.02(\pi \times 0.02 \text{ m})} + \frac{1}{(\pi \times 0.02 \text{ m})1} + 2\left[\frac{1 - 0.02}{0.02(\pi \times 0.035 \text{ m})}\right] + \frac{1}{(\pi \times 0.035 \text{ m})1} + \frac{1 - 0.05}{0.05(\pi \times 0.05 \text{ m})}\right\}$$

$$R_{tot} = \frac{1}{L}(779.9 + 15.9 + 891.3 + 9.1 + 121.0) = \frac{1817}{L} \text{ m}^{-2}$$

Hence, the heat rate *with the radiation shield* is

$$q'_w = \frac{q_w}{L} = \frac{5.67 \times 10^{-8} \text{ W/m}^2 \cdot \text{K}^4[(77 \text{ K})^4 - (300 \text{ K})^4]}{1817 \text{ m}^{-1}} = -0.25 \text{ W/m} \lhd$$

The percentage change in the heat transfer to the cryogenic fluid is then

$$\frac{q'_w - q'_{wo}}{q'_{wo}} \times 100 = \frac{(-0.25 \text{ W/m}) - (-0.50 \text{ W/m})}{-0.50 \text{ W/m}} \times 100 = -50\% \lhd$$

### 18.7.3 The Three-Surface Enclosure with a Reradiating Surface

*reradiating surface*

We can use the *network representation* of Fig. 18.25 for a *single surface* to construct the network corresponding to a *three-surface enclosure.* We will consider the special case (Fig. 18.28) where one of the surfaces is perfectly insulated on the backside, with negligible radiation (and convection) on the enclosure side. Termed a ***reradiating surface,*** the idealized surface is characterized by *zero net radiation transfer* ($q_i = 0$). This situation is common in many industrial applications, especially furnaces and ovens where radiation is the dominant mode of heat transfer.

The three-surface enclosure, for which the third surface $R$ is reradiating, is shown in Fig. 18.28*a*, and the corresponding network is shown in Fig. 18.28*b*. Since surface $R$ is presumed to be well insulated with negligible convection effects, the net radiation transfer must be zero. That is, $q_R = 0$, and from an *overall energy balance on the enclosure,* it follows that $q_1 = -q_2$.

Since $q_R = 0$, according to Eq. 18.51, the driving potential for the surface radiative resistance element must be zero. Hence, the blackbody emissive power of the reradiating surface must equal its radiosity

$$E_{bR} = J_R$$

If the radiosity of a reradiating surface, $J_R$, is known, then its temperature is readily determined. In such an enclosure, *the temperature of the reradiating surface is determined by its interaction with the other surfaces,* and is *independent of the emissivity of the reradiating surface.*

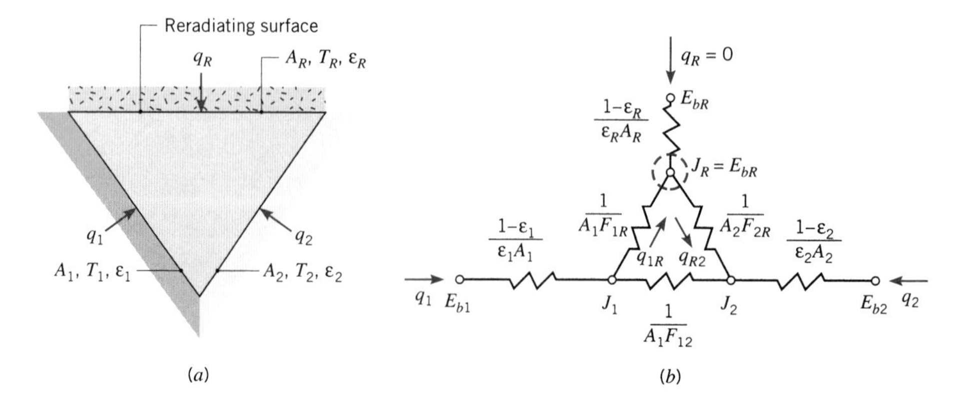


*Figure 18.28* A three-surface enclosure with one surface reradiating. (*a*) Schematic. (*b*) Network representation with three *surface* and three *space* radiative resistances.

The network representing the enclosure, Fig. 18.26*b*, is a simple series-parallel arrangement, and from its analysis it is readily shown that the *net radiation transfer rate* is

$$q_1 = \frac{E_{b1} - E_{b2}}{\dfrac{1-\varepsilon_1}{\varepsilon_1 A_1} + \dfrac{1}{A_1F_{12} + [(1/A_1F_{1R}) + (1/A_2F_{2R})]^{-1}} + \dfrac{1-\varepsilon_2}{\varepsilon_2 A_2}} \tag{18.63}$$

Knowing $q_1 = -q_2$, Eq. 18.51 can be applied to surfaces 1 and 2 to determine their radiosities $J_1$ and $J_2$. Knowing $J_1$, $J_2$, and the *geometrical resistances,* the radiosity of the reradiating surface $J_R$ can be determined from the radiation balance on node $R$ ($q_{1R} - q_{2R} = 0$)

$$\frac{J_1 - J_R}{(1/A_1F_{1R})} - \frac{J_R - J_2}{(1/A_2F_{2R})} = 0 \tag{18.64}$$

The ***temperature of the reradiating surface,*** $T_R$, can then be determined from the requirement that $\sigma T_R^4 = J_R$.

*temperature of reradiating surface*

As you have seen for the two- and three-surface enclosure (with one reradiating surface), the network representation is convenient for setting up an analysis. For complicated enclosures, a more direct approach involves working with the energy balance relations, Eqs. 18.51 and 18.55. For an $N$-surface enclosure with $N$-unknown radiosities (or a combination of $N$ radiosities and temperatures), the analysis requires simultaneously solving the system of $N$-energy balance equations. The methods for performing such analysis are provided in more advanced heat transfer texts.

## *Example 18.10* Three-Surface Enclosure Analysis: Paint Baking Oven

A paint baking oven consists of a long, triangular duct in which a heated surface is maintained at 1200 K and another surface is insulated. Painted panels, which are maintained at 500 K, occupy the third surface. The triangle is of width $W = 1$ m on a side, and the heated and insulated surfaces have an emissivity of 0.8. The emissivity of the panels is 0.4. **(a)** During steady-state operation, at what rate must energy be supplied to the heated side per unit length of the duct to maintain its temperature at 1200 K? **(b)** What is the temperature of the insulated surface?

### Solution

***Known:*** Surface properties of a long triangular duct that is insulated on one side and heated and cooled on the other sides.

***Find:***

**(a)** Rate at which energy must be supplied per unit length of duct.

**(b)** Temperature of the insulated surface.

***Schematic and Given Data:***

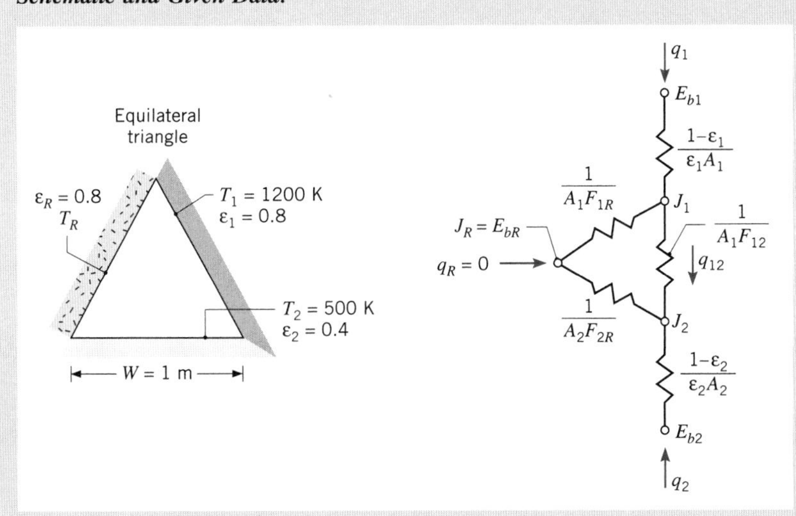


***Assumptions:***

1. Steady-state conditions exist.
2. All surfaces are opaque, diffuse, gray, and of uniform radiosity.
3. Convection effects are negligible.
4. Surface $R$ is reradiating.
5. End effects are negligible.

*Figure E18.10*

***Analysis:***

**(a)** The system may be modeled as a three-surface enclosure with one surface reradiating. The rate at which energy must be supplied to the heated surface can be obtained from Eq. 18.63

$$q_1 = \frac{E_{b1} - E_{b2}}{\dfrac{1 - \varepsilon_1}{\varepsilon_1 A_1} + \dfrac{1}{A_1 F_{12} + [(1/A_1 F_{1R}) + (1/A_2 F_{2R})]^{-1}} + \dfrac{1 - \varepsilon_2}{\varepsilon_2 A_2}}$$

From symmetry considerations, $F_{12} = F_{1R} = F_{2R} = 0.5$. Also, $A_1 = A_2 = W \cdot L$, where $L$ is the duct length. Substituting numerical values, find the heat transfer rate to the duct per unit length as

$$q_1' = \frac{q_1}{L} = \frac{5.67 \times 10^{-8}\ \text{W/m}^2 \cdot \text{K}^4 (1200^4 - 500^4)\ \text{K}^4}{\dfrac{1 - 0.8}{0.8 \times 1\ \text{m}} + \dfrac{1}{1\ \text{m} \times 0.5 + (2 + 2)^{-1}\ \text{m}} + \dfrac{1 - 0.4}{0.4 \times 1\ \text{m}}}$$

or

$$q_1' = 37\ \text{kW/m} \ \lhd$$

**(b)** The temperature of the insulated surface can be obtained from the requirement that $J_R = E_{bR}$, where $J_R$ is dertermined from Eq. 18.64. However, to use this expression $J_1$ and $J_2$ must be known. Applying the *surface energy balance,* Eq. 18.51, to surfaces 1 and 2, and recognizing from the network, Fig. E18.10, that $q_2' = -q_1'$, it follows that

$$J_1 = E_{b1} - \frac{1 - \varepsilon_1}{\varepsilon_1 W} q_1' = 5.67 \times 10^{-8}\ \text{W/m}^2 \cdot \text{K}^4 (1200\ \text{K})^4 - \frac{1 - 0.8}{0.8 \times 1\ \text{m}} \times 37{,}000\ \text{W/m} = 108{,}323\ \text{W/m}^2$$

$$J_2 = E_{b2} - \frac{1 - \varepsilon_2}{\varepsilon_2 W} q_2' = 5.67 \times 10^{-8}\ \text{W/m}^2 \cdot \text{K}^4 (500\ \text{K})^4 - \frac{1 - 0.4}{0.4 \times 1\ \text{m}} (-37{,}000\ \text{W/m}) = 59{,}043\ \text{W/m}^2$$

From the energy balance for the reradiating surface, Eq. 18.64, it follows that

$$\frac{108{,}323 - J_R}{\dfrac{1}{W \times L \times 0.5}} - \frac{J_R - 59{,}043}{\dfrac{1}{W \times L \times 0.5}} = 0$$

Hence, the radiosity of the reradiating surface is

$$J_R = 83{,}683\ \text{W/m}^2$$

Since $J_R = E_{bR} = \sigma T_R^4$ for the reradiating surface, its temperature is

$$T_R = \left(\frac{J_R}{\sigma}\right)^{1/4} = \left(\frac{83{,}683\ \text{W/m}^2}{5.67 \times 10^{-8}\ \text{W/m}^2 \cdot \text{K}^4}\right)^{1/4} = 1102\ \text{K} \ \lhd$$

## 18.8 Chapter Summary and Study Guide

In this chapter we studied *radiation processes* and *properties,* and we applied these fundamentals to methods for determining radiative transfer from *spectrally selective surfaces* and between *diffuse-gray surfaces* comprising an enclosure.

We described the nature of thermal radiation and then introduced the *radiation processes* of emission and irradiation. The concept of the *blackbody,* the perfect absorber and ideal emitter, provides a basis for our understanding of the *spectral distribution* of radiation as a function of wavelength and temperature. *Radiation properties,* defined in terms of *blackbody* behavior, describe the interaction between radiation processes and *real* surfaces. In our treatment, we considered *diffuse* surfaces, for which the equality of the *spectral* properties applies, $\varepsilon_\lambda = \alpha_\lambda$. For the *diffuse-gray* surface, a useful model in many engineering applications, the equality on a *total* basis applies, $\varepsilon = \alpha$. The equality does not apply, however, to *spectrally-selective* surfaces, which have spectral properties that are different in the wavelength ranges associated with the emission and irradiation processes.

The diffuse and gray surface assumptions allow for expressing the geometrical features of radiative exchange between surfaces of an enclosure in terms of the *view factor*. Based upon energy balance relations, we introduced a *network representation* for the *two-surface enclosure,* treating *radiation shields* as a special case, and for *three-surface enclosures with one surface reradiating*.

Many new concepts and terms were introduced in this chapter, so careful reading of the material will be required to make you more comfortable with their application. Review the terms summarized in Table 18.1, the characteristics of the blackbody listed in Sec. 18.3, and property interrelationships summarized in Sec. 18.4.4.

The following check list provides a study guide for this chapter. When your study of the text and end-of-chapter problems has been completed, you should be able to

- write out the meanings of the terms listed in the margins throughout the chapter and understand each of the related concepts. The subset of key terms listed here are particularly important.
- describe the nature of radiation and the important features that characterize radiation.
- define the *spectral* and *total emissive powers,* and explain the role the latter plays in a surface energy balance.
- define the *total irradiation* and *total radiosity,* and explain the role they play in a surface energy balance.
- list the characteristics of a *blackbody,* and explain the principal role of blackbody behavior in radiation analysis.
- describe the *Planck distribution* and explain the use of *Wien's displacement law,* the *Stefan-Boltzmann law,* and the *band emission fraction* in problem solving.
- list the important characteristics of the *spectrally selective* and *gray* surfaces and explain what is a *diffuse* surface.
- explain the concept of a view factor and use of the *reciprocity relation* and the *summation rule.*
- apply the network representation to calculate net radiant exchange in a two surface, diffuse-gray enclosure.
- explain the use of a *radiation shield* and whether it is advantageous for the shield to have a high surface absorptivity or reflectivity.

***thermal radiation***
***emissive power***
***irradiation***
***radiosity***
***blackbody***
***Planck spectral distribution***
***Wien's displacement law***
***Stefan–Boltzmann law***
***spectral, total properties***
***spectrally selective surface***
***view factor***
***networks for enclosures***
***radiation shield***
***reradiating surface***

## Problems

**Note:** Unless otherwise indicated in the problem statement, use values of the required thermophysical properties given in appropriate tables of Appendix HT when solving these problems.

### Radiation Quantities and Processes

**18.1** The spectral distribution of the radiation emitted by a diffuse surface may be approximated as shown in Fig. P18.1. The surface has a total irradiation of 1500 W/m$^2$, 70% of which is absorbed and 30% reflected.

(a) What is the total emissive power?
(b) What is the radiosity?
(c) What is the *net* radiative heat flux *leaving* the surface, $q''_{rad}$? Show your surface energy balance schematically and label the radiation processes.

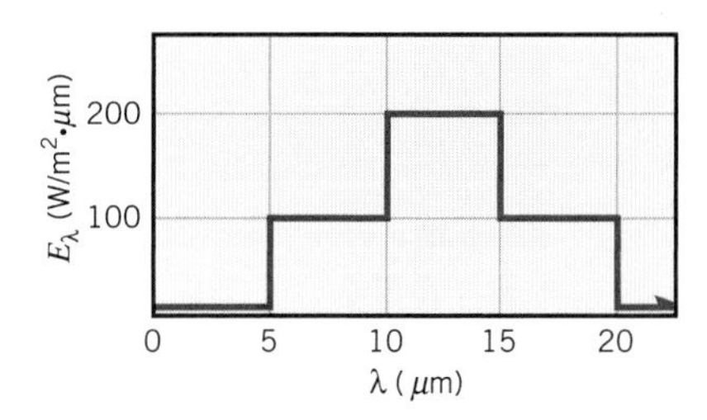

*Figure P18.1*

**18.2** A surface is subjected to the spectral irradiation shown in Fig. P18.2. The surface reflects 40% of the irradiation and has an emissive power of 600 W/m$^2$.

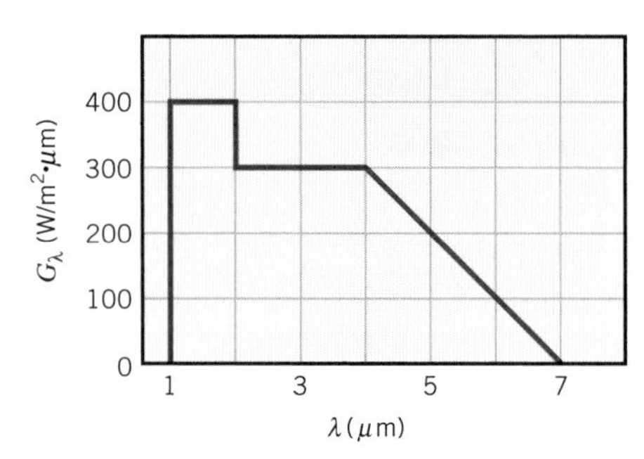

*Figure P18.2*

(a) What is the irradiation?
(b) What is the radiosity?
(c) What is the *net* radiative heat flux *leaving* the surface, $q''_{\text{rad}}$? Represent the surface energy balance schematically and label the radiation processes.

**18.3** Consider a plate that is well insulated on its back side and maintained at 200°C by imbedded electrical resistance elements. The exposed surface has an emissive power of 1200 W/m$^2$, irradiation of 2500 W/m$^2$, and reflectivity of 30%. The exposed surface experiences air flow having a free stream temperature of 20°C with a convection coefficient of 15 W/m$^2$ · K.
(a) What is the radiosity, $J$?
(b) Determine the *net* radiation heat flux *leaving* the surface, $q''_{\text{rad,net}}$, in terms of the radiosity and irradiation.
(c) Determine the *combined* convection and *net* radiation heat flux *leaving* the surface.
(d) Represent a surface energy balance schematically, and label all the radiation processes.
(e) What is the electrical power requirement, $P''_e$ (W/m$^2$), to maintain the plate under these conditions?

### Blackbody Radiation

**18.4** A spherical aluminum shell of inside diameter $D = 2$ m is evacuated and is used as a radiation test chamber. If the inner surface is coated with carbon black and maintained at 600 K, what is the irradiation on a small test surface placed in the chamber? If the inner surface were not coated, but still maintained at 600 K, what would the irradiation be?

**18.5** An enclosure has an inside area of 100 m$^2$, and its inside surface is black and is maintained at a constant temperature. A small opening in the enclosure has an area of 0.02 m$^2$. The radiant power emitted from this opening is 70 W. What is the temperature of the interior enclosure wall? If the interior surface is maintained at this temperature, but is now polished, what will be the value of the radiant power emitted from the opening?

**18.6** The energy flux associated with solar radiation incident on the outer surface of the earth's atmosphere has been accurately measured and is known to be 1353 W/m$^2$. The diameters of the sun and earth are $1.39 \times 10^9$ and $1.29 \times 10^7$ m, respectively, and the distance between the sun and the earth is $1.5 \times 10^{11}$ m.
(a) What is the emissive power of the sun?
(b) Approximating the sun's surface as black, what is its temperature?
(c) At what wavelength is the spectral emissive power of the sun a maximum?
(d) Assuming the earth's surface to be black and the sun to be the only source of energy for the earth, estimate the earth's surface temperature.

**18.7** Estimate the wavelength corresponding to maximum blackbody emission from each of the following surfaces: the sun, a tungsten filament at 2500 K, a heated metal at 1500 K, human skin at 305 K, and a cryogenically cooled metal surface at 60 K. Estimate the fraction of the solar emission that is in the following spectral regions: the ultraviolet, the visible, and the infrared.

**18.8** A 100-W light source consists of a filament that is in the form of a thin rectangular strip, 5 mm long by 2 mm wide, and radiates as a blackbody at 2900 K.
(a) Assuming that the glass bulb transmits all incident visible radiation, what is its efficiency? The efficiency is defined as the ratio of the visible radiant power to the consumed electrical power.
(b) Determine the efficiency as a function of filament temperature for the range from 1300 to 3300 K.

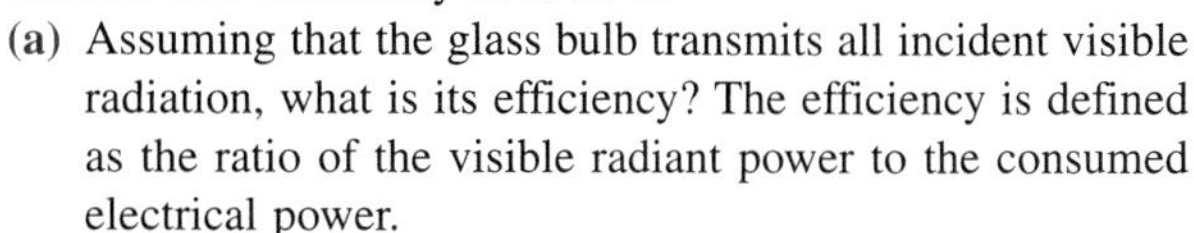

**18.9** (CD-ROM)

### Properties: Emissivity

**18.10** The spectral emissivity of tungsten may be approximated by the distribution shown in Fig. P18.10. Consider a cylindrical tungsten filament that is of diameter $D = 0.8$ mm and length $L = 20$ mm. The filament is enclosed in an evacuated bulb and is heated by an electrical current to a steady-state temperature of 2900 K.

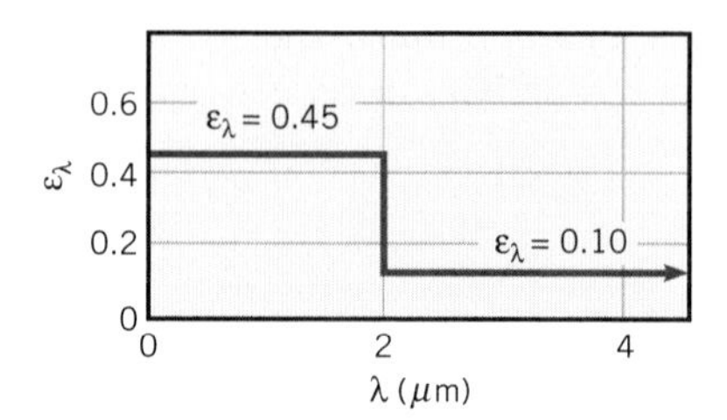

*Figure P18.10*

(a) What is the total emissivity when the filament temperature is 2900 K?
(b) Generate a plot of the emissivity as a function of filament temperature for $1300 \leq T \leq 2900$ K.

**18.11** (CD-ROM)

**18.12** The spectral emissivity of a diffuse material at 2000 K has the distribution shown in Fig. P18.12.

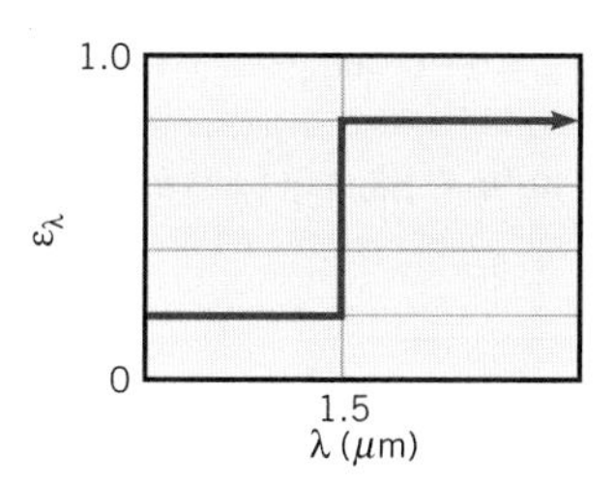


*Figure P18.12*

(a) Determine the total emissivity at 2000 K.
(b) Determine the emissive power over the spectral range 0.8 to 2.5 μm.

**18.13** For materials A and B, whose spectral emissivities vary with wavelength as shown below, how does the total emissivity vary with temperature? Explain briefly.

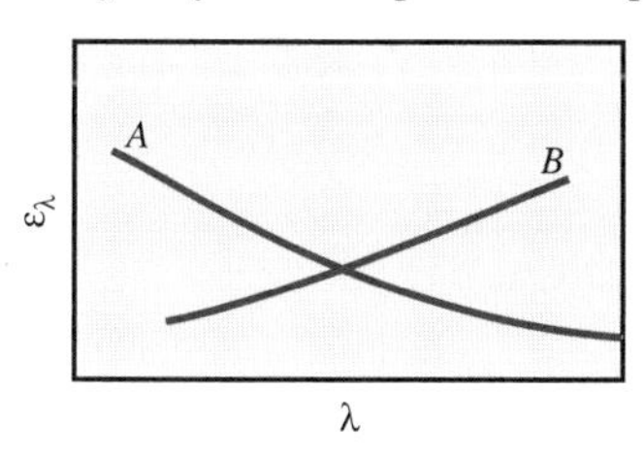


*Figure P18.13*

### Properties and Energy Balances

**18.14** An opaque surface with the prescribed spectral reflectivity distribution is subjected to the spectral irradiation in Fig. P18.14.

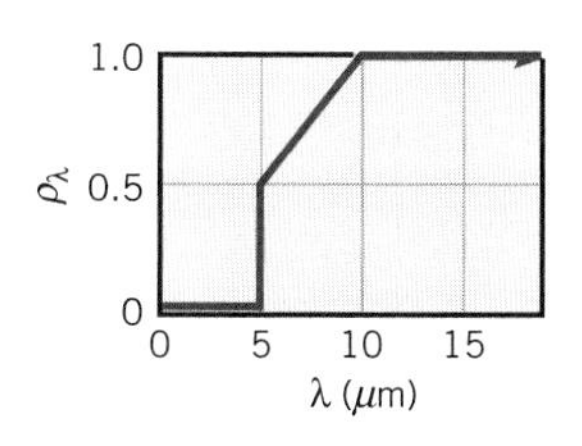


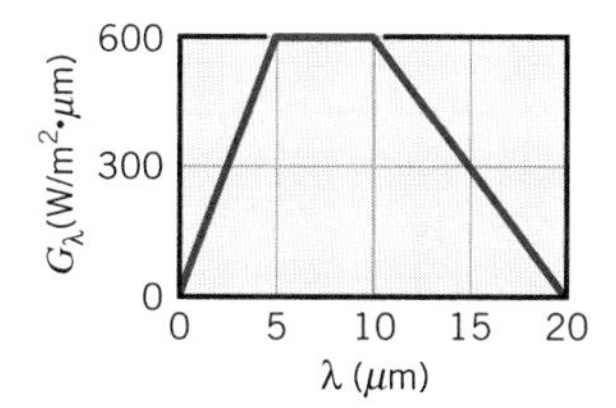


*Figure P18.14*

(a) Sketch the spectral absorptivity distribution.
(b) Determine the total irradiation on the surface.
(c) Determine the radiant flux that is absorbed by the surface.
(d) What is the total absorptivity of this surface?

**18.15** An opaque surface, 2 m by 2 m, is maintained at 400 K and is simultaneously exposed to solar irradiation with $G$ = 1200 W/m$^2$. The surface is diffuse and its spectral absorptivity is $\alpha_\lambda = 0, 0.8, 0,$ and 0.9 for $0 \leq \lambda \leq 0.5$ μm, 0.5 μm $< \lambda \leq 1$ μm, 1 μm $< \lambda \leq 2$ μm, and $\lambda > 2$ μm, respectively. Determine the absorbed irradiation, emissive power, radiosity, and net radiation heat transfer from the surface.

**18.16** The spectral absorptivity of an opaque surface is shown in Fig. P18.16.

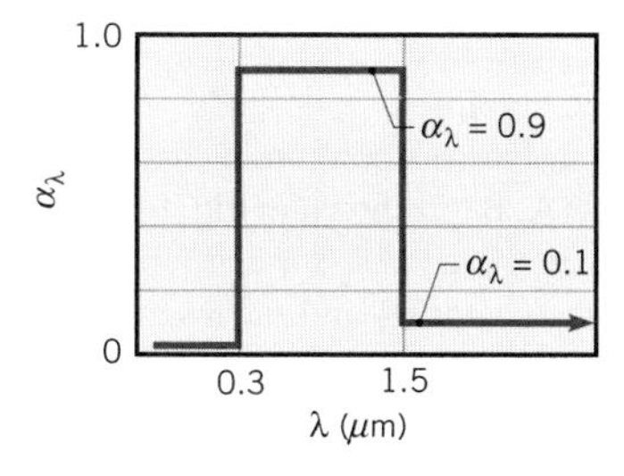


*Figure P18.16*

What is the solar absorptivity, $\alpha_S$? If it is assumed that $\varepsilon_\lambda = \alpha_\lambda$ and that the surface is at a temperature of 340 K, what is its total emissivity?

**18.17** The spectral absorptivity of an opaque surface and the spectral distribution of radiation incident on the surface are depicted below.

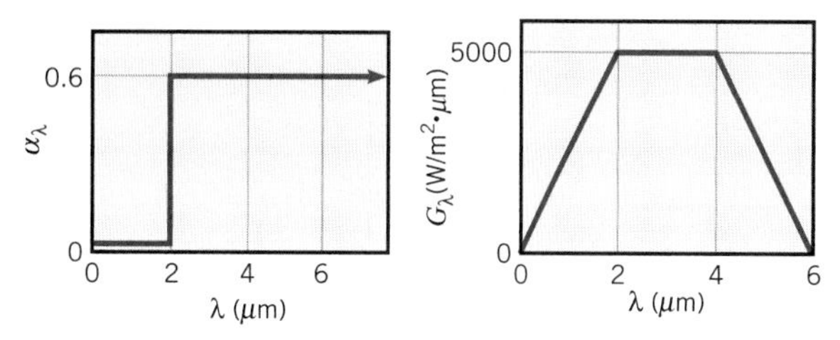


*Figure P18.17*

What is the total absorptivity of the surface? If it is assumed that $\varepsilon_\lambda = \alpha_\lambda$ and that the surface is at 1000 K, what is its total emissivity? What is the net radiant heat flux to the surface?

**18.18** (CD-ROM)

**18.19** The spectral emissivity of an opaque, diffuse surface is as shown.

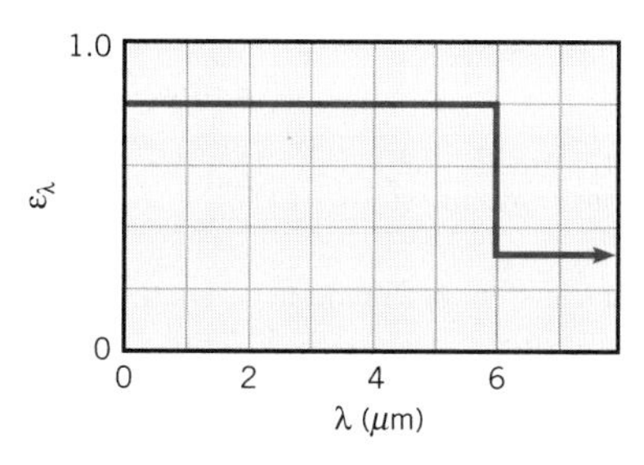


*Figure P18.19*

(a) If the surface is maintained at 1000 K, what is the total emissivity?
(b) What is the total absorptivity of the surface when irradiated by large surroundings of emissivity 0.8 and temperature 1500 K?
(c) What is the radiosity of the surface when it is maintained at 1000 K and subjected to the irradiation prescribed in part (b)?
(d) Determine the *net* radiation flux *leaving* the surface for the conditions of part (c).

**18.20** (CD-ROM)

### Energy Balance Applications

**18.21** An opaque, horizontal flat plate has a top surface area of 3 m$^2$, and its edges and lower surface are well insulated. The plate is uniformly irradiated at its top surface at a rate of 1300 W. Consider steady-state conditions for which 1000 W of the incident radiation is absorbed, the plate temperature is 500 K, and heat transfer by convection from the surface is 300 W. Determine the irradiation $G$, emissive power $E$, radiosity $J$, absorptivity $\alpha$, reflectivity $\rho$, and emissivity $\varepsilon$.

**18.22** A small workpiece is placed in a large oven having isothermal walls at $T_f = 1000$ K with an emissivity of $\varepsilon_f = 0.5$. The workpiece experiences convection with moving air at 600 K and a convection coefficient of $h = 60$ W/m$^2 \cdot$ K. The surface of the workpiece has a spectrally selective coating for which the emissivity has the following spectral distribution:

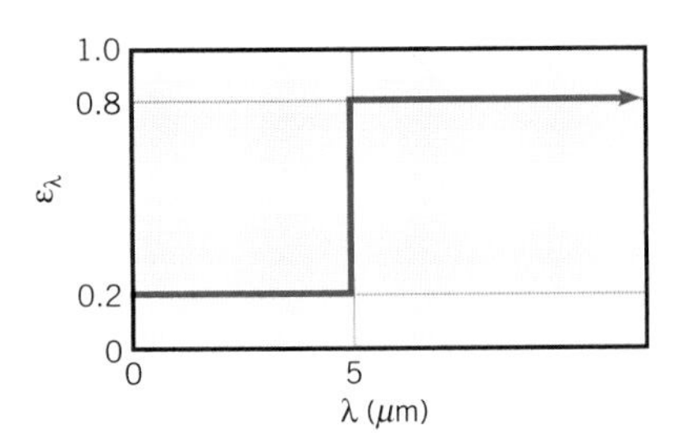


*Figure P18.22*

(a) Beginning with identification of all relevant processes for a control surface about the workpiece, perform an energy balance on the workpiece and determine its steady-state temperature, $T_s$.

(b) Plot the surface temperature $T_s$ as a function of the convection coefficient for $10 \leq h \leq 120$ W/m$^2 \cdot$ K. On the same plot, show the surface temperature as a function of the convection coefficient for diffuse, gray surfaces with emissivities of 0.2 and 0.8.

**18.23** A thermocouple whose surface is diffuse and gray with an emissivity of 0.6 indicates a temperature of 180°C when used to measure the temperature of a gas flowing through a large duct whose walls have an emissivity of 0.85 and a uniform temperature of 450°C.

(a) If the convection heat transfer coefficient between the thermocouple and the gas stream is $\bar{h} = 125$ W/m$^2 \cdot$ K and there are negligible conduction losses from the thermocouple, determine the temperature of the gas.

(b) Consider a gas temperature of 125°C. Compute and plot the thermocouple *measurement error* as a function of the convection coefficient for $10 \leq \bar{h} \leq 1000$ W/m$^2 \cdot$ K. What are the implications of your results?

**18.24** Solar irradiation of 1100 W/m$^2$ is incident on a large, flat, horizontal metal roof on a day when the wind blowing over the roof causes a convection heat transfer coefficient of 25 W/m$^2 \cdot$ K. The outside air temperature is 27°C, the metal surface absorptivity for incident solar radiation is 0.60, the metal surface emissivity is 0.20, and the roof is well insulated from below.

(a) Estimate the roof temperature under steady-state conditions.

(b) Explain qualitatively the effect of changes in the absorptivity, emissivity, and convection coefficient on the steady-state temperature.

**18.25** (CD-ROM)

**18.26** (CD-ROM)

**18.27** Square plates freshly sprayed with an epoxy paint must be cured at 140°C for an extended period of time. The plates are located in a large enclosure and heated by a bank of infrared lamps as illustrated in Fig. P18.27. The top surface of each plate has an emissivity of $\varepsilon = 0.8$ and experiences convection with a ventilation air stream that is at $T_\infty = 27$°C and provides a convection coefficient of $h = 20$ W/m$^2 \cdot$ K. The irradiation from the enclosure walls is estimated to be $G_{wall} = 450$ W/m$^2$, for which the plate absorptivity is $\alpha_{wall} = 0.7$.

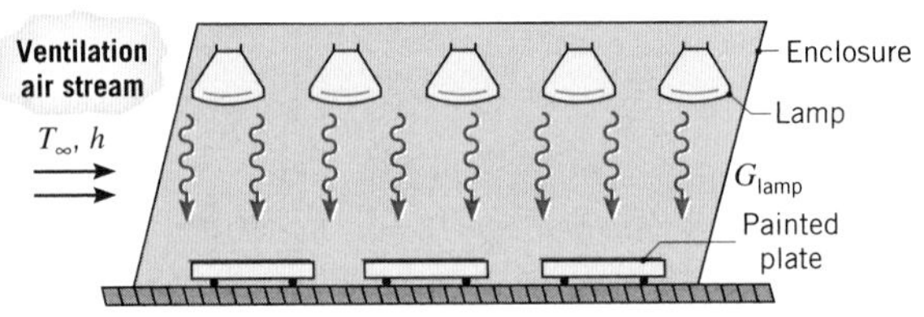


*Figure P18.27*

Determine the irradiation that must be provided by the lamps, $G_{lamp}$. The absorptivity of the plate surface for this irradiation is $\alpha_{lamp} = 0.6$.

**18.28** (CD-ROM)

### Environmental and Space Radiation

**18.29** A contractor must select a roof covering material from the two diffuse, opaque coatings with $\alpha_\lambda(\lambda)$ as in Fig. P18.29. Which of the two coatings would result in a lower roof temperature? Which is preferred for summer use? For winter use? Sketch the spectral distribution of $\alpha_\lambda$ that would be ideal for summer use. For winter use.

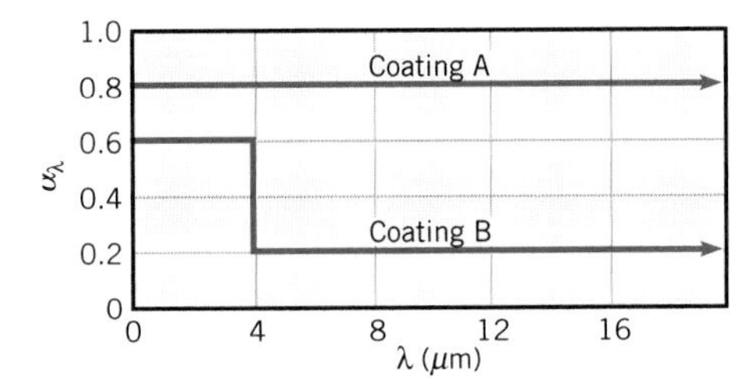


*Figure P18.29*

**18.30** A radiator on a proposed satellite solar power station must transfer by radiation to deep space the electrical power dissipated within the satellite. The radiator surface has a solar absorptivity of 0.5 and an emissivity of 0.95. What is the equilibrium surface temperature when the solar irradiation is 1000 W/m$^2$ and the electrical power dissipation is 1500 W/m$^2$?

**18.31** The exposed surface of a power amplifier for an earth satellite receiver of area 130 mm by 130 mm has a diffuse, gray, opaque coating with an emissivity of 0.5. For typical amplifier operating conditions, the surface temperature is 58°C under the following environmental conditions: air temperature, $T_\infty = 27$°C; sky temperature, $T_{sky} = -20$°C; convection coefficient, $h = 15$ W/m$^2 \cdot$ K; and solar irradiation, $G_S = 800$ W/m$^2$.

(a) For the above conditions, determine the electrical power being dissipated within the amplifier.

(b) It is desired to reduce the surface temperature by applying one of the diffuse coatings (A, B, C) shown in Fig. P18.31. Which coating will result in the coolest surface temperature for the same amplifier operating and environmental conditions?

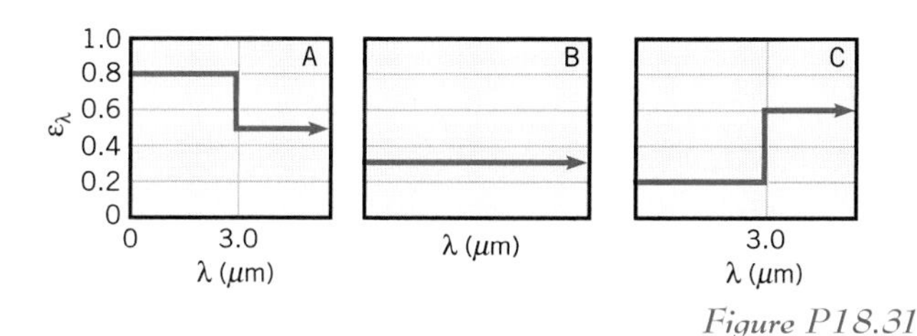


*Figure P18.31*

**18.32** It is not uncommon for the night sky temperature in desert regions to drop to $-40°C$. If the ambient air temperature is 20°C and the convection coefficient for still air conditions is approximately 5 $W/m^2 \cdot K$, can a shallow pan of water freeze?

### View Factors

**18.33** Determine $F_{12}$ and $F_{21}$ for the configurations shown in Fig. P18.33 using the reciprocity theorem and other basic shape factor relations. Do not use tables or charts.

(a) Long duct

(b) Small sphere of area $A_1$ under a concentric hemisphere of area $A_2 = 2A_1$

(c) Long duct. What is $F_{22}$ for this case?

(d) Long inclined plates (point $B$ is directly above the center of $A_1$)

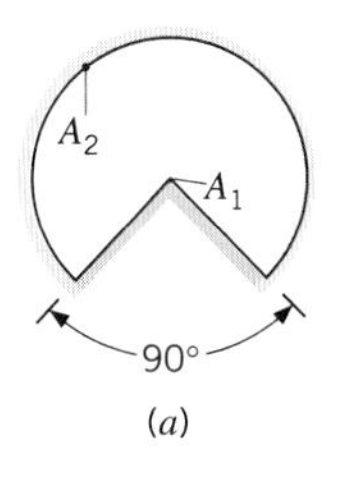


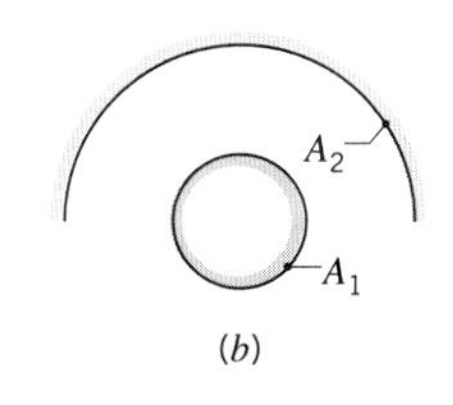


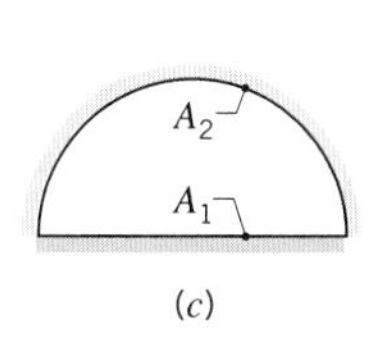


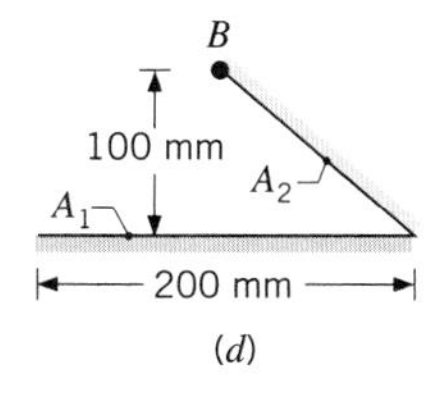


*Figure P18.33*

**18.34** Consider the following grooves, each of width $W$, that have been machined from a solid block of material.

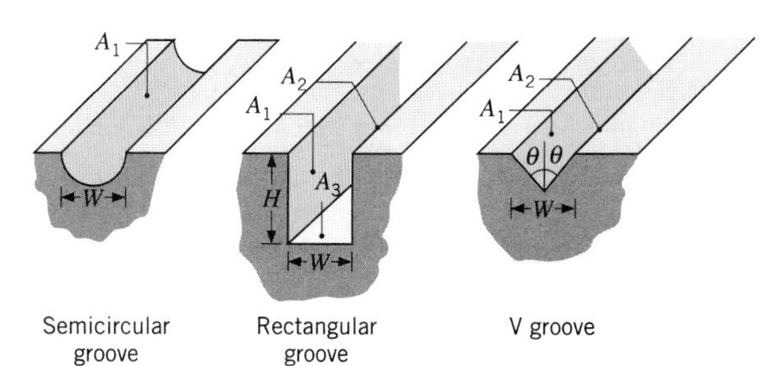


*Figure P18.34*

(a) For each case obtain an expression for the view factor of the groove with respect to the surroundings outside the groove.

(b) For the $V$ groove, obtain an expression for the view factor $F_{12}$, where $A_1$ and $A_2$ are opposite surfaces.

(c) If $H = 2W$ in the rectangular groove, what is the view factor $F_{12}$?

**18.35** Calculate all the shape factors associated with (a) a regular tetrahedron, whose sides are in the shape of an equilateral triangle and (b) a cubical enclosure.

**18.36** Consider the long concentric cylinders with diameters $D_1$ and $D_2$ and surface areas $A_1$ and $A_2$.

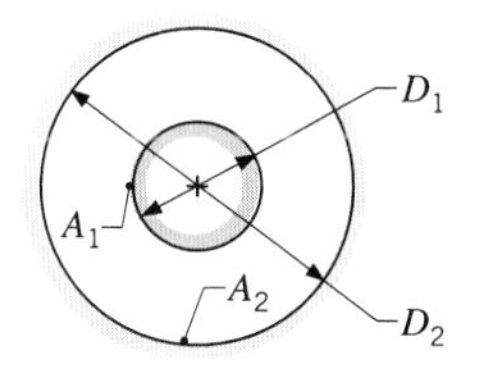


*Figure P18.36*

(a) What is the view factor $F_{12}$?

(b) Obtain expressions for the view factors $F_{22}$ and $F_{21}$ in terms of the cylinder diameters.

**18.37** Consider the two coaxial disks having diameters $D$ = 250 mm that are separated a distance $L$ = 150 mm. The upper disk has a 125 mm hole. Determine the view factor $F_{12}$.

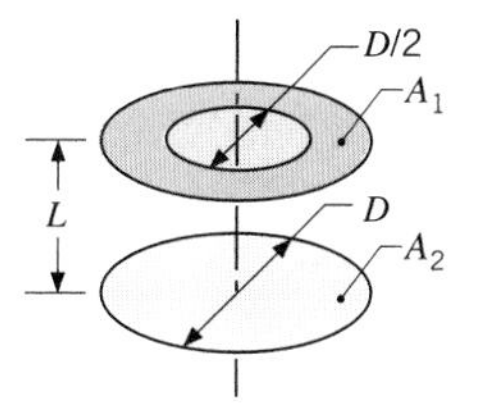


*Figure P18.37*

**18.38** Consider the perpendicular rectangles shown schematically in Fig. P18.38. Determine the shape factor $F_{12}$.

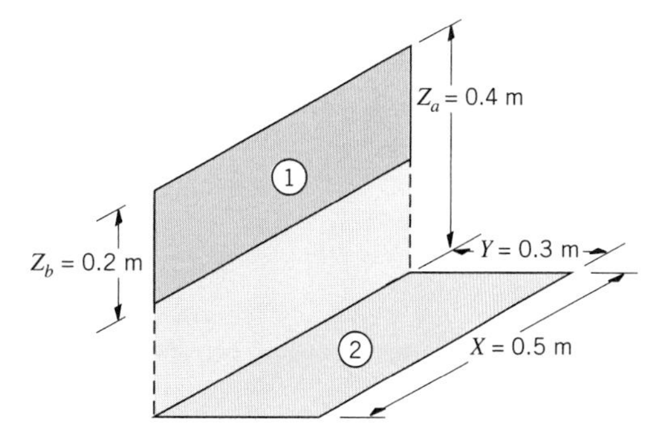


*Figure P18.38*

**18.39** (CD-ROM)

### Blackbody Radiation Exchange

**18.40** A drying oven consists of a long semicircular duct of diameter $D$ = 1 m as shown in Fig. P18.40.

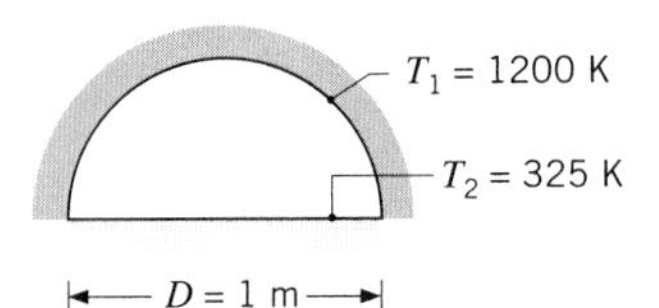


Figure P18.40

Materials to be dried cover the base of the oven, while the wall is maintained at 1200 K. What is the drying rate per unit length of the oven (kg/s · m) if a water-coated layer of material is maintained at 325 K during the drying process? Blackbody behavior may be assumed for the water surface and for the oven wall.

**18.41** Consider the arrangement of the three black surfaces shown in Fig. P18.41, where $A_1$ is small compared to $A_2$ or $A_3$.

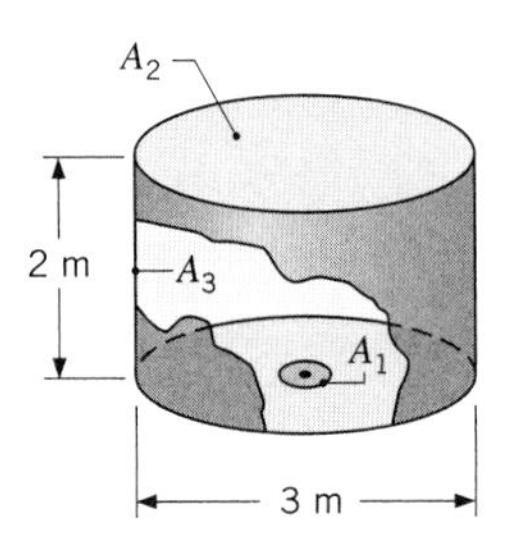


Figure P18.41

Determine the value of $F_{13}$. Calculate the net radiation heat transfer from $A_1$ to $A_3$ if $A_1 = 0.05\ \text{m}^2$, $T_1 = 1000$ K, and $T_3 = 500$ K.

**18.42** A circular disk of diameter $D_1 = 20$ mm is located at the base of an enclosure that has a cylindrical sidewall and a hemispherical dome. The enclosure is of diameter $D = 0.5$ m, and the height of the cylindrical section is $L = 0.3$ m. The disk and the enclosure surface are black and at temperatures of 1000 and 300 K, respectively.

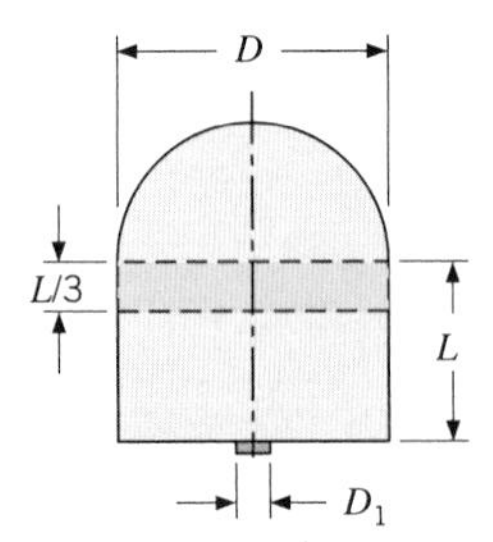


Figure P18.42

What is the net rate of radiation exchange between the disk and the hemispherical dome? What is the net rate of radiation exchange between the disk and the top one-third portion of the cylindrical section?

**18.43** Consider coaxial, parallel, black disks separated a distance of 0.20 m as shown in Fig. P18.43. The lower disk of diameter 0.40 m is maintained at 500 K and the surroundings are at 300 K. What temperature will the upper disk of diameter 0.20 m achieve if electrical power of 17.5 W is supplied to the heater on the back side of the disk?

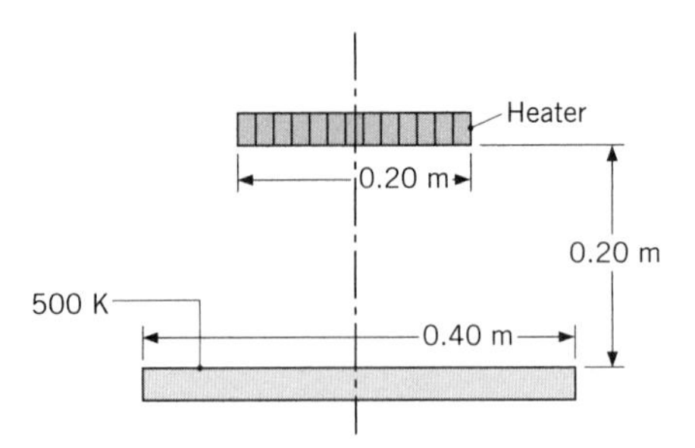


Figure P18.43

**18.44** (CD-ROM)

**18.45** Consider the very long, inclined black surfaces ($A_1$, $A_2$) maintained at uniform temperatures of $T_1 = 1000$ K and $T_2 = 800$ K.

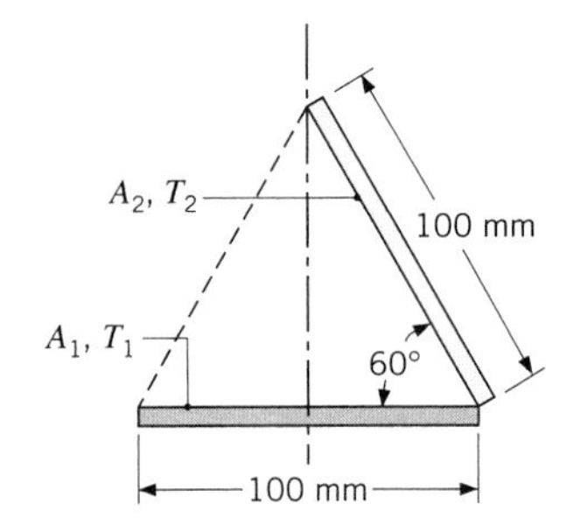


Figure P18.45

(a) Determine the net radiation exchange between the surfaces per unit length of the surfaces.
(b) Consider the configuration when a black surface ($A_3$), whose back side is insulated, is positioned along the dashed line shown. Calculate the net radiation transfer to surface $A_2$ per unit length of the surface and determine the temperature of the insulated surface $A_3$.

**18.46** (CD-ROM)

### Two-Surface Enclosures

**18.47** Consider two very large parallel plates with diffuse, gray surfaces.

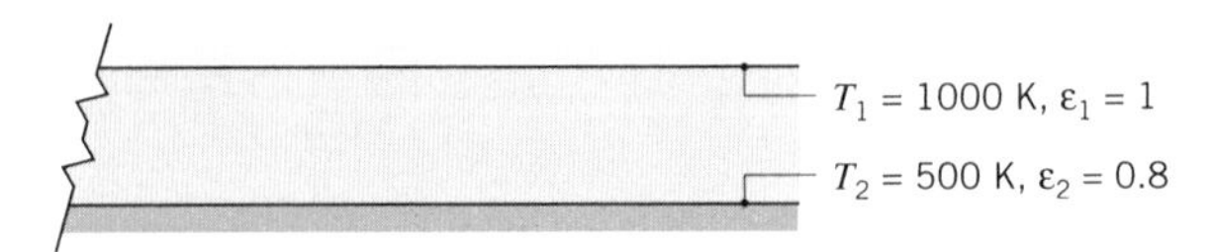


Figure P18.47

Determine the irradiation and radiosity for the upper plate. What is the radiosity for the lower plate? What is the net radiation exchange between the plates per unit area of the plates?

**18.48** A flat-bottomed hole 6 mm in diameter is drilled to a depth of 24 mm in a diffuse, gray material having an emissivity of 0.8 and a uniform temperature of 1000 K.
(a) Determine the radiant power leaving the opening of the cavity.
(b) The effective emissivity $\varepsilon_e$ of a cavity is defined as the ratio of the radiant power leaving the cavity to that from a

blackbody having the area of the cavity opening and a temperature of the inner surfaces of the cavity. Calculate the effective emissivity of the cavity described above.

(c) If the depth of the hole were increased, would $\varepsilon_e$ increase or decrease? What is the limit of $\varepsilon_e$ as the depth increases?

**18.49** (CD-ROM)

**18.50** (CD-ROM)

**18.51** A very long electrical conductor 10 mm in diameter is concentric with a cooled cylindrical tube 50 mm in diameter whose inner surface is diffuse and gray with an emissivity of 0.9 and temperature of 27°C. The electrical conductor has a diffuse, gray surface with an emissivity of 0.6 and is dissipating 6.0 W per meter of length. Assuming that the space between the two surfaces is evacuated, calculate the surface temperature of the conductor.

**18.52** Liquid oxygen is stored in a thin-walled, spherical container 0.8 m in diameter, which is enclosed within a second thin-walled, spherical container 1.2 m in diameter. The opaque, diffuse, gray container surfaces have an emissivity of 0.05 and are separated by an evacuated space. If the outer surface is at 280 K and the inner surface is at 95 K, what is the mass rate of oxygen lost due to evaporation? (The heat of vaporization of oxygen is $h_{fg} = 2.13 \times 10^5$ J/kg.)

**18.53** Two concentric spheres of diameters $D_1 = 0.8$ m and $D_2 = 1.2$ m are separated by an air space and have surface temperatures of $T_1 = 400$ K and $T_2 = 300$ K.

(a) If the surfaces are black, what is the net rate of radiation exchange between the spheres?

(b) What is the net rate of radiation exchange between the surfaces if they are diffuse and gray with $\varepsilon_1 = 0.5$ and $\varepsilon_2 = 0.05$?

(c) What is the net rate of radiation exchange if $D_2$ is increased to 20 m, with $\varepsilon_2 = 0.05$, $\varepsilon_1 = 0.5$, and $D_1 = 0.8$ m? What error would be introduced by assuming blackbody behavior for the outer surface ($\varepsilon_2 = 1$), with all other conditions remaining the same?

## Radiation Shields

**18.54** Determine the steady-state temperatures of two radiation shields placed in the evacuated space between two infinite planes at temperatures of 600 and 325 K. All the surfaces are diffuse and gray with emissivities of 0.7.

**18.55** Consider two large, diffuse, gray, parallel surfaces separated by a small distance. If the surface emissivities are 0.8, what emissivity should a thin radiation shield have to reduce the radiation heat transfer rate between the two surfaces by a factor of 10?

**18.56** (CD-ROM)

**18.57** In free space, the end of a cylindrical liquid cryogenic propellant tank is to be protected from external (solar) radiation by placing a thin metallic shield in front of the tank as shown in Fig. P18.57. Assume the view factor $F_{ts}$ between the tank and the shield is unity; all surfaces are diffuse and gray, and the surroundings are at 0 K. Find the temperature of the shield $T_s$ and the heat flux (W/m$^2$) to the end of the tank.

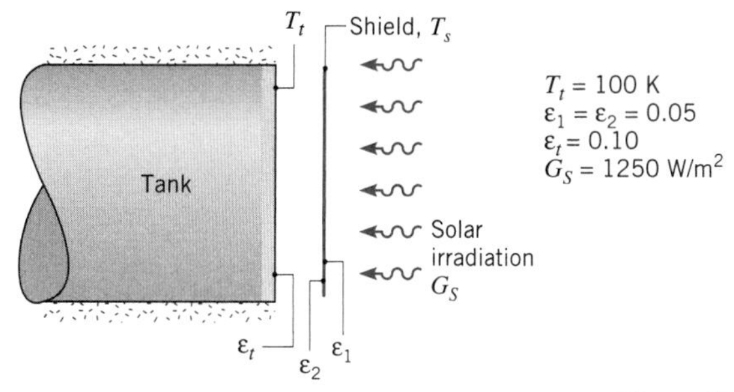


*Figure P18.57*

**18.58** At the bottom of a very large vacuum chamber whose walls are at 300 K, a black panel 0.1 m in diameter is maintained at 77 K. To reduce the heat transfer to this panel, a radiation shield of the same diameter $D$ and an emissivity of 0.05 is placed very close to the panel. Calculate the net heat transfer to the panel.

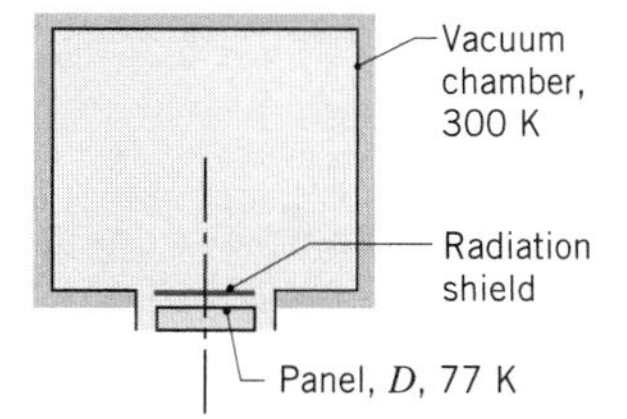


*Figure P18.58*

**18.59** A diffuse, gray radiation shield of 60-mm diameter and emissivities of $\varepsilon_{2,i} = 0.01$ and $\varepsilon_{2,o} = 0.1$ on the inner and outer surfaces, respectively, is concentric with a long tube transporting a hot process fluid. The tube outer surface is black with a diameter of 20 mm. The region interior to the shield is evacuated. The exterior surface of the shield is exposed to a large room whose walls are at 17°C and experiences convection with air at 27°C and a convection heat transfer coefficient of 10 W/m$^2\cdot$K. Determine the operating temperature for the inner tube if the shield temperature is maintained at 42°C.

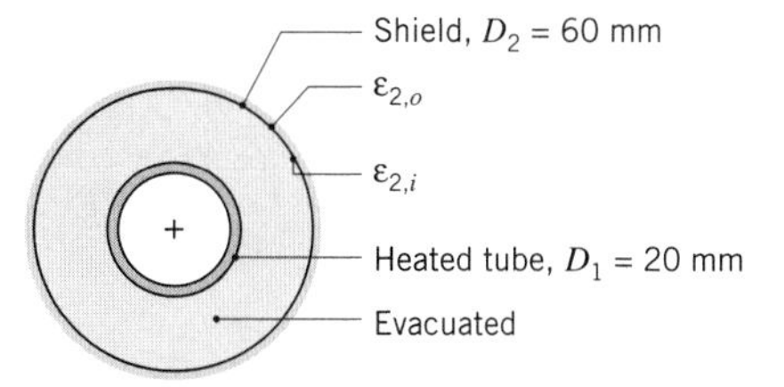


*Figure P18.59*

## Enclosures with a Reradiating Surface

**18.60** Two parallel, coaxial disks, 0.4 m in diameter and separated by 0.1 m, are located in a large room whose walls are maintained at 300 K. One of the disks is maintained at a uniform temperature of 500 K with an emissivity of 0.6, while the back side of the second disk is well insulated. If the disks are diffuse-gray surfaces, determine the temperature of the insulated disk.

**18.61** Consider two aligned, parallel, square planes (0.4 m × 0.4 m) spaced 0.8 m apart and maintained at $T_1$ = 500 K and $T_2$ = 800 K. Calculate the *net* radiative heat transfer *from surface 1* for the following special conditions:

(a) Both planes are black, and the surroundings are at 0 K.

(b) Both planes are black with connecting, reradiating walls.

(c) Both planes are diffuse and gray with $\varepsilon_1 = 0.6$, $\varepsilon_2 = 0.8$, and the surroundings at 0 K.

(d) Both planes are diffuse-gray ($\varepsilon_1 = 0.6$ and $\varepsilon_2 = 0.8$) with connecting, reradiating walls.

**18.62** (CD-ROM)

**18.63** (CD-ROM)

# Index to Property Tables and Figures



[1]The convention used to present numerical values is illustrated by this example:

| $T$ (K) | $\nu \cdot 10^7$ (m²/s) | $k \cdot 10^3$ (W/m · K) |
|---|---|---|
| 300 | 0.349 | 521 |

where $\nu = 0.349 \times 10^{-7}$ m²/s and $k = 521 \times 10^{-3}$ W/m · K at 300K.

[2]The convention used to present numerical values of the specific volume of liquids in the SI tables is illustrated by this example:

| Temp. (°C) | $v_f \times 10^3$ (m³/kg) |
|---|---|
| 20 | 1.0018 |

where $v_f = 1.0018 \times 10^{-3}$ m³/kg.

## CD-ROM

### Tables



### Figures

**Table FM-1** Properties of Common Fluids

(a) Approximate Physical Properties of Some Common Fluids (SI Units)

| | Temperature (°C) | Density, $\rho$ (kg/m$^3$) | Specific Weight, $\gamma$ (kN/m$^3$) | Dynamic Viscosity, $\mu$ (N · s/m$^2$) | Kinematic Viscosity, $\nu$ (m$^2$/s) |
|---|---|---|---|---|---|
| Liquids | | | | | |
| Carbon tetrachloride | 20 | 1,590 | 15.6 | $9.58 \times 10^{-4}$ | $6.03 \times 10^{-7}$ |
| Ethyl alcohol | 20 | 789 | 7.74 | $1.19 \times 10^{-3}$ | $1.51 \times 10^{-6}$ |
| Gasoline | 15.6 | 680 | 6.67 | $3.1 \times 10^{-4}$ | $4.6 \times 10^{-7}$ |
| Glycerin | 20 | 1,260 | 12.4 | $1.50 \times 10^{+0}$ | $1.19 \times 10^{-3}$ |
| Mercury | 20 | 13,600 | 133 | $1.57 \times 10^{-3}$ | $1.15 \times 10^{-7}$ |
| SAE 30 oil | 15.6 | 912 | 8.95 | $3.8 \times 10^{-1}$ | $4.2 \times 10^{-4}$ |
| Seawater | 15.6 | 1,030 | 10.1 | $1.20 \times 10^{-3}$ | $1.17 \times 10^{-6}$ |
| Water | 15.6 | 999 | 9.80 | $1.12 \times 10^{-3}$ | $1.12 \times 10^{-6}$ |
| Gases at Standard Atmospheric Pressure[1] | | | | | |
| Air (standard) | 15 | 1.23 | 12.0 | $1.79 \times 10^{-5}$ | $1.46 \times 10^{-5}$ |
| Carbon dioxide | 20 | 1.83 | 18.0 | $1.47 \times 10^{-5}$ | $8.03 \times 10^{-6}$ |
| Helium | 20 | 0.166 | 1.63 | $1.94 \times 10^{-5}$ | $1.15 \times 10^{-4}$ |
| Hydrogen | 20 | 0.0838 | 0.822 | $8.84 \times 10^{-6}$ | $1.05 \times 10^{-4}$ |
| Methane (natural gas) | 20 | 0.667 | 6.54 | $1.10 \times 10^{-5}$ | $1.65 \times 10^{-5}$ |
| Nitrogen | 20 | 1.16 | 11.4 | $1.76 \times 10^{-5}$ | $1.52 \times 10^{-5}$ |
| Oxygen | 20 | 1.33 | 13.0 | $2.04 \times 10^{-5}$ | $1.53 \times 10^{-5}$ |

(b) Approximate Physical Properties of Some Common Fluids (Other Units)

| | Temperature (°F) | Density, $\rho$ (slug/ft$^3$) | Specific Weight, $\gamma$ (lbf/ft$^3$) | Dynamic Viscosity, $\mu$ (lbf · s/ft$^2$) | Kinematic Viscosity, $\nu$ (ft$^2$/s) |
|---|---|---|---|---|---|
| Liquids | | | | | |
| Carbon tetrachloride | 68 | 3.09 | 99.5 | $2.00 \times 10^{-5}$ | $6.47 \times 10^{-6}$ |
| Ethyl alcohol | 68 | 1.53 | 49.3 | $2.49 \times 10^{-5}$ | $1.63 \times 10^{-5}$ |
| Gasoline | 60 | 1.32 | 42.5 | $6.5 \times 10^{-6}$ | $4.9 \times 10^{-6}$ |
| Glycerin | 68 | 2.44 | 78.6 | $3.13 \times 10^{-2}$ | $1.28 \times 10^{-2}$ |
| Mercury | 68 | 26.3 | 847 | $3.28 \times 10^{-5}$ | $1.25 \times 10^{-6}$ |
| SAE 30 oil | 60 | 1.77 | 57.0 | $8.0 \times 10^{-3}$ | $4.5 \times 10^{-3}$ |
| Seawater | 60 | 1.99 | 64.0 | $2.51 \times 10^{-5}$ | $1.26 \times 10^{-5}$ |
| Water | 60 | 1.94 | 62.4 | $2.34 \times 10^{-5}$ | $1.21 \times 10^{-5}$ |
| Gases at Standard Atmospheric Pressure[1] | | | | | |
| Air (standard) | 59 | $2.38 \times 10^{-3}$ | $7.65 \times 10^{-2}$ | $3.74 \times 10^{-7}$ | $1.57 \times 10^{-4}$ |
| Carbon dioxide | 68 | $3.55 \times 10^{-3}$ | $1.14 \times 10^{-1}$ | $3.07 \times 10^{-7}$ | $8.65 \times 10^{-5}$ |
| Helium | 68 | $3.23 \times 10^{-4}$ | $1.04 \times 10^{-2}$ | $4.09 \times 10^{-7}$ | $1.27 \times 10^{-3}$ |
| Hydrogen | 68 | $1.63 \times 10^{-4}$ | $5.25 \times 10^{-3}$ | $1.85 \times 10^{-7}$ | $1.13 \times 10^{-3}$ |
| Methane (natural gas) | 68 | $1.29 \times 10^{-3}$ | $4.15 \times 10^{-2}$ | $2.29 \times 10^{-7}$ | $1.78 \times 10^{-4}$ |
| Nitrogen | 68 | $2.26 \times 10^{-3}$ | $7.28 \times 10^{-2}$ | $3.68 \times 10^{-7}$ | $1.63 \times 10^{-4}$ |
| Oxygen | 68 | $2.58 \times 10^{-3}$ | $8.31 \times 10^{-2}$ | $4.25 \times 10^{-7}$ | $1.65 \times 10^{-4}$ |

[1]For gases at atmospheric pressure, the ideal gas model (Sec. 4.5) applies, and $\rho = p/RT$.

## Table HT-1 Thermophysical Properties of Selected Technical Materials

| Composition | Melting Point (K) | Properties at 300 K | | | | Properties at Various Temperatures (K) $k$ (W/m · K) / $c_p$ (J/kg · K) | | | | | | | | | |
|---|---|---|---|---|---|---|---|---|---|---|---|---|---|---|---|
| | | $\rho$ (kg/m$^3$) | $c_p$ (J/kg · K) | $k$ (W/m · K) | $\alpha \cdot 10^6$ (m$^2$/s) | 100 | 200 | 400 | 600 | 800 | 1000 | 1200 | 1500 | 2000 | 2500 |
| *Metallic Solids* | | | | | | | | | | | | | | | |
| Aluminum | | | | | | | | | | | | | | | |
| Pure | 933 | 2702 | 903 | 237 | 97.1 | 302 | 237 | 240 | 231 | 218 | | | | | |
| | | | | | | 482 | 798 | 949 | 1033 | 1146 | | | | | |
| Alloy 2024-T6 | 775 | 2770 | 875 | 177 | 73.0 | 65 | 163 | 186 | 186 | | | | | | |
| | | | | | | 473 | 787 | 925 | 1042 | | | | | | |
| Beryllium | 1550 | 1850 | 1825 | 200 | 59.2 | 990 | 301 | 161 | 126 | 106 | 90.8 | 78.7 | | | |
| | | | | | | 203 | 1114 | 2191 | 2604 | 2823 | 3018 | 3227 | 3519 | | |
| Copper | | | | | | | | | | | | | | | |
| Pure | 1358 | 8933 | 385 | 401 | 117 | 482 | 413 | 393 | 379 | 366 | 352 | 339 | | | |
| | | | | | | 252 | 356 | 397 | 417 | 433 | 451 | 480 | | | |
| Cartridge brass (70% Cu, 30% Zn) | 1188 | 8530 | 380 | 110 | 33.9 | 75 | 95 | 137 | 149 | | | | | | |
| | | | | | | | 360 | 395 | 425 | | | | | | |
| Germanium | 1211 | 5360 | 322 | 59.9 | 34.7 | 232 | 96.8 | 43.2 | 27.3 | 19.8 | 17.4 | 17.4 | | | |
| | | | | | | 190 | 290 | 337 | 348 | 357 | 375 | 395 | | | |
| Gold | 1336 | 19,300 | 129 | 317 | 127 | 327 | 323 | 311 | 298 | 284 | 270 | 255 | | | |
| | | | | | | 109 | 124 | 131 | 135 | 140 | 145 | 155 | | | |
| Iron | | | | | | | | | | | | | | | |
| Pure | 1810 | 7870 | 447 | 80.2 | 23.1 | 134 | 94.0 | 69.5 | 54.7 | 43.3 | 32.8 | 28.3 | 32.1 | | |
| | | | | | | 216 | 384 | 490 | 574 | 680 | 975 | 609 | 654 | | |
| Plain carbon steel | | 7854 | 434 | 60.5 | 17.7 | | | 58.7 | 48.8 | 39.2 | 31.3 | | | | |
| | | | | | | | | 487 | 559 | 685 | 1168 | | | | |
| AISI 1010 | | 7832 | 434 | 63.9 | 18.8 | | | 58.7 | 48.8 | 39.2 | 31.3 | | | | |
| | | | | | | | | 487 | 559 | 685 | 1168 | | | | |
| Stainless steel AISI 316 | | 8238 | 468 | 13.4 | 3.48 | | | 15.2 | 18.3 | 21.3 | 24.2 | | | | |
| | | | | | | | | 504 | 550 | 576 | 602 | | | | |
| Molybdenum | 2894 | 10,240 | 251 | 138 | 53.7 | 179 | 143 | 134 | 126 | 118 | 112 | 105 | 98 | 90 | 86 |
| | | | | | | 141 | 224 | 261 | 275 | 285 | 295 | 308 | 330 | 380 | 459 |
| Nickel | | | | | | | | | | | | | | | |
| Pure | 1728 | 8900 | 444 | 90.7 | 23.0 | 164 | 107 | 80.2 | 65.6 | 67.6 | 71.8 | 76.2 | 82.6 | | |
| | | | | | | 232 | 383 | 485 | 592 | 530 | 562 | 594 | 616 | | |
| Platinum | 2045 | 21,450 | 133 | 71.6 | 25.1 | 77.5 | 72.6 | 71.8 | 73.2 | 75.6 | 78.7 | 82.6 | 89.5 | 99.4 | |
| | | | | | | 100 | 125 | 136 | 141 | 146 | 152 | 157 | 165 | 179 | |
| Silicon | 1685 | 2330 | 712 | 148 | 89.2 | 884 | 264 | 98.9 | 61.9 | 42.2 | 31.2 | 25.7 | 22.7 | | |
| | | | | | | 259 | 556 | 790 | 867 | 913 | 946 | 967 | 992 | | |

**Table HT-1** *(Continued)*

| Composition | Melting Point (K) | Properties at 300 K $\rho$ (kg/m$^3$) | $c_p$ (J/kg · K) | $k$ (W/m · K) | $\alpha \cdot 10^6$ (m$^2$/s) | Properties at Various Temperatures (K) $k$ (W/m · K) / $c_p$ (J/kg · K) 100 | 200 | 400 | 600 | 800 | 1000 | 1200 | 1500 | 2000 | 2500 |
|---|---|---|---|---|---|---|---|---|---|---|---|---|---|---|---|
| Silver | 1235 | 10,500 | 235 | 429 | 174 | 444 | 430 | 425 | 412 | 396 | 379 | 361 | | | |
| | | | | | | 187 | 225 | 239 | 250 | 262 | 277 | 292 | | | |
| Tantalum | 3269 | 16,600 | 140 | 57.5 | 24.7 | 59.2 | 57.5 | 57.8 | 58.6 | 59.4 | 60.2 | 61.0 | 62.2 | 64.1 | 65.6 |
| | | | | | | 110 | 133 | 144 | 146 | 149 | 152 | 155 | 160 | 172 | 189 |
| Thorium | 2023 | 11,700 | 118 | 54.0 | 39.1 | 59.8 | 54.6 | 54.5 | 55.8 | 56.9 | 56.9 | 58.7 | | | |
| | | | | | | 99 | 112 | 124 | 134 | 145 | 156 | 167 | | | |
| Titanium | 1953 | 4500 | 522 | 21.9 | 9.32 | 30.5 | 24.5 | 20.4 | 19.4 | 19.7 | 20.7 | 22.0 | 24.5 | | |
| | | | | | | 300 | 465 | 551 | 591 | 633 | 675 | 620 | 686 | | |
| Tungsten | 3660 | 19,300 | 132 | 174 | 68.3 | 208 | 186 | 159 | 137 | 125 | 118 | 113 | 107 | 100 | 95 |
| | | | | | | 87 | 122 | 137 | 142 | 145 | 148 | 152 | 157 | 167 | 176 |
| *Nonmetallic solids* | | | | | | | | | | | | | | | |
| Aluminum oxide, polycrystalline | 2323 | 3970 | 765 | 36.0 | 11.9 | 133 | 55 | 26.4 | 15.8 | 10.4 | 7.85 | 6.55 | 5.66 | 6.00 | |
| | | | | | | — | — | 940 | 1110 | 1180 | 1225 | — | — | — | |
| Aluminum oxide, sapphire | 2323 | 3970 | 765 | 46 | 15.1 | 450 | 82 | 32.4 | 18.9 | 13.0 | 10.5 | | | | |
| | | | | | | — | — | 940 | 1110 | 1180 | 1225 | | | | |
| Beryllium oxide | 2725 | 3000 | 1030 | 272 | 88.0 | | | 196 | 111 | 70 | 47 | 33 | 21.5 | 15 | |
| | | | | | | | | 1350 | 1690 | 1865 | 1975 | 2055 | 2145 | 2750 | |
| Graphite, pyrolytic | 2273 | 2210 | | | | | | | | | | | | | |
| $k$, ∥ to layers | | | | 1950 | | 4970 | 3230 | 1390 | 892 | 667 | 534 | 448 | 357 | 262 | |
| $k$, ⊥ to layers | | | | 5.70 | | 16.8 | 9.23 | 4.09 | 2.68 | 2.01 | 1.60 | 1.34 | 1.08 | 0.81 | |
| $c_p$ | | | 709 | | | 136 | 411 | 992 | 1406 | 1650 | 1793 | 1890 | 1974 | 2043 | |
| Pyroceram, Corning 9606 | 1623 | 2600 | 808 | 3.98 | 1.89 | 5.25 | 4.78 | 3.64 | 3.28 | 3.08 | 2.96 | 2.87 | 2.79 | | |
| | | | | | | — | — | 908 | 1038 | 1122 | 1197 | 1264 | 1498 | | |
| Silicon carbide | 3100 | 3160 | 675 | 490 | 230 | | | — | — | — | 87 | 58 | 30 | | |
| | | | | | | | | 880 | 1050 | 1135 | 1195 | 1243 | 1310 | | |
| Silicon dioxide, polycrystalline (fused silica) | 1883 | 2220 | 745 | 1.38 | 0.834 | 0.69 | 1.14 | 1.51 | 1.75 | 2.17 | 2.87 | 4.00 | | | |
| | | | | | | — | — | 905 | 1040 | 1105 | 1155 | 1195 | | | |
| Silicon nitride | 2173 | 2400 | 691 | 16.0 | 9.65 | — | — | 13.9 | 11.3 | 9.88 | 8.76 | 8.00 | 7.16 | 6.20 | |
| | | | | | | — | 578 | 778 | 937 | 1063 | 1155 | 1226 | 1306 | 1377 | |
| Thorium dioxide | 3573 | 9110 | 235 | 13 | 6.1 | | | 10.2 | 6.6 | 4.7 | 3.68 | 3.12 | 2.73 | 2.5 | |
| | | | | | | | | 255 | 274 | 285 | 295 | 303 | 315 | 330 | |

## Table HT-2 Thermophysical Properties of Selected Common Materials

| | Typical Properties at 300 K | | |
|---|---|---|---|
| Description/Composition | Density, ρ (kg/m$^3$) | Thermal Conductivity, $k$ (W/m · K) | Specific Heat, $c_p$ (J/kg · K) |
| *Insulating Materials and Systems* | | | |
| Blanket and Batt | | | |
| Glass fiber, paper faced | 16 | 0.046 | — |
| | 28 | 0.038 | — |
| | 40 | 0.035 | — |
| Board and Slab | | | |
| Cellular glass | 145 | 0.058 | 1000 |
| Glass fiber, organic bonded | 105 | 0.036 | 795 |
| Polystyrene, expanded | | | |
| Extruded (R-12) | 55 | 0.027 | 1210 |
| Molded beads | 16 | 0.040 | 1210 |
| Loose Fill | | | |
| Glass fiber, poured or blown | 16 | 0.043 | 835 |
| Vermiculite, flakes | 80 | 0.068 | 835 |
| | 160 | 0.063 | 1000 |
| Formed/Foamed-in-Place | | | |
| Polyvinyl acetate cork mastic; sprayed or troweled | — | 0.100 | — |
| Urethane, two-part mixture; rigid foam | 70 | 0.026 | 1045 |
| Reflective | | | |
| Aluminum foil separating fluffy glass mats; 10–12 layers, evacuated; for cryogenic applications (150 K) | 40 | 0.00016 | — |
| Aluminum foil and glass paper laminate; 75–150 layers; evacuated; for cryogenic application (150 K) | 120 | 0.000017 | — |
| Typical silica powder, evacuated | 160 | 0.0017 | — |
| *Structural Building Materials* | | | |
| Building Boards | | | |
| Gypsum or plaster board | 800 | 0.17 | — |
| Hardboard, siding | 640 | 0.094 | 1170 |
| Particle board, low density | 590 | 0.078 | 1300 |
| Particle board, high density | 1000 | 0.170 | 1300 |
| Plywood | 545 | 0.12 | 1215 |
| Woods | | | |
| Hardwoods (oak, maple) | 720 | 0.16 | 1255 |
| Softwoods (fir, pine) | 510 | 0.12 | 1380 |
| Masonry Materials | | | |
| Brick, common | 1920 | 0.72 | 835 |
| Concrete (stone mix) | 2300 | 1.4 | 880 |

## Table HT-2 Thermophysical Properties of Selected Common Materials (*Continued*)

| Description/ Composition | Temperature (K) | Density, ρ (kg/m$^3$) | Thermal Conductivity, $k$ (W/m · K) | Specific Heat, $c_p$ (J/kg · K) |
|---|---|---|---|---|
| *Other Materials* | | | | |
| Asphalt | 300 | 2115 | 0.062 | 920 |
| Coal, anthracite | 300 | 1350 | 0.26 | 1260 |
| Cotton | 300 | 80 | 0.06 | 1300 |
| Foodstuffs | | | | |
| Apple, red (75% water) | 300 | 840 | 0.513 | 3600 |
| Cake, batter | 300 | 720 | 0.223 | — |
| Cake, fully baked | 300 | 280 | 0.121 | — |
| Chicken meat, white | 198 | — | 1.60 | — |
| (74.4% water content) | 273 | | 0.476 | |
| Glass | | | | |
| Plate (soda lime) | 300 | 2500 | 1.4 | 750 |
| Pyrex | 300 | 2225 | 1.4 | 835 |
| Ice | 273 | 920 | 1.88 | 2040 |
| | 253 | — | 2.03 | 1945 |
| Leather (sole) | 300 | 998 | 0.159 | — |
| Paper | 300 | 930 | 0.180 | 1340 |
| Paraffin | 300 | 900 | 0.240 | 2890 |
| Rock | | | | |
| Granite, Barre | 300 | 2630 | 2.79 | 775 |
| Marble, Halston | 300 | 2680 | 2.80 | 830 |
| Sandstone, Berea | 300 | 2150 | 2.90 | 745 |
| Rubber, vulcanized | | | | |
| Soft | 300 | 1100 | 0.13 | 2010 |
| Hard | 300 | 1190 | 0.16 | — |
| Sand | 300 | 1515 | 0.27 | 800 |
| Soil | 300 | 2050 | 0.52 | 1840 |
| Snow | 273 | 110 | 0.049 | — |
| | | 500 | 0.190 | — |
| Teflon | 300 | 2200 | 0.35 | — |
| | 400 | | 0.45 | — |
| Tissue, human | | | | |
| Skin | 300 | — | 0.37 | — |
| Fat layer (adipose) | 300 | — | 0.2 | — |
| Muscle | 300 | — | 0.41 | — |
| Wood, cross grain | | | | |
| Fir | 300 | 415 | 0.11 | 2720 |
| Oak | 300 | 545 | 0.17 | 2385 |
| Wood, radial | | | | |
| Oak | 300 | 545 | 0.19 | 2385 |
| Fir | 300 | 420 | 0.14 | 2720 |

## Table HT-3 Thermophysical Properties of Gases at Atmospheric Pressure[1]

| $T$ (K) | $\rho$ (kg/m$^3$) | $c_p$ (kJ/kg · K) | $\mu \cdot 10^7$ (N · s/m$^2$) | $\nu \cdot 10^6$ (m$^2$/s) | $k \cdot 10^3$ (W/m · K) | $\alpha \cdot 10^6$ (m$^2$/s) | $Pr$ |
|---|---|---|---|---|---|---|---|
| **Air** | | | | | | | |
| 100 | 3.5562 | 1.032 | 71.1 | 2.00 | 9.34 | 2.54 | 0.786 |
| 150 | 2.3364 | 1.012 | 103.4 | 4.426 | 13.8 | 5.84 | 0.758 |
| 200 | 1.7458 | 1.007 | 132.5 | 7.590 | 18.1 | 10.3 | 0.737 |
| 250 | 1.3947 | 1.006 | 159.6 | 11.44 | 22.3 | 15.9 | 0.720 |
| 300 | 1.1614 | 1.007 | 184.6 | 15.89 | 26.3 | 22.5 | 0.707 |
| 350 | 0.9950 | 1.009 | 208.2 | 20.92 | 30.0 | 29.9 | 0.700 |
| 400 | 0.8711 | 1.014 | 230.1 | 26.41 | 33.8 | 38.3 | 0.690 |
| 450 | 0.7740 | 1.021 | 250.7 | 32.39 | 37.3 | 47.2 | 0.686 |
| 500 | 0.6964 | 1.030 | 270.1 | 38.79 | 40.7 | 56.7 | 0.684 |
| 550 | 0.6329 | 1.040 | 288.4 | 45.57 | 43.9 | 66.7 | 0.683 |
| 600 | 0.5804 | 1.051 | 305.8 | 52.69 | 46.9 | 76.9 | 0.685 |
| 650 | 0.5356 | 1.063 | 322.5 | 60.21 | 49.7 | 87.3 | 0.690 |
| 700 | 0.4975 | 1.075 | 338.8 | 68.10 | 52.4 | 98.0 | 0.695 |
| 750 | 0.4643 | 1.087 | 354.6 | 76.37 | 54.9 | 109 | 0.702 |
| 800 | 0.4354 | 1.099 | 369.8 | 84.93 | 57.3 | 120 | 0.709 |
| 850 | 0.4097 | 1.110 | 384.3 | 93.80 | 59.6 | 131 | 0.716 |
| 900 | 0.3868 | 1.121 | 398.1 | 102.9 | 62.0 | 143 | 0.720 |
| 950 | 0.3666 | 1.131 | 411.3 | 112.2 | 64.3 | 155 | 0.723 |
| 1000 | 0.3482 | 1.141 | 424.4 | 121.9 | 66.7 | 168 | 0.726 |
| 1100 | 0.3166 | 1.159 | 449.0 | 141.8 | 71.5 | 195 | 0.728 |
| **Helium (He)** | | | | | | | |
| 100 | 0.4871 | 5.193 | 96.3 | 19.8 | 73.0 | 28.9 | 0.686 |
| 120 | 0.4060 | 5.193 | 107 | 26.4 | 81.9 | 38.8 | 0.679 |
| 140 | 0.3481 | 5.193 | 118 | 33.9 | 90.7 | 50.2 | 0.676 |
| 180 | 0.2708 | 5.193 | 139 | 51.3 | 107.2 | 76.2 | 0.673 |
| 220 | 0.2216 | 5.193 | 160 | 72.2 | 123.1 | 107 | 0.675 |
| 260 | 0.1875 | 5.193 | 180 | 96.0 | 137 | 141 | 0.682 |
| 300 | 0.1625 | 5.193 | 199 | 122 | 152 | 180 | 0.680 |
| 400 | 0.1219 | 5.193 | 243 | 199 | 187 | 295 | 0.675 |
| 500 | 0.09754 | 5.193 | 283 | 290 | 220 | 434 | 0.668 |
| 700 | 0.06969 | 5.193 | 350 | 502 | 278 | 768 | 0.654 |
| 1000 | 0.04879 | 5.193 | 446 | 914 | 354 | 1400 | 0.654 |

[1]For gases at atmospheric pressure, the ideal gas model (Sec. 4.5) applies, and $\rho = p/RT$.

## Table HT-4 Thermophysical Properties of Saturated Liquids

*Saturated Liquids*

| $T$ (K) | $\rho$ (kg/m$^3$) | $c_p$ (kJ/kg · K) | $\mu \cdot 10^2$ (N · s/m$^2$) | $\nu \cdot 10^6$ (m$^2$/s) | $k \cdot 10^3$ (W/m · K) | $\alpha \cdot 10^7$ (m$^2$/s) | $Pr$ | $\beta \cdot 10^3$ (K$^{-1}$) |
|---|---|---|---|---|---|---|---|---|
| **Engine Oil (Unused)** | | | | | | | | |
| 273 | 899.1 | 1.796 | 385 | 4280 | 147 | 0.910 | 47,000 | 0.70 |
| 280 | 895.3 | 1.827 | 217 | 2430 | 144 | 0.880 | 27,500 | 0.70 |
| 290 | 890.0 | 1.868 | 99.9 | 1120 | 145 | 0.872 | 12,900 | 0.70 |
| 300 | 884.1 | 1.909 | 48.6 | 550 | 145 | 0.859 | 6400 | 0.70 |
| 310 | 877.9 | 1.951 | 25.3 | 288 | 145 | 0.847 | 3400 | 0.70 |
| 320 | 871.8 | 1.993 | 14.1 | 161 | 143 | 0.823 | 1965 | 0.70 |
| 330 | 865.8 | 2.035 | 8.36 | 96.6 | 141 | 0.800 | 1205 | 0.70 |
| 340 | 859.9 | 2.076 | 5.31 | 61.7 | 139 | 0.779 | 793 | 0.70 |
| 350 | 853.9 | 2.118 | 3.56 | 41.7 | 138 | 0.763 | 546 | 0.70 |
| 360 | 847.8 | 2.161 | 2.52 | 29.7 | 138 | 0.753 | 395 | 0.70 |
| 370 | 841.8 | 2.206 | 1.86 | 22.0 | 137 | 0.738 | 300 | 0.70 |
| 380 | 836.0 | 2.250 | 1.41 | 16.9 | 136 | 0.723 | 233 | 0.70 |
| 390 | 830.6 | 2.294 | 1.10 | 13.3 | 135 | 0.709 | 187 | 0.70 |
| 400 | 825.1 | 2.337 | 0.874 | 10.6 | 134 | 0.695 | 152 | 0.70 |
| 410 | 818.9 | 2.381 | 0.698 | 8.52 | 133 | 0.682 | 125 | 0.70 |
| 420 | 812.1 | 2.427 | 0.564 | 6.94 | 133 | 0.675 | 103 | 0.70 |
| 430 | 806.5 | 2.471 | 0.470 | 5.83 | 132 | 0.662 | 88 | 0.70 |
| **Ethylene Glycol [$C_2H_4(OH)_2$]** | | | | | | | | |
| 273 | 1130.8 | 2.294 | 6.51 | 57.6 | 242 | 0.933 | 617 | 0.65 |
| 280 | 1125.8 | 2.323 | 4.20 | 37.3 | 244 | 0.933 | 400 | 0.65 |
| 290 | 1118.8 | 2.368 | 2.47 | 22.1 | 248 | 0.936 | 236 | 0.65 |
| 300 | 1114.4 | 2.415 | 1.57 | 14.1 | 252 | 0.939 | 151 | 0.65 |
| 310 | 1103.7 | 2.460 | 1.07 | 9.65 | 255 | 0.939 | 103 | 0.65 |
| 320 | 1096.2 | 2.505 | 0.757 | 6.91 | 258 | 0.940 | 73.5 | 0.65 |
| 330 | 1089.5 | 2.549 | 0.561 | 5.15 | 260 | 0.936 | 55.0 | 0.65 |
| 340 | 1083.8 | 2.592 | 0.431 | 3.98 | 261 | 0.929 | 42.8 | 0.65 |
| 350 | 1079.0 | 2.637 | 0.342 | 3.17 | 261 | 0.917 | 34.6 | 0.65 |
| 360 | 1074.0 | 2.682 | 0.278 | 2.59 | 261 | 0.906 | 28.6 | 0.65 |
| 370 | 1066.7 | 2.728 | 0.228 | 2.14 | 262 | 0.900 | 23.7 | 0.65 |
| 373 | 1058.5 | 2.742 | 0.215 | 2.03 | 263 | 0.906 | 22.4 | 0.65 |
| **Glycerin [$C_3H_5(OH)_3$]** | | | | | | | | |
| 273 | 1276.0 | 2.261 | 1060 | 8310 | 282 | 0.977 | 85,000 | 0.47 |
| 280 | 1271.9 | 2.298 | 534 | 4200 | 284 | 0.972 | 43,200 | 0.47 |
| 290 | 1265.8 | 2.367 | 185 | 1460 | 286 | 0.955 | 15,300 | 0.48 |
| 300 | 1259.9 | 2.427 | 79.9 | 634 | 286 | 0.935 | 6780 | 0.48 |
| 310 | 1253.9 | 2.490 | 35.2 | 281 | 286 | 0.916 | 3060 | 0.49 |
| 320 | 1247.2 | 2.564 | 21.0 | 168 | 287 | 0.897 | 1870 | 0.50 |

## Table HT-5 Thermophysical Properties of Saturated Water[1]

| Temperature, $T$ (K) | Specific Heat (kJ/kg · K) | | Viscosity (N · s/m$^2$) | | Thermal Conductivity (W/m · K) | | Prandtl Number | | Expansion Coefficient, $\beta_f \cdot 10^6$ ($K^{-1}$) |
|---|---|---|---|---|---|---|---|---|---|
| | $c_{p,f}$ | $c_{p,g}$ | $\mu_f \cdot 10^6$ | $\mu_g \cdot 10^6$ | $k_f \cdot 10^3$ | $k_g \cdot 10^3$ | $Pr_f$ | $Pr_g$ | |
| 273.15 | 4.217 | 1.854 | 1750 | 8.02 | 569 | 18.2 | 12.99 | 0.815 | −68.05 |
| 275 | 4.211 | 1.855 | 1652 | 8.09 | 574 | 18.3 | 12.22 | 0.817 | −32.74 |
| 280 | 4.198 | 1.858 | 1422 | 8.29 | 582 | 18.6 | 10.26 | 0.825 | 46.04 |
| 285 | 4.189 | 1.861 | 1225 | 8.49 | 590 | 18.9 | 8.81 | 0.833 | 114.1 |
| 290 | 4.184 | 1.864 | 1080 | 8.69 | 598 | 19.3 | 7.56 | 0.841 | 174.0 |
| 295 | 4.181 | 1.868 | 959 | 8.89 | 606 | 19.5 | 6.62 | 0.849 | 227.5 |
| 300 | 4.179 | 1.872 | 855 | 9.09 | 613 | 19.6 | 5.83 | 0.857 | 276.1 |
| 305 | 4.178 | 1.877 | 769 | 9.29 | 620 | 20.1 | 5.20 | 0.865 | 320.6 |
| 310 | 4.178 | 1.882 | 695 | 9.49 | 628 | 20.4 | 4.62 | 0.873 | 361.9 |
| 315 | 4.179 | 1.888 | 631 | 9.69 | 634 | 20.7 | 4.16 | 0.883 | 400.4 |
| 320 | 4.180 | 1.895 | 577 | 9.89 | 640 | 21.0 | 3.77 | 0.894 | 436.7 |
| 325 | 4.182 | 1.903 | 528 | 10.09 | 645 | 21.3 | 3.42 | 0.901 | 471.2 |
| 330 | 4.184 | 1.911 | 489 | 10.29 | 650 | 21.7 | 3.15 | 0.908 | 504.0 |
| 335 | 4.186 | 1.920 | 453 | 10.49 | 656 | 22.0 | 2.88 | 0.916 | 535.5 |
| 340 | 4.188 | 1.930 | 420 | 10.69 | 660 | 22.3 | 2.66 | 0.925 | 566.0 |
| 345 | 4.191 | 1.941 | 389 | 10.89 | 665 | 22.6 | 2.45 | 0.933 | 595.4 |
| 350 | 4.195 | 1.954 | 365 | 11.09 | 668 | 23.0 | 2.29 | 0.942 | 624.2 |
| 355 | 4.199 | 1.968 | 343 | 11.29 | 671 | 23.3 | 2.14 | 0.951 | 652.3 |
| 360 | 4.203 | 1.983 | 324 | 11.49 | 674 | 23.7 | 2.02 | 0.960 | 697.9 |
| 365 | 4.209 | 1.999 | 306 | 11.69 | 677 | 24.1 | 1.91 | 0.969 | 707.1 |
| 370 | 4.214 | 2.017 | 289 | 11.89 | 679 | 24.5 | 1.80 | 0.978 | 728.7 |
| 373.15 | 4.217 | 2.029 | 279 | 12.02 | 680 | 24.8 | 1.76 | 0.984 | 750.1 |

[1]See Table T-2 for specific volume, $v_f$ and $v_g$.

**Table T-1 and Table T-1E** Atomic or Molecular Weights and Critical Properties of Selected Elements and Compounds

| Substance | Chemical Formula | $M$ (kg/kmol) (lb/lbmol) | SI $T_c$ (K) | SI $p_c$ (bar) | E $T_c$ (°R) | E $p_c$ (atm) |
|---|---|---|---|---|---|---|
| Acetylene | $C_2H_2$ | 26.04 | 309 | 62.8 | 556 | 62 |
| Air (equivalent) | — | 28.97 | 133 | 37.7 | 239 | 37.2 |
| Ammonia | $NH_3$ | 17.03 | 406 | 112.8 | 730 | 111.3 |
| Argon | Ar | 39.94 | 151 | 48.6 | 272 | 47.97 |
| Benzene | $C_6H_6$ | 78.11 | 563 | 49.3 | 1013 | 48.7 |
| Butane | $C_4H_{10}$ | 58.12 | 425 | 38.0 | 765 | 37.5 |
| Carbon | C | 12.01 | — | — | — | — |
| Carbon dioxide | $CO_2$ | 44.01 | 304 | 73.9 | 548 | 72.9 |
| Carbon monoxide | CO | 28.01 | 133 | 35.0 | 239 | 34.5 |
| Copper | Cu | 63.54 | — | — | — | — |
| Ethane | $C_2H_6$ | 30.07 | 305 | 48.8 | 549 | 48.2 |
| Ethyl alcohol | $C_2H_5OH$ | 46.07 | 516 | 63.8 | 929 | 63.0 |
| Ethylene | $C_2H_4$ | 28.05 | 283 | 51.2 | 510 | 50.5 |
| Helium | He | 4.003 | 5.2 | 2.3 | 9.33 | 2.26 |
| Hydrogen | $H_2$ | 2.016 | 33.2 | 13.0 | 59.8 | 12.8 |
| Methane | $CH_4$ | 16.04 | 191 | 46.4 | 344 | 45.8 |
| Methyl alcohol | $CH_3OH$ | 32.04 | 513 | 79.5 | 924 | 78.5 |
| Nitrogen | $N_2$ | 28.01 | 126 | 33.9 | 227 | 33.5 |
| Octane | $C_8H_{18}$ | 114.22 | 569 | 24.9 | 1025 | 24.6 |
| Oxygen | $O_2$ | 32.00 | 154 | 50.5 | 278 | 49.8 |
| Propane | $C_3H_8$ | 44.09 | 370 | 42.7 | 666 | 42.1 |
| Propylene | $C_3H_6$ | 42.08 | 365 | 46.2 | 657 | 45.6 |
| Refrigerant 12 | $CCl_2F_2$ | 120.92 | 385 | 41.2 | 693 | 40.6 |
| Refrigerant 22 | $CHClF_2$ | 86.48 | 369 | 49.8 | 665 | 49.1 |
| Refrigerant 134a | $CF_3CH_2F$ | 102.03 | 374 | 40.7 | 673 | 40.2 |
| Sulfur dioxide | $SO_2$ | 64.06 | 431 | 78.7 | 775 | 77.7 |
| Water | $H_2O$ | 18.02 | 647.3 | 220.9 | 1165 | 218.0 |

*Sources:* Adapted from *International Critical Tables* and L. C. Nelson and E. F. Obert, Generalized Compressibility Charts, *Chem. Eng., 61:* 203 (1954).

**Table T-2** Properties of Saturated Water (Liquid–Vapor): Temperature Table

| | | Specific Volume $m^3/kg$ | | Internal Energy kJ/kg | | Enthalpy kJ/kg | | | Entropy kJ/kg · K | | |
|---|---|---|---|---|---|---|---|---|---|---|---|
| Temp. °C | Press. bar | Sat. Liquid $v_f \times 10^3$ | Sat. Vapor $v_g$ | Sat. Liquid $u_f$ | Sat. Vapor $u_g$ | Sat. Liquid $h_f$ | Evap. $h_{fg}$ | Sat. Vapor $h_g$ | Sat. Liquid $s_f$ | Sat. Vapor $s_g$ | Temp. °C |
| .01 | 0.00611 | 1.0002 | 206.136 | 0.00 | 2375.3 | 0.01 | 2501.3 | 2501.4 | 0.0000 | 9.1562 | .01 |
| 4 | 0.00813 | 1.0001 | 157.232 | 16.77 | 2380.9 | 16.78 | 2491.9 | 2508.7 | 0.0610 | 9.0514 | 4 |
| 5 | 0.00872 | 1.0001 | 147.120 | 20.97 | 2382.3 | 20.98 | 2489.6 | 2510.6 | 0.0761 | 9.0257 | 5 |
| 6 | 0.00935 | 1.0001 | 137.734 | 25.19 | 2383.6 | 25.20 | 2487.2 | 2512.4 | 0.0912 | 9.0003 | 6 |
| 8 | 0.01072 | 1.0002 | 120.917 | 33.59 | 2386.4 | 33.60 | 2482.5 | 2516.1 | 0.1212 | 8.9501 | 8 |
| 10 | 0.01228 | 1.0004 | 106.379 | 42.00 | 2389.2 | 42.01 | 2477.7 | 2519.8 | 0.1510 | 8.9008 | 10 |
| 11 | 0.01312 | 1.0004 | 99.857 | 46.20 | 2390.5 | 46.20 | 2475.4 | 2521.6 | 0.1658 | 8.8765 | 11 |
| 12 | 0.01402 | 1.0005 | 93.784 | 50.41 | 2391.9 | 50.41 | 2473.0 | 2523.4 | 0.1806 | 8.8524 | 12 |
| 13 | 0.01497 | 1.0007 | 88.124 | 54.60 | 2393.3 | 54.60 | 2470.7 | 2525.3 | 0.1953 | 8.8285 | 13 |
| 14 | 0.01598 | 1.0008 | 82.848 | 58.79 | 2394.7 | 58.80 | 2468.3 | 2527.1 | 0.2099 | 8.8048 | 14 |
| 15 | 0.01705 | 1.0009 | 77.926 | 62.99 | 2396.1 | 62.99 | 2465.9 | 2528.9 | 0.2245 | 8.7814 | 15 |
| 16 | 0.01818 | 1.0011 | 73.333 | 67.18 | 2397.4 | 67.19 | 2463.6 | 2530.8 | 0.2390 | 8.7582 | 16 |
| 17 | 0.01938 | 1.0012 | 69.044 | 71.38 | 2398.8 | 71.38 | 2461.2 | 2532.6 | 0.2535 | 8.7351 | 17 |
| 18 | 0.02064 | 1.0014 | 65.038 | 75.57 | 2400.2 | 75.58 | 2458.8 | 2534.4 | 0.2679 | 8.7123 | 18 |
| 19 | 0.02198 | 1.0016 | 61.293 | 79.76 | 2401.6 | 79.77 | 2456.5 | 2536.2 | 0.2823 | 8.6897 | 19 |
| 20 | 0.02339 | 1.0018 | 57.791 | 83.95 | 2402.9 | 83.96 | 2454.1 | 2538.1 | 0.2966 | 8.6672 | 20 |
| 21 | 0.02487 | 1.0020 | 54.514 | 88.14 | 2404.3 | 88.14 | 2451.8 | 2539.9 | 0.3109 | 8.6450 | 21 |
| 22 | 0.02645 | 1.0022 | 51.447 | 92.32 | 2405.7 | 92.33 | 2449.4 | 2541.7 | 0.3251 | 8.6229 | 22 |
| 23 | 0.02810 | 1.0024 | 48.574 | 96.51 | 2407.0 | 96.52 | 2447.0 | 2543.5 | 0.3393 | 8.6011 | 23 |
| 24 | 0.02985 | 1.0027 | 45.883 | 100.70 | 2408.4 | 100.70 | 2444.7 | 2545.4 | 0.3534 | 8.5794 | 24 |
| 25 | 0.03169 | 1.0029 | 43.360 | 104.88 | 2409.8 | 104.89 | 2442.3 | 2547.2 | 0.3674 | 8.5580 | 25 |
| 26 | 0.03363 | 1.0032 | 40.994 | 109.06 | 2411.1 | 109.07 | 2439.9 | 2549.0 | 0.3814 | 8.5367 | 26 |
| 27 | 0.03567 | 1.0035 | 38.774 | 113.25 | 2412.5 | 113.25 | 2437.6 | 2550.8 | 0.3954 | 8.5156 | 27 |
| 28 | 0.03782 | 1.0037 | 36.690 | 117.42 | 2413.9 | 117.43 | 2435.2 | 2552.6 | 0.4093 | 8.4946 | 28 |
| 29 | 0.04008 | 1.0040 | 34.733 | 121.60 | 2415.2 | 121.61 | 2432.8 | 2554.5 | 0.4231 | 8.4739 | 29 |
| 30 | 0.04246 | 1.0043 | 32.894 | 125.78 | 2416.6 | 125.79 | 2430.5 | 2556.3 | 0.4369 | 8.4533 | 30 |
| 31 | 0.04496 | 1.0046 | 31.165 | 129.96 | 2418.0 | 129.97 | 2428.1 | 2558.1 | 0.4507 | 8.4329 | 31 |
| 32 | 0.04759 | 1.0050 | 29.540 | 134.14 | 2419.3 | 134.15 | 2425.7 | 2559.9 | 0.4644 | 8.4127 | 32 |
| 33 | 0.05034 | 1.0053 | 28.011 | 138.32 | 2420.7 | 138.33 | 2423.4 | 2561.7 | 0.4781 | 8.3927 | 33 |
| 34 | 0.05324 | 1.0056 | 26.571 | 142.50 | 2422.0 | 142.50 | 2421.0 | 2563.5 | 0.4917 | 8.3728 | 34 |
| 35 | 0.05628 | 1.0060 | 25.216 | 146.67 | 2423.4 | 146.68 | 2418.6 | 2565.3 | 0.5053 | 8.3531 | 35 |
| 36 | 0.05947 | 1.0063 | 23.940 | 150.85 | 2424.7 | 150.86 | 2416.2 | 2567.1 | 0.5188 | 8.3336 | 36 |
| 38 | 0.06632 | 1.0071 | 21.602 | 159.20 | 2427.4 | 159.21 | 2411.5 | 2570.7 | 0.5458 | 8.2950 | 38 |
| 40 | 0.07384 | 1.0078 | 19.523 | 167.56 | 2430.1 | 167.57 | 2406.7 | 2574.3 | 0.5725 | 8.2570 | 40 |
| 45 | 0.09593 | 1.0099 | 15.258 | 188.44 | 2436.8 | 188.45 | 2394.8 | 2583.2 | 0.6387 | 8.1648 | 45 |
| 50 | 0.1235 | 1.0121 | 12.032 | 209.32 | 2443.5 | 209.33 | 2382.7 | 2592.1 | .7038 | 8.0763 | 50 |
| 55 | 0.1576 | 1.0146 | 9.568 | 230.21 | 2450.1 | 230.23 | 2370.7 | 2600.9 | .7679 | 7.9913 | 55 |
| 60 | 0.1994 | 1.0172 | 7.671 | 251.11 | 2456.6 | 251.13 | 2358.5 | 2609.6 | .8312 | 7.9096 | 60 |
| 65 | 0.2503 | 1.0199 | 6.197 | 272.02 | 2463.1 | 272.06 | 2346.2 | 2618.3 | .8935 | 7.8310 | 65 |
| 70 | 0.3119 | 1.0228 | 5.042 | 292.95 | 2469.6 | 292.98 | 2333.8 | 2626.8 | .9549 | 7.7553 | 70 |
| 75 | 0.3858 | 1.0259 | 4.131 | 313.90 | 2475.9 | 313.93 | 2321.4 | 2635.3 | 1.0155 | 7.6824 | 75 |
| 80 | 0.4739 | 1.0291 | 3.407 | 334.86 | 2482.2 | 334.91 | 2308.8 | 2643.7 | 1.0753 | 7.6122 | 80 |
| 85 | 0.5783 | 1.0325 | 2.828 | 355.84 | 2488.4 | 355.90 | 2296.0 | 2651.9 | 1.1343 | 7.5445 | 85 |
| 90 | 0.7014 | 1.0360 | 2.361 | 376.85 | 2494.5 | 376.92 | 2283.2 | 2660.1 | 1.1925 | 7.4791 | 90 |
| 95 | 0.8455 | 1.0397 | 1.982 | 397.88 | 2500.6 | 397.96 | 2270.2 | 2668.1 | 1.2500 | 7.4159 | 95 |

**Table T-2** *(Continued)*

| Temp. °C | Press. bar | Specific Volume m³/kg Sat. Liquid $v_f \times 10^3$ | Sat. Vapor $v_g$ | Internal Energy kJ/kg Sat. Liquid $u_f$ | Sat. Vapor $u_g$ | Enthalpy kJ/kg Sat. Liquid $h_f$ | Evap. $h_{fg}$ | Sat. Vapor $h_g$ | Entropy kJ/kg · K Sat. Liquid $s_f$ | Sat. Vapor $s_g$ | Temp. °C |
|---|---|---|---|---|---|---|---|---|---|---|---|
| 100 | 1.014 | 1.0435 | 1.673 | 418.94 | 2506.5 | 419.04 | 2257.0 | 2676.1 | 1.3069 | 7.3549 | 100 |
| 110 | 1.433 | 1.0516 | 1.210 | 461.14 | 2518.1 | 461.30 | 2230.2 | 2691.5 | 1.4185 | 7.2387 | 110 |
| 120 | 1.985 | 1.0603 | 0.8919 | 503.50 | 2529.3 | 503.71 | 2202.6 | 2706.3 | 1.5276 | 7.1296 | 120 |
| 130 | 2.701 | 1.0697 | 0.6685 | 546.02 | 2539.9 | 546.31 | 2174.2 | 2720.5 | 1.6344 | 7.0269 | 130 |
| 140 | 3.613 | 1.0797 | 0.5089 | 588.74 | 2550.0 | 589.13 | 2144.7 | 2733.9 | 1.7391 | 6.9299 | 140 |
| 150 | 4.758 | 1.0905 | 0.3928 | 631.68 | 2559.5 | 632.20 | 2114.3 | 2746.5 | 1.8418 | 6.8379 | 150 |
| 160 | 6.178 | 1.1020 | 0.3071 | 674.86 | 2568.4 | 675.55 | 2082.6 | 2758.1 | 1.9427 | 6.7502 | 160 |
| 170 | 7.917 | 1.1143 | 0.2428 | 718.33 | 2576.5 | 719.21 | 2049.5 | 2768.7 | 2.0419 | 6.6663 | 170 |
| 180 | 10.02 | 1.1274 | 0.1941 | 762.09 | 2583.7 | 763.22 | 2015.0 | 2778.2 | 2.1396 | 6.5857 | 180 |
| 190 | 12.54 | 1.1414 | 0.1565 | 806.19 | 2590.0 | 807.62 | 1978.8 | 2786.4 | 2.2359 | 6.5079 | 190 |
| 200 | 15.54 | 1.1565 | 0.1274 | 850.65 | 2595.3 | 852.45 | 1940.7 | 2793.2 | 2.3309 | 6.4323 | 200 |
| 210 | 19.06 | 1.1726 | 0.1044 | 895.53 | 2599.5 | 897.76 | 1900.7 | 2798.5 | 2.4248 | 6.3585 | 210 |
| 220 | 23.18 | 1.1900 | 0.08619 | 940.87 | 2602.4 | 943.62 | 1858.5 | 2802.1 | 2.5178 | 6.2861 | 220 |
| 230 | 27.95 | 1.2088 | 0.07158 | 986.74 | 2603.9 | 990.12 | 1813.8 | 2804.0 | 2.6099 | 6.2146 | 230 |
| 240 | 33.44 | 1.2291 | 0.05976 | 1033.2 | 2604.0 | 1037.3 | 1766.5 | 2803.8 | 2.7015 | 6.1437 | 240 |
| 250 | 39.73 | 1.2512 | 0.05013 | 1080.4 | 2602.4 | 1085.4 | 1716.2 | 2801.5 | 2.7927 | 6.0730 | 250 |
| 260 | 46.88 | 1.2755 | 0.04221 | 1128.4 | 2599.0 | 1134.4 | 1662.5 | 2796.6 | 2.8838 | 6.0019 | 260 |
| 270 | 54.99 | 1.3023 | 0.03564 | 1177.4 | 2593.7 | 1184.5 | 1605.2 | 2789.7 | 2.9751 | 5.9301 | 270 |
| 280 | 64.12 | 1.3321 | 0.03017 | 1227.5 | 2586.1 | 1236.0 | 1543.6 | 2779.6 | 3.0668 | 5.8571 | 280 |
| 290 | 74.36 | 1.3656 | 0.02557 | 1278.9 | 2576.0 | 1289.1 | 1477.1 | 2766.2 | 3.1594 | 5.7821 | 290 |
| 300 | 85.81 | 1.4036 | 0.02167 | 1332.0 | 2563.0 | 1344.0 | 1404.9 | 2749.0 | 3.2534 | 5.7045 | 300 |
| 320 | 112.7 | 1.4988 | 0.01549 | 1444.6 | 2525.5 | 1461.5 | 1238.6 | 2700.1 | 3.4480 | 5.5362 | 320 |
| 340 | 145.9 | 1.6379 | 0.01080 | 1570.3 | 2464.6 | 1594.2 | 1027.9 | 2622.0 | 3.6594 | 5.3357 | 340 |
| 360 | 186.5 | 1.8925 | 0.006945 | 1725.2 | 2351.5 | 1760.5 | 720.5 | 2481.0 | 3.9147 | 5.0526 | 360 |
| 374.14 | 220.9 | 3.155 | 0.003155 | 2029.6 | 2029.6 | 2099.3 | 0 | 2099.3 | 4.4298 | 4.4298 | 374.14 |

*Source:* Tables T-2 through T-5 are extracted from J. H. Keenan, F. G. Keyes, P. G. Hill, and J. G. Moore, *Steam Tables,* Wiley, New York, 1969.

**Table T-3** Properties of Saturated Water (Liquid–Vapor): Pressure Table

| Press. bar | Temp. °C | Specific Volume m³/kg Sat. Liquid $v_f \times 10^3$ | Sat. Vapor $v_g$ | Internal Energy kJ/kg Sat. Liquid $u_f$ | Sat. Vapor $u_g$ | Enthalpy kJ/kg Sat. Liquid $h_f$ | Evap. $h_{fg}$ | Sat. Vapor $h_g$ | Entropy kJ/kg · K Sat. Liquid $s_f$ | Sat. Vapor $s_g$ | Press. bar |
|---|---|---|---|---|---|---|---|---|---|---|---|
| 0.04 | 28.96 | 1.0040 | 34.800 | 121.45 | 2415.2 | 121.46 | 2432.9 | 2554.4 | 0.4226 | 8.4746 | 0.04 |
| 0.06 | 36.16 | 1.0064 | 23.739 | 151.53 | 2425.0 | 151.53 | 2415.9 | 2567.4 | 0.5210 | 8.3304 | 0.06 |
| 0.08 | 41.51 | 1.0084 | 18.103 | 173.87 | 2432.2 | 173.88 | 2403.1 | 2577.0 | 0.5926 | 8.2287 | 0.08 |
| 0.10 | 45.81 | 1.0102 | 14.674 | 191.82 | 2437.9 | 191.83 | 2392.8 | 2584.7 | 0.6493 | 8.1502 | 0.10 |
| 0.20 | 60.06 | 1.0172 | 7.649 | 251.38 | 2456.7 | 251.40 | 2358.3 | 2609.7 | 0.8320 | 7.9085 | 0.20 |
| 0.30 | 69.10 | 1.0223 | 5.229 | 289.20 | 2468.4 | 289.23 | 2336.1 | 2625.3 | 0.9439 | 7.7686 | 0.30 |
| 0.40 | 75.87 | 1.0265 | 3.993 | 317.53 | 2477.0 | 317.58 | 2319.2 | 2636.8 | 1.0259 | 7.6700 | 0.40 |
| 0.50 | 81.33 | 1.0300 | 3.240 | 340.44 | 2483.9 | 340.49 | 2305.4 | 2645.9 | 1.0910 | 7.5939 | 0.50 |
| 0.60 | 85.94 | 1.0331 | 2.732 | 359.79 | 2489.6 | 359.86 | 2293.6 | 2653.5 | 1.1453 | 7.5320 | 0.60 |
| 0.70 | 89.95 | 1.0360 | 2.365 | 376.63 | 2494.5 | 376.70 | 2283.3 | 2660.0 | 1.1919 | 7.4797 | 0.70 |

**Table T-3** *(Continued)*

| | | Specific Volume $m^3/kg$ | | Internal Energy kJ/kg | | Enthalpy kJ/kg | | | Entropy kJ/kg · K | | |
|---|---|---|---|---|---|---|---|---|---|---|---|
| Press. bar | Temp. °C | Sat. Liquid $v_f \times 10^3$ | Sat. Vapor $v_g$ | Sat. Liquid $u_f$ | Sat. Vapor $u_g$ | Sat. Liquid $h_f$ | Evap. $h_{fg}$ | Sat. Vapor $h_g$ | Sat. Liquid $s_f$ | Sat. Vapor $s_g$ | Press. bar |
| 0.80 | 93.50 | 1.0380 | 2.087 | 391.58 | 2498.8 | 391.66 | 2274.1 | 2665.8 | 1.2329 | 7.4346 | 0.80 |
| 0.90 | 96.71 | 1.0410 | 1.869 | 405.06 | 2502.6 | 405.15 | 2265.7 | 2670.9 | 1.2695 | 7.3949 | 0.90 |
| 1.00 | 99.63 | 1.0432 | 1.694 | 417.36 | 2506.1 | 417.46 | 2258.0 | 2675.5 | 1.3026 | 7.3594 | 1.00 |
| 1.50 | 111.4 | 1.0528 | 1.159 | 466.94 | 2519.7 | 467.11 | 2226.5 | 2693.6 | 1.4336 | 7.2233 | 1.50 |
| 2.00 | 120.2 | 1.0605 | 0.8857 | 504.49 | 2529.5 | 504.70 | 2201.9 | 2706.7 | 1.5301 | 7.1271 | 2.00 |
| 2.50 | 127.4 | 1.0672 | 0.7187 | 535.10 | 2537.2 | 535.37 | 2181.5 | 2716.9 | 1.6072 | 7.0527 | 2.50 |
| 3.00 | 133.6 | 1.0732 | 0.6058 | 561.15 | 2543.6 | 561.47 | 2163.8 | 2725.3 | 1.6718 | 6.9919 | 3.00 |
| 3.50 | 138.9 | 1.0786 | 0.5243 | 583.95 | 2546.9 | 584.33 | 2148.1 | 2732.4 | 1.7275 | 6.9405 | 3.50 |
| 4.00 | 143.6 | 1.0836 | 0.4625 | 604.31 | 2553.6 | 604.74 | 2133.8 | 2738.6 | 1.7766 | 6.8959 | 4.00 |
| 4.50 | 147.9 | 1.0882 | 0.4140 | 622.25 | 2557.6 | 623.25 | 2120.7 | 2743.9 | 1.8207 | 6.8565 | 4.50 |
| 5.00 | 151.9 | 1.0926 | 0.3749 | 639.68 | 2561.2 | 640.23 | 2108.5 | 2748.7 | 1.8607 | 6.8212 | 5.00 |
| 6.00 | 158.9 | 1.1006 | 0.3157 | 669.90 | 2567.4 | 670.56 | 2086.3 | 2756.8 | 1.9312 | 6.7600 | 6.00 |
| 7.00 | 165.0 | 1.1080 | 0.2729 | 696.44 | 2572.5 | 697.22 | 2066.3 | 2763.5 | 1.9922 | 6.7080 | 7.00 |
| 8.00 | 170.4 | 1.1148 | 0.2404 | 720.22 | 2576.8 | 721.11 | 2048.0 | 2769.1 | 2.0462 | 6.6628 | 8.00 |
| 9.00 | 175.4 | 1.1212 | 0.2150 | 741.83 | 2580.5 | 742.83 | 2031.1 | 2773.9 | 2.0946 | 6.6226 | 9.00 |
| 10.0 | 179.9 | 1.1273 | 0.1944 | 761.68 | 2583.6 | 762.81 | 2015.3 | 2778.1 | 2.1387 | 6.5863 | 10.0 |
| 15.0 | 198.3 | 1.1539 | 0.1318 | 843.16 | 2594.5 | 844.84 | 1947.3 | 2792.2 | 2.3150 | 6.4448 | 15.0 |
| 20.0 | 212.4 | 1.1767 | 0.09963 | 906.44 | 2600.3 | 908.79 | 1890.7 | 2799.5 | 2.4474 | 6.3409 | 20.0 |
| 25.0 | 224.0 | 1.1973 | 0.07998 | 959.11 | 2603.1 | 962.11 | 1841.0 | 2803.1 | 2.5547 | 6.2575 | 25.0 |
| 30.0 | 233.9 | 1.2165 | 0.06668 | 1004.8 | 2604.1 | 1008.4 | 1795.7 | 2804.2 | 2.6457 | 6.1869 | 30.0 |
| 35.0 | 242.6 | 1.2347 | 0.05707 | 1045.4 | 2603.7 | 1049.8 | 1753.7 | 2803.4 | 2.7253 | 6.1253 | 35.0 |
| 40.0 | 250.4 | 1.2522 | 0.04978 | 1082.3 | 2602.3 | 1087.3 | 1714.1 | 2801.4 | 2.7964 | 6.0701 | 40.0 |
| 45.0 | 257.5 | 1.2692 | 0.04406 | 1116.2 | 2600.1 | 1121.9 | 1676.4 | 2798.3 | 2.8610 | 6.0199 | 45.0 |
| 50.0 | 264.0 | 1.2859 | 0.03944 | 1147.8 | 2597.1 | 1154.2 | 1640.1 | 2794.3 | 2.9202 | 5.9734 | 50.0 |
| 60.0 | 275.6 | 1.3187 | 0.03244 | 1205.4 | 2589.7 | 1213.4 | 1571.0 | 2784.3 | 3.0267 | 5.8892 | 60.0 |
| 70.0 | 285.9 | 1.3513 | 0.02737 | 1257.6 | 2580.5 | 1267.0 | 1505.1 | 2772.1 | 3.1211 | 5.8133 | 70.0 |
| 80.0 | 295.1 | 1.3842 | 0.02352 | 1305.6 | 2569.8 | 1316.6 | 1441.3 | 2758.0 | 3.2068 | 5.7432 | 80.0 |
| 90.0 | 303.4 | 1.4178 | 0.02048 | 1350.5 | 2557.8 | 1363.3 | 1378.9 | 2742.1 | 3.2858 | 5.6772 | 90.0 |
| 100. | 311.1 | 1.4524 | 0.01803 | 1393.0 | 2544.4 | 1407.6 | 1317.1 | 2724.7 | 3.3596 | 5.6141 | 100. |
| 110. | 318.2 | 1.4886 | 0.01599 | 1433.7 | 2529.8 | 1450.1 | 1255.5 | 2705.6 | 3.4295 | 5.5527 | 110. |
| 120. | 324.8 | 1.5267 | 0.01426 | 1473.0 | 2513.7 | 1491.3 | 1193.6 | 2684.9 | 3.4962 | 5.4924 | 120. |
| 130. | 330.9 | 1.5671 | 0.01278 | 1511.1 | 2496.1 | 1531.5 | 1130.7 | 2662.2 | 3.5606 | 5.4323 | 130. |
| 140. | 336.8 | 1.6107 | 0.01149 | 1548.6 | 2476.8 | 1571.1 | 1066.5 | 2637.6 | 3.6232 | 5.3717 | 140. |
| 150. | 342.2 | 1.6581 | 0.01034 | 1585.6 | 2455.5 | 1610.5 | 1000.0 | 2610.5 | 3.6848 | 5.3098 | 150. |
| 160. | 347.4 | 1.7107 | 0.009306 | 1622.7 | 2431.7 | 1650.1 | 930.6 | 2580.6 | 3.7461 | 5.2455 | 160. |
| 170. | 352.4 | 1.7702 | 0.008364 | 1660.2 | 2405.0 | 1690.3 | 856.9 | 2547.2 | 3.8079 | 5.1777 | 170. |
| 180. | 357.1 | 1.8397 | 0.007489 | 1698.9 | 2374.3 | 1732.0 | 777.1 | 2509.1 | 3.8715 | 5.1044 | 180. |
| 190. | 361.5 | 1.9243 | 0.006657 | 1739.9 | 2338.1 | 1776.5 | 688.0 | 2464.5 | 3.9388 | 5.0228 | 190. |
| 200. | 365.8 | 2.036 | 0.005834 | 1785.6 | 2293.0 | 1826.3 | 583.4 | 2409.7 | 4.0139 | 4.9269 | 200. |
| 220.9 | 374.1 | 3.155 | 0.003155 | 2029.6 | 2029.6 | 2099.3 | 0 | 2099.3 | 4.4298 | 4.4298 | 220.9 |

**Table T-4** Properties of Superheated Water Vapor

| $T$ °C | $v$ m³/kg | $u$ kJ/kg | $h$ kJ/kg | $s$ kJ/kg · K | $v$ m³/kg | $u$ kJ/kg | $h$ kJ/kg | $s$ kJ/kg · K | $v$ m³/kg | $u$ kJ/kg | $h$ kJ/kg | $s$ kJ/kg · K |
|---|---|---|---|---|---|---|---|---|---|---|---|---|
| | $p$ = 0.06 bar = 0.006 MPa ($T_{sat}$ = 36.16°C) | | | | $p$ = 0.35 bar = 0.035 MPa ($T_{sat}$ = 72.69°C) | | | | $p$ = 0.70 bar = 0.07 MPa ($T_{sat}$ = 89.95°C) | | | |
| Sat. | 23.739 | 2425.0 | 2567.4 | 8.3304 | 4.526 | 2473.0 | 2631.4 | 7.7158 | 2.365 | 2494.5 | 2660.0 | 7.4797 |
| 80 | 27.132 | 2487.3 | 2650.1 | 8.5804 | 4.625 | 2483.7 | 2645.6 | 7.7564 | 2.434 | 2509.7 | 2680.0 | 7.5341 |
| 120 | 30.219 | 2544.7 | 2726.0 | 8.7840 | 5.163 | 2542.4 | 2723.1 | 7.9644 | 2.571 | 2539.7 | 2719.6 | 7.6375 |
| 160 | 33.302 | 2602.7 | 2802.5 | 8.9693 | 5.696 | 2601.2 | 2800.6 | 8.1519 | 2.841 | 2599.4 | 2798.2 | 7.8279 |
| 200 | 36.383 | 2661.4 | 2879.7 | 9.1398 | 6.228 | 2660.4 | 2878.4 | 8.3237 | 3.108 | 2659.1 | 2876.7 | 8.0012 |
| 240 | 39.462 | 2721.0 | 2957.8 | 9.2982 | 6.758 | 2720.3 | 2956.8 | 8.4828 | 3.374 | 2719.3 | 2955.5 | 8.1611 |
| 280 | 42.540 | 2781.5 | 3036.8 | 9.4464 | 7.287 | 2780.9 | 3036.0 | 8.6314 | 3.640 | 2780.2 | 3035.0 | 8.3162 |
| 320 | 45.618 | 2843.0 | 3116.7 | 9.5859 | 7.815 | 2842.5 | 3116.1 | 8.7712 | 3.905 | 2842.0 | 3115.3 | 8.4504 |
| 360 | 48.696 | 2905.5 | 3197.7 | 9.7180 | 8.344 | 2905.1 | 3197.1 | 8.9034 | 4.170 | 2904.6 | 3196.5 | 8.5828 |
| 400 | 51.774 | 2969.0 | 3279.6 | 9.8435 | 8.872 | 2968.6 | 3279.2 | 9.0291 | 4.434 | 2968.2 | 3278.6 | 8.7086 |
| 440 | 54.851 | 3033.5 | 3362.6 | 9.9633 | 9.400 | 3033.2 | 3362.2 | 9.1490 | 4.698 | 3032.9 | 3361.8 | 8.8286 |
| 500 | 59.467 | 3132.3 | 3489.1 | 10.1336 | 10.192 | 3132.1 | 3488.8 | 9.3194 | 5.095 | 3131.8 | 3488.5 | 8.9991 |
| | $p$ = 1.0 bar = 0.10 MPa ($T_{sat}$ = 99.63°C) | | | | $p$ = 1.5 bar = 0.15 MPa ($T_{sat}$ = 111.37°C) | | | | $p$ = 3.0 bar = 0.30 MPa ($T_{sat}$ = 133.55°C) | | | |
| Sat. | 1.694 | 2506.1 | 2675.5 | 7.3594 | 1.159 | 2519.7 | 2693.6 | 7.2233 | 0.606 | 2543.6 | 2725.3 | 6.9919 |
| 100 | 1.696 | 2506.7 | 2676.2 | 7.3614 | | | | | | | | |
| 120 | 1.793 | 2537.3 | 2716.6 | 7.4668 | 1.188 | 2533.3 | 2711.4 | 7.2693 | | | | |
| 160 | 1.984 | 2597.8 | 2796.2 | 7.6597 | 1.317 | 2595.2 | 2792.8 | 7.4665 | 0.651 | 2587.1 | 2782.3 | 7.1276 |
| 200 | 2.172 | 2658.1 | 2875.3 | 7.8343 | 1.444 | 2656.2 | 2872.9 | 7.6433 | 0.716 | 2650.7 | 2865.5 | 7.3115 |
| 240 | 2.359 | 2718.5 | 2954.5 | 7.9949 | 1.570 | 2717.2 | 2952.7 | 7.8052 | 0.781 | 2713.1 | 2947.3 | 7.4774 |
| 280 | 2.546 | 2779.6 | 3034.2 | 8.1445 | 1.695 | 2778.6 | 3032.8 | 7.9555 | 0.844 | 2775.4 | 3028.6 | 7.6299 |
| 320 | 2.732 | 2841.5 | 3114.6 | 8.2849 | 1.819 | 2840.6 | 3113.5 | 8.0964 | 0.907 | 2838.1 | 3110.1 | 7.7722 |
| 360 | 2.917 | 2904.2 | 3195.9 | 8.4175 | 1.943 | 2903.5 | 3195.0 | 8.2293 | 0.969 | 2901.4 | 3192.2 | 7.9061 |
| 400 | 3.103 | 2967.9 | 3278.2 | 8.5435 | 2.067 | 2967.3 | 3277.4 | 8.3555 | 1.032 | 2965.6 | 3275.0 | 8.0330 |
| 440 | 3.288 | 3032.6 | 3361.4 | 8.6636 | 2.191 | 3032.1 | 3360.7 | 8.4757 | 1.094 | 3030.6 | 3358.7 | 8.1538 |
| 500 | 3.565 | 3131.6 | 3488.1 | 8.8342 | 2.376 | 3131.2 | 3487.6 | 8.6466 | 1.187 | 3130.0 | 3486.0 | 8.3251 |
| | $p$ = 5.0 bar = 0.50 MPa ($T_{sat}$ = 151.86°C) | | | | $p$ = 7.0 bar = 0.70 MPa ($T_{sat}$ = 164.97°C) | | | | $p$ = 10.0 bar = 1.0 MPa ($T_{sat}$ = 179.91°C) | | | |
| Sat. | 0.3749 | 2561.2 | 2748.7 | 6.8213 | 0.2729 | 2572.5 | 2763.5 | 6.7080 | 0.1944 | 2583.6 | 2778.1 | 6.5865 |
| 180 | 0.4045 | 2609.7 | 2812.0 | 6.9656 | 0.2847 | 2599.8 | 2799.1 | 6.7880 | | | | |
| 200 | 0.4249 | 2642.9 | 2855.4 | 7.0592 | 0.2999 | 2634.8 | 2844.8 | 6.8865 | 0.2060 | 2621.9 | 2827.9 | 6.6940 |
| 240 | 0.4646 | 2707.6 | 2939.9 | 7.2307 | 0.3292 | 2701.8 | 2932.2 | 7.0641 | 0.2275 | 2692.9 | 2920.4 | 6.8817 |
| 280 | 0.5034 | 2771.2 | 3022.9 | 7.3865 | 0.3574 | 2766.9 | 3017.1 | 7.2233 | 0.2480 | 2760.2 | 3008.2 | 7.0465 |
| 320 | 0.5416 | 2834.7 | 3105.6 | 7.5308 | 0.3852 | 2831.3 | 3100.9 | 7.3697 | 0.2678 | 2826.1 | 3093.9 | 7.1962 |
| 360 | 0.5796 | 2898.7 | 3188.4 | 7.6660 | 0.4126 | 2895.8 | 3184.7 | 7.5063 | 0.2873 | 2891.6 | 3178.9 | 7.3349 |
| 400 | 0.6173 | 2963.2 | 3271.9 | 7.7938 | 0.4397 | 2960.9 | 3268.7 | 7.6350 | 0.3066 | 2957.3 | 3263.9 | 7.4651 |
| 440 | 0.6548 | 3028.6 | 3356.0 | 7.9152 | 0.4667 | 3026.6 | 3353.3 | 7.7571 | 0.3257 | 3023.6 | 3349.3 | 7.5883 |
| 500 | 0.7109 | 3128.4 | 3483.9 | 8.0873 | 0.5070 | 3126.8 | 3481.7 | 7.9299 | 0.3541 | 3124.4 | 3478.5 | 7.7622 |
| 600 | 0.8041 | 3299.6 | 3701.7 | 8.3522 | 0.5738 | 3298.5 | 3700.2 | 8.1956 | 0.4011 | 3296.8 | 3697.9 | 8.0290 |

**Table T-4** *(Continued)*

| $T$ °C | $v$ m³/kg | $u$ kJ/kg | $h$ kJ/kg | $s$ kJ/kg · K | $v$ m³/kg | $u$ kJ/kg | $h$ kJ/kg | $s$ kJ/kg · K | $v$ m³/kg | $u$ kJ/kg | $h$ kJ/kg | $s$ kJ/kg · K |
|---|---|---|---|---|---|---|---|---|---|---|---|---|
| | $p$ = 15.0 bar = 1.5 MPa ($T_{sat}$ = 198.32°C) | | | | $p$ = 20.0 bar = 2.0 MPa ($T_{sat}$ = 212.42°C) | | | | $p$ = 30.0 bar = 3.0 MPa ($T_{sat}$ = 233.90°C) | | | |
| Sat. | 0.1318 | 2594.5 | 2792.2 | 6.4448 | 0.0996 | 2600.3 | 2799.5 | 6.3409 | 0.0667 | 2604.1 | 2804.2 | 6.1869 |
| 200 | 0.1325 | 2598.1 | 2796.8 | 6.4546 | | | | | | | | |
| 240 | 0.1483 | 2676.9 | 2899.3 | 6.6628 | 0.1085 | 2659.6 | 2876.5 | 6.4952 | 0.0682 | 2619.7 | 2824.3 | 6.2265 |
| 280 | 0.1627 | 2748.6 | 2992.7 | 6.8381 | 0.1200 | 2736.4 | 2976.4 | 6.6828 | 0.0771 | 2709.9 | 2941.3 | 6.4462 |
| 320 | 0.1765 | 2817.1 | 3081.9 | 6.9938 | 0.1308 | 2807.9 | 3069.5 | 6.8452 | 0.0850 | 2788.4 | 3043.4 | 6.6245 |
| 360 | 0.1899 | 2884.4 | 3169.2 | 7.1363 | 0.1411 | 2877.0 | 3159.3 | 6.9917 | 0.0923 | 2861.7 | 3138.7 | 6.7801 |
| 400 | 0.2030 | 2951.3 | 3255.8 | 7.2690 | 0.1512 | 2945.2 | 3247.6 | 7.1271 | 0.0994 | 2932.8 | 3230.9 | 6.9212 |
| 440 | 0.2160 | 3018.5 | 3342.5 | 7.3940 | 0.1611 | 3013.4 | 3335.5 | 7.2540 | 0.1062 | 3002.9 | 3321.5 | 7.0520 |
| 500 | 0.2352 | 3120.3 | 3473.1 | 7.5698 | 0.1757 | 3116.2 | 3467.6 | 7.4317 | 0.1162 | 3108.0 | 3456.5 | 7.2338 |
| 540 | 0.2478 | 3189.1 | 3560.9 | 7.6805 | 0.1853 | 3185.6 | 3556.1 | 7.5434 | 0.1227 | 3178.4 | 3546.6 | 7.3474 |
| 600 | 0.2668 | 3293.9 | 3694.0 | 7.8385 | 0.1996 | 3290.9 | 3690.1 | 7.7024 | 0.1324 | 3285.0 | 3682.3 | 7.5085 |
| 640 | 0.2793 | 3364.8 | 3783.8 | 7.9391 | 0.2091 | 3362.2 | 3780.4 | 7.8035 | 0.1388 | 3357.0 | 3773.5 | 7.6106 |
| | $p$ = 40 bar = 4.0 MPa ($T_{sat}$ = 250.4°C) | | | | $p$ = 60 bar = 6.0 MPa ($T_{sat}$ = 275.64°C) | | | | $p$ = 80 bar = 8.0 MPa ($T_{sat}$ = 295.06°C) | | | |
| Sat. | 0.04978 | 2602.3 | 2801.4 | 6.0701 | 0.03244 | 2589.7 | 2784.3 | 5.8892 | 0.02352 | 2569.8 | 2758.0 | 5.7432 |
| 280 | 0.05546 | 2680.0 | 2901.8 | 6.2568 | 0.03317 | 2605.2 | 2804.2 | 5.9252 | | | | |
| 320 | 0.06199 | 2767.4 | 3015.4 | 6.4553 | 0.03876 | 2720.0 | 2952.6 | 6.1846 | 0.02682 | 2662.7 | 2877.2 | 5.9489 |
| 360 | 0.06788 | 2845.7 | 3117.2 | 6.6215 | 0.04331 | 2811.2 | 3071.1 | 6.3782 | 0.03089 | 2772.7 | 3019.8 | 6.1819 |
| 400 | 0.07341 | 2919.9 | 3213.6 | 6.7690 | 0.04739 | 2892.9 | 3177.2 | 6.5408 | 0.03432 | 2863.8 | 3138.3 | 6.3634 |
| 440 | 0.07872 | 2992.2 | 3307.1 | 6.9041 | 0.05122 | 2970.0 | 3277.3 | 6.6853 | 0.03742 | 2946.7 | 3246.1 | 6.5190 |
| 500 | 0.08643 | 3099.5 | 3445.3 | 7.0901 | 0.05665 | 3082.2 | 3422.2 | 6.8803 | 0.04175 | 3064.3 | 3398.3 | 6.7240 |
| 540 | 0.09145 | 3171.1 | 3536.9 | 7.2056 | 0.06015 | 3156.1 | 3517.0 | 6.9999 | 0.04448 | 3140.8 | 3496.7 | 6.8481 |
| 600 | 0.09885 | 3279.1 | 3674.4 | 7.3688 | 0.06525 | 3266.9 | 3658.4 | 7.1677 | 0.04845 | 3254.4 | 3642.0 | 7.0206 |
| 640 | 0.1037 | 3351.8 | 3766.6 | 7.4720 | 0.06859 | 3341.0 | 3752.6 | 7.2731 | 0.05102 | 3330.1 | 3738.3 | 7.1283 |
| 700 | 0.1110 | 3462.1 | 3905.9 | 7.6198 | 0.07352 | 3453.1 | 3894.1 | 7.4234 | 0.05481 | 3443.9 | 3882.4 | 7.2812 |
| 740 | 0.1157 | 3536.6 | 3999.6 | 7.7141 | 0.07677 | 3528.3 | 3989.2 | 7.5190 | 0.05729 | 3520.4 | 3978.7 | 7.3782 |
| | $p$ = 100 bar = 10.0 MPa ($T_{sat}$ = 311.06°C) | | | | $p$ = 120 bar = 12.0 MPa ($T_{sat}$ = 324.75°C) | | | | $p$ = 140 bar = 14.0 MPa ($T_{sat}$ = 336.75°C) | | | |
| Sat. | 0.01803 | 2544.4 | 2724.7 | 5.6141 | 0.01426 | 2513.7 | 2684.9 | 5.4924 | 0.01149 | 2476.8 | 2637.6 | 5.3717 |
| 320 | 0.01925 | 2588.8 | 2781.3 | 5.7103 | | | | | | | | |
| 360 | 0.02331 | 2729.1 | 2962.1 | 6.0060 | 0.01811 | 2678.4 | 2895.7 | 5.8361 | 0.01422 | 2617.4 | 2816.5 | 5.6602 |
| 400 | 0.02641 | 2832.4 | 3096.5 | 6.2120 | 0.02108 | 2798.3 | 3051.3 | 6.0747 | 0.01722 | 2760.9 | 3001.9 | 5.9448 |
| 440 | 0.02911 | 2922.1 | 3213.2 | 6.3805 | 0.02355 | 2896.1 | 3178.7 | 6.2586 | 0.01954 | 2868.6 | 3142.2 | 6.1474 |
| 480 | 0.03160 | 3005.4 | 3321.4 | 6.5282 | 0.02576 | 2984.4 | 3293.5 | 6.4154 | 0.02157 | 2962.5 | 3264.5 | 6.3143 |
| 520 | 0.03394 | 3085.6 | 3425.1 | 6.6622 | 0.02781 | 3068.0 | 3401.8 | 6.5555 | 0.02343 | 3049.8 | 3377.8 | 6.4610 |
| 560 | 0.03619 | 3164.1 | 3526.0 | 6.7864 | 0.02977 | 3149.0 | 3506.2 | 6.6840 | 0.02517 | 3133.6 | 3486.0 | 6.5941 |
| 600 | 0.03837 | 3241.7 | 3625.3 | 6.9029 | 0.03164 | 3228.7 | 3608.3 | 6.8037 | 0.02683 | 3215.4 | 3591.1 | 6.7172 |
| 640 | 0.04048 | 3318.9 | 3723.7 | 7.0131 | 0.03345 | 3307.5 | 3709.0 | 6.9164 | 0.02843 | 3296.0 | 3694.1 | 6.8326 |
| 700 | 0.04358 | 3434.7 | 3870.5 | 7.1687 | 0.03610 | 3425.2 | 3858.4 | 7.0749 | 0.03075 | 3415.7 | 3846.2 | 6.9939 |
| 740 | 0.04560 | 3512.1 | 3968.1 | 7.2670 | 0.03781 | 3503.7 | 3957.4 | 7.1746 | 0.03225 | 3495.2 | 3946.7 | 7.0952 |

**Table T-4** *(Continued)*

| $T$ °C | $v$ m³/kg | $u$ kJ/kg | $h$ kJ/kg | $s$ kJ/kg · K | $v$ m³/kg | $u$ kJ/kg | $h$ kJ/kg | $s$ kJ/kg · K |
|---|---|---|---|---|---|---|---|---|
| | $p$ = 160 bar = 16.0 MPa ($T_{sat}$ = 347.44°C) | | | | $p$ = 180 bar = 18.0 MPa ($T_{sat}$ = 357.06°C) | | | |
| Sat. | 0.00931 | 2431.7 | 2580.6 | 5.2455 | 0.00749 | 2374.3 | 2509.1 | 5.1044 |
| 360 | 0.01105 | 2539.0 | 2715.8 | 5.4614 | 0.00809 | 2418.9 | 2564.5 | 5.1922 |
| 400 | 0.01426 | 2719.4 | 2947.6 | 5.8175 | 0.01190 | 2672.8 | 2887.0 | 5.6887 |
| 440 | 0.01652 | 2839.4 | 3103.7 | 6.0429 | 0.01414 | 2808.2 | 3062.8 | 5.9428 |
| 480 | 0.01842 | 2939.7 | 3234.4 | 6.2215 | 0.01596 | 2915.9 | 3203.2 | 6.1345 |
| 520 | 0.02013 | 3031.1 | 3353.3 | 6.3752 | 0.01757 | 3011.8 | 3378.0 | 6.2960 |
| 560 | 0.02172 | 3117.8 | 3465.4 | 6.5132 | 0.01904 | 3101.7 | 3444.4 | 6.4392 |
| 600 | 0.02323 | 3201.8 | 3573.5 | 6.6399 | 0.02042 | 3188.0 | 3555.6 | 6.5696 |
| 640 | 0.02467 | 3284.2 | 3678.9 | 6.7580 | 0.02174 | 3272.3 | 3663.6 | 6.6905 |
| 700 | 0.02674 | 3406.0 | 3833.9 | 6.9224 | 0.02362 | 3396.3 | 3821.5 | 6.8580 |
| 740 | 0.02808 | 3486.7 | 3935.9 | 7.0251 | 0.02483 | 3478.0 | 3925.0 | 6.9623 |
| | $p$ = 200 bar = 20.0 MPa ($T_{sat}$ = 365.81°C) | | | | $p$ = 240 bar = 24.0 MPa | | | |
| Sat. | 0.00583 | 2293.0 | 2409.7 | 4.9269 | | | | |
| 400 | 0.00994 | 2619.3 | 2818.1 | 5.5540 | 0.00673 | 2477.8 | 2639.4 | 5.2393 |
| 440 | 0.01222 | 2774.9 | 3019.4 | 5.8450 | 0.00929 | 2700.6 | 2923.4 | 5.6506 |
| 480 | 0.01399 | 2891.2 | 3170.8 | 6.0518 | 0.01100 | 2838.3 | 3102.3 | 5.8950 |
| 520 | 0.01551 | 2992.0 | 3302.2 | 6.2218 | 0.01241 | 2950.5 | 3248.5 | 6.0842 |
| 560 | 0.01689 | 3085.2 | 3423.0 | 6.3705 | 0.01366 | 3051.1 | 3379.0 | 6.2448 |
| 600 | 0.01818 | 3174.0 | 3537.6 | 6.5048 | 0.01481 | 3145.2 | 3500.7 | 6.3875 |
| 640 | 0.01940 | 3260.2 | 3648.1 | 6.6286 | 0.01588 | 3235.5 | 3616.7 | 6.5174 |
| 700 | 0.02113 | 3386.4 | 3809.0 | 6.7993 | 0.01739 | 3366.4 | 3783.8 | 6.6947 |
| 740 | 0.02224 | 3469.3 | 3914.1 | 6.9052 | 0.01835 | 3451.7 | 3892.1 | 6.8038 |
| 800 | 0.02385 | 3592.7 | 4069.7 | 7.0544 | 0.01974 | 3578.0 | 4051.6 | 6.9567 |
| | $p$ = 280 bar = 28.0 MPa | | | | $p$ = 320 bar = 32.0 MPa | | | |
| 400 | 0.00383 | 2223.5 | 2330.7 | 4.7494 | 0.00236 | 1980.4 | 2055.9 | 4.3239 |
| 440 | 0.00712 | 2613.2 | 2812.6 | 5.4494 | 0.00544 | 2509.0 | 2683.0 | 5.2327 |
| 480 | 0.00885 | 2780.8 | 3028.5 | 5.7446 | 0.00722 | 2718.1 | 2949.2 | 5.5968 |
| 520 | 0.01020 | 2906.8 | 3192.3 | 5.9566 | 0.00853 | 2860.7 | 3133.7 | 5.8357 |
| 560 | 0.01136 | 3015.7 | 3333.7 | 6.1307 | 0.00963 | 2979.0 | 3287.2 | 6.0246 |
| 600 | 0.01241 | 3115.6 | 3463.0 | 6.2823 | 0.01061 | 3085.3 | 3424.6 | 6.1858 |
| 640 | 0.01338 | 3210.3 | 3584.8 | 6.4187 | 0.01150 | 3184.5 | 3552.5 | 6.3290 |
| 700 | 0.01473 | 3346.1 | 3758.4 | 6.6029 | 0.01273 | 3325.4 | 3732.8 | 6.5203 |
| 740 | 0.01558 | 3433.9 | 3870.0 | 6.7153 | 0.01350 | 3415.9 | 3847.8 | 6.6361 |
| 800 | 0.01680 | 3563.1 | 4033.4 | 6.8720 | 0.01460 | 3548.0 | 4015.1 | 6.7966 |
| 900 | 0.01873 | 3774.3 | 4298.8 | 7.1084 | 0.01633 | 3762.7 | 4285.1 | 7.0372 |

**Table T-5** Properties of Compressed Liquid Water

| $T$ °C | $v \times 10^3$ m³/kg | $u$ kJ/kg | $h$ kJ/kg | $s$ kJ/kg · K | $v \times 10^3$ m³/kg | $u$ kJ/kg | $h$ kJ/kg | $s$ kJ/kg · K |
|---|---|---|---|---|---|---|---|---|
| | $p$ = 25 bar = 2.5 MPa ($T_{sat}$ = 223.99°C) | | | | $p$ = 50 bar = 5.0 MPa ($T_{sat}$ = 263.99°C) | | | |
| 20 | 1.0006 | 83.80 | 86.30 | .2961 | .9995 | 83.65 | 88.65 | .2956 |
| 40 | 1.0067 | 167.25 | 169.77 | .5715 | 1.0056 | 166.95 | 171.97 | .5705 |
| 80 | 1.0280 | 334.29 | 336.86 | 1.0737 | 1.0268 | 333.72 | 338.85 | 1.0720 |
| 100 | 1.0423 | 418.24 | 420.85 | 1.3050 | 1.0410 | 417.52 | 422.72 | 1.3030 |
| 140 | 1.0784 | 587.82 | 590.52 | 1.7369 | 1.0768 | 586.76 | 592.15 | 1.7343 |
| 180 | 1.1261 | 761.16 | 763.97 | 2.1375 | 1.1240 | 759.63 | 765.25 | 2.1341 |
| 200 | 1.1555 | 849.9 | 852.8 | 2.3294 | 1.1530 | 848.1 | 853.9 | 2.3255 |
| 220 | 1.1898 | 940.7 | 943.7 | 2.5174 | 1.1866 | 938.4 | 944.4 | 2.5128 |
| Sat. | 1.1973 | 959.1 | 962.1 | 2.5546 | 1.2859 | 1147.8 | 1154.2 | 2.9202 |
| | $p$ = 75 bar = 7.5 MPa ($T_{sat}$ = 290.59°C) | | | | $p$ = 100 bar = 10.0 MPa ($T_{sat}$ = 311.06°C) | | | |
| 20 | .9984 | 83.50 | 90.99 | .2950 | .9972 | 83.36 | 93.33 | .2945 |
| 40 | 1.0045 | 166.64 | 174.18 | .5696 | 1.0034 | 166.35 | 176.38 | .5686 |
| 80 | 1.0256 | 333.15 | 340.84 | 1.0704 | 1.0245 | 332.59 | 342.83 | 1.0688 |
| 100 | 1.0397 | 416.81 | 424.62 | 1.3011 | 1.0385 | 416.12 | 426.50 | 1.2992 |
| 140 | 1.0752 | 585.72 | 593.78 | 1.7317 | 1.0737 | 584.68 | 595.42 | 1.7292 |
| 180 | 1.1219 | 758.13 | 766.55 | 2.1308 | 1.1199 | 756.65 | 767.84 | 2.1275 |
| 220 | 1.1835 | 936.2 | 945.1 | 2.5083 | 1.1805 | 934.1 | 945.9 | 2.5039 |
| 260 | 1.2696 | 1124.4 | 1134.0 | 2.8763 | 1.2645 | 1121.1 | 1133.7 | 2.8699 |
| Sat. | 1.3677 | 1282.0 | 1292.2 | 3.1649 | 1.4524 | 1393.0 | 1407.6 | 3.3596 |
| | $p$ = 150 bar = 15.0 MPa ($T_{sat}$ = 342.24°C) | | | | $p$ = 200 bar = 20.0 MPa ($T_{sat}$ = 365.81°C) | | | |
| 20 | .9950 | 83.06 | 97.99 | .2934 | .9928 | 82.77 | 102.62 | .2923 |
| 40 | 1.0013 | 165.76 | 180.78 | .5666 | .9992 | 165.17 | 185.16 | .5646 |
| 80 | 1.0222 | 331.48 | 346.81 | 1.0656 | 1.0199 | 330.40 | 350.80 | 1.0624 |
| 100 | 1.0361 | 414.74 | 430.28 | 1.2955 | 1.0337 | 413.39 | 434.06 | 1.2917 |
| 140 | 1.0707 | 582.66 | 598.72 | 1.7242 | 1.0678 | 580.69 | 602.04 | 1.7193 |
| 180 | 1.1159 | 753.76 | 770.50 | 2.1210 | 1.1120 | 750.95 | 773.20 | 2.1147 |
| 220 | 1.1748 | 929.9 | 947.5 | 2.4953 | 1.1693 | 925.9 | 949.3 | 2.4870 |
| 260 | 1.2550 | 1114.6 | 1133.4 | 2.8576 | 1.2462 | 1108.6 | 1133.5 | 2.8459 |
| 300 | 1.3770 | 1316.6 | 1337.3 | 3.2260 | 1.3596 | 1306.1 | 1333.3 | 3.2071 |
| Sat. | 1.6581 | 1585.6 | 1610.5 | 3.6848 | 2.036 | 1785.6 | 1826.3 | 4.0139 |
| | $p$ = 250 bar = 25 MPa | | | | $p$ = 300 bar = 30.0 MPa | | | |
| 20 | .9907 | 82.47 | 107.24 | .2911 | .9886 | 82.17 | 111.84 | .2899 |
| 40 | .9971 | 164.60 | 189.52 | .5626 | .9951 | 164.04 | 193.89 | .5607 |
| 100 | 1.0313 | 412.08 | 437.85 | 1.2881 | 1.0290 | 410.78 | 441.66 | 1.2844 |
| 200 | 1.1344 | 834.5 | 862.8 | 2.2961 | 1.1302 | 831.4 | 865.3 | 2.2893 |
| 300 | 1.3442 | 1296.6 | 1330.2 | 3.1900 | 1.3304 | 1287.9 | 1327.8 | 3.1741 |

**Table T-6** Properties of Saturated Refrigerant 134a (Liquid–Vapor): Temperature Table

| | | Specific Volume $m^3/kg$ | | Internal Energy kJ/kg | | Enthalpy kJ/kg | | | Entropy kJ/kg · K | | |
|---|---|---|---|---|---|---|---|---|---|---|---|
| Temp. °C | Press. bar | Sat. Liquid $v_f \times 10^3$ | Sat. Vapor $v_g$ | Sat. Liquid $u_f$ | Sat. Vapor $u_g$ | Sat. Liquid $h_f$ | Evap. $h_{fg}$ | Sat. Vapor $h_g$ | Sat. Liquid $s_f$ | Sat. Vapor $s_g$ | Temp. °C |
| −40 | 0.5164 | 0.7055 | 0.3569 | −0.04 | 204.45 | 0.00 | 222.88 | 222.88 | 0.0000 | 0.9560 | −40 |
| −36 | 0.6332 | 0.7113 | 0.2947 | 4.68 | 206.73 | 4.73 | 220.67 | 225.40 | 0.0201 | 0.9506 | −36 |
| −32 | 0.7704 | 0.7172 | 0.2451 | 9.47 | 209.01 | 9.52 | 218.37 | 227.90 | 0.0401 | 0.9456 | −32 |
| −28 | 0.9305 | 0.7233 | 0.2052 | 14.31 | 211.29 | 14.37 | 216.01 | 230.38 | 0.0600 | 0.9411 | −28 |
| −26 | 1.0199 | 0.7265 | 0.1882 | 16.75 | 212.43 | 16.82 | 214.80 | 231.62 | 0.0699 | 0.9390 | −26 |
| −24 | 1.1160 | 0.7296 | 0.1728 | 19.21 | 213.57 | 19.29 | 213.57 | 232.85 | 0.0798 | 0.9370 | −24 |
| −22 | 1.2192 | 0.7328 | 0.1590 | 21.68 | 214.70 | 21.77 | 212.32 | 234.08 | 0.0897 | 0.9351 | −22 |
| −20 | 1.3299 | 0.7361 | 0.1464 | 24.17 | 215.84 | 24.26 | 211.05 | 235.31 | 0.0996 | 0.9332 | −20 |
| −18 | 1.4483 | 0.7395 | 0.1350 | 26.67 | 216.97 | 26.77 | 209.76 | 236.53 | 0.1094 | 0.9315 | −18 |
| −16 | 1.5748 | 0.7428 | 0.1247 | 29.18 | 218.10 | 29.30 | 208.45 | 237.74 | 0.1192 | 0.9298 | −16 |
| −12 | 1.8540 | 0.7498 | 0.1068 | 34.25 | 220.36 | 34.39 | 205.77 | 240.15 | 0.1388 | 0.9267 | −12 |
| −8 | 2.1704 | 0.7569 | 0.0919 | 39.38 | 222.60 | 39.54 | 203.00 | 242.54 | 0.1583 | 0.9239 | −8 |
| −4 | 2.5274 | 0.7644 | 0.0794 | 44.56 | 224.84 | 44.75 | 200.15 | 244.90 | 0.1777 | 0.9213 | −4 |
| 0 | 2.9282 | 0.7721 | 0.0689 | 49.79 | 227.06 | 50.02 | 197.21 | 247.23 | 0.1970 | 0.9190 | 0 |
| 4 | 3.3765 | 0.7801 | 0.0600 | 55.08 | 229.27 | 55.35 | 194.19 | 249.53 | 0.2162 | 0.9169 | 4 |
| 8 | 3.8756 | 0.7884 | 0.0525 | 60.43 | 231.46 | 60.73 | 191.07 | 251.80 | 0.2354 | 0.9150 | 8 |
| 12 | 4.4294 | 0.7971 | 0.0460 | 65.83 | 233.63 | 66.18 | 187.85 | 254.03 | 0.2545 | 0.9132 | 12 |
| 16 | 5.0416 | 0.8062 | 0.0405 | 71.29 | 235.78 | 71.69 | 184.52 | 256.22 | 0.2735 | 0.9116 | 16 |
| 20 | 5.7160 | 0.8157 | 0.0358 | 76.80 | 237.91 | 77.26 | 181.09 | 258.36 | 0.2924 | 0.9102 | 20 |
| 24 | 6.4566 | 0.8257 | 0.0317 | 82.37 | 240.01 | 82.90 | 177.55 | 260.45 | 0.3113 | 0.9089 | 24 |
| 26 | 6.8530 | 0.8309 | 0.0298 | 85.18 | 241.05 | 85.75 | 175.73 | 261.48 | 0.3208 | 0.9082 | 26 |
| 28 | 7.2675 | 0.8362 | 0.0281 | 88.00 | 242.08 | 88.61 | 173.89 | 262.50 | 0.3302 | 0.9076 | 28 |
| 30 | 7.7006 | 0.8417 | 0.0265 | 90.84 | 243.10 | 91.49 | 172.00 | 263.50 | 0.3396 | 0.9070 | 30 |
| 32 | 8.1528 | 0.8473 | 0.0250 | 93.70 | 244.12 | 94.39 | 170.09 | 264.48 | 0.3490 | 0.9064 | 32 |
| 34 | 8.6247 | 0.8530 | 0.0236 | 96.58 | 245.12 | 97.31 | 168.14 | 265.45 | 0.3584 | 0.9058 | 34 |
| 36 | 9.1168 | 0.8590 | 0.0223 | 99.47 | 246.11 | 100.25 | 166.15 | 266.40 | 0.3678 | 0.9053 | 36 |
| 38 | 9.6298 | 0.8651 | 0.0210 | 102.38 | 247.09 | 103.21 | 164.12 | 267.33 | 0.3772 | 0.9047 | 38 |
| 40 | 10.164 | 0.8714 | 0.0199 | 105.30 | 248.06 | 106.19 | 162.05 | 268.24 | 0.3866 | 0.9041 | 40 |
| 42 | 10.720 | 0.8780 | 0.0188 | 108.25 | 249.02 | 109.19 | 159.94 | 269.14 | 0.3960 | 0.9035 | 42 |
| 44 | 11.299 | 0.8847 | 0.0177 | 111.22 | 249.96 | 112.22 | 157.79 | 270.01 | 0.4054 | 0.9030 | 44 |
| 48 | 12.526 | 0.8989 | 0.0159 | 117.22 | 251.79 | 118.35 | 153.33 | 271.68 | 0.4243 | 0.9017 | 48 |
| 52 | 13.851 | 0.9142 | 0.0142 | 123.31 | 253.55 | 124.58 | 148.66 | 273.24 | 0.4432 | 0.9004 | 52 |
| 56 | 15.278 | 0.9308 | 0.0127 | 129.51 | 255.23 | 130.93 | 143.75 | 274.68 | 0.4622 | 0.8990 | 56 |
| 60 | 16.813 | 0.9488 | 0.0114 | 135.82 | 256.81 | 137.42 | 138.57 | 275.99 | 0.4814 | 0.8973 | 60 |
| 70 | 21.162 | 1.0027 | 0.0086 | 152.22 | 260.15 | 154.34 | 124.08 | 278.43 | 0.5302 | 0.8918 | 70 |
| 80 | 26.324 | 1.0766 | 0.0064 | 169.88 | 262.14 | 172.71 | 106.41 | 279.12 | 0.5814 | 0.8827 | 80 |
| 90 | 32.435 | 1.1949 | 0.0046 | 189.82 | 261.34 | 193.69 | 82.63 | 276.32 | 0.6380 | 0.8655 | 90 |
| 100 | 39.742 | 1.5443 | 0.0027 | 218.60 | 248.49 | 224.74 | 34.40 | 259.13 | 0.7196 | 0.8117 | 100 |

*Source:* Tables T-6 through T-8 are calculated based on equations from D. P. Wilson and R. S. Basu, "Thermodynamic Properties of a New Stratospherically Safe Working Fluid—Refrigerant 134a," *ASHRAE Trans.*, Vol. 94, Pt. 2, 1988, pp. 2095–2118.

**Table T-7** Properties of Saturated Refrigerant 134a (Liquid–Vapor): Pressure Table

| | | Specific Volume $m^3/kg$ | | Internal Energy kJ/kg | | Enthalpy kJ/kg | | | Entropy $kJ/kg \cdot K$ | | |
|---|---|---|---|---|---|---|---|---|---|---|---|
| Press. bar | Temp. °C | Sat. Liquid $v_f \times 10^3$ | Sat. Vapor $v_g$ | Sat. Liquid $u_f$ | Sat. Vapor $u_g$ | Sat. Liquid $h_f$ | Evap. $h_{fg}$ | Sat. Vapor $h_g$ | Sat. Liquid $s_f$ | Sat. Vapor $s_g$ | Press. bar |
| 0.6 | −37.07 | 0.7097 | 0.3100 | 3.41 | 206.12 | 3.46 | 221.27 | 224.72 | 0.0147 | 0.9520 | 0.6 |
| 0.8 | −31.21 | 0.7184 | 0.2366 | 10.41 | 209.46 | 10.47 | 217.92 | 228.39 | 0.0440 | 0.9447 | 0.8 |
| 1.0 | −26.43 | 0.7258 | 0.1917 | 16.22 | 212.18 | 16.29 | 215.06 | 231.35 | 0.0678 | 0.9395 | 1.0 |
| 1.2 | −22.36 | 0.7323 | 0.1614 | 21.23 | 214.50 | 21.32 | 212.54 | 233.86 | 0.0879 | 0.9354 | 1.2 |
| 1.4 | −18.80 | 0.7381 | 0.1395 | 25.66 | 216.52 | 25.77 | 210.27 | 236.04 | 0.1055 | 0.9322 | 1.4 |
| 1.6 | −15.62 | 0.7435 | 0.1229 | 29.66 | 218.32 | 29.78 | 208.19 | 237.97 | 0.1211 | 0.9295 | 1.6 |
| 1.8 | −12.73 | 0.7485 | 0.1098 | 33.31 | 219.94 | 33.45 | 206.26 | 239.71 | 0.1352 | 0.9273 | 1.8 |
| 2.0 | −10.09 | 0.7532 | 0.0993 | 36.69 | 221.43 | 36.84 | 204.46 | 241.30 | 0.1481 | 0.9253 | 2.0 |
| 2.4 | −5.37 | 0.7618 | 0.0834 | 42.77 | 224.07 | 42.95 | 201.14 | 244.09 | 0.1710 | 0.9222 | 2.4 |
| 2.8 | −1.23 | 0.7697 | 0.0719 | 48.18 | 226.38 | 48.39 | 198.13 | 246.52 | 0.1911 | 0.9197 | 2.8 |
| 3.2 | 2.48 | 0.7770 | 0.0632 | 53.06 | 228.43 | 53.31 | 195.35 | 248.66 | 0.2089 | 0.9177 | 3.2 |
| 3.6 | 5.84 | 0.7839 | 0.0564 | 57.54 | 230.28 | 57.82 | 192.76 | 250.58 | 0.2251 | 0.9160 | 3.6 |
| 4.0 | 8.93 | 0.7904 | 0.0509 | 61.69 | 231.97 | 62.00 | 190.32 | 252.32 | 0.2399 | 0.9145 | 4.0 |
| 5.0 | 15.74 | 0.8056 | 0.0409 | 70.93 | 235.64 | 71.33 | 184.74 | 256.07 | 0.2723 | 0.9117 | 5.0 |
| 6.0 | 21.58 | 0.8196 | 0.0341 | 78.99 | 238.74 | 79.48 | 179.71 | 259.19 | 0.2999 | 0.9097 | 6.0 |
| 7.0 | 26.72 | 0.8328 | 0.0292 | 86.19 | 241.42 | 86.78 | 175.07 | 261.85 | 0.3242 | 0.9080 | 7.0 |
| 8.0 | 31.33 | 0.8454 | 0.0255 | 92.75 | 243.78 | 93.42 | 170.73 | 264.15 | 0.3459 | 0.9066 | 8.0 |
| 9.0 | 35.53 | 0.8576 | 0.0226 | 98.79 | 245.88 | 99.56 | 166.62 | 266.18 | 0.3656 | 0.9054 | 9.0 |
| 10.0 | 39.39 | 0.8695 | 0.0202 | 104.42 | 247.77 | 105.29 | 162.68 | 267.97 | 0.3838 | 0.9043 | 10.0 |
| 12.0 | 46.32 | 0.8928 | 0.0166 | 114.69 | 251.03 | 115.76 | 155.23 | 270.99 | 0.4164 | 0.9023 | 12.0 |
| 14.0 | 52.43 | 0.9159 | 0.0140 | 123.98 | 253.74 | 125.26 | 148.14 | 273.40 | 0.4453 | 0.9003 | 14.0 |
| 16.0 | 57.92 | 0.9392 | 0.0121 | 132.52 | 256.00 | 134.02 | 141.31 | 275.33 | 0.4714 | 0.8982 | 16.0 |
| 18.0 | 62.91 | 0.9631 | 0.0105 | 140.49 | 257.88 | 142.22 | 134.60 | 276.83 | 0.4954 | 0.8959 | 18.0 |
| 20.0 | 67.49 | 0.9878 | 0.0093 | 148.02 | 259.41 | 149.99 | 127.95 | 277.94 | 0.5178 | 0.8934 | 20.0 |
| 25.0 | 77.59 | 1.0562 | 0.0069 | 165.48 | 261.84 | 168.12 | 111.06 | 279.17 | 0.5687 | 0.8854 | 25.0 |
| 30.0 | 86.22 | 1.1416 | 0.0053 | 181.88 | 262.16 | 185.30 | 92.71 | 278.01 | 0.6156 | 0.8735 | 30.0 |

**Table T-8** Properties of Superheated Refrigerant 134a Vapor

| $T$ °C | $v$ m³/kg | $u$ kJ/kg | $h$ kJ/kg | $s$ kJ/kg · K | $v$ m³/kg | $u$ kJ/kg | $h$ kJ/kg | $s$ kJ/kg · K | $v$ m³/kg | $u$ kJ/kg | $h$ kJ/kg | $s$ kJ/kg · K |
|---|---|---|---|---|---|---|---|---|---|---|---|---|
| | $p$ = 0.6 bar = 0.06 MPa ($T_{sat}$ = −37.07°C) | | | | $p$ = 1.0 bar = 0.10 MPa ($T_{sat}$ = −26.43°C) | | | | $p$ = 1.4 bar = 0.14 MPa ($T_{sat}$ = −18.80°C) | | | |
| Sat. | 0.31003 | 206.12 | 224.72 | 0.9520 | 0.19170 | 212.18 | 231.35 | 0.9395 | 0.13945 | 216.52 | 236.04 | 0.9322 |
| −20 | 0.33536 | 217.86 | 237.98 | 1.0062 | 0.19770 | 216.77 | 236.54 | 0.9602 | | | | |
| −10 | 0.34992 | 224.97 | 245.96 | 1.0371 | 0.20686 | 224.01 | 244.70 | 0.9918 | 0.14549 | 223.03 | 243.40 | 0.9606 |
| 0 | 0.36433 | 232.24 | 254.10 | 1.0675 | 0.21587 | 231.41 | 252.99 | 1.0227 | 0.15219 | 230.55 | 251.86 | 0.9922 |
| 10 | 0.37861 | 239.69 | 262.41 | 1.0973 | 0.22473 | 238.96 | 261.43 | 1.0531 | 0.15875 | 238.21 | 260.43 | 1.0230 |
| 20 | 0.39279 | 247.32 | 270.89 | 1.1267 | 0.23349 | 246.67 | 270.02 | 1.0829 | 0.16520 | 246.01 | 269.13 | 1.0532 |
| 30 | 0.40688 | 255.12 | 279.53 | 1.1557 | 0.24216 | 254.54 | 278.76 | 1.1122 | 0.17155 | 253.96 | 277.97 | 1.0828 |
| 40 | 0.42091 | 263.10 | 288.35 | 1.1844 | 0.25076 | 262.58 | 287.66 | 1.1411 | 0.17783 | 262.06 | 286.96 | 1.1120 |
| 50 | 0.43487 | 271.25 | 297.34 | 1.2126 | 0.25930 | 270.79 | 296.72 | 1.1696 | 0.18404 | 270.32 | 296.09 | 1.1407 |
| 60 | 0.44879 | 279.58 | 306.51 | 1.2405 | 0.26779 | 279.16 | 305.94 | 1.1977 | 0.19020 | 278.74 | 305.37 | 1.1690 |
| 70 | 0.46266 | 288.08 | 315.84 | 1.2681 | 0.27623 | 287.70 | 315.32 | 1.2254 | 0.19633 | 287.32 | 314.80 | 1.1969 |
| 80 | 0.47650 | 296.75 | 325.34 | 1.2954 | 0.28464 | 296.40 | 324.87 | 1.2528 | 0.20241 | 296.06 | 324.39 | 1.2244 |
| 90 | 0.49031 | 305.58 | 335.00 | 1.3224 | 0.29302 | 305.27 | 334.57 | 1.2799 | 0.20846 | 304.95 | 334.14 | 1.2516 |

| $T$ °C | $v$ m³/kg | $u$ kJ/kg | $h$ kJ/kg | $s$ kJ/kg · K | $v$ m³/kg | $u$ kJ/kg | $h$ kJ/kg | $s$ kJ/kg · K | $v$ m³/kg | $u$ kJ/kg | $h$ kJ/kg | $s$ kJ/kg · K |
|---|---|---|---|---|---|---|---|---|---|---|---|---|
| | $p$ = 1.8 bar = 0.18 MPa ($T_{sat}$ = −12.73°C) | | | | $p$ = 2.0 bar = 0.20 MPa ($T_{sat}$ = −10.09°C) | | | | $p$ = 2.4 bar = 0.24 MPa ($T_{sat}$ = −5.37°C) | | | |
| Sat. | 0.10983 | 219.94 | 239.71 | 0.9273 | 0.09933 | 221.43 | 241.30 | 0.9253 | 0.08343 | 224.07 | 244.09 | 0.9222 |
| −10 | 0.11135 | 222.02 | 242.06 | 0.9362 | 0.09938 | 221.50 | 241.38 | 0.9256 | | | | |
| 0 | 0.11678 | 229.67 | 250.69 | 0.9684 | 0.10438 | 229.23 | 250.10 | 0.9582 | 0.08574 | 228.31 | 248.89 | 0.9399 |
| 10 | 0.12207 | 237.44 | 259.41 | 0.9998 | 0.10922 | 237.05 | 258.89 | 0.9898 | 0.08993 | 236.26 | 257.84 | 0.9721 |
| 20 | 0.12723 | 245.33 | 268.23 | 1.0304 | 0.11394 | 244.99 | 267.78 | 1.0206 | 0.09399 | 244.30 | 266.85 | 1.0034 |
| 30 | 0.13230 | 253.36 | 277.17 | 1.0604 | 0.11856 | 253.06 | 276.77 | 1.0508 | 0.09794 | 252.45 | 275.95 | 1.0339 |
| 40 | 0.13730 | 261.53 | 286.24 | 1.0898 | 0.12311 | 261.26 | 285.88 | 1.0804 | 0.10181 | 260.72 | 285.16 | 1.0637 |
| 50 | 0.14222 | 269.85 | 295.45 | 1.1187 | 0.12758 | 269.61 | 295.12 | 1.1094 | 0.10562 | 269.12 | 294.47 | 1.0930 |
| 60 | 0.14710 | 278.31 | 304.79 | 1.1472 | 0.13201 | 278.10 | 304.50 | 1.1380 | 0.10937 | 277.67 | 303.91 | 1.1218 |
| 70 | 0.15193 | 286.93 | 314.28 | 1.1753 | 0.13639 | 286.74 | 314.02 | 1.1661 | 0.11307 | 286.35 | 313.49 | 1.1501 |
| 80 | 0.15672 | 295.71 | 323.92 | 1.2030 | 0.14073 | 295.53 | 323.68 | 1.1939 | 0.11674 | 295.18 | 323.19 | 1.1780 |
| 90 | 0.16148 | 304.63 | 333.70 | 1.2303 | 0.14504 | 304.47 | 333.48 | 1.2212 | 0.12037 | 304.15 | 333.04 | 1.2055 |
| 100 | 0.16622 | 313.72 | 343.63 | 1.2573 | 0.14932 | 313.57 | 343.43 | 1.2483 | 0.12398 | 313.27 | 343.03 | 1.2326 |

| $T$ °C | $v$ m³/kg | $u$ kJ/kg | $h$ kJ/kg | $s$ kJ/kg · K | $v$ m³/kg | $u$ kJ/kg | $h$ kJ/kg | $s$ kJ/kg · K | $v$ m³/kg | $u$ kJ/kg | $h$ kJ/kg | $s$ kJ/kg · K |
|---|---|---|---|---|---|---|---|---|---|---|---|---|
| | $p$ = 2.8 bar = 0.28 MPa ($T_{sat}$ = −1.23°C) | | | | $p$ = 3.2 bar = 0.32 MPa ($T_{sat}$ = 2.48°C) | | | | $p$ = 4.0 bar = 0.40 MPa ($T_{sat}$ = 8.93°C) | | | |
| Sat. | 0.07193 | 226.38 | 246.52 | 0.9197 | 0.06322 | 228.43 | 248.66 | 0.9177 | 0.05089 | 231.97 | 252.32 | 0.9145 |
| 0 | 0.07240 | 227.37 | 247.64 | 0.9238 | | | | | | | | |
| 10 | 0.07613 | 235.44 | 256.76 | 0.9566 | 0.06576 | 234.61 | 255.65 | 0.9427 | 0.05119 | 232.87 | 253.35 | 0.9182 |
| 20 | 0.07972 | 243.59 | 265.91 | 0.9883 | 0.06901 | 242.87 | 264.95 | 0.9749 | 0.05397 | 241.37 | 262.96 | 0.9515 |
| 30 | 0.08320 | 251.83 | 275.12 | 1.0192 | 0.07214 | 251.19 | 274.28 | 1.0062 | 0.05662 | 249.89 | 272.54 | 0.9837 |
| 40 | 0.08660 | 260.17 | 284.42 | 1.0494 | 0.07518 | 259.61 | 283.67 | 1.0367 | 0.05917 | 258.47 | 282.14 | 1.0148 |
| 50 | 0.08992 | 268.64 | 293.81 | 1.0789 | 0.07815 | 268.14 | 293.15 | 1.0665 | 0.06164 | 267.13 | 291.79 | 1.0452 |
| 60 | 0.09319 | 277.23 | 303.32 | 1.1079 | 0.08106 | 276.79 | 302.72 | 1.0957 | 0.06405 | 275.89 | 301.51 | 1.0748 |
| 70 | 0.09641 | 285.96 | 312.95 | 1.1364 | 0.08392 | 285.56 | 312.41 | 1.1243 | 0.06641 | 284.75 | 311.32 | 1.1038 |
| 80 | 0.09960 | 294.82 | 322.71 | 1.1644 | 0.08674 | 294.46 | 322.22 | 1.1525 | 0.06873 | 293.73 | 321.23 | 1.1322 |
| 90 | 0.10275 | 303.83 | 332.60 | 1.1920 | 0.08953 | 303.50 | 332.15 | 1.1802 | 0.07102 | 302.84 | 331.25 | 1.1602 |
| 100 | 0.10587 | 312.98 | 342.62 | 1.2193 | 0.09229 | 312.68 | 342.21 | 1.2076 | 0.07327 | 312.07 | 341.38 | 1.1878 |
| 110 | 0.10897 | 322.27 | 352.78 | 1.2461 | 0.09503 | 322.00 | 352.40 | 1.2345 | 0.07550 | 321.44 | 351.64 | 1.2149 |
| 120 | 0.11205 | 331.71 | 363.08 | 1.2727 | 0.09774 | 331.45 | 362.73 | 1.2611 | 0.07771 | 330.94 | 362.03 | 1.2417 |

**Table T-8** *(Continued)*

| $T$ °C | $v$ m³/kg | $u$ kJ/kg | $h$ kJ/kg | $s$ kJ/kg · K | $v$ m³/kg | $u$ kJ/kg | $h$ kJ/kg | $s$ kJ/kg · K | $v$ m³/kg | $u$ kJ/kg | $h$ kJ/kg | $s$ kJ/kg · K |
|---|---|---|---|---|---|---|---|---|---|---|---|---|
| | $p$ = 5.0 bar = 0.50 MPa ($T_{sat}$ = 15.74°C) | | | | $p$ = 6.0 bar = 0.60 MPa ($T_{sat}$ = 21.58°C) | | | | $p$ = 7.0 bar = 0.70 MPa ($T_{sat}$ = 26.72°C) | | | |
| Sat. | 0.04086 | 235.64 | 256.07 | 0.9117 | 0.03408 | 238.74 | 259.19 | 0.9097 | 0.02918 | 241.42 | 261.85 | 0.9080 |
| 20 | 0.04188 | 239.40 | 260.34 | 0.9264 | | | | | | | | |
| 30 | 0.04416 | 248.20 | 270.28 | 0.9597 | 0.03581 | 246.41 | 267.89 | 0.9388 | 0.02979 | 244.51 | 265.37 | 0.9197 |
| 40 | 0.04633 | 256.99 | 280.16 | 0.9918 | 0.03774 | 255.45 | 278.09 | 0.9719 | 0.03157 | 253.83 | 275.93 | 0.9539 |
| 50 | 0.04842 | 265.83 | 290.04 | 1.0229 | 0.03958 | 264.48 | 288.23 | 1.0037 | 0.03324 | 263.08 | 286.35 | 0.9867 |
| 60 | 0.05043 | 274.73 | 299.95 | 1.0531 | 0.04134 | 273.54 | 298.35 | 1.0346 | 0.03482 | 272.31 | 296.69 | 1.0182 |
| 70 | 0.05240 | 283.72 | 309.92 | 1.0825 | 0.04304 | 282.66 | 308.48 | 1.0645 | 0.03634 | 281.57 | 307.01 | 1.0487 |
| 80 | 0.05432 | 292.80 | 319.96 | 1.1114 | 0.04469 | 291.86 | 318.67 | 1.0938 | 0.03781 | 290.88 | 317.35 | 1.0784 |
| 90 | 0.05620 | 302.00 | 330.10 | 1.1397 | 0.04631 | 301.14 | 328.93 | 1.1225 | 0.03924 | 300.27 | 327.74 | 1.1074 |
| 100 | 0.05805 | 311.31 | 340.33 | 1.1675 | 0.04790 | 310.53 | 339.27 | 1.1505 | 0.04064 | 309.74 | 338.19 | 1.1358 |
| 110 | 0.05988 | 320.74 | 350.68 | 1.1949 | 0.04946 | 320.03 | 349.70 | 1.1781 | 0.04201 | 319.31 | 348.71 | 1.1637 |
| 120 | 0.06168 | 330.30 | 361.14 | 1.2218 | 0.05099 | 329.64 | 360.24 | 1.2053 | 0.04335 | 328.98 | 359.33 | 1.1910 |
| 130 | 0.06347 | 339.98 | 371.72 | 1.2484 | 0.05251 | 339.38 | 370.88 | 1.2320 | 0.04468 | 338.76 | 370.04 | 1.2179 |
| 140 | 0.06524 | 349.79 | 382.42 | 1.2746 | 0.05402 | 349.23 | 381.64 | 1.2584 | 0.04599 | 348.66 | 380.86 | 1.2444 |
| | $p$ = 8.0 bar = 0.80 MPa ($T_{sat}$ = 31.33°C) | | | | $p$ = 9.0 bar = 0.90 MPa ($T_{sat}$ = 35.53°C) | | | | $p$ = 10.0 bar = 1.00 MPa ($T_{sat}$ = 39.39°C) | | | |
| Sat. | 0.02547 | 243.78 | 264.15 | 0.9066 | 0.02255 | 245.88 | 266.18 | 0.9054 | 0.02020 | 247.77 | 267.97 | 0.9043 |
| 40 | 0.02691 | 252.13 | 273.66 | 0.9374 | 0.02325 | 250.32 | 271.25 | 0.9217 | 0.02029 | 248.39 | 268.68 | 0.9066 |
| 50 | 0.02846 | 261.62 | 284.39 | 0.9711 | 0.02472 | 260.09 | 282.34 | 0.9566 | 0.02171 | 258.48 | 280.19 | 0.9428 |
| 60 | 0.02992 | 271.04 | 294.98 | 1.0034 | 0.02609 | 269.72 | 293.21 | 0.9897 | 0.02301 | 268.35 | 291.36 | 0.9768 |
| 70 | 0.03131 | 280.45 | 305.50 | 1.0345 | 0.02738 | 279.30 | 303.94 | 1.0214 | 0.02423 | 278.11 | 302.34 | 1.0093 |
| 80 | 0.03264 | 289.89 | 316.00 | 1.0647 | 0.02861 | 288.87 | 314.62 | 1.0521 | 0.02538 | 287.82 | 313.20 | 1.0405 |
| 90 | 0.03393 | 299.37 | 326.52 | 1.0940 | 0.02980 | 298.46 | 325.28 | 1.0819 | 0.02649 | 297.53 | 324.01 | 1.0707 |
| 100 | 0.03519 | 308.93 | 337.08 | 1.1227 | 0.03095 | 308.11 | 335.96 | 1.1109 | 0.02755 | 307.27 | 334.82 | 1.1000 |
| 110 | 0.03642 | 318.57 | 347.71 | 1.1508 | 0.03207 | 317.82 | 346.68 | 1.1392 | 0.02858 | 317.06 | 345.65 | 1.1286 |
| 120 | 0.03762 | 328.31 | 358.40 | 1.1784 | 0.03316 | 327.62 | 357.47 | 1.1670 | 0.02959 | 326.93 | 356.52 | 1.1567 |
| 130 | 0.03881 | 338.14 | 369.19 | 1.2055 | 0.03423 | 337.52 | 368.33 | 1.1943 | 0.03058 | 336.88 | 367.46 | 1.1841 |
| 140 | 0.03997 | 348.09 | 380.07 | 1.2321 | 0.03529 | 347.51 | 379.27 | 1.2211 | 0.03154 | 346.92 | 378.46 | 1.2111 |
| 150 | 0.04113 | 358.15 | 391.05 | 1.2584 | 0.03633 | 357.61 | 390.31 | 1.2475 | 0.03250 | 357.06 | 389.56 | 1.2376 |
| 160 | 0.04227 | 368.32 | 402.14 | 1.2843 | 0.03736 | 367.82 | 401.44 | 1.2735 | 0.03344 | 367.31 | 400.74 | 1.2638 |
| 170 | 0.04340 | 378.61 | 413.33 | 1.3098 | 0.03838 | 378.14 | 412.68 | 1.2992 | 0.03436 | 377.66 | 412.02 | 1.2895 |
| 180 | 0.04452 | 389.02 | 424.63 | 1.3351 | 0.03939 | 388.57 | 424.02 | 1.3245 | 0.03528 | 388.12 | 423.40 | 1.3149 |
| | $p$ = 12.0 bar = 1.20 MPa ($T_{sat}$ = 46.32°C) | | | | $p$ = 14.0 bar = 1.40 MPa ($T_{sat}$ = 52.43°C) | | | | $p$ = 16.0 bar = 1.60 MPa ($T_{sat}$ = 57.92°C) | | | |
| Sat. | 0.01663 | 251.03 | 270.99 | 0.9023 | 0.01405 | 253.74 | 273.40 | 0.9003 | 0.01208 | 256.00 | 275.33 | 0.8982 |
| 50 | 0.01712 | 254.98 | 275.52 | 0.9164 | | | | | | | | |
| 60 | 0.01835 | 265.42 | 287.44 | 0.9527 | 0.01495 | 262.17 | 283.10 | 0.9297 | 0.01233 | 258.48 | 278.20 | 0.9069 |
| 70 | 0.01947 | 275.59 | 298.96 | 0.9868 | 0.01603 | 272.87 | 295.31 | 0.9658 | 0.01340 | 269.89 | 291.33 | 0.9457 |
| 80 | 0.02051 | 285.62 | 310.24 | 1.0192 | 0.01701 | 283.29 | 307.10 | 0.9997 | 0.01435 | 280.78 | 303.74 | 0.9813 |
| 90 | 0.02150 | 295.59 | 321.39 | 1.0503 | 0.01792 | 293.55 | 318.63 | 1.0319 | 0.01521 | 291.39 | 315.72 | 1.0148 |
| 100 | 0.02244 | 305.54 | 332.47 | 1.0804 | 0.01878 | 303.73 | 330.02 | 1.0628 | 0.01601 | 301.84 | 327.46 | 1.0467 |
| 110 | 0.02335 | 315.50 | 343.52 | 1.1096 | 0.01960 | 313.88 | 341.32 | 1.0927 | 0.01677 | 312.20 | 339.04 | 1.0773 |
| 120 | 0.02423 | 325.51 | 354.58 | 1.1381 | 0.02039 | 324.05 | 352.59 | 1.1218 | 0.01750 | 322.53 | 350.53 | 1.1069 |
| 130 | 0.02508 | 335.58 | 365.68 | 1.1660 | 0.02115 | 334.25 | 363.86 | 1.1501 | 0.01820 | 332.87 | 361.99 | 1.1357 |
| 140 | 0.02592 | 345.73 | 376.83 | 1.1933 | 0.02189 | 344.50 | 375.15 | 1.1777 | 0.01887 | 343.24 | 373.44 | 1.1638 |
| 150 | 0.02674 | 355.95 | 388.04 | 1.2201 | 0.02262 | 354.82 | 386.49 | 1.2048 | 0.01953 | 353.66 | 384.91 | 1.1912 |
| 160 | 0.02754 | 366.27 | 399.33 | 1.2465 | 0.02333 | 365.22 | 397.89 | 1.2315 | 0.02017 | 364.15 | 396.43 | 1.2181 |
| 170 | 0.02834 | 376.69 | 410.70 | 1.2724 | 0.02403 | 375.71 | 409.36 | 1.2576 | 0.02080 | 374.71 | 407.99 | 1.2445 |
| 180 | 0.02912 | 387.21 | 422.16 | 1.2980 | 0.02472 | 386.29 | 420.90 | 1.2834 | 0.02142 | 385.35 | 419.62 | 1.2704 |

**Table T-9** Ideal Gas Properties of Air

$T$(K), $h$ and $u$(kJ/kg), $s^\circ$ (kJ/kg · K)

| | | | | when $\Delta s = 0$[1] | | | | | | when $\Delta s = 0$ | |
|---|---|---|---|---|---|---|---|---|---|---|---|
| $T$ | $h$ | $u$ | $s^\circ$ | $p_r$ | $v_r$ | $T$ | $h$ | $u$ | $s^\circ$ | $p_r$ | $v_r$ |
| 200 | 199.97 | 142.56 | 1.29559 | 0.3363 | 1707. | 600 | 607.02 | 434.78 | 2.40902 | 16.28 | 105.8 |
| 210 | 209.97 | 149.69 | 1.34444 | 0.3987 | 1512. | 610 | 617.53 | 442.42 | 2.42644 | 17.30 | 101.2 |
| 220 | 219.97 | 156.82 | 1.39105 | 0.4690 | 1346. | 620 | 628.07 | 450.09 | 2.44356 | 18.36 | 96.92 |
| 230 | 230.02 | 164.00 | 1.43557 | 0.5477 | 1205. | 630 | 638.63 | 457.78 | 2.46048 | 19.84 | 92.84 |
| 240 | 240.02 | 171.13 | 1.47824 | 0.6355 | 1084. | 640 | 649.22 | 465.50 | 2.47716 | 20.64 | 88.99 |
| 250 | 250.05 | 178.28 | 1.51917 | 0.7329 | 979. | 650 | 659.84 | 473.25 | 2.49364 | 21.86 | 85.34 |
| 260 | 260.09 | 185.45 | 1.55848 | 0.8405 | 887.8 | 660 | 670.47 | 481.01 | 2.50985 | 23.13 | 81.89 |
| 270 | 270.11 | 192.60 | 1.59634 | 0.9590 | 808.0 | 670 | 681.14 | 488.81 | 2.52589 | 24.46 | 78.61 |
| 280 | 280.13 | 199.75 | 1.63279 | 1.0889 | 738.0 | 680 | 691.82 | 496.62 | 2.54175 | 25.85 | 75.50 |
| 285 | 285.14 | 203.33 | 1.65055 | 1.1584 | 706.1 | 690 | 702.52 | 504.45 | 2.55731 | 27.29 | 72.56 |
| 290 | 290.16 | 206.91 | 1.66802 | 1.2311 | 676.1 | 700 | 713.27 | 512.33 | 2.57277 | 28.80 | 69.76 |
| 295 | 295.17 | 210.49 | 1.68515 | 1.3068 | 647.9 | 710 | 724.04 | 520.23 | 2.58810 | 30.38 | 67.07 |
| 300 | 300.19 | 214.07 | 1.70203 | 1.3860 | 621.2 | 720 | 734.82 | 528.14 | 2.60319 | 32.02 | 64.53 |
| 305 | 305.22 | 217.67 | 1.71865 | 1.4686 | 596.0 | 730 | 745.62 | 536.07 | 2.61803 | 33.72 | 62.13 |
| 310 | 310.24 | 221.25 | 1.73498 | 1.5546 | 572.3 | 740 | 756.44 | 544.02 | 2.63280 | 35.50 | 59.82 |
| 315 | 315.27 | 224.85 | 1.75106 | 1.6442 | 549.8 | 750 | 767.29 | 551.99 | 2.64737 | 37.35 | 57.63 |
| 320 | 320.29 | 228.42 | 1.76690 | 1.7375 | 528.6 | 760 | 778.18 | 560.01 | 2.66176 | 39.27 | 55.54 |
| 325 | 325.31 | 232.02 | 1.78249 | 1.8345 | 508.4 | 770 | 789.11 | 568.07 | 2.67595 | 41.31 | 53.39 |
| 330 | 330.34 | 235.61 | 1.79783 | 1.9352 | 489.4 | 780 | 800.03 | 576.12 | 2.69013 | 43.35 | 51.64 |
| 340 | 340.42 | 242.82 | 1.82790 | 2.149 | 454.1 | 790 | 810.99 | 584.21 | 2.70400 | 45.55 | 49.86 |
| 350 | 350.49 | 250.02 | 1.85708 | 2.379 | 422.2 | 800 | 821.95 | 592.30 | 2.71787 | 47.75 | 48.08 |
| 360 | 360.58 | 257.24 | 1.88543 | 2.626 | 393.4 | 820 | 843.98 | 608.59 | 2.74504 | 52.59 | 44.84 |
| 370 | 370.67 | 264.46 | 1.91313 | 2.892 | 367.2 | 840 | 866.08 | 624.95 | 2.77170 | 57.60 | 41.85 |
| 380 | 380.77 | 271.69 | 1.94001 | 3.176 | 343.4 | 860 | 888.27 | 641.40 | 2.79783 | 63.09 | 39.12 |
| 390 | 390.88 | 278.93 | 1.96633 | 3.481 | 321.5 | 880 | 910.56 | 657.95 | 2.82344 | 68.98 | 36.61 |
| 400 | 400.98 | 286.16 | 1.99194 | 3.806 | 301.6 | 900 | 932.93 | 674.58 | 2.84856 | 75.29 | 34.31 |
| 410 | 411.12 | 293.43 | 2.01699 | 4.153 | 283.3 | 920 | 955.38 | 691.28 | 2.87324 | 82.05 | 32.18 |
| 420 | 421.26 | 300.69 | 2.04142 | 4.522 | 266.6 | 940 | 977.92 | 708.08 | 2.89748 | 89.28 | 30.22 |
| 430 | 431.43 | 307.99 | 2.06533 | 4.915 | 251.1 | 960 | 1000.55 | 725.02 | 2.92128 | 97.00 | 28.40 |
| 440 | 441.61 | 315.30 | 2.08870 | 5.332 | 236.8 | 980 | 1023.25 | 741.98 | 2.94468 | 105.2 | 26.73 |
| 450 | 451.80 | 322.62 | 2.11161 | 5.775 | 223.6 | 1000 | 1046.04 | 758.94 | 2.96770 | 114.0 | 25.17 |
| 460 | 462.02 | 329.97 | 2.13407 | 6.245 | 211.4 | 1020 | 1068.89 | 776.10 | 2.99034 | 123.4 | 23.72 |
| 470 | 472.24 | 337.32 | 2.15604 | 6.742 | 200.1 | 1040 | 1091.85 | 793.36 | 3.01260 | 133.3 | 22.39 |
| 480 | 482.49 | 344.70 | 2.17760 | 7.268 | 189.5 | 1060 | 1114.86 | 810.62 | 3.03449 | 143.9 | 21.14 |
| 490 | 492.74 | 352.08 | 2.19876 | 7.824 | 179.7 | 1080 | 1137.89 | 827.88 | 3.05608 | 155.2 | 19.98 |
| 500 | 503.02 | 359.49 | 2.21952 | 8.411 | 170.6 | 1100 | 1161.07 | 845.33 | 3.07732 | 167.1 | 18.896 |
| 510 | 513.32 | 366.92 | 2.23993 | 9.031 | 162.1 | 1120 | 1184.28 | 862.79 | 3.09825 | 179.7 | 17.886 |
| 520 | 523.63 | 374.36 | 2.25997 | 9.684 | 154.1 | 1140 | 1207.57 | 880.35 | 3.11883 | 193.1 | 16.946 |
| 530 | 533.98 | 381.84 | 2.27967 | 10.37 | 146.7 | 1160 | 1230.92 | 897.91 | 3.13916 | 207.2 | 16.064 |
| 540 | 544.35 | 389.34 | 2.29906 | 11.10 | 139.7 | 1180 | 1254.34 | 915.57 | 3.15916 | 222.2 | 15.241 |
| 550 | 554.74 | 396.86 | 2.31809 | 11.86 | 133.1 | 1200 | 1277.79 | 933.33 | 3.17888 | 238.0 | 14.470 |
| 560 | 565.17 | 404.42 | 2.33685 | 12.66 | 127.0 | 1220 | 1301.31 | 951.09 | 3.19834 | 254.7 | 13.747 |
| 570 | 575.59 | 411.97 | 2.35531 | 13.50 | 121.2 | 1240 | 1324.93 | 968.95 | 3.21751 | 272.3 | 13.069 |
| 580 | 586.04 | 419.55 | 2.37348 | 14.38 | 115.7 | 1260 | 1348.55 | 986.90 | 3.23638 | 290.8 | 12.435 |
| 590 | 596.52 | 427.15 | 2.39140 | 15.31 | 110.6 | 1280 | 1372.24 | 1004.76 | 3.25510 | 310.4 | 11.835 |

1. $p_r$ and $v_r$ data for use with Eqs. 7.32 and 7.33, respectively.

**Table T-9** *(Continued)*

$T$(K), $h$ and $u$(kJ/kg), $s°$ (kJ/kg · K)

| | | | | when $\Delta s = 0$ | | | | | | when $\Delta s = 0$ | |
|---|---|---|---|---|---|---|---|---|---|---|---|
| $T$ | $h$ | $u$ | $s°$ | $p_r$ | $v_r$ | $T$ | $h$ | $u$ | $s°$ | $p_r$ | $v_r$ |
| 1300 | 1395.97 | 1022.82 | 3.27345 | 330.9 | 11.275 | 1600 | 1757.57 | 1298.30 | 3.52364 | 791.2 | 5.804 |
| 1320 | 1419.76 | 1040.88 | 3.29160 | 352.5 | 10.747 | 1620 | 1782.00 | 1316.96 | 3.53879 | 834.1 | 5.574 |
| 1340 | 1443.60 | 1058.94 | 3.30959 | 375.3 | 10.247 | 1640 | 1806.46 | 1335.72 | 3.55381 | 878.9 | 5.355 |
| 1360 | 1467.49 | 1077.10 | 3.32724 | 399.1 | 9.780 | 1660 | 1830.96 | 1354.48 | 3.56867 | 925.6 | 5.147 |
| 1380 | 1491.44 | 1095.26 | 3.34474 | 424.2 | 9.337 | 1680 | 1855.50 | 1373.24 | 3.58335 | 974.2 | 4.949 |
| 1400 | 1515.42 | 1113.52 | 3.36200 | 450.5 | 8.919 | 1700 | 1880.1 | 1392.7 | 3.5979 | 1025 | 4.761 |
| 1420 | 1539.44 | 1131.77 | 3.37901 | 478.0 | 8.526 | 1750 | 1941.6 | 1439.8 | 3.6336 | 1161 | 4.328 |
| 1440 | 1563.51 | 1150.13 | 3.39586 | 506.9 | 8.153 | 1800 | 2003.3 | 1487.2 | 3.6684 | 1310 | 3.944 |
| 1460 | 1587.63 | 1168.49 | 3.41247 | 537.1 | 7.801 | 1850 | 2065.3 | 1534.9 | 3.7023 | 1475 | 3.601 |
| 1480 | 1611.79 | 1186.95 | 3.42892 | 568.8 | 7.468 | 1900 | 2127.4 | 1582.6 | 3.7354 | 1655 | 3.295 |
| 1500 | 1635.97 | 1205.41 | 3.44516 | 601.9 | 7.152 | 1950 | 2189.7 | 1630.6 | 3.7677 | 1852 | 3.022 |
| 1520 | 1660.23 | 1223.87 | 3.46120 | 636.5 | 6.854 | 2000 | 2252.1 | 1678.7 | 3.7994 | 2068 | 2.776 |
| 1540 | 1684.51 | 1242.43 | 3.47712 | 672.8 | 6.569 | 2050 | 2314.6 | 1726.8 | 3.8303 | 2303 | 2.555 |
| 1560 | 1708.82 | 1260.99 | 3.49276 | 710.5 | 6.301 | 2100 | 2377.4 | 1775.3 | 3.8605 | 2559 | 2.356 |
| 1580 | 1733.17 | 1279.65 | 3.50829 | 750.0 | 6.046 | 2150 | 2440.3 | 1823.8 | 3.8901 | 2837 | 2.175 |
| | | | | | | 2200 | 2503.2 | 1872.4 | 3.9191 | 3138 | 2.012 |
| | | | | | | 2250 | 2566.4 | 1921.3 | 3.9474 | 3464 | 1.864 |

*Source:* Tables T-9 are based on J. H. Keenan and J. Kaye, *Gas Tables,* Wiley, New York, 1945.

**Table T-10** Ideal Gas Specific Heats of Some Common Gases (kJ/kg · K)

| | $c_p$ | $c_v$ | $c_p$ | $c_v$ | $c_p$ | $c_v$ | $c_p$ | $c_v$ | $c_p$ | $c_v$ | $c_p$ | $c_v$ | |
|---|---|---|---|---|---|---|---|---|---|---|---|---|---|
| Temp. K | Air | | Nitrogen, $N_2$ | | Oxygen, $O_2$ | | Carbon Dioxide, $CO_2$ | | Carbon Monoxide, CO | | Hydrogen, $H_2$ | | Temp. K |
| 250 | 1.003 | 0.716 | 1.039 | 0.742 | 0.913 | 0.653 | 0.791 | 0.602 | 1.039 | 0.743 | 14.051 | 9.927 | 250 |
| 300 | 1.005 | 0.718 | 1.039 | 0.743 | 0.918 | 0.658 | 0.846 | 0.657 | 1.040 | 0.744 | 14.307 | 10.183 | 300 |
| 350 | 1.008 | 0.721 | 1.041 | 0.744 | 0.928 | 0.668 | 0.895 | 0.706 | 1.043 | 0.746 | 14.427 | 10.302 | 350 |
| 400 | 1.013 | 0.726 | 1.044 | 0.747 | 0.941 | 0.681 | 0.939 | 0.750 | 1.047 | 0.751 | 14.476 | 10.352 | 400 |
| 450 | 1.020 | 0.733 | 1.049 | 0.752 | 0.956 | 0.696 | 0.978 | 0.790 | 1.054 | 0.757 | 14.501 | 10.377 | 450 |
| 500 | 1.029 | 0.742 | 1.056 | 0.759 | 0.972 | 0.712 | 1.014 | 0.825 | 1.063 | 0.767 | 14.513 | 10.389 | 500 |
| 550 | 1.040 | 0.753 | 1.065 | 0.768 | 0.988 | 0.728 | 1.046 | 0.857 | 1.075 | 0.778 | 14.530 | 10.405 | 550 |
| 600 | 1.051 | 0.764 | 1.075 | 0.778 | 1.003 | 0.743 | 1.075 | 0.886 | 1.087 | 0.790 | 14.546 | 10.422 | 600 |
| 650 | 1.063 | 0.776 | 1.086 | 0.789 | 1.017 | 0.758 | 1.102 | 0.913 | 1.100 | 0.803 | 14.571 | 10.447 | 650 |
| 700 | 1.075 | 0.788 | 1.098 | 0.801 | 1.031 | 0.771 | 1.126 | 0.937 | 1.113 | 0.816 | 14.604 | 10.480 | 700 |
| 750 | 1.087 | 0.800 | 1.110 | 0.813 | 1.043 | 0.783 | 1.148 | 0.959 | 1.126 | 0.829 | 14.645 | 10.521 | 750 |
| 800 | 1.099 | 0.812 | 1.121 | 0.825 | 1.054 | 0.794 | 1.169 | 0.980 | 1.139 | 0.842 | 14.695 | 10.570 | 800 |
| 900 | 1.121 | 0.834 | 1.145 | 0.849 | 1.074 | 0.814 | 1.204 | 1.015 | 1.163 | 0.866 | 14.822 | 10.698 | 900 |
| 1000 | 1.142 | 0.855 | 1.167 | 0.870 | 1.090 | 0.830 | 1.234 | 1.045 | 1.185 | 0.888 | 14.983 | 10.859 | 1000 |

*Source:* Tables T-10 are adapted from K. Wark, *Thermodynamics,* 4th ed., McGraw-Hill, New York, 1983, as based on "Tables of Thermal Properties of Gases," NBS Circular 564, 1955.

**Table T-11** Ideal Gas Properties of Selected Gases

$T$(K), $\bar{h}$ and $\bar{u}$ (kJ/kmol), $\bar{s}^\circ$ (kJ/kmol · K)

| | Carbon Dioxide, $CO_2$ | | | Carbon Monoxide, CO | | | Water Vapor, $H_2O$ | | | Oxygen, $O_2$ | | | Nitrogen, $N_2$ | | | |
|---|---|---|---|---|---|---|---|---|---|---|---|---|---|---|---|---|
| $T$ | $\bar{h}$ | $\bar{u}$ | $\bar{s}^\circ$ | $\bar{h}$ | $\bar{u}$ | $\bar{s}^\circ$ | $\bar{h}$ | $\bar{u}$ | $\bar{s}^\circ$ | $\bar{h}$ | $\bar{u}$ | $\bar{s}^\circ$ | $\bar{h}$ | $\bar{u}$ | $\bar{s}^\circ$ | $T$ |
| 220 | 6,601 | 4,772 | 202.966 | 6,391 | 4,562 | 188.683 | 7,295 | 5,466 | 178.576 | 6,404 | 4,575 | 196.171 | 6,391 | 4,562 | 182.638 | 220 |
| 230 | 6,938 | 5,026 | 204.464 | 6,683 | 4,771 | 189.980 | 7,628 | 5,715 | 180.054 | 6,694 | 4,782 | 197.461 | 6,683 | 4,770 | 183.938 | 230 |
| 240 | 7,280 | 5,285 | 205.920 | 6,975 | 4,979 | 191.221 | 7,961 | 5,965 | 181.471 | 6,984 | 4,989 | 198.696 | 6,975 | 4,979 | 185.180 | 240 |
| 250 | 7,627 | 5,548 | 207.337 | 7,266 | 5,188 | 192.411 | 8,294 | 6,215 | 182.831 | 7,275 | 5,197 | 199.885 | 7,266 | 5,188 | 186.370 | 250 |
| 260 | 7,979 | 5,817 | 208.717 | 7,558 | 5,396 | 193.554 | 8,627 | 6,466 | 184.139 | 7,566 | 5,405 | 201.027 | 7,558 | 5,396 | 187.514 | 260 |
| 270 | 8,335 | 6,091 | 210.062 | 7,849 | 5,604 | 194.654 | 8,961 | 6,716 | 185.399 | 7,858 | 5,613 | 202.128 | 7,849 | 5,604 | 188.614 | 270 |
| 280 | 8,697 | 6,369 | 211.376 | 8,140 | 5,812 | 195.173 | 9,296 | 6,968 | 186.616 | 8,150 | 5,822 | 203.191 | 8,141 | 5,813 | 189.673 | 280 |
| 290 | 9,063 | 6,651 | 212.660 | 8,432 | 6,020 | 196.735 | 9,631 | 7,219 | 187.791 | 8,443 | 6,032 | 204.218 | 8,432 | 6,021 | 190.695 | 290 |
| 300 | 9,431 | 6,939 | 213.915 | 8,723 | 6,229 | 197.723 | 9,966 | 7,472 | 188.928 | 8,736 | 6,242 | 205.213 | 8,723 | 6,229 | 191.682 | 300 |
| 310 | 9,807 | 7,230 | 215.146 | 9,014 | 6,437 | 198.678 | 10,302 | 7,725 | 190.030 | 9,030 | 6,453 | 206.177 | 9,014 | 6,437 | 192.638 | 310 |
| 320 | 10,186 | 7,526 | 216.351 | 9,306 | 6,645 | 199.603 | 10,639 | 7,978 | 191.098 | 9,325 | 6,664 | 207.112 | 9,306 | 6,645 | 193.562 | 320 |
| 330 | 10,570 | 7,826 | 217.534 | 9,597 | 6,854 | 200.500 | 10,976 | 8,232 | 192.136 | 9,620 | 6,877 | 208.020 | 9,597 | 6,853 | 194.459 | 330 |
| 340 | 10,959 | 8,131 | 218.694 | 9,889 | 7,062 | 201.371 | 11,314 | 8,487 | 193.144 | 9,916 | 7,090 | 208.904 | 9,888 | 7,061 | 195.328 | 340 |
| 350 | 11,351 | 8,439 | 219.831 | 10,181 | 7,271 | 202.217 | 11,652 | 8,742 | 194.125 | 10,213 | 7,303 | 209.765 | 10,180 | 7,270 | 196.173 | 350 |
| 360 | 11,748 | 8,752 | 220.948 | 10,473 | 7,480 | 203.040 | 11,992 | 8,998 | 195.081 | 10,511 | 7,518 | 210.604 | 10,471 | 7,478 | 196.995 | 360 |
| 370 | 12,148 | 9,068 | 222.044 | 10,765 | 7,689 | 203.842 | 12,331 | 9,255 | 196.012 | 10,809 | 7,733 | 211.423 | 10,763 | 7,687 | 197.794 | 370 |
| 380 | 12,552 | 9,392 | 223.122 | 11,058 | 7,899 | 204.622 | 12,672 | 9,513 | 196.920 | 11,109 | 7,949 | 212.222 | 11,055 | 7,895 | 198.572 | 380 |
| 390 | 12,960 | 9,718 | 224.182 | 11,351 | 8,108 | 205.383 | 13,014 | 9,771 | 197.807 | 11,409 | 8,166 | 213.002 | 11,347 | 8,104 | 199.331 | 390 |
| 400 | 13,372 | 10,046 | 225.225 | 11,644 | 8,319 | 206.125 | 13,356 | 10,030 | 198.673 | 11,711 | 8,384 | 213.765 | 11,640 | 8,314 | 200.071 | 400 |
| 410 | 13,787 | 10,378 | 226.250 | 11,938 | 8,529 | 206.850 | 13,699 | 10,290 | 199.521 | 12,012 | 8,603 | 214.510 | 11,932 | 8,523 | 200.794 | 410 |
| 420 | 14,206 | 10,714 | 227.258 | 12,232 | 8,740 | 207.549 | 14,043 | 10,551 | 200.350 | 12,314 | 8,822 | 215.241 | 12,225 | 8,733 | 201.499 | 420 |
| 430 | 14,628 | 11,053 | 228.252 | 12,526 | 8,951 | 208.252 | 14,388 | 10,813 | 201.160 | 12,618 | 9,043 | 215.955 | 12,518 | 8,943 | 202.189 | 430 |
| 440 | 15,054 | 11,393 | 229.230 | 12,821 | 9,163 | 208.929 | 14,734 | 11,075 | 201.955 | 12,923 | 9,264 | 216.656 | 12,811 | 9,153 | 202.863 | 440 |
| 450 | 15,483 | 11,742 | 230.194 | 13,116 | 9,375 | 209.593 | 15,080 | 11,339 | 202.734 | 13,228 | 9,487 | 217.342 | 13,105 | 9,363 | 203.523 | 450 |
| 460 | 15,916 | 12,091 | 231.144 | 13,412 | 9,587 | 210.243 | 15,428 | 11,603 | 203.497 | 13,535 | 9,710 | 218.016 | 13,399 | 9,574 | 204.170 | 460 |
| 470 | 16,351 | 12,444 | 232.080 | 13,708 | 9,800 | 210.880 | 15,777 | 11,869 | 204.247 | 13,842 | 9,935 | 218.676 | 13,693 | 9,786 | 204.803 | 470 |
| 480 | 16,791 | 12,800 | 233.004 | 14,005 | 10,014 | 211.504 | 16,126 | 12,135 | 204.982 | 14,151 | 10,160 | 219.326 | 13,988 | 9,997 | 205.424 | 480 |
| 490 | 17,232 | 13,158 | 233.916 | 14,302 | 10,228 | 212.117 | 16,477 | 12,403 | 205.705 | 14,460 | 10,386 | 219.963 | 14,285 | 10,210 | 206.033 | 490 |
| 500 | 17,678 | 13,521 | 234.814 | 14,600 | 10,443 | 212.719 | 16,828 | 12,671 | 206.413 | 14,770 | 10,614 | 220.589 | 14,581 | 10,423 | 206.630 | 500 |
| 510 | 18,126 | 13,885 | 235.700 | 14,898 | 10,658 | 213.310 | 17,181 | 12,940 | 207.112 | 15,082 | 10,842 | 221.206 | 14,876 | 10,635 | 207.216 | 510 |
| 520 | 18,576 | 14,253 | 236.575 | 15,197 | 10,874 | 213.890 | 17,534 | 13,211 | 207.799 | 15,395 | 11,071 | 221.812 | 15,172 | 10,848 | 207.792 | 520 |
| 530 | 19,029 | 14,622 | 237.439 | 15,497 | 11,090 | 214.460 | 17,889 | 13,482 | 208.475 | 15,708 | 11,301 | 222.409 | 15,469 | 11,062 | 208.358 | 530 |
| 540 | 19,485 | 14,996 | 238.292 | 15,797 | 11,307 | 215.020 | 18,245 | 13,755 | 209.139 | 16,022 | 11,533 | 222.997 | 15,766 | 11,277 | 208.914 | 540 |

**Table T-11** *(Continued)*

| | $T$(K), $\bar{h}$ and $\bar{u}$ (kJ/kmol), $\bar{s}^\circ$ (kJ/kmol · K) | | | | | | | | | | | | | | | |
|---|---|---|---|---|---|---|---|---|---|---|---|---|---|---|---|---|
| | Carbon Dioxide, $CO_2$ | | | Carbon Monoxide, CO | | | Water Vapor, $H_2O$ | | | Oxygen, $O_2$ | | | Nitrogen, $N_2$ | | | |
| $T$ | $\bar{h}$ | $\bar{u}$ | $\bar{s}^\circ$ | $\bar{h}$ | $\bar{u}$ | $\bar{s}^\circ$ | $\bar{h}$ | $\bar{u}$ | $\bar{s}^\circ$ | $\bar{h}$ | $\bar{u}$ | $\bar{s}^\circ$ | $\bar{h}$ | $\bar{u}$ | $\bar{s}^\circ$ | $T$ |
| 550 | 19,945 | 15,372 | 239.135 | 16,097 | 11,524 | 215.572 | 18,601 | 14,028 | 209.795 | 16,338 | 11,765 | 223.576 | 16,064 | 11,492 | 209.461 | 550 |
| 560 | 20,407 | 15,751 | 239.962 | 16,399 | 11,743 | 216.115 | 18,959 | 14,303 | 210.440 | 16,654 | 11,998 | 224.146 | 16,363 | 11,707 | 209.999 | 560 |
| 570 | 20,870 | 16,131 | 240.789 | 16,701 | 11,961 | 216.649 | 19,318 | 14,579 | 211.075 | 16,971 | 12,232 | 224.708 | 16,662 | 11,923 | 210.528 | 570 |
| 580 | 21,337 | 16,515 | 241.602 | 17,003 | 12,181 | 217.175 | 19,678 | 14,856 | 211.702 | 17,290 | 12,467 | 225.262 | 16,962 | 12,139 | 211.049 | 580 |
| 590 | 21,807 | 16,902 | 242.405 | 17,307 | 12,401 | 217.693 | 20,039 | 15,134 | 212.320 | 17,609 | 12,703 | 225.808 | 17,262 | 12,356 | 211.562 | 590 |
| 600 | 22,280 | 17,291 | 243.199 | 17,611 | 12,622 | 218.204 | 20,402 | 15,413 | 212.920 | 17,929 | 12,940 | 226.346 | 17,563 | 12,574 | 212.066 | 600 |
| 610 | 22,754 | 17,683 | 243.983 | 17,915 | 12,843 | 218.708 | 20,765 | 15,693 | 213.529 | 18,250 | 13,178 | 226.877 | 17,864 | 12,792 | 212.564 | 610 |
| 620 | 23,231 | 18,076 | 244.758 | 18,221 | 13,066 | 219.205 | 21,130 | 15,975 | 214.122 | 18,572 | 13,417 | 227.400 | 18,166 | 13,011 | 213.055 | 620 |
| 630 | 23,709 | 18,471 | 245.524 | 18,527 | 13,289 | 219.695 | 21,495 | 16,257 | 214.707 | 18,895 | 13,657 | 227.918 | 18,468 | 13,230 | 213.541 | 630 |
| 640 | 24,190 | 18,869 | 246.282 | 18,833 | 13,512 | 220.179 | 21,862 | 16,541 | 215.285 | 19,219 | 13,898 | 228.429 | 18,772 | 13,450 | 214.018 | 640 |
| 650 | 24,674 | 19,270 | 247.032 | 19,141 | 13,736 | 220.656 | 22,230 | 16,826 | 215.856 | 19,544 | 14,140 | 228.932 | 19,075 | 13,671 | 214.489 | 650 |
| 660 | 25,160 | 19,672 | 247.773 | 19,449 | 13,962 | 221.127 | 22,600 | 17,112 | 216.419 | 19,870 | 14,383 | 229.430 | 19,380 | 13,892 | 214.954 | 660 |
| 670 | 25,648 | 20,078 | 248.507 | 19,758 | 14,187 | 221.592 | 22,970 | 17,399 | 216.976 | 20,197 | 14,626 | 229.920 | 19,685 | 14,114 | 215.413 | 670 |
| 680 | 26,138 | 20,484 | 249.233 | 20,068 | 14,414 | 222.052 | 23,342 | 17,688 | 217.527 | 20,524 | 14,871 | 230.405 | 19,991 | 14,337 | 215.866 | 680 |
| 690 | 26,631 | 20,894 | 249.952 | 20,378 | 14,641 | 222.505 | 23,714 | 17,978 | 218.071 | 20,854 | 15,116 | 230.885 | 20,297 | 14,560 | 216.314 | 690 |
| 700 | 27,125 | 21,305 | 250.663 | 20,690 | 14,870 | 222.953 | 24,088 | 18,268 | 218.610 | 21,184 | 15,364 | 231.358 | 20,604 | 14,784 | 216.756 | 700 |
| 710 | 27,622 | 21,719 | 251.368 | 21,002 | 15,099 | 223.396 | 24,464 | 18,561 | 219.142 | 21,514 | 15,611 | 231.827 | 20,912 | 15,008 | 217.192 | 710 |
| 720 | 28,121 | 22,134 | 252.065 | 21,315 | 15,328 | 223.833 | 24,840 | 18,854 | 219.668 | 21,845 | 15,859 | 232.291 | 21,220 | 15,234 | 217.624 | 720 |
| 730 | 28,622 | 22,552 | 252.755 | 21,628 | 15,558 | 224.265 | 25,218 | 19,148 | 220.189 | 22,177 | 16,107 | 232.748 | 21,529 | 15,460 | 218.059 | 730 |
| 740 | 29,124 | 22,972 | 253.439 | 21,943 | 15,789 | 224.692 | 25,597 | 19,444 | 220.707 | 22,510 | 16,357 | 233.201 | 21,839 | 15,686 | 218.472 | 740 |
| 750 | 29,629 | 23,393 | 254.117 | 22,258 | 16,022 | 225.115 | 25,977 | 19,741 | 221.215 | 22,844 | 16,607 | 233.649 | 22,149 | 15,913 | 218.889 | 750 |
| 760 | 30,135 | 23,817 | 254.787 | 22,573 | 16,255 | 225.533 | 26,358 | 20,039 | 221.720 | 23,178 | 16,859 | 234.091 | 22,460 | 16,141 | 219.301 | 760 |
| 770 | 30,644 | 24,242 | 255.452 | 22,890 | 16,488 | 225.947 | 26,741 | 20,339 | 222.221 | 23,513 | 17,111 | 234.528 | 22,772 | 16,370 | 219.709 | 770 |
| 780 | 31,154 | 24,669 | 256.110 | 23,208 | 16,723 | 226.357 | 27,125 | 20,639 | 222.717 | 23,850 | 17,364 | 234.960 | 23,085 | 16,599 | 220.113 | 780 |
| 790 | 31,665 | 25,097 | 256.762 | 23,526 | 16,957 | 226.762 | 27,510 | 20,941 | 223.207 | 24,186 | 17,618 | 235.387 | 23,398 | 16,830 | 220.512 | 790 |
| 800 | 32,179 | 25,527 | 257.408 | 23,844 | 17,193 | 227.162 | 27,896 | 21,245 | 223.693 | 24,523 | 17,872 | 235.810 | 23,714 | 17,061 | 220.907 | 800 |
| 810 | 32,694 | 25,959 | 258.048 | 24,164 | 17,429 | 227.559 | 28,284 | 21,549 | 224.174 | 24,861 | 18,126 | 236.230 | 24,027 | 17,292 | 221.298 | 810 |
| 820 | 33,212 | 26,394 | 258.682 | 24,483 | 17,665 | 227.952 | 28,672 | 21,855 | 224.651 | 25,199 | 18,382 | 236.644 | 24,342 | 17,524 | 221.684 | 820 |
| 830 | 33,730 | 26,829 | 259.311 | 24,803 | 17,902 | 228.339 | 29,062 | 22,162 | 225.123 | 25,537 | 18,637 | 237.055 | 24,658 | 17,757 | 222.067 | 830 |
| 840 | 34,251 | 27,267 | 259.934 | 25,124 | 18,140 | 228.724 | 29,454 | 22,470 | 225.592 | 25,877 | 18,893 | 237.462 | 24,974 | 17,990 | 222.447 | 840 |
| 850 | 34,773 | 27,706 | 260.551 | 25,446 | 18,379 | 229.106 | 29,846 | 22,779 | 226.057 | 26,218 | 19,150 | 237.864 | 25,292 | 18,224 | 222.822 | 850 |
| 860 | 35,296 | 28,125 | 261.164 | 25,768 | 18,617 | 229.482 | 30,240 | 23,090 | 226.517 | 26,559 | 19,408 | 238.264 | 25,610 | 18,459 | 223.194 | 860 |
| 870 | 35,821 | 28,588 | 261.770 | 26,091 | 18,858 | 229.856 | 30,635 | 23,402 | 226.973 | 26,899 | 19,666 | 238.660 | 25,928 | 18,695 | 223.562 | 870 |
| 880 | 36,347 | 29,031 | 262.371 | 26,415 | 19,099 | 230.227 | 31,032 | 23,715 | 227.426 | 27,242 | 19,925 | 239.051 | 26,248 | 18,931 | 223.927 | 880 |
| 890 | 36,876 | 29,476 | 262.968 | 26,740 | 19,341 | 230.593 | 31,429 | 24,029 | 227.875 | 27,584 | 20,185 | 239.439 | 26,568 | 19,168 | 224.288 | 890 |

**Table T-11** *(Continued)*

$T$(K), $\bar{h}$ and $\bar{u}$ (kJ/kmol), $\bar{s}^\circ$ (kJ/kmol · K)

| | Carbon Dioxide, $CO_2$ | | | Carbon Monoxide, CO | | | Water Vapor, $H_2O$ | | | Oxygen, $O_2$ | | | Nitrogen, $N_2$ | | | |
|---|---|---|---|---|---|---|---|---|---|---|---|---|---|---|---|---|
| $T$ | $\bar{h}$ | $\bar{u}$ | $\bar{s}^\circ$ | $\bar{h}$ | $\bar{u}$ | $\bar{s}^\circ$ | $\bar{h}$ | $\bar{u}$ | $\bar{s}^\circ$ | $\bar{h}$ | $\bar{u}$ | $\bar{s}^\circ$ | $\bar{h}$ | $\bar{u}$ | $\bar{s}^\circ$ | $T$ |
| 900 | 37,405 | 29,922 | 263.559 | 27,066 | 19,583 | 230.957 | 31,828 | 24,345 | 228.321 | 27,928 | 20,445 | 239.823 | 26,890 | 19,407 | 224.647 | 900 |
| 910 | 37,935 | 30,369 | 264.146 | 27,392 | 19,826 | 231.317 | 32,228 | 24,662 | 228.763 | 28,272 | 20,706 | 240.203 | 27,210 | 19,644 | 225.002 | 910 |
| 920 | 38,467 | 30,818 | 264.728 | 27,719 | 20,070 | 231.674 | 32,629 | 24,980 | 229.202 | 28,616 | 20,967 | 240.580 | 27,532 | 19,883 | 225.353 | 920 |
| 930 | 39,000 | 31,268 | 265.304 | 28,046 | 20,314 | 232.028 | 33,032 | 25,300 | 229.637 | 28,960 | 21,228 | 240.953 | 27,854 | 20,122 | 225.701 | 930 |
| 940 | 39,535 | 31,719 | 265.877 | 28,375 | 20,559 | 232.379 | 33,436 | 25,621 | 230.070 | 29,306 | 21,491 | 241.323 | 28,178 | 20,362 | 226.047 | 940 |
| 950 | 40,070 | 32,171 | 266.444 | 28,703 | 20,805 | 232.727 | 33,841 | 25,943 | 230.499 | 29,652 | 21,754 | 241.689 | 28,501 | 20,603 | 226.389 | 950 |
| 960 | 40,607 | 32,625 | 267.007 | 29,033 | 21,051 | 233.072 | 34,247 | 26,265 | 230.924 | 29,999 | 22,017 | 242.052 | 28,826 | 20,844 | 226.728 | 960 |
| 970 | 41,145 | 33,081 | 267.566 | 29,362 | 21,298 | 233.413 | 34,653 | 26,588 | 231.347 | 30,345 | 22,280 | 242.411 | 29,151 | 21,086 | 227.064 | 970 |
| 980 | 41,685 | 33,537 | 268.119 | 29,693 | 21,545 | 233.752 | 35,061 | 26,913 | 231.767 | 30,692 | 22,544 | 242.768 | 29,476 | 21,328 | 227.398 | 980 |
| 990 | 42,226 | 33,995 | 268.670 | 30,024 | 21,793 | 234.088 | 35,472 | 27,240 | 232.184 | 31,041 | 22,809 | 243.120 | 29,803 | 21,571 | 227.728 | 990 |
| 1000 | 42,769 | 34,455 | 269.215 | 30,355 | 22,041 | 234.421 | 35,882 | 27,568 | 232.597 | 31,389 | 23,075 | 243.471 | 30,129 | 21,815 | 228.057 | 1000 |

$T >$ 1000 K (CD-ROM)

*Source*: Tables T-11 are based on the JANAF Thermochemical Tables, NSRDS-NBS-37, 1971.

**Table T-2E** Properties of Saturated Water (Liquid–Vapor): Temperature Table

| | | Specific Volume $ft^3/lb$ | | Internal Energy Btu/lb | | Enthalpy Btu/lb | | | Entropy Btu/lb · °R | | |
|---|---|---|---|---|---|---|---|---|---|---|---|
| Temp. °F | Press. $lbf/in.^2$ | Sat. Liquid $v_f$ | Sat. Vapor $v_g$ | Sat. Liquid $u_f$ | Sat. Vapor $u_g$ | Sat. Liquid $h_f$ | Evap. $h_{fg}$ | Sat. Vapor $h_g$ | Sat. Liquid $s_f$ | Sat. Vapor $s_g$ | Temp. °F |
| 32 | 0.0886 | 0.01602 | 3305 | −.01 | 1021.2 | −.01 | 1075.4 | 1075.4 | −.00003 | 2.1870 | 32 |
| 35 | 0.0999 | 0.01602 | 2948 | 2.99 | 1022.2 | 3.00 | 1073.7 | 1076.7 | 0.00607 | 2.1764 | 35 |
| 40 | 0.1217 | 0.01602 | 2445 | 8.02 | 1023.9 | 8.02 | 1070.9 | 1078.9 | 0.01617 | 2.1592 | 40 |
| 45 | 0.1475 | 0.01602 | 2037 | 13.04 | 1025.5 | 13.04 | 1068.1 | 1081.1 | 0.02618 | 2.1423 | 45 |
| 50 | 0.1780 | 0.01602 | 1704 | 18.06 | 1027.2 | 18.06 | 1065.2 | 1083.3 | 0.03607 | 2.1259 | 50 |
| 52 | 0.1917 | 0.01603 | 1589 | 20.06 | 1027.8 | 20.07 | 1064.1 | 1084.2 | 0.04000 | 2.1195 | 52 |
| 54 | 0.2064 | 0.01603 | 1482 | 22.07 | 1028.5 | 22.07 | 1063.0 | 1085.1 | 0.04391 | 2.1131 | 54 |
| 56 | 0.2219 | 0.01603 | 1383 | 24.08 | 1029.1 | 24.08 | 1061.9 | 1085.9 | 0.04781 | 2.1068 | 56 |
| 58 | 0.2386 | 0.01603 | 1292 | 26.08 | 1029.8 | 26.08 | 1060.7 | 1086.8 | 0.05159 | 2.1005 | 58 |
| 60 | 0.2563 | 0.01604 | 1207 | 28.08 | 1030.4 | 28.08 | 1059.6 | 1087.7 | 0.05555 | 2.0943 | 60 |
| 62 | 0.2751 | 0.01604 | 1129 | 30.09 | 1031.1 | 30.09 | 1058.5 | 1088.6 | 0.05940 | 2.0882 | 62 |
| 64 | 0.2952 | 0.01604 | 1056 | 32.09 | 1031.8 | 32.09 | 1057.3 | 1089.4 | 0.06323 | 2.0821 | 64 |
| 66 | 0.3165 | 0.01604 | 988.4 | 34.09 | 1032.4 | 34.09 | 1056.2 | 1090.3 | 0.06704 | 2.0761 | 66 |
| 68 | 0.3391 | 0.01605 | 925.8 | 36.09 | 1033.1 | 36.09 | 1055.1 | 1091.2 | 0.07084 | 2.0701 | 68 |
| 70 | 0.3632 | 0.01605 | 867.7 | 38.09 | 1033.7 | 38.09 | 1054.0 | 1092.0 | 0.07463 | 2.0642 | 70 |
| 72 | 0.3887 | 0.01606 | 813.7 | 40.09 | 1034.4 | 40.09 | 1052.8 | 1092.9 | 0.07839 | 2.0584 | 72 |
| 74 | 0.4158 | 0.01606 | 763.5 | 42.09 | 1035.0 | 42.09 | 1051.7 | 1093.8 | 0.08215 | 2.0526 | 74 |
| 76 | 0.4446 | 0.01606 | 716.8 | 44.09 | 1035.7 | 44.09 | 1050.6 | 1094.7 | 0.08589 | 2.0469 | 76 |
| 78 | 0.4750 | 0.01607 | 673.3 | 46.09 | 1036.3 | 46.09 | 1049.4 | 1095.5 | 0.08961 | 2.0412 | 78 |
| 80 | 0.5073 | 0.01607 | 632.8 | 48.08 | 1037.0 | 48.09 | 1048.3 | 1096.4 | 0.09332 | 2.0356 | 80 |
| 82 | 0.5414 | 0.01608 | 595.0 | 50.08 | 1037.6 | 50.08 | 1047.2 | 1097.3 | 0.09701 | 2.0300 | 82 |
| 84 | 0.5776 | 0.01608 | 559.8 | 52.08 | 1038.3 | 52.08 | 1046.0 | 1098.1 | 0.1007 | 2.0245 | 84 |
| 86 | 0.6158 | 0.01609 | 527.0 | 54.08 | 1038.9 | 54.08 | 1044.9 | 1099.0 | 0.1044 | 2.0190 | 86 |
| 88 | 0.6562 | 0.01609 | 496.3 | 56.07 | 1039.6 | 56.07 | 1043.8 | 1099.9 | 0.1080 | 2.0136 | 88 |
| 90 | 0.6988 | 0.01610 | 467.7 | 58.07 | 1040.2 | 58.07 | 1042.7 | 1100.7 | 0.1117 | 2.0083 | 90 |
| 92 | 0.7439 | 0.01611 | 440.9 | 60.06 | 1040.9 | 60.06 | 1041.5 | 1101.6 | 0.1153 | 2.0030 | 92 |
| 94 | 0.7914 | 0.01611 | 415.9 | 62.06 | 1041.5 | 62.06 | 1040.4 | 1102.4 | 0.1189 | 1.9977 | 94 |
| 96 | 0.8416 | 0.01612 | 392.4 | 64.05 | 1041.2 | 64.06 | 1039.2 | 1103.3 | 0.1225 | 1.9925 | 96 |
| 98 | 0.8945 | 0.01612 | 370.5 | 66.05 | 1042.8 | 66.05 | 1038.1 | 1104.2 | 0.1261 | 1.9874 | 98 |
| 100 | 0.9503 | 0.01613 | 350.0 | 68.04 | 1043.5 | 68.05 | 1037.0 | 1105.0 | 0.1296 | 1.9822 | 100 |
| 110 | 1.276 | 0.01617 | 265.1 | 78.02 | 1046.7 | 78.02 | 1031.3 | 1109.3 | 0.1473 | 1.9574 | 110 |
| 120 | 1.695 | 0.01621 | 203.0 | 87.99 | 1049.9 | 88.00 | 1025.5 | 1113.5 | 0.1647 | 1.9336 | 120 |
| 130 | 2.225 | 0.01625 | 157.2 | 97.97 | 1053.0 | 97.98 | 1019.8 | 1117.8 | 0.1817 | 1.9109 | 130 |
| 140 | 2.892 | 0.01629 | 122.9 | 107.95 | 1056.2 | 107.96 | 1014.0 | 1121.9 | 0.1985 | 1.8892 | 140 |
| 150 | 3.722 | 0.01634 | 97.0 | 117.95 | 1059.3 | 117.96 | 1008.1 | 1126.1 | 0.2150 | 1.8684 | 150 |
| 160 | 4.745 | 0.01640 | 77.2 | 127.94 | 1062.3 | 127.96 | 1002.2 | 1130.1 | 0.2313 | 1.8484 | 160 |
| 170 | 5.996 | 0.01645 | 62.0 | 137.95 | 1065.4 | 137.97 | 996.2 | 1134.2 | 0.2473 | 1.8293 | 170 |
| 180 | 7.515 | 0.01651 | 50.2 | 147.97 | 1068.3 | 147.99 | 990.2 | 1138.2 | 0.2631 | 1.8109 | 180 |
| 190 | 9.343 | 0.01657 | 41.0 | 158.00 | 1071.3 | 158.03 | 984.1 | 1142.1 | 0.2787 | 1.7932 | 190 |
| 200 | 11.529 | 0.01663 | 33.6 | 168.04 | 1074.2 | 168.07 | 977.9 | 1145.9 | 0.2940 | 1.7762 | 200 |

**Table T-2E** *(Continued)*

| Temp. °F | Press. lbf/in.$^2$ | Specific Volume ft$^3$/lb Sat. Liquid $v_f$ | Sat. Vapor $v_g$ | Internal Energy Btu/lb Sat. Liquid $u_f$ | Sat. Vapor $u_g$ | Enthalpy Btu/lb Sat. Liquid $h_f$ | Evap. $h_{fg}$ | Sat. Vapor $h_g$ | Entropy Btu/lb · °R Sat. Liquid $s_f$ | Sat. Vapor $s_g$ | Temp. °F |
|---|---|---|---|---|---|---|---|---|---|---|---|
| 210 | 14.13 | 0.01670 | 27.82 | 178.1 | 1077.0 | 178.1 | 971.6 | 1149.7 | 0.3091 | 1.7599 | 210 |
| 212 | 14.70 | 0.01672 | 26.80 | 180.1 | 1077.6 | 180.2 | 970.3 | 1150.5 | 0.3121 | 1.7567 | 212 |
| 220 | 17.19 | 0.01677 | 23.15 | 188.2 | 1079.8 | 188.2 | 965.3 | 1153.5 | 0.3241 | 1.7441 | 220 |
| 230 | 20.78 | 0.01685 | 19.39 | 198.3 | 1082.6 | 198.3 | 958.8 | 1157.1 | 0.3388 | 1.7289 | 230 |
| 240 | 24.97 | 0.01692 | 16.33 | 208.4 | 1085.3 | 208.4 | 952.3 | 1160.7 | 0.3534 | 1.7143 | 240 |
| 250 | 29.82 | 0.01700 | 13.83 | 218.5 | 1087.9 | 218.6 | 945.6 | 1164.2 | 0.3677 | 1.7001 | 250 |
| 260 | 35.42 | 0.01708 | 11.77 | 228.6 | 1090.5 | 228.8 | 938.8 | 1167.6 | 0.3819 | 1.6864 | 260 |
| 270 | 41.85 | 0.01717 | 10.07 | 238.8 | 1093.0 | 239.0 | 932.0 | 1170.9 | 0.3960 | 1.6731 | 270 |
| 280 | 49.18 | 0.01726 | 8.65 | 249.0 | 1095.4 | 249.2 | 924.9 | 1174.1 | 0.4099 | 1.6602 | 280 |
| 290 | 57.53 | 0.01735 | 7.47 | 259.3 | 1097.7 | 259.4 | 917.8 | 1177.2 | 0.4236 | 1.6477 | 290 |
| 300 | 66.98 | 0.01745 | 6.472 | 269.5 | 1100.0 | 269.7 | 910.4 | 1180.2 | 0.4372 | 1.6356 | 300 |
| 310 | 77.64 | 0.01755 | 5.632 | 279.8 | 1102.1 | 280.1 | 903.0 | 1183.0 | 0.4507 | 1.6238 | 310 |
| 320 | 89.60 | 0.01765 | 4.919 | 290.1 | 1104.2 | 290.4 | 895.3 | 1185.8 | 0.4640 | 1.6123 | 320 |
| 330 | 103.00 | 0.01776 | 4.312 | 300.5 | 1106.2 | 300.8 | 887.5 | 1188.4 | 0.4772 | 1.6010 | 330 |
| 340 | 117.93 | 0.01787 | 3.792 | 310.9 | 1108.0 | 311.3 | 879.5 | 1190.8 | 0.4903 | 1.5901 | 340 |
| 350 | 134.53 | 0.01799 | 3.346 | 321.4 | 1109.8 | 321.8 | 871.3 | 1193.1 | 0.5033 | 1.5793 | 350 |
| 360 | 152.92 | 0.01811 | 2.961 | 331.8 | 1111.4 | 332.4 | 862.9 | 1195.2 | 0.5162 | 1.5688 | 360 |
| 370 | 173.23 | 0.01823 | 2.628 | 342.4 | 1112.9 | 343.0 | 854.2 | 1197.2 | 0.5289 | 1.5585 | 370 |
| 380 | 195.60 | 0.01836 | 2.339 | 353.0 | 1114.3 | 353.6 | 845.4 | 1199.0 | 0.5416 | 1.5483 | 380 |
| 390 | 220.2 | 0.01850 | 2.087 | 363.6 | 1115.6 | 364.3 | 836.2 | 1200.6 | 0.5542 | 1.5383 | 390 |
| 400 | 247.1 | 0.01864 | 1.866 | 374.3 | 1116.6 | 375.1 | 826.8 | 1202.0 | 0.5667 | 1.5284 | 400 |
| 410 | 276.5 | 0.01878 | 1.673 | 385.0 | 1117.6 | 386.0 | 817.2 | 1203.1 | 0.5792 | 1.5187 | 410 |
| 420 | 308.5 | 0.01894 | 1.502 | 395.8 | 1118.3 | 396.9 | 807.2 | 1204.1 | 0.5915 | 1.5091 | 420 |
| 430 | 343.3 | 0.01909 | 1.352 | 406.7 | 1118.9 | 407.9 | 796.9 | 1204.8 | 0.6038 | 1.4995 | 430 |
| 440 | 381.2 | 0.01926 | 1.219 | 417.6 | 1119.3 | 419.0 | 786.3 | 1205.3 | 0.6161 | 1.4900 | 440 |
| 450 | 422.1 | 0.01943 | 1.1011 | 428.6 | 1119.5 | 430.2 | 775.4 | 1205.6 | 0.6282 | 1.4806 | 450 |
| 460 | 466.3 | 0.01961 | 0.9961 | 439.7 | 1119.6 | 441.4 | 764.1 | 1205.5 | 0.6404 | 1.4712 | 460 |
| 470 | 514.1 | 0.01980 | 0.9025 | 450.9 | 1119.4 | 452.8 | 752.4 | 1205.2 | 0.6525 | 1.4618 | 470 |
| 480 | 565.5 | 0.02000 | 0.8187 | 462.2 | 1118.9 | 464.3 | 740.3 | 1204.6 | 0.6646 | 1.4524 | 480 |
| 490 | 620.7 | 0.02021 | 0.7436 | 473.6 | 1118.3 | 475.9 | 727.8 | 1203.7 | 0.6767 | 1.4430 | 490 |
| 500 | 680.0 | 0.02043 | 0.6761 | 485.1 | 1117.4 | 487.7 | 714.8 | 1202.5 | 0.6888 | 1.4335 | 500 |
| 520 | 811.4 | 0.02091 | 0.5605 | 508.5 | 1114.8 | 511.7 | 687.3 | 1198.9 | 0.7130 | 1.4145 | 520 |
| 540 | 961.5 | 0.02145 | 0.4658 | 532.6 | 1111.0 | 536.4 | 657.5 | 1193.8 | 0.7374 | 1.3950 | 540 |
| 560 | 1131.8 | 0.02207 | 0.3877 | 548.4 | 1105.8 | 562.0 | 625.0 | 1187.0 | 0.7620 | 1.3749 | 560 |
| 580 | 1324.3 | 0.02278 | 0.3225 | 583.1 | 1098.9 | 588.6 | 589.3 | 1178.0 | 0.7872 | 1.3540 | 580 |
| 600 | 1541.0 | 0.02363 | 0.2677 | 609.9 | 1090.0 | 616.7 | 549.7 | 1166.4 | 0.8130 | 1.3317 | 600 |
| 620 | 1784.4 | 0.02465 | 0.2209 | 638.3 | 1078.5 | 646.4 | 505.0 | 1151.4 | 0.8398 | 1.3075 | 620 |
| 640 | 2057.1 | 0.02593 | 0.1805 | 668.7 | 1063.2 | 678.6 | 453.4 | 1131.9 | 0.8681 | 1.2803 | 640 |
| 660 | 2362 | 0.02767 | 0.1446 | 702.3 | 1042.3 | 714.4 | 391.1 | 1105.5 | 0.8990 | 1.2483 | 660 |
| 680 | 2705 | 0.03032 | 0.1113 | 741.7 | 1011.0 | 756.9 | 309.8 | 1066.7 | 0.9350 | 1.2068 | 680 |
| 700 | 3090 | 0.03666 | 0.0744 | 801.7 | 947.7 | 822.7 | 167.5 | 990.2 | 0.9902 | 1.1346 | 700 |
| 705.4 | 3204 | 0.05053 | 0.05053 | 872.6 | 872.6 | 902.5 | 0 | 902.5 | 1.0580 | 1.0580 | 705.4 |

*Source:* Tables T-2E through T-5E are extracted from J. H. Keenan, F. G. Keyes, P. G. Hill, and J. G. Moore, *Steam Tables,* Wiley, New York, 1969.

**Table T-3E** Properties of Saturated Water (Liquid–Vapor): Pressure Table

| | | Specific Volume ft³/lb | | Internal Energy Btu/lb | | Enthalpy Btu/lb | | | Entropy Btu/lb · °R | | | |
|---|---|---|---|---|---|---|---|---|---|---|---|---|
| Press. lbf/in.² | Temp. °F | Sat. Liquid $v_f$ | Sat. Vapor $v_g$ | Sat. Liquid $u_f$ | Sat. Vapor $u_g$ | Sat. Liquid $h_f$ | Evap. $h_{fg}$ | Sat. Vapor $h_g$ | Sat. Liquid $s_f$ | Evap. $s_{fg}$ | Sat. Vapor $s_g$ | Press. lbf/in.² |
| 0.4 | 72.84 | 0.01606 | 792.0 | 40.94 | 1034.7 | 40.94 | 1052.3 | 1093.3 | 0.0800 | 1.9760 | 2.0559 | 0.4 |
| 0.6 | 85.19 | 0.01609 | 540.0 | 53.26 | 1038.7 | 53.27 | 1045.4 | 1098.6 | 0.1029 | 1.9184 | 2.0213 | 0.6 |
| 0.8 | 94.35 | 0.01611 | 411.7 | 62.41 | 1041.7 | 62.41 | 1040.2 | 1102.6 | 0.1195 | 1.8773 | 1.9968 | 0.8 |
| 1.0 | 101.70 | 0.01614 | 333.6 | 69.74 | 1044.0 | 69.74 | 1036.0 | 1105.8 | 0.1327 | 1.8453 | 1.9779 | 1.0 |
| 1.2 | 107.88 | 0.01616 | 280.9 | 75.90 | 1046.0 | 75.90 | 1032.5 | 1108.4 | 0.1436 | 1.8190 | 1.9626 | 1.2 |
| 1.5 | 115.65 | 0.01619 | 227.7 | 83.65 | 1048.5 | 83.65 | 1028.0 | 1111.7 | 0.1571 | 1.7867 | 1.9438 | 1.5 |
| 2.0 | 126.04 | 0.01623 | 173.75 | 94.02 | 1051.8 | 94.02 | 1022.1 | 1116.1 | 0.1750 | 1.7448 | 1.9198 | 2.0 |
| 3.0 | 141.43 | 0.01630 | 118.72 | 109.38 | 1056.6 | 109.39 | 1013.1 | 1122.5 | 0.2009 | 1.6852 | 1.8861 | 3.0 |
| 4.0 | 152.93 | 0.01636 | 90.64 | 120.88 | 1060.2 | 120.89 | 1006.4 | 1127.3 | 0.2198 | 1.6426 | 1.8624 | 4.0 |
| 5.0 | 162.21 | 0.01641 | 73.53 | 130.15 | 1063.0 | 130.17 | 1000.9 | 1131.0 | 0.2349 | 1.6093 | 1.8441 | 5.0 |
| 6.0 | 170.03 | 0.01645 | 61.98 | 137.98 | 1065.4 | 138.00 | 996.2 | 1134.2 | 0.2474 | 1.5819 | 1.8292 | 6.0 |
| 7.0 | 176.82 | 0.01649 | 53.65 | 144.78 | 1067.4 | 144.80 | 992.1 | 1136.9 | 0.2581 | 1.5585 | 1.8167 | 7.0 |
| 8.0 | 182.84 | 0.01653 | 47.35 | 150.81 | 1069.2 | 150.84 | 988.4 | 1139.3 | 0.2675 | 1.5383 | 1.8058 | 8.0 |
| 9.0 | 188.26 | 0.01656 | 42.41 | 156.25 | 1070.8 | 156.27 | 985.1 | 1141.4 | 0.2760 | 1.5203 | 1.7963 | 9.0 |
| 10 | 193.19 | 0.01659 | 38.42 | 161.20 | 1072.2 | 161.23 | 982.1 | 1143.3 | 0.2836 | 1.5041 | 1.7877 | 10 |
| 14.696 | 211.99 | 0.01672 | 26.80 | 180.10 | 1077.6 | 180.15 | 970.4 | 1150.5 | 0.3121 | 1.4446 | 1.7567 | 14.696 |
| 15 | 213.03 | 0.01672 | 26.29 | 181.14 | 1077.9 | 181.19 | 969.7 | 1150.9 | 0.3137 | 1.4414 | 1.7551 | 15 |
| 20 | 227.96 | 0.01683 | 20.09 | 196.19 | 1082.0 | 196.26 | 960.1 | 1156.4 | 0.3358 | 1.3962 | 1.7320 | 20 |
| 25 | 240.08 | 0.01692 | 16.31 | 208.44 | 1085.3 | 208.52 | 952.2 | 1160.7 | 0.3535 | 1.3607 | 1.7142 | 25 |
| 30 | 250.34 | 0.01700 | 13.75 | 218.84 | 1088.0 | 218.93 | 945.4 | 1164.3 | 0.3682 | 1.3314 | 1.6996 | 30 |
| 35 | 259.30 | 0.01708 | 11.90 | 227.93 | 1090.3 | 228.04 | 939.3 | 1167.4 | 0.3809 | 1.3064 | 1.6873 | 35 |
| 40 | 267.26 | 0.01715 | 10.50 | 236.03 | 1092.3 | 236.16 | 933.8 | 1170.0 | 0.3921 | 1.2845 | 1.6767 | 40 |
| 45 | 274.46 | 0.01721 | 9.40 | 243.37 | 1094.0 | 243.51 | 928.8 | 1172.3 | 0.4022 | 1.2651 | 1.6673 | 45 |
| 50 | 281.03 | 0.01727 | 8.52 | 250.08 | 1095.6 | 250.24 | 924.2 | 1174.4 | 0.4113 | 1.2476 | 1.6589 | 50 |
| 55 | 287.10 | 0.01733 | 7.79 | 256.28 | 1097.0 | 256.46 | 919.9 | 1176.3 | 0.4196 | 1.2317 | 1.6513 | 55 |
| 60 | 292.73 | 0.01738 | 7.177 | 262.1 | 1098.3 | 262.2 | 915.8 | 1178.0 | 0.4273 | 1.2170 | 1.6443 | 60 |
| 65 | 298.00 | 0.01743 | 6.647 | 267.5 | 1099.5 | 267.7 | 911.9 | 1179.6 | 0.4345 | 1.2035 | 1.6380 | 65 |
| 70 | 302.96 | 0.01748 | 6.209 | 272.6 | 1100.6 | 272.8 | 908.3 | 1181.0 | 0.4412 | 1.1909 | 1.6321 | 70 |
| 75 | 307.63 | 0.01752 | 5.818 | 277.4 | 1101.6 | 277.6 | 904.8 | 1182.4 | 0.4475 | 1.1790 | 1.6265 | 75 |
| 80 | 312.07 | 0.01757 | 5.474 | 282.0 | 1102.6 | 282.2 | 901.4 | 1183.6 | 0.4534 | 1.1679 | 1.6213 | 80 |
| 85 | 316.29 | 0.01761 | 5.170 | 286.3 | 1103.5 | 286.6 | 898.2 | 1184.8 | 0.4591 | 1.1574 | 1.6165 | 85 |
| 90 | 320.31 | 0.01766 | 4.898 | 290.5 | 1104.3 | 290.8 | 895.1 | 1185.9 | 0.4644 | 1.1475 | 1.6119 | 90 |
| 95 | 324.16 | 0.01770 | 4.654 | 294.5 | 1105.0 | 294.8 | 892.1 | 1186.9 | 0.4695 | 1.1380 | 1.6075 | 95 |
| 100 | 327.86 | 0.01774 | 4.434 | 298.3 | 1105.8 | 298.6 | 889.2 | 1187.8 | 0.4744 | 1.1290 | 1.6034 | 100 |
| 110 | 334.82 | 0.01781 | 4.051 | 305.5 | 1107.1 | 305.9 | 883.7 | 1189.6 | 0.4836 | 1.1122 | 1.5958 | 110 |
| 120 | 341.30 | 0.01789 | 3.730 | 312.3 | 1108.3 | 312.7 | 878.5 | 1191.1 | 0.4920 | 1.0966 | 1.5886 | 120 |
| 130 | 347.37 | 0.01796 | 3.457 | 318.6 | 1109.4 | 319.0 | 873.5 | 1192.5 | 0.4999 | 1.0822 | 1.5821 | 130 |
| 140 | 353.08 | 0.01802 | 3.221 | 324.6 | 1110.3 | 325.1 | 868.7 | 1193.8 | 0.5073 | 1.0688 | 1.5761 | 140 |
| 150 | 358.48 | 0.01809 | 3.016 | 330.2 | 1111.2 | 330.8 | 864.2 | 1194.9 | 0.5142 | 1.0562 | 1.5704 | 150 |
| 160 | 363.60 | 0.01815 | 2.836 | 335.6 | 1112.0 | 336.2 | 859.8 | 1196.0 | 0.5208 | 1.0443 | 1.5651 | 160 |

**Table T-3E** *(Continued)*

| | | Specific Volume ft³/lb | | Internal Energy Btu/lb | | Enthalpy Btu/lb | | | Entropy Btu/lb · °R | | | |
|---|---|---|---|---|---|---|---|---|---|---|---|---|
| Press. lbf/in.² | Temp. °F | Sat. Liquid $v_f$ | Sat. Vapor $v_g$ | Sat. Liquid $u_f$ | Sat. Vapor $u_g$ | Sat. Liquid $h_f$ | Evap. $h_{fg}$ | Sat. Vapor $h_g$ | Sat. Liquid $s_f$ | Evap. $s_{fg}$ | Sat. Vapor $s_g$ | Press. lbf/in.² |
| 170 | 368.47 | 0.01821 | 2.676 | 340.8 | 1112.7 | 341.3 | 855.6 | 1196.9 | 0.5270 | 1.0330 | 1.5600 | 170 |
| 180 | 373.13 | 0.01827 | 2.553 | 345.7 | 1113.4 | 346.3 | 851.5 | 1197.8 | 0.5329 | 1.0223 | 1.5552 | 180 |
| 190 | 377.59 | 0.01833 | 2.405 | 350.4 | 1114.0 | 351.0 | 847.5 | 1198.6 | 0.5386 | 1.0122 | 1.5508 | 190 |
| 200 | 381.86 | 0.01839 | 2.289 | 354.9 | 1114.6 | 355.6 | 843.7 | 1199.3 | 0.5440 | 1.0025 | 1.5465 | 200 |
| 250 | 401.04 | 0.01865 | 1.845 | 375.4 | 1116.7 | 376.2 | 825.8 | 1202.1 | 0.5680 | 0.9594 | 1.5274 | 250 |
| 300 | 417.43 | 0.01890 | 1.544 | 393.0 | 1118.2 | 394.1 | 809.8 | 1203.9 | 0.5883 | 0.9232 | 1.5115 | 300 |
| 350 | 431.82 | 0.01912 | 1.327 | 408.7 | 1119.0 | 409.9 | 795.0 | 1204.9 | 0.6060 | 0.8917 | 1.4977 | 350 |
| 400 | 444.70 | 0.01934 | 1.162 | 422.8 | 1119.5 | 424.2 | 781.2 | 1205.5 | 0.6218 | 0.8638 | 1.4856 | 400 |
| 450 | 456.39 | 0.01955 | 1.033 | 435.7 | 1119.6 | 437.4 | 768.2 | 1205.6 | 0.6360 | 0.8385 | 1.4745 | 450 |
| 500 | 467.13 | 0.01975 | 0.928 | 447.7 | 1119.4 | 449.5 | 755.8 | 1205.3 | 0.6490 | 0.8154 | 1.4644 | 500 |
| 550 | 477.07 | 0.01994 | 0.842 | 458.9 | 1119.1 | 460.9 | 743.9 | 1204.8 | 0.6611 | 0.7941 | 1.4451 | 550 |
| 600 | 486.33 | 0.02013 | 0.770 | 469.4 | 1118.6 | 471.7 | 732.4 | 1204.1 | 0.6723 | 0.7742 | 1.4464 | 600 |
| 700 | 503.23 | 0.02051 | 0.656 | 488.9 | 1117.0 | 491.5 | 710.5 | 1202.0 | 0.6927 | 0.7378 | 1.4305 | 700 |
| 800 | 518.36 | 0.02087 | 0.569 | 506.6 | 1115.0 | 509.7 | 689.6 | 1199.3 | 0.7110 | 0.7050 | 1.4160 | 800 |
| 900 | 532.12 | 0.02123 | 0.501 | 523.0 | 1112.6 | 526.6 | 669.5 | 1196.0 | 0.7277 | 0.6750 | 1.4027 | 900 |
| 1000 | 544.75 | 0.02159 | 0.446 | 538.4 | 1109.9 | 542.4 | 650.0 | 1192.4 | 0.7432 | 0.6471 | 1.3903 | 1000 |
| 1100 | 556.45 | 0.02195 | 0.401 | 552.9 | 1106.8 | 557.4 | 631.0 | 1188.3 | 0.7576 | 0.6209 | 1.3786 | 1100 |
| 1200 | 567.37 | 0.02232 | 0.362 | 566.7 | 1103.5 | 571.7 | 612.3 | 1183.9 | 0.7712 | 0.5961 | 1.3673 | 1200 |
| 1300 | 577.60 | 0.02269 | 0.330 | 579.9 | 1099.8 | 585.4 | 593.8 | 1179.2 | 0.7841 | 0.5724 | 1.3565 | 1300 |
| 1400 | 587.25 | 0.02307 | 0.302 | 592.7 | 1096.0 | 598.6 | 575.5 | 1174.1 | 0.7964 | 0.5497 | 1.3461 | 1400 |
| 1500 | 596.39 | 0.02346 | 0.277 | 605.0 | 1091.8 | 611.5 | 557.2 | 1168.7 | 0.8082 | 0.5276 | 1.3359 | 1500 |
| 1600 | 605.06 | 0.02386 | 0.255 | 616.9 | 1087.4 | 624.0 | 538.9 | 1162.9 | 0.8196 | 0.5062 | 1.3258 | 1600 |
| 1700 | 613.32 | 0.02428 | 0.236 | 628.6 | 1082.7 | 636.2 | 520.6 | 1156.9 | 0.8307 | 0.4852 | 1.3159 | 1700 |
| 1800 | 621.21 | 0.02472 | 0.218 | 640.0 | 1077.7 | 648.3 | 502.1 | 1150.4 | 0.8414 | 0.4645 | 1.3060 | 1800 |
| 1900 | 628.76 | 0.02517 | 0.203 | 651.3 | 1072.3 | 660.1 | 483.4 | 1143.5 | 0.8519 | 0.4441 | 1.2961 | 1900 |
| 2000 | 636.00 | 0.02565 | 0.188 | 662.4 | 1066.6 | 671.9 | 464.4 | 1136.3 | 0.8623 | 0.4238 | 1.2861 | 2000 |
| 2250 | 652.90 | 0.02698 | 0.157 | 689.9 | 1050.6 | 701.1 | 414.8 | 1115.9 | 0.8876 | 0.3728 | 1.2604 | 2250 |
| 2500 | 668.31 | 0.02860 | 0.131 | 717.7 | 1031.0 | 730.9 | 360.5 | 1091.4 | 0.9131 | 0.3196 | 1.2327 | 2500 |
| 2750 | 682.46 | 0.03077 | 0.107 | 747.3 | 1005.9 | 763.0 | 297.4 | 1060.4 | 0.9401 | 0.2604 | 1.2005 | 2750 |
| 3000 | 695.52 | 0.03431 | 0.084 | 783.4 | 968.8 | 802.5 | 213.0 | 1015.5 | 0.9732 | 0.1843 | 1.1575 | 3000 |
| 3203.6 | 705.44 | 0.05053 | 0.0505 | 872.6 | 872.6 | 902.5 | 0 | 902.5 | 1.0580 | 0 | 1.0580 | 3203.6 |

**Table T-4E** Properties of Superheated Water Vapor

| $T$ °F | $v$ $ft^3/lb$ | $u$ Btu/lb | $h$ Btu/lb | $s$ Btu/lb · °R | $v$ $ft^3/lb$ | $u$ Btu/lb | $h$ Btu/lb | $s$ Btu/lb · °R | $v$ $ft^3/lb$ | $u$ Btu/lb | $h$ Btu/lb | $s$ Btu/lb · °R |
|---|---|---|---|---|---|---|---|---|---|---|---|---|
| | $p$ = 1 lbf/in.$^2$ ($T_{sat}$ = 101.7°F) | | | | $p$ = 5 lbf/in.$^2$ ($T_{sat}$ = 162.2°F) | | | | $p$ = 10 lbf/in.$^2$ ($T_{sat}$ = 193.2°F) | | | |
| Sat. | 333.6 | 1044.0 | 1105.8 | 1.9779 | 73.53 | 1063.0 | 1131.0 | 1.8441 | 38.42 | 1072.2 | 1143.3 | 1.7877 |
| 150 | 362.6 | 1060.4 | 1127.5 | 2.0151 | | | | | | | | |
| 200 | 392.5 | 1077.5 | 1150.1 | 2.0508 | 78.15 | 1076.0 | 1148.6 | 1.8715 | 38.85 | 1074.7 | 1146.6 | 1.7927 |
| 250 | 422.4 | 1094.7 | 1172.8 | 2.0839 | 84.21 | 1093.8 | 1171.7 | 1.9052 | 41.95 | 1092.6 | 1170.2 | 1.8272 |
| 300 | 452.3 | 1112.0 | 1195.7 | 2.1150 | 90.24 | 1111.3 | 1194.8 | 1.9367 | 44.99 | 1110.4 | 1193.7 | 1.8592 |
| 400 | 511.9 | 1147.0 | 1241.8 | 2.1720 | 102.24 | 1146.6 | 1241.2 | 1.9941 | 51.03 | 1146.1 | 1240.5 | 1.9171 |
| 500 | 571.5 | 1182.8 | 1288.5 | 2.2235 | 114.20 | 1182.5 | 1288.2 | 2.0458 | 57.04 | 1182.2 | 1287.7 | 1.9690 |
| 600 | 631.1 | 1219.3 | 1336.1 | 2.2706 | 126.15 | 1219.1 | 1335.8 | 2.0930 | 63.03 | 1218.9 | 1335.5 | 2.0164 |
| 700 | 690.7 | 1256.7 | 1384.5 | 2.3142 | 138.08 | 1256.5 | 1384.3 | 2.1367 | 69.01 | 1256.3 | 1384.0 | 2.0601 |
| 800 | 750.3 | 1294.4 | 1433.7 | 2.3550 | 150.01 | 1294.7 | 1433.5 | 2.1775 | 74.98 | 1294.6 | 1433.3 | 2.1009 |
| 900 | 809.9 | 1333.9 | 1483.8 | 2.3932 | 161.94 | 1333.8 | 1483.7 | 2.2158 | 80.95 | 1333.7 | 1483.5 | 2.1393 |
| 1000 | 869.5 | 1373.9 | 1534.8 | 2.4294 | 173.86 | 1373.9 | 1534.7 | 2.2520 | 86.91 | 1373.8 | 1534.6 | 2.1755 |
| | $p$ = 14.7 lbf/in.$^2$ ($T_{sat}$ = 212.0°F) | | | | $p$ = 20 lbf/in.$^2$ ($T_{sat}$ = 228.0°F) | | | | $p$ = 40 lbf/in.$^2$ ($T_{sat}$ = 267.3°F) | | | |
| Sat. | 26.80 | 1077.6 | 1150.5 | 1.7567 | 20.09 | 1082.0 | 1156.4 | 1.7320 | 10.50 | 1093.3 | 1170.0 | 1.6767 |
| 250 | 28.42 | 1091.5 | 1168.8 | 1.7832 | 20.79 | 1090.3 | 1167.2 | 1.7475 | | | | |
| 300 | 30.52 | 1109.6 | 1192.6 | 1.8157 | 22.36 | 1108.7 | 1191.5 | 1.7805 | 11.04 | 1105.1 | 1186.8 | 1.6993 |
| 400 | 34.67 | 1145.6 | 1239.9 | 1.8741 | 25.43 | 1145.1 | 1239.2 | 1.8395 | 12.62 | 1143.0 | 1236.4 | 1.7606 |
| 500 | 38.77 | 1181.8 | 1287.3 | 1.9263 | 28.46 | 1181.5 | 1286.8 | 1.8919 | 14.16 | 1180.1 | 1284.9 | 1.8140 |
| 600 | 42.86 | 1218.6 | 1335.2 | 1.9737 | 31.47 | 1218.4 | 1334.8 | 1.9395 | 15.69 | 1217.3 | 1333.4 | 1.8621 |
| 700 | 46.93 | 1256.1 | 1383.8 | 2.0175 | 34.47 | 1255.9 | 1383.5 | 1.9834 | 17.20 | 1255.1 | 1382.4 | 1.9063 |
| 800 | 51.00 | 1294.4 | 1433.1 | 2.0584 | 37.46 | 1294.3 | 1432.9 | 2.0243 | 18.70 | 1293.7 | 1432.1 | 1.9474 |
| 900 | 55.07 | 1333.6 | 1483.4 | 2.0967 | 40.45 | 1333.5 | 1483.2 | 2.0627 | 20.20 | 1333.0 | 1482.5 | 1.9859 |
| 1000 | 59.13 | 1373.7 | 1534.5 | 2.1330 | 43.44 | 1373.5 | 1534.3 | 2.0989 | 21.70 | 1373.1 | 1533.8 | 2.0223 |
| 1100 | 63.19 | 1414.6 | 1586.4 | 2.1674 | 46.42 | 1414.5 | 1586.3 | 2.1334 | 23.20 | 1414.2 | 1585.9 | 2.0568 |
| | $p$ = 60 lbf/in.$^2$ ($T_{sat}$ = 292.7°F) | | | | $p$ = 80 lbf/in.$^2$ ($T_{sat}$ = 312.1°F) | | | | $p$ = 100 lbf/in.$^2$ ($T_{sat}$ = 327.8°F) | | | |
| Sat. | 7.17 | 1098.3 | 1178.0 | 1.6444 | 5.47 | 1102.6 | 1183.6 | 1.6214 | 4.434 | 1105.8 | 1187.8 | 1.6034 |
| 300 | 7.26 | 1101.3 | 1181.9 | 1.6496 | | | | | | | | |
| 350 | 7.82 | 1121.4 | 1208.2 | 1.6830 | 5.80 | 1118.5 | 1204.3 | 1.6476 | 4.592 | 1115.4 | 1200.4 | 1.6191 |
| 400 | 8.35 | 1140.8 | 1233.5 | 1.7134 | 6.22 | 1138.5 | 1230.6 | 1.6790 | 4.934 | 1136.2 | 1227.5 | 1.6517 |
| 500 | 9.40 | 1178.6 | 1283.0 | 1.7678 | 7.02 | 1177.2 | 1281.1 | 1.7346 | 5.587 | 1175.7 | 1279.1 | 1.7085 |
| 600 | 10.43 | 1216.3 | 1332.1 | 1.8165 | 7.79 | 1215.3 | 1330.7 | 1.7838 | 6.216 | 1214.2 | 1329.3 | 1.7582 |
| 700 | 11.44 | 1254.4 | 1381.4 | 1.8609 | 8.56 | 1253.6 | 1380.3 | 1.8285 | 6.834 | 1252.8 | 1379.2 | 1.8033 |
| 800 | 12.45 | 1293.0 | 1431.2 | 1.9022 | 9.32 | 1292.4 | 1430.4 | 1.8700 | 7.445 | 1291.8 | 1429.6 | 1.8449 |
| 900 | 13.45 | 1332.5 | 1481.8 | 1.9408 | 10.08 | 1332.0 | 1481.2 | 1.9087 | 8.053 | 1331.5 | 1480.5 | 1.8838 |
| 1000 | 14.45 | 1372.7 | 1533.2 | 1.9773 | 10.83 | 1372.3 | 1532.6 | 1.9453 | 8.657 | 1371.9 | 1532.1 | 1.9204 |
| 1100 | 15.45 | 1413.8 | 1585.4 | 2.0119 | 11.58 | 1413.5 | 1584.9 | 1.9799 | 9.260 | 1413.1 | 1584.5 | 1.9551 |
| 1200 | 16.45 | 1455.8 | 1638.5 | 2.0448 | 12.33 | 1455.5 | 1638.1 | 2.0130 | 9.861 | 1455.2 | 1637.7 | 1.9882 |

**Table T-4E** *(Continued)*

| $T$ °F | $v$ ft³/lb | $u$ Btu/lb | $h$ Btu/lb | $s$ Btu/lb · °R | $v$ ft³/lb | $u$ Btu/lb | $h$ Btu/lb | $s$ Btu/lb · °R | $v$ ft³/lb | $u$ Btu/lb | $h$ Btu/lb | $s$ Btu/lb · °R |
|---|---|---|---|---|---|---|---|---|---|---|---|---|
| | $p$ = 120 lbf/in.² ($T_{sat}$ = 341.3°F) | | | | $p$ = 140 lbf/in.² ($T_{sat}$ = 353.1°F) | | | | $p$ = 160 lbf/in.² ($T_{sat}$ = 363.6°F) | | | |
| Sat. | 3.730 | 1108.3 | 1191.1 | 1.5886 | 3.221 | 1110.3 | 1193.8 | 1.5761 | 2.836 | 1112.0 | 1196.0 | 1.5651 |
| 350 | 3.783 | 1112.2 | 1196.2 | 1.5950 | | | | | | | | |
| 400 | 4.079 | 1133.8 | 1224.4 | 1.6288 | 3.466 | 1131.4 | 1221.2 | 1.6088 | 3.007 | 1128.8 | 1217.8 | 1.5911 |
| 450 | 4.360 | 1154.3 | 1251.2 | 1.6590 | 3.713 | 1152.4 | 1248.6 | 1.6399 | 3.228 | 1150.5 | 1246.1 | 1.6230 |
| 500 | 4.633 | 1174.2 | 1277.1 | 1.6868 | 3.952 | 1172.7 | 1275.1 | 1.6682 | 3.440 | 1171.2 | 1273.0 | 1.6518 |
| 600 | 5.164 | 1213.2 | 1327.8 | 1.7371 | 4.412 | 1212.1 | 1326.4 | 1.7191 | 3.848 | 1211.1 | 1325.0 | 1.7034 |
| 700 | 5.682 | 1252.0 | 1378.2 | 1.7825 | 4.860 | 1251.2 | 1377.1 | 1.7648 | 4.243 | 1250.4 | 1376.0 | 1.7494 |
| 800 | 6.195 | 1291.2 | 1428.7 | 1.8243 | 5.301 | 1290.5 | 1427.9 | 1.8068 | 4.631 | 1289.9 | 1427.0 | 1.7916 |
| 900 | 6.703 | 1330.9 | 1479.8 | 1.8633 | 5.739 | 1330.4 | 1479.1 | 1.8459 | 5.015 | 1329.9 | 1478.4 | 1.8308 |
| 1000 | 7.208 | 1371.5 | 1531.5 | 1.9000 | 6.173 | 1371.0 | 1531.0 | 1.8827 | 5.397 | 1370.6 | 1530.4 | 1.8677 |
| 1100 | 7.711 | 1412.8 | 1584.0 | 1.9348 | 6.605 | 1412.4 | 1583.6 | 1.9176 | 5.776 | 1412.1 | 1583.1 | 1.9026 |
| 1200 | 8.213 | 1454.9 | 1637.3 | 1.9679 | 7.036 | 1454.6 | 1636.9 | 1.9507 | 6.154 | 1454.3 | 1636.5 | 1.9358 |
| | $p$ = 180 lbf/in.² ($T_{sat}$ = 373.1°F) | | | | $p$ = 200 lbf/in.² ($T_{sat}$ = 381.8°F) | | | | $p$ = 250 lbf/in.² ($T_{sat}$ = 401.0°F) | | | |
| Sat. | 2.533 | 1113.4 | 1197.8 | 1.5553 | 2.289 | 1114.6 | 1199.3 | 1.5464 | 1.845 | 1116.7 | 1202.1 | 1.5274 |
| 400 | 2.648 | 1126.2 | 1214.4 | 1.5749 | 2.361 | 1123.5 | 1210.8 | 1.5600 | | | | |
| 450 | 2.850 | 1148.5 | 1243.4 | 1.6078 | 2.548 | 1146.4 | 1240.7 | 1.5938 | 2.002 | 1141.1 | 1233.7 | 1.5632 |
| 500 | 3.042 | 1169.6 | 1270.9 | 1.6372 | 2.724 | 1168.0 | 1268.8 | 1.6239 | 2.150 | 1163.8 | 1263.3 | 1.5948 |
| 550 | 3.228 | 1190.0 | 1297.5 | 1.6642 | 2.893 | 1188.7 | 1295.7 | 1.6512 | 2.290 | 1185.3 | 1291.3 | 1.6233 |
| 600 | 3.409 | 1210.0 | 1323.5 | 1.6893 | 3.058 | 1208.9 | 1322.1 | 1.6767 | 2.426 | 1206.1 | 1318.3 | 1.6494 |
| 700 | 3.763 | 1249.6 | 1374.9 | 1.7357 | 3.379 | 1248.8 | 1373.8 | 1.7234 | 2.688 | 1246.7 | 1371.1 | 1.6970 |
| 800 | 4.110 | 1289.3 | 1426.2 | 1.7781 | 3.693 | 1288.6 | 1425.3 | 1.7660 | 2.943 | 1287.0 | 1423.2 | 1.7301 |
| 900 | 4.453 | 1329.4 | 1477.7 | 1.8174 | 4.003 | 1328.9 | 1477.1 | 1.8055 | 3.193 | 1327.6 | 1475.3 | 1.7799 |
| 1000 | 4.793 | 1370.2 | 1529.8 | 1.8545 | 4.310 | 1369.8 | 1529.3 | 1.8425 | 3.440 | 1368.7 | 1527.9 | 1.8172 |
| 1100 | 5.131 | 1411.7 | 1582.6 | 1.8894 | 4.615 | 1411.4 | 1582.2 | 1.8776 | 3.685 | 1410.5 | 1581.0 | 1.8524 |
| 1200 | 5.467 | 1454.0 | 1636.1 | 1.9227 | 4.918 | 1453.7 | 1635.7 | 1.9109 | 3.929 | 1453.0 | 1634.8 | 1.8858 |
| | $p$ = 300 lbf/in.² ($T_{sat}$ = 417.4°F) | | | | $p$ = 350 lbf/in.² ($T_{sat}$ = 431.8°F) | | | | $p$ = 400 lbf/in.² ($T_{sat}$ = 444.7°F) | | | |
| Sat. | 1.544 | 1118.2 | 1203.9 | 1.5115 | 1.327 | 1119.0 | 1204.9 | 1.4978 | 1.162 | 1119.5 | 1205.5 | 1.4856 |
| 450 | 1.636 | 1135.4 | 1226.2 | 1.5365 | 1.373 | 1129.2 | 1218.2 | 1.5125 | 1.175 | 1122.6 | 1209.5 | 1.4901 |
| 500 | 1.766 | 1159.5 | 1257.5 | 1.5701 | 1.491 | 1154.9 | 1251.5 | 1.5482 | 1.284 | 1150.1 | 1245.2 | 1.5282 |
| 550 | 1.888 | 1181.9 | 1286.7 | 1.5997 | 1.600 | 1178.3 | 1281.9 | 1.5790 | 1.383 | 1174.6 | 1277.0 | 1.5605 |
| 600 | 2.004 | 1203.2 | 1314.5 | 1.6266 | 1.703 | 1200.3 | 1310.6 | 1.6068 | 1.476 | 1197.3 | 1306.6 | 1.5892 |
| 700 | 2.227 | 1244.0 | 1368.3 | 1.6751 | 1.898 | 1242.5 | 1365.4 | 1.6562 | 1.650 | 1240.4 | 1362.5 | 1.6397 |
| 800 | 2.442 | 1285.4 | 1421.0 | 1.7187 | 2.085 | 1283.8 | 1418.8 | 1.7004 | 1.816 | 1282.1 | 1416.6 | 1.6844 |
| 900 | 2.653 | 1326.3 | 1473.6 | 1.7589 | 2.267 | 1325.0 | 1471.8 | 1.7409 | 1.978 | 1323.7 | 1470.1 | 1.7252 |
| 1000 | 2.860 | 1367.7 | 1526.5 | 1.7964 | 2.446 | 1366.6 | 1525.0 | 1.7787 | 2.136 | 1365.5 | 1523.6 | 1.7632 |
| 1100 | 3.066 | 1409.6 | 1579.8 | 1.8317 | 2.624 | 1408.7 | 1578.6 | 1.8142 | 2.292 | 1407.8 | 1577.4 | 1.7989 |
| 1200 | 3.270 | 1452.2 | 1633.8 | 1.8653 | 2.799 | 1451.5 | 1632.8 | 1.8478 | 2.446 | 1450.7 | 1621.8 | 1.8327 |
| 1300 | 3.473 | 1495.6 | 1688.4 | 1.8973 | 2.974 | 1495.0 | 1687.6 | 1.8799 | 2.599 | 1494.3 | 1686.8 | 1.8648 |

**Table T-4E** *(Continued)*

| $T$ °F | $v$ ft³/lb | $u$ Btu/lb | $h$ Btu/lb | $s$ Btu/lb · °R | $v$ ft³/lb | $u$ Btu/lb | $h$ Btu/lb | $s$ Btu/lb · °R | $v$ ft³/lb | $u$ Btu/lb | $h$ Btu/lb | $s$ Btu/lb · °R |
|---|---|---|---|---|---|---|---|---|---|---|---|---|
| | $p$ = 450 lbf/in.² ($T_{sat}$ = 456.4°F) | | | | $p$ = 500 lbf/in.² ($T_{sat}$ = 467.1°F) | | | | $p$ = 600 lbf/in.² ($T_{sat}$ = 486.3°F) | | | |
| Sat. | 1.033 | 1119.6 | 1205.6 | 1.4746 | 0.928 | 1119.4 | 1205.3 | 1.4645 | 0.770 | 1118.6 | 1204.1 | 1.4464 |
| 500 | 1.123 | 1145.1 | 1238.5 | 1.5097 | 0.992 | 1139.7 | 1231.5 | 1.4923 | 0.795 | 1128.0 | 1216.2 | 1.4592 |
| 550 | 1.215 | 1170.7 | 1271.9 | 1.5436 | 1.079 | 1166.7 | 1266.6 | 1.5279 | 0.875 | 1158.2 | 1255.4 | 1.4990 |
| 600 | 1.300 | 1194.3 | 1302.5 | 1.5732 | 1.158 | 1191.1 | 1298.3 | 1.5585 | 0.946 | 1184.5 | 1289.5 | 1.5320 |
| 700 | 1.458 | 1238.2 | 1359.6 | 1.6248 | 1.304 | 1236.0 | 1356.7 | 1.6112 | 1.073 | 1231.5 | 1350.6 | 1.5872 |
| 800 | 1.608 | 1280.5 | 1414.4 | 1.6701 | 1.441 | 1278.8 | 1412.1 | 1.6571 | 1.190 | 1275.4 | 1407.6 | 1.6343 |
| 900 | 1.752 | 1322.4 | 1468.3 | 1.7113 | 1.572 | 1321.0 | 1466.5 | 1.6987 | 1.302 | 1318.4 | 1462.9 | 1.6766 |
| 1000 | 1.894 | 1364.4 | 1522.2 | 1.7495 | 1.701 | 1363.3 | 1520.7 | 1.7371 | 1.411 | 1361.2 | 1517.8 | 1.7155 |
| 1100 | 2.034 | 1406.9 | 1576.3 | 1.7853 | 1.827 | 1406.0 | 1575.1 | 1.7731 | 1.517 | 1404.2 | 1572.7 | 1.7519 |
| 1200 | 2.172 | 1450.0 | 1630.8 | 1.8192 | 1.952 | 1449.2 | 1629.8 | 1.8072 | 1.622 | 1447.7 | 1627.8 | 1.7861 |
| 1300 | 2.308 | 1493.7 | 1685.9 | 1.8515 | 2.075 | 1493.1 | 1685.1 | 1.8395 | 1.726 | 1491.7 | 1683.4 | 1.8186 |
| 1400 | 2.444 | 1538.1 | 1741.7 | 1.8823 | 2.198 | 1537.6 | 1741.0 | 1.8704 | 1.829 | 1536.5 | 1739.5 | 1.8497 |
| | $p$ = 700 lbf/in.² ($T_{sat}$ = 503.2°F) | | | | $p$ = 800 lbf/in.² ($T_{sat}$ = 518.3°F) | | | | $p$ = 900 lbf/in.² ($T_{sat}$ = 532.1°F) | | | |
| Sat. | 0.656 | 1117.0 | 1202.0 | 1.4305 | 0.569 | 1115.0 | 1199.3 | 1.4160 | 0.501 | 1112.6 | 1196.0 | 1.4027 |
| 550 | 0.728 | 1149.0 | 1243.2 | 1.4723 | 0.615 | 1138.8 | 1229.9 | 1.4469 | 0.527 | 1127.5 | 1215.2 | 1.4219 |
| 600 | 0.793 | 1177.5 | 1280.2 | 1.5081 | 0.677 | 1170.1 | 1270.4 | 1.4861 | 0.587 | 1162.2 | 1260.0 | 1.4652 |
| 700 | 0.907 | 1226.9 | 1344.4 | 1.5661 | 0.783 | 1222.1 | 1338.0 | 1.5471 | 0.686 | 1217.1 | 1331.4 | 1.5297 |
| 800 | 1.011 | 1272.0 | 1402.9 | 1.6145 | 0.876 | 1268.5 | 1398.2 | 1.5969 | 0.772 | 1264.9 | 1393.4 | 1.5810 |
| 900 | 1.109 | 1315.6 | 1459.3 | 1.6576 | 0.964 | 1312.9 | 1455.6 | 1.6408 | 0.851 | 1310.1 | 1451.9 | 1.6257 |
| 1000 | 1.204 | 1358.9 | 1514.9 | 1.6970 | 1.048 | 1356.7 | 1511.9 | 1.6807 | 0.927 | 1354.5 | 1508.9 | 1.6662 |
| 1100 | 1.296 | 1402.4 | 1570.2 | 1.7337 | 1.130 | 1400.5 | 1567.8 | 1.7178 | 1.001 | 1398.7 | 1565.4 | 1.7036 |
| 1200 | 1.387 | 1446.2 | 1625.8 | 1.7682 | 1.210 | 1444.6 | 1623.8 | 1.7526 | 1.073 | 1443.0 | 1621.7 | 1.7386 |
| 1300 | 1.476 | 1490.4 | 1681.7 | 1.8009 | 1.289 | 1489.1 | 1680.0 | 1.7854 | 1.144 | 1487.8 | 1687.3 | 1.7717 |
| 1400 | 1.565 | 1535.3 | 1738.1 | 1.8321 | 1.367 | 1534.2 | 1736.6 | 1.8167 | 1.214 | 1533.0 | 1735.1 | 1.8031 |

**Table T-4E** *(Continued)*

| $T$ °F | $v$ ft³/lb | $u$ Btu/lb | $h$ Btu/lb | $s$ Btu/lb · °R | $v$ ft³/lb | $u$ Btu/lb | $h$ Btu/lb | $s$ Btu/lb · °R | $v$ ft³/lb | $u$ Btu/lb | $h$ Btu/lb | $s$ Btu/lb · °R |
|---|---|---|---|---|---|---|---|---|---|---|---|---|
| | $p$ = 1000 lbf/in.² ($T_{sat}$ = 544.7°F) | | | | $p$ = 1200 lbf/in.² ($T_{sat}$ = 567.4°F) | | | | $p$ = 1400 lbf/in.² ($T_{sat}$ = 587.2°F) | | | |
| Sat. | 0.446 | 1109.0 | 1192.4 | 1.3903 | 0.362 | 1103.5 | 1183.9 | 1.3673 | 0.302 | 1096.0 | 1174.1 | 1.3461 |
| 600 | 0.514 | 1153.7 | 1248.8 | 1.4450 | 0.402 | 1134.4 | 1223.6 | 1.4054 | 0.318 | 1110.9 | 1193.1 | 1.3641 |
| 650 | 0.564 | 1184.7 | 1289.1 | 1.4822 | 0.450 | 1170.9 | 1270.8 | 1.4490 | 0.367 | 1155.5 | 1250.5 | 1.4171 |
| 700 | 0.608 | 1212.0 | 1324.6 | 1.5135 | 0.491 | 1201.3 | 1310.2 | 1.4837 | 0.406 | 1189.6 | 1294.8 | 1.4562 |
| 800 | 0.688 | 1261.2 | 1388.5 | 1.5665 | 0.562 | 1253.7 | 1378.4 | 1.5402 | 0.471 | 1245.8 | 1367.9 | 1.5168 |
| 900 | 0.761 | 1307.3 | 1448.1 | 1.6120 | 0.626 | 1301.5 | 1440.4 | 1.5876 | 0.529 | 1295.6 | 1432.5 | 1.5661 |
| 1000 | 0.831 | 1352.2 | 1505.9 | 1.6530 | 0.685 | 1347.5 | 1499.7 | 1.6297 | 0.582 | 1342.8 | 1493.5 | 1.6094 |
| 1100 | 0.898 | 1396.8 | 1562.9 | 1.6908 | 0.743 | 1393.0 | 1557.9 | 1.6682 | 0.632 | 1389.1 | 1552.8 | 1.6487 |
| 1200 | 0.963 | 1441.5 | 1619.7 | 1.7261 | 0.798 | 1438.3 | 1615.5 | 1.7040 | 0.681 | 1435.1 | 1611.4 | 1.6851 |
| 1300 | 1.027 | 1486.5 | 1676.5 | 1.7593 | 0.853 | 1483.8 | 1673.1 | 1.7377 | 0.728 | 1481.1 | 1669.6 | 1.7192 |
| 1400 | 1.091 | 1531.9 | 1733.7 | 1.7909 | 0.906 | 1529.6 | 1730.7 | 1.7696 | 0.774 | 1527.2 | 1727.8 | 1.7513 |
| 1600 | 1.215 | 1624.4 | 1849.3 | 1.8499 | 1.011 | 1622.6 | 1847.1 | 1.8290 | 0.865 | 1620.8 | 1844.8 | 1.8111 |
| | $p$ = 1600 lbf/in.² ($T_{sat}$ = 605.1°F) | | | | $p$ = 1800 lbf/in.² ($T_{sat}$ = 621.2°F) | | | | $p$ = 2000 lbf/in.² ($T_{sat}$ = 636.0°F) | | | |
| Sat. | 0.255 | 1087.4 | 1162.9 | 1.3258 | 0.218 | 1077.7 | 1150.4 | 1.3060 | 0.188 | 1066.6 | 1136.3 | 1.2861 |
| 650 | 0.303 | 1137.8 | 1227.4 | 1.3852 | 0.251 | 1117.0 | 1200.4 | 1.3517 | 0.206 | 1091.1 | 1167.2 | 1.3141 |
| 700 | 0.342 | 1177.0 | 1278.1 | 1.4299 | 0.291 | 1163.1 | 1259.9 | 1.4042 | 0.249 | 1147.7 | 1239.8 | 1.3782 |
| 800 | 0.403 | 1237.7 | 1357.0 | 1.4953 | 0.350 | 1229.1 | 1345.7 | 1.4753 | 0.307 | 1220.1 | 1333.8 | 1.4562 |
| 900 | 0.466 | 1289.5 | 1424.4 | 1.5468 | 0.399 | 1283.2 | 1416.1 | 1.5291 | 0.353 | 1276.8 | 1407.6 | 1.5126 |
| 1000 | 0.504 | 1338.0 | 1487.1 | 1.5913 | 0.443 | 1333.1 | 1480.7 | 1.5749 | 0.395 | 1328.1 | 1474.1 | 1.5598 |
| 1100 | 0.549 | 1385.2 | 1547.7 | 1.6315 | 0.484 | 1381.2 | 1542.5 | 1.6159 | 0.433 | 1377.2 | 1537.2 | 1.6017 |
| 1200 | 0.592 | 1431.8 | 1607.1 | 1.6684 | 0.524 | 1428.5 | 1602.9 | 1.6534 | 0.469 | 1425.2 | 1598.6 | 1.6398 |
| 1300 | 0.634 | 1478.3 | 1666.1 | 1.7029 | 0.561 | 1475.5 | 1662.5 | 1.6883 | 0.503 | 1472.7 | 1659.0 | 1.6751 |
| 1400 | 0.675 | 1524.9 | 1724.8 | 1.7354 | 0.598 | 1522.5 | 1721.8 | 1.7211 | 0.537 | 1520.2 | 1718.8 | 1.7082 |
| 1600 | 0.755 | 1619.0 | 1842.6 | 1.7955 | 0.670 | 1617.2 | 1840.4 | 1.7817 | 0.602 | 1615.4 | 1838.2 | 1.7692 |

**Table T-4E** *(Continued)*

| T °F | v ft³/lb | u Btu/lb | h Btu/lb | s Btu/lb · °R | v ft³/lb | u Btu/lb | h Btu/lb | s Btu/lb · °R |
|---|---|---|---|---|---|---|---|---|
| | $p$ = 2500 lbf/in.² ($T_{sat}$ = 668.3°F) | | | | $p$ = 3000 lbf/in.² ($T_{sat}$ = 695.5°F) | | | |
| Sat. | 0.1306 | 1031.0 | 1091.4 | 1.2327 | 0.0840 | 968.8 | 1015.5 | 1.1575 |
| 700 | 0.1684 | 1098.7 | 1176.6 | 1.3073 | 0.0977 | 1003.9 | 1058.1 | 1.1944 |
| 750 | 0.2030 | 1155.2 | 1249.1 | 1.3686 | 0.1483 | 1114.7 | 1197.1 | 1.3122 |
| 800 | 0.2291 | 1195.7 | 1301.7 | 1.4112 | 0.1757 | 1167.6 | 1265.2 | 1.3675 |
| 900 | 0.2712 | 1259.9 | 1385.4 | 1.4752 | 0.2160 | 1241.8 | 1361.7 | 1.4414 |
| 1000 | 0.3069 | 1315.2 | 1457.2 | 1.5262 | 0.2485 | 1301.7 | 1439.6 | 1.4967 |
| 1100 | 0.3393 | 1366.8 | 1523.8 | 1.5704 | 0.2772 | 1356.2 | 1510.1 | 1.5434 |
| 1200 | 0.3696 | 1416.7 | 1587.7 | 1.6101 | 0.3086 | 1408.0 | 1576.6 | 1.5848 |
| 1300 | 0.3984 | 1465.7 | 1650.0 | 1.6465 | 0.3285 | 1458.5 | 1640.9 | 1.6224 |
| 1400 | 0.4261 | 1514.2 | 1711.3 | 1.6804 | 0.3524 | 1508.1 | 1703.7 | 1.6571 |
| 1500 | 0.4531 | 1562.5 | 1772.1 | 1.7123 | 0.3754 | 1557.3 | 1765.7 | 1.6896 |
| 1600 | 0.4795 | 1610.8 | 1832.6 | 1.7424 | 0.3978 | 1606.3 | 1827.1 | 1.7201 |
| | $p$ = 3500 lbf/in.² | | | | $p$ = 4000 lbf/in.² | | | |
| 650 | 0.0249 | 663.5 | 679.7 | 0.8630 | 0.0245 | 657.7 | 675.8 | 0.8574 |
| 700 | 0.0306 | 759.5 | 779.3 | 0.9506 | 0.0287 | 742.1 | 763.4 | 0.9345 |
| 750 | 0.1046 | 1058.4 | 1126.1 | 1.2440 | 0.0633 | 960.7 | 1007.5 | 1.1395 |
| 800 | 0.1363 | 1134.7 | 1223.0 | 1.3226 | 0.1052 | 1095.0 | 1172.9 | 1.2740 |
| 900 | 0.1763 | 1222.4 | 1336.5 | 1.4096 | 0.1462 | 1201.5 | 1309.7 | 1.3789 |
| 1000 | 0.2066 | 1287.6 | 1421.4 | 1.4699 | 0.1752 | 1272.9 | 1402.6 | 1.4449 |
| 1100 | 0.2328 | 1345.2 | 1496.0 | 1.5193 | 0.1995 | 1333.9 | 1481.6 | 1.4973 |
| 1200 | 0.2566 | 1399.2 | 1565.3 | 1.5624 | 0.2213 | 1390.1 | 1553.9 | 1.5423 |
| 1300 | 0.2787 | 1451.1 | 1631.7 | 1.6012 | 0.2414 | 1443.7 | 1622.4 | 1.5823 |
| 1400 | 0.2997 | 1501.9 | 1696.1 | 1.6368 | 0.2603 | 1495.7 | 1688.4 | 1.6188 |
| 1500 | 0.3199 | 1552.0 | 1759.2 | 1.6699 | 0.2784 | 1546.7 | 1752.8 | 1.6526 |
| 1600 | 0.3395 | 1601.7 | 1831.6 | 1.7010 | 0.2959 | 1597.1 | 1816.1 | 1.6841 |
| | $p$ = 4400 lbf/in.² | | | | $p$ = 4800 lbf/in.² | | | |
| 650 | 0.0242 | 653.6 | 673.3 | 0.8535 | 0.0237 | 649.8 | 671.0 | 0.8499 |
| 700 | 0.0278 | 732.7 | 755.3 | 0.9257 | 0.0271 | 725.1 | 749.1 | 0.9187 |
| 750 | 0.0415 | 870.8 | 904.6 | 1.0513 | 0.0352 | 832.6 | 863.9 | 1.0154 |
| 800 | 0.0844 | 1056.5 | 1125.3 | 1.2306 | 0.0668 | 1011.2 | 1070.5 | 1.1827 |
| 900 | 0.1270 | 1183.7 | 1287.1 | 1.3548 | 0.1109 | 1164.8 | 1263.4 | 1.3310 |
| 1000 | 0.1552 | 1260.8 | 1387.2 | 1.4260 | 0.1385 | 1248.3 | 1317.4 | 1.4078 |
| 1100 | 0.1784 | 1324.7 | 1469.9 | 1.4809 | 0.1608 | 1315.3 | 1458.1 | 1.4653 |
| 1200 | 0.1989 | 1382.8 | 1544.7 | 1.5274 | 0.1802 | 1375.4 | 1535.4 | 1.5133 |
| 1300 | 0.2176 | 1437.7 | 1614.9 | 1.5685 | 0.1979 | 1431.7 | 1607.4 | 1.5555 |
| 1400 | 0.2352 | 1490.7 | 1682.3 | 1.6057 | 0.2143 | 1485.7 | 1676.1 | 1.5934 |
| 1500 | 0.2520 | 1542.7 | 1747.6 | 1.6399 | 0.2300 | 1538.2 | 1742.5 | 1.6282 |
| 1600 | 0.2681 | 1593.4 | 1811.7 | 1.6718 | 0.2450 | 1589.8 | 1807.4 | 1.6605 |

**Table T-5E** Properties of Compressed Liquid Water

| T<br>°F | v<br>ft³/lb | u<br>Btu/lb | h<br>Btu/lb | s<br>Btu/lb · °R | v<br>ft³/lb | u<br>Btu/lb | h<br>Btu/lb | s<br>Btu/lb · °R |
|---|---|---|---|---|---|---|---|---|
| | $p$ = 500 lbf/in.$^2$ ($T_{sat}$ = 467.1°F) | | | | $p$ = 1000 lbf/in.$^2$ ($T_{sat}$ = 544.7°F) | | | |
| 32 | 0.015994 | 0.00 | 1.49 | 0.00000 | 0.015967 | 0.03 | 2.99 | 0.00005 |
| 50 | 0.015998 | 18.02 | 19.50 | 0.03599 | 0.015972 | 17.99 | 20.94 | 0.03592 |
| 100 | 0.016106 | 67.87 | 69.36 | 0.12932 | 0.016082 | 67.70 | 70.68 | 0.12901 |
| 150 | 0.016318 | 117.66 | 119.17 | 0.21457 | 0.016293 | 117.38 | 120.40 | 0.21410 |
| 200 | 0.016608 | 167.65 | 169.19 | 0.29341 | 0.016580 | 167.26 | 170.32 | 0.29281 |
| 300 | 0.017416 | 268.92 | 270.53 | 0.43641 | 0.017379 | 268.24 | 271.46 | 0.43552 |
| 400 | 0.018608 | 373.68 | 375.40 | 0.56604 | 0.018550 | 372.55 | 375.98 | 0.56472 |
| Sat. | 0.019748 | 447.70 | 449.53 | 0.64904 | 0.021591 | 538.39 | 542.38 | 0.74320 |
| | $p$ = 1500 lbf/in.$^2$ ($T_{sat}$ = 596.4°F) | | | | $p$ = 2000 lbf/in.$^2$ ($T_{sat}$ = 636.0°F) | | | |
| 32 | 0.015939 | 0.05 | 4.47 | 0.00007 | 0.015912 | 0.06 | 5.95 | 0.00008 |
| 50 | 0.015946 | 17.95 | 22.38 | 0.03584 | 0.015920 | 17.91 | 23.81 | 0.03575 |
| 100 | 0.016058 | 67.53 | 71.99 | 0.12870 | 0.016034 | 67.37 | 73.30 | 0.12839 |
| 150 | 0.016268 | 117.10 | 121.62 | 0.21364 | 0.016244 | 116.83 | 122.84 | 0.21318 |
| 200 | 0.016554 | 166.87 | 171.46 | 0.29221 | 0.016527 | 166.49 | 172.60 | 0.29162 |
| 300 | 0.017343 | 267.58 | 272.39 | 0.43463 | 0.017308 | 266.93 | 273.33 | 0.43376 |
| 400 | 0.018493 | 371.45 | 376.59 | 0.56343 | 0.018439 | 370.38 | 377.21 | 0.56216 |
| 500 | 0.02024 | 481.8 | 487.4 | 0.6853 | 0.02014 | 479.8 | 487.3 | 0.6832 |
| Sat. | 0.02346 | 605.0 | 611.5 | 0.8082 | 0.02565 | 662.4 | 671.9 | 0.8623 |
| | $p$ = 3000 lbf/in.$^2$ ($T_{sat}$ = 695.5°F) | | | | $p$ = 4000 lbf/in.$^2$ | | | |
| 32 | 0.015859 | 0.09 | 8.90 | 0.00009 | 0.015807 | 0.10 | 11.80 | 0.00005 |
| 50 | 0.015870 | 17.84 | 26.65 | 0.03555 | 0.015821 | 17.76 | 29.47 | 0.03534 |
| 100 | 0.015987 | 67.04 | 75.91 | 0.12777 | 0.015942 | 66.72 | 78.52 | 0.12714 |
| 150 | 0.016196 | 116.30 | 125.29 | 0.21226 | 0.016150 | 115.77 | 127.73 | 0.21136 |
| 200 | 0.016476 | 165.74 | 174.89 | 0.29046 | 0.016425 | 165.02 | 177.18 | 0.28931 |
| 300 | 0.017240 | 265.66 | 275.23 | 0.43205 | 0.017174 | 264.43 | 277.15 | 0.43038 |
| 400 | 0.018334 | 368.32 | 378.50 | 0.55970 | 0.018235 | 366.35 | 379.85 | 0.55734 |
| 500 | 0.019944 | 476.2 | 487.3 | 0.6794 | 0.019766 | 472.9 | 487.5 | 0.6758 |
| Sat. | 0.034310 | 783.5 | 802.5 | 0.9732 | | | | |

**Table T-6E** Properties of Saturated Refrigerant 134a (Liquid–Vapor): Temperature Table

| | | Specific Volume ft³/lb | | Internal Energy Btu/lb | | Enthalpy Btu/lb | | | Entropy Btu/lb · °R | | |
|---|---|---|---|---|---|---|---|---|---|---|---|
| Temp. °F | Press. lbf/in.² | Sat. Liquid $v_f$ | Sat. Vapor $v_g$ | Sat. Liquid $u_f$ | Sat. Vapor $u_g$ | Sat. Liquid $h_f$ | Evap. $h_{fg}$ | Sat. Vapor $h_g$ | Sat. Liquid $s_f$ | Sat. Vapor $s_g$ | Temp. °F |
| −40 | 7.490 | 0.01130 | 5.7173 | −0.02 | 87.90 | 0.00 | 95.82 | 95.82 | 0.0000 | 0.2283 | −40 |
| −30 | 9.920 | 0.01143 | 4.3911 | 2.81 | 89.26 | 2.83 | 94.49 | 97.32 | 0.0067 | 0.2266 | −30 |
| −20 | 12.949 | 0.01156 | 3.4173 | 5.69 | 90.62 | 5.71 | 93.10 | 98.81 | 0.0133 | 0.2250 | −20 |
| −15 | 14.718 | 0.01163 | 3.0286 | 7.14 | 91.30 | 7.17 | 92.38 | 99.55 | 0.0166 | 0.2243 | −15 |
| −10 | 16.674 | 0.01170 | 2.6918 | 8.61 | 91.98 | 8.65 | 91.64 | 100.29 | 0.0199 | 0.2236 | −10 |
| −5 | 18.831 | 0.01178 | 2.3992 | 10.09 | 92.66 | 10.13 | 90.89 | 101.02 | 0.0231 | 0.2230 | −5 |
| 0 | 21.203 | 0.01185 | 2.1440 | 11.58 | 93.33 | 11.63 | 90.12 | 101.75 | 0.0264 | 0.2224 | 0 |
| 5 | 23.805 | 0.01193 | 1.9208 | 13.09 | 94.01 | 13.14 | 89.33 | 102.47 | 0.0296 | 0.2219 | 5 |
| 10 | 26.651 | 0.01200 | 1.7251 | 14.60 | 94.68 | 14.66 | 88.53 | 103.19 | 0.0329 | 0.2214 | 10 |
| 15 | 29.756 | 0.01208 | 1.5529 | 16.13 | 95.35 | 16.20 | 87.71 | 103.90 | 0.0361 | 0.2209 | 15 |
| 20 | 33.137 | 0.01216 | 1.4009 | 17.67 | 96.02 | 17.74 | 86.87 | 104.61 | 0.0393 | 0.2205 | 20 |
| 25 | 36.809 | 0.01225 | 1.2666 | 19.22 | 96.69 | 19.30 | 86.02 | 105.32 | 0.0426 | 0.2200 | 25 |
| 30 | 40.788 | 0.01233 | 1.1474 | 20.78 | 97.35 | 20.87 | 85.14 | 106.01 | 0.0458 | 0.2196 | 30 |
| 40 | 49.738 | 0.01251 | 0.9470 | 23.94 | 98.67 | 24.05 | 83.34 | 107.39 | 0.0522 | 0.2189 | 40 |
| 50 | 60.125 | 0.01270 | 0.7871 | 27.14 | 99.98 | 27.28 | 81.46 | 108.74 | 0.0585 | 0.2183 | 50 |
| 60 | 72.092 | 0.01290 | 0.6584 | 30.39 | 101.27 | 30.56 | 79.49 | 110.05 | 0.0648 | 0.2178 | 60 |
| 70 | 85.788 | 0.01311 | 0.5538 | 33.68 | 102.54 | 33.89 | 77.44 | 111.33 | 0.0711 | 0.2173 | 70 |
| 80 | 101.37 | 0.01334 | 0.4682 | 37.02 | 103.78 | 37.27 | 75.29 | 112.56 | 0.0774 | 0.2169 | 80 |
| 85 | 109.92 | 0.01346 | 0.4312 | 38.72 | 104.39 | 38.99 | 74.17 | 113.16 | 0.0805 | 0.2167 | 85 |
| 90 | 118.99 | 0.01358 | 0.3975 | 40.42 | 105.00 | 40.72 | 73.03 | 113.75 | 0.0836 | 0.2165 | 90 |
| 95 | 128.62 | 0.01371 | 0.3668 | 42.14 | 105.60 | 42.47 | 71.86 | 114.33 | 0.0867 | 0.2163 | 95 |
| 100 | 138.83 | 0.01385 | 0.3388 | 43.87 | 106.18 | 44.23 | 70.66 | 114.89 | 0.0898 | 0.2161 | 100 |
| 105 | 149.63 | 0.01399 | 0.3131 | 45.62 | 106.76 | 46.01 | 69.42 | 115.43 | 0.0930 | 0.2159 | 105 |
| 110 | 161.04 | 0.01414 | 0.2896 | 47.39 | 107.33 | 47.81 | 68.15 | 115.96 | 0.0961 | 0.2157 | 110 |
| 115 | 173.10 | 0.01429 | 0.2680 | 49.17 | 107.88 | 49.63 | 66.84 | 116.47 | 0.0992 | 0.2155 | 115 |
| 120 | 185.82 | 0.01445 | 0.2481 | 50.97 | 108.42 | 51.47 | 65.48 | 116.95 | 0.1023 | 0.2153 | 120 |
| 140 | 243.86 | 0.01520 | 0.1827 | 58.39 | 110.41 | 59.08 | 59.57 | 118.65 | 0.1150 | 0.2143 | 140 |
| 160 | 314.63 | 0.01617 | 0.1341 | 66.26 | 111.97 | 67.20 | 52.58 | 119.78 | 0.1280 | 0.2128 | 160 |
| 180 | 400.22 | 0.01758 | 0.0964 | 74.83 | 112.77 | 76.13 | 43.78 | 119.91 | 0.1417 | 0.2101 | 180 |
| 200 | 503.52 | 0.02014 | 0.0647 | 84.90 | 111.66 | 86.77 | 30.92 | 117.69 | 0.1575 | 0.2044 | 200 |
| 210 | 563.51 | 0.02329 | 0.0476 | 91.84 | 108.48 | 94.27 | 19.18 | 113.45 | 0.1684 | 0.1971 | 210 |

*Source:* Tables T-6E through T-8E are calculated based on equations from D. P. Wilson and R. S. Basu, "Thermodynamic Properties of a New Stratospherically Safe Working Fluid—Refrigerant 134a," *ASHRAE Trans.*, Vol. 94, Pt. 2, 1988, pp. 2095–2118.

**Table T-7E** Properties of Saturated Refrigerant 134a (Liquid–Vapor): Pressure Table

| | | Specific Volume ft³/lb | | Internal Energy Btu/lb | | Enthalpy Btu/lb | | | Entropy Btu/lb · °R | | |
|---|---|---|---|---|---|---|---|---|---|---|---|
| Press. lbf/in.² | Temp. °F | Sat. Liquid $v_f$ | Sat. Vapor $v_g$ | Sat. Liquid $u_f$ | Sat. Vapor $u_g$ | Sat. Liquid $h_f$ | Evap. $h_{fg}$ | Sat. Vapor $h_g$ | Sat. Liquid $s_f$ | Sat. Vapor $s_g$ | Press. lbf/in.² |
| 5 | −53.48 | 0.01113 | 8.3508 | −3.74 | 86.07 | −3.73 | 97.53 | 93.79 | −0.0090 | 0.2311 | 5 |
| 10 | −29.71 | 0.01143 | 4.3581 | 2.89 | 89.30 | 2.91 | 94.45 | 97.37 | 0.0068 | 0.2265 | 10 |
| 15 | −14.25 | 0.01164 | 2.9747 | 7.36 | 91.40 | 7.40 | 92.27 | 99.66 | 0.0171 | 0.2242 | 15 |
| 20 | −2.48 | 0.01181 | 2.2661 | 10.84 | 93.00 | 10.89 | 90.50 | 101.39 | 0.0248 | 0.2227 | 20 |
| 30 | 15.38 | 0.01209 | 1.5408 | 16.24 | 95.40 | 16.31 | 87.65 | 103.96 | 0.0364 | 0.2209 | 30 |
| 40 | 29.04 | 0.01232 | 1.1692 | 20.48 | 97.23 | 20.57 | 85.31 | 105.88 | 0.0452 | 0.2197 | 40 |
| 50 | 40.27 | 0.01252 | 0.9422 | 24.02 | 98.71 | 24.14 | 83.29 | 107.43 | 0.0523 | 0.2189 | 50 |
| 60 | 49.89 | 0.01270 | 0.7887 | 27.10 | 99.96 | 27.24 | 81.48 | 108.72 | 0.0584 | 0.2183 | 60 |
| 70 | 58.35 | 0.01286 | 0.6778 | 29.85 | 101.05 | 30.01 | 79.82 | 109.83 | 0.0638 | 0.2179 | 70 |
| 80 | 65.93 | 0.01302 | 0.5938 | 32.33 | 102.02 | 32.53 | 78.28 | 110.81 | 0.0686 | 0.2175 | 80 |
| 90 | 72.83 | 0.01317 | 0.5278 | 34.62 | 102.89 | 34.84 | 76.84 | 111.68 | 0.0729 | 0.2172 | 90 |
| 100 | 79.17 | 0.01332 | 0.4747 | 36.75 | 103.68 | 36.99 | 75.47 | 112.46 | 0.0768 | 0.2169 | 100 |
| 120 | 90.54 | 0.01360 | 0.3941 | 40.61 | 105.06 | 40.91 | 72.91 | 113.82 | 0.0839 | 0.2165 | 120 |
| 140 | 100.56 | 0.01386 | 0.3358 | 44.07 | 106.25 | 44.43 | 70.52 | 114.95 | 0.0902 | 0.2161 | 140 |
| 160 | 109.56 | 0.01412 | 0.2916 | 47.23 | 107.28 | 47.65 | 68.26 | 115.91 | 0.0958 | 0.2157 | 160 |
| 180 | 117.74 | 0.01438 | 0.2569 | 50.16 | 108.18 | 50.64 | 66.10 | 116.74 | 0.1009 | 0.2154 | 180 |
| 200 | 125.28 | 0.01463 | 0.2288 | 52.90 | 108.98 | 53.44 | 64.01 | 117.44 | 0.1057 | 0.2151 | 200 |
| 220 | 132.27 | 0.01489 | 0.2056 | 55.48 | 109.68 | 56.09 | 61.96 | 118.05 | 0.1101 | 0.2147 | 220 |
| 240 | 138.79 | 0.01515 | 0.1861 | 57.93 | 110.30 | 58.61 | 59.96 | 118.56 | 0.1142 | 0.2144 | 240 |
| 260 | 144.92 | 0.01541 | 0.1695 | 60.28 | 110.84 | 61.02 | 57.97 | 118.99 | 0.1181 | 0.2140 | 260 |
| 280 | 150.70 | 0.01568 | 0.1550 | 62.53 | 111.31 | 63.34 | 56.00 | 119.35 | 0.1219 | 0.2136 | 280 |
| 300 | 156.17 | 0.01596 | 0.1424 | 64.71 | 111.72 | 65.59 | 54.03 | 119.62 | 0.1254 | 0.2132 | 300 |
| 350 | 168.72 | 0.01671 | 0.1166 | 69.88 | 112.45 | 70.97 | 49.03 | 120.00 | 0.1338 | 0.2118 | 350 |
| 400 | 179.95 | 0.01758 | 0.0965 | 74.81 | 112.77 | 76.11 | 43.80 | 119.91 | 0.1417 | 0.2102 | 400 |
| 450 | 190.12 | 0.01863 | 0.0800 | 79.63 | 112.60 | 81.18 | 38.08 | 119.26 | 0.1493 | 0.2079 | 450 |
| 500 | 199.38 | 0.02002 | 0.0657 | 84.54 | 111.76 | 86.39 | 31.44 | 117.83 | 0.1570 | 0.2047 | 500 |

**Table T-8E** Properties of Superheated Refrigerant 134a Vapor

| $T$ °F | $v$ ft³/lb | $u$ Btu/lb | $h$ Btu/lb | $s$ Btu/lb · °R | $v$ ft³/lb | $u$ Btu/lb | $h$ Btu/lb | $s$ Btu/lb · °R | $v$ ft³/lb | $u$ Btu/lb | $h$ Btu/lb | $s$ Btu/lb · °R |
|---|---|---|---|---|---|---|---|---|---|---|---|---|
| | $p$ = 10 lbf/in.² ($T_{sat}$ = −29.71°F) | | | | $p$ = 15 lbf/in.² ($T_{sat}$ = −14.25°F) | | | | $p$ = 20 lbf/in.² ($T_{sat}$ = −2.48°F) | | | |
| Sat. | 4.3581 | 89.30 | 97.37 | 0.2265 | 2.9747 | 91.40 | 99.66 | 0.2242 | 2.2661 | 93.00 | 101.39 | 0.2227 |
| −20 | 4.4718 | 90.89 | 99.17 | 0.2307 | | | | | | | | |
| 0 | 4.7026 | 94.24 | 102.94 | 0.2391 | 3.0893 | 93.84 | 102.42 | 0.2303 | 2.2816 | 93.43 | 101.88 | 0.2238 |
| 20 | 4.9297 | 97.67 | 106.79 | 0.2472 | 3.2468 | 97.33 | 106.34 | 0.2386 | 2.4046 | 96.98 | 105.88 | 0.2323 |
| 40 | 5.1539 | 101.19 | 110.72 | 0.2553 | 3.4012 | 100.89 | 110.33 | 0.2468 | 2.5244 | 100.59 | 109.94 | 0.2406 |
| 60 | 5.3758 | 104.80 | 114.74 | 0.2632 | 3.5533 | 104.54 | 114.40 | 0.2548 | 2.6416 | 104.28 | 114.06 | 0.2487 |
| 80 | 5.5959 | 108.50 | 118.85 | 0.2709 | 3.7034 | 108.28 | 118.56 | 0.2626 | 2.7569 | 108.05 | 118.25 | 0.2566 |
| 100 | 5.8145 | 112.29 | 123.05 | 0.2786 | 3.8520 | 112.10 | 122.79 | 0.2703 | 2.8705 | 111.90 | 122.52 | 0.2644 |
| 120 | 6.0318 | 116.18 | 127.34 | 0.2861 | 3.9993 | 116.01 | 127.11 | 0.2779 | 2.9829 | 115.83 | 126.87 | 0.2720 |
| 140 | 6.2482 | 120.16 | 131.72 | 0.2935 | 4.1456 | 120.00 | 131.51 | 0.2854 | 3.0942 | 119.85 | 131.30 | 0.2795 |
| 160 | 6.4638 | 124.23 | 136.19 | 0.3009 | 4.2911 | 124.09 | 136.00 | 0.2927 | 3.2047 | 123.95 | 135.81 | 0.2869 |
| 180 | 6.6786 | 128.38 | 140.74 | 0.3081 | 4.4359 | 128.26 | 140.57 | 0.3000 | 3.3144 | 128.13 | 140.40 | 0.2922 |
| 200 | 6.8929 | 132.63 | 145.39 | 0.3152 | 4.5801 | 132.52 | 145.23 | 0.3072 | 3.4236 | 132.40 | 145.07 | 0.3014 |

| $T$ °F | $v$ ft³/lb | $u$ Btu/lb | $h$ Btu/lb | $s$ Btu/lb · °R | $v$ ft³/lb | $u$ Btu/lb | $h$ Btu/lb | $s$ Btu/lb · °R | $v$ ft³/lb | $u$ Btu/lb | $h$ Btu/lb | $s$ Btu/lb · °R |
|---|---|---|---|---|---|---|---|---|---|---|---|---|
| | $p$ = 30 lbf/in.² ($T_{sat}$ = 15.38°F) | | | | $p$ = 40 lbf/in.² ($T_{sat}$ = 29.04°F) | | | | $p$ = 50 lbf/in.² ($T_{sat}$ = 40.27°F) | | | |
| Sat. | 1.5408 | 95.40 | 103.96 | 0.2209 | 1.1692 | 97.23 | 105.88 | 0.2197 | 0.9422 | 98.71 | 107.43 | 0.2189 |
| 20 | 1.5611 | 96.26 | 104.92 | 0.2229 | | | | | | | | |
| 40 | 1.6465 | 99.98 | 109.12 | 0.2315 | 1.2065 | 99.33 | 108.26 | 0.2245 | | | | |
| 60 | 1.7293 | 103.75 | 113.35 | 0.2398 | 1.2723 | 103.20 | 112.62 | 0.2331 | 0.9974 | 102.62 | 111.85 | 0.2276 |
| 80 | 1.8098 | 107.59 | 117.63 | 0.2478 | 1.3357 | 107.11 | 117.00 | 0.2414 | 1.0508 | 106.62 | 116.34 | 0.2361 |
| 100 | 1.8887 | 111.49 | 121.98 | 0.2558 | 1.3973 | 111.08 | 121.42 | 0.2494 | 1.1022 | 110.65 | 120.85 | 0.2443 |
| 120 | 1.9662 | 115.47 | 126.39 | 0.2635 | 1.4575 | 115.11 | 125.90 | 0.2573 | 1.1520 | 114.74 | 125.39 | 0.2523 |
| 140 | 2.0426 | 119.53 | 130.87 | 0.2711 | 1.5165 | 119.21 | 130.43 | 0.2650 | 1.2007 | 118.88 | 129.99 | 0.2601 |
| 160 | 2.1181 | 123.66 | 135.42 | 0.2786 | 1.5746 | 123.38 | 135.03 | 0.2725 | 1.2484 | 123.08 | 134.64 | 0.2677 |
| 180 | 2.1929 | 127.88 | 140.05 | 0.2859 | 1.6319 | 127.62 | 139.70 | 0.2799 | 1.2953 | 127.36 | 139.34 | 0.2752 |
| 200 | 2.2671 | 132.17 | 144.76 | 0.2932 | 1.6887 | 131.94 | 144.44 | 0.2872 | 1.3415 | 131.71 | 144.12 | 0.2825 |
| 220 | 2.3407 | 136.55 | 149.54 | 0.3003 | 1.7449 | 136.34 | 149.25 | 0.2944 | 1.3873 | 136.12 | 148.96 | 0.2897 |
| 240 | | | | | 1.8006 | 140.81 | 154.14 | 0.3015 | 1.4326 | 140.61 | 153.87 | 0.2969 |
| 260 | | | | | 1.8561 | 145.36 | 159.10 | 0.3085 | 1.4775 | 145.18 | 158.85 | 0.3039 |

| $T$ °F | $v$ ft³/lb | $u$ Btu/lb | $h$ Btu/lb | $s$ Btu/lb · °R | $v$ ft³/lb | $u$ Btu/lb | $h$ Btu/lb | $s$ Btu/lb · °R | $v$ ft³/lb | $u$ Btu/lb | $h$ Btu/lb | $s$ Btu/lb · °R |
|---|---|---|---|---|---|---|---|---|---|---|---|---|
| | $p$ = 60 lbf/in.² ($T_{sat}$ = 49.89°F) | | | | $p$ = 70 lbf/in.² ($T_{sat}$ = 58.35°F) | | | | $p$ = 80 lbf/in.² ($T_{sat}$ = 65.93°F) | | | |
| Sat. | 0.7887 | 99.96 | 108.72 | 0.2183 | 0.6778 | 101.05 | 109.83 | 0.2179 | 0.5938 | 102.02 | 110.81 | 0.2175 |
| 60 | 0.8135 | 102.03 | 111.06 | 0.2229 | 0.6814 | 101.40 | 110.23 | 0.2186 | | | | |
| 80 | 0.8604 | 106.11 | 115.66 | 0.2316 | 0.7239 | 105.58 | 114.96 | 0.2276 | 0.6211 | 105.03 | 114.23 | 0.2239 |
| 100 | 0.9051 | 110.21 | 120.26 | 0.2399 | 0.7640 | 109.76 | 119.66 | 0.2361 | 0.6579 | 109.30 | 119.04 | 0.2327 |
| 120 | 0.9482 | 114.35 | 124.88 | 0.2480 | 0.8023 | 113.96 | 124.36 | 0.2444 | 0.6927 | 113.56 | 123.82 | 0.2411 |
| 140 | 0.9900 | 118.54 | 129.53 | 0.2559 | 0.8393 | 118.20 | 129.07 | 0.2524 | 0.7261 | 117.85 | 128.60 | 0.2492 |
| 160 | 1.0308 | 122.79 | 134.23 | 0.2636 | 0.8752 | 122.49 | 133.82 | 0.2601 | 0.7584 | 122.18 | 133.41 | 0.2570 |
| 180 | 1.0707 | 127.10 | 138.98 | 0.2712 | 0.9103 | 126.83 | 138.62 | 0.2678 | 0.7898 | 126.55 | 138.25 | 0.2647 |
| 200 | 1.1100 | 131.47 | 143.79 | 0.2786 | 0.9446 | 131.23 | 143.46 | 0.2752 | 0.8205 | 130.98 | 143.13 | 0.2722 |
| 220 | 1.1488 | 135.91 | 148.66 | 0.2859 | 0.9784 | 135.69 | 148.36 | 0.2825 | 0.8506 | 135.47 | 148.06 | 0.2796 |
| 240 | 1.1871 | 140.42 | 153.60 | 0.2930 | 1.0118 | 140.22 | 153.33 | 0.2897 | 0.8803 | 140.02 | 153.05 | 0.2868 |
| 260 | 1.2251 | 145.00 | 158.60 | 0.3001 | 1.0448 | 144.82 | 158.35 | 0.2968 | 0.9095 | 144.63 | 158.10 | 0.2940 |
| 280 | 1.2627 | 149.65 | 163.67 | 0.3070 | 1.0774 | 149.48 | 163.44 | 0.3038 | 0.9384 | 149.32 | 163.21 | 0.3010 |
| 300 | 1.3001 | 154.38 | 168.81 | 0.3139 | 1.1098 | 154.22 | 168.60 | 0.3107 | 0.9671 | 154.06 | 168.38 | 0.3079 |

**Table T-8E** *(Continued)*

| T °F | v ft³/lb | u Btu/lb | h Btu/lb | s Btu/lb · °R | v ft³/lb | u Btu/lb | h Btu/lb | s Btu/lb · °R | v ft³/lb | u Btu/lb | h Btu/lb | s Btu/lb · °R |
|---|---|---|---|---|---|---|---|---|---|---|---|---|
| | $p$ = 90 lbf/in.² ($T_{sat}$ = 72.83°F) | | | | $p$ = 100 lbf/in.² ($T_{sat}$ = 79.17°F) | | | | $p$ = 120 lbf/in.² ($T_{sat}$ = 90.54°F) | | | |
| Sat. | 0.5278 | 102.89 | 111.68 | 0.2172 | 0.4747 | 103.68 | 112.46 | 0.2169 | 0.3941 | 105.06 | 113.82 | 0.2165 |
| 80 | 0.5408 | 104.46 | 113.47 | 0.2205 | 0.4761 | 103.87 | 112.68 | 0.2173 | | | | |
| 100 | 0.5751 | 108.82 | 118.39 | 0.2295 | 0.5086 | 108.32 | 117.73 | 0.2265 | 0.4080 | 107.26 | 116.32 | 0.2210 |
| 120 | 0.6073 | 113.15 | 123.27 | 0.2380 | 0.5388 | 112.73 | 122.70 | 0.2352 | 0.4355 | 111.84 | 121.52 | 0.2301 |
| 140 | 0.6380 | 117.50 | 128.12 | 0.2463 | 0.5674 | 117.13 | 127.63 | 0.2436 | 0.4610 | 116.37 | 126.61 | 0.2387 |
| 160 | 0.6675 | 121.87 | 132.98 | 0.2542 | 0.5947 | 121.55 | 132.55 | 0.2517 | 0.4852 | 120.89 | 131.66 | 0.2470 |
| 180 | 0.6961 | 126.28 | 137.87 | 0.2620 | 0.6210 | 125.99 | 137.49 | 0.2595 | 0.5082 | 125.42 | 136.70 | 0.2550 |
| 200 | 0.7239 | 130.73 | 142.79 | 0.2696 | 0.6466 | 130.48 | 142.45 | 0.2671 | 0.5305 | 129.97 | 141.75 | 0.2628 |
| 220 | 0.7512 | 135.25 | 147.76 | 0.2770 | 0.6716 | 135.02 | 147.45 | 0.2746 | 0.5520 | 134.56 | 146.82 | 0.2704 |
| 240 | 0.7779 | 139.82 | 152.77 | 0.2843 | 0.6960 | 139.61 | 152.49 | 0.2819 | 0.5731 | 139.20 | 151.92 | 0.2778 |
| 260 | 0.8043 | 144.45 | 157.84 | 0.2914 | 0.7201 | 144.26 | 157.59 | 0.2891 | 0.5937 | 143.89 | 157.07 | 0.2850 |
| 280 | 0.8303 | 149.15 | 162.97 | 0.2984 | 0.7438 | 148.98 | 162.74 | 0.2962 | 0.6140 | 148.63 | 162.26 | 0.2921 |
| 300 | 0.8561 | 153.91 | 168.16 | 0.3054 | 0.7672 | 153.75 | 167.95 | 0.3031 | 0.6339 | 153.43 | 167.51 | 0.2991 |
| 320 | 0.8816 | 158.73 | 173.42 | 0.3122 | 0.7904 | 158.59 | 173.21 | 0.3099 | 0.6537 | 158.29 | 172.81 | 0.3060 |
| | $p$ = 140 lbf/in.² ($T_{sat}$ = 100.56°F) | | | | $p$ = 160 lbf/in.² ($T_{sat}$ = 109.55°F) | | | | $p$ = 180 lbf/in.² ($T_{sat}$ = 117.74°F) | | | |
| Sat. | 0.3358 | 106.25 | 114.95 | 0.2161 | 0.2916 | 107.28 | 115.91 | 0.2157 | 0.2569 | 108.18 | 116.74 | 0.2154 |
| 120 | 0.3610 | 110.90 | 120.25 | 0.2254 | 0.3044 | 109.88 | 118.89 | 0.2209 | 0.2595 | 108.77 | 117.41 | 0.2166 |
| 140 | 0.3846 | 115.58 | 125.54 | 0.2344 | 0.3269 | 114.73 | 124.41 | 0.2303 | 0.2814 | 113.83 | 123.21 | 0.2264 |
| 160 | 0.4066 | 120.21 | 130.74 | 0.2429 | 0.3474 | 119.49 | 129.78 | 0.2391 | 0.3011 | 118.74 | 128.77 | 0.2355 |
| 180 | 0.4274 | 124.82 | 135.89 | 0.2511 | 0.3666 | 124.20 | 135.06 | 0.2475 | 0.3191 | 123.56 | 134.19 | 0.2441 |
| 200 | 0.4474 | 129.44 | 141.03 | 0.2590 | 0.3849 | 128.90 | 140.29 | 0.2555 | 0.3361 | 128.34 | 139.53 | 0.2524 |
| 220 | 0.4666 | 134.09 | 146.18 | 0.2667 | 0.4023 | 133.61 | 145.52 | 0.2633 | 0.3523 | 133.11 | 144.84 | 0.2603 |
| 240 | 0.4852 | 138.77 | 151.34 | 0.2742 | 0.4192 | 138.34 | 150.75 | 0.2709 | 0.3678 | 137.90 | 150.15 | 0.2680 |
| 260 | 0.5034 | 143.50 | 156.54 | 0.2815 | 0.4356 | 143.11 | 156.00 | 0.2783 | 0.3828 | 142.71 | 155.46 | 0.2755 |
| 280 | 0.5212 | 148.28 | 161.78 | 0.2887 | 0.4516 | 147.92 | 161.29 | 0.2856 | 0.3974 | 147.55 | 160.79 | 0.2828 |
| 300 | 0.5387 | 153.11 | 167.06 | 0.2957 | 0.4672 | 152.78 | 166.61 | 0.2927 | 0.4116 | 152.44 | 166.15 | 0.2899 |
| 320 | 0.5559 | 157.99 | 172.39 | 0.3026 | 0.4826 | 157.69 | 171.98 | 0.2996 | 0.4256 | 157.38 | 171.55 | 0.2969 |
| 340 | 0.5730 | 162.93 | 177.78 | 0.3094 | 0.4978 | 162.65 | 177.39 | 0.3065 | 0.4393 | 162.36 | 177.00 | 0.3038 |
| 360 | 0.5898 | 167.93 | 183.21 | 0.3162 | 0.5128 | 167.67 | 182.85 | 0.3132 | 0.4529 | 167.40 | 182.49 | 0.3106 |
| | $p$ = 200 lbf/in.² ($T_{sat}$ = 125.28°F) | | | | $p$ = 300 lbf/in.² ($T_{sat}$ = 156.17°F) | | | | $p$ = 400 lbf/in.² ($T_{sat}$ = 179.95°F) | | | |
| Sat. | 0.2288 | 108.98 | 117.44 | 0.2151 | 0.1424 | 111.72 | 119.62 | 0.2132 | 0.0965 | 112.77 | 119.91 | 0.2102 |
| 140 | 0.2446 | 112.87 | 121.92 | 0.2226 | | | | | | | | |
| 160 | 0.2636 | 117.94 | 127.70 | 0.2321 | 0.1462 | 112.95 | 121.07 | 0.2155 | | | | |
| 180 | 0.2809 | 122.88 | 133.28 | 0.2410 | 0.1633 | 118.93 | 128.00 | 0.2265 | 0.0965 | 112.79 | 119.93 | 0.2102 |
| 200 | 0.2970 | 127.76 | 138.75 | 0.2494 | 0.1777 | 124.47 | 134.34 | 0.2363 | 0.1143 | 120.14 | 128.60 | 0.2235 |
| 220 | 0.3121 | 132.60 | 144.15 | 0.2575 | 0.1905 | 129.79 | 140.36 | 0.2453 | 0.1275 | 126.35 | 135.79 | 0.2343 |
| 240 | 0.3266 | 137.44 | 149.53 | 0.2653 | 0.2021 | 134.99 | 146.21 | 0.2537 | 0.1386 | 132.12 | 142.38 | 0.2438 |
| 260 | 0.3405 | 142.30 | 154.90 | 0.2728 | 0.2130 | 140.12 | 151.95 | 0.2618 | 0.1484 | 137.65 | 148.64 | 0.2527 |
| 280 | 0.3540 | 147.18 | 160.28 | 0.2802 | 0.2234 | 145.23 | 157.63 | 0.2696 | 0.1575 | 143.06 | 154.72 | 0.2610 |
| 300 | 0.3671 | 152.10 | 165.69 | 0.2874 | 0.2333 | 150.33 | 163.28 | 0.2772 | 0.1660 | 148.39 | 160.67 | 0.2689 |
| 320 | 0.3799 | 157.07 | 171.13 | 0.2945 | 0.2428 | 155.44 | 168.92 | 0.2845 | 0.1740 | 153.69 | 166.57 | 0.2766 |
| 340 | 0.3926 | 162.07 | 176.60 | 0.3014 | 0.2521 | 160.57 | 174.56 | 0.2916 | 0.1816 | 158.97 | 172.42 | 0.2840 |
| 360 | 0.4050 | 167.13 | 182.12 | 0.3082 | 0.2611 | 165.74 | 180.23 | 0.2986 | 0.1890 | 164.26 | 178.26 | 0.2912 |
| 380 | | | | | 0.2699 | 170.94 | 185.92 | 0.3055 | 0.1962 | 169.57 | 184.09 | 0.2983 |

**Table T-9E** Ideal Gas Properties of Air

$T$(°R), $h$ and $u$(Btu/lb), $s°$(Btu/lb · °R)

| | | | | when $\Delta s = 0$[1] | | | | | | when $\Delta s = 0$ | |
|---|---|---|---|---|---|---|---|---|---|---|---|
| $T$ | $h$ | $u$ | $s°$ | $p_r$ | $v_r$ | $T$ | $h$ | $u$ | $s°$ | $p_r$ | $v_r$ |
| 360 | 85.97 | 61.29 | 0.50369 | 0.3363 | 396.6 | 1480 | 363.89 | 262.44 | 0.85062 | 53.04 | 10.34 |
| 380 | 90.75 | 64.70 | 0.51663 | 0.4061 | 346.6 | 1520 | 374.47 | 270.26 | 0.85767 | 58.78 | 9.578 |
| 400 | 95.53 | 68.11 | 0.52890 | 0.4858 | 305.0 | 1560 | 385.08 | 278.13 | 0.86456 | 65.00 | 8.890 |
| 420 | 100.32 | 71.52 | 0.54058 | 0.5760 | 270.1 | 1600 | 395.74 | 286.06 | 0.87130 | 71.73 | 8.263 |
| 440 | 105.11 | 74.93 | 0.55172 | 0.6776 | 240.6 | 1650 | 409.13 | 296.03 | 0.87954 | 80.89 | 7.556 |
| 460 | 109.90 | 78.36 | 0.56235 | 0.7913 | 215.33 | 1700 | 422.59 | 306.06 | 0.88758 | 90.95 | 6.924 |
| 480 | 114.69 | 81.77 | 0.57255 | 0.9182 | 193.65 | 1750 | 436.12 | 316.16 | 0.89542 | 101.98 | 6.357 |
| 500 | 119.48 | 85.20 | 0.58233 | 1.0590 | 174.90 | 1800 | 449.71 | 326.32 | 0.90308 | 114.0 | 5.847 |
| 520 | 124.27 | 88.62 | 0.59172 | 1.2147 | 158.58 | 1850 | 463.37 | 336.55 | 0.91056 | 127.2 | 5.388 |
| 537 | 128.34 | 91.53 | 0.59945 | 1.3593 | 146.34 | 1900 | 477.09 | 346.85 | 0.91788 | 141.5 | 4.974 |
| 540 | 129.06 | 92.04 | 0.60078 | 1.3860 | 144.32 | 1950 | 490.88 | 357.20 | 0.92504 | 157.1 | 4.598 |
| 560 | 133.86 | 95.47 | 0.60950 | 1.5742 | 131.78 | 2000 | 504.71 | 367.61 | 0.93205 | 174.0 | 4.258 |
| 580 | 138.66 | 98.90 | 0.61793 | 1.7800 | 120.70 | 2050 | 518.61 | 378.08 | 0.93891 | 192.3 | 3.949 |
| 600 | 143.47 | 102.34 | 0.62607 | 2.005 | 110.88 | 2100 | 532.55 | 388.60 | 0.94564 | 212.1 | 3.667 |
| 620 | 148.28 | 105.78 | 0.63395 | 2.249 | 102.12 | 2150 | 546.54 | 399.17 | 0.95222 | 233.5 | 3.410 |
| 640 | 153.09 | 109.21 | 0.64159 | 2.514 | 94.30 | 2200 | 560.59 | 409.78 | 0.95868 | 256.6 | 3.176 |
| 660 | 157.92 | 112.67 | 0.64902 | 2.801 | 87.27 | 2250 | 574.69 | 420.46 | 0.96501 | 281.4 | 2.961 |
| 680 | 162.73 | 116.12 | 0.65621 | 3.111 | 80.96 | 2300 | 588.82 | 431.16 | 0.97123 | 308.1 | 2.765 |
| 700 | 167.56 | 119.58 | 0.66321 | 3.446 | 75.25 | 2350 | 603.00 | 441.91 | 0.97732 | 336.8 | 2.585 |
| 720 | 172.39 | 123.04 | 0.67002 | 3.806 | 70.07 | 2400 | 617.22 | 452.70 | 0.98331 | 367.6 | 2.419 |
| 740 | 177.23 | 126.51 | 0.67665 | 4.193 | 65.38 | 2450 | 631.48 | 463.54 | 0.98919 | 400.5 | 2.266 |
| 760 | 182.08 | 129.99 | 0.68312 | 4.607 | 61.10 | 2500 | 645.78 | 474.40 | 0.99497 | 435.7 | 2.125 |
| 780 | 186.94 | 133.47 | 0.68942 | 5.051 | 57.20 | 2550 | 660.12 | 485.31 | 1.00064 | 473.3 | 1.996 |
| 800 | 191.81 | 136.97 | 0.69558 | 5.526 | 53.63 | 2600 | 674.49 | 496.26 | 1.00623 | 513.5 | 1.876 |
| 820 | 196.69 | 140.47 | 0.70160 | 6.033 | 50.35 | 2650 | 688.90 | 507.25 | 1.01172 | 556.3 | 1.765 |
| 840 | 201.56 | 143.98 | 0.70747 | 6.573 | 47.34 | 2700 | 703.35 | 518.26 | 1.01712 | 601.9 | 1.662 |
| 860 | 206.46 | 147.50 | 0.71323 | 7.149 | 44.57 | 2750 | 717.83 | 529.31 | 1.02244 | 650.4 | 1.566 |
| 880 | 211.35 | 151.02 | 0.71886 | 7.761 | 42.01 | 2800 | 732.33 | 540.40 | 1.02767 | 702.0 | 1.478 |
| 900 | 216.26 | 154.57 | 0.72438 | 8.411 | 39.64 | 2850 | 746.88 | 551.52 | 1.03282 | 756.7 | 1.395 |
| 920 | 221.18 | 158.12 | 0.72979 | 9.102 | 37.44 | 2900 | 761.45 | 562.66 | 1.03788 | 814.8 | 1.318 |
| 940 | 226.11 | 161.68 | 0.73509 | 9.834 | 35.41 | 2950 | 776.05 | 573.84 | 1.04288 | 876.4 | 1.247 |
| 960 | 231.06 | 165.26 | 0.74030 | 10.61 | 33.52 | 3000 | 790.68 | 585.04 | 1.04779 | 941.4 | 1.180 |
| 980 | 236.02 | 168.83 | 0.74540 | 11.43 | 31.76 | 3050 | 805.34 | 596.28 | 1.05264 | 1011 | 1.118 |
| 1000 | 240.98 | 172.43 | 0.75042 | 12.30 | 30.12 | 3100 | 820.03 | 607.53 | 1.05741 | 1083 | 1.060 |
| 1040 | 250.95 | 179.66 | 0.76019 | 14.18 | 27.17 | 3150 | 834.75 | 618.82 | 1.06212 | 1161 | 1.006 |
| 1080 | 260.97 | 186.93 | 0.76964 | 16.28 | 24.58 | 3200 | 849.48 | 630.12 | 1.06676 | 1242 | 0.9546 |
| 1120 | 271.03 | 194.25 | 0.77880 | 18.60 | 22.30 | 3250 | 864.24 | 641.46 | 1.07134 | 1328 | 0.9069 |
| 1160 | 281.14 | 201.63 | 0.78767 | 21.18 | 20.29 | 3300 | 879.02 | 652.81 | 1.07585 | 1418 | 0.8621 |
| 1200 | 291.30 | 209.05 | 0.79628 | 24.01 | 18.51 | 3350 | 893.83 | 664.20 | 1.08031 | 1513 | 0.8202 |
| 1240 | 301.52 | 216.53 | 0.80466 | 27.13 | 16.93 | 3400 | 908.66 | 675.60 | 1.08470 | 1613 | 0.7807 |
| 1280 | 311.79 | 224.05 | 0.81280 | 30.55 | 15.52 | 3450 | 923.52 | 687.04 | 1.08904 | 1719 | 0.7436 |
| 1320 | 322.11 | 231.63 | 0.82075 | 34.31 | 14.25 | 3500 | 938.40 | 698.48 | 1.09332 | 1829 | 0.7087 |
| 1360 | 332.48 | 239.25 | 0.82848 | 38.41 | 13.12 | 3550 | 953.30 | 709.95 | 1.09755 | 1946 | 0.6759 |
| 1400 | 342.90 | 246.93 | 0.83604 | 42.88 | 12.10 | 3600 | 968.21 | 721.44 | 1.10172 | 2068 | 0.6449 |
| 1440 | 353.37 | 254.66 | 0.84341 | 47.75 | 11.17 | 3650 | 983.15 | 732.95 | 1.10584 | 2196 | 0.6157 |

[1] $p_r$ and $v_r$ data for use with Eqs. 7.32 and 7.33, respectively.

**Table T-9E** *(Continued)*

| $T(°R)$, $h$ and $u$(Btu/lb), $s°$(Btu/lb · °R) | | | | | | | | | | | |
|---|---|---|---|---|---|---|---|---|---|---|---|
| | | | | when $\Delta s = 0$ | | | | | | when $\Delta s = 0$ | |
| $T$ | $h$ | $u$ | $s°$ | $p_r$ | $v_r$ | $T$ | $h$ | $u$ | $s°$ | $p_r$ | $v_r$ |
| 3700 | 998.11 | 744.48 | 1.10991 | 2330 | .5882 | 4200 | 1148.7 | 860.81 | 1.14809 | 4067 | .3826 |
| 3750 | 1013.1 | 756.04 | 1.11393 | 2471 | .5621 | 4300 | 1179.0 | 884.28 | 1.15522 | 4513 | .3529 |
| 3800 | 1028.1 | 767.60 | 1.11791 | 2618 | .5376 | 4400 | 1209.4 | 907.81 | 1.16221 | 4997 | .3262 |
| 3850 | 1043.1 | 779.19 | 1.12183 | 2773 | .5143 | 4500 | 1239.9 | 931.39 | 1.16905 | 5521 | .3019 |
| 3900 | 1058.1 | 790.80 | 1.12571 | 2934 | .4923 | 4600 | 1270.4 | 955.04 | 1.17575 | 6089 | .2799 |
| 3950 | 1073.2 | 802.43 | 1.12955 | 3103 | .4715 | 4700 | 1300.9 | 978.73 | 1.18232 | 6701 | .2598 |
| 4000 | 1088.3 | 814.06 | 1.13334 | 3280 | .4518 | 4800 | 1331.5 | 1002.5 | 1.18876 | 7362 | .2415 |
| 4050 | 1103.4 | 825.72 | 1.13709 | 3464 | .4331 | 4900 | 1362.2 | 1026.3 | 1.19508 | 8073 | .2248 |
| 4100 | 1118.5 | 837.40 | 1.14079 | 3656 | .4154 | 5000 | 1392.9 | 1050.1 | 1.20129 | 8837 | .2096 |
| 4150 | 1133.6 | 849.09 | 1.14446 | 3858 | .3985 | 5100 | 1423.6 | 1074.0 | 1.20738 | 9658 | .1956 |
| | | | | | | 5200 | 1454.4 | 1098.0 | 1.21336 | 10539 | .1828 |
| | | | | | | 5300 | 1485.3 | 1122.0 | 1.21923 | 11481 | .1710 |

**Table T-10E** Ideal Gas Specific Heats of Some Common Gases (Btu/lb · °R)

| Temp. °F | $c_p$ | $c_v$ | $c_p$ | $c_v$ | $c_p$ | $c_v$ | $c_p$ | $c_v$ | $c_p$ | $c_v$ | $c_p$ | $c_v$ | Temp. °F |
|---|---|---|---|---|---|---|---|---|---|---|---|---|---|
| | Air | | Nitrogen, $N_2$ | | Oxygen, $O_2$ | | Carbon Dioxide, $CO_2$ | | Carbon Monoxide, CO | | Hydrogen, $H_2$ | | |
| 40 | 0.240 | 0.171 | 0.248 | 0.177 | 0.219 | 0.156 | 0.195 | 0.150 | 0.248 | 0.177 | 3.397 | 2.412 | 40 |
| 100 | 0.240 | 0.172 | 0.248 | 0.178 | 0.220 | 0.158 | 0.205 | 0.160 | 0.249 | 0.178 | 3.426 | 2.441 | 100 |
| 200 | 0.241 | 0.173 | 0.249 | 0.178 | 0.223 | 0.161 | 0.217 | 0.172 | 0.249 | 0.179 | 3.451 | 2.466 | 200 |
| 300 | 0.243 | 0.174 | 0.250 | 0.179 | 0.226 | 0.164 | 0.229 | 0.184 | 0.251 | 0.180 | 3.461 | 2.476 | 300 |
| 400 | 0.245 | 0.176 | 0.251 | 0.180 | 0.230 | 0.168 | 0.239 | 0.193 | 0.253 | 0.182 | 3.466 | 2.480 | 400 |
| 500 | 0.248 | 0.179 | 0.254 | 0.183 | 0.235 | 0.173 | 0.247 | 0.202 | 0.256 | 0.185 | 3.469 | 2.484 | 500 |
| 600 | 0.250 | 0.182 | 0.256 | 0.185 | 0.239 | 0.177 | 0.255 | 0.210 | 0.259 | 0.188 | 3.473 | 2.488 | 600 |
| 700 | 0.254 | 0.185 | 0.260 | 0.189 | 0.242 | 0.181 | 0.262 | 0.217 | 0.262 | 0.191 | 3.477 | 2.492 | 700 |
| 800 | 0.257 | 0.188 | 0.262 | 0.191 | 0.246 | 0.184 | 0.269 | 0.224 | 0.266 | 0.195 | 3.494 | 2.509 | 800 |
| 900 | 0.259 | 0.191 | 0.265 | 0.194 | 0.249 | 0.187 | 0.275 | 0.230 | 0.269 | 0.198 | 3.502 | 2.519 | 900 |
| 1000 | 0.263 | 0.195 | 0.269 | 0.198 | 0.252 | 0.190 | 0.280 | 0.235 | 0.273 | 0.202 | 3.513 | 2.528 | 1000 |
| 1500 | 0.276 | 0.208 | 0.283 | 0.212 | 0.263 | 0.201 | 0.298 | 0.253 | 0.287 | 0.216 | 3.618 | 2.633 | 1500 |
| 2000 | 0.286 | 0.217 | 0.293 | 0.222 | 0.270 | 0.208 | 0.312 | 0.267 | 0.297 | 0.226 | 3.758 | 2.773 | 2000 |

**Table T-11E** Ideal Gas Properties of Selected Gases

$T$(°R), $\bar{h}$ and $\bar{u}$ (Btu/lbmol), $\bar{s}^\circ$ (Btu/lbmol · °R)

| | Carbon Dioxide, $CO_2$ | | | Carbon Monoxide, CO | | | Water Vapor, $H_2O$ | | | Oxygen, $O_2$ | | | Nitrogen, $N_2$ | | | |
|---|---|---|---|---|---|---|---|---|---|---|---|---|---|---|---|---|
| $T$ | $\bar{h}$ | $\bar{u}$ | $\bar{s}^\circ$ | $\bar{h}$ | $\bar{u}$ | $\bar{s}^\circ$ | $\bar{h}$ | $\bar{u}$ | $\bar{s}^\circ$ | $\bar{h}$ | $\bar{u}$ | $\bar{s}^\circ$ | $\bar{h}$ | $\bar{u}$ | $\bar{s}^\circ$ | $T$ |
| 300 | 2108.2 | 1512.4 | 46.353 | 2081.9 | 1486.1 | 43.223 | 2367.6 | 1771.8 | 40.439 | 2073.5 | 1477.8 | 44.927 | 2082.0 | 1486.2 | 41.695 | 300 |
| 320 | 2256.6 | 1621.1 | 46.832 | 2220.9 | 1585.4 | 43.672 | 2526.8 | 1891.3 | 40.952 | 2212.6 | 1577.1 | 45.375 | 2221.0 | 1585.5 | 42.143 | 320 |
| 340 | 2407.3 | 1732.1 | 47.289 | 2359.9 | 1684.7 | 44.093 | 2686.0 | 2010.8 | 41.435 | 2351.7 | 1676.5 | 45.797 | 2360.0 | 1684.4 | 42.564 | 340 |
| 360 | 2560.5 | 1845.6 | 47.728 | 2498.8 | 1783.9 | 44.490 | 2845.1 | 2130.2 | 41.889 | 2490.8 | 1775.9 | 46.195 | 2498.9 | 1784.0 | 42.962 | 360 |
| 380 | 2716.4 | 1961.8 | 48.148 | 2637.9 | 1883.3 | 44.866 | 3004.4 | 2249.8 | 42.320 | 2630.0 | 1875.3 | 46.571 | 2638.0 | 1883.4 | 43.337 | 380 |
| 400 | 2874.7 | 2080.4 | 48.555 | 2776.9 | 1982.6 | 45.223 | 3163.8 | 2369.4 | 42.728 | 2769.1 | 1974.8 | 46.927 | 2777.0 | 1982.6 | 43.694 | 400 |
| 420 | 3035.7 | 2201.7 | 48.947 | 2916.0 | 2081.9 | 45.563 | 3323.2 | 2489.1 | 43.117 | 2908.3 | 2074.3 | 47.267 | 2916.1 | 2082.0 | 44.034 | 420 |
| 440 | 3199.4 | 2325.6 | 49.329 | 3055.0 | 2181.2 | 45.886 | 3482.7 | 2608.9 | 43.487 | 3047.5 | 2173.8 | 47.591 | 3055.1 | 2181.3 | 44.357 | 440 |
| 460 | 3365.7 | 2452.2 | 49.698 | 3194.0 | 2280.5 | 46.194 | 3642.3 | 2728.8 | 43.841 | 3186.9 | 2273.4 | 47.900 | 3194.1 | 2280.6 | 44.665 | 460 |
| 480 | 3534.7 | 2581.5 | 50.058 | 3333.0 | 2379.8 | 46.491 | 3802.0 | 2848.8 | 44.182 | 3326.5 | 2373.3 | 48.198 | 3333.1 | 2379.9 | 44.962 | 480 |
| 500 | 3706.2 | 2713.3 | 50.408 | 3472.1 | 2479.2 | 46.775 | 3962.0 | 2969.1 | 44.508 | 3466.2 | 2473.2 | 48.483 | 3472.2 | 2479.3 | 45.246 | 500 |
| 520 | 3880.3 | 2847.7 | 50.750 | 3611.2 | 2578.6 | 47.048 | 4122.0 | 3089.4 | 44.821 | 3606.1 | 2573.4 | 48.757 | 3611.3 | 2578.6 | 45.519 | 520 |
| 540 | 4056.8 | 2984.4 | 51.082 | 3750.3 | 2677.9 | 47.310 | 4282.4 | 3210.0 | 45.124 | 3746.2 | 2673.8 | 49.021 | 3750.3 | 2678.0 | 45.781 | 540 |
| 560 | 4235.8 | 3123.7 | 51.408 | 3889.5 | 2777.4 | 47.563 | 4442.8 | 3330.7 | 45.415 | 3886.6 | 2774.5 | 49.276 | 3889.5 | 2777.4 | 46.034 | 560 |
| 580 | 4417.2 | 3265.4 | 51.726 | 4028.7 | 2876.9 | 47.807 | 4603.7 | 3451.9 | 45.696 | 4027.3 | 2875.5 | 49.522 | 4028.7 | 2876.9 | 46.278 | 580 |
| 600 | 4600.9 | 3409.4 | 52.038 | 4168.0 | 2976.5 | 48.044 | 4764.7 | 3573.2 | 45.970 | 4168.3 | 2976.8 | 49.762 | 4167.9 | 2976.4 | 46.514 | 600 |
| 620 | 4786.6 | 3555.6 | 52.343 | 4307.4 | 3076.2 | 48.272 | 4926.1 | 3694.9 | 46.235 | 4309.7 | 3078.4 | 49.993 | 4307.1 | 3075.9 | 46.742 | 620 |
| 640 | 4974.9 | 3704.0 | 52.641 | 4446.9 | 3175.9 | 48.494 | 5087.8 | 3816.8 | 46.492 | 4451.4 | 3180.4 | 50.218 | 4446.4 | 3175.5 | 46.964 | 640 |
| 660 | 5165.2 | 3854.6 | 52.934 | 4586.6 | 3275.8 | 48.709 | 5250.0 | 3939.3 | 46.741 | 4593.5 | 3282.9 | 50.437 | 4585.8 | 3275.2 | 47.178 | 660 |
| 680 | 5357.6 | 4007.2 | 53.225 | 4726.2 | 3375.8 | 48.917 | 5412.5 | 4062.1 | 46.984 | 4736.2 | 3385.8 | 50.650 | 4725.3 | 3374.9 | 47.386 | 680 |
| 700 | 5552.0 | 4161.9 | 53.503 | 4866.0 | 3475.9 | 49.120 | 5575.4 | 4185.3 | 47.219 | 4879.3 | 3489.2 | 50.858 | 4864.9 | 3474.8 | 47.588 | 700 |
| 720 | 5748.4 | 4318.6 | 53.780 | 5006.1 | 3576.3 | 49.317 | 5738.8 | 4309.0 | 47.450 | 5022.9 | 3593.1 | 51.059 | 5004.5 | 3574.7 | 47.785 | 720 |
| 740 | 5946.8 | 4477.3 | 54.051 | 5146.4 | 3676.9 | 49.509 | 5902.6 | 4433.1 | 47.673 | 5167.0 | 3697.4 | 51.257 | 5144.3 | 3674.7 | 47.977 | 740 |
| 760 | 6147.0 | 4637.9 | 54.319 | 5286.8 | 3777.5 | 49.697 | 6066.9 | 4557.6 | 47.893 | 5311.4 | 3802.2 | 51.450 | 5284.1 | 3774.9 | 48.164 | 760 |
| 780 | 6349.1 | 4800.1 | 54.582 | 5427.4 | 3878.4 | 49.880 | 6231.7 | 4682.7 | 48.106 | 5456.4 | 3907.5 | 51.638 | 5424.2 | 3875.2 | 48.345 | 780 |
| 800 | 6552.9 | 4964.2 | 54.839 | 5568.2 | 3979.5 | 50.058 | 6396.9 | 4808.2 | 48.316 | 5602.0 | 4013.3 | 51.821 | 5564.4 | 3975.7 | 48.522 | 800 |
| 820 | 6758.3 | 5129.9 | 55.093 | 5709.4 | 4081.0 | 50.232 | 6562.6 | 4934.2 | 48.520 | 5748.1 | 4119.7 | 52.002 | 5704.7 | 4076.3 | 48.696 | 820 |
| 840 | 6965.7 | 5297.6 | 55.343 | 5850.7 | 4182.6 | 50.402 | 6728.9 | 5060.8 | 48.721 | 5894.8 | 4226.6 | 52.179 | 5845.3 | 4177.1 | 48.865 | 840 |
| 860 | 7174.7 | 5466.9 | 55.589 | 5992.3 | 4284.5 | 50.569 | 6895.6 | 5187.8 | 48.916 | 6041.9 | 4334.1 | 52.352 | 5985.9 | 4278.1 | 49.031 | 860 |
| 880 | 7385.3 | 5637.7 | 55.831 | 6134.2 | 4386.6 | 50.732 | 7062.9 | 5315.3 | 49.109 | 6189.6 | 4442.0 | 52.522 | 6126.9 | 4379.4 | 49.193 | 880 |
| 900 | 7597.6 | 5810.3 | 56.070 | 6276.4 | 4489.1 | 50.892 | 7230.9 | 5443.6 | 49.298 | 6337.9 | 4550.6 | 52.688 | 6268.1 | 4480.8 | 49.352 | 900 |
| 920 | 7811.4 | 5984.4 | 56.305 | 6419.0 | 4592.0 | 51.048 | 7399.4 | 5572.4 | 49.483 | 6486.7 | 4659.7 | 52.852 | 6409.6 | 4582.6 | 49.507 | 920 |
| 940 | 8026.8 | 6160.1 | 56.536 | 6561.7 | 4695.0 | 51.202 | 7568.4 | 5701.7 | 49.665 | 6636.1 | 4769.4 | 53.012 | 6551.2 | 4684.5 | 49.659 | 940 |
| 960 | 8243.8 | 6337.4 | 56.765 | 6704.9 | 4798.5 | 51.353 | 7738.0 | 5831.6 | 49.843 | 6786.0 | 4879.5 | 53.170 | 6693.1 | 4786.7 | 49.808 | 960 |

**Table T-11E** *(Continued)*

$T$(°R), $\bar{h}$ and $\bar{u}$ (Btu/lbmol), $\bar{s}^\circ$ (Btu/lbmol · °R)

| | Carbon Dioxide, $CO_2$ | | | Carbon Monoxide, CO | | | Water Vapor, $H_2O$ | | | Oxygen, $O_2$ | | | Nitrogen, $N_2$ | | | |
|---|---|---|---|---|---|---|---|---|---|---|---|---|---|---|---|---|
| $T$ | $\bar{h}$ | $\bar{u}$ | $\bar{s}^\circ$ | $\bar{h}$ | $\bar{u}$ | $\bar{s}^\circ$ | $\bar{h}$ | $\bar{u}$ | $\bar{s}^\circ$ | $\bar{h}$ | $\bar{u}$ | $\bar{s}^\circ$ | $\bar{h}$ | $\bar{u}$ | $\bar{s}^\circ$ | $T$ |
| 980 | 8462.2 | 6516.1 | 56.990 | 6848.4 | 4902.3 | 51.501 | 7908.2 | 5962.0 | 50.019 | 6936.4 | 4990.3 | 53.326 | 6835.4 | 4889.3 | 49.955 | 980 |
| 1000 | 8682.1 | 6696.2 | 57.212 | 6992.2 | 5006.3 | 51.646 | 8078.9 | 6093.0 | 50.191 | 7087.5 | 5101.6 | 53.477 | 6977.9 | 4992.0 | 50.099 | 1000 |
| 1020 | 8903.4 | 6877.8 | 57.432 | 7136.4 | 5110.8 | 51.788 | 8250.4 | 6224.8 | 50.360 | 7238.9 | 5213.3 | 53.628 | 7120.7 | 5095.1 | 50.241 | 1020 |
| 1040 | 9126.2 | 7060.9 | 57.647 | 7281.0 | 5215.7 | 51.929 | 8422.4 | 6357.1 | 50.528 | 7391.0 | 5325.7 | 53.775 | 7263.8 | 5198.5 | 50.380 | 1040 |
| 1060 | 9350.3 | 7245.3 | 57.861 | 7425.9 | 5320.9 | 52.067 | 8595.0 | 6490.0 | 50.693 | 7543.6 | 5438.6 | 53.921 | 7407.2 | 5302.2 | 50.516 | 1060 |
| 1080 | 9575.8 | 7431.1 | 58.072 | 7571.1 | 5426.4 | 52.203 | 8768.2 | 6623.5 | 50.854 | 7696.8 | 5552.1 | 54.064 | 7551.0 | 5406.2 | 50.651 | 1080 |
| 1100 | 9802.6 | 7618.1 | 58.281 | 7716.8 | 5532.3 | 52.337 | 8942.0 | 6757.5 | 51.013 | 7850.4 | 5665.9 | 54.204 | 7695.0 | 5510.5 | 50.783 | 1100 |
| 1120 | 10030.6 | 7806.4 | 58.485 | 7862.9 | 5638.7 | 52.468 | 9116.4 | 6892.2 | 51.171 | 8004.5 | 5780.3 | 54.343 | 7839.3 | 5615.2 | 50.912 | 1120 |
| 1140 | 10260.1 | 7996.2 | 58.689 | 8009.2 | 5745.4 | 52.598 | 9291.4 | 7027.5 | 51.325 | 8159.1 | 5895.2 | 54.480 | 7984.0 | 5720.1 | 51.040 | 1140 |
| 1160 | 10490.6 | 8187.0 | 58.889 | 8156.1 | 5851.5 | 52.726 | 9467.1 | 7163.5 | 51.478 | 8314.2 | 6010.6 | 54.614 | 8129.0 | 5825.4 | 51.167 | 1160 |
| 1180 | 10722.3 | 8379.0 | 59.088 | 8303.3 | 5960.0 | 52.852 | 9643.4 | 7300.1 | 51.630 | 8469.8 | 6126.5 | 54.748 | 8274.4 | 5931.0 | 51.291 | 1180 |
| 1200 | 10955.3 | 8572.3 | 59.283 | 8450.8 | 6067.8 | 52.976 | 9820.4 | 7437.4 | 51.777 | 8625.8 | 6242.8 | 54.879 | 8420.0 | 6037.0 | 51.413 | 1200 |
| 1220 | 11189.4 | 8766.6 | 59.477 | 8598.8 | 6176.0 | 53.098 | 9998.0 | 7575.2 | 51.925 | 8782.4 | 6359.6 | 55.008 | 8566.1 | 6143.4 | 51.534 | 1220 |
| 1240 | 11424.6 | 8962.1 | 59.668 | 8747.2 | 6284.7 | 53.218 | 10176.1 | 7713.6 | 52.070 | 8939.4 | 6476.9 | 55.136 | 8712.6 | 6250.1 | 51.653 | 1240 |
| 1260 | 11661.0 | 9158.8 | 59.858 | 8896.0 | 6393.8 | 53.337 | 10354.9 | 7852.7 | 52.212 | 9096.7 | 6594.5 | 55.262 | 8859.3 | 6357.2 | 51.771 | 1260 |
| 1280 | 11898.4 | 9356.5 | 60.044 | 9045.0 | 6503.1 | 53.455 | 10534.4 | 7992.5 | 52.354 | 9254.6 | 6712.7 | 55.386 | 9006.4 | 6464.5 | 51.887 | 1280 |
| 1300 | 12136.9 | 9555.3 | 60.229 | 9194.6 | 6613.0 | 53.571 | 10714.5 | 8132.9 | 52.494 | 9412.9 | 6831.3 | 55.508 | 9153.9 | 6572.3 | 51.001 | 1300 |
| 1320 | 12376.4 | 9755.0 | 60.412 | 9344.6 | 6723.2 | 53.685 | 10895.3 | 8274.0 | 52.631 | 9571.6 | 6950.2 | 55.630 | 9301.8 | 6680.4 | 52.114 | 1320 |
| 1340 | 12617.0 | 9955.9 | 60.593 | 9494.8 | 6833.7 | 53.799 | 11076.6 | 8415.5 | 52.768 | 9730.7 | 7069.6 | 55.750 | 9450.0 | 6788.9 | 52.225 | 1340 |
| 1360 | 12858.5 | 10157.7 | 60.772 | 9645.5 | 6944.7 | 53.910 | 11258.7 | 8557.9 | 52.903 | 9890.2 | 7189.4 | 55.867 | 9598.6 | 6897.8 | 52.335 | 1360 |
| 1380 | 13101.0 | 10360.5 | 60.949 | 9796.6 | 7056.1 | 54.021 | 11441.4 | 8700.9 | 53.037 | 10050.1 | 7309.6 | 55.984 | 9747.5 | 7007.0 | 52.444 | 1380 |
| 1400 | 13344.7 | 10564.5 | 61.124 | 9948.1 | 7167.9 | 54.129 | 11624.8 | 8844.6 | 53.168 | 10210.4 | 7430.1 | 56.099 | 9896.9 | 7116.7 | 52.551 | 1400 |
| 1420 | 13589.1 | 10769.2 | 61.298 | 10100.0 | 7280.1 | 54.237 | 11808.8 | 8988.9 | 53.299 | 10371.0 | 7551.1 | 56.213 | 10046.6 | 7226.7 | 52.658 | 1420 |
| 1440 | 13834.5 | 10974.8 | 61.469 | 10252.2 | 7392.6 | 54.344 | 11993.4 | 9133.8 | 53.428 | 10532.0 | 7672.4 | 56.326 | 10196.6 | 7337.0 | 52.763 | 1440 |
| 1460 | 14080.8 | 11181.4 | 61.639 | 10404.8 | 7505.4 | 54.448 | 12178.8 | 9279.4 | 53.556 | 10693.3 | 7793.9 | 56.437 | 10347.0 | 7447.6 | 52.867 | 1460 |
| 1480 | 14328.0 | 11388.9 | 61.800 | 10557.8 | 7618.7 | 54.522 | 12364.8 | 9425.7 | 53.682 | 10855.1 | 7916.0 | 56.547 | 10497.8 | 7558.7 | 52.969 | 1480 |
| 1500 | 14576.0 | 11597.2 | 61.974 | 10711.1 | 7732.3 | 54.665 | 12551.4 | 9572.7 | 53.808 | 11017.1 | 8038.3 | 56.656 | 10648.0 | 7670.1 | 53.071 | 1500 |
| 1520 | 14824.9 | 11806.4 | 62.138 | 10864.9 | 7846.4 | 54.757 | 12738.8 | 9720.3 | 53.932 | 11179.6 | 8161.1 | 56.763 | 10800.4 | 7781.9 | 53.171 | 1520 |
| 1540 | 15074.7 | 12016.5 | 62.302 | 11019.0 | 7960.8 | 54.858 | 12926.8 | 9868.6 | 54.055 | 11342.4 | 8284.2 | 56.869 | 10952.2 | 7893.9 | 53.271 | 1540 |
| 1560 | 15325.3 | 12227.3 | 62.464 | 11173.4 | 8075.4 | 54.958 | 13115.6 | 10017.6 | 54.117 | 11505.4 | 8407.4 | 56.975 | 11104.3 | 8006.4 | 53.369 | 1560 |
| 1580 | 15576.7 | 12439.0 | 62.624 | 11328.2 | 8190.5 | 55.056 | 13305.0 | 10167.3 | 54.298 | 11668.8 | 8531.1 | 57.079 | 11256.9 | 8119.2 | 53.465 | 1580 |
| 1600 | 15829.0 | 12651.6 | 62.783 | 11483.4 | 8306.0 | 55.154 | 13494.4 | 10317.6 | 54.418 | 11832.5 | 8655.1 | 57.182 | 11409.7 | 8232.3 | 53.561 | 1600 |
| 1620 | 16081.9 | 12864.8 | 62.939 | 11638.9 | 8421.8 | 55.251 | 13685.7 | 10468.6 | 54.535 | 11996.6 | 8779.5 | 57.284 | 11562.8 | 8345.7 | 53.656 | 1620 |
| 1640 | 16335.7 | 13078.9 | 63.095 | 11794.7 | 8537.9 | 55.347 | 13877.0 | 10620.2 | 54.653 | 12160.9 | 8904.1 | 57.385 | 11716.4 | 8459.6 | 53.751 | 1640 |
| 1660 | 16590.2 | 13293.7 | 63.250 | 11950.9 | 8654.4 | 55.411 | 14069.2 | 10772.7 | 54.770 | 12325.5 | 9029.0 | 57.484 | 11870.2 | 8573.6 | 53.844 | 1660 |

**Table T-11E** *(Continued)*

$T$(°R), $\bar{h}$ and $\bar{u}$ (Btu/lbmol), $\bar{s}^\circ$ (Btu/lbmol · °R)

| | Carbon Dioxide, $CO_2$ | | | Carbon Monoxide, CO | | | Water Vapor, $H_2O$ | | | Oxygen, $O_2$ | | | Nitrogen, $N_2$ | | | |
|---|---|---|---|---|---|---|---|---|---|---|---|---|---|---|---|---|
| $T$ | $\bar{h}$ | $\bar{u}$ | $\bar{s}^\circ$ | $\bar{h}$ | $\bar{u}$ | $\bar{s}^\circ$ | $\bar{h}$ | $\bar{u}$ | $\bar{s}^\circ$ | $\bar{h}$ | $\bar{u}$ | $\bar{s}^\circ$ | $\bar{h}$ | $\bar{u}$ | $\bar{s}^\circ$ | $T$ |
| 1680 | 16845.5 | 13509.2 | 63.403 | 12107.5 | 8771.2 | 55.535 | 14261.9 | 10925.6 | 54.886 | 12490.4 | 9154.1 | 57.582 | 12024.3 | 8688.1 | 53.936 | 1680 |
| 1700 | 17101.4 | 13725.4 | 63.555 | 12264.3 | 8888.3 | 55.628 | 14455.4 | 11079.4 | 54.999 | 12655.6 | 9279.6 | 57.680 | 12178.9 | 8802.9 | 54.028 | 1700 |
| 1720 | 17358.1 | 13942.4 | 63.704 | 12421.4 | 9005.7 | 55.720 | 14649.5 | 11233.8 | 55.113 | 12821.1 | 9405.4 | 57.777 | 12333.7 | 8918.0 | 54.118 | 1720 |
| 1740 | 17615.5 | 14160.1 | 63.853 | 12579.0 | 9123.6 | 55.811 | 14844.3 | 11388.9 | 55.226 | 12986.9 | 9531.5 | 57.873 | 12488.8 | 9033.4 | 54.208 | 1740 |
| 1760 | 17873.5 | 14378.4 | 64.001 | 12736.7 | 9241.6 | 55.900 | 15039.8 | 11544.7 | 55.339 | 13153.0 | 9657.9 | 57.968 | 12644.3 | 9149.2 | 54.297 | 1760 |
| 1780 | 18132.2 | 14597.4 | 64.147 | 12894.9 | 9360.0 | 55.990 | 15236.1 | 11701.2 | 55.449 | 13319.2 | 9784.4 | 58.062 | 12800.2 | 9265.3 | 54.385 | 1780 |
| 1800 | 18391.5 | 14816.9 | 64.292 | 13053.2 | 9478.6 | 56.078 | 15433.0 | 11858.4 | 55.559 | 13485.8 | 9911.2 | 58.155 | 12956.3 | 9381.7 | 54.472 | 1800 |
| 1820 | 18651.5 | 15037.2 | 64.435 | 13212.0 | 9597.7 | 56.166 | 15630.6 | 12016.3 | 55.668 | 13652.5 | 10038.2 | 58.247 | 13112.7 | 9498.4 | 54.559 | 1820 |
| 1840 | 18912.2 | 15258.2 | 64.578 | 13371.0 | 9717.0 | 56.253 | 15828.7 | 12174.7 | 55.777 | 13819.6 | 10165.6 | 58.339 | 13269.5 | 9615.5 | 54.645 | 1840 |
| 1860 | 19173.4 | 15479.7 | 64.719 | 13530.2 | 9836.5 | 56.339 | 16027.6 | 12333.9 | 55.884 | 13986.8 | 10293.1 | 58.428 | 13426.5 | 9732.8 | 54.729 | 1860 |

$T >$ 1860 °R (CD-ROM)

# Index